D1452327

The Geology of North America
Volume I-2

The Atlantic Continental Margin: U.S.

Edited by

Robert E. Sheridan
Department of Geological Sciences
Busch Campus
Rutgers University
New Brunswick, New Jersey 08903

John A. Grow
U.S. Geological Survey
MS 960, Box 25046
Denver Federal Center
Denver, Colorado 80225

1988

Acknowledgment

Publication of this volume, one of the synthesis volumes of *The Decade of North American Geology Project* series, has been made possible by members and friends of the Geological Society of America, corporations, and government agencies through contributions to the Decade of North American Geology fund of the Geological Society of America Foundation.

Following is a list of individuals, corporations, and government agencies giving and/or pledging more than $50,000 in support of the DNAG Project:

ARCO Exploration Company
Chevron Corporation
Conoco, Inc.
Diamond Shamrock Exploration
  Corporation
Exxon Production Research Company
Getty Oil Company
Gulf Oil Exploration and Production
  Company
Paul V. Hoovler
Kennecott Minerals Company
Kerr McGee Corporation
Marathon Oil Company
McMoRan Oil and Gas Company
Mobil Oil Corporation
Pennzoil Exploration and Production
  Company

Phillips Petroleum Company
Shell Oil Company
Caswell Silver
Sohio Petroleum Corporation
Standard Oil Company of Indiana
Sun Exploration and Production Company
Superior Oil Company
Tenneco Oil Company
Texaco, Inc.
Union Oil Company of California
Union Pacific Corporation and
  its operating companies:
    Union Pacific Resources Company
    Union Pacific Railroad Company
    Upland Industries Corporation
U.S. Department of Energy

© 1988 by The Geological Society of America, Inc.
All rights reserved.

All materials subject to this copyright and included in this volume may be photocopied for the noncommercial purpose of scientific or educational advancement.

Copyright is not claimed on any material prepared by government employees within the scope of their employment.

Published by the Geological Society of America, Inc.
3300 Penrose Place, P.O. Box 9140, Boulder, Colorado 80301

Printed in U.S.A.

Front Cover: Selected Mesozoic structural and geophysical features on the U.S. Atlantic continental margin. Structural features include Mesozoic rift basins (black-exposed; yellow-buried), major marginal sedimentary basins (dark green), marginal platforms, and embayments as labeled, and oceanic fracture zones. Geophysical lineaments include the East Coast Magnetic Anomaly and Blake Spur Magnetic Anomaly (orange), seafloor spreading magnetic lineations (M-0 to M-25), and the Appalachian gravity gradient (asterisks). From Klitgord and others (this volume, Plate 2C) and Grow and Sheridan (this volume, Fig. 1). Graphics by P. Forrestel, J. Zwinakis, and D. Blackwood.

Library of Congress Cataloging-in-Publication Data

The Atlantic continental margin / edited by Robert E. Sheridan, John A. Grow
    p.   cm. — (The Geology of North America ; v. I-2)
    Includes bibliographies and index.
    ISBN 0-8137-5204-3
    1. Geology—Atlantic Coast (North America)   2. Continental margins—Atlantic Coast (North America) I. Sheridan, Robert E., 1940- . II. Grow, John A.   III. Geological Society of America. IV. Series.
QE71.G48 1986 vol. I-2
557 s—dc19
[557.4]
                           87-33849
                           CIP

Contents

GEOLOGICAL RESOURCES

ENVIRONMENTAL HAZARDS

CONCLUSIONS

Plates
(in accompanying slipcase)

Contents

Plates (continued)

Preface

The Geology of North America series has been prepared to mark the Centennial of The Geological Society of America. It represents the cooperative efforts of more than 1,000 individuals from academia, state and federal agencies of many countries, and industry to prepare syntheses that are as current and authoritative as possible about the geology of the North American continent and adjacent oceanic regions.

This series is part of the Decade of North American Geology (DNAG) Project which also includes eight wall maps at a scale of 1:5,000,000 that summarize the geology, tectonics, magnetic and gravity anomaly patterns, regional stress fields, thermal aspects, seismicity and neotectonics of North America and its surroundings. Together, the synthesis volumes and maps are the first coordinated effort to integrate all available knowledge about the geology and geophysics of a crustal plate on a regional scale.

The products of the DNAG Project present the state of knowledge of the geology and geophysics of North America in the 1980s, and they point the way toward work to be done in the decades ahead.

In addition to the contributions from organizations and individuals acknowledged at the front of this book, major support has been provided to the editors of this volume by the Departments of Geology (University of Delaware) and Geological Sciences (Rutgers) (Sheridan) and the U.S. Geological Survey (Grow).

A. R. Palmer
General Editor for the volumes
published by the Geological
Society of America

J. O. Wheeler
General Editor for the volumes
published by the Geological
Survey of Canada

Foreword

Our goal in producing this volume of the DNAG Series on *The Geology of North America* is simply stated: To provide a synthesis of our current thinking on the geology and geophysics of the Atlantic continental margin of the United States. Since our knowledge of this subject is ever growing, and rapidly so in the recent past, the word current is somewhat relative. Consequently, the volume does not include some data and ideas that have become available in the last two years while the volume was in production. Our thinking on this region not only involves data, but theoretical interpretations, both historical and modern. Thus, we have included some chapters on these theoretical matters as well. Also, because this region straddles the coastline of the most densely populated part of the United States, man's use and abuse of the geological resources of the region are also considered.

Management and direction of the project must be credited to Allison R. (Pete) Palmer, who, as general editor for GSA, guided us and the contributors of the chapters through several critical steps. With the able assistance of Jean Davis of GSA, Pete Palmer organized a preliminary workshop in Woods Hole, Massachusetts, which gathered the potential chapter authors. There the contents of the book and chapters were hotly debated. Even the book outline generated controversy. Should the book have a thread leading from the known to the unknown? For the coastal plain and offshore, where our geological knowledge is largely subsurface, this thread would have gone from top to bottom of sections, from the ocean floor to the subbottom, from the youngest, more factually known paleoenvironments to the older, more interpreted ones. But from a stratigraphic perspective, discussion of the geology of a region should proceed from the older history and rocks to the younger, to reveal evolutionary change. Given the maturity of exploration and the extensive, albeit not conclusive, data in the region, the more stratigraphic and evolutionary approach was accepted by the editors and authors.

A second workshop at Woods Hole gathered the actual authors and their preliminary chapters for review and critique. This meeting really established the form of the book. For example, two proposed chapters were eliminated, one on rifting theory and one on geo-chemistry, when it was found that these subjects would be covered within other chapters on mechanics of subsidence and mineral resources, respectively.

Palmer also assisted in the organization of a symposium on the Atlantic continental magin at the 1985 Annual GSA meeting in Orlando, Florida, and a poster session of some of the plates that accompany this volume. This effort induced the chapter authors to complete their chapters and offered the community a "coming-attraction" glimpse of the volume.

The U.S. Atlantic continental margin is the archetype passive margin, and as such it has geology of interest to many scientists and resources of interest to many government agencies and commerical companies. It is thus understandable that the contributors to this volume are geologists and geophysicsts from academia, government, and industry. This reflects the wealth of data and ideas contributed, especially over the last two decades, by commercial and scientific exploration. All these chapter authors are volunteers; no remuneration for their efforts was directly paid. Their time and expertise were freely given, and production of the chapters was supported by their institutions, agencies, and companies, where the subjects were directly related to their sponsored research. The contributors must be acknowledged and thanked for their generous extracurricular efforts on this project.

One of our authors, Francis Kohout of the U.S.G.S., died before the publication of the volume. Fran gathered a group of experts on the groundwater of the margin, and they wrote an excellent, comprehensive, and informative chapter on the subject reflecting his insight as a leading authority. As an indication of Fran's professionalism, he worked on this chapter in his Woods Hole home to within days of his death, and made his deadlines as previously committed. We admire his dedication very much and will miss his personal friendship.

As editors, we encouraged joint authorship of chapters, especially on the regional syntheses. We hoped to have the "groups" of experts reach a consensus, or if consensus was not possible, that conflicting opinions would be given equal exposure. In some cases, we could not get obvious competitors to become joint authors, and we made special efforts to make certain competing views were mentioned in chapters through the review process. We hope we were successful; if not, the chapters might be viewed as just one opinion of those particular authors. In this case, we hope we have selected the current experts on the chapter subjects, so the best opinions are presented.

In some chapters, there are interpretations in direct conflict with interpretations of the same geology in other chapters. This is the nature of our science and to be expected. We tried to resolve these conflicts in the workshops, but where they were unresolvable, we identify these conflicts in the chapters themselves, and in the final chapter.

Final production steps were ably managed by Pete Palmer at GSA headquarters in Boulder, Colorado. These efforts included the hounding of late authors to make their deadlines, the computer listing of manuscript status, and the review of final galley proofs and plates. The efforts of the GSA staff on these matters is greatly appreciated.

For us, one of the rewards of the project is the finished product: the collection in one place of the current thinking and status of data on the U.S. Atlantic continental margin. In the process of doing this project, we recognized holes in the data base, conflicts of interpretation, and inabilities to resolve controversial new ideas on the evolution of the margin because of poor or lacking data. With these observations as editors, we were afforded the luxury to suggest future research directions for the region. While these suggestions reflect our bias, they also reflect our insights. Perhaps they will be found valuable in the future.

Robert E. Sheridan
New Brunswick, New Jersey

John A. Grow
Denver, Colorado

Chapter 1

U.S. Atlantic Continental Margin; A typical Atlantic-type or passive continental margin

John A. Grow
U.S. Geological Survey, Box 25046, Denver Federal Center, Denver, Colorado 80225
Robert E. Sheridan
Department of Geological Sciences, Rutgers, The State University, New Brunswick, New Jersey 08903

INTRODUCTION

The conceptual revolutions of sea-floor spreading and plate tectonics between 1961 and 1968 (Dietz, 1961; Hess, 1962; Vine and Matthews, 1963; Wilson, 1965; McKenzie and Parker, 1967; Heirtzler and others, 1968; Le Pichon, 1968; Morgan, 1968; Isacks and others, 1968) led quickly to new models for continental margin evolution starting with an initial rifting phase (Dewey and Bird, 1970). The separation of Africa and North America in the post-Triassic to form the central North Atlantic Ocean and its adjacent rifted continental margins was soon considered the type example of this tectonic process (Dewey and Bird, 1970; Le Pichon and Fox, 1971; Pitman and Talwani, 1972; Le Pichon and Sibuet, 1981; Sclater and others, 1977; Klitgord and others, Chapter 3).

During the last two decades, the collection of marine multichannel seismic reflection data and the completion of deep drill holes for both scientific and commercial exploration have greatly increased our understanding of the U.S. Atlantic Continental Margin and passive or Atlantic-type continental margins in general (Bally, 1981). Early single-channel seismic reflection systems were capable of penetrating 1 to 2 km of sediments in the deep basins and the continental rise (i.e., water depths greater than 3,000 m), but were not capable of penetrating beneath the continental shelf and slope where water-bottom multiples severely contaminate the subbottom reflectors (Emery and others, 1970). Proprietary multichannel seismic reflection profiles were collected on the U.S. Atlantic Continental Margin in the late 1960's by the petroleum industry, and although never formally published, interpretations of the margin structure soon appeared indicating up to 10 km of sedimentary rocks beneath the continental shelf of New Jersey, suggesting significant potential for oil and gas exploration (Emery and Uchupi, 1972; Sheridan, 1974; Mattick and others, 1974). Although the 1973 international oil crisis accelerated the demand by the petroleum industry to open the U.S. Atlantic Continental Shelf for oil and gas exploration, many state and local governments as well as numerous environ-mental groups opposed leasing that might lead to a damaging oil spill, such as occurred off California near Santa Barbara in 1969.

In response to the debate in 1973 between industry advocates favoring accelerated offshore oil exploration and environmental groups opposing exploration, U.S. academic research institutions and government agencies initiated numerous regional geophysical, geological, and environmental studies of the U.S. Atlantic Continental Margin and the Bahamas. Between 1973 and 1978, over 30,000 km of state-of-the-art multichannel seismic reflection data and a high-resolution aeromagnetic survey, supplemented by numerous marine-magnetic and marine-gravity data, were obtained along the margin. These helped to define five major Mesozoic and Cenozoic basins off the U.S. Atlantic Coast and beneath the Bahama Banks (Fig. 1, Plate 2C) (Folger and others, 1979; Klitgord and Behrendt, 1979; Sheridan and others, 1981). The structure and evolution of these five offshore basins, the adjacent Coastal Plain onshore to the west, and the adjacent part of the deep-water basin of the western Atlantic Ocean on the east will be the focus of this volume.

Scientific drilling along the continental margin began in 1965 with the JOIDES (Joint Oceanographic Institutions for Deep Earth Sampling) drilling on the Blake Plateau (JOIDES, 1965). Subsequent DSDP (Deep Sea Drilling Project) drilling on Leg 11 (Hollister and Ewing, 1972), Leg 44 during 1975 (Benson and Sheridan, 1978), the USGS AMCOR (Atlantic Margin Coring) shallow drilling project (305 m) on the Continental Shelf in 1975 (Hathaway and others, 1979), IPOD (International Phase of Ocean Drilling) Leg 76 during 1980 (Sheridan and Gradstein, 1983), and IPOD Leg 95 during 1983 (Poag and Watts, 1987) have provided an enormous new data base for the shallow subsurface portion (less than 1,600 m) of the U.S. Atlantic Margin. Concurrently, five deep Continental Offshore Stratigraphic Test (COST) wells (up to 6.7 km below sea level) were completed by the petroleum industry prior to oil and gas lease sales between 1976 and 1979 (COST B-2: Scholle, 1977; COST

Grow, J. A., and Sheridan, R. E., 1988, U.S. Atlantic Continental Margin; A typical Atlantic-type or passive continental margin, *in* Sheridan, R. E., and Grow, J. A., eds., The Atlantic Continental Margin, U.S.: Geological Society of America, The Geology of North America, v. I-2.

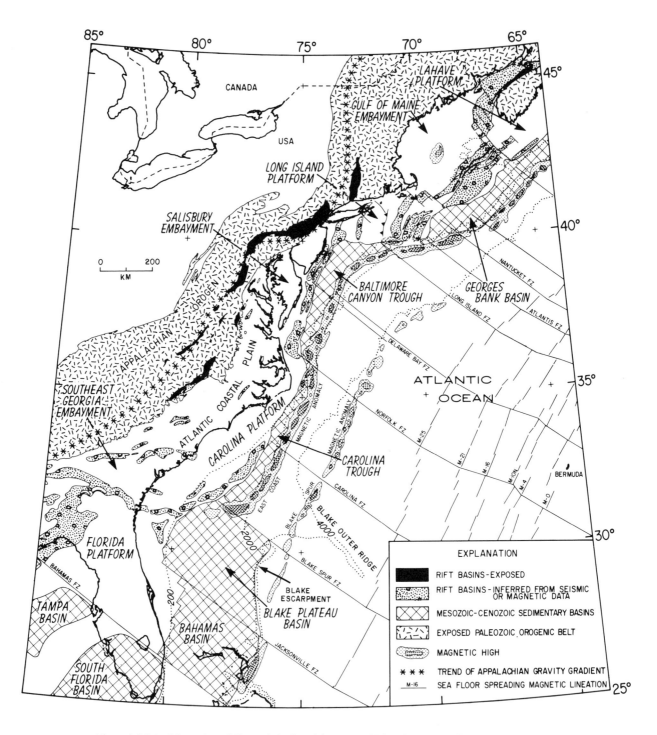

Figure 1. Major Mesozoic and Cenozoic basins of the eastern United States (modified from Klitgord and Behrendt, 1979; Hutchinson and others, 1986; Klitgord and others, Chapter 3 and Plate 2C). The most intensively studied of these basins is the Baltimore Canyon Trough. A schematic cross-section of this basin (Fig. 2) illustrates the primary structural elements of an Atlantic-type or passive continental margin.

GE-1: Scholle, 1979; COST B-3, Scholle, 1980; COST G-1 and G-2: Scholle and Wenkem, 1982). Forty-eight oil and gas exploration wells have been drilled along the U.S. Atlantic Margin to date (Mattick and Libby-French, Chapter 22), although no deep drill holes have yet tested the Carolina Trough or Blake Plateau Basin. While no commercial quantities of oil or gas have yet been discovered and current prices have shut down industry exploration, there is a vast amount of new data from both scientific and commercial drilling in the last two decades. The regional characteristics of these test wells have been summarized (Poag and Valentine, Chapter 5), but a comprehensive analysis and integration of all the offshore wells has yet to be completed and is beyond the scope of the present volume. However, highlights of these test wells and the existing geophysical data base allow an excellent opportunity for a regional synthesis.

MAJOR BASINS AND STRUCTURAL ELEMENTS OF THE U.S. ATLANTIC MARGIN

The Carolina Trough, Baltimore Canyon Trough, and Georges Bank Basin are classical Atlantic-type marginal basins (Fig. 1). In these basins, the separation between North America and Africa appears to be almost purely extensional; rift grabens and deep marginal basins are parallel or subparallel to the overall trends within the earlier Appalachian Mountains (Klitgord and others, Chapter 3; Grow and others, Chapter 13; and Schlee and Klitgord, Chapter 12; Dillon and Popenoe, Chapter 14). Deep seismic refraction and gravity models indicate that transitional or rift-stage crust of intermediate thickness (20 to 30 km thick, not including the postrift sediments) lies beneath these basins, with 30- to 40-km-thick continental crust to the northwest and 5- to 10-km-thick oceanic crust to the southeast (Fig. 2) (Grow and others, 1979; Hutchinson and others, 1983; Klitgord and others, Chapter 3, Fig. 16; Sheridan and others, Chapter 9). The intermediate thickness of the rift-stage crust is thought to be caused primarily by extensional thinning or stretching of the pre-existing continental crust, but mafic intrusions (or magmatic underplating) (LASE Study Group, 1986; and Diebold and others, Chapter 17; Klitgord and others, Chapter 3) and deep crustal metamorphism (Falvey and Middleton, 1981) are certainly important secondary processes during the transition from rifting to drifting (i.e., the initiation of seafloor spreading).

The two southern basins of the U.S. margin are not as well known as the three to the north. The Blake Plateau Basin is over 400 km wide, and it is inferred that rifting occurred over a wider zone for a longer period than the basins to the north and was probably accompanied by widespread mafic intrusions (Klitgord and others, Chapter 3; Dillon and Popenoe, Chapter 14). The Bahamas Basin is the most poorly known because of the complex structure of the shallow-water reefs, and it has not been systematically surveyed like the four basins to the north (Sheridan and others, Chapter 15). Besides the massive carbonate reefs, the structure of the Bahamas Basin is complicated by a shear zone

that connected the eastern Gulf of Mexico with the central Atlantic during Middle and Late Jurassic time (Klitgord and Schouten, 1986; Klitgord and others, Chapter 3). The Bahamas Basin may have further structural complications on its southern flank due to the formation of strike-slip and compressional zones north of Cuba during the Cretaceous and Cenozoic (Sheridan and others, Chapter 15). While the five basins vary from simple to very complex, the primary extensional, subsidence, and thermal processes are common to all passive or Atlantic-type margins (Bally, 1981; Steckler and others, Chapter 18; Sawyer, Chapter 19).

The schematic cross-section through Baltimore Canyon Trough (Fig. 2) illustrates most of the major structural elements observed on the U.S. Atlantic Margin and other Atlantic-type margins (Bally, 1981; Klitgord and others, Chapter 3). The most unusual aspect of the U.S. margin, and Baltimore Canyon Trough, in particular, is the enormous thickness of postrift sedimentary rocks (i.e., those deposited after the initiation of sea-floor spreading; see Plate 2D). Up to 13 km of postrift sedimentary rocks are present in the deepest part of Baltimore Canyon Trough (Fig. 2). The extremely thick postrift deposits along the U.S. Atlantic margin are probably due to the fact that the rifting and sea-floor spreading occurred while the Appalachian Mountains were still young and could supply a large quantity of erosional debris. The thickest sediment accumulations occur between the "hinge zone" and the East Coast Magnetic Anomaly (ECMA). The ECMA is a prominent positive magnetic high which marks the landward edge of the oceanic crust from the Carolina Trough northeast into the Canadian margin (Fig. 1 and 2; see Klitgord and others, Chapter 3, for a discussion of magnetic modeling studies of the ECMA). Recent deep seismic velocity studies have mapped a 7.1- to 7.5-km/s layer from the base of the oceanic crust on the east side of the ECMA up to 40 km west of the ECMA along the base of the rift-stage crust (LASE Study Group, 1986; Diebold and others, Chapter 17). This layer has been interpreted as evidence for massive intrusions or underplating of the rift-stage crust.

Because of the great amount of postrift sedimentary cover, seismic resolution is poor within the synrift deposits in the deepest part of Baltimore Canyon Trough (Plate 4). They are inferred to be nonmarine arkoses and lacustrine deposits intruded and interbedded with mafic volcanics, analogous to the synrift deposits in the onshore basins (Manspeizer and Cousminer, Chapter 10). Evaporite deposits are inferred to form in the latest stage of rifting when marine seaways begin to intermittently invade the basin. These later formed salt diapirs, which are also common along the ECMA in the Carolina Trough (Dillon and Popenoe, Chapter 14) and numerous other Atlantic-type margins.

Within the postrift sedimentary deposits, the shelf edge may prograde and retreat significant distances depending on the sediment supply and the rate of subsidence (Fig. 2). Carbonate bank and reef deposits are associated with the prograding shelf edge in Baltimore Canyon Trough (Grow and others, Chapter 13) and are also found in Georges Bank Basin (Schlee and Klitgord, Chapter 12), the Blake Plateau Basin (Dillon and Popenoe,

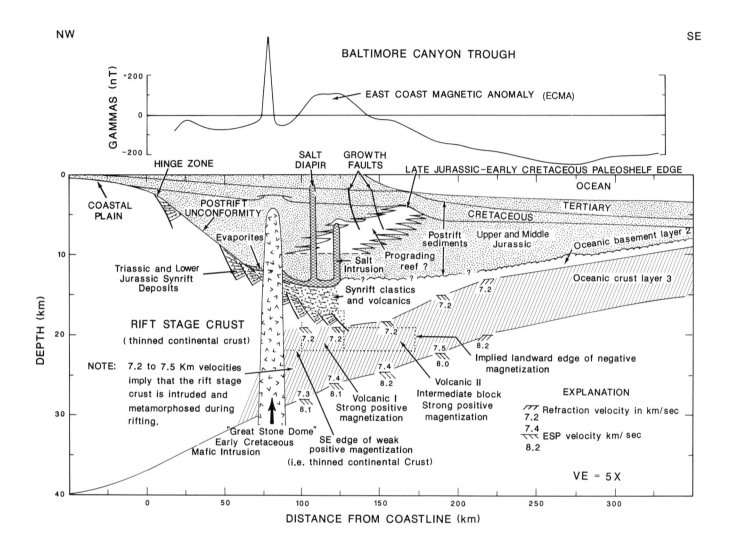

Figure 2. Cross-section through Baltimore Canyon Trough, a typical Atlantic-type continental margin. This is a region with extremely thick postrift sedimentary fill (i.e., deposited after initiation of sea-floor spreading). Prograding and retreat of shelf edges is common to all the Mesozoic and Cenozoic basins, and carbonate bank and reef formation was common near the continental shelf edge during the Jurassic and Early Cretaceous. Salt deposited during the late stage of rifting is indicated by numerous diapirs. Secondary volcanism is rare along the eastern continental margin, and the "Great Stone Dome" in Baltimore Canyon is the only large intrusive to penetrate high into the postrift sedimentary section. However, seamounts were formed during the Cretaceous south of Georges Bank, and deeper intrusions are inferred within Georges Bank Basin. Deep seismic measurements in Baltimore Canyon Trough observed (1) a 7.1- to 7.5-km/s velocity in the lower oceanic crust, which extends up to 40 km landward of the ECMA, and (2) gradual slope of the Moho beneath the ECMA (Diebold and others, Chapter 17). The magnetic boundaries are from Alsop and Talwani (1984). See Klitgord and others (Chapter 3) for additional discussion of the magnetic models. See Grow and others (Chapter 13) for a complete discussion of Baltimore Canyon Trough.

Chapter 14), and the Bahamas Basin (Sheridan and others, Chapter 15).

The Coastal Plain strata did not begin to onlap the eroded Appalachian surface west of the hinge zone until the Late Jurassic or Early Cretaceous (Fig. 2). At maximum transgression, however, these strata extended more than 200 km landward of the present coastline during periods of high sea level, like the Late Cretaceous (Olsson and others, Chapter 6; Gohn, Chapter 7).

The large mafic intrusion of Early Cretaceous age, the Great Stone Dome (Fig. 2), is the only major intrusion to penetrate high up into the postrift deposits on the U.S. margin, although much deeper intrusions are suspected in Georges Bank Basin and appear as Cretaceous seamounts on the continental rise south of Georges Bank Basin (Schlee and Klitgord, Chapter 12).

This discussion of Figures 1 and 2 is intended as a brief preview of much more detailed analyses and interpretations given by Klitgord and others (Chapter 3) on the structural and tectonic framework and by other chapters on individual basins.

OVERVIEW OF THE ATLANTIC CONTINENTAL MARGIN VOLUME

The volume is organized into six sections. The introductory section consists of four chapters, including this one. This chapter presents a simplified tectonic map and one representative crustal cross-section through Baltimore Canyon Trough (Figs. 1 and 2) in order to point out the major features of Atlantic-type margins and outline the topics that we will address in the volume. Shor and McClennen (Chapter 2) present the regional bathymetry or physiography of the margin (Plate 1A) and discuss the processes which are thought to control sedimentation and erosion in the modern oceans. Klitgord, Hutchinson, and Schouten (Chapter 3) review the structural and tectonic framework of the margin in terms of its rifting, sea-floor-spreading history and the general crustal structure of the five marginal basins (Plates 2C and 2D) that formed during the Mesozoic and Cenozoic. Drake (Chapter 4) gives a historical review of the scientific progress from the early 18th century through the post–World War II expansion of marine geological and geophysical research programs.

The second section of the volume contains four chapters which concern the stratigraphy, depositional processes, and depositional history of the postrift sedimentary rocks on the margin. Poag and Valentine (Chapter 5 and Plate 3) summarize the Mesozoic and Cenozoic lithostratigraphy and chronostratigraphy from the deep exploration wells on the Continental Shelf and Slope. The portion of postrift sedimentary rocks that continues landward of the coastline forms the emerged Coastal Plain (Figs. 1 and 2). The emerged Coastal Plain north of Cape Hatteras is summarized by Olsson, Gibson, and Hansen (Chapter 6), while that south of Hatteras is presented by Gohn (Chapter 7). Riggs and Belknap (Chapter 8) review the geologic processes, environments, sea-level changes, and climatic fluctuations that have been discovered from recent studies of the upper Cenozoic or Neogene sediments on the margin.

The third section of the volume contains eight chapters (9 through 16) concerning basin syntheses. Sheridan, Grow, and Klitgord (Chapter 9) review the geophysical data cited in subsequent chapters on individual basins. This includes location maps for seismic reflection profiles (Plate 1B), seismic refraction profiles (Plate 1C), magnetic total field and second vertical derivative maps (Plates 2A and 2B, by Behrendt and Grim), and seismic displays of four representative profiles (Plate 4 by Grow, Klitgord, Schlee, and Dillon). Manzpeizer and Cousminer (Chapter 10) synthesize all of the onshore Mesozoic basins containing Upper Triassic–Lower Jurassic synrift sedimentary rocks and correlate them to the offshore in Georges Bank Basin, Canada and Morocco. Mesozoic and Cenozoic magmatism is reviewed by De-Boer, McHone, Puffer, Ragland, and Whittington (Chapter 11). Schlee and Klitgord (Chapter 12) describe in detail Georges Bank Basin, the northernmost marginal basin along the U.S. Continental Margin where salt was encountered—at a depth of 6.7 km in the bottom of the COST G-2 well. This is the deepest penetrating offshore well on the U.S. Atlantic margin to date. Disagreement exists with respect to age of the deeper sedimentary rocks in the COST G-2 well. Manspeizer and Cousminer have found Late Triassic pollen in a core at 4.4 km depth, while Schlee and Klitgord (Chapter 12) and Poag and Valentine (Chapter 5) argue that the Triassic core material is reworked and the bottom of the COST G-2 well is no older than late Early Jurassic, based on seismic correlations to other wells on Georges Bank. A more detailed discussion of the disagreement is given by Klitgord and others (Chapter 3).

The section on basin syntheses continues with the Baltimore Canyon Trough (Chapter 13, by Grow, Klitgord and Schlee) where the thickest sediment accumulation on the margin is found. Northern Baltimore Canyon Trough is also the only place on the margin where the Jurassic shelf edge prograded seaward onto the oceanic crust (Fig. 2). The Carolina Trough and Blake Plateau Basin are the narrowest and widest of the marginal basins (Dillon and Popenoe, Chapter 14). The Carolina Trough has a linear chain of 23 salt diapirs which are aligned along the axis of the East Coast Magnetic Anomaly (ECMA); the diapirs are thought to have risen from late synrift evaporites at 11 km depth. Sheridan, Mullins, Austin, Ball, and Ladd (Chapter 15 and Plate 5) synthesize the geology of the Bahamas Basin where thick carbonate bank and reef sedimentation has continued into the Holocene. Schlee, Manspeizer, and Riggs summarize the paleoenvironments for the offshore Atlantic margin in Chapter 16 and Plate 6, the final chapter in the section on basin syntheses.

The fourth section of the volume contains five papers of a more theoretical nature concerning deep crustal structure, rifting, and subsidence theory (Chapters 17 to 21). Diebold, Stoffa, and the LASE Study Group (Chapter 17) summarize recent deep seismic velocity studies in Baltimore Canyon Trough that have discovered a 7.1- to 7.5-km/s layer at the base of the crust on both sides of the ECMA. Steckler, Watts, and Thorne (Chapter 18) review subsidence modeling of the Atlantic margin, while Sawyer (Chapter 19) discusses thermal modeling of the margin.

Pitman and Golovchenko (Chapter 20) address the inter-reaction of fluctuating sealevels and sediment supply with subsidence. The final chapter in this section concerns the estimation of eustatic sealevel fluctuations from a well-calibrated seismic profile off New Jersey (Chapter 21 and Plate 7, by Greenlee, Schroeder, and Vail).

The fifth section of the volume presents five chapters concerning geological resources (Chapters 22 to 26). Mattick and Libby-French review the petroleum geology and recent exploration activity that has yet to discover any oil or gas in commercial quantities (Chapter 22). Kohout, Meisler, Meyer, Johnston, Leve, and Wait (Chapter 23) discuss the fresh-water hydrogeology of the margin, including evidence for a relict Pleistocene fresh-water lens that extends most of the distance across the Continental Shelf off New Jersey. Duane and Stubblefield (Chapter 24) review the large quantity of sand and gravel resources on the inner Continental Shelf. Riggs and Manheim review the phosphate and other mineral resources on the continental shelf in Chapter 25. Finally, Costain and Speer analyze the potential for geothermal resources of the onshore Coastal Plain in Chapter 26.

The sixth section of the volume addresses environmental hazards to the margin due to both natural processes and the activities of man (Chapters 27 to 31). Folger reviews the research studies conducted in the last decade to determine the effects of offshore oil and gas exploration on the Atlantic margin (Chapter 27). Pilkey and Neal address the effects of coastal erosion and the great difficulties man encounters combating a continuing gradual rise in sea level (Chapter 28). Cretaceous and Cenozoic tectonism along the margin are revealed by numerous faults which have

been identified in the Coastal Plain (Prowell, Chapter 29 and Plate 8). Seeber and Armbruster review the destructive 1886 Charleston earthquake and the potential for future intraplate earthquakes along the Atlantic margin (Chapter 30 and Plate 8). Finally, Palmer summarizes the methods used in the disposal of waste products generated by man (Chapter 31).

The seventh and final section of the volume concludes with a single chapter by Sheridan and Grow (Chapter 32) which summarizes some of the recent advances in knowledge of the Atlantic margin, points out some of the continuing controversies, and suggests future directions for solving the major unanswered questions.

This volume attempts to integrate a very broad spectrum of geological disciplines represented by approximately 60 different contributors from more than 20 institutions. The tremendous increase in new information about the margin that has occurred in the last two decades has been impossible for any one scientist to master, and the opportunity to synthesize these new discoveries into a single volume has been a considerable challenge. The exchange of new concepts among the authors and editors during the preparation of this volume has been an invaluable experience to the participants. We hope the reader will share the excitement of exploration which these research studies have given to the participants. Finally, we hope that this synthesis volume will benefit students and scholars from diverse backgrounds and provide a valuable guide to the next generation of earth scientists interested in the Atlantic margin and in the geologic processes active along Atlantic-type margins.

REFERENCES CITED

Alsop, L. E., and Talwani, M., 1984, The East Coast magnetic anomaly: Science, v. 226, p. 1189–1191.

Bally, A. W., 1981, Atlantic-type margins, *in* Bally, A. W., ed., Geology of passive continental margins: American Association of Petroleum Geologists Education Course Notes, series 19, p. 1-1–1-48.

Benson, W. E., and Sheridan, R. E., eds., 1978, Initial reports of the Deep Sea Drilling Project: Washington, D.C., U.S. Government Printing Office, v. 44, 1005 p.

Dewey, J. F., and Bird, J. M., 1970, Mountain Belts and the New Global Tectonics: Journal of Geophysical Research, v. 75, p. 2625–2647.

Dietz, R. S., 1961, Continent and ocean basin evolution by spreading of the sea floor: Nature, v. 190, p. 854–857.

Emery, K. O., and Uchupi, E., 1972, Western North Atlantic Ocean: Topography, rocks, structure, water, life, and sediments: American Association of Petroleum Geologists Memoir 17, 532 p.

Emery, K. O., Uchupi, E., Phillips, J. D., Bowin, C. O., Bunce, E. T., and Knott, S. T., 1970, Continental rise of eastern North America: American Association of Petroleum Geologists Bulletin, v. 54, p. 44–108.

Falvey, D. A., and Middleton, M. F., 1981, Passive continental margins; Evidence for a prebreakup deep crustal metamorphic subsidence mechanism: Proceeding, 26th International Geologic Congress, Geology of Continental Margins Symposium, Paris, 1980, Oceanologica Acta, p. 103–114.

Folger, D. W., Dillon, W. P., Grow, J. A., Klitgord, K. D., and Schlee, J. S., 1979, Evolution of the Atlantic Continental Margin of the United States, *in* Talwani, M., Hay, W., and Ryan, W.B.F., eds., Deep drilling results in the

Atlantic Ocean, continental margins, and paleoenvironment: American Geophysical Union Maurice Ewing Series, v. 3, p. 87–108.

Grow, J. A., Bowin, C. O., and Hutchinson, D. R., 1979, The gravity field of the U.S. Atlantic Continental Margin: Tectonophysics, v. 59, p. 27–52.

Hathaway, J. C., and 8 others, 1979, U.S. Geological Survey Core Drilling on the Atlantic Shelf: Science, v. 206, p. 515–527.

Hess, H. H., 1962, History of the ocean basins, *in* Engel, A.E.J., James, H. L., and Leonard, B. P., eds., Petrologic studies: Geological Society of America A. F. Buddington volume, p. 599–620.

Heirtzler, J. R., Dickson, G. O., Herron, E. M., Pitman, W. C., III, and Le Pichon, X., 1968, Marine magnetic anomalies, geomagnetic field reversals, and motion of the ocean floor and continents: Journal of Geophysical Research, v. 73, p. 2119–2136.

Hollister, C. D., and Ewing, J. I., eds., 1972, Initial report of the Deep Sea Drilling Project: Washington, D.C., U.S. Government Printing Office, v. 11, 1077 p.

Hutchinson, D. R., Grow, J. A., Klitgord, K. D., and Swift, B. A., 1983, Deep structure and evolution of the Carolina Trough, *in* Watkins, J. S., and Drake, C. L., eds., Studies in continental margin geology: American Association of Petroleum Geologists Memoir 34, p. 129–152.

Hutchinson, D. R., Klitgord, K. D., and Detride, R. S., 1986, Rift basins of the Long Island platform: Geological Society of America Bulletin, v. 97, p. 688–702.

Isacks, G. L., Oliver, J., and Sykes, L. R., 1968, Seismology and the new global tectonics: Journal of Geophysical Research, v. 73, p. 5855–5899.

JOIDES, 1965, Ocean drilling on the continental margin: Science, v. 150, p. 709–716.

Klitgord, K. D., and Behrendt, J. C., 1979, Basin structure on the U.S. Atlantic margin, *in* Watkins, J. S., Montadert, L., and Dickerson, P., eds., Geological and geophysical investigations of continental margins: American Association of Petroleum Geologists Memoir 29, p. 85–112.

Klitgord, K. D., and Schouten, H., 1986, Plate kinematics of the central Atlantic, *in* Vogt, P. R., and Tucholke, B. E., eds., The western North Atlantic region: Boulder, Colorado, Geological Society of America, The Geology of North America, v. M, p. 351–378.

LASE Study Group, 1986, Deep structure of the U.S. East Coast passive margin from large aperture seismic experiments (LASE): Marine Petroleum Geology, v. 3, p. 234–242.

Le Pichon, X., 1968, Sea-floor spreading and continental drift: Journal of Geophysical Research, v. 73, p. 3661–3697.

Le Pichon, X., and Fox, P. J., 1971, Marginal offsets, fracture zones, and the early opening of the North Atlantic: Journal of Geophysical Research, v. 76, p. 6294–6308.

Le Pichon, X., and Sibuet, J.-C., 1981, Passive margins: A model of formation: Journal of Geophysical Research, v. 86, p. 3708–3720.

Mattick, R. E., Foote, R. Q., Weaver, N. L., and Grim, M. S., 1974, Structural framework of the U.S. Atlantic Outer Continental Shelf north of Cape Hatteras: American Association of Petroleum Geologists Bulletin, v. 58, p. 1179–1190.

McKenzie, D. P., and Parker, R. L., 1967, The North Pacific; An example of tectonics on a sphere: Nature, v. 216, p. 1276–1280.

Morgan, W. J., 1968, Rises, trenches, great faults, and crustal blocks: Journal of Geophysical Research, v. 78, p. 1959–1982.

Pitman, W. C., III, and Talwani, M., 1972, Seafloor spreading in the North Atlantic: Geological Society of America Bulletin, v. 83, p. 619–646.

Poag, C. W., and Watts, A. B., eds., 1987, Initial reports of the Deep Sea Drilling Project: Washington, D.C., U.S. Government Printing Office, v. 95, (in press).

Scholle, P. A., 1977, Geological studies on the COST no. B-2 well, U.S. Mid-Atlantic Outer Continental Shelf area: U.S. Geological Survey Circular 750, 71 p.

—— , 1979, Geological studies of the COST GE-1 well, U.S. South Atlantic Continental Shelf area: U.S. Geological Survey Circular 800, 114 p.

—— , ed., 1980, Geological Studies of the COST No. B-3 well, United States mid-Atlantic Continental Slope area: U.S. Geological Survey Circular 833, 182 p.

Scholle, P. A., and Wenkam, C. R., 1982, Geological studies of the COST G-1 and G-2 wells, U.S. North Atlantic Continental Shelf area: U.S. Geological Survey Circular, 193 p.

Sclater, J. G., Hellinger, S., and Tapscott, C., 1977, The paleobathymetry of the Atlantic Ocean from the Jurassic to the present: Journal of Geology, v. 85, p. 509–552.

Sheridan, R. E., 1974, Atlantic Continental Margin of North America, *in* Burk, C. A., and Drake, C. L., eds., The geology of continental margins: New York, Springer-Verlag, p. 391–407.

Sheridan, R. E., Crosby, J. T., Bryan, G. M., and Stoffa, P. L., 1981, Stratigraphy and structure of the southern Blake Plateau, northern Florida Straits and northern Bahama Platform from recent multichannel seismic reflection data: American Association of Petroleum Geologists Bulletin, v. 65, p. 2571–2593.

Sheridan, R. E., and Gradstein, F. M., eds., 1983, Initial reports of the Deep Sea Drilling Project: Washington, D.C., U.S. Government Printing Office, v. 76, 947 p.

Vine, F. J., and Matthews, D. H., 1963, Magnetic anomalies over oceanic ridges: Nature, v. 199, p. 947–949.

Wilson, J. T., 1965, Convection currents and continental drift: Philosophical Transactions of the Royal Society of London, ser. A, v. 258, p. 145–147.

MANUSCRIPT ACCEPTED BY THE SOCIETY OCTOBER 13, 1987

ACKNOWLEDGMENTS

We thank Kim D. Klitgord for providing a simplified version of his tectonic map (Plate 2C) which we have used for our Figure 1. This manuscript has benefited from reviews by Kim D. Klitgord and William P. Dillon. We also thank Patty Forrestel and Tom Kostick for drafting the figures and Pat Worl for typing the manuscript.

Printed in U.S.A.

The Geology of North America
Vol. I-2, The Atlantic Continental Margin: U.S.
The Geological Society of America, 1988

Chapter 2

Marine physiography of the U.S. Atlantic margin

Alexander N. Shor*
Lamont-Doherty Geological Observatory of Columbia University, Palisades, NY 10964
Charles E. McClennen
Department of Geology, Colgate University, Hamilton, NY 13346

INTRODUCTION

Although the study of Atlantic continental margin physiography (Plate 1a) started with lead-line sounding data centuries ago, it was the development and automation of precision echo sounding in the second quarter of the twentieth century that fostered modern investigation and interpretation (Veatch and Smith, 1939; Emery and Uchupi, 1972). Bottom samples gathered in the last four decades have strongly influenced the explanations for the development of submarine topography (Field and others, 1979; Hollister, 1973; Knebel, 1981; Milliman and others, 1972; Schlee and Pratt, 1970). High-resolution seismic profiling, side-scan sonar, bottom photography, current meters, and submersibles have allowed more detailed examination and interpretation of selected areas. Key recent articles cited in this chapter, from the vast literature about the Atlantic margin, provide references for an up-to-date understanding of the major physiographic features of the shelf, slope, and rise. In addition, controversies and unresolved questions have been identified. For readers less familiar with the location of physiographic features discussed in the text, the map provided should be of considerable assistance.

CONTINENTAL SHELF

Because the nature of the shelf physiography and the processes responsible for creating the submarine landforms differ markedly along the U.S. Atlantic shelf (Burk and Drake, 1974; Nairn and Stehli, 1974; Emery and Uchupi, 1972; and Uchupi, 1968), six areas are recognized. From north to south they are: 1) Gulf of Maine, 2) Georges Bank, 3) Southern New England, 4) Mid-Atlantic from Rhode Island to Cape Hatteras, 5) South-Atlantic from Cape Hatteras to Southern Florida, and 6) the Bahamas. A seventh section introduces the recent literature on microtopography and related sediment transport.

Gulf of Maine

The Gulf of Maine is bounded by Nova Scotia, Coastal New

*Present address: Hawaii Institute of Geophysics, University of Hawaii at Manoa, Honolulu, HI 96822.

England south to Cape Cod, and Georges Bank to the south and east, an area roughly 225 by 400 km in extent. It is noted for the numerous basins and banks (ledges, swells, ridges, knolls, and rocks) that have local relief of up to a couple hundred meters and a maximum depth of nearly 400 m (Emery, 1966; Uchupi, 1968). The formation of the topography was controlled, to a large extent, by Triassic rift structures (Ballard and Uchupi, 1975). Glacial scouring of fluvial valleys and interfluves of Tertiary to perhaps Mesozoic age, followed by glacial retreat and deposition as well as post-glacial marine reworking and deposition, account for many of the physiographic features (Emery, 1966; Uchupi, 1968, 1970; Oldale and Uchupi, 1970; Schlee and Pratt, 1970; Denny, 1982). The nearly flat and fine-grained nature of the basin floors indicates the extent of post-glacial sediment deposition (Emery and Uchupi, 1972).

Georges Bank

Georges Bank, which lies southeast of the Gulf of Maine and southwest of the Northeast Channel and Brown's Bank off Nova Scotia (about 400 by 170 km in extent), exhibits three physiographic patterns. The first feature, on the northwest crest, is Georges Shoal (less than 20 m depth) where strong rotary semidiurnal tidal currents continually rework dynamic sand waves superimposed on sand ridges (Belderson and others, 1978; Uchupi, 1968; Twichell, 1983b). To the southwest, beyond Great South Channel, Nantucket Shoals (100 by 100 km in area) display similar sand waves. Southeast of Georges Shoal and Nantucket Shoal is the deeper, smoother, and more extensive, but less mobile, shelf. Surface samples and cores indicate that the sediments there are reworked glacial moraine and outwash deposits (Bothner and Spiker, 1980; Oldale and Uchupi, 1970; Schlee and Pratt, 1970).

The third pattern occurs along the eastern and southern sides of Georges Bank at the shelf break, typically at 140 to 160 m water depths. Here, submarine canyons are common and often extensive, displaying heads as shallow as 120 m. The characteris-

Shor, A. N., and McClennen, C. E., 1988, Marine physiography of the U.S. Atlantic margin; *in* Sheridan, R. E., and Grow, J. A., eds., The Geology of North America, Volume I-2, The Atlantic Continental Margin, U.S.: Geological Society of America.

tics of these canyons on the slope and their continuation out onto the rise are discussed in subsequent sections of this chapter. Because no North American rivers can be tied to the canyons on this part of the margin (due to the intervening Gulf of Maine), mass wasting and/or glacial meltwaters and associated sediments probably played key roles in canyon formation. West of Georges Bank and south to Cape Hatteras, submarine canyons on the slope are similarly well documented and some, such as the Hudson, are clearly related to adjacent rivers. Knebel (1981) reviews the processes controlling many of the characteristics of the surficial sand sheet along the outer continental shelf, whereas Southard and Stanley (1976) review the nature of the shelf break, including the processes related to canyon head formation.

Southern New England

South of Nantucket, Martha's Vineyard, and Rhode Island, the shelf seaward of the 55 to 65 m isobath (about 90 by 180 km extent) displays an anomalously smooth topography and exceptionally fine-grained surface sediment (Garrison and McMaster, 1966, and Twichell and others, 1981). Although somewhat smaller in area than the Gulf of Maine and Georges Bank, this shelf region is worthy of special attention because very fine Holocene deposits, derived dominantly from the winnowing of the glacial deposits on Georges Bank and Nantucket Shoals, have buried and smoothed the initial ridge and swale morphology formed during the initial stages of this last transgression (Twichell and others, 1981). Although the present-day rate of deposition is quite low, sedimentation continues, because westward-moving mean currents transport fines from Georges Bank to the zone of weaker currents and sediment deposition south of New England (Bothner and others, 1981).

Mid-Atlantic—Rhode Island to Cape Hatteras

From Rhode Island to Cape Hatteras, the shelf (about 650 km long by 20 to 150 km wide) has been intensely studied in terms of its physiography and related sedimentary processes (Emery and Uchupi, 1972; Knebel, 1981; Swift and others, 1972; Uchupi, 1968 and 1970). The area contains an abundance of elongate ridges and nearly parallel depressions (or swales) that generally trend parallel to the adjacent coast (Veatch and Smith, 1939; Stearns and Garrison, 1967; Uchupi, 1968). Similarly, relict shorelines (or "scarps") were identified (Veatch and Smith, 1939) and subsequently used to evaluate glacio-tectonic influence (Dillon and Oldale, 1978). Cross shelf channels (or valleys) and deltas (or aprons), that in some places extend out to the shelf break (at 40 to 180 m water depth), were identified and associated with Pleistocene and/or Holocene stands of lowered sea level (Emery and Uchupi, 1972). Freeland and others (1981) and Swift and others (1980) compare the Hudson Shelf Valley morphology to that of the other valleys (now mostly sediment filled), noting the contrasting depositional histories and shelf morphological development.

Shelf valleys and their sediment filled or partially filled precursors have been traced as cross-shelf fluvial pathways to adjacent major canyons in complex and changing courses as follows: Block Shelf Valley/Block Canyon (McMaster and Ashraf, 1973), Hudson Channel/Hudson Canyon (sometimes) (Freeland and others, 1981), Great Egg Valley/Wilmington Canyon (McClennen, 1973a), and Delaware Shelf Valley/Wilmington Canyon (Twichell and others, 1977). Surface valleys are typically not congruent with the routes of the buried valleys, which can be explained in part by the estuary-mouth-retreat model (Swift and others, 1980) and by shifting river direction (Knebel and others, 1979).

More recently, terms such as estuary-mouth or shoal-retreat massif and cape-retreat massif have been added to describe clusters of ridges and swales located on somewhat elevated shelf areas associated with the adjacent coastal features for which they are named (see Stanley and Swift, 1976, Chapters 10, 14, and 15; Swift and others, 1977). Considerable debate exists as to the degree to which the ridge and swale topographic features are relict barrier-beach and lagoonal pairs (fossil shorelines) or post-transgressive shore-face-connected ridges that are still active under storm conditions (Emery and Uchupi, 1972; Uchupi, 1968; Swift and others, 1973; Sheridan and others, 1974; Stubblefield and Swift, 1976; Field, 1980; Swift, 1980; Swift and Field, 1981; Twichell and others, 1981; Stubblefield and others, 1983).

South Atlantic—Cape Hatteras to Southern Florida

The Cape Hatteras to Southern Florida margin is distinctive in that in addition to a long (about 1200 km), relatively narrow (1 to 130 km), and shallow shelf (edge at 20 to 100 m), there is the parallel but much deeper Blake Plateau (350 by 800 km extent) along all but southern Florida (Uchupi, 1968; Emery and Uchupi, 1972). On the shelf the linear sand ridges display radial patterns concentrated on shoals at coastal capes and river mouths (Hoyt and Henry, 1971; Field and Duane, 1976). The increasing abundance of carbonate sediments and reefs strongly influences the physiography of the margin (MacIntyre and Milliman, 1970; Meisburger and Duane, 1971; Ginsburg and James, 1974; MacIntyre and others, 1975), as have the location and strength of the Florida Current/Gulf Stream and eustatic changes in sea level (Henry and Hoyt, 1968). Reefs and hardgrounds are estimated at less than 10% of the shelf area between Cape Hatteras and Jacksonville, Florida (Henry and Foley, 1981). Terraces and a discontinuous ridge are prominent on the extremely narrow Florida shelf, reflecting active reef development (Miocene?) at times of lowered sea level as well as Gulf Stream erosion (Emery and Uchupi, 1972; Mullins and Neumann, 1979b).

The Bahamas

Little Bahama Bank and Great Bahama Bank are very shallow (5–15 m) continuations of the carbonate platforms of the Blake Plateau and Florida (Sheridan and others, 1981). The nearly flat (and in places sand wave-covered) banks indicate the importance of tidal currents, wind driven currents, and waves on the unconsolidated sands between the numerous islands (Hine and others, 1981). The exceptionally steep slopes (in places essentially vertical) along the Straits of Florida, the Atlantic margin, and the indentations of Exuma Sound, Tongue of the Ocean, and other canyons and channels are explained by former carbonate reef growth and submarine lithification of carbonate sediment sufficient to keep up with subsidence of the possibly faulted platform basement (Emery and Uchupi, 1972; Meyerhoff and Hatten, 1974; and Mullins and Lynts, 1977). Much of the banks is composed of eolianite consolidated during times of glacially lowered sea level (Emery and Uchupi, 1972).

Microtopographic Features

Microtopographic features, widely distributed on the shelf, have been discovered through the use of side-scan sonar, bottom photography, television cameras, and submersible dives (Sanders and others, 1969; McKinney and others, 1974; Henry and Foley, 1981). Belderson and others (1982) reviewed the significance of bedforms found on the shelf, while Swift and others (1979) focused on megaripple distribution and their relation to wind driven flow on the North Atlantic Shelf. Twichell (1983a) provided detail on bedforms in and around the heads of Lydonia and Oceanographer Canyons of Georges Bank. Knebel and Folger (1976) discussed the significance of asymmetric sand waves on the shelf around Wilmington Canyon, and Knebel (1979) interpreted the anomalous rough topography around Hudson Canyon as a relict erosion surface. Twichell and others (1985) interpreted the hummocky topography around Hudson Submarine Canyon as erosion due to tile fish and associated crustaceans. Ripples, megaripples, sand stringers, animal burrows and mounds, and bottom trawl scars have been used to infer the amount, direction, and frequency of sediment erosion, transport, and deposition on the shelf (Butman and others, 1979; Swift and others, 1979; Twichell, 1983b, 1985; Vincent and others, 1981). Near bottom, current-meter data have been gathered and used to estimate the nature of modern shelf sediment transport (Butman and others, 1979; Gadd and others, 1978; Lavelle and others, 1978; McClennen, 1973b; Vincent and others, 1981). Shinn and others (1981) provided a reexamination of the nature of spur and groove features associated with carbonate reefs.

CONTINENTAL SLOPE

The first comprehensive survey of the continental slope between Cape Hatteras and Georges Bank was conducted during the 1930s by the U.S. Coast and Geodetic Survey (Veatch and Smith, 1939). Based on early versions of the acoustic echo sounder to determine depths, and navigated by radio-acoustic ranging techniques with anchored buoys for reference, the contour charts published by Veatch and Smith (1939) were the first accurate charts of the continental slope. They were also the first

illustration of the intense dissection of this slope by submarine canyons. According to Scanlon (1984), the Veatch and Smith (1939) charts portray the pinnate drainage morphology of the slope more accurately than more recent bathymetric charts published by the U.S. National Ocean Survey (1:250,000 Scale Bathymetric Chart Series, Eastern United States Continental Margin, 1978–1980).

The eastern U.S. slope is divided morphologically into three major segments: 1) the section between Cape Lookout and Georges Bank, 2) the Florida-Hatteras Slope, and 3) the slope seaward of Blake Plateau. In addition, the Blake Plateau and the slopes off the Bahama Platform are described.

Cape Lookout to Northeast Channel

The depth of the shelf/slope break decreases toward the south, ranging from more than 120 m northeast of Hudson Canyon to approximately 60 m off Cape Lookout. Dillon and Oldale (1978) describe a hinge zone on the margin near 39°N, off the coast of New Jersey. Outer shelf Wisconsinan shorelines (Nichols and Franklin Shores) each deepen by about 50 m over 150 km toward the northeast, suggesting subsidence of this part of the margin in post-Wisconsinan time.

The width of the slope north of Cape Lookout is narrowest (less than 20 km) off northern Georges Bank. The average width between Cape Cod and Cape Lookout is 40 km. The base of the slope typically lies between 2,000 and 2,200 m, marked by a substantial decrease in gradient on the upper rise.

Knebel (1984), following earlier work by Uchupi (1968) and Emery and Uchupi (1972), identifies "average" gradients on the upper slope as 1 to 2 degrees, and on the lower slope as 3 to 8 degrees. He places the boundary between upper and lower slope between 400 and 1,000 m. This range of gradients is reasonable for intercanyon sections of the slope on a regional scale. As pointed out by Twichell and Roberts (1982), however, a major part of the upper slope north of Cape Lookout is, in fact, part of a series of canyon drainage systems. Within canyons, wall gradients commonly reach 30° (and are occasionally nearly vertical) based on submersible observations (e.g., Ryan and others, 1978).

Mass wasting processes on the slope also give rise to locally steep scarps and ponding. Slumps, slide scars, and debris flow deposits locally modify the inter-canyon slope. The largest debris flow observed on the U.S. margin heads on the slope between 70° and 71°W. The deposit from this failure covers 40,000 km^2 of the lower slope and rise (Embley, 1980; Vassallo and others, 1984). Numerous smaller failures are mapped on the slope (Vassallo and others, 1984.)

The slope between Cape Lookout and Northeast Channel is dominated by submarine canyons, as originally described by Veatch and Smith (1939). Long-range side-scan sonar (GLORIA) surveys carried out in 1979 by the U.S. Geological Survey and the Institute of Oceanographic Sciences in England (Twichell and Roberts, 1982) identified 51 canyons cutting the upper slope between the regions of the Hudson and Baltimore canyons. Scanlon (1982, 1984), reporting on the northern part of the same survey, illustrates more than 80 additional upper slope canyons along the Georges Bank slope between Northeast Channel and Alvin Canyon. All told, the upper slope between Northeast Channel and Cape Hatteras is incised by approximately 200 discrete canyons over about 1,200 km, or one every 6 km. The longest undissected slope segment is 40 km, located immediately southwest of Hudson Canyon.

Fewer canyons are described from the lower slope in the reports of the GLORIA surveys. In part this is due to canyon coalescence, with convergence toward the rise, as described by McGregor (1985) in the Lydonia Canyon region. Side-scan sonar images there show that seven slope canyons merge to form only three channels on the upper rise. Other canyons do not reach the lower slope (Scanlon, 1984). However, detailed side-scan sonar and bathymetric mapping of the lower slope in the vicinity of Carteret Canyon show a number of lower slope canyons ½ to 1 km wide, not indicated on the GLORIA interpretations, and which are not connected to canyons on the upper slope (Farre, 1985). These are interpreted as incipient slope canyons formed by mass wasting of the lower slope. Their recognition supports the "headward growth" hypothesis of canyon development described by Twichell and Roberts (1982), McGregor and others (1982), and Farre and others (1983).

Individual canyons vary widely in size. The larger, shelf-indenting canyons can be as wide as 5 km or more, with flat floors 500 to 1,500 m across, and relief that locally exceeds 1,000 m. In contrast, the smaller canyons are commonly only ½ to 2 km wide, and often do not exceed 200 to 300 m relief.

Along the Georges Bank margin, many of the larger canyons indent the shelf/slope break, including Corsair, Georges, Lydonia, Hydrographer, Oceanographer, Veatch, Atlantis, and Alvin canyons. In contrast, only five of the canyons further to the south have eroded significantly landward of the shelf edge (Hudson, Wilmington, Baltimore, Washington, and Norfolk canyons). Shelf channels (some now buried) link two of these southern canyons with fluvial systems (Hudson and Delaware rivers, as described previously). Shelf-indenting canyons off Georges Bank and New England are not directly related to existing river systems, except possibly Block Canyon (McMaster and Ashraf, 1973).

Florida-Hatteras Slope

Unlike the slope further north, the Florida-Hatteras slope is largely devoid of submarine canyons. The slope drops from the shallow shelf edge (60 m) onto the Blake Plateau and (farther south) the Straits of Florida. The depth of the Blake Plateau is 600 to 800 m at the base of the slope. The upper slope (above 600 m) is generally smooth, and the underlying strata are interpreted as a series of Neogene and older progradational wedges (Paull and Dillon, 1980b). Slope erosion by the Gulf Stream is evident off Georgia between 600 and 800 m (Paull and Dillon, 1980b).

Blake Plateau and Blake Escarpment

Blake Plateau is a terrace approximately 800 m deep, bounded to the west by the Florida-Hatteras Slope. The Blake Plateau slope diverges from the Florida-Hatteras slope south of Cape Lookout, until the plateau gradually widens to more than 300 km off northern Florida. South of Blake Spur the slope steepens to more than 10°; this section is called the Blake Escarpment. The gradient north of Blake Spur is more typical of other lower slope segments along the margin, ranging from 2 to 6° on average (Knebel, 1984).

The surface of Blake Plateau has been, and continues to be, eroded by the Gulf Stream, and the Tertiary sediment cover is both thin and irregular (Pinet and Popenoe, 1985a, b; Pinet and others, 1981). Relief of 50 to 100 meters is common on this eroded surface beneath the modern Gulf Stream, and shifting pre-Holocene Gulf Stream pathways are evident as well (Pinet and Popenoe, 1985a). Deep water coral mounds and erosional pits are both common on the surface (Pratt, 1971).

The Blake Escarpment and the contiguous Bahama Escarpment to the south are the steepest sections of the slope along the eastern U.S. margin. The Blake Escarpment is interpreted as an outer reef, drowned in Albian time, whereas the Bahama Escarpment survives as an active carbonate platform margin (Mullins and Neumann, 1979a). Freeman-Lynde and others (1981) deduced 1 to 5 km of erosion of Bahama Escarpment based on exposed backreef facies along the scarp. Paull and Dillon (1980a) interpret the origin of the Blake Escarpment as a process of abyssal current erosion with mass failure along the scarp face due to undercutting. However, all slump blocks ascribed to this process are now buried.

Both the slope north of Blake Spur and the Blake Escarpment north of 27°30'N are devoid of major canyons. With few exceptions, both of these slope segments lack slump features as well, exhibiting a smooth surface morphology. South of 27°30'N, platform-indenting canyons dissect the escarpment, including Great Abaco Canyon off of Blake Plateau. Great Bahama Canyon and Exuma Sound incise the Bahama Banks.

Bahamas Slope

Little Bahama Bank is bounded by the 600- to 800-m deep Blake Plateau and Straits of Florida to the north and west, and by the Bahama Escarpment, dropping steeply into the 4,500-m deep Blake Basin to the east. It is separated from Great Bahama Bank by Great Bahama Canyon. Great Bahama Bank is similarly bounded by the Straits of Florida and the Bahama Escarpment.

Mullins and Neumann (1979a) summarize the average gradients off Little Bahama Bank. Slopes to the north and west average 4 to 6 degrees except off the northwest corner, where they are only 1 to 1.5 degrees. The gradient into Great Bahama Canyon to the south is 6 to 12 degrees, and the extremely steep Bahama Escarpment is 20 to 40 degrees on the average. Morphologic characteristics of the bank slopes vary substantially according to a variety of environmental and tectonic processes. Mullins and Neumann (1979a) identify seven different slope types around Little Bahama Bank alone.

The extreme slope of the Bahama Escarpment results largely from a thick reef buildup. Structural control on the location of the escarpment is discussed elsewhere in this volume (Sheridan and others, Chapter 15); it has been interpreted as the transition line between oceanic and continental crust (Mullins and Lynts, 1977).

The Bahama Bank carbonate slopes exhibit a variety of both constructional and erosional structures, some of which are analogous to features on terrigenous margins. Others, such as lithoherms (biologic structures) are unique to the carbonate environment.

Slope Processes: Models for Morphologic Development

In terms of continental slope processes, the most exciting recent discoveries have been based on detailed geomorphology of the submarine drainage systems, which enable comparisons with subaerial drainage systems. While the Veatch and Smith (1939) concept of stream erosion cutting slope canyons during times of subaerial exposure has been rejected, recent studies using long- and medium-range side-scan sonar (McGregor and others, 1982, 1984; Farre and others, 1983; Twichell and Roberts, 1982; Scanlon, 1982, 1984) have demonstrated that the landforms developed in submerged canyons have notable similarities to those of fluvial systems. These side-scan sonar studies have given rise to growth models for submarine canyon development, which include mass failures, debris flows, and biologic erosion. One implication of this work is that turbidity currents play a smaller role in slope erosion than previously thought, particularly in canyon initiation and canyon wall dissection. A further implication of this recent work is that continental slope physiography is largely the result of purely submarine processes, and that the subaerial exposure of large segments of the continental shelf (as in Pleistocene low sea level stands) is probably not required to initiate canyon development.

Among the important observations that have led to recent re-examination of canyon growth and slope erosion are the following:

1. Gully systems feed into most of the slope canyons between Cape Hatteras and Georges Bank. Gullies, which are often branched, form "pinnate" drainage systems that do not head at the shelf break, but rather incise the intercanyon slope (Twichell and Roberts, 1982; McGregor and others, 1982; Farre and others, 1983; Scanlon, 1984). In many instances the canyon spacing is sufficiently close that only a sharp-crested ridge separates adjacent canyons, with gullying extending right up to the ridge crest.

2. Studies in the vicinity of Wilmington Canyon suggest that two distinct classes of canyons are present on the slope. The larger, shelf-indenting canyons are classified as "mature." They tend to meander and to have lower gradients than canyons that head on the slope. These latter canyons typically have steep walls, straight down-slope trends, and often chutes headed by well-

developed scarps on the floor. They have been described as "juvenile," and have led to the development of the concept of "headward erosion" (Twichell and Roberts, 1982; Farre and others, 1983).

3. Both medium-range side-scan sonar images and submersible observations of canyon floors indicate that at least some of the shelf-indenting canyons are flat-floored, with well-defined, often meandering, thalweg channels (McGregor and others, 1982; Stubblefield and others, 1982). The absence of debris aprons at the junction between gullies and the canyon floor has been used as evidence of probable Holocene turbidity current flows (McGregor and others, 1982).

4. Long-range side-scan sonar studies (GLORIA II) have shown that "when the gullies are included as part of the canyon systems, little of the upper and middle slope remains outside the canyon systems" between Cape Hatteras and Georges Bank (Twichell and Roberts, 1982).

The model emerging from these studies is one in which canyon initiation ("headward erosion") and expansion (gullying) result largely from the prolonged effects of local mass failure of continental slope deposits. This model implies that the modern slope is largely an erosional surface incised by numerous adjacent canyon systems. Turbidity currents may be responsible for maintaining unobstructed channel floors, particularly in the "mature" shelf-indenting canyons, which are thought to develop their unique characteristics by serving as a conduit for river sediment, particularly during times of lowered sea level. Slumping is a more likely cause of most canyon morphology, however. Processes causing slumps may include undercutting by turbidity currents (Malahoff and others, 1980), biologic erosion (Rowe and Haedrich, 1979), and ground water flow (Robb, 1984) in addition to seismic activity and inherent sediment instability (MacIlvaine and Ross, 1979). Field observations of erosion are difficult to make, and assessing the relative importance of various mechanisms may ultimately rest in examination of the nature of the deposits on the continental rise. It seems likely that canyons are polygenetic in the origin, however, and it is worth remembering the caution of Shepard (1981, p. 1076), who summarized more than 50 years of experience in the study of submarine canyons: "[I]t seems unwise for authors to continue to offer simple, unified explanation(s) for these huge valleys of the sea floor."

A key question that remains unanswered concerning the erosion of the continental slope is whether modern geologic processes are responsible for slope dissection, or whether the slope is instead largely a relict feature resulting from more energetic erosion during periods of lower sea level. A related question is whether modern slope morphology is an appropriate analog for interpreting slope erosion processes in the ancient geologic record.

Holocene erosion has been convincingly demonstrated locally on the slope (e.g. Stanley and others, 1984), and erosive mechanisms apparently independent of sea level and of sufficient magnitude to incise deep canyons have been proposed. Some

recent studies have indicated that only minor erosion has occurred during the Holocene and that the modern slope surface is
primarily relict (e.g., in the vicinity of Lindenkohl to South Toms
canyons: Robb and others, 1981). Prior and others (1984,
p. 928), based on high resolution study of the slope near Carteret
Canyon, state that "new data show that this ancient, relict
landscape, apparently unaltered for several thousand years, is
unlikely to provide significant constraints for industrial development as a result of modern geologic processes." What is still
lacking, however, is a quantitative comparison of slope erosion
during periods of high and low sea level stand, and the incorporation in slope-erosion models of both canyon and non-canyon
slope sections. Stanley and others (1984) examined structures and
ages of sediments in 18 sediment cores from the slope and upper
rise in the vicinity of Wilmington and Spencer Canyons. They
conclude that slope erosion processes and sediment displacement
continue during Holocene time, with "meaningful differences in
petrology and depositional rates between cores taken from different environments." (Stanley and others, 1984, p. 127). Much
further work will be necessary, however, before we can critically
evaluate models such as that proposed by Vail and others (1977),
which assume that slope erosion is predominantly a result of
lowered sea level, and use slope depositional sequences and their
bounding unconformities as a basis for the interpretation of sea
level fluctuation, worldwide continental margin growth processes
and age correlations.

CONTINENTAL RISE

In contrast to the continental slope, the continental rise has
been dominantly a depositional regime since at least Miocene
time. The gross morphology was first described by Heezen and
others (1959), who distinguished an upper and lower rise. The
upper rise, with a broadly convex surface (gradient typically
1:100) extends from the base of the continental slope, onto which
it onlaps, to approximately 4,000 m water depth. It has been
interpreted as a "contourite" deposit, composed primarily of fine-
grained sediments deposited from the slope-parallel Western
Boundary Undercurrent (Hollister and Heezen, 1972). Mass
wasting processes are clearly important as well (e.g. Embley,
1980). In contrast, the lower rise is a relatively flat region with
only a slight seaward slope (less than 1:300). In the vicinity of
Hudson and Wilmington submarine channels, the rise consists of
turbidites ponded behind the Hatteras Outer Ridge in the post-
Miocene period (Tucholke and Laine, 1982).

The seaward edge of the continental rise, below 4,200 m,
separates the margin sequence from the Hatteras Abyssal Plain,
which lies below 5,200 m. A continuous sediment drift, or
"Outer Ridge," extends from north of Hudson Fan to south of
Blake Spur. It consists of the partially eroded Gulf Stream Outer
Ridge on the north, the Hatteras Outer Ridge (Tucholke and
Laine, 1982), and the arcuate Blake-Bahama Outer Ridge system
to the south and west. The Outer Ridges are mantled by sediment
waves characteristic of regions swept by contour currents, par-

ticularly on the seaward flanks. The waves on the northern part of
the margin are called the Lower Continental Rise Hills (Asquith,
1979; Ayers and Cleary, 1980), and typically have amplitudes of
50 to 100 m and wavelengths of 1 to 2 km. In some areas these
waves have been dissected by channel systems. Waves on the
Blake-Bahama Outer Ridge are more variable in size, but waves
in both regions decrease in both amplitude and wavelength from
the crest of the Outer Ridge toward the abyssal plain (Hollister
and others, 1974). Additional smaller-scale bedforms are developed in conjunction with sediment waves, including abyssal
furrows and current ripples (Flood, 1983).

Rise Channels and "Submarine Fans"

Three major submarine channels incise the continental rise
along the eastern U.S. margin; all are located north of Cape
Hatteras. The submarine Hudson Valley extends more than 600
kilometers seaward from the base of the continental slope to the
Hatteras Abyssal Plain.* It connects upslope with Hudson
Canyon, and hence with Hudson Shelf Valley (Ewing and others,
1963) and the Pleistocene Hudson River drainage system.
Hudson Valley is slightly sinuous and has eroded more than 500
meters into the sediments of the middle rise. On the upper rise,
the surface southwest of the channel is elevated by more than
100 m relative to the northeastern surface. A stratified sequence
of levee-like sediments has been deposited on the southwestern
side through interaction of the along-slope Western Boundary
Undercurrent flow with the fine-grained fraction of Hudson Canyon turbidity currents.

Hudson Valley is floored by a flat thalweg, from which
coarse sediments, including both gravel and large boulders (to >1
meter diameter), have been sampled (Hanselman and Ryan,
1983). The source of these coarse sediments, including Eocene
chalk boulders, is apparently the continental slope. Eocene samples on the channel floor are reported from 4,000 meters depth.
Although Cacchione and others (1978) reported Paleocene or
Eocene outcrops exposed in the walls of Hudson Valley near
3,000 meters, more recent seismic reflection studies strongly suggest this material was not in place (Mountain and Tucholke,
1985). The presence of such large boulders so far from the slope
is striking evidence for the competence of the turbidity currents
that are responsible for eroding the rise channel. Their exposure at
the seabed suggests that such flows are no older than Wisconsinan, and possibly even Holocene.

Hudson Valley changes from a single leveed erosional
channel, nearly 4,000 m in depth, to a distributary system of three
or more shallow (30–50 m deep) channels that cross the surface
of the sediment pond behind Hatteras Outer Ridge. Several of
these channels rejoin near 35°55′N 69°45′W, where a 200-m-
deep channel has eroded through the crest of Hatteras Outer
Ridge, allowing turbidity currents to reach the abyssal plain
(Asquith, 1979; Ayers and Cleary, 1980).

---

*Discussion of Hudson Valley based on 1984 surveys by Shor and others (in
preparation) in addition to cited references.

The overall morphology of Wilmington and Carstens Valleys is similar to the Hudson system, but their bathymetry has not been investigated in comparable detail. Wilmington Valley may have a more complex history, as it drains not only Wilmington Canyon, but also a broad region of the Mid-Atlantic continental slope, including several of the larger shelf-incising canyons (Washington, Baltimore, and Hatteras). As with the Hudson Valley, the Wilmington Valley extends far across the rise, emptying onto the southern end of the sediment pond behind the Hatteras Outer Ridge and spilling across the Lower Continental Rise Hills onto Hatteras Abyssal Plain. This section of the rise has been called the "Wilmington Fan" by Ayers and Cleary (1980).

Carstens Valley, which lies immediately northeast of Hudson Valley and south of Block Canyon, is of similar size and extent over the middle rise, but has no known landward continuation. Heezen and others (1959) refer to it as the "Eastern Branch" of the Hudson Valley, although the two valleys do not directly connect. The landward end of the channel may be buried beneath a broad slide and debris flow complex mapped by Embley (1980). Scarps west of Alvin Canyon on the slope (Scanlon, 1984) may be the source of this material.

Strikingly absent on the eastern U.S. margin is the classic submarine fan morphology described by Normark (1978) and others along most of the margins of the world. Normark (1974) and Ayers and Cleary (1980) suggest that this results from redistribution of "fan" turbidites by slope-parallel contour currents. It seems likely, however, that the Miocene Outer Ridge system offshore of the eastern United States has formed a barrier that has not allowed the development of convex fan morphology, Instead, the turbidites of the channel systems have filled the broad "pond" behind the outer ridges, forming the gently seaward-sloping bench that makes up the lower rise. Due to the youth of the system, fan growth has only just begun, as the pond has only recently filled.

CONCLUSION

As seen from the literature review in this chapter, the marine physiography of the U.S. Atlantic shelf, slope, and rise displays significant local and regional variations. Our present understanding of margin geomorphology is highly variable. Recent research emphasis has been on the origin and processes responsible for evolution of physiographic features; the development of swath imaging technology and improved navigation and sampling equipment has inspired new views of these incompletely explored regions. While most of the large features of the shelf and slope have been mapped, many smaller-scale features have yet to be mapped or even discovered. Similarly, morphologic characterization of many of the larger canyons remain to be investigated; without understanding the details of canyon morphology, we remain uncertain about the relative influence of proposed canyon-cutting processes. The decreasing resolution of the physiography with increasing depth reflects both the amount of research effort expended and the increasing difficulty of working at greater depths. Accordingly, additional studies of shelf physiography will be required to resolve the few, albeit important, existing debates and explain any yet-to-be-discovered detailed features. Considerable exploration and detailed mapping and interpretation is in progress for the slope and rise regions, and we can anticipate that the next decade of research will produce new views on margin growth processes as well as refinements of existing ideas. Improvements in our understanding of continental slope stability during the Holocene sea level high stand will be of particular interest to petroleum exploration on the deeper margin, but will also be important as a basis for understanding processes active on this and other margins earlier in geologic time.

REFERENCES CITED

Asquith, S. M., 1979, Nature and origin of the lower continental rise hills off the east coast of the United States: Marine Geology, v. 32, p. 165–190.

Ayers, M. W., and Cleary, W. J., 1980, Wilmington Fan; Mid-Atlantic lower rise development: Journal of Sedimentary Petrology, v. 50, p. 235–245.

Ballard, R. D., and Uchupi, E., 1975, Triassic rift structure in Gulf of Maine: American Association of Petroleum Geologists Bulletin, v. 59, p. 1041–1072.

Belderson, R. H., Johnson, M. A., and Stride, A. H., 1978, Bed-load partings and convergences at the entrance to the White Sea, U.S.S.R., and between Cape Cod and Georges Bank, U.S.A.: Marine Geology, v. 28, p. 65–75.

Belderson, R. H., Johnson, M. A., and Kenyon, N. H., 1982, Bedforms, *in* Stride, A. H., ed., Offshore Tidal Sands; Processes and Deposits: New York, Chapman and Hall Ltd., p. 27–57.

Bothner, M. H., and Spiker, E. C., 1980, Upper Wisconsinan till recovered on the continental shelf southeast of New England: Science, v. 210, p. 423–425.

Bothner, M. H., Spiker, E. C., Johnson, P. P., Rendigs, R. R., and Aruscavage, P. J., 1981, Geochemical evidence for modern sediment accumulation on the continental shelf off southern New England: Journal of Sedimentary Petrology, v. 51, p. 281–292.

Burk, C. A., and Drake, C. L., 1974, The geology of continental margins: New York, Springer-Verlag, 1009 p.

Butman, B., Noble, M., and Folger, D. W., 1979, Long-term observations of bottom currents and bottom sediment movement on the mid-Atlantic continental shelf: Journal of Geophysical Research, v. 84, p. 1187–1205.

Cacchione, D. A., Rowe, G. T., and Malahoff, A., 1978, Submersible investigation of outer Hudson Submarine Canyon, *in* Stanley, D. I., and Kelling, G., eds., Sedimentation in Submarine Canyons, Fans, and Trenches: Dowden, Hutchinson, and Ross, p. 42–50.

Denny, C. S., 1982, Geomorphology of New England: U.S. Geological Survey Professional Paper 1208, 18 p.

Dillon, W. P., and Oldale, R. N., 1978, Late Quaternary sea-level curve; Reinterpretation based on glaciotectonic influence: Geology, v. 6, p. 56–60.

Embley, R. W., 1980, The role of mass transport in the distribution and character of deep-ocean sediments with special reference to the North Atlantic: Marine Geology, v. 38, p. 23–50.

Emery, K. O., 1966, Atlantic continental shelf and slope of the United States; Geologic background: U.S. Geological Survey Professional Paper 529-A, p. A1–A23.

Emery, K. O., and Uchupi, E., 1972, Western north Atlantic Ocean; Topography, rocks, structure, water, life, and sediments: American Association of Petroleum Geologists Memoir 17, 532 p.

Ewing, J., LePichon, X., and Ewing, M., 1963, Upper stratification of Hudson Apron region: Journal of Geophysical Research, v. 68, p. 6303–6316.

Farre, J. A., 1985, The importance of mass wasting processes on the continental slope [thesis]: Columbia University, 229 p.

Farre, J. A., McGregor, B. A., Ryan, W.B.F., and Robb, J. M., 1983, Breaching the shelf break; Passing from youthful to mature phase in submarine canyon evolution, *in* Stanley, D. J., and Moore, G. T., eds., The Shelfbreak; A Critical Interface on Continental Margins: Society of Economic Paleontologists and Mineralogists Special Publication 33, p. 25–39.

Field, M. E., 1980, Sand bodies on coastal plain shelves; Holocene record of the U.S. Atlantic inner shelf off Maryland: Journal of Sedimentary Petrology, v. 50, p. 505–528.

Field, M. E., and Duane, D. B., 1976, Post Pleistocene history of the United States inner continental shelf; Significance to origin of barrier islands: Geological Society of America Bulletin, v. 87, p. 691–702.

Field, M. E., Meisburger, E. P., Stanley, E. A., and Williams, S. J., 1979, Upper Quaternary peat deposits on the Atlantic inner shelf of the United States: Geological Society of America Bulletin, v. 90, p. 618–628.

Flood, R. D., 1983, Classification of sedimentary furrows and a model for furrow initiation and evolution: Geological Society of America Bulletin, v. 94, p. 630–639.

Freeland, G. L., Stanley, D. J., Swift, D.J.P., and Lambert, D. N., 1981, The Hudson shelf valley; Its role in shelf sediment transport: Marine Geology, v. 42, p. 399–428.

Freeman-Lynde, R. P., Cita, M. B., Jadoul, F., Miller, E. L., and Ryan, W.B.F., 1981, Marine geology of the Bahama Escarpment: Marine Geology, v. 44, p. 119–156.

Gadd, P. E., Lavelle, J. W., and Swift, D.J.P., 1978, Estimates of sand transport on the New York shelf using near-bottom current observations: Journal of Sedimentary Petrology, v. 48, p. 239–252.

Garrison, L. E., and McMaster, R. L., 1966, Sediments and geomorphology of the continental shelf of southern New England: Marine Geology, v. 4, p. 273–289.

Ginsburg, R. N., and James, N. P., 1974, Holocene carbonate sediments of continental shelves, *in* Burke, C. A., and Drake, C. L., eds., The Geology of Continental Margins: New York, Springer-Verlag, p. 137–155.

Hanselman, D. H., and Ryan, W.B.F., 1983, 1978 Atlantic 3800-meter radioactive waste disposal site survey; Sedimentary micromorphologic and geophysical analysis: Environmental Protection Agency 520/1-83-017.

Heezen, B. C., Tharp, M., and Ewing, M., 1959, The floors of the Oceans, 1. The North Atlantic: Geological Society of America Special Paper 65, 122 p.

Henry, V. J., and Foley, F. D., 1981, Bottom morphology and shallow subbottom sedimentary structures on U.S. continental shelf between Cape Hatteras, North Carolina, and Jacksonville, Florida: American Association of Petroleum Geologists Bulletin, v. 65, p. 1663.

Henry, V. J., and Hoyt, J. H., 1968, Quaternary paralic and shelf sediments of Georgia: Southeastern Geology, v. 9, no. 4, p. 195–214.

Hine, A. C., Wilber, R. J., Bane, J. M., Neuman, A. C., and Lorenson, K. R., 1981, Offbank transport of carbonate sands along open, leeward bank margins; Northern Bahamas: Marine Geology, v. 42, p. 327–348.

Hollister, C. D., 1973, Atlantic continental shelf and slope of the United States; Texture of surface sediments from New Jersey to Southern Florida: U.S. Geological Survey Professional Paper 529M, 23 p.

Hollister, C. D., and Heezen, B. C., 1972, Geologic effects of ocean bottom currents; Western North Atlantic, *in* Gordon, A. L., ed., Studies in Physical Oceanography: London, Gordon and Breach, p. 37–66.

Hollister, C. D., Flood, R. D., Johnson, D. A., Lonsdale, P. F., and Southard, J. B., 1974, Abyssal furrows and hyperbolic echo traces on the Bahama Outer Ridge: Geology, v. 2, p. 395–400.

Hoyt, J. H., and Henry, V. J., 1971, Origin of capes and shoals along the southeastern coast of the United States: Geological Society of American Bulletin, v. 82, p. 59–66.

Knebel, H. J., 1979, Anomalous topography on the continental shelf around Hudson Canyon: Marine Geology, v. 33, M67–M75.

Knebel, H. J., 1981, Processes controlling the characteristics of the surficial sand sheets, U.S. Atlantic outer continental shelf: Marine Geology, v. 42, p. 349–368.

—— , 1984, Sedimentary processes on the Atlantic continental slope of the United States: Marine Geology, v. 61, p. 43–74.

Knebel, H. J., and Folger, D. W., 1976, Large sand waves on the Atlantic outer continental shelf around Wilmington Canyon, off eastern United States: Marine Geology, v. 22, p. M7–M15.

Knebel, H. J., Wood, S. A., and Spiker, E., 1979, Hudson River; Evidence for extensive migration on the exposed continental shelf during the Pleistocene: Geology, v. 7, p. 254–258.

Lavelle, J. W., Swift, D.J.P., Gadd, P. E., Stubblefield, W. L., Case, F. N., Brashear, H. R., and Haff, K. W., 1978, Fair weather and storm sand transport on the Long Island, N.Y., inner shelf: Sedimentology, v. 25, p. 823–842.

MacIlvaine, J. C., and Ross, D. A., 1979, Sedimentary processes on the continental slope of New England: Journal of Sedimentary Petrology, v. 49, p. 563–574.

MacIntyre, I. G., and Milliman, J. D., 1970, Physiographic features on the outer shelf and upper slope, Atlantic continental margin, southeastern United States: Geological Society of America, v. 80, p. 2577–2598.

MacIntyre, I. G., Blackwelder, B. W., Land, L. S., and Stuckenrath, R., 1975, North Carolina shelf-edge sandstone; Age, environment of origin, and relationship to pre-existing sea levels: Geological Society of America Bulletin, v. 86, p. 1073–1078.

Malahoff, A., Embley, R. W., Perry R. B., and Fefe, C., 1980, Submarine masswasting of sediments on the continental slope and upper rise south of Baltimore Canyon: Earth and Planetary Science Letters, v. 49, p. 1–7.

McClennen, C. E., 1973a, Great Egg buried channel on the New Jersey Continental Shelf; A Possible Continuation of the Pleistocene Schuylkill River to Wilmington Canyon: Geological Society of America Abstracts with Programs, v. 5, no. 2, p. 194–195.

—— 1973b, New Jersey continental shelf near bottom current meter records and recent sediment activity: Journal of Sedimentary Petrology, v. 43, p. 371–380.

McGregor, B. A., 1985, Role of submarine canyons in shaping the rise between Lydonia and Oceanographer canyons, Georges Bank: Marine Geology, v. 62, p. 277–293.

McGregor, B. A., Stubblefield, W. L., Ryan, W.B.F., and Twichell, D. C., 1982, Wilmington Submarine Canyon; A marine fluvial-like system: Geology, v. 10, no. 1, p. 27–30.

McGregor, B. A., Nelsen, T. A., Stubblefield, W. L., and Merrill, G. F., 1984, The role of canyons in late Quaternary deposition on the United States mid-Atlantic continental rise, *in* Stow, D.A.V., and Piper, D.J.W., eds., Fine-Grained Sediments; Deep-Water Processes and Facies: Geological Society Special Publication 15, p. 319–330.

McKinney, T. F., Stubblefield, W. L., and Swift, D.J.P., 1974, Large scale current lineations on the central New Jersey shelf, investigations by sidescan sonar: Marine Geology, v. 17, p. 79–102.

McMaster, R. L., and Ashraf, A., 1973, Drowned and buried valleys on the southern New England continental shelf: Marine Geology, v. 15, p. 249–268.

Meisburger, E. P., Duane, D. B., 1971, Geomorphology and sediments of the inner continental shelf, Palm Beach to Cape Kennedy, Florida: U.S. Army Coastal Engineering Research Center Technical Memorandum 34, 42 p.

Meyerhoff, A. A., and Hatten, C. W., 1974, Bahamas salient of North America, *in* Burke, C. A., and Drake, C. L., eds., The Geology of Continental Margins: New York, Springer-Verlag, p. 429–446.

Milliman, J. D., Pilkey, O. H., Ross, D. A., 1972, Sediments of the continental margin off the eastern United States: Geological Society of America Bulletin, v. 83, p. 1315–1334.

Mountain, G. S., and Tucholke, B. E., 1985, Mesozoic and Cenozoic geology of the U.S. Atlantic continental slope and rise, *in* Poag, C. W., ed., Geologic Evolution of the United States Atlantic Margin: Van Nostrand Reinhold Co., Inc., p. 293–341.

Mullins, H. T., and Lynts, G. W., 1977, Origin of the northwestern Bahama Platform; Review and reinterpretation: Geological Society of America Bulletin, v. 88, p. 1447–1461.

Mullins, H. T., and Neumann, A. C., 1979a, Deep carbonate bank margin structure and sedimentation in the northern Bahamas, *in* Geology of Continental

Slopes: Society of Economic Paleontologists and Mineralogists Special Publication 27, p. 165–192.

——, 1979b, Geology of the Miami terrace and its paleo-oceanographic implications: Marine Geology, v. 30, p. 205–232.

Nairn, A.E.M., and Stehli, F. G., 1974, The Ocean Basins and Margins, Volume II, The North Atlantic: New York, Plenum Press, 598 p.

Normark, W. R., 1974, Submarine canyons and fan valleys; Factors affecting growth patterns of deep-sea fans, *in* Modern and Ancient Geosynclinal Sedimentation: Society of Economic Paleontologists and Mineralogists Special Publication 19, p. 56–68.

——, 1978, Fan valleys, channels, and depositional lobes on modern submarine fans; Characters for recognition of sandy turbidite environments: American Association of Petroleum Geologists Bulletin, v. 62, p. 912–931.

Oldale, R. N., and Uchupi, E., 1970, The glaciated shelf off northeastern United States: U.S. Geological Survey Professional Paper 700-B, p. B167–B173.

Paull, C. K., and Dillon, W. P., 1980a, Erosional origin of the Blake Escarpment; An alternative hypothesis: Geology, v. 8, p. 538–542.

——, 1980b, Structure, stratigraphy, and geologic history of Florida-Hatteras Shelf and Inner Blake Plateau: American Association of Petroleum Geologists Bulletin, v. 64, no. 3, p. 339–358.

Pinet, P. R., and Popenoe, P., 1985a, A scenario of Mesozoic–Cenozoic ocean circulation over the Blake Plateau and its environs: Geological Society of American Bulletin, v. 96, p. 618–626.

——, 1985b, Shallow seismic stratigraphy and post-Albian geologic history of the northern and central Blake Plateau: Geological Society of America Bulletin, v. 96, p. 627–638.

Pinet, P. R., and Popenoe, P., Nellingham, D. F., 1981, Gulf stream; Reconstruction of Cenozoic flow patterns over the Blake Plateau: Geology, v. 9, p. 266–270.

Pratt, R. M., 1971, Lithology of rocks dredged from the Blake Plateau: Southeastern Geology, v. 13, p. 19–38.

Prior, D. B., Coleman, J. M., and Doyle, E. H., 1984, Antiquity of the continental slope along the middle-Atlantic margin of the United States: Science, v. 223, p. 926–928.

Robb, J. M., 1984, Spring sapping on the lower continental slope, offshore New Jersey: Geology, v. 12, p. 278–282.

Robb, J. M., Hampson, J. C., Jr., and Twichell, D. C., 1981, Geomorphology and sediment stability of a segment of the U.S. continental slope off New Jersey: Science, v. 211, p. 935–937.

Rowe, G. T., and Haedrich, R. L., 1979, The biota and biological processes of the continental slope, *in* Doyle, L. J., and Pilkey, O. H., Jr., eds., Geology of Continental Slopes: Society of Economic Paleontologists and Mineralogists Special Publication 27, p. 49–60.

Ryan, W.B.F., Cita, M. B., Miller, E. L., Hanselman, D., Nesteroff, W. D., Hecker, B., and Nibbelink, M., 1978, Bedrock geology in New England submarine canyons: Oceanologica Acta, v. 1, no. 2, p. 233–254.

Sanders, J. E., Emery, K. O., and Uchupi, E., 1969, Microtopography of five small areas of the continental shelf by side-scanning sonar: Geological Society of America Bulletin, v. 80, p. 561–572.

Scanlon, K. M., 1982, Geomorphic features of the western North Atlantic continental slope between Northeast Channel and Alvin Canyon as interpreted from Gloria II long-range sidescan-sonar data: U.S. Geological Survey Open-File Report.

——, 1984, The continental slope off New England; A long-range sidescan-sonar perspective: Geo-Marine Letters, v. 4, p. 1–4.

Schlee, J., and Pratt, R. M., 1970, Atlantic continental shelf and slope of the United States; Gravels of the northeastern part: U.S. Geological Survey Professional Paper 529H, p. H1–H39.

Shepard, F. P., 1981, Submarine canyons; Multiple causes and long-time persistence: American Association of Petroleum Geologists Bulletin, v. 65, p. 1062–1077.

Sheridan, R. E., Dill, C. E., and Kraft, J. C., 1974, Holocene sedimentary environment of the Atlantic inner shelf off Delaware: Geological Society of American Bulletin, v. 85, p. 1319–1328.

Sheridan, R. E., Crosby, J. T., Kent, K. M., Dillon, W. P., and Paul, C. K., 1981, The geology of the Blake Plateau and Bahamas region: Canadian Society of Petroleum Geologists Memoir 7, p. 487–502.

Shinn, E. A., Hudson, J. H., Robbin, D. M., and Lidz, B., 1981, Spurs and grooves revisited; Construction versus erosion Looe Key Reef, Florida, *in* Gomez, E. D., Birkeland, C. E., Buddlemeier, R. W., Johannes, R. E., Marsh, J. A., and Tsuda, R. T., eds., Reef and Man; Proceedings of the Fourth International Coral Reef Symposium, Volume I: Quezon City, Marine Sciences Center, University of the Philippines, p. 475–484.

Southard, J. B., and Stanley, D. J., 1976, Shelf break processes and sedimentation, *in* Stanley, D. J., and Swift, D.J.P., eds., Marine Sediment Transport and Environmental Management: New York, John Wiley and Sons, p. 351–377.

Stanley, D. J., and Swift, D.J.P., 1976, eds., Marine Sediment Transport and Environmental Management: New York, John Wiley and Sons, 602 p.

Stanley, D. J., Nelsen, T. A., and Stuckenrath, R., 1984, Recent sedimentation on the New Jersey slope and rise: Science, v. 226, p. 125–133.

Stearns, F., and Garrison, L. E., 1967, Bathymetric maps, middle Atlantic U.S. continental shelf: Rockville, Maryland, National Oceanic and Atmospheric Administration, National Ocean Survey, scale 1:125,000.

Stubblefield, W. L., and Swift, D.J.P., 1976, Ridge development as revealed by subbottom profiles on the central New Jersey shelf: Marine Geology, v. 20, p. 315–334.

Stubblefield, W. L., McGregor, B. A., Forde, E. B., Lambert, D. N., and Merrill, G. F., 1982, Reconnaissance in DSRV *Alvin* of a "fluvial-like" meander system in Wilmington Canyon and slump features in South Wilmington Canyon: Geology, v. 10, no. 1, p. 31–36.

Stubblefield, W. L., Kersey, D. G., and McGrail, D. W., 1983, Development of middle continental shelf ridges, New Jersey: American Association of Petroleum Geologists Bulletin, v. 67, p. 817–831.

Swift, D.J.P., 1980, Shoreline periodicities and linear offshore shoals; A discussion: Journal of Geology, v. 88, p. 365–368.

Swift, D.J.P., and Field, M. E., 1981, Evolution of a classic sand ridge field; Maryland sector, North American inner shelf: Sedimentology, v. 80, p. 461–482.

Swift, D.J.P., Duane, D. B., and Pilkey, O. H., eds., 1972, Shelf Sediment Transport; Process and Pattern: Stroudsburg, Pennsylvania, Dowden, Hutchinson, and Ross, Inc., 656 p.

Swift, D.J.P., Duane, D. B., and McKinney, T. F., 1973, Ridge and swale topography of the Middle Atlantic Bight, North America; Secular response to the Holocene hydraulic regime: Marine Geology, v. 15, p. 227–247.

Swift, D.J.P., Nelson, T., McHone, J., Holliday, B., Palmer, H., and Shideler, G., 1977, Holocene evolution of the inner shelf off southern Virginia: Journal of Sedimentary Petrology, v. 47, p. 1454–1474.

Swift, D.J.P., Freeland, G. L., and Young, R. A., 1979, Time and space distribution of megaripples and associated bedforms, Middle Atlantic Bight, North American Atlantic Shelf: Sedimentology, v. 26, p. 389–406.

Swift, D.J.P., Moir, R., and Freeland, G. L., 1980, Quaternary rivers on the New Jersey shelf; Relation of seafloor to buried valleys: Geology, v. 8, p. 276–280.

Tucholke, B. E., and Laine, E. P., 1982, Neogene and Quaternary development of the lower continental rise off the central U.S. east coast, *in* Watkins, J. S., and Drake, C. L., eds., Studies in Continental Margin Geology: American Association of Petroleum Geologists Memoir 34, p. 295–305.

Twichell, D. C., 1983a, Geology of the head of Lydonia Canyon, U.S. Atlantic outer continental shelf: Marine Geology, v. 54, p. 91–108.

——, 1983b, Bedform distribution and inferred sand transport on Georges Bank, U.S. Atlantic continental shelf: Sedimentology, v. 30, p. 695–710.

Twichell, D. C., and Roberts, D. G., 1982, Morphology, distribution, and development of submarine canyons on the United States Atlantic continental slope between Hudson and Baltimore Canyons: Geology, v. 10, no. 8, p. 408–412.

Twichell, D. C., Knebel, H. J., and Folger, D. W., 1977, Delaware River; Evidence for its former extension to Wilmington Submarine Canyon: Science,

v. 195, p. 483–485.

Twichell, D. C., McClennen, C. E., and Butman, B., 1981, Morphology and processes associated with the accumulation of the fine-grained sediment deposit on the southern New England shelf: Journal of Sedimentary Petrology, v. 51, p. 269–280.

Twichell, D. C., Grimes, C. B., Jones, R. S., and Able, K. W., 1985, The role of erosion by fish in shaping topography around Hudson Submarine Canyon: Journal of Sedimentary Petrology, v. 55, p. 712–719.

Uchupi, E., 1968, Atlantic continental shelf and slope of the United States; Physiography: U.S. Geological Survey Professional Paper 529C, p. C1–C30.

Uchupi, E., 1970, Atlantic continental shelf and slope of the United States; Shallow structure: U.S. Geological Survey Professional Paper 529-I, p. I1–I44.

Vail, P. R., Mitchum, R. M., Jr., Todd, R. G., Widmier, J. M., Thompson, S., III, Sangree, J. B., Bubb, J. N., and Hattlelid, W. G., 1977, Seismic stratigraphy and global changes of sea level: American Association of Petroleum Geologists Memoir 26, p. 49–50.

Vassallo, K., Jacobi, R. D., and Shor, A. N., 1984, Echo character, microphysiography, and geologic hazards, *in* Ewing, J. I., and Rabinowitz, P. D., eds., Eastern North American Continental Margin and Adjacent Ocean Floor: Woods Hole, Massachusetts, Ocean Margin Drilling Program, Regional Atlas Series, Atlas 4, Marine Science International.

Veatch, A. C., and Smith, P. A., 1939, Atlantic submarine valleys of the United States and the Congo submarine valley: Geological Society of America Special Paper 7, 101 p.

Vincent, C. E., Swift, D.J.P., and Hilliard, B., 1981, Sediment transport in the New York Bight, North American Atlantic Shelf: Marine Geology, v. 42, p. 369–398.

MANUSCRIPT ACCEPTED BY THE SOCIETY JULY 18, 1986

ACKNOWLEDGMENTS

We appreciate the thoughtful reviews provided by W.B.F. Ryan, G. Mountain, R. Flood, H. Knebel and D. Twichell. Reviews of earlier drafts by B. McGregor, R. Bennett, J. Farre, and S. Lewis were also helpful. Support for A. Shor for studies of Hudson Fan was provided by NSF grant OCE82-08826. This is contribution #4000 of Lamont-Doherty Geological Observatory.

Chapter 3

U.S. Atlantic continental margin; Structural and tectonic framework

Kim D. Klitgord and Deborah R. Hutchinson
U.S. Geological Survey, Woods Hole, Massachusetts 02543
Hans Schouten
Woods Hole Oceanographic Institution, Woods Hole, Massachusetts 02543

INTRODUCTION

The U.S. Atlantic continental margin (Plate 2C, Fig. 1) is one of the best studied passive (Atlantic-type) continental margins. As the U.S. margin evolved through compressional, extensional (rifting), and vertical (subsidence) tectonic phases, a distinctive set of deep crustal structures, basement structures, and sedimentary features was created. A series of Paleozoic orogenies created the Appalachian mountains and formed large thrust faults, terrane boundaries, and magmatic structures that would control the locus of crustal fracturing during the subsequent extensional phase. During the rifting phase, as the African plate started to break away from the North American plate, the margin was an active plate boundary. Only during the subsidence phase is an Atlantic-type margin actually a passive continental margin (i.e., not an active plate boundary). The very thick sedimentary wedge that overlies crystalline basement on the margin limits our knowledge of basement and underlying crustal structures, but it nevertheless provides a detailed record of the subsidence phase of margin evolution. Distinctive magnetic-anomaly and gravity-anomaly lineations, discontinuities, and characteristic patterns also developed during the evolution of this margin. These geophysical anoomalies provide the basis for inferring crustal structures and crustal types in lieu of more direct seismic or sample information.

Models for the evolution of Atlantic-type continental margins have been developed from studies of other extensional-tectonic regimes, such as the southern Australian margin (Falvey and Mutter, 1981), Biscay margin of France (Montadert and others, 1979; LePichon and Barbier, 1987), the North Sea (Sclater and Christie, 1980), the Red Sea (Cochran and others, 1986), the Ligurian Tethys (Lemoine and others, 1986), and the Basin and Range Province (Anderson and others, 1983; Wernicke, 1985). Applications of subsidence models to the U.S. Atlantic margin (Watts, 1982; Steckler and Watts, 1982; Sawyer and others, 1983; Steckler and others, this volume) have enhanced our ability to interpret the sedimentary record of the postrift evolution of the margin. Models for the mechanical de-

formation of the crust during rifting and early subsidence phases are just now being more closely examined (McKenzie, 1978; Bally, 1981; LePichon and Sibuet, 1981; Falvey and Middleton, 1981; Beaumont and others, 1982; Foucher and others, 1982; Hellinger and Sclater, 1983; Wernicke, 1985; Lister and others, 1986; LePichon and Barbier, 1987).

A considerable volume of information now exists from which inferences can be made on the crustal structure of the U.S. Atlantic margin. Summaries of structural studies, as they apply to individual segments of the margin, can be found in subsequent chapters. In this chapter we examine the geophysical framework, general structural framework, deep crustal framework, and plate-tectonic framework of the Atlantic margin. We outline the important structural elements and geophysical lineaments that have been identified and mapped along the margin. We conclude with a discussion of margin evolution that includes this structural information as well as models of rift margin evolution.

Age Constraints

The development of a tectonic evolution history for the U.S. margin incorporates the distribution of structural features and age constraints on when they formed. Isotopic age data for the Mesozoic are restricted to the dikes and sills onshore (McHone and Butler, 1984; Sutter, 1985), one sill within Georges Bank Basin and igneous rocks at the base of the South Florida Basin, the New England Seamount Chain and White Mountain Volcanic Series, and a few samples of oceanic basement in the deep sea. Biochronology provides the primary basis for developing a geochronology for the synrift and postrift sedimentation. A summary for the synrift and early postrift age control is given by Manspeizer and Cousminer (this volume), and the postrift age controls are given by Poag and Valentine (this volume). A sharp discrepancy occurs, however, in the late synrift and early postrift age assignments for the offshore basins in these two summaries. Based on biostratigraphic and seismic-stratigraphic studies tied into petro-

Klitgord, K. D., Hutchinson, D. R., and Schouten, H., 1988, U.S. Atlantic continental margin; Structural and tectonic framework, *in* Sheridan, R. E., and Grow, J. A., eds., The Atlantic Continental Margin, U.S.: Geological Society of America, The Geology of North America, v. I-2.

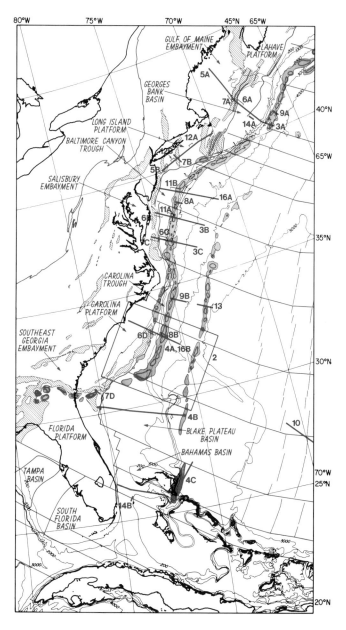

Figure 1. Location map for seismic-reflection profiles and regional area shown in subsequent figures. Numbers refer to figure numbers. Basic Mesozoic structural and geophysical features on the U.S. Atlantic margin are shown, including marginal basins, platforms, and embayments; rift-basins; the basement hinge zone; Brunswick Magnetic Anomaly (BMA); East Coast Magnetic Anomaly (ECMA); Blake Spur Magnetic Anomaly (BSMA); and selected Mesozoic sea-floor spreading lineations and fracture zones (Klitgord and Schouten, 1986). See Plate 2C for addition information.

leum industry wells along the U.S. and Canadian Atlantic margins, there is a reasonable consensus on the age assignments down to the Mohican Formation (Bathonian-Bajocian?; early Middle Jurassic) (Given, 1977; Barss and others, 1979; Hurtubise and others, 1987; Manspeizer and Cousminer, this volume; Poag and Valentine, this volume). The discrepancy arises from the poor preservation and/or sparse occurrence of palynomorph, foraminiferal, and ostracod assemblages of early Middle Jurassic and Early Jurassic age within the lower strata (Iroquois Formation) penetrated by wells in Georges Bank Basin (COST G-1 and COST G-2) and Scotian Basin (e.g., Shell Mohican I-100). The equivalent deep strata have not been penetrated in any of the other U.S. Atlantic marginal basins.

Initial age assignment for the deep Iroquois Formation in Georges Bank Basin by Poag and others was latest Triassic(?) (Rhaetian) to Early Jurassic(?), based on seismic-sequence analyses correlated with sea-level fluctuation curves and on correlation with equivalent stratigraphic units in the Scotian Basin (Poag, 1982). No biostratigraphic age control was available. Subsequent re-examination of the COST G-1 well by Ascoli (1983) yielded Bathonian assemblages for the Mohican Formation and Bajocian assemblages at the base of the Iroquois Formation. Re-examination of the COST G-2 well by Cousminer and others (1984) revealed reworked Late Triassic palynomorphs above the Mohican Formation. They interpreted marine Late Triassic (Norian) palynomorphs near the top of the Iroquois Formation, however, as an in situ assemblage (Manspeizer and Cousminer, this volume). They also point out that the Iroquois Formation interval in the Mohican I-100 well (dated as late Early Jurassic by Barss and others, 1979) does not contain age-restricted palynomorphs to support this age interpretation. The salt at the base of the Mohican I-100 well has been dated, however, as early Early Jurassic (Hettangian-Sinemurian) (Barss and others, 1979). There is disagreement not about the age-assemblage of the palynomorphs but about their biostratigraphic interpretation. Because Manspeizer and Cousminer (this volume) assert that the Late Triassic palynomorphs in the upper Iroquois Formation are in situ, they conclude that, except for perhaps a thin Lower Jurassic section (less than 300 m), the entire lower strata (below 4,344 m at COST G-2) in Georges Bank Basin (and therefore probably in the other marginal basins) are Late Triassic in age. They place the breakup (postrift) unconformity at the top of the Iroquois Formation; this is similar to the interpretation of Given (1977) for the Scotian Basin, but it disagrees with that of Haworth and Keen (1979) and Schlee and Jansa (1981), who place the unconformity at the base of the Iroquois Formation.

In contrast, Poag and Valentine (this volume) and Schlee and Klitgord (this volume) assert that the Iroquois Formation is mostly Middle Jurassic and perhaps late Early Jurassic in age. This conclusion is based on seismic stratigraphic studies combined with the latest biochronology of Ascoli (1983). The Georges Bank cross section of Manspeizer and Cousminer (this volume, Fig. 2) suggests a stratigraphic relationship between the COST G-1 and COST G-2 wells for these re-examined zones that is not correct. Merging of this well information onto the composite seismic-reflection profile between these two wells (Amato and Simonis, 1980, Plate 4) establishes that the Bajocian unit in the COST G-1 well is stratigraphically below (nearly 1,500 m) the unit from which Cousminer and others (1984) reported the Late Triassic palynomorphs. Poag and Valentine (this volume) and we conclude, therefore, that all the Late Triassic assemblages in the COST G-2 well cited by Manspeizer and Cousminer (this volume) must have been reworked. These strata containing Late Triassic palynomorphs onlap the hinge zone region of Georges Bank Basin at the subcrop of a rift basin, Franklin Basin. Franklin Basin would be a likely source for the reworked Triassic palynomorphs, although its sedimentary units have never been drilled. Poag and Valentine (this volume) and Schlee and Klitgord (this volume) placed the breakup (postrift) unconformity just above the salt layer, making the oldest postrift sedimentary units in the lowest 500 m of section in the COST G-2 well late Early Jurassic in age.

In this chapter we use the age assignments of Poag and Valentine (this volume). As in Georges Bank Basin, we have indicated thick Middle Jurassic sections in all the other basins. The lowest strata of the postrift sections are assumed to be late Early Jurassic, but as mentioned above, age control for this later assumption is minimal.

GEOPHYSICAL FRAMEWORK

Identification and mapping of structural features on the U.S. Atlantic Coastal Plain and Continental Margin are based primarily on seismic-reflection, seismic-refraction, gravity, and magnetic studies integrated with a sparse set of rock samples from drill holes or dredges. Each of these data sets, by itself, is of limited use because it provides information about only a few specific properties of the rocks under the seafloor. When they are taken as complementary data sets along with reasonable geologic inferences, their combined information is a powerful data base for examining buried geologic features. A summary of geophysical data bases on the U.S. Atlantic margin is given by Sheridan and others (Ch. 9, this volume). A brief outline is given below of how these data sets are used in this paper.

Seismic-reflection data

Seismic images from multichannel seismic-reflection data (Fig. 1, Plate 1B) are used to identify structural features within sedimentary units, crystalline basement, and deeper in the crust. With the data set used here, acoustic basement usually coincides with crystalline basement (base of the sedimentary fill) in regions of less than 6 to 8 km of sediment thickness. Within the thicker sedimentary sequences (>10 km) of the marginal sedimentary basins, acoustic basement usually is above crystalline basement, and deeper structures must be inferred from the overlying sedimentary structures or from the magnetic, gravity, and seismic-refraction data. Moho is usually observed only on profiles over slightly thinned continental crust or oceanic crust.

Wide-angle seismic data

The most useful crustal information has been derived from combinations of seismic reflection with wide-angle reflection and seismic-refraction studies across the margin (Plate 1C) (LASE Study Group, 1986; Hutchinson and others, 1987a, 1987b; Trehu and others, 1987; Sheridan and others, Ch. 9, this volume; Diebold and others, this volume). Both horizontal and vertical variations in velocity structure are essential for understanding crustal structures imaged by seismic-reflection techniques. Variations in velocity structure across the margin are related to changes in depth of distinctive crustal layers and compositional changes within these layers. These velocity variations provide an important parameter for evaluating crustal modification across and along the margin, in response to the various tectonic processes.

Estimates of average crustal velocity structure are still being evaluated. Sheridan and others (1979) estimated average velocity structures seaward of the marginal basins by dividing the region into two crustal zones, whereas Trehu and others (1987) divided the region into three crustal zones plus the rift-stage crust in marginal basins. Anomalous velocity structures associated with fracture zones (White and others, 1984; Purdy and Ewing, 1986) also can bias velocity averages for normal oceanic crust. Trehu and others' more detailed segregation of data and elimination of data sets clearly associated with fracture zones result in a regrouping of data sets published by Sheridan and others (1979, Figs. 9 and 10) that makes a significant change in the assigned "average" velocity structures. In particular, Sheridan and others (1979) had concluded that the velocity structure of the inner and outer magnetic quiet zones contained a normal oceanic crustal layer 3a (6.3–6.9 km/sec) underlain by a higher-velocity lower crustal layer 3b (7.1–7.6 km/sec). The regrouping places the stations with thicker crust and a higher-velocity lower crustal layer (layer 3b) almost exclusively within rift-stage crust and in the zone with poorly defined oceanic basement just seaward of the East Coast Magnetic Anomaly (Sheridan and others, this volume, Ch. 9, Fig. 2). Stations with thinner crust and a higher velocity lower (upper mantle?) layer are located near fracture zones (White and others, 1984).

Gravity anomaly data

Gravity data in conjunction with seismic-velocity data are used to identify and map deep crustal structures in regions lacking seismic-refraction control. The structural units of the margin are characterized by different free-air gravity signatures (Grow and others, 1979, Fig. 2; Sheridan and others, this volume, Ch. 9, Fig. 8). There are broad gravity lows over oceanic crust near the margin and over the broad marginal embayments, whereas the marginal platforms are marked by shorter-wavelength, higher-amplitude gravity highs and lows. The most prominent free-air gravity anomaly is located over the present shelf edge, but it also overlies one of the most important deep crustal boundaries on the margin—the continent-ocean boundary. It is a large (75 to 150 mgal peak to trough; 75 to 125 km wide) free-air gravity anomaly and has a large variation in amplitude and shape along the margin. Seafloor relief causes a significant part of this anomaly; the remainder is associated with horizontal density variations in the crust and relief on the crust-mantle boundary (Karner and Watts, 1982). Significant horizontal variations in density are found within the carbonate-bank paleoshelf-edge complex beneath the outer shelf and slope, but they are poorly constrained (Hutchinson and others, 1983). A summary of the gravity field and models on the U.S. Atlantic margin is given by Karner and Watts (1982) and Sheridan and others (Ch. 9, this volume).

Magnetic anomaly data

Four important evaluation aspects of the magnetic data (Plate 2A) are used in our study: (1) depth-to-basement estimations, (2) basement rock and crustal-type estimations, (3) identification of crustal boundary markers, and (4) identification of discontinuities (e.g., faults) within crustal units.

Depth-to-basement estimations. Magnetic depth-to-basement estimation provides a quantitative value for the depth to the base of sedimentary fill. Detailed discussions of the depth-to-basement technique and results are given by Klitgord and Behrendt (1979) and Behrendt and Klitgord (1980). It is important to recognize the uncertainties associated with such estimates that are dependent upon magnetic source assumptions. Depth-to-basement estimations are used primarily to interpolate known structures between seismic lines. Many of the small, linear magnetic anomalies within and seaward of the marginal basins are associated with small changes in basement depth at faults (Fig. 2, Plate 2A) (Klitgord and Behrendt, 1979; Klitgord and Grow, 1980; Klitgord and others, 1982).

Basement rock and crustal type estimations. Source analyses of magnetic signatures have been applied to the platform areas landward of the marginal basins (e.g., Higgins and Zietz, 1983; Hinze, 1985). For example, broad-wavelength magnetic and gravity anomaly lows and small-amplitude short-wavelength magnetic anomalies are associated with calc-alkaline granitic batholiths. Comparisons between geologic features and magnetic-field patterns within the exposed Appalachian mountain belt have demonstrated that long, narrow sinuous belts of magnetic lineations are associated with metamorphic zones of thrust faults and shear zones (Hatcher and others, 1977; Robinson and others, 1985). In contrast, a more chaotic pattern of short magnetic lineations is associated with foliated rock units. In the Coastal Plain and offshore regions, where there is a paucity of drill holes into crystalline rock, it is at least possible to identify large crustal units by using these signatures. A variety of signal-processing techniques can be applied to the magnetic (e.g., Plate 2B) and gravity data in order to quantify some of these inferences (Behrendt and Grim, 1985; Hinze, 1985).

Crustal boundary magnetic lineaments. Three large-amplitude magnetic-anomaly lineations on the U.S. Atlantic margin (Fig. 2, Plates 2A and 2C) have been recognized for

several decades as obviously important markers of geologic boundaries: the East Coast Magnetic Anomaly (ECMA), Brunswick Magnetic Anomaly (BMA), and Blake Spur Magnetic Anomaly (BSMA) (Taylor and others, 1968; Vogt, 1973; Rabinowitz, 1974; Klitgord and Behrendt, 1979) (see Figs. 3 and 4). The origins of these anomalies are still being debated, but their usefulness as crustal markers is generally accepted.

The *East Coast Magnetic Anomaly* is a large magnetic high (300 to 500 nT peak-to-trough) that is located near the continental shelf edge from offshore Nova Scotia to South Carolina. The ECMA terminates at its intersection with the Blake Spur Fracture Zone where the anomaly becomes a large peak-trough pair that straddles the fracture zone (peak to north). The general shape and amplitude of the ECMA vary in a segmented manner along the margin. A large-scale segmentation (300 to 400 km) corresponds to the lengths of individual basin or platform segments of the margin, and a small-scale segmentation (50 to 100 km) is associated with fracture-zone spacings. Contoured magnetic-anomaly data show that the small-scale variation in the ECMA is a series of elongate highs separated by narrow zones of lower-amplitude anomalies (Fig. 2) (Klitgord and Behrendt, 1979; Klitgord and Schouten, 1987a, 1987b). These narrow zones of lower-amplitude anomalies are where the oceanic fracture zones intersect the margin.

The *Brunswick Magnetic Anomaly* is a paired anomaly peak and trough located about 100 km landward of the ECMA between Cape Hatteras and offshore Georgia. Although the character of this peak-trough pair changes along the length of the BMA, the magnetic trough is always west or northwest of the magnetic peak. Between the Norfolk Fracture Zone (36°N) and the Carolina Fracture Zone (33.5°N), the magnetic signature is dominated by the peak. Between the Carolina Fracture Zone and the Blake Spur Fracture Zone (31.5°N), the magnetic trough dominates the signature, and the peak is almost nonexistent. This dominance is manifest in the lack of a large magnetic low separating the BMA from the ECMA, and the two anomalies seem to merge. Southwest of the Blake Spur Fracture Zone, the BMA swings westward across the continental shelf, and the magnetic peak increases markedly in amplitude. Both the magnetic high and magnetic low

---

Figure 2. Magnetic anomaly contour map and selected basement and sedimentary features on swath across the U.S. Atlantic margin in the Carolina Trough region. The Brunswick Magnetic Anomaly (BMA), East Coast Magnetic Anomaly (ECMA), and Blake Spur Magnetic Anomaly (BSMA) are indicated. Basement structures include normal faults (lines with hachures on down thrown side), the upper edge of the basement hinge zone (H), fracture zones (FZ-heavy dash lines), and basement peaks (triangles). Sedimentary features include the location of the paleoshelf edge at the end of the Jurassic (heavy line with slanted hachures) and salt diapirs (dots). The landward limit of well-defined oceanic crust coincides with the J_3 scarp. Magnetic contour intervals are 10nT offshore and 50nT onshore. Modified from Klitgord and Schouten (1987a). See Figure 1 for location.

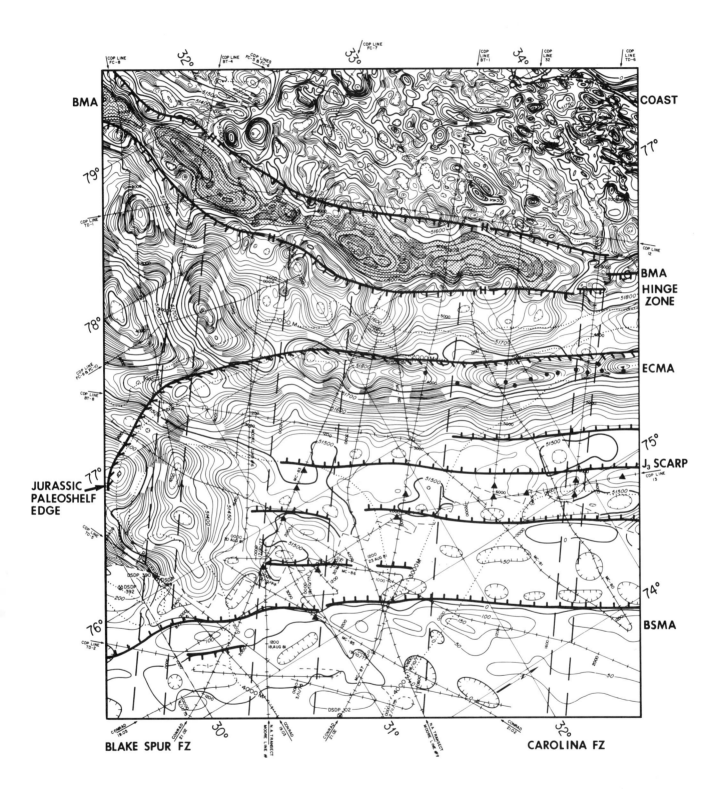

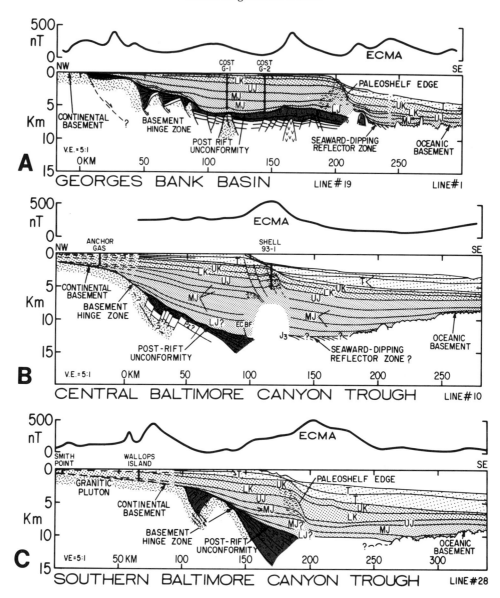

Figure 3. Cross sections of northern U.S. Atlantic margin displaying sedimentary and basement structures. Magnetic anomaly profile is shown across the top of each section. Major magnetic lineaments and structural elements discussed in text are labeled. Ages of sedimentary units are indicated (T = Tertiary; UK = Upper Cretaceous; LK = Lower Cretaceous; UJ = Upper Jurassic; MJ = Middle Jurassic; LJ = Lower Jurassic). Synrift deposits are shaded in red. A. Cross section of Georges Bank Basin along USGS seismic line 19. Magnetic depth estimates to basement are also shown. From Klitgord and Hutchinson (1985, Fig. 9.6A). B. Cross section of central Baltimore Canyon Trough along USGS seismic line 10. C. Cross section of southern Baltimore Canyon Trough along USGS seismic line 28. From Klitgord and Hutchinson (1985, Fig. 9.7). See Figure 1 for locations.

equally dominate the magnetic signature of the BMA until just after it crosses onshore near Brunswick, Georgia. The onshore segment of the BMA (called the Altamaha Anomaly by Higgins and Zietz [1983]) is dominated by a magnetic high, with a broader-wavelength, lower-amplitude magnetic low than is associated with the offshore segment.

The *Blake Spur Magnetic Anomaly* is located 150 to 250 km seaward of the ECMA and within oceanic crust. It resembles the ECMA in that it is a series of elongate magnetic highs separated by narrow zones of reduced anomaly amplitude (Fig. 2). The BSMA is a magnetic high that can be traced clearly from the Long Island Fracture Zone (about 38°N, 69°W) to the Bahamas

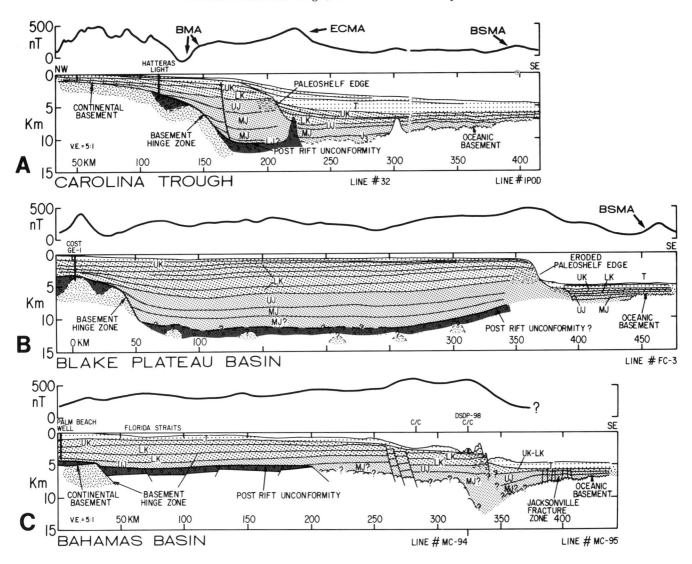

Figure 4. Cross sections of southern U.S. Atlantic margin. A. Cross section of Carolina Trough along USGS seismic line 32. Modified from Grow and others (1983, Fig. 4). B. Cross section of Blake Plateau Basin along USGS seismic line FC-3. Modified from Dillon and others (1979, Fig. 5). C. Cross section of Bahamas Basin along Lamont-Doherty Geological Observatory (LDGO) seismic lines MC-94 and MC-95 projected along trend of 110°. Modified from Sheridan and others (1981a, Fig. 14). See Figure 1 for location and Figure 3 for explanation.

Fracture Zone, with major disruptions or offsets at the Carolina, Blake Spur, and Jacksonville Fracture Zones (Klitgord and Schouten, 1986). It is a much lower-amplitude magnetic anomaly (50 to 100 nT) than the ECMA or BMA. The BSMA is located just east of a distinctive scarp in basement, and seismic-reflection and refraction data indicate that it is flanked on both sides by typical oceanic crust (Sheridan and others, 1979; Klitgord and Grow, 1980, Trehu and others, 1987).

Magnetic lineation discontinuities. Numerous disruptions are present in the magnetic-lineation patterns on the sea-

ward side of the margin. Within the Mesozoic seafloor-spreading magnetic lineations east of the BSMA, these disruptions are associated with oceanic fracture zones, and their locations can be predicted by flow lines calculated from poles of motion for the Mesozoic evolution of the central Atlantic (Schouten and Klitgord, 1982; Klitgord and Schouten, 1986). A consistent set of these discontinuities can be traced across the magnetic quiet zone from seafloor-spreading lineation M-25 to the ECMA (Fig. 2). This pattern of discontinuities was used by Klitgord and Schouten (1986) to calculate the poles of rotation for the Middle to Late

USGS LINE IA GULF OF MAINE

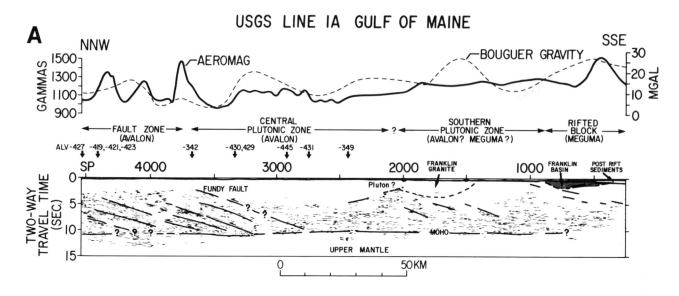

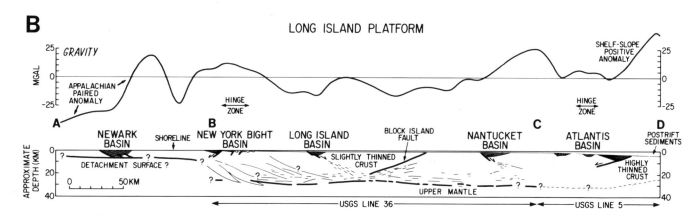

Figure 5. Cross sections of the northern U.S. Atlantic margin indicating major crustal structures and reflective character found landward of the basement hinge zone. A. Cross section of the Gulf of Maine based on the USGS deep-crustal seismic line 1-1a. Magnetic and Bouguer gravity anomaly profiles and major crustal zones are shown. From Hutchinson and others (1987b, Fig. 11). B. Cross section of the Long Island Platform based on the COCORP New England line onshore (Ando and others, 1983) and USGS seismic lines 36 and 5 offshore. Bouguer gravity anomaly profile is shown. From Hutchinson and Klitgord (1987, Fig. 4). See Figure 1 for location and Figure 3 for explanation.

Jurassic evolution of the central Atlantic and to tie the Middle Jurassic evolution for the Gulf of Mexico with the evolution of the central Atlantic.

A key point in the above discussion is the *consistent* match of discontinuities with a *predicted* trend. Examination of the magnetic-anomaly contour maps for any region will quickly suggest that numerous trends are plausible. Within the extensional regime of spreading centers in the deep ocean basins, the theory of plate tectonics predicts that a single pole of motion should describe the motion between two plates along the entire length of their common extensional boundary at any given time. Therefore the trace of a fracture zone (a flowline) and its associated discon-

tinuities in the magnetic-anomaly pattern should be predictable within oceanic crust of any particular age. Along the same two-plate boundary, a plausible trend in magnetic discontinuities for one section of oceanic crust along the margin must be consistent (as a flow line) with trends in the same age crust on other parts of the margin. The set of fracture zones mapped by Klitgord and Schouten (1986, 1987a, 1987b) from anomaly M-25 to the BSMA to the ECMA (Fig. 2) is based on estimates of the most consistent set of trends that would fit the magnetic discontinuity and basement relief records of fracture zones from the Grand Banks to the Bahamas. Notice in Figure 2 how these predicted fracture-zone locations intersect both the BSMA and ECMA at

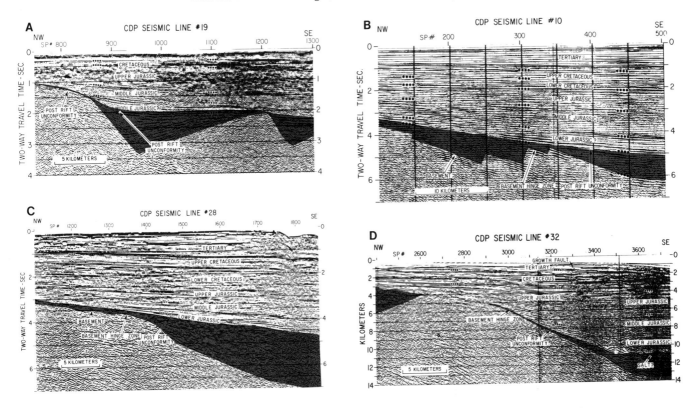

Figure 6. Seismic-reflection records showing examples of the postrift unconformity (PRU) and rift basins within or just seaward of the hinge zone. A. USGS seismic line 19 across the Georges Bank hinge zone. From Klitgord and Hutchinson (1985, Fig. 9.6B). B. USGS seismic line 10 across the central Baltimore Canyon Trough hinge zone. C. USGS seismic line 28 across southern Baltimore Canyon Trough hinge zone showing thick synrift wedge sequence. D. USGS seismic line 32 across the Carolina Trough hinge zone. See Figure 1 for location and Figure 3 for explanation.

the narrow zones of reduced magnetic amplitude and match the discontinuities and other anomaly trends in the intervening crustal section.

STRUCTURAL FRAMEWORK

The basic structural units of the U.S. Atlantic margin are sediment-filled marginal basins, marginal platforms, and marginal embayments (Fig. 1, Plate 2C) that are buried beneath the Coastal Plain and continental shelf, slope, and upper rise (Plate 1A) (Sheridan, 1974; Klitgord and Behrendt, 1979; Uchupi, 1983; Uchupi and others, 1983a, 1983b; Emery and Uchupi, 1984). Deep-sea sedimentary basins developed on oceanic crust along the seaward edge of the marginal basins. Within these basic structural units are common structural elements (e.g., hinge zones, rift basins, postrift unconformity) (Figs. 3–5) that can be mapped along the margin. These elements provide the vertical and horizontal boundaries that are used to segment the margin into major structural units and crustal layers. We also use them to help characterize the internal structure of different marginal units. They provide the primary constraints for models of the tectonic

evolution and formation of the margin. Distinctive variations along the margin in the individual structural characteristics of these elements are important indicators of the tectonic phases which influenced their formation.

Margin structural elements: Sedimentary and Basement Features

Basement Hinge zone. A basement hinge zone (Figs. 3 and 4) forms the landward edge of the marginal sedimentary basins. It is a zone where basement deepens rapidly in a seaward direction from about 2 to 4 km depth to over 8 km depth. Structures associated with the hinge zone vary along the margin and include half-graben structures with seaward-dipping border faults (Fig. 6A), tilted blocks bounded by landward-dipping faults (Fig. 6B), and thick sedimentary(?)-wedge sequences (Fig. 6C; Dillon and Popenoe, this volume, Fig. 4). These fault-bounded structures represent small crustal fragments that probably developed during the rifting processes. The rotated block structures under Georges Bank (Fig. 6A) closely resemble faulted structures on the Irish margin (de Graciansky and others, 1985, Fig. 2), North Biscay

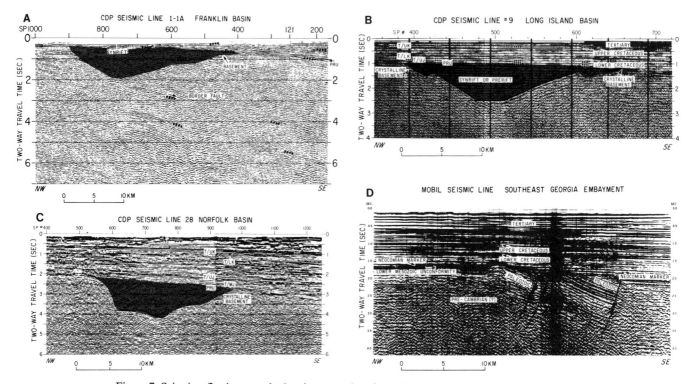

Figure 7. Seismic-reflection records showing examples of postrift unconformity (PRU) and rift basins just landward of basement hinge zone. A. USGS deep-crustal seismic line 1-1a across Franklin Basin landward of Georges Bank Basin. From Hutchinson and others (1987b, Fig. 8). B. USGS seismic line 9 across the Long Island Basin on the Long Island Platform. From Hutchinson and others (1986b, Fig. 3). C. USGS seismic line 28 across Norfolk Basin landward of southern Baltimore Canyon Trough. From Klitgord and Hutchinson (1985, Fig. 9.8B). D. Seismic line across rift basin just landward of Blake Plateau Basin in Southeast Georgia Embayment. From Crutcher (1983). See Figure 1 for location and Figure 3 for explanation.

margin (Montadert and others, 1979, Fig. 4), and western Ligurian Tethys margin (Lemoine and others, 1986, Fig. 11). The thick sedimentary wedges are seaward of the large fault block structures and resemble a broad set of fan deposits (Figs. 3B, 3C). On line 28 (Fig. 6C), the wedge has the acoustic character of a single packet of seaward-dipping reflectors with a sharp base. Internal reflectors are subparallel to this base and are continuous across the width of the wedge. We have interpreted it as a clastic wedge (probably volcaniclastic) that developed across the hinge zone during the rifting phase.

Rift basins. Rift basins are found along the entire margin (Plate 2C, Figs. 6 and 7) and demonstrate the chaotic nature associated with crustal failure in an extensional tectonic regime. The sedimentary fill has not been sampled in any of the U.S. offshore rift basins, but the fanning pattern of deposition, as exemplified in Figure 6A, does indicate a significant synrift component (Klitgord and Hutchinson, 1985; Hutchinson and others, 1986b; Hutchinson and Klitgord, 1987). The character of these basins changes markedly across the hinge zone. Those basins within or seaward of the hinge zone are much more completely preserved (e.g., Fig. 6A) than the basins landward of the hinge

zone, which are deeply eroded, with only lower sections remaining (Fig. 7). These latter basins closely resemble the rift basins found in the onshore region (Robinson and Froelich, 1985; Manspeizer and Cousminer, this volume). Distinctive magnetic anomaly lows are associated with most of the rift basins landward of the hinge zone (Popenoe and Zietz, 1977; Daniels and others, 1983; Klitgord and Hutchinson, 1985), and these anomalies have been used to map the areal extent of the buried rift basins.

Carbonate-bank, paleoshelf-edge complex. A carbonate-bank, paleoshelf-edge complex (Fig. 8) developed along the seaward edge of the marginal basins and separated shelf sediment deposition from slope-rise deposition (Schlee and others, 1979; Schlee and Grow, 1980; Poag, 1985; Poag and Valentine, this volume). The Middle Jurassic to mid-Cretaceous shelf edge was a carbonate-bank complex that built upward, prograded seaward, or was eroded back at various times. The upper surface of this carbonate-bank complex is a very strong acoustic reflector, whereas the underlying carbonate complex absorbs (attenuates or scatters) energy and masks the deeper seismic structures on most seismic-reflection profiles across it. The initial location of this shelf edge is important for evaluating basin geometries at the time

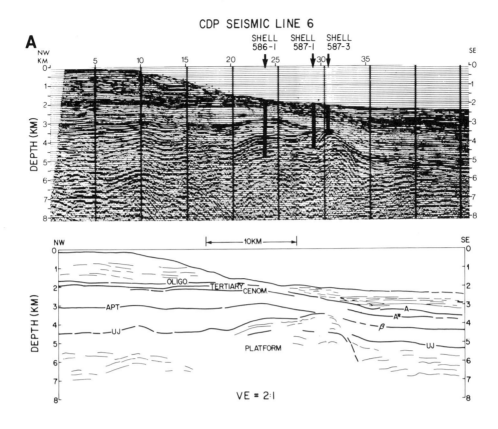

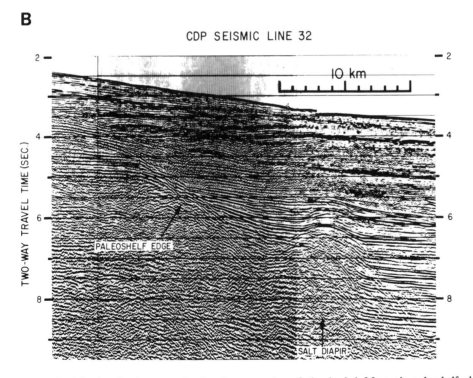

Figure 8. Seismic-reflection records showing examples of the buried Mesozoic paleoshelf-edge carbonate-bank complex. A. USGS seismic line 6 across the northern Baltimore Canyon Trough. Locations of exploration wells by Shell Oil Co. are indicated. From Schlee and Klitgord (1986, Fig. 10). B. USGS seismic line 32 across the southern Carolina Trough. From Dillon and others (1983b, Fig. 10). See Figure 1 for location and Figure 3 for explanation.

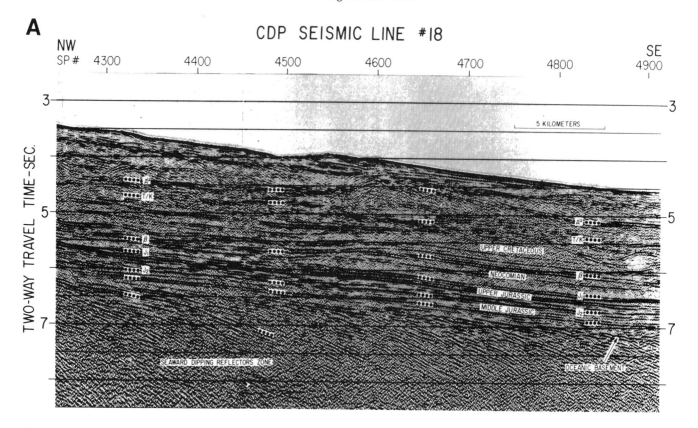

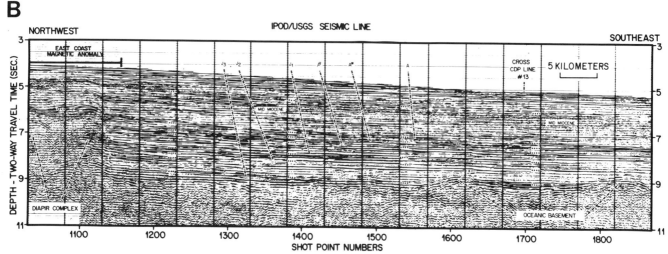

Figure 9. Seismic-reflection records across crust just seaward of ECMA. A. USGS seismic line 18 showing seaward-dipping reflector zone east of Georges Bank Basin. B. Seismic line IPOD/USGS showing block-faulted basement, acoustic reflector J_3, and salt diapirs just east of Carolina Trough. From Klitgord and Grow (1980, Fig. 5B). See Figure 1 for location and Figure 3 for explanation.

when the margin was in transition from an active rift zone to a passive margin. Its subsequent position controlled the distribution of sediment loading and related subsidence through time, an important constraint on the postrift depositional history.

 Salt diapirs. Salt diapirs (Figs. 8B and 9B) form a second important set of markers of the seaward edges of the northern three marginal basins. Nearly every salt diapir found on the U.S. Atlantic margin is located beneath the axis of the ECMA (Grow, 1980; Dillon and others, 1983b; Klitgord and Schlee, 1987). These diapirs are just seaward of the Jurassic paleoshelf edge

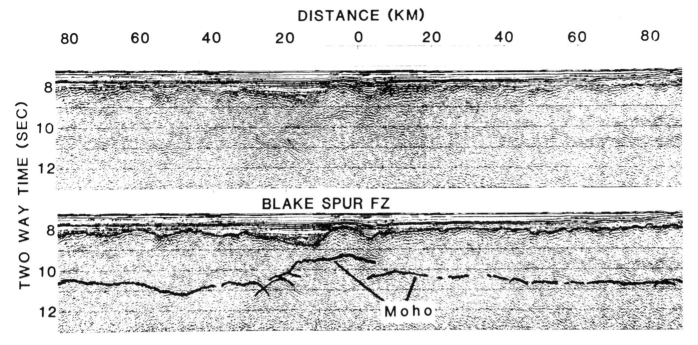

Figure 10. Seismic-reflection record showing basement and Moho structure on seismic line NAT-15 across Blake Spur Fracture Zone. From Mutter and others (1984, Fig. 2). See Figure 1 for location.

along the sections of the margin where the shelf edge grew vertically. They penetrate the carbonate bank complex in the regions where the shelf edge prograded seaward. Other examples of salt diapirs are shown in Grow and others (this volume, Fig. 11) and Dillon and others (1983b, Fig. 14).

Seaward-dipping seismic reflectors. Seaward-dipping seismic reflectors (Fig. 9A) characterize basement in a 25-km-wide zone just seaward of the salt diapirs offshore of Georges Bank, north of the New England Seamounts. The acoustic character of this dipping-reflector zone is distinctly different from the packet of seaward-dipping reflectors associated with volcaniclastic wedges between the basement hinge zone and the ECMA (Fig. 6C). Individual reflectors can be traced only part way across the zone. They all merge upward into a conformable unit that blankets the zone; the reflectors disappear at depth, and there is no distinctive base to the zone. Seaward-dipping reflectors can be seen on seismic profiles just seaward of the ECMA at a few places along the rest of the U.S. margin, but most of the profiles have the very reflective sedimentary horizon J_3 (Fig. 9B; Klitgord and Grow, 1980) as acoustic basement in this region. Similar zones of seaward-dipping reflectors have been found at the very edge of oceanic crust on other continental margins (Hinz, 1981; Mutter, 1985). The crust containing similar dipping reflectors has been drilled on the Norwegian and Rockall margins; it is composed primarily of basaltic flows and dikes, interbedded with small amounts of sediment (Roberts and others, 1985; Eldholm and others, 1986). The origin of this zone has been attributed to the earliest formation of oceanic crust or final formation of rift-stage crust (Mutter, 1985).

Seamounts. Seamounts and intrusive igneous bodies are scattered along the margin. The New England Seamount Chain (Uchupi and others, 1970) represents the bulk of these features and intersects the margin near the southern edge of Georges Bank Basin. Three intrusive bodies found within Georges Bank Basin (Schlee and Klitgord, this volume) and the White Mountain volcanic series onshore (McHone and Butler, 1984) may mark the landward extension of this seamount chain. The only other clearly intrusive body on the margin is the Great Stone Dome in the northern Baltimore Canyon Trough (Grow and others, this volume). Middle Jurassic and Early Cretaceous igneous sills have been drilled in Georges Bank Basin (Hurtubise and others, 1987), and a Middle Jurassic sill drilled near Charleston has been traced offshore across the southern Carolina Platform (Dillon and others, 1983a).

Fracture zones. Fracture zones are prominent structural and geophysical features in the crust seaward of the ECMA. Detailed examinations of the fracture zones in the western Atlantic (Klitgord and Grow, 1980; Detrick and Purdy, 1980; Schouten and Klitgord, 1982; Tucholke and others, 1982; Mutter and others, 1984) have identified distinctive seismic basement and Moho structures (Fig. 10). They form long ridges and troughs that are approximately orthogonal to the margin. Hudson Ridge (Plate 2C; Grow and others, this volume, Fig. 13) represents one of the best examples of such a structure close to the margin.

Margin structural elements: Crustal faults

Large faults within the postrift sedimentary section and small faults in the synrift section are found seaward of the hinge zone. Most of the large faults in the postrift section are growth faults associated with salt flow (Fig. 6D) (Dillon and others, 1983b) or with differential compaction and subsidence in the vicinity of the paleoshelf edge (Fig. 11A, Plate 3B) (Poag, 1987, Fig. 20). The deepest and most landward of these large faults within the Baltimore Canyon Trough is along the landward edge of the ECMA (Fig. 11B, Plate 2C). This fault has been called the East Coast Boundary Fault (ECBF) and interpreted as a basement fault (Alsop and Talwani, 1984; LASE Study Group, 1986). It is imaged at a depth of about 13 km (6 seconds two-way travel time) as a curved set of diffraction reflectors (Fig. 11A) that migrate back to a sharp edge (Alsop and Talwani, 1984, Fig. 1). A closer examination, however, shows that the ECBF is not a basement feature; it is the juncture of two sets of closely spaced faults—one dipping west into synrift deposits and one dipping east within postrift deposits (Fig. 11B). Only the upper part of the west-dipping fault set is imaged. These faults penetrate a zone of seaward-dipping sedimentary units similar to those shown on USGS seismic line 10 (Figs. 3B and 6B) and seismic line 28 (Figs. 3C and 6C). The ECBF is the most landward of the large east-dipping faults, and it cuts the postrift sedimentary section seaward of shot point 2050 on line 25 (Fig. 11B). This fault appears to sole into the postrift unconformity surface near shot point 2200. The large offsets (~1 km) in the Bajocian(?) unit and postrift unconformity across this fault zone indicate significant synrift and early postrift differential subsidence across it. No clear basement fault is actually imaged on USGS seismic line 25, but there may be a large drop in the underlying basement structure at this point. A large Early Cretaceous intrusion, the Great Stone Dome (Grow, 1980; Grow and others, this volume), was intruded at this fault boundary.

Faults that penetrate deep into the crust are the most poorly recorded structural features on the margin. Deep penetration faults on the U.S. margin have been imaged only landward of the hinge zone. Only the upper ends of the faults have been identified on seismic profiles across the marginal basins and the adjacent deep ocean basin. Basement and subbasement structures across the hinge zone and in the marginal basins are poorly imaged on the seismic data south of the Baltimore Canyon Trough.

Deep crustal-penetrating faults.

Deep crustal-penetrating faults have been imaged offshore on the Carolina Platform (Behrendt and others, 1983; Behrendt, 1986), on the Long Island Platform (Figs. 5B, 12A) (Hutchinson and others, 1985, 1986b; Phinney, 1986), and within the Gulf of Maine (Fig. 5A) (Hutchinson and others, 1987a, 1987b). Some of these faults can be traced to where they intersect Moho (e.g., the Block Island Fault), whereas other faults disappear at mid-crustal depths or are antithetic to deeper crustal-penetrating faults (e.g., New Shoreham Fault, Fig. 12A). These faults are often characterized by a narrow zone of parallel, strong reflectors that separates two crustal blocks

containing different reflective patterns (see Hutchinson and others, 1985; Ratcliffe and others, 1986). They have been interpreted as being part of the system of large faults imaged onshore in the eastern part of the Appalachian orogenic belt (Cook and others, 1981; Ando and others, 1983; Hamilton and others, 1983; Stewart and others, 1986; Nelson and others, 1987). Some of these faults bound rift basins and may be Paleozoic thrust faults reactivated during Mesozoic rifting (Cook and others, 1981; Klitgord and others, 1983; Hutchinson and others, 1985, 1986b; Ratcliffe and others, 1986; Behrendt, 1986; Swanson, 1986). These sets of faults and associated rift basins are similar to the sets of faults found around the British Isles that are associated with Paleozoic thrust faults reactivated by Mesozoic rifting (Brewer and others, 1983; BIRPS and ECORS, 1986).

Basement-cutting faults within the hinge zone are the edges of small crustal fragments (10 to 20 km wide) and generally are associated with the formation of rift basins. The character of this faulting differs significantly along the margin, but there are long segments (~200 km) in which a consistent pattern is found. In eastern Georges Bank Basin, these faults are east-dipping features that form the border faults of half-graben structures (Fig. 6A). In contrast, the faults along the western edge of the Baltimore Canyon Trough (Fig. 6B) are landward dipping, and they appear to bound a chaotic set of fault blocks covered by a layer of synrift deposits. Some of these faults penetrate upward a small distance into postrift sedimentary units (Fig. 11B). The depth extent of these faults and base of these crustal fragments are not imaged on most of the seismic data. Therefore, we cannot determine the geometries by which these faults merge into deeper surfaces within the crust.

Landward-facing scarps.

A series of landward-facing scarps, interpreted as faults, has been mapped in the oceanic crust between the BSMA and the ECMA (Klitgord and Grow, 1980; Sheridan and others, 1981a). The most seaward scarp parallels and lies along the landward edge of the BSMA (Figs. 2 and 13). At least three more faults can be traced for long distances within oceanic crust landward of the BSMA and roughly parallel to the margin (Fig. 2). The most landward of these scarps, which we call the J_3 Scarp, coincides with the position where the highly reflective horizon J_3 onlaps basement and marks the landward limit of well-defined oceanic crust (Klitgord and Schouten, 1986; 1987a, 1987b). Between the J_3 Scarp and ECMA, the J_3 reflector masks the deeper structures on most seismic profiles south of the New England Seamount Chain (e.g., Fig. 9B).

Crustal Layer Boundaries

Crustal modification can occur at different levels in the crust as well as at different spatial locations. Horizontal layering of the crust provides one of the characteristic signatures by which crustal types can be distinguished. We have defined five primary crustal surfaces: (1) top of the sedimentary deposits that blanket the margin (topographic elevations of the Coastal Plain and bathymetric depths of the seafloor—see Plate 1A), (2) the boundary

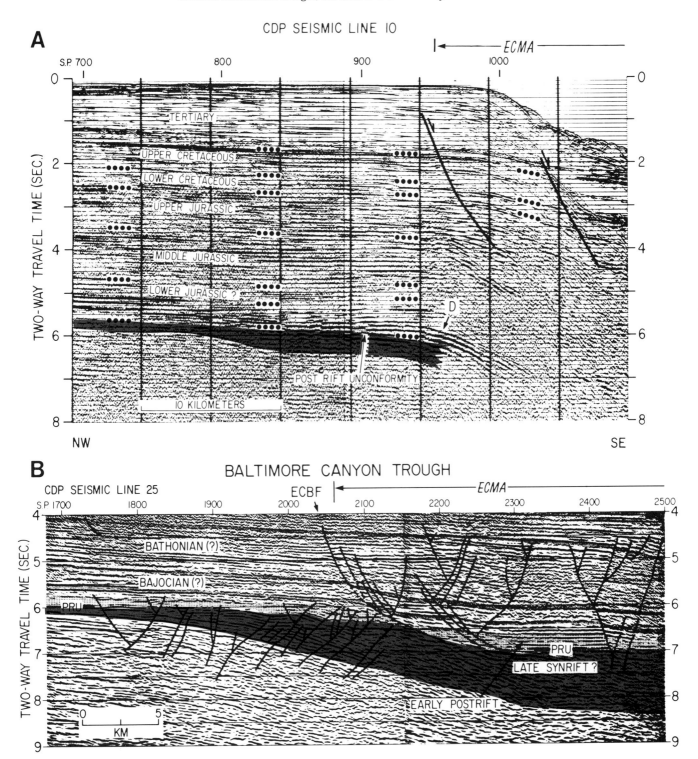

Figure 11. Seismic-reflection records across crust just landward of ECMA showing the East Coast Boundary Fault (ECBF). A. USGS seismic line 10 (unmigrated) across the central Baltimore Canyon Trough. Diffraction pattern (D) is associated with the ECBF between shotpoints 950 and 1000 and a depth of 6–7 sec. B. USGS seismic line 25 (migrated) across northern Baltimore Canyon Trough. Note that the ECBF is at the juncture of two series of faults. The primary discontinuity is actually at the boundary between small west-dipping faults and the large east-dipping faults. Sedimentary units beneath the postrift unconformity are shaded red. Upper bounds of the Bathonian(?) and Bajocian(?) units are also shaded to highlight the structural patterns. See Figure 1 for location and Figure 3 for explanation.

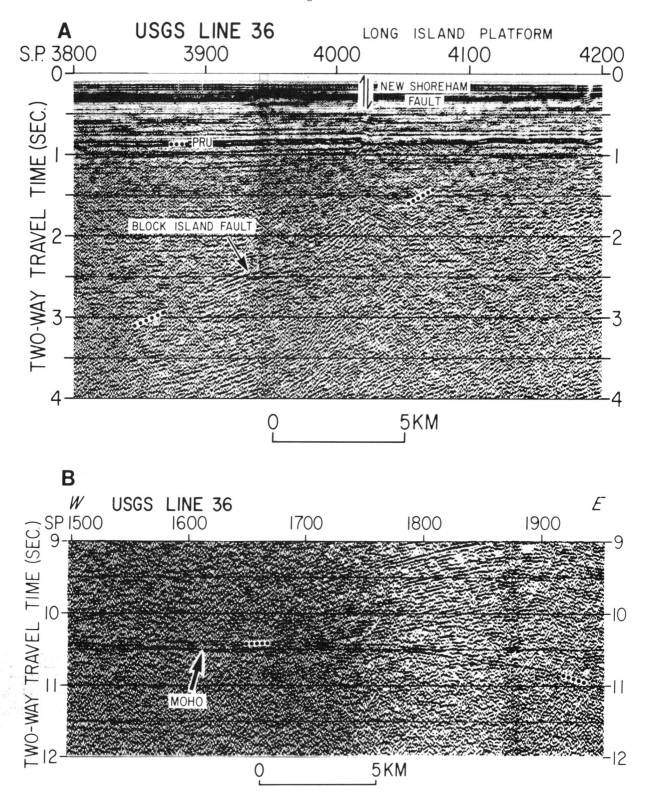

Figure 12. Seismic-reflection records from USGS seismic line 36 showing deep crustal reflectors on the Long Island Platform. A. Upper end of Block Island Fault. From Hutchinson and others (1985, Fig. 3). B. Moho under the west end of the Long Island Platform. From Hutchinson and others (1986a, Fig. 4). See Figure 1 for location and Figure 3 for explanation.

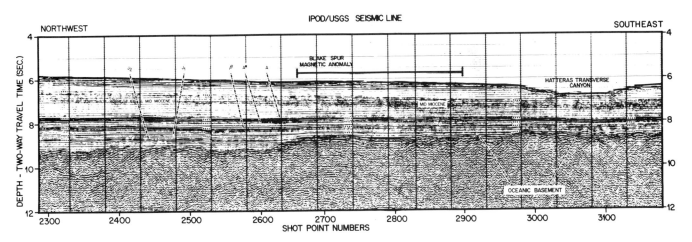

Figure 13. Seismic-reflection records showing examples of oceanic basement and fault scarps in this basement on seismic line IPOD/USGS across the Blake Spur Magnetic Anomaly. From Klitgord and Grow (1980, Fig. 5A). See Figure 1 for location.

between postrift and synrift-prerift rocks, (3) base of synrift deposits (generally the top of metamorphic or crystalline basement), (4) a mid-crustal boundary (7.2 km/sec layer), and (5) base of the crust.

Postrift unconformity. The base of postrift sedimentary rocks, the postrift unconformity (PRU), is one of the most important erosional-depositional sedimentary surface on the continental margin. This boundary represents the deepest mappable surface across the margin (Plate 2D), and it separates sedimentary units deposited in very different tectonic environments (Figs. 6, 7, 11, 14). The PRU is a time-transgressive surface that merges with younger unconformities across and landward of the hinge zone. Within the northern four marginal basins, the PRU is associated with the end of rifting (late Early Jurassic or early Middle Jurassic) and corresponds to the breakup unconformity of Falvey (1974); it is a disconformity that separates latest Lower Jurassic and Middle Jurassic marine carbonate and anhydrite deposits from early Lower Jurassic and Upper Triassic evaporites (including salt), nonmarine clastics, and volcanics (Fig. 14A) (Schlee and Fritsch, 1983; Poag and Valentine, this volume; Schlee and Klitgord, this volume). This places the PRU at the base of the Iroquois Formation instead of at the top as proposed by Manspeizer and Cousminer (this volume). In the Bahamas and South Florida basins, the PRU (Fig. 14B) separates Upper Jurassic carbonate platform deposits from Lower Jurassic(?) or Triassic(?) clastic deposits (Sheridan and others, this volume). Postrift depositional units onlap this surface across the hinge zone. The oldest sedimentary units found above the PRU are Middle Jurassic in age near the hinge zone and Late Jurassic just landward of the rift basins along the upper end of the hinge zone. Within the embayments, the PRU is covered only by Cretaceous and younger sediments.

The PRU marks an important change in the margin evolu-

tion record. Synrift deposits accumulated when the active plate boundary was within the marginal basin. This would have been a period of active crustal stretching, block faulting, graben formation, igneous intrusion, and large amounts of differential vertical motion across the margin. These deposits are often faulted, rotated, and otherwise deformed. Postrift sediments, by contrast, accumulated on the passive margin after the plate boundary moved seaward of the margin. The primary tectonic activity on the margin was then one of regional subsidence caused by lithospheric cooling and sediment loading over a broad region. This caused successive sedimentary units to be relatively conformable with each other. We have identified the PRU in the regions of the hinge zone where it is an erosional unconformity and then traced it across the marginal basin, where it is a more conformable surface. Differential subsidence across the margin has created subsequent relief on the postrift unconformity in the form of small faults or dips in the surface (Figs. 6A, 11B; Klitgord and others, 1982, Fig. 79). A clear distinction between late rift and early postrift faults is not always possible (e.g., Fig. 11B), because postrift sediment loading and compaction can re-activate older faults or create new ones.

Crystalline basement. Crystalline basement is the upper surface of crystalline rock, either igneous or metamorphic, and the lower surface of the sedimentary fill. Landward of the hinge zone on the U.S. margin, crystalline basement corresponds to the PRU except within the rift basins (Klitgord and others, 1982; Klitgord and others, 1983; Hutchinson and others, 1986b). Near the hinge zone, crystalline basement is most often imaged as the top of fault blocks (Figs. 6A, 6B, 7). Although acoustic basement often coincides with the boundary surface between Mesozoic sedimentary rocks and the underlying Paleozoic or older metamorphic and igneous rocks, Paleozoic (prerift) sedimentary rocks probably lie beneath acoustic basement in some areas (Hutchin-

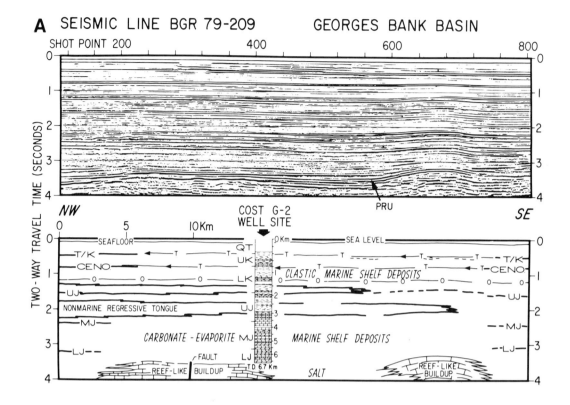

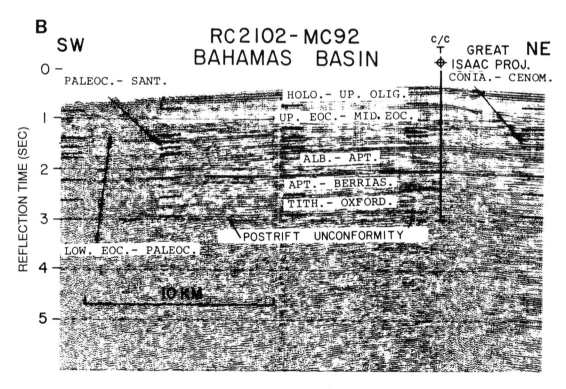

Figure 14. Seismic-reflection records showing examples of the postrift unconformity and underlying rift sequences within marginal sedimentary basins. A. Seismic line BGR79-209 across Georges Bank Basin. From Schlee and Fritsch (1983, Fig. 4). B. Seismic line MC-92 of Lamont-Doherty Geological Observatory across the central Bahamas Basin. Modified from Sheridan and others (1981a, Fig. 8). See Figure 1 for location and Figure 3 for explanation.

son and others, 1986b). On sediment-starved margins, such as those of western Europe (de Graciansky and others, 1985), the base of synrift deposits is known to be prerift sedimentary rock, and the same probably is true on the U.S. Atlantic margin. Hence, the conformable layering seen in the tilted fault blocks on seismic line 10 (Fig. 6B) and seismic line 9 (Fig. 7B) may be caused by Paleozoic sedimentary units on top of continental crustal fragments. The tilted conformable layers within the updip section of the rift basin landward of the Blake Plateau Basin (Fig. 7D) are Paleozoic sedimentary strata that have been drilled for petroleum exploration (Crutcher, 1983).

Seaward of the hinge zone in the marginal basins, crystalline basement is rarely imaged with seismic-reflection data. Although this may change with improved acquisition and processing techniques, at present it is not possible to map crystalline basement or estimate synrift sediment thicknesses within the marginal basins on the U.S. margin. Even seismic-refraction measurements are limited because of the similarity in velocity structure of deeply buried marine carbonate and clastic sedimentary rocks and the underlying crystalline rock (e.g., Keen and Cordsen, 1981).

Oceanic basement. Seaward of the marginal basins, crystalline basement is oceanic basement, the upper surface of oceanic crustal layer 2. Oceanic basement in the western Atlantic is often characterized by a series of hyperbolic echos (Fig. 13). On most of the multichannel seismic-reflection profiles, there is a landward limit of well-defined oceanic crust (based on its acoustic character) at the J_3 Scarp (Klitgord and Grow, 1980; Klitgord and Schouten, 1987a, 1987b). Off Georges Bank between this landward limit and the Jurassic paleoshelf edge, basement is the zone of seaward-dipping reflectors; we have mapped the upper surface of this zone on Plate 2D. South of the New England Seamount Chain, the J_3 reflector usually forms acoustic basement and is the surface mapped on Plate 2D. Lack of a distinctive acoustic reflector below horizon J_3 makes it difficult to identify crystalline basement between the J_3 Scarp and the ECMA, even using seismic-refraction data.

Moho. Moho, at the base of the crust, is the deepest of the crustal surfaces. Recent improvements in seismic-reflection techniques have resulted in the imaging of this surface beneath continental crust. This surface has now been traced beneath rifted-continental crust in the Gulf of Maine Embayment (Hutchinson and others, 1987a, 1987b), on the Long Island Platform (Fig. 12B) (Hutchinson and others, 1986a), and on the Carolina Platform (Behrendt and others, 1983). Even with the very best of available data and techniques, this surface has not been imaged beneath the U.S. Atlantic marginal basins (LASE Study Group, 1986), but it has been imaged on profiles across sediment-starved margins (Montadert and others, 1979; LePichon and Barbier, 1987). This surface is clearly identifiable beneath oceanic crust (Fig. 10) (Grow and Markl, 1977; Mutter and others, 1985).

Intermediate crustal layers just above Moho are now being recognized on the margin, and their surfaces may be important crustal boundaries. For example, Mutter and others (1985) have identified a lower oceanic crustal layer that may be related to magma-chamber geometries. There is a 7.2 km/sec seismic-velocity layer just above Moho beneath the marginal basins and the landward edge of oceanic crust (Sheridan and others, 1979; Keen and Cordsen, 1981; LASE Study Group, 1986; Trehu and others, 1986; White and others, 1987; Diebold and others, this volume) that may be a characteristic product of rift-stage or initial oceanic magmatism.

Margin structural units

In terms of basic structural units, the U.S. Atlantic margin can be divided into six marginal basins, four platforms, and three embayments (Fig. 1, Plate 2C). *Marginal basins* evolved along the margin between continental and oceanic crust. They are the successor basins to rift basins and contain more than 6 to 8 km of sediment. *Marginal basement platforms and embayments* are the crustal regions modified slightly by rifting and located immediately landward of the marginal basins. Their crust has been faulted, intruded, and perhaps thinned by rifting, but otherwise it resembles continental crust. They are covered by up to 5 km of postrift sedimentary rock.

We have made a distinction between platforms and embayments because differences between their sedimentary cover, structural elements, and geophysical signatures are sufficient to infer that they have undergone a slightly different tectonic evolution. Marginal platforms protrude partially across the margin and form the along-margin (shear zone) and landward (extensional) boundaries of marginal basins. Marginal embayments are recessed into the margin and form only the landward edge of marginal basins. The embayments are covered by a thin layer of synrift deposits (Brown and others, 1972; Daniels and others, 1983; Klitgord and others, 1983). Synrift sediments on the platforms are found only within narrow grabens. The embayments also contain more extensive onlapping wedges of Cretaceous and Cenozoic sediments than the platforms. Both the Salisbury Embayment and Southeast Georgia Embayment were identified on the basis of thickening Upper Cretaceous units (Maher, 1971; Hansen, 1978; Olsson and others, this volume; Gohn, this volume). The platforms are cut by numerous deep-penetration crustal faults and contain many of the clearly identified Mesozoic rift basins. Deep faults and rift basins rim the embayments, but their central regions appear to be devoid of structures on seismic data. These central regions are characterized by magnetic and gravity anomalies of broader wavelength and lower amplitude than the anomalies found over the platforms (Klitgord and others, 1983). This feature served as the main criterion upon which we have outlined the embayments in Plate 2C.

Marginal basins. The six marginal basins along the U.S. Atlantic margin are indicated in Figure 1, and cross sections of the northern five basins are presented in Figures 3 and 4. Georges Bank Basin, Baltimore Canyon Trough, Carolina Trough, and Blake Plateau Basin contain 8 to 13 km of sedimentary fill (Klitgord and Behrendt, 1979; Grow and Sheridan, 1981). The Bahamas Basin and the South Florida Basin overlie the boundary between the Atlantic and Gulf of Mexico continental margins and contain only 6 or 7 km of sedimentary fill (Sheridan and others, 1981a, 1981b; Klitgord and others, 1984). The Scotian Basin on the Canadian margin is similar to the four northern basins (Haworth and Keen, 1979).

Georges Bank Basin is the most northern of the U.S. Atlantic marginal basins (Austin and others, 1980; Schlee and Klitgord, this volume). It is bounded to the northeast by a broad shear-zone edge of the LaHave Platform and to the northwest by extensional structures along the edge of the Gulf of Maine Embayment and Long Island Platform. The basin narrows to the southwest and ends near 40°N, 70°W where the Atlantis Fracture Zone intersects the margin. A major change in hinge-zone basement structure occurs at the Nantucket Fracture Zone (Plate 2C); seaward-dipping faults bound rift basins to the northeast, whereas landward-dipping faults are found to the southwest.

In cross section (Fig. 3A), the basin provides an ideal representation of the general characteristic elements of a U.S. Atlantic marginal basin. Northeast of the Nantucket Fracture Zone, the landward edge is a hinge zone formed by a series of half-graben structures that have seaward-dipping border faults (Fig. 6A). There is a small rift graben (Franklin Basin) just landward of the hinge zone adjacent to the Gulf of Maine (Fig. 7A). The Atlantis Rift Basin has a landward-dipping border fault and is located astride the hinge zone between Georges Bank Basin and the Long Island Platform (Hutchinson and others, 1986b). A prominent carbonate-bank shelf-edge complex forms the seaward edge of the basin. In this area, the shelf edge appears to have been constructed straight upward and is located at the landward edge of the ECMA. There are salt diapirs just seaward of the paleoshelf edge and directly beneath the ECMA. The zone of these diapirs broadens to the north, where they form the "sedimentary ridge" province along the seaward edge of the Scotian Basin (Haworth and Keen, 1979; Uchupi and Austin, 1979). Near Northeast Channel, a few diapirs also intrude the marginal basin between the hinge zone and paleoshelf edge. A 25-km-wide zone of seaward-dipping reflectors (Fig. 9A) forms acoustic basement between the salt diapir zone and oceanic basement to the east of Georges Bank Basin. The broadened diapir zone along the east edge of the Scotian Basin is located on the northern extension of the seaward-dipping reflector zone.

Baltimore Canyon Trough is the deepest marginal basin and the one in which most of the petroleum exploration has been concentrated for the U.S. Atlantic margin (Grow and others, this volume). It is separated by strike-slip shear zones from the Long Island Platform to the northeast and the Carolina Platform to the southwest. These two shear zones are coincident with the Long Island and Norfolk fracture zones. The northwestern edge of the basin borders the Salisbury Embayment and is a steeply dipping hinge zone. The basin has a distinctive segmented structure that includes a major change in basin width and basement depth at its intersection with the Delaware Bay Fracture Zone and at the fracture zone just north of the Norfolk Basin; it narrows and becomes shallower to the south.

In cross section (Figs. 3B, 3C; Grow and others, this volume, Fig. 5A), Baltimore Canyon Trough has basic structural features that are distinctly different from those of Georges Bank Basin. In the northern and central Baltimore Canyon Trough, basement at the hinge zone drops rapidly to a depth of more than 7 km, where there is a series of tilted fault blocks with landward-dipping border faults (Fig. 6B). The top of the hinge zone is near the coast line and is not crossed by most of our seismic lines. A set of shallow-penetration reflection profiles at the mouth of Delaware Bay (Benson and others, 1986) indicates a chaotic fault structure at the shallow end of the hinge zone (Benson, 1984). A broad rift basin with no distinctive border fault developed over several tilted blocks at the deeper end of the hinge zone in the northern and central sections (Fig. 6B). In contrast, a single large rift basin (Norfolk Basin) with an east-dipping border fault developed at the upper end of the hinge zone in the southern section (Fig. 7C).

A seaward-thickening wedge of synrift deposits developed across the marginal basin (e.g., shot points 500–1000 on seismic line 10, Fig. 3B, and shot points 1300–2300 on seismic line 28, Fig. 3C). This seaward-thickening wedge masks deeper structures, and it is cut by landward-dipping faults west of the East Coast boundary fault system (Fig. 11B). The PRU forms the upper surface of this sedimentary wedge and shallows progressively from 10 to 13 km in the north to 5 to 10 km in the south (Plate 2D). Within the southern segment of the Baltimore Canyon Trough, the top of the wedge (Figs. 3C and 6C) has shallowed to 5 km, but it deepens (to 10 km) more rapidly to the east than in the northern and central zones. Hinz (1981) has interpreted this wedge as the same type of crustal feature that is found off the Norwegian margin—the result of a late-rifting phase or an early phase of oceanic crust generation with massive igneous flows. As discussed previously, however, the wedge on seismic line 28 (Fig. 6C) has a distinctly different acoustic pattern from the seaward-dipping reflectors found just offshore Georges Bank seaward of the ECMA (Fig. 9A).

The Jurassic paleoshelf edge in the northern Baltimore Canyon Trough (Fig. 8A) prograded seaward more than 40 km and masks a broad region of deeper structures beneath it (Grow, 1981). Salt diapirs that intrude far enough upward into the sediment pile to be imaged are all located beneath the ECMA (Grow, 1980; Grow and others, this volume). The Jurassic paleoshelf-edge in the southern region appears to have been constructed straight upward and is located at the landward edge of the ECMA. No salt diapirs have been found along the seaward edge of the southern Baltimore Canyon Trough. Crystalline basement is obscured just seaward of the paleoshelf edge along the entire basin by the very reflective sedimentary surface, horizon J_3.

Carolina Trough is the narrowest of the marginal basins (Dillon and Popenoe, this volume). It is bounded at its northern end by the Norfolk Fracture Zone and at the southern end by the Blake Spur Fracture Zone. The western edge of the basin borders the Carolina Platform.

In cross section (Fig. 4A), the Carolina Trough resembles the northern Baltimore Canyon Trough. There is a rapid drop in basement at the hinge zone (Fig. 6D), there are no obvious half-graben structures within the hinge zone, and a possible rift basin is located just landward of the hinge zone. The PRU forms acoustic basement for the entire basin from the upper end of the hinge zone to the Jurassic paleoshelf edge. Just seaward of the hinge zone, the PRU is a flat surface that Dillon and others (1983b) have postulated to coincide with a salt layer. A large growth fault, which penetrates nearly the entire postrift section, soles onto the PRU and may be caused by salt withdrawal to the east. The carbonate-bank paleoshelf edge in this region (Fig. 8B) was constructed straight upward and perhaps eroded back slightly during the Jurassic and Early Cretaceous. It is located along the landward edge of the ECMA. The Late Cretaceous and Cenozoic shelf edge was a clastic wedge that retreated nearly 100 km landward during the Cenozoic. Numerous salt diapirs have been mapped along the seaward front of the carbonate-bank paleoshelf edge, directly beneath the axis of the ECMA (Dillon and others, 1983b). The deep reflective surface, horizon J_3, obscures basement just east of the ECMA on most of the seismic lines.

Deep structures are poorly defined landward of the hinge zone. There appears to be a rift basin, the Brunswick Graben, just landward of the hinge zone on line 32 (Hutchinson and others, 1983, Figure 4), but graben structures are not found on other lines across this part of the margin. A Middle Jurassic basalt layer covers much of the southern Carolina Platform out to the hinge zone (Dillon and others, 1983a; Klitgord and others, 1983) and masks deeper structures.

Blake Plateau Basin is the widest of the marginal basins (Dillon and Popenoe, this volume). It is separated from the Carolina Trough by the Blake Spur Fracture Zone and from the Bahamas Basin by the Jacksonville Fracture Zone. The northwestern edge of the basin is the hinge zone that separates it from the Southeast Georgia Embayment (Dillon and others, 1985, Fig. 6).

In cross section (Fig. 4B), the Blake Plateau Basin is a broad basin bounded by a block-faulted hinge zone on the west and a steep escarpment on the east. Presently, the Blake Plateau is at a water depth of 1 km, is separated from the Florida shelf by a narrow slope above the basement hinge zone, and is separated from the deep sea floor by the Blake Escarpment. The rift basin just landward of the hinge zone (Fig. 7D) is filled with early Mesozoic synrift and Paleozoic prerift sediments (Crutcher, 1983; Dillon and Popenoe, this volume, Fig. 16). The primary border fault is on the southeast side of the rift basin, and the oldest sedimentary units that are truncated by the postrift unconformity are in the north end of the basin. The seaward edge of the Blake Plateau Basin is a massive carbonate escarpment that may have been eroded back a few kilometers during the Jurassic (Dillon and others, 1983a, 1985). There is no ECMA in this region, and the BSMA is located just seaward of the escarpment. Basement is not imaged in the southern part of the basin, but the pre–Upper Jurassic sedimentary sequence thickens rapidly across the Jacksonville Fracture Zone, suggesting a large drop in basement from the Bahamas Basin into the Blake Plateau Basin. Upper Jurassic

and Cretaceous sedimentary units are about the same thicknesses across this boundary, but the thick Middle Jurassic or older section is missing in the Bahamas Basin.

Bahamas Basin is the most southern marginal basin that can be clearly assigned to the U.S. Atlantic margin (Sheridan and others, this volume). The basin is bounded on the northeast by the Jacksonville Fracture Zone and on the southwest by the Bahamas Fracture Zone. There is a much thinner postrift sediment section in the Bahamas Basin than in the basins to the north. The shallow section of the entire basin is constructed of carbonate platform rocks, making the deeper basin structures difficult to image (Sheridan and others, 1981a, 1981b).

In cross section (Fig. 4C), it can be seen that the basin's seaward and landward edges are both very different from those of the basins to the north. Whereas basement in the other basins deepens westward across the seaward edge into the basin, the opposite occurs for the Bahamas Basin. At its landward edge, there is only a small basement drop at the hinge zone. There also is a major difference in postrift sediment fill within the basins; the Bahamas Basin does not contain the thick Middle Jurassic deposits found in the basins to the north. It contains Oxfordian sedimentary units directly over synrift deposits (Fig. 14B).

South Florida Basin straddles the U.S. Atlantic and Gulf of Mexico margins (Klitgord and others, 1984). It is separated from the Bahamas Basin by the Bahamas Fracture Zone and is bounded to the north and northwest by the Florida and West Florida shelf platforms. There are no publicly available seismic-reflection profiles with deep structural information across the basin, but drill-hole, magnetic, and gravity data indicate that it has a deep crustal structure that is probably the same as that of the Bahamas Basin. It also has a thin postrift sedimentary section in which the Middle Jurassic section is missing.

Marginal platforms. There are four marginal platforms along the U.S. Atlantic margin: the LaHave, Long Island, Carolina, and Florida Platforms (Fig. 1, Plate 2C).

LaHave Platform is on the Canadian Atlantic margin, but it does form the northeast end of Georges Bank Basin. It is underlain by Paleozoic crystalline rock and deepens slowly to the southeast and southwest. This region includes a large fragment of the African plate (the Meguma Terrane) that was accreted to the North American plate during the middle Paleozoic and then stranded there by the Mesozoic rifting and seafloor spreading (Williams and Hatcher, 1983; Keppie, 1985). In general geophysical character, it closely resembles the Long Island Platform. The southwestern end of the platform becomes highly fractured and merges into a block-faulted zone called the Yarmouth Arch and Yarmouth Sag (Klitgord and others, 1982). Insufficient data are available to determine if the crustal rocks of the Meguma Terrane in Nova Scotia (Keppie, 1985) continue southwestward across the entire platform and descend beneath Georges Bank Basin. The Scotian Basin formed along the seaward edge of this platform and merges southward into the Georges Bank Basin in the vicinity of Northeast Channel.

Long Island Platform bounds Georges Bank Basin and Baltimore Canyon Trough. Its northern edge may coincide with a major geologic shear zone boundary along the New England coast, whereas its southern and southeastern boundary is the basement hinge zone. The western end of the Long Island Platform includes the Newark Basin. Detailed discussions of the structure and geophysical signature of the Long Island Platform can be found in Hutchinson and others (1985, 1986a, 1986b), Klitgord and Hutchinson (1985), and Phinney (1986). The platform is covered by 2 to 5 km of postrift sedimentary material. Its underlying crust is fragmented by deep, crustal-penetrating faults with rift basins at the upper end of several of these faults (Figs. 5B, 7B, 12A). These faults appear to have been formed as paired structures of large antithetic faults and deep detachment surfaces (Hutchinson and Klitgord, 1987). A thin, layered set of reflectors at a depth of 10.5 sec (two-way travel time) is interpreted as Moho (Fig. 12B). Numerous short-wavelength magnetic anomalies characterize the magnetic field over the platform (Hutchinson and others, 1986b, Fig. 13). Several northeast-trending large-amplitude magnetic highs and broad magnetic lows are associated with the deep crustal-penetrating faults and rift basins, respectively. A major side lobe in the ECMA is located over the southern edge of the platform near 40°N, 70°–72°W

Carolina Platform extends from beneath the Coastal Plain to offshore North and South Carolina (Klitgord and Behrendt, 1979; Hutchinson and others, 1983; Klitgord and others, 1983). It is separated from the Baltimore Canyon Trough and Blake Plateau Basin by the Norfolk and Blake Spur Fracture Zones and is adjacent to the Carolina Trough. The best deep seismic data offshore across this platform are near Charleston, South Carolina, where Behrendt and others (1983) have identified a major deep crustal-penetrating fault (the Helena Banks Fault) and a laminated lower crust, similar to that found on the Long Island Platform. The magnetic field over most of the platform is characterized by large-amplitude (100 to 300 nT), short-wavelength (<20 km) magnetic anomalies (Plate 2A). The exception is the broad magnetic (and gravity) low just offshore Charleston that may be associated with a large granitic batholith (Klitgord and others, 1983). This batholith is bordered on the east by a linear chain of mafic bodies (a north trending set of magnetic highs), on the north by the Helena Banks Fault, on the west by a series of magnetic highs, and on the southeast by a small rift basin at 32°N, 79°W.

Florida Platform formed the southeastern corner of the North American plate during the Jurassic (Klitgord and others, 1984). It is bounded on its northeast side by the BMA and on its southeast by the hinge zone of the Blake Plateau and Bahamas Basins. The Bahamas Fracture Zone and Gulf Rim Fault Zone separate it from the South Florida Basin and the rift basins and arches that underlie the West Florida Shelf. This region is another fragment of old African crust (Suwannee Terrane) that has been left on the North American margin by Mesozoic rifting and sea-floor spreading (Williams and Hatcher, 1983). The western edge of the platform, the Apalachicola Embayment, was fractured by Mesozoic rifting and is part of the South Georgia Rift system. The

magnetic field over the Florida Platform is characterized by high-amplitude, short-wavelength anomalies, similar to those over the other platforms. Detailed summaries of this region are given by Klitgord and others (1983, 1984), Daniels and others (1983), and Sheridan and others (1981a, 1981b).

Marginal embayments. There are three major embayments along the U.S. Atlantic margin: Gulf of Maine, Salisbury, and Southeast Georgia Embayments (Fig. 1, Plate 2C).

Southeast Georgia Embayment crosses southern Georgia and merges with the Apalachicola Embayment in the Florida Panhandle and southwestern Georgia (Maher, 1971; Klitgord and others, 1983, 1984; Daniels and others, 1983; Chowns and Williams, 1983). Its boundaries are best defined by the BMA to the south, the Riddleville Basin fault zone and Charleston block to the west and north, and the large granitic batholith offshore Charleston (see Klitgord and others, 1983, Plate 1; Higgins and Zietz, 1983, Fig. 1). Broad magnetic and gravity lows cover most of the embayment (Klitgord and others, 1983; Long and Dainty, 1985).

The South Georgia Rift of Daniels and others (1983) covers the entire Southeast Georgia Embayment, but the rift basins imaged on seismic profiles and regions of thick (>2 km) synrift sediments are located only around the edges of the embayment. COCORP lines across the Southeast Georgia Embayment (Nelson and others, 1985a, 1987) show that the embayment is bounded on the northwest by a southeast-dipping, crustal-penetrating fault (see COCORP line GA-14, Nelson and others, 1985a) that also forms the border fault for the Riddleville Basin (see COCORP line GA-5, Cook and others, 1981; Long and Dainty, 1985). A rift basin is imaged on COCORP line GA-19 across the magnetic low of the BMA but not on the adjacent lines (GA-13/14 and GA-16). Drill-hole data (Chowns and Williams, 1983) and faint northwest-dipping reflectors on line GA-13/14 are consistent with an elongate rift basin beneath the western onshore part of the BMA low. Line GA-16 crosses the eastern onshore part of the BMA low just west of where Paleozoic volcanic tuffs were sampled beneath about 1.5 km of post-Jurassic sediment (Chowns and Williams, 1983). The central zone of the embayment is more of a broad regional sag in which a thin unit of early Mesozoic sedimentary and igneous material accumulated (Klitgord and others, 1983). Basement samples are sparse but indicate that the region is underlain by granitic or other felsic rocks (Daniels and others, 1983; Williams and Chowns, 1983; Klitgord and others, 1983) at depths less than 2 km. This region coincides with a major Cretaceous depocenter landward of the Blake Plateau Basin (Dillon and others, 1983a, 1985).

Salisbury Embayment is recessed into the U.S. Atlantic margin landward of the central and southern Baltimore Canyon Trough (Hansen, 1978). Its western boundary is best defined by a linear chain of elongate magnetic and gravity highs that may be associated with a Paleozoic suture zone (Lefort and Van der Voo, 1981) and by a set of west-verging thrust sheets (Pratt and others, 1986, 1987). The low-gradient magnetic and gravity signatures of the Salisbury Embayment resemble those of the Southeast

Georgia Embayment (Klitgord and others, 1983). The boundary between the embayment and the Baltimore Canyon Trough is marked by a change in wavelength of the magnetic and gravity fields (shorter wavelengths to the west of the hinge zone). Rift basins are only found around the rim of the embayment. It was a broad depocenter in which a thin layer of Triassic and Cretaceous sediment accumulated (Maher, 1971; Hansen, 1978).

Gulf of Maine Embayment is recessed into the Atlantic margin between Cape Cod and Nova Scotia (Ballard and Uchupi, 1975; Hutchinson and others, 1987a, 1987b). The Gulf of Maine contains several deep crustal-cutting faults, several small rift basins, and at least one major granitic batholith (Fig. 5A). It is the only embayment on the U.S. Atlantic margin not covered by a veneer of Mesozoic sedimentary rock; it may be that much of this material was removed by glacial erosion. The Gulf of Maine Fault Zone forms the western edge of the embayment; it merges onshore to the southwest with the Clinton-Newbury Fault System and to the northeast with the Fundian Fault System. The Gulf of Maine Fault Zone is a series of southeast-dipping surfaces, but it lacks the distinctive pattern of antithetic faults and rift basins found on the Long Island Platform. This indicates that it probably was not a major zone of Mesozoic extension. The Fundy Basin, a Mesozoic rift basin in the Bay of Fundy, does not cross the Gulf of Maine Embayment; it swings southward toward Northeast Channel where it merges with Franklin Basin. Franklin Rift Basin (Fig. 7A), which we have inferred to be an early Mesozoic rift basin (Klitgord and Hutchinson, 1985; Hutchinson and others, 1987b), developed along the southeastern boundary of the embayment with Georges Bank Basin. The southern half of the embayment (41°–42.5°N) is characterized by low-gradient magnetic and gravity fields, similar to the fields in the other embayments. Just landward of the Franklin Rift Basin there is a distinctively coherent low-gradient zone in the magnetic and gravity fields associated with a crustal feature that we have called the Franklin Batholith. This batholith has not been sampled, but its geophysical signature is similar to that of the South Mountain granite batholith (Late Devonian age) that covers much of Nova Scotia (Keppie, 1985).

DEEP CRUSTAL FRAMEWORK

The U.S. continental margin is the boundary between two major crustal units: continental crust and oceanic crust. Often this boundary is discussed as if it is a line on the map: the continent-ocean boundary. In reality it is a broad zone over which crustal modification has created a hybrid crust, usually referred to as transitional or rift-stage crust. The tectonic evolution of a passive margin is partially the evolution of crustal modification, starting with continental crust and eventually creating oceanic crust. Thus, variation of crustal structure across and along the margin is an important record of this evolution.

Margin crustal zones

To summarize the crustal framework of the Atlantic margin,

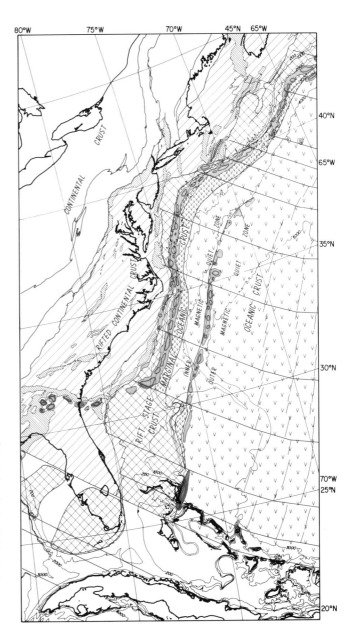

Figure 15. Distribution of crustal types and locations of important crustal boundary markers on the U.S. Atlantic Margin. Prominent geophysical lineaments are the East Coast Magnetic anomaly (ECMA), Brunswick Magnetic Anomaly (BMA), and Blake Spur Magnetic Anomaly (BSMA). Selected other features as in Plate 2C.

we have divided the margin into crustal zones (Fig. 15): continental crust, rifted-continental crust, rift-stage crust, marginal oceanic crust, and oceanic crust. Velocity structures on the U.S. margin for these crustal zones are summarized in Figure 16 and in Sheridan and others (this volume, Ch. 9, Fig. 2).

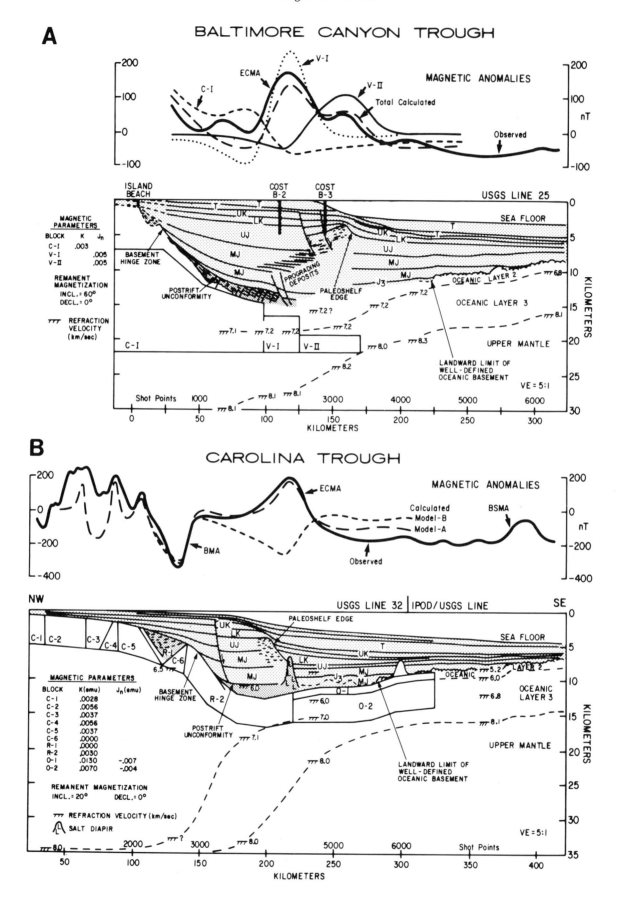

Continental crust. Continental crust on the U.S. margin refers to the thick (>30 km) crust that existed on the North American–African boundary prior to breakup of the megacontinent of Pangea. This crust was far from uniform because some of it had been formed during the Paleozoic by the accretion of terranes and collision of plates (Williams and Hatcher, 1983). Therefore, the continental crust that was to be modified by Mesozoic rifting was quite variable along the margin. West-verging and east-verging thrust sheets, separated by the remnants of magmatic arcs and suture zones, created a variety of crustal structures, which probably failed differently when subjected to extension.

Rifted-continental crust. Rifted-continental crust is crust that has been stretched, faulted, and thinned slightly by rifting but that is otherwise recognizable as continental crust. The marginal platforms and embayments, which contain early Mesozoic rift basins, are of this type of crust. Variability in the velocity structure of continental crust precludes identifying rifted-continental crust based on velocity distributions. The most western deep crustal-penetrating faults associated with Mesozoic rift basins mark the landward limit of this crust. The basement hinge zone marks the eastward limit.

Rift-stage or transitional crust. Rift-stage or transitional crust is continental crust that has been intensely modified by rifting. Changes in nearly all of the parameters that usually characterize continental crust make this a distinctively different type of crust. It is thinner than rifted-continental crust. It most likely contains fragments of the original continental crust from which it evolved, but there also are significant components of new igneous material (Sheridan, 1974; LASE Study Group, 1986; Lister and others, 1986). The block-faulted synrift and prerift structures landward of the ECMA (Figs. 6B, 11B) are the most distinctive evidence of these rifted crustal fragments. This type of crust is found beneath the marginal sedimentary basins and is transitional from continental to oceanic crust. For the seaward boundary of rift-stage crust, we have used the landward edge of the ECMA and the East Coast Boundary Fault. The great variety of sediment and deeper crustal-layer thicknesses and the sparse set of velocity-structure data have precluded establishing characteristic crustal

properties. One characteristic, however, is becoming evident—that of a 7.2 km/sec seismic-velocity layer just above Moho (LASE Study Group, 1986; Trehu and others, 1986; White and others, 1987) (Fig. 16). Velocities actually vary from 7.1 to 7.6 km/sec in this lower crustal layer, and its thickness ranges from at least 9 to 15 km.

Oceanic crust. The most seaward part of the margin is underlain by oceanic crust. This crust, which has a fairly distinctive velocity structure (Purdy and Ewing, 1986), is much thinner than either continental or rift-stage crust. Oceanic layer 2 (4.5–6.0 km/sec) varies in thickness from 1 to 2 km in the area of well-defined oceanic basement of older oceanic crust. Oceanic layer 3 (6.0–7.0 km/sec) is 5 to 6 km thick in zones of well-defined oceanic basement landward and seaward of the BSMA (this is layer 3A of Sheridan and others, 1979). Oceanic basement has a seismic-reflection signature that is usually characterized by hyperbolic echos. Oceanic crust has a segmented character that is reflected in both the morphologic and magnetic signatures (Schouten and Klitgord, 1982; Schouten and others, 1985). Based on all these characteristics, oceanic crust can be traced at least as far west as a position we refer to as the landward limit of well-defined oceanic crust. Most of the continental-rise sediments on the U.S. Atlantic margin were deposited on oceanic crust. This contrasts with sediment-starved margins where slope-rise deposits are found on rift-stage crust (de Graciansky and others, 1985). The oceanic crust east of the ECMA is separated into an inner and outer magnetic quiet zone by the BSMA (Vogt, 1973; Klitgord and Grow, 1980). The inner magnetic quiet zone includes both the marginal oceanic crust and oceanic crust. East of the BSMA is the zone of oceanic crust with the outer magnetic quiet zone and Mesozoic seafloor-spreading magnetic lineations.

Marginal oceanic crust. There may be a fifth crustal type, located along the landward edge of oceanic crust, that we have for simplicity called marginal oceanic crust. It was the first new oceanic crust to form as the tectonic activity evolved from rifting to sea-floor spreading. This crust includes the zones of seaward-dipping reflectors east of the ECMA and the crust directly beneath the ECMA; it may extend as far east as the J_3 Scarp and the landward limit of well-defined oceanic crust. It has a crustal thickness (~10 km) that is slightly greater than that of oceanic crust and includes the 7.2 km/sec layer (oceanic layer 3B of Sheridan and others, 1979) (LASE Study Group, 1986; Trehu and others, 1986, 1987). Oceanic layer 2 is not resolvable in this region of the U.S. margin where basement is not imaged. The 5.6 to 6.1 km/sec layer at ESP-5 on the LASE transect (LASE Study Group, 1986, Fig. 6) includes the sedimentary unit below horizon J_3 plus the landward extension of oceanic layer 2. Moho (top of a 7.8–8.2 km/sec layer) gradually deepens from 13 to 20 km in a westward direction across the magnetic quiet zone and then drops rapidly near the ECMA to a depth of 30 km.

There is still a debate as to whether the seaward-dipping reflector zones were formed as part of the seafloor-spreading process or whether the basaltic layers erupted through and formed on top of continental crust (Hinz, 1981; Mutter, 1985;

Figure 16. Magnetic models for the origin of the East Coast Magnetic Anomaly (ECMA). A. Magnetic model of Alsop and Talwani (1984) along USGS seismic line 25, northern Baltimore Canyon trough. Components of the magnetic field associated with this model are shown. Seismic structure from Grow and others (this volume, Fig. 5A). B. Magnetic model of Hutchinson and others (1983) along USGS Seismic line 32, Carolina trough. Modified from Hutchinson and others (1983, Fig. 12). Continental crustal blocks (C) and rift-stage crustal blocks (R) have induced magnetizations with susceptibilities (K) and magnetic field parameters as indicated. Oceanic crustal blocks (O) and intrusive volcanic blocks (V) have remanent magnetizations (J_n) as indicated. The dashed calculated anomaly in B is assuming all induced magnetizations for the blocks. Superimposed is the deep crustal velocity structure (in km/sec) from the LASE experiment (LASE Study Group, 1986) and other refraction stations (Trehu and others, 1987).

White and others, 1987). The formation of seafloor-spreading magnetic anomalies within the seaward-dipping reflector zone on the Norwegian margin (anomaly 24) (Mutter, 1985) and the southwest African margin (anomalies M4-M9) (Austin and Uchupi, 1982) clearly shows that at least on these margins there was crustal generation by the sea-floor spreading process. Because the elongate, segmented pattern of the ECMA closely matches the segmented pattern of the BSMA (Fig. 2) and of the sea-floor spreading magnetic lineations to the east (Klitgord and Schouten, 1986, 1987a, 1987b), we conclude that the source layers for all of these anomalies on the U.S. margin have a similar origin—the sea-floor spreading process. Keen and others (LASE Study Group, 1986) have suggested that the 7.2 km/sec layer could be caused by melt migration and underplating, producing a new lower crustal layer that is continuous from rift-stage crust to oceanic crust. Mutter (1985, p. 117–118) contemplated identifying this region as a separate crustal zone. Nicolas (1985) specifically calls it a new type of crust, the metasedimentary crust. There are sufficient differences in the crustal structure of this zone for it to be included as a separate crustal type in terms of crustal evolution models.

Marginal crustal boundary markers

Three large magnetic anomaly lineations that parallel the margin, the BMA, ECMA, and BSMA, have been interpreted as markers of major crustal boundaries. Although we have used them as such, there still is considerable debate as to their origin.

East Coast Magnetic Anomaly. The ECMA is usually associated with the continent-ocean crustal boundary, an elusive but critical element in margin crustal evolution (Keen, 1969; Emery and others, 1970; Rabinowitz, 1974; Klitgord and Behrendt, 1979; Hutchinson and others, 1983; Alsop and Talwani, 1984; LASE Study Group, 1986). It is situated within the zone of marginal oceanic crust, a region where crystalline basement is not imaged. As mentioned previously, the segmented character of the ECMA suggests that the ECMA source crust may have been created by the sea-floor spreading tectonic and/or magmatic process rather than by the rifting process. Synrift and prerift sedimentary units are faulted landward of this zone, indicating that continental crustal fragments exist at least as far east as the East Coast Boundary Fault (Fig. 11B). The large change in crustal structures at the ECBF indicates that it probably overlies a major crustal discontinuity. Basement just east of the ECBF must be at a depth of at least 15 km, whereas the top of oceanic basement at the J_3 Scarp (100 km to the east) is at a depth of 10.5 km (Fig. 16A). At the paleoshelf edge (50 to 60 km to the east of the ECBF), basement must be deeper than 12 km (depth of J_3 horizon) and shallower than 15 km (top of 7.2 km/sec layer) at ESP 5 on the LASE line (LASE Study Group, 1986, Fig. 6). Therefore, unless there is sediment directly on top of the 7.2 km/sec layer, there probably is a drop in basement of 1 to 2 km, down to the west, between the Jurassic paleoshelf edge and the axis of the ECMA (near shotpoint 2500 in Fig. 11B).

Two other important features associated with the ECMA are the paleoshelf edge and the salt diapirs. The initial location of the paleoshelf edge during the Middle Jurassic was along the landward edge of the ECMA. This initial location has been imaged only in the northern Baltimore Canyon Trough, where the shelf edge prograded seaward more than 50 km during the Jurassic and the initial paleoshelf is not overlain by thick carbonate-bank deposits. The seismic pattern of prograding carbonate shelf edge deposits (Grow and others, this volume, Fig. 7) indicates that the initial shelf edge was just landward of the ECBF. Structural relief at this fault zone may have provided the local highs upon which a carbonate bank developed. It is possible that the initial shelf-edge location coincides with that of the ECBF along the entire margin. Most of the salt diapirs are seaward of the Middle Jurassic paleoshelf edge, beneath the axis of the ECMA. If the salt diapirs are from a salt unit just below the PRU (Klitgord and others, 1982; Dillon and others, 1983b), then the horizontal flow may have been controlled by sediment loading behind the paleoshelf edge, with diapirism the result of deep structural or sedimentation changes at the ECMA.

Magnetic source models for the ECMA generally fall into two categories, edges and dikes, and include an intrusive dike, a magnetized ridge, the landward edge of reversely magnetized oceanic crust, the seaward edge of continental crust, or a mixture of these models. Most of the detailed magnetic model studies, however, were carried out over only one of the margin segments of the ECMA; each of the different models can explain the ECMA within that segment of the margin. For instance, in the Hutchinson and others (1983) model for the Carolina Trough region (Fig. 16B), the primary source is the landward edge of a reversely magnetized (inclination = 20°) oceanic crust adjacent to rift-stage crust that has a weaker induced magnetization. In contrast, in the Alsop and Talwani (1984) model for the ECMA along the Baltimore Canyon Trough (Fig. 16A), the primary source is an intrusive body (V-I and V-II in Fig. 16A) that has a steeply dipping remanent magnetization (inclination = 60°) adjacent to the edge of weakly magnetized continental crust. Because both models have the same weakly magnetized layer of induced magnetization (K = 0.003) landward of the ECMA, the principal difference is between the positively magnetized intrusive body and the edge of a reversely magnetized oceanic crust layer. Seismic-reflection data provide strong evidence that the oceanic crustal layer (source of seafloor-spreading magnetic lineations) continues almost to the ECMA and must be at least a partial source for the ECMA. Intrusive bodies have not been imaged beneath the ECMA, but low-angle detachment models for margin evolution (e.g., Lister and others, 1986) indicate a high likelihood of such intrusives in this region. The thick zones of seaward-dipping volcanic strata along the margins of the North Atlantic (Mutter, 1985) are indicative of a major intrusive phase near the transition from rifting to sea-floor spreading.

Our preferred interpretation for the ECMA is that its primary source is the landward edge of reversely magnetized marginal oceanic crust, above the 7.2 km/sec layer. We interpret the gradual depth increase and low relief on the J_3 horizon as an indication that a major depth increase for the top of oceanic crust or marginal oceanic crust does not occur until landward of the Upper Jurassic paleoshelf edge but seaward of the axis of the ECMA. This is consistent with all of the seismic data. Dikes of Early Jurassic age onshore have remanent inclinations of 20 to 40° in the Carolinas and New Jersey (DeBoer and Snider, 1979), which is significantly different from the inclination used by Alsop and Talwani (1984). Most of the intrusive source postulated by Alsop and Talwani (1984) is located within the 7.2 layer, which the LASE results suggest is quite uniform across the margin beneath the ECMA (LASE Study Group, 1986); their source layer is at a depth (~20 km) where the thermal regime would significantly reduce any remanent magnetization. A magnetized oceanic crustal layer that is nearly 5 km shallower than the source bodies of Alsop and Talwani easily accounts for the seaward gradient and peak of the ECMA. The steep landward gradient of the ECMA in the Baltimore Canyon Trough requires an additional small change in magnetization (decreasing to the west) beneath the East Coast Boundary Fault. This is consistent with the evidence of a crustal change beneath this fault and may represent an "igneous dike" component to the source of the ECMA—perhaps a thickened volcanic layer at the landward edge of oceanic crust.

Brunswick Magnetic Anomaly. As with the ECMA, the source of the BMA is attributed to a major crustal boundary. In the offshore region, we have noted that the BMA marks the hinge zone along the landward edge of the marginal sedimentary basins (Klitgord and Behrendt, 1979; Hutchinson and others, 1983). We also have associated the magnetic low with a rift basin just landward of the hinge zone, but evidence for such a rift basin is found on only one of five multichannel seismic-reflection profiles offshore (seismic line 32) crossing the magnetic low.

The onshore segment of the anomaly has been interpreted as the signature of a suture zone boundary between a fragment of old African crust in the Florida region and North American crust to the north (Popenoe and Zietz, 1977; Daniels and others, 1983; Klitgord and others, 1983, 1984). Recent seismic-reflection profiling across the BMA by COCORP (Nelson and others, 1985a, 1985b, 1987) seems to confirm this interpretation of the anomaly source as a major crustal boundary. To the north, in the Salisbury Embayment, a possible suture zone is associated with mafic rocks that are marked by a large-amplitude, elongate set of magnetic highs with broader magnetic lows marking the adjacent granitic and gneissic terranes (Lefort and Van der Voo, 1981; Klitgord and others, 1983; Pratt and others, 1986). A similar association is found in the Southeast Georgia Embayment, where magnetic highs of the onshore segment of the BMA are located over the axis of the possible suture zone and where granitic and rhyolitic rocks have been recovered just onshore at a subsurface depth of 1.5 km beneath the BMA low (Klitgord and others, 1983; Chowns and Williams, 1983). Nelson and others (1985b, 1987) have proposed that the suture-zone origin applies to both the onshore and offshore segments of this anomaly, whereas we have argued that it is very likely that they are separate sources that

coincide at an intersection of crustal boundaries (Klitgord and others, 1983). The coherent part of the BMA, which can be traced from landward of the hinge zone to along the hinge zone, is the distinctive magnetic trough; it is not the magnetic high associated with mafic material and the suture zone model. Rift basins have been identified both onshore and offshore beneath the magnetic low of the BMA. Roussel and Liger (1983) have postulated that the offshore segment of the suture zone is actually on the African margin in the Senegal Basin.

Blake Spur Magnetic Anomaly. Origin of the BSMA, although poorly defined in terms of actual source, is the least controversial of the three lineations. Vogt (1973) attributed its origin to a spreading-center jump eastward to the BSMA location, which was against the African margin at that time (Middle Jurassic). Subsequent analyses and studies have supported this conclusion (Sheridan, 1974; Klitgord and Behrendt, 1979; Sheridan and others, 1979; Klitgord and Grow, 1980). The BSMA is the most western isochron marker in the Atlantic deep ocean basin (Plate 2C), and it has been used along with the other seafloor-spreading magnetic lineations to determine the poles of rotation for reconstructing paleo-positions of continents bordering the Atlantic (Klitgord and Schouten, 1986).

Crustal Evolution Models

Thermo-mechanical models of crustal modification are essential tools for converting a variety of geologic, structural, and geophysical observations into a crustal evolution history of passive margin development (Sawyer, this volume; Steckler and others, this volume). Simple extensional models based on symmetric extension that involves listric faulting in a brittle upper crust and plastic flow in a ductile lower lithosphere (McKenzie, 1978; Le-Pichon and Sibuet, 1981) have been partially successful in explaining margin evolution. Ductile thinning of the lower crust and upper mantle occurring directly beneath the zone of brittle failure is an assumption used in many mechanical models for continental rift evolution (Fig. 17) (Hellinger and Sclater, 1983). These models predict an initial tectonic subsidence, with block-faulting brittle failure of the upper crust and heating of the lower crust, during the rifting phase. This is followed by a cooling and sediment-loading subsidence phase when the zone of extension has moved seaward to a spreading-center system. Variations in lithospheric strength and flexural rigidity, both vertically and with time (see Watts, 1982; LePichon and Sibuet, 1981; Hellinger and Sclater, 1983), are added to the models in order to improve the agreement of predicted initial subsidence with observations. These models usually predict the postrift subsidence history very well, but significant discrepancies remain for the initial tectonic subsidence predictions on the U.S. Atlantic margin (Sawyer and others, 1983). More importantly, these symmetric extension models do not predict the observed wide variation in gross margin architecture, crustal thinning, and continental uplift (Lister and others, 1986).

Studies of continental extensional regimes, such as the Basin

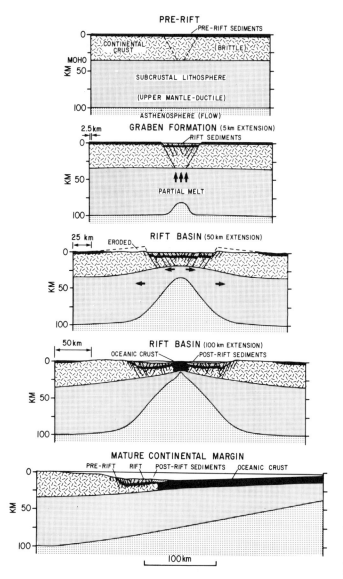

Figure 17. Crustal evolution model for Atlantic-type margins based on pure-shear deformation. When tensional stress is applied, the crust fails by brittle fracture, faulting, and shearing. The upper mantle fails by necking. The future fault zones in the crust are shown on the prerift stage of the model. A graben forms as extension takes place, and the prerift sediments are faulted down and preserved. The isostatic disequilibrium caused by crustal extension leads to compensating rise of the astheno-sphere and some regional uplift. Necking in the upper mantle allows the asthenosphere to flow up, implacing a source of heat. The lower part of the upper mantle is raised to a decreasing pressure regime, and a partial melt is generated that leads to volcanism and further transfer of heat upward. Further extension thins the crust more (rift basin, 50-km exten-sion). The uplifted shoulders have become deeply eroded, and more sediments fill the basin. Extension of the continental crust finally stops (rift basin, 100-km extension), and extension of the lithosphere continues by generation of oceanic crust. As the rifted basin evolves into a mature continental margin, the lithosphere is thickened by accretion from below. From Hellinger and Sclater (1983, Fig. 1).

and Range Province (Anderson and others, 1983), have led to the recent development of a different type of thermo-mechanical model for extension involving simple-shear deformation (Bally, 1981; Wernicke, 1985; Lister and others, 1986; LePichon and Barbier, 1987). Crustal failure in this type of model is not sym-metric but takes place along low-angle detachment faults or low-angle crustal shear zones (Fig. 18). This asymmetric model predicts the development of distinctly different margin crustal structures and subsidence histories for conjugate margins. The separating continental margins evolve as upper-plate or lower-plate margins (Lister and others, 1986) and have different initial crustal properties and different tectonic/magmatic processes modifying their crust and upper mantle. The lower-plate margin has the more complex structure; it comprises rocks originally below the detachment surface plus highly faulted remnants of the upper plate over a bowed-up detachment surface. The upper-plate margin has a relatively simple fault structure and comprises rocks originally above the detachment surface plus underplated material.

Specific structural and deep crustal features on the U.S. Atlantic margin are compatible with the asymmetric detachment model. The Paleozoic thrust faults provide the low-angle surfaces (zones of weakness) that could be reactivated during rifting (e.g., Block Island Fault, Fig. 12A). The broad zone of rift basins west of the basement hinge zone represents either numerous reacti-vated detachment surfaces or the stranded crustal fragments on a low-angle detachment surface (Hutchinson and Klitgord, 1987). The change in character along the margin of rift structures at the hinge zone could be related to the existence of large rift transfer faults (Bally, 1981), with some parts of the U.S. Atlantic margin forming as an upper-plate margin and some parts as a lower-plate margin. For example, the Baltimore Canyon Trough and Georges Bank Basin regions have very different crustal structures (includ-ing sediments) and are located on either side of a major shear zone, the western continuation of the South Atlas Fault. The Baltimore Canyon Trough margin south of this shear zone ap-pears to be a lower-plate margin. The Paleozoic thrust faults are east-dipping, and the fault-block structures (Fig. 3B) could be crustal fragments left on the detachment surface. In contrast, the half-graben structures along the Georges Bank Basin hinge zone (Fig. 3A) are similar to the Bay of Biscay structures that LePichon and Barbier (1987, Fig. 10) locate on an upper-plate margin. The asymmetric distribution of early Mesozoic rift basins on the flanks of these two marginal basins (Fig. 19) is suggestive of a non-symmetric rifting process; numerous rift basins are preserved west of Baltimore Canyon Trough, whereas a similar broad distribu-tion of rift basins is found to the east of Georges Bank Basin in Morocco.

PLATE TECTONIC FRAMEWORK AND STRUCTURAL EVOLUTION

The U.S. Atlantic margin has evolved through a series of tectonic phases, each of which has left its imprint on the structural

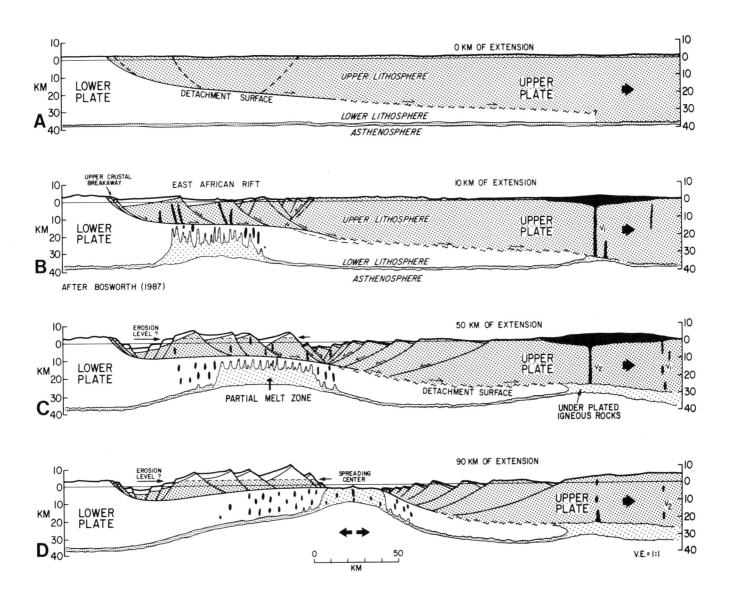

Figure 18. Crustal evolution model for Atlantic-type margins based on simple shear along low-angle detachment faults and the development of lower-plate (left) and upper-plate (right) margins (Lister and others, 1986). A. Initial fracturing of the crust is along a low-angle surface that extends at least as deep as the brittle-ductile boundary in the crust. As the two plates separate along the detachment surface, the lower plate must warp upward as the load formerly exerted by the upper plate is removed, and a broad arch or culmination will develop. B. The second section, a model of the tectonic configuration after about 10 km of extension, is based on the model for the East African Rift system proposed by Bosworth (1987). C. The third section is a model of the tectonic configuration after about 50 km of extension. Active faulting in the upper plate migrates to the right and leaves fragments of the upper plate on the lower plate. This causes the locus of partial melting and uplift in the lower plate to also migrate to the right. Uplift and erosion (or gravitational sliding) denudes the upper crustal fragments on the lower plate. D. The fourth section is a model of the tectonic configuration just after seafloor spreading has started. Cross sections based on model cross sections of Bosworth (1987, Fig. 3), Wernicke (1985, Figs. 3, 12), Lister and others (1986, Fig. 2), and LePichon and Barbier (1987, Fig. 10).

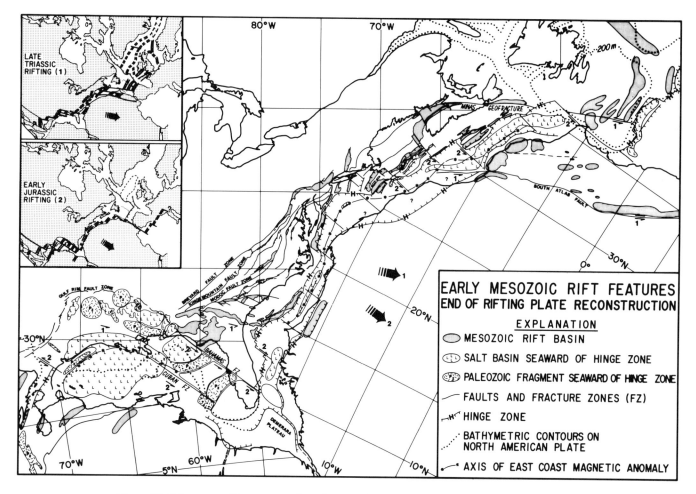

Figure 19. Paleogeographic reconstruction of the continents bordering the Atlantic at the end of the rifting phase in the late Early Jurassic. The minimum-closure reconstruction poles-of-rotation of Klitgord and Schouten (1986) were used for the North American, South American, and African plate positions. Rift basin distribution is taken from Klitgord and others (1982, 1983, 1984), Daniels and others (1983), Long and Dainty (1985), Robinson and Froelich (1985), and Hutchinson and others (1986b, 1987b). Selected Paleozoic faults are taken from Hatcher and Zietz (1980), Williams and Hatcher (1983), Haworth and Jacobi (1983), and Hutchinson and others (1987b). Salt distribution is from Dillon and others (1983b), Jansa (1986), and Salvador (1987). Plate motions and major fault motion trends are indicated for the early synrift (1) and late synrift (2) phases. Insets show Late Triassic rift configuration (1) with the North Atlantic—Arctic system and the Early Jurassic rift configuration (2) with the Tethys system.

framework outlined in the previous sections. Sediment deposition patterns on the margin were influenced by the interaction between tectonic activity, continental configurations, and prevailing paleoenvironment during each phase (Jansa, 1986; Manspeizer and Cousminer, this volume; Schlee and others, this volume). Important aspects of any reconstruction of continental configuration are the data sets used to construct the model and the age constraints on the period when the model is valid (Klitgord and Schouten, 1986). The age constraints must be considered because some reconstructions may represent successive periods, whereas others may represent modified models for the same time period. The reconstruction models of Lefort (1983), Van der Voo and others (1976), and Klitgord and Schouten (1986) are successive

time models because they are based on data sets of different geologic ages (late Paleozoic, Triassic, and post Triassic, respectively). In contrast, the Bullard and others (1965), LePichon and others (1977), and Klitgord and Schouten (1986) closure models are for the same geologic time period, just prior to initiation of sea-floor spreading. The Bullard and others (1965) reconstruction was the initial attempt at a closure fit based on matching isobaths, and it now has been superseded by newer reconstruction models.

Late Paleozoic collision phase

A number of terranes, microcontinents, and continents were accreted to the North American plate throughout the Paleozoic

(Williams and Hatcher, 1983). Construction of the megacontinent Pangea was finally completed in the Late Paleozoic during the Alleghanian Orogeny. Many of the details of this final suturing between Gondwana and Laurasia are still under active investigation, including the location of the suture zones (e.g., Hatcher and Zietz, 1980; Lefort and Van der Voo, 1981; Haworth and Jacobi, 1983; Lefort, 1983; Klitgord and others, 1984; Hutchinson and others, 1987b; Nelson and others, 1987).

Paleogeography of the Atlantic margin at the end of the Paleozoic (Klitgord and Schouten, 1986, Fig. 9, Plate 9A) represents the continental configuration at the end of the collision event. This reconstruction places Cap Blanc (West Africa) adjacent to the Salisbury Embayment, the Carolina Platform adjacent to the Senegal Basin, and the northern margin of South America within the Gulf of Mexico. Three data sets help to constrain this Atlantic closure reconstruction: Late Paleozoic geologic and geophysical markers (Lefort, 1983; Haworth and Jacobi, 1983), late Paleozoic and Triassic paleomagnetic data (Van der Voo and others, 1976; Lefort and Van der Voo, 1981), and Mesozoic sea-floor spreading data (Klitgord and Schouten, 1986). The maximum closure reconstruction of Klitgord and Schouten (1986) is nearly identical to the reconstructions based on late Paleozoic markers and paleomagnetic data. Each of these closure reconstructions is based on data sets for a different geologic age, yet the differences are too small to be significant.

Numerous large thrust faults were generated during the Paleozoic collision phase (Cook and others, 1981; Ando and others, 1983; Stewart and others, 1986; Hutchinson and others, 1985, 1987b; Phinney, 1986; Pratt and others, 1987). There also is considerable evidence for dextral strike-slip shearing along this same thrust fault system during the Middle to Late Carboniferous (Lefort and Van der Voo, 1981; Bradley, 1982; Gates and others, 1986; Manspeizer and Cousminer, this volume). Carboniferous rift basins such as the Narragansett Basin developed within this shear zone.

Distribution of possible late Paleozoic thrust and strike-slip shear faults is indicated on Plate 2C. The faults beneath the Coastal Plain and offshore are known only in coarse detail based on magnetic-anomaly patterns and seismic-reflection profiling. This fault distribution includes both a zone of thin-skinned tectonics where splay faults sole onto a major decollement (Cook and others, 1981) and a zone of "thick-skinned" tectonics where faults cut the entire crust (Hutchinson and others, 1985, 1987b; Pratt and others, 1987). The zone of late Paleozoic thin-skinned tectonics is over the North American craton, whereas the deep penetration faults are within the accreted terranes and fractured eastern edge of the North American craton. These faults all represented potential failure zones that could be reactivated under extension.

Early rifting phase (Late Triassic)

Tectonic activity switched from compression to extension during the late Permian when the Arctic–North Atlantic rift sys-

tem (Ziegler, 1982) propagated southward along the North American–African (central Atlantic) boundary (Manspeizer and others, 1978). An initial phase of uplift and crustal thinning was postulated by Manspeizer and others (1978) to explain the absence of late Permian to Middle Triassic sediments along the boundary. Rift basins formed during the Late Triassic over a broad (300-km-wide) region from the Grand Banks to the Gulf of Mexico (Fig. 19) (see Manspeizer and Cousminer, this volume). The coastal New England plutonic complexes were emplaced during this rifting phase (McHone and Butler, 1984). In some basins, Paleozoic thrust faults were reactivated and formed the border faults; in other basins, the border faults are antithetic faults that merged with these older thrust faults (Cook and others, 1981; Brewer and others, 1983; Hutchinson and others, 1986b). During this early rifting phase, there was a major zone of rifting in the Bay of Fundy and the South Georgia Rift, and the two major African fragments (the Meguma and Suwannee Terranes) were still partially attached to the African plate.

Major left-lateral strike-slip shear zones connected the central Atlantic rift system with the North Atlantic rift zone and the western Gulf of Mexico rift zone. The northern shear zone is bounded by the Minas Geofracture (Cobequid-Chedabucto Fault Zone) and the South Atlas Fault Zone. The South Atlas Fault Zone intersects the North American margin in the vicinity of the Long Island Platform, but there still is uncertainty as to whether it coincides with faults along the north side of the platform or with the hinge zone along the south side (the 40°N Fault). The southern shear zone is bounded by the Gulf Rim Fault Zone plus the western segment of the BMA and by second fault zone south of the Cuban Fracture Zone. All of the northern Gulf Basin salt basins and the southern half of the South Georgia Rift are within this shear zone.

There is only minimal control on the actual direction of rifting. The stress regime of an east–west-trending, left-lateral shear couple postulated by Manspeizer and Cousminer (this volume, Fig. 10) is a reasonable estimate for the early rift phase (Late Triassic). The predicted fault trends are consistent with the trends (trend 1 in Fig. 19) of the Minas Geofracture (Keppie, 1985; Haworth and Jacobi, 1983), the South Atlas Fault (Manspeizer, 1985), and a possible shear zone along the Gulf Rim Fault Zone and the onshore segment of the BMA (Klitgord and others, 1984).

There is no evidence to suggest large amounts of northeast strike-slip motion. Swanson's (1982) postulated existence of large dextral shear during this phase is based on the use of out-of-date plate reconstruction models. He assumed that it was necessary to move the African plate position from the Van der Voo and others (1976) closure to that of Bullard and others (1965), a rotation of 20° about a pole in the Sahara. Klitgord and Schouten (1980, 1986) have pointed out that the Bullard and others (1965) model should no longer be used because it has been replaced by better constrained closure models for the early Mesozoic. All three of the most recent models of continental reconstructions for the latest Paleozoic to early Mesozoic (Van der Voo and others,

1976; Lefort and Van der Voo, 1981; Klitgord and Schouten, 1986) are very similar to each other and indicate very little motion during this period. Therefore, the plate-tectonic models do not require major amounts of Triassic and Jurassic shear movement as Swanson (1982) has suggested.

Late Rifting Phase (Early Jurassic)

Distinct changes in the distribution of rifting and in the direction of crustal extension (trend 1 to trend 2 in Fig. 19) mark a second phase of rifting. Rifting in the central Atlantic became linked to the Ligurian Tethys rift system (Lemoine and others, 1986) rather than to the Arctic–North Atlantic rift system. As rifting continued, it was focused within the zone of marginal basins, and the basement hinge zone developed. Rifting in the onshore basins was waning at this time, whereas the large offshore marginal basins were still to undergo major amounts of tectonic subsidence. For the offshore region, we have attributed the uplift of the basement hinge zone to the beginning of the late rifting phase (Hutchinson and others, 1986b; Hutchinson and Klitgord, 1987). Increased depth of erosion of synrift deposits toward the hinge zone is the primary indicator of uplift after the rift basins had formed on the Long Island Platform. Rifting ceased in the Fundy and South Georgia Rifts and shifted primarily to the Scotian and Bahamas Basins, leaving the two African fragments, the Meguma and Suwannee Terranes, attached to the North American plate. In the paleogeography, this zone of activity is indicated by the regions seaward of the hinge zone on Figure 19. Mechanical failure and subsidence of the brittle upper crust would continue to limit deposition to small rift basins and deform existing synrift deposits. The maximum extension during this rifting phase, seaward of the hinge zone, is only about 100 km, based on the 100-km width of the Baltimore Canyon Trough and a stretching β-factor of >5 (Sawyer and others, 1983).

Direction of rifting shifted more to the southeast (Africa relative to North America; trend 2 in Fig. 19), closer to the subsequent seafloor-spreading trend. The Newfoundland-Gibraltar Fault system connected the central Atlantic rift system to the Ligurian Tethys system (Klitgord and Schouten, 1986, Fig. 10; Lemoine and others, 1986), and the Bahamas-Cuban Fault system connected the central Atlantic rift system to the western Gulf of Mexico rift system (Klitgord and others, 1984). Major offsets in the basement hinge zone at the Jacksonville, Blake Spur, and Norfolk Fracture Zones (Klitgord and Behrendt, 1979) and offsets between the Nantucket, Franklin, and Fundy Rift Basins (Klitgord and Hutchinson, 1985; Hutchinson and others, 1986b, 1987b) are approximately parallel to these two major fault systems.

Age constraints on the start and end of the late-rifting phase are still only speculative. The sedimentation record in the Western Alps indicates that rifting started there at the beginning of the Early Jurassic (Lemoine and others, 1986). Oldest sediments seaward of the hinge zone in the Central Atlantic are early Early Jurassic in age (see discussion in Introduction). In contrast, the youngest sediments in rift basins onshore North America are earliest Early Jurassic, except in the Hartford-Deerfield Basin (Manspeizer and Cousminer, this volume, Fig. 7). Thus, we assume that the late rifting phase began in the earliest Early Jurassic. The best estimates for the end of rifting come from the age of the last major pulse of igneous activity onshore (191 Ma; Sutter, 1985), from the age of earliest postrift sedimentary units (late Early Jurassic), and from the sea-floor spreading record on younger oceanic crust (Klitgord and Schouten, 1986). The oldest sedimentary rock recovered in wells seaward of the basement hinge zone and above the PRU are latest Early Jurassic to earliest Middle Jurassic in the Scotian and Georges Bank Basins (see discussion in Introduction) and mid–Middle Jurassic (Bathonian) in the Ligurian Tethys (Lemoine and others, 1986) and in the Gulf of Mexico (Salvador, 1987). A possible age limitation of 175 Ma was derived by Klitgord and Schouten (1986) based on the sea-floor spreading record and assumption of a constant spreading rate prior to Tithonian. The 175-Ma also coincides with the final igneous event in the onshore rift basins, although Sutter (1985) has indicated that these dates are suspect. The end of the rifting phase is not a synchronous event along the margin, but it is sometime between the late Early Jurassic and the early Middle Jurassic.

Earliest sea-floor spreading phase

The central Atlantic spreading center system began to produce oceanic crust seaward of the continental margin during the early Middle Jurassic (Klitgord and Schouten, 1986). The diffuse plate boundary beneath the continental margin was replaced by a narrow ridge-axis spreading center. Extension within the marginal basins would have effectively ended when a spreading center became the primary focus of extensional strain release. The shift from a rift magmatic process to a sea-floor spreading magmatic process was not synchronous along the margin. Rifting was still active in the Blake Plateau and Bahamas Basins when to the north sea-floor spreading was creating typical oceanic crust east of the Carolina to Scotian Basins. During this initial stage of the seafloor-spreading phase, it is likely that the zone of seaward-dipping reflectors and marginal oceanic crust was created east of the ECMA. After a brief period of normal oceanic crust generation, there was the ridge jump eastward to the Blake Spur Magnetic Anomaly (Vogt, 1973). By this time (170 Ma), oceanic crust was being generated in the Ligurian Tethys, along the entire central Atlantic spreading system (Newfoundland to Bahamas Fracture Zones), and in the Gulf of Mexico (Klitgord and Schouten, 1986, Fig. 10).

By 170 Ma, the central Atlantic Ocean basin was over 200 km wide. Continental shelves already rimmed the ocean basin and basin-wide water circulation patterns began to develop (Jansa, 1986). The plate boundary shift at 170 Ma ended the restricted water circulation between the Tethys and the central Atlantic and led to a change in shelf sediment deposition from anhydrite-dolomite units to limestone, with the construction of the carbonate-bank shelf-edge complex (Schlee and Fritsch, 1983; Jansa, 1986). Circulation was still restricted in the Gulf of Mexico, but the central Atlantic was open to the Tethys via the straits between Africa and the Grand Banks–Iberia platform. The

margin was cooling and subsiding, and substantial sediment was accumulating on the shelf behind the paleoshelf edge. This was the start of the period of rapid subsidence on the margin as the spreading center and active plate boundary shifted farther and farther to the east.

REFERENCES CITED

Alsop, L., and Talwani, M., 1984, The east coast magnetic anomaly: Science, v. 226, p. 1189–1191.

Amato, R. W., and Simonis, E. K., 1980, Geologic and operational summary, COST No. G-2 well, Georges Bank area, north Atlantic OCS: U.S. Geological Survey Open File Report 80-269, 120 p.

Anderson, R. E., Zoback, M. L., and Thompson, G. A., 1983, Implications of selected subsurface data on the structural form and evolution of some basins in the northern Basin and Range province, Nevada and Utah: Geological Society of America Bulletin, v. 94, p. 1055–1072.

Ando, C. J., Cook, F. A., Oliver, J. E., Brown, L. D., and Kaufman, S., 1983, Crustal geometry of the Appalachian orogen from seismic reflection studies, *in* Hatcher, R. D., Jr., Williams, H., and Zietz, I., eds., Contributions to the tectonics and geophysics of mountain chains: Geological Society of America Memoir 158, p. 83–101.

Ascoli, P., 1983, Summary report on the biostratigraphy (Foraminifera and Ostracoda) and depositional environments of the Atlantic COST G-1 well, Georges Bank, from 10,000 to 14,800′: Geological Survey of Canada Report EPGS-PAL. 21-83PA, 6 p.

Austin, J. A., Jr., and Uchupi, E., 1982, Continental-oceanic crustal transition off southwest Africa: American Association of Petroleum Geologists Bulletin, v. 66, p. 1328–1347.

Austin, J. A., Jr., Uchupi, E., Shaughnessy, D. R., III, and Ballard, R. D., 1980, Geology of New England passive margin: American Association of Petroleum Geologists Bulletin, v. 64, p. 501–526.

Ballard, R. D., and Uchupi, E., 1975, Triassic rift structures in Gulf of Maine: American Association of Petroleum Geologists Bulletin, v. 59, p. 1041–1072.

Bally, A. W., 1981, Atlantic-type margins, *in* Bally, A. W., ed., Geology of passive continental margins: American Association of Petroleum Geologists Education Course Note Series no. 19, p. 11 to 148.

Barss, M. S., Bujak, J. P., and Williams, G. L., 1979, Palynological zonations and correlations of sixty-seven wells, Eastern Canada: Geological Survey of Canada Paper 78-24, 118 p.

Beaumont, C., Keen, C. E., and Boutilier, R., 1982, A comparison of foreland and rift margin sedimentary basins: Philosophical Transactions of the Royal Society of London, v. A305, p. 295–317.

Behrendt, J. C., 1986, Structural interpretation of multichannel seismic reflection profiles crossing the southeastern United States and adjacent continental margin; décollements, faults, Triassic(?) basins and Moho reflections, *in* Barazangi, M., and Brown, L., eds., Reflection Seismology; The Continental Crust: American Geophysical Union Geodynamics Series, v. 14, p. 201–213.

Behrendt, J. C., and Grim, M. S., 1985, Structure of the U.S. Atlantic continental margin from derivative and filtered maps of the magnetic field, *in* Hinze, W. J., ed., The Utility of Regional Gravity and Magnetic Anomaly Maps: Tulsa, Oklahoma, Society of Exploration Geophysicists, p. 325–338.

Behrendt, J. C., and Klitgord, K. D., 1980, High-sensitivity aeromagnetic survey of the U.S. Atlantic continental margin: Geophysics, v. 45, p. 1813–1846.

Behrendt, J. C., Hamilton, R. M., Ackerman, H. D., Henry, V. J., and Bayer, K. C., 1983, Marine multichannel seismic-reflection evidence for Cenozoic faulting and deep crustal structure near Charleston, S.C., *in* Gohn, G. S., ed., Studies related to the Charleston, South Carolina, earthquake of 1886; Tectonics and seismicity: U.S. Geological Survey Professional Paper 1313, p. J1–J29.

Benson, R. N., 1984, Structure contour map of pre-Mesozoic basement, landward

margin of Baltimore Canyon Trough: Delaware Geological Survey Miscellaneous Map Series no. 2, scale 1:500,000.

Benson, R. N., Andres, A. S., Roberts, J. H., and Woodruff, K. D., 1986, Seismic stratigraphy along three multichannel seismic reflection profiles off Delaware's coast: Delaware Geological Survey Miscellaneous Map Series no. 4, 1 sheet.

BIRPS and ECORS, 1986, Deep seismic reflection profiling between England, France, and Ireland: Geological Society of London Journal, v. 143, p. 45–52.

Bosworth, W., 1987, Off-axis volcanism in the Gregory Rift, east Africa; Implications for models of continental rifting: Geology, v. 15, p. 397–400.

Bradley, D. C., 1982, Subsidence in Late Paleozoic basins in the northern Appalachians: Tectonics, v. 1, p. 107–123.

Brewer, J. A., Matthews, D. H., Warner, M. R., Hall, J. R., Smythe, D. K., and Whittington, R. J., 1983, BIRPS deep seismic reflection studies of the British Caledonides: Nature, v. 305, p. 206–210.

Brown, P. M., Miller, J. A., and Swain, F. M., 1972, Structural and stratigraphic framework, and spatial distribution of permeability of the Atlantic Coastal Plain, North Carolina to New York: U.S. Geological Survey Professional Paper 796, 79 p.

Bullard, E. C., Everett, J. E., and Smith, A. G., 1965, Fit of the continents around the Atlantic, *in* Blackett, P.M.S., Bullard, E. C., and Runcorn, K. S., eds., A symposium on continental drift: Philosophical Transactions of the Royal Society of London, v. A258, p. 41–51.

Chowns, T. M., and Williams, C. T., 1983, Pre-Cretaceous rocks beneath the Georgia coastal plain; Regional implications, *in* Gohn, G. S., ed., Studies related to the Charleston, South Carolina, earthquake of 1886; Tectonics and seismicity: U.S. Geological Survey Professional Paper 1313, p. L1–L42.

Cochran, J. R., Martinez, F., Steckler, M. S., and Hobart, M. A., 1986, Conrad Deep; A new northern Red Sea deep; Origin and implications for continental rifting: Earth and Planetary Science Letters, v. 78, p. 18–32.

Cook, F. A., Brown, L. D., Kaufman, S., Oliver, J. E., and Peterson, T. A., 1981, COCORP seismic profiling of the Appalachian orogen beneath the Coastal Plain of Georgia: Geological Society of America Bulletin, pt. I, v. 92, p. 738–748.

Cousminer, H. L., Steinkraus, W. E., and Hall, R. E., 1984, Biostratigraphic restudy documents Triassic/Jurassic section in Georges Bank COST G-2 well [abs.]: American Association of Petroleum Geologists Bulletin, v. 68, p. 466.

Crutcher, T. D., 1983, Southeast Georgia Embayment, *in* Bally, A. W., ed., Seismic expression of structural styles: American Association of Petroleum Geologists Studies in Geology Series no. 15, v. 2, p. 2.2.3-27–2.2.3-29.

Daniels, D. L., Zietz, I., and Popenoe, P., 1983, Distribution of subsurface lower Mesozoic rocks in the southeastern United States as interpreted from regional aeromagnetic and gravity maps, *in* Gohn, G. S., ed., Studies related to the Charleston, South Carolina, Earthquake of 1886; Tectonics and Seismicity: U.S. Geological Survey Professional Paper 1313, p. K1–K24.

deBoer, J., and Snider, F. G., 1979, Magnetic and chemical variations of Mesozoic diabase dikes from eastern North America; Evidence for a hot spot in the Carolinas?: Geological Society of America Bulletin, v. 90, p. 185–198.

de Graciansky, P. C., Poag, C. W., and others, 1985, The Goban Spur transect; Geologic evolution of a sediment-starved passive continental margin: Geological Society of America Bulletin, v. 96, p. 58–76.

Detrick, R. S., and Purdy, G. M., 1980, The crustal structure of the Kane fracture zone from seismic refraction studies: Journal of Geophysical Research, v. 85, p. 3759–3777.

Dillon, W. P., Paull, C. L., Dahl, A. G., and Patterson, W. C., 1979, Structure of the continental margin near the COST GE-1 well site from a common depth point seismic reflection profile, *in* Scholle, P. A., ed., Geological studies of the COST GE-1 well, United States South-Atlantic Outer Continental Shelf area: U.S. Geological Survey Circular 800, p. 97–107.

Dillon, W. P., Klitgord, K. D., and Paull, C. K., 1983a, Mesozoic development and structure of the continental margin off South Carolina, *in* Gohn, G. S., ed., Studies related to the Charleston, South Carolina, earthquake of 1886; Tectonics and seismicity: U.S. Geological Survey Professional Paper 1313, p. N1–N16.

Dillon, W. P., Popenoe, P., Grow, J. A., Klitgord, K. D., Swift, B. A., Paull, C. K., and Cashman, K. V., 1983b, Growth faulting and salt diapirism; Their relationship and control in the Carolina Trough, eastern North America, *in* Watkins, J. S., and Drake, C. L., eds., Studies in continental margin geology: American Association of Petroleum Geologists Memoir 34, p. 21–46.

Dillon, W. P., Paull, C. K., and Gilbert, L. E., 1985, History of the Atlantic continental margin off Florida; The Blake Plateau Basin, *in* Poag, C. W., ed., Geologic evolution of the United States Atlantic margin: New York, Van Nostrand Reinhold, p. 189–215.

Eldholm, O., and ODP Leg 104 Scientists, 1986, Formation of the Norwegian Sea: Nature, v. 319, p. 360–361.

Emery, K. O., and Uchupi, E., 1984, The Geology of the Atlantic Ocean: New York, Spring-Verlag, 1050 p. and 23 charts.

Emery, K. O., Uchupi, E., Phillips, J. D., Bowin, C. O., Bunce, E. T., and Knott, S. T., 1970, Continental rise off eastern North America: American Association of Petroleum Geologists Bulletin, v. 54, p. 44–108.

Falvey, D. A., 1974, The development of continental margins in plate tectonic theory: Australian Petroleum Exploration Journal, v. 14, p. 95–106.

Falvey, D. A., and Middleton, M. F., 1981, Passive continental margins; Evidence for a prebreakup deep crustal metamorphic subsidence mechanism, *in* Proceedings of 26th International Congress, Geology of continental margins symposium, Paris, 1980: Oceanologica Acta, no. SP, p. 143–153.

Falvey, D. A., and Mutter, J. C., 1981, Regional plate tectonics and the evolution of Australia's passive continental margins: BMR Journal of Australian Geology and Geophysics, v. 6, p. 1–29.

Foucher, J.-P., LePichon, X., and Sibuet, J.-C., 1982, The ocean-continent transition in the uniform lithospheric stretching model; Role of partial melting in the mantle: Philosophical Transactions of the Royal Society of London, v. A305, p. 27–43.

Gates, A. E., Simpson, C., and Glover, L., III, 1986, Appalachian Carboniferous dextral strike-slip faults; An example from Brookneal, Virginia: Tectonics, v. 5, p. 119–133.

Given, M. M., 1977, Mesozoic and early Cenozoic geology of offshore Nova Scotia: Canadian Petroleum Geology Bulletin, v. 25, p. 63–91.

Grow, J. A., 1980, Deep structure and evolution of the Baltimore Canyon Trough in the vicinity of the Cost No. B-3 well, *in* Scholle, P. A., ed., Geological studies of the COST No. B-3 Well, United States Mid-Atlantic Continental Slope area: U.S. Geological Survey Circular 833, p. 117–132.

——, 1981, The Atlantic margin of the United States, *in* Bally, A. W., ed., Geology of passive Continental margins; History, structure, and sedimentologic record (with special emphasis on the Atlantic margin): American Association of Petroleum Geologists Education Course Note Series no. 19, p. 3-0 to 3-41.

Grow, J. A., and Markl, R. G., 1977, IPOD-USGS multichannel seismic reflection profile from Cape Hatteras to the mid-Atlantic Ridge: Geology, v. 5, p. 625–630.

Grow, J. A., and Sheridan, R. E., 1981, Deep structure and evolution of the continental margin off eastern United States, *in* Proceedings of 26th International Geological Congress, Geology of continental margins symposium, Paris, 1980: Oceanologica Acta, no. SP, p. 11–19.

Grow, J. A., Bowin, C. O., and Hutchinson, D. R., 1979, The gravity field of the U.S. Atlantic continental margin: Tectonophysics, v. 59, p. 27–52.

Grow, J. A., Hutchinson, D. R., Klitgord, K. D., Dillon, W. P., and Schlee, J. S., 1983, Representative multichannel seismic profiles over the U.S. Atlantic margin, *in* Bally, A. W., ed., Seismic Expression of Structural Styles: American Association of Petroleum Geologists Studies in Geology Series no. 15, v. 2, p. 2.2.3-1–2.2.3-19.

Hamilton, R. M., Behrendt, J. C., and Ackermann, H. D., 1983, Land multichannel seismic-reflection evidence for tectonic features near Charleston, South Carolina, *in* Gohn, G. S., ed., Studies related to the Charleston, South Carolina, earthquake of 1886; Tectonics and seismicity: U.S. Geological Survey Professional Paper 1313, p. I1–I18.

Hansen, H. J., 1978, Upper Cretaceous (Senonian) and Paleocene (Danian) pinchouts on the south flank of the Salisbury Embayment, Maryland, and their relationship to antecedent basement structures: Maryland Geological Survey Report of Investigations no. 29, 36 p.

Hatcher, R. D., Jr., and Zietz, I., 1980, Tectonic implications of regional aeromagnetic data from the southern Appalachians, *in* Wones, D. R., ed., The Caledonides in the U.S.A., Proceedings of the second symposium of the International Geological Correlation Program, Caledonide Orogen Project: Blacksburg, Virginia Polytechnic Institute and State University Memoir 2, p. 235–244.

Hatcher, R. D., Jr., Howell, D. E., and Talwani, P., 1977, Eastern Piedmont Fault System; Speculation on its extent: Geology, v. 5, p. 636–640.

Haworth, R. T., and Jacobi, R. D., 1983, Geophysical correlation between the geological zonation of Newfoundland and the British Isles, *in* Hatcher, R. D., Jr., Williams, H., and Zietz, I., eds., Contributions to the tectonics and geophysics of mountain chains: Geological Society of America Memoir 158, p. 25–32.

Haworth, R. T., and Keen, C. E., 1979, The Canadian Atlantic margin; A passive continental margin encompassing an active past: Tectonophysics, v. 59, p. 83–126.

Hellinger, S. J., and Sclater, J. G., 1983, Some comments on two-layer extensional models for the evolution of sedimentary basins: Journal of Geophysical Research, v. 88, p. 8251–8269.

Higgins, M., and Zietz, I., 1983, Geologic interpretation of geophysical maps of the Pre-Cretaceous "basement" beneath the Coastal Plain of the southeastern United States, *in* Hatcher, R. D., Jr., Williams, H., and Zietz, I., eds., Contributions to the tectonics and geophysics of mountain chains: Geological Society of America Memoir 158, p. 125–130.

Hinz, K., 1981, A hypothesis on terrestrial catastrophes; Wedges of oceanward dipping subacoustic basement reflections; Their origin and paleoenvironmental consequences: Geologisches Jahrbuch Reihe E, Geophysik, v. 22, p. 3–28.

Hinze, W. J., ed., 1985, The utility of regional gravity and magnetic anomaly maps: Tulsa, Oklahoma, Society of Exploration Geophysicists, 454 p.

Hurtubise, D. O., Puffer, J. H., and Cousminer, H. L., 1987, An offshore Mesozoic igneous sequence, Georges Bank basin, North Atlantic: Geology, v. 98, p. 430–438.

Hutchinson, D. R., and Klitgord, K. D., 1987, Deep structure of rift basins from the continental margin around New England: U.S. Geological Survey Bulletin 2776 (in press).

Hutchinson, D. R., Grow, J. A., Klitgord, K. D., and Swift, B. A., 1983, Deep structure and evolution of the Carolina trough, *in* Watkins, J. S., and Drake, C. L., eds., Studies in Continental Margin Geology: American Association of Petroleum Geologists Memoir 34, p. 129–152.

Hutchinson, D. R., Klitgord, K. D., and Detrick, R. S., Jr., 1985, Block Island fault; A Paleozoic crustal boundary on the Long Island platform: Geology, v. 13, p. 875–879.

Hutchinson, D. R., Grow, J. A., Klitgord, K. D., and Detrick, R. S., 1986a, Moho reflections from the Long Island platform, eastern United States, *in* Barazangi, M., and Brown, L., eds., Reflection Seismology; The Continental Crust: American Geophysical Union Geodynamics Series, v. 14, p. 173–187.

Hutchinson, D. R., Klitgord, K. D., and Detrick, R. S., Jr., 1986b, Rift basins of the Long Island platform: Geological Society of America Bulletin, v. 97, p. 688–702.

Hutchinson, D. R., Klitgord, K. D., and Trehu, A. M., 1987a, Structure of the lower crust beneath the Gulf of Maine: Geophysical Journal of the Royal Astronomical Society, v. 89, p. 189–194.

Hutchinson, D. R., Klitgord, K. D., Lee, M. W., and Trehu, A. M., 1987b, U.S.G.S. deep seismic reflection profile across the Gulf of Maine: Geological Society of America Bulletin, v. 99, (in press).

Jansa, L. F., 1986, Paleoceanography and evolution of the North Atlantic Ocean Basin during the Jurassic, *in* Vogt, P. R., and Tucholke, B. E., eds., The Western North Atlantic Region: Boulder, Colorado, Geological Society of America, The Geology of North America, v. M, p. 603–616.

Karner, G. D., and Watts, A. B., 1982, On isostasy at Atlantic-type continental margins: Journal of Geophysical Research, v. 87, p. 2923–2948.

Keen, C. E., and Cordsen, A., 1981, Crustal structure, seismic stratigraphy, and rift processes of the continental margin off eastern Canada; Ocean bottom seismic refraction results off Nova Scotia: Canadian Journal of Earth Sciences, v. 18, p. 1523–1538.

Keen, M. J., 1969, Possible edge effect to explain magnetic anomalies off the eastern seaboard of the U.S.: Nature, v. 222, p. 72–74.

Keppie, J. D., 1985, Geology and tectonics of Nova Scotia, *in* Appalachian geotraverse (Canadian mainland): Geological Association of Canada–Mineralogical Association of Canada, Fredericton, New Brunswick, excursion 1, chapter 2, p. 23–108.

Klitgord, K. D., and Behrendt, J. C., 1979, Basin structure of the U.S. Atlantic margin, *in* Watkins, J. S., Montadert, L., and Dickerson, P. W., eds., Geological and geophysical investigations of continental margins: American Association Petroleum Geologists Memoir 29, p. 85–112.

Klitgord, K. D., and Grow, J. A., 1980, Jurassic seismic stratigraphy and basement structure of western Atlantic magnetic quiet zone: American Association Petroleum Geologists Bulletin, v. 64, p. 1658–1680.

Klitgord, K. D., and Hutchinson, D. R., 1985, Distribution and geophysical signatures of early Mesozoic rift basins beneath the U.S. Atlantic continental margin, *in* Robinson, G. R., Jr., and Froelich, A. J., eds., Proceedings of the Second U.S. Geological Survey Workshop on the Early Mesozoic basins of the eastern United States: U.S. Geological Survey Circular 946, p. 45–53.

Klitgord, K. D., and Schlee, J. S., 1987, Georges Bank subsurface geology, *in* Backus, R. H., ed., Georges Bank: Cambridge, MIT Press, p. 40–51.

Klitgord, K. D., and Schouten, H., 1980, Mesozoic evolution of the Atlantic, Caribbean, and Gulf of Mexico, *in* Pilger, R. H., Jr., ed., The origin of the Gulf of Mexico and the early opening of the central North Atlantic; Proceedings of a symposium, March 3–5, 1980: Baton Rouge, Louisiana State University, p. 100–101.

—— , 1986, Plate kinematics of the Central Atlantic, *in* Vogt, P. R., and Tucholke, B. E., The Western North Atlantic Region: Boulder, Colorado, Geological Society of America, The Geology of North America, v. M, p. 351–378.

—— , 1987a, Tectonic and magnetic features; Carolina Trough and adjacent magnetic quiet zone: U.S. Geological Survey Miscellaneous Field Studies Map, scale 1:1,000,000 (in press).

—— , 1987b, Tectonic and magnetic features: Baltimore Canyon Trough and adjacent magnetic quiet zone: U.S. Geological Survey Miscellaneous Field Studies Map, scale 1:1,000,000 (in press).

Klitgord, K. D., Schlee, J. S., and Hinz, K., 1982, Basement structure, sedimentation, and tectonic history of the Georges Bank basin, *in* Scholle, P. A., and Wenkam, C. R., eds., Geological Studies of the COST Nos. G-1 and G-2 Wells, United States North Atlantic Outer Continental Shelf: U.S. Geological Survey Circular 861, p. 160–186.

Klitgord, K. D., Dillon, W. P., and Popenoe, P., 1983, Mesozoic tectonics of the southeastern United States Coastal Plain and Continental Margin, *in* Gohn, G. S., ed., Studies related to the Charleston, South Carolina, earthquake of 1886; Tectonics and seismicity: U.S. Geological Survey Professional Paper 1313, p. P1–P15.

Klitgord, K. D., Popenoe, P., and Schouten, H., 1984, Florida; A Jurassic transform plate boundary: Journal of Geophysical Research, v. 89, p. 7753–7772.

LASE Study Group, 1986, Deep structure of the U.S. East Coast passive margin from large aperture seismic experiments (LASE): Marine and Petroleum Geology, v. 3, p. 234–242.

Lefort, J. P., 1983, A new geophysical criterion to correlate the Acadian and Hercynian orogenies of western Europe and eastern America, *in* Hatcher, R. D., Jr., Wililams, H., and Zietz, I., eds., Contributions to the tectonics and geophysics of mountain chains: Geological Society of America Memoir 158, p. 3–18.

Lefort, J. P., and Van der Voo, R., 1981, A kinematic model for the collision and complete suturing between Gondwanaland and Laurussia in the Carboniferous: Journal of Geology, v. 89, p. 537–550.

Lemoine, M., and others, 1986, The continental margin of the Mesozoic Tethys in the Western Alps: Marine and Petroleum Geology, v. 3, p. 179–199.

LePichon, X., and Barbier, F., 1987, Passive margin formation by low-angle faulting within the upper crust; The northern Bay of Biscay margin: Tectonics, v. 6, p. 133–150.

LePichon, X., and Sibuet, J. C., 1981, Passive margins; A model of formation: Journal of Geophysical Research, v. 86, p. 3708–3720.

LePichon, X., Sibuet, J. C., and Francheteau, J., 1977, The fit of the continents around the North Atlantic Ocean: Tectonophysics, v. 38, p. 169–209.

Lister, G. S., Etheridge, M. A., and Symonds, P. A., 1986, Detachment faulting and the evolution of passive continental margins: Geology, v. 14, p. 246–250.

Long, L. T., and Dainty, A. M., 1985, Studies of gravity anomalies in Georgia and adjacent areas of the southeastern United States, *in* Hinze, W. J., ed., The Utility of Regional Gravity and Magnetic Anomaly Maps: Tulsa, Oklahoma, Society of Exploration Geophysicists, p. 308–319.

McHone, J. G., and Butler, J. R., 1984, Mesozoic igneous provinces of New England and the opening of the North Atlantic Ocean: Geological Society of America Bulletin, v. 95, p. 75–765.

McKenzie, D., 1978, Some remarks on the development of sedimentary basins: Earth and Planetary Science Letters, v. 40, p. 25–32.

Maher, J. C., 1971, Geologic framework and petroleum potential of the Atlantic Coastal Plain and continental shelf: U.S. Geological Survey Professional Paper 659, 98 p.

Manspeizer, W., 1985, Early Mesozoic history of the Atlantic passive margin, *in* Poag, C. W., ed., Geologic evolution of the United States Atlantic Margin: New York, Van Nostrand Reinhold Co., p. 1–23.

Manspeizer, W., Puffer, J. H., and Cousminer, H. L., 1978, Separation of Morocco and eastern North America, A Triassic-Liassic stratigraphic record: Geological Society of America Bulletin, v. 89, p. 901–920.

Montadert, L., Roberts, D. G., DeCharpal, O., and Guennoc, P., 1979, Rifting and subsidence of the northern continental margin of the Bay of Biscay, *in* Montadert, L., and Roberts, D. G., eds., Initial reports of the Deep Sea Drilling Project: Washington, D.C., U.S. Government Printing Office, v. 48, p. 1025–1060.

Mutter, J. C., 1985, Seaward dipping reflectors and the continent-ocean boundary at passive continental margins: Tectonophysics, v. 114, p. 117–131.

Mutter, J. C., and North Atlantic Transect Study Group, 1985, Multichannel seismic images of the oceanic crust's internal structure; Evidence for a magma chamber beneath the Mesozoic Mid-Atlantic Ridge: Geology, v. 13, p. 629–632.

Mutter, J. C., Detrick, R. S., and North Atlantic Transect Study Group, 1984, Multichannel seismic evidence for anomalously thin crust at Blake Spur fracture zone: Geology, v. 12, p. 534–537.

Nelson, K. D., Arnow, J. A., McBride, J. H., Willemin, J. H., Huang, J., Zheng, L., Oliver, J. E., Brown, L. D., and Kaufman, S., 1985a, New COCORP profiling in the southeastern United States. Part I, Late Paleozoic suture and Mesozoic rift basin: Geology, v. 13, p. 714–718.

Nelson, K. D., McBride, J. H., Arnow, J. A., Oliver, J. E., Brown, L. D., and Kaufman, S., 1985b, New COCORP profiling in the southeastern United States. Part II, Brunswick and East Coast Magnetic anomalies; Opening of the north-central Atlantic Ocean: Geology, v. 13, p. 718–721.

Nelson, K. D., McBride, J. H., Arnow, J. A., Wille, L. D., Oliver, J. E., and Kaufman, S., 1987, Results of recent COCORP profiling in the southeastern

United States: Geophysical Journal of the Royal Astronomical Society, v. 89, p. 141–146.

Nicolas, A., 1985, Novel type of crust produced during continental rifting: Nature, v. 315, p. 112–115.

Phinney, R. A., 1986, A seismic cross section of the New England Appalachians; The orogen exposed, *in* Barazangi, M., and Brown, L., eds., Reflection Seismology; The continental crust: American Geophysical Union Geodynamics Series, v. 14, p. 157–172.

Poag, C. W., 1982, Stratigraphic reference section for Georges Bank Basin; Depositional model for New England passive margin: American Association of Petroleum Geologists, v. 66, p. 1021–1041.

—— , ed., 1985, Geologic evolution of the United States Atlantic Margin: New York, Van Nostrand Reinhold Co., 383 p.

—— , 1987, The New Jersey transect; Stratigraphic framework and depositional history of a sediment-rich passive margin, *in* Poag, C. W., and Watts, A. B., eds., Initial Reports of the Deep Sea Drilling Project: Washington, D.C., U.S. Government Printing Office, v. 95, p. 763–817.

Popenoe, P., and Zietz, I., 1977, The nature of the geophysical basement beneath the Coastal Plain of South Carolina and Northeastern Georgia, *in* Rankin, D. W., ed., Studies related to the Charleston, South Carolina, earthquake of 1886; A preliminary report: U.S. Geological Survey Professional Paper 1028-I, p. 119–137.

Pratt, T. L., Coruh, C., Costain, J. K., Glover, L., III, and Gates, A. E., 1986, Preliminary interpretation of reprocessed USGS I-64 seismic reflection line in central Virginia: Geological Society of America Abstracts with Programs, v. 18, no. 6, p. 722.

Pratt, T. L., Coruh, C., and Costain, J. K., 1987, Lower crustal reflections in central Virginia: Geophysical Journal of the Royal Astronomical Society, v. 89, p. 163–170.

Purdy, G. M., and Ewing, J., 1986, Seismic Structure of the ocean crust, *in* Vogt, P. R., and Tucholke, B. E., eds., The Western North Atlantic Region: Geological Society of America, The Geology of North America, v. M, p. 313–330.

Rabinowitz, P., 1974, The boundary between oceanic and continental crust in the western North Atlantic, *in* Burk, C. A., and Drake, C. L., eds., The Geology of continental margins: New York, Springer-Verlag, p. 67–84.

Ratcliffe, N. M., Burton, W. C., D'Angelo, R. M., and Costain, J. K., 1986, Low-angle extensional faulting, reactivated mylonites, and seismic reflection geometry of the Newark basin margin in Pennsylvania: Geology, v. 14, p. 766–770.

Roberts, D. G., Morton, A. C., and Backman, J., 1985, Late Paleocene–Early Eocene volcanic events in the northern North Atlantic Ocean, *in* Roberts, D. G., and Schnitker, D., eds., Initial Reports of the Deep Sea Drilling Project: Washington, D.C., U.S. Government Printing Office, v. 81, p. 913–923.

Robinson, E. S., Poland, P. V., Glover, L., III, and Speer, J. A., 1985, Some effects of regional metamorphism and geologic structure on magnetic anomalies over the Carolina Slate Belt near Roxboro, North Carolina, *in* Hinze, W. J., ed., The Utility of Regional Gravity and Magnetic Anomaly Maps: Tulsa, Oklahoma, Society of Exploration Geophysicists, p. 320–324.

Robinson, G. R., Jr., and Froelich, A. J., eds., 1985, Proceedings of the Second U.S. Geological Survey Workshop on the Early Mesozoic Basins of the eastern United States: U.S. Geological Survey Circular 946, 147 p.

Roussel, J., and Liger, J. L., 1983, A review of deep structure and ocean-continent transition in the Senegal Basin (West Africa): Tectonophysics, v. 91, p. 183–211.

Salvador, A., 1987, Late Triassic-Jurassic paleogeography and origin of the Gulf of Mexico Basin: American Association of Petroleum Geologists Bulletin, v. 71, p. 419–451.

Sawyer, D. S., Toksoz, M. N., Sclater, J. G., and Swift, B. A., 1983, Thermal evolution of the Baltimore Canyon Trough and Georges Bank Basin: American Association of Petroleum Geologists Memoir 34, p. 743–762.

Schlee, J. S., and Fritsch, J., 1983, Seismic stratigraphy of the Georges Bank basin complex, offshore New England, *in* Watkins, J. S., and Drake, C. L., eds.,

Studies in Continental Margin Geology: American Association of Petroleum Geologists Memoir 34, p. 223–251.

Schlee, J. S., and Grow, J. A., 1980, Buried carbonate shelf edge beneath the Atlantic continental slope: Oil and Gas Journal, 25 Feb. 1980, p. 148–159.

Schlee, J. S., and Jansa, L. F., 1981, The paleoenvironment and development of the eastern North American continental margin: Oceanologica Acta, no. SP, p. 71–80.

Schlee, J. S., and Klitgord, K. D., 1986, Structure of the North American Atlantic continental margin: Journal of Geological Education, v. 34, p. 72–89.

Schlee, J. S., Dillon, W. P., and Grow, J. A., 1979, Structure of the continental slope off the eastern United States, *in* Doyle, L. J., and Pikey, O. H., eds., Geology of continental slopes: Society of Economic Paleontologist and Mineralogists Special Publication 27, p. 95–118.

Schouten, H., and Klitgord, K. D., 1982, The memory of the accreting plate boundary and the continuity of fracture zones: Earth and Planetary Science Letters, v. 59, p. 255–266.

Schouten, H., Klitgord, K. D., and Whitehead, J. A., 1985, Segmentation of mid-ocean ridges: Nature, v. 317, p. 225–229.

Sclater, J. G., and Christie, P. A., 1980, Continental stretching; An explanation of the post Mid-Cretaceous subsidence of the Central North Sea basin: Journal of Geophysical Research, v. 85, p. 3711–3739.

Sheridan, R. E., 1974, Atlantic Continental Margin of North America, *in* Burk, C. A., and Drake, C. L., eds., The Geology of Continental Margins: New York, Springer-Verlag, p. 391–407.

Sheridan, R. E., Grow, J. A., Behrendt, J. C., and Bayer, K. C., 1979, Seismic refraction study of the continental edge off the eastern United States: Tectonophysics, v. 59, p. 1–26.

Sheridan, R. E., Crosby, J. T., Bryan, G. M., and Stoffa, P. L., 1981a, Stratigraphy and structure of southern Blake Plateau, northern Florida Straits, and northern Bahama Platform from multichannel seismic reflection data: American Association of Petroleum Geologists Bulletin, v. 65, p. 2571–2593.

Sheridan, R. E., Crosby, J. T., Kent, K. M., Dillon, W. P., and Paull, C. K., 1981b, The geology of the Blake Plateau and Bahamas region, *in* Kerr, J. W., Fergusson, A. J. and Machan, L. C., eds., Geology of the North American Borderlands: Canadian Society of Petroleum Geologists Memoir 7, p. 487–502.

Steckler, M. S., and Watts, A. B., 1982, Subsidence history and tectonic evolution of Atlantic-type continental margins, *in* Scrutton, R. A., ed., Dynamics of passive margins: American Geophysical Union Geodynamics Series, v. 6, p. 184–196.

Stewart, D. B., and others, 1986, The Quebec–Western Maine seismic reflection profile—setting and first year results, *in* Barazangi, M., and Brown, L., eds., Reflection Seismology; The continental crust: American Geophysical Union Geodynamics Series, v. 14, p. 189–199.

Sutter, J. F., 1985, Progress on geochronology of Mesozoic diabases and basalts: U.S. Geological Survey Circular 946, p. 110–114.

Swanson, M. T., 1982, Preliminary model for an early transform history in central Atlantic rifting: Geology, v. 10, p. 317–320.

—— , 1986, Pre-existing fault control for Mesozoic basin formation in eastern North America: Geology, v. 14, p. 419–422.

Taylor, P. T., Zietz, I., and Dennis, L. S., 1968, Geologic implications of aeromagnetic data for the eastern continental margin of the United States: Geophysics, v. 33, p. 755–780.

Trehu, A. M., Muller, G. K., Gettrust, J. F., Ballard, A., Dorman, L. M., and Schreiner, T., 1986, Deep crustal structure beneath the Carolina Trough [abs.]: EOS Transactions of the American Geophysical Union, v. 67, no. 44, p. 1103.

Trehu, A. M., Klitgord, K. D., Sawyer, D. S., and Buffler, R. T., 1987, Regional investigations of the crust and upper mantle, Atlantic and Gulf of Mexico continental margins, *in* Pakiser, L., and Mooney, W., eds., Geophysical framework of the continental United States: Geological Society of America Memoir (in press).

Tucholke, B. E., Houtz, R. E., and Ludwig, W. J., 1982, Sediment thickness and depth to basement in western North Atlantic ocean basin: American Association of Petroleum Geologists Bulletin, v. 66, p. 1384–1395.

Uchupi, E., 1983, Tectonic features, *in* Uchupi, E., and Shor, A. N., eds., Eastern North American Continental Margin and adjacent ocean floor, 39° to 46°N and 64° to 74°W: Ocean Margin Drilling Program, Regional Atlas Series 3, sheet 27.

Uchupi, E., and Austin, J. A., 1979, The geologic history of the passive margin off New England and the Canadian Maritime Provinces: Tectonophysics, v. 59, p. 53–69.

Uchupi, E., Phillips, J. D., and Prada, K. E., 1970, Origin and structure of the New England Seamount Chain: Deep Sea Research, v. 17, p. 483–494.

Uchupi, E., Bolmer, S. T., Jr., Eusden, J. D., Jr., Ewing, J. I., Costain, J. K., Gleason, R. J., and Glover, L. III, 1983a, Tectonic features, *in* Shor, A. N., and Uchupi, E., eds., Eastern North American Continental Margin and adjacent ocean floor, 34° to 41°N and 68° to 78°W: Ocean Margin Drilling Program, Regional Atlas Series 4, sheet 28.

Uchupi, E., Crosby, J. T., Bolmer, S. T., Jr., Eusden, J. D., Jr., Ewing, J. I., Costain, J. K., Gleason, R. J., and Glover, L., III, 1983b, Tectonic features, *in* Bryan, G. M., and Heirtzler, J. R., eds., Eastern North American Continental Margin and adjacent ocean floor, 28° to 36°N and 70° to 82°W: Ocean Margin Drilling Program, Regional Atlas Series 5, sheet 36.

Van der Voo, R., Mauk, F. J., and French, R. B., 1976, Permian-Triassic continental configurations and the origin of the Gulf of Mexico: Geology, v. 4, p. 177–180.

Vogt, P. R., 1973, Early events in the opening of the North Atlantic, *in* Tarling, D. H., and Runcorn, S. K., eds., Implications of continental drift to the Earth Sciences, v. 2: London, Academic Press, p. 693–712.

Watts, A. B., 1982, Tectonic subsidence, flexure, and global changes of sea level: Nature, v. 297, p. 469–474.

Wernicke, B., 1985, Uniform-sense normal simple shear of the continental lithosphere: Canadian Journal of Earth Sciences, v. 22, p. 108–125.

White, R. S., Detrick, R. S., Sinha, M. C., and Cormier, M. H., 1984, Anomalous seismic crustal structure of oceanic fracture zones: Geophysical Journal of the Royal Astronomical Society, v. 79, p. 779–798.

White, R. S., Westbrook, G. K., Fowler, S. R., Spence, G. D., Barton, P. J., Joppen, M., Morgan, J., Bowen, A. N., Prestcott, C., and Bott, M.H.P., 1987, Hatton Bank (northwest U.K.) continental margin structure: Geophysical Journal of the Royal Astronomical Society, v. 89, p. 265–272.

Williams, H., and Hatcher, R. D., Jr., 1983, Appalachian suspect terranes, *in* Hatcher, R. D., Jr., Williams, H., and Zietz, I., eds., Contributions to the tectonics and geophysics of mountain chains: Geological Society of America Memoir 158, p. 33–53.

Ziegler, P. A., 1982, Faulting and graben formation in western and central Europe: Royal Society of London, Philosophical Transactions, v. A305, p. 113–143.

MANUSCRIPT ACCEPTED BY THE SOCIETY SEPTEMBER 17, 1987
WOODS HOLE OCEANOGRAPHIC INSTITUTION CONTRIBUTION #6533

ACKNOWLEDGMENTS

Numerous field programs and discussions with our colleagues at Woods Hole (U.S. Geological Survey and Woods Hole Oceanographic Institution) contributed to substantial amounts of the work summarized here. In particular, we thank J. Behrendt, M. Ball, C. Bowin, W. Dillon, K. O. Emery, J. Ewing, D. Folger, J. Grow, W. Poag, P. Popenoe, M. Purdy, J. Schlee, A. Trehu, B. Tucholke, and E. Uchupi. Field acquisition and processing support were provided by T. O'Brien, D. Nichols, G. Miller, D. Taylor, and R. Wise. We also acknowledge fruitful discussions with J. Austin, W. Bothner, L. Glover, C. Keen, W. Manspeizer, J. Mutter, R. Sheridan, D. Sawyer, and M. Talwani and careful critiques of the final manuscript by W. Dillon, J. Grow, J. Mutter, and W. Manspeizer. We thank M. C. Mons-Wengler and E. Winget for manuscript technical assistance and P. Forrestel, J. Zwinakis, and D. Blackwood for graphics assistance.

Printed in U.S.A.

The Geology of North America
Vol. I-2, The Atlantic Continental Margin: U.S.
The Geological Society of America, 1988

Chapter 4

History of studies of the Atlantic margin of the United States

Charles L. Drake
Earth Sciences Department, Dartmouth College, Hanover, New Hampshire 03755

INTRODUCTION

Early views on the continental margin of the eastern United States were derived from studies of land geology, since there was little information available at sea. The parallelism of the Appalachian mountains and the continental margin led to speculations about the relationship between the two. In the middle of the 19th century, and indeed, well into the early part of the 20th, much of the crystalline part of the Appalachian system was regarded as pre-Paleozoic in age and discussion of mountain building concentrated on the folding of the sedimentary rock sequences of the Valley and Ridge Province rather than upon the whole system.

In 1846, James Dana suggested that early in the history of the earth the continents cooled first and maintained their elevations while the oceanic areas, of volcanic character, continued to cool and subside through time. As a consequence, "Ruptures, elevations, foldings and contortions of strata have been produced in the course of contraction. The greater subsidence of the oceanic parts would necessarily occasion that lateral pressure required for the rise and various foldings of the Appalachians and like regions."

Chamberlin and Salisbury (1907) held similar views about the antiquity of continents and oceans, but pictured the oceanic areas as sinking because of their greater specific gravity to cause shear forces at the continental margins.

James Hall (1843) recognized that thickening of the Paleozoic sandstones to the east required a source to the southeast. Dana (1856) pictured a "great reef or sand bank, partly hemming in a large lagoon . . ." later to be raised as the folded Appalachians and separating this lagoon from the open Atlantic, but he could not accept that the rocks of the continent had been supplied from any such region now sunk beneath the seas. He called upon the land to the north as a source and on the Labrador Current as the mechanism for supplying those sediments. By 1890 new knowledge caused him to change his views and he concluded that the source for the Appalachian sediments was an Archaean land mass now beneath the continental margin (Dana, 1890). This view was supported by Walcott (1891) on the basis of earliest Paleozoic stratigraphy: "It is not improbable that the area of the great Coastal Plain of the Atlantic was then an elevated portion of the continent and that much of the sediment deposited during

Cambrian and later Paleozoic was washed from it into the seas immediately to the west." Claypole (1891) noted that the sediment source problem was not limited to North America:

The immense Paleozoic deposits in eastern North America and northwestern Europe are far too large to have been supplied from any existing Paleozoic land. The increasing thickness of the former to the eastward and northward, and of the latter to the west and northwest, are strong indications that their origin must be sought in that direction. Accordingly, in this region—that is in the area of the present North Atlantic—not a few geologists are inclined to place a Paleozoic continent which has since disappeared, but which was then the quarry from whence came the material that has built up these massive strata on both sides of the Atlantic."

This disappearing continent ran counter to the concept of permanence, but there were doubts whether there was sufficient space beneath the coastal plain and the continental shelf for the proposed Paleozoic landmass that supplied sediments to the Appalachian system (Gilbert, 1893). The idea of a landmass to the east of the folded Appalachians, named Appalachia by Williams (1897), found favor and became an integral part of the borderland concept championed particularly by Schuchert (1910, 1923) (Fig. 1). This view of the continental margin of the eastern United States was the ruling hypothesis until the 1940s, although not totally unquestioned (van der Gracht, 1931).

Geophysical measurements on the continental margin were limited before 1930. Soundings and, after 1921, echo soundings had begun to define the major bathymetric features, and the propagation of earthquake waves was used to investigate the differences between continental and oceanic structure. These studies were interpreted to suggest that the crustal layers characteristic of the continents were absent beneath the Pacific but present, though thinner, beneath the Atlantic and Indian Oceans (Gutenberg and Richter, 1925). Only a few gravity measurements existed in the offshore part of the margin (Vening Meinesz and Wright, 1930); too few to be useful in determining whether the transition across the margin followed the model of Helmert (1909). Wegener (1924), from the few available gravity measurements of limited accuracy, suggested that the oceanic

Drake, C. L., 1988, History of studies of the Atlantic margin in the United States; *in* Sheridan, R. E., and Grow, J. A., eds., The Geology of North America, Volume I-2, The Atlantic Continental Margin, U.S.: Geological Society of America.

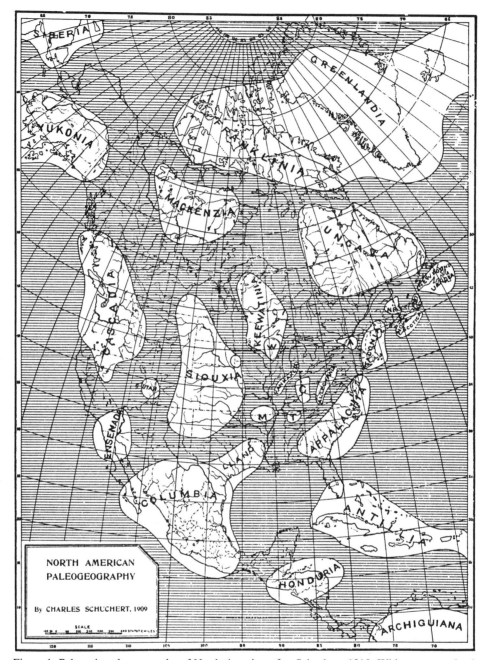

Figure 1. Paleozoic paleogeography of North America, after Schuchert, 1910. White areas are land, horizontally ruled areas are epicontinental or deep seas. Appalachia, named by Williams (1897) was proposed as a source for the thick clastic sediments of the folded Appalachians.

crust was qualitatively different from that of the continents and lacked the granitic, or sialic, layer.

EARLY OFFSHORE WORK

The first chart of the coast—the New York area—was published in 1835. By the 1870s the shelf edge had been identified and was found to be notched by submarine canyons. The Blake Plateau had been discovered and was surveyed in some detail in

the 1880s. Maury's (1858) map of the North Atlantic basin, based on a limited number of wire soundings, showed a gradual transition from the shelf to the abyss.

Prevailing opinion had developed that sediments should become progressively finer toward the shelf edge despite an abundant record from bottom samples, taken for navigation purposes, that contradicted it (Pourtales, 1872). It was known (Murray, 1885; Agassiz, 1888) that the Gulf Stream reached to the sea floor on the Blake Plateau and that beneath it the bottom, for the most part, was hard with little unconsolidated sediment. However, other currents or wave actions were perceived to be of insufficient strength to move coarse sediments except in nearshore areas or on shallow fishing banks.

Maury (1858) attributed these submerged banks off New England to the accumulation of iceberg rafted debris. Based on studies of the materials from the banks, Thoulet (1889) concluded that they came from neighboring shores rather than the far north, brought by shore ice, rather than icebergs, and moved by rivers instead of the Labrador Current.

In 1840, MacLaren noted that the formation of large ice sheets would result in a lowering of sea level. Evidence of this lower sea level and the subsequent rise following glacial retreat was found by Mudge (1858) who observed that the salt marshes of Lynn, Massachusetts, to depths of 8 ft (2.4 m) were composed of supratidal vegetation. He attributed this to local subsidence, but Johnson (1925) subsequently concluded that it was the result of slow submergence in post-glacial time.

The origin of submarine canyons was a difficult problem. Although acknowledging that there must be deep currents in the ocean, Maury (1858) did not credit these or any other agents with the power to erode or transport sediments: "In the deep sea there are no abrading processes at work; neither frosts or rain are felt there, and the force of gravitation is so paralyzed that it cannot use half its power, as on dry land, in tearing the overhanging rock from the precipice and casting it down into the valley below."

Consequently, early concepts about the formation of topographic features on the shelf and of submarine canyons cut into the slope focused on subaerial processes acting during lower relative sea level. From fragments of Upper Tertiary rock recovered in fishing nets, Upham (1894) concluded: "The Fishing Banks are thus to be accounted, like the fjords of our northern coasts, the submerged continuation of the Hudson River channel, and the similar very deep submarine valleys off the coast of California near Cape Mendocino . . . as evidence of a great epeirogenic uplift of the northern part of the continent preceding and producing the ice age."

Spencer (1903) concluded from "scanty" soundings that the Hudson Canyon could be inferred to extend to a depth of between 11,400 and 15,000 ft (3,500 to 4,600 m) (Fig. 2). He concluded that the submarine canyons had been produced subaerially, but could not explain what became of the waters or what caused the great continental movements during the period of erosion.

Sands and gravels had been recovered from the deep sea as early as the *Porcupine* expedition in 1869 (Thomson, 1874) and were usually attributed to ice rafting or "the ordinary action of marine currents." Agassiz was astonished by the amount of shallow water and terrestrial material—logs, twigs, decaying vegetable matter—from below 2000 fathoms. Many of these materials could be explained by the above mechanisms or by gradual sinking or downslope movements, but some could not. Prevailing opinion followed Maury who wrote (1855): "It is an established fact that there is no running water at the bottom of the deep

sea. . . .The agents which disturb the equilibrium of the sea, giving violence to its waves and force to its currents, all reside near or above its surface; none of them have its home in its depths."

By 1858 Maury recognized that there must be deep undercurrents in the Atlantic: "I may here remark, that there seems to be a larger flow of polar waters into the Atlantic than of other waters from it, and I cannot account for the preservation of equilibrium of this ocean by any other hypothesis than that which calls in the aid of undercurrents. They, I have no doubt, bear an important part in the system of ocean circulation." He did not attribute to these deep undercurrents enough strength to erode or transport sediments as evidenced by his comment that a submarine cable on the deep sea floor would "lie in cold obstruction, without anything to fret, chafe, or wear, save only the tooth of time."

NASCO AND AGU

The National Academy of Sciences formed a Committee on Oceanography (NASCO) in 1927. Its recommendations led to a series of grants from the Rockefeller Foundation and the Carnegie Corporation, which were used to establish the Woods Hole Oceanographic Institution (WHOI). The steel ketch, *Atlantis,* was built for WHOI and began research on the geology of the continental margin. In 1932 the American Geophysical Union (AGU) organized a symposium on ocean basins and margins, chaired by Professor R. M. Field of Princeton, to "further emphasize the importance of this phase of oceanographic research [geology and geophysics] not only from the point of view of the relation of geophysical methods to marine geophysical problems, but also the development of a reasonably sound method of attack" (Field, 1932).

The symposium discussed the application of echo sounding, gravity measurements, and both earthquake and explosion seismology to determine the structure and properties of the ocean basins and their margins and led to the establishment of a Committee on the Geophysical and Geological Study of the Ocean Basins and their Margins to promote further research. At the same meeting, in the Seismology Section, a paper was presented (de Smitt, 1932) about submarine cables that broke following the 1929 Grand Banks earthquake. Large portions of the cables were buried or carried away and the cables broke progressively according to their distance from the epicenter. The last break, some 300 miles distant and 13 hours after the earthquake, was on sea floor sloping "less than a degree."

Shepard (1932) noted that samples collected by the Coast and Geodetic Survey did not support the concept of sediments fining seaward and that there was a lack of sedimentation, even evidence of erosion, related to the last glacial epoch. Further sampling of the shelf (Shepard and Cohee, 1936; Shepard and others, 1934; Shepard, 1939; Stetson, 1934, 1938, 1939) led to the conclusion that the surficial sediments on the shelf were relict; the sediments were related to an earlier glacial environment and were only moderately modified by modern processes. Shepard

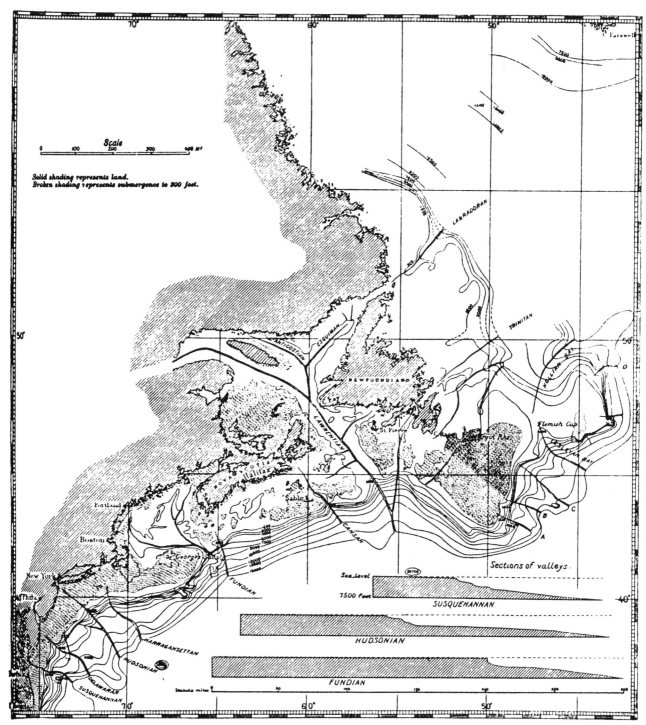

Figure 2. Submarine channels and canyons of eastern North America, after Spencer, 1903. On the basis of very scanty data Spencer concluded that these canyons extended to depths of 12,000 to 15,000 ft. (3658 to 4572 m).

noted the resemblance of the shelf sediments to beach sands from New Jersey and Long Island that had been studied earlier (McCarthy, 1931).

Stetson was less convinced of this near shore origin, and noted a varied sediment distribution. He called attention to the erosive powers of the Gulf Stream and suggested that they might result in the anomalous situation of a sand being laid down in deep water contemporaneously with future shales in much shallower water. Stetson (1936, 1937) also dredged sediments of Cretaceous and Late Tertiary age from the walls of submarine

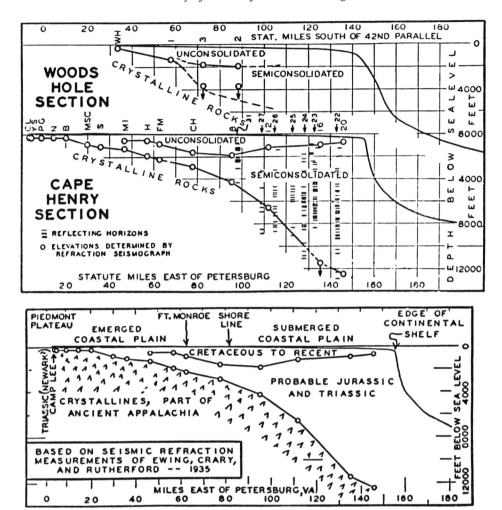

Figure 3. Top: Results of the first seismic refraction experiments on the continental shelf, after Ewing and others, 1937. Circles indicate depth determinations; horizontal bars denote reflecting horizons. Bottom: Geological Interpretation of seismic results, after Miller, 1937. The crystalline rock surface was interpreted as the subsided remnant of Appalachia.

canyons off Georges Bank, the exposure of which was an indication of erosion. Measurements of (tidal) currents indicated that they were similar to those on the adjacent shelf. The currents flowed in pre-existing cuts but did not cause them. Although Stetson indicated that the floor of the Gulf of Maine must have stood some 7000 ft (2100 m) higher with respect to sea level at the time of the canyon cutting, he regarded the question of subaerial erosion as open.

Daly (1936) found the concept of major sea level changes as an explanation of canyon cutting hard to accept and argued for density currents produced by mud-laden storm waters. Bucher (1940) suggested that tsunamis might contribute, perhaps in combination with other phenomena. Smith, however (Veatch and Smith, 1939), was taken by the resemblances of the slope topography contoured from detailed Coast and Geodetic Survey data to erosional features on land and concluded that the evidence was overwhelmingly in favor of subaerial erosion to depths of at least 1200 fathoms.

After tests of techniques in New Jersey and aboard the USCGS vessel *Oceanographer* in the Spring of 1935, M. Ewing and his colleagues made the first seismic refraction measurements on the Atlantic shelf off Woods Hole and Virginia (Ewing and others, 1937) (Fig. 3). The basement surface on the outer end of the Virginia line was found to be some 12,000 ft. (3658 m) below sea level. In the accompanying geological interpretation Miller (1937) concluded that this surface, "the remnants of Ancient Appalachia," had sunk by downwarping rather than profound faulting. Woollard (1940), who combined the seismic refraction results with land gravity measurements and a few marine measurements by Vening Meinesz, suggested alternatively that there might be eastward thinning of the continental crust or possibly a density transition in the crust at the continental border. Gunn (1944), on the other hand, concluded that a significant part of the gravity anomalies was due to finite strength of the lithosphere.

Two more seismic sections were shot across the coastal plain in New Jersey in the late 1930s (Ewing and others, 1939, 1940), and experiments were carried out to test methods for doing refraction experiments in deep water (Ewing, 1938; Ewing and Vine, 1938).

THE POSTWAR PERIOD: 1946–1963

In the years immediately after the war, Richards (1945, 1948) summarized the stratigraphy of the Atlantic coastal plain sediments, and petroleum exploration led to drilling of a number of [dry] holes. Marshall Kay (1951) had become convinced that the Paleozoic borderland, Appalachia, had been a great geosyncline and he divided the Appalachian orthogeosyncline into an inner non-volcanic miogeosyncline and an outer eugeosyncline that lay in a belt of active volcanism in association with relatively rapid subsidence. In this model the sediments are essentially indigenous; from the volcanic island arc or from welts raised tectonically within the geosyncline. Eardley (1947) suggested that there might be paleogeographic analogies to modern volcanic archipelagoes, a concept supported by Kay.

Kay noted the similarity of the thick sedimentary accumulations along the Atlantic and Gulf margins to the miogeosynclines, but in the absence of a seaward belt of active deformation and volcanism he concluded that these regions represented "a new cycle of subsidence in a region that had passed through an orogenic and plutonic history after having been an eugeosynclinal belt of the same general trend."

Dietz (1952) suggested that the continental shelf, because it is swept by strong currents, and the continental slope, because of its steep topography, accumulate little permanent sediment; most of the sediment ends up on the continental rise. As the sediments build up, the underlying crust is isostatically depressed and this downwarping controls the deposition on the shelf itself.

From studies of the sediments and the cable breaks, Heezen and Ewing (1952) proposed that the propagating disturbance that broke the cables following the Grand Banks earthquake of 1929 was a scaled-up example of the high-density turbidity currents produced experimentally by Kuenen (1948). In subsequent papers the ability of these currents to produce graded beds of the kind found in the graywackes of geosynclines and in deep sea sediments, to actively erode submarine canyons, and to carry shallow water sediments and biota to the deep sea floor was demonstrated.

The first deep sea seismic refraction measurements (Ewing and others, 1950) indicated that the "granitic" crust was absent and the mantle was very shallow. Seismic refraction measurements were continued across the continental margin by many investigators from the Woods Hole Oceanographic Institution and Columbia University; the results are summarized in Drake and others (1959). In that paper a comparison was made between the continental margin and the Appalachian system (Fig. 4, A, B). The sedimentary trough beneath the shelf, containing shallow water sediments, was compared with the Appalachian miogeosyncline and the trough beneath the slope and rise, presumably

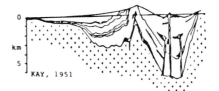

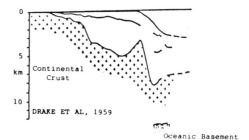

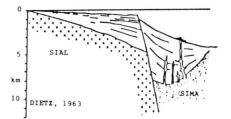

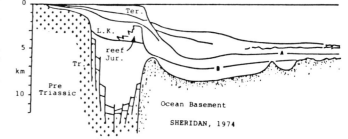

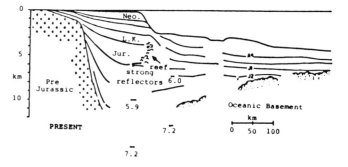

Figure 4. Post World War II subsidence of the eastern continental margin, after Steckler, 1981. A: Early Paleozoic reconstruction of the Appalachian system, after Kay, 1951. B: Structure of the continental margin off Cape May, New Jersey, from seismic measurements, after Drake and others 1959. C: Proposed structure of the eastern continental margin based on a sediment loading model, after Dietz, 1963. D: Reinterpretation of seismic refraction data across the New Jersey continental margin, after Sheridan, 1974. Seismic horizon interpreted as basement in section B is now considered to represent Mesozoic sediments. E: Recent section based on seismic reflection and refraction and borehole data, after Steckler, 1974. Configuration of the basement surface is not clear, but the basement is deep and no basement ridge is apparent.

filled with turbidites, was likened to the Appalachian eugeosyncline. A ridge near the shelf edge was identified as basement and compared with the Precambrian axis that separates the two geosynclinal troughs of the Appalachians. Seismic velocities in the layer identified as (Paleozoic) basement decreased seaward and this was attributed to a lessening of the degree of metamorphism and a change in character of the basement from crystallines to sediments, pyroclastics, and volcanic flows. Although there was no active volcanism in the deep water trough identified with the eugeosyncline, it was presumed that volcanism would occur when deformation of the geosyncline began. When deformation elevated the rise area above sea level, it would serve as a source for sedimentation in the shallow waters to the west as suggested by Kay (1951).

The sediments in the deep water trough were thought to be similar to graywackes, and many examples of graywackes in orogenic belts have been cited (Pettijohn, 1949). However, Rich (1950) proposed a continental slope environment for the Silurian rocks of Wales, marked by a scarcity of fossils, evenness of bedding, flow markings, and small scale cross bedding. He found that, within his experience, it is difficult to find thin, evenly-bedded alternations of siltstone or shale that did not show flowage phenomena. These were not the stuff of turbidity currents; another current type must be responsible for their deposition.

Ripple marks began to show up in underwater photographs at depths of over 3,000 fathoms (Elmendorf and Heezen, 1957), indicating the presence of deep currents. Currents with velocities of 5–15 cm sec-1 were measured in the North Atlantic Western Boundary Current (Swallow and Worthington, 1957). Further study (Heezen and Hollister, 1964; Heezen and others, 1966; Heezen and Schneider, 1968) demonstrated that these currents were of sufficient strength to erode, transport, and deposit sediment in the silt and fine sand size range. The sediments deposited by these currents following the bathymetric contours were thin laminated silts containing fine scale crossbedding. The continental rise was shaped not only by turbidity currents, but by the more continuous Western Boundary Undercurrent as well.

South of Cape Hatteras, both surface and buried sediments show an increase in carbonate content culminating in the thick carbonate platforms of Florida and the Bahamas. It proved difficult to identify basement by seismic refraction measurements beneath the thick limestone section (Sheridan and others 1966), but it was concluded that the structure of the Blake Plateau was similar to that to the north; a north-south trough with a basement ridge at its seaward edge. Massive subsidence was indicated by Lower Cretaceous algal limestone dredged from near the base of the Blake escarpment (Heezen and Sheridan, 1966). The difference in topography was attributed to the change from clastic to carbonate sedimentation. Drilling on the Blake Plateau (JOIDES, 1965) supported the conclusion that the thinness of the post-Paleocene sediments was due to the strong currents that swept across it.

Murray (1961) synthesized the results of the geological and geophysical investigations of the Atlantic and Gulf coastal plains and margins, and Maher (1965) published correlations of the subsurface Cenozoic and Mesozoic rocks.

INTRODUCTION OF PLATE TECTONICS

Plate tectonics provided a genetic model for the transition from continental to oceanic crust that could be tested against the data from this area and other marginal areas created by similar processes, but in different stages of development. DuToit (1937) suggested that many continental margins, created by the splitting of Pangaea, were fault line coasts, so this was not new; what was new was that there was now general acceptance of a rift origin for the Atlantic margin of the United States. Plate tectonics also provided a mechanism for deforming the continental margin and turning it into a mountain system (Dietz, 1963). But even with this new model, it was difficult to find agreement on where continental crust ended or ocean crust began, or in what manner the transition took place (see for example, Burk and Drake, 1974). Inferences were drawn from gravity and magnetic data and a few long refraction profiles, but the basement surface and the mantle configuration were poorly defined.

Dewey (1969a) and Bird and Dewey (1970) noted that plate tectonics provided a mechanism through which a thick sedimentary prism such as that off the eastern United States could be converted into the deformed, metamorphosed, and intruded sequences of orogenic belts, and showed that Appalachian geology was consistent with such an interpretation. How subduction is initiated is still a mystery, but with subduction it became possible to conceive of conversion of a margin of the Atlantic type into an orogen of the Andean type and, with continental collision, into one of Appalachian character.

In 1967 investigators from Woods Hole Oceanographic Institution made detailed studies of the continental rise (Emery and others, 1970) and concluded that the basement ridge described by Drake and others (1959) acted as a sediment trap for land-derived sediments through Mesozoic time, but it was overtopped in the Eocene and the present continental rise was built atop a Mesozoic abyssal plain. At a GSA Penrose conference in 1972, industry data on the Scotia shelf were cited to suggest that the seismic horizon identified with Paleozoic basement beneath the U.S. outer shelf to the south was probably indurated Mesozoic sediments (Sheridan, 1974) (Fig. 4D). A basement ridge was still included in the sections but, on the basis of modelling of the East Coast Magnetic Anomaly (Taylor and others, 1968; Emery and others, 1970), it was now considered as oceanic rather than continental crust. In 1974 multichannel seismic reflection measurements taken for the USGS supported the premise that early Mesozoic sediments of some thickness lay beneath the shelf and showed little evidence of a basement ridge on a line running seaward from Cape Hatteras (Grow and Markl, 1977). Airborne magnetic data were acquired by the USGS in 1975 and while they indicated no basement ridge on the Cape Hatteras line, magnetic depth-to-basement calculations suggested that a ridge was present to the north (Klitgord and Behrendt, 1979). The ridge was still present, though shrinking, in the interpretations of reflection profiles in 1976 (Schlee and others, 1976), but in the most recent interpretations of the data from LASE (Large Aperture Seismic Experiment) it has virtually disappeared. As Steckler (1981) pointed out, there has been rapid subsidence of the eastern continental margin since World War II.

The Triassic fault basins of the eastern United States were included in the original definition of geosynclines by Dana (1873) and were called taphrogeosynclines by Kay (1945). The rough parallelism of these basins to Appalachian structures created a tendency to think of them as a late continuation of an earlier orogeny (Emerson, 1917), but Bucher (1933) emphasized the discordance of these structures and concluded that they "must represent a new start dynamically which disrupted the old structural lines." Subsequently, Triassic fault basins were found in the subsurface beneath the coastal plain and beneath the shelf (Drake and others, 1959; Ballard and Uchupi, 1972). These fault troughs, with their associated red beds, lacustrine sediments, volcanics, and evaporites, have now been associated with the initial rifting that led to opening of the Atlantic ocean. The original basement ridge identified by Drake and others (1959) is now interpreted as a prograding carbonate reef ranging in age from Jurassic to Lower Cretaceous (Folger and others, 1979). This fits the pattern of sedimentation to be expected on rifted margins following more open access to the sea (Brice and others, 1982; Ponte and Asmus, 1976).

To the south, in the Blake-Bahama region, carbonates form a platform sequence on basement presumed to be transitional between continental and oceanic (Folger and others, 1979). Here, too, the basement ridge has been reinterpreted as a buried reef. The subsidence history of the Blake Plateau appears to be similar to that of the Mid-Atlantic region (Sheridan and Enos, 1979) and its greater depth is attributed to Gulf Stream action in preventing sedimentation or even eroding (JOIDES, 1965). Drilling on the Blake Plateau and subsequently in deep water showed that the thick sedimentary pile that makes up the Blake Outer Ridge can be explained in terms of the combined actions of the Gulf Stream and the Western Boundary Undercurrent (Bryan, 1970; Markl and others, 1970). Subsidence of the eastern continental margin is due in part to loading by sediments and water, but more important is the thinning and cooling of the lithosphere that resulted from the rifting and subsequent drifting of the margin away from the active spreading center (Watts and Ryan, 1976; Watts and Steckler, 1979; Steckler, 1981). Syntheses of the investigations described above may be found in the comprehensive reviews of Emery and Uchupi (1972, 1984).

CONCLUDING REMARKS

Charles Darwin commented in *Descent of Man*: "False facts are highly injurious to the progress of science, for they often endure long; but false views, if supported by some evidence, do little harm, for everyone takes a salutary pleasure in proving their falseness; and when this is done, one path towards error is closed and the road to truth is often at the same time opened."

Our views about the nature and development of the Atlantic continental margin of the United States have been the evolutionary product of data, ideas, and common wisdom. Ideas have always been plentiful, uninhibited initially by much data, but constrained by common wisdom. We found it hard to accept moving continents or to believe that deep subsea currents had the power to erode and transport coarse grained sediments, but the data brought us around. We sometimes ignored data, such as the sediment samples taken on the shelf during mapping, although these would have demonstrated that some of our ideas were flawed. But eventually the data win and as we apply new tools to examination of the margin, accumulate more and better data, generate new ideas, and test common wisdom, we come closer to opening the door to truth.

REFERENCES CITED

Agassiz, A., 1888, Three Cruises of the United States Coast and Geodetic Survey Steamer "Blake": Harvard University Museum of Comparative Zoology, v. 14, 314 p.; v. 15, 220 p.

Ballard, R. D., and Uchupi, E., 1972, Carboniferous and Triassic Rifting: A Preliminary Outline of the Gulf of Maine's Tectonic History: Geological Society of America Bulletin, v. 83, p. 2285–2302.

Bird, J. M., and Dewey, J. F., 1970, Lithosphere Plate–Continental Margin Tectonics and the Evolution of the Appalachian Orogen: Geological Society of America Bulletin, v. 81, p. 1031–1060.

Brice, S. E., Cochran, M. D., Pardo, G., and Edwards, A. D., 1982, Tectonics and Sedimentation of the South Atlantic Rift Sequence: Cabinda, Angola, *in* Watkins, J. S., and Drake, C. L., eds., Continental Margin Geology: American Association of Petroleum Geologists Memoir 34, p. 5–50.

Bryan, G. H., 1970, Hydrodynamic Model for the Blake Outer Ridge: Journal of Geophysical Research, v. 75, p. 4530–4537.

Bucher, W. H., 1933, Deformation of the Earth's crust: Princeton University Press, 518 p.

—— 1940, Submarine Valleys and Related Geological Problems of the North Atlantic: Geological Society of America Bulletin, v. 51, p. 489–512.

Burk, C. A., and Drake, C. L., eds., 1974, The Geology of Continental Margins: New York, Springer-Verlag, 1009 p.

Chamberlin, T. C., and Salisbury, R. D., 1907, Geology, v. II, Earth History: New York, Henry Holt & Co., 692 p.

Claypole, E. W., 1891, Continents and the Deep Seas [abs.]: Geological Society of America Bulletin, v. 2, p. 10–16.

Daly, R. A., 1936, Origin of Submarine "Canyons": American Journal of Science, 5th ser., v. 31, p. 401–420.

Dana, J., 1846, Volcanoes of the Moon: American Journal of Science, 2nd ser., v. 2, p. 381–382, 353–355.

—— 1856, On the Plan of Development in the Geological History of North America: American Journal of Science, v. 22, p. 335–349.

—— 1873, On Some Results of the Earth's Contraction from Cooling Including a Discussion of the Origin of Mountains and the Nature of the Earth's Interior: American Journal of Science, v. 5, p. 423–443.

—— 1890, Archaean Axes of Eastern North America: American Journal of Science, v. 40, p. 181–196.

Darwin, C., 1871, Descent of Man, 1st edition, 2 volumes, London.

de Smitt, V. P., 1932, Earthquakes in the North Atlantic as Related to Submarine Cables: Transactions of the American Geophysical Union, 13th Annual Meeting, Washington, D.C., p. 103–109.

Dewey, J. F., 1969a, Continental Margins: A model for Conversion of Atlantic Type to Andean Type: Earth and Planetary Science Letters, v. 6, p. 189–197.

—— 1969b, Evolution of the Appalachian/Caledonian Orogen: Nature, v. 222, p. 124–129.

Dietz, R. S., 1952, Geomorphic Evolution of the Continental Terrace (Continental Shelf and Slope): American Association of Petroleum Geologists Bulletin, v. 36, p. 1802–1819.

—— 1963, Collapsing Continental Rises: An Actualistic Concept of Geosynclines and Mountain Building: Journal of Geology, v. 71, p. 314–333.

Drake, C. L., Ewing, M., and Sutton, G. H., 1959, Continental Margins and Geosynclines: The East Coast of North America North of Cape Hatteras. Physics and Chemistry of the Earth, v. 3: London, Pergamon Press, p. 110–198.

DuToit, A. L., 1937, Our Wandering Continents: Edinburgh, Oliver and Boyd, 336 p.

Eardley, A. J., 1947, Paleozoic Cordilleran Geosyncline and Related Orogeny: Journal of Geology, v. 55, p. 309–342.

Elmendorf, C. H., and Heezen, B. C., 1957, Oceanographic Information for Engineering Submarine Cable Systems: Bell System Technical Journal XXXVI, p. 1047–1093.

Emerson, B. K., 1917, Geology of Massachusetts and Rhode Island: U.S. Geological Survey Bulletin 597.

Emery, K. O., and Uchupi, E., 1972, Western North Atlantic Ocean: Topography, Rocks, Structure, Water, Life, and Sediments: American Association of Petroleum Geologists Memoir 17, 532 p.

—— 1984, Geology of the Atlantic: New York, Springer-Verlag, 1050 p.

Emery, K. O., Uchupi, E., Phillips, J. D., Bowin, C. O., Bunce, E. T., and Knott, S. D., 1970, Continental Rise of Eastern North America: American Association of Petroleum Geologists Bulletin, v. 54, p. 44–108.

Erickson, D. B., Ewing, M., and Heezen, B. C., 1951, Deep Sea Sands and Submarine Canyons: Geological Society of America Bulletin, v. 62, p. 961–965.

—— 1952, Turbidity Currents and Sediments in the North Atlantic: American Association of Petroleum Geologists Bulletin, v. 36, p. 489–511.

Ewing, M., 1938, Sub-Oceanic Seismology; (2) Report on the Application of Seismic Reflection and Refraction Methods: Advance Report of the Commission on Continental Margins and Ocean Structure, Part 5, to the Washington General Assembly of IUGG, p. 50–51.

Ewing, M., Crary, A. P., and Rutherford, H. M., 1937, Geophysical Measurements in the Emerged and Submerged Atlantic Coastal Plain: Pt. I. Methods and Results: Geological Society of America Bulletin, v. 48, p. 753–801.

Ewing, M., and Vine, A. C., 1938, Deep Sea Measurements without Wires or Cables: Transactions of the American Geophysical Union, 19th Annual Meeting, p. 248–251.

Ewing, M., Woollard, G. P., and Vine, A. C., 1939, Geophysical Investigations in the Emerged and Submerged Atlantic Coastal Plain, Part III, Barnegat Bay, New Jersey, Section: Geological Society of America Bulletin, v. 50, p. 257–296.

—— 1940, Geophysical Investigations in the Emerged and Submerged Atlantic Coastal Plain, Pt. IV, Cape May, New Jersey, Section: Geological Society of America Bulletin, v. 51, p. 1821–1840.

Ewing, M., Woollard, G. P., Vine, A. C., and Worzel, J. L., 1946, Recent Results in Submarine Geophysics: Geological Society of America Bulletin, v. 57, p. 909–934.

Ewing, M., Worzel, J. L., Hersey, J. B., Press, F., and Hamilton, G. R., 1950, Seismic Refraction Measurements in the Atlantic Ocean Basin, Part I: Seismicological Society of America Bulletin, v. 40, p. 233–242.

Field, R. M., 1932, Symposium on the Application of Geophysics to the Ocean Basins and Margins: Transactions of the American Geophysical Union, 13th

Annual Meeting, Washington, D.C., p. 11–12.

—— 1935, Report of Special Committee on Geophysical and Geological Study of Ocean Basins: Transactions of the American Geophysical Union, 16th Annual Meeting, Washington, D.C., p. 6–9.

Folger, D. W., Dillon, W. P., Grow, J. A., Klitgord, K. D., and Schlee, J. S., 1979, Evolution of the Atlantic Continental Margin of the United States: American Geophysical Union, Maurice Ewing Series, v. 3, p. 87–208.

Gilbert, G. K., 1893, Continental Problems: Geological Society of America Bulletin, v. 4, p. 179–190.

Grow, J. A., and Markl, R. G., 1977, IPOD-USGS Multichannel Reflection Profile from Cape Hatteras to the Mid-Atlantic Ridge: Geology, v. 5, p. 625–630.

Gunn, R., 1944, A Quantitative Study of the Lithosphere and Gravity Anomalies along the Atlantic Coast: Journal of the Franklin Institute, v. 237, p. 139–154.

Gutenberg, B., and Richter, C. F., 1925, Bearbeitung von Aufzeichnungen einiger Weltbeben: Senckenberg naturf. Gesell. Abh. 40, p. 57.

Hall, James, 1843, Geology of New York, Part IV, Comprising the Survey of the Fourth Geological District.

Heezen, B. C., and Ewing, M., 1952, Turbidity Currents and Submarine Slumps, and the Grand Banks Earthquake: American Journal of Science, v. 250, p. 849–873.

Heezen, B. C., and Hollister, C., 1964, Deep Current Evidence from Abyssal Sediments: Marine Geology, v. 1, p. 141–174.

Heezen, B. C., and Sheridan, R. E., 1966, Lower Cretaceous Rocks (Neocomian-Albian) Dredged from the Blake Escarpment: Science, v. 154, p. 1644–1647.

Heezen, B. C., and Schneider, E. D., 1968, The Shaping and Sediment Stratification of the Continental Rise [abs.]: Mar. Tech. Soc. Nat. Sym., Ocean Science and Engineering of the Atlantic Shelf Trans., p. 279–280.

Heezan, B. C., Hollister, C., and Ruddiman, W. F., 1966, Shaping of the Continental Rise by Deep Geostrophic Contour Currents: Science, v. 152, p. 502–508.

Helmert, F. R., 1909, Die Tiefe der Ausgleichflache bei des Prattschen Hypothese fur das Gleichgewicht der Erdkruste und der Verlauf der Schwerstorung vom Innern der Kontinente und Ozeane nach den Kusten: Sitzber. d. Kgl. Preusz. Akad. d. Wiss., v. 18, p. 1192–1198.

Johnson, D. W., 1925, The New England–Acadian Shoreline: New York, John Wiley & Sons, 608 p.

—— 1939, Origin of Submarine Canyons: Journal of Geomorphology, v. 2, p. 133–158, 213–236.

JOIDES, 1965, Ocean Drilling on the Continental Margin: Science, v. 150, p. 709–716.

Kay, M., 1945, North American Geosynclines—Their Classification [abs.]: Geological Society of America Bulletin, v. 56, p. 1172.

—— 1951, North American Geosynclines: Geological Society of America Memoir 48, 143 p.

Klitgord, K. D., and Behrendt, J. C., 1979, Basin Structures of the U.S. Atlantic Margin, *in* Watkins, J. S., Montadert, L., and Dickerson, P. W., eds., Geological and Geophysical Investigations of Continental Margins: American Association of Petroleum Geologists Memoir 29, p. 85–112.

Kuenen, Ph. H., 1948, Turbidity Currents of High Density; 18th International Geological Congress.

Kuenen, Ph. H., and Migliorini, C. I., 1950, Turbidity Currents as a Cause of Graded Bedding: Journal of Geology, v. 58, p. 1–127.

McCarthy, G. R., 1931, Coastal Sands of the Eastern United States: American Journal of Science, 5th series, v. 22, p. 35–50.

Maher, J. C., 1965, Correlations of Subsurface Mesozoic and Cenozoic Rocks along the Atlantic Coast: American Association of Petroleum Geologists, 18 p.

Markl, R. G., Bryan, G. H., and Ewing, J. I., 1970, Structure of the Blake-Bahama Outer Ridge: Journal of Geophysical Research, v. 75, p. 4539–4555.

Maury, M. F., 1855, Physical Geography of the Sea (1st edition); 1858 (5th edition): New York, Harper & Brothers.

Miller, B. L., 1937, Geophysical Investigations in the Emerged and Submerged Atlantic Coastal Plain, Part II, Geological Significance of the Geophysical Data: Geological Society of America Bulletin, v. 48, p. 803–812.

Mudge, B. F., 1858, The Salt Marshes of Lynn: Proc. Essex. Inst., v. 2, p. 117–119.

Murray, G. E., 1961, Geology of the Atlantic and Gulf Coastal Province of North America: Harper & Brothers, 692 p.

Murray, J., 1885, Report on the Specimens of Bottom Deposits (collected by the U.S. Coast Survey steamer "Blake" 1877–1880). Harvard University Museum of Comparative Zoology Bulletin, v. 12, no. 2, p. 1430.

Pettijohn, F. J., 1949, Archean Sedimentation: Geological Society of America Bulletin, v. 54, p. 925–972.

Ponte, F. C., and Asmus, H. E., 1976, The Brazilian Marginal Basins: Current State of Knowledge, in de Almeida, F.F.M., ed., Continental Margins of Atlantic Type: Annales Academie Brasilias Ciencias, v. 48, p. 215–240.

Pourtales, L. F., 1872, The Characteristics of the Atlantic Sea Bottom off the Coast of the United States: Report by the Superintendent of the U.S. Coast Survey for 1869, App. 11, p. 220–225.

Rich, J. L., 1950, Flowmarkings, Groovings, and Intra-stratal Crumplings as Criteria for recognition of Slope Deposits with Illustrations from Silurian Rocks of Wales: American Association of Petroleum Geologists Bulletin, v. 34, p. 717–741.

Richards, H. G., 1945, Subsurface Stratigraphy of Atlantic Coastal Plain between New Jersey and Georgia: American Association of Petroleum Geologists Bulletin, v. 29, p. 885–955.

—— 1948, Studies of the Subsurface Geology and Paleontology of the Atlantic Coastal Plain: Acad. Nat. Sci. Phil., v. 100, p. 39–76.

Schlee, J., Behrendt, J. C., Grow, J. A., Mattick, R. E., Taylor, P. T., and Lawson, B. J., 1976, Regional Geological Framework off Northeastern United States: American Association of Petroleum Geologists Bulletin, v. 60, p. 926–951.

Schuchert, C., 1910, Paleogeography of North America: Geological Society of America Bulletin, v. 20, p. 427–606.

—— 1923, Sites and Natures of the North American Geosynclines: Geological Society of America Bulletin, v. 34, p. 151–229.

Shepard, F. P., 1932, Sediments of the Continental Shelves: Geological Society of America Bulletin, v. 43, p. 1017–1039.

—— 1939, Continental Shelf Sediments, in Trask, P. D., ed., Recent Marine Sediments: American Association of Petroleum Geologists, p. 219–229.

Shepard, F. P., and Cohee, G. V., 1936, Continental Shelf Sediments off the Mid-Atlantic States: Geological Society of America Bulletin, v. 47, p. 441–458.

Shepard, F. P., Trefethen, J. M., and Cohee, G. V., 1934, Origin of Georges Bank: Geological Society of America Bulletin, v. 45, p. 281–302.

Sheridan, R. E., 1974, Atlantic Continental Margin of North America, in Burke, C. A., and Drake, C. L., eds., The Geology of Continental Margins: New York, Springer-Verlag, p. 391–408.

Sheridan, R. E., Drake, C. L., Nafe, J. E., and Hennion, J., 1966, Seismic Refraction Study of Continental Margin East of Florida: American Association of Petroleum Geologists Bulletin, v. 50, p. 1972–1991.

Sheridan, R. E., and Enos, P., 1979, Stratigraphic Evolution of the Blake Plateau after a Decade of Scientific Drilling: American Geophysical Union, Maurice Ewing Series, v. 3, p. 109–122.

Spencer, J. W., 1903, Submarine Valleys off the American Coast and in the North Atlantic: Geological Society of America Bulletin, v. 14, p. 207–226.

Steckler, M. S., 1981, The Thermal and Mechanical Evolution of Atlantic-Type Continental Margins [Ph.D. Thesis]: Columbia University, 261 p.

Stetson, H. C., 1934, Origin and Limits of a Zone of Rounded Quartz Sand off the Southern New England Coast: Journal of Sedimentary Petrology, v. 4, no. 3, p. 152–153.

—— 1936, Dredge-Samples from the Submarine Canyons Between the Hudson Gorge and Chesapeake Bay: Transactions of the American Geophysical Union, 17th Annual Meeting, Washington, D.C., p. 223–225.

—— 1937, Current Measurements in the Georges Bank Canyons: Transactions of the American Geophysical Union, 18th Annual Meeting, Washington, D.C., p. 216–219.

—— 1938, The Sediments of the Continental Shelf off the Eastern Coast of the United States: Papers in Phys. Ocean. and Met., Massachusetts Institute of Technology and Woods Hole Oceanographic Institute, v. 5, no. 4.

—— 1939, Summary of Sedimentary Conditions on the Continental Shelf off the East Coast of the United States, in Trask, P. D., ed., Recent Marine Sediments: American Association of Petroleum Geologists, p. 230–244.

Swallow, J. C., and Worthington, L. V., 1957, Measurements of Deep Currents in the Western North Atlantic: Nature, v. 179, p. 1183.

Taylor, P. T., Zietz, I., and Dennis, L. S., 1968, Geological Implications of Aeromagnetic Data for the Eastern Continental Margin of the United States: Geophysics, v. 33, p. 755–780.

Thomson, C. W., 1874, The Depths of the Sea: London, MacMillan & Co., 527 p.

Thoulet, J., 1889, Considerations sur la Structure et la Genese des Bancs de Terre-Neuve. Geological Society of Paris Bulletin, 7th ser. v. 10, p. 203–241.

Upham, W., 1894, The Fishing Banks between Cape Cod and Newfoundland: American Journal of Science, 3rd ser., v. 47, p. 123–129.

van der Gracht, W.J.A.M. van W., 1931, Permo-Carboniferous Orogeny in the South Central United States: American Association of Petroleum Geologists Bulletin, v. 15, p. 991–1057.

Veatch, A. C., and Smith, P. A., 1939, Atlantic Submarine Valleys of the United States and the Congo Submarine Valley. Geological Society of America Special Paper 7, 101 p.

Vening Meinesz, F. A., and Wright, F. E., 1930, The Gravity Measuring Cruise of the U.S. Submarine S-21: Pub. Nav. Obs., 2nd Ser., v. XIII—App. I, 94 p.

Walcott, A., 1891, Correlation Papers; Cambrian: U.S. Geological Survey Bulletin 81.

Watts, A. B., and Ryan, W.B.F., 1976, Flexure of the Lithosphere and Continental Margin Basins: Tectonophysics, v. 36, p. 25–44.

Watts, A. B., and Steckler, M. S., 1979, Subsidence and Eustasy at the Continental Margin of Eastern North America: American Geophysical Union, Maurice Ewing Series, v. 3, p. 218–234.

Wegener, A., 1924, The Origin of Continents and Oceans (translated from the 3rd German edition by J.G.A. Skerl): E. P. Dutton & Co., 212 p.

Williams, H. S., 1897, On the Southern Devonian Formations: American Journal of Science, v. 3, p. 393–403.

Woollard, G. P., 1940, Gravitational Determination of Deep-Seated Crustal Structure of Continental Borders (Structural Interpretations of Gravity Observations): Transactions of the American Geophysical Union, Symposium on the Surface and Subsurface Exploration of Continental Margins, Richmond, Virginia, December 1938, p. 808–815.

Worzel, J. L., and Ewing, M., 1948, Explosion Sounds in Shallow Water, in Propagation of Sound in the Ocean: Geological Society of America Memoir 27, 53 p.

Wust, G., 1936, Schichtung und Zirkulation des Atlantischen Ozeans. Das Bodenwasser und die Stratosphare: Wiss. Erg. Deutsch. Atlant. Exp. "Meteor" 1925-27, Berlin VI, (2), 420 p.

Manuscript Accepted by the Society May 20, 1985

Printed in U.S.A.

The Geology of North America
Vol. I-2, The Atlantic Continental Margin: U.S.
The Geological Society of America, 1988

Chapter 5

Mesozoic and Cenozoic stratigraphy of the United States Atlantic continental shelf and slope

C. Wylie Poag and Page C. Valentine
U.S. Geological Survey, Woods Hole, Massachusetts 02543

INTRODUCTION

Geological and geophysical knowledge of the U.S. Atlantic margin has advanced rapidly during the last ten years, spurred by commercial interest in offshore petroleum exploration. The U.S. Geological Survey (U.S.G.S.), in particular, has been instrumental in providing thousands of line-kilometers of seismic reflection, geomagnetic, and gravity surveys. Several academic research institutions, such as Lamont-Doherty Geological Observatory, Woods Hole Oceanographic Institution, University of Rhode Island, Duke University, University of Georgia, University of Miami, and University of Texas have also contributed heavily to this geophysical data base. Geologic ground truth has been acquired, at the same time, from 5 deep (4,000 ± m) Continental Offshore Stratigraphic Test (COST) wells, 46 exploratory wildcat wells, 41 intermediate depth (300 m) coreholes (Joint Oceanographic Institutions for Deep Earth Sampling [JOIDES]; Atlantic Slope Project [ASP]; Atlantic Margin Coring Project [AMCOR]), a few Deep Sea Drilling Project (DSDP) core holes, and several hundred shallow-penetration (3–10 m) cores, seabed grab samples, and submersible samples (see Poag, 1978; 1982a,b; 1985a,b,c).

Four major offshore basins and troughs have been identified as a result of these research efforts. We summarize the stratigraphic framework and depositional history of the three best studied features (Georges Bank basin, Baltimore Canyon trough, and Blake Plateau basin), emphasizing their main features with a cross section through each basin. For additional regional compilations the reader should consult Bryan and Heirtzler (1984), Ewing and Rabinowitz (1984), Uchupi and Shor (1984) and Poag (1985c).

GEORGES BANK BASIN

Previous Studies and Geologic Setting

An estimated 6,300 line-km of multichannel and single-channel seismic reflection profiles are the basis for seismostratigraphic interpretations of Georges Bank basin (e.g., Austin and others, 1980; Ballard and Uchupi, 1975; Emery and Uchupi, 1965; Klitgord and others, 1982; Mattick and others, 1974; Schlee and Fritsch, 1983; Schlee and others, 1976; Schlee and others, 1985; Uchupi and Shor, 1984; Uchupi and others, 1977; Schlee and Klitgord, this volume; Fig. 1). Two deep stratigraphic tests and eight exploratory wildcat wells have been drilled on Georges Bank (Giordano and others, 1983; Scholle and Wenkam, 1982), but prior to this summary, data from only the two stratigraphic tests (COST G-1 and G-2) had been released. Here, we incorporate data from two of the exploratory wells (Exxon 133-1 and Shell 410-1); the other exploratory well data are still proprietary (see Fig. 1; Table 1). A few shallow borings (Atlantic Slope Project [ASP] and Atlantic Margin Coring Project [AMCOR]), dredgings, and manned submersible samplings provide supplementary stratigraphic data (Arthur, 1982; Gibson and others, 1968; Hathaway and others, 1979; Poag, 1978, 1982a,b, 1985c; Valentine, 1982b; Valentine and others, 1980; Trumbull and Hathaway, 1968).

A stratigraphic integration of paleontologic, lithologic, and seismic data for the Georges Bank basin was presented by Poag, (1982a, b). Poag projected the stratigraphic columns of the G-1 and G-2 wells onto USGS seismic line 19 (a dip section), and recommended that Line 19 serve as the standard stratigraphic reference section for the basin. We also use Line 19, projecting to it the stratigraphic data from the two additional wells (Plate 3A, in pocket). These new data are derived from oil company reports released through the National Geophysical Data Center, Boulder, Colorado. The projected locations and total depths of the unreleased wells are also shown on the cross section (Plate 3A, in pocket).

Georges Bank basin underlies the New England Continental Shelf (Fig. 1). Its topographic expression is a broad, flat-topped high, known as George Bank. The basin is surrounded on three sides by structural platforms: the LaHave platform to the northeast; Gulf of Maine platform to the northwest; and Long Island platform to the southwest. The structural floor of the basin is composed of a series of subbasins (grabens and half-grabens)

Poag, C. W., and Valentine, P. C., 1988, Mesozoic and Cenozoic stratigraphy of the United States Atlantic continental shelf and slope; *in* Sheridan, R. E., and Grow, J. A., eds., The Geology of North America, Volume I-2, The Atlantic Continental Margin, U.S.: Geological Society of America.

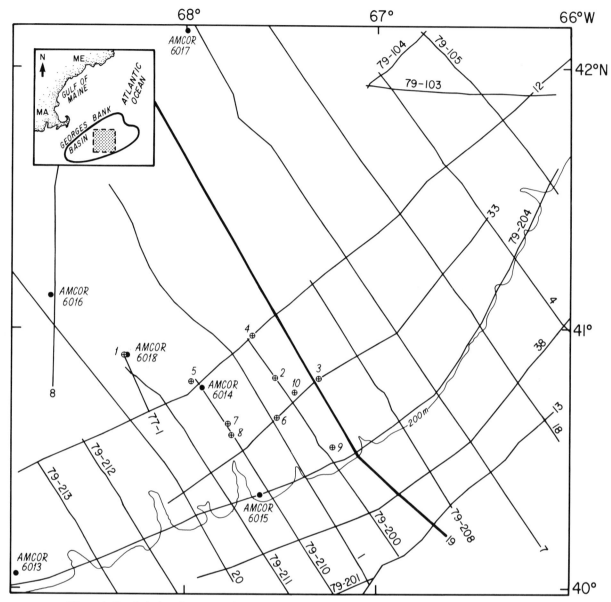

Figure 1. Borehole locations and principal multichannel seismic reflection profiles for Georges Bank basin, offshore Massachusetts and Maine. See Table 1 for further identification of numbered boreholes. Interpretation of Line 19 illustrated in Plate 3A.

formed during the rifting phase of Pangaean continental breakup (Schlee and Fritsch, 1983; Schlee and Klitgord, 1982, this volume; Uchupi, 1984).

At the termination of rifting, tectonic activity ceased in the individual fault blocks and broad downwarping began to form an elongate depression encompassing all the subbasins. This elongate basin filled with a thick postrift prism of Jurassic and younger sedimentary rocks. Maximum total sediment thickness in the depocenter (main basin) exceeds 16 km (Schlee and Fritsch, 1983; Schlee and Klitgord, 1982, this volume). At least five buried seamounts or large igneous intrusions are inferred to be present within Georges Bank basin on the basis of distinctive

magnetic anomaly patterns (Schlee and Klitgord, 1982; Uchupi, 1984). The southeast margin of the basin is marked by a prominent linear magnetic anomaly (East Coast Magnetic Anomaly), which is thought to be produced by edge effects at the junction of continental and oceanic crust (Grow and others, 1979; Klitgord and others, 1982).

Stratigraphy

Chronostratigraphic interpretation of the COST G-1 and G-2 wells, projected to Line 19, reveals that about 75% of the sedimentary section drilled on Georges Bank is of Jurassic age

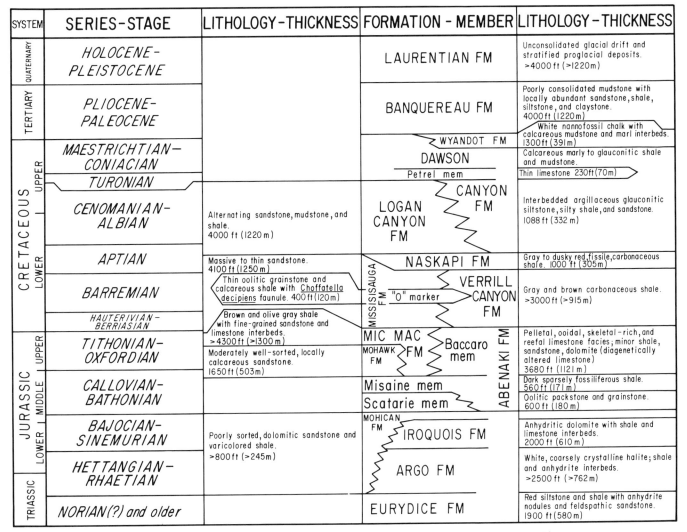

SYSTEM	SERIES-STAGE	LITHOLOGY-THICKNESS	FORMATION - MEMBER		LITHOLOGY-THICKNESS
QUATERNARY	HOLOCENE-PLEISTOCENE		LAURENTIAN FM		Unconsolidated glacial drift and stratified proglacial deposits. >4000 ft (>1220 m)
TERTIARY	PLIOCENE-PALEOCENE		BANQUEREAU FM		Poorly consolidated mudstone with locally abundant sandstone, shale, siltstone, and claystone. 4000 ft (1220 m)
			WYANDOT FM		White nannofossil chalk with calcareous mudstone and marl interbeds. 1300 ft (391 m)
CRETACEOUS UPPER	MAESTRICHTIAN-CONIACIAN		DAWSON	CANYON FM	Calcareous marly to glauconitic shale and mudstone.
	TURONIAN		Petrel mem		Thin limestone 230 ft (70 m)
	CENOMANIAN-ALBIAN	Alternating sandstone, mudstone, and shale. 4000 ft (1220 m)	LOGAN CANYON FM	LOGAN CANYON FM	Interbedded argillaceous glauconitic siltstone, silty shale, and sandstone. 1088 ft (332 m)
CRETACEOUS LOWER	APTIAN	Massive to thin sandstone. 4100 ft (1250 m)	NASKAPI FM	VERRILL CANYON FM	Gray to dusky red, fissile, carbonaceous shale. 1000 ft (305 m)
	BARREMIAN	Thin oolitic grainstone and calcareous shale with *Choffatella decipiens* faunule. 400 ft (120 m)	"O" marker		Gray and brown carbonaceous shale. >3000 ft (>915 m)
	HAUTERIVIAN-BERRIASIAN	Brown and olive gray shale with fine-grained sandstone and limestone interbeds. >4300 ft (>1300 m)			
JURASSIC UPPER	TITHONIAN-OXFORDIAN	Moderately well-sorted, locally calcareous sandstone. 1650 ft (503 m)	MIC MAC FM MOHAWK FM	Baccaro mem	Pelletal, ooidal, skeletal-rich, and reefal limestone facies; minor shale, sandstone, dolomite (diagenetically altered limestone) 3680 ft (1121 m)
JURASSIC MIDDLE	CALLOVIAN-BATHONIAN		Misaine mem		Dark sparsely fossiliferous shale. 560 ft (171 m)
			Scatarie mem		Oolitic packstone and grainstone. 600 ft (180 m)
JURASSIC LOWER	BAJOCIAN-SINEMURIAN	Poorly sorted, dolomitic sandstone and varicolored shale. >800 ft (>245 m)	MOHICAN FM	IROQUOIS FM	Anhydritic dolomite with shale and limestone interbeds. 2000 ft (610 m)
	HETTANGIAN-RHAETIAN		ARGO FM		White, coarsely crystalline halite; shale and anhydrite interbeds. >2500 ft (>762 m)
TRIASSIC	NORIAN(?) and older		EURYDICE FM		Red siltstone and shale with anhydrite nodules and feldspathic sandstone. 1900 ft (580 m)

Figure 2. Lithostratigraphic nomenclature applied in Scotian basin, Georges Bank basin, and Baltimore Canyon trough (from Poag, 1982b; after Eliuk, 1978; Given, 1977; Jansa and Wade, 1975; McIver, 1972). Thickness shown is maximum drilled.

(Poag, 1982a,b); ca. 20% is Cretaceous; ca. 5% is Cenozoic. We used seismic facies analysis to extrapolate the lithostratigraphic units drilled at the four well sites; the results are shown in Plate 3A (in pocket). Formation names are taken from the adjacent Scotian basin (Eliuk, 1978; Given, 1977; Jansa and Wade, 1975; McIver, 1972; Poag, 1982a,b; Fig. 2). Poag (1982b) discusses the basis for and limitations of applying this lithostratigraphic nomenclature.

Basement Rocks

Seismic reflection profiles combined with magnetic and gravity data are used to infer the nature of basement rocks beneath the Georges Bank basin (Austin and others, 1980; Ballard and Uchupi, 1975; Grow and Schlee, 1976; Klitgord and Behrendt, 1979; Klitgord and others, 1982; Mattick and others, 1974;

Schlee and others, 1976, 1979). Two conceptually different types of basement can be recognized: (1) crystalline basement, a surface of igneous or metamorphic rocks upon which the synrift and postrift sedimentary rocks have accumulated; and (2) seismic basement, which is the surface of maximum interpretable acoustic penetration (Klitgord and others, 1982). On the shoreward and seaward flanks of the basin, seismic and crystalline basement commonly coincide (as in the COST G-1 well). However, in the deeper parts of the main basin, crystalline basement is too deep in places to be detected on available seismic profiles. Seismic basement in these areas is formed by stratal contacts within the Lower Jurassic-Upper Triassic sequences.

At the updip end of Line 19 (Plate 3A), seismic basement coincides with crystalline basement, which probably consists of blockfaulted, metamorphosed sedimentary rocks similar to the black graphitic slate, schist, phyllite, and sericitic metadolomite

TABLE 1. COMMERCIAL WILDCAT AND STRATIGRAPHIC TEST WELLS DRILLED ON THE UNITED STATES ATLANTIC COASTAL SHELF AND SLOPE

Georges Bank Basin

Number on Figs.	1	2	3	4	5	6	7	8	9	10
Operator	Ocean Production	Ocean Production	Conoco	Exxon	Exxon	Mobil	Mobil	Shell	Shell	Tenneco
Block	COST	COST	145	975	133	273	312	357	410	187
Well no.	G-1	G-2	1	1	1	1	1	1	1	1
Water depth (m)	48	83	91	64	69	92	76	81	138	91
Total Depth BKB (m)	4899	6667	4420	4452	4398	4749	6196	5921	4745	5525
Oldest sedimentary Rx	Early? Jurassic	Triassic?	Early? Jurassic	Early? Jurassic	Early? Jurassic	Early? Jurassic	Early? Jurassic	Early? Jurassic	Early? Jurassic	Early? Jurassic

TABLE 1. (CONTINUED)

Baltimore Canyon Trough

Number on Figs.	1	2	3	4	5	6	7	8	9	10
Operator	Ocean Production	Chevron	Conoco	Exxon	Exxon	Exxon	Exxon	Exxon	Exxon	Exxon
Block	COST	COST	590	500	599	684	684	728	816	902
Well no.	B-2	B-3	1	1	1	1	2	1	1	1
Water depth (m)	91	819	70	62	135	126	126	132	143	132
Total Depth BKB (m)	4863	4822	3658	3735	5219	5371	5121	4635	5411	4867
Oldest sedimentary Rx	Late Jurassic	Late Jurassic	Early Cretaceous	Late Jurassic	Late Jurassic	Late Jurassic	Late Jurassic	Late Jurassic	Late Jurassic	Late Jurassic

BKB = Below Kelly Bushing on rig floor; distance from KB to sea surface varies from rig to rig.

drilled in the COST G-1 well. Downdip from shot point 2300, the crystalline basement plunges beneath seismic basement in the main basin (Klitgord and others, 1982). Seismic basement here is near the top of the Triassic graben-fill and may be associated with a salt layer (Klitgord and others, 1982; Poag, 1982a,b). A high block of crystalline basement appears to be present between shot points 2800 and 2900. Between shot points 3050 and 3400, crystalline basement comprises a large intrusive igneous body that produces a magnetic high (Klitgord and others, 1982). Between shot points 3400 and 3900, crystalline basement descends again below resolving depth of the seismic profiles, and seismic basement appears to be a stratal boundary near the Lower Jurassic-Upper Triassic transition. Neither type of basement can be recognized beneath the shelf-edge reefal bank, but seaward of shot point 4300, crystalline basement, in the form of basaltic oceanic crust, is indicated by hyperbolic reflectors (Klitgord and others, 1982; Poag, 1982a,b; Schlee and others, 1985).

Synrift Sedimentary Rocks

The synrift deposits of Georges Bank basin are poorly known. However, in the COST G-1 well, a 221 m section of reddish conglomerate and arkosic sandstone, containing occasional interbeds of limestone, dolomite, and anhydrite, may represent the upper part of the Eurydice Formation, which fills a Triassic half-graben adjacent to the COST G-1 site (Arthur, 1982; Poag, 1982a,b). By analogy with coastal plain exposures (Manspeizer, 1982) and the Scotian basin (Jansa and Wade, 1975), we expect that chiefly nonmarine siliciclastics, evaporites, and volcanic intrusions fill the synrift subbasins.

TABLE 1. (CONTINUED)

Baltimore Canyon Trough (continued)

Number on Figs.	11	12	13	14	15	16	17	18	19	20	21
Operator	Gulf	Gulf	Houston Oil & Minerals	Houston Oil & Minerals	Mobil	Mobil	Mobil	Mobil	Murphy	Shell	Shell
Block	718	857	676	855	17	17	544	544	106	93	272
Well no.	1	1	1	1	1	2	1	2	1	1	1
Water depth (m)	58	102	67	89	85	79	65	66	125	1527	66
Total Depth BKB (m)	3906	5655	3810	5336	366	4265	5319	2611	5610	5409	4115
Oldest sedimentary Rx	Late Jurassic	Late Jurassic	Late Jurassic	Late Jurassic	Middle Miocene	Late Jurassic	Late Jurassic	Early Creta- ceous	Late Jurassic	Late Jurassic	Late Jurassic

TABLE 1. (CONTINUED)

Baltimore Canyon Trough (continued)

Number on Figs.	22	23	24	25	26	27	28	29	30	31
Operator	Shell	Shell	Shell	Shell	Shell	Tenneco	Tenneco	Tenneco	Texaco	Texaco
Block	273	372	586	587	632	495	642	642	598	598
Well no.	1	1	1	1	1	1	2	3	1	2
Water depth (m)	72	2116	1779	1965	64	108	139	137	130	128
Total Depth BKB (m)	5334	3546	4880	4420	4267	5578	5609	5022	4580	5398
Oldest sedimentary Rx	Late Jurassic	Late Jurassic	Late Jurassic	Late Jurassic	Late Jurassic	Late Jurassic	Late Jurassic	Late Jurassic	Late Jurassic	Late Jurassic

TABLE 1. (CONTINUED)

	Baltimore Canyon Trough (continued)			Southeast Georgia Embayment						
Number on Figs.	32	33	34	1	2	3	4	5	6	7
Operator	Texaco	Texaco	Texaco	Ocean Production	Exxon	Exxon	Getty	Tenneco	Tenneco	Transco
Block	598	598	642	COST	472	564	913	208	427	1005
Well no.	3	4	1	GE-1	1	1	1	1	1	1
Water depth (m)	129	129	139	42	32	139	35	35	30	40
Total Depth BKB (m)	4968	4892	5428	4040	2310	3921	2134	2365	2278	3546
Oldest sedimentary Rx	Late Jurassic	Late Jurassic	Late Jurassic	Early Creta- ceous	Early Creta- ceous?	Early Creta- ceous?	Early Creta- ceous?	Early Creta- ceous?	Early Creta- ceous?	Early Creta- ceous

BKB = Below Kelly Bushing on rig floor; distance from KB to sea surface varies from rig to rig.

Postrift Sedimentary Rocks

As rifting ceased, regional subsidence allowed deposition across the tops of upthrown basement blocks. Restricted marine paleoenvironments developed across the outer parts of Georges Bank basin, as evidenced by halite in the bottom of the COST G-2 well (Arthur, 1982; Poag, 1982a,b; Plate 3A). A series of mounded reflections on line 19 may represent small carbonate buildups developed downdip from the COST G-1 well. At the G-2 site however, equivalent early postrift strata assignable to the Mohican Formation are nonmarine and shallow marine sandstones and red and brown shales containing occasional interbeds of conglomerate and dolomite. Similar strata are present at the bottom of the Exxon 133-1 well. The age of these unfossiliferous early postrift beds is not definitely known. Poag (1982a,b) suggested that the salt at the bottom of the G-2 well could be equivalent to the Argo Formation of the Scotian basin, representing part of the transition from Late Triassic to Early Jurassic time (Rhaetian-Hettangian age).

When seafloor spreading began in the New England region during the Early to Middle Jurassic, shallow marine deposition became more widespread in Georges Bank basin. The Iroquois Formation, characterized by dolomite containing anhydrite interbeds, was drilled at all four control wells (COST G-1, G-2; Exxon 133-1; Shell 410-1). Several groups of mounded reflections presumed to represent carbonate buildups are present in the Iroquois Formation and one near shot point 3600 (Line 19) may have formed a shelf-edge barrier during Iroquois deposition. Evidence of igneous activity during Iroquois time comes from the Exxon 133-1 well, which encountered a diabase sill at approximately 4,054 m below the seafloor (Plate 3A). There is disagreement as to the age of the Iroquois section in Georges Bank basin (Ascoli, written communication, 1983; Cousminer and others, 1984; Poag, 1982a,b; Manspeizer and Cousminer, this volume). Triassic and Middle Jurassic microfossils have been found, and seismic sequence analysis led Poag (1982a,b) to suggest that Lower Jurassic rocks are also present here. We show Lower and Middle Jurassic rocks in the Iroquois interval, as inferred from Ascoli'a (unpublished) micropaleontological analysis and Poag's revised seismostratigraphic analysis. We believe that the Triassic microfossils reported by Cousminer and others (1984) in the G-2 well are reworked into the Jurassic section.

Following deposition of the Iroquois dolomite, continued seafloor spreading allowed normal marine conditions to pervade the outer shelf segment of the Georges Bank basin, and limestones of the Abenaki Formation (including the Scatarie and Baccaro members) were widespread (Plate 3A). These carbonate strata prograded seaward as the siliciclastic Mohawk and Mic Mac Formations accumulated in the inner and middle parts of the basin. A prominent reefal bank formed at the shelf edge during Abenaki time (shot point 4000 on Line 19; Plate 3A). These strata contain richer microfossil assemblages than older beds and can be confidently assigned to the Middle and Late Jurassic (Poag, 1982a,b; Ascoli, unpublished data).

During the Early Cretaceous, a major phase of siliciclastic progradation brought coarse sandstones (Missisauga Formation) as far eastward as the Shell 410-1 well (Plate 3A). Carbonate deposition in the basin was curtailed as the shelf-edge reefal bank was overtopped by Missisauga clastic materials and significant volumes of terrigenous debris were supplied to the deep oceanic basin (Hatteras Formation; Jansa and others, 1979; Poag, 1982a,b). Siliciclastic deposition continued into the Late Cretaceous, but rising sea levels (Poag, 1982a,b; Vail and others, 1977) brought finer grained deposits of the Logan Canyon and Dawson Canyon formations into the Georges Bank basin. Chiefly silty shales are present in the Upper Cretaceous section at Shell 410-1, and calcareous siltstones have been recovered in this interval from canyon-side outcrops and borings (Gibson and others, 1968; Poag, 1978, 1982a; Ryan and others, 1978; Stetson, 1949; Trumbull and Hathaway, 1968; Valentine and others, 1980).

Most Paleogene rocks in Georges Bank basin appear to be Eocene strata, including porcellanite (Schlee and Cheetham, 1967), glauconitic clay and hard gray limestone (Hathaway and others, 1979), clayey silt and greensand (Folger and others, 1978), and white biosiliceous chalks and clays (Gibson and others, 1968). These strata have been sampled at the COST G-1 and G-2 wells, at ASP 17 and AMCOR 6019, from coreholes on Nantucket and Martha's Vineyard, and in deep water dredgings and submersible samplings (Gibson and others, 1968; Folger and others, 1978; Hall and others, 1980; Hathaway and others, 1979; Poag, 1978; Ryan and others, 1978; Schlee and Cheetham, 1967).

Sparsely sampled Paleocene and Oligocene strata in Georges Bank basin are known chiefly from deep-water dredgings, which comprise silty and sandy clays (Gibson and others, 1968; Trumbull and Hathaway, 1968). Drill holes on Nantucket and Martha's Vineyard sampled Paleocene strata composed of clayey silt (Folger and others, 1978; Hall and others, 1980).

Neogene rocks, sampled at a few sites in the Georges Bank basin, come from deep water dredgings, outcrops in submarine canyons, and a well on Nantucket Island (Poag, 1978; 1984). Variable lithologies range from silty clays to indurated pebbly sandstones.

Quaternary rocks, including glaciomarine deposits, form a thick (200 m or more) sedimentary cap on Georges Bank and have been extensively sampled. Coarse shelly sands are widespread but a considerable variety of coarser and finer grained lithologies have been encountered (Hathaway and others, 1979; Poag, 1978, 1982a,b; Valentine and others, 1980).

BALTIMORE CANYON TROUGH

Previous Studies and Geologic Setting

On the basis of seismic refraction and multichannel seismic reflection surveys and magnetic and gravity data, at least three subbasins have been recognized on the basement surface of Baltimore Canyon trough (Bayer and Mattick, 1980; Grow and

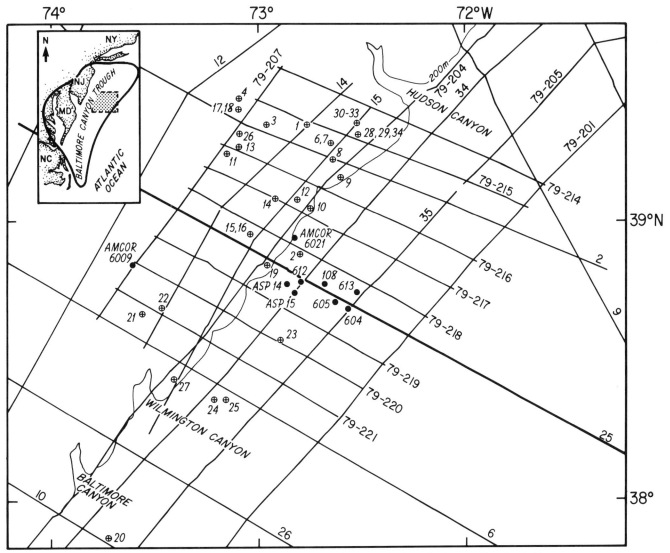

Figure 3. Borehole locations and principal multichannel seismic reflection profiles for Baltimore Canyon trough, offshore New Jersey to North Carolina. See Table 1 for further identification of numbered boreholes. Interpretation of Line 25 illustrated in Plate 3B.

Klitgord, 1980; Klitgord and Behrendt, 1979). The deepest and widest subbasin (sediment depocenter), located off New Jersey, contains at least 18 km of sedimentary rocks (Fig. 3; Bayer and Mattick, 1980; Grow, 1980; Poag, 1985a; Schlee, 1981; Grow and others, this volume). Several igneous intrusions are indicated by magnetic depth-to-basement maps (Klitgord and Behrendt, 1979), the largest of which is the Great Stone Dome (Grow, 1980; Grow and others, this volume). At least three diapiric structures, lacking measurable magnetic and gravity anomalies, are thought to be salt intrusions (Grow, 1980). Salt was encountered on the flank of the Great Stone Dome in the Houston Oil and Minerals (HOM) 676-1 well (Fig. 3; Grow and Klitgord, 1980). The Baltimore Canyon trough is bounded to the northeast by the structurally high Long Island platform and to the south-

west by the Carolina platform (Klitgord and Behrendt, 1979). The seaward margin is marked by a prominent shelf-edge reefal bank of Mesozoic age (Grow, 1980; Poag, 1980a, 1985a; Schlee and others, 1976). In this trough, the reefal bank extends well seaward of the East Coast Magnetic Anomaly, which marks the presumed junction of oceanic and continental crust (Grow, 1980). The Baltimore Canyon trough extends northwestward beneath the coastal plain as a series of embayed basement depressions (Poag, 1985a).

A series of two deep stratigraphic test wells (COST B-2 and B-3) and 28 exploratory wildcat wells constitute the chief geologic data base for the offshore part of the Baltimore Canyon trough (Fig. 3; Plate 3B; Table 1; Libby-French, 1981, 1984; Poag, 1980a,b, 1985a; Scholle, 1977, 1980). Four additional exploratory wells have been drilled on the lower continental slope, but data from them are not available at this writing. In addition, the Atlantic Slope Project (ASP) drilled seven shallow (maximum 305 m penetration) coreholes in Baltimore Canyon trough (Poag, 1978, 1985a) and the Atlantic Margin Coring Project (AMCOR) drilled nine coreholes (maximum penetration 305 m; Hathaway and others, 1979; Poag, 1978, 1985a; Fig. 3). On the seaward side of the trough, the Deep Sea Drilling Project (DSDP) has drilled five core sites (Poag, 1985b). Several gravity and piston cores and dredgings have been recovered on the continental slope and rise (Northrop and Heezen, 1951; Poag, 1985a; Robb, Hampson and Twichell, 1981; Robb, Hampson, and others, 1981; Robb and others, 1983).

The stratigraphy of the onshore part of the trough has been studied from outcrops and subsurface borings (e.g., Brown and others, 1972; Charletta, 1980; Maher, 1971; Olsson, 1964, 1978; Olsson and Nyong, 1984; Olsson and others, 1980; Owens and Gohn, 1985; Owens and Sohl, 1969; Petters, 1976; Poag, 1985a; Ward and Strickland, 1985; Youssefnia, 1978; Olsson and others, this volume).

In addition to the geological data, a dense network of seismic reflection profiles (ca. 10,000 line-km) provides seismostratigraphic information (Grow, 1980; Poag, 1985a; Schlee, 1981; Schlee and others, 1976). Seismostratigraphic analyses allow regional extrapolation of the borehole data and inference of stratigraphic relationships in the deeper, undrilled parts of the trough.

Stratigraphy

Poag (1985a) summarized the details of the stratigraphic framework of the Baltimore Canyon trough, using USGS multichannel seismic reflection Line 25 (Grow, 1980; Grow and others, 1983) as the standard reference section. We also use Line 25 to illustrate the general stratigraphy of Baltimore Canyon trough (Plate 3B).

Basement Rocks

Basement rocks have been drilled in numerous wells on the coastal plain (Brown and others, 1972), but have not been sampled offshore. Basement sections onshore are chiefly granitic and metasedimentary rocks of Paleozoic age. From the seaward side of the trough, hyperbolic echoes of oceanic (basaltic) basement can be traced on seismic reflection profiles from the deep sea to the continental slope (Grow, 1980; Mountain and Tucholke, 1985). The continental-oceanic crust boundary is inferred to be a zone of "transitional" igneous crust where the fractured continental edge has been intruded by numerous basaltic dikes and sills (Poag, 1985a; Schlee and Jansa, 1981; see Grow and others, this volume, for further discussion).

Synrift Sedimentary Rocks

The basement offshore is overlain by a thick wedge of sea-ward thickening strata (>5 km maximum thickness) that accumulated in grabens and half grabens during continental rifting. By analogy with the Scotian basin, Georges Bank basin, and Triassic basins of the coastal plain, we infer that synrift sediments consist of chiefly terrigenous siliciclastics formed by rapid erosion of the high-standing basement blocks. They probably include reddish-brown mudstones, petromict conglomerates, arkosic sandstones, gray to black lacustrine shales, evaporites, and coal lenses (Manspeizer, 1982, 1985; Manspeizer and Cousminer, this volume; van Houten, 1969). Tholeiitic lava flows, sills and dikes are found extensively in the coastal plain synrift basins, and some high-amplitude, continuous reflections within the Triassic sequence on Line 25 may represent such volcanic layers (Grow, 1980; Poag, 1985a). Poag (1985a) provisionally applied the term Eurydice Formation of the Scotian and Georges Bank basins to the Triassic sequence of Baltimore Canyon trough (Plate 3B).

Postrift Sedimentary Rocks

Following the cessation of block-faulting, which characterized the rifting phase of continental breakup, marine waters presumably entered Baltimore Canyon trough, depositing evaporitic rocks in the arid Early Jurassic climates (Manspeizer, 1985; Poag, 1985a). We infer that these strata are represented by the deepest sub-horizontal reflections in the trough, which mark a major angular unconformity relative to the more steeply dipping synrift strata below. Several presumed salt diapirs appear to have arisen from these early postrift evaporites (Grow, 1980). The term Argo Formation is applied to the basal evaporite sequence, by analogy with the Scotian and Georges Bank basins.

The seismic facies analysis of Poag (1985a) reveals a system of presumed Lower and Middle Jurassic facies analogous to those of the Scotian and Georges Bank basins. Nearly continuous, high-amplitude reflections in the middle to outer parts of the trough are interpreted to represent carbonate deposits (dolomites and anydrites overlain by shallow-water limestones) of the Iroquois and Abenaki Formations (including the Scatarie limestone member). Shoreward, discontinuous, variable-amplitude reflections are interpreted to reflect marine and non-marine siliciclastic deposits assignable to the Mohican, Mohawk, and Mic Mac Formations. A general seaward progradation of the carbonate facies took place throughout this Early-Middle Jurassic period and several small reef-like carbonate buildups have been identified (Plate 3B; Poag, 1985a). A large shelf-edge carbonate build-up (barrier reef?) appears to have formed at the end of Middle Jurassic time.

Seaward progradation continued during the Late Jurassic, and a large reefal bank system formed a precipitous shelf-edge barrier about 20 km farther southeastward than the Middle Jurassic barrier (Plate 3B). Several exploratory wells have penetrated the Upper Jurassic sediments, revealing the presence of Mohawk-Mic Mac siliciclastics (including gray shale, white sandstone and thin coal beds) as far downdip as the COST B-3 well. Overlying limestones of the Baccaro member of the Abenaki

Formation have been identified as far updip as the COST B-2 well, and appear to extend shoreward as far as shot point 1300 on Line 25 (Figure 5, Plate 3; Poag, 1985a). Within the Abenaki section, bioclastic debris derived from sponges, corals, echinoderms, bryozoans, bivalves, calcareous algae, and stromatoporoids has been reported from near the crest of the shelf-edge reefal bank (Edson, 1985). The Baccaro member represents the last major pulse of Mesozoic carbonate deposition in this trough, although minor carbonate deposits are present in the younger Cretaceous interval.

Siliciclastics of the Missisauga Formation (medium to coarse sandstones and silty shales in the COST B-3 well) buried the shelf-edge barrier in the Early Cretaceous, as they did in Georges Bank basin. This initiated a period of terrigenous sediment accumulation throughout the basin that lasted, with only a few exceptions, throughout the rest of Cretaceous time. However, Upper Cretaceous strata are characteristically finer grained (silty shales and mudstones in the COST B-3 well) than Lower Cretaceous strata and contain deeper water microfossil assemblages (Poag, 1985a).

As in the Georges Bank basin, Paleogene rocks in the Baltimore Canyon trough are chiefly carbonates (calcareous shales, chalks, and limestones) of Eocene age. These limey strata, extending from the New Jersey coastal plain (e.g., Island Beach and other wells) to the continental rise (e.g., DSDP Sites 605, 613), were deposited during a major marine transgression (Charletta, 1980; Olsson, 1978; Poag, 1985a,b). Siliceous microfossils (chiefly radiolarians and diatoms) are conspicuous elements in the slope and rise facies of the Eocene (DSDP sites; Poag, 1985b). Paleocene and Oligocene strata are also present in the trough, but are less persistent, often being completely missing or only partly represented (e.g., COST B-2 and B-3 wells, DSDP Site 612; Poag, 1985a).

Neogene rocks are particularly thick in the Baltimore Canyon trough, reaching more than 1,000 m in the depocenter (Poag, 1980; 1985a). Most of this section is a series of prograding, middle Miocene, delta lobes enriched in organic carbon and siliceous microfossils (diatoms and radiolarians). Neogene lithologies range from pebbly and shelly sands at the coastline to silty clays, glauconitic sands and sandy pebbly conglomerates on the continental slope and rise.

More than 300 m of Pleistocene sandy and silty clay is present near the shelf edge of Baltimore Canyon trough. Farther shoreward, unconsolidated shelly-to-barren sands and gravels are common in the Quaternary section.

BLAKE PLATEAU BASIN

Previous Studies and Geologic Setting

Prior to 1970 little was known of the stratigraphy of the Blake Plateau region, although the more accessible coastal plain had been studied for many years through drilling and geologic mapping. Since 1970 the Blake Plateau basin has been investigated by geophysical surveying and deep drilling.

Only a few studies have analyzed the composition and topography of the basement surface and most have been directed at the onshore region. Basement rock composition is discussed by Applin (1951), Chowns and Williams (1983), Gohn (1983), Milton (1972), and Milton and Hurst (1965); and basement topography by Bonini and Woollard (1960), Daniels and others (1983), and Klitgord and Behrendt (1979). The subsurface stratigraphy of the North Carolina, South Carolina, Georgia, and northern Florida coastal plains is known from many wells and has been discussed by Applin (1955); Applin and Applin (1965, 1967); Brown and others (1972, 1979); Cramer (1974); Gohn and others (1977; 1980; 1982); Gohn, Bybell, and others (1978); Gohn, Christopher and others (1978); Gohn, Gottfried, and others (1978); Hazel and others (1977); Herrick (1961); Herrick and Vorhis (1963); Maher (1971); Marsalis (1970); Rankin (1977); Vorhis (1974); Valentine (1979a, 1982b, 1984).

Offshore, there are relatively few boreholes, and samples of exposed older rocks are rare. The stratigraphy of offshore boreholes has been studied by Bunce and others (1965), Benson, Sheridan, and others (1978), Charm and others (1969), Hathaway and others (1979), Hollister, Ewing, and others (1972), Scholle (1979), Sheridan, Gradstein, and others (1983), Poag and Hall (1979), and Valentine (1979b). Six exploratory wells and one deep stratigraphic test have been drilled into the Southeast Georgia embayment and the results are available from the National Geophysical Data Center (NGDC), Boulder, CO. Rocks as old as Early Cretaceous have been collected from the Blake Escarpment by dredging and from submersible dives (Dillon and others, 1981; Heezen and Sheridan, 1966; Sheridan and others, 1969).

Offshore geophysical studies incorporating borehole stratigraphic information have been published by Buffler and others (1978, 1979); Dillon and McGinnis (1983); Dillon and Paull (1978); Dillon and others, (1983, 1985); Dillon, Paull and others (1979); Dillon, Poag, and others (1979); Emery and Zarudski (1967); Ewing and others (1966); Grow and Markl (1977); Hersey and others (1959); Hutchinson and others (1982); Klitgord and others (1983); Paull and Dillon (1980a,b); Popenoe (1985); Schlee (1977); Schlee and others (1979); Sheridan and Osburn (1975); Sheridan and others (1979); Shipley and others (1978), (see Byran and Heirtzler, 1984, for a regional compilation of these data).

In the broadest sense, the Blake Plateau region extends from Cape Hatteras in the north to the Bahamas in the south (Fig. 4). It is delimited on the west by the Fall Line where sedimentary rocks of the coastal plain lap onto igneous and metamorphic rocks of the Piedmont Province, and on the east by the boundary between continental and oceanic basement rocks, which generally lies seaward of the present continental slope.

Rifting between North America and Africa in Late Triassic and Early Jurassic time initiated the formation of two large sedimentary basins in the Blake Plateau region. To the north, the deep, elongate Carolina trough developed off the coasts of North and South Carolina (Hutchinson and others, 1982). This trough,

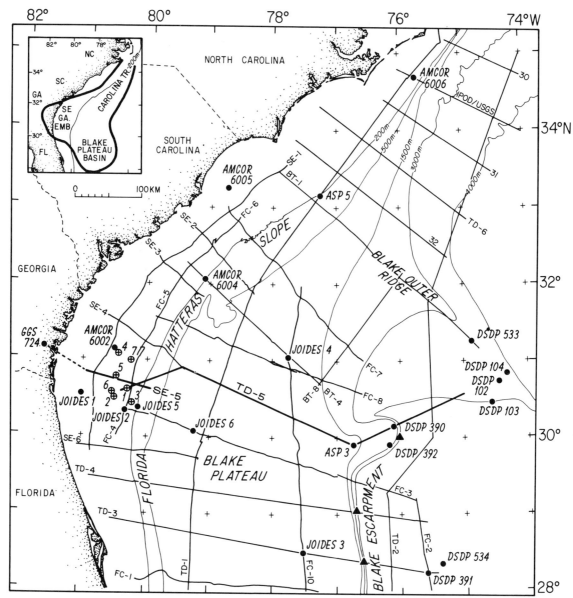

Figure 4. Borehole locations and principal multichannel seismic reflection profiles for Blake Plateau basin and Southeast Georgia embayment, offshore Florida, Georgia, and the Carolinas. Seismic lines SE-5 and TD-5 are discussed in text; DSDP wells (Benson and Sheridan, 1978); ASP wells (Dillon et al., 1979; Poag, 1978); AMCOR wells (Hathaway and others, 1979; Poag, 1978); JOIDES wells (Charm and others, 1969; Poag, 1978; Poag and Hall, 1979). See Table 1 for further identification of numbered exploratory boreholes 1-7. Interpretation of Lines SE-5 and TD-5 illustrated in Plate 3C and D.

approximately 450 km long and 90 km wide, contains up to 11 km of sedimentary strata beneath the continental slope (Dillon and others, this volume). The stratigraphy of the Carolina trough is poorly known, as no deep wells have been drilled there, and it is not discussed further here. A steep basement gradient separates the Carolina trough from the Carolina platform to the west. This platform is overlain by Upper Cretaceous and Cenozoic sedimentary rocks, which are only 1.5 km thick beneath the outer shelf; they thin toward the Fall Line.

South of the Carolina trough, the submerged continental margin widens considerably and encompasses the Blake Plateau basin, a large, subcircular feature approximately 300 km in diameter and 13 km deep (Dillon, Paull, and others, 1979; Dillon and others, 1985 and this volume). This basin is centered beneath the Blake Plateau and is separated from the deep ocean by the Blake Escarpment, a near-vertical feature, 2,500 m high, which formed, at least in part, by erosional retreat of a former continental slope. A shoreward extension of the Blake Plateau basin is known beneath the continental shelf and the South Carolina and Georgia coastal plains as the Southeast Georgia embayment. In its deepest part in southeastern Georgia, the embayment contains only about 1.4 km of Cretaceous and Cenozoic sedimentary rocks (Valentine 1979a, 1982b).

Stratigraphy

Basement and Synrift Rocks. Basement rocks in the Blake Plateau region have been sampled almost exclusively by drilling. Metamorphic and plutonic rocks characteristic of the Piedmont Province are present in a narrow band beneath the landward edge of the coastal plain in Georgia and South Carolina (Chowns and Williams, 1983). In Georgia, they are bounded on the east by a region of Late Triassic and Early Jurassic synrift sedimentary red beds and diabase intrusions. These rocks lie in the South Georgia rift, which extends northeast into South Carolina where coeval red beds, diabase sills, and basalt flows have been sampled by drilling and inferred from geophysical studies (Daniels and others, 1983). In southeastern Georgia the basement is composed of felsic volcanic rocks and granitic plutons of Proterozoic to Early Paleozoic(?) age that abut sedimentary rocks of Ordovician to Devonian age, which extend southward into Florida (Chowns and Williams, 1983).

Offshore, the COST GE-1 well drilled into Devonian argillite, quartzite, and metamorphosed trachyte, and the Transco 1005-1 well penetrated Silurian shale and Ordovician quartzite (unpublished data, NGDC; Fig. 4). On the basis of seismostratigraphic interpretations, the basalt flows of the South Carolina subsurface extend seaward beneath the continental shelf off South Carolina and northern Georgia. Farther offshore, in the deep parts of the Carolina trough and the Blake Plateau basin, basement rocks are inferred from seismic data to compose a complex terrane of rifted blocks of Paleozoic igneous, metamorphic, and sedimentary rocks. These blocks form synrift basins that contain mafic igneous intrusives and volcanics of Late Triassic and Early Jurassic age (Dillon and others, 1983).

Postrift Sedimentary Rocks. *Southeast Georgia Embayment.* Beneath the coastal plain, the Southeast Georgia embayment is bounded on the northeast by the Cape Fear arch in South Carolina, on the southwest by the Florida platform, and on the west by the Piedmont Province in Georgia. The embayment deepens from 1,400 m in southeastern Georgia to 3,300 m at the COST GE-1 well (Fig. 4). In the deepest part of the Southeast Georgia embayment onshore, the oldest postrift rocks are undated nonmarine to marginal marine clastic rocks about 50 m thick, confined to eastern and southern Georgia. These rocks are underlain by an Upper Cretaceous section that includes 200 m of marine clastic strata overlain by marine carbonate beds 350 m thick. The carbonate facies dominates the Upper Cretaceous section on the Florida platform to the south, whereas the section is dominantly clastic northward toward the Cape Fear arch. Paleocene, Eocene, and Oligocene strata comprise 640 m of marine carbonates in southeastern Georgia and northern Florida, which

thin to the northeast. They apparently are absent on the crest of the Cape Fear arch.

Sedimentary rocks in the offshore part of the Southeast Georgia embayment are known from seven deep wells. The most thoroughly studied of these is the COST GE-1 well, which was drilled to basement (Scholle, 1979). It penetrated 4,040 m of Cenozoic and Cretaceous sedimentary rocks and Devonian metasedimentary and metavolcanic rocks (Plate 3C and D).

The oldest sedimentary section in the GE-1 well comprises 1,066 m of red sandstone, siltstone, and shale of probable Early Cretaceous age, including some gray shale and anhydrite and fragments of wood and lignite that suggest a non-marine environment of deposition. The lower 400 m of the red beds are barren of fossils, but dark siltstone and shale in the middle of the section contain sporomorph assemblages of Barremian to Aptian age; others contain Aptian dinoflagellates that suggest a shallow marine paleoenvironment. The upper 350 m of the red bed interval contains sporomorph and dinoflagellate assemblages indicative of fluctuating non-marine and shallow marine conditions. Above the red beds, a 473-m section of limestone, calcareous sand, and minor amounts of dolomite, oolite, and anhydrite, represents shallow marine depositional environments. These strata contain dinoflagellates, poor assemblages of calcareous nannofossils, and agglutinated foraminifers; sporomorphs suggest an Aptian to Albian age. Between this Aptian-Albian section and a thick succession of Upper Cretaceous marine limestone, a 50-m interval contains argillaceous limestone, calcareous sandstone, and assemblages of sporomorphs, dinoflagellates, and agglutinated foraminifers. These fossils are not good biozone indicators, but they and the associated lithology indicate a shallow marine paleoenvironment.

In contrast, the overlying Upper Cretaceous rocks in the COST GE-1 section are argillaceous chalks of Turonian to Maestrichtian age. This 671-m interval contains rich assemblages of calcareous nannofossils, foraminifers, and dinoflagellates deposited in sublittoral and bathyal paleoenvironments. Above this, a 787-m sequence of Paleocene, Eocene, and Oligocene limestones contains calcareous nannofossils, foraminifers, and ostracodes of sublittoral origin. However, an additional 92 m of Oligocene strata and the overlying 54 m of Miocene limestones and shelly carbonate sands contain microfossil assemblages indicative of bathyal deposition. A post-Miocene sequence, approximately 95 m thick, consists of calcareous and quartzose sand deposited in a sublittoral paleoenvironment.

In the Southeast Georgia embayment, the stratigraphy beneath the coastal plain is quite different from that found offshore. Seismic line SE-5 (Plate 3C) crosses the shelf about 8 km north of COST GE-1 and intersects multichannel profile TD-5 (Plate 3D), which crosses the GE-1 well. The stratigraphic units present at COST GE-1 can be traced landward along multichannel seismic line SE-5 to within 73 km of the GGS 724 well in the deep part of the embayment onshore (Dillon, Poag, and others, 1979; Dillon and others, 1985).

The basement surface rises from about 3,300 m beneath the

seafloor at COST GE-1 to about 2,180 m at the western end of line SE-5. This surface is uneven in some areas, presumably representing blocks of igneous and metamorphic Paleozoic rocks. Elsewhere the basement surface is represented by a high-amplitude reflector; this reflector may arise from an erosional unconformity truncating the synrift section of Triassic and Jurassic sedimentary and volcanic rocks. From the western end of line SE-5, the basement continues to rise to about 1,400 m at GGS 724 onshore.

Lower Cretaceous rocks at COST GE-1 are about, 1,600 m thick. They thin shoreward and may be correlative with the thin interval of undated rocks at GGS 724, which lies between the Upper Cretaceous section and the basement.

The 671 m of Upper Cretaceous chalks at COST GE-1 thins to about 359 m at the western end of line SE-5, but then thickens to almost 600 m at GGS 724. The upward change from carbonates to clastics within the Campanian strata of GGS 724 is not seen at COST GE-1, where the Upper Cretaceous section is entirely carbonate. The transition shown on line SE-5 is inferred from seismic facies analysis. Clastic facies are dominant throughout the Upper Cretaceous of the coastal plain but their extent offshore is unknown.

Paleocene limestone is widespread in the Southeast Georgia embayment in a persistent layer that ranges in thickness from 45 m at COST GE-1 to 130 m at GGS 724. However, the Paleocene section thickens seaward beneath the Florida-Hatteras Slope. Eocene and Oligocene carbonate rocks are also widespread, thinning landward from 834 m at COST GE-1 to about 509 m at GGS 724; this section thickens seaward from the GE-1 site. Post-Oligocene strata are limestones and calcareous sands of Miocene age overlain by Pliocene and Pleistocene calcareous and quartzose sands. These deposits are no thicker than 150–200 m throughout the region except, perhaps, beneath the Florida-Hatteras Slope.

Depositional styles in the Southeast Georgia embayment have varied throughout Late Cretaceous and Cenozoic time (Paull and Dillon, 1980a). Sedimentation on an almost level marine platform that extended beneath the present shelf and Blake Plateau took place in pre-Coniacian-Santonian time and again during the Campanian, Maestrichtian, and Paleocene. In contrast, the shelf prograded during the Santonian and Coniacian and also from Eocene time to the present, constructing the modern shelf and Florida-Hatteras Slope. Eocene and younger strata thin at the foot of the Florida-Hatteras Slope and seaward of this position, due to erosion by the Gulf Stream.

Blake Plateau Basin. The Blake Plateau basin lies beneath the Blake Plateau and contains 13 km of sedimentary rock in its deepest part (Fig. 4; Dillon and others, this volume). Drilling has been limited to a few, scattered shallow holes that penetrated only Cenozoic rocks (ASP 3, Joides 3, 4, 5, 6). Two shallow coreholes on the Blake Spur terminated in Lower Cretaceous rocks (DSDP sites 390, 392). One can trace seismic reflectors seaward across the basin from the COST GE-1 well and landward from the DSDP holes and from dated outcrops on the Blake Escarpment.

However, only the upper 3–4 km of the postrift sedimentary sequence has been dated. The age and depositional environments of the remaining 9–10 km of sedimentary strata are interpreted from seismostratigraphic analyses (Dillon, Paull, and others, 1979; Dillon, Poag, and others, 1979; Dillon and others, 1985).

On the seaward margin of the basin, DSDP sites 390 and 392 were drilled in water depths of 2,601 and 2,665 m. Drilling revealed a thin layer of lower Tertiary and Upper Cretaceous nannofossil ooze (79 to 135 m thick) overlying Albian, Aptian, and Barremian nannofossil ooze and Barremian or older shallow-water limestone that is poorly fossiliferous (Benson, Sheridan, and others, 1978).

Additional stratigraphic information comes from a series of samples collected on the Blake Escarpment with the research submersible Alvin (Dillon and others, 1981; 1985). Three sampling transects were undertaken with Alvin; one at the Blake spur and two farther south (Fig. 4). Predominantly shallow-water limestones from present depths of 4,000 to 1,450 m contain calcareous nannofossil assemblages ranging in age from Valanginian-Hauterivian to Albian. Although the three transects are separated by distances of 72 and 119 km, the stratigraphic succession along the escarpment is consistent. Valanginian-Hauterivian strata are present from 4,000 to 3,300 m; a relatively thin Barremian interval extends from 3,300 to 3,100 m; and Aptian-Albian strata are present up to 1,500 m.

The structure and stratigraphy of the Blake Plateau basin are best revealed by multichannel seismic line TD-5 (Plate 3D), which has the best stratigraphic control of any multichannel line in the region. Line TD-5 crosses the COST GE-1 well on the outer shelf, the ASP 3 well on the other Blake Plateau, and DSDP site 390 on the Blake Spur. The most recent interpretations of the stratigraphy of the Blake Plateau basin are by Dillon, Paull, and others (1979) and Dillon and others (1985); they are the basis for much of the following discussion.

Seaward of the Blake Escarpment, oceanic basement lies approximately 2 km below the seafloor, but descends to about 5 or 6 km at the base of the escarpment. Beneath the Blake Plateau, the basement surface is not well defined on line TD-5, but appears to be about 13 km below the seafloor in the center of the basin. The basement is better defined as it rises to the west beneath the inner Blake Plateau and the Florida-Hatteras Slope. Basement blocks and several small synrift basins beneath the Florida-Hatteras Slope are separated by an apparent erosional unconformity from overlying postrift sedimentary strata of Early Cretaceous age or older.

Basement rocks beneath the Blake Plateau basin have not been sampled, but are interpreted to represent a "transitional" mixture of continental and oceanic crust formed during an episode of continental breakup. Neither have the oldest sedimentary rocks in the basin been sampled, but they probably were deposited in the Late Triassic and Early Jurassic during the early stages of rifting and basin subsidence.

The oldest sedimentary rocks dated with certainty are Early Cretaceous in age. Valanginian-Hauterivian shallow water limestone is exposed on the Blake Escarpment, from where seismic reflectors can be traced landward across the basin to the COST GE-1 well. These strata crop out on the escarpment as deep as 4,000 m below sea level, and are well above basement in the center of the basin. Landward they appear to terminate against basement rocks starting at shotpoint 2601 on line TD-5, and they reach the COST GE-1 site in the undated, unfossiliferous, non-marine interval just above Paleozoic basement.

Barremian, Aptian, and Albian shallow-water carbonate rocks thicken landward from the Blake Escarpment, and a reef-like structure of Aptian-Albian age is present at shotpoint 1041 beneath the outer Blake Plateau near ASP 3. These strata maintain their thickness across the basin, but thin as basement rises beneath the inner Blake Plateau. Barremian-Albian strata in the COST GE-1 well are non-marine at the base, but are shallow-marine carbonates in the upper part. The boundary between the non-marine and marine strata seaward of the G3-1 well is defined by a change in seismic facies along line TD-5.

Upper Cretaceous carbonates of Turonian to Maestrichtian age are well represented at COST GE-1 and seaward beneath the Florida-Hatteras Slope and inner Blake Plateau. These carbonates mark the onset of deeper-water paleoenvironments and much slower rates of sedimentation in the Blake Plateau basin. The Upper Cretaceous section is thin across most of the basin and seismic reflectors are difficult to trace seaward of shotpoint 4501. The ASP 3 borehole was drilled at shotpoint 1001 on the landward edge of the Aptian-Albian reef. This borehole terminated in Paleocene strata 150 m below the sea floor, constraining the Upper Cretaceous section to a thin interval between the Paleocene and the top of the reef. On the Blake Spur, only about 30 m of Campanian and Maestrichtian nannofossil ooze is present at DSDP sites 390 and 392. Cenozoic strata also are thickest beneath the outer shelf and Florida-Hatteras Slope. They are very thin beneath the inner Blake Plateau but thicken somewhat across the basin.

REGIONAL DEPOSITIONAL TRENDS

General Depositional History

Borehole data and seismostratigraphic interpretations show that most of the sedimentary fill in the offshore basins of the U.S. Atlantic margin is of Mesozoic age. Triassic strata have been sparsely recovered, having been drilled in situ only in the bottom of the COST G-1 and G-2 wells in the Georges Bank basin (Poag, 1980a,b). Triassic microfossils from higher in the G-2 well (Cousminer and others, 1984) appear to have been redeposited. These Triassic strata are composed of conglomerates, red and brown shales, arkosic sandstones, dolomite, and evaporites, which accumulated chiefly in grabens and half-grabens during the rifting stage of continental breakup. As much as 8 km of Triassic strata may be present in the Georges Bank basin (Mattick and others, 1981; Schlee and Fritsch, 1983); 5 km in the Baltimore Canyon trough (Schlee, 1981; Poag, 1985a) and 3 km in the

Blake Plateau basin (Dillon, Poag, and others, 1979; Dillon and others, 1985).

Seafloor spreading appears to have begun in the Early Jurassic, initiating a period of evaporite deposition, followed by carbonate deposition on the middle and outer parts of the Atlantic paleoshelf. Terrigenous siliciclastics accumulated updip, prograding steadily seaward as Jurassic time progressed. An elongate, discontinuous, reefal bank developed at the paleoshelf edge during the Middle and Late Jurassic and prevailed into the Early Cretaceous. To the north (Georges Bank basin and Baltimore Canyon trough), terrigenous deposits buried the shelf-edge reef in the middle of the Early Cretaceous (Hauterivian-Barremian), but in the Blake Plateau basin a shelf-edge reef system persisted into Aptian-Albian time.

Siliciclastic deposition continued to dominate these margin basins throughout the remaining Cretaceous, although biogenous chalks began to accumulate in some areas (especially to the south) during the Late Cretaceous highstand of sea level. Following a major period of erosion and nondeposition at the end of the Cretaceous, the Cenozoic gave rise to another regime of carbonate deposition spanning the U.S. margin from the Blake Plateau basin to Georges Bank. Particularly high sea levels during the Eocene resulted in thick carbonate deposits of deep-water origin in the Baltimore Canyon trough, which are characterized by a relatively high biosilica in the downdip facies. However, in the Southeast Georgia embayment high sediment accumulation rates maintained a thick column of shallow-water carbonate deposits during the Eocene. Siliciclastic deposition returned all along the margin during the Oligocene and Miocene and has persisted generally to the present.

Depositional Sequences and Unconformities

The biostratigraphic record of Upper Cretaceous and Cenozoic strata contains abundant planktonic microfossils that allow finer resolution of biozones and their associated depositional sequences than in Lower Cretaceous and older rocks. Poag (1980a,b; 1985a,b) and Poag and Schlee (1984) have examined this biostratigraphic record and compared it with paleobathymetric trends, which are inferred on the basis of microfossils. They concluded that a series of regional depositional sequences can be recognized and can be related to sea-level changes, following the Vail model (Vail and others, 1977; Vail and Hardenbol, 1979; Vail and Todd, 1981).

The Middle to Upper Cretaceous strata constitute six depositional sequences that are bounded by stratigraphic gaps or unconformities (Fig. 5). Above an especially notable gap at the Cretaceous/Tertiary contact, six Paleogene depositional sequences, bounded by unconformities, can be recognized. These sequences appear to represent third order depositional cycles encompassed by the second order supercycles Kb, Ta, Tb, Tc, and Td of the Vail model. The stratigraphic gaps are inferred to coincide with major periods of erosion and nondeposition caused by sea-level falls.

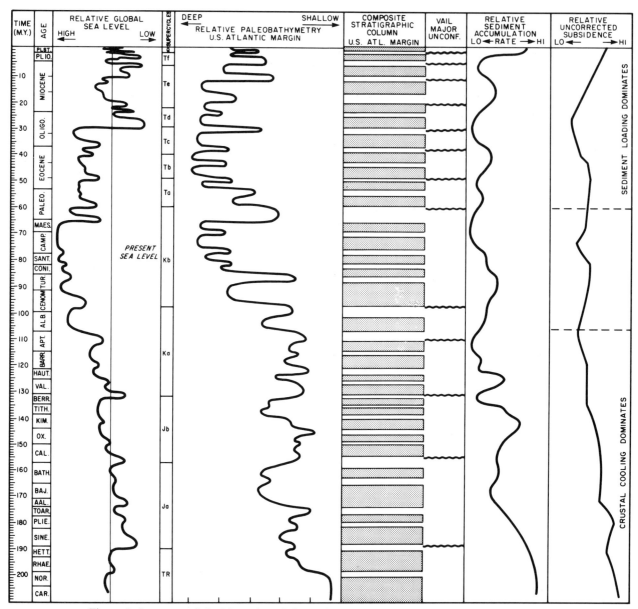

Figure 5. Summary of the Mesozoic and Cenozoic development of the U.S. Atlantic margin, as illustrated by composite curves of relative paleobathymetry, sediment accumulation rate, and basin subsidence compared to a composite stratigraphic column, and a curve of relative global sea-level. Sea-level curve modified from Vail and Hardenbol (1979) and Vail and Todd (1981).

The Neogene-Quaternary section is composed of five depositional sequences separated by unconformities (Fig. 5). These sequences appear to represent third order cycles encompassed by three second order supercycles (Te, Tf, Q) of the Vail model, and the unconformities include the major ones of the Vail model.

The stratigraphic column older than Cenomanian lacks a reliable planktonic biozonation for the facies we studied, and has been divided into depositional sequences chiefly by calibrating seismic sequences with the Vail model. Poag and Schlee (1984) and Poag (1985a) believed they could recognize at least 16 unconformity-bound depositional cycles in this pre-Cenomanian

section and that they represent cycles and supercycles of the Vail model (Fig. 5). However, because the seismostratigraphic interpretations are highly subjective (Poag and Schlee, 1984), the chronostratigraphic positions of the cycles must be considered tentative until the inferred geochronology can be corroborated by other methods.

Sediment Accumulation Rates and Basin Subsidence

Poag and Schlee (1984) calculated average rates of sediment accumulation (uncorrected for sediment compaction) for each basin, which we have composited in Figure 5 as a curve of relative accumulation rate. There is considerable variability from basin to basin, but the regional trends shown by the composite curves appear to be substantive. The terrigenous filling of the Triassic basins in the rifting phase took place at relatively high rates, perhaps higher than in any following period. These rates declined to a low point at the beginning of the Middle Jurassic (Bajocian), when carbonate deposition was widespread in each basin. Later, peaks of high accumulation rate occurred during the Kimmeridgian and Valanginian, apparently associated with lowered sea levels (Fig. 5). However, subsequent high peaks in the Turonian, Eocene and middle Miocene are associated with high sea levels. The Miocene peak is especially notable in the Baltimore Canyon trough, where it resulted from a major period of deltaic progradation. Calculated Quaternary rates are higher than any in the Cretaceous or Tertiary, being in part the result of multiple sea-level falls, glaciomarine sedimentation, and an uncompacted sediment column.

Relative subsidence rates were also calculated for each basin by Poag and Schlee (1984), using a modified version of the geohistory analysis originally described by van Hinte (1978; Poag, 1985a; see Poag and Hall, 1979). In Figure 5 we show a composite subsidence rate uncorrected for sediment compaction. The early post-rift history of the three basins was characterized by rapid subsidence dominated by the effects of crustal cooling (Watts, 1981; Watts and Steckler, 1979). A long period of gradually declining rates ended in the Aptian-Albian. This was followed by an increased rate in the Cenomanian-middle Campanian interval. This surge may have been related to plate tectonic adjustments or the increased height of the water column, as the low rate of sediment accumulation would rule out sediment loading effects.

In the Cenozoic, however, sediment loading appears to have been a more effective agent for subsidence. Pulses of increased subsidence rates in the Eocene, and Miocene-Pleistocene are associated with significantly increased sediment accumulation rates (see also Heller and others, 1982).

SUMMARY

Integration of lithostratigraphic, biostratigraphic, and seismostratigraphic data from three major offshore sedimentary basins of the U.S. Atlantic margin (Georges Bank basin, Baltimore Canyon trough, Blake Plateau basin) reveals common aspects of their stratigraphic framework and depositional history. The oldest sedimentary rocks are synrift graben-fill deposits inferred to comprise chiefly coarse terrigenous siliciclastics of Triassic age. Maximum thickness of the synrift section reaches more than 5 km (Baltimore Canyon trough). Following a major erosional hiatus, restricted marine carbonates and evaporites (including bedded halite) were laid down as the initial postrift deposits during an Early Jurassic transition from rifting to seafloor spreading. As seafloor spreading proceeded, shallow-water limestones and shelf-edge reefs characterized the outer parts of the basins, culminating in the Late Jurassic-Early Cretaceous as a discontinuous, margin-rimming reefal bank. Jurassic strata constitute most of the postrift deposits, reaching a maximum thickness of more than 12 km (Baltimore Canyon trough).

Deposition of terrigenous siliciclastics terminated this carbonate regime during the Early and Middle Cretaceous, burying the shelf-edge barrier. Upper Cretaceous deposits accumulated as increasingly finer-grained sandstones, siltstones, mudstones, shales, and chalks, beneath a steadily deepening shelf-sea. Maximum thickness of the Cretaceous sediment column is more than 2 km (Baltimore Canyon trough). Cretaceous deposition was terminated by an especially widespread interval of erosion and nondeposition, followed by renewed carbonate deposition in the Paleogene. Neogene deposition was chiefly siliciclastic, characterized by deltaic progradation. Quaternary deposits also are dominantly siliciclastic and include glaciomarine facies in the Georges Bank basin. Total Cenozoic sediment thickness is maximum in the Baltimore Canyon trough, where more than 2 km accumulated beneath the present shelf.

The offshore sedimentary column of the U.S. Atlantic margin can be divided into at least 32 depositional sequences, each of which is bounded above and below by unconformities. The stratigraphic positions of these sequences and their relationships to paleobathymetric cycles are strongly correlative with cycles and supercycles of the Vail depositional model and appear to be controlled chiefly by sea-level change.

REFERENCES CITED

Applin, E. R., 1955, A biofacies of Woodbine age in the southeastern Gulf Coast region: U.S. Geological Survey Professional Paper 264-I, p. 187–197.

Applin, P. L., 1951, Preliminary report on buried pre-Mesozoic rocks in Florida and adjacent states: U.S. Geological Survey Circular 91, 28 p.

Applin, P. L., and Applin, E. R., 1965, The Comanche Series and associated rocks in the subsurface in central and south Florida: U.S. Geological Survey Professional Paper 447, 84 p.

—— , 1967, The Gulf Series in the subsurface in northern Florida and southern Georgia: U.S. Geological Survey Professional Paper 524-G, 35 p.

Arthur, M. A., 1982, Lithology and petrology of the COST Nos. G-1 and G-2 wells, *in* Scholle, P. A., and Wenkam, C. R., eds., Geological studies of the COST Nos. G-1 and G-2 wells, United States North Atlantic Outer Continental Shelf: U.S. Geological Survey Circular 861, p. 11–33.

Austin, J. A., Jr., Uchupi, E., and Schaughnessy, D. R. III., 1980, Geology of New England passive margin: American Association of Petroleum Geologists Bulletin, v. 64, p. 501–526.

Ballard, R. D., and Uchupi, E., 1975, Triassic rift structure in Gulf of Maine: American Association of Petroleum Geologists Bulletin, v. 59, p. 1041–1072.

Bayer, K. C., and Mattick, R. E., 1980, Geologic setting, *in* Mattick, R. E., and Hennessy, J. L., eds., Structural framework, stratigraphy, and petroleum geology of the area of oil and gas lease sale No. 49 on the U.S. Atlantic Continental Shelf and Slope: U.S. Geological Survey Circular 812, p. 6–8.

Benson, W. E., Sheridan, R. E., and others, 1978, Initial reports of the Deep Sea Drilling Project: Washington, D.C., U.S. Government Printing Office, v. 44, 1005 p.

Bonini, W. E., and Woollard, G. P., 1960, Subsurface geology of North Carolina–South Carolina Coast Plain from seismic data: American Association of Petroleum Geologists Bulletin, v. 44, no. 3, p. 298–315.

Brown, P. M., Miller, J. A., and Swain, F. M., 1972, Structural and stratigraphic framework and spatial distribution of permeability of the Atlantic Coastal Plain, North Carolina to New York: U.S. Geological Survey Professional Paper 796, 70 p.

Brown, P. M., Brown, D. C., Reid, M. S., and Lloyd, O. B., 1979, Evaluation of the geologic and hydrologic factors related to the waste-storage potential of Mesozoic aquifers in the southern part of the Atlantic Coastal Plain, South Carolina and Georgia: U.S. Geological Survey Professional Paper 1088, 37 p.

Bryan, G. M., and Heirtzler, J. R., eds., 1984, Eastern North American continental margin and adjacent ocean floor, 28° to 36°N and 70° to 82°W: Ocean Margin Drilling Program regional atlas series, Woods Hole, Massachusetts, Marine Science International, Atlas 5, 49 p.

Buffler, R. T., Shipley, T. H., and Watkins, J. S., 1978, Blake continental margin seismic secton: American Association of Petroleum Geologists Seismic Section No. 2.

Buffler, R. T., Watkins, J. S., and Dillon, W. P., 1979, Geology of the offshore Southeast Georgia embayment, U.S. Atlantic continental margin, based on multichannel seismic reflection profiles, in Watkins, J. S., Montadert, L., and Dickerson, P. W., eds., Geological and geophysical investigations of continental margins: American Association of Petroleum Geologists Memoir 29, p. 11–25.

Bunce, E. T., Emery, K. O., Gerard, R. D., Knott, S. T., Lidz, L., Saito, T., and Schlee, J., 1965, Ocean drilling on the continental margin: Science, v. 150, p. 709–716.

Charletta, A. C., 1980, Eocene benthic foraminiferal paleoecology and paleobathymetry of the New Jersey continental margin [Ph.D. thesis]: Rutgers University, 83 p.

Charm, W. B., Nesterhoff, W. D., and Valdes, S., 1969, Detailed stratigraphic description of the JOIDES cores on the continental margin off Florida: U.S. Geological Survey Professional Paper 581-D, 13 p.

Chowns, T. M., and Williams, C. T., 1983, Pre-Cretaceous rocks beneath the Georgia Coastal Plain; Regional implications, in Gohn, G. S., ed., Studies related to the Charleston, South Carolina earthquake of 1886; Tectonics and seismicity: U.S. Geological Survey Professional Paper 1313-L, 42 p.

Cousminer, H. L., Steinkraus, W. E., and Hall, R. E., 1984, Biostratigraphic restudy documents Triassic/Jurassic section in Georges Bank COST G-2 well: American Association of Petroleum Geologists Bulletin, v. 68, p. 466.

Cramer, H. R., 1974, Isopach and lithofacies analyses of the Cretaceous and Cenozoic rocks of the Coastal Plain of Georgia: Georgia Geological Survey Bulletin 87, p. 21–44.

Daniels, D. L., Zietz, I., and Popenoe, P., 1983, Distribution of subsurface Lower Mesozoic rocks in the southeastern United States, as interpreted from regional aeromagnetic and gravity maps, in Gohn, G. S., ed., Studies related to the Charleston, South Carolina earthquake of 1886; Tectonics and seismicity: U.S. Geological Survey Professional Paper 1313-K, 24 p.

Dillon, W. P., and McGinnis, C. D., 1983, Basement structure indicated by seismic refraction measurements offshore from South Carolina and adjacent areas, in Gohn, G. S., ed., Studies related to the Charleston, South Carolina, earthquake of 1886; Tectonics and seismicity: U.S. Geological Survey Professional Paper 1313-O, 7 p.

Dillon, W. P., and Paull, C. K., 1978, Interpretation of multichannel seismic reflection profiles of the Atlantic Continental Margin off the coasts of South Carolina and Georgia: U.S. Geological Survey Miscellaneous Field Studies Map MF-936.

Dillon, W. P., Paull, C. K., Buffler, R. T., and Fail, J. P., 1979, Structure and development of the Southeast Georgia Embayment and northern Blake Plateau, preliminary analysis, in Watkins, J. S., Montadert, L., and Dickerson, P. W., eds., Geological and geophysical investigations of continental margins: American Association of Petroleum Geologists Memoir 29, p. 27–41.

Dillon, W. P., Poag, C. W., Valentine, P. C., and Paull, C. K., 1979, Structure, biostratigraphy and seismic stratigraphy along a common-depth-point seismic profile through three drill holes on the continental margin off Jackson-

ville: U.S. Geological Survey Miscellaneous Field Studies Map MF-1090.

Dillon, W. P., Paull, C. K., Valentine, P. C., Ball, M. M., Arthur, M. A., Shinn, E., and Kent, K. M., 1981, The Blake Escarpment platform edge; Conclusions based on observations and sampling from a research submersible: American Association of Petroleum Geologists Bulletin, v. 65, p. 918.

Dillon, W. P., Klitgord, K. D., and Paull, C. K., 1983, Mesozoic development and structure of the continental margin off South Carolina, in Gohn, G. S., ed., Studies related to the Charleston, South Carolina, earthquake of 1886; Tectonics and seismicity: U.S. Geological Survey Professional Paper 1313-N, 16 p.

Dillon, W. P., Paull, C. K., and Gilbert, L. E., 1985, History of the Atlantic continental margin off Florida; The Blake Plateau basin, in Poag, C. W., ed., Geologic evolution of the United States Atlantic margin: New York, Van Nostrand Reinhold, p. 189–215.

Edson, G. M., 1985, The Mid-Atlantic Mesozoic paleoshelf edge; Carbonate buildup or reef?: American Association of Petroleum Geologists Bulletin, v. 69, p. 1436.

Eliuk, L. S., 1978, The Abenaki Formation, Nova Scotia shelf, Canada; A depositional and diagenetic model for a Mesozoic carbonate platform: Bulletin of Canadian Petroleum Geology, v. 26, p. 424–514.

Emery, K. O., 1965, Geology of the continental margin off eastern United States, in Whittard, W. F., and Bradshaw, R., eds., Submarine geology and geophysics: London, Butterworths, p. 1–20.

Emery, K. O., and Uchupi, E., 1965, Structure of Georges Bank: Marine Geology, v. 3, p. 349–358.

Emery, K. O., and Zarudski, E.F.K., 1967, Seismic reflection profiles along the drill holes on the continental margin off Florida: U.S. Geological Survey Professional Paper 581-A, 8 p.

Ewing, J. I., and Rabinowitz, P. D., eds., 1984, Eastern North American continental margin and adjacent ocean floor, 34° to 41°N and 68° to 78°W: Ocean margin Drilling Program regional atlas series, Woods Hole, Massachusetts, Marine Science International, Atlas 4, 40 p.

Ewing, J. I., Ewing, M., and Leyden, R., 1966, Seismic-profiler survey of Blake Plateau: American Association of Petroleum Geologists Bulletin, v. 50, p. 1948–1971.

Folger, D. W., Hathaway, J. C., Christopher, R. A., Valentine, P. C., and Poag, C. W., 1978, Stratigraphic test well, Nantucket Island, Massachusetts: U.S. Geological Survey Circular 773, 28 p.

Gibson, T. G., Hazel, J. E., and Mello, J. F., 1968, Fossiliferous rocks from submarine canyons off the northeastern United States: U.S. Geological Survey Professional Paper 600-D, p. 222–230.

Giordano, A. C., Carpenter, G. B., and Amato, R. V., 1983, Oil and gas developments in Atlantic Coastal Plain and outer Continental Shelf: American Association of Petroleum Geologists Bulletin, v. 67, p. 1566–1569.

Given, M. M., 1977, Mesozoic and early Cenozoic geology of offshore Nova Scotia: Bulletin of Canadian Petroleum Geology, v. 25, p. 63–91.

Gohn, G. S., 1983, Geology of the basement rocks near Charleston, South Carolina; Data from detrital rock fragments in Lower Mesozoic(?) rocks in Clubhouse Crossroads test hole #3, in Gohn, G. S., ed., Studies related to the Charleston, South Carolina, earthquake of 1886; Tectonics and seismicity: U.S. Geological Survey Professional Paper 1313-E, 22 p.

Gohn, G. S., Higgins, B. B., Smith, C. C., and Owens, J. P., 1977, Lithostratigraphy of the deep corehole (Clubhouse Crossroads Corehole 1) near Charleston, South Carolina: U.S. Geological Survey Professional Paper 1028-E, p. 59–70.

Gohn, G. S., Bybell, L. M., Smith, C. C., and Owens, J. P., 1978, Preliminary stratigraphic cross sections of Atlantic Coastal Plain sediments of the southeastern United States; Cenozoic sediments along the South Carolina coastal margin: U.S. Geological Survey Miscellaneous Field Studies Map MF-1015-B.

Gohn, G. S., Christopher, R. A., Smith, C. C., and Owens, J. P., 1978, Preliminary stratigraphic cross sections of Atlantic Coastal Plain sediments of the southeastern United States; Cretaceous sediments along the South Carolina coastal margin: U.S. Geological Survey Miscellaneous Field Studies Map

MF-1015-A.

Gohn, G. S., Gottfried, D., Lamphere, M. A., and Higgins, B. B., 1978, Regional implications of Triassic or Jurassic age for basalt and sedimentary red beds in the South Carolina Coastal Plain: Science, v. 202, no. 4370, p. 887–889.

Gohn, G. S., Smith, C. C., Christopher, R. A., and Owens, J. P., 1980, Preliminary stratigraphic cross sections of Atlantic Coastal Plain sediments of the southeastern United States; Cretaceous sediments along the Georgia coastal margin: U.S. Geological Survey Miscellaneous Field Studies Map MF-1015-C.

Gohn, G. S., Bybell, L. M., Christopher, R. A., Owens, J. P., and Smith, C. C., 1982, A stratigraphic framework for Cretaceous and Paleogene sediments along the South Carolina and Georgia coastal margins, *in* Arden, D. D., Beck, B. F., and Morrow, E., eds., Second symposium on the geology of the Southeastern Coastal Plain: Georgia Geological Survey Information Circular 53, p. 64–74.

Grow, J. A., 1980, Deep structure and evolution of the Baltimore Canyon trough in the vicinity of the COST No. B-3 well, *in* Scholle, P. A., ed., Geological studies of the COST No. B-3 well, United States Mid-Atlantic Continental Slope area: U.S. Geological Survey Circular 833, p. 117–125.

Grow, J. A., and Klitgord, K. D., 1980, Structural framework, *in* Mattick, R. E., and Hennessy, J. L., eds., Structural framework stratigraphy and petroleum geology of the area of oil and gas lease sale No. 49 on the U.S. Atlantic Continental Shelf and Slope: U.S. Geological Survey Circular 812, p. 8–35.

Grow, J. A., and Markl, R. G., 1977, IPOD-USGS multichannel seismic reflection profile from Cape Hatteras to the Mid-Atlantic Ridge: Geology, v. 5, p. 625–630.

Grow, J. A., and Schlee, J. S., 1976, Interpretation and velocity analysis of U.S. Geological Survey multichannel reflection profiles 4, 5, and 6, Atlantic continental margin: U.S. Geological Survey Miscellaneous Field Studies Map MF-808.

Grow, J. A., Mattick, R. E., and Schlee, J. S., 1979, Multichannel seismic depth sections and interval velocities over continental shelf and upper continental slope between Cape Hatteras and Cape Cod, *in* Watkins, J. S., Montadert, L., and Dickerson, P. W., eds., Geological and geophysical investigations of continental margins: American Association of Petroleum Geologists Memoir 29, p. 65–83.

Grow, J. A., Hutchinson, D. B., Klitgord, K. D., Dillon, W. P., and Schlee, J. S., 1983, Representative multichannel seismic reflection profiles over the U.S. Atlantic continental margin, *in* Bally, A. W., ed., Seismic expression of structural styles: American Association of Petroleum Geologists, Studies in Geology, Ser. 15, pt. 2, p. 2.2.3-1–2.2.3-19.

Hall, R. E., Poppe, L. J., and Ferrebee, W. M., 1980, A stratigraphic test well, Martha's Vineyard, Massachusetts: U.S. Geological Survey Bulletin No. 1488, 19 p.

Hathaway, J. C., Poag, C. W., Valentine, P. C., Miller, R. E., Schultz, D. M., Manheim, F. T., Kohout, F. A., Bothner, M. H., and Sangree, D. A., 1979, U.S. Geological Survey core drilling on the Atlantic Shelf: Science, v. 206, p. 515–527.

Hazel, J. E., Bybell, L. M., Christopher, R. A., Fredericksen, N. O., May, F. E., McLean, D. M., Poore, R. Z., Smith, C. C., Sohl, N. F., Valentine, P. C., and Witmer, R. J., 1977, Biostratigraphy of the deep corehole (clubhouse crossroads corehole 1) near Charleston, South Carolina: U.S. Geological Survey Professional Paper 1028-F, p. 71–89.

Heezen, B. C., and Sheridan, R. E., 1966, Lower Cretaceous rocks (Neocomian-Albian) dredged from the Blake Escarpment: Science, v. 154, p. 1644–1647.

Heller, P. L., Wentworth, C. M., and Poag, C. W., 1982, Episodic post-rift subsidence of the U.S. Atlantic continental margin: Geological Society of America Bulletin, v. 93, p. 379–390.

Herrick, S. M., 1961, Well logs of the Coastal Plain of Georgia: Georgia Geological Survey Bulletin 70, 461 p.

Herrick, S. M., and Vorhis, R. C., 1963, Subsurface geology of the Georgia Coastal Plain: Georgia Geophysical Survey Information Circular 25, 78 p.

Hersey, J. B., Bunce, E. J., Wyrick, R. F., and Dietz, F. T., 1959, Geophysical investigation of the continental margin between Cape Henry, Virginia, and Jacksonville, Florida: Geological Society of America Bulletin, v. 70, p. 437–466.

Hollister, C. D., Ewing, J. I., and others, 1972, Initial reports of the Deep Sea Drilling Project: Washington, D.C., U.S. Government Printing Office, v. 11, 1077 p.

Hutchinson, D. R., Grow, J. A., Klitgord, K. D., and Swift, B. A., 1982, Deep structure and evolution of the Carolina Trough, *in* Watkins, J. S., and Drake, C. L., eds., Studies in continental margin geology: American Association of Petroleum Geologists Memoir 34, p. 129–152.

Jansa, L. F., and Wade, J. A., 1975, Geology of the continental margin off Nova Scotia and Newfoundland: Geological Survey of Canada Paper 74-30, p. 51–105.

Jansa, L. F., Enos, P., Tucholke, B. E., Gradstein, F. M., and Sheridan, R. E., 1979, Mesozoic-Cenozoic sedimentary formations of the North American Basin; western North Atlantic, *in* Talwani, M., Hay, W. W., and Ryan, W.B.F., eds., Deep drilling results in the Atlantic Ocean; Continental margins and paleoenvironments: Maurice Ewing Series 3, American Geophysical Union, p. 1–57.

Klitgord, K. D., and Behrendt, J. C., 1979, Basin structure of the U.S. Atlantic margin, *in* Watkins, J. S., Montadert, L., and Dickerson, P. W., eds., Geological and geophysical investigations of continental margins: American Association of Petroleum Geologists Memoir 29, p. 85–112.

Klitgord, K. D., Behrendt, J. C., Schlee, J. S., and Hinz, K., 1982, Basement structure, sedimentation, and tectonic history of the Georges Bank basin, *in* Scholle, P. A., and Wenkham, C. R., eds., Geological studies of the COST Nos. G-1 and G-2 wells, United States North Atlantic Outer Continental Shelf: U.S. Geological Survey Circular 861, p. 160–186.

Klitgord, K. D., Dillon, W. P., and Popenoe, P., 1983, Mesozoic tectonics of the Southeastern United States Coastal Plain and continental margins, *in* Gohn, G. S., ed., Studies related to the Charleston, South Carolina, earthquake of 1886; Tectonics and seismicity: U.S. Geological Survey Professional Paper 1313-P, 15 p.

Libby-French, J., 1981, Lithostratigraphy of the Shell 272-1 and 273-1 wells; Implications as to depositional history of the Baltimore Canyon trough, mid-Atlantic OCS: American Association of Petroleum Geologists Bulletin, v. 65, p. 1476–1484.

—— , 1984, Stratigraphic framework and petroleum potential of northeastern Baltimore Canyon trough, mid-Atlantic Outer Continental Shelf: American Association of Petroleum Geologists Bulletin, v. 68, p. 50–73.

Maher, J. C., 1971, Geologic framework and petroleum potential of the Atlantic Coastal Plain and Continental Shelf: U.S. Geological Survey Professional Paper 695, 98 p.

Manspeizer, W., 1982, Early Mesozoic basins of the central Atlantic passive margins, *in* Bally, A. W., ed., Geology of passive continental margins; History, structure, and sedimentologic record: American Association of Petroleum Geologists, Education Course Note Series No. 19, Art. 4, p. 1–60.

—— , 1985, Early Mesozoic history of the Atlantic passive margin, *in* Poag, C. W., ed., Geologic evolution of the United States Atlantic margin: New York, Van Nostrand Reinhold, p. 1–23.

Marsalis, W. E., 1970, Petroleum exploration in Georgia: Georgia Geological Survey Information Circular 38, 52 p.

Mattick, R. E., Foote, R. Q., Weaver, N. L., and Grim, M. S., 1974, Structural framework of United States Atlantic Outer Continental Shelf north of Cape Hatteras: American Association of Petroleum Geologists Bulletin, v. 58, p. 1179–1190.

Mattick, R. E., Schlee, J. S., and Bayer, K. C., 1981, The geology and hydrocarbon potential of the Georges Bank–Baltimore Canyon trough area: Canadian Society of Petrology Geologists Memoir 7, p. 461–486.

McIver, N. L., 1972, Cenozoic and Mesozoic stratigraphy of the Nova Scotia Shelf: Canadian Journal of Earth Science, v. 9, p. 54–70.

Milton, C., 1972, Igneous and metamorphic basement rocks of Florida: Florida Bureau of Geology Bulletin 55, 125 p.

Milton, C., and Hurst, V. J., 1965, Subsurface "basement" rocks of Georgia: Georgia Geological Survey Bulletin 76, 56 p.

Mountain, G. S., and Tucholke, B. E., 1985, Mesozoic and Cenozoic geology of the U.S. Atlantic Continental Slope and Rise, *in* Poag, C. W., ed., Geologic evolution of the United States Atlantic margin: New York, Van Nostrand Reinhold, p. 293–341.

Northrop, J., and Heezen, B. C., 1951, An outcrop of Eocene sediment on the continental slope: Journal of Geology, v. 59, p. 396–399.

Olsson, R. K., 1964, Late Cretaceous planktonic foraminifera from New Jersey and Delaware: Micropaleontology, v. 10, p. 157–188.

——, 1978, Summary of lithostratigraphy and biostratigraphy of Atlantic Coastal Plain (northern part), *in* Benson, W. E., and Sheridan, R. E., eds., Initial reports of the Deep Sea Drilling Project: Washington, D.C., U.S. Government Printing Office, v. 44, p. 941–947.

Olsson, R. K., and Nyong, E. E., 1984, A paleoslope model for Campanian-lower Maestrichtian foraminifera of New Jersey and Delaware: Journal of Foraminiferal Research, v. 14, p. 50–68.

Olsson, R. K., Miller, K. G., and Ungrady, T. E., 1980, Late Oligocene transgression of middle Atlantic Coastal Plain: Geology, v. 8, p. 549–554.

Owens, J. P., and Gohn, G. S., 1985, Depositional history of the Cretaceous System in the United States Atlantic Coastal Plain; Stratigraphy, paleoenvironments, and basin evolution, *in* Poag, C. W., ed., Geologic evolution of the United States Atlantic margin: New York, Van Nostrand Reinhold, p. 25–86.

Owens, J. P., and Sohl, N. F., 1969, Shelf and deltaic paleoenvironments in the Cretaceous-Tertiary formations of the New Jersey Coastal Plain, *in* Subitzky, S., ed., Geology of selected areas in New Jersey and eastern Pennsylvania: New Brunswick, New Jersey, Rutgers University Press, p. 235–278.

Paull, C. K., and Dillon, W. P., 1980a, Structure, stratigraphy, and geologic history of Florida-Hatteras Shelf and inner Blake Plateau: American Association of Petroleum Geologists Bulletin, v. 64, p. 339–358.

——, 1980b, Erosional origin of the Blake Escarpment: An alternative hypothesis: Geology, v. 8, p. 538–542.

Petters, S. W., 1976, Upper Cretaceous subsurface stratigraphy of Atlantic Coastal Plain of New Jersey: American Association of Petroleum Geologists Bulletin, v. 60, p. 87–107.

Poag, C. W., 1978, Stratigraphy of the Atlantic Continental Shelf and Slope of the United States: Annual Review of Earth and Planetary Sciences, v. 6, p. 251–280.

——, 1980a, Foraminiferal stratigraphy and paleoecology, *in* Mattick, R. E., and Hennessy, J. L., eds., Structural framework, stratigraphy, and petroleum geology of the area of oil and gas lease sale No. 49 on the U.S. Atlantic Continental Shelf and Slope: U.S. Geological Survey Circular 812, p. 35–48.

——, 1980b, Foraminiferal stratigraphy, paleoenvironments, and depositional cycles in the outer Baltimore Canyon trough, *in* Scholle, P. A., ed., Geological studies of the COST No. B-3 well, United States Mid-Atlantic Continental Slope area: U.S. Geological Survey Circular 833, p. 44–65.

——, 1982a, Foraminiferal and seismic stratigraphy, paleoenvironments, and depositional cycles in the Georges Bank basin, *in* Scholle, P. A., and Wenkham, C. R., ed., Geological studies of the COST Nos. G-1 and G-2 wells, United States North Atlantic Outer Continental Shelf: U.S. Geological Survey Circular 861, p. 43–91.

——, 1982b, Stratigraphic reference section for Georges Bank basin–depositional model for New England passive margin: American Association of Petroleum Geologists Bulletin, v. 66, p. 1021–1041.

——, 1984, Neogene stratigraphy of the submerged U.S. Atlantic margin: Palaeogeography, Palaeoclimatology, Palaeoecology, v. 47, p. 103–127.

——, 1985a, Depositional history and stratigraphic reference section for central Baltimore Canyon trough, *in* Poag, C. W., ed., Geologic evolution of the United States Atlantic margin: New York, Van Nostrand Reinhold, p. 217–264.

——, 1985b, Cenozoic and Upper Cretaceous sedimentary facies and depositional systems of the New Jersey Slope and Rise, *in* Poag, C. W., ed., Geologic evolution of the United States Atlantic margin: New York, Van Nostrand Reinhold, p. 343–365.

——, 1985c, Geologic evolution of the United States Atlantic margin: New

York, Van Nostrand Reinhold, 383 p.

Poag, C. W., and Hall, R. E., 1979, Foraminiferal biostratigraphy, paleoecology, and sediment accumulation rates, *in* Scholle, P. A., ed., Geological studies of the COST GE-1 well, United States South Atlantic Outer Continental Shelf area: U.S. Geological Survey Circular 800, pp. 49–63.

Poag, C. W., and Schlee, J. S., 1984, Depositional sequences and stratigraphic gaps on submerged U.S. Atlantic margin, *in* Schlee, J. S., ed., Interregional unconformities and hydrocarbon accumulation: American Association of Petroleum Geologists Memoir 36, p. 165–182.

Popenoe, P., 1985, Cenozoic depositional and structural history of the North Carolina margin from seismic-stratigraphic analysis, *in* Poag, C. W., ed., Geologic evolution of the United States Atlantic margin: New York, Van Nostrand Reinhold, p. 125–187.

Rankin, D. W., ed., 1977, Studies related to the Charleston, South Carolina, Earthquake of 1886; A preliminary report: U.S. Geological Survey Professional Paper 1028, 204 p.

Robb, J. M., Hampson, J. C., Jr., Kirby, J. R., and Twichell, D. C., 1981, Geology and potential hazards of the continental slope between Lindenkohl and South Toms Canyons, offshore mid-Atlantic United States: U.S. Geological Survey Open-File Report 81-600, 38 p.

Robb, J. M., Hampson, J. C., Jr., and Twichell, D. C., 1981, Geomorphology and sediment stability of a segment of the U.S. Continental Slope off New Jersey: Science, v. 211, p. 935–937.

Robb, J. M., Kirby, J. R., Hampson, J. C., Jr., Gibson, P. R., and Hecker, B., 1983, Furrowed outcrops of Eocene chalk on the lower continental slope offshore New Jersey: Geology, v. 11, p. 182–186.

Ryan, W.B.F., Cita, M. B., Miller, E. L., Hanselman, D., Nesteroff, W. D., Hecker, B., and Nibbelink, M., 1978, Bedrock geology in New England submarine canyons: Oceanologica Acta, v. 1, p. 233–254.

Schlee, J. S., 1977, Stratigraphy and Tertiary development of the continental margin east of Florida: U.S. Geological Survey Professional Paper 581-F, 25 p.

——, 1981, Seismic stratigraphy of Baltimore Canyon trough: American Association of Petroleum Geologists Bulletin, v. 65, p. 26–53.

Schlee, J. C., and Cheetham, A. H., 1967, Rocks of Eocene age on Fippennies Ledge, Gulf of Maine: Geological Society of America Bulletin, v. 78, p. 681–684.

Schlee, J. S., and Fritsch, J., 1983, Seismic stratigraphy of the Georges Bank complex, offshore New England, *in* Watkins, J. S., and Drake, C. L., eds., Studies in continental margin geology: American Association of Petroleum Geologists Memoir 34, p. 223–251.

Schlee, J. S., and Jansa, L. F., 1981, The paleoenvironment and development of the eastern North American continental margin; Proceedings 26th International Geological Congress, Paris, 1980. Geology of continental margins symposium: Oceanologica Acta, No. Sp. p. 71–80.

Schlee, J. S., and Klitgord, K. D., 1982, Geologic setting of the Georges Bank basin, *in* Scholle, P. A., and Wenkam, C. R., eds., Geologic studies of the COST Nos. G-1 and G-2 wells, United States North Atlantic outer continental shelf: U.S. Geological Survey Circular 861, p. 4–10.

Schlee, J. S., Behrendt, J. C., Grow, J. A., Robb, J. M., Mattick, R. E., Taylor, P.T., and Lawson, B. J., 1976, Regional geologic framework off northeastern United States: American Association of Petroleum Geologists Bulletin, v. 60, p. 926–951.

Schlee, J. S., Dillon, W. P., and Grow, J. A., 1979, Structure of the continental slope off the eastern United States: Society of Economic Paleontologists and Mineralogists Special Publication No. 27, p. 95–117.

Schlee, J. C., Poag, C. W., and Hinz, K., 1985, Seismic stratigraphy of the continental slope and rise seaward of Georges Bank, *in* Poag, C. W., ed., Geologic evolution of the United States Atlantic margin: New York, Van Nostrand Reinhold, p. 265–292.

Scholle, P. A., ed., 1977, Geological studies on the COST No. B-2 well, U.S. Mid-Atlantic Outer Continental Shelf area: U.S. Geological Survey Circular 750, 71 p.

——, 1979, Geological studies of the COST GE-1 well, United States South

Atlantic Outer Continental Shelf area: U.S. Geological Survey Circular 800, 114 p.

——, 1980, Geological studies of the COST No. B-3 well, United States Mid-Atlantic Continental Slope area: U.S. Geological Survey Circular 833, 132 p.

Scholle, P. A., and Wenkam, C. R., eds., 1982, Geological studies of the COST Nos. G-1 and G-2 wells, United States North Atlantic outer continental shelf: U.S. Geological Survey Circular 861, 193 p.

Sheridan, R. E., and Osburn, W. L., 1975, Marine geological and geophysical studies of the Florida–Blake Plateau–Bahamas area: Canadian Society of Petroleum Geologists Memoir 4, p. 9–32.

Sheridan, R. E., Smith, J. D., and Garnder, J., 1969, Rock dredges from Blake Escarpment near Great Abaco Canyon: American Association of Petroleum Geologists Bulletin, v. 53, p. 2551–2558.

Sheridan, R. E., Windisch, C. C., Ewing, J. I., and Stoffa, P. L., 1979, Structure and stratigraphy of the Blake Escarpment based on seismic reflection profiles, *in* Watkins, J. S., Montadert, L., and Dickerson, P. W., eds., Geological and geophysical investigations of continental margins: American Association of Petroleum Geologists Memoir 29, p. 177–186.

Sheridan, R. E., Gradstein, F. M., and others, 1983, Initial Reports of the Deep Sea Drilling Project: Washington, D.C., U.S. Government Printing Office, v. 76, 947 p.

Shipley, T. H., Buffler, R. J., and Watkins, J. S., 1978, Seismic stratigraphy and geologic history of Blake Plateau and adjacent western Atlantic continental margin: American Association of Petroleum Geologists Bulletin, v. 62, p. 792–812.

Stetson, H. C., 1949, The sediments and stratigraphy of the east coast continental margin–Georges Bank to Norfolk Canyon: Massachusetts Institute of Technology and Woods Hole Oceanographic Institution, Papers in Physical Oceanography and Meteorology, v. 11, no. 2, p. 1–60.

Trumbull, J.V.A., and Hathaway, J. C., 1968, Further exploration of Oceanographer Canyon: Woods Hole Oceanographic Institution, Reference No. 68-37, 57 p.

Uchupi, E., 1984, Tectonic features, *in* Uchupi, E., and Shor, A. N., eds., Eastern North American continental margin and adjacent ocean floor, 39° to 47°N and 64° to 74°W, Ocean Margin Drilling Program regional atlas series: Woods Hole, Massachusetts, Marine Science International, Atlas 3, p. 27.

Uchupi, E., and Shor, A. N., eds., 1984, Eastern North American continental margin and adjacent ocean floor, 39° to 46°N and 64° to 74°W; Ocean Margin Drilling Program regional atlas series: Woods Hole, Massachusetts, Marine Science International, Atlas 3, 38 p.

Uchupi, E., Ballard, R. D., and Ellis, J. P., 1977, Continental slope and upper rise off western Nova Scotia and Georges Bank: American Association of Petroleum Geologists Bulletin, v. 61, p. 1483–1492.

Vail, P. R., and Hardenbol, J., 1979, Sea-level changes during Tertiary: Oceanus, v. 22, p. 71–79.

Vail, P. R., and Todd, R. G., 1981, Northern North Sea Jurassic unconformities, chronostratigraphy, and sea-level changes from seismic stratigraphy, *in* Illing, L. V., and Hobson, G. D., eds., Petroleum geology of the continental shelf of northwest Europe: London, Institute of Petroleum, p. 216–235.

Vail, P. R., Mitchum, R. M., Jr., Todd, R. G., Widmier, J. M., Thompson III, S., Sangree, J. B., Bubb, J. N., and Hatlelid, W. G., 1977, Seismic stratigraphy and global changes of sea level, *in* Payton, C. E., ed., Seismic stratigraphy; Applications to hydrocarbon exploration: American Association of Petroleum Geologist Memoir 26, p. 49–212.

Valentine, P. C., 1979a, Regional stratigraphy and structure of the Southeast Georgia Embayment, *in* Scholle, P. A., ed., Geological studies of the COST GE-1 well, United States South Atlantic Outer Continental Shelf area: U.S.

Geological Survey Circular 800, p. 7–17.

——, 1979b, Calcareous nannofossil biostratigraphy and paleoenvironmental interpretation, *in* Scholle, P. A., ed., Geological studies of the COST GE-1 well, United States South Atlantic Outer Continental Shelf area: U.S. Geological Survey Circular 800, p. 64–70.

——, 1982a, Calcareous nannofossil biostratigraphy and paleoenvironment of the COST Nos. G-1 and G-2 wells in the Georges Bank basin, *in* Scholle, P. A., and Wenkam, C. R., eds., Geological studies of the COST Nos. G-1 and G-2 wells, United States North Atlantic Outer Continental Shelf: U.S. Geological Survey Circular 861, p. 34–42.

——, 1982b, Upper Cretaceous subsurface stratigraphy and structure of coastal Georgia and South Carolina: U.S. Geological Survey Professional Paper 1222, 33 p.

——, 1984, Turonian (Eaglefordian) stratigraphy of the Atlantic Coastal Plain and Texas: U.S. Geological Survey Professional Paper 1315, 21 p.

Valentine, P. C., Uzmann, J. R., and Cooper, R. A., 1980, Geology and biology of Oceanographer submarine canyon: Marine Geology, v. 38, p. 283–312.

van Hinte, J. E., 1978, Geohistory analysis; Application of micropaleontology in exploration geology: American Association of Petroleum Geologists Bulletin, v. 62, p. 201–222.

van Houten, F. B., 1969, Late Triassic Newark Group, north central New Jersey and adjacent Pennsylvania and New York, *in* Subitzky, S., ed., Geology of selected areas in New Jersey and eastern Pennsylvania: New Brunswick, New Jersey, Rutgers University Press, p. 314–347.

Vorhis, R. C., 1974, Structural patterns on sediments of the Georgia Coastal Plain, *in* Stafford, L. P., ed., Symposium on the petroleum geology of the Georgia Coastal Plain: Georgia Geological Survey Bulletin 87, p. 87–97.

Ward, L. W., and Strickland, G. L., 1985, Outline of the Tertiary stratigraphy and depositional history of the U.S. Atlantic Coastal Plain, *in* Poag, C. W., ed., Geologic evolution of the United States Atlantic Margin: New York, Van Nostrand Reinhold, p. 87–123.

Watts, A. B., 1981, The U.S. Atlantic continental margin; Subsidence history, crustal structure, and thermal evolution, *in* Bally, A. W., ed., Geology of passive continental margins; History, structure, and sedimentologic record: American Association of Petroleum Geologists Education Course Note Series No. 19, p. 2-i-2-70.

Watts, A. B., and Steckler, M. S., 1979, Subsidence and eustacy at the continental margin of eastern North America, *in* Talwani, M., Hay, W. W., and Ryan, W.B.F., eds., Deep drilling results in the Atlantic Ocean; Continental margins and paleoenvironments: American Geophysical Union, Maurice Ewing Symposium, Series 3, p. 273–310.

Youssefnia, I., 1978, Paleocene benthonic foraminiferal paleoecology of Atlantic Coastal Plain: Journal of Foraminiferal Research, v. 8, p. 114–126.

MANUSCRIPT ACCEPTED BY THE SOCIETY JANUARY 21, 1986

ACKNOWLEDGMENTS

We thank our colleagues at Woods Hole, especially Jack Hathaway, Bill Dillon, John Schlee, Kim Klitgord, John Grow, and Jim Robb for their invaluable contributions to the stratigraphic framework now established for the Atlantic offshore region. Doris Low, Ray Hall, Tom Gibson, Joe Hazel, Ruth Todd, Ken Miller, Dick Olsson, Bill Abbott, Paul Huddlestun, Paul Belanger, Ray Christopher, Charlie Smith, and Norman Fredricksen have provided biostratigraphic data and advice during various phases of our studies. We are grateful to John Grow and Bob Sheridan for the invitation to participate in this monumental DNAG project.

Printed in U.S.A.

The Geology of North America
Vol. I-2, The Atlantic Continental Margin: U.S.
The Geological Society of America, 1988

Chapter 6

Geology of the northern Atlantic coastal plain:
Long Island to Virginia

Richard K. Olsson
Department of Geological Sciences, Rutgers University, New Brunswick, New Jersey 08903
Thomas G. Gibson
U.S. Geological Survey, U.S. National Museum, Washington, D.C. 20560
Harry J. Hansen
Maryland Geological Survey, Baltimore, Maryland 21211
James P. Owens
U.S. Geological Survey, Reston, Virginia 22092

INTRODUCTION

The northern Atlantic coastal plain forms the western margin of the Baltimore Canyon Trough, a large sedimentary basin that underlies the continental shelf along the Middle Atlantic states (Fig. 1). The coastal plain narrows northeastwardly from Virginia to Long Island where it plunges beneath the Atlantic Ocean; small exposures of coastal plain sediments occur on Block, Marthas Vineyard, and Nantucket islands east of Long Island.

Deposition in the coastal region is related to the development of the Baltimore Canyon Trough, which took place during the postrift phase of the opening of the Atlantic Ocean. The coastal plain is composed of unconsolidated and semi-consolidated sediments of Cretaceous and Cenozoic age. Sediments of late Jurassic age possibly lie beneath the eastern edge of the coastal plain but this has not been clearly documented. In outcrop, the coastal plain is divided into an inner belt of Cretaceous and early Tertiary formations and an outer belt of younger Tertiary and Quaternary formations. The coastal plain sediments thicken eastwardly into the Baltimore Canyon Trough as a series of basin fills that vary in thickness, along strike. In general the section thins northeastwardly. Near Salisbury, Maryland, the sedimentary section is approximately 2165 m in thickness, whereas at Long Island the thickness is less than 625 m. The variation in thickness along strike is related to structural highs and lows of the underlying basement rocks.

Sedimentation began on the coastal plain during Early Cretaceous (Neocomian) or possibly Late Jurassic time with deposition of fluvial sands, gravels, and variegated clays. The Lower Cretaceous stratigraphic sequence is composed almost entirely of sediments of continental (fluvial) origin. In the distal downdip parts of the coastal plain, marine fossils (molluscs, dinoflagellates,

and foraminifers) occur in the Lower Cretaceous section in some wells. This suggests that the coastal plain was influenced from time to time by marginal marine incursions.

A major cycle of sea level rise, which began during Albian time in the Baltimore Canyon Trough, led to the spread of seas over the coastal plain for the first time in the Cenomanian. Following this rise, until Pleistocene time, marine deposition influenced the coastal plain. The sediments consist of coastal deposits of beach, lagoon, marsh, and related deposits; inner shelf sediments of shoreface sands and related delta front facies of micaceous clay and silty, thinly bedded fine sands; mid and outer shelf clayey glauconite sands and glauconitic clays, often extensively burrowed; and slope deposits composed of calcareous clays and silts. The formations of these various facies were deposited during major cycles of sea level change and they are genetically linked to most of the sedimentary units found further offshore in the Baltimore Canyon Trough.

The geology of the coastal plain figured prominently in the early stages of the development of the science of geology. Initial studies began in the 1700s as attention focused on the relationship of rock units in the United States to those recognized in Europe. Study and description of coastal plain formations became a priority when state geological surveys were organized in the 1800s. Emphasis was on the classification of coastal plain units and their placement in the emerging geologic rock column. No less a prominent figure at that time as Charles Lyell visited the coastal plain and assigned various "beds" to the new classification. He also attempted correlations with formations in Europe.

By the early 1900s most of the formations exposed in the coastal plain had been described and classified, a number of paleontologic reports had been prepared on the richly fossiliferous strata in the coastal plain, and detailed surface mapping had been completed. Commercial use of coastal plain materials was

Olsson, R. K., Gibson, T. G., Hansen, H. J., and Owens, J. P., 1988, Geology of the northern Atlantic coastal plain: Long Island to Virginia; *in* Sheridan, R. E., and Grow, J. A., eds., The Geology of North America, Volume I-2, The Atlantic Continental Margin, U.S.: Geological Society of America.

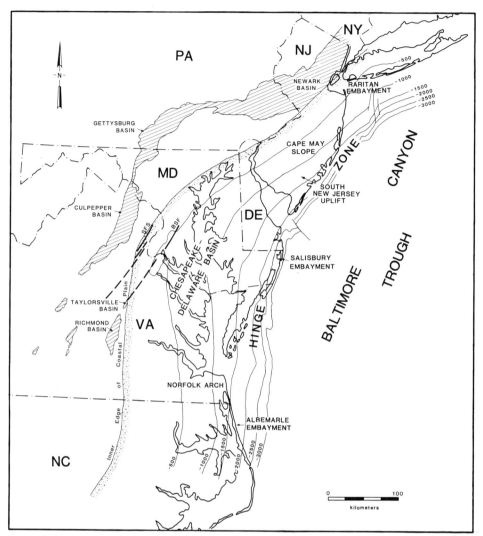

Figure 1. Structure contour map of pre-Mesozoic basement of coastal plain showing major structural features and location of Triassic-Jurassic synrift basins (after Benson, 1984). BFS—Brandywine fault system and gravity gradient (dashed line). SFS—Stafford fault system and aeromagnetic lineament (dashed line).

directed at the greensand (glauconite) marls, which were dug for fertilizer, and at clay, sand and gravel deposits. Much valuable fossil material (vertebrate and invertebrate) was retrieved from the marl pits but, unfortunately, it was not always adequately labeled. In the 1920s the first exploration wells for oil and gas were drilled to shallow depths, but without success. Subsurface geology was poorly understood except in regions where water-bearing formations were utilized.

An important phase in understanding the geology of the coastal plain in the subsurface took place during 1944 to 1947 with the drilling of deep exploration wells in Maryland and North Carolina by petroleum companies. Five wells were drilled, three of which reached basement (Fig. 1). These wells and two others drilled to basement in the early 1960s in New Jersey provide the most up-to-date information on the geology of the coastal plain. They have allowed identification of the major structural highs

and lows of the coastal plain basement (Fig. 1) and they have given valuable information on the stratigraphic development of the coastal plain.

Additional data on the subsurface have come from studies of samples from gas storage test wells drilled in 1951 in New Jersey, from more recent water test wells in Delaware, New Jersey, and Maryland, from a deep geothermal test well in Maryland (Hansen, 1982; Lambiase and others, 1980; Doyle, 1982), and from a corehole in Virginia (Hazel and others, 1977). More recent data on the stratigraphy of outcropping units has come from mapping programs and hydrologic projects of the various state geological surveys and the U.S. Geological Survey.

STRUCTURAL SETTING

The coastal plain sediments were deposited on a pre-

Mesozoic basement complex consisting of gneiss, schist, granite, and other crystalline rocks. Along its western edge, the coastal plain sediments overlap onto the crystalline metamorphic Piedmont Province from Virginia into New Jersey. In the central part of New Jersey the sediments onlap sedimentary rocks of the Triassic-Jurassic Newark Basin.

Triassic-Jurassic rift basins probably lie buried beneath the coastal plain. A K/Ar date of 169 ± 8 m.y. on a diabase core from a basement test well in St. Marys County, Maryland indicates that early to middle Jurassic igneous activity affected the coastal plain basement (Hansen and Wilson, 1984). Also in the southern Maryland coastal plain, steeply dipping sandstones and shales have been encountered in wells beneath unconsolidated sediments (Jacobeen, 1972). On strike in Virginia, a down-faulted basin buried beneath coastal plain sediments has been suggested by a magnetic survey (LeVan and Pharr, 1963) and confirmed by drilling. These indurated rocks are interpreted to be on extension of the late Triassic(?) Richmond and Taylorsville rift basins, which outcrop near the coastal plain onlap in Virginia (Hansen, 1978; Mixon and Newell, 1977).

Shale and sandstone similar to the continental sediments present in the Newark rift basin are encountered in wells in eastern Maryland (Richards, 1967), but Hansen (1978) regards them as strata of possible late Jurassic age. If so, they are related to the post-rifting phase of the stratigraphic development of the Atlantic margin. Triassic grabens have been tentatively identified in seismic reflection lines east of the coastal plain beneath the Baltimore Canyon Trough (Grow, 1980). A structure contour map prepared by Benson (1984) shows north-south aligned grabens and half-grabens, which may contain Triassic-Jurassic rift sediments.

A western extension of the hinge zone of the Baltimore Canyon Trough forms the Salisbury Embayment of Delaware, Maryland, and Virginia (Fig. 1). The hinge zone (Fig. 1) north and south of the Salisbury Embayment underlies the continental shelf (Benson, 1984). Beneath the coastal regions the basement rises from greater than 2165 m near Salisbury, Maryland to 1187 m at Island Beach, New Jersey. The rise of the basement in southern New Jersey was termed the Cape May Slope by Richards (1967). A shallow embayment was recognized by Owens and others (1968) in northern New Jersey and Long Island, to which they gave the name Raritan Embayment. Between this feature and the Chesapeake-Delaware Basin (Fig. 1) they defined a northwest-southeast trending uplift, the South New Jersey Uplift (Normandy Arch of Brown and others, 1972). Bounding the Chesapeake-Delaware Basin on the south is the Fort Monroe High.

Brown and others (1972) interpret the coastal plain to have a wrench-fault tectonic framework controlled by lateral compressive stress. In their view, fault troughs controlled the structural-sedimentary geometry of 17 chronostratigraphic units recognized by them. Periodic rotational realignment of structural axes (Fig. 2) caused crustal segments in the coastal plain to subside differentially so that the depositional alignments of the majority

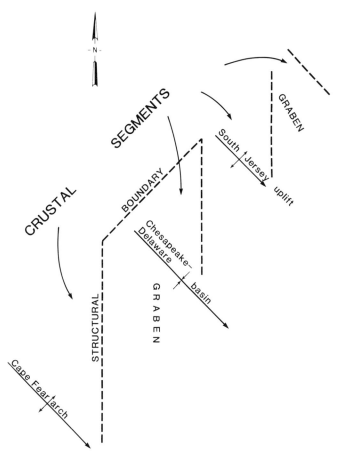

Figure 2. Brown and others' (1972) interpretation of regional system of crustal segments and structural alignments in the coastal plain.

of these units were not concordant with the present basement topography. On the other hand, basin subsidence studies (Steckler and others, this volume) indicate that downward flexure, leading to progressive overlapping of older sediments by younger sediments, is the dominant force in the evolution of the coastal plain.

Two major faults have been postulated to cross the New Jersey coastal plain. The Cornwall-Kelvin fault of the Atlantic basin is hypothesized to cross through the Raritan Bay region between New Jersey and Long Island and to link up with the eastward curvature of the Appalachians (Drake and Woodward, 1963). Another fault is implied on aeromagnetic data to trend westwardly through Cape May (Taylor and others, 1968). The pre-Mesozoic basement map of Benson (1984) shows a westward trending fault with a left-lateral displacement in this region.

The Stafford fault system (Fig. 1), a series of high-angle reverse faults that parallel the inner edge of the coastal plain in Virginia, is the most extensive zone of faulting mapped in the coastal plain (Mixon and Newell, 1977). Up to 100 m of displacement is associated with this fault zone, which was active in Mesozoic and Cenozoic time. Arched Paleocene strata occur just east of the Stafford system (Mixon and Powars, 1984). About 25 km to the northeast in Maryland, Jacobeen (1972) recognized the

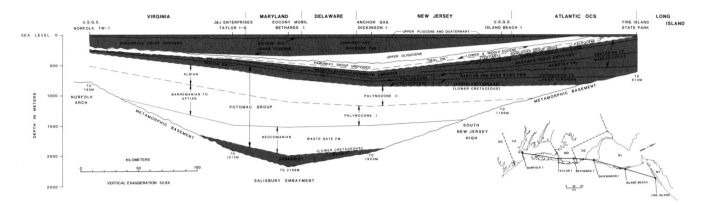

Figure 3. Cross section drawn parallel to strike of coastal plain from Norfolk 1 Well northward to Fire Island Well. The Jurassic(?) section is in red, the Lower Cretaceous is in white, the Upper Cretaceous is in red, the Paleogene is in white, and the Neogene is in red.

Brandywine fault system, a basement system of high angle up-to-the-SE reverse faults. On strike and connecting the Brandywine fault system with the Richmond Basin border fault is an east-dipping gravity gradient that is interpreted to be the faulted northwest edge of a rift basin terrane (Mixon and Newell, 1977). This buried rift basin belt is defined by a relatively flat magnetic gradient and has been called the Richmond-Brandywine trend (Hansen, 1978). At the southeastern boundary of the Richmond-Brandywine trend, stratigraphic pinchouts of Senonian and Danian strata may be related to the Hillville Fault, a high angle up-to-the-SE fault recorded on a seismic record section (Hansen, 1978). This fault has an apparent offset at top-of-basement of 76 m.

Vibratory reflection seismic surveys have identified other small faults in the basement in Delaware (Spoljaric, 1973) and in Maryland (Hansen, 1978). Small scale earthquakes in parts of the coastal plain from time to time attest to the presence of faults, but little is known about the basement in these areas.

Sedimentation on the coastal plain was controlled by major structural features. Subsidence in the Chesapeake-Delaware Basin during Early Cretaceous and later Jurassic time allowed more than 1500 m of continental to nearshore sediments to accumulate (Fig. 3). To the north, on the South New Jersey High, less than 500 m of sediment was deposited. Thinning and lapping out of the older sections onto this structural high is also suggested by paleontological criteria (Fig. 3).

The Upper Cretaceous section and early Tertiary section, which is largely marine in origin, is thickest on the South New Jersey High where it reaches a thickness of 366 m. It thins into the Chesapeake-Delaware Basin and, on the south flank of the basin, the upper part (Senonian and Paleocene) pinches out (Hansen, 1978). The pinchout of section on the south is apparently related to basement structures. Thickness of the upper Tertiary section increases into the Chesapeake-Delaware Basin. Thus, during the development of the coastal plain, an early phase of nonmarine to marginal marine sedimentation was concentrated in the Salisbury Embayment. The embayment continued to sub-side during a subsequent marine phase with outer shelf and slope deposition, but structural movements on its southern flank caused pinchouts to occur. On the north the Upper Cretaceous and lower Tertiary sections of the New Jersey coastal plain were deposited. The embayment again controlled deposition of upper Tertiary sediments, which are of shallow marine to continental origin.

In the coastal plain the Upper Jurassic strata are overlapped by Lower Cretaceous strata, which in turn are overlapped by Upper Cretaceous strata. Vail and others (1977) ascribe the overlapping pattern or coastal onlap along passive margins to eustasy. Watts (1982) and Karner and Watts (1982) believe that the supercycles of sea level change identified by Vail and others (1977) are due to tectonic synchroneity rather than eustasy. However, they regarded short-term changes of sea level (<1 m.y.) as due to factors other than flexure. Although the initial stages of coastal onlap may be caused by flexure, the Cretaceous flooding of continents is regarded as a major eustatic event in earth history. Along the Atlantic margin the Cretaceous rise in sea level is evident in the coastal plain in Cenomanian, lower Turonian, Santonian, Campanian, and Maestrichtian marine units, which extend over much of the Lower Cretaceous nonmarine sediments. A major upper Turonian/Coniacian hiatus interrupts the sequence.

LITHOSTRATIGRAPHY

Jurassic

Upper Jurassic marine and nonmarine rocks have been identified in wells drilled in the Baltimore Canyon Trough (Scholle, 1980). These rocks, which formed during an early part of the drifting stage of the Atlantic margin, consist of limestones, sandstones, shales, and coal. It is clear on seismic reflection lines (Schlee and Grow, 1980) that these formations extend shoreward toward the coastal plain. Nonmarine sandstones and shales encountered in the bottoms of deep wells in Maryland (Fig. 4) lie unconformably beneath rocks of Neocomian age and have been tentatively placed in the Jurassic (Hansen, 1982; Doyle, 1982). Perry and others (1975) also suggested the presence of Jurassic rocks in the basal parts of deep coastal plain wells. Thus, it seems reasonable to conclude that in places such as the Salisbury Embayment, upper Jurassic nonmarine sediments onlap the basement.

Lower Cretaceous

The basal sequence of sediments in the coastal plain outcrop belt is the Potomac Group of Early Cretaceous age (Fig. 5). The Potomac Group consists of white, grey, and red interbedded variegated silts, clays, and quartzose sands. In outcrop the total thickness of the Potomac Group in Maryland is about 230 m, but it thins and is overlapped by younger sediments in New Jersey and Virginia.

In the subsurface the Potomac Group thickens considerably (Figs. 3, 4, 6). In Maryland it reaches a maximum thickness of over 1370 m, and in New Jersey, in the Island Beach Well, it is about 460 m thick. Thus, about half of the stratigraphic section in the coastal plain is taken up by the Potomac Group.

The Potomac Group is interpreted as a complex of fluvial-deltaic lithofacies of a major axial river system centered in Maryland (Hansen, 1969). Lithofacies include braided and meandering stream sands and gravels, upper floodplain, lower floodplain, and fringing swamp deposits. In deep wells marine fossils in the Potomac interval (Anderson, 1948; Richards, 1967; Hansen, 1982) indicate proximity of these lithofacies to nearshore environments of deposition.

Palynological zonation of the Potomac (Brenner, 1963; Doyle, 1969; Doyle and Robbins, 1977) indicates a Barremanian to Albian age (Palynozone I), an Albian age (Palynozone II), and an early Cenomanian age (Palynozone III). Separation of the Potomac in the coastal plain subsurface is difficult because of lithologic similarity of adjacent units, but Palynozones I-III are used to identify Potomac age sediments (Hansen, 1982; Perry and others, 1975).

Beneath the Potomac and above questionable Jurassic sediments in the subsurface of Maryland, Hansen (1982) has recognized a new formation, the Waste Gate Formation (Figs. 3, 4). This unit contains pre-Palynozone I palynomorphs and has been dated as mid-Berriasian in age (Doyle, 1982). The Waste Gate, which Hansen places in the Potomac Group, consists chiefly of unconsolidated to moderately lithified arkosic sandstones and laminated silty shales and clays. The formation is up to 460 m thick and is continental in origin.

Upper Cretaceous

The Upper Cretaceous is characterized by numerous marine units. The marine section is best exposed in New Jersey where a sequence of transgressive and regressive units can be traced from outcrop into the subsurface. South and north of New Jersey the cycles of deposition are less well defined, and on the south flank of the Chesapeake-Delaware Basin the marine sequence pinches out (Hansen, 1978).

The Upper Cretaceous sediments, and the Lower Tertiary as well, contain large quantities of glauconite, which in some units are concentrated in large amounts. Glauconite is generally believed to be of fecal pellet origin (Pryor, 1975). Fecal pellets, which are composed mostly of clay, alter to glauconite under microreducing conditions within the pellet. Glauconite concentrates in marine environments where sedimentation rates are very low. Filter-feeding organisms are the principal mechanism of sedimentation in the form of biogenic pelletization of suspended clay. In the coastal plain, glauconitic units accumulated under slow rates of deposition. They exhibit extensive bioturbation and homogenization of sediment, which results from filter-feeding activity. Very slow rates of subsidence in the coastal plain during Late Cretaceous and Early Tertiary time created environments that favored formation of abundant glauconite.

The first extensive marine deposition in the coastal plain occurred during the late Cenomanian to early Turonian. Throughout the subsurface of the New Jersey coastal plain, marine clayey silts (Fig. 6) reach a maximum thickness of 122 m. They are replaced updip by sands and variegated clays of largely fluvial origin. These marine and fluvial deposits are separated from the underlying nonmarine lower Cenomanian deposits by an unconformity. Deposition occurred during a world-wide transgression that had begun in Albian time. This transgression is clearly observed in the Baltimore Canyon Trough in the shoreward overlapping of nonmarine Lower Cretaceous sediments by Albian marine sandstones and shales. Fully marine environments did not reach the coastal plain until late Cenomanian time. The transgression ended during the Turonian.

The late Cenomanian-Turonian fluvial facies (Raritan Formation) thins southwestwardly along the outcrop belt and disappears north of Trenton, New Jersey. This unit is distinguished in the subsurface of the coastal plain from lithologically similar sediments (Potomac Group) on the basis of palynology, as it falls within Palynozone IV of late Cenomanian-Turonian age (Perry and others, 1975).

Marine strata of late Cenomanian-Turonian age have not been recognized in Delaware and Maryland, but in southeastern Virginia marine strata have been identified in wells (Brown and others, 1972). In Maryland, Palynozone IV sediments with rare molluscan remains suggest marginal marine environments were present in this area.

Unconformably overlying the late Cenomanian-Turonian unit is the Magothy Formation, which is Santonian to possibly earliest Campanian in age (Figs. 4, 6). The Magothy is the basal unit of a Santonian to Campanian transgression, which established marine environments for the first time over the entire coastal plain (Perry and others, 1975; Petters, 1976; Spoljaric, 1972).

The Magothy, which consists of cross-bedded sands with thin beds of clay, silt, and scattered lignite, formed as a beach-

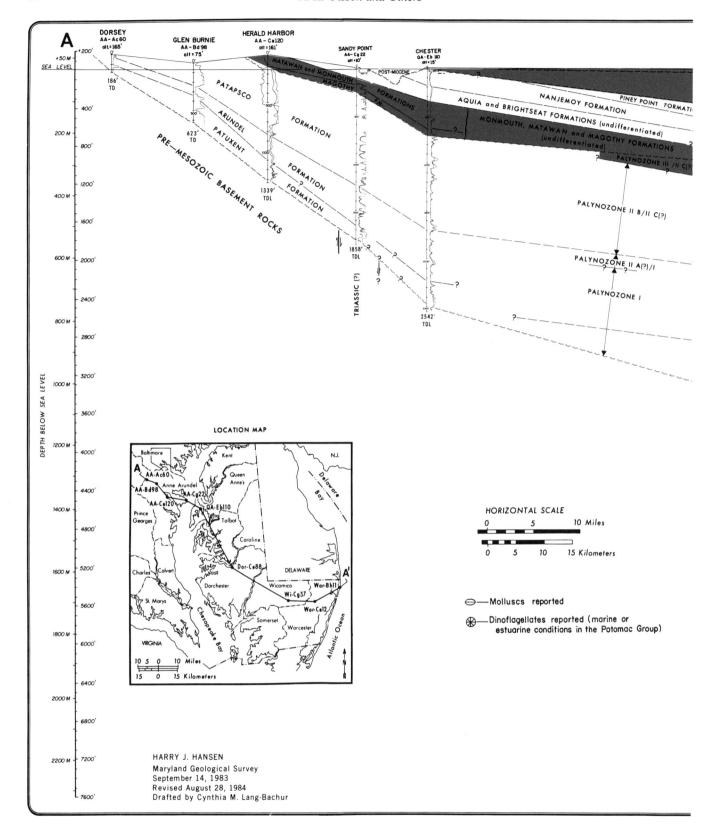

Figure 4. Cross section drawn parallel to dip direction in Maryland. Jurassic (?) section is in red, the Lower Cretaceous is in white, the Upper Cretaceous is in red, the Paleogene is in white, and the Neogene is in red.

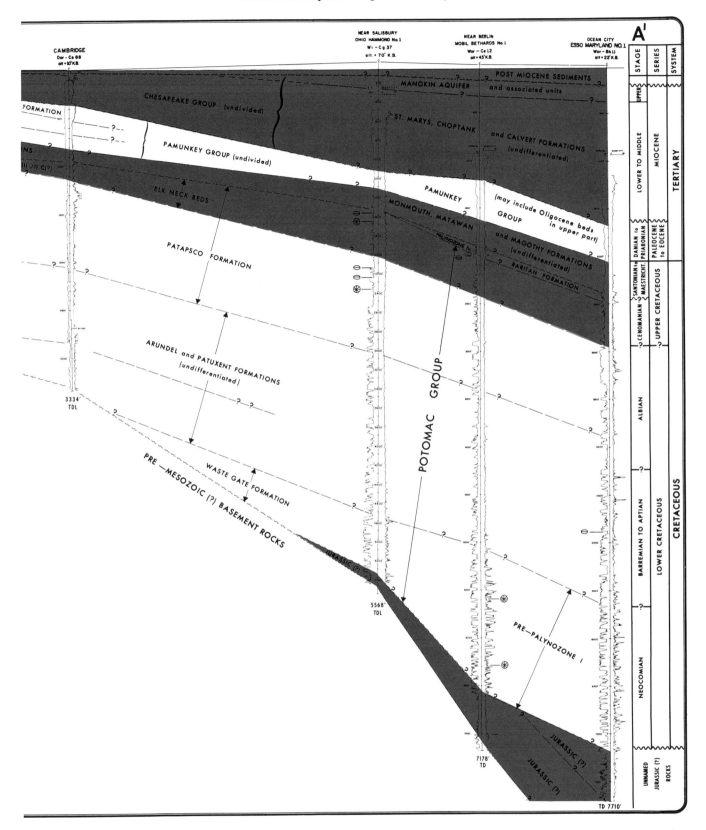

Figure 4. (continued).

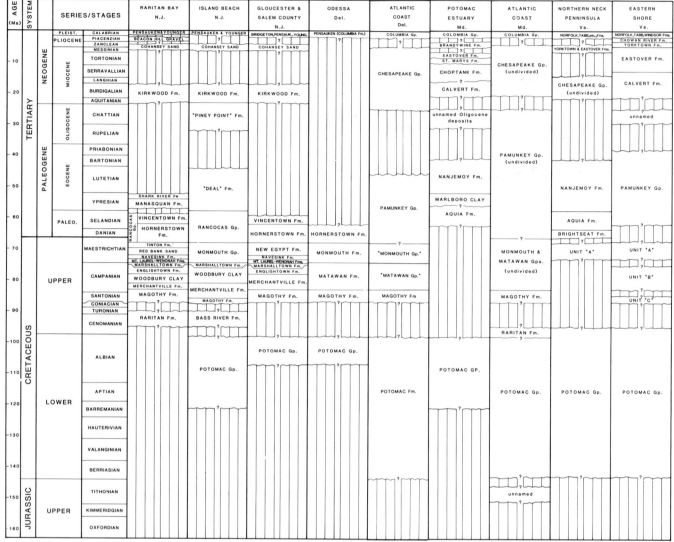

Figure 5. Correlation chart of coastal plain formation. Adapted from COSUNA Correlation Chart for Atlantic Coastal Plain (Jordan and Smith, 1983).

zone complex with related fluvial and estuarine facies along the strandline of the advancing Santonian sea. The Magothy is thickest under Long Island where it is about 245 m thick. In outcrop it thins along strike to about 15 m in southern New Jersey; it thins in Delaware and Maryland and disappears in southern Maryland.

In the subsurface the Magothy is present as a thin persistent unit from New Jersey to Maryland, but it is absent in southern Maryland and Virginia and also in the Dickinson I Well in New Jersey. Christopher (1979) placed the Magothy in pollen zone V, which he considers Santonian to early Campanian in age. His correlation of the Magothy agrees with the interpretation of the Magothy as a facies of the transgressive marine Merchantville Formation. Petters (1976) demonstrated that the Merchantville becomes older in its lower part as it is traced into the subsurface. He dated the Merchantville as Santonian to early Campanian on the basis of planktonic foraminifera, thus documenting on bio-

stratigraphic data the transgressive facies relationship between the Magothy and the Merchantville.

The Merchantville Formation, a glauconitic sand to a micaceous silty clay, is a thin (15 m thick) unit in outcrop. Like the Magothy, it thins and disappears south of Maryland. In Long Island, and in Maryland in the subsurface, the Merchantville is not differentiated within the Matawan Group (Figs. 3, 4). The Merchantville is the second of two major transgressions, which established marine environments of deposition in the coastal plain. Thus, in two steps, marine deposition on the Atlantic margin advanced from a relatively narrow shelf province to a very broad shelf margin.

A regressive pulse led to the deposition of the Woodbury Formation, a micaceous, chloritic, silty clay. The Woodbury is an inner shelf deposit that grades into and is replaced in the subsurface by the deeper shelf sediments of the Merchantville (Olsson

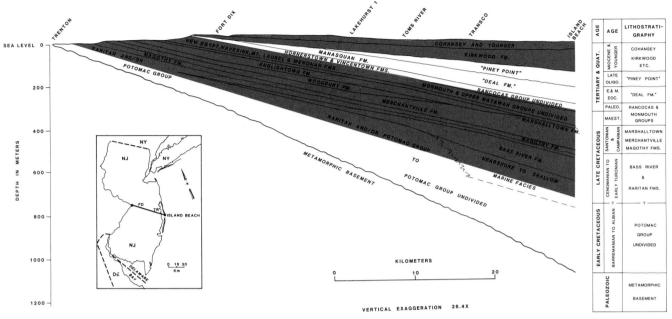

Figure 6. Cross section drawn parallel to dip direction in New Jersey. Lower Cretaceous section is in white, the Upper Cretaceous is in red, the Paleogene is in white, and the Neogene is in red.

and Nyong, 1984; Petters, 1976). The Woodbury also grades into the overlying sands of the Englishtown Formation, which it in turn replaces in the subsurface (Fig. 6). The Woodbury thins southward in outcrop; in Delaware it is replaced by the Merchantville. On the north, in the subsurface of Long Island, the formation is not differentiated within the Matawan Group (Fig. 5). The Woodbury appears to be a shoaling facies developed north of the Chesapeake-Delaware Basin. The shoaling can also be recognized in the Baltimore Canyon Trough (Nyong and Olsson, 1984).

The Woodbury is Campanian in age. The age of the Woodbury in outcrop is based on molluscan evidence (Stephenson and others, 1942). More definitive dating of the Woodbury is based on abundant foraminifera found in the subsurface (Petters, 1976). In Delaware micropaleontologic data indicates that the upper part of the Merchantville is the age equivalent of the Woodbury (Houlik and others, 1983).

The Englishtown Formation marks the maximum extent of the Woodbury regressive pulse. It consists of cross-stratified quartz sands, silty sands, and silts, which were deposited in an innermost shelf to shoreface environment (Houlik and others, 1983; Olsson and Nyong, 1984).

The Englishtown is thickest in the northern part of the coastal plain in New Jersey where it is about 37 m thick. It thins southward along strike, and in southern New Jersey and Delaware it becomes a micaceous silt to a very fine sand (Minard, 1965; Owens and others, 1970; Pickett, 1970). Farther south in Maryland the Englishtown is not recognized. Its equivalent there is the Matawan Formation. Downdip in New Jersey the Englishtown is replaced by the Woodbury (Petters, 1976). It also disap-

pears in the subsurface of Delaware, but there is no data on whether it has a facies relationship with the Merchantville. In Long Island the Englishtown is undifferentiated within the Matawan Group (Fig. 5).

The Marshalltown Formation represents a return to transgressive conditions following the regressive pulse of the Woodbury and the Englishtown. Like other transgressive units in the coastal plain, it is a thin, easily recognized formation. The Marshalltown is a highly glauconitic, very silty fine sand that is extensively burrowed. In places in outcrop it contains a rich calcareous molluscan assemblage characterized by abundant *Exogyra ponderosa*. In other places only internal molds attest to a formerly abundant assemblage. In outcrop the formation is 3 to 4 m thick, whereas in the subsurface it thickens to about 14 m. On the north in Long Island the formation is not differentiated within the Matawan Group. On the south in Maryland the Marshalltown interval would fall within the Matawan Formation (Fig. 5).

The Marshalltown is latest Campanian in age and contains the uppermost Campanian planktonic foraminiferal *Globotruncana calcarata* Zone (Houlik and others, 1983; Olsson, 1964; Petters, 1976). The Marshalltown was deposited under middle shelf to upper slope environments (Olsson and Nyong, 1984). The deepening trend of Marshalltown deposition is also evident in the offshore Baltimore Canyon Trough and thus signals a sea level rise event in this region during latest Campanian time.

The Maestrichtian sequence of formations in the New Jersey coastal plain is characterized by a number of interesting facies relationships. These relationships occur between regressive sand deposits and they are related to shifting environments of deposition. Some of the facies have been interpreted differently by different workers, but biostratigraphic data have clarified some of the relationships.

The Wenonah Formation and the Mount Laurel Sand were deposited during a regressive pulse that followed the Marshalltown transgression. The Wenonah, a clayey, silty, slightly glauconitic fine quartz sand, is gradational with the underlying Marshalltown and with the overlying Mount Laurel Sand. The Wenonah is thickest (21 m) in outcrop in the central part of the New Jersey coastal plain and thins on the south and north. It is not present beyond New Jersey (Fig. 5). In southern New Jersey it coarsens and becomes indistinguishable from the Mount Laurel (Owens and Sohl, 1969).

The Mount Laurel is more variable in lithology than the Wenonah. Lithofacies include thinly bedded clays and sands, massive sand beds, and thin pebbly sands. Cross-bedding is common in the thin-bedded sequences. Rounded pebbles, glauconite in-filling of burrows and abraded fossil molds at the top of this formation suggest that it is a deposit related to the overlying transgressive Navesink Formation.

Deposition of the Wenonah and Mount Laurel occurred in inner shelf to shoreface environments. The Wenonah contains molds of molluscs and *Asterosoma* and zoophycus-like trace fossils (Martino, 1976). Sparse foraminiferal assemblages (Olsson and Nyong, 1984) also indicate an inner shelf environment. The Mount Laurel contains abundant trace fossils of *Ophiomorpha* and *Asterosoma*. A massive sand facies in the Mount Laurel is interpreted as a nonemergent offshore bar complex (Martino, 1976). In Delaware and New Jersey the Mount Laurel ranges from 6 to 26 m in thickness. Foraminifera in the lower part indicate a mid- to outer-shelf environment (Olsson and Nyong, 1984).

In New Jersey the Monmouth Group includes the formations above the Wenonah Formation (Fig. 5). In outcrop and in the shallow subsurface, these units are easily differentiated. They lose their identity further downdip and along the outcrop belt as they are traced southward through Delaware. In Maryland and Virginia, glauconitic sands and clays are called Monmouth Formation. In the subsurface there, the Monmouth merges with the Matawan Formation; it is difficult to separate the two (Fig. 4). In the New Jersey subsurface, Monmouth Group is used where the formations are not easily separated. In Long Island the Monmouth Group is also used for sediments that cannot be separated into distinct units.

A return to mid-shelf conditions following deposition of the Mt. Laurel and Wenonah formations resulted in accumulation of the clayey glauconite sands of the transgressive Navesink Formation. The Navesink is rich in skeletal fossil content. The most prominent megafossils are the oysters *Exogyra, Pycnodonte,* and *Ostrea*; the brachiopod *Choristothyris*; and the belemnite *Belemnitella*. In addition, molds of various molluscs are common, microfossils (foraminifers, ostracodes, coccoliths, dinoflagellates, epibiont bryozoans) are abundant, and fish and reptilian remains are present.

The Navesink is recognized only in New Jersey; it thins along strike on the south and disappears north of Delaware. In the subsurface the Navesink glauconites blend with similar younger sediments, and it becomes difficult to separate the Navesink as a formation.

In the northern part of the New Jersey coastal plain, micaceous, feldspathic quartz sands of the Redbank Formation lie above the Navesink. They are of limited geographic extent, thin rapidly, and disappear north of the central part of the outcrop belt as well as in the shallow subsurface. On the south and downdip, the Redbank is replaced by the clayey glauconite sands of the New Egypt Formation.

Large scale cross-bedding in the upper part of the Redbank indicates a transport southwardly of a prograding sand from a northern source (Martino, 1976). The paleocurrent direction and the association of the trace fossils *Ophiomorpha* and *Asterosoma* suggest a nonemergent offshore bar complex origin for the upper sands of the Redbank. The lower, more clayey and silty beds of the Redbank are interpreted as inner to middle shelf deposits (Olsson, 1963; Owens and Sohl, 1969).

The New Egypt Formation has been considered a more glauconitic facies of the lower Redbank (Owens and Sohl, 1969) but biostratigraphic data (Koch and Olsson, 1977) show that it is equivalent to the entire Redbank, and the Tinton Formation as well. The New Egypt is a shelf facies marginal to these formations. It overlies the Navesink Formation and in turn is overlain by the Hornerstown Formation.

The Tinton Formation is the only indurated unit in the Upper Cretaceous section of New Jersey. It is very thin and is more limited in extent than the Redbank Formation upon which it lies. It is an argillaceous, medium to coarse, quartz and glauconite sandstone interbedded with layers and lenses of gray claystone. Molds of molluscs, crab claws, and the trace fossil *Ophiomorpha* are common in places. The formation is interpreted as an inner shelf facies related to the regressive Redbank facies.

The Hornerstown Formation is almost a pure glauconite sand, containing little fine-grained matrix. This gives it a distinctive deep-green color. It is a very thin (6 m thick), persistent unit that can be traced along the entire outcrop belt in New Jersey and southward into Maryland. This massive and extensively burrowed facies originated in an inner- to mid-shelf environment. The Hornerstown possibly represents an environment where most suspended clay was utilized by filter-feeding organisms and sedimented as pellets that glauconitized rapidly. In the subsurface increasing amounts of clay matrix are present and it gradually loses its lithologic characteristics in the far downdip. A varied assortment of fossil remains of invertebrate megafossils, microfossils (foraminifers, ostracods, coccoliths, dinoflagellates), and vertebrates (fish, reptiles, birds) is found in the formation. A thin shell bed consisting of the brachiopod *Oleneothyris* and the oyster *Pycnodonte* occurs at the top of the formation.

The Hornerstown has been regarded as the basal formation in the Tertiary of Delaware and New Jersey. However, paleontological data (Baird, 1964; Richards, and others, 1973; Richards and Gallagher, 1974; Jordan, 1976; Koch and Olsson, 1977) indicate that the basal beds are Cretaceous in age. The upper beds of the Hornerstown range in age from early Paleocene to late Paleocene (Olsson, 1970).

Lower Tertiary (Paleogene)

Tertiary strata in the coastal plain can be separated into two main lithologic groups, a Paleogene group dominated by moderately to highly glauconitic sand, silt, and clay, and a Neogene group of non-glauconitic sand, silt, and clay. The Tertiary strata can be separated geographically into units that occur only to the south or north of the Delaware River. As a consequence, stratigraphic terminology is almost entirely different between these two areas (Fig. 5). This presumably reflects control of sediment input by the shape and location of tectonic features in the coastal plain area.

In New Jersey the Paleogene formations include the Hornerstown, Vincentown, Manasquan, Shark River, and "Piney Point" Formations (Fig. 5). The Hornerstown, which was discussed above, and the Vincentown are the only formations that can be traced in outcrop south of the Delaware River—the Hornerstown into northern Maryland and the Vincentown into Delaware. The Vincentown, an inner to middle shelf deposit of late Paleocene age, contains two prominent facies: a massive quartz sand facies, and a quartz calcarenite facies rich in bryozoans and foraminifera. The formation thins rapidly and is replaced by a silt facies in the shallow subsurface.

The Lower Eocene Manasquan Formation unconformably overlies the Vincentown. The Manasquan contains a glauconite rich lower member and an upper clayey, sand to silt member. The upper member thickens considerably in the subsurface and becomes the dominant lithology of the Eocene sediments there, also replacing the overlying Middle Eocene Shark River Formation, a glauconitic sand and mudstone. The lower and middle Eocene sediments are separated by an unconformity. Deposition during the early Eocene occurred in middle to outer shelf environments, whereas during the middle Eocene, inner to middle shelf environments prevailed.

An extensive beveled erosional surface on the Eocene can be traced from the subsurface of New Jersey southward into the subsurface of Maryland (Olsson and others, 1980). This surface transgresses lower to middle Eocene rocks in the coastal plain to upper Eocene rocks offshore in the Baltimore Canyon Trough (Olsson and Miller, 1979). In the subsurface of New Jersey (Fig. 6) glauconitic silt and sand containing reworked weathered Eocene lithoclasts lie on this surface. This unit was identified and correlated with the Piney Point Formation of Delaware and Maryland by Richards (1967). Olsson and others (1980) dated this unit on planktonic foraminifera as late Oligocene, and correlated it with sediments in the subsurface of Maryland identified as the Piney Point Formation.

The term Piney Point is also used for an aquifer system present in the Maryland, Delaware, and in the southern New Jersey subsurface. Brown and others (1972) reported middle Eocene ostracods from the Piney Point in its type well in Maryland, but the water-bearing horizons of the aquifer include not only middle Eocene sands but also the upper Oligocene sands. Separation of the Eocene and Oligocene is difficult where the sands occur adjacent to one another in the shallow updip of the coastal plain of Maryland and Delaware.

The Oligocene part of the "Piney Point" is an inner to middle shelf facies deposited during a late Oligocene transgression. For the most part, lower Oligocene strata and upper Eocene strata are missing in the coastal plain.

The Paleogene units in Virginia and Maryland comprise the Pamunkey Group and include, in ascending order, the Brightseat Formation, Aquia Formation, Marlboro Clay, Nanjemoy Formation, and the Chickahominy Formation (Fig. 5). The formations represent one, and in some cases several, depositional cycles. Disconformities, recognized by burrowed surfaces, separate each cycle. Deposition during each cycle occurred in inner to middle shelf environments (Gibson and others, 1980). The formations are generally widespread both along strike and downdip; most units are present in the surface and subsurface with similar lithologies and thicknesses, although there are notable exceptions. The hiatuses between the formational units in the Paleocene to middle Eocene part of the section are of relatively short duration; considerably longer hiatuses are present in the middle Eocene to upper Oligocene part of the section and indicate that only a few marine incursions of relatively short duration occurred in this area during the later part of the Paleogene. The dominant lithology of the Paleogene units is glauconitic clayey sand with the amount of glauconite varying considerably among the units, but being as high as 70 percent (Glaser, 1971). In outcrop, the formations are, in general, less than 30 m thick and some are commonly less than 6 m thick. They generally increase in thickness downdip to the east; however, the Aquia and Nanjemoy, along with possibly others, initially thicken in the subsurface (Fig. 7) but then thin eastward in Virginia and are absent from the well sections toward the coast (Shifflett, 1948; Brown and others, 1972).

The earliest Tertiary record in the southern part of the Chesapeake-Delaware Basin begins with the Brightseat Formation, a micaceous, slightly glauconitic, clayey and silty sand that is placed in calcareous nannoplankton zone NP3 (Bybell and Govoni, 1977). Outcrops of Brightseat are thin and discontinuous but extend as far west as Washington, D.C., where inner to middle shelf environments of deposition are interpreted from the foraminiferal assemblages (Nogan, 1964). In Virginia, the formation becomes less glauconitic and is only documented as far south as the Oak Grove corehole in the northern part of the Virginia Coastal Plain (Gibson and others, 1980).

The upper Paleocene Aquia Formation was deposited in a northeast trending basin extending northward from the James River in southern Virginia; to the south the formation disappears against the Norfolk Arch. The formation consists of shelly glauconitic sand, reaching as much as 70 percent glauconite. The Aquia is approximately 60 m thick in the subsurface of southern

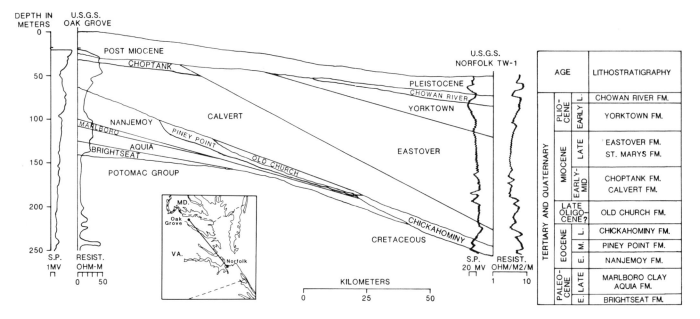

Figure 7. Cross section drawn parallel to dip direction in Virginia. Cross section connects to the Norfolk 1 Well.

Maryland, but it thins in outcrop there and southward into Virginia to around 30 m. The unit was deposited in a very shallow shelf environment, and probably represents a general upward shoaling in the basin. Downdip, the Aquia changes to a fine sandy to clayey outer shelf facies that is difficult to differentiate from the underlying Brightseat Formation in well logs (Hansen, 1974). This regression continued with the deposition of the silty clay of the overlying Marlboro Clay, which is a marginal marine deposit. The Marlboro may be entirely of late Paleocene age or may represent deposition across the Paleocene-Eocene boundary (Frederiksen and others, 1982).

The Nanjemoy Formation, a shelly glauconitic clayey sand, was deposited during several transgressive cycles following the Aquia-Marlboro regression. The formation is early Eocene in age. The Nanjemoy thickens northward from the James River, reaching 30 m in the subsurface in Virginia and in the outcrop belt in Maryland. In the subsurface of southern Maryland, thicknesses of as much as 75 m are found. In both areas shallow marine, inner shelf environments dominate.

Above the Nanjemoy Formation, the sedimentary record for the remainder of the Paleogene is much more fragmentary. Unconformably overlying the Nanjemoy is the Piney Point Formation of middle Eocene age (DiMarzio, 1984; Ward, 1984). This unit consists of clayey, highly glauconitic, fossiliferous sand. It is a 15 m (50 ft) thick subsurface unit in its type area in southern Maryland and extends southward to the Norfolk arch in southern Virginia, outcropping in central Virginia with a thickness of about 7.6 m (25 ft).

Uppermost Eocene deposition is found in the Chickahominy Formation. This glauconitic clay is only known from the subsurface, being found in eastern Virginia and in places in the eastern part of the basin (Brown, and others; 1972; Gibson, 1970). The middle shelf environments of deposition indicated by the foraminiferal faunas suggest a more extensive coverage before Oligocene erosion.

A thin fossiliferous quartz sand unit of latest Oligocene to earliest Miocene age, called the Old Church Formation (Ward, 1984) is found in only one part of the Paleogene outcrop area, in south-central Virginia, but it is commonly recognized in the subsurface from southern Virginia northward (Fig. 7).

Upper Tertiary (Neogene)

The Miocene and Pliocene sedimentary record in the southern part of the Chesapeake-Delaware Basin in Maryland and Virginia and in the subsurface of Delaware begins with thin units composed largely of biogenic silica, but it quickly changes upward into a thicker and increasingly clastic mode of deposition. The clastic input, largely of quartz sand, silt, and clay, commonly carbonaceous, appears earlier in the northern part of the basin and moves southward during the middle and late Miocene. It dominates the entire Pliocene record. Glauconitic units are of minor importance. Numerous depositional cycles of relatively short duration are present; six formations are presently named, of which several probably contain more than one transgressive-regressive cycle as seen in the outcrop belt. These formations comprise the Chesapeake Group (Fig. 5). Along the western shore of the Chesapeake Bay in Maryland, the lower three formations, the Calvert, Choptank, and St. Marys, total approximately 90 m thick, whereas downdip, the overall thickness of these three units increases to 240 m (Fig. 7).

The lowest Miocene deposits in the basin consist of diatomaceous clay and quartz sand of the Calvert Formation. Southward, the diatomaceous clay becomes interbedded with phosphatic sand and clay in the southeastern part of the basin near Norfolk. These phosphatic strata are the northernmost extension of the phosphate deposits characteristic of the Pungo River Formation in the Albermarle Embayment in North Carolina. The diatomaceous sediments were deposited in inner- to possibly middle-shelf environments of deposition. Radiolarians occurring in these strata suggest that there was upwelling of cool, deeper oceanic water coming into the basin (Palmer, 1982). In Maryland and northern Virginia the diatomaceous beds are overlain by a series of shelly, largely clayey sands. In northeastern Maryland, in outcrop and the subsurface, these sediments contain interlaminated carbonaceous clay and sand and are considered to be prodelta in origin, reflecting input from New Jersey and easternmost Pennsylvania (Isphording, 1970; Gibson, 1982).

The pattern seen in the Miocene and Pliocene strata, starting with the Calvert Formation, is a higher proportion of clastics initially in Maryland followed by a southward movement of clastic facies into Virginia, and ultimately into the Albermarle Embayment in North Carolina. This pattern continues in the overlying Choptank Formation, where sand of inner shelf origin dominates the formation in Maryland and northernmost Virginia, but coeval deposits farther south in Virginia consist of shallow marine diatomaceous clay. The overlying St. Marys Formation, composed largely of sand and clay, is found both in Maryland and northern Virginia; environments of deposition range from very shallow shelf to marginal marine. The St. Marys Formation is not known from southern Virginia. After deposition of the St. Marys Formation, the depocenter in this part of the basin shifted southward into Virginia where strata of the Eastover and Yorktown Formations were deposited (Stephenson and MacNeil, 1954; Newell and Radar, 1982).

Deposition of the Eastover occurred mainly in the southern part of the basin during a transgressive cycle in the late Miocene. The formation thickens considerably in southern Virginia, from 3 m in the western part of the outcrop belt to over 90 m in the subsurface at Norfolk (Fig. 7).

The Chesapeake Group thins and disappears northward in Delaware. In the subsurface of Delaware and Maryland, it contains several important aquifers.

The Yorktown Formation is also developed primarily in the southern part of the basin and is widespread over east-central and southeastern Virginia. Marine deposits of the Yorktown are not known to the north in Maryland. The extensive transgressive character is seen in southern Virginia where the formation rests upon shallowly buried Piedmont rocks and crosses the Norfolk Arch into the Albemarle Embayment in North Carolina. The highly fossiliferous strata of this formation suggest deposition in shallow shelf seas for the most part.

In New Jersey the Neogene consists of the Kirkwood Formation, Cohansey Formation, and several formations of fluvial origin that occur in isolated exposures on the coastal plain surface

(Fig. 5). The Kirkwood is the thickest of the Neogene units, increasing in thickness from about 30 m in outcrop to near 210 m in the subsurface. In outcrop it can be divided into three members based upon clay, silt, and sand content. In the subsurface, the lower part is composed of sand and silt and the upper part is a diatomaceous sandy silt. The Kirkwood is probably a composite of strata deposited during two, possibly three, marine cycles. Foraminifera and nannofossils in the lower part indicate an early Miocene age (NN4), whereas the Tortonian *Globorotalia acostaensis* Zone occurs in the upper part (Melillo and Olsson, 1981). The age of the overlying Cohansey Sand is uncertain as it contains sparse fossils. Pollen studies on a basal lignite bed suggest a late Miocene-early Pliocene age (Rachele, 1976). Deposition of the Kirkwood occurred in inner to middle shelf environments. The Cohansey was deposited along an emergent shoreline (Carter, 1978).

The latest Tertiary was a period of deposition and dissection of much of the emerged coastal plain. At this time a series of coarse clasts spread across the upper part of the Chesapeake-Delaware Basin. In the north, in New Jersey, the Bridgeton Formation was deposited; in Delmarva, the Pensauken (Columbia); and to the northwest, in southern Maryland, the Brandywine Formation. Coarse upland clasts are known to exist between Fredericksburg and Richmond, Virginia, along the western edge of the coastal plain, but precise age relationships are poorly understood.

In general, the coarse clasts were deposited in an upper delta plain environment, which in some places interfingers with marginal marine deposits downdip (lower delta plain–delta front), and less commonly with prodelta–inner shelf beds. A major controversy has long existed as to the time of emplacement of these deposits, with many scientists favoring a Pleistocene age.

Because the coarse clasts were first studied in New Jersey, it was widely assumed that the Bridgeton and Pensauken were Pleistocene in age because of their proximity to the known glacial deposits to the north. They were interpreted as glacial melt water deposits. Such an explanation for the Brandywine and upland deposits in Virginia, however, is less likely. Another line of evidence used in New Jersey was that the Bridgeton-Pensauken post-dated the Cohansey Sand, which was presumed to be Pliocene. As a point of fact, the Cohansey has not been precisely dated.

An important development in coastal plain stratigraphy has been the documentation of widespread Pliocene beds (the Yorktown Formation in Virginia [Hazel, 1971]). A pre-Pleistocene, post Yorktown unit named the Chowan Formation has been recognized by Blackwelder (1981) in southern Virginia and North Carolina. The Chowan, which overlies the Yorktown unconformably, consists of inner shelf deposits. Farther northward, in lower Delmarva Peninsula, post-Yorktown–pre-Pleistocene marginal marine deposits are present (R. Mixon, written communication). These beds are replaced northward by the coarse feldspathic sands of the Beaverdam Formation, a largely fluviatile (deposit) with some marginal marine tongues. Pollen data suggest that the Beaverdam is pre-Pleistocene in age. In New Jersey little is known about the equivalents of the latest Tertiary.

The Beaverdam is particularly important because it unconformably overlies the Pensauken (Columbia) and the Upper Miocene aquifer complex (Manokin, Ocean City, and Pocomoke aquifers; Hansen, 1981). Owens and Denny (1979) interpret the Pensauken as the upper delta plain equivalents of the Upper Miocene aquifers, which they interpret as lower delta plain to delta front deposits. In this case the Pensauken may be, at least in part, Late Miocene in age, but Hansen (1981) considers the Pensauken younger in age and separated from the Upper Miocene aquifer complex by a channeled unconformity.

In New Jersey the Bridgeton lies upon the Cohansey Sand and is overlain unconformably by the Pensauken. Based on regional stratigraphic relationships, Owens and Denny (1979) place a late Miocene age on the Cohansey. Based solely on topographic position, they correlate the Cohansey of New Jersey with the Brandywine Formation of southern Maryland. The age of the Brandywine is conjectural, but recent mapping in this area (McCartan, personal communication) suggests that it may be equivalent, at least in part, to marine units occurring in the upper part of the Chesapeake Group.

The oldest dated marine Pleistocene unit is 200,000 years old (variously called Omar-Accomack or Shirley [Norfolk]). This unit, which crops out from southeastern Virginia to the Delmarva Peninsula, consists largely of barrier to back barrier facies and minor inner shelf lithofacies in the coastal region. Extensive estuarine-fluvial facies interfinger with this unit along Delaware and Chesapeake Bays.

The 200,000 year old deposit was extensively eroded during the emplacement of a younger 70,000 year old Pleistocene unit. This unit lies at about 8 m above sea level and is the Pamlico Formation of previous investigators. As presently used, local beds of this age include the Tabb (Norfolk) Formation in southeastern Virginia, the Ironshire Formation in the Delmarva, and the Cape May Formation in New Jersey. These beds are best exposed and most widespread in southeastern Virginia. In general the Tabb Formation is similar to the 200,000 year old deposits in that it consists of a barrier to back barrier sequence in the coastal areas and interfingers with estuarine-fluvial facies along the Delaware and Chesapeake Bays.

The Sinepuxent Formation (Owens and Denny, 1979) is a third Pleistocene unit that is present in the lower Delmarva Peninsula. The Sinepuxent, which has surface elevations about 3 to 4.5 m above sea level, lies adjacent to the Ironshire Formation. Where best preserved, the bulk of the Sinepuxent consists of dark gray, massive, silty, micaceous fine to medium sand. Fossils typically dominated by *Mulinia* are common in the basal 3 m. These massive sands are capped in the type area by thick, very peaty fresh water swamp deposits. Pollen studies in both the swamp deposits and Sinepuxent identify a cool temperate assemblage (for example, spruce is common to abundant). Carbon 14 determinations on wood yielded ages ranging from 40,000 to 28,000 years old.

The west side of the Delmarva Peninsula, mainly the Maryland portion, is bordered by a broad flat surface, the Kent Island, which ranges from about 6 m in elevation in the upper bay to about 3 m in the lower (Owens and Denny, 1979). This surface is underlain by older units, suggesting that most of the surface was wave cut. Locally some of the younger sediments yield radiocarbon ages about 30,000 years B.P. and contain a common spruce microflora. These beds appear to be the estuarine equivalent of the Sinepuxent Formation.

Dunal deposits mapped as the Parsonsburg Sand are widespread south of Salisbury, Maryland (Sirkin, 1977). Radioactive ages from peats at the base of the dunes range from 13,000 to 30,000 years and, like the Sinepuxent and Kent Island, typically contain a cool temperate microfloral assemblage. It is probable that at least part of the Parsonsburg is coeval with the Sinepuxent and Kent Island deposition.

Based on these data, it is suggested by Owens and Denny (1979) that a marine event occurred in the Chesapeake Bay region post–70,000 years ago in which sea level rose to nearly its present position. Both the oceanic temperatures and terrestrial conditions were cooler than present.

Holocene

In general the northeastern emerged Atlantic Coastal Plain is undergoing a relatively rapid Holocene transgression (Cinquemani and Newman, 1981). This transgression appears to have removed or overlapped much of the youngest Pleistocene marine formations in coastal Delmarva and in the lower Chesapeake and Delaware Bays. In a few areas, notably coastal Delaware, Kraft (1971) has studied the transgression in great detail.

The character and general distribution of part of the Holocene fill in the major estuaries is discussed by Owens and others (1974). As outlined, the major estuaries are aggrading as a result of flooding during the Holocene transgression.

BIOSTRATIGRAPHY

Biostratigraphic zonation (Figs. 8, 9) in the coastal plain has been based principally on studies of planktonic foraminifera and palynomorphs. Planktonic foraminiferal zonation has been applied to the marine strata of the Upper Cretaceous and Lower Tertiary (Olsson, 1964, 1975; Olsson and others, 1980; Petters, 1976), but palynostratigraphy has been applied principally to the nonmarine strata of the Lower Cretaceous and part of the Upper Cretaceous (Doyle, 1969, 1982; Christopher, 1979, 1982). In addition to these microfossil groups, dinoflagellates (Aurisano, 1980; McLean, 1969), ostracodes (Hazel, 1971), diatoms (Andrews, 1978), and calcareous nannofossils (Gibson and others, 1980) have been utilized in biostratigraphy of the coastal plain.

Planktonic Foraminifera

At the base of the marine section in the subsurface of New

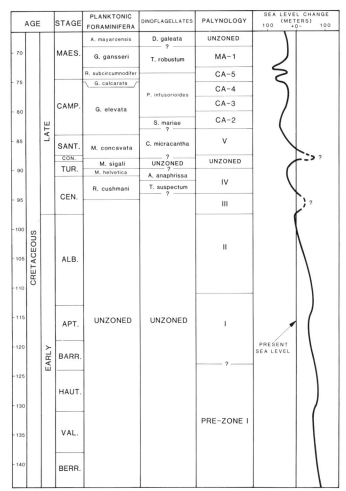

Figure 8. Cretaceous Biostratigraphy of coastal plain showing zonation based on planktonic foraminifera, dinoflagellates, and pollen. Paleodepth curve is shown in the right column.

Jersey, a late Cenomanian age for the Bass River Formation (Figures 5, 8) is recognized by the species *Rotalipora cushmani* (Petters, 1976). The presence of the species *Marginotruncana helvetica* in the upper part of the Bass River Formation places this interval in the lower Turonian *M. helvetica* Zone. These two zones have not been recognized southward in the coastal plain. The upper Turonian and Coniacian is absent in most of the coastal plain except in the Dickinson I well where Petters (1976) has recognized parts of the upper Turonian and the Coniacian on the occurrence of *M. schneegansi, M. sigali,* and *M. concavata.*

Santonian to Maestrichtian planktonic foraminiferal markers include *M. concavata* for the Santonian, *Globotruncana elevata* and *G. calcarata* for the Campanian, and *G. gansseri* and *G. contusa* for the Maestrichtian (Fig. 8). The uppermost Maestrichtian zonal species *Abathomphalus mayaroensis* has not been identified in the coastal plain. Nevertheless, the occurrence in New Jersey of the uppermost Maestrichtian calcareous nannofossil species *Micula muris* and *Nephrolithus frequens* indicates that the uppermost Cretaceous beds are contained within the *Abathomphalus mayaroensis* Zone (Fig. 7).

Although the uppermost part of the Maestrichtian appears to be documented, the lowermost part of the Tertiary has not been identified. Its absence or presence is not proven conclusively because closely spaced fossiliferous samples across the Cretaceous Tertiary boundary have not been available for study. The earliest Tertiary zones identified are the *Subbotina pseudobulloides* Zone and the calcareous nannofossil zones NP 2 and NP 3 (Fig. 9). Thus the foraminiferal *"Globigerina" eugubina* Zone and the NP 1 zones have so far not been identified but the unsampled interval immediately above the top of the Cretaceous allows the possibility of their eventual recognition.

In the Paleogene, the Paleocene and lower Eocene sections can be zoned utilizing low latitude or tethyan zonations (Fig. 9). All of the Paleocene zones have been identified, but the *Morozovella uncinata* and the *M. velascoensis* zones are only partially present due to hiatuses. The zones recognized in the lower Eocene also compare favorably with tethyan zones except for the *Morozovella edgari* Zone, which can not be differentiated, possibly due to inadequate sampling.

Biostratigraphic zonation in the Middle Eocene (Fig. 9) compares favorably with the temperature *Globorotalia cerroazulansis* zones, which are very useful not only in the coastal plain but in the Atlantic Basin as well. For the most part, the Upper Eocene is missing in the coastal plain but is present in the Baltimore Canyon Trough where it is zoned on species of the *G. cerroazulensis* group.

Lower Oligocene strata are missing in the coastal plain but upper Oligocene strata are identified on the occurrence of the *Globorotalia opima* and *Globigerina ciperoesis* Zones (Fig. 9).

The Neogene is the least understood biostratigraphically due to unconformities and shallow marine to fluvial environments of deposition. In the Miocene, the Aquitanian appears to be missing or represented by a very thin interval because only the *Catapsdrax dissimilis* Zone can be recognized above the Oligocene (Fig. 9). This zone is based on the temperate zonation of Kennett (1973) and indicates cooler ocean temperatures in this area in the early Miocene. In the upper part of the lower Miocene and the lower part of the middle Miocene, zonation can be equated with the tropical zones of *Globigerinatella insueta, Praeorbulina glomerosa,* and *Globorotalia fohsi peripheroronda.* Above, only the *Neogloboquadrina acostaensis* Zone has been recognized (Fig. 9).

Calcareous Nannofossils

Calcareous nannofossil biostratigraphy (Fig. 9) has been applied in some parts of the Tertiary with the most complete zonation recognized being in the Paleocene and the Eocene (Gibson and others, 1980). In the Cretaceous certain intervals in the Turonian, Campanian, and Maestrichtian have been dated by calcareous nannofossils (Valentine, 1982; Sissingh, 1977; Worsley and Werle, 1981).

Palynology

Palynostratigraphy has been most important in clarifying

Hanley Library
University of Pittsburgh
Bradford Campus

stratigraphic relationships in the nonmarine Lower and Upper Cretaceous (Fig. 8). Five zones are presently recognized (Zones I to V) but their correlation with marine zones has not yet been fully established. The zones encompassing the Barremanian to the early Campanian (Christoffer, 1979, 1982) are very important in subsurface stratigraphy. A pre-Zone 1 interval of Berriasian to Barremanian age has been recognized in the subsurface of Maryland (Doyle, 1982).

SEA LEVEL EVENTS

The coastal plain has long been recognized for its sequence of transgressive and regressive formations. This sequence can be conveniently divided into three parts: a Lower Cretaceous and possibly Upper Jurassic, largely nonmarine section of sand and clay with some gravel; Upper Cretaceous to Eocene glauconitic marine section of alternative sand, silt, and clay; and an upper Oligocene to Holocene section of nonmarine to shallow shelf sand and silt. This division expresses a long term cycle of sea level rise that influenced the coastal plain in the Late Cretaceous and early Tertiary. Within these broad divisions, numerous sea level events are recognized. Many of these sea level events compare favorably with the coastal onlap-offlap record (Vail and others, 1977; Vail and Hardenbol, 1979).

The estimates of sea-level elevations for the maximum rise in the Late Cretaceous by Pitman (1978) and Vail and others (1977) show over 300 m above present level. Watts and Steckler (1979) estimated a maximum rise of about 100 m for the continental margin of eastern North America. These estimates are not based directly on paleontological criteria but on onlap (Vail and others, 1977) or subsidence calculations (Watts and Steckler, 1979). Paleontological estimates based on foraminiferal models (Olsson and Nyong, 1984) show sea level at maximum high stand in the coastal plain at about 80 m above present level in the Campanian. The sea level curves in Figures 8 and 9 show paleodepths for the coastal plain based on this and other studies of foraminifera.

In the subsurface, thin intervals of marine or marginal marine sediments encountered within the Potomac Group indicate that, periodically, sea level rise events influenced the coastal plain during the early Cretaceous (Fig. 8). Offshore in the Baltimore Canyon Trough, intervals of marine sandstone and shale in the Lower Cretaceous (Poag, 1980) may be partially equivalent to marine intervals in the coastal plain, but they are not well constrained to geologic stage.

In the upper Cretaceous, sea-level events are in most cases biostratigraphically well constrained. The first extensive marine deposition in the coastal plain occurred in the late Cenomanian to early Turonian (Fig. 8). The apparent absence of marine early Cenomanian in the coastal plain may be due to the sea-level fall event shown on the onlap-offlap curve of Vail and others (1977).

A major fall in sea level caused a prominent Upper Turonian to Coniacian unconformity in the coastal plain, but offshore in the Baltimore Canyon Trough, nonmarine to inner shelf sands, shales, and lignites are encountered (Poag, 1980). After the Coni-

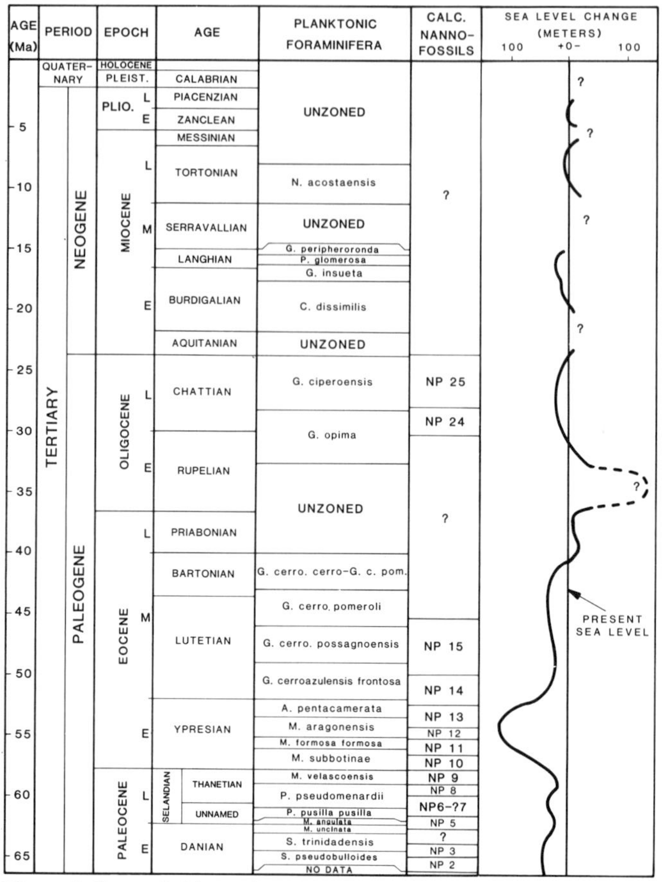

Figure 9. Cenozoic biostratigraphy of coastal plain showing zonation of planktonic foraminifera and calcareous nannofossils. Paleodepth curve is shown in the right column.

acian fall in sea level, a Santonian to early Campanian rise event, which falls within the *Marginotruncana concavata* and *Globotruncana elevata* Zones (Fig. 8), led to deposition in the coastal plain of the Magothy and Merchantville Formations. Rises in sea level (Fig. 8), separated by a fall in sea level, occur at the end of the Campanian (*G. calcarata* Zone) and in the early Maestrichtian (*Rugotruncana subcircumnodifier* Zone). The Campanian sea-level changes of the coastal plain arc also recognized in the Baltimore Canyon (Nyong and Olsson, 1984). A slight rise in sea level occurs in the coastal plain at the end of the Cretaceous.

An unconformity in the Baltimore Canyon Trough with missing Paleocene and Maestrichtian sections contrasts sharply with the coastal plain, where inner to middle shelf sands, silts, and glauconitic clays of these ages occur. Thus the unconformity does not appear related to a single sea-level event and it is difficult to explain its origin.

Sea-level events are well constrained biostratigraphically in the Paleogene (Fig. 8). Except in the lower Oligocene, the general pattern of these events compares favorably with the curve of Vail and others (1977) and Vail and Hardenbol (1979). Unconformities have been noted at the base of the Paleocene, in the mid-Paleocene, between the Upper Paleocene and Lower Eocene, and between the Lower and Middle Eocene. The cycles in the Paleocene and Eocene are similar in magnitude to those of the Santonian to Maestrichtian with relatively minor changes in sea level except in the Early Eocene. The Early Eocene rise in sea level in the coastal plain appears to have been a large magnitude event (Fig. 9). An unconformity separates the Middle Eocene from the Upper Oligocene. Upper Eocene sediments are encountered offshore in the Baltimore Canyon Trough, but lower Oligocene sediments are absent in the coastal plain and in the offshore basin (Hazel and others, 1984; Olsson and others, 1980; Poag, 1980). The curve of Vail and others shows a major rise in sea level during the early Oligocene followed by a rapid fall in sea level. Miller and others (1985), compared the lower Oligocene unconformity in the coastal plain with the Irish continental margin where a similar unconformity is present. The unconformity, which is associated with an enrichment of $\delta^{18}O$ in the isotopic record, is explained by a rapid glacio-eustatic fall in sea level (Miller and others, 1985).

The Neogene record of sea-level events in the coastal plain is not well constrained biostratigraphically (Fig. 9). At least five unconformities are recognized in the Miocene (Fig. 5). Kidwell (1984) identified numerous additional unconformities in the Miocene of Maryland. She correlates three of the unconformities in the Maryland section with interregional unconformities recognized by Vail and Hardenbol (1979). The chronostratigraphic position of Pliocene marine strata in Virginia and the various "Pliocene-Pleistocene" sand and gravel deposits of the coastal plain are not well known.

SUMMARY

Many advances have been made in understanding the geologic history of the coastal plain, but many problems remain unresolved. Future advances can be projected from current directions of research. They emphasize increased stratigraphic resolution, sea-level history, and basement tectonics.

Application of marine and nonmarine microfossil biostratigraphy to achieve increased resolution is allowing more refined analyses of sea-level events in the coastal plain and the recognition of important unconformities. Ultimately, the coastal onlap-offlap model (Vail and others, 1977; Vail and Hardenbol, 1979) will be more thoroughly tested, and coastal plain events will be compared to events recognized in the Baltimore Canyon Basin and the North Atlantic Basin. Increased biostratigraphic resolution will also lead to closer constraints of chronostratigraphic units so that coastal plain basement tectonics can be better understood, and ideas such as those of Brown and others (1972) tested.

Stratigraphic test holes and increased use of geophysical techniques such as seismic profiling will provide more rigid analyses of postrift and synrift sedimentary packages and will provide an understanding of the relationship between coastal plain and hinge zone tectonics. This will also provide more knowledge of the economic value of the coastal plain and its adjacent offshore zone.

REFERENCES CITED

Anderson, J. L., 1948, Cretaceous and Tertiary subsurface geology: Maryland Department of Geology, Mines, and Water Resources, Bulletin 2, 456 p.

Andrews, G. W., 1978, Marine diatom sequence in Miocene strata of the Chesapeake Bay region, Maryland: Micropalentology, v. 24, p. 371–406.

Aurisano, R. W., 1980, Upper Cretaceous subsurface dinoflagellate stratigraphy and paloecology of the Atlantic Coastal Plain of New Jersey [Ph.D. thesis]: Rutgers University, 204 p.

Baird, D., 1964, A Fossil Sea-Turtle from New Jersey; New Jersey State Museum Investigation No. 1, 26 p.

Benson, R. N., 1984, Structure contour map of pre-Mesozoic basement, landward margin of Baltimore Canyon Trough: Delaware Geological Survey, Miscellaneous Map Series No. 2.

Blackwelder, B. W., 1981, Stratigraphy of upper Pliocene and lower Pleistocene marine and estuarine deposits of northeastern North Carolina and southeastern Virginia: U.S. Geological Survey Bulletin 1502-B, 16 p.

Brenner, G. J., 1963, The spores and pollen of the Potomac Group of Maryland: Maryland Department of Geology, Mines, and Water Resources Bulletin 27, p. 1–215.

Brown, P. M., Miller, J. A., and Swain, F. M., 1972, Structural and stratigraphic framework, and spatial distribution of permeability of the Atlantic coastal plain, North Carolina to New York: U.S. Geological Survey Professional Paper 796, p. 1–79.

Bybell, L. M., and Govoni, D. L., 1977, Preliminary calcareous nannofossil zonation of Brightseat and Aquia Formations (Paleocene) of Maryland and Virginia—stratigraphic implications: American Association of Petroleum Geologists Bulletin, v. 61, p. 773–774.

Carter, C. H., 1978, A regressive barrier and barrier-protected deposit: Depositional environments and geographic setting of the Tertiary Cohansey Sand:

Journal of Sedimentary Petrology, v. 48, p. 933–950.

Christopher, R. A., 1979, Normapolles and triporate pollen assemblages from the Raritan and Magothy formations (Upper Cretaceous) of New Jersey; Palynology, v. 3, p. 73–122.

——1982, The occurrence of the *Complexiopollis-Atlantipollis* Zone (palynomorphs) in the Eagle Ford Group (Upper Cretaceous) of Texas: Journal of Paleontology, v. 56, p. 525–541.

Cinquemani, L. J., and Newman, W. S., 1981, Holocene sea-level changes in vertical movements along the eastern coast of the U.S.: A preliminary report, in Colquhoun, D. J., ed., Holocene sea level fluctuations, magnitude and causes: University of South Carolina, p. 13–33.

DiMarzio, J. A., 1984, Calcareous nannofossils from the Piney Point Formation, Pamunkey River, Virginia, in Ward, L. W., and Krafft, K., eds., Stratigraphy and paleontology of the outcropping Tertiary beds in the Pamunkey River region, central Virginia Coastal Plain: Guidebook for Atlantic Coastal Plain Geological Association 1984 field trip, p. 111–116.

Doyle, J. A., 1969, Angiosperm pollen evolution and biostratigraphy of the basal Cretaceous formations of Maryland, Delaware, and New Jersey: Geological Society of America Abstracts with Programs, Annual Meeting, Atlantic City, N.J., p. 51.

——1982, Waste Gate Formation: Part II Palynology of continental Cretaceous sediments, Crisfield geothermal test well, eastern Maryland: Maryland Geological Survey, Open File Report, p. 51–87.

Doyle, J. A., and Robbins, E. I., 1977, Angiosperm pollen zonation of the continental Cretaceous of the Atlantic Coastal Plain and its application to deep wells in the Salisbury embayent: Palynology, v. 1, p. 43–78.

Drake, C., and Woodward, H., 1963, Appalachian curvature, wrench faulting and offshore structures: Transactions of the New York Academy of Science,

Series II, v. 26, p. 48–63.

Frederiksen, N. O., Gibson, T. G., and Bybell, L. M., 1982, Paleocene-Eocene boundary in the eastern Gulf Coast: Transactions of the Gulf Coast Association of Geological Societies, v. 32, p. 289–294.

Gibson, T. G., 1970, Late Mesozoic-Cenozoic Tectonic Aspects of the Atlantic Coastal Margin: Geological Society of America Bulletin, v. 81, p. 1813–1822.

—— 1982, Depositional framework and paleoenvironments of Miocene strata from North Carolina to Maryland, *in* Scott, T. M., and Upchurch, S. B., eds., Miocene of the southern United States: Florida Bureau of Geology Special Publication 25, p. 1–22.

Gibson, T. G., Andrews, G. W., Bybell, L. M., Fredericksen, N. O., Hansen, T., Hazel, J. E., McLean, D. M., Witmer, R. J., and Van Nieuwenhuise, D. S., 1980, Geology of the Oak Grove Core, Part 2: Biostratigraphy of the Tertiary Strata of the Core: Virginia Division of Mineral Resources Publication, v. 20, p. 14–30.

Glaser, J. D., 1971, Geology and Mineral Resources of Southern Maryland: Maryland Geological Survey, Report of Investigations, v. 15, 84 p.

Grow, J. A., 1980, Deep structure and evolution of the Baltimore Canyon Trough in the vicinity of the COST No. B-3 Well: U.S. Geological Survey Circular 833, p. 117–132.

Hansen, H. J., 1969, Depositional environments of subsurface Potomac Group in southern Maryland: American Association of Petroleum Geologists Bulletin, v. 53, p. 1923–1937.

—— 1974, Sedimentary facies of the Aquia Formation in the subsurface of the Maryland coastal plain: Maryland Geological Survey, Report of Investigations No. 21, p. 1–47.

—— 1978, Upper Cretaceous (Senonian) and Paleocene (Danian) pinchouts on the south flanks of the Salisbury Embayment, Maryland, and their relationship to antecedent basement structures: Maryland Geological Survey, Report of Investigations No. 29, p. 1–36.

—— 1981, Stratigraphic discussion in support of a major unconformity separating the Columbia Group from the underlying upper Miocene aquifer complex in eastern Maryland: Southeastern Geology, v. 22, p. 123–138.

—— 1982, Waste Gate Formation: Part I Hydrologic framework and potential utilization of the brine aquifers of the Waste Gate Formation, a new unit of the Potomac Group underlying the Delmarva Peninsula: Maryland Geological Survey, Open File Report, p. 1–50.

Hansen, J. H., and Wilson, J. M., 1984, Summary of hydrologic data from a deep (2,678 ft.) well at Lexington Park, St. Mary's County, Maryland: Maryland Geological Survey, Open File Report No. 84-02-1, p. 1–61.

Hazel, J. E., 1971, Ostracode biostratigraphy of the Yorktown Formation (Upper Miocene and Lower Pliocene) of Virginia and North Carolina: U.S. Geological Survey Professional Paper 704, p. 1–13.

Hazel, J. E., Bybell, L. M., Christopher, R. A., Fredericksen, N. O., May, F. E., McLean, D. M., Poore, R. C., Smith, C. C., Sohl, N. F., Valentine, P. C., and Witmer, R. V., 1977, Biostratigraphy of the deep corehole (Clubhouse crossroads Corehole 1) near Charleston, South Carolina, *in* Rankin, D. W., ed., Studies related to the Charleston South Carolina Earthquake of 1886: U.S. Geological Survey Professional Paper 1028, p. 71–89.

Hazel, J. E., Edwards, L. E., and Bybell, L. M., 1984, Significant unconformities and the hiatuses represented by them in the Paleogene of the Atlantic and Gulf Coastal Province, *in* Schlee, J. S., ed., Interregional unconformities and hydrocarbon accumulation: American Association of Petroleum Geologists Memoir 36, p. 59–66.

Houlik, C. W., Jr., Olsson, R. K., and Aurisano, R. W., 1983, Upper Cretaceous (Campanian-Maestrichtian) marine strata in the subsurface of northern Delaware: Southeastern Geology, v. 24, p. 57–65.

Isphording, W. C., 1970, Petrology, stratigraphy and re-definition of the Kirkwood Formation (Miocene) of New Jersey: Journal of Sedimentary Petrology, v. 40, p. 986–997.

Jacobeen, F. H., 1972, Seismic evidence for high angle reverse faulting in the coastal plain of Prince George's and Charles Counties, Maryland: Maryland Geological Survey, Information Circular No. 13, 21 p.

Jordan, R. R., 1976, The Cretaceous-Tertiary boundary in Delaware, *in* Guidebook to the stratigraphy of the Atlantic Coastal Plain in Delaware: Petroleum Exploration Society of New York, p. 74–80.

—— 1983, Stratigraphic nomenclature of nonmarine Cretaceous rocks of inner margin of coastal plain in Delaware and adjacent states: Delaware Geological Survey, Report of Investigations No. 37, p. 1–43.

Jordan, R. R., and Smith, R. V., 1983, Atlantic coastal plain: Correlation of stratigraphic units of North American (COSUNA) Project: American Association of Petroleum Geologists, CSD #120.

Karner, G. D., and Watts, A. B., 1982, On isostasy at Atlantic-Type Continental Margins: Journal of Geophysical Research, v. 87, p. 2923–2948.

Kennett, J. P., 1973, Middle and late Cenozoic planktonic foraminiferal biostratigraphy of the southwest Pacific—DSDP Leg 21, *in* Burnes, R. E., et al., eds., Initial Reports of the Deep Sea Drilling Project, v. 21, p. 575–640.

Kidwell, S. M., 1984, Outcrop features and origin of basin margin unconformities in the lower Chesapeake Group (Miocene), Atlantic coastal plain, *in* Schlee, J. S., ed., Interregional unconformities and hydrocarbon accumulation: American Association of Petroleum Geologists Memoir 36, p. 37–58.

Koch, R. C., and Olsson, R. K., 1977, Dinoflagellate and planktonic foraminiferal biostratigraphy of the uppermost Cretaceous of New Jersey: Journal of Paleontology, v. 51, p. 480–491.

Kraft, J. C., 1971, Sedimentary environment, facies patterns, and geologic history of a Holocene marine transgression: Geological Society of America Bulletin, v. 82, p. 2131–2158.

Lambiase, J. J., Daskevsky, S. S., Costain, J. K., Gleason, R. J., and McClung, W. S., 1980, Moderate-temperature geothermal resource potential of the northern Atlantic Coastal Plain: Geology, v. 8, p. 447–449.

LeVan, D. C., and Pharr, R. F., 1963, A magnetic survey of the coastal plain in Virginia: Virginia Division of Mineral Resources, Report of Investigations, no. 4, 17 p.

Martino, R. L., 1976, Sedimentology and paleoenvironments of the Maestrichtian Monmouth Group in the northern and central New Jersey coastal plain [M.S. thesis]: Rutgers University, 70 p.

McLean, D. M., 1969, Organic-walled phytoplankton from the lower Tertiary Pamunkey Group of Virginia and Maryland [Ph.D. thesis]: Stanford University, 165 p.

Melillo, A. J., and Olsson, R. K., 1981, Late Miocene (late Tortonian) sea level event of Maryland-New Jersey coastal plain: Geological Society of America Abstracts with Programs, v. 13, no. 3, p. 166.

Miller, K. G., Mountain, G. S., and Tucholke, B. E., 1985, Oligocene glacioeustary and erosion on the margins of the North Atlantic: Geology, v. 13, p. 10–13.

Minard, J. P., 1965, Geologic map of the Woodstown Quadrangle, Gloucester and Salem Counties, New Jersey: U.S. Geological Survey, Geological Quadrangle Maps, U.S., Map GQ-404.

Mixon, R. B., and Newell, W. L., 1977, Strafford fault system: Structures documenting Cretaceous and Tertiary deformation along the Fall Line in northeastern Virginia: Geology, v. 5, p. 437–440.

Mixon, R. B., and Powars, D. S., 1984, Folds and faults in the inner coastal plain of Virginia and Maryland: Their effect on distribution and thickness of Tertiary rock units and local geomorphic history, *in* Frederiksen, N. O., and Krafft, K., eds., Cretaceous and Tertiary stratigraphy, paleontology, and structure, southwestern Maryland and northeastern Virginia: American Association of Stratigraphic Palynologists, Inc., Field Trip and Guidebook, p. 112–122.

Newell, W. L., and Radar, E. K., 1982, Tectonic control of cyclic sedimentation in the Chesapeake Group of Virginia and Maryland, *in* Lyttle, P. T., ed., Central Appalachian geology: Geological Society of America NE-SE Field Trip Guidebook, American Geological Institute, p. 1–27.

Norgan, D. S., 1964, Foraminifera, stratigraphy, and paleoecology of the Aquia Formation of Maryland and Virginia: Cushman Foundation Foraminiferal Research, Special Publication No. 7, p. 1–50.

Nyong, E. E., and Olsson, R. K., 1984, A paleoslope model of Campanion to lower Maestrichtian foraminifera in the North American Basin and adjacent continental margin: Marine Micropaleontology, v. 8, p. 437–477.

Olsson, R. K., 1963, Latest Cretaceous and Earliest Tertiary Stratigraphy of New Jersey Coastal Plain: American Association of Petroleum Geologists Bulletin, v. 57, p. 643–665.

—— 1964, Late Cretaceous planktonic foraminifera from New Jersey and Delaware: Micropaleontology, v. 10, p. 157–188.

—— 1970, Paleocene planktonic foraminiferal biostratigraphy and paleozoogeography of New Jersey: Journal of Paleontology, v. 44, p. 589–597.

Olsson, R. K., and Miller, K. G., 1979, Oligocene transgressive sediments of New Jersey continental margin: American Association of Petroleum Geologists Bulletin, v. 63, p. 505.

Olsson, R. K., Miller, K. G., and Ungrady, T. E., 1980, Late Oligocene transgression of middle Atlantic coastal plain: Geology, v. 8, p. 549–554.

Olsson, R. K., and Nyong, E. E., 1984, A paleoslope model for Campanian-lower Maestrichtian foraminifera of New Jersey and Delaware: Journal of Foraminiferal Research, v. 14, p. 50–68.

Owens, J. P., and Denny, C. S., 1979, Upper Cenozoic deposits of the Central Delmarva Peninsula, Maryland and Delaware: U.S. Geological Survey Professional Paper 1067-A, p. 1–27.

Owens, J. P., Minard, J. P., and Shol, N. F., 1968, Cretaceous deltas in the northern New Jersey coastal plain, *in* Guidebook to field excursions: New York State Geological Association, 40th Annual Meetings, p. 31–48.

Owens, J. P., Minard, J. P., Shol, N. F., and Mellow, J., 1970, Stratigraphy of the outcropping of post-Magothy Upper Cretaceous formations in New Jersey and northern Delmarva Peninsula, Delaware and Maryland: U.S. Geological Survey, Professional Paper 674, 60 p.

Owens, J. P., and Sohl, N. F., 1969, Shelf and deltaic paleoenvironments in the Cretaceous-Tertiary formations of the New Jersey coastal plain, *in* Subitsky, S., ed., Geology of selected areas in New Jersey and eastern Pennsylvania and guidebook of excursions: New Brunswick, N.J., Rutgers University Press, p. 235–278.

Owens, J. P., Stefansson, K., and Sirkin, L. A., 1974, Chemical, mineralogic, and palynologic character of the upper Wisconsinan-lower Holocene fill in parts of Hudson, Delaware, and Chesapeake estuaries: Journal of Sedimentary Petrology, v. 44, p. 390–408.

Palmer, A. A., 1982, Depositional environments of Miocene diatomaceous sediments of the United States middle Atlantic coastal plain: Geological Society of America Abstracts with Programs, v. 14, no. 7, p. 582.

Perry, W. J., Jr., Minard, J. P., Weed, E.G.A., Robbins, E. I., and Rhodehamel, E. C., 1975, Stratigraphy of Atlantic Coastal margin of United States north of Cape Hatteras—Brief Survey: American Association of Petroleum Geologists Bulletin, v. 59, p. 1529–1548.

Petters, S. W., 1976, Upper Cretaceous subsurface stratigraphy of Atlantic Coastal Plain of New Jersey: American Association of Petroleum Geologists Bulletin, v. 60, no. 1, p. 87–107.

Pickett, T. E., 1970, Geology of the Chesapeake and Delaware Canal area, Delaware: Delaware Geological Survey Map Series No. 1.

Pitman, W. C., 1978, Relationship between eustacy and stratigraphic sequences of passive continental margins: Geological Society of America Bulletin, v. 89, p. 1389–1403.

Poag, C. W., 1980, Foraminiferal stratigraphy, paleoenvironments, and depositional cycles in the outer Baltimore Canyon Trough, *in* Scholle, P. A., ed., Geological studies of the COST No. B-3 Well, United States Mid-Atlantic continental slope area: Geological Survey Circular 833, p. 44–66.

Pryor, W. A., 1975, Biogenic sedimentation and alteration of argillaceous sediments in shallow marine environments: Geological Society of America Bulletin, v. 86, p. 1244–1254.

Rachele, L. D., 1976, Palynology of the Legler lignite: A deposit in the Tertiary Cohansey Formation of New Jersey, U.S.A.: Review of Palaeobotany and Palynology, v. 22, p. 225–252.

Richards, H. G., 1948, Studies on the subsurface geology and paleontology of the Atlantic coastal plain: Philadelphia Academy of Natural Science Proceedings, v. 100, p. 39–76.

—— 1967, Stratigraphy of Atlantic Coastal Plain between Long Island and Georgia: Review: American Association of Petroleum Geologists Bulletin, v. 51, p. 2400–2429.

Richards, H. G., and Gallagher, W., 1974, The problem of the Cretaceous-Tertiary boundary in New Jersey: Notulae Naturae, no. 449, p. 1–6.

Richards, H. G., White, R. S., Jr., Madden, K., 1973, Upper Cretaceous Geology and Paleontology at Sewell, New Jersey: Geological Society of America Abstracts with Programs, v. 5, no. 2, p. 212.

Schlee, J. S., and Grow, J. A., 1980, Seismic stratigraphy in the vicinity of the COST No. B-3 Well: U.S. Geological Survey Circular 833, p. 111–116.

Scholle, P. A., ed., 1980, Geological Studies of the COST No. B-3 Well, United States Mid-Atlantic Continental Slope Area: U.S. Geological Survey Circular 833, 132 p.

Shifflett, E., 1948, Eocene stratigraphy and foraminifera of the Aquia Formation: Maryland Natural Resources Board Bulletin, v. 3, p. 1–93.

Sirkin, L. A., 1977, Late Pleistocene environment of the central Delmarva Peninsula, Delaware-Maryland: Geological Society of America Bulletin, v. 88, p. 139–142.

Sissingh, W., 1977, Biostratigraphy of Cretaceous calcareous nannoplankton: Geologie en Mijnbouw, v. 56, p. 37–65.

Spoljaric, N., 1972, Upper Cretaceous marine transgression in northern Delaware: Southeastern Geology, v. 14, no. 1, p. 25–37.

—— 1973, Normal faults in basement rocks of the northern coastal plain, Delaware: Geological Society of America Bulletin, v. 84, p. 2781–2784.

Stephenson, L. W., King, P. B., Monroe, W. H., and Imlay, R. W., 1942, Correlation of the outcropping Cretaceous rocks of the Atlantic and Gulf coastal plain and Trans-Pecos region: Geological Society of America Bulletin, v. 53, p. 435–448.

Stephenson, L. W., and MacNeil, F. S., 1954, Extension of Yorktown Formation (Miocene) of Virginia into Maryland: Geological Society of America Bulletin, v. 65, p. 733–738.

Taylor, P. I., Zietz, I., and Dennis, L. S., 1968, Geologic implication of aeromagnetic data for the eastern continental margin of the U.S.: Geophysics, v. 33, p. 755–780.

Vail, P. R., and Hardenbol, J., 1979, Sea-level change during the Tertiary: Oceanus, v. 22, p. 71–79.

Vail, P. R., Mitchum, R. M., Jr., and Thompson, S., III, 1977, Seismic stratigraphy and global changes of sea level, Part 4: Global cycles of relative changes of sea level, *in* Payton, C. E., ed., Seismic stratigraphy—applications to hydrocarbon exploration: American Association of Petroleum Geologists Memoir 26, p. 83–97.

Valentine, P. C., 1982, Upper Cretaceous subsurface stratigraphy and structure of coastal Georgia and South Carolina: U.S. Geological Survey Professional Paper 1222, p. 1–33.

Ward, L. W., 1984, Stratigraphy of outcropping Tertiary beds along the Pamukey River—central Virginia Coastal Plain, *in* Ward, L. W., and Krafft, K., eds., Stratigraphy and paleontology of the outcropping Tertiary beds in the Pamunkey region, central Virginia Coastal Plain: Guidebook for Atlantic Coastal Plain Geological Association 1984 field trip, p. 11–77.

Watts, A. B., 1982, Tectonic subsidence, flexure, and global changes of sea level: Nature, v. 297, p. 469–474.

Watts, A. B., and Steckler, M. S., 1979, Subsidence and eustasy at the continental margin of eastern North America, *in* Talwani, M., Hay, W., and Ryan, W.B.F., eds., Deep Drilling Results in the Atlantic Ocean: Continental Margins and Paleoenvironment: Washington, D.C., American Geophysical Union, Maurice Ewing Ser., v. 3, p. 218–234.

Worlsey, T., and Werle, K. V., 1981, Onshore calcareous nannofossil biostratigraphy of Atlantic margin Cretaceous and Paleogene: Abstracts American Association of Petroleum Geologists, eastern sectional meeting, Atlantic City, p. 52–53.

MANUSCRIPT ACCEPTED BY THE SOCIETY JULY 15, 1985

Printed in U.S.A.

Chapter 7

Late Mesozoic and early Cenozoic geology of the Atlantic Coastal Plain: North Carolina to Florida

Gregory S. Gohn
U.S. Geological Survey, 926 National Center, Reston, Virginia 22092

INTRODUCTION

The Atlantic Coastal Plain is a physiographic surface of low relief that is underlain by a mildly deformed, seaward-dipping wedge of Mesozoic and Cenozoic sediments. It constitutes a major segment of the Atlantic continental margin of the United States. In this chapter, the Mesozoic and early Cenozoic geologic history of the southern part of the Atlantic Coastal Plain, recorded within the Coastal Plain sediments, is reviewed within the broader context of the development of the entire continental margin.

The study area consists of the Atlantic Coastal Plain in North Carolina, South Carolina, eastern Georgia, and peninsular Florida (Fig. 1). The boundaries of this area are the Fall Line, the modern coastline, the Virginia/North Carolina border, and an arbitrary northwest-trending line in central Georgia that separates the Atlantic and Gulf Coastal Plains. The Fall Line (Fig. 1) separates Precambrian and Paleozoic crystalline rocks in the Piedmont province of the Appalachian orogen to the west from the relatively undeformed younger sediments of the low-lying Coastal Plain to the east. With the major exceptions of sedimentary and basaltic igneous rocks in exposed early Mesozoic rift basins (Fig. 1), upper Cenozoic fluvial deposits on Piedmont stream terraces, and local upper Cenozoic colluvial deposits in the Piedmont, Mesozoic and Cenozoic deposits in the Atlantic margin are restricted to the Coastal Plain east of the Fall Line.

In this chapter, the geology of the southern Atlantic Coastal Plain is summarized through integration of surface and subsurface structural and lithologic information for biostratigraphically determined time intervals of the Mesozoic and Cenozoic Eras. This approach stresses the historical development of the southern Atlantic Coastal Plain on a broad scale and requires that chronostratigraphic units and their contained lithofacies be emphasized, although formal lithostratigraphic units are also listed. Summaries for individual basins are presented through use of stratigraphic charts and a cross section. The general tectonic framework of this large sedimentary province is given before the chronostratigraphic summary.

The data base for this chapter varies considerably in size and detail throughout the study area. Although geologic studies in the area can be traced back to the 1700s, detailed field studies and maps are not abundant, and paleontologic data are sparse, in part because of the poorly fossiliferous nature of the sections in some areas. The subsurface part of the province was not explored in any detail until the 1940s (see, for example, Spangler, 1950; Applin and Applin, 1944, 1947), and hydrocarbon exploration has been minimal. Although a few stratigraphic test holes have been drilled, the subsurface data base in many areas consists primarily of geophysical logs and drill cuttings from water wells. Summaries of the province, or major parts thereof, include Chen (1965), Applin and Applin (1965, 1967), Maher and Applin (1971), Brown and others (1972, 1979), Swain (1977), Valentine (1979, 1982, 1984), Cramer and Arden (1980), Gohn and others (1982), Carter (1983; includes an extensive bibliography), Owens (1983), Colquhoun and others (1983), Jordan and Smith (1984), Ward and Strickland (1985), and Owens and Gohn (1985).

The locally small data base required generalizations and extrapolations of existing data for some areas described in this chapter. In addition, continuing uncertainty about such fundamental points as stratigraphic correlations required difficult choices with regard to representatives of conflicting viewpoints in a short chapter of broad scope. Although multiple interpretations are mentioned where the implications are of regional extent, relevant reports may have been inadvertently omitted.

TECTONIC HISTORY AND FRAMEWORK

Postrift Tectonic Processes

The tectonic history of the Atlantic continental margin began with continental fragmentation and rifting in the early Mesozoic and continued with continental drifting and the opening of the modern Atlantic Ocean in the late Mesozoic and Cenozoic. Onshore evidence of the rifting and postrifting history of the margin includes the postrift sedimentary wedge of the

Gohn, G. S., 1988, Late Mesozoic and early Cenozoic geology of the Atlantic Coastal Plain: North Carolina to Florida; *in* Sheridan, R. E., and Grow, J. A., eds., The Geology of North America, Volume I-2, The Atlantic Continental Margin, U.S. Geological Society of America.

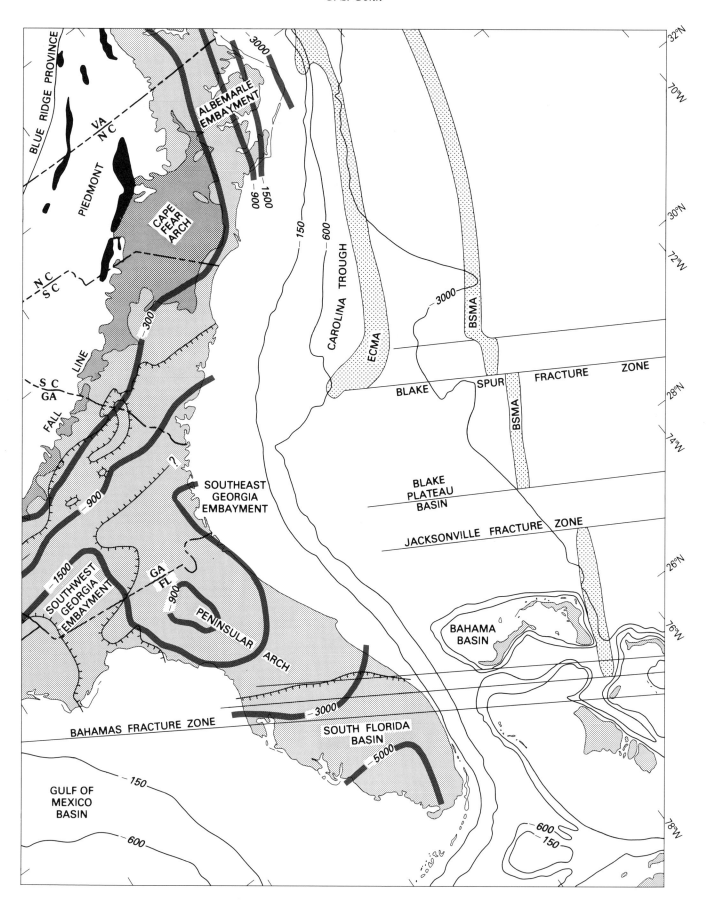

southern Atlantic Coastal Plain, which is the main focus of this chapter, and several exposed and subsurface rift-stage basins (Fig. 1). The geology of the northern part of the Coastal Plain is discussed by Olsson and others (this volume), and the geology of the offshore postrift section is discussed by Poag and Valentine (this volume).

Postrift tectonism (subsidence and uplift) in the Atlantic Coastal Plain is closely linked to subsidence along the outer continental margin (Steckler and others, and Sawyer, this volume). Outer margin subsidence resulted from extension and thinning of the crust during early Mesozoic rifting and subsequent postrift thermal contraction of the cooling lithosphere and from the loading of sediments and the oceanic water column on the lithosphere. The deepest sedimentary basins of the margin, for example the Carolina trough (Fig. 1), developed on this altered (rift-stage or transitional) crust at the seaward edge of the margin. Landward, less altered and thicker continental crust subsided at a decreasing rate away from the oceanic edge of the margin. Because of the differential subsidence, a major sedimentary hinge zone marks the change from continental to rift-stage crust. With the significant exception of the South Florida basin (Fig. 1), the Atlantic Coastal Plain is located entirely landward of this major hinge zone and lies entirely on continental crust. The effect of outer margin subsidence on subsidence in the Coastal Plain was complex, due to thermal and loading effects on the continental lithosphere. Variations in the age, composition, and thermal conductivity of the continental crust below the Coastal Plain probably produced local differences in flexural strength and heat flow that added to the complexity of its subsidence history.

However, not all tectonism recorded by Atlantic Coastal Plain sediments can be attributed directly to outer continental margin subsidence. Subsidence due to lithospheric cooling decreases exponentially with time, and subsidence due to sediment loading must have been much greater in the Jurassic and Early Cretaceous (when most of the sediments now present on the margin were deposited) than in the Late Cretaceous and Cenozoic. Therefore, the presence of Late Cretaceous and Cenozoic compressional faults and differences in the thickness of Upper

Cretaceous and Cenozoic stratigraphic units suggest that intraplate stress fields and resultant fault systems largely unrelated to continental margin subsidence played a role in Coastal Plain tectonism (Brown and others, 1972; Heller and others, 1982; Prowell, this volume). In the wrench-fault model of Brown and others (1972), sedimentary hingelines in the Coastal Plain section are inferred from changes in basement slope and the orientation of physiographic boundaries, as well as from vertical and lateral changes in lithofacies and sediment thickness. The hingeline system is interpreted to result from a regional crustal fault system that includes strike-slip and tensional faults produced by lateral compressive stress. Vertical movement on the various faults, which apparently changed direction with periodic rotations of the stress system, produced differential subsidence of fault-bounded blocks and, hence, variable geometries for the sedimentary units deposited on the blocks.

Major Postrift Structures

The major postrift basins and highs of the southern Atlantic Coastal Plain are outlined by the structure contours drawn on the top of pre-Upper Jurassic rocks in Figure 1. The basins include, from north to south, the Albemarle (or Hatteras) embayment, the Southeast Georgia embayment, and the South Florida basin. The highs are the Cape Fear arch, which is the only extensive outcrop belt of Cretaceous sediments in the southern Atlantic Coastal Plain, and the Florida Peninsular arch and its extension northward into Georgia. A broad area of gently dipping beds in southern South Carolina is called the Carolina platform by some authors. The Southwest Georgia embayment (Fig. 1) is a major basin of the eastern Gulf Coastal Plain.

As expected, the postrift sedimentary sections are thinner and less complete on the arches than in the basins (Figs. 1, 2). In particular, the older stratigraphic units (Upper Jurassic, Berriasian-Aptian) are confined to the deeper parts of the basins, and even middle Cretaceous beds (Aptian-Cenomanian) are missing from the Cape Fear arch and the highest parts of the Peninsular arch. Santonian beds are the oldest sediments to completely cover the Cape Fear and Peninsular arches and extend to the Fall Line throughout the study area.

The Southeast Georgia embayment is the shallowest of the onshore basins (Fig. 2) and contains little pre-Upper Cretaceous sediment and only moderately thick sections of Upper Cretaceous and Cenozoic sediments (maximum of about 1.5 km). However, this basin opens and deepens seaward to merge with the Blake Plateau basin, which contains more than 12 km of postrift sediments (Dillon and others, this volume; Klitgord and Behrendt, 1979).

The Albemarle embayment contains 3 km of sediment in the Hatteras Light #1 well on Cape Hatteras in which Upper Jurassic sediments were encountered above granitic rocks (Fig. 2). This embayment is not a typical structural basin with partial closure of structure contours, but rather is a roughly homoclinal sequence of beds that dip seaward at an increasing angle from the

---

◄————————————————

Figure 1 (facing page). Generalized geologic map showing the location of the southern Atlantic Coastal Plain along the United States Atlantic continental margin (adapted from Plate 4 and Figure 2 of Maher and Applin, 1971). Structure contours (m; in red) drawn on the top of pre-Upper Jurassic rocks outline the major postrift Coastal Plain basins and arches (metric contours modified from contours in feet on Plate 4 of Maher and Applin, 1971, using data in Brown and others, 1972; Popenoe and Zietz, 1977; Gohn and others, 1977; Applegate and others, 1981; Chowns and Williams, 1983). Cretaceous (dark shading) and Cenozoic (light shading) outcrop belts are indicated, as are exposed early Mesozoic rift basins (black) in the Piedmont. The boundaries of subsurface early Mesozoic rift basins are shown by hachured lines (data from Barnett, 1975; Chowns and Williams, 1983; Daniels and others, 1983). The locations of offshore Mesozoic-Cenozoic basins, the East Coast (ECMA) and Blake Spur (BSMA) magnetic anomalies, and selected oceanic fracture zones are from Klitgord and Behrendt (1979) and Klitgord and others (1983). Bathymetric contours are in meters.

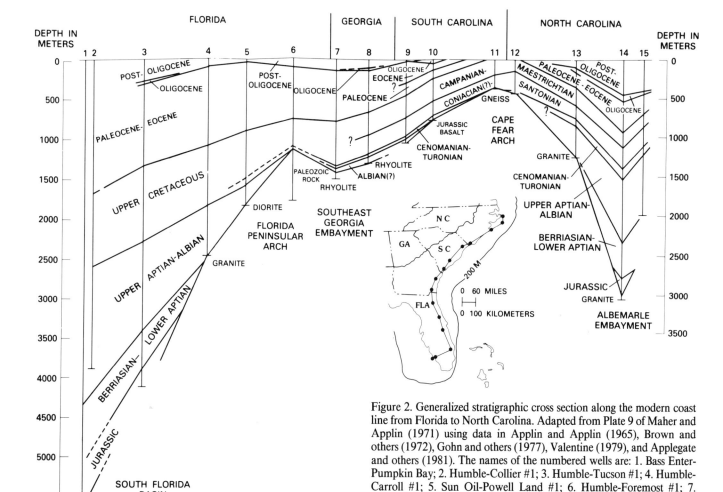

Figure 2. Generalized stratigraphic cross section along the modern coast line from Florida to North Carolina. Adapted from Plate 9 of Maher and Applin (1971) using data in Applin and Applin (1965), Brown and others (1972), Gohn and others (1977), Valentine (1979), and Applegate and others (1981). The names of the numbered wells are: 1. Bass Enter-Pumpkin Bay; 2. Humble-Collier #1; 3. Humble-Tucson #1; 4. Humble-Carroll #1; 5. Sun Oil-Powell Land #1; 6. Humble-Foremost #1; 7. California-Buie #1; 8. Larue-Jelks and Rodgers #1; 9. U.S.-Parris Island #2; 10. USGS-Clubhouse Crossroads #1; 11. USGS-Brittons Neck #1; 12. USGS-Calabash #1; 13. Karston-Laughton #1; 14. Standard Oil-Hatteras Light #1; 15. Standard Oil-Esso #2.

Fall Line to the coast (Fig. 1). The concept of an Albemarle "embayment" or "basin" has resulted primarily because progradation of the modern North Carolina coast line nearly to the edge of the continental shelf has provided easy access for drilling in thick sections near the edge of the continental margin. If the modern coastline were located farther inland, the Albemarle area probably would not be viewed as an embayment. However, because of its widespread usage in the literature, the term "Albemarle embayment" is used in this report.

The South Florida basin contains the thickest postrift section in the study area. Wells in the Florida Keys bottom in postrift beds of Late Jurassic or Early Cretaceous age at depths of about 3.5 to 4.7 km (Applin and Applin, 1965; Winston, 1971a). In a well in the deepest part of the South Florida basin (Fig. 2), sediments of probable Late Jurassic age lie above rhyolite of probable Early Jurassic age at a depth of 5.66 km (Applegate and others, 1981). Basement below this basin likely consists of rift-stage crust south of the Bahamas fracture zone in central Florida (Fig. 1) (Griffin and others, 1977; Klitgord and others, 1983; Sheridan, 1983; Mueller and Porch, 1983), which accounts for the deep subsidence of this area. The presence of numerous smaller troughs and highs within the South Florida basin (Winston, 1971a) indicates a considerable internal structural complexity to this embayment.

Other smaller but significant faults, basins, and arches have been defined throughout the southern Atlantic Coastal Plain. Some of these features may be erosional in nature and not structural, and some appear to have been active only at certain times. A few of them, for example the Suwanee channel, are mentioned below with regard to their effect on specific parts of the stratigraphic section. However, other features of this scale cannot be treated in detail in this report, and the reader is referred to articles by Applin and Applin (1965, 1967), Winston (1971a), Popenoe and Zietz (1977), Baum and others (1979), Harris and others

(1979), Cramer and Arden (1980), Brown and others (1982), and Prowell (1983, this volume).

DEPOSITIONAL HISTORY AND STRATIGRAPHIC FRAMEWORK

Synrift and Transitional Synrift-to-Postrift Geology

Although continental separation eventually occurred along a central axial rift near the eastern edge of the modern continental margin, peripheral fault-bounded troughs were formed by extension of continental crust across a broad area (Fig. 1) during the Late Triassic to Early or Middle Jurassic. These troughs, which are known in outcrop (Piedmont) and the subsurface (Coastal Plain), were filled by continental sediments (Newark Supergroup in outcrop; Froelich and Olsen, 1984). Rifting was accompanied by intrusion of diabase stocks and sills and extrusion of basalt within several of the rift basins and by intrusion of diabase dikes within these basins and across most of the continental crust east of the Blue Ridge province (Fig. 1; Daniels and others, 1983; Ragland and others, 1983; deBoer and others, this volume). The lower boundary of the rift basins is the rift-onset unconformity of Falvey (1974). Onshore, their upper boundary is the postrift or breakup unconformity of Falvey (1974), as used in most reports on the Atlantic continental margin. Tectonic, stratigraphic, magmatic, and other aspects of the onshore and offshore rift-stage basins are discussed at length in this volume by Manspeizer and Cousminer, deBoer and others, Dillon and Popenoe, Schlee and Klitgord, Grow and others, and Sheridan and others (Chapter 15).

The transition from continental rifting to the onset of seafloor spreading and the early production of oceanic crust probably occurred during the Early or Middle Jurassic in the corridor bounded by the East Coast and Blake Spur magnetic anomalies (Fig. 1; Sheridan, 1983). Sedimentation during this transitional period was concentrated in this corridor underlain by newly formed rift-stage crust. Initial marine flooding of this main axial rift before and during continental separation produced a juvenile Atlantic Ocean that was probably similar to the modern Gulf of Elat and Red Sea, where terrigenous sediments, salt, and anhydritic and reefal carbonate sediments accumulated under at least occasionally restricted conditions (Falvey, 1974; Epstein and Friedman, 1983; Cochran, 1983). At the modern continental edge, the top of the transitional sedimentary section would approximate the postrift or breakup unconformity in the original sense of Falvey (1974, Fig. 8). The salt-bearing lowest parts of the thick Jurassic sections in the major offshore Atlantic basins probably represent this phase (Grow and others, this volume; Dillon and Popenoe, this volume).

Because of their genetic restriction to the outer edge of the continental margin, transitional sediments would not be expected to occur in the southern Atlantic Coastal Plain. The possible exception to this premise is the South Florida basin, where basal sediments of Jurassic age overlie rift-stage crust. However, the absence in south Florida of a major salt-anhydrite unit similar to

those of the outer Atlantic-margin basins or the Middle and Upper Jurassic Louann Salt of the Gulf of Mexico basin suggests that the basal Florida sediments were not deposited in a highly restricted, small-ocean setting. On this limited basis, the entire south Florida sedimentary section is considered to represent postrift deposition.

Early Postrift Deposition

In the late Middle Jurassic, the spreading center shifted eastward from a position between the East Coast and Blake Spur anomalies to a position marked in the modern western Atlantic Ocean by the Blake Spur anomaly (Fig. 1; Sheridan, 1983; Gradstein and Sheridan, 1983). As the continental margin moved away from this spreading center during the Late Jurassic and Early Cretaceous, the zone of crustal subsidence began to progress landward due to lithospheric cooling, and sediment deposition spread across a progressively wider expanse of the newly submerging Continental Shelf.

The earliest postrift sediments in the southern Atlantic Coastal Plain are of Late Jurassic (Tithonian) and Early Cretaceous (Berriasian-Aptian) age (Fig. 3). Although these units do not crop out, sediments of this approximate age are present in several deep wells in the South Florida basin and in the Albemarle embayment near Cape Hatteras (Fig. 2).

South Florida Basin. In south Florida, a basal Coastal Plain section of carbonate sediments, anhydrite, and terrigenous sediments (Figs. 4, 5a) has been assigned to the Fort Pierce Formation along the basin's northeastern edge (Applin and Applin, 1965) and to the Wood River, Bone Island, and Pumpkin Bay Formations in the basin's center (Applegate and others, 1981). The foraminifer *Choffatella decipiens* Schlumberger is common near the top of the Fort Pierce (Fig. 4; Applin and Applin, 1965; Maher and Applin, 1971) and indicates a Berriasian to early Aptian age (Ascoli, 1976; Gradstein and Sheridan, 1983). Another foraminifer, *Anchispirocyclina lusitanica* (Egger) (reported as *Anchispirocyclina henbesti* Jordan and Applin by Maher and Applin, 1971, Table 3), occurs in one core from the upper part of the Fort Pierce and indicates a Late Jurassic (Tithonian) age (Jansa and others, 1980; Ascoli and others, 1984).

Basinward projection of the *Anchispirocyclina lusitanica* occurrence to the top of the Pumpkin Bay Formation would require as much as 1.3 km of Jurassic sediments in the center of the South Florida basin (Fig. 4). Alternatively, Applegate and others (1981) suggested that the Pumpkin Bay and Bone Island Formations may be Berriasian-Aptian (Coahuilan) in age and that only the Wood River Formation may be Jurassic in age (Figs. 2, 3).

Along the basin margin, the Fort Pierce Formation consists of three lithofacies (Applin and Applin, 1965); from base to top, these are a terrigenous red-bed facies, a mixed carbonate-terrigenous clastic facies, and a carbonate facies. The relatively thin, unfossiliferous red-bed facies consists of reddish arkosic sandstone, calcareous sandstone, and red or varicolored shale

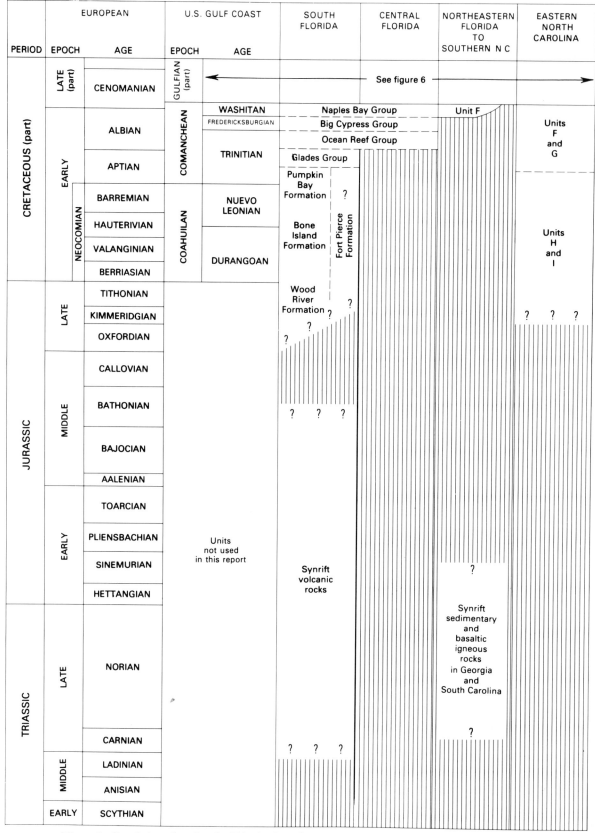

Figure 3. Correlation chart for the Triassic, Jurassic, and Lower Cretaceous units of the southern Atlantic Coastal Plain. Data sources for each unit are given in the text.

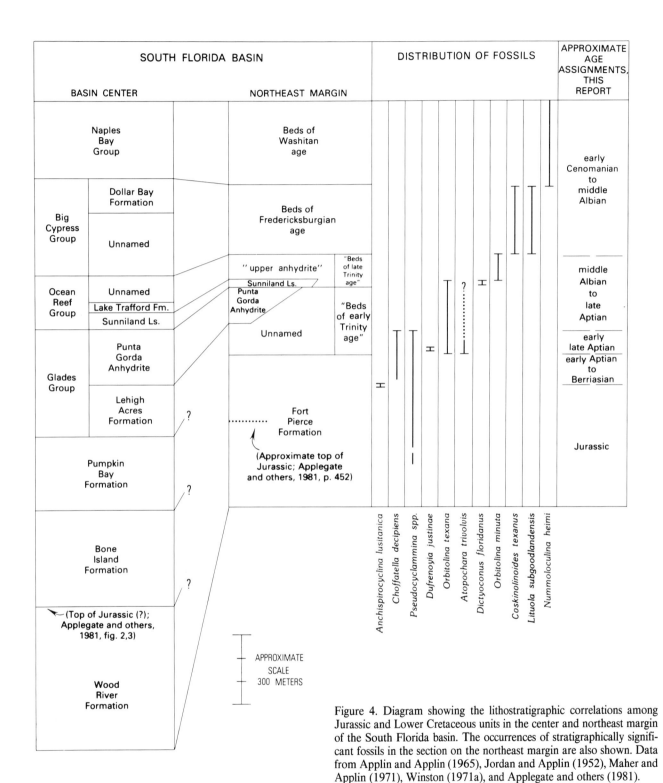

Figure 4. Diagram showing the lithostratigraphic correlations among Jurassic and Lower Cretaceous units in the center and northeast margin of the South Florida basin. The occurrences of stratigraphically significant fossils in the section on the northeast margin are also shown. Data from Applin and Applin (1965), Jordan and Applin (1952), Maher and Applin (1971), Winston (1971a), and Applegate and others (1981).

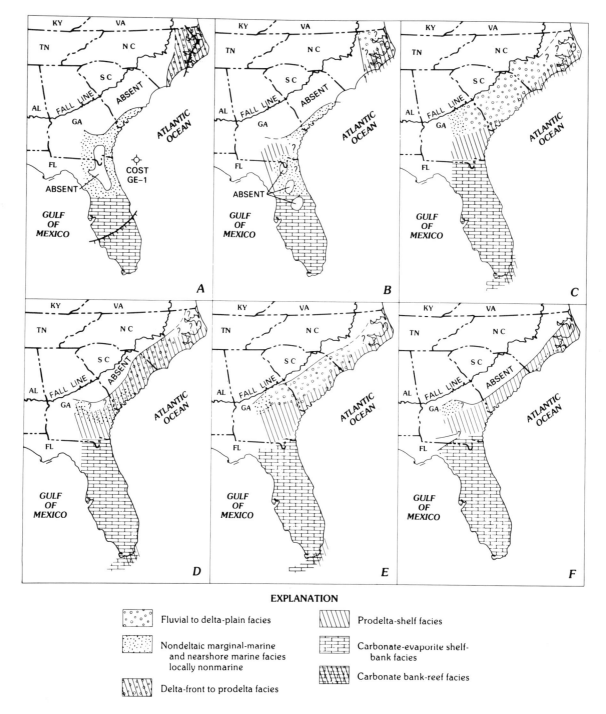

EXPLANATION

Fluvial to delta-plain facies	Prodelta-shelf facies
Nondeltaic marginal-marine and nearshore marine facies locally nonmarine	Carbonate-evaporite shelf-bank facies
Delta-front to prodelta facies	Carbonate bank-reef facies

Figure 5. A. Distribution of Upper Jurassic and Lower Cretaceous sediments in the southern Atlantic Coastal Plain. Updip limits of Jurassic plus Berriasian-lower Aptian sediments are shown by hachured lines. Lithofacies-paleoenvironment symbols are for upper Aptian-lower Cenomanian units. Modified from Figure 2-8 of Owens and Gohn (1985). B. Distribution of upper Cenomanian-lower Turonian sediments in the southern Atlantic Coastal Plain. Modified from Figure 2-14 of Owens and Gohn (1985). C. Distribution of Coniacian(?)-Santonian sediments in the southern Atlantic Coastal Plain. Adapted from Figures 2-15A and 2-15B of Owens and Gohn (1985). D. Distribution of lower Campanian sediments in the southern Atlantic Coastal Plain. Adapted from Figure 2-17 of Owens and Gohn (1985). E. Distribution of the upper Campanian and lower Maestrichtian sediments in the southern Atlantic Coastal Plain. Modified from Figure 2-19 of Owens and Gohn (1985). F. Distribution of middle and upper Maestrichtian sediments in the southern Atlantic Coastal Plain. Modified from Figure 2-21 of Owens and Gohn (1985).

with minor dolomite and anhydrite. The mixed facies consists primarily of dolomite with common anhydrite (as cement and nodules in carbonate sediments and probably as bedded anhydrite) and laminated gray shale and minor sandstone. Sparse ostracodes, benthic foraminifers, and algal boundstones are the principal biogenic elements of this facies. The carbonate facies contains little shale and little or no sandstone; it consists primarily of cemented bioclastic or oolitic limestone. Fossils are abundant in the carbonate facies (Applin and Applin, 1965; Maher and Applin, 1971). In the center of the basin, a basal 40-m-thick, terrigenous clastic sequence in the Wood River Formation underlies a thick carbonate-evaporite sequence in the upper Wood River, Bone Island, and Pumpkin Bay Formations (Applegate and others, 1981).

Albemarle Embayment. A thick section of Late Jurassic and Early Cretaceous (Berriasian-Aptian) age occurs in the deepest onshore part of the Albemarle embayment (Figs. 2, 3, 5a). These sediments have been assigned to chronostratigraphic units I (Late Jurassic?) and H (Late Jurassic? to Berriasian-Aptian) of Brown and others (1972).

Unit I, known only from the Hatteras Light No. 1 well, consists of unfossiliferous, feldspathic, coarse-grained, and conglomeratic sandstone interbedded with varicolored shale (Spangler, 1950; Swain, 1952; Brown and others, 1972). This unit resembles but is thicker than known sections of the basal terrigenous clastic facies of the Fort Pierce and Wood River Formations in Florida.

Unit H in the Hatteras well consists of a lower unit of carbonate sediments with minor anhydrite, sand, and clay; a middle coarse quartz sand; and an upper carbonate unit (Swain, 1952; Brown and others, 1972). The lower carbonate unit in the Hatteras Light No. 1 well contains *Anchispirocyclina lusitanica* (Egger) (reported as *Anchispirocyclina henbesti* Jordan and Applin) in a core near its base (Jordan and Applin, 1952; Maher and Applin, 1971). *Choffatella decipiens* Schlumberger occurs in cuttings from higher in this unit in several wells (Brown and others, 1972; Maher and Applin, 1971). The presence of *A. lusitanica* suggests a Late Jurassic age, as does the ostracode fauna reported by Swain (1952; Swain and Brown, 1972). Several species of ostracodes and *C. decipiens* in the upper part of unit H indicate a Berriasian-Aptian age for that section (Swain, 1952; Brown and others, 1972). A general chronostratigraphic equivalence of units H and I with the Fort Pierce Formation and its correlative units in Florida is indicated.

Early Postrift Depositional Trends. There is a general upward trend in the Albemarle and South Florida sections from basal terrigenous clastic sediments and overlying evaporitic carbonate sediments to open-marine, dominantly carbonate sediments. This trend probably represents the large-scale initial transgression of the Continental Shelf that reached parts of the Coastal Plain during the Late Jurassic and early Early Cretaceous.

The terrigenous sediments were restricted to nearshore environments and occur primarily as a basal transgressive sheet, as

predicted by the general model of Falvey (1974, p. 102). Deposition of the open-marine sediments primarily involved aggradation of shelfal carbonate sections behind a shelf-edge reef system (Dillon and others, 1985).

Late Postrift Deposition

During the late Early Cretaceous, Late Cretaceous, and Cenozoic, epicontinental sedimentation was widespread in the southern Atlantic Coastal Plain. Postrift sediments covered progressively larger parts of the continental margin from about the Aptian until the Santonian, at which time delta plain or barrier-strandplain deposition occurred near the modern Fall Line in virtually all of the southern Atlantic Coastal Plain. All exposed sediments in the study area were deposited during this late postrift period.

The transition from early to late postrift deposition took place during the Early Cretaceous as the average rate of crustal subsidence slowed. During late postrift time, delta systems periodically built out across the continental shelf, and effects on the sedimentary record from eustatic and/or tectonically driven sea-level changes became more obvious due to the reduced subsidence rate. Therefore, despite the temporal and geographic variability in the subsidence rate (Brown and others, 1972; Poag and Schlee, 1984), certain Late Cretaceous and Cenozoic transgressive unconformities can be traced entirely across the Coastal Plain (where preserved) and constitute a basis for dividing the late postrift section into transgressive-regressive cycles, which represent the progradation and ultimate truncation of major delta systems in the Carolinas and Georgia. For example, Owens and Gohn (1985) divided the Upper Cretaceous Series of the Atlantic Coastal Plain into five chronostratigraphic sequences that are widely traceable packages of genetically and temporally related sediments bounded by physical unconformities.

Following the large-scale trend of progressive landward subsidence and sediment overlap, Lower Cretaceous deltas are located primarily offshore, whereas Upper Cretaceous and Paleocene deltas are located onshore in the Carolinas. Carbonate sedimentation continued in the Florida peninsula throughout the Cretaceous and early Tertiary because it was far from the inland sources of terrigenous sediment. The transition zone between the carbonate province and the deltaic province was located in southern Georgia throughout most of the Early and Late Cretaceous and Paleocene epochs. During the Eocene and Oligocene, this transition zone shifted to the north, and many deposits of these ages in the southern Atlantic Coastal Plain consist of carbonate sediments. Depositional patterns of Neogene and Quaternary sediments of the southern Atlantic Coastal Plain are discussed by Riggs and Belknap (this volume).

Cretaceous System. **Lower Cretaceous (Aptian-Albian) Series.** Aptian, Albian, and associated basal Upper Cretaceous (lower Cenomanian) units are the oldest postrift beds to cover a major part of the southern Atlantic Coastal Plain (Fig. 5a). They occur in the subsurface throughout most of Florida, southern and eastern Georgia, southern South Carolina, and eastern North Carolina.

In peninsular Florida, the dominantly carbonate upper Aptian-Albian-lower Cenomanian section has been divided into four groups in the center of the South Florida basin and into equivalent informal units elsewhere in the state, as shown in Figures 3 and 4 (Applin and Applin, 1965; Winston, 1971a; Applegate and others, 1981). These sediments constitute the Comanchean Series of Applin and Applin (1965) and, together with the Fort Pierce Formation, constitute the Marquesas Supergroup of Meyerhoff and Hatten (1974).

Applin and Applin (1965) reported the ammonite *Dufrenoya texana* Burckhardt from the lower part of their "beds of early Trinity age" (Fig. 4). This report may represent the early late Aptian species *Dufrenoyia justinae* (Hill) (see synonymy in Young, 1974), a common ammonite in the middle Trinitian Cow Creek Limestone of Texas. The presence of the foraminifers *Choffatella decipiens* Schlumberger and *Orbitolina texana* (Roemer) with the charophyte *Atopochara trivolvis* Peck in this unit (Applin and Applin, 1965) supports the Aptian age for the lower "beds of early Trinity age" and for the largely equivalent Lehigh Acres Formation of the Glades Group (Fig. 4). The upper "beds of early Trinity age" and the "beds of late Trinity age" (equivalent to the Ocean Reef Group of Winston, 1971a) contain the late Aptian-middle Albian foraminifers *Orbitolina minuta* Douglass and *O. texana*. The common marker foraminifers for the Big Cypress Group of Winston (1971b) (middle Albian; Fredericksburgian) are *Lituola subgoodlandensis* (Vanderpool) and *Coskinolinoides texanus* Keijzer. The abundance of the foraminifer *Nummoloculina heimi* Bonet in the Naples Bay Group of Winston (1971a) suggests a middle Albian to early Cenomanian (Washitan) age (Swain, 1982).

The Aptian-Albian section consists of limestone, dolomite, anhydrite, and terrigenous clastic beds in sedimentary cycles at several scales. Applin and Applin (1965) recognized three major lithofacies within the Florida Aptian-Albian section: a basin-marginal terrigenous clastic facies, a facies of mixed carbonate and siliciclastic sediments, and a carbonate-evaporite facies. The marginal clastic and mixed facies are the nearshore (updip) units of each stratigraphic group. Within the carbonate-evaporite sections, Winston (1971a,b) recognized numerous cyclic repetitions of anhydrite, skeletal and oolitic calcarenitic limestone, chalky limestone, and anhydritic and argillaceous dolomite on scales of a few meters to 120 meters.

The Florida Aptian-Albian section represents a long-term, fundamentally transgressive sequence lacking major progradational deltaic wedges. Along any given time line, relatively thin sections of marginal-marine terrigenous clastic sediments are restricted to the edge of the basin and grade laterally basinward into aggradational shelfal carbonate sediments. Winston (1971b) interpreted the carbonate-anhydrite cycles to represent deposition in generally shallow but variable water depths on a locally restricted shelf behind a shelf-edge reef system. The presence of

marine, brackish-water, and perhaps freshwater ostracodes in the Florida Aptian-Albian sections also suggests variability in the salinity and depositional environments of these sediments (Swain and Miller, 1979; Swain, 1982).

A distinct transition from the dominantly calcareous Aptian-Albian section of south Florida to a northern terrigenous clastic province occurs in east-central Florida. Offshore on the Atlantic Continental Shelf, the COST GE-1 test hole (Fig. 5a) contains a thick (about 1200 m) Barremian (or older)-Aptian-Albian section of coal-bearing continental sediments that likely represents delta-plain and related environments of a major delta (data in Amato and Bebout, 1978, and Scholle, 1979). This deltaic unit is apparently not represented onshore, although its sediments must have been derived from the Georgia-South Carolina area. A younger Albian carbonate-shale section in the GE-1 well likely represents an extension of the carbonate facies of the Naples Bay Group.

Onshore, a northward extension of the terrigenous clastic beds of the Naples Bay Group occurs above pre-Cretaceous rocks in southern Georgia and southern South Carolina. This thin section of unfossiliferous sands and clays has been assigned to the lower member of the Atkinson Formation (Applin and Applin, 1947), the lower part of the lower member of the Atkinson (Applin, 1955; Applin and Applin, 1965, 1967), unit F (Brown and others, 1979), and unit K$_1$ (Gohn and others, 1978, 1980).

North of the Cape Fear arch, the upper Aptian through lower Cenomanian section is represented by chronostratigraphic units G and F of Brown and others (1972). Brown and others (1972) and Swain and Brown (1972) assigned a middle to late Trinitian (late Aptian to early Albian) age to unit G on the basis of its ostracode fauna and the presence of the charophytes *Atopochara trivolvis* Peck and *Clavator harrisi* Peck. Unit G is restricted to the subsurface in eastern North Carolina (Brown and others, 1972). Primarily on the basis of ostracodes, Brown and others (1972; Swain and Brown, 1972) assigned a middle to late Albian and early Cenomanian (Frederickburgian-Washitan) age to their unit F. Valentine (1982, 1984) and Owens and Gohn (1985) have suggested, on the basis of additional fossil data, that unit F as originally defined also includes outcropping beds of Coniacian(?)/Santonian age in North Carolina and beds of late Cenomanian and/or early Turonian age in the subsurface of South Carolina and Georgia. In this report, unit F (restricted sense; Albian-early Cenomanian) is considered to occur as a subsurface unit in eastern North Carolina (beds containing *Fossocytheridea lenoirensis* Swain and Brown and associated ostracodes that are below a marine unit containing late Cenomanian-Turonian fossils) and as the thin Cretaceous unit described above from the subsurface of southern South Carolina and eastern Georgia. Unit F (restricted) in North Carolina is broadly correlative with the Big Cypress and Naples Bay Groups of Florida.

In North Carolina, units G and F (restricted) consist of interbedded sands and clays and minor limestones (Spangler, 1950; Swain, 1952; Brown and others, 1972). Owens and Gohn

(1985) interpreted coarsening-upward clay-sand cycles in units G and F (restricted) in North Carolina to represent major deltaic (delta-plain and delta-front) sequences and intervening prodelta-shelf sequences.

Upper Cretaceous Series. The Upper Cretaceous depositional sequences of Owens and Gohn (1985; their sequences 2 through 6) and chronostratigraphic units A through E of Brown and others (1972, 1979; Swain and Brown, 1972) provide a basis for summarizing Late Cretaceous deposition in the southern Atlantic Coastal Plain. The stratigraphic positions of units A through E, as originally related to the Gulf Coast provincial stages by Brown and others (1972), are shown in Figure 6. My interpretation of these units and the chronostratigraphic units of Gohn and others (1978, 1980) also are shown in Figure 6 (also see Valentine, 1982, 1984).

Cenomanian-Turonian sediments: Cenomanian and Turonian sediments have a relatively limited distribution and, with the probable exception of certain beds in northern North Carolina, do not crop out in the southern Atlantic Coastal Plain. Sediments of this age are present, however, in the subsurface of eastern North Carolina, southern South Carolina, Georgia, and Florida (Fig. 5b). Stratigraphic units of Cenomanian and/or Turonian age include (Fig. 6) the fossiliferous part of the lower member of the Atkinson Formation (Applin and Applin, 1965, 1967; equals middle member of the Atkinson of Applin and Applin, 1947), subsurface unit K_2 of Gohn and others (1978, 1980), the fossiliferous part of the subsurface Cape Fear Formation of Gohn and others (1977; Hazel and others, 1977; this South Carolina unit is of Cenomanian-Turonian age and is incorrectly correlated with the younger outcropping Cape Fear Formation), and the subsurface beds of Eaglefordian age of Swain (1952). Collectively, these units form depositional sequence 2 of Owens and Gohn (1985) in the southern Atlantic Coastal Plain. Sequence 2 is of late Cenomanian to early Turonian age (middle Eaglefordian) as indicated by the planktic faunas in several drill holes as well as the ostracode *Rehacythereis eaglefordensis* (Alexander) and pollen species of the *Complexiopollis-Atlantopollis* Zone (Swain, 1952; Applin, 1955; Swain and Brown, 1964; Hazel, 1969; Brown and others, 1972; Hazel and others, 1977; Hattner and Wise, 1980; Christopher, 1982a,b; Valentine, 1982, 1984).

Sediments of middle Cenomanian (Woodbinian) and late Turonian (late Eaglefordian) ages likely are absent in the southern Atlantic Coastal Plain (Valentine, 1984). Brown and others (1972, 1979) assigned a middle Cenomanian (Woodbinian) age to their subsurface unit E (Fig. 6); however, the widespread occurrence of *Rehacythereis eaglefordensis* in unit E (Brown and others, 1972; Swain and Brown, 1964) suggests a late Cenomanian (middle Eaglefordian) age for at least part of that unit (Hazel, 1969). Primarily on the basis of calcareous nannofossils, Valentine (1984) assigned an early Turonian age to unit E and locally to part of unit F in several South Carolina and Georgia wells. Owens and Gohn (1985) included unit E and part of unit F (discussed above) in their sequence 2.

Several subsurface units of proposed late Cenomanian-Turonian (Eaglefordian) age are interpreted herein to be Coniacian(?)-Santonian in age. Applin and Applin (1964, 1967) assigned an Eaglefordian age to their upper member of the Atkinson Formation in Georgia. However, this unit contains the mollusk *Ostrea cretacea* Morton, the ostracode *Veenia quadrialira* (Swain), and other fossils that indicate a Coniacian(?)-Santonian (or slightly younger) age in several wells (references in Owens and Gohn, 1985). Unit D of Brown and others (1972, 1979) was assigned a Cenomanian to Turonian (middle to late Eaglefordian) age by those authors; however, the position of unit D above the Cenomanian-Turonian unit E and the presence of a Coniacian-Santonian nannoflora in unit D in one well led Valentine (1982, 1984) to assign a Coniacian-Santonian age to unit D. Unit K_3 of Gohn and others (1978, 1980) is broadly correlative with unit D and is Coniacian(?)-Santonian in age (Valentine, 1982, 1984) rather than Cenomanian as originally assigned.

The Cenomanian-Turonian section consists of a variety of marine and marginal-marine sediments (Fig. 5b). Little-studied marginal-marine beds that crop out near the Fall Line in northern North Carolina (Owens and Gohn, 1985) probably represent delta-plain environments located updip from fine-grained pro-delta-shelf deposits in the subsurface of the Albemarle embayment (Swain, 1952; Brown and others, 1972). Marine and marginal-marine sections consist of terrigenous clastic sediments in coastal South Carolina, southern Georgia, and northern Florida and carbonate sediments in peninsular Florida; major deltaic sections apparently are absent south of the Cape Fear arch.

Coniacian(?)-Santonian sediments: Sediments of Coniacian(?)-Santonian (early to middle Austinian) age occur in virtually all areas of the southern Atlantic Coastal Plain (Fig. 5c). Owens and Gohn (1985) assigned these beds to their depositional sequence 3, which correlates with units D (as discussed above) and C (part) of Brown and others (1972, 1979). Owens and Gohn (1985) recognized two distinct subsequences of their sequence 3 that individually include a major deltaic system centered in South Carolina, laterally equivalent nondeltaic nearshore deposits in the northern Georgia Coastal Plain, and shelf deposits in southern Georgia (terrigenous) and Florida (carbonate).

Litho- and chronostratigraphic units assigned to the lower subsequence include (Fig. 6) the outcropping Cape Fear Formation (Swift and Heron, 1969; Christopher and others, 1979) of the Cape Fear arch (equivalent to the lower part of the outcropping unit F of Brown and others, 1972, in that area) and its subsurface equivalents (see discussion above)—unit K_3, unit D, the upper member of the Atkinson Formation, and updip subsurface beds of unit UK_1 of Prowell and others (1985). Units assigned to the upper subsequence include (Fig. 6) the outcropping Middendorf Formation of the Cape Fear arch (Swift and Heron, 1969; Woollen and Colquhoun, 1977a) and its subsurface equivalents—unit K_4 of Gohn and others (1978, 1980), unit C (part) of Brown and others (1972, 1979), the subsurface beds of Austinian age (part) of Applin and Applin (1964, 1967) and Maher and Applin (1971), and updip subsurface beds of unit UK_2 of Prowell and others (1985) in Georgia and western South Carolina. The

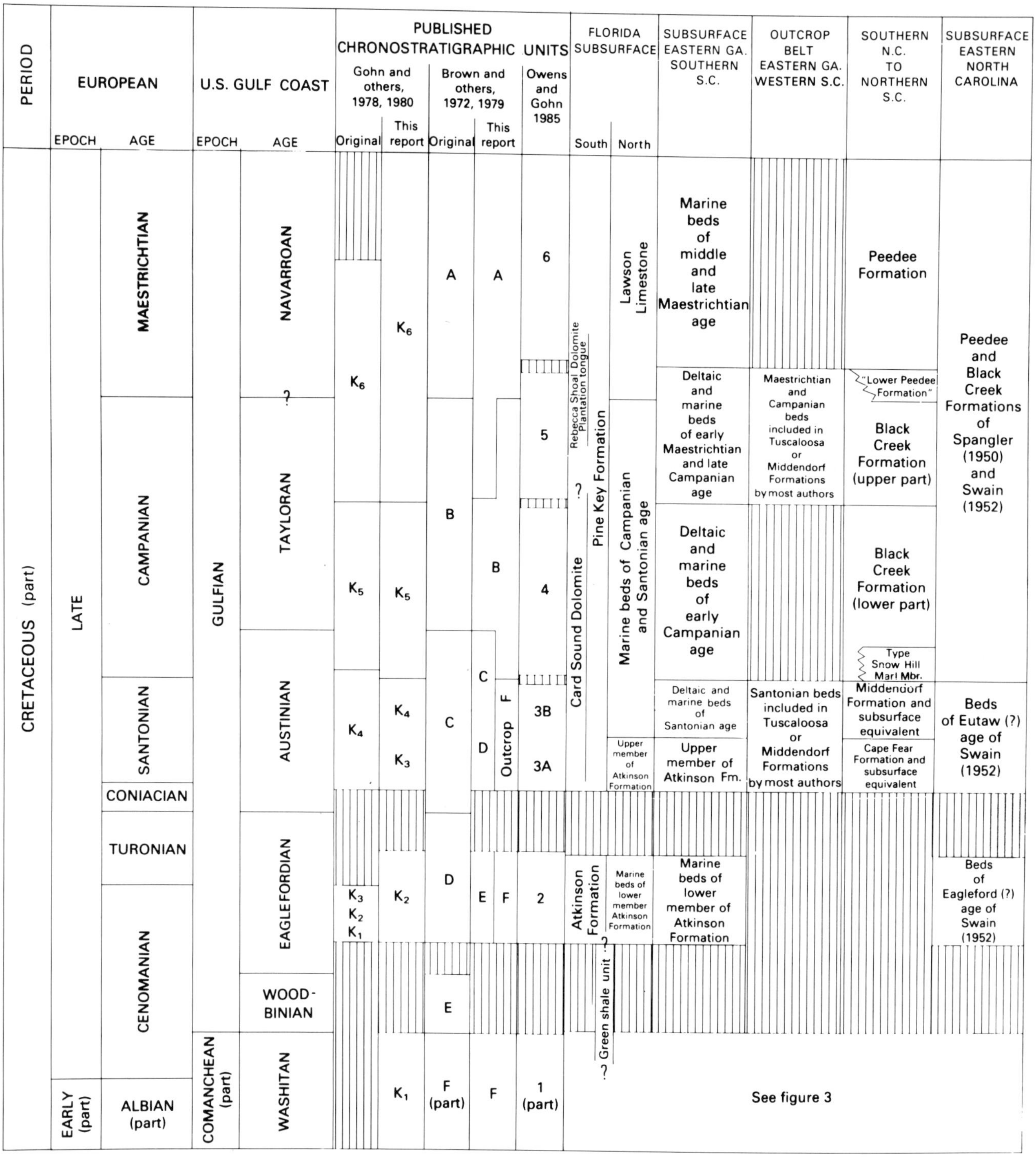

Figure 6. Correlation chart for the Upper Cretaceous units of the southern Atlantic Coastal Plain. Data sources for each unit are given in the text.

subsurface beds of Eutaw(?) age of Swain (1952) in coastal North Carolina probably represent both the lower and upper subsequences, as do Coniacian(?)-Santonian parts of the Pine Key Formation and Card Sound Dolomite in Florida (Winston, 1971a; Meyerhoff and Hatten, 1974).

Calcareous fossils reported from these units indicate a Coniacian(?)-Santonian age (see references for individual units), and nonmarine facies of both subsequences contain pollen species indicative of the *Complexiopollis exigua-Santalacites minor, Pseudoplicapollis longiannulata-Plicapollis incisa,* and *?Pseudoplicapollis cuneata-Semioculopollis verrucosa* zones of Coniacian(?)-Santonian age (collectively pollen zone V of earlier workers; Christopher, 1977, 1979, 1982a; Christopher and others, 1979).

Both subsequences are characterized by large and probably composite delta systems. In the lower subsequence (Owens and Gohn, 1985), fluvial to upper-delta-plain environments are represented from southern North Carolina to Georgia (Fig. 5c). Lower-delta-plain and delta-front deposits occur in southern Georgia and eastern North Carolina and prodelta-shelf deposits are present in southern Georgia and northern Florida. Much of the subtidal part of this large delta system is probably located offshore in the Southeast Georgia-Blake Plateau basin. A similar pattern of paleoenvironments is shown by the upper subsequence, except that the deltaic units are smaller and carbonate sedimentation extended farther north. Away from the Coniacian(?)-Santonian deltas, relatively deep-water chalks of the Pine Key Formation accumulated in peninsular Florida between a reef-bank system, represented onshore by the Card Sound Dolomite in the Florida Keys (Winston, 1971a; Meyerhoff and Hatten, 1974), and the shallow-shelf sediments to the north (Fig. 5c).

Lower Campanian sediments: The Coniacian(?)-Santonian deposits are separated from overlying Campanian beds by a regionally extensive transgressive unconformity. Although the stratigraphic position of this unconformity is not precisely known, it must occur very close to the Santonian-Campanian boundary (Owens and Gohn, 1985). Above the unconformity, lower Campanian (upper Austinian-lower Tayloran) sections in the Carolinas and Gerogia consist of prodelta-shelf sections and overlying regressive deltaic sands or delta-margin barrier complexes (Fig. 5d), which are truncated above by a middle Campanian transgressive unconformity. Owens and Gohn (1985) assigned this lower Campanian cycle to their sequence 4.

Lower Campanian beds crop out in central and southern North Carolina and northern South Carolina (Fig. 7), where they are assigned to the lower part of the Black Creek Formation. The detailed stratigraphic relationships and depositional history of the Black Creek Formation (delta plain-delta front facies) and associated Peedee Formation (prodelta-shelf facies) of the Cape Fear arch area are controversial (Brown, 1957; Swift, 1966; Benson, 1969; Swift and Heron, 1969; Woollen and Colquhoun, 1977b; Sohl and Christopher, 1983). This controversy can be resolved by recognizing that these units represent deposition in deltaic systems containing considerable lateral variation and vertical repetition of lithofacies and by acquiring sufficient biostratigraphic information.

On the Cape Fear arch, delta-plain and delta-front deposits of the lower part of the Black Creek Formation contain pollen indicative of an early Campanian age (Christopher and others, 1979, amended by R. A. Christopher, oral communication, 1982). At central North Carolina outcrops (Fig. 7), the Snow Hill Marl Member of the Black Creek represents delta-front or open-bay to prodelta environments on the flank of the lower Black Creek delta (Fig. 5d); the ostracode fauna from the Snow Hill (Brown, 1957) contains early Campanian species (Hazel and Brouwers, 1982). Fine-grained, lower Campanian marine beds in the subsurface of eastern North Carolina (Swain, 1952; Brown and others, 1972) represent prodelta-shelf deposition seaward of the lower Black Creek delta.

The lower part of the Black Creek Formation can also be traced into the subsurface in South Carolina, where it is mapped as units B (part) and C (part) of Brown and others (1979) and as unit K_5 of Gohn and others (1978, 1980). These units contain early Campanian pollen, ostracodes, calcareous nannofossils, and planktic foraminifers (Hazel and others, 1977; Christopher, 1978; Gohn and others, 1978, 1982; Hattner and Wise, 1980; Valentine, 1982). In this area, the lower Campanian section consists of coarsening-upward, delta-front cycles that overlie a basal prodelta-shelf sequence (Fig. 5d). From central South Carolina westward into Georgia, this delta-front section changes into a barrier-strandplain system overlying marginal prodelta and shelf deposits (Reinhardt, 1982; Owens and Gohn, 1985). These non-deltaic shoreline sequences may have received their sediments through strike-parallel, littoral movement of sand from the lower Black Creek delta. Terrigenous prodelta-shelf deposition occurred seaward of the delta front and barrier system in southern Georgia (Applin and Applin, 1967), and deposition of carbonate sediments in the Pine Key-Card Sound system continued in Florida (Winston, 1971a).

Upper Campanian-lower Maestrichtian sediments: Upper Campanian and lower Maestrichtian (upper Tayloran-lower Navarroan) sediments constitute depositional sequence 5 of Owens and Gohn (1985). This sequence is bounded below by a middle Campanian transgressive unconformity and above by a similar unconformity that occurs at or just below the lower Maestrichtian-middle Maestrichtian boundary. A third localized transgressive unconformity occurs at the Campanian-Maestrichtian boundary in North Carolina. The facies pattern for sequence 5 is similar to that of the lower Campanian section; a major deltaic section is centered in South Carolina, and prodelta-shelf deposits are present in eastern North Carolina, southern Georgia, and Florida (Fig. 5e).

Upper Campanian-lower Maestrichtian sediments crop out along their inland margin from North Carolina to eastern Georgia. On the Cape Fear arch (Fig. 7), lower Maestrichtian beds contain diagnostic pollen floras and mollusks (Sohl and Christopher, 1983). In North Carolina, sections of this age consist of homogeneous prodelta-shelf sediments and are assigned to the

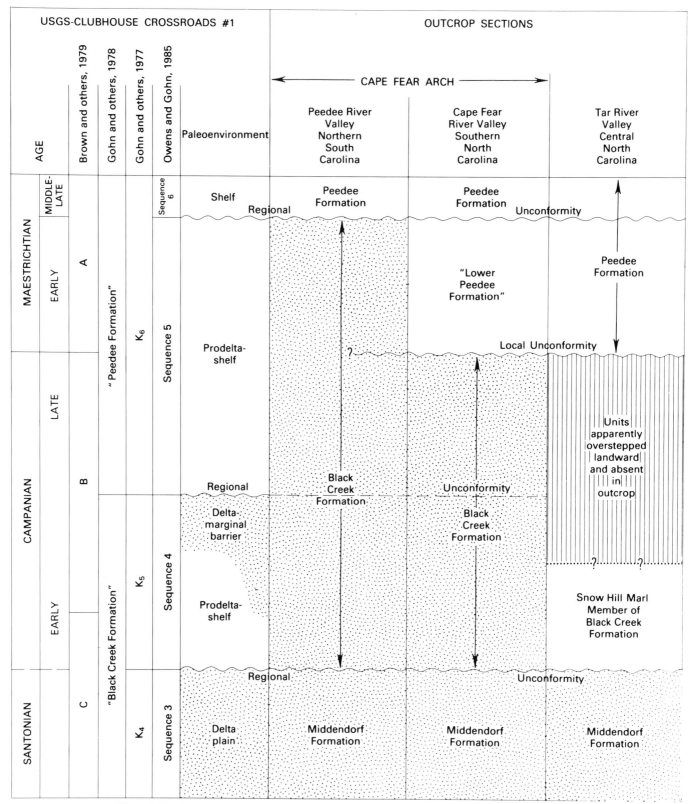

Figure 7. Stratigraphic and sedimentologic relationships of the outcropping Campanian and Maestrichtian sediments of the Cape Fear arch area. Stratigraphic units used for the Campanian-Maestrichtian sediments in the USGS-Clubhouse Crossroads #1 core (near Charleston, South Carolina) by several authors are also shown. Stippled units consist of delta-front and delta-plain sediments. Unpatterned units consist of prodelta-shelf sediments. Adapted from Figure 2-20 of Owens and Gohn (1985).

Peedee Formation (Brown, 1957; Sohl and Christopher, 1983). However, in northern South Carolina (Fig. 7), the outcropping lower Maestrichtian section consists of delta-plain and delta-front sediments and is assigned to the Black Creek Formation (Van Nieuwenhuise and Kanes, 1976; Woollen and Colquhoun, 1977b). The Campanian part of sequence 5 throughout the Cape Fear arch area also consists of delta-front and delta-plain sediments that are assigned to the Black Creek Formation and contain Tayloran (Campanian) mollusks and a Campanian pollen assemblage (Sohl and Christopher, 1983). In the Cretaceous outcrop belt of western South Carolina and eastern Georgia, units UK_4 and UK_5 of Prowell and others (1985) are of middle Campanian to early Maestrichtian age. These units consist of clayey gravels and coarse sands and probably represent deposition in delta-plain environments (Fig. 5e).

Subsurface upper Campanian and lower Maestrichtian deposits are widespread in the Carolinas, Georgia, and Florida (Figs. 2, 6). In South Carolina and eastern Georgia, the lower part of unit K_6 of Gohn and others (1978, 1980), units B (part) and A (part) of Brown and others (1979), and the beds of Tayloran age (part) of Applin and Applin (1967) contain late Campanian and early Maestrichtian mollusks, nannofossils, and ostracodes (Hazel and others, 1977; Hattner and Wise, 1980; Valentine, 1982). Units B and A also occur in the subsurface in eastern North Carolina (Brown and others, 1972). These units consist of progradational delta-front sediments in most of South Carolina and southern North Carolina and prodelta-shelf sediments in eastern North Carolina (data in Brown and others, 1972; Swain, 1952; Owens and Gohn, 1985) and Georgia (Fig. 5e). In Florida, the facies pattern established in the Santonian-Coniacian(?) continued with the deposition of chalky limestones (Pine Key Formation) and reef-bank carbonate sediments (Card Sound Dolomite or Rebecca Shoal Dolomite) (Winston, 1971a, 1978; Meyerhoff and Hatten, 1974).

Middle and upper Maestrichtian sediments: Fossiliferous marine beds of middle and late Maestrichtian age (middle to late Navarroan) are widespread in the subsurface of the Carolinas, eastern Georgia, and the Florida peninsula (Fig. 5f). They crop out, however, only in the river sections of the Cape Fear arch where they constitute the Peedee Formation (Fig. 7; Brown, 1957; Swift and Heron, 1969; Woollen and Colquhoun, 1977b). In the subsurface of North Carolina, South Carolina, and eastern Georgia, they constitute depositional sequence 6 of Owens and Gohn (1985), as well as the upper part of unit K_6 of Gohn and others (1978, 1980) and the upper part of unit A of Brown and others (1972, 1979). In Florida, most or all of the Lawson Limestone of Applin and Applin (1964, 1967) and Maher and Applin (1971) and the equivalent uppermost part of the Pine Key Formation, and perhaps part of the Plantation Tongue of the Rebecca Shoal Dolomite, represent this interval (Winston, 1971a, 1978; Meyerhoff and Hatten, 1974).

A widespread transgressive unconformity that occurs at or just below the lower-middle Maestrichtian boundary (first appearance datum of *Globotruncana gansseri* Bolli) throughout the Atlantic Coastal Plain was used by Owens and Gohn (1985) as the lower boundary of their sequence 6. This contact is exposed at the lectostratotype (Burches Ferry, Peedee River) of the Peedee Formation in South Carolina (Van Nieuwenhuise and Kanes, 1976; Hattner and Wise, 1980) and at Black Rock Landing on the Cape Fear River in southern North Carolina (N. F. Sohl, oral communication, 1982). Above this contact, planktic faunas and floras in outcrop and subsurface Maestrichtian sections contain *Globotruncana gansseri* Bolli, *Globotruncana aegyptiaca* Nakkady, *Trinitella scotti* Bronnimann, *Racemiguembelina fructicosa* (Egger), and *Micula murus* (Martini) (Hazel and others, 1977; Hattner and Wise, 1980; Valentine, 1984; C. C. Smith cited in Owens and Gohn, 1985).

Unlike the Santonian and Campanian sections, the middle-to-upper Maestrichtian unit does not contain any discernible deltaic sections in the Carolinas. Except along the coast in the Cape Fear arch area, the unit consists almost entirely of bioturbated marine beds. In southern North Carolina, the Rocky Point Member of the Peedee Formation (Scotts Hill Member of Ward and Blackwelder, 1979) consists of quartz sands and quartzose calcarenites that represent a shallow marine to nearshore cape-shoal complex at the top of the unit (Wheeler and Curan, 1974; Harris, 1978). In contrast, however, Prowell and others (1985) recognized a subsurface deltaic unit (UK_6) of this age in updip areas of Georgia that may be more closely associated with sediment sources in the Gulf Coastal Plain than in the Atlantic Coastal Plain. In Florida, the equivalent interval consists of peritidal to subtidal carbonate sediments of the Lawson Limestone in the north and subtidal chalks of the Pine Key Formation in the peninsula (Applin and Applin, 1967; Winston, 1971a, 1978; Meyerhoff and Hatten, 1974).

In the area between the carbonate and terrigenous provinces along the Georgia-Florida boundary, the middle-to-late Maestrichtian section is thin or locally absent (Applin and Applin, 1964, 1967). This unusual pattern, which continued into the Tertiary, has been referred to as the Suwanee saddle, strait, or channel. Its origin is most frequently ascribed to slow deposition or nondeposition/erosion in a bathymetric depression that resulted from its location far from the continental sediment source (starved basin; carbonate-suppression model) and/or from the action of oceanic currents (Applin and Applin, 1967; Chen, 1965; Pinet and Popenoe, 1985; McKinney, 1984). However, similar differences in the thickness of Tertiary units in Georgia have been ascribed to differential erosion of units on various fault-bounded blocks (Cramer and Arden, 1980; Gelbaum and Howell, 1982).

Tertiary (Paleogene) System. Paleocene Series. The Paleocene Series occurs throughout the southern Atlantic Coastal Plain except for the Cape Fear arch and updip areas of the Carolinas (Fig. 8a). Paleocene sediments crop out in discontinuous belts along their inner margin in North Carolina and Georgia and in a broader belt in east-central South Carolina. The general pattern of sedimentary facies is similar to that of the Upper Cretaceous section in which carbonate sediments are restricted to

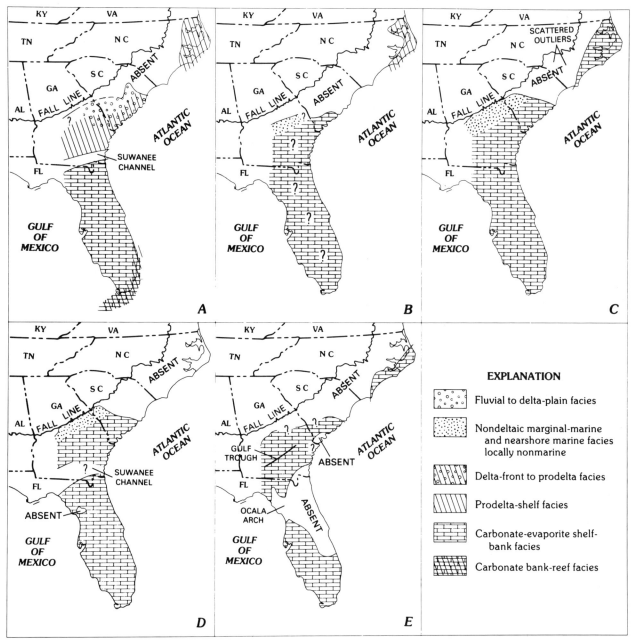

Figure 8. Distribution of sediments in the southern Atlantic Coastal Plain. A) Paleocene sediments, B) lower Eocene sediments, C) middle Eocene sediments, D) upper Eocene sediments, and E) Oligocene sediments.

peninsular Florida and southern Georgia and terrigenous clastic deposits dominate elsewhere.

In eastern North Carolina, the outcropping Beaufort Formation (Fig. 9) consists of a Danian Jericho Run Member (zone P1; early Midwayan) and a Selandian unnamed member (lower zone P4; late Midwayan) (Brown, 1958; Brown and others, 1972, 1982; Harris and Baum, 1977). (Cenozoic alphanumeric fossil zones used in this report follow or approximate the zonations of Blow [1979; P zones] and Martini [1971; NP zones].) The sub-

surface Midwayan unit of Brown and others (1972; also see Brown, 1958; Swain, 1951) in coastal areas of North Carolina is approximately equivalent to the Beaufort Formation. In the same area, the subsurface Sabinian unit of Brown and others (1972) may include upper Paleocene as well as lower Eocene sediments.

The Beaufort Formation and its subsurface equivalent consist of fossiliferous cristobalitic mudstones and fossiliferous, glauconitic clayey sands that likely represent prodelta and shelf environments. Because marine sediments are present at the landward edge of the Paleocene outcrop belt, and because associated nearshore sediments are apparently absent, it is assumed that the Paleocene section originally extended farther to the west than its present distribution. The source of these clastic sediments therefore may have been located in the North Carolina Piedmont, or it may be represented by the major delta located south of the arch (Fig. 8a).

In east-central South Carolina, outcropping and subsurface Paleocene sediments are assigned to the Rhems Formation and overlying Williamsburg Formation of the Black Mingo Group (Van Nieuwenhuise and Colquhoun, 1982). In the continuously cored section in USGS-Clubhouse Crossroads #1 near Charleston, South Carolina, the Rhems Formation is assigned a Danian to early Selandian age (zones NP2-4, P1-3; early and middle Midwayan), and the Williamsburg Formation is assigned a late Selandian age (NP5 or 6 to NP9, P4-5; late Midwayan to early Sabinian) (Hazel and others, 1977, 1984b). Van Nieuwenhuise and Colquhoun (1982) assigned similar ages to these formations throughout South Carolina.

In downdip areas of South Carolina, the lower members of the Rhems and the Williamsburg Formations consist of fine-grained, fossiliferous, prodelta-shelf sediments (data in Gohn and others, 1977). The upper members of both formations consist of well-sorted sand bodies, shell beds, and sequences of thinly inter-bedded sands and dark clays that represent nearshore and marginal-marine deposition. The presence of nearshore sands gradationally above prodelta-shelf sections in both formations, and the presence of progradational sequences within both near-shore sand sections, suggests that the Black Mingo Group consists of two deltaic sequences. Alternatively, Colquhoun and others (1983) interpreted these downdip units to represent shallow to deep shelf environments.

In updip areas of northwestern South Carolina and eastern Georgia, outcropping Paleocene sections consist of sparingly fos-siliferous sands, kaolinite beds, and lignitic clays assigned to the lower part of the Huber Formation (Buie, 1978, 1980; Nystrom and Willoughby, 1982), which contains Tertiary beds of several ages but definitely includes Paleocene sediments (Tschudy and Patterson, 1975). Updip subsurface Paleocene beds consist of a variety of sparingly fossiliferous and highly lignitic lithologies similar to the Huber Formation (lithostratigraphic units P_1 and P_2 of Prowell and others, 1985). Parts of these updip subsurface beds have been assigned previously to the Ellenton Formation (Siple, 1967; Prowell and others, 1986) and the "unnamed unit of Paleocene age" of Siple (1975). Colquhoun and others (1983) assigned these beds to the Lang Syne and Sawdust Landing Members of the Rhems Formation and the Chicora Member of the Williamsburg Formation. The Huber (Lang Syne-Sawdust Landing-updip Chicora) section consists of the fluvial and delta-plain parts of the two Black Mingo deltas (Fig. 8a).

In downdip areas of eastern Georgia, dominantly marine sections of Paleocene sediments are typically assigned to the un-divided Midway and Wilcox Groups or their constituent forma-tions (Applin and Applin, 1944; Herrick, 1961; Cramer and

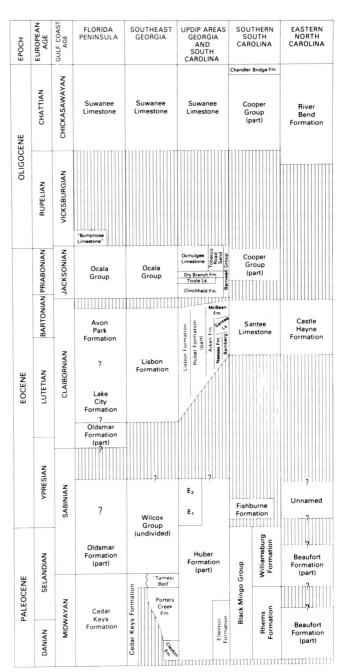

Figure 9. Correlation chart for the Paleogene (lower Tertiary) units of the southern Atlantic Coastal Plain. Data sources for each unit are given in the text.

Arden, 1980). Fossil data for these sections are relatively sparse except for one interval that contains the "Tamesi fauna" of Applin and Applin (1944, 1964; Applin and Jordan, 1945). The "Tamesi fauna" occurs in glauconitic marine beds above an un-conformity and consists of planktic foraminifers that indicate as-signment to the lower part of zone P4 of Selandian (late Midwayan) age. Except near the Florida border, older Paleocene beds occur between beds containing the "Tamesi fauna" and Cretaceous beds. This intervening unit is typically referred to the Porters Creek Formation and/or the Clayton Formation of the Midway Group and is assigned a Danian age, as indicated by the presence of *Globoconusa daubjergensis* (Bronnimann) and Danian ostracodes (Ogren, 1970, as reported in Cramer and Arden, 1980). In wells near the Florida-Georgia border, beds containing the "Tamesi fauna" lie directly on Cretaceous beds (Applin and Applin, 1944, 1964). Chen (1965) recognized this area of missing lower Paleocene sediments and decided that it marked the location of the early Paleocene Suwanee channel. Chen (1965) also recognized a Sabinian Suwanee channel in about the same location (probably Paleocene in part on the basis of the age of the "Tamesi fauna"), although thinning of the Sabinian section is not obvious on his cross sections or those of Applin and Applin (1944) or Cramer and Arden (1980).

In contrast to the terrigenous units of the Carolinas and Georgia, the lower Tertiary section of the Florida peninsula (Fig. 2) consists almost entirely of carbonate and evaporite sediments. The stratigraphic units originally defined for this section by Applin and Applin (1944), Cole (1944), and others were primarily biostratigraphic. Although Winston (1977) subsequently provided cotype sections and lithologic criteria (also see Chen, 1965) for recognition of these formations, their identification in most areas remains difficult because of vertical and lateral repetitions of similar lithologies and a strong overprint of diagenetic facies (for example, see Southeastern Geological Society, 1976). The traditional formational units of the lower Tertiary of Florida are used in the following discussion despite the difficulties in recognizing them.

The oldest Paleocene unit in peninsular Florida is the Cedar Keys Formation (Fig. 9), which is typically assigned a Midwayan age although the available fossil information is relatively sparse. The unit is characterized by the foraminifers *Borelis gunteri* Cole and *B. floridanus* Cole, whose presence was originally used to define the top of the formation (Cole, 1944). Applin and Applin (1944) and Chen (1965) correlated the Cedar Keys with terrigenous Midwayan sediments to the north, including the beds containing the "Tamesi fauna"; hence the Cedar Keys Formation is likely of Danian and early Selandian age.

Above the Cedar Keys Formation, the Oldsmar Formation is generally considered to be of early Eocene age (Applin and Applin, 1944; Chen, 1965; Maher and Applin, 1971). However, examination of the available data suggests that some, and perhaps most, of the Oldsmar is late Paleocene in age. Applin and Applin (1944) recognized four faunal zones (benthic foraminifers) in the Oldsmar. The second youngest of these zones (zone II), the "Salt Mountain faunal unit," contains *Discocyclina blanpiedi* Vaughan, a characteristic foraminifer of the Salt Mountain Limestone of Alabama. Because the Salt Mountain Limestone is of Selandian age (NP7; Siesser and others, 1985), it is possible that much of the Oldsmar is also of late Paleocene age.

The Cedar Keys Formation consists primarily of dolomite, evaporitic carbonate rocks, and evaporite beds that locally consti-

tute more than 30 percent of the section. The Oldsmar Formation consists of dolomite, limestone, and sparse evaporite beds. One additional facies, the Paleocene Tavernier Tongue of the Rebecca Shoal Dolomite, occurs in the Florida Keys (Winston, 1978). Chen's (1965) lithofacies map for the Paleocene (Midwayan) suggests that this unit may extend some distance along Florida's Atlantic coast (Fig. 8a). The Rebecca Shoal Dolomite likely represents a shelf-edge reef complex along a bathymetric break in slope (paleo-Florida straits?) that grades northward and westward into the evaporitic carbonate bank represented by the Cedar Keys and Oldsmar Formations.

Eocene Series. *Lower Eocene sediments:* Lower Eocene sediments probably do not crop out in the southern Atlantic Coastal Plain, and their distribution in the subsurface is uncertain in several areas (Fig. 8b). Sediments of this age are absent on the Cape Fear arch and probably are absent in most updip areas of the Carolinas. Swain (1951) identified a subsurface lower Eocene section in several wells in eastern North Carolina. However, his tentative correlation of this unit with the Paleocene Black Mingo and Aquia (Virginia-Maryland) Formations suggests the possibility of a Paleocene rather than Eocene age for at least part of this unit. Similarly, the subsurface Sabinian unit of Brown and others (1972) in North Carolina contains the foraminifers *Subbotina inaequispira* (Subbotina), *Morozovella wilcoxensis* (Cushman and Ponton), and *Pseudohastigerina wilcoxensis* (Cushman and Ponton) and may include upper Paleocene and middle Eocene as well as lower Eocene sediments. The distribution of the Sabinian unit of Brown and others (1972) in North Carolina is shown in Figure 8b. In southern South Carolina, the subsurface lower Eocene Fishburne Formation (Gohn and others, 1983; Hazel and others, 1984b) is assigned to zone NP10-11 and to the *Morozovella subbotinae* Zone of Stainforth and others (1975) (approximately P6b).

Lower Eocene sediments do not occur in the Eocene outcrop belt of western South Carolina and eastern Georgia (Huddlestun and others, 1974) unless they exist as an unrecognized part of the Huber Formation (Buie, 1978; Tschudy and Patterson, 1975; Nystrom and Willoughby, 1982). However, immediately downdip of the outcrop belt, subsurface units E_1 and E_2 of Prowell and others (1985) are early Eocene in age. Further downdip, the Wilcox Group (undivided) and the Oldsmar Formation are traditionally assigned early Eocene ages, although parts of these units are almost certainly older or younger.

The known or suspected lower Eocene units in the southern Atlantic Coastal Plain consist of carbonate (South Carolina-eastern Georgia-Florida) and terrigenous (North Carolina) shelf deposits and minor nearshore terrigenous deposits (updip Georgia). The absence of nearshore deposits in the Carolinas and the limited geographic distribution of these units suggest that lower Eocene deposits were significantly eroded before the deposition of overlapping middle Eocene deposits.

Middle Eocene sediments: Middle Eocene units occur throughout the southern Atlantic Coastal Plain except in the area of the Cape Fear arch and in updip areas of North Carolina

(Fig. 8c). these units crop out along the entire length of their inland margin, where they have been studied extensively.

In North Carolina, the outcropping Castle Hayne Formation (Fig. 9) has been assigned to zones NP16-17 and P12-13 (Berggren and Aubry, 1984; Jones, 1982; Hazel and others, 1984a) and apparently is of middle Eocene age, although a late Eocene age for at least part of the unit has been suggested (Harris and Zullo, 1980, 1982; Worsley and Turco, 1979; Zullo, 1979). In South Carolina, the Santee Limestone (Moultrie and overlying Cross Members) as defined by Ward and others (1979) is of middle Eocene age (approximately zones NP16-17 and P11-14; Hazel and others, 1984b), whereas the same interval as described by Baum and others (1980) and Powell and Baum (1982) consists of the middle Eocene Santee Limestone (restricted) and overlying upper Eocene Cross Formation. In this report, the practice of assigning all parts of the Castle Hayne and Santee (including Cross) limestones to the middle Eocene is followed (Fig. 9).

In updip areas of South Carolina and eastern Georgia (outcrop and shallow subsurface), terrigenous clastic units of middle Eocene age also have a long and complex history of study (Cooke and MacNeil, 1952; Huddlestun, 1982). Most recent reports on these units depict a fluvial to marginal-marine sand-clay facies along the Fall Line that is typically assigned to the upper part of the Huber Formation (Buie, 1978, 1980; Tschudy and Patterson, 1975; Colquhoun and others, 1983). Seaward of this unit, shallow- to deep-water, siliceous to calcareous shelf deposits have been assigned to the "Aiken," "Neeses" (may equal the Chapel Branch Member of the Santee of Powell, 1984), "Bamberg" (may equal the Caw Caw Member of the Santee of Powell and Baum, 1982), and McBean Formations in South Carolina (Colquhoun and others, 1983) and the Lisbon Formation in Georgia (Huddlestun, 1982).

Middle Eocene limestones (Santee and Castle Hayne formations) are present downdip in eastern North Carolina (Brown and others, 1972), southern South Carolina, and eastern Georgia (Cramer and Arden, 1980; Gohn and others, 1982). In Florida, the middle Eocene section traditionally is assigned to the Lake City Limestone and the overlying Avon Park Limestone (Chen, 1965; Winston, 1977; Cramer and Arden, 1980). Huddlestun and others (1974) assigned these units to zones P11 (Lake City) and P13 (Avon Park).

As discussed above, the Oldsmar Formation is traditionally assigned an early Eocene age. However, the upper part of the Oldsmar assigned to Zone I (*Helicostegina gyralis* zone) of Applin and Applin (1944) is probably middle Eocene (Lutetian?) in age, as the zonal marker *Helicostegina gyralis* Barker and Grimsdale is restricted to the middle Eocene in Caribbean sections (Cole and Applin, 1964).

The depositional pattern for the middle Eocene sediments is reminiscent of the pattern described for the Lower Cretaceous sections of Florida. During the middle Eocene, relatively small volumes of terrigenous clastic sediments were deposited in fluvial and marginal-marine environments near the modern Fall Line, whereas shallow-marine carbonate sediments were deposited

seaward of this nearshore terrigenous province (Banks, 1977; Powell, 1984). The truncation of the middle Eocene sediments at a high angle to their depositional strike in South Carolina (Fig. 8c), and their absence throughout much of the Carolinas, suggests that post-middle Eocene erosion has removed much of the middle Eocene section. In the Georgia Coastal Plain, Chen (1965) mapped a middle Eocene Suwanee channel similar to the Maestrichtian and Paleocene channels in that area. However, no evidence for a thin, "starved-basin" section of the type representative of the older channels is apparent on his cross sections or those of Cramer and Arden (1980).

Upper Eocene sediments: Upper Eocene deposits occur in outcrop and in the subsurface throughout eastern Georgia and southern South Carolina, where they extend landward to the Fall Line (Fig. 8d). As noted above, Eocene units are absent on the Cape Fear arch. Curiously, upper Eocene sediments apparently are also missing in the Eocene outcrop belt (Ward and others, 1978) and the subsurface (Brown and others, 1972; Swain, 1951) of eastern North Carolina. However, Baum and others (1978, 1980; Baum, 1981; Powell and Baum, 1982) have assigned a late Eocene age to their New Bern Formation, which crops out in North Carolina; this unit includes beds that Ward and Blackwelder (1980; Ward and others, 1978) assigned to their Spring Garden Member of the middle Eocene Castle Hayne Formation or their Oligocene River Bend Formation. Hence, the recognition of upper Eocene sediments in North Carolina outcrops remains equivocal, and their presence in the subsurface of that state remains undocumented.

The depositional pattern of upper Eocene sediments in South Carolina and eastern Georgia is similar to that of the middle Eocene section (Fig. 8c-d). A relatively narrow band of terrigenous sediments occurs along the modern Fall Line and grades downdip into carbonate sediments. Facies relationships in the updip sections and in the transitional interval between the terrigenous and carbonate provinces are complex, as is the stratigraphic nomenclature that has been applied to these units (Huddlestun, 1982). In this report, Huddlestun and Hetrick's (1978, 1979) concept of an upper Eocene Barnwell Group (Fig. 9) is followed. The Barnwell Group consists of nearshore calcareous sand units (Riggins Mill Member and undivided sections of the Clinchfield Formation; Irwinton Sand and Griffins Landing Sand Members of the Dry Branch Formation; Tobacco Road Sand) that grade and interfinger downdip directly into carbonate shelf deposits (Treadwell and Utley Limestone Members of the Clinchfield Formation; Sandersville Limestone Member of the Tobacco Road Sand; Tivola Limestone; Ocmulgee Limestone) or through fine-grained siliciclastic marine units (Albion Member of the Clinchfield Formation; Twiggs Clay Member of the Dry Branch Formation) into the carbonate sections.

The widespread upper Eocene Ocala Limestone (Group) consists primarily of macrofossiliferous limestone that accumulated in shallow shelf environments in Georgia and South Carolina and a shallow-water, tropical carbonate bank in Florida (Chen, 1965; Cramer and Arden, 1980; Cheetham, 1963). A different carbonate facies occurs in east-central South Carolina (Fig. 8d), where microfossiliferous, fine-grained limestones of the Harleyville and Parkers Ferry Formations of the Cooper Group (Ward and others, 1979; Weems and Lemon, 1984) replace the Ocala Group. These fine-grained units are relatively deep-water sediments that could represent the northeastern end of the late Eocene Suwanee channel in Georgia (Cheetham, 1963; Chen, 1965).

Oligocene Series. Lower Oligocene sediments are not known from the study area except in south Florida, where Cheetham (1963) recognized a thin unit he called the Bumpnose Limestone. Upper Oligocene sediments occur in coastal areas of North Carolina and some parts of southern South Carolina, eastern Georgia, and Florida (Fig. 8e). Generally, the Oligocene sections consist of fossiliferous limestone, although the subsurface sections in Georgia and Florida locally consist of recrystallized limestone and dolomite (Gelbaum and Howell, 1982; Herrick, 1961). Oligocene sediments are absent throughout updip areas, the Cape Fear arch, and a large part of the Florida peninsula (Fig. 8e). However, small outliers and boulders of silicified limestone containing Oligocene fossils located inland of the main outcrop-subcrop belt in Georgia and western South Carolina (Huddlestun and others, 1974; Willoughby and others, 1984) and in the upper Eocene outcrop belt (Ocala arch) of peninsular Florida (Puri and others, 1967), suggest that Oligocene sediments originally were more extensive in these areas.

Ward and others (1978) assigned Oligocene limestones in North Carolina to the River Bend Formation, which probably correlates with the Oligocene Trent Marl and Miocene Belgrade Formation (part) of Baum and others (1978) in the same area (but not with the Miocene Belgrade Formation of Ward and others, 1978; also see Zullo and others, 1982). Ward and others (1978) assigned a middle to late Oligocene (late Vicksburgian-Chickasawhayan) age to the River Bend on the basis of its ostracode and mollusk faunas. Baum (1981) interpreted the North Carolina Oligocene section as a facies sequence of intertidal to subtidal deposits.

Oligocene sediments of South Carolina are assigned to the Ashley Formation of the Cooper Group (Ward and others, 1979; Weems and Lemon, 1984). The Ashley Formation is assigned to zones NP24 and P21-22 and also contains late Oligocene ostracodes, pollen, and dinoflagellates (Oligocene section described in Hazel and others, 1977). The Ashley consists of microfossiliferous, phosphatic, clayey fine (quartz) sand or sandy fine-grained limestone that was deposited in an outer shelf environment (Gohn and others, 1977). Relief on the basal contact of the Ashley can be as great as 60 m, and this unit appears to fill broad (5- to 10-km wide) channels (Ward and others, 1979). The very thin and highly discontinuous Chandler Bridge Formation of latest Oligocene age also occurs in the Charleston, South Carolina, area (Sanders and others, 1982).

The Oligocene section in eastern Georgia and Florida is traditionally assigned to the Suwannee Limestone (Applin and Applin, 1944; Yon and Hendry, 1972; Huddlestun and others, 1974; Gelbaum and Howell, 1982) of Chattian age. This section of peritidal and shallow subtidal carbonate sediments is highly variable in its thickness and is locally absent (Cramer and Arden, 1980; Gelbaum and Howell, 1982). The thickest sections occur in the narrow northeast-trending Gulf trough of Georgia (Fig. 8e), which is generally considered to be a fault-controlled basin (Cramer and Arden, 1980; Gelbaum and Howell, 1982).

CONCLUSIONS

Although locally limited in quantity and detail, the geological data base for the southern Atlantic Coastal Plain is now adequate for describing the general tectonic and sedimentary history of this large province. As discussed in this chapter, the spatial and temporal distribution of sedimentary units is reasonably well known throughout the area, and generalized depositional models for genetically related, time-equivalent lithofacies can be postulated. The biostratigraphic positions of regional unconformities are now defined with varying levels of precision, and our understanding of the ages of formations and the ages and durations of unconformities will doubtless improve as paleontologic data continue to accumulate. In the near future, it will thus be possible for Coastal Plain geologists to address more completely the fundamental questions of sedimentation rates and sources, subsidence rates and causes, the effects of local continental tectonism, and the effects of eustatic sea-level changes. Ultimately, integration of presently evolving offshore studies of these questions (for example, Poag and Schlee, 1984) with the onshore geologic record should produce a complete picture of the sedimentary history of the United States Atlantic continental margin.

REFERENCES CITED

Amato, R. V., and Bebout, J. W., eds., 1978, Geological and operational summary, COST no. GE-1 well, Southeast Georgia embayment area, South Atlantic OCS: U.S. Geological Survey Open-File Report 78-668, 122 p.

Applegate, A. S., Winston, G. O., and Palacas, J. G., 1981, Subdivision and regional stratigraphy of the pre-Punta Gorda rocks (lowermost Cretaceous-Jurassic?) in south Florida: Supplement to Transactions, Gulf Coast Association of Geological Societies, 31st Annual Meeting, Corpus Christi, 1981, p. 447–453.

Applin, E. R., 1955, A biofacies of Woodbine age in the southeastern Gulf Coast region: U.S. Geological Survey Professional Paper 264-I, p. 187–197.

Applin, E. R., and Applin, P. L., 1964, Logs of selected wells in the Coastal Plain of Georgia: Georgia Geological Survey Bulletin 74, 229 p.

Applin, E. R., and Jordan, L., 1945, Diagnostic foraminifera from subsurface formations in Florida: Journal of Paleontology, v. 19, no. 2, p. 129–148.

Applin, P. L., and Applin, E. R., 1944, Regional subsurface stratigraphy and structure of Florida and central Georgia: American Association of Petroleum Geologists Bulletin, v. 28, p. 1673–1753.

Applin, P. L., and Applin, E. R., 1947, Regional subsurface stratigraphy, structure, and correlation of middle and early Upper Cretaceous rocks in Alabama, Georgia, and north Florida: U.S. Geological Survey Oil and Gas Investigations Preliminary Chart 26, 3 sheets.

Applin, P. L., and Applin, E. R., 1965, The Comanche Series and associated rocks in the subsurface in central and south Florida: U.S. Geological Survey Professional Paper 447, 84 p.

Applin, P. L., and Applin, E. R., 1967, The Gulf Series in the subsurface in northern Florida and southern Georgia: U.S. Geological Survey Professional Paper 524-G, 34 p.

Ascoli, P., 1976, Foraminiferal and ostracod biostratigraphy of the Mesozoic-Cenozoic Scotian shelf, Atlantic Canada: 1st International Symposium on Benthonic Foraminifera of Continental Margins, Maritime Sediments Special Publication 1, p. 653–771.

Ascoli, P., Poag, C. W., and Remane, J., 1984, Microfossil zonation across the Jurassic-Cretaceous boundary on the Atlantic margin of North America, *in* Westermann, G.E.G., Jurassic-Cretaceous Biochronology and Paleogeography of North America: Geological Association of Canada Special Paper 27, p. 33–48.

Banks, R. S., 1977, Stratigraphy of the Eocene Santee Limestone in three quarries of the Coastal Plain of South Carolina: South Carolina Geological Survey Geologic Notes, v. 21, no. 3, p. 85–149.

Barnett, R. S., 1975, Basement structure of Florida and its tectonic implications: Gulf Coast Association of Geological Societies Transactions, v. 25, p. 122–142.

Baum, G. R., 1981, Lithostratigraphy, depositional environments, and tectonic framework of the Eocene New Bern Formation and Oligocene Trent Formation, North Carolina: Southeastern Geology, v. 22, no. 4, p. 177–191.

Baum, G. R., Harris, W. B., and Zullo, V. A., 1978, Stratigraphic revision of the exposed middle Eocene and lower Miocene formations of North Carolina: Southeastern Geology, v. 20, p. 1–19.

—— , 1979, Structural and stratigraphic framework for the Coastal Plain of North Carolina: Field Trip Guidebook for Carolina Geological Society and Atlantic Coastal Plain Geological Society, North Carolina Department of Natural Resources and Development, 111 p.

Baum, G. R., Collins, J. S., Jones, R. M., Madlinger, B. A., and Powell, R. J., 1980, Correlation of the Eocene strata of the Carolinas: South Carolina Geology, v. 24, no. 1, p. 19–27.

Benson, P. H., 1969, Evidence against a large scale disconformity between the Upper Cretaceous Black Creek and Peedee Formations in South Carolina: Geologic Notes, v. 13, no. 2, p. 47–50.

Berggren, W. A., and Aubrey, M.-P, 1984, Rb-Sr glauconite isochron of the Eocene Castle Hayne Limestone, North Carolina; Further discussion: Geological Society of America Bulletin, v. 95, no. 3, p. 364–370.

Blow, W. H., 1979, The Cainozoic Globigerinida: Leiden, E. J. Brill, v. 1–3, 1413 p.

Brown, P. M., 1957, Upper Cretaceous Ostracoda from North Carolina: North Carolina Department of Conservation and Development Bulletin 70, 28 p.

—— , 1958, Well logs from the Coastal Plain of North Carolina: North Carolina Department of Conservation and Development Bulletin 72, 68 p.

Brown, P. M., Miller, J. A., and Swain, F. M., 1972, Structural and stratigraphic framework, and spatial distribution of permeability of the Atlantic Coastal Plain, North Carolina to New York: U.S. Geological Survey Professional Paper 796, p. 1–79.

Brown, P. M., Brown, D. L., Reid, M. S., and Lloyd, O. B., Jr., 1979, Evaluation of the geologic and hydrologic factors related to the waste-storage potential of Mesozoic aquifers in the southern part of the Atlantic Coastal Plain, South Carolina and Georgia: U.S. Geological Survey Professional Paper 1088, 37 p.

Brown, P. M., Shufflebarger, T. E., Jr., and Smith, S. R., 1982, Structural-stratigraphic framework and geomorphic signature of the Graingers wrench zone, North Carolina Coastal Plain: Atlantic Coastal Plain Geological Association Field Trip Guidebook, New Bern, 1982, 39 p.

Buie, B. F., 1978, The Huber Formation in eastern central Georgia, *in* Short contributions to the geology of Georgia: Georgia Geological Survey Bulletin 93, p. 1–7.

—— , 1980, Kaolin deposits and the Cretaceous-Tertiary boundary in east-central Georgia, *in* Frey, R. W., ed., Excursions in Southeastern Geology, Vol. II: Geological Society of America Field Trip Guidebook, 1980 Annual Meeting, Atlanta, p. 311–322.

Carter, J. G., 1983, Biostratigraphy Newsletter (Gulf and Atlantic Coasts of North America), Supplement No. 1, October, 1983, Chapel Hill, 118 p.

Cheetham, A. H., 1963, Late Eocene zoogeography of the eastern Gulf Coast region: Geological Society of America Memoir 91, 113 p.

Chen, Chih Shan, 1965, The regional lithostratigraphic analysis of Paleocene and Eocene rocks of Florida: Florida Geological Survey Geological Bulletin 45, 105 p.

Chowns, T. M., and Williams, C. T., 1983, Pre-Cretaceous rocks beneath the Georgia Coastal Plain; Regional implications, *in* Gohn, G. S., ed., Studies related to the Charleston, South Carolina, earthquake of 1886; Tectonics and seismicity: U.S. Geological Survey Professional Paper 1313, p. L1–L42.

Christopher, R. A., 1977, Selected Normapolles pollen genera and the age of the Raritan and Magothy Formations (Upper Cretaceous) of northern New Jersey, *in* Owens, J. P., Sohl, N. F., and Minard, J. P., eds., A field guide to Cretaceous and lower Tertiary beds of the Raritan and Salisbury embayments, New Jersey, Delaware, and Maryland: American Association of Petroleum Geologists-Society of Economic Paleontologists and Mineralogists Guidebook, Annual meeting, Washington, D.C., June, 1977, p. 58–68.

—— , 1978, Quantitative palynologic correlation of three Campanian and Maestrichtian sections (Upper Cretaceous) from the Atlantic Coastal Plain: Palynology, v. 2, p. 1–27.

—— , 1979, Normapolles and triporate pollen assemblages from the Raritan and Magothy Formations (Upper Cretaceous) of New Jersey: Palynology, v. 3, p. 73–121.

—— 1982a, Palynostratigraphy of the basal Cretaceous units of the eastern Gulf and southern Atlantic Coastal Plains, *in* Arden, D. D., Beck, B. F., and Morrow, E., Second symposium on the geology of the southeastern Coastal Plain: Georgia Geologic Survey Information Circular 53, p. 10–23.

—— , 1982b, The occurrence of the *Complexiopollis-Atlantopollis* Zone (palynomorphs) in the Eagle Ford Group (Upper Cretaceous) of Texas: Journal of Paleontology, v. 56, no. 2, p. 525–541.

Christopher, R. A., Owens, J. P., and Sohl, N. F., 1979, Late Cretaceous palynomorphs from the Cape Fear Formation of North Carolina: Southeastern Geology, v. 20, no. 3, p. 145–159.

Cochran, J. R., 1983, A model for development of Red Sea: American Association of Petroleum Geologists Bulletin, v. 67, p. 41–69.

Cole, W. S., 1944, Stratigraphic and paleontologic studies of wells in Florida—

no. 3: Florida Geological Survey Geological Bulletin no. 26, 168 p.

Cole, W. S., and Applin, E. R., 1964, Problems of the geographic and stratigraphic distribution of American middle Eocene larger Foraminifera: Bulletin of American Paleontology, v. 47, no. 212, p. 5–48.

Colquhoun, D. J., Woollen, I. D., Van Nieuwenhuise, D. S., Padgett, G. G., Oldham, R. W., Boylan, D. C., Bishop, J. W., and Howell, P. D., 1983, Surface and subsurface stratigraphy, structure, and aquifers of the South Carolina Coastal Plain: Columbia, State of South Carolina, Report to the Department of Health and Environmental Control, 78 p.

Cooke, C. W., and MacNeil, F. S., 1952, Tertiary stratigraphy of South Carolina: U.S. Geological Survey Professional Paper 243-B, p. 19–29.

Cramer, H. R., and Arden, D. D., 1980, Subsurface Cretaceous and Paleogene geology of the Coastal Plain of Georgia: Georgia Geological Survey Open-File Report 80-8, 184 p.

Daniels, D. L., Zietz, I., and Popenoe, P., 1983, Distribution of subsurface lower Mesozoic rocks in the Southeastern United States as interpreted from regional aeromagnetic and gravity maps, *in* Gohn, G. S., ed., Studies related to the Charleston, South Carolina, earthquake of 1886; Tectonics and seismicity: U.S. Geological Survey Professional Paper 1313, p. K1–K24.

Dillon, W. P., Paull, C. K., and Gilbert, L. E., 1985, History of the Atlantic Continental margin off Florida; The Blake Plateau basin, *in* Poag, C. W., ed., Geologic Evolution of the United States Atlantic Margin: New York, Van Nostrand Reinhold, p. 189–215.

Epstein, S. A., and Friedman, G. M., 1983, Depositional and diagenetic relationships between Gulf of Elat (Aqaba) and Mesozoic of United States east coast offshore: American Association of Petroleum Geologists Bulletin, v. 67, no. 6, p. 953–962.

Falvey, D. A., 1974, The development of continental margins in plate tectonic theory: Australian Petroleum Exploration Association Journal, v. 14, p. 95–106.

Froelich, A. J., and Olsen, P. E., 1984, Newark Supergroup, A revision of the Newark Group in eastern North America: U.S. Geological Survey Bulletin 1537-A, p. A55–A58.

Gelbaum, C., and Howell, J., 1982, The geohydrology of the Gulf trough, *in* Arden, D. D., Beck, B. F., and Morrow, E., eds., Second symposium on the geology of the southeastern Coastal Plain: Georgia Geologic Survey Information Circular 53, p. 140–153.

Gohn, G. S., Higgins, B. B., Smith, C. C., and Owens, J. P., 1977, Lithostratigraphy of the deep corehole (Clubhouse Crossroads corehole 1) near Charleston, South Carolina, *in* Rankin, D. W., ed., Studies related to the Charleston, South Carolina, earthquake of 1886; A preliminary report: U.S. Geological Survey Professional Paper 1028, p. 59–70.

Gohn, G. S., Christopher, R. A., Smith, C. C., and Owens, J. P., 1978, Preliminary stratigraphic cross sections of Atlantic Coastal Plain sediments of the southeastern United States; Cretaceous sediments along the South Carolina coastal margin: U.S. Geological Survey Miscellaneous Field Studies Map MF-1015-A, 2 sheets.

Gohn, G. S., Smith, C. C., Christopher, R. A., and Owens, J. P., 1980, Preliminary stratigraphic cross sections of Atlantic Coastal Plain sediments of the southeastern United States; Cretaceous sediments along the Georgia coastal margin: U.S. Geological Survey Miscellaneous Field Studies Map MF-1015-C, 2 sheets.

Gohn, G. S., Bybell, L. M., Christopher, R. A., Owens, J. P., and Smith, C. C., 1982, A stratigraphic framework for Cretaceous and Paleogene sediments along the South Carolina and Georgia coastal margins, *in* Arden, D. D., Beck, B. F., and Morrow, E., eds., Second symposium on the geology of the Southeastern Coastal Plain: Georgia Geologic Survey Information Circular 53, p. 64–74.

Gohn, G. S., Hazel, J. E., Bybell, L. M., and Edwards, L. E., 1983, The Fishburne Formation (lower Eocene), a newly defined subsurface unit in the South Carolina Coastal Plain: U.S. Geological Survey Bulletin 1537-C, 16 p.

Gradstein, F. M., and Sheridan, R. E., 1983, On the Jurassic Atlantic Ocean and a synthesis of results of Deep Sea Drilling Project Leg 76: Initial Reports of the Deep Sea Drilling Project, v. LXXVI, p. 913–943.

Griffin, G. M., Reel, D. A., and Pratt, R. W., 1977, Heat flow in Florida oil test holes and indications of oceanic crust beneath the southern Florida-Bahamas platform: Florida Bureau of Geology Special Publication No. 21, p. 43–63.

Harris, W. B., 1978, Stratigraphic and structural framework of the Rocky Point Member of the Cretaceous Peedee Formation, North Carolina: Southeastern Geology, v. 19, no. 4, p. 207–229.

Harris, W. B., and Baum, G. R., 1977, Foraminifera and Rb-Sr glauconite ages of a Paleocene Beaufort Formation outcrop in North Carolina: Geological Society of America Bulletin, v. 88, p. 869–872.

Harris, W. B., and Zullo, V. A., 1980, Rb-Sr glauconite isochron of the Eocene Castle Hayne Limestone, North Carolina: Geological Society of America Bulletin, Part I, v. 91, no. 10, p. 587–592.

——, 1982, Rb-Sr glauconite isochron of the Eocene Castle Hayne Limestone, North Carolina; Discussion and reply: Geological Society of America Bulletin, v. 93, no. 2, p. 179–182.

Harris, W. B., Zullo, V. A., and Baum, G. R., 1979, Tectonic effects on Cretaceous, Paleogene, and early Neogene sedimentation, North Carolina: Carolina Geological Society Field Trip Guidebook, p. 17–29.

Hattner, J. G., and Wise, S. W., Jr., 1980, Upper Cretaceous calcareous nannofossil biostratigraphy of South Carolina: South Carolina Geology, v. 24, no. 2, p. 41–117.

Hazel, J. E., 1969, *Cythereis eaglefordensis* Alexander, 1929; A guide fossil for deposits of latest Cenomanian age in the Western Interior and Gulf Coast regions of the United States: U.S. Geological Survey Professional Paper 650-D, p. D155–D158.

Hazel, J. E., and Brouwers, E. M., 1982, Biostratigraphic and chronostratigraphic distribution of ostracodes in the Coniacian-Maestrichtian (Austinian-Navarroan) in the Atlantic and Gulf Coastal province, *in* Maddocks, R. F., ed., Texas Ostracoda; Guidebook of excursions and related papers for the Eighth International Symposium on Ostracoda: Houston, Texas, University of Houston, Department of Geosciences, p. 166–198.

Hazel, J. E., Bybell, L. M., Christopher, R. A., Frederiksen, N. O., May, F. E., McLean, D. M., Poore, R. Z., Smith, C. C., Sohl, N. F., Valentine, P. C., and Witmer, R. J., 1977, Biostratigraphy of the deep corehole (Clubhouse Crossroads corehole 1) near Charleston, South Carolina, *in* Rankin, D. W., ed., Studies related to the Charleston, South Carolina, earthquake of 1886; A preliminary report: U.S. Geological Survey Professional Paper 1028-F, p. 71–89.

Hazel, J. E., Bybell, L. M., Edwards, L. E., Jones, G. D., and Ward, L. W., 1984a, Age of the Comfort Member of the Castle Hayne Formation, North Carolina: Geological Society of America Bulletin, v. 95, no. 9, p. 1040–1044.

Hazel, J. E., Edwards, L. E., and Bybell, L. M., 1984b, Significant unconformities and the hiatuses represented by them in the Paleogene of the Atlantic and Gulf Coastal Province, *in* Schlee, J. S., ed., Interregional Unconformities and Hydrocarbon Accumulations: American Association of Petroleum Geologists Memoir 36, p. 59–66.

Heller, P. L., Wentworth, C. M., and Poag, C. W., 1982, Episodic post-rift subsidence of the United States Atlantic continental margin: Geological Society of America Bulletin, v. 93, no. 5, p. 379–390.

Herrick, S. M., 1961, Well logs of the Coastal Plain of Georgia: Georgia Department of Mines, Mining, and Geology Bulletin 70, 462 p.

Huddlestun, P. F., 1982, The development of the stratigraphic terminology of the Claibornian and Jacksonian marine deposits of western South Carolina and eastern Georgia, *in* Nystrom, P. G., Jr., and Willoughby, R. H., eds., Geological investigations related to the stratigraphy in the kaolin mining district, Aiken County, South Carolina: Carolina Geological Society Field Trip Guidebook, 1982, p. 21–23.

Huddlestun, P. F., and Hetrick, J. H., 1978, Stratigraphy of the Tobacco Road Sand; A new formation, *in* Shorter Contributions to the Geology of Georgia: Georgia Geologic Survey Bulletin 93, p. 56–77.

——, 1979, The stratigraphy of the Barnwell Group of Georgia: Georgia Geologic Survey Open-File Report 80-1, 89 p.

Huddlestun, P. F., Marsalis, W. E., and Pickering, S. M., Jr., 1974, Tertiary stratigraphy of the central Georgia Coastal Plain, *in* Geological Society of American Southeastern Section Annual Meeting Guidebook 12: Georgia Geological Survey, p. 2-1–2-35.

Jansa, L. F., Remane, J., and Ascoli, P., 1980, Calpionellid and foraminiferal-ostracod biostratigraphy at the Jurassic-Cretaceous boundary, offshore eastern Canada: Revista Italiana de Paleontologia e Stratigrafia, v. 86, p. 67–126.

Jones, G. D., 1982, Rb-Sr glauconite isochron of the Eocene Castle Hayne Limestone, North Carolina; Discussion and reply: Geological Society of America Bulletin, v. 93, no. 2, p. 179–182.

Jordan, L., and Applin, E. R., 1952, *Choffatella* in the Gulf Coast regions of the United States and description of *Anchispirocyclina* n. gen.: Contributions from the Cushman Foundation for Foraminiferal Research, v. III, part 1, p. 1–5.

Jordan, R. R., and Smith, R. V., 1984, Notes on COSUNA correlation chart for Atlantic Coastal Plain: Southeastern Geology, v. 24, no. 4, p. 195–205.

Klitgord, K. D., and Behrendt, J. C., 1979, Basin structure of the U.S. Atlantic margin, *in* Watkins, J. S., Montadert, L., and Dickerson, P. W., eds., Geological and geophysical investigations of continental margins: American Association of Petroleum Geologists Memoir 29, p. 85–112.

Klitgord, K. D., Dillon, W. P., and Popenoe, P., 1983, Mesozoic tectonics of the Southeastern United States Coastal Plain and continental margin, *in* Gohn, G. S., ed., Studies related to the Charleston, South Carolina, earthquake of 1886; Tectonics and seismicity: U.S. Geological Survey Professional Paper 1313, p. P1–P15.

Maher, J. C., and Applin, E. R., 1971, Stratigraphy, *in* Maher, J. C., Geologic framework and petroleum potential of the Atlantic Coastal Plain and Continental Shelf: U.S. Geological Survey Professional Paper 659, p. 26–56.

Martini, E., 1971, Standard Tertiary and Quaternary calcareous nannoplankton zonation, *in* Farinacci, A., ed., Proceedings, Second Planktonic Conference: Rome, Edizioni Tecnoscienza, v. 2, p. 739–785.

McKinney, M. L., 1984, Suwanee channel of the Paleogene Coastal Plain; Support for the "carbonate suppression" model of basin formation: Geology, v. 12, no. 6, p. 343–345.

Meyerhoff, A. A., and Hatten, C. W., 1974, Bahamas salient of North America; Tectonic framework, stratigraphy, and petroleum potential: American Association of Petroleum Geologists Bulletin, v. 56, no. 6, part II, p. 1201–1239.

Mueller, P. A., and Porch, J. W., 1983, Tectonic implications of Paleozoic and Mesozoic igneous rocks in the subsurface of Peninsular Florida: Transactions, Gulf Coast Association of Geological Societies, v. XXXIII, p. 169–173.

Nystrom, P. G., and Willoughby, R. H., eds., 1982, Geological investigations related to the stratigraphy in the Kaolin Mining District, Aiken County, South Carolina: Carolina Geological Society Field Trip Guidebook, Oct. 9-12, 1982, South Carolina Geological Survey, 183 p.

Owens, J. P., 1983, The northwestern Atlantic Ocean margin, *in* The Phanerozoic Geology of the World II, The Mesozoic B: Elsevier, New York, p. 33–60.

Owens, J. P., and Gohn, G. S., 1985, Depositional history of the Cretaceous Series in the U.S. Atlantic Coastal Plain; Stratigraphy, paleoenvironments, and tectonic controls of sedimentation, *in* Poag, C. W., ed., Geologic Evolution of the United States Atlantic Margin: New York, Van Nostrand Reinhold, p. 25–86.

Pinet, P. R., and Popenoe, P., 1985, A scenario of Mesozoic-Cenozoic ocean circulation over the Blake Plateau and its environs: Geological Society of America Bulletin, v. 96, p. 618–626.

Poag, C. W., and Schlee, J. S., 1984, Depositional sequences and stratigraphic gaps on submerged United States Atlantic margin, *in* Schlee, J. S., ed., Interregional unconformities and hydrocarbon accumulation: American Association of Petroleum Geologists Memoir 36, p. 165–182.

Popenoe, P., and Zietz, I., 1977, The nature of the geophysical basement beneath the Coastal Plain of South Carolina and northeastern Georgia, *in* Rankin, D. W., ed., Studies related to the Charleston, South Carolina, earthquake of 1886; A preliminary report: U.S. Geological Survey Professional Paper 1028, p. 119–137.

Powell, R. J., 1984, Lithostratigraphy, depositional environment, and sequence framework of the middle Eocene Santee Limestone, South Carolina Coastal Plain: Southeastern Geology, v. 25, p. 79–100.

Powell, R. J., and Baum, G. R., 1982, Eocene biostratigraphy of South Carolina and its relationship to Gulf Coastal Plain zonations and global changes of coastal onlap: Geological Society of America Bulletin, v. 93, no. 11, p. 1099–1108.

Prowell, D. C., 1983, Index of faults of Cretaceous and Cenozoic age in the eastern United States: U.S. Geological Survey Map MF-1269, 3 plates.

Prowell, D. C., Christopher, R. A., Edwards, L. E., Bybell, L. M., and Gill, H. E., 1985, Geologic section of the updip Coastasl Plain from central Georgia to western South Carolina: U.S. Geological Survey Miscellaneous Field Studies Map 1737, in press.

Prowell, D. C., Edwards, L. E., and Frederiksen, N. O., 1986, The Ellenton Formation in South Carolina: A revised age designation from Cretaceous to Paleocene: U.S. Geological Survey Bulletin 1605, in press.

Puri, H. S., Yon, J. W., Jr., and Oglesby, W. R., 1967, Geology of Dixie and Gilchrist Counties, Florida: Florida Division of Geology Bulletin No. 49, 155 p.

Ragland, P. C., Hatcher, R. D., Jr., and Whittington, David, 1983, Juxtaposed Mesozoic diabase dike sets from the Carolinas; A preliminary assessment: Geology, v. 11, no. 7, p. 394–399.

Reinhardt, J., 1982, Lithofacies and depositional cycles in Upper Cretaceous rocks, central Georgia to eastern Alabama, *in* Arden, D. D., Beck, B. F., and Morrow, E., eds., Second Symposium on the Geology of the Southeastern Coastal Plain: Georgia Geologic Survey Information Circular 53, p. 89–96.

Sanders, A. E., Weems, R. E., and Lemon, E. M., Jr., 1982, Chandler Bridge Formation; A new Oligocene stratigraphic unit in the lower Coastal Plain of South Carolina: U.S. Geological Survey Bulletin 1529-H, p. H105–H124.

Scholle, P. A., ed., 1979, Geological studies of the COST GE-1 well, United States South Atlantic Outer Continental Shelf area: U.S. Geological Survey Circular 800, 114 p.

Sheridan, R. E., 1983, Phenomena of pulsation tectonics related to the breakup of the eastern North American continental margin: Initial Reports of the Deep Sea Drilling Project, Leg 76, v. LXXVI, p. 897–908.

Siesser, W. G., Fitzgerald, B. G., and Kronman, D. J., 1985, Correlation of Gulf Coast Provincial Paleogene stages with European standard stages: Geological Society of America Bulletin, v. 96, p. 827–831.

Siple, G. E., 1967, Geology and ground water of the Savannah River Plant and vicinity: U.S. Geological Survey Water-Supply Paper 1841, 113 p.

——, 1975, Groundwater resources of Orangeburg County, South Carolina: South Carolina State Development Board, Division of Geology Bulletin 36, 59 p.

Sohl, N. F., and Christopher, R. A., 1983, The Black Creek-Peedee formational contact (Upper Cretaceous) in the Cape Fear River region of North Carolina: U.S. Geological Survey Professional Paper 1285, 37 p.

Southeastern Geological Society, 1976, Mid-Tertiary carbonates, Citrus, Levy, and Marion Counties, west-central Florida: Field Trip Guide Book 18, Twentieth Field Conference, Gainesville, 1976, 102 p.

Spangler, W. B., 1950, Subsurface geology of Atlantic Coastal Plain of North Carolina: American Association of Petroleum Geologists Bulletin, v. 34, no. 1, p. 100–132.

Stainforth, R. M., Lamb, J. L., Luterbacher, H., Beard, J. H., and Jeffords, R. M., 1975, Cenozoic planktonic foraminiferal zonation and characteristics of index forms: University of Kansas Paleontological Contributions, Article 62, 425 p.

Swain, F. M., 1951, Ostracoda from wells in North Carolina, Part 1, Cenozoic Ostracoda: U.S. Geological Survey Professional Paper 234-A, 58 p.

——, 1952, Ostracoda from wells in North Carolina; Part 2, Mesozoic Ostracoda: U.S. Geological Survey Professional Paper 234-B, p. 59–93.

——, ed., 1977, Stratigraphic micropaleontology of Atlantic Basin and borderlands: Developments in Paleontology and Stratigraphy, v. 6, Amsterdam, Elsevier, 603 p.

——, 1982, Marine and brackish water Cretaceous Ostracoda from wells in central and southern Florida: Journal of Micropaleontology, v. 1,

p. 115–128.

Swain, F. M., and Brown, P. M., 1964, Cretaceous ostracods from wells in the southeastern United States: North Carolina Division of Mineral Resources Bulletin no. 78, 42 p.

——, 1972, Lower Cretaceous, Jurassic(?), and Triassic Ostracoda from the Atlantic coastal region: U.S. Geological Survey Professional Paper 795, 55 p.

Swain, F. M., and Miller, J. A., 1979, Biostratigraphy of Early and earliest Late Cretaceous Ostracoda from peninsular Florida: American Association of Petroleum Geologists Bulletin, v. 63, p. 536.

Swift, D.J.P., 1966, The Black Creek-Peedee contact in South Carolina: South Carolina State Development Board, Division of Geology, Geologic Notes, v. 10, no. 2, p. 17–36.

Swift, D.J.P., and Heron, S. D., Jr., 1969, Stratigraphy of the Carolina Cretaceous: Southeastern Geology, v. 10, p. 201–245.

Tschudy, R. R., and Patterson, S. H., 1975, Palynological evidence for Late Cretaceous, Paleocene, and early and middle Eocene ages for strata in the kaolin belt, central Georgia: U.S. Geological Survey Journal of Research, v. 3, no. 4, p. 437–445.

Valentine, P. C., 1979, Regional stratigraphy and structure of the Southeast Georgia embayment, in Scholle, P. A., ed., Geological studies of the COST GE-1 well, United States South Atlantic Outer Continental Shelf area: U.S. Geological Survey Circular 800, p. 7–17.

——, 1982, Upper Cretaceous subsurface stratigraphy and structure of coastal Georgia and South Carolina: U.S. Geological Survey Professional Paper 1222, 33 p.

——, 1984, Turonian (Eaglefordian) stratigraphy of the Atlantic Coastal Plain and Texas: U.S. Geological Survey Professional Paper 1315, 21 p.

Van Nieuwenhuise, D. S., and Colquhoun, D. J., 1982, The Paleocene-lower Eocene Black Mingo Group of the east central Coastal Plain of South Carolina: South Carolina Geology, v. 16, no. 2, p. 47–67.

Van Nieuwenhuise, D. S., and Kanes, W. H., 1976, Lithology and ostracod assemblages of the Peedee Formation at Burches Ferry, South Carolina: South Carolina State Development Board, Division of Geology, Geologic Notes, v. 20, no. 3, p. 73–87.

Ward, L. W., and Blackwelder, B. W., 1979, Scotts Hill Member (new name) of the Cretaceous Peedee Formation of southeasternmost North Carolina and east-central South Carolina, in Sohl, N. F., and Wright, W. B., Changes in stratigraphic nomenclature by the U.S. Geological Survey, 1978: U.S. Geological Survey Bulletin 1482-A, p. A87–A88.

——, 1980, Stratigraphy of Eocene, Oligocene, and lower Miocene Formations—Coastal Plain of the Carolinas, in DuBar, J. R., and others, Cenozoic biostratigraphy of the Carolinas outer Coastal Plain: Excursions in Southeastern Geology, Frey, R. W., ed., Geological Society of America 1980 Annual Meeting Guidebook, Atlanta, v. 1, p. 190–208.

Ward, L. W., and Strickland, G. L., 1985, Outline of Tertiary stratigraphy and depositional history of the U.S. Atlantic Coastal Plain, in Poag, C. W., ed., Geologic Evolution of the United States Atlantic Margin: New York, Van Nostrand Reinhold, p. 87–123.

Ward, L. W., Lawrence, D. R., and Blackwelder, B. W., 1978, Stratigraphic revision of the middle Eocene, Oligocene, and lower Miocene—Atlantic Coastal Plain of North Carolina: U.S. Geological Survey Bulletin 1457-F, 23 p.

Ward, L. W., Blackwelder, B. W., Gohn, G. S., and Poore, R. Z., 1979, Stratigraphic revision of Eocene, Oligocene, and lower Miocene formations of South Carolina: South Carolina Geological Survey Geologic Notes, v. 23, no. 1, p. 2–32.

Weems, R. E., and Lemon, E. M., Jr., 1984, Geologic map of the Stallsville Quadrangle, Dorchester and Charleston Counties, South Carolina: U.S. Geological Survey Geological Quadrangle Map GQ 1581.

Wheeler, W. H., and Curan, H. A., 1974, Relation of Rocky Point Member (Peedee Formation) to Cretaceous-Tertiary boundary in North Carolina: American Association of Petroleum Geologists Bulletin, v. 58, no. 9, p. 1751–1757.

Willoughby, R. H., Zullo, V. A., Edwards, L. E., Nystrom, P. G., Jr., Prowell, D. C., Kite, L. E., and Colquhoun, D. J., 1984, Oligocene (to Miocene?) marine deposits in Aiken County, South Carolina: Geological Society of America Abstracts with Programs, v. 16, no. 3, p. 205.

Winston, G. O., 1971a, Regional structure, stratigraphy, and oil possibilities of the South Florida basin: Gulf Coast Association of Geological Societies Transactions, v. XXI, p. 15–29.

——, 1971b, The Dollar Bay Formation of Lower Cretaceous (Fredericksburg) age in south Florida, its stratigraphy and petroleum possibilities: Florida Bureau of Geology Special Publication 15, 99 p.

——, 1977, Cotype wells for the five classic formations in peninsular Florida: Gulf Coast Association of Geological Societies Transactions, v. XXVII, p. 421–427.

——, 1978, Rebecca Shoal reef complex (Upper Cretaceous and Paleocene) in south Florida: American Association of Petroleum Geologists Bulletin, v. 62, no. 1, p. 121–127.

Woollen, I. D., and Colquhoun, D. J., 1977a, The Black Creek and Middendorf Formations in Darlington and Chesterfield Counties, South Carolina, their type areas: South Carolina Geological Survey Geologic Notes, v. 21, no. 4, p. 164–197.

——, 1977b, The Black Creek-Peedee contact in Florence County, South Carolina: South Carolina Geological Survey Geologic Notes, v. 21, no. 1, p. 20–41.

Worsley, T. R., and Turco, K. P., 1979, Calcareous nannofossils from the lower Tertiary of North Carolina, in Baum, G. R., Harris, W. B., and Zullo, V. A., eds., Structural and stratigraphic framework for the Coastal Plain of North Carolina: Carolina Geological Society Field Trip Guidebook, p. 73–86.

Yon, J. W., Jr., and Hendry, C. W., Jr., 1972, Suwanee Limestone in Hernando and Pasco Counties, Florida: Florida Bureau of Geology Bulletin no. 54, part I, p. 1–42.

Young, Keith, 1974, Lower Albian and Aptian (Cretaceous) ammonites of Texas: Geoscience and Man, v. VIII, p. 175–228.

Zullo, V. A., 1979, Biostratigraphy of Eocene through Miocene Cirripedia, North Carolina Coastal Plain, in Baum, G. R., Harris, W. B., and Zullo, V. A., eds., Structural and stratigraphic framework for the Coastal Plain of North Carolina, Carolina Geological Society and Atlantic Coastal Plain Geological Association Field Trip Guidebook, Oct. 19-21, Wrightsville Beach: Raleigh, North Carolina, Geological Survey Section, Department of Natural Resources and Community Development, p. 73–86.

Zullo, V. A., Willoughby, R. H., and Nystrom, P. G., Jr., 1982, A late Oligocene or early Miocene age for the Dry Branch Formation and Tobacco Road Sand in Aiken County, South Carolina, in Nystrom, P. G., Jr., and Willoughby, R. H., eds., Geological investigations related to stratigraphy in the kaolin mining district, Aiken, South Carolina, Carolina Geological Society Field Trip Guidebook, Oct. 9-10, 1982: Columbia, South Carolina Geological Survey, p. 34–35.

Manuscript Accepted by the Society July 18, 1986

ACKNOWLEDGMENTS

I thank R. E. Mattick, D. C. Prowell, and L. W. Ward (U.S. Geological Survey), P. M. Brown (North Carolina Geological Survey), and F. M. Swain (University of Delaware) for their reviews of the first draft of this manuscript. Because I could not accommodate all of their suggestions for modification of the stratigraphy used in this report, some stratigraphic correlations presented herein may not represent the views of individual reviewers. A subsequent draft was reviewed by N. F. Sohl and R. Z. Poore (U.S. Geological Survey). Numerous colleagues have contributed to my understanding of Atlantic Coastal Plain geology; among them, I particularly wish to acknowledge J. P. Owens, N. F. Sohl, J. E. Hazel, J. Reinhardt, D. C. Prowell, R. A. Christopher, L. W. Ward, C. C. Smith, L. M. Bybell, and W. A. Bryant for their many informative discussions.

Printed in U.S.A.

Chapter 8

Upper Cenozoic processes and environments of continental margin sedimentation: eastern United States

Stanley R. Riggs*
Department of Geology, East Carolina University, Greenville, North Carolina 27834
Daniel F. Belknap*
Department of Geological Sciences, University of Maine, Orono, Maine 04469

INTRODUCTION

Most early studies of the U.S. Atlantic continental margin were dominated by the concept of 'layer-cake' stratigraphy with disruptions in continuity often explained by 'yo-yo' processes of basin faulting. Recent studies now demonstrate that these two concepts are not totally satisfactory in explaining the stratigraphic patterns of the past 24-million-year history of the Atlantic margin.

During the past two decades, increasing sophistication of such tools as high-resolution seismic stratigraphy, biostratigraphic time zonations, and absolute dating techniques have provided a detailed basis for interregional correlations and environmental interpretations of upper Cenozoic lithostratigraphic units. The coastal plain and continental shelf are now recognized as parts of a coherent geologic province on a passive plate margin that have responded as an integral unit to complex sets of rapidly changing environmental conditions. The resulting upper Cenozoic sediment record is characterized by extremely variable lithologies with complex geometries and which are extensively dissected by unconformities.

The large-scale structural framework produced by the early tectonic history of the Atlantic continental margin controlled subsequent regional patterns of upper Cenozoic sedimentation. However, detailed depositional histories have been more complexly interwoven with climatic change and climatically controlled processes of glaciation, deglaciation, and rapid glacio-eustatic-sea-level fluctuations than with regional tectonism and structural

processes. This realization necessitates the increased recognition of erosional processes as equal counterparts to depositional processes and as a major controlling factor in determining the final distribution and character of the partially preserved regional and local stratigraphic records.

Extreme climatic fluctuations and associated continental glaciation and deglaciation resulted in rapid changes in paleoceanographic processes throughout the upper Cenozoic (Vail and others, 1977; Matthews, 1984). These changes include: 1) fluctuation in global sea level, which causes oceanic waters to move on and off the continental margin; 2) interaction of changing sea level with continental margin bathymetry, which causes modification of the path, intensity, and processes of the Gulf Stream and associated currents; 3) different oceanographic water masses, which have varying physical and chemical characteristics, alternately dominate the margin; and 4) conditions of sediment deposition and accumulation that alternate with erosion and truncation by both submarine currents and subaerial processes. Changing climatic conditions also produce extensive latitudinal migration of climatic zones; modification of vegetative cover, weathering, and fluvial processes; and alternating episodes dominated by siliciclastic sedimentation with episodes of authigenic sedimentation and minor siliciclastic dilution.

Thus, continental margin basin development and sediment accumulation during the upper Cenozoic is largely the result of cyclical paleoclimatic and paleoceanographic conditions and processes that interact with the geometry of the margin and regional patterns of differential thermal subsidence through time. The shelf-slope geometry determines whether sediment erosion or deposition takes place and, if sediment is deposited, whether it accumulates as vertical accretion on the shelf or lateral progradation off the shelf-edge. Also, the modern sedimentary environments and sedimentary processes presently active within the

*With contributions from H. W. Borns, Department of Geological Sciences, University of Maine, Orono, Maine 04469. L. J. Doyle and A. C. Hine, Department of Marine Science, University of South Florida, St. Petersburg, Florida 33701. J. T. Kelley, Department of Geological Sciences, University of Maine, Orono, Maine 04469. H. T. Mullins, Department of Geology, Syracuse University, Syracuse, New York 13210. S. W. Snyder, Department of Geology, East Carolina University, Greenville, North Carolina 27834. S.W.P. Snyder, Department of Marine Science, University of South Florida, St. Petersburg, Florida 33701. D.J.P. Swift, 1127 Chesterton Drive, Richardson, Texas 75080.

Riggs, S. R., and Belknap, D. F., 1988, Upper Cenozoic processes and environments of continental margin sedimentation: eastern United States; *in* Sheridan, R. E., and Grow, J. A., eds., The Geology of North America, Volume I-2, The Atlantic Continental Margin, U.S., Geological Society of America.

province are similar to portions of the ancestral upper Cenozoic systems; however, they only represent momentary time slices within a continuously changing sequence of paleoenvironments.

CLIMATIC AND SEA-LEVEL FLUCTUATIONS DURING THE UPPER CENOZOIC

Causes of Sea-Level Fluctuation

The fact that the stratigraphic section represents a sensitive record of global, regional, and local fluctuations in relative sea level is self-evident. Also, the relationship of sea-level fluctuations during the Quaternary to major change in global climate and resulting glaciation and deglaciation is an established fact. However, only recently have extensive research efforts from many different avenues generated a data base that demonstrates a similar sea-level cyclicity throughout the Neogene. Thus, initiation of rapid climatic changes, development of polar glaciation, and the resulting patterns of sedimentation during the Neogene represent an important difference in geologic histories of the lower and upper Cenozoic (see chapter 16, this volume).

Explanations for climatic changes that resulted in the development of polar glaciation can be grouped into two categories of theories (Flint, 1971; Bowen, 1978; Gates, 1976). Theories based upon extraterrestrial causes include solar variability, cosmic dust clouds or supernovae, and variations in Earth's orbital and rotational characteristics. Theories based upon terrestrial causes include volcanic dust and CO_2 input to the atmosphere, lateral and vertical crustal movements, and changes in atmospheric and oceanic circulation. Many of these causes are interrelated.

Recent studies have focused on the Milankovitch (1941) theory as the prime triggering mechanism for glacial-interglacial cycles (Imbrie and Imbrie, 1980; Berger, 1982). This theory invokes periodicities in the geometry of the earth's axial orientation and orbital parameters to influence variations in isolation at given latitudes as the link to climatic change. Through spectral analysis of astronomical data, Imbrie and Imbrie (1980) identified similar periodic variations for the past 730,000 years. The Milankovitch theory is also supported by a significant correlation between the earth's periodicity and oxygen isotopic cycles occurring within the last 1 million years of depositional history in deep sea sediment sequences (Emiliani, 1966, 1970, 1978; Shackleton and Opdyke, 1973, 1976).

The Neogene of the Atlantic continental margin represents a period of major change in depositional regimes (see chapter 16, this volume). By the end of the Paleogene, extensive carbonate sedimentation characteristic of the Eocene and Oligocene had diminished. The Neogene was dominated by siliciclastic sedimentation intermixed with extensive authigenic phosphate, dolomite, glauconite, diatomaceous muds, and only minor and local carbonate sedimentation. The Neogene represents a transition zone into the well-known Quaternary glaciation. Increasing faunal and floral evidence suggests that global climates were becoming less stable with declining temperatures by the end of the Oligocene in

Europe (Nilsson and Persson, 1983) and the United States (Dorf, 1964).

Based upon oxygen isotope data, Matthews and Poore (1980) and Matthews (1984) argued that glacio-eustatic sea-level fluctuations have been an important process in marine sedimentation at least since the Eocene and possibly since the Cretaceous. Recent oxygen isotopic studies of deep-sea sediments (Miller and others, 1985; Keigwin and Corliss, 1986) suggest that permanent ice-building was initiated in Antarctica during the Eocene. During the early Miocene, about 25 to 20 Ma, additional global cooling followed the opening of the Drake Passage, which established the deep circumpolar current and led to thermal isolation of Antarctica (Keller and Barron, 1983). They recognized eight distinct deep-sea dissolution hiatuses within the Neogene. These hiatuses represent periods of cooling and intensification of bottom-water circulation, which increases the corrosion of calcareous fauna. Keller and Barron (1983) believe that the dissolution zones represent unstable climatic conditions characterized by cooling at high latitudes or increased glaciation. By the middle Miocene, between 16 and 13 Ma, Woodruff and others (1981) and Woodruff (1985) found that the climate had become extremely unstable with oxygen isotope records suggesting similarities to the Pleistocene with periodicities on the order of 100,000 years. Woodruff (1985) found that changes in distribution of deep-sea benthic foraminifera, as well as evolutionary originations and extinctions, were concomitant with paleoclimatic and paleoceanographic changes presumably related to Antarctic glacier expansion and cooling of the deep Pacific Ocean between 16 and 13 Ma. The Antarctic ice sheet continued to experience major expansion episodes between 13 to 11.5 Ma (Keigwin, 1979; Woodruff and others, 1981; Keller and Barron, 1983) and again between 10 to 8 Ma (Woodruff, 1985). Keller and Barron (1983) found strong correlations between the Neogene deep-sea hiatal surfaces with onshore unconformities produced by eustatic lowering of sea level, both reflecting periods of intensified polar glaciation. Some of these events correlate with severe downward shifts of coastal onlap and low eustatic sea level as shown in Figure 1 and in Chapter 16 Figure 15 (Fig. 15, this volume).

Vail and others (1977) defined second-order cycles of relative sea-level change with durations of 10 to 80 million years and third-order cycles of 1 to 10 million years (Fig. 1). They believed that some second-order cycles may be of sufficient magnitude and duration that a suggestion of a response to geotectonic mechanisms is warranted; however, some second-order and many third-order cycles, particularly during the Neogene, are the result of glaciation and deglaciation. Vail and others (1977) and Vail and Mitchum (1979) delineated a series of global highstands (when sea level is above the shelf edge in most regions of the world with widespread deposition of marine sediments on continental margins) and global lowstands (when sea level is below the shelf edge with widespread erosion and nondeposition on continental margins producing interregional unconformities).

Stratigraphic analysis of Neogene continental margin sedimentation in North Carolina has documented fourth-order cycli-

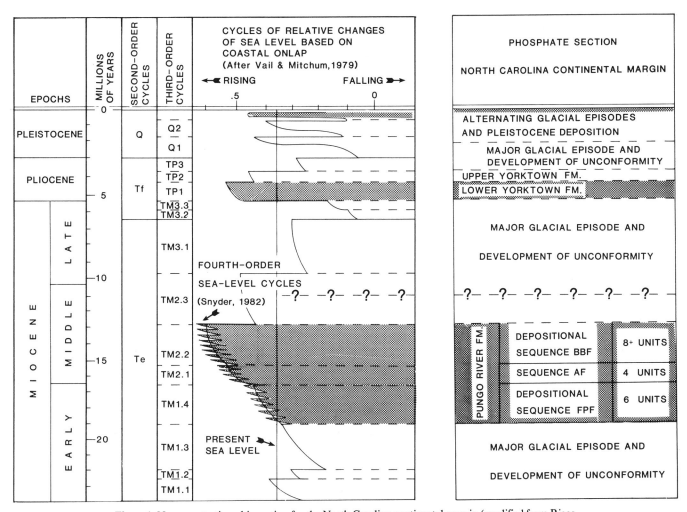

Figure 1. Neogene stratigraphic section for the North Carolina continental margin (modified from Riggs, 1984) superimposed upon the second- and third-order coastal onlap curves of Vail and Mitchum (1979). Fourth-order sea-level cycles during the Miocene are based upon the detailed high-resolution seismic mapping of depositional sequences within the Pungo River Formation by Snyder (1982). Deposition of the 18 fourth-order depositional sequences of the Pungo River Formation began about 19 Ma and continued to at least 13 Ma and possibly to 10.5 Ma. Stratigraphic units containing anomalous concentrations of phosphate are stippled.

city (Fig. 1) believed to be associated with sea-level and climatic fluctuations (Riggs, 1984). Snyder (1982) identified 18 depositional sequences within the Pungo River Formation. Each sequence is bound by unconformities that were traced throughout the North Carolina continental margin. These regional unconformities and the sequences they bound are interpreted to be depositional products of sea-level fluctuations on the scale of 10^5 years duration as direct responses to glaciation and deglaciation during the early and middle Miocene (Riggs and others, 1982; Snyder and others, 1982b; Snyder and others, 1983; Riggs and others, 1985). Kidwell (1984) described similar cyclicity in the middle Miocene Calvert and Choptank Formations of Maryland and Virginia, partial stratigraphic equivalents of the Pungo River Formation.

Changes in Quaternary sea levels are dominated by rapid exchanges between oceans and ice sheets on scales of 10^2 meters in 10^4 year time ranges. Longer-term changes in ocean volume resulting from variations in ocean ridge spreading rates are much slower (Pitman, 1979). Volume estimates of water and ice exchanges are best accomplished with oxygen isotopes. Using $\delta^{18}O$ records and neglecting isostatic adjustments, Shackleton (1977) and Fairbanks and Matthews (1978) respectively, estimated a full-glacial to interglacial difference in sea level of between 155 and 160 meters. Figure 2 shows a typical fluctuating $\delta^{18}O$ record of ice volume (Shackleton and Opdyke, 1973) compared to the Imbrie and Imbrie (1980) model of ice volume response to Milankovitch climatic cycles over the past 500,000 years. This ice-volume curve seems to be a reasonable representation of

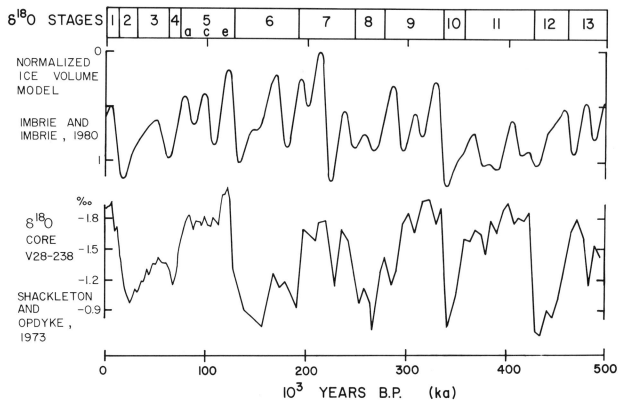

Figure 2. Model of global ice volume in response to Milankovitch cyclicity (from Imbrie and Imbrie, 1980) compared to deep sea oxygen isotope data from core V-28-238, an equitorial Pacific foraminifera ooze (from Shackleton and Opdyke, 1973).

glacio-eustatic sea-volume changes on continental margins where tectonic and isostatic effects have not warped the margin and associated basins (Belknap, 1979; Cronin and others, 1981).

A study of timing of glacial events based upon worldwide deep sea isochronous records in the Atlantic (van Donk, 1976) defined 21 glacial-interglacial cycles through the Pleistocene. This record is certainly more realistic and representative of the complexity of climatic change than the classic view of four Pleistocene 'ice ages.'

Recent modeling studies carried out for land and sea-surface temperatures since the glacial maximum 18,000 years B.P. (CLIMAP, 1976; McIntyre and others, 1976; Webb and Andrews, 1976; Wright, 1981) have led to the reconstruction of climatic conditions associated with the latest glacial cycle. Since the Holocene record of glaciation is the best known and preserved stratigraphic section, it allows extrapolation to the many older and less well known glacial-interglacial stages of the Quaternary and Neogene.

Regional Variations Upon Sea-Level Fluctuations

In spite of the fact that the Quaternary record is dominated by global cycles of climatic change driving glacial eustatic and isostatic variations, there are many important modifying processes that determine local relative sea-level records. The three most important regional variables include glacial isostasy, differential thermal subsidence, and hydro-isostatic warping; of lesser importance are such long-term effects as sedimentation and erosion isostasy (Menard, 1983) and short-term effects of minor and regional climatic changes (Lisitzin, 1974).

Glacial isostasy involves the subsidence and rebound of the land in response to the addition and removal of the mass of glacial ice. Extent and rate of adjustment laterally and vertically due to loading and rebound are still actively debated (Mörner, 1980; Denton and Hughes, 1981), but it is clear that geoidal deformation occurs worldwide during glaciation. For example, Cathles (1980) found about 150 m of tilt over a distance of 10,000 km at 6,000 years B.P. General rebound from major deformation under the ice and inboard of the forebulge was compensated for rapidly in an exponentially decreasing manner. Up to 90% of the rebound is accomplished in 6,000 to 8,000 years after the ice has melted off the land and within 300 to 500 km of the ice margin. More distant areas experience subsidence of a margin forebulge, decreasing in amplitude asymptotically away from the former ice front.

Emery and Uchupi (1972) interpreted Pleistocene bathymetric features on the mid-Atlantic outer continental shelf as low-sea-level shorelines. Deformations of presumed horizontal planes through these features were interpreted as responses to glacial rebound off Long Island and subsidence off Delaware and New Jersey as a glacial forebulge relaxed. Dillon and Oldale (1978) supported this idea with seismic reflection profiling data, and suggested a strong inflection point in tilting off central New Jersey, possibly related to an oceanic fracture zone. These data suggest that warping of the margin can occur both in a coast-normal or ice-front-normal direction in response to isostatic adjustments.

As demonstrated elsewhere in this volume, tectonic influences on sea level are obvious in the long-term history of continental margins, especially across basin boundaries on different crustal blocks. Klitgord and Behrendt (1979) and Grow and Sheridan (1981) defined the first-order basin structure of the margin and related its origin to the rift and early drift stage of continental breakup during the Mesozoic. Brown and others (1972) utilized tectonic and structural processes to describe the continuous evolution of Cenozoic sedimentary basins on the coastal plain. Extensive high-resolution seismic profiling on the adjacent continental shelf since the late 1970s, has generally demonstrated a decreasing importance of contemporaneous structural deformation due to tectonic processes during upper Cenozoic sedimentation (Snyder, 1982; Snyder and others, 1982b; Dillon, 1983; Dillon and others, 1983; Popenoe, 1985; Riggs and others, 1985). Recent releveling studies (Brown, 1978), however, do demonstrate that some neotectonic movement, related both to rebound (Newman and others, 1971) and to deep structure, is locally deforming the Atlantic coast today.

Mörner (1980) believed that isostatic readjustment due to weight differences in the change from solid ice on land to liquid in the oceans, and vice versa, is important. Sea level is interpreted to have dropped 130 m on the Atlantic margin during peak Wisconsin glaciation (Milliman and Emery, 1968); part of this drop is probably due to hydro-isostatic loading of the shelf (Bloom, 1967; Belknap and Kraft, 1977). Belknap and Kraft noted a possible 40 m per 100 km seaward tilt of isochrons derived from radiocarbon dates obtained on samples taken normal to the Delaware shoreline.

Evidence for Sea-Level Cyclicity in the Patterns of Sedimentation

Reviews of development of glacial theory and mapping of glaciated terrain on the Atlantic continental margin are provided by Schafer and Hartshorn (1965), Flint (1971), Emery and Uchupi (1972), Black and others (1973), and Denton and Hughes (1981). In glaciated areas, direct evidence recording climatic change and the resulting glacial and deglacial sedimentary environments and associated sea-level fluctuations includes: fairly obvious and distinctive landforms, sediments, fossils, and isotopic signals preserved in the lithostratigraphic record. These features reflect rapid and complex glacial-interglacial and stadial-interstadial fluctuations. Interglacial sediments represent normal

depositional conditions and by themselves are virtually impossible to identify; however, they do occur within the cyclical sequences representing the changing climatic and oceanographic conditions.

Evidence for glaciation and deglaciation in sedimentary environments peripheral to actual glaciation becomes much less obvious. Oldale (1982a) utilized ice wedges, dry valleys, and tundra pollen to interpret periglacial environments. Changing climatic conditions on land environments remote to glaciation can only be recognized indirectly through studies of pollen (Sirkin and others, 1977) or fossil soil profiles (Owens and others, 1983); both reflect patterns of moisture and temperature over wide areas.

In the shallow marine environments of the nonglaciated mid-Atlantic continental margin, sea-level events resulting from glaciation and deglaciation are reflected in complex lithofacies patterns of rapidly changing depositional regimes and subsequent erosional events (Riggs, 1984). High energy coastal systems migrate rapidly back and forth across the low-sloping continental shelf. Except in basinal areas with active subsidence, the coastal system repeatedly reworks and destroys much of what was previously deposited, leaving complex partial sections behind (Belknap and Kraft, 1981, 1985; Hine and Snyder, 1985). Fossil mollusk and micro-organism zonations reflect these major paleoclimatic changes along the continental margin (Blackwelder, 1981; Cronin, 1980; and Cronin and others, 1981). Thus, the stratigraphic record shows irregular patterns of deposition in which similar, but highly discontinuous lithostratigraphic units with irregular geometries are stacked in complex lithic sequences. The resulting complex stratigraphy is the product of nondepositional/erosional modification where only portions of each cycle are preserved; those sediments that are preserved are often diagenetically altered.

On the continental shelf, major deposition occurs during transgressive phases and high sea levels; nondeposition, weathering, and erosion result from regression and low sea levels. Thus, stratigraphic analysis of coastal plain formations represent dominantly peak sea-level transgressive to regressive transition deposits and are only partial records of full sea-level cycles (Demarest and others, 1981). The submerged margin carries evidence of low stand shorelines of sea-level cycles (Emery and Uchupi, 1972; Dillon and Oldale, 1978).

In shallow water carbonate environments of the southern Atlantic continental margin, sea-level fluctuations did not produce rapidly changing types of lithofacies as in the mid-Atlantic region. Rather, these fluctuations produced complex carbonate facies punctuated by extensive dissolution and cementation surfaces that formed on exposed carbonates during low sea-level stands. These indurated carbonates form the cores of many of the islands of South Florida and the Bahamas (Ball, 1967; Halley and others, 1983; McKee and Ward, 1983).

In the deep sea, changes in ice volume and oceanic temperatures are recorded directly by the changing populations of microfossils and their preservation (Keller and Barron, 1983), as well as

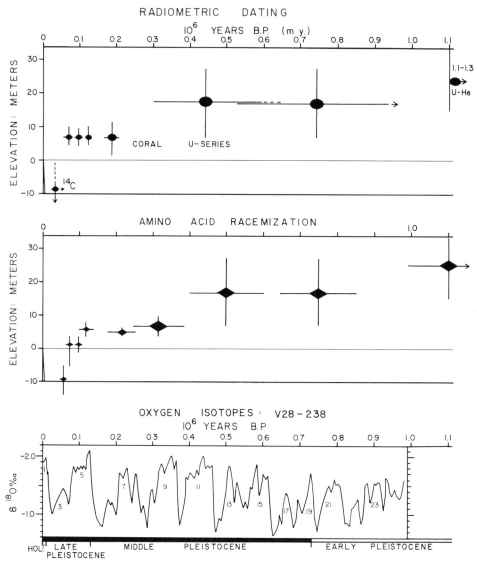

Figure 3. Comparison of approximate sea-level positions on the continental margin with the deep sea oxygen isotope record for the past million years. Radiometric dating is from Cronin (1980) and Cronin and others (1981), amino acid racemization dating is from Belknap (1979), and oxygen isotope data is from Shackleton and Opdyke (1973, 1976).

the isotopic shifts in the sediments. Organisms such as foraminifera that precipitate calcium carbonate tests record the isotopic composition of oxygen in water, as well as a further fractionation, which is temperature dependent (Emiliani, 1966, 1970, 1978; Emiliani and Shackleton, 1974; Shackleton and Opdyke, 1973, 1976). Thus, the $\delta^{18}O$ record in foraminifera (Keigwin, 1979; Woodruff and others, 1981) can be used to reconstruct complex signals of climatic change through geologic time (Fig. 2).

Sea-Level Curves

Despite pioneering work by Fairbridge (1961) to produce a global eustatic sea-level curve for the Quaternary, there are great world-wide variations in sea-level curves. A great deal more data are necessary to clarify the details of worldwide geoidal warping. Belknap and Kraft (1977) emphasized that the data base must consist of numerous local relative sea-level curves.

Construction of local relative glacio-eustatic sea-level curves must take into account only true age and depth of deposition and subsequent comparison of all relative sea-level indicators. The first factor, the age range of dating techniques, determines the nature of Quaternary sea-level curves. Pleistocene curves (Fig. 3) are based on U-series dating of corals (0 to 400 ka) or amino acid dating of mollusks (0 to 800 ka or more) with limitations based on availability of specimens and questions as to the reliability of various techniques (McCarten and others, 1982; Wehmiller and

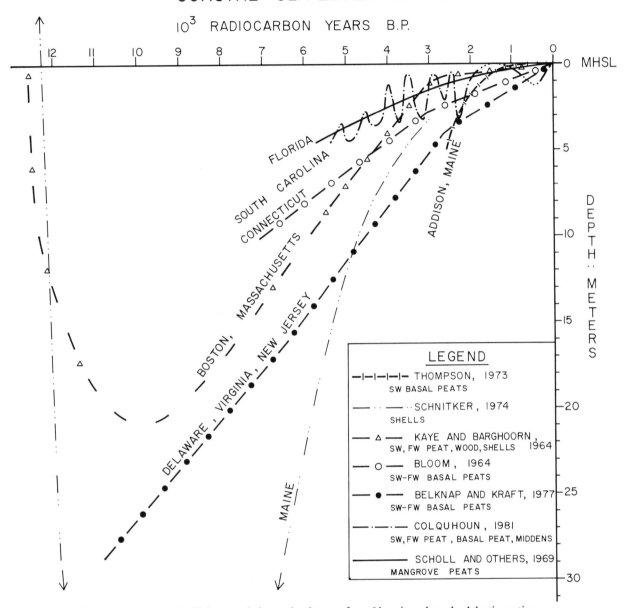

Figure 4. Representative Holocene relative sea-level curves from 6 locations along the Atlantic continental margin.

Belknap, 1982). The second factor, water depth of the depositional environment, is generally based on marine organisms with vertical ranges of 1 to 10 m, most of which have a high degree of uncertainty in their depth below sea level. Cronin and others (1981) found that error ranges of about 6 m are probably more typical of the best available precision for paleo-sea-level maxima determinations. Dating of low sea-level events on the Atlantic margin remains an elusive objective.

Late Quaternary sea-level curve construction is almost in-variably based on radiocarbon analysis. The practical range of ^{14}C dating is about 40 ka, despite recent advances in enrichment techniques (Grootes, 1978; Stuiver and others, 1978) and accelerator techniques (Bennett and others, 1977; Muller and others, 1978). However, conflicts between radiocarbon data and interpretations of marine oxygen isotopic data suggest that contamination of geologic samples generally limits the reliable application of this technology to samples within the past 25,000 to 35,000 years (Thom, 1973; Bowen, 1978). Figure 4 presents representative sea-level curves for the last 12,000 years B.P. from several Atlantic coast localities. Each is plotted as originally presented and demonstrates the along-coast variations in relative sea level. This is probably related to glacio-isostatic processes and other factors previously discussed, as well as author variation in interpreting rapid fluctuations on a scale of 1 m/yr. Lisle (1982) produced an annotated bibliography of sea-level change along the Atlantic and Gulf coasts.

NEOGENE GEOLOGY

Introduction

The upper Cenozoic stratigraphy of the U.S. mid- and south-Atlantic continental margin records major shifts in patterns of sedimentation from the Paleogene to the Quaternary. The Neogene section is dominated by cyclical episodes of phosphogenesis producing extensive sequences of phosphate and related authigenic sediments (Riggs, 1984). These sediments represent a transition between the deposition of predominantly carbonate sediments during the Eocene and Oligocene and predominantly siliciclastic sediments during the Quaternary (see chapter 16, this volume).

Throughout the mid- and south-Atlantic continental margin, the Neogene section, particularly the Miocene portion, is characterized by the following (Riggs, 1984): 1) abnormally high concentrations of authigenic minerals including phosphate, dolomite, opaline silica, glauconite, Mg-rich clay, and clinoptilolite; 2) regional patterns of phosphate, carbonate, and siliciclastic sedimentation; and 3) cyclical sediment units that reflect changing lithologies and processes of deposition and erosion. The changing paleoenvironmental conditions represented in this complex Neogene sediment sequence are dramatic responses to changing physical, chemical, and biological conditions associated with changing global paleoclimatic conditions and development of polar glaciation with the resultant fluctuations in paleoceanographic processes along the Atlantic continental margin (Katrosh and Snyder, 1982; Riggs and others, 1982, 1985; Snyder, 1982; Snyder and others, 1982b; Riggs, 1984; Hine and Snyder, 1985; Waters and Snyder, 1986).

The southern coastal plain and continental shelf of North Carolina contain a typical Neogene section, which has the most extensive and detailed stratigraphic and sedimentologic data base of the entire U.S. Atlantic margin. Consequently, it can serve as a valuable model for sedimentation along other portions of the U.S. Atlantic margin and similar passive margins in other areas of the world. Also, most of the basic processes of deposition and erosion, patterns of sediment distribution, and environmental responses to fluctuating paleoclimatic and paleoceanographic conditions are directly applicable to understanding the Quaternary. Thus, much of the following discussion focuses on the North Carolina stratigraphic section as a sensitive record of changing climatic patterns with their resulting oceanographic responses of sea level and current regimes interacting with each

other and with regional geologic structures to affect shelf/slope sedimentation along the Atlantic margin.

Tectonic, Structural, and Stratigraphic Framework of the Neogene

General patterns of Neogene sedimentation throughout the U.S. Atlantic continental margin were controlled and defined by the pre-Neogene structural framework and erosional topography (see Chapters 1 and 3, this volume). Continental break-up occurred along an irregular boundary forming a series of structural promontories or platforms and re-entrants or basins (Klitgord and Behrendt, 1979; Dillon and others, 1979; Grow and others, 1979; Schlee, 1982). Deposited upon this complex rifted terrain is a Mesozoic/Cenozoic sedimentary wedge that is over 12 km thick in the basins and thins and disappears as it onlaps onto the adjacent platforms, which have experienced minimal subsidence. Subsidence processes decreased exponentially with time such that about 95% of the total subsidence occurred by the end of the Cretaceous (Parsons and Sclater, 1977), which explains the relatively thin Cenozoic section, as well as the decreasing influence of structural deformation resulting from tectonism through the Cenozoic; deformation reached a minimum in the Neogene.

These large-scale structures and topographic features controlled the general location, size, and regional patterns of Neogene sediment sequences on the U.S. Atlantic margin (Fig. 16 in Schlee and others, this volume). These first-order structural features include, from south to north, the Florida, Carolina, and Long Island platforms and the adjacent South Florida and Southeast Georgia embayments, the Baltimore Canyon Trough, and the Georges Bank Basin (Dillon and others, 1979; Klitgord and Behrendt, 1979). Superimposed on this regional structural pattern is a series of smaller-scale or second-order structural or topographic highs and adjacent entrapment basins that define the detailed geometry and lithofacies of each sediment depocenter (Fig. 16 in Schlee and others, and Fig. 1 in Riggs and Manheim, this volume). Geographic locations of second-order features within the first-order structural framework change through time in response to changing patterns of deposition and erosion.

Within the Tertiary covering of the mid- and south-Atlantic continental shelf, only rare faults of possible basement tectonic origin have been recognized (e.g., the Helena Banks Fault of Behrendt and others, 1981). Other faults and folds are small and of non-tectonic origin, such as those associated with collapse over limestone solution cavities and compaction/subsidence over buried, irregular topographic highs and shelf-edge features (Snyder and others, 1982b; Dillon, 1983; Dillon and others, 1983; Popenoe and others, 1982). Consequently, it appears that the Neogene depositional system has had only minimal and local influence of contemporaneous structural deformation controlling sedimentation. This understanding is in contrast to other interpretations of coastal plain stratigraphy, which based depositional patterns largely on extensive basin faulting throughout the Neogene (Brown and others, 1972; Gibson, 1982, 1983; Miller, 1982; Ward and Strickland, 1985).

Neogene stratigraphy of the mid-Atlantic continental margin has been strongly influenced by fluctuations in eustatic sea level and by regional paleogeographic setting. Widespread sedimentary units were deposited in response to two second-order marine transgressions (Vail and Mitchum, 1979), one in late early to middle Miocene and the other in middle early to late Pliocene. Vertical facies variations developed in response to smaller scale, third- and fourth-order sea-level fluctuations; lateral facies were controlled by latitude, location of the first-order structural features previously described, and smaller-scale paleotopographic highs that delineated the intervening depositional embayments.

The major Miocene transgression, which began in the early Miocene and culminated in the middle Miocene (Fig. 1), produced deposits of anomalous sediments all along the mid- and south-Atlantic coast (Blackwelder, 1981; Gibson, 1982, 1983; Ward and Strickland, 1985). The most notable include the phosphatic, dolomitic, and Mg-rich clayey Hawthorn Formation of Florida through South Carolina (Riggs, 1979b, 1979c, 1984); the phosphatic, clastic-rich Pungo River Formation that was deposited in the Onslow and Aurora Embayments of North Carolina (Miller, 1982; Riggs and others, 1982, 1985); and the glauconite and diatom-rich sediments of the time-equivalent Calvert and lower portion of the Choptank Formations, which accumulated in the Salisbury Embayment of Virginia through Delaware (Gibson, 1982, 1983; Ward and Strickland, 1985). Marine clayey sands of the St. Marys Formation in the Salisbury Embayment of Virginia and Maryland represents the end of middle Miocene deposition (Ward and Strickland, 1985).

Late Miocene deposits occur on a more local basis along portions of the continental margin and are often controversial in their age assignments. The biostratigraphy of many late Miocene units is based upon molluscan zonations; where planktonic foraminiferal work has been done, the same units are often considered to be early or middle Miocene. Clastic marine deposits of the Eastover Formation, interpreted as late Miocene on the basis of molluscan zonations, accumulated in Maryland, Virginia, and into northern North Carolina (Ward and Strickland, 1985). Throughout much of North and South Carolina, there was little or no deposition. The poorly understood, interbedded sands, clays, and limestones of the Marks Head, Coosawhatchie, and Screven Formations were deposited through southern South Carolina and Georgia (Carter, 1984). Laterally equivalent units in Peninsular Florida include the Choctawhatchee and lower portion of the Tamiami Formations (Blackwelder, 1980) and the Bone Valley Member of the Hawthorn Formation (Scott, 1986).

The latest Miocene and earliest Pliocene interval was a time of nondeposition along the entire Atlantic margin (Blackwelder, 1980; Ward and Strickland, 1985). Widespread deposition resumed during the major Pliocene marine transgression (Fig. 1), which inundated pre-existing paleotopographic highs and produced deposits that were not restricted to distinct embayments. The Yorktown Formation occurs from Maryland through northern North Carolina (Blackwelder, 1980; Gibson, 1983). Age-equivalent strata in southern North Carolina, South Carolina, and Georgia are assigned to the Duplin Formation with minor variations in names on a local basis; they become the Jackson Bluff Formation in northwest Florida and the upper portion of the Tamiami Formation in Peninsular Florida (Blackwelder, 1980; Scott, 1986).

Regression during the late Pliocene produced a widespread unconformity throughout the Atlantic margin (Blackwelder, 1980). By the end of the Pliocene, there was a general resumption of marine deposition marking the beginning of very complex depositional patterns that characterize the Pleistocene sediment units: thin and highly variable lithologic sequences of regional deposits characterized by extensive erosion and truncation. This resulted in an exceedingly complex and often confusing stratigraphic nomenclature due to difficulties in identifying and correlating a complex multiplicity of similar Pleistocene sediment units.

North Carolina Continental Margin—A Model System

Seismic Stratigraphy. The dominant first-order basement structure within the Carolina sediment province (Fig. 16 in chap. 16, this volume) is the Carolina Platform. It is a positive feature located between subsiding margin basins including the Baltimore Canyon Trough to the northeast, the Carolina Trough on the seaward side, and the Blake Plateau Basin with its landward extension, the Southeast Georgia Embayment, to the south. The Cape Fear Arch, or Mid-Carolina Platform High, is the highest and most stable portion of the Carolina Platform as a result of being located away from the rapidly subsiding (during Mesozoic) transitional or rift-stage crust (Popenoe and others, 1982). It is becoming increasingly clear that the Carolina Platform has significantly affected or controlled regional sedimentation processes since the Cretaceous (Sheridan, 1974; Dillon and others, 1979; Klitgord and Behrendt, 1979; Grow and Sheridan, 1981; Popenoe and others, 1982; Snyder, 1982; Snyder and others, 1982b; Dillon, 1983; Dillon and others, 1983; Riggs, 1984; Popenoe, 1985). This feature has provided the major, underlying topography over which sea-level fluctuation and oceanographic currents have interacted to produce complex stratigraphy and lithology.

Miocene sediments accumulated on the east flank of the Mid-Carolina Platform High in a large Neogene basin (Fig. 5). An east-west oriented second-order erosional feature, known as the Cape Lookout High (Scarborough and others, 1982; Snyder and others, 1982b), divides the Neogene basin into two local embayments. North of the Cape Lookout High is the Aurora Embayment, which contains the Aurora phosphate district. South of the Cape Lookout High is the Onslow Embayment, which contains the newly discovered Onslow Bay phosphate deposits (Riggs and others, 1985). Here Miocene sediments occur in shallow subcrop as a series of apparent offlapping erosional units around the eastern side of the Mid-Carolina Platform High, and in the subsurface at the shelf edge (Blackwelder and others, 1982; Snyder and others, 1982b; Riggs, 1984; Riggs and others, 1985).

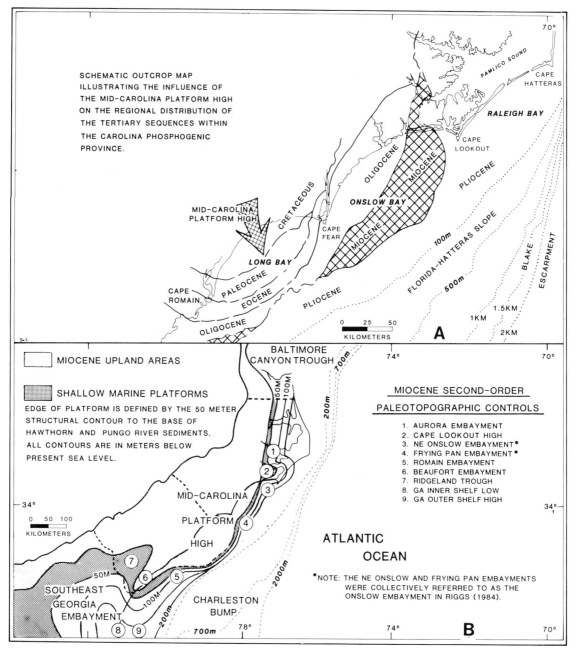

Figure 5. Maps of the middle Atlantic continental margin in eastern United States (from Riggs and others, 1985). Map A illustrates the influence of the Mid-Carolina Platform High on the regional distribution of Tertiary units within the Carolina Phosphogenic Province. Map B shows the general Miocene shoreline, bathymetry, and major first-order structural and second-order paleotopographic features that controlled Neogene sedimentation.

Single channel, high-resolution seismic profiles provide the basis for determining the Neogene evolution of the lower coastal plain, continental shelf, and upper slope off North Carolina (Riggs and others, 1982, 1985; Snyder and others, 1982b; Popenoe, 1985). Popenoe (1985) obtained regional coverage of deep penetration seismic reflection data while Snyder (1982), Snyder and others (1982b), and Matteucci (1984) obtained a detailed

network of higher-frequency, higher-resolution seismic data across the Carolina continental margin. These studies, utilizing different levels of seismic resolution, compliment one another and provide perhaps the most detailed high-resolution seismic coverage along any passive continental margin.

Stratigraphic control is provided by drill holes, rock dredgings, and vibracores on the continental shelf and numerous drill holes on the adjacent coastal plain. Structure contour and isopach maps provide the spatial and temporal view necessary to reconstruct Neogene geologic and paleoceanographic history.

Popenoe (1985) defined seven major unconformities and six seismic sequences in the Neogene on the North Carolina margin. These sequences show remarkable chronological agreement with established third-order cycles within the second-order supercycles (Fig. 1) of Vail and others (1977); the only disagreement is minor and concerns the magnitude of low stand events (Popenoe, 1985). As explained by Popenoe, the global, eustatic curve was reconstructed by recognizing and determining the chronology of shifts in position of the Gulf Stream responding to sea-level fluctuations (Pinet and others, 1981; Pinet and Popenoe, 1982), rather than by developing a local coastal onlap curve. The primary axioms are that most of the unconformities result from Gulf Stream submarine erosion and not subaerial exposure, and that lateral Gulf Stream shifts are induced by sea-level fluctuations.

More detailed work on the continental shelf of Onslow Bay, North Carolina (Snyder, 1982; Snyder and others, 1982b; Matteucci, 1984; Riggs and others, 1985), indicates that the third-order late Burdigalian, Langhian, and Serravellian seismic sequences of the Miocene crop out directly on the sea floor (Fig. 6) and are covered by Pliocene and Quaternary strata to the east and south along the shelf-slope margin. Based upon very high-resolution seismic data with vibracores for biostratigraphic control, the three third-order seismic sequences (durations of between 1 to 10 m.y.) of the Miocene stratigraphic section were subdivided into at least 18 fourth-order seismic sequences with durations of less than 1 m.y. (Figs. 1 and 6). These data reveal an even more complex history of higher frequency sea-level fluctuations and associated lateral shifts in the Gulf Stream, which combined to form multiple unconformities and the intervening depositional sequences (Fig. 6). These smaller-scale, fourth-order sea-level fluctuations controlled the spatial and temporal variation in lithofacies.

Seismic stratigraphy has demonstrated that major changes in sediment thickness and abrupt appearance or disappearance of seismic sequences are due to paleoceanographic and eustatic effects, and not due to faulting or folding (Snyder, 1982; Snyder and others, 1982b; Popenoe, 1985; Riggs and others, 1985). Regional differential subsidence as well as local collapse or sag structures do play a role in sequence geometry and stratal deformation. However, the depositional and erosional processes, operating during periods of varying duration and frequency, have controlled Neogene seismic stratigraphy along the North Carolina continental margin.

Lithostratigraphy. Detailed lithofacies studies of Neogene sediments in the Aurora phosphate district, North Carolina, (Riggs and others, 1982) have led to the interpretation of the Pungo River and Yorktown Formations as depositional products of two second-order sea-level cycles (supercycles of Vail and Mitchum, 1979) (Fig. 6). Bounding unconformities represent measurable periods of geologic time and are characterized by: 1) topographic relief up to 15 m, 2) hardgrounds of moldic sand-

stone or limestone that supported extensive populations of hardrock boring infauna, and 3) local pavements of laminated and burrowed phosphate mud that were subsequently indurated, bored, and locally fragmented to produce phosphate intraclast pebbles. Regional erosion associated with each of the three surfaces has, in part, determined the final occurrence and distribution of the underlying lithofacies.

A series of smaller-scale cyclic depositional units occur within each formation (Riggs and others, 1982, 1985). Within the Lee Creek Mine, the Pungo River Formation consists of four sediment units (units A through D) and the Yorktown consists of two units (lower and upper), each separated by minor unconformities that mark abrupt lithologic changes between units. Immediately below unconformities are highly burrowed, fossiliferous carbonates, which were exposed as submarine or subaerial surfaces during sediment bypass and nondeposition. During these brief periods, the carbonates were often indurated, the shells leached to produce undeformed fossil molds, and the surfaces bored by hard-rock boring infauna.

The sediment units are characterized by a distinct vertical succession of three main sediment components (siliciclastics, phosphate, and carbonate), and the units are cyclically repeated through time (Riggs and others, 1982). Within a typical sediment unit, clayey phosphorite quartz sands grade upward into phosphorite sands, and culminate in carbonate sediments containing minor siliciclastic and phosphatic sediments. Thus, 1) there is a strong inverse relationship between the siliciclastic and phosphate components, both of which are inversely related to the carbonate component; 2) each unit represents variations on an idealized lithologic cycle, which begins with dominantly siliciclastic sediments that grade upward through a zone that is relatively enriched in phosphate or other similar authigenic minerals into dominantly carbonate sediments; and 3) the major pattern of deposition is repeated within each of the four Pungo River units.

Sediments within the Pungo River Formation are laterally persistent throughout the Aurora phosphate district with minor lithologic variations (Riggs and others, 1982); however, regional changes across Aurora and Onslow Embayments are significant (Lewis and others, 1982; Miller, 1982; Scarborough and others, 1982; Snyder, S. W., and others, 1982; Riggs, 1984). Generally, the dominantly siliciclastic sand units grade into clay facies seaward; the dominantly phosphorite sand units grade seaward into diatomaceous clay facies; the development of the carbonate facies increases onto the Cape Lookout High (Fig. 2 in Riggs and Manheim, this volume).

Vibracoring and seismic profiling of the Pungo River sediments in Onslow and Long Bays, North Carolina, have demonstrated that the sediment units in the Aurora phosphate district represent only a portion of a larger, more complex depositional regime operating during the Miocene (Lewis and others, 1982; Snyder, 1982; Snyder and others, 1982b; Waters, 1983; Riggs, 1984; Riggs and others, 1985). These workers have found the following additinal factors.

1. Miocene Pungo River sediments occur in Onslow Bay on

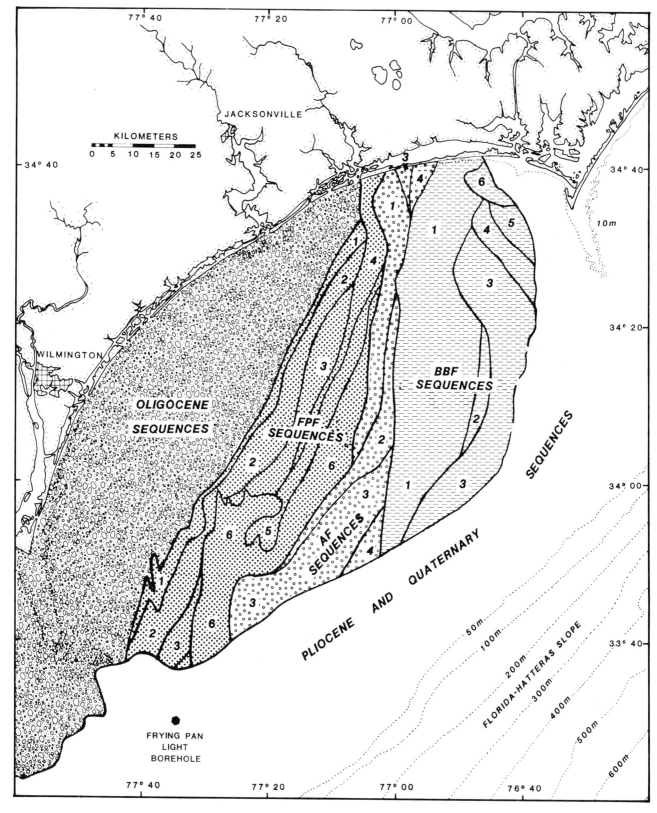

Figure 6. Geologic map of Onslow Bay showing the complex outcrop distribution of three third-order depositional seismic sequences (FPF, AF, and BBF in Fig. 1) and 16 fourth-order depositional seismic sequences of the Miocene Pungo River Formation (after Snyder, 1982). At least 2 more fourth-order seismic units have been mapped in the subsurface but do not crop out in Onslow Bay.

the east flank of the Mid-Carolina Platform where they crop out in a NE–SW trending belt more than 150 km long by 25 to 50 km wide. The western sediment limit is an erosional pinchout with remnants preserved in small flexure basins higher on the Platform. The sediments extend into the subsurface to the east and southeast and thicken rapidly to over 350 m (Fig. 2 in chap. 25, this volume).

2. Minor quantities of phosphate (<2%) occur within all Pungo River lithofacies; however, numerous depositional units represent periods of increased phosphate formation and accumulation (>2% and up to 75%). Pungo River phosphate formation began at least by late early Miocene (about 19 Ma) and continued cyclically through late middle Miocene (to at least 13 Ma and possibly to about 10.5 Ma) (Fig. 1).

3. Four regional unconformities subdivide the Pungo River section into three major depositional sequences (Figs. 1 and 6) (late Burdigalian = FPF Sequence; Langhian = AF Sequence; and early Serravallian = BBF Sequence), which reflect third-order sea-level cycles of 1 to 10 million years duration (Vail and others, 1977). These three third-order sequences are composed of at least 18 smaller-scale depositional sequences (Figs. 1 and 6) similar to those in the Aurora Area. Each sequence was deposited in response to fourth-order sea-level cycles of 100,000 to 1 million years duration. The major sediment types consist of various combinations of quartz sand, clay, phosphate, dolosilt, calcite mud, chert, and bioclastic hash.

4. Between each of the 18 depositional sequences there is an erosional surface, which is often characterized by major changes in lithology. However, following the deposition of some of the sediment sequences, extensive erosion and channeling during periods of low sea level truncated large portions of previously deposited sediments, producing severely dissected surfaces upon which the subsequent sequence of Pungo River sediments were deposited. Consequently, entire depositional cycles with an upper carbonate facies are often only locally preserved due to complex patterns of deposition and erosion of each sediment unit.

5. When the carbonate facies is preserved, it is generally semi-indurated to indurated with a well developed and corroded hardground marking the hiatal surface of nondeposition. The mineralogy and chemistry of underlying sediments often have diagenetic signatures that decrease downward from each hiatal surface. However, upper portions of the unit, including the carbonate facies and associated hardground, have often been truncated by erosion leaving only a subtle diagenetic profile between similar lithologies of successive depositional units as evidence of both cyclic deposition and presence of a hiatus. Vertical profiles of diagenetic alteration associated with hardgrounds and erosional surfaces includes the following: a) foraminiferal recrystallization, dissolution, and mineralization; b) diagenetic growth of such minerals as dolomite, siderite-magnesite, opal CT, clinoptilolite, pyrite, glauconite, and possibly some clay minerals; c) dissolution, precipitation, and trace element modification of phosphate grains; and d) alteration of organic carbon components.

Many Miocene depositional units, if not partially or totally

removed by subsequent erosion, are characterized by a diagenetic profile in which foraminifera occur as molds in the upper portion and grade down-section from the hiatal surface with increasing test preservation (Waters, 1983). Internal chambers of foraminiferal tests form important microenvironments in which various diagenetic minerals are precipitated. Seven different minerals have been recognized forming within the tests during various stages of diagenesis; they are a) calcite, b) carbonate fluorapatite, c) clinoptilotite, d) dolomite, e) glauconite, f) pyrite, and g) siderite or magnesite (Riggs, 1984). Most of these minerals did not form together; rather, the occurrence of each appears to be linked to specific diagenetic processes and conditions associated with the nondepositional or erosional period, following each depositional cycle.

Diagenetic profiles displayed by foraminiferal tests are related to Miocene lithologic cycles and appear to be the result of alteration during subsequent nondepositional or erosional periods. Thus, they are 1) important indicators of degree and type of geochemical processes that have affected more complex authigenic sediment components, such as phosphate; 2) a basis for mapping diagenetic facies within depositional units; and 3) guides to post-depositional history of each sediment unit, including the extent of sediment removal by erosion.

The Pliocene Yorktown Formation is an extensive and significant sediment unit only within the Aurora Embayment, however, thin units of possible Pliocene age do occur in eastern portions of the Onslow Embayment (Snyder, 1982). In the Aurora phosphate district, the Yorktown consists of a lower phosphorite gravelly sand unit and an upper very fossiliferous clayey sand unit without any phosphate. Phosphate in the lower Yorktown is generally most abundant and coarsest at the base, and becomes finer grained upward. Phosphate gravels have been interpreted in the past to be 'obviously reworked' from the underlying Pungo River Formation (Brown, 1958; Gibson, 1967; Miller, 1982), however, similar types of phosphate occur only rarely and locally within Pungo River sediments and always associated with unconformities. Riggs and others (1982) suggested that the lower Yorktown phosphate gravels are related to development of phosphorite pavements on Pungo River carbonate hardgrounds during initial stages of Pliocene transgression and thus represent primary phosphorite sedimentation at that time. In addition, preliminary petrographic and chemical analyses of the lower Yorktown suggest that all of this phosphate may represent primary deposition during a Pliocene phosphogenic episode rather than reworked phosphate formed during a previous episode.

Biostratigraphy and Paleoecology. The biostratigraphic framework of Neogene sediments of the North Carolina continental margin is based upon analyses of faunal and floral assemblages of the Pungo River and Yorktown Formations from exposures in the Lee Creek Mine and extensive drilling throughout the Aurora and Onslow Embayments. Within the Lee Creek Mine, Riggs and others (1982) described four cyclic depositional units that comprise the Pungo River Formation. Only the upper two units (C and D) have yielded sufficient planktonic foraminifera to permit reliable biostratigraphic interpretation. This plank-tonic assemblage was correlated with Blow Zone N8 by Katrosh and Snyder (1982) and Zones N8 and early N9 (Langhian) by Gibson (1983). Diatomaceous clay units in the upper part of the Pungo River Formation from cores approximately 10 km north-east of the Lee Creek Mine contain two diatom assemblages. Based on standard diatom zonation, the older is age-equivalent to Blow Zones N8 and N9; the younger is equivalent to Zone N11 (early Serravallian) (Abbott and Ernissee, 1983). Although a Langhian/early Serravallian age is indicated for the upper part of the formation, no unequivocal age assignment is yet available for the lower two units (A and B) in the Pungo River sediments of the Aurora area.

Lithologic changes within each of the four depositional units appear to reflect paleoclimatic and paleoceanographic fluctua-tions. Clayey sediments in the lower portions of the cycles are diatom-rich and appear to be related to cooler-water conditions (Abbott and Ernissee, 1983). The phosphorite sands within the lower two units (A and B) contain low diversity benthic foramin-iferal and diatom assemblages that are numerically dominated by species favoring high nutrient, oxygen deficient conditions associated with cold-water upwelling (Katrosh and Snyder, 1982; Snyder, S. W. and others, 1982). Phosphorite sediment within Unit C contains a higher diversity foraminiferal fauna, reflecting a more open marine, middle shelf environment. In the carbonates on top of Units C and D, dissolution has severely altered the foraminiferal faunas. The rich, diverse, moldic megafauna of the unit C carbonate caprock reflects subtropical conditions (Carter, unpublished data; Blackwelder, 1980; Blackwelder and others, 1982).

Within the Pungo River Formation in the Aurora area, the changing patterns in composition and diversity of benthic species and the abundance of planktonic specimens suggest multiple ma-rine transgressions with regressive pulses creating minor uncon-formities between successive units. The units recognized at Aurora grade eastward into deeper water deposits with increasing concentrations of diatomaceous muds (Miller, 1982); southward they thin and interfinger with shelly sands across the Cape Look-out High (Scarborough and others, 1982).

The Pungo River Formation in the Onslow Embayment, separated from the Aurora Embayment by the Cape Lookout High (Fig. 2 in chapter 25, this volume), is characterized by at least 18 seismic sequences that are lithologically similar to the four depositional units in the Aurora area. The Onslow Bay units are characterized by complex patterns of deposition and erosion, which are interpreted to be associated with sea-level fluctuations. Major accumulation of phosphate occurred in the Frying Pan area of southern Onslow Bay (Riggs and others, 1985). Here, phosphorite sands are overlain, in succession, by phosphatic sandy mud and muddy foraminiferal quartz sand.

Moderate values of species diversity and faunal predomi-nance typify the rich benthic assemblages from the entire Pungo River section in the Frying Pan area. Foraminiferal assemblages from each unit are distinctive in terms of species composition.

They indicate that all units were deposited on the middle shelf to upper slope, with faunal changes directly reflecting changes in water mass properties and substrate type (Waters, 1983). Phos-phorite sands and phosphatic muds are numerically dominated by species associated with high nutrient levels and low oxygen conditions that characterize upwelling zones. Species adapted to organic-rich fine sediments become prominent secondary faunal elements in muds. Abundant planktonic species predominate in clean quartz sands with negligible phosphate; associated benthic faunas indicate well oxygenated and more open marine environ-ments (Waters, 1983).

Rich planktonic foraminiferal assemblages from the Frying Pan area have produced correlations of the Pungo River sequence with Blow Zones N6 to early N7 (late Burdigalian). The lower phosphorite sands and overlying muds have been assigned to Zone N6; the foraminiferal quartz sands are latest N6 and early N7 (Waters and Snyder, 1986). Calcareous nannofloral assem-blages corroborate this biostratigraphic interpretation (Steinmetz, unpublished data). Pungo River sediments from the Frying Pan area are older than the dated units (Units C and D) exposed at Aurora.

Pungo River sediments recovered from northern and central portions of Onslow Bay represent deposits that include the late Burdigalian, Langhian, and early Serravallian. Planktonic fora-minifera occur in only a few of these sediments, but biostrati-graphic age assignments (Snyder and Moore, 1985) are supported by evidence from calcareous nannofloras (Steinmetz, unpublished data) and seismic correlation with the southern portion of Onslow Bay. Benthic foraminifera are common to abundant. The species associated with high nutrient conditions are most abun-dant in sediments with the highest phosphate concentration; how-ever, these species are generally less abundant than in the Aurora and Frying Pan areas. The predominant benthic taxa indicate that the nutrient-rich conditions noted in those areas were less-well developed in northern Onslow Bay. Lower primary productivity and greater dilution with siliciclastic sediments, possibly derived from the vicinity of Cape Lookout High, resulted in minor phos-phate accumulation.

Age of the Yorktown Formation has now been firmly estab-lished as Pliocene (Snyder, S.W., and others, 1983), based upon planktonic foraminifera from the Lee Creek Mine. Concurrent range zones of the taxa indicate Blow Zones N19 and N20 (mid-dle early through late Pliocene). Hazel (1983) equated ostracode faunas from the Yorktown Formation at the Lee Creek Mine with planktonic foraminiferal Zones N19 and N20. A broader regional study by Gibson (1983) resulted in similar interpreta-tions with a tentative assignment of the uppermost Yorktown strata to Zone N21.

Pliocene clastic sediments are geographically very wide-spread. In southern North Carolina, the Pliocene units contain a warmer water fauna and are referred to as the Duplin Formation, whereas time-correlative sediments in the northern portion of North Carolina are called the Yorktown Formation. At Aurora, the very fossiliferous Yorktown Formation grades upward from muddy phosphorite sand to very muddy quartz sand to sandy mud to slightly muddy quartz sand (Snyder, S. W., and others, 1983). The changing patterns of foraminifera indicate a major decrease in organic content from the lower phosphatic and muddy portions to the non-phosphatic upper portion of the formation. Changes in benthic species composition, diversity values, and the abundance of planktonic specimens upward through the formation indicate a transgressive phase with middle shelf environments followed by a regressive phase with inner shelf environments (Snyder, S. W., and others, 1983).

Model of Neogene Continental Margin Sedimentation

Patterns of Sedimentation. Spatial controls of sedimentation result from the regional tectonic setting (i.e. the major large-scale structural framework and local smaller second-order structural or topographic features), while temporal controls result from the cyclicity of major paleoclimatic and paleoceanographic events (i.e. sea-level oscillations, changing current patterns and water masses, etc.). Thus, the type, intensity, and duration of each oceanographic event dramatically affect the environmental conditions of the continental margin and determine the resulting sediments and associated biota that are deposited as follows.

Neogene sediments on the North Carolina continental margin are characterized by cyclical depositional patterns representing at least three different time scales. Large-scale sediment sequences were deposited in response to the relatively long second-order Miocene sea-level cycle and the subsequent shorter, second-order Pliocene cycle of Vail and Mitchum (1979) (Fig. 1). Deposition of multiple smaller-scale sequences occurred within the Miocene and Pliocene events in response to established third-order cycles of eustatic sea-level change. At a still smaller scale, each third-order depositional sequence is composed of multiple depositional units associated with fourth-order sea-level cycles. Local erosional events were produced by changing patterns of Gulf Stream processes. Regression and low stand portions of sea-level cycles were characterized by nondeposition, erosional scouring and channeling, and diagenetic alteration. These destructional processes severely complicate the depositional patterns of Neogene sediments.

Internal reflection geometries within the Neogene depositional sequences exhibit depositional and erosional features, which are similar in seismic facies to modern and Quaternary shelf-edge sequences. This suggests that processes now operating along the continental margin were active throughout the Neogene. The modern outer shelf extends seaward onto the upper slope along a series of protruding bulges separated by embayments that appear to be zones of relative non-deposition. These sediment bulges consist of internal, seaward-oriented prograding clinoforms as seen in seismic profiles (Matteucci, 1984). Subsequently, tops of the bulges may be eroded by Gulf Stream activity, truncating the clinoforms to produce terrace and scarp morphologies with the formation of hiatal surfaces, hardgrounds, and mineral pavements. Cape Fear Terrace is the best of numerous examples that occur on the North and South Carolina slope.

Along specific sections of the modern Carolina outer shelf, sediments are selectively transported seaward in response to interactions between the outer shelf and major oceanographic currents. Physical oceanographic models show that strong off-shelf currents are generated by dynamic interactions between the Gulf Stream and the continental shelf (L. Pietrafesa, unpublished data). Large, seaward oriented, asymmetric sand lobes and sand waves demonstrate the sediment response to such flows (Matteucci, 1984). High frequency sea-level cyclicity throughout the upper Cenozoic would have accentuated this depositional pattern of shelf progradation along the paleoshelf-slope boundary.

Major lithologic components in Neogene sediments on the North Carolina continental margin are characterized by distinctive regional and vertical distribution patterns (Riggs, 1984). An idealized depositional sequence is as follows. Siliciclastic sedimentation was dominant during early stages of each transgression and within shallower, landward environments; carbonate sedimentation was dominant during late stages of each transgression and within deeper, seaward environments. If phosphogenesis occurred, it began on the upper slope and outer shelf in the early-stages and migrated across the shelf with the transgression, continuing through the mid-stages of transgression. Phosphates formed in local environments within the transition zone between dominantly siliciclastic and dominantly carbonate sedimentation, both laterally and vertically (see Chapter 16, this volume).

Each fourth-order sea-level cycle could have produced a sediment unit characterized by a complete and idealized lithologic sequence or some variation on this theme as determined by local structure, topography, and siliciclastic sediment supply. Preservation of all or part of each depositional unit would be dependent upon contemporaneous erosional activity of the Gulf Stream, and extent and duration of subsequent regressions. Both processes could have severely eroded portions of a lithologic sequence or diagenetically altered the included mineral components. Submarine erosion occurred on the seaward side of the margin during low sea-level stands between each fourth-order depositional unit. Subaerial channeling and truncation of large portions of the section occurred simultaneously on the landward side of the margin following deposition of each major third-order unit, as well as some of the fourth-order units.

Detailed structural contour and isopach maps of each Neogene depositional unit, based on high-resolution seismic data (Snyder, 1982; Matteucci, 1984; Popenoe, 1985), demonstrate that Neogene sedimentation actually took place along a generally northeast-southwest shelf-edge, defined by the Oligocene sediment sequence, which was between 80 to 100 km landward of the present shelf-edge. No sedimentary record of the Aquitanian and early Burdigalian is preserved on the North Carolina continental margin (Fig. 1). During this time sea-level was generally low, which allowed fluvial processes to erode the continental shelf, defining the Aurora and Onslow Embayments and intervening Cape Lookout High.

During the ensuing second-order Miocene sea-level rise, the

late Burdigalian, Langhian, and early Serravallian depositional units of the Pungo River Formation completely filled the Onslow Embayment and a portion of the Aurora Embayment (Figs. 1 and 6). The latter remained as a shelf margin basin during much of the late Miocene and early Pliocene, until it was totally infilled with a thick sequence of Pliocene and Pleistocene siliciclastic sediments from both the north and south directions (Popenoe, 1985). Only the eastern and southern edges of the Onslow Embayment received Pliocene and Pleistocene deposition where an onlapping shelf-edge cover lies unconformably over the Miocene sequences (Fig. 6).

Neogene Phosphogenesis. Neogene phosphorites in North Carolina represent a small portion of a series of major deposits (see Chapter 25, this volume), which formed in response to oceanographic episodes of global extent (Riggs, 1986). During the Miocene, anomalous concentrations of phosphate formed contemporaneously, along with a suite of other aberrant authigenic minerals (see Chapter 16, this volume), throughout an extensive portion of the world's continental margins. Most of these deposits occur in the subsurface below the continental shelf, occasionally cropping out on the seafloor where the phosphate is locally reworked into the Holocene sediments (Baturin, 1982). Some of these phosphate deposits extend onto the emerged coastal plain where they are often economic or subeconomic in value. The known Miocene phosphate deposits have the following general distribution below 45° latitude (Riggs, 1986): 1) east Atlantic continental margin (southwest Europe, northwest Africa through South Africa, and Agulhas Bank); 2) west Atlantic continental margin (North Carolina through Florida, Cuba, and Venezuela); 3) east Pacific continental margin (California, Baja California Mexico, and Peru through Chile); and 4) west Pacific continental margins (East Australia, Chatham Rise east of New Zealand, Indonesia, Sea of Japan, and Sakhalin Island). The stratigraphy, sedimentology, and structural relationships of most of these deposits are similar. Thus, on the basis of temporal and geological relationships, these deposits should be products of similar sets of processes.

Formation and subsequent deposition of Neogene phosphorites throughout the southeastern U.S. continental margin were controlled and defined by two scales of structures and topographic features (Riggs, 1984). Large-scale features defined the regional setting and determined location, size, and geometry of the phosphogenic provinces. Maximum concentrations of phosphate formed and accumulated in shelf environments around the nose and flanks of the Ocala and Mid-Carolina Platform Highs, producing the Florida and Carolina phosphogenic provinces, respectively (Fig. 16, Schlee and others, this volume). Phosphate formation decreased away from the structural noses to minimums in adjacent large-scale embayments.

Superimposed upon this regional pattern of phosphate distribution are a series of smaller-scale structural or topographic highs and adjacent entrapment basins that define the various phosphate districts (Riggs, 1984) (Fig. 5; see also Fig. 2 in Riggs and Manheim, this volume). The geographic location of each

small-scale feature within the larger structural framework determines the amount and type of phosphate formed and type of associated sediment components. Specific size and geometry of each district is dictated by the genesis of the basin where the phosphates accumulated (i.e., structural deformation, primary depositional processes, subaerial or submarine erosion, etc.) and the post-depositional erosional history of the basin.

During Miocene time, an extremely large volume of phosphate sediment formed in the southeastern U.S., as well as in many other portions of the world (Riggs, 1979c, 1984; Riggs and others, 1982, 1985). Minor concentrations of phosphate grains (<2%) occur within all Miocene lithofacies. However, specific lithofacies represent periods of increased phosphate formation and accumulation (up to 75% phosphate grains). Miocene phosphate formation began at least by late early Miocene, about 19 Ma, and continued cyclically into late middle Miocene, between 13 and 10.5 Ma (Fig. 1). The Pliocene was characterized by dramatically decreased volumes of phosphate sedimentation during the much shorter second-order sea-level cycle between 5 and 4 Ma. Local concentrations of Pleistocene/Holocene phosphorite on the North Carolina shelf are only partially reworked from the Pungo River Formation; some phosphate appears to be primary and formed during one or more brief Pleistocene phosphogenic episodes, which coincided with the Pleistocene transgressive cycles.

The volume of phosphate formed during the major episodes was primarily dependent upon their duration, which decreased from approximately 6 million years for the Miocene, to about 1 million years for the Pliocene, and probably 10,000s of years for the Pleistocene episodes. A second factor could be the depletion of the deep-ocean phosphorus sink as a result of Miocene phosphogenesis, as proposed by Sheldon (1980). Riggs (1984) suggested that cooler climates with considerably lower sea levels, coupled with frequent but brief glacial episodes during the Neogene and Pleistocene, could have significantly limited phosphorus renewal in ocean sinks. Consequently, the potential intensity of phosphogenic episodes associated with successive sea-level transgressions through the Neogene and Pleistocene would have decreased due to smaller volumes of available phosphorus, lower organic productivity, and inadequate duration of time for the production and accumulation of significant volumes of phosphate sediments.

Stratigraphic evidence (Riggs, 1984; Riggs and others, 1982, 1985) demonstrates that the major episodes of phosphogenesis occurred in the early to mid stages of transgression associated with sea-level cyclic events resulting from major glaciation and deglaciation, as defined by Vail and others (1977). Regional structural highs determined the location of the phosphogenic provinces. Smaller-scale structural or topographic highs controlled specific sites of phosphate deposition by interacting with oceanic boundary currents along the continental margin. This resulted in cold, nutrient-rich, deeper waters being upwelled into adjacent embayments where high organic production produced disaerobic environments essential for phosphate formation. During transgression, phosphate depocenters migrated farther onto the continental shelf of regional embayments as the oceanic current system increasingly impinged on the shelf.

QUATERNARY GEOLOGY

Introduction

The history of Quaternary investigations on the Atlantic continental margin reflects the disparate approaches between those studies being done on the coastal plain compared to those on the continental shelf, and those studies concerning Pleistocene to early Holocene deposits compared to those of modern sediments and processes. The coastal plain Quaternary was studied by generalists in the 1800s, with little detailed development of stratigraphic models. From the early 1900s, Cooke (1930) championed the 'terrace-formation' concept of Shattuck (1901). This work, stimulated by the topographic mapping of the U.S. Geological Survey, considered terraces to be Pleistocene shorelines that could be correlated along the coast by elevation.

By the 1950s, it became clear that detailed stratigraphic and sedimentologic studies were required to separate marine from non-marine deposits and to correlate units throughout the coastal plain. This new approach was coincident with development of radiometric and other chronological tools that provided a better understanding of age relationships of units. Recent research has become progressively more interdisciplinary in nature, integrating lithostratigraphic, biostratigraphic, and chronological information into detailed chronostratigraphic correlation schemes (Oaks and DuBar, 1974; Wehmiller and Belknap, 1982; McCarten and others, 1982; Oldale and others, 1982).

A similar history of research of the submerged part of the margin was detailed by Uchupi (1968). This research was strongly tied to developments in marine technology such as in situ sampling and remote sensing, emergence of major oceanographic institutions, development of the National Science Foundation research vessel fleet, and growth of marine science programs in universities in most coastal states. Much of the initial new work concentrated on modern sedimentary processes and distribution of sediments, but new techniques for coring on the margin in consort with high-resolution seismic profiling have begun to elucidate Quaternary shelf stratigraphy. Recognition of the complexities of the shelf, slope, and rise have increased with each new research tool and area investigated. The net results are major stratigraphic advances that are beginning to force the reevaluation of more conventional stratigraphic interpretations based upon standard coastal plain drilling and outcrop mapping.

Modern sedimentary environments are typically discussed by type rather than region, such as continental shelf (Swift and others, 1972a), barrier islands (Leatherman, 1979), and estuaries (Lauff, 1967). These studies have generally followed a course of development from early descriptive geomorphic studies (Johnson, 1919), to more rigorous, often quantitative approaches. This

evolution indicates the maturation of modern sedimentology as a scientific discipline.

Pleistocene Systems

Basis of Pleistocene Stratigraphy. Two aims of stratigraphy are to place rocks in a chronological sequence and to correlate them from place to place. This requires basic sedimentologic data, as well as correlation tools such as biostratigraphic and absolute dating techniques. Geologic interpretations of the Pleistocene record are based on different types and quality of information for emergent and submergent portions of the continental margin. Also, geomorphic, lithostratigraphic, biostratigraphic, and chronostratigraphic approaches require different data bases, which are usually either not carried out simultaneously or are not equally available for use in unraveling complex relationships.

Regional studies of scattered outcrops and quarries form the data base of most geological studies on the coastal plain. However, due to the cyclical patterns of Pleistocene sedimentation in consort with subtle tectonic movements and complex processes of deposition and erosion, interregional correlation is extremely difficult. The result is that similar lithologies are repeated through time with complex geometric relationships. Consequently, it becomes essential to invoke tools that have more than local significance, such as absolute dating, biostratigraphy, and seismic stratigraphy.

On the submerged margin, the understanding of Pleistocene geologic systems lags behind that of the coastal plain and is further complicated by the increased remoteness. Evolution in geomorphic thought is tied to recent bathymetric mapping (Emery and Uchupi, 1972), compilation of physiographic studies (Uchupi, 1968), and development of remote sidescan sonar (Laughton, 1981) for mapping detailed sedimentologic systems (Prior and others, 1979; McGregor and others, 1982). Lithostratigraphic studies are often limited to surface samples, shallow cores, and an occasional deep drill hole (Hathaway and others, 1979). Only during the past decade or so have detailed efforts begun to extend coastal plain mapping onto the submerged shelf. Mixon and Pilkey (1976) and Blackwelder and others (1982) made preliminary extrapolation of the surface geology into Onslow Bay, North Carolina. Detailed stratigraphic mapping of the Carolina margin utilizing a multidisciplinary analysis of high-resolution seismics and vibracores extended the subsurface coastal plain data base seaward (Lewis and others, 1982; Snyder and others, 1982b; Hine and Snyder, 1985; Popenoe, 1985; Riggs and others, 1985).

Pleistocene biostratigraphic information comes from three major sources, mollusks, microfauna, and pollen. Blackwelder (1981) utilized mollusks, which are abundant and well preserved, to define three biozones from the central and southern coastal plain (Fig. 7). Ostracoda are the dominant microfossil group used to date (Valentine, 1971; Hazel, 1977; Cronin, 1980; Cronin and others, 1981). Cronin (1980) defined three ostracode assemblage biozonations on the coastal plain (Fig. 7). Planktonic foramini-

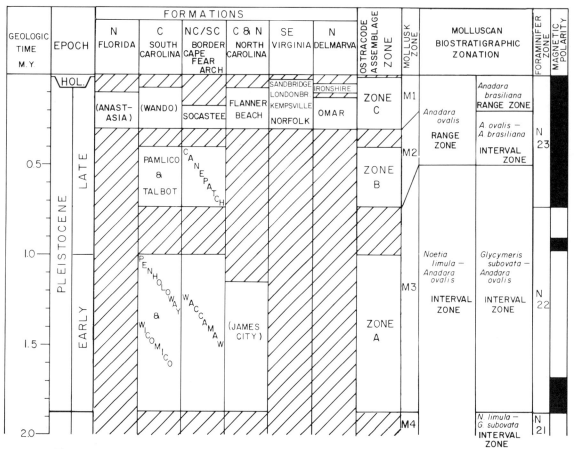

Figure 7. Correlation of Pleistocene stratigraphic units of the Atlantic continental margin based on biozonations of molluscs (Blackwelder, 1981), ostracodes (Cronin, 1980), planktonic foraminifera (Berggren and van Couvering, 1974), and standard magnetic polarity chrons.

fera indicators for the Pleistocene are referred to Berggren and van Couvering's (1974) world-wide planktonic foraminifera zones N22 and N23 (Fig. 7). Pollen is used primarily in paleovegetation reconstructions (Delcourt and Delcourt, 1981).

Absolute dating techniques used for Quaternary sediments of the Atlantic margin include radiometric (^{14}C and U-series dating) and non-radiometric methods (amino acid racemization). Other sequential dating techniques used for relative ages include magnetostratigraphy (Liddicoat and others, 1981, 1982), soils (Owens and others, 1983), and pollen (Cronin and others, 1981) that allow dating only if complete, uninterrupted sediment sequences are available. These methods provide binary information useful in stratigraphy such as glacial or nonglacial and reverse or normal polarity.

Radiocarbon dating may be useful on materials containing organic carbon that are less than 40,000 years old, if the limitations are understood and contamination can be avoided (Stuckenrath, 1977). Uranium-series dating measures the ratios of various parent and daughter elements in a variety of decay sequences of uranium to thorium. If the sample represents a closed

system with respect to all parents, daughters, and intermediate decay products, then the technique is useful for dating samples that range from 0 to 300 or 400 ka (Cronin and others, 1981; Mixon and others, 1982; McCarten and others, 1982). Uranium-series dating is useful in unrecrystallized marine calcium carbonates such as corals, because carbonate contains appreciable uranium but essentially no thorium when deposited.

Amino acid racemization is a method of dating mollusk shells and other organic materials based on changes in stereochemistry of preserved amino acids (Wehmiller, 1982). This method has a dating range up to 2 million years and has been successfully used as a relative stratigraphic dating tool in the Atlantic coastal plain (Mitterer, 1974, 1975; Belknap, 1979; Belknap and Wehmiller, 1980; Wehmiller and Belknap, 1978, 1982). Problems with the amino acid racemization dating technique are primarily due to temperature sensitivity; models of along-coast temperatures through time must be constructed for accurate chronologic correlations (Wehmiller and Belknap, 1978, 1982).

Ideally, stratigraphic correlations are made with the use of multiple chronostratigraphic techniques. Absolute age dating techniques are playing an increasingly more important role as interregional correlations and become critical to understanding the complex continental margin depositional patterns. Future Quaternary stratigraphic studies will, of necessity, become more dependent upon interdisciplinary approaches utilizing multiple stratigraphic techniques.

Depositional Environments During Glaciation. Reviews of glacial theory development and mapping of glaciated terrain on the Atlantic continental margin are abundant (Schafer and Hartshorn, 1965; Flint, 1971; Emery and Uchupi, 1972; White, 1973; Black and others, 1973; Borns, 1973; Denton and Hughes, 1981). Early work concluded that the primary Quaternary event in New England was a major overriding by the Laurentide Ice Sheet during the latest glacial stage (Wisconsinian). The pre-Wisconsin history of New England was a matter of serious debate after Fuller (1914) placed the Gardiner's Clay of Long Island in the Yarmouth Interglacial. In 1964, Kaye argued for multiple glaciations based upon preserved glacial deposits on Martha's Vineyard and Nantucket. Recent work has demonstrated that units such as Gardiner's Clay contain warm water mollusks, ostracodes, and pollen, which suggest interglacial climates (Gustavson, 1972, 1976; Oldale and others, 1982). Radiocarbon dates (Sirkin and Stuckenrath, 1980) indicate mid-Wisconsin events, but amino acid racemization and U-series coral dates (Belknap, 1979; Oldale and others, 1982) suggest that the same units are at least 125,000 years old and that there are at least two separate pre-Wisconsin interglacials represented on Long Island. Thus, it now appears that the Quaternary history of southern New England is more completely preserved than previously thought.

During the past decade, new understandings of New England's glacial geology have resulted from increased emphasis in four areas. These study areas include 1) mechanisms of glaciation and deglaciation (Mayeski and others, 1981), 2) production of glacial sediments and processes of glacier-front sedimentation (Koteff, 1974; Koteff and Pessl, 1981; Smith, 1982; Thompson, 1982), 3) use of absolute dating techniques (Stuiver and Borns, 1975; Sirkin and Stuckenrath, 1980; Oldale and others, 1982), and 4) integration of mapping, detailed sedimentologic and stratigraphic site studies, and glacial modeling (Larson and Stone, 1982). For example, when marine-based glaciers are grounded below sea level, frontal sediments are deposited directly into sea water and on top of normal basal tills (Hughes, 1981).

Ice-flow directions of the Laurentide Ice Sheet across the New England coastal zone and Gulf of Maine (Fig. 8) have been summarized by Schlee and Pratt (1970). They used bedrock and sedimentary landform flow indicators on land and an interpretation of sedimentary characteristics and supposed sediment provenance for the now submerged shelf. However, control from absolute age dating is now essential to determine if these sediments resulted from multiple glaciations.

Proximal to glaciated areas, extreme changes in climate cause changes in surficial processes related to frost activity that increase mass-wasting and mechanical weathering and decrease chemical weathering (Flint, 1971; Oldale, 1982a). Cooler and drier conditions during glaciation leads to increased aeolian activity such as deposition of the Parsonsburg Sand over central Delmarva Peninsula between 30 ka and 13 ka (Denny, 1982). Then, during interglacials, soil development reflects major climatic shifts with increased chemical weathering (Owens and others, 1983).

Glacial History. It is generally accepted that the Laurentide Ice Sheet reached its maximum (Fig. 8) in the northeast 21,000 to 18,000 years ago (Sirkin and Stuckenrath, 1980). It terminated at the Ronkonkoma Moraine on Long Island and at its correlatives, the Vineyard Moraine and Nantucket Moraine in Massachusetts, and on submerged portions of the continental shelf (Fig. 8) (Schlee and Pratt, 1970; Flint, 1971; Trumbull, 1972; Borns, 1973; Schlee, 1973; Sirkin and Stuckenrath, 1980; Mayewski and others, 1981). This moraine system has been extended eastward, based upon submerged gravels, to Georges Bank and across the Northeast Channel to the shelf break off Nova Scotia (Schlee and Pratt, 1970; Trumbull, 1972; Schlee, 1973; Tucholke and Hollister, 1973; Mayewski and others, 1981). To the west, maximum ice position reached into northern New Jersey where it was surrounded by Glacial Lake Passaic (Conally and Sirkin, 1973). Clear evidence now exists suggesting that multiple glacial and interglacial stages are present on Long Island (Belknap, 1979), Nantucket (Oldale and others, 1982) and in New Jersey, where Wisconsin drift overlies older Kansan and Illinoisan drift (NRC, 1959).

Subsequently, the glacial margin retreated inland to the southern coastal area of New England leaving a series of end moraines (Fig. 8), including Harbor Hill Moraine of Long Island, New York, Sandwich Moraine on Cape Cod of Massachusetts and Ledyard Moraine of Connecticut (Flint, 1971; Oldale, 1982b). The eastern glacial margin receded across the Gulf of Maine, mostly in contact with rising sea level. The Fundian Moraine system was constructed 17,000 to 16,000 years ago (Fader and others, 1977) and a belt of end moraines was constructed along the present coastal zone of Maine and New Hampshire between 14,000 and 12,500 years ago (Borns, 1973) (Fig. 8). Georges Bank and the Scotian Shelf on the outer continental shelf were ultimately drowned by rising sea level. Contemporaneously, the southern ice margin retreated north of the present coast of Connecticut (Borns, 1973; Connally and Sirkin, 1973) and formed a topographically controlled, irregular, terrestrial ice margin across southern Vermont and New Hampshire, which was grounded and ablating (Mayhewski and others, 1981).

As the eastern ice margin retreated across Maine, ice was afloat on the sea and receded by calving and ablation (Stuiver and Borns, 1975). As the ice receded, the isostatically depressed central portion of Maine was flooded, at which time the widespread glaciomarine muds of the Presumpscot Formation were deposited (Bloom, 1963). As deglacial isostatic uplift exceeded eustatic sea-level rise, the shoreline withdrew from its high stand about 140 m above mean sea level to its present position along the Maine coast by about 12,000 years ago. The sea continued to

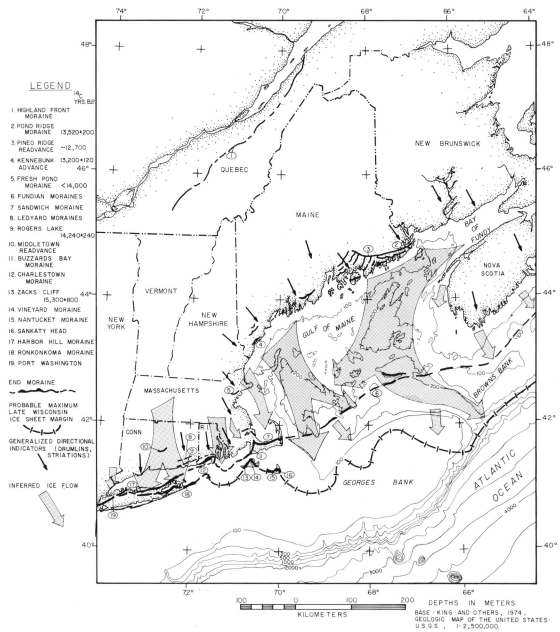

Figure 8. Schematic map of the geologic history of the Northeast Atlantic continental margin during the Wisconsin glaciation (after Schlee and Pratt, 1970; Borns, 1973).

withdraw to a lowstand of about 47 m below mean sea level by 9,000 years ago off Massachusetts and New Hampshire in response to deglacial uplift (Schnitker, 1974; Oldale and others, 1983; Birch, 1984). At this time the Merrimack, Kennebec, and Machias Rivers carried coarse sediments and built deltas on the shelf (Oldale and others, 1983; Belknap and others, 1986). By late Holocene time, rising sea level had drowned a terrain of moderately high relief and produced the present highly irregular shoreline of northern New England.

The character of the present coastline in the northeastern

United States is determined by location of the zero isobase of emergence, which passes through Boston and into the Gulf of Maine and Nova Scotia (Flint, 1971). Interaction between glacial rebound and eustatic sea-level rise has resulted in the emergence of glacio-marine deposits in the northern New England coastal zone and partial submergence of terrestrial deposits in southern New England and New York. The emerged belt increases from zero width at sea level in the Boston area, northward into Maine where it is up to 60 km wide and up to 140 m above sea level (Thompson, 1979). In Maine, ice-marginal deposits within this belt consist of submarine stratified end moraines, small washboard end moraines, submarine outwash fans and aprons, and ice-contact marine deltas (Smith, 1982), all of which interfinger with or are stratigraphically overlain by a blanket of glaciomarine silts and clays of the Presumpscot Formation (Bloom, 1963).

Sedimentary Processes in Nonglaciated Environments.
The Pleistocene nonglaciated province is bounded on the north by glacial end moraines in New Jersey and Long Island; it extends southward through the Florida Keys. Comparison between continental margin morphologies within and south of the glaciated area shows profound differences. The glaciated portion of the Gulf of Maine consists of several large closed basins surrounded by broad areas of irregular topography produced by glacial bedrock scouring (Schlee and Pratt, 1970). The basins contain basal glacial sediments capped by shallow marine muds deposited immediately after ice retreat from the Gulf of Maine (Lougee, 1953; Schnitker, 1975). The margin south of the glaciated area has no large closed basins, is characterized primarily by sand, and was subjected to subaerial exposure and subsequent coastal sedimentary processes during glaciation and deglaciation.

In nonglaciated areas, nonmarine environments were altered by systematic migrations of climatic and biogeographic zones south and north along the margin in response to episodes of glaciation and deglaciation. Pollen records present strong evidence for these climatic changes through the late Pleistocene glaciation (Delcourt and Delcourt, 1981). Figure 9 shows reconstructions of climatic zones and sea-surface temperature for 18,000 radiocarbon years B.P. and emphasizes the southern shift and narrowing of vegetation zones with steeper latitudinal temperature gradients than at present (CLIMAP, 1976; Delcourt and Delcourt, 1981).

Sedimentary environments and their associated processes are controlled by temperature and amount of rainfall, which determine the vegetative cover, type and degree of weathering, and volume of sediment delivered to rivers (Meade, 1969, 1982; Milliman and Meade, 1983). The underfit character of modern coastal plain streams (Flint, 1971) and extensive post-Miocene gravel deposits that occur throughout the upper coastal plain from Maryland (Schlee, 1957) and Delaware (Jordan, 1974) through North Carolina (Daniels and Gamble, 1974) suggest the presence of braided fluvial systems associated with glacial periods. Lowered sea levels during glaciations would have brought subaerial processes along with abundant fluvial sediment across the continental shelf, as evidenced by extensive Quaternary channeling (Snyder and others, 1982b; Hine and Snyder, 1985; Riggs and others, 1985) and occurrence of freshwater peat and mammoth teeth (Emery, 1968; Edwards and Emery, 1977). Climatic patterns also control incident wave energy and associated coastal processes. For example, storm tracks were diverted southward to follow the ice margin during glacial stages (Ruddiman and McIntyre, 1979; Cronin and others, 1981) causing increased storm-related coastal sediment redistribution.

The nonglaciated portion of the margin records multiple

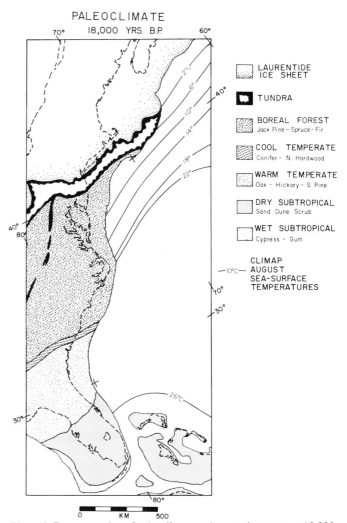

Figure 9. Reconstruction of paleoclimate and vegetation patterns 18,000 years ago based upon pollen records and planktonic foraminifera data (after Delcourt and Delcourt, 1981, Fig. 5). Sea-surface temperature reconstruction for the same period is based upon CLIMAP (1976).

Pleistocene sea-level fluctuations, which have left estuarine, barrier, and marine shelf deposits preserved throughout the coastal plain. However, on most parts of the present inner and middle continental shelves, the Pleistocene record is thin and poorly preserved because of destruction by migrating littoral erosion zones. The thickness of Pleistocene sediments increases to over 100 m northward towards the southern limit of glaciation, suggesting glacial sources (Knott and Hoskins, 1968; Emery and Uchupi, 1972), and increases along the outer shelf and upper slope where wedge-shaped units up to 150 m and 500 m thick cap the shelf-edge off Cape Fear (Matteucci, 1984) and Cape Hatteras, North Carolina, respectively (Popenoe, 1985). The latter suggests either sediment bypassing the shelf during lower sea stands (Swift and others, 1972b) or increased preservation potential due to shorter exposure to shoreline erosion during low stands. The Pleistocene record on the shelf is very poorly known in comparison to the coastal plain.

During lowered sea level, fluvial sediments were carried across the shelf in their fluvial channels. Near the shelf edge, these sediments entered littoral transport until they were intercepted by submarine canyons (Shepard and Dill, 1966; Normark, 1974; Farre and others, 1983); some Pleistocene paleochannels emptied directly into submarine canyons, such as the Delaware River paleochannel, which emptied into the Wilmington Canyon (Twichell and others, 1977; McGregor and others, 1982) and the Hudson River shelf valley follows its paleochannel to the Hudson Canyon, the largest such feature on the east coast (Emery and Uchupi, 1972). Thus, during Pleistocene low sea-level events, fluvial sediment bypassed the shelf and slope to be deposited on deep-sea fans associated with the continental rise. Also, slumps and other forms of mass movement transported unstable Pleistocene sediments down the continental slope, producing areas of outcropping Tertiary sediments on the slope (Robb and others, 1981; May and others, 1983).

During transgression, only Pleistocene sediments occurring within local areas of active subsidence or within abundant, deeply scoured fluvial and tidal inlet channels crossing the continental shelf will be preserved. These paleochannels have been backfilled with fluvial, estuarine, littoral, and shelf sediments (Swift and others, 1972b; Belknap and Kraft, 1981; Riggs, 1984; Hine and Snyder, 1985). The ancestral Delaware River valley was incised into pre-Holocene sediments in a distinctly separate channel system from the modern shelf valley, which is incised entirely into Holocene sands (Sheridan and others, 1974; Twichell and others, 1977). Other examples of paleochannels have been documented off North Carolina (Hine and Snyder, 1985; Riggs and others, 1985) and New Jersey (Swift and others, 1980).

The primary process affecting the long-term preservation potential of the Pleistocene record on the continental shelf is shoreface erosion. After each glacial episode, sea-level transgression produces a ravinement surface, which migrates landward and erodes large portions of the previously deposited sediments (Fischer, 1961; Bruun, 1962; Swift, 1968, 1975; Belknap and Kraft, 1981, 1985; Hine and Snyder, 1985). Important factors in preservation potential include balances between wave energy, sediment budget, rates of relative sea-level rise, and antecedent topography.

During high sea-level stands, the subaerially eroded Pleistocene sediments deposited during the previous high stand now receive a new sequence of estuarine, barrier island, and inner shelf sediments emplaced over an unconformity, which cuts into previously deposited sequences. The result is a complex set of multiple stacked deposits of similar lithologies whose geometries have been totally modified by subsequent erosional processes. Extensive sequences of complex Pleistocene sediment units have been described in Delaware (Demarest and others, 1981), Virginia (Oaks and others, 1974), northeastern North Carolina (Eames, 1983), and South Carolina (Colquhoun, 1974).

In summary, Quaternary shorelines migrating rapidly across

the shelf have destroyed much of the depositional record of multiple glaciations and deglaciations. However, peak high and low sea-level stands associated with each transgression and regression have left complexly stacked sequences of similar lithologies on the coastal plain and at the shelf edge, both of which have been further complicated by subsequent erosional processes. DSDP drill holes have demonstrated a more complete record of Pleistocene sections on the continental rise and abyssal plains. Thus, the Pleistocene record has been largely erased on the continental shelf and is primarily preserved on the landward and oceanic edges of the continental margin; the landward coastal plain portion of the record is poorly legible and the oceanic edge is accessible with great difficulty and cost.

Stratigraphy of Nonglaciated Environments. Terrigenous clastic materials occurring within marine, paralic, and continental sedimentary facies dominate the Pleistocene stratigraphic units in the mid-Atlantic region. Southward, the terrigenous clastics become increasingly interbedded with carbonate units. In South Florida most of the Pleistocene units are dominated by carbonates such as in the Miami and Key Largo Limestones.

Pleistocene formations in the mid- and south-Atlantic regions were initially correlated by the geomorphic terrace concept of Cooke (1930). Hoyt and Hails (1974) refined the concept and defined six terrace formations in Georgia. These occur at decreasing elevations ranging from 30 m above present sea level on the inland side to 1.5 m on the seaward side and include the following, respectively: Wicomoco, Penholoway, Talbot, Pamlico, Princess Anne, and Silver Bluff Formations. Winker and Howard (1977) traced these shoreline elevations laterally and demonstrated tectonic deformation along the coastal plain. More recent detailed studies supply evidence of complex shoreline reoccupations, overrunning of older terraces, and large hiatuses in the record. The advent of new absolute and relative dating techniques has helped with time correlations of units. Table 1 is a compilation of common formation names, age assignments, and major references for the mid- and south-Atlantic regions.

Figure 10 is a generalized geologic map of Pleistocene marine formations and features of the Atlantic continental margin. Due to lack of precise controls, absolute dating and correlation of locally defined lithologic units by fauna is tentative in some cases. Late Pleistocene deposits are preserved on the coastal plain in the Southeast Georgia, Hatteras, and Salisbury Embayments. Early and middle Pleistocene deposits are only preserved over the Carolina Platform. Up to 10 m of gentle differential warping and 25 m of absolute uplift have deformed the relatively stable coastal plain (Winker and Howard, 1977; Blackwelder, 1981; Cronin and others, 1981). Similar warping is expected on the submerged margin.

Holocene Systems

Sedimentary Processes Associated with the Holocene Transgression. The Holocene is the latest interglacial half-cycle, or transgressive event, of the highly cyclical patterns of sedimen-

TABLE 1. SUMMARY OF MAJOR PLEISTOCENE MARINE STRATIGRAPHIC UNITS OCCURRING
ON THE ATLANTIC COASTAL PLAIN.

Formation	Age Estimate (ka = 10^3 years)	Major Information Sources[*]
New Jersey		
Cape May Fm.	100-140 ka	Richards (1967); Wehmiller and Belknap (1982)
Delaware and Maryland		
Sinepuxent Fm.	<40 ka (^{14}C)	Mixon and others (1982).
Ironshire Fm.	70 ka (U-Th)	Mixon and others (1982).
Omar-Accomac Fm.	190 ka (U-Th)	Mixon and others (1982).
Omar Fm.	Early and Mid-Pleistocene	Demarest (1981); Belknap and Wehmiller (1980); Belknap (1979); Jordan (1974); Wehmiller and Belknap (1982); Owens and Denny (1979).
Virginia		
Kempsville Fm.	60-80 ka (U-Th)	Oaks and others (1974); Johnson
Norfolk Fm.	Mid-Late Pleistocene	(1976); Belknap (1979).
Tabb Fm.	220-350 ka (AAR)	
Windsor Fm.	Early Pleistocene	
North Carolina (North and Central)		
Diamond City Fm.	<40 ka (^{14}C) Early, Middle, and Late Pleistocene	Moslow and Heron (1978); Belknap (1982); Snyder and others (1982a).
Atlantic Sand	Late Pleistocene	Mixon and Pilkey (1976).
Core Creek Sand	Mid-Late Pleistocene	Mixon and Pilkey (1976).
Flanner Beach Fm.	Middle Pleistocene	DuBar and others (1974).
James City Fm.	Early Pleistocene	DuBar and others (1974).
North Carolina (Cape Fear Arch)		
Socastee Fm.	Middle Pleistocene	McCarten and others (1982).
Canepatch Fm.	Middle Pleistocene	Wehmiller and Belknap (1982).
Waccamaw Fm.	Early Pleistocene	Liddicoat and others (1981); Cronin and others (1981); Blackwelder (1981); DuBar (1974).
South Carolina		
Wando Fm.	Late Pleist.(U-Th)	McCarten and others (1980; 1982).
Princess Anne Fm.	Mid- Late Pleist.(AAR)	Wehmiller and Belknap (1982).
Silver Bluff Fm.	Mid- Late Pleist.(AAR)	Wehmiller and Belknap (1982).
Socastee Fm.	Middle Pleistocene	Colquhoun (1974).
Pamlico Fm.	Middle Pleistocene	Colquhoun (1974).
Talbot Fm.	Middle Pleistocene	Colquhoun (1974).
Penholoway Fm.	Early Pleistocene	Colquhoun (1974).
Wicomoco Fm.	Early Pleistocene	Colquhoun (1974).
Georgia		
Silver Bluff Fm.	<40 ka (^{14}C)	Hoyt and Hails (1974).
Princess Anne Fm.	Late Pleistocene	Winker and Howard (1977).
Pamlico Fm.	400-600 ka (AAR)	Belknap (1979).
Talbot Fm.	400-600 ka (AAR)	Belknap (1979).
Penholoway Fm.		
Wicomico Fm.		
Florida		
Anastasia Fm.	Late Pleistocene	DuBar (1974); Blackwelder (1981); Bender (1973); Mitterer (1974, 1975).
Miami Limestone	Late Pleistocene	Broecker and Thurber (1965); Osmond and others (1970); Wehmiller and Belknap (1978, 1982).
Bermont Fm.	Early Pleistocene	Broecker and Thurber (1965); Osmond and others (1970).
Caloosahatchee Fm.	Early Pleistocene	Broecker and Thurber (1965); Osmond and others (1970); Wehmiller and Belknap (1978, 1982).

[*]For greater detail, see Oaks and DuBar (1974) and Jordan and Smith (1983).

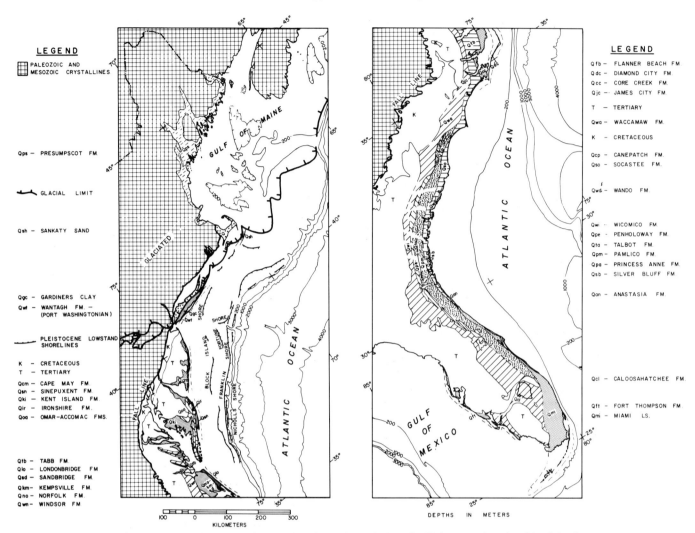

Figure 10. Generalized geologic map of marine and marginal marine Pleistocene deposits of the Atlantic continental margin. Correlations based on relationships shown in Table 1 and Figure 7. Map is based on data from Oaks and DuBar (1974), Stuiver and Borns (1975), Mixon and Pilkey (1976), Winker and Howard (1977), Owens and Denny (1979), and Mixon and others (1982). Base maps are from the 1:2,500,000 map by King and others (1974) and from the 1:1,750,000 glacial map by NRC (1959).

tation that has dominated the Atlantic continental margin throughout the Neogene and Quaternary. The Modern sediment system is an instantaneous time slice through a small segment of a fourth- or fifth-order sea-level cycle within the complex pattern of multiple cyclic events. Considering the past 24 million years of earth history, it is highly probable that the global cycles controlling climatic variations will re-establish glacial conditions along with rapid evolutionary changes of the modern sedimentary environments.

During the Holocene, the Atlantic continental margin has been dominated by post-glacial rise of sea level and resultant marine transgression. The shorelines and associated paralic sedimentary environments formed near the shelf break about 18,000 years B.P. and migrated across the shelf to its present position

following the sea-level curves delineated in Figure 4. Transgression continues today along most sections of the coast.

Changing climatic conditions resulting from deglaciation during the Holocene are readily apparent in the preserved sediment sequence in northeastern North Carolina (Copeland and others, 1983). During maximum glacial development and lowstand, approximately 18,000 years B.P., cool and semiarid climatic conditions supported sparse vegetative cover with enough precipitation to maximize sediment erosion. The resulting sediment-choked streams discharged coarse terrigenous sediments to the subaerially exposed continental shelf, as evidenced by basal gravels and coarse sands in preserved fluvial channels. As climates moderated and glaciers receded, first boreal and then temperate vegetation developed along with extensive wetland pocosins (Whitehead, 1981). Increased vegetative cover decreased the volume and size of fluvial sediment regimes to predominantly a suspended load of silt and clay (Copeland and others, 1983). The leading edge of the transgression flooded topographic lows, forming the present embayed estuarine system; old fluvial channels in the estuaries filled first with coarse fluvial sand and gravel followed by thick accumulations of organic-rich, estuarine mud. Modern rivers are delivering predominantly suspended sediments (Fig. 11), which are being discharged to the continental shelf, producing a thin, but ephemeral mud blanket on top of older sediment along portions of the inner shelf (Swift, 1976). Fluvial sands have not been transported into the modern estuaries since flooding began; sands within the modern estuarine system have been derived either internally from estuarine shoreline erosion or from the sand system in adjacent barrier islands (Copeland and others, 1983). Sandy carbonates and calcareous quartz sandstones of Holocene age are scattered across the middle and outer shelf in Onslow Bay and grade into contemporaneous coralline algal limestones along the outer shelf-edge (Blackwelder and others, 1982).

Thus, within the mid- and south-Atlantic regions, major north-south and cross-shelf shifts occur between terrigenous and carbonate sediments through each sea-level cycle resulting from glaciation and deglaciation. Other sections of this chapter and Chapter 16 (this volume) demonstrate that most of the Quaternary and Neogene sediment units of the southeastern U.S. continental margin are characterized by similar lithologic patterns (Riggs, 1984).

Variations in the interaction between rising sea level and sediment supply will determine whether stability, regression, or transgression will take place (Curray, 1965; Swift, 1975). The present Atlantic coast is dominated by transgressive shorelines (Kraft, 1971; Pierce and Colquhoun, 1970; Rampino and Sanders, 1981). However, local portions of the coast have a regressive stratigraphy, due to local effects of sediment supply or wave energy overwhelming relative sea-level rise (Kraft and others, 1978; Hayes, 1979; Hine and Snyder, 1985).

Bruun (1962) suggested that in a rising sea level, the equilibrium profile of erosion and deposition would rise as well, and shoreface erosion would supply sediment for deposition offshore. Since equilibrium profiles are most probably related to and controlled by the energy regime of both seasonal and individual storm events, the time framework becomes important when considering the null line between erosion and deposition. Studies of erosional shoreface retreat on the middle Atlantic coast suggest that retreat is a continuous process at a time scale of a century or greater (Niedoroda and others, 1985).

High-resolution seismic profiling, in consort with detailed vibracoring, has led to a new understanding of Holocene shelf sedimentation, and to preservation potential of coastal systems on the continental shelf following shoreline migration. Belknap and Kraft (1981, 1985) suggested that preservation potential is dependent upon depth of shoreface erosion, which in turn is a function of incident wave energy, sediment supply, rate of local

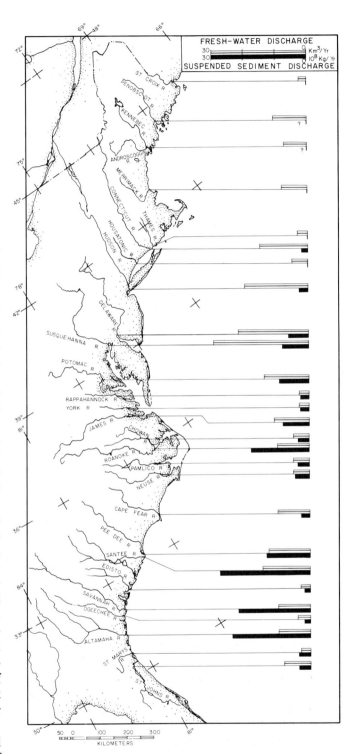

Figure 11. Suspended sediment load associated with fresh-water discharges of coastal plain streams along the U.S. Atlantic margin (after Meade, 1969).

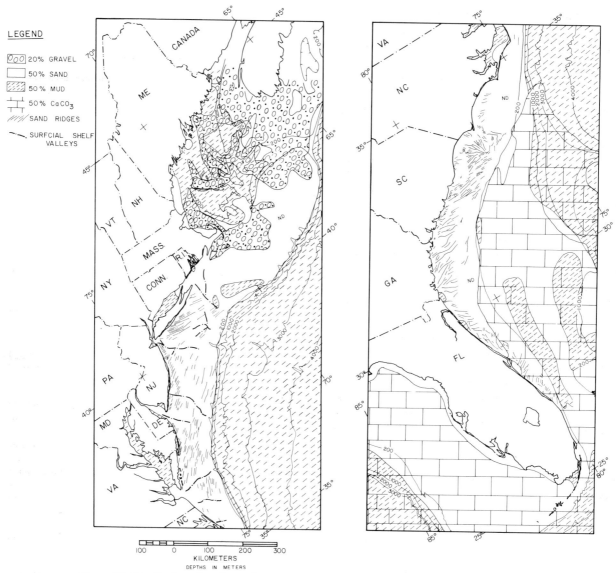

Figure 12. Distribution of marine sediment types on the Atlantic continental margin. Data are from Schlee and Pratt (1970), Emery and Uchupi (1972), Milliman and others (1972), Trumbull (1972), Hollister (1973), and Schlee (1973).

relative sea-level change, and antecedent topography. Their model shows enhanced preservation potential on the outer shelf during rapid transgression, near zero potential on the inner shelf during slow transgression, and maximum potential during peak high sea-stands as systems switch from transgression to regression. Most preserved Pleistocene formations on the Atlantic coastal plain are peak sea-level deposits (Demarest and others, 1981).

Without active subsidence or major sediment input, coastal lithosome preservation on the middle and inner continental shelf is minimal (Hine and Snyder, 1985). Only the lowest portion of the coastal lithosome has a significant potential for preservation. The preserved lithosome includes lower estuarine facies, bottom portions of deeper inlet channels, and fluvial facies infilling the

deep, cross-shelf fluvial channels (Swift and others, 1972b; Kumar and Sanders, 1974; Sheridan and others, 1974; Field and others, 1979; Field 1980; Belknap and Kraft, 1981, 1985; Hine and Snyder, 1985). In local shelf areas, sediments representing more extensive coastal lithosomes have survived transgression (Sanders and Kumar, 1975; Rampino and Sanders, 1981; Stubblefield and others, 1983). Sanders and Kumar (1975) suggested that if sea-level rise was not smooth, barrier island sediments might have been preserved on the inner shelf.

Emery and Uchupi (1972) summarized the general distribution of modern sediment types on the Atlantic continental margin (Fig. 12). The shelf is predominated by sands composed of detrital quartz and feldspar with low organic content ($<0.5\%$ C).

Terrigenous gravels are generally confined to the glaciated margin, such as Georges Bank and the Gulf of Maine (see Chapter 25, this volume). Organic-rich mud (>0.5% C) occurs in deep basins of the Gulf of Maine, estuaries and lagoons along the coast, and on the slope and rise north of Georgia. Minor phosphate occurs on the shelf and slope terraces from the Carolinas through Florida, with extensive phosphate and manganese sediments occurring on the Blake Plateau. Calcium carbonate precipitation and accumulation constitutes 100% of the sediment in the Bahama Banks, is the predominant sediment component in southern Florida, and constitutes more than 25% of sediments on the Atlantic slope and rise where foraminiferal tests are predominant.

Emery (1968) placed the shelf sediment distribution into a simple climatic model, which shifts latitudinally with each glacial/interglacial cycle. Glacial detritus predominates in northern portions of the temperate zone, while water-transported terrigenous clastics predominate in middle and southern portions of the temperate zone. Biogenic sediments predominate in equatorial tropical and subtropical zones. Maximum volumes of detritus were transported to the shelf by streams and glaciers during glaciation and temporarily stored in fluvial and coastal systems. Terrigenous sediments are continuously supplied to modern environments by reworking these temporary storage deposits as coastal systems migrate across the shelf in response to sea-level changes.

Processes active in modern coastal and marine environments are dominated by the present post-glacial transgression and reworking of sediments deposited during the previous glacial episode. Rates of sediment accumulation are highly variable in time and space on the continental margin. However, Emery and Uchupi (1972) estimated that long-term sedimentation rates during the Quaternary were 4 cm/1,000 years on the shelf, 30 cm/1,000 years on the slope, and 5 cm/1,000 years in the deep sea. Multiple events of nondeposition and erosion are probably as important on the shelf as times of sedimentation. The combined effects result in abnormally low, long-term rates of accumulation on the shelf and very high, long-term accumulation rates on the adjacent slope environments. Thus, regressions create temporary storage of new sediment on the shelf, while transgressions extensively rework these sediments, carrying some shoreward with the coastal system and losing some seaward to the slope. In contrast to the shelf, the slope and rise are generally zones of more continuous sedimentation, interrupted locally by mass movement and by strong erosional currents of the Gulf Stream and Western Boundary Undercurrent (Pinet and Popenoe, 1982, 1985; Popenoe, 1985).

Clastic Sedimentary Environments. Fluvial Systems.

Early research on the geology of streams focused on relationships between climate, as measured by precipitation and vegetation, and stream sediment yields (Langbein and Schumm, 1958). Sediment yield increases rapidly as a function of precipitation, from a minimum in deserts to a maximum in grasslands; this yield decreases rapidly when rainfall is sufficient to produce a forest cover, which reverses the process due to higher efficiency in retaining runoff (Leopold and others, 1964; Gordon, 1979).

Recent work on the U.S. Atlantic coast has focused on differences in discharge and sediment yield between northern (glaciated) and southern (nonglaciated) streams. Figure 11 demonstrates that northern rivers deliver greater volumes of fresh water but less total sediment to the ocean than do southern rivers (Meade, 1969). Also, northern fluvial systems have more deeply incised valleys, created by downcutting of glacial meltwaters, which are filling more slowly than their southern counterparts (White, 1978).

Judson (1968) estimated that prior to human impact, the overall denudation rate of land was 24 mm/100 years. Meade (1972b, 1982) emphasized that essentially all resulting sediment being transported by fluvial systems is moved through the Atlantic coastal plain and is temporarily trapped in marshes of downstream estuarine systems; probably less than 5% escapes directly to shelf and deeper water environments. In a longer-term view, coastal erosion and shore-face retreat in response to ongoing sea-level rise ultimately bring these temporarily stored sediments into the marine realm.

Meade (1969) pointed out the importance of interpreting modern denudation rates and stream sediment yields in terms of human impacts on fluvial systems. Although early deforestation for agricultural purposes released huge quantities of sediment to rivers, subsequent reforestation and better farming practices have greatly reduced the amount of erosion. Material eroded in colonial days may still be moving through fluvial systems due to trapping capabilities of dams on many streams (Meade, 1972b, 1982). A growing body of literature concerning human impact on fluvial systems is summarized by Meade (1982) and Toy (1982).

Another factor to consider concerning modern streams is the meaning of short-term measurements in regard to geological time scales. Ritter (1974) documented the extraordinary effects of a single storm event, Hurricane Agnes, on a large fluvial system; that storm equaled decades of 'normal' sedimentary processes in the Susquehanna River system (Schubel, 1974). Modern observations do not supply an adequate basis for understanding fluvial conditions on the Atlantic margin during Quaternary glacial events. In nonglaciated terrain, sea-level fluctuations have resulted in alternating periods of down-cutting and alluviation. To the north, drainage systems were completely rejuvenated following the recent removal of glacial ice and are slowly reaching a new equilibrium with ambient conditions (Denny, 1982).

Estuarine Systems. Early workers treated Atlantic coast estuaries as drowned extensions of their riverine sources and produced isolated studies on the texture, mineralogy, and distribution of bottom sediments (Ryan, 1953; Powers, 1957). By the mid-1960s, multidisciplinary research focused on estuaries as unique transitional environments between fresh water river systems and the marine environment. Resulting symposia publications marked the beginning of modern estuarine research (Lauff, 1967; Nelson, 1972).

Pritchard (1967) defined an estuary as a semi-enclosed coastal body of water with a salinity measurably diluted by fresh water from land drainage. Along the U.S. Atlantic coast, these transitional water bodies can be categorized on the basis of their origins. The northern region is predominated by drowned river valleys, deeply scoured by glaciers and their meltwaters (Upson and Spencer, 1964; Meade, 1972b; White, 1978). The mid-Atlantic region contains mixed estuarine systems produced by flooding deeply scoured fluvial valleys in conjunction with the development of a system of coastal barrier islands. South of Cape Fear, North Carolina, the estuaries are largely barrier-fronted lagoons. Estuaries in southernmost Florida are associated with organic developments, including coral reefs and mangrove islands. Tectonic basins are absent on the Atlantic coast; however, Belknap and Kraft (1977) and Brown (1978) stated that downwarping can be an important process in modifying estuaries. This is demonstrated by low relief shorelines around Delaware Bay and the eastern shores of Chesapeake Bay, which contrast with the high bluffs and incised drainage along the western shores of Chesapeake Bay.

Estuarine systems along the Atlantic margin are characterized by extreme variation in both dynamic processes and sedimentary environments. Four variables determine general estuarine environments and resulting sediments: 1) regional geologic setting controls geometry and sediment supply; 2) latitude controls climate and river discharge; 3) oceanic influence controls tidal fluctuation, currents, and chemical characteristics; and 4) development within the associated drainage basin and adjacent coastal system determines the degree of human impact.

Processes of sediment deposition and distribution in estuaries are controlled by physical, chemical, and biological factors that are highly variable from estuary to estuary (Postma, 1967; Schubel, 1971; Biggs, 1978). Estuarine circulation controls primary physical and chemical processes within the estuaries. In deeper estuaries, which are dominated by a large river discharge, density stratification occurs in which salt and fresh water mix along a near-horizontal interface. Most shallow estuaries and those without large river discharges are partially to completely mixed by advection, turbulent tidal currents, and winds that break down density stratification to ultimately produce vertically homogeneous water columns; however, horizontal salinity gradients still persist.

Flocculation, the agglomeration of clay particles where turbid fresh water mixes with more saline water (Zabawa, 1978), is one of the physical/chemical process controlling fine-grained sediment distribution within estuaries. Flocculation creates composite particles, which settle more rapidly than the individual finer-grained components (Einstein and Krone, 1962; Kranck, 1975). In stratified estuaries, flocculation is most prevalent along the interface of fresh and salt water masses and it has been cited as one of the causes of turbidity maxima in such estuaries (Schubel, 1969, 1971). Coagulation or binding of dissolved and complexed organic matter and heavy metals with clays is another important process that occurs near turbidity maxima (Zabawa, 1978; Sholkovitz and Copland, 1981).

Primary biological processes also act as agglomerating

mechanisms to bind and deposit fine-grained sediment in estuaries (Zabawa, 1978). In coastal environments, biological processes probably overwhelm the purely physical/chemical processes of flocculation. Algal mucus (Avnimelech and others, 1982) and bacterial slime (Paerl, 1973) bind most particles. In addition, many filter-feeding benthic organisms ingest suspended detritus, pelletize the sediment, and excrete it as larger composite fecal pellets and pseudofeces, which settle more rapidly than the detritus (Haven and Morales-Alamo, 1972). Benthic and nektonic organisms are also important in reworking and redistributing deposited sediments to produce the final textural characteristics and sediment structures (Visher and Howard, 1974; Dorjes and Howard, 1975).

Resulting sediments in bay head or riverine portions of typical east coast estuaries are organic-rich (>2%), silty clays, which consist of vermiculite, smectite, and heavy minerals derived from soils in nearby uplands (Folger, 1972; Hathaway, 1972; Neiheisal, 1973). Sediments in the bay mouth are generally organic-poor (<1%), illite- and chlorite-rich muddy sands derived from the barrier islands and shelf. Estuarine circulation traps riverine sands in upper estuaries, deposits marine sands in lower estuarine reaches (Meade, 1969, 1972b), and flocculates fine-grained sediment within the estuary (Schubel, 1969, 1971). Thus, specific zones and patterns of sediment accumulation in estuaries are determined by estuarine geometry, patterns of circulation, and mixing and effectiveness of resuspension by wave and current action.

Rates of short-term, modern sediment accumulation have been evaluated utilizing ^{210}Pb and ^{137}Cs profiles (Goldberg and others, 1977, 1978; Benninger, 1978; Knebel and others, 1981; Yarbro and others, 1983). This work demonstrates that accumulation within estuarine systems has ranged from 1.5 mm/year to 7.9 mm/year, with changes in sediment-accumulation patterns occurring during historic times. However, construction of estuarine sediment budgets has met with mixed success, partly because of difficulties in obtaining representative measurements over meaningful time scales (Guilcher, 1967; Oostdam and Jordan, 1972; Boon, 1975; Bokenoweiz and others, 1976; Ward, 1981; Kelley, 1983).

With the advent of radiometric dating methods, vibracoring, and high-resolution seismic profiling in the late 1960s, a three-dimensional understanding of the Holocene history of many estuaries began to emerge (Weil, 1977; Kraft, 1978; Knebel and others, 1981; Eames, 1983). For example, such work in Delaware Bay (Moody and Van Reenan, 1967; Weil, 1977) has led to a three-dimensional stratigraphic model, which emphasizes preservation potential of lithosomes (Belknap and Kraft, 1985). With this understanding, it is increasingly apparent that sediment accumulation within estuarine systems represents only temporary sediment storage, particularly when considered in light of long-term transgression.

Barrier Island Systems. Clastic sediments of the present eastern U.S. coastline consist of two contrasting regimes. The regime extending northward from New York is the result of interaction between crystalline and older sedimentary rocks of the Appalachian orogenic belt, glacial rebound, and eustatic sea-level rise. The regime extending southward from New Jersey to Miami, Florida was remote to the glacial ice masses. Consequently, the development and evolution is the result of interaction between the young sedimentary rocks of the coastal plain province and the indirect effects of deglaciation such as sea-level rise and major paleoclimatic changes.

The zero isobase of emergence passes offshore in the area of Boston, Massachusetts (Flint, 1971), resulting in partial submergence of terrestrial deposits in southern New England and New York, and a contrasting belt of emerged glacio-marine deposits in northern New England. The northern New England shoreline is composed largely of rocky headlands associated with the Appalachian orogenic belt with local strandplains of coarse sands and gravels derived either from associated crystalline rocks or reworked from ice-marginal and glacial-marine deposits. From the Boston area southward to New York the coastline consists of scattered rocky headlands with extensive strandplains and local barrier islands consisting of sediments derived from the erosion and reworking of partially submerged terrestrial deposits.

Barrier islands and associated back-barrier estuaries are characteristic sedimentary environments throughout the open-ocean facing coastal system extending from New Jersey southward to Miami, Florida. Multiple barrier sand ridges on the coastal plain (Cooke, 1930; Winker and Howard, 1977) demonstrate that similar environments existed throughout the Quaternary, at least during interglacial periods.

Schwartz (1973), Swift (1975), Leatherman (1979), and Oertel and Leatherman (1985) discussed the origins, processes, and evolution of barrier islands. They summarized four major theories of origin: 1) buildup of an offshore bar; 2) longshore spit growth and subsequent breaching; 3) drowning of antecedent topography; and 4) formation on the outer shelf with subsequent migration landward. Schwartz (1971) and Zenkovitch (1967) emphasized the probability of multiple origins of barrier islands in varying geologic settings.

Geomorphology as a process-response (morphodynamics) on barriers has been intensely studied for many years. Hayes (1979) summarized a geomorphic model for barrier-inlet systems based on tidal range and wave energy: macrotidal systems have few barrier islands; tide-dominated mesotidal systems have short barriers with many large inlets; and wave-dominated microtidal systems have long, narrow barriers with few, small inlets. Vegetation interacts with aeolian and overwash processes to form and maintain the superstructure of barriers (Rosen, 1979; Godfrey and others, 1979). Migrating tidal inlets and their remnant flood-tide deltas play important roles in formation and maintenance of back-barrier systems, including development of a base for barrier island migration and substrate for marsh growth (Oertel, 1976; Moslow and Heron, 1978; FitzGerald and FitzGerald, 1977; O'Brien, 1969). On mesotidal barriers in Georgia and South Carolina, wave refraction by ebb-tidal deltas in adjacent inlets causes increased erosion along mid-island segments and prograd-

ing beach ridge plains on the ends of islands, producing depositional lobes, resulting in a 'drumstick-shaped' island (Hayes and Kana, 1976). Prograding segments also occur at the downdrift end of longshore drift systems on wave-dominated microtidal barriers, such as Cape Henlopen (Kraft and others, 1978) and Fire Island Inlet (Kumar and Sanders, 1974; McCormick and Toscano, 1981).

Hoyt (1967), Hoyt and Henry (1967), and Halsey (1979) demonstrated that modern coastal geometries and processes on barrier islands are also linked to drowning of antecedent geologic features. For example, Halsey (1979) has shown that short barriers with wide inlets and lagoons on the south side of Delmarva Peninsula and long barriers with narrow lagoons on the north side of the Peninsula are controlled by pre-Holocene relief on the transgressed surface. Belknap and Kraft (1981, 1985) stated that antecedent topography and sediment supply determine the location and geometry of barriers, inlets, lagoons, and headland beaches. Preservation potential and detailed evolution during marine transgression is then strongly tied to incident wave energy and rate of local relative sea-level change. Deep valleys and abundant sedient supply enhance preservation potential while paucity of sediment and shallow interfluves decrease preservation potential. Detailed stratigraphic analysis of barrier island systems is ultimately the most important data for unravelling their complex evolutionary history.

Swift and others (1985) and Niedoroda and others (1985) have shown that the dynamical shoreface surface is a key morphologic element in the migrating barrier system. Sand released by erosional shoreface retreat cycles tank-tread fashion through the barrier system. Aeolian action, storm washover, and inlet migration (Leatherman, 1979) move sand to the back barrier flat where it is buried and subsequently re-exhumed at the shoreface as the barrier passes over it. Not all of the sand released at the eroding shoreface takes part in this recycling process. Downwelling coastal storm flows sweep some sand down the shoreface and onto the adjacent shelf floor where it is deposited on the leading edge of the thin, discontinuous transgressive sand sheet forming on the inner shelf surface. Thus, the migrating barrier system steadily loses sand to the shelf floor. However, barrier sand supplies are continuously replenished through the process of shoreface retreat, 'blowing out' old inlets or forming new inlets, and by planation of shoal massifs and protruding headlands. If the sands are not replenished, the barrier will starve and be overstepped. The nearly continuous sheet of back barrier sediments on the North Atlantic shelf (Sheridan and others, 1974; Stahl and others, 1974) suggests that overstepping has not been a common process during Holocene barrier island evolution.

Continental Shelf Systems. Modern Atlantic shelf sediments can be divided into a series of depositional subsystems (Fig. 12). The Gulf of Maine, north of Cape Cod, Massachusetts, is characterized by several large closed basins with relief in excess of 200 m resulting from pre-Holocene fluvial and glacial erosion of crystalline rocks of the Appalachian orogenic belt (Uchupi, 1968; Schlee and Pratt, 1970). These crystalline rocks extend under the Gulf of Maine to the western edge of Nantucket Shoals, Georges and Browns Banks, and the eastern side of Nova Scotia. The crystalline rock floor has a thin veneer of coarse glacial gravels; marine muds have been accumulating in the basins since the ice retreated (Lougee, 1953; Schnitker, 1975). Nantucket Shoals and Georges Banks are characterized by sands controlled by strong tidal flows (Twichell, 1983).

South of Georges Bank, the New England shelf contains no large closed basins. This area was subjected to subaerial exposure immediately after ice retreat and has been subsequently dominated by coastal and marine sedimentary processes associated with the Holocene transgression. Sediments in this shelf area are predominately sands transported southward from Georges Bank (Twichell, 1983) with local patches (Fig. 12) of fine-grained sediment on the outer shelf south of New England (Twichell and others, 1981).

The Middle Atlantic Bight, from Montauk Point, N.Y. to Cape Hatteras, N.C., consists of shelf valley complexes and associated sand shoal massifs (Fig. 12) separated by broad, plateau-like interfluves (Swift and others, 1972b; Swift and Sears, 1974). Shelf valley complexes are river valleys excavated during previous Quaternary low stands of the sea and subsequently filled with estuarine sediments during sea-level rise. Continued sea-level rise with an intense wave climate caused erosional shoreface retreat of river forelands or estuary mouths to produce widely spaced cape retreat massifs with broad intervening plateau-like interfluves containing extensive fields of sand ridges (Fig. 12). In some cases, modern shelf valleys have been incised into previous valley fill (Sheridan and others, 1974).

Vibracores collected on interfluve areas within the Middle Atlantic Bight indicate a three-fold stratigraphy deposited by landward barrier migration during Holocene sea-level rise (Niedoroda and others, 1985). When fully developed, it consists of a lower sequence of basal gravels overlain by estuarine silts and clays, a middle sequence of backbarrier sands with intercalated beds of shell and mud, and an upper sequence of shelf sediments consisting of a basal gravel and well sorted sands. Dune, beach, and shoreface deposits have been destroyed by the erosional retreat process, leaving a thin and irregular (1 to 10 m thick) Holocene shelf sand sheet with a sublittoral molluscan fauna (Swift, 1976; Niedoroda and others, 1985). The basal layer of the surficial sand sheet is often a thin (<1 m) discontinuous gravel consisting of shell hash rich in beach species and rock lithoclasts (Swift, 1976). In high interfluve areas or areas with poorly developed estuarine strata, the retreating shoreface will incise into the underlying Pleistocene or Tertiary units supplying sand, lithoclasts, and other minerals such as phosphate (Riggs and others, 1985) to the surficial sand sheet.

Within the Middle Atlantic Bight, the hydraulic climate is sufficiently intense to produce modern textural and morphologic patterns (Swift, 1976; Swift and others, 1979). The shelf surface is characterized by a pervasive ridge and swale topography. This topography has formed locally at the foot of the retreating shoreface in response to coastal boundary flow and in some areas has

developed more or less spontaneously further out on the shelf surface. Current lineations are abundant and coarse sand lags occur on highs with finer sands on downcurrent slopes and in adjacent lows. These bedforms have a post-transgression origin and formed in response to storm flows (Swift, 1976; Swift and Field, 1981).

Most large-scale subaerial morphologic elements now occurring on the shelf have been subtly but pervasively modified by the erosional retreat of the shoreface during the Holocene transgression, while most small-scale elements have been totally destroyed (Swift, 1976). Niedoroda and others (1985) and Swift (1976) believed that erosional shoreface retreat is the immediate source of most sand for barrier islands, as well as for many shelf sand bodies such as sand shoal massifs, cape extension shoals, and shoreface-connected sand ridges. Figure 13 shows the distribution of sediment types in the Middle Atlantic Bight as a result of shoreface retreat.

Between Cape Hatteras, North Carolina and Cape Canaveral, Florida, the retreat of cuspate forelands has generated extensive cape-associated, sand shoal massifs on the shelf. As the spacing of the cuspate forelands decreases southwards into the Georgia Bight (between Cape Romain, South Carolina, and Cape Canaveral), increased tidal ranges, milder wave climates, and closely spaced estuarine-dominated river mouths have generated a Holocene retreat sand blanket consisting of coalesced shelf valley complexes and shoal retreat massifs (Swift, 1976).

Carbonate sediments north of Cape Canaveral are mixed with siliciclastic sands and consist predominantly of invertebrate skeletal material derived from benthic communities on the shelf and adjacent coastal zone (Longman, 1981). Many of these organisms form living veneers on submarine outcrops (Riggs, 1979a; Crowson, 1980; Henry, 1981; Blackwelder and others, 1982). These outcrops are Tertiary and Pleistocene hardgrounds of various types of dolomite, coquina, oolitic calcarenite, and calcareous sandstones (Pilkey and others, 1969; Macintyre and Milliman, 1970; Mixon and Pilkey, 1976). Carbonate sediment production from these mostly temperate shelf environments is generally poorly understood. South of Cape Canaveral, shelf sedimentation is dominated by carbonates (Fig. 12); this is discussed in a following section.

Sedimentation in the three Atlantic shelf subsystems is controlled by the relationship between rates of sea-level rise, sediment input, and fluid power expenditure. Sea-level rise on the Atlantic shelf averages 10^{-3}m/yr and fluvial sediment input is about 6.2×10^3 metric tons/yr/km of shoreline (Meade, 1972a). Mean annual fluid power expenditure per m^2 of sea floor is not easily calculated. Tidal currents are major agents of sedimentation in tidal inlets, most estuaries, on offshore banks such as Georges Bank, and in the Georgia Bight, where spring tides of 3 m occur. Wave climate ranges from moderate off-Florida and Georgia to extreme off-North Carolina (Dolan and others, 1973; Nummedal and others, 1977). On the upper shoreface of the open shelf, wave orbital currents dominate the fluid power spectrum. Seaward of the 5 m isobath, however, a wind-driven unidi-

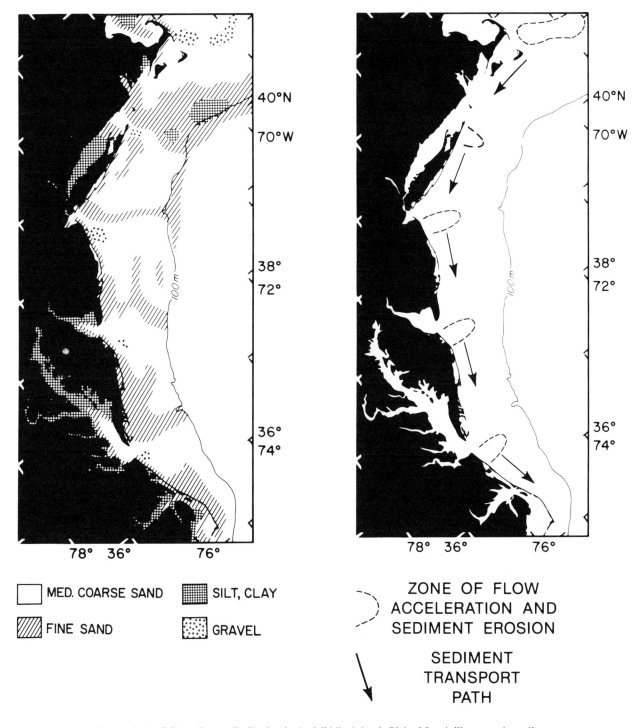

Figure 13. Surficial sediment distribution in the Middle Atlantic Bight. Map A illustrates the sediment distribution (after Milliman and others, 1972; Trumbull, 1972; Schlee, 1973; Hollister, 1973). Map B illustrates zones of seafloor erosion and paths of sediment transport that are inferred to have created the observed pattern (from Swift, 1986).

rectional component of flow plays a role equal to, or greater than that of wave orbital currents in the mean annual energy budget (Beardsley and others, 1976).

The three process variables, sea-level change, sediment supply, and fluid power interact in a manner that may be described as follows. Sea level is rising so rapidly relative to rates of the other two variables that an estuary-lagoon coast has been created; fluvial sediment is either trapped in the estuaries or is bypassed seaward over the relatively high energy shelf surface. In this setting the modern sedimentary regime is autochthonous (Swift, 1976) and erosional shoreface retreat, not river input, is the dominant source of sediment for the shelf surface.

Sediment transport on the Atlantic shelf floor is event-dominated (Butman and others, 1979; Vincent and others, 1981; Swift and others, 1986). Wind-induced currents arise in response to mid-latitude low pressure systems, which cross the Atlantic shelf with a periodicity of 3 to 10 days. Most have a relatively minor effect on the seafloor; major sediment transport is accomplished by three to five major storms per year (Vincent and others, 1981). The most efficient storms are those which, during their passage, generate northeasterly winds that parallel the shoreline. During such occasions the sea surface is set up against the coast by as much as 50 cm, and the resulting internal pressure field in the water column drives a powerful geostrophic flow southeastward along the coast (Beardsley and others, 1976). Peak mean velocities tend to correlate with peak wave orbital currents, since both are of storm origin (Vincent and others, 1981).

Holocene continental shelf deposits formed by this storm-derived flow regime consist primarily of medium gray to yellowish brown sand with a negligible admixture of silt and clay. Zones of gravel occur in areas where storm current velocities increase in a down current direction. Here the sedimentary regime is largely erosional and basal gravels of the Holocene transgression are exposed at the sea floor. In the Middle Atlantic Bight, such erosional zones tend to occur on the down current sides of major estuaries (Fig. 13). In response to expansion and deceleration of flow during storm events, sands move southeast (down current) with decreasing grain size, covering the gravels. Locally, these down-current grain size gradients end in mud, such as the 'mud patch' of the southern New England shelf (Fig. 12) (Twichell and others, 1981).

The Holocene sand sheet consists of stacked sand strata 20 to 50 cm thick. About 40% of the beds exhibit an upward decrease in shell content or quartz grain size, or both. Horizontal laminations are the most prevalent primary structure in the finer-grained, upward fining beds, and appear to reflect a highly turbulent combined regime that develops during peak storm events (Swift and others, 1972a, 1979, 1986). Beds containing sands coarser than two phi tend to exhibit high angle cross-stratification. The remaining 60% of the beds are indistinctly mottled to massive as a consequence of bioturbation, and become more abundant offshore. Stratification characteristics of much of the Holocene sand sheet suggest that they are deposited by successive storm events.

The character of Holocene sediments and sedimentation on the shelf is partially a prooduct of inherited geologic controls and partially due to presently active forces (Swift and others, 1986). Sediment supply is inherited from fluvial and glacial deposition on the shelf during periods of lower sea level. Reworking of this material has occurred during Holocene shoreline retreat, and the upper several decimeters is presently being reworked during frequent high energy events. Rather than the simplistic fining-offshore pattern suggested during the first half of this century, or the quiet relict shelf envisioned during the 1960s, the shelf is now known to be a dynamic area of active sediment redistribution.

Continental Slope and Rise Systems. Transition from shelf to slope occurs at the shelf break, which represents a critical interface between traction-dominated processes of the shelf and gravity-dominated processes of the slope (Vanney and Stanley, 1983). The shelf break is a zone that may migrate through geologic time by either progradation or erosion or may be fixed by reefs or tectonic activity (Stanley and Moore, 1983).

Schlee and others (1979) summarized the regional structural framework of the Atlantic continental slope and concluded that the present slope formed over a complex zone of thinned continental, transitional, and oceanic crust covered by a thick sequence of Mesozoic sediments. A thick Tertiary sequence prograded the continental slope seaward with alternating periods of deposition and erosion. During the Quaternary, position of the Atlantic shelf break has been relatively stable except for active slumping and canyon cutting (Farre and others, 1983). The continental rise consists largely of clastic sediments derived from eroded Tertiary and Quaternary units on the shelf.

Pinet and others (1981), Paull and Dillon (1980), and Pinet and Popenoe (1982) demonstrated that the Gulf Stream, or its predecessor, has been operational since the late Paleogene. Axes of the Gulf Stream and its associated southward-flowing Western Boundary Undercurrent have migrated in response to major fluctuations in global sea level, causing shifts in depocenters and creating erosional unconformities with associated channel networks up to 100 m deep (Popenoe, 1985). During low stands of the sea, both the Gulf Stream and Western Boundary Undercurrent were deflected seaward away from the continental margin. During periods of high sea level, current flow shifted landward against the Florida-Hatteras Slope causing extensive erosion of the slope and rise and producing major unconformities within the continental margin sediment sequence. Migration patterns of these currents appear to be directly correlative with published sea-level curves, and the currents have controlled the type and distribution of sediment facies, as well as location and geometry of sediment depocenters throughout the past 24 million years (Riggs, 1984; Popenoe, 1985).

Eustatic sea-level fluctuations are equally important in influencing where, how much, and what kinds of sediments are delivered to the slope. During low stands, the coastal zone is at or on the continental slope. Rivers discharge sediment, including sand-sized fractions, directly into canyon heads that deliver sediment directly to the continental rise. During high stands, the coastal zone migrates westward across the shelf, canyon systems become relatively inactive, and inter-canyon areas with steeper gradients become more active loci of erosion and downslope sediment transport (McGregor, 1984).

Regional changes in both physiography and sediment composition of the continental slope and rise system occur off Cape Hatteras in both the Recent and Pleistocene. Between Cape Hatteras and Labrador, the slope is cut by more than 190 submarine canyons (Emery and Uchupi, 1972), while the slope south of the Cape is essentially devoid of canyons. North of Cape Hatteras, the shelf break has been breached by some mature submarine canyons, with a resultant redeposition of shelf sediment on the rise and abyssal plains by episodic turbulent suspensions (Farre and others, 1983; May and others, 1983).

Emery and Uchupi (1972) related the plethora of canyons north of Hatteras to the proximity of Pleistocene ice sheets, which fed large quantities of sediment to rivers and hence to canyon systems during lowered stands of sea level. South of Cape Hatteras, the shelf break throughout the Quaternary was generally characterized by constructive carbonate reef developments (Blackwelder and others, 1982) associated with the warm water Gulf Stream. Holocene carbonate reefs still persist northward to Cape Hatteras (Macintyre and Pilkey, 1969; Macintyre and Milliman, 1970).

The continental slope is veneered with Pleistocene to Recent sediment ranging from thin surficial skims up to several hundred m (Doyle and others, 1979). North of Cape Hatteras, slope sediments are predominantly terrigenous muds with a high silt content and less than 10% calcium carbonate. Low amounts of terrigenous sand occur as distinct layers and contain hornblende, garnet, and opaques in the detrital heavy mineral fraction. The clay mineral suite is predominated by illite-mica (Wall, 1981). Surface sediments on the slope may contain over 2% organic carbon by weight and accumulate at rates between 22 cm/1,000 years (Doyle and others, 1981) and 30 cm/1,000 years or more (Emery and Uchupi, 1972). South of the Cape, quartz sands predominate, with calcium carbonate generally composing more than 30% of the sediment. The clay mineral suite is predominated by smectites (Wall, 1981), while the heavy mineral suite is largely hornblende and epidote (Doyle and others, 1979).

Regional differences in mineral composition of slope sediments reflect basic changes in sediments of adjacent shelves and land provenance areas. Differences in textural composition result from the increased role of the Gulf Stream south of Cape Hatteras in scouring the slope, winnowing sediment, and limiting deposition of finer grain sizes (Doyle and others, 1979). North of the Cape, where there is decreasing influence by the Gulf Stream, hemipelagic sedimentation is dominant. Off-shelf sediment transport occurs during storm events and low sea-level stands. Doyle and others (1979) and Robb and others (1981) have shown that active gravitationally induced reworking and mass movement downslope occur in both regions in the form of submarine creep, sliding, and slumping; however, their importance appears quite variable (Nelsen and Stanley, 1986).

The mudline is defined as the water depth below which deposition prevails and shelf and slope sediments contain substantially increased mud contents (Stanley and others, 1983). Position of the mudline on the continental margin is quite variable. In areas with major mud input, such as the Mississippi River delta, the mudline is high on the shelf. In areas with high current activity, as between Florida and Cape Hatteras, the mudline is near the bottom of the slope. North of the Cape, decreased interaction between currents and the slope allows mud to accumulate on the slope to just below the shelf break. Stanley and others (1983) believed that the mudline represents a good tracer of changing energy conditions through the sea-level cycles.

Carbonate Sedimentary Environments. Modern sedimentary environments of the South Florida/Bahama Banks province mantle one of the largest zones of limestone accumulation in earth history. These platforms are up to 14 km thick, are structurally tied together, and represent the modern component of an enormous carbonate bank system that once stretched from Nova Scotia southward into the Gulf of Mexico and lower Caribbean Basin during early and mid-rifting stages of the North Atlantic (Sheridan, 1974; Worzel and Burke, 1978; Buffler and others, 1979; Grow and others, 1979; Schlee and others, 1979; Ryan and Miller, 1981; Sheridan and others, 1981). Passive margins with subsidence and without significant siliciclastic influx have an extensive 'carbonate phase' and resultant thick carbonate platforms (Schneider, 1972; Schlee and others, 1979; Watts and Steckler, 1981). The Bahama Banks and South Florida are modern remnants of that once extensive carbonate system.

The now restricted Florida-Bahamas zone of Holocene carbonate sedimentation has resulted from: 1) continuity of subtropical to tropical environments since rifting began—carbonate platforms of Nova Scotia terminated due to slow climatic change, partly as a result of moving into higher latitudes (Schlager, 1981); 2) no siliciclastic influx or increased turbidity from river discharge; 3) no uplift terminating bank development; and 4) long-term subsidence creating wide, broad, flat shelves and platforms on which shallow marine carbonate environments could continue to flourish.

Shallow-Water Carbonates. Distribution of shallow Holocene carbonate sedimentary environments of South Florida and the Bahamas has been controlled by 1) rate of sea-level rise and depth of flooding, 2) antecedent topography, and 3) climatic variations in temperature and duration and magnitude of physical energy flux including prevailing winds, storms, waves, and currents. All of these factors are closely linked and interact with each other to control island formation, reef development and demise, sediment production and transport, tidal flat development, mud bank and ooid sand body formation, and establishment of broad shelf and interior lagoons.

Carbonate-producing environments are highly sensitive to slight fluctuations of sea level (Neumann and Moore, 1975; Lighty and others, 1978; Schlager, 1981; Hine and others, 1981a, 1981b; Kendall and Schlager, 1981; Hine, 1983; Hine and Steinmetz, 1984). History of Holocene sea level rise over the

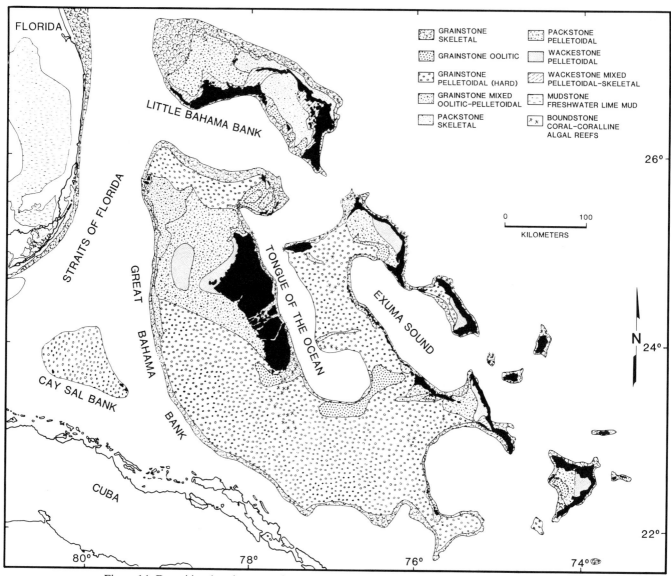

Figure 14. Depositional environments in south Florida and the Bahamas showing the distribution of carbonate sediment types (from Enos, 1974).

exposed Pleistocene carbonate terrace had three stages. First, older topographic highs on the bank margins became sites for new reefs (Enos and Perkins, 1977; Hine and Neumann, 1977); widespread, interior blankets of peloids, aggregate grains, and ooids with large-scale bedforms formed in response to waves, tides, and storm-generated currents (Winland and Matthews, 1974; Hine, 1983); distribution of antecedent islands and rock ridges along bank margins controlled sediment transport both onto and off of the banks (Hine and others, 1981a, 1981b). Second, as water depth increased, these barriers became less effective; tidal flows were accelerated on certain bank-margins resulting in large bank-normal, tidal bar belts of ooid sands (Ball, 1967; Dravis, 1979; Harris, 1979, 1983; Halley and others, 1983); and

leesides of major islands developed broad wind-tidal flats, which were dominated by storm events, tidal creek activity, and significant early diagenetic processes (Hardie, 1977). Third, the widespread, active sand blanket rimming the margins was unable to build vertically and maintain itself close to sea level with continued flooding of the banks; much of it became relict and is now stabilized beneath a cover of benthic flora. Zones of active sand transport and ooid formation became restricted to the most energetic portions of open-shelf environments on the Bahama Banks. In contrast, sand shoals consisting of only skeletal sands (no ooids) occur along the Florida reef tract (Enos and Perkins, 1977).

Carbonate producing environments respond to changing

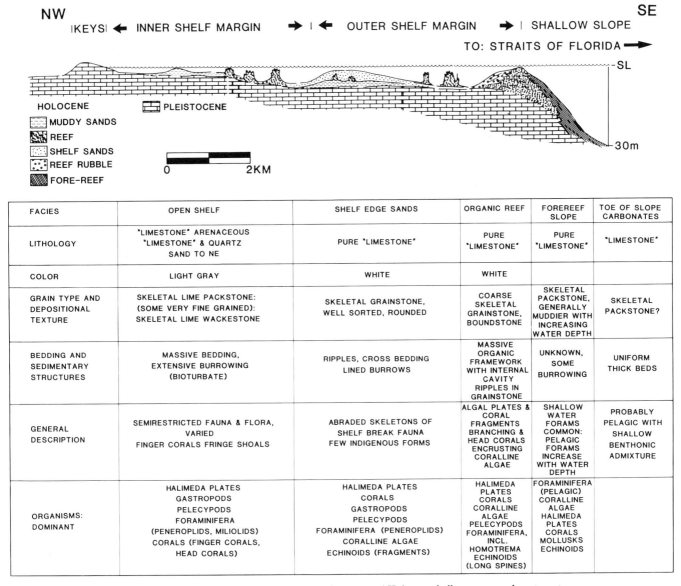

Figure 15. Typical geologic section across the Pleistocene and Holocene shallow-water carbonate system of the Florida Keys illustrating changes in morphology, lithofacies, biofacies, and sedimentary environments (modified from Enos and Perkins, 1977).

climatic effects including seasonal temperature changes, waves, and storm-generated tides and currents. The South Florida/Bahama Banks province (Figs. 14 and 15) lies within the Trade Wind Belt with prevailing winds from the east. As a result, pronounced differences occur between those shallow bank margins facing east (windward margins) and those facing west (leeward margins) (Ginsburg and Shinn, 1964; Hine and Neumann, 1977; Hine and others, 1981a). Windward margins are reef-dominated systems in front of major islands. These margins produce reefal boundstones in association with sporadic and discontinuous active reefs and high energy grainstones (oolites and skeletal grains). Leeward of islands are low energy lagoons and tidal flats that

produce peloidal and skeletal wackestones. Mud-dominant environments of Florida Bay are similarly situated behind the Pleistocene Florida Keys (Ginsburg, 1956; Enos and Perkins, 1977). Within open shelf-lagoons and away from protective islands, the dominant sediments are peloidal grainstones that grade into peloidal packstones where large islands occur on the margins. Leeward margins are generally open without rock barriers; have efficient exchanges of water on and off the banks; and develop thick (up to 25 m) Holocene sand bodies with relatively high concentrations of non-skeletal components including ooids, peloids, and aggregates (Purdy, 1963; Wilber, 1981). These open, leeward margins export significant quantities of sands to adjacent deep flanks (Mullins and Neumann, 1979; Mullins and others, 1980; Hine and others, 1981b).

Holocene reef systems of the South Florida/Bahama Banks are poorly developed in comparison to well-known shallow, fringing, and barrier reefs of other tropical seas. Many Florida/Bahama margin reefs are deep (down to –30 m), have low relief, and do not generally support dense, active, healthy communities of corals (Lighty and others, 1978; Goldberg, 1983; Hine and Steinmetz, 1984). These are not climax reefs; rather, they are in start-up or catch-up phases of bank margin development whereby the rapid rate of sea-level flooding coupled with environmental stress have not allowed reefed rims to maintain vertical growth with sea level (Schlager, 1981; Kendall and Schlager, 1981; Hine and Steinmetz, 1984). Environmental stress occurs as turbidity and burial by offbank sediment transport and as export of cold bank-top waters across the reefs. Such surface waters may be cooled down to 13°C by winter extra-tropical frontal systems (Shinn, 1976; Hudson and others, 1976; Schlager, 1981; Roberts and others, 1982).

Deep-Water Carbonates. Pleistocene and Holocene deep-water carbonates of the Bahamas occur in greater than 200 m of water and consist of diverse sediment facies (see Fig. 22 in chapter 15, this volume) (Mullins and Neumann, 1979; Cook and Mullins, 1983).

Periplatform sands consist of a coarse, poorly sorted mixture of carbonates derived from both the adjacent shallow-water platforms and overlying water column. These sands occur in narrow (1 to 3 km) belts along the base of the Bahama marginal escarpment and are particularly well developed on the leeward margins where there is extensive off-bank transport of carbonate sediment (Mullins and Neumann, 1979; Hine and others, 1981b).

Periplatform oozes are fine-grained carbonates that consist of a blend of bank-derived aragonite and magnesian calcite, and pelagic-derived calcite (Schlager and James, 1978). In Northwest Providence Channel, more than 80% of the fine-grained Holocene carbonates are produced on, and derived from adjacent platforms (Neumann and Land, 1975; Boardman and Neumann, 1984). In cores, these oozes typically display cyclic variations of carbonate mineralogy that appear to be controlled eustatically (Kier and Pilkey, 1971; Droxler and others, 1983). Periplatform oozes are distributed ubiquitously, but are volumetrically important only where they are not winnowed by bottom currents, diluted by gravity flows, or deposited below carbonate compensation depths (Mullins and Neumann, 1979).

Carbonate gravity flow deposits are mostly biolithoclastic turbidites and debris flows (Bornhold and Pilkey, 1971; Crevello and Schlager, 1980) that are deposited on lower slopes as wedge-shaped aprons that parallel adjacent shelf/slope breaks (Mullins and others, 1984). Carbonate contourites are present along current-swept open seaways, such as the Florida Straits (Mullins and others, 1980). Hemiconical sediment drifts, up to 100 km long and 60 km wide, have formed since the middle Miocene in the 'oceanographic lee' of the Great and Little Bahama Banks, where there is a confluence of strong bottom currents. Sediments on these drifts are coarse, well-sorted sands that are highly susceptible to submarine cementation (Mullins and others, 1980).

Ahermatypic coral 'reefs' are also found in areas of strong bottom currents in water depths from 600 to 1,300 m (Neumann and others, 1977; Mullins and others, 1981). Individual ecologically zoned mounds, up to 50 m in relief and 200 m long, extend more than 200 km along the eastern side of the northern Straits of Florida, making it one of the largest coral 'reefs' in the Caribbean region (Mullins and Neumann, 1979). Surprisingly diverse ahermatypic coral mounds also occur north of Little Bahama Bank where they appear to build vertically and laterally via in situ accumulation of skeletal debris and baffling/trapping of finer-grained sediment (Mullins and others, 1981).

SUMMARY

The upper Cenozoic record of the Atlantic continental margin is characterized by rapidly changing sedimentary processes and environments of deposition and erosion as compared to the Paleogene. First, there were the extreme differences between glacial and interglacial episodes, as well as the transitions into and out of both extremes. Second, there were strong latitudinal shifts of climatic zones and environments that affected oceanographic conditions, biological zonations, and weathering processes associated with sediment production and soil formation. Third, major fluctuations in sea level caused rapid migrations of the shoreline and associated nonmarine and marine environments and processes across the continental margin. In glaciated regions, processes were dominated by glacial ice erosion and deposition and by glacial isostasy. In non-glaciated regions, the continental margin was dominated by alternating marine and nonmarine environments in which the stratigraphy was further complicated by interacting processes of deposition, erosion, and diagenesis. These changes have taken place on a stage with structural and stratigraphic underpinnings related to rifting and post-rift subsidence, which have continuously impacted large-scale patterns of sedimentation.

Knowledge of the upper Cenozoic stratigraphic record resides in several data bases: geomorphic, lithologic, biostratigraphic, seismic, and geochronologic information. Unfortunately, investigators find it difficult to remain current with new information in all these fields. Integration of different data bases remains a challenge, in part due to increasing specialization and sophistication of methods employed, and in part due to conflicts raised between different approaches and regions. Nevertheless, correlation of complex sequences such as the Pleistocene units along the coastal plain is reaching beyond the tentative stages. Older views of simple 'bathtub ring' terraces and shorelines, correlated by elevation, have been supplanted by a more rigorous stratigraphic approach, which recognizes gentle warping of the coastal plain. In general, the coastal plain is much better understood than the submerged margin, where high-resolution seismic profiling and detailed coring has only begun.

The future will see continued growth of interdisciplinary

studies integrating results of new methods of stratigraphic analysis into regional syntheses. Critical issues for the next decade include reconciliation of various dating techniques with each other by 1) controlled interlaboratory comparison, 2) careful examination of assumptions, 3) better control of stratigraphic sequence models, and 4) improved correlations through the use of better lithologic, paleontologic, and chronologic techniques. However, field description and analysis of new and old exposures, coupled with remote sensing and sampling, will continue to be the backbone of the science and allow stratigraphic concepts to evolve.

For pre-Modern geologic systems, stratigraphy and correlation will be further clarified by integration among many disciplines. In both the Holocene and Pleistocene, an important issue is preservation potential of units. Can shoreline deposits be efficiently preserved only during stillstands, or can rapid transgressions and regressions leave mappable units? Also, what is the role of long-term subsidence or emergence of the margin as a means of preservation of littoral and marine units? Can eustatic and tectonic events be differentiated? Finally, how can Modern, Pleistocene, and Neogene models of continental margin sedimentation be used in understanding ancient continental margins?

Holocene sedimentary environments on the Atlantic continental margin are known to various confidence levels. Coastal systems, including fluvial, estuarine, and barrier island environments, have been intensely studied and, despite active debate about details, are generally well understood from process and stratigraphic points of view. Interdisciplinary research on interactions within these complex environmental systems is increasing as their importance to man increases. Less well known are submerged, offshore portions of the continental margin; our degree of understanding is roughly proportional to water depth in the marine environment. Limiting factors include our dependence upon remote sensing and remote sampling, technological limitations of navigation and location systems, and cost of marine operations.

Many problems related to the sedimentary environments of the continental margin remain to be solved, but only a few can be mentioned in the space available. Relatively little is known about continental slopes and processes that affect them. What are the rates, mechanisms, and importance of mass-movement processes? What is the origin of submarine canyons, when are they active, and why are they distributed as they are? Are continental rises the result of downslope movement and submarine fan deposition associated with canyon systems, or are they contour current deposits? Somewhat more is known about the geology of the Atlantic continental shelf. However, questions concerning its growth and development include understanding processes of vertical aggradation versus horizontal progradation, mechanisms and pathways of sediment transport, and changing modes of deposition and erosion in response to rapid changes in paleoclimatic and paleoceanographic conditions. The limited Quaternary stratigraphic record is not fully understood; a large part of this problem is associated with the concept of preservation potential and processes of transgression and regression. Answers to these questions will only come with detailed work including high-resolution

seismic profiling, side-scan sonar mapping, and analysis and dating of cored sediment sequences.

Within the coastal zone, some important problems include the following. Concerning sediment budgets, little is known about the short-term and long-term volumes, timing, mechanisms, and pathways of sediment transport interacting between fluvial, estuarine, barrier, and offshore systems. The many process-response models must be refined with respect to interrelationships of energy regimes, sediment budgets, and paleoclimatic and paleoceanographic changes. Development of regional Holocene relative sea-level curves is important in separating isostatic and neotectonic along-coast land-level changes from eustatic sea level changes; this is important in understanding the role of fluctuating sea level in controlling coastal processes. For all of the continental margin environments, it is imperative that we resolve the problem of importance of extreme versus average energy events. What is a 'normal' event? How much of the stratigraphic record is represented by extreme events and small fractions of total elapsed geologic time? Finally, future investigations must continue to involve detailed field measurements and observations, but there is an increasing need for this work to be done in consort with interdisciplinary cooperation, including analytical work utilizing the most advanced techniques and instrumentation.

In addition to the dynamics of natural geologic systems as previously discussed, we must recognize that man produces profound impacts upon geologic processes. Burning fossil fuels may be changing climates and accelerating sea-level rise. Changing land-use patterns, such as deforestation, urbanization, and poor agricultural practices have increased soil erosion and sediment loads in streams, accelerated desertification, and decreased groundwater recharge. Direct human interference with geologic processes, such as damming and channelization of rivers, construction of shoreline structures, and dredging in coastal systems have altered natural sedimentation and erosion patterns, often resulting in increased economic losses from flood and wave attack. Pollution is also a major concern in altering ecological systems and in directly affecting human health. Future geologic research will be tied to problems associated with alteration of natural systems, prediction of future changes, and in planning for rational use of resources and adjustment to uncontrollable forces such as climatic changes and sea-level fluctuations.

Clearly, we are now in an age of interdisciplinary research, with crossovers from other fields within and outside geology essential in contributing to the solution of complex stratigraphic and sedimentologic problems. Application of rapidly expanding technology, growth of critical data bases, and increased sophistication in abilities to model stratigraphic and sedimentologic systems, both on continents and in the oceans, will continue to expand capabilities for resource and environmental management within the dynamic Atlantic continental margin. This may be critical as the hypothesized 'greenhouse effect,' resulting from increased CO_2 concentration in the atmosphere due to burning of fossil fuels, modifies global climates and accelerates the rate of future sea-level rise (Hansen and others, 1981).

REFERENCES CITED

Abbott, W. H., and Ernissee, J. J., 1983, Biostratigraphy and paleoecology of a diatomaceous clay unit in the Miocene Pungo River Formation of Beaufort County, North Carolina, *in* Ray, C. E., ed., Geology and Paleontology of the Lee Creek Mine, North Carolina: Smithsonian Contributions to Paleobiology, no. 53, p. 287–353.

Avnimelech, Y., Troeger, B. W., and Reed, L. W., 1982, Mutual flocculation of algae and clay; Evidence and implications: Science, v. 216, p. 63–65.

Ball, M. M., 1967, Carbonate sand bodies of Florida and the Bahamas: Journal of Sedimentary Petrology, v. 37, p. 556–591.

Baturin, G. N., 1982, Phosphorites on the Sea Floor: Amsterdam, Elsevier, 343 p.

Bearsdley, R. C., Biocourt, W. C., and Hansen, D. V., 1976, Physical oceanography of the Middle Atlantic Bight: American Society of Limnology and Oceanography Special Symposium 2, p. 20–33.

Behrendt, J. C., Hamilton, R. M., Ackerman, H. D., and Henry, V. J., 1981, Cenozoic faulting in the vicinity of the Charleston, South Carolina, 1886 earthquake: Geology, v. 9, p. 117–122.

Belknap, D. F., 1979, Applicaton of amino acid geochronology to stratigraphy of late Cenozoic emerged marine units of the Atlantic coastal plain [Ph.D. thesis]: Newark, Department of Geology, University of Delaware, 550 p.

——— , 1982, Amino acid racemization from C-14 dated "Mid-Wisconsin" mollusks of the Atlantic coastal plain: Geological Society of America Abstracts with Programs, v. 14, p. 4.

Belknap, D. F., and Kraft, J. C., 1977, Holocene relative sea-level changes and coastal stratigraphic units on the northwest flank of the Baltimore Canyon Trough geosyncline: Journal of Sedimentary Petrology, v. 47, p. 610–629.

——— , 1981, Preservation potential of transgressive coastal lithosomes on the U.S. Atlantic shelf: Marine Geology, v. 42, p. 429–442.

——— , 1985, Influence of antecedent geology on evolution of Delaware's barrier system: Marine Geology, v. 63, p. 235–262.

Belknap, D. F., and Wehmiller, J. F., 1980, Application of amino acid racemization to geochronology of coastal plain mollusks; Delmarva Peninsula and shores of Chesapeake Bay, *in* Hare, P. E., Hoering, T. C., and King, K., eds., Biogeochemistry of Amino Acids: New York, John Wiley & Sons, p. 401–414.

Belknap, D. F., Kelley, J. T., and Shipp, R. C., 1986, Quaternary stratigraphy of representative Maine estuaries; initial examination by high resolution seismic reflection profiling, *in* FitzGerald, D. M. and Rosen, P. S., eds., Treatise on Glaciated Coasts: New York, Academic Press, (in press).

Bender, M. L., 1973, Helium-uranium dating of corals: Geochimica et Cosmochimica Acta, v. 37, p. 1229–1247.

Bennett, C. L., Beukins, R. P., Clover, M. R., Gove, H. E., Liebert, R. B., Litherland, A. E., Purser, K. H., and Sondheim, W. E., 1977, Radiocarbon dating by using electrostatic accelerators; Negative ions provide the key: Science, v. 198, p. 508–510.

Benninger, L. K., 1978, Lead-210 balance in Long Island Sound: Geochimica et Cosmochima Acta, v. 42, p. 1165–1174.

Berger, W. H., 1982, Climate steps in ocean history-lessons from the Pleistocene, *in* Berger, W. H. and Crowell, J. C., eds., Climate in Earth History: National Research Council Geophysical Study Committee, Washington, D.C., National Academy Press, p. 43–54.

Berggren, W. A., and van Couvering, J. A., 1974, The Late Neogene; Biostratigraphy, geochronology, and paleoclimatology of the last 15 million years in marine and continental sequences: Paleogeography, Paleoclimatology, and Paleoecology, v. 16, p. 1–216.

Biggs, R. B., 1978, Coastal bays, *in* Davis, R. A., Jr., ed., Coastal Sedimentary Environments: New York, Springer-Verlag, p. 69–99.

Birch, F. S., 1984, Seismic sedimentary units of the inner continental shelf of New Hampshire: Geological Society of America Abstracts with Programs, v. 16, p. 3.

Black, R. F., Goldthwait, R. P., and Willman, H. B., 1973, The Wisconsinian Stage: Geological Society of America Memoir 136, 334 p.

Blackwelder, B. W., 1980, Late Cenozoic marine deposition in the United States Atlantic coastal plain related to tectonism and global climate: Paleogeography, Paleoclimatology, Paleoecology, v. 34, p. 87–114.

——— , 1981, Late Cenozoic stages and molluscan zones of the U.S. middle Atlantic coastal plain: The Paleontological Society Memoir 12, Journal of Paleontology, v. 55, no. 5, 34 p.

Blackwelder, B. W., Pilkey, O. H., and Howard, J. D., 1979, Late Wisconsinan sea levels on the southeast U.S. Atlantic shelf based on in-place shoreline indicators: Science, v. 204, p. 618–620.

Blackwelder, B. W., Macintyre, I. G., and Pilkey, O. H., 1982, Geology of continental shelf, Onslow Bay, North Carolina, as revealed by submarine outcrops: American Association of Petroleum Geologists Bulletin, v. 66, p. 44–66.

Bloom, A. L., 1963, Late Pleistocene fluctuations of sea-level and post glacial crustal rebound in coastal Maine: American Journal of Science, v. 261, p. 862–879.

——— , 1964, Peat accumulation and compaction in a Connecticut coastal marsh: Journal of Sedimentary Petrology, v. 34, p. 599–603.

——— , 1967, Pleistocene shorelines: A new test of isostasy: Geological Society of America Bulletin, v. 78, p. 1477–1494.

Boardman, M. R., and Neumann, A. C., 1984, Sources of periplatform carbonates; Northwest Providence Channel, Bahamas: Journal of Sedimentary Petrology, v. 54, p. 1110–1123.

Bokenowiez, H. J., Gebert, J., and Gordan, R. B., 1976, Sediment mass balance of a large estuary, Long Island Sound: Estuarine and Coastal Marine Science, v. 4, p. 523–536.

Boon, J. D., 1975, Tidal discharge asymmetry in a salt marsh drainage system: Limnology and Oceanography, v. 20, p. 71–80.

Bornhold, B. D., and Pilkey, O. H., 1971, Bioclastic turbidite sedimentation in Columbus Basin, Bahamas: Geological Society of America Bulletin, v. 82, p. 1341–1354.

Borns, H. W., Jr., 1973, Late Wisconsin fluctuations of the Laurentide ice sheet in southern and eastern New England, *in* Black, R. F., Goldthwait, R. P., and Willman, H. B., eds., The Wisconsinan Stage: Geological Society of America Memoir 136, p. 37–45.

Bowen, D. Q., 1978, Quaternary Geology: Elmsford, New York, Pergamon Press, Inc., 221 p.

Broecker, W. S., and Thurber, D. L., 1965, Uranium-series dating of corals from Bahaman and Florida Key limestones: Science, v. 149, p. 58–60.

Brown, L. D., 1978, Recent crustal movement along the east coast of the United States: Tectonophysics, v. 44, p. 205–231.

Brown, P. M., 1958, The relation of phosphorites to ground water in Beaufort County, North Carolina: Economic Geology, v. 53, no. 1, p. 85–101.

Brown, P. M., Miller, J. A., and Swain, F. M., 1972, Structural and stratigraphic framework, and spatial distribution of permeability of Atlantic coastal plain, North Carolina to New York: U.S. Geological Survey Professional Paper 796, 79 p.

Brunn, P., 1962, Sea level rise as a cause of shore erosion: Journal of the Waterways and Harbors Division, American Society of Civil Engineers Proceedings, v. 88, p. 117–130.

Buffler, R. T., Watkins, J. S., and Dillon, W. P., 1979, Geology of the offshore Southeast Georgia Embayment, U.S. Atlantic Continental Margin, based on multichannel seismic reflection profiles, *in* Watkins, J. S., Montadert, L., and Dickerson, P. W., eds., Geological and Geophysical Investigations of Continental Margins: American Association of Petroleum Geologists Memoir 29, p. 11–75.

Butman, B., Noble, M., and Folger, D. W., 1979, Long-term observations of bottom current and bottom sediment movement on the mid-Atlantic continental shelf: Journal of Geophysical Research, v. 84, no. C3, p. 1187–1205.

Carter, J. G., 1984, Summary of lithostratigraphy and biostratigraphy for the coastal plain of the southeastern United States: Chapel Hill, University of North Carolina, Biostratigraphy Newsletter, No. 2.

Cathles, L. M., 1980, Interpretation of postglacial isostatic adjustment phenomena in terms of mantle rheology, *in* Mörner, N. A., ed., Earth Rheology, Isostasy, and Eustasy: New York, John Wiley & Sons, p. 11–44.

CLIMAP, 1976, The surface of the ice-age Earth: Science, v. 191, p. 1131–1137.

Colquhoun, D. J., 1974, Cyclic surficial stratigraphic units of the middle and lower coastal plains, central South Carolina, *in* Oaks, R. Q., Jr., and DuBar, J. R., eds., Post-Miocene Stratigraphy Central and Southern Atlantic Coastal Plain: Logan, Utah State University Press, p. 179–190.

—— , 1981, Variation in sea level on the South Carolina coastal plain: UNESCO International Geological Correlation Program No. 61, The Sea Level Program, International Quaternary Association Quaternary Shoreline, Holocene, and Neotectonics Commissions, Columbia, Department of Geology, University of South Carolina, 181 p.

Connally, G. G., and Sirkin, L. A., 1973, Wisconsinian history of the Hudson–Champlain Lobe, *in* Black, R. F., Goldthwait, R. P., and Willman, H. B., eds., The Wisconsinian Stage: Geological Society of America Memoir 136, p. 47–69.

Cook, H. E., and Mullins, H. T., 1983, Basin margin environment, *in* Scholle, P. A., Bebout, D. G., and Moore, C. H., eds., Carbonate Depositional Environments: American Association of Petroleum Geologists Memoir 33, p. 539–617.

Cooke, C. W., 1930, Correlation of coastal terraces: Journal of Geology, v. 38, p. 577–589.

Copeland, B. J., Hodson, R. G., Riggs, S. R., and Easley, J. E., 1983, Ecology of Albemarle Sound North Carolina; An Estuarine Profile: Washington, D.C., U.S. Department of the Interior, Fish and Wildlife Service, FWS/OBS-83-1, 68 p.

Crevello, P. D., and Schlager, W., 1980, Carbonate debris sheets and turbidites, Exuma Sound, Bahamas: Journal of Sedimentary Petrology, v. 50, p. 1121–1148.

Cronin, T. M., 1980, Biostratigraphic correlation of Pleistocene marine deposits and sea levels, Atlantic coastal plain of the southeastern United States: Quaternary Research, v. 13, p. 213–229.

Cronin, T. M., Szabo, B. J., Ager, T. A., Hazel, J. E., and Owens, J. P., 1981, Quaternary climates and sea levels of the U.S. Atlantic coastal plain: Science, v. 211, p. 233–240.

Crowson, R. A., 1980, Nearshore rock exposures and their relationships to modern shelf sedimentation, Onslow Bay, North Carolina [M.S. thesis]: Greenville, North Carolina, East Carolina University, 128 p.

Curray, J. R., 1965, late Quaternary history, continental shelves of the United States, *in* Wright, H. E., Jr., and Grey, D. G., eds., The Quaternary of the United States: Princeton, New Jersey, Princeton University Press, p. 723–735.

Daniels, R. B., and Gamble, E. E., 1974, Surficial deposits of the Neuse–Cape Fear divide above the Surry Scarp, North Carolina, *in* Oaks, R. Q., Jr., and DuBar, J. R., eds., Post-Miocene Stratigraphy Central and Southern Atlantic Coastal Plain: Logan, Utah, Utah State University Press, p. 88–101.

Delcourt, P. A., and Delcourt, H. R., 1981, Vegetation maps for North America; 40,000 yr. B.P. to the present, *in* Romans, R. C., ed., Geobotany II: New York, Plenum Publishing Company, p. 123–165.

Demarest, J. M., II, 1981, Genesis and preservation of Quaternary paralic deposits on Delmarva peninsula [Ph.D. thesis]: Newark, Department of Geology, University of Delaware, 240 p.

Demarest, J. M., II. Biggs, R. B., and Kraft, J. C., 1981, Time-stratigraphic concepts of a formation; Interpretation of surficial Pleistocene deposits by analogy with Holocene paralic deposits, southeastern Delaware: Geology, v. 9, p. 360–365.

Denny, C. S., 1982, The geomorphology of New England: U.S. Geological Survey Professional 1208, 18 p.

Denton, G. H., and Hughes, T. J., eds., 1981, The Last Great Ice Sheets: New York, John Wiley & Sons, 484 p.

Dillon, W. P., 1983, Geology report for proposed oil and gas lease sale No. 90; Continental margin off the southeastern United States: U.S. Geological Survey Open-File Report 83-1896, 125 p.

Dillon, W. P., and Oldale, R. N., 1978, Late Quaternary sea-level curve; Reinterpretation based on glaciotectonic influence: Geology, v. 6, p. 56–60.

Dillon, W. P., Paull, C. K., Buffler, R. T., and Fail, J. P., 1979, Structure and development of the Southeast Georgia Embayment and northern Blake Plateau; Preliminary analysis, *in* Watkins, J. S., Montadert, L., and Dickerson, P. W., eds., Geologic and geophysical investigations of continental margins: American Association of Petroleum Geologists Memoir 29, p. 27–46.

Dillon, W. P., Popenoe, P., Grow, J. A., Klitgord, K. D., Swift, B. A., Paull, C. K., and Cashman, K. V., 1983, Growth faulting and salt diapirism; Their relationship and control in the Carolina Trough, eastern North America, *in* Watkins, J. S., and Drake, C. C., eds., Studies in Continental Margin Geology: American Association of Petroleum Geologists Memoir 34, p. 21–46.

Dolan, R., Godfrey, P. J., and Odum, W. E., 1973, Man's impact on the barrier islands of North Carolina: American Science, v. 61, no. 2, p. 152–162.

Dorf, E., 1964, The use of fossil plants in paleoclimatic interpretations, *in* Narin, A.E.M., ed., Problems in Paleoclimatology: p. 13–31 and 46–48.

Dorjes, J., and Howard, J. D., 1975, Estuaries of the Georgia coast, USA; Sedimentology and biology. IV. Fluvial-marine transition indicators in an estuarine environment, Ogeechee River–Ossabaw Sound: Senckenbergiana Maritima, v. 7, p. 137–179.

Doyle, L. J., Pilkey, O. H., and Woo, C. C., 1979, Sedimentation on the eastern United States continental slope, *in* Doyle, L. J. Pilkey, O. H., eds., Geology of Continental Slopes: Society of Economic Paleontologists and Mineralogists Special Publication No. 27, p. 119–130.

Doyle, L. J., Wall, F. M., and Schroeder, P. D., 1981, Sediments and sedimentary processes as interpreted from piston cores and grab samples from the continental slope of the southeastern United States, *in* Popenoe, P., ed., Environmental geologic studies on the southeastern Atlantic outer continental shelf, 1977–78: U.S. Geological Survey Open File Report 81-582, 691 p.

Dravis, J., 1979, Rapid and widespread generation of recent oolite hardgrounds on a high energy Bahamian platform, Eleuthera Bank, Bahamas: Journal of Sedimentary Petrology, v. 49, p. 195–208.

Droxler, A. W., Schlager, W., and Whallon, C. C., 1983, Quaternary aragonite cycles and oxygen-isotope record in Bahamian carbonate ooze: Geology, v. 11, p. 235–239.

DuBar, J. R., 1974, Summary of the Neogene stratigraphy of southern Florida, *in* Oaks, R. Q., Jr., and DuBar, J. R., eds., Post-Miocene Stratigraphy Central and Southern Atlantic Coastal Plain: Logan, Utah State University Press, p. 206–231.

DuBar, J. R., Solliday, J. R., and Howard, J. F., 1974, Stratigraphy and morphology of Neogene deposits, Neuse River Estuary, North Carolina, *in* Oaks, R. Q., Jr., and DuBar, J. R., eds., Post-Miocene Stratigraphy Central and Southern Atlantic Coastal Plain: Logan, Utah State University Press, p. 88–101.

Eames, G. B., 1983, The late Quaternary seismic stratigraphy, lithostratigraphy, and geology history of a shelf-barrier-estuarine system, Dare County, North Carolina [M.S. thesis]: Greenville, North Carolina, Department of Geology, East Carolina University, 196 p.

Edwards, R. L., and Emery, K. O., 1977, Man on the continental shelf, *in* Newman, W. S., and Salwen, B., eds., Amerinds and their Paleoenvironments in Eastern North America: Annals of the New York Academy of Sciences, v. 288, p. 245–256.

Einstein, H. A., and Krone, R. B., 1962, Experiments to determine modes of cohesive sediment transport in salt water: Journal of Geophysical Research, v. 62, p. 1451–1461.

Emery, K. O., 1968, Relict sediments on continental shelves of the world: American Association of Petroleum Geologists Bulletin, v. 52, p. 445–464.

Emery, K. O., and Uchupi, E., 1972, Western North Atlantic Ocean; Topography, rocks, strucure, water, life, and sediments: American Association of Petroleum Geologists Memoir 17, 532 p.

Emiliani, C., 1966, Paleotemperature analysis of Caribbean cores P6304-8 and P6304-9 and a generalized temperature curve for the past 425,000 years: Journal of Geology, v. 74, p. 109–126.

—— , 1970, Pleistocene paleotemperatures: Science, v. 168, p. 822–825.

—— , 1978, The cause of the ice ages: Earth and Planetary Science Letters, v. 37,

p. 349–352.

Emiliani, C., and Shackleton, N. J., 1974, The Brunhes epoch; Isotopic paleo-temperatures and geochronology: Science, v. 183, p. 511–514.

Enos, P., 1974, Reefs, platforms, and basins of Middle Cretaceous in northeast Mexico, *in* Comparative Sedimentology of Carbonates Symposium: American Association of Petroleum Geologists Bulletin, v. 58, no. 5, p. 800–809.

Enos, P., and Perkins, R. D., 1977, Quaternary sedimentation in south Florida: Geological Society of America Memoir 147, 198 p.

Fader, G. B., King, L. H., and MacLean, B., 1977, Surficial geology of the eastern Gulf of Maine and Bay of Fundy: Geological Survey of Canada Paper 76-17, 23 p.

Fairbanks, R. G., and Matthews, R. K., 1978, The marine oxygen-isotope record in Pleistocene corals, Barbados, West Indies: Quaternary Research, v. 10, p. 181–196.

Fairbridge, R. W., 1961, Eustatic changes in sea level: Physics and Chemistry of the Earth, v. 14, p. 99–185.

Farre, J. A., McGregor, B. A., Ryan, W.B.F., and Robb, J. M., 1983, Breaching the shelfbreak; Passage from youthful to mature phase in canyon evolution, *in* Stanley, D. J., and Moore, G. T., eds., The Shelfbreak; Critical Interface on the Continental Margin: Society of Economic Paleontologists and Mineralogists Special Publication 33, p. 25–39.

Field, M. E., 1980, Sand bodies on coastal plain shelves: Holocene record of the U.S. Atlantic inner shelf off Maryland: Journal of Sedimentary Petrology, v. 50, p. 505–528.

Field, M. E., Meisburger, E. P., Stanley, E. A., and Williams, S. J., 1979, Upper Quaternary peat deposits on the Atlantic inner shelf of the United States: Geological Society of America Bulletin, v. 90, p. 618–628.

Fischer, A. G., 1961, Stratigraphic record of transgressing seas in light of sedimentation on Atlantic coast of New Jersey: American Association of Petroleum Geologists Bulletin, v. 45, p. 1656–1666.

FitzGerald, D. M., and FitzGerald, S. A., 1977, Factors influencing tidal inlet throat geometry, *in* Coastal Sediments '77, Fifth Symposium of the Waterway, Port, Coastal, and Ocean Division of the American Society of Civil Engineers, Charleston, South Carolina, p. 563–581.

Flint, R. F., 1971, Glacial and Quaternary Geology: New York, John Wiley and Sons, 892 p.

Folger, D. W., 1972, Texture and organic carbon content of bottom sediments in some estuaries of the United States: Geological Society of America Memoir 133, p. 391–408.

Fuller, M. L., 1914: The geology of Long Island, New York: U.S. Geological Survey Professional Paper 82, 231 p.

Gates, W. L., 1976, Modeling the ice-age climate: Science, v. 191, p. 1138–1144.

Gibson, T. C., 1967, Stratigraphy and paleoenvironment of the phosphatic Miocene strata of North Carolina: Geological Society of America Bulletin, v. 78, p. 631–649.

——, 1982, Depositional framework and paleoenvironments of Miocene strata from North Carolina to Maryland, *in* Upchurch, S. B., and Scott, T. M., eds., Miocene of the Southeastern United States: Florida Bureau of Geology Special Publication No. 25, p. 1–22.

——, 1983, Stratigraphy of Miocene through lower Pleistocene strata of the United States Central Atlantic Coastal Plain, *in* Ray, C. E., ed., Geology and Paleontology of the Lee Creek Mine, North Carolina: Smithsonian Contributions to Paleobiology, no. 53, p. 35–80.

Ginsburg, R. N., 1956, Environmental relationships of grain size and constituent particles in some south Florida carbonate sediments: American Association of Petroleum Geologists Bulletin, v. 40, p. 2384–2427.

Ginsburg, R. N., and Shinn, E. A., 1964, Distribution of the reef-building community in Florida and the Bahamas: American Association of Petroleum Geologists Bulletin, v. 48, p. 527.

Godfrey, P. J., Leatherman, S. P., and Zaremba, R., 1979, A geobotanical approach to classification of barrier beach systems, *in* Leatherman, S. P., ed., Barrier Islands from the Gulf of St. Lawrence to the Gulf of Mexico: New York, Academic Press, p. 185–210.

Goldberg, E. D., Gamble, E., Griffon, J. J., and Koide, M., 1977, Pollution

history of Narragansett Bay as recorded in its sediments: Estuarine and Coastal Marine Science, v. 5, p. 549–561.

Goldberg, E. D., Hodge, V., Koide, M., Griffen, J. J., Gamble, E., Bricker, O. P., Matisoff, G., and Holdren, G. R., Jr., 1978, A pollution history of Chesapeake Bay: Geochimica et Cosmochimica Acta, v. 42, p. 1413–1425.

Goldberg, W. M., 1983, Cay Sal Bank; Biologically impoverished, physically controlled environment: Atoll Research Bulletin, 271 p.

Gordon, R. B., 1979, Denudation rate of central New England determined from estuarine sedimentation: American Journal of Science, v. 279, p. 632–642.

Grootes, P. M., 1978, Carbon-14 time scale extended; Comparison of chronologies: Science, v. 200, p. 11–15.

Grow, J. A., and Sheridan, R. E., 1981, Deep structure and evolution of the continental margin off the eastern United States: Oceanologica Acta, Proceedings 26th International Geological Congress, Geology of Continental Margins Symposium, Paris, July 7-17, 1980, p. 11–19.

Grow, J. A., Mattick, R. E., and Schlee, J. S., 1979, Multichannel seismic depth sections and interval velocities over outer continental shelf and upper continental slope between Cape Hatteras and Cape Cod, *in* Watkins, J. S., Montadert, L., and Dickerson, P. W., eds., Geological and Geophysical Investigations of Continental Margins: American Association of Petroleum Geologists Memoir 29, p. 65–83.

Guilcher, A., 1967, Origin of sediments in estuaries, *in* Lauff, G. H., ed. Estuaries: American Association for the Advancement of Science Publication no. 83, p. 149–157.

Gustavson, T. C., 1972, A warm-water Pleistocene fauna from the Gardiners Clay of eastern Long Island: Journal of Paleontology, v. 46, p. 447–459.

——, 1976, Paleotemperature analysis of the marine Pleistocene of Long Island, New York, and Nantucket Island, Massachusetts: Geological Society of America Bulletin, v. 87, p. 1–8.

Halley, R. B., Harris, P. M., and Hine, A. C., 1983, Bank margin environments, *in* Scholle, P. A., Bebout, D. G., and Moore, C. H., eds., Carbonate Depositional Environments: American Association of Petroleum Geologists Memoir 33, p. 463–506.

Halsey, S. D., 1979, Nexus; New model of barrier island development, *in* Leatherman, S. P., ed., Barrier Islands from the Gulf of St. Lawrence to the Gulf of Mexico: New York, Academic Press, p. 185–210.

Hansen, J., Johnson, D., Lacis, A., Lebedeff, S., Lee, P., Rind, D., and Russell, G., 1981, Climate impact of increasing atmospheric carbon dioxide: Science, v. 213, p. 957–966.

Hardie, L. A., 1977, Sedimentation on the modern carbonate tidal flats of northwest Andros Island, Bahamas: The Johns Hopkins University Studies in Geology, No. 22, Baltimore, The Johns Hopkins University Press, 202 p.

Harris, P. M., 1979, Facies anatomy and diagenesis of a Bahamian ooid shoal: Sedimentation No. VII, The comparative sedimentology laboratory, Florida, University of Miami, 163 p.

——, 1983, The Joulters ooid shoal, Great Bahama Bank, *in* Peryt, T., ed., Coated Grains: New York, Springer-Verlag, p. 132–141.

Hathaway, J. C., 1972, Regional clay mineral facies in estuaries and continental margin of the United States east coast: Geological Society of America Memoir 133, p. 293–316.

Hathaway, J. C., Poag, C. W., Valentine, P. C., Miller, R. E., Schultz, D. M., Manheim, F. T., Kohout, F. A., Bothner, M. H., and Sangrey, D. A., 1979, U.S. Geological Survey core drilling on the Atlantic Shelf: Science, v. 206, p. 515–527.

Haven, D. S., and Morales-Alamo, R., 1972, Biodeposition as a factor in sedimentation of fine suspended solids in estuaries: Geological Society of America Memoir 133, p. 121–130.

Hayes, M. O., 1979, Barrier island morphology as a function of tidal and wave regime, *in* Leatherman, S. P., ed., Barrier Islands from the Gulf of St. Lawrence to the Gulf of Mexico: New York, Academic Press, p. 1–27.

Hayes, M. O., and Kana, T. W., eds., 1976, Terrigenous clastic depositional environments: Columbia, University of South Carolina, Coastal Research Division, Technical Report No. 11-CRD, 315 p.

Hazel, J. E., 1977, Distribution of some biostratigraphically diagnostic ostracodes in the Pliocene and lower Pleistocene of Virginia and northern North Carolina: Journal of Research of the U.S. Geological Survey, v. 5, p. 373–388.

——, 1983, Age and correlation of the Yorktown (Pliocene) and Croatan (Pliocene and Pleistocene) Formations at the Lee Creek Mine, *in* Ray, C. E., ed., Geology and Paleontology of the Lee Creek Mine, North Carolina: Smithsonian Contributions to Paleobiology, no. 53, p. 81–199.

Henry, V. J., 1981, Ocean bottom survey of the Georgia Bight, *in* Popenoe, P., ed., Environmental geologic studies on the southeastern Atlantic outer continental shelf: U.S. Geologic Survey Open-file Report 81-582-A, chapter 6.

Hine, A. C., 1983, Relict sand bodies and bedforms of the northern Bahamas; Evidence of extensive early Holocene sand transport, *in* Peryt, T. M., ed., Coated Grains: Heidelberg, Springer-Verlag, p. 116–131.

Hine, A. C., and Neumann, A. C., 1977, Shallow carbonate bank margin growth and structure, Little Bahama Bank: American Association of Petroleum Geologists Bulletin, v. 61, p. 376–406.

Hine, A. C., and Snyder, S. W., 1985, Coastal lithosome preservation; Evidence from the shoreface and continental shelf off Bogue Banks, North Carolina: Marine Geology, v. 63, p. 307–330.

Hine, A. C., and Steinmetz, J. S., 1984, Cay Sal Bank; A partially drowned carbonate platform: Marine Geology, v. 59, p. 135–164.

Hine, A. C., Wilber, R. J., and Neumann, A. C., 1981a, Carbonate sand bodies along contrasting shallow bank margins facing open seaways in northern Bahamas: American Association of Petroleum Geologists Bulletin, v. 65, p. 261–190.

Hine, A. C., Wilber, R. J., Bane, J. M., Neumann, A. C., and Lorenson, K. R., 1981b, Offbank transport of carbonate sands along open leeward bank margins, northern Bahamas: Marine Geology, v. 42, p. 327–348.

Hollister, C. D., 1973, Atlantic continental shelf and slope of the United States; Texture of the sediments from northern New Jersey to southern Florida: U.S. Geological Survey Professional Paper 529, 23 p.

Hoyt, J. H., 1967, Barrier island formation: Geological Society of America Bulletin, v. 79, p. 1125–1136.

Hoyt, J. H., and Hails, J. R., 1974, Pleistocene stratigraphy of southeast Georgia, *in* Oakes, R. Q., and DuBar, J. R., eds., Post-Miocene Stratigraphy Central and Southern Atlantic Coastal Plain: Logan, Utah State University Press, p. 191–205.

Hoyt, J. H., and Henry, V. J., 1967, Influence of island migration on barrier-island sedimentation: Geological Society of America Bulletin, v. 78, p. 77–86.

Hudson, J. H., Shinn, E. A., Halley, R. B., and Lidz, B., 1976, Sclerochronology; A tool for interpreting past environments: Geology, v. 4, p. 361–364.

Hughes, J. T., 1981, Numerical reconstruction of paleo-ice sheets, *in* Denton, G. H., and Highes, T. J., eds., The Last Great Ice Sheets: New York, John Wiley & Sons, p. 222–261.

Imbrie, J., and Imbrie, J. Z., 1980, Modeling the climatic response to orbital variations: Science, v. 207, p. 943–953.

Johnson, D. W., 1919, Shore Processes and Shoreline Development: New York, Hafner Publishing Company, Facsimile Edition 1972, 584 p.

Johnson, G. H., 1976, Geology of the Mulberry Island, Newport News North, and Hampton Quadrangles, Virginia: Charlottesville, Virginia Division of Mineral Resources, Report of Investigations no. 41, 72 p.

Jordan, R. R., 1974, Pleistocene deposits of Delaware, *in* Oaks, R. Q., Jr. and DuBar, J. R., eds., Post-Miocene Stratigraphy Central and Southern Atlantic Coastal Plain: Logan, Utah State University Press, p. 30–52.

Jordan, R. R., and Smith, R. V., 1983, Atlantic Coastal Plain, Correlation of Stratigraphic Units of North America (COSUNA) Project: American Association of Petroleum Geologists, 1 Chart.

Judson, S., 1968, Erosion of the land; What's happening to our continents?: American Scientist, v. 56, p. 356–374.

Katrosh, M. R., and Snyder, S. W., 1982, Diagnostic foraminifera and paleoecology of the Pungo River Formation, central coastal plain of North Carolina: Southeastern Geology, v. 23, p. 217–232.

Kaye, C. A., 1964, Illinoian and early Wisconsin moraines of Martha's Vineyard, Massachusetts: U.S. Geological Survey Professional Paper 501-C,

p. C140–C143.

Kaye, C. A., and Barghoorn, E. S., 1964, Late Quaternary sea-level change and crustal rise at Boston, Massachusetts, with notes on the autocompaction of peat: Geological Society of America Bulletin, v. 75, p. 63–80.

Keigwin, L. D., 1979, Late Cenozoic stable isotope stratigraphy and paleoceanography of DSDP site from nearby Pacific Ocean and Caribbean Sea cores: Geology, v. 6, p. 630–634.

Keigwin, L. D., and Corliss, B. H., 1986, Stable isotopes in middle late Eocene to Oligocene foraminifera: Geological Society of America Bulletin, v. 97, p. 335–345.

Keller, G., and Barron, J. A., 1983, Paleoceanographic implications of Miocene deep-sea hiatuses: Geological Society of America Bulletin, v. 94, p. 590–613.

Kelley, J. T., 1983, Composition and origin of the inorganic fraction of southern New Jersey coastal mud deposits: Geological Society of America Bulletin, v. 94, p. 689–699.

Kendall, C. G. St. G., and Schlager, W., 1981, Carbonates and relative changes in sea level: Marine Geology, v. 44, p. 181–212.

Kidwell, S. M., 1984, Outcrop features and origin of basin margin unconformities in the Lower Chesapeake Group (Miocene), Atlantic coastal plain, *in* Schlee, J. S., ed., Interregional Unconformities and Hydrocarbon Accumulation: American Association of Petroleum Geologists Memoir 36, p. 37–58.

Kier, J. S., and Pilkey, O. H., 1971, The influence of sea-level changes on sediment carbonate mineralogy, Tongue of the Ocean, Bahamas: Marine Geology, v. 11, p. 189–200.

King, P. B., Beikman, H. M., and Edmonstron, G. J., 1974, Geologic map of the United States: U.S. Geological Survey, scale 1:2,500,000.

Klitgord, K. D., and Behrendt, J. C., 1979, Basin structure of the United States Atlantic Continental Margin, *in* Watkins, J. S., Montadert, L., and Dickerson, P. W., eds., Geological and Geophysical Investigations of Continental Margins: American Association of Petroleum Geologists Memoir 29, p. 85–112.

Knebel, H. J., Martin, E. A., Glenn, J. L., and Needell, S. W., 1981, Sedimentary framework of the Potomac River estuary, Maryland: Geological Society of America Bulletin, v. 92, p. 578–589.

Knott, S. T., and Hoskins, H., 1968, Evidence of Pleistocene events in the structure of the continental shelf off the northeastern United States: Marine Geology, v. 6, p. 5–26.

Koteff, C., 1974, The morphologic sequence concept and deglaciation of southern New England, *in* Coates, D. R., ed., Glacial Geomorphology: Binghamton, New York, State University of New York Publications in Geomorphology, p. 121–144.

Koteff, C., and Pessl, F., Jr., 1981, Systematic ice retreat in New England: U.S. Geological Survey Professional Paper 1179, 20 p.

Kraft, J. C., 1971, Sedimentary facies patterns and geologic history of a Holocene marine transgression: Geological Society of America Bulletin, v. 82, p. 2131–2158.

——, 1978, Coastal stratigraphic sequences, *in* Davis, R. A., Jr., ed., Coastal Sedimentary Environments: New York, Springer-Verlag, p. 361–381.

Kraft, J. C., Allen, E. A., and Maurmeyer, E. M., 1978, The geological and paleogeomorphological evolution of a spit system and its associated coastal environments, Cape Henlopen Spit, coastal Delaware: Journal of Sedimentary Petrology, v. 48, p. 211–226.

Kranck, K., 1975, Sediment deposition from flocculated suspensions: Sedimentology, v. 22, p. 111–123.

Kumar, N., and Sanders, J. E., 1974, Inlet sequence; A vertical succession of sedimentary structures and textures created by the lateral migration of tidal inlets: Sedimentology, v. 21, p. 491–532.

Langbein, W. B., and Schumm, S. A., 1958, Yield of sediment in relation to mean annual precipitation: American Geophysical Union Transactions, v. 39, p. 1075–1084.

Larson, G. J., and Stone, B. D., eds., 1982, Late Wisconsinian Glaciation of New England: Dubuque, Iowa, Kendall/Hunt Publishing Company, 242 p.

Lauff, G. H., ed., 1967, Estuaries: American Association for the Advancement of Science Publication No. 83, 757 p.

Laughton, A. S., 1981, The first decade of GLORIA: Journal of Geophysical Research, v. 86, p. 11511–11534.

Leatherman, S. P., ed., 1979, Barrier Islands from the Gulf of St. Lawrence to the Gulf of Mexico: New York, Academic Press, 325 p.

Leopold, L. B., Wolman, M. G., and Miller, J. P., 1964, Fluvial processes in geomorphology: San Francisco, W. H. Freeman Co., 522 p.

Lewis, D. W., Riggs, S. R., Hine, A. C., Snyder, S.W.P., Snyder, S. W., and Waters, V., 1982, Preliminary stratigraphic report on the Pungo River Formation of the Atlantic continental shelf, Onslow Bay, North Carolina, *in* Scott, T. M., and Upchurch, S. B., eds., Miocene of the Southeastern United States: Florida Bureau of Geology Special Publication no. 25, p. 122–137.

Liddicoat, J. C., McCartan, L., Weems, R. E., and Lemon, E. M., Jr., 1981, Paleomagnetic investigations of Pliocene and Pleistocene sediments in the Charleston, South Carolina area: Geological Society of America Abstracts with Programs, v. 13, p. 12–13.

Liddicoat, J. C., Belknap, D. F., and Wehmiller, J. F., 1982, Paleomagnetic and amino acid dating of sediment in the Atlantic coastal plain: Geological Society of America Abstracts with Programs, v. 14, p. 546–547.

Lighty, R. G., Macintyre, I. G., and Stuckenrath, R., 1978, Submerged early Holocene barrier reef, southeast Florida Shelf: Nature, v. 275, p. 50–60.

Lisitzin, E., 1974, Sea-level Changes: New York, Elsevier Scientific Publishing Company, 286 p.

Lisle, L. D., 1982, Annotated bibliography of sea level changes along the Atlantic and Gulf coasts of North America: Shore and Beach, v. 50, no. 7, p. 24–33.

Longman, M. W., 1981, A process approach to recognizing facies of reef complexes, *in* Toomey, D. F., ed., European Fossil Reef Models: Society of Economic Paleontologists and Mineralogists Special Publication 30, p. 9–40.

Lougee, R. J., 1953, A chronology of post-glacial time in eastern North America: Scientific Monthly, v. 76, p. 259–276.

McCarten, L., Weems, R. E., and Lemon, E. M., Jr., 1980, The Wando Formation (upper Pleistocene) in the Charleston, South Carolina area: U.S. Geological Survey Bulletin 1502A, p. A110–A116.

McCarten, L., Owens, J. P., Blackwelder, B. W., Szabo, B. J., Belknap, D. F., Kriausakul, N., Mitterer, R. M., and Wehmiller, J. F., 1982, Comparison of amino acid racemization geochronometry with lithostratigraphhy, biostratigraphy, uranium-series coral dating, and magnetostratigraphy in the Atlantic coastal plain of the southeastern United States: Quaternary Research, v. 18, p. 337–359.

McCormick, C. L., and Toscano, M. A., 1981, Origin of the barrier island system of Long Island, New York: Northeastern Geology, v. 3, p. 230–234.

McGregor, B. A., 1984, Environmental geologic studies on the United States mid- and north-Atlantic outer continental shelf area 1980–1982; Executive summary: U.S. Geological Survey Open-File Report, Volume I, 54 p.

McGregor, B., Stubblefield, W. L., Ryan, W.B.F., and Twichell, D. C., 1982, Wilmington Submarine Canyon; A marine fluvial-like system: Geology, v. 10, p. 27–30.

McIntyre, A., Kipp, N. G., Be, A.W.H., Crowley, T., Kellogg, T., Gardner, J. V., Prell, W., and Ruddiman, W. F., 1976, Glacial North Atlantic 18,000 years ago; A CLIMAP reconstruction, *in* Investigation of Late Quaternary Paleoceanography and Paleoclimatology: Geological Society of America Memoir 145, p. 43–76.

McKee, E. D., and Ward, W. C., 1983, Eolian environment, *in* Scholle, P. A., Bebout, D. G. and Moore, C. H., eds., Carbonate Depositional Environments: American Association of Petroleum Geologists Memoir 33, p. 131–170.

Macintyre, I. G., and Milliman, J. D., 1970, Physiographic features on the outer shelf and upper slope, Atlantic continental margin, southeastern United States: Geological Society of America Bulletin, v. 81, p. 2577–2597.

Macintyre, I. G., and Pilkey, O. H., 1969, Tropical reef corals; Tolerance of low temperature of the North Carolina continental shelf: Science, v. 166, p. 374–375.

Matteucci, T. D., 1984, High-resolution seismic stratigraphy of the North Carolina continental margin–the Cape Fear Region; Sea-level cyclicity, paleobathymetry, and Gulf Stream dynamics [M.S. thesis]: St. Petersburg, Florida, University of South Florida, 151 p.

Matthews, R. K., 1984, Oxygen-isotope record of ice-volume history; 100 million years of glacio-eustatic sea-level fluctuation, *in* Schlee, J. S., ed., Interregional Unconformities and Hydrocarbon Accumulation: American Association of Petroleum Geology Memoir 36, p. 97–107.

Matthews, R. K., and Poore, R. Z., 1980, Tertiary $\delta^{18}O$ record and glacio–eustatic sea level fluctuations: Geology, v. 8, p. 501–504.

May, J. A., Warme, J. E., and Slater, R. A., 1983, Role of submarine canyons on shelfbreak erosion and sedimentation; Modern and ancient examples, *in* Stanley, D. J., and Moore, G. T., eds., The Shelfbreak; Critical Interface on the Continental Margin: Society of Economic Paleontologists and Mineralogists Special Publication 33, p. 315–332.

Mayewski, P. A., Denton, G. H., and Hughes, T. J., 1981, Late Wisconsin ice sheets of North America, *in* Denton, G. H., and Hughes, T. J., eds., The Last Great Ice Sheets: New York, John Wiley & Sons, p. 67–178.

Meade, R. H., 1969, Landward transport of bottom sediments in estuaries of the Atlantic coastal plain: Journal of Sedimentary Petrology, v. 39, p. 222–234.

—— , 1972a, Sources and sinks of suspended matter on continental shelves, *in* Swift, D.J.P., Duane, D. B., and Pilkey, O. H., Jr., eds., Shelf Sediment Transport; Process and Pattern: Stroudsburg, Pennsylvania, Dowden, Hutchinson and Ross, p. 249–262.

—— , 1972b, Transport and deposition of sediments in estuaries: Geological Society of America Memoir 133, p. 91–120.

—— , 1982, Sources, sinks, and storage of river sediment in the Atlantic drainage of the United States: Journal of Geology, v. 90, p. 235–252.

Menard, H. W., 1983, Insular erosion, isostasy, and subsidence: Science, v. 220, p. 913–918.

Milankovitch, M., 1941, Kanon der Erdbestrahlung und seine Anwendung auf des Eiszeit problem: Academie Royal de Serbe, Edition Special, v. 133, Section de Science, Mathematiques, et Nature, v. 33, 633 p.

Miller, J. A., 1982, Stratigraphy, structure, and phosphate deposits of the Pungo River Formation of North Carolina: North Carolina Department of Natural Resources and Community Development, Geological Survey Bulletin 87, 32 p.

Miller, K. G., Aubry, M. P., Khan, M. J., Melillo, A. J., Kent, D. V., and Berggren, W. A., 1985, Oligocene-Miocene biostratigraphy, magnetostratigraphy, and isotopic stratigraphy of the western North Atlantic: Geology, v. 13, p. 257–261.

Milliman, J. D., and Emery, K. O., 1968, Sea levels during the past 35,000 years: Science, v. 162, p. 1121–1123.

Milliman, J. D., and Meade, R. H., 1983, World-wide delivery of river sediment to the oceans: Journal of Geology, v. 91, p. 1–22.

Milliman, J. D., Pilkey, O. H., and Ross, D. A., 1972, Sediments of the continental margin off the eastern United States: Geological Society of America Bulletin, v. 83, p. 1315–1334.

Mitterer, R. M., 1974, Pleistocene stratigraphy in southern Florida based on amino acid diagenesis in fossil *Mercenaria:* Geology, v. 2, p. 425–428.

—— , 1975, Ages and diagenetic temperatures of Pleistocene deposits of Florida based on isoleucine epimerization in *Merceneria:* Earth and Planetary Science Letters, v. 28, p. 275–282.

Mixon, R. B., and Pilkey, O. H., 1976, Reconnaissance geology of the submerged and emerged coastal plain province, Cape Lookout area, North Carolina: U.S. Geological Survey Professional Paper 859, 45 p.

Mixon, R. B., Szabo, B. J., and Owens, J. P., 1982, Uranium-setting dating of mollusks and corals, and age of Pleistocene deposits, Chesapeake Bay area, Virginia and Maryland: U.S. Geological Survey Professional Paper 1067-E, 18 p.

Moody, D., and van Reenan, E., 1967, High-resolution subbottom seismic profiles of the Delaware estuary and its bay mouth: U.S. Geological Survey Professional Paper 575-D, p. D247–D252.

Mörner, N. A., ed., 1980, Earth Rheology, Isostasy, and Eustasy: New York, John Wiley & Sons, 599 p.

Moslow, T. F., and Heron, S. D., Jr., 1978, Relict inlets; Preservation and occurrence in the Holocene stratigraphy of southern Core Banks, North

Carolina: Journal of Sedimentary Petrology, v. 48, p. 1275–1286.

Muller, R. A., Stephanson, E. J., and Mast, T. S., 1978, Radioisotope dating with an accelerator; A blind measurement: Science, v. 201, p. 347–348.

Mullins, H. T., and Neumann, A. C., 1979, Deep carbonate bank margin structure and sedimentation in the northern Bahamas, *in* Doyle, L. J. and Pilkey, O. H., eds., Geology of Continental Slopes: Society of Economic Paleontologists and Mineralogists Special Publication No. 27, p. 165–192.

Mullins, H. T., Neumann, A. C., Wilber, R. J., Hine, A. C., and Chinburg, S. J., 1980, Carbonate sediment drifts in the northern Straits of Florida: American Association of Petroleum Geologists Bulletin, v. 64, p. 1701–1717.

Mullins, H. T., Newton, C. R., Heath, K., and Van Buren, H. M., 1981, Modern deep-water coral mounds north of Little Bahama Bank; Criteria for recognition of deep-water coral bioherms in the rock record: Journal of Sedimentary Petrology, v. 51, p. 999–1031.

Mullins, H. T., Heath, K., Van Buren, H. M., and Newton, C. R., 1984, Anatomy of a modern open-ocean carbonate slope; Northern Little Bahama Bank: Sedimentology, v. 31, p. 141–168.

Neiheisel, J., 1973, Source of detrital heavy minerals in estuaries of the Atlantic coastal plain [Ph.D. thesis]: Atlanta, Georgia Institute of Technology, 112 p.

Nelsen, T. A., and Stanley, D. J., 1986, Variable deposition rates on the slope and rise off the Mid-Atlantic states: Geo-Marine Letters (in press).

Nelson, B. W., ed., 1972, Environmental framework of coastal plain estuaries: Geological Society of America Memoir 133, 619 p.

Neumann, A. C., and Land, L. S., 1975, Lime mud deposition and calcareous algae in the Bight of Abaco, Bahamas: Journal of Sedimentary Petrology, v. 45, p. 763–768.

Neumann, A. C., and Moore, W. S., 1975, Sea level events and Pleistocene coral ages in the northern Bahamas: Quaternary Research, v. 5, p. 215–224.

Neumann, A. C., Kofoed, J. W., and Keller, G. H., 1977, Lithoherms in the Straits of Florida: Geology, v. 5, p. 4–10.

Newman, W. S., Fairbridge, R. W., and March, S., 1971, Marginal subsidence of glaciated areas; United States, Baltic, and North Sea: Etudes sur le Quaternaire dans le monde, Association Francais pour l'Etude Quaternaire Bulletin Supplemental No. 4, v. 2, p. 795–801.

Niedoroda, A. W., Swift, D.J.P., Figueiredo, A. G., and Freeland, G. L., 1985, Barrier island evolution, Middle Atlantic Shelf, U.S.A. Part II; Evidence from the shelf floor: Marine Geology, v. 63, p. 363–396.

Nilsson, S. T., and Persson, S., 1983, Tree-pollen spectra in the Stockholm region (Sweden) [abs.]: International Palynological Conference, no. 5, p. 288.

Normark, W. R., 1974, Submarine canyons and fan valleys; Factors affecting growth patterns of deep-sea fans, *in* Dott, R. H., Jr., and Shaver, R. H., eds., Modern and Ancient Geosynclinal Sedimentation: Society of Economic Paleontologists and Mineralogists Special Publication No. 19, p. 56–68.

NRC, 1959, Glacial Map of the United States East of the Rocky Mountains; Division of Earth Sciences, R. F. Flint, Chairman: Geological Society of America, scale 1:1,750,000.

Nummedal, D., Oertal, G. F., Hubbard, D. K., and Hine, A. C., 1977, Tidal inlet variability; Cape Hatteras to Cape Canaveral, *in* Coastal Sediments 1977: New York, American Society of Civil Engineers, p. 543–562.

Oaks, R. Q., Jr., and Dubar, J. R., eds., 1974, Post-Miocene Stratigraphy Central and Southern Atlantic Coastal Plain: Logan, Utah State University Press, 275 p.

Oaks, R. Q., Coch, N. K., Sanders, J. E., and Flint, R. F., 1974, Post-Miocene shorelines and sea levels, southeastern Virginia, *in* Oaks, R. Q., and DuBar, J. R., eds., Post-Miocene Stratigraphy Central and Southern Atlantic Coastal Plain: Logan, Utah State University Press, p. 63–87.

O'Brien, M. P., 1969, Equilibrium flow areas of inlets on sandy coasts: Journal of the Waterways and Harbors Division, Proceedings of the American Society of Civil Engineers, WW195, p. 43–52.

Oertel, G. F., 1976, Characteristics of suspended sediments in estuaries and nearshore waters of Georgia: Southeastern Geology, v. 18, p. 107–118.

Oertel, G. F., and Leatherman, S. P., eds., 1985, Barrier islands: Marine Geology Special Issue, v. 63, p. 1–369.

Oldale, R. N., 1982a, Permafrost in the northeastern United States coastal plain: American Quaternary Association 7th Biennial Conference Program and Abstracts, p. 150.

——, 1982b, Pleistocene stratigraphy of Nantucket, Martha's Vineyard, the Elizabeth Islands, and Cape Cod, Massachusetts, *in* Larson, G. J., and Stone, B. D., eds., Late Wisconsinan Glaciation of New England: Dubuque, Iowa, Kendall/Hunt Publishing Company, p. 1–34.

Oldale, R. N., Valentine, P. C., Cronin, T. M., Spiker, E. C., Blackwelder, B. W., Belknap, D. F., Wehmiller, J. W., and Szabo, B. J., 1982, Stratigraphy, structure, absolute age, and paleontology of the upper Pleistocene deposits of Sankaty Head, Nantucket Island, Massachusetts: Geology, v. 10, p. 246–252.

Oldale, R. N., Wommack, L. E., and Whitney, A. B., 1983, Evidence for postglacial low relative sea-level stand in the drowned delta of the Merrimack River, western Gulf of Maine: Quaternary Research, v. 33, p. 325–336.

Oostdam, B. L., and Jordan, R. R., 1972, Suspended sediment transport in Delaware Bay: Geological Society of America Memoir 133, p. 143–150.

Osmond, J. K., May, J. P., and Tanner, W. F., 1970, Age of the Cape Kennedy barrier-and-lagoon complex: Journal of Geophysical Research, v. 75, p. 469–479.

Owens, J. P., and Denny, C. S., 1979, Upper Cenozoic deposits of the central Delmarva Peninsula, Maryland and Delaware: U.S. Geological Survey Professional Paper 1067-A, 28 p.

Owens, J. P., Hess, M. M., Denny, C. S., and Dwarnik, E. J., 1983, Postdepositional alteration of surface and near-surface minerals in selected coastal plain formations of the middle Atlantic states: U.S. Geological Survey Professional Paper 1067-F, 45 p.

Paerl, H. W., 1973, Detritus in Lake Tahoe; Structural modification by attached microflora: Science, v. 180, p. 496–498.

Parsons, B., and Sclater, J. G., 1977, An analysis of ocean floor bathymetry and heat flow with age: Journal of Geophysical Research, v. 82, p. 803–827.

Paull, C. K., and Dillon, W. P., 1980, Structure, stratigraphy, and geologic history of Florida-Hatteras Shelf and inner Blake Plateau: American Association of Petroleum Geologists Bulletin 64, p. 339–358.

Pierce, J. W., and Colquhoun, D. J., 1970, Holocene evolution of a portion of the North Carolina coast: Geological Society of America Bulletin, v. 81, p. 3697–3714.

Pilkey, O. H., Blackwelder, B. W., Doyle, L. J., Estes, E., and Terlecky, P. M., 1969, Aspects of carbonate sedimentation on the Atlantic continental shelf off the southern United States: Journal of Sedimentary Petrology, v. 39, p. 744–768.

Pinet, P. R., and Popenoe, P., 1982, Blake Plateau; Control of Miocene sedimentation patterns by large-scale shifts of the Gulf Stream axis: Geology, v. 10, p. 257–259.

——, 1985, A scenario of Mesozoic-Cenozoic ocean circulation over the Blake Plateau and its environs: Geological Society of America Bulletin, v. 96, p. 618–626.

Pinet, P. R., Popenoe, P., and Nelligan, D. F., 1981, Gulf Stream; Reconstruction of Cenozoic flow patterns over the Blake Plateau: Geology, v. 9, p. 266–270.

Pitman, W. C., II., 1979, The effect of eustatic sea level changes on stratigraphic sequences at Atlantic Margins, *in* Watkins, J. S., Montadert, L., and Dickerson, P. W., eds., Geological and Geophysical Investigations of Continental Margins: American Association of Petroleum Geologists Memoir 29, p. 453–460.

Popenoe, P., 1985, Cenozoic depositional and structural history of the North Carolina margin from seismic stratigraphic analysis, *in* Poag, C. W., ed., Geologic evolution of the United States Atlantic Margin: New York, Van Nostrand Rheinhold, p. 125–188.

Popenoe, P., Coward, E. L., and Cashman, K. V., 1982, A regional assessment of potential environmental hazards to and limitations on petroleum development of the SE United States Atlantic continental shelf, slope, and rise, offshore, North Carolina: U.S. Geological Survey Open-File Report 82-136, 67 p.

Postma, H., 1967, Sediment transport and sedimentation in the estuarine environment, *in* Lauff, G. H., ed., Estuaries: American Association for the Advancement of Science Publication No. 83, p. 158–178.

Powers, M. C., 1957, Adjustment of land-derived clays to the marine environment: Journal of Sedimentary Petrology, v. 27, p. 355–372.

Prior, D. B., Coleman, J. M., and Garrison, L. E., 1979, Digitally acquired undistorted side-scan sonar images of submarine landslides, Mississippi River delta: Geology, v. 7, p. 423–425.

Pritchard, D. W., 1967, Observations of circulation in coastal plain estuaries, *in* Lauff, G. H., ed., Estuaries: American Association for the Advancement of Science Publication No. 83, p. 37–44.

Purdy, E. G., 1963, Recent calcium carbonate facies of the Great Bahama Bank; Sedimentary facies: Journal of Geology, v. 71, p. 472–497.

Rampino, M. R., and Sanders, J. E., 1981, Episodic growth of Holocene tidal marshes in the northeastern United States; A possible indicator of eustatic sea-level fluctuations: Geology, v. 9, p. 63–67.

Richards, H. G., 1967, Stratigraphy of the Atlantic coastal plain between Long Island and Georgia; A review: American Association of Petroleum Geologists Bulletin, v. 51, p. 2400–2429.

Riggs, S. R., 1979a, A geologic profile of the North Carolina coastal-inner continental shelf system, *in* Langfelder, J., ed., Ocean Outfall Wastewater Disposal Feasibility and Planning: Raleigh, North Carolina State University Press, p. 90–113.

—— , 1979b, Petrology of the Tertiary phosphorite system of Florida: Economic Geology, v. 74, p. 195–220.

—— , 1979c, Phosphorite sedimentation in Florida; A model phosphogenic system: Economic Geology, v. 74, p. 285–314.

—— , 1984, Paleoceanographic model of Neogene phosphorite deposition, U.S. Atlantic continental margin: Science, v. 223, no. 4632, p. 123–131.

—— , 1986, Model of Tertiary phosphorites on the World's continental margins, *in* Teleki, P. G., ed., Marine minerals; Resource assessment strategies: Dordricht, Holland, Reidel Publishing Co. (in press).

Riggs, S. R., Lewis, D. W., Scarborough, A. K., and Snyder, S. W., 1982, Cyclic deposition of the Neogene phosphorites in the Aurora area, North Carolina, and their possible relationship to global sea-level fluctuations: Southeastern Geology, v. 23, no. 4, p. 189–204.

Riggs, S. R., Snyder, S.W.P., Hine, A. C., Snyder, S. W., Ellington, M. D., and Mallette, P. M., 1985, Geologic framework of phosphate resources in Onslow Bay, North Carolina continental shelf: Economic Geology, v. 80, p. 716–738.

Ritter, J. R., 1974, The effects of Hurricane Agnes flood on channel geometry and sediment discharge of selected streams in the Susquehanna River basin, Pennsylvania: Journal of Research of the U.S. Geological Survey, v. 2, p. 753–762.

Robb, J. M., Hampson, J. C., Jr., and Twichell, D. C., 1981, Geomorphology and sediment stability of a segment of the U.S. continental slope off New Jersey: Science, v. 211, p. 935–937.

Roberts, H. H., Rouse, L. J., Walker, N. D., and Hudson, J. H., 1982, Coldwater stress in Florida Bay and northern Bahamas; A product of winter cold-air outbreaks: Journal of Sedimentary Petrology, v. 52, p. 145–155.

Rosen, P. S., 1979, Aeolian dynamics of a barrier island system, *in* Leatherman, S. P., ed., Barrier Islands from the Gulf of St. Lawrence to the Gulf of Mexico: New York, Academic Press, p. 81–98.

Ruddiman, W. F., and McIntyre, A., 1979, Warmth of the subpolar North Atlantic ocean during northern hemisphere ice sheet growth: Science, v. 204, p. 173–175.

Ryan, J. D., 1953, The sediments of Chesapeake Bay: Maryland Board of Natural Resources, Department of Mines and Water Resources, Bulletin 12, 125 p.

Ryan, W.B.F., and Miller, E. L., 1981, Evidence of a carbonate platform beneath Georges Bank: Marine Geology, v. 44, p. 213–228.

Sanders, J. E., and Kumar, N., 1975, Evidence of shoreface retreat and in-place "drowning" during Holocene submergence of barriers, shelf off Fire Island, New York: Geological Society of America Bulletin, v. 86, no. 1, p. 65–76.

Scarborough, A. K., Riggs, S. R., and Snyder, S. W., 1982, Stratigraphy and petrology of the Pungo River Formation, central coastal plain, North Carolina: Southeastern Geology, v. 23, no. 4, p. 205–216.

Schafer, J. P., and Hartshorn, J. H., 1965, The Quaternary of New England, *in*

Wright, H. E., and Frey, D. G., eds., The Quaternary of the United States: Princeton, New Jersey, Princeton University Press, p. 113–128.

Schlager, W., 1981, The paradox of drowned reefs and carbonate platforms: Geological Society of America Bulletin, v. 92, p. 197–211.

Schlager, W., and James, N. P., 1978, Low-magnesian calcite limestones forming at the deep-sea floor, Tongue of the Ocean, Bahamas: Sedimentology, v. 25, p. 675–702.

Schlee, J., 1957, Upland gravels of southern Maryland: Geological Society of America Bulletin, v. 68, p. 1371–1410.

—— , 1973, Atlantic continental shelf and slope of the United States; sediment texture of the northeastern part: U.S. Geological Survey Professional Paper 529L, 64 p.

—— , 1982, Summary report of the sediments, structural framework, petroleum potential, and environmental conditions of the United States middle and northern continental margin in area of proposed oil and gas lease sale No. 82: U.S. Geological Survey, Open-File Report No. 81-1353, 133 p.

Schlee, J. S., and Pratt, R. M., 1970, Atlantic continental shelf and slope of the United States; Gravels of the northeastern part: U.S. Geological Survey Professional Paper 529-H, 39 p.

Schlee, J. S., Dillon, W. P., and Grow, J. A., 1979, Structure of the continental slope off the eastern United States, *in* Doyle, L. J., and Pilkey, O. H., eds., Geology of Continental Slopes: Society of Economic Paleontologists and Mineralogists Special Publication 27, p. 95–117.

Schneider, E. D., 1972, Sedimentary evolution of rifted continental margins: Geological Society of America Memoir 137, p. 109–118.

Schnitker, D., 1974, Postglacial emergence of the Gulf of Maine: Geological Society of America Bulletin, v. 85, p. 491–494.

—— , 1975, Late glacial to Recent paleoceanography of the Gulf of Maine; 1st International Symposium on Continental Margin Benthonic Foraminifera, Part B; Paleoecology and Biostratigraphy: Maritime Sediments Special Publication 1, p. 385–392.

Scholl, D. W., Craighead, F. C., and Stuiver, M., 1969, Florida submergence curve revisited; Its relation to coastal sedimentation rates: Science, v. 163, p. 562–564.

Schubel, J. R., 1969, Size distribution of the suspended particles of the Chesapeake Bay turbidity maximum: Netherlands Journal of Sea Research, v. 4, p. 283–309.

—— , 1971, Estuarine circulation and sedimentation, American Geological Institute Lecture Notes: Washington, D.C., American Geological Institute, p. V1–V7.

—— , 1974, Effects of tropical Storm Agnes on the suspended solids of the northern Chesapeake Bay, *in* Gibbs, R. J., ed., Suspended Solids in Water: New York, Plenum Press, p. 113–133.

Schwartz, M. L., 1971, The multiple causality of barrier islands: Journal of Geology, v. 79, p. 91–94.

Schwartz, M. L., ed., 1973, Barrier Islands: Stroudsburg, Pennsylvania, Dowden, Hutchinson and Ross, Benchmark Paper in Geology, v. 9, 451 p.

Scott, T. M., 1986, Lithostratigraphy of the Hawthorn Group (Miocene) of Florida: Florida Geological Survey Publication (in press).

Shackleton, N. J., 1977, The ocean oxygen isotope record; Stratigraphic tool and paleoglacial record: Abstracts of X'th International Quaternary Association Congress, p. 415.

Shackleton, N. J., and Opdyke, N. D., 1973, Oxygen isotope and paleomagnetic stratigraphy of equatorial Pacific core V28-238; Oxygen isotope temperatures and ice volumes on a 10^5 and 10^6 year scale: Quaternary Research, v. 3, p. 39–55.

—— , 1976, Oxygen isotope and paleomagnetic stratigraphy of Pacific core V28-239 late Pliocene to latest Pleistocene, *in* Cline, R. M., and Hays, J. D., eds., Investigation of late Quaternary Paleoceanography and Paleoclimatology: Geological Society of America Memoir 145, p. 449–464.

Shattuck, G. B., 1901, The Pleistocene problem of the north Atlantic coastal plain: Johns Hopkins University Circular, v. 20, no. 152, p. 67–75.

Sheldon, R. P., 1980, Episodicity of phosphate deposition and deep ocean circulation; An hypothesis, *in* Bentor, Y. K., ed., Marine Phosphorites: Society of

Economic Paleontologists and Mineralogists Special Publication 29, p. 239–248.

Shepard, F. P., and Dill, R. F., 1966, Submarine Canyons and Other Sea Valleys: Chicago, Rand McNally, 381 p.

Sheridan, R. E., 1974, Atlantic continental margin of North America, *in* Burk, C. A., and Drake, C. L., eds., The Geology of Continental Margins: New York, Springer-Verlag, p. 391–407.

Sheridan, R. E., Dill, C. E., Jr., and Kraft, J. C., 1974, Holocene sedimentary environments of the Atlantic Inner Shelf off Delaware: Geological Society of America Bulletin, v. 85, p. 1319–1328.

Sheridan, R. E., Crosby, J. T., Bryan, G. M., and Stoffa, P. L., 1981, Stratigraphy and structure of southern Blake Plateau, northern Florida Straits, and northern Bahama platform from multichannel seismic reflection data: American Association of Petroleum Geologists Bulletin, v. 65, p. 2571–2593.

Shinn, E. A., 1976, Coral reef recovery in Florida and the Persian Gulf: Environmental Geology, v. 1, p. 241–254.

Sholkovitz, E. R., and Copland, D., 1981, The coagulation, solubitility, and absorbtion properties of Fe, Mn, Cu, Ni, Cd, Co, and humic acids in river water: Geochimica et Cosmochimica Acta, v. 45, p. 181–189.

Sirkin, L. A., and Stuckenrath, R., 1980, The Portwashingtonian warm interval in the northern Atlantic coastal plain: Geological Society of America Bulletin, v. 19, p. 332–336.

Sirkin, L. A., Denny, C. S., and Rubin, M., 1977, Late Pleistocene environment of the central Delmarva Peninsula, Delaware–Maryland: Geological Society of America Bulletin, v. 88, p. 139–142.

Smith, G. W., 1982, End moraines and the pattern of last ice retreat from central and south coastal Maine, *in* Larson, G. J., and Stone, B. D., eds., Late Wisconsinan glaciation of New England: Dubuque, Iowa, Kendall/Hunt Publishing Company, p. 195–210.

Snyder, S.W.P., 1982, Seismic stratigraphy within the Miocene Carolina Phosphogenic Province; Chronostratigraphy, paleotopographic controls, sea-level cyclicity, Gulf Stream dynamics, and the resulting depositional framework [M.S. thesis]: Chapel Hill, University of North Carolina, 183 p.

Snyder, S.W.P., Belknap, D. F., Hine, A. C., and Steele, G. A., 1982a, Seismic stratigraphy, lithostratigraphy, and amino acid racemization of the Diamond City Formation; Reinterpretation of a reported Mid-Wisconsin high sea-level indicator from the North Carolina coastal plain: Geological Society of America Abstracts with Programs, v. 14, p. 84.

Snyder, S.W.P., Hine, A. C., and Riggs, S. R., 1982b, Miocene seismic stratigraphy, structural framework, and sea-level cyclicity; North Carolina continental shelf: Southeastern Geology, v. 23, p. 247–266.

Snyder, S. W., Riggs, S. R., Katrosh, M. R., Lewis, D. W., and Scarborough, A. K., 1982, Synthesis of phosphatic sediment-faunal relationships within the Pungo River Formation; Paleoenvironmental implications: Southeastern Geology, v. 23, no. 4, p. 233–246.

Snyder, S. W., Mauger, L. L., and Akers, W. H., 1983, Planktonic foraminifera and biostratigraphy of the Yorktown Formation, Lee Creek Mine, *in* Ray, C. E., ed., Geology and paleontology of the Lee Creek Mine, North Carolina: Smithsonian Contributions to Paleobiology, No. 53, p. 455–482.

Snyder, S. W., and Moore, T. L., 1985, Planktonic foraminiferal biostratigraphy of cyclical Miocene deposits in central and northern Onslow Bay, North Carolina continental shelf: Geological Society of America Abstracts with Programs, v. 17, no. 2, p. 135.

Stahl, L., Koczan, J., and Swift, D., 1974, Anatomy of a shore-face-connected ridge system on the New Jersey shelf; Implications for genesis of the shelf surficial sand sheet: Geology, v. 2, p. 117–120.

Stanley, D. J., Addy, S. K., and Behrens, E. W., 1983, The mudline; Variability of its position relative to shelfbreak, *in* Stanley, D. J., and Moore, G. T., eds., The Shelfbreak; Critical Interface on the Continental Margin: Society of Economic Paleontologists and Mineralogists Special Publication 33, p. 279–298.

Stanley, D. J., and Moore, G. T., 1983, The shelfbreak; Critical interface on continental margins: Society of Economic Paleontologists and Mineralogists Special Publication 33, 467 p.

Stubblefield, W. L., Kersey, D. G., and McGrail, D. W., 1983, Development of middle continental shelf sand ridges, New Jersey: American Association of Petroleum Geologists Bulletin, v. 67, p. 817–830.

Stuckenrath, R., 1977, Radiocarbon; Some notes from Merlin's diary, *in* Newman, W. S., and Salwen, B., eds., Amerinds and their Paleoenvironments in Northeastern North America: Annals of the New York Academy of Sciences, v. 288, p. 181–188.

Stuiver, M., and Borns, H. W., Jr., 1975, Late Quaternary marine invasion in Maine; Its chronology and associated crustal movement: Geological Society of America Bulletin, v. 86, p. 99–104.

Stuiver, M., Heusser, C. J., and Yang, I. C., 1978, North American glacial history extended to 75,000 years ago: Science, v. 200, p. 16–21.

Swift, D.J.P., 1968, Coastal erosion and transgressive stratigraphy: Journal of Geology, v. 76, p. 444–456.

—— , 1975, Barrier island genesis; Evidence from the central Atlantic shelf, eastern U.S.A.: Sedimentary Geology, v. 14, p. 1–43.

—— , 1976, Coastal sedimentation, *in* Stanley, D. J., and Swift, D.J.P., eds., Marine Sediment Transport and Environmental Management: New York, John Wiley and Sons, p. 255–310.

Swift, D.J.P., and Field, M. F., 1981, Evolution of a classic ridge field, Maryland sector, North American inner shelf: Sedimentology, v. 28, p. 461–482.

Swift, D.J.P., and Sears, P., 1974, Estuaries and littoral depositional patterns in the surficial sand sheet, central and southern Atlantic shelf of North America, *in* Internal Symposium on Interrelationship of Estuarine and Continental Shelf Sedimentation: Institute of Geology Du Rossin D'aquitaine, Memoir 7, p. 171–189.

Swift, D.J.P., Duane, D. B. and Pilkey, O. H., eds., 1972a, Shelf sediment transport-process and pattern: Stroudsburg, Pennsylvania, Dowden, Hutchinson and Ross, 656 p.

Swift, D.J.P., Kofoed, J. W., Saulsbury, F. J., and Sears, P., 1972b, Holocene evolution of the shelf surface, central and southern Atlantic shelf of North America, *in* Swift, D.J.P., Duane, D. B., and Pilkey, O. H., Jr., Shelf sediment transport-process and pattern: Stroudsburg, Pennsylvania, Dowden Hutchinson and Ross, p. 499–574.

Swift, D.J.P., Freeland, G. L., and Young, R. A., 1979, Time and space distributions of megaripples and associated bedforms, Middle Atlantic Bight, North American Atlantic Shelf: Sedimentology, v. 26, p. 389–706.

Swift, D.J.P., Moir, R., and Freeland, G. L., 1980, Quaternary rivers on the New Jersey shelf; Relation of sea floor to buried valleys: Geology, v. 8, p. 176–280.

Swift, D.J.P., Thorne, J. A., and Oertel, G., 1986, Fluid process and seafloor response on a modern storm-dominated shelf; Middle Atlantic Bight of North America. Part II; Response of the shelf floor, *in* Knight, J., ed., Shelf Sands and Sandstone Reservoirs, A Symposium: Canadian Society of Petroleum Geologists, Symposium Volume (in press).

Thom, B. G., 1973, The dilemma of high interstadial sea levels during the last glaciation: Progress in Geography, v. 5, p. 170–246.

Thompson, S. N., 1973, Sea-level rise along the Maine coast during the last 3000 years [M.S. thesis]: Orono, University of Maine, 78 p.

Thompson, W. B., 1979, Surficial geology handbook for coastal Maine: Maine Geological Survey, 68 p.

—— , 1982, Recession of the Late Wisconsinan ice sheet in coastal Maine, *in* Larson, G. J., and Stone, B. D., eds., Late Wisconsinan Glaciation of New England: Dubuque, Iowa, Kendall/Hunt Publishing Company, p. 211–228.

Toy, T. J., 1982, Accelerated erosion; Process, problems, prognosis: Geology, v. 10, p. 524–529.

Trumbull, J.V.A., 1972, Atlantic continental shelf and slope of the United States; Sand sized fraction of bottom sediments, New Jersey to Nova Scotia: U.S. Geological Survey Professional paper, 529K, 45 p.

Tucholke, B. E., and Hollister, C. D., 1973, Late Wisconsin glaciation of the southwestern Gulf of Maine; New evidence from the marine environment: Geological Society of America Bulletin, v. 84, p. 3279–3296.

Twichell, D. C., 1983, Bedform distribution and inferred sand transport on Georges Bank, United States Atlantic Continental shelf: Sedimentology,

v. 30, p. 695–710.

Twichell, D. C., Knebel, H. J., and Folger, D. W., 1977, Delaware River, evidence for its former extension to Wilmington Submarine Canyon: Science, v. 195, p. 483–484.

Twichell, D. C., McClennen, C. E., an Butma, B., 1981, Fine grained sediment accumulation, southern New England shelf: Journal of Sedimentary Petrology, v. 51, p. 269–280.

Uchupi, E., 1968, The Atlantic continental shelf and slope of the United States-physiography: U.S. Geological Survey Professional Paper 529-C, 30 p.

Upson, J. E., and Spencer, C. W., 1964, Bedrock valleys of the New England coast as related to fluctuations of sea level: U.S. Geological Survey Professional Paper 454-M, p. M1–M41.

Vail, P. R., and Mitchum, R. M., Jr., 1979, Global cycles of relative changes of sea level from seismic stratigraphy, *in* Watkins, J. S., Montadert, L., and Dickerson, P. W., eds., Geological and Geophysical Investigations of Continental Margins: American Association of Petroleum Geologists Memoir 29, p. 469–472.

Vail, P. R., Mitchum, R. M., Todd, R. G., Widemier, J. M., Thompson, S., Sangree, J. B., Bubb, J. H., and Hatelid, W. G., 1977, Seismic stratigraphy and global changes of sea level, *in* Payton, C. E., ed., Seismic stratigraphy; Applications to hydrocarbon exploration: American Association of Petroleum Geologists Memoir 26, p. 49–212.

Valentine, P. C., 1971, Climatic implications of a late Pleistocene ostracode assemblage from southeastern Virginia: U.S. Geological Survey Professional Paper 683-D, 26 p.

van Donk, J., 1976, δO^{18} record of the Atlantic Ocean for the entire Pleistocene Epoch, *in* Cline, R. M., and Hays, J. D., eds., Investigation of Late Quaternary Paleoceanography and Paleoclimatology: Geological Society of America Memoir 145, p. 267–299.

Vanney, J. R., and Stanley, D. J., 1983, Shelfbreak physiography; An overview, *in* Stanley, D. J., and Moore, G. T., eds., The Shelfbreak; Critical Interface on the Continental Margin: Society of Economic Paleontologists and Mineralogists Special Publication 33, p. 1–24.

Vincent, C. E., Swift, D.J.P., and Hilliard, B., 1981, Transport in the New York Bight, North American Atlantic Shelf: Marine Geology, v. 42, p. 201–232.

Visher, G. S., and Howard, J. D., 1974, Dynamic relationship between hydraulics and sedimentation in the Altamaha estuary: Journal of Sedimentary Petrology, v. 74, p. 502–521.

Wall, F. M., 1981, The clay mineralogy of continental slope and rise sediments off the Eastern United States [M.S. thesis]: St. Petersburg, University of South Florida, 81 p.

Ward, L. G., 1981, Suspended material transport in marsh tidal channels, Kiawah Island, South Carolina: Marine Geology, v. 40, p. 139–154.

Ward, L. W., and Strickland, G. L., 1985, Outline of Tertiary stratigraphy and depositional history of the U.S. Atlantic coastal plain, *in* Poag, C. W., ed., Geologic evolution of the United States Atlantic margin: New York, Von Nostrand Reinhold Co., p. 87–124.

Waters, V. J., 1983, Foraminiferal paleoecology and biostratigraphy of the Pungo River Formation, southern Onslow Bay, North Carolina continental shelf [M.S. thesis]: Greenville, North Carolina, East Carolina University, 186 p.

Waters, V. J., and Snyder, S. W., 1986, Planktonic foraminiferal biostratigraphy of the Pungo River Formation, southern Onslow Bay, North Carolina continental shelf: Journal of Foraminiferal Research, v. 16, p. 9–23.

Watts, A. B., and Steckler, M. S., 1981, Subsidence and tectonics of Atlantic-type continental margins: Oceanologica Acta, Proceedings, 26th International Geological Congress, Geology of Continental Margins Symposium, Paris,

p. 143–153.

Webb, T., III, and McAndrews, J. H., 1976, Corresponding patterns of contemporary pollen and vegetation in central North America, *in* Cline, R. M., and Hays, J. D., eds., Investigation of Late Quaternary Paleoceanography and Paleoclimatology: Geological Society of America Memoir 145, p. 267–299.

Wehmiller, J. F., 1982, A review of amino acid racemization studies in Quaternary mollusks; Stratigraphic and chronologic applications in coastal and interglacial sites, Pacific and Atlantic coasts, United States: Quaternary Research, v. 18, p. 311–336.

Wehmiller, J. F., and Belknap, D. F., 1978, Alternative kinetic models for the interpretation of amino acid racemization ratios, in Pleistocene mollusks; examples from California, Washington, and Florida: Quaternary Research, v. 9, p. 330–348.

—— , 1982, Amino acid age estimates, Quaternary Atlantic coastal plain; Comparison with U-series dates, biostratigraphy, and paleomagnetic control: Quaternary Research, v. 18, p. 311–336.

Weil, C. B., 1977, Sediments, structural framework, and evolution of Delaware Bay, a transgressing estuarine delta: College of Marine Studies, University of Delaware, Publication No. DEL-SG-4-77, 199 p.

White, G. W., 1973, History of investigation and classification of Wisconsinan drift in north-central United States, *in* Black, R. F., Goldthwait, R. P., and Willman, H. B., eds., The Wisconsinan Stage: Geological Society of America Memoir 136, p. 3–36.

White, W. A., 1978, Influence of glacial meltwater in the Atlantic coastal plain: Southeastern Geology, v. 19, p. 139–156.

Whitehead, D. R., 1981, Late-Pleistocene vegetational changes in northeastern North Carolina: Ecological Monographs, v. 51, p. 451–471.

Wilber, R. J., 1981, Late Quaternary history of a leeward carbonate bank margin; A chronostratigraphic approach [Ph.D. thesis]: Chapel Hill, University of North Carolina, 290 p.

Winker, C. D., and Howard, J. D., 1977, Correlation of tectonically deformed shorelines on the southeastern Atlantic coastal plain: Geology, v. 5, p. 125–127.

Winland, H. D., and Matthews, R. K., 1974, Origin and significance of grapestone, Bahama Islands: Journal of Sedimentary Petrology, v. 44, p. 921–927.

Woodruff, D., Savin, S. M., and Douglas, R. G., 1981, Miocene stable isotope record; A detailed deep Pacific Ocean study and its paleoclimatic implications: Science, v. 212, no. 4495, p. 665–668.

Woodruff, F., 1985, Changes in Miocene deep-sea benthic foraminiferal distribution in the Pacific Ocean; Relationship to paleoceanography, *in* Kennett, J. P., ed., The Miocene Ocean; Paleoceanography and Biogeography: Geological Society of America Memoir 163, p. 131–176.

Worzel, J. C., and Burke, C. A., 1978, Margins of Gulf of Mexico: American Association of Petroleum Geologists, v. 62, p. 2290–2307.

Wright, H. E., Jr., 1981, Vegetation east of the Rocky Mountains 18,000 years ago: Quaternary Research, v. 15, p. 113–125.

Yarbro, L. A., Carlson, P. R., Fisher, T. R., Chanton, J. P., and Kemp, W. M., 1983, A sediment budget for the Choptank River estuary in Maryland, USA: Estuarine and Coastal Shelf Science, v. 17, p. 555–570.

Zabawa, C. F., 1978, Microstructure of agglomerated suspended sediments in the northern Chesapeake Bay estuary: Science, v. 202, p. 49–51.

Zenkovitch, V. P., 1967, Processes of coastal development: Edinburgh, Oliver and Boyd, 738 p.

MANUSCRIPT ACCEPTED BY THE SOCIETY MAY 6, 1986

Chapter 9

Geophysical data

Robert E. Sheridan
Department of Geological Sciences, Rutgers University, New Brunswick, New Jersey 08903
John A. Grow
U.S. Geological Survey, MS 913, Box 25046, Denver Federal Center, Denver, Colorado 80225
Kim D. Klitgord
U.S. Geological Survey, Woods Hole, Massachusetts 02543

INTRODUCTION

Understanding the geology of the U.S. Atlantic margin is based on a sparse set of drilling and dredge data integrated with a much larger base of geophysical data. Geophysical data provide the primary basis for interpreting structural features and extrapolating from the sparse geological control. The purpose of this chapter is to present the scope of geophysical data coverage of the Atlantic continental margin in a general review format. The discussion will touch on the kinds of data produced by different technologies, the historical reasons for the use of various methods, and the improvement of methods.

There are two basic classes of geophysical data used in margin studies: seismic data and potential field data. Seismic techniques utilize acoustic energy sources and receivers to examine seismic-wave propagation characteristics of the subsurface rocks. This information includes variation in acoustic reflectivity and velocity within the rock. Near-vertical incidence reflection profiling is used to map structures on a regional basis. Wide-angle seismic-reflection and seismic-refraction profiling are used to examine seismic-velocity structures. Magnetic and gravity potential field anomaly data sets complement the seismic information, providing a means for extrapolating it over broad regions. The comparatively low cost of magnetic and gravity surveying has resulted in the acquisition of fairly dense regional data sets (2 to 10 km line spacing) compared with more coarse regional seismic grids (20 to 30 km line spacing).

The main features of crustal seismic velocities and density models are reviewed as an introduction to the more detailed disucssions of the interpretation of these features in later chapters. A discussion of the prominent magnetic and gravity features is given in Klitgord and others (this volume).

SEISMIC-REFRACTION DATA

Seismic-refraction studies of the U.S. Atlantic continental margin have been extensive (Table 1). There has been an evolu-

tion of deep crustal seismic techniques from short-offset two-ship methods, to large-offset sonobuoy methods, to large-offset continuous coverage two-ship methods. Nearly 300 profiles have been acquired, including those on the coastal plain (Ewing and others, 1937, 1939; Woollard and others, 1957), and those recorded from the 1930s to the 1970s in the offshore (Ewing and others, 1937, 1939, 1940, 1950; Hersey and others, 1959; Drake and others, 1959; Ewing and Ewing, 1959; Houtz and Ewing, 1963, 1964; Sheridan and others, 1966, 1979; Emery and others, 1970). Most of these profiles were shot with explosive charges at discrete, separate shot points with variable (2 to 5 km) spacing, and with two ships. Source-receiver offsets varied as the ships moved apart with maximum offsets of over 70 km. Later, short-offset (up to 20 km) sonobuoy profiles with denser shot spacing (0.05 to 0.4 km) were made with explosives and air guns. A few large-offset (up to 80 km) sonobuoy profiles shot with air guns and explosives have been done, but with widely spaced shot points (3 to 4 km) at extreme range.

Not included in this discussion of refraction data are the large offset experiments using two ships and long multichannel streamers and ocean bottom seismometers (see section on LASE below and chapter by Diebold and others, this volume). Improvements both in seismic sources and receivers have led most recently to deep-crustal seismic experiments that combine large offset with more continuous, closely spaced (0.05 km) coverage.

The seismic-refraction data through the 1970s is presented in summary fashion as six columnar cross sections (Figs. 1–6) with the velocity layers subdivided by ranges in compressional wave velocities. The profiles used in the cross sections are identified as to reference in Table 1, and their locations are indicated on Plate 1C.

The general distribution of seismic compressional wave velocities based on those data is determined by the offset used and by the geology of the Atlantic continental margin. Seismic velocities within sedimentary rocks generally increase with depth and

Sheridan, R. E., Grow, J. A., and Klitgord, K. D., 1988, Geophysical data, *in* Sheridan, R. E., and Grow, J. A., eds., Continental Margin, U.S.: Geological Society of America, The Geology of North America, v. I-2.

R. E. Sheridan and Others

TABLE 1. SEISMIC-REFRACTION PROFILE REFERENCES
(Locations of mapped profiles are on Plate 1C)

Mapped Profile	Reference Profile	Reference* Source	Mapped Profile	Reference Profile	Reference* Source
1	1	9	51	7-NY & 8NY	7
2	1-56A	8	52	8	12
3	1-56B	8	53	8-56	8
4	1-56C	8	54	8-CP	13
5	1A & 3.2	7	55	8-GB	1
6	1-CP	13	56	8-NA	10
7	1-GB	1	57	8-NY & 9-NY	7
8	1-GM	1	58	8-SC	13
9	1-NC	13	59	9	12
10	2-3	9	60	9-54	8
11	2-56	8	61	9-CP	13
12	2-CP	13	62	9-GB	1
13	2-GM	1	63	9-NA	10
14	2-NC	13	64	9-NY & 10-NY	7
15	3-54	8	65	9-SC	13
16	3-56	8	66	10	12
17	3A & 3.2	7	67	10-54	8
18	3-CP	13	68	10-CP	13
19	3-GB	1	69	10-NY & 11-NY	7
20	30-GM	1	70	10-SC	13
21	3-NC	13	71	11-54	8
22	3R	9	72	11-55	8
23	4	12	73	11-CP	13
24	4.3 & 5	7	74	11-NA	10
25	4-5	9	75	11-NY & 12NY	7
26	4-54	8	76	11-SC	13
27	4-55	8	77	12-54	8
28	4-CP	13	78	12-55	8
29	4-GM	1	79	12-CP	13
30	4-NC	13	80	12-NA	10
31	5 & 6	7	81	12-SC	13
32	5-54	8	82	13 & 14	7
33	5-55	8	83	13-54	8
34	5-56	8	84	13-55	8
35	5-CP	13	85	13-CP	13
36	5-GM	1	86	13-NA	10
37	5-NC	13	87	14 & N15	7
38	5R-6	9	88	14-54	8
39	6-55	8	89	14-55	8
40	6-56	8	90	14-CP	13
41	6-CP	13	91	14-G	13
42	6-GB	1	92	15 & 16	7
43	6-NC	13	93	15-54	8
44	6R	9	94	15-55	8
45	7	12	95	15-CP	13
46	7-8	9	96	15-G	13
47	7-56	8	97	15-NA	10
48	7-CP	13	98	16-CP	13
49	7-GB	1	99	16-G	13
50	7-NC	13	100	16-NA	10

*See References at end of Table 1.

TABLE 1 (CONTINUED)

Mapped Profile	Reference Profile	Reference* Source	Mapped Profile	Reference Profile	Reference* Source
101	17-55	8	151	39-55	8
102	17-CP	13	152	39-CP	13
103	17-NA	10	153	39-PM & 40-PM	1
104	18-55	8	154	40-CP	13
105	18-CP	13	155	41-CP	13
106	18-NA	10	156	41-PM & 42-PM	1
107	19	12	157	42-CP	13
108	19-CP	13	158	43-CP	13
109	19-NA	10	159	43-PM & 44-PM	1
110	20	12	160	44-CP	13
111	20-CP	13	161	44-PM & 45-PM	1
112	20-NA	10	162	45-CP	13
113	21	12	163	46-CP	13
114	21-CP	13	164	46-PM	1
115	21-NA	10	165	47-CP	13
116	22	12	166	47-PM	1
117	22-CP	13	167	48-CP	13
118	22-NA	10	168	48-PM	1
119	23	12	169	49-CP	13
120	23-CP	13	170	49-PM	1
121	23-NA	10	171	50-CP	13
122	24	12	172	50-PM	1
123	24-CP	13	173	51-CP	13
124	29-55	8	174	51-PM & 52-PM	1
125	29-CP	13	175	52-CP	13
126	29-NA	10	176	52-PM & 53-PM	1
127	30-55	8	177	53-CP	13
128	30-CP	13	178	54-CP	13
129	30-NA	10	179	54-PM	1
130	30-PM	1	180	55-CP	13
131	31-55	8	181	55-PM & 56-PM	1
132	31-CP	13	182	56-CP	13
133	31-NA	10	183	57-CP	13
134	32-PM	1	184	57-PM & 58-PM	1
135	33-55	8	185	58-CP	13
136	33-CP	13	186	59-CP	13
137	33-NA	10	187	59-PM & 60-PM	1
138	33-PM	1	188	60-CP	13
139	34-55	8	189	61-PM	1
140	34-CP	13	190	62-PM & 63-PM	1
141	34-CH & 35-WH	1	191	64-GB	1
142	35-55	8	192	64-PM	1
143	35-CP	13	193	65-GB	1
144	36-55	8	194	65-PM	1
145	36-CP	13	195	66-GB	1
146	37-55	8	196	66-PM & 67-PM	1
147	37-CP	13	197	67-GB	1
148	37-PM & 38-PM	1	198	68-GB	1
149	38-55	8	199	68-PM & 69-GB	1
150	38-CP	13	200	69-PM	1

*See References at end of Table 1.

TABLE 1 (CONTINUED)

Mapped Profile	Reference Profile	Reference* Source	Mapped Profile	Reference Profile	Reference* Source
201	70-GB	1	251	G15	3
202	70-PM & 71-PM	1	252	G16	3
203	71-PM & 72-PM	1	253	GS-2	11
204	74-WH & 75-WH	1	254	GS-3	11
205	76-WH & 77-WH	1	255	GS-4	11
206	77-WH & 78-WH	1	256	ME	1
207	78-WH & 79-WH	1	257	32-NA	10
208	80-WH & 81-WH	1	258	31-PM	1
209	82-WH	1	259	32-55	8
210	82-NY & 83-NY	1	260	25-CP	13
211	86-NY & 85-NY	1	261	26-CP	13
212	92	12	262	27-CP	13
213	93	12	263	28-CP	13
214	94	12	264	32-CP	13
215	95	12	265	BRIDGEPORT	6
216	96	12	266	PROSPECT	6
217	97	12	267	SWEDESBORO	6
218	98	12	268	LINCOLN	6
219	99	12	269	PITTSGROVE	6
220	100	12	270	ELMER	6
221	102	12	271	NORMA	6
222	119	12	272	MILLVILLE	6
223	120	12	273	PORT ELIZABETH	6
224	121	12	274	WOODBINE	6
225	122	12	275	SEAVILLE	6
226	123	12	276	AVALON	6
227	124	12	277	PLAINSBORO	5
228	125	12	278	HIGHTSTOWN	5
229	126	12	279	DIS BROW'S HILL	5
230	127	12	280	CHARLESTON SPRING	5
231	267	2	281	JACKSON'S MILL	5
232	272	2	282	LAKEWOOD	5
233	273-4	2	283	CEDAR BRIDGE	5
234	273-5	2	284	SILVERTON	5
235	273-6	2	285	CAMP LEE	4
236	274	2	286	YOUNGBLOOD'S STORE	4
237	275-8	2	287	PRINCE GEORGE	4
238	275-9	2	288	NEWVILLE	4
239	275-10	2	289	BRANDON	4
240	276-11	2	290	MERCY SEAT CHURCH	4
241	277-12	2	291	SURRY	4
242	277-13	2	292	MULBERRY ISLAND	4
243	277-14	2	293	HAMPTON	4
244	A-156-28	3	294	FORT MONROE	4
245	A-157-29	3	295	CAPE HENRY	4
246	A-157-30	3	296	STATION 8	4
247	A-173-4	3	297	STATION 12	4
248	G12	3	298	STATION 16	4
249	G13	3	299	STATION 20	4
250	G14	3			

*References:

1. Drake and others, 1959	6. Ewing and others, 1940
2. Emery and others, 1970	5. Ewing and others, 1950
3. Ewing and Ewing, 1959	8. Hersey and others, 1959
4. Ewing and others, 1937	9. Houtz and Ewing, 1963
5. Ewing and others, 1939	

10. Houtz and Ewing, 1964
11. Sheridan and others, 1979
12. Sheridan and others, 1966
13. Woolard and others, 1957

CROSS SECTION A

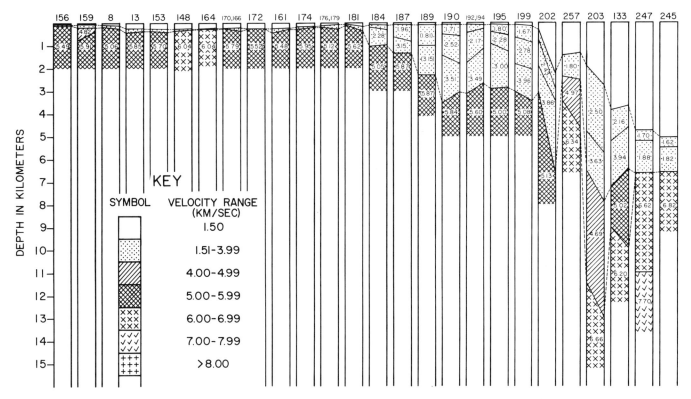

Figure 1. Refraction cross-section summary A—Georges Bank. Insert shows symbols for velocity range. Profiles located in Plate 1C.

age as a result of burial and lithification. Velocities range from approximately the velocity of water, 1.5 km/s, to more than 5 km/s. This velocity increase can be correlated with density increase using the Nafe and Drake (1957) curves. At these higher velocities, however, the velocities of sedimentary and crystalline rocks overlap. For example, the velocities observed at 1 to 2 km depths under the coastal plain are in the range of 5 to 6 km/s, and they correspond with granitic continental basement rocks. Under the continental shelf, similar high velocities of 5 to 6 km/s were found at 4 to 5 km depths, but these are ambiguous. These high velocities were originally, and understandably, correlated with the granitic basement under the coastal plain (Ewing and others, 1937, 1950; Drake and others, 1959), except off Florida and the Bahamas where they were correlated with limestone and dolomites (Sheridan and others, 1966). Subsequently, with the benefit of deep-penetration seismic-reflection data, these high seismic velocities have been attributed to carbonate rocks all along the margin, from Georges Bank to the Bahamas (Figs. 1–6).

Velocities greater than 6.5 km/s are generally associated with deeper crustal rocks. A few higher velocities in the 7-km/s

range were measured at shallow depth (2–7 km) under the continental shelf off the Carolinas and under the Blake Plateau. Off the Carolinas, there is evidence in the seismic-reflection, gravity, and magnetic data for relatively young gabbroic intrusions, and these could explain the high seismic velocities. On the Blake Plateau, deeply buried (10 km) dolomites might reach velocities of 7 km/s. Under the continental slope and rise, layers with higher velocity in the 7-km/s range at depths of 12 to 15 km were measured. These were correlated with oceanic crustal layers until the 1970s. However, later large-offset data (see LASE section below) have been used to reinterpret these velocities.

Only under the deeper water of the continental rise were good mantle velocities, greater than 8 km/s, measured in layers at 15 to 25 km depths. This depth to Moho is compatible with gravity modeling and with the correlation of the overlying velocity layers with normal oceanic crust, which has been depressed by the sediment load of the continental rise prism (Sheridan and others, 1979). However, other interpretations are possible (see LASE section below).

Postcritical seismic-refraction data did not determine well

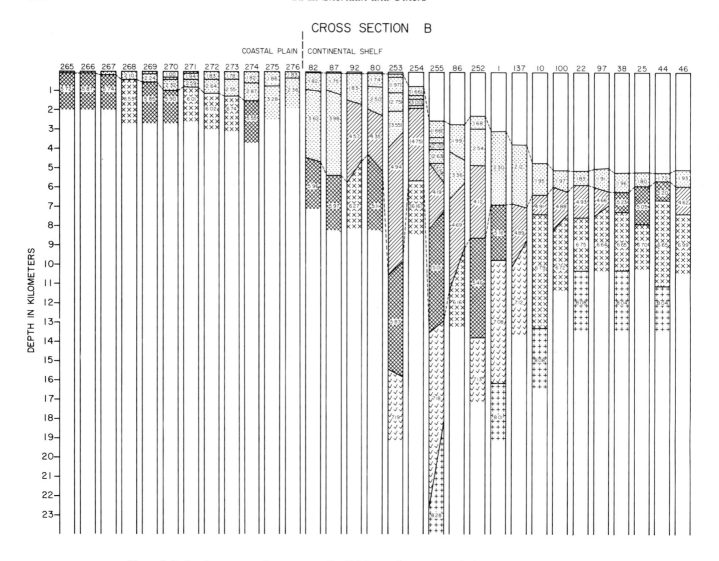

Figure 2. Refraction cross-section summary B—Baltimore Canyon Trough. Same symbols for velocity range as Figure 1.

the detailed velocity structure of the shallow coastal plain and continental margin. Features such as velocity inversions and velocity gradients could only be determined by the subcritical wide-angle reflection profiles discussed in the next sections. However, velocity increases in the range of 1.6 to 5.5 km/s with depth and higher velocities and velocity/depth gradients in carbonate rocks were recognized on the U.S. Atlantic margin just from refraction data (Sheridan and others, 1966).

SEISMIC-REFLECTION DATA

Following the international oil crisis in 1973, an accelerated schedule of offshore oil and gas leasing was planned that required geological and geophysical data for geologic framework and en-

vironmental hazard research. On the U.S. Atlantic continental margin, approximately 25,000 km of multichannel seismic-reflection profiles were acquired by the U.S. Geological Survey between 1973 and 1979 (Fig. 7 and Plate 1B). Although most of the multichannel seismic-reflection profiles were acquired and processed by contract geophysical companies, some of the lines were obtained in cooperation with the University of Texas, the Institut Francais du Petrole (France), and Bundesanstalt Geowissenshaften und Rohstoffe (West Germany). Most of the lines were acquired with source arrays of airguns with volumes ranging from 1,400 to 2,200 in³ (2,000 psi) fired on 50-m shot intervals; data were received on hydrophone streamers composed of 24 to 48 channels totaling 2,400 to 3,600 m in length. The data were recorded from 6 to 12 s on 48-channel digital recorders. Seismic processing included true-amplitude recovery, common-depth-point gather, prestack deconvolution, velocity analysis, normal move-out correction, CDP stack, poststack deconvolution, time-variant filtering, and time-variant scaling. Additional multichannel profiles were acquired by other academic institutions (Lamont-Doherty Geological Observatory, Woods Hole Oceanographic Institution, and University of Texas) as part of locally focused research projects.

The first publication of these multichannel seismic-reflection lines ushered in a new era of understanding for the U.S. Atlantic continental margin (Behrendt and others, 1974; Schlee and others, 1976). Dozens of reports and publications have come out based on U.S. Geological Survey seismic lines in the last decade plus additional studies based on academic institution profiles and industry data (e.g., Shipley and others, 1978; Buffler and others, 1979; Klitgord and Behrendt, 1979; Tucholke and Mountain, 1979; Dillon and others, 1979; Austin and others, 1980; Watts, 1981; Sheridan and others, 1981; Poag, 1982; Schlee and Fritsch, 1983; Grow and others, 1979a, 1983; Crutcher, 1983; Lippert, 1983; Morgan and Dowdall, 1983; and Gamboa and others, 1985). One of the great values of the reflection data was the abundant, closely spaced velocity data on the shallow structure that became available (Grow and others, 1979a). These data showed details on horizontal and vertical gradients in the sedimentary rocks.

LASE AND OTHER DEEP-CRUSTAL SEISMIC STUDIES

A major deficiency of the marine seismic-refraction studies of the Atlantic continental margin through the 1970s was the lack of deep-velocity determinations under the coastal plain and continental shelf (Figs. 1–6, Table 1). In the 1960s, several onshore-offshore experiments utilizing very large explosive charges (1 to 6 tons) were made (Steinhart and others, 1962; Hales and others, 1968). For the North Carolina coastal plain the East Coast Onshore Offshore Experiment (ECOOE), lines showed mantle with velocities of 8.13 km/s at approximately 30 km beneath a typical granitic continental crust of 5 to 6 km/s velocities (Hales and others, 1968). In Maine and the Gulf of Maine, Steinhart and others (1962) found a two-layer crust with velocities of 6.05 and 6.8 km/s over a Moho at 8.1-km/s velocity at 34 km depth.

The most recent, and startling, seismic experiments have been in the offshore Baltimore Canyon Trough and Carolina Trough where long-offset continuous-coverage data revealed the depth to Moho and surprising velocity structures. The Large Aperture Seismic Experiment (LASE) (Diebold and others, this volume) used expanding-spread, two-ship-long streamer multichannel data to measure the detailed velocities on five long (80 km) strike-parallel profiles. The long offset of precritical reflection data allowed the detection of velocity inversion beneath the high-velocity carbonate lid within the sedimentary section (Diebold and others, this volume). Another intriguing result is that the 7+ km/s velocity range layer, previously observed just under and

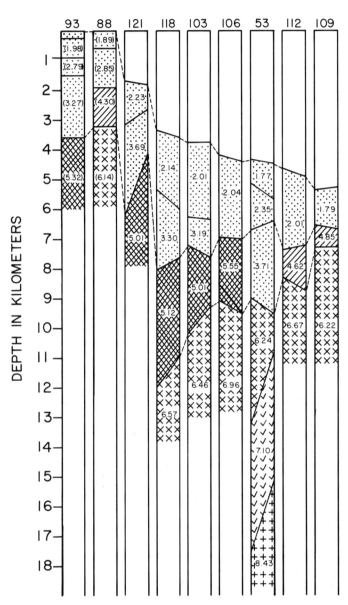

Figure 3. Refraction cross-section summary C—Cape Hatteras. Same symbols for velocity range as Figure 1.

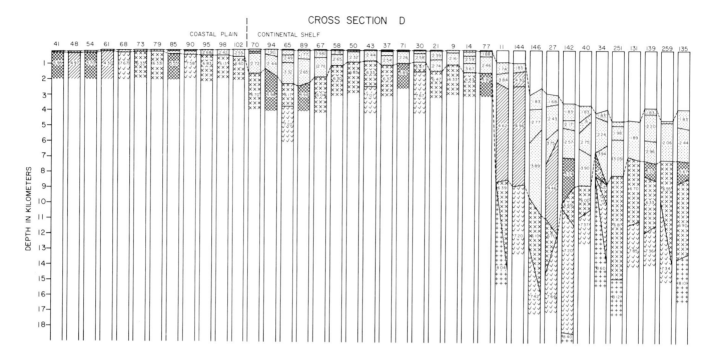

Figure 4. Refraction cross-section summary D—Carolina Trough. Same symbols for velocity range as Figure 1.

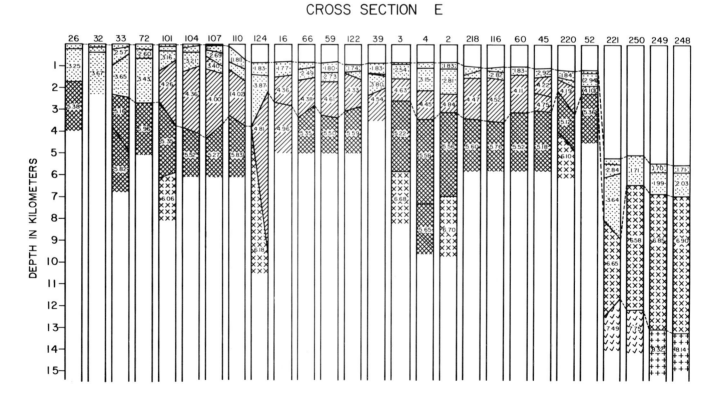

Figure 5. Refraction cross-section summary E—Blake Plateau. Same symbols for velocity range as Figure 1.

CROSS SECTION F

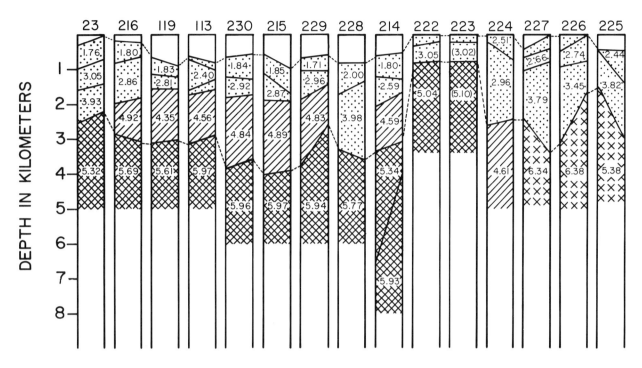

Figure 6. Refraction cross-section summary F—Florida Bahamas. Same symbols for velocity range as Figure 1.

seaward of the ECMA, was now found at comparable depths and thickness landward of the ECMA. Several possible explanations are possible for this, but one put forward by the LASE group (LASE Study Group, 1986; Diebold and others, this volume) is that some mafic intrusions underplate both the continental and oceanic crust near the locus of breakup.

Continuous-coverage long-offset profiles shot with explosives and airguns to ocean bottom seismometers (OBS) in the Carolina Trough appear to show the same occurrence. A 7$^+$ km/s lower crustal layer occurs landward of the ECMA, but seaward of the hinge zone (Trehu and others, 1987; Klitgord and others, this volume, Fig. 16). Thus, this layer, whatever its interpretation, is found in at least two places on the Atlantic continental margin. Its possible distribution on the U.S. Atlantic margin is discussed by Klitgord and others (this volume). A similar velocity structure has been reported on the Norwegian margin (Mutter and others, 1984; Mutter, 1985). Indeed the creation of this layer may be common to all passive margins.

MAGNETIC DATA

Magnetic surveys now cover nearly the entire U.S. Atlantic margin and provide comprehensive data used for mapping buried

geologic features over a broad region. Early shipborne and airborne magnetic surveys of the U.S. Atlantic continental margin were sparsely spaced, but they did discover a positive linear magnetic anomaly along the trend of the continental slope from off the Carolinas to Nova Scotia (Drake and others, 1963). The first major aeromagnetic survey of the U.S. margin by the U.S. Navy in 1964–66 (Project Magnet) extended the profile coverage (8 km line spacing) all along the margin. Besides mapping the regional extent of the very prominent magnetic lineament, known as the East Coast Magnetic Anomaly (ECMA), numerous other magnetic anomalies were discovered and broad geologic interpretations made (Taylor and others, 1968). A 185,000-km high-sensitivity digital aeromagnetic survey was acquired by the USGS and industry in 1975 with line densities of 2.4 to 4.8 km (Klitgord and Behrendt, 1979). Large-scale black and white maps (1:250,000) and color maps (Behrendt and Klitgord, 1980—1:1,000,000; Behrendt and Grim, 1983—1:2,500,000) have been published. Behrendt and Grim (1983, 1985) present both the total magnetic field and the second vertical derivative of the magnetic field; the latter display highlights many of the more subtle trends such as the hinge zones and buried Mesozoic basins (see Plate 2A and 2B for a 1:5,000,000 reduction of the Behrendt and Grim maps). A series of aeromagnetic surveys of the U.S.

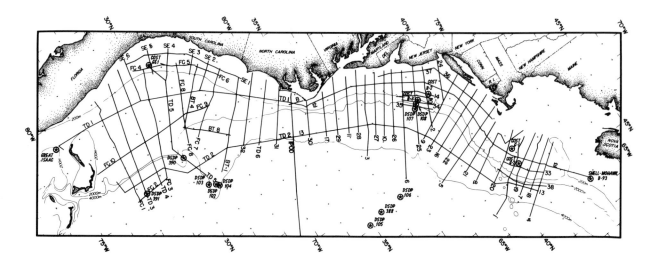

Figure 7. USGS multichannel seismic-reflection profiles from 1973 to 1978, from Folger and others (1979). See Plate 1B for a complete summary.

Coastal Plain acquired by the U.S.G.S. complement the detailed offshore magnetic data (Zietz, 1980).

The digital aeromagnetic data has been used for magnetic depth to basement studies in conjunction with the multichannel seismic-reflection profiles to outline the regional tectonic framework (Klitgord and Behrendt, 1979; Klitgord and others, this volume). Very detailed studies of the relationship between the magnetic field anomalies and tectonic features have recently been compiled by Klitgord and Schouten (1987a, b). Interpretations of the ECMA and other magnetic features along the margin are discussed in detail in Klitgord and others (this volume) and in chapters on individual basins.

GRAVITY DATA

During 1975 and 1976, the USGS conducted a number of geophysical cruises along the U.S. Atlantic continental margin and collected approximately 39,000 km of new offshore gravity data. At approximately the same time, the U.S. Navy declassified a large quantity of gravity data over the continental shelf. These new data were combined with previously available data from academic institutions to produce a 10-mgal free-air anomaly map of the margin (Fig. 8). Lyons and others (1982) have subsequently produced a 1:2,500,000-scale gravity map of the U.S. (5-mgal contour interval), which included additional U.S. Navy data farther offshore.

The free-air gravity field typically has a positive (0 to +70 mgal) along the edge of the continental shelf and a negative (−20 to −140 mgal) along the base of the continental slope. The abrupt change in water depth and depth to Moho causes significant edge effects at the shelf edge and lower continental slope, which can be

larger than +50 and −30 mgal, respectively, even if the margin were in perfect isostatic balance (Fig. 9). Abrupt changes in the free-air anomalies along the length of the margin are frequently correlated with the major fracture zones that intersect the margin (Fig. 8) and are associated with segmentation of the margins deep structure into marginal basins and platforms (Klitgord and others, this volume).

Because of the large edge effects along the shelf edge and continental slope, direct interpretation of the free-air anomalies near the shelf edge and slope is extremely difficult. The utilization of two-dimensional, local Airy isostatic anomalies across continental margins has been recommended to remove the edge effects (Talwani and Eldholm, 1972; Rabinowitz and LaBreque, 1977). Summary cross-sections of isostatic anomalies along eight USGS multichannel seismic-reflection profiles showed that values were typically between −10 and −30 mgal over the major shelf basins and 0 to +20 mgal over the platforms, with local positives of 10 to 20 mgal over the hinge zones and paleoshelf edges and no consistent positive or negative near the ECMA (Grow and others, 1979b). Recent large-scale isostatic anomaly maps of the U.S. confirm this observation (Simpson and others, 1986; Jachens and others, 1986).

Two-dimensional gravity modeling using density estimates based on seismic refraction or reflection horizons provides another interpretation technique. Although not unique, modeling can be instructive if good refraction and/or reflection data are available to provide strong constraints on the models. Seismic-velocity information is used to estimate density in the shallow sedimentary layers using the Nafe and Drake (1957) curves. Gravity models along eight USGS multichannel seismic-reflection profiles have been calculated where the upper 10 to 13 km of the crust is

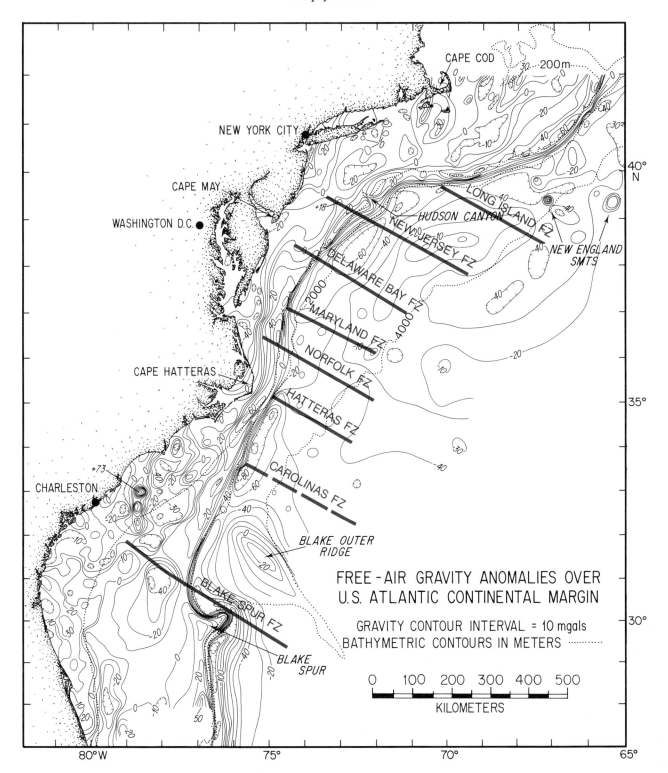

Figure 8. Free-air gravity map for the U.S. Atlantic continental margin from Grow and others (1979b). Gravity models from eight multichannel seismic-reflection profiles are summarized in Figures 9 through 13. Fracture zone locations are from Klitgord and others (this volume Plate 2C). More recent large-scale compilations (1:2,500,000) of the free-air anomalies offshore and Bouguer anomalies onshore have been published by the Society of Exploration Geophysicists (Lyons and others, 1982). Similar scale isostatic anomaly maps have also been published recently (Jachens and others, 1986). Smaller-scale color isostatic anomaly maps are also now available (Simpson and others, 1986).

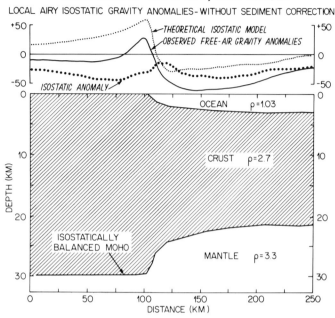

Figure 9. Illustration of edge effects due to changing water depth and rising depth to Moho assuming local Airy isostatic compensation. The theoretical curve from the isostatic model, assuming a crust with a density of 2.7 g/cc overlying a mantle with a density of 3.3 g/cc, predicts an anomaly of +58 mgal at the shelf edge and –30 mgal along the base of the continental slope. The isostatic anomaly is obtained by substracting the theoretical curve from the observed free-air gravity. Rabinowitz (1974) has shown that the sediment corrections to the isostatic anomalies are small. From Grow and others (1979b, Fig. 4).

constrained, although deep refraction control is not available beneath the marginal basins except for very recent LASE results in Baltimore Canyon Trough (LASE Study Group, 1986; Diebold and others, this volume), the landward end of the Georges Bank Basin (Hutchinson and others, 1987), and preliminary ocean-bottom seismography results in the Caroline Trough (Trehu and others, 1987). The eight gravity models all assume a three-layer structure beneath the sediments, defined by the reflection data. The upper crust, lower crust, and upper mantle are assumed to have densities of 2.7, 3.0, and 3.3 g/cc, respectively. The boundary between the 2.7 and 3.0 g/cc layers is assumed to be generally flat or gently dipping toward the continent and is assumed to be the boundary between the oceanic layer 2 and oceanic layer 3 on the seaward side of the ECMA. The boundary between the 3.0 and 3.3 g/cc layers is assumed to be the Moho; it is a free variable over most of the models except the oceanic crust of Baltimore Canyon Trough where refraction control was available (Sheridan

and others, 1979). One of the most significant uncertainties in the gravity modeling is the horizontal density variation within the sedimentary layers near the shelf edge (Hutchinson and others, 1983). In this region, a buried carbonate platform of high density grades landward into shelf clastics and seaward into slope rise deposits.

Gravity models across Georges Bank Basin are available along USGS seismic lines 4, 5 and 19 (Fig. 10). The postrift sedimentary rocks have a maximum thickness of approximately 7 km in the central part of the basin (line 19), but became narrow and thin to the northeast (line 4) and southwest (line 5). The distance from the hinge zone to the ECMA is narrowest on line 5, and the calculated depth to Moho decreases from more than 30 km beneath the hinge zone to less than 18 km near the ECMA (Fig. 10A). Recent deep seismic studies in the Gulf of Maine (Hutchinson and others, 1987) indicate that Moho is at a depth of 28 km beneath the landward edge of Georges Bank Basin. The basin is wider at lines 4 and 19, and the depth to the calculated Moho stays nearly 30 km deep until the ECMA, where it rises abruptly to approximately 15 km depth where thin oceanic crust occurs seaward of the ECMA. The Moho appears to deepen very slightly landward of the hinge zone at lines 4 and 19 (Fig. 10B and 10C). In Georges Bank Basin the primary crustal transition, based on gravity modeling, is at the ECMA.

Two gravity models have been calculated across the Baltimore Canyon Trough along USGS seismic line 6 and 25 (Fig. 11). Seismic line 6 has a shallow acoustic and magnetic basement on the seaward side of the ECMA, while better seismic penetration along line 25 suggests little or no basement ridge. The line 25 gravity model indicates a major thinning near the ECMA. A slightly simpler model across the northern Baltimore Canyon Trough by Watts (1981) also indicated a similar two-stage thinning of the crust beneath the hinge zone and the ECMA, and Karner and Watts (1982) suggest that two abrupt zones of thinning could be expected if local Airy isostatic subsidence were more dominant than flexural loading of the lithosphere. However, more recent LASE deep-crustal velocity suggests a more gradual depth transition beneath the ECMA than the line 25 gravity model (LASE Study Group, 1986; Diebold and others, this volume).

Two gravity models have been calculated in the Carolina Trough along the International Phase of Ocean Drilling (IPOD)/USGS line and USGS line 32 (Fig. 12). Both lines indicate abrupt changes in crustal thickness beneath the hinge zone and ECMA with thick-rifted continental landward of the hinge zone, transitional or "rift-stage" crust between the hinge zone and the ECMA, and thin oceanic crust seaward of the ECMA. No basement ridge is required near the ECMA for the gravity or magnetic models in the Carolina Trough (Hutchinson and others, 1983). Preliminary results of three recent refraction lines in the Carolina Trough using ocean-bottom seismographs suggest a more gradual transition across the ECMA than was used in the gravity models and appear consistent with the LASE results in Baltimore Canyon Trough (Trehu and others, 1987).

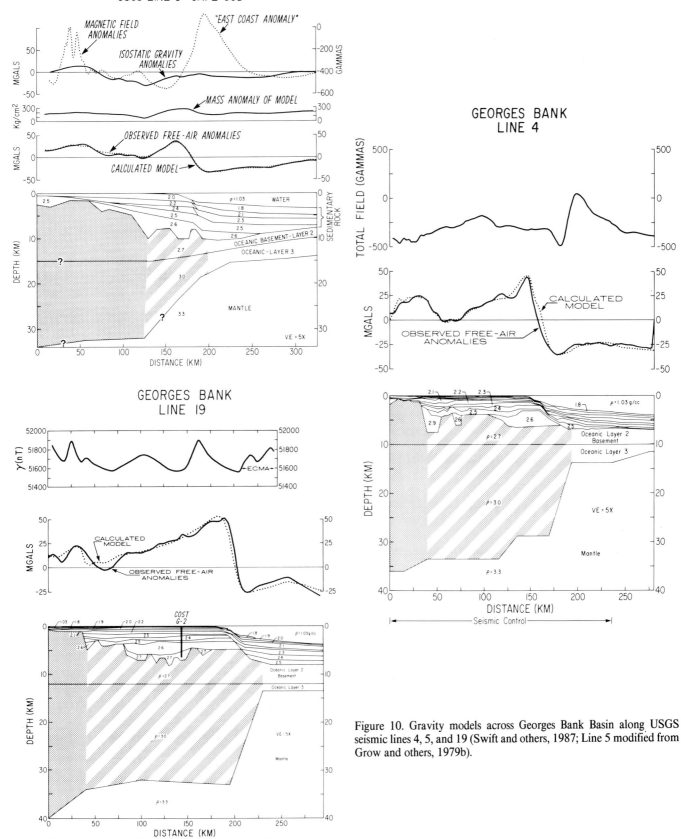

Figure 10. Gravity models across Georges Bank Basin along USGS seismic lines 4, 5, and 19 (Swift and others, 1987; Line 5 modified from Grow and others, 1979b).

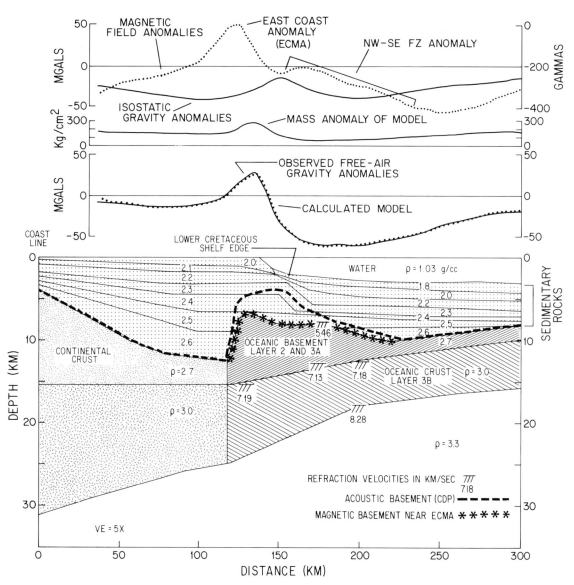

Figure 11 (this and facing page). Gravity models across Baltimore Canyon Trough along USGS seismic lines 6 and 25 (Grow and others, 1979b, 1983). The ridge of thickened oceanic crust interpreted in line 6 was not found on line 25, which had much deeper seismic penetration and control than line 6 (Grow and others, 1983).

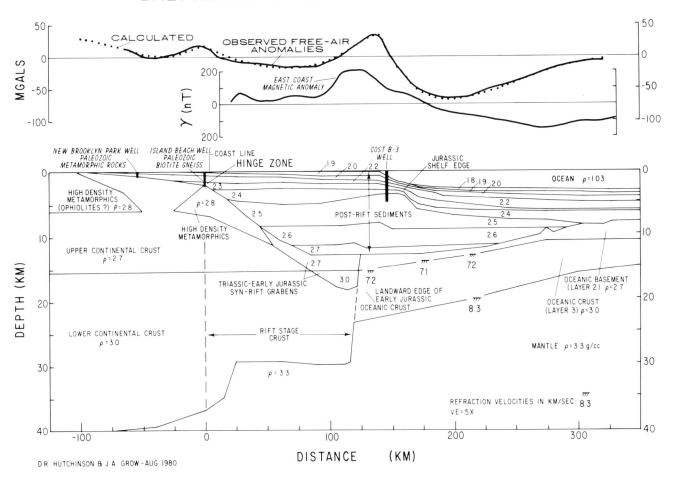

USGS LINE 25 GRAVITY MODEL
BALTIMORE CANYON TROUGH

D R HUTCHINSON & J A GROW - AUG 1980

USGS LINE 32 GRAVITY MODEL, CAPE FEAR

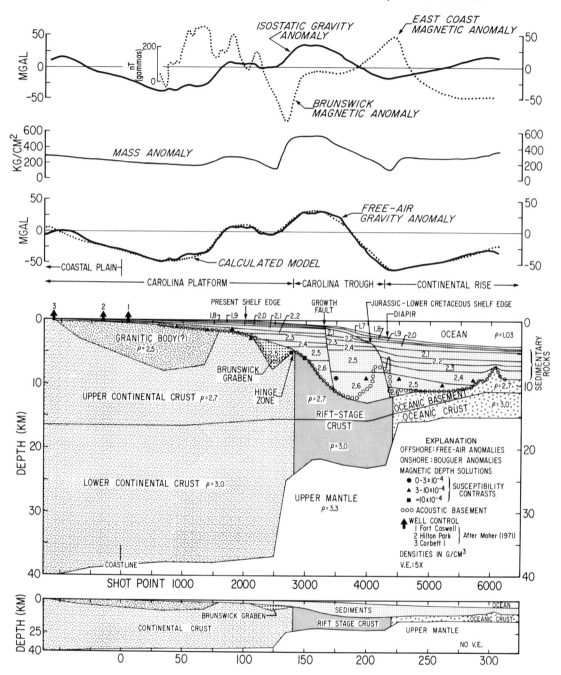

Figure 12 (this and facing page). Gravity models across the Carolina Trough along the IPOD/USGS line and USGS line 32 (Hutchinson and others, 1983).

LINE IPOD/USGS GRAVITY MODEL, CAPE HATTERAS

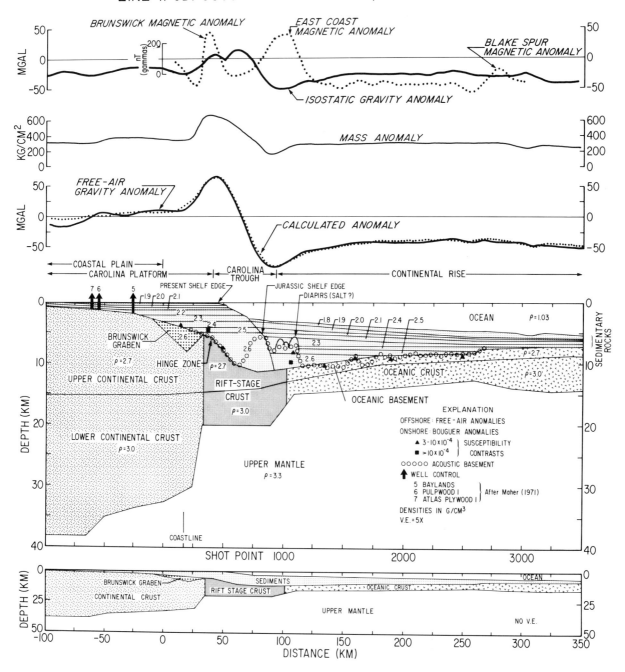

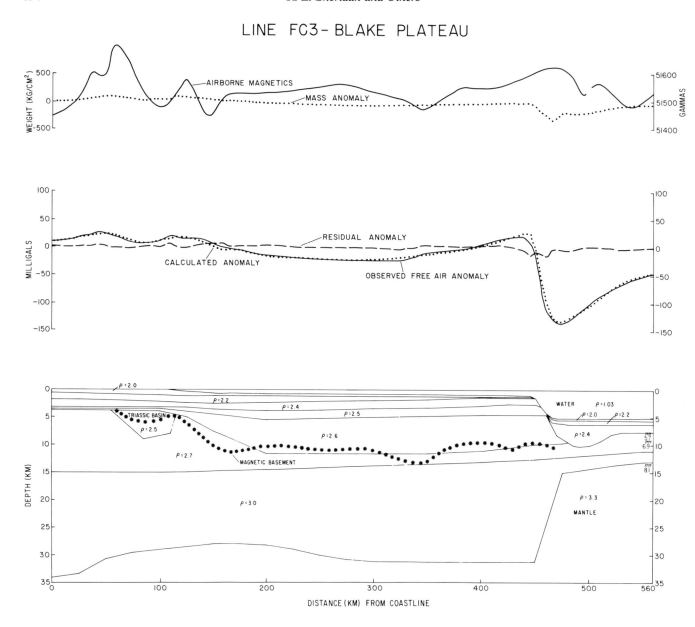

Figure 13. Gravity model across the Blake Plateau Basin along seismic line FC-3 (modified from Kent and others, 1979).

Finally, one gravity model is available across the Blake Plateau Basin along seismic line FC-3 (Fig. 13). The model suggests that the transitional crust is extremely wide in this basin and that a very abrupt change in crustal thickness occurs beneath the Blake Escarpment. The formation of transitional or rift-stage is inferred to have continued later than in the three basins to the north and involved a spreading center jump at the time of the Blake Spur Anomaly (Klitgord and Schouten, 1986; Klitgord and others, this volume).

The eight gravity models, briefly discussed above, all need additional deep seismic-refraction or LASE control for the lower crust and upper mantle, but they still allow an overall comparison between the basins and should aid in the design of deep seismic experiments.

REFERENCES CITED

Austin, J. A., Jr., Uchupi, E., Shaughnessy, D. R., III, and Ballard, R. D., 1980, Geology of the New England passive margin: American Association of Petroleum Geologists Bulletin, v. 64, p. 501–526.

Behrendt, J. C., and Grim, M. S., 1983, Structural elements of the U.S. Atlantic margin delineated by the second vertical derivative of the aeromagnetic data: U.S. Geological Survey Geophysical Investigations Map GP–956, scale 1:2,500,000.

—— , 1985, Structure of the U.S. Atlantic continental margin from derivative and filtered maps of the magnetic field, *in* Hinze, W. J., ed., The utility of regional gravity and magnetic anomaly maps: Tulsa, Oklahoma, Society of Exploration Geophysicists, p. 325–338.

Behrendt, J. C., and Klitgord, K. D., 1980, High-sensitivity aero-magnetic survey of the U.S. Atlantic continental margin: Geophysics, v. 45, p. 1813–1846.

Behrendt, J. C., Schlee, J., and Foote, R. Q., 1974, Seismic evidence of a thick section of sedimentary rock on the Atlantic continental shelf and slope of the United States: EOS Transactions of the American Geophysical Union, v. 55, p. 278.

Buffler, R. T., Watkins, J. S., and Dillon, W. P., 1979, Geology of the offshore Southeast Georgia Embayment, U.S. Atlantic continental margin, based on multichannel seismic reflection profiles: American Association of Petroleum Geologists Memoir 29, p. 11–25.

Crutcher, T. D., 1983, Southeast Georgia Embayment, *in* Bally, A. W., ed., Seismic expression of structural styles: American Association of Petroleum Geologists Studies in Geology Series no. 15, v. 2, p. 2.2.3-27–2.2.3-29.

Dillon, W. P., Paull, C. L., Dahl, A. G., and Patterson, W. C., 1979, Structure of the continental margin near the COST GE–1 well site from a common depth point seismic reflection profile, *in* Scholle, P. A., ed., Geologic studies of the COST GE–1 well: U.S. Geological Survey Circular 800, p. 97–107.

Drake, C. L., Ewing, M., and Sutton, G. H., 1959, Continental margins and geosyncline of the East Coast of North America north of Cape Hatteras: Physics and Chemistry of the Earth, v. 3, p. 110–198.

Drake, C. L., Heirtzler, J., and Hirshman, J., 1963, Magnetic anomalies off eastern North America: Journal of Geophysical Research, v. 68, p. 5259–5275.

Emery, K. O., and five others, 1970, Continental rise off eastern North America: American Association of Petroleum Geologists Bulletin, v. 54, p. 44–108.

Ewing, J. I., and Ewing, M., 1959, Seismic refraction measurements in the Atlantic Ocean basins, in the Mediterranean Sea, on the Mid-Atlantic Ridge, and in the Norwegian Sea: Geological Society of America Bulletin, v. 70, p. 291–318.

Ewing, M., Cary, A. P., and Rutherford, H. M., 1937, Geophysical investigations in the emerged and submerged Atlantic Coastal Plain, Part I: Geological Society of America Bulletin, v. 48, p. 753–802.

Ewing, M., Woolard, G. P., and Vine, A. C., 1939, Geophysical investigations in the merged and submerged Atlantic Coastal Plain, Part III: Geological Society of America Bulletin, v. 50, p. 257–296.

—— , 1940, Geophysical investigations in the emerged and submerged Atlantic Coastal Plain, Part IV: Geological Society of America Bulletin, v. 51, p. 1821–1840.

Ewing, M., Worzel, J. L., Steenland, N. C., and Press, F., 1950, Geophysical investigations in the emerged and submerged Atlantic Coastal Plain, Part V: Geological Society of America Bulletin, v. 61, p. 877–892.

Folger, D. W., Dillon, W. P., Grow, J. A., Klitgord, K. D., and Schlee, J. S., 1979, Evolution of the Atlantic Continental Margin of the U.S., *in* Talwani, M., Hay, W., and Ryan, W.B.F., eds., Deep Drilling Results in the Atlantic Ocean—Continental Margins and Paleoenvironment: American Geophysical Union Maurice Ewing Series, v. 3, p. 87–108.

Gamboa, L. A., Truchan, M., and Stoffa, P. L., 1985, Middle and Upper Jurassic depositional environments at outer shelf and slope of Baltimore Canyon Trough: American Association of Petroleum Geologists Bulletin, v. 69, p. 610–621.

Grow, J. A., Mattick, R. E., and Schlee, J. S., 1979a, Multichannel seismic depth sections and interval velocities over outer continental shelf and upper continental slope between Cape Hatteras and Cape Cod, *in* Watkins, J. S., Montadert, L., and Dickerson, P. W., eds., Geological and geophysical investigations of continental margins: American Association of Petroleum Geologists Memoir 29, p. 65–83.

Grow, J. A., Bowin, C. O., and Hutchinson, D. R., 1979b, The gravity field of the U.S. Atlantic continental margin: Tectonophysics, v. 59, p. 27–52.

Grow, J. A., Hutchinson, D. H., Klitgord, K. D., Dillon, W. P., and Schlee, J. S., 1983, Representative multichannel seismic profiles over the U.S. Atlantic margin, *in* Bally, A. W., ed., Seismic expression of structural styles: American Association of Petroleum Geologists Studies in Geology Series no. 15, v. 2, p. 2.2.3-1–2.2.3-19.

Hales, A. L., Helsey, C. E., Dowling, J. J., and Nation, J. B., 1968, The east coast onshore-offshore experiment; I, The first arrival phases: Seismological Society of America Bulletin, v. 58, p. 757–783.

Hersey, J. B., Bunce, E. T., Wyrick, R. F., and Dietz, F. T., 1959, Geophysical investigation of the continental margin between Cape Henry, Virginia, and Jacksonville, Florida: Geological Society of America Bulletin, v. 70, p. 437–466.

Houtz, R. E., and Ewing, J. I., 1963, Detailed sedimentary velocities from seismic refraction profiles in the western North Atlantic: Journal of Geophysical Research, v. 68, p. 5233–5258.

—— , 1964, Sedimentary velocities of the western North Atlantic margin: Seismological Society of America Bulletin, v. 54, p. 867–895.

Hutchinson, D. R., Grow, J. A., Klitgord, K. D., and Swift, B. A., 1983, Deep structure and evolution of the Carolina Trough, *in* Watkins, J. S., and Drake, C. L., eds., Studies in continental margin geology: American Association of Petroleum Geologists Memoir 34, p. 129–152.

Hutchinson, D. R., Klitgord, K. D., Lee, M. W., and Trehu, A. M., 1987, USGS deep seismic reflection profile across the Gulf of Maine: Geological Society of America Bulletin, v. 98, (in press).

Jachens, R. C., Simpson, R. W., Laltus, R. W., and Blakely, R. J., 1986, Isotatic residual gravity anomaly map of the United States (excluding Alaska and Hawaii): Boulder, Colorado, National Oceanic and Atmospheric Administration Geophysical Data Center, scale 1:2,500,000.

Karner, G. D., and Watts, A. B., 1982, On isostasy at Atlantic-type continental margins: Journal of Geophysical Research, v. 87, p. 2923–2948.

Kent, K. M., Grow, J. A., and Dillon, W. P., 1979, Gravity studies of the continental margin off northern Florida: Geological Society of America Abstracts with Programs, v. 11, no. 4, p. 184.

Klitgord, K. D., and Behrendt, J. C., 1979, Basin structure of the U.S. Atlantic margin, *in* Watkins, J. S., Montadert, L., and Dickerson, P. W., eds., Geological and geophysical investigations of continental margins: American Association of Petroleum Geologists Memoir 29, p. 85–112.

Klitgord, K. D., and Schouten, H., 1986, Plate kinematics of the central Atlantic, *in* Vogt, P. R., and Tucholke, B. E., eds., The western North Atlantic region: Boulder, Colorado, Geological Society of America, Geology of North America, v. M., p. 351–378.

—— , 1987a, Tectonic and magnetic features, Carolina Trough and adjacent magnetic quiet zone: U.S. Geological Survey Miscellaneous Field Studies Map MF–XX, scale 1:1,000,000.

—— , 1987b, Tectonic and magnetic features, Baltimore Canyon Trough and adjacent magnetic quiet zone: U.S. Geological Survey Miscellaneous Field Studies Map MF–XX, scale 1:1,000,000.

LASE Study Group, 1986, Deep structure of the U.S. East Coast passive margin from large aperture seismic experiments (LASE): Marine Petroleum Geology, v. 3, p. 234–242.

Lippert, R. H., 1983, The 'Great Stone Dome'; A compaction structure, *in* Bally, A. W., ed., Seismic expression of structural styles: American Association of Petroleum Geologists Studies in Geology Series no. 15, p. 1.3-1–1.3-4.

Lyons, P. L., O'Hara, N. W., and others, 1982, Gravity anomaly map of the United States: Society of Exploration Geophysicists, scale 1:2,500,000,

2 sheets.

Morgan, L., and Dowdall, W., 1983, The Atlantic continental margin, *in* Bally, A. W., ed., Seismic expression of structural styles: American Association of Petroleum Geologists Studies in Geology Series no. 15, v. 2, p. 2.2.3.30–2.2.3.35.

Mutter, J. C., Talwani, M., and Stoffa, P. L., 1984, Evidence for a thick oceanic crust adjacent to the Norwegian margin: Journal for Geophysical Research, v. 89, p. 483–503.

Mutter, J. C., and North Atlantic Transect Study Group, 1985, Multi-channel seismic evidence for anomalously thin crust at Blake Spur fracture zone: Geology, v. 12, p. 534–537.

Nafe, J. E., and Drake, C. D., 1957, Variations with depth in shallow and deep water marine sediments of porosity, density, and the velocities of compressional and shear waves: Geophysics, v. 22, p. 523–552.

Poag, C. W., 1982, Stratigraphic reference section for Georges Bank Basin; Depositional model for New England passive margin: American Association of Petroleum Geologists Bulletin, v. 66, p. 1021–1041.

Rabinowitz, P. D., 1974, The boundary between oceanic and continental crust in the western North Atlantic, *in* Burk, C. A., and Drake, C. L., eds., The geology of continental margins: New York, Springer-Verlag, p. 67–84.

Rabinowitz, P. D., and LaBrecque, J. L., 1977, The isostatic gravity anomaly; Key to the evolution of the ocean-continent boundary at passive continental margins: Earth and Planetary Science Letters, v. 35, p. 145–150.

Schlee, J. .S, and Fritsch, J., 1983, Seismic stratigraphy of the Georges Bank Basin complex, offshore New England, *in* Watkins, J. S., and Drake, C. D., eds., Studies in continental margin geology: American Association of Petroleum Geologists Memoir 34, p. 223–251.

Schlee, J. S., and six others, 1976, Regional geologic framework off northeastern United States: American Association of Petroleum Geologists Bulletin, v. 60, p. 926–951.

Sheridan, R. E., Drake, C. L., Nafe, J. E., and Hennion, J., 1966, Seismic refraction study of the continental margin east of Florida: American Association of Petroleum Geologists Bulletin, v. 60, p. 1972–1991.

Sheridan, R. E., Grow, J. A., Behrendt, J. C., and Bayer, K. C., 1979, Seismic refraction study of the continental edge off the eastern United States: Tectonophysics, v. 59, p. 1–26.

Sheridan, R. E., Crosby, J. T., Bryan, G. M., and Stoffa, P. L., 1981, Stratigraphy and structure of southern Blake Plateau, northern Florida Straits, and north Bahama Platform from multichannel seismic reflection data: American Association of Petroleum Geologists Bulletin, v. 65, p. 2571–2593.

Shipley, T. A., Buffler, R. T., and Watkins, J. S., 1978, Seismic stratigraphy and

geologic history of the Blake Plateau and adjacent western Atlantic continental margin: American Association of Petroleum Geologists Bulletin, v. 62, p. 792–812.

Simpson, R. W., Jachens, R. C., Blakely, R. J., and Saltus, R. W., 1986, A new isostatic residual gravity map of the conterminous United States with a discussion on the significance of isostatic residual anomalies: Journal of Geophysical Research, v. 91, p. 8348–8372.

Steinhart, J. S., and five others, 1962, The Maine seismic experiment: Washington, D.C., Carnegie Institution Yearbook, v. 63, p. 221–231.

Swift, B. A., Sawyer, D. D., Grow, J. A., and Klitgord, K. D., 1987, Subsidence, crustal structure, and thermal evolution of Georges Bank Basin: American Association of Petroleum Geologists Bulletin, v. 71, p. 702–718.

Talwani, M., and Eldholm, O., 1972, Continental margin off Norway; A geophysical study: Geological Society of America Bulletin, v. 83, p. 3606–3675.

Taylor, P. T., Zietz, I., and Dennis, L. S., 1968, Geological implications of aeromagnetic data for the eastern continental margin of the United States: Geophysics, v. 33, p. 755–780.

Trehu, A. M., Klitgord, K. D., Sawyer, D. S., and Buffler, R. T., 1987, Regional investigations of crust and upper mantle; Atlantic and Gulf of Mexico continental margins, *in* Pakiser, L., and Mooney, W., eds., Geophysical framework of the continental United States: Geological Society of America Memoir (in press).

Tucholke, B. E., and Mountain, G. S., 1979, Seismic stratigraphy, lithostratigraphy, and paleosedimentation patterns in the North American Basin, *in* Talwani, M., Hayes, W., and Ryan, W.B.F., eds., Deep drilling results in the Atlantic Ocean: American Geophysical Union Maurice Ewing Series, v. 3, p. 58–86.

Watts, A. B., 1981, The U.S. Atlantic continental margin; Subsidence history, crustal structure, and thermal evolution, *in* Bally, A. W., ed., Geology of passive continental margins; History, structure, and sedimentologic record (with special emphasis on the Atlantic margin): American Association of Petroleum Geologists Educational Course Notes Series #19, p. 2-1–2-75.

Woollard, G. P., Bonini, W. E., and Meyer, R. P., 1957, A seismic refraction study of the sub-surface geology of the Atlantic Coastal Plain and Continental Shelf between Virginia and Florida: University of Wisconsin Technical Report, 128 p.

Zietz, I., 1980, Exploration of the continental crust using aeromagnetic data, *in* Studies in geophysics; Continental tectonics: Washington, D. C., National Academy of Sciences, p. 127–138.

MANUSCRIPT ACCEPTED BY THE SOCIETY AUGUST 26, 1987

The Geology of North America
Vol. I-2, The Atlantic Continental Margin: U.S.
The Geological Society of America, 1988

Chapter 10

Late Triassic–Early Jurassic synrift basins of the U.S. Atlantic margin

Warren Manspeizer
Geology Department, Rutgers University, Newark, New Jersey 07102
Harold L. Cousminer
Minerals Management Service, U.S. Department of the Interior, Vienna, Virginia 22180

INTRODUCTION

The present Atlantic passive margin of North America has been the site of recurrent plate activity over the course of geologic time. Different tectonic styles were superimposed over previously deformed basement rocks, so that today's margin is a collage of ancient converging, diverging, and strike slip plate boundaries. Tectonic suturing towards the middle and close of the Paleozoic Era produced a vast continental landmass extending from the Appalachians eastward through the Mauretanides and Hercynides (Arthaud and Matte, 1977; Ziegler, 1982).

As the Late Triassic land mass drifted north, perhaps over a field of hotspots (Morgan, 1980) or across a tensional stress field (Bedard, 1985), its crust was pulled apart along older fracture zones, sutures, and transforms. The landscape was marked by high-standing coastal ranges with synrift depositional basins that bordered a narrow band of salt flats, which were overlain by shallow, hypersaline waters derived from the Tethys Ocean to the east and/or from arctic Canada to the north (Fig. 1). Continued displacement along east-west trending transforms, coupled with extension along the proto-Atlantic axis, finally broke the narrow platform into multiple basins (e.g., Georges Bank and the Baltimore Canyon Trough) near the end of the Triassic and beginning of the Jurassic, thus bringing to a close the early rifting stage of the passive margins. Intermittent episodes of rifting and drifting followed into the Early Jurassic. Since the Middle Jurassic, however, the U.S. margin has been dominated by regional subsidence (drifting stage), so that today the record of Triassic-Liassic rifting is buried under a vast prism of postrift or drift sediment.

Because the offshore Triassic-Jurassic synrift data base is limited primarily to geophysical data that are presented by others (Poag and Valentine, this volume; Schlee and Klitgord, this volume), this paper will focus on the rocks of the offshore and onshore basins, for they provide additional data yielding some answers to vexing questions surrounding the origin of synrift basins and the history of the U.S. Atlantic passive margin.

OFFSHORE DATA BASE

The Triassic-Jurassic data base for the U.S. margin relies heavily on multichannel seismic reflection profiles that have been supplemented by magnetic and gravity studies, and by the COST wells in Georges Bank. Even the offshore drill hole data from Triassic strata of the Canadian margin are quite limited. Of 23 deep wells on the Scotian Shelf (Barss and others, 1979), only two may reach uppermost Triassic strata. The Mohican I-100, which is commonly used as a standard for correlation to the U.S. margin (see Poag, 1982, Fig. 24), does not even penetrate the Jurassic-Triassic systemic boundary (Barss and others, 1979). Only the Sandpiper 2J-77, the Osprey H-84, and the Spoonbill C-30 wells of Grand Banks penetrate virtually the entire Upper Triassic section (Barss and others, 1979; see Fig. 1).

The COST G-2 well off Georges Bank is the deepest and most important stratigraphic well on the U.S. margin (Figs. 2, 3 and 4). It was drilled to a depth of 6,667 m (i.e., 1,769 m deeper than the COST G-1 well), and penetrated a thick Upper Triassic section of dolomite with limestone and anhydrite, bottoming in Upper Triassic salt. Palynomorphs, in both cores (Cousminer, 1983) and cuttings (Cornet, written communications, 1983, 1984), indicate that the post-rift unconformity occurs within an attenuated (less than 300 m) Liassic section of carbonates and evaporites, and at the base (4,153 m) of the Middle Jurassic Mohican Formation (Fig. 3), which according to Given (1977) represents the first sands to transgress the newly formed margin after the breakup of the North American and Afro-European plates. In addition, Cousminer (1983) reports the presence of phytoplankton in core number 5 at 4,441 m, indicating that intermittent marine conditions were present on Georges Bank as early as the Late Triassic (Fig. 3). Based on seismic data, Poag (1982) tentatively correlated the basal salt, which occurs at the 6,667 m level in the well with the Early Jurassic Argo Salt of the Scotian Shelf. However, the palynologic data from over 2,200 m above the salt indicate that intermittent marine conditions were

Manspeizer, W., and Cousminer, H. L., 1988, Late Triassic–Early Jurassic synrift basins of the U.S. Atlantic margin; *in* Sheridan, R. E., and Grow, J. A., eds., The Geology of North America, Volume I-2, The Atlantic Continental Margin, U.S.: Geological Society of America.

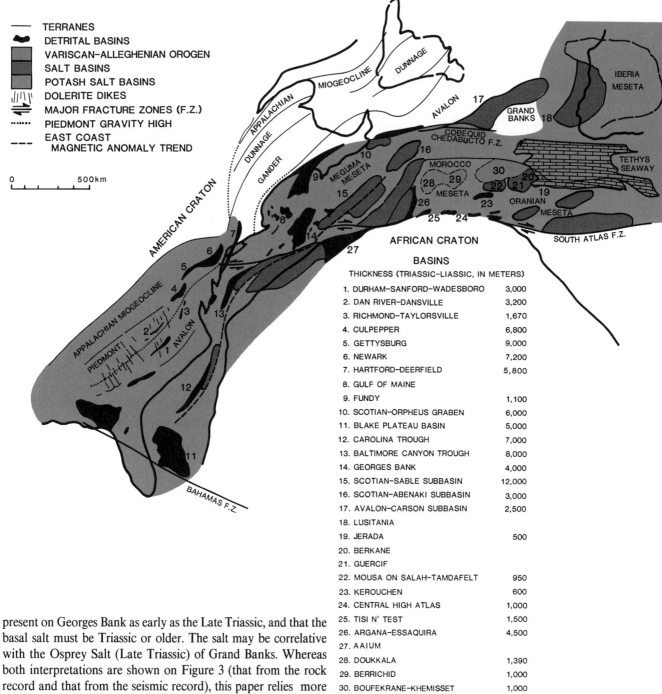

TERRANES
DETRITAL BASINS
VARISCAN-ALLEGHENIAN OROGEN
SALT BASINS
POTASH SALT BASINS
DOLERITE DIKES
MAJOR FRACTURE ZONES (F.Z.)
PIEDMONT GRAVITY HIGH
EAST COAST
  MAGNETIC ANOMALY TREND

0   500km

BASINS

THICKNESS (TRIASSIC-LIASSIC, IN METERS)

1. DURHAM–SANFORD–WADESBORO	3,000
2. DAN RIVER–DANSVILLE	3,200
3. RICHMOND–TAYLORSVILLE	1,670
4. CULPEPPER	6,800
5. GETTYSBURG	9,000
6. NEWARK	7,200
7. HARTFORD–DEERFIELD	5,800
8. GULF OF MAINE	
9. FUNDY	1,100
10. SCOTIAN–ORPHEUS GRABEN	6,000
11. BLAKE PLATEAU BASIN	5,000
12. CAROLINA TROUGH	7,000
13. BALTIMORE CANYON TROUGH	8,000
14. GEORGES BANK	4,000
15. SCOTIAN–SABLE SUBBASIN	12,000
16. SCOTIAN–ABENAKI SUBBASIN	3,000
17. AVALON–CARSON SUBBASIN	2,500
18. LUSITANIA	
19. JERADA	500
20. BERKANE	
21. GUERCIF	
22. MOUSA ON SALAH–TAMDAFELT	950
23. KEROUCHEN	600
24. CENTRAL HIGH ATLAS	1,000
25. TISI N' TEST	1,500
26. ARGANA–ESSAQUIRA	4,500
27. AAIUM	
28. DOUKKALA	1,390
29. BERRICHID	1,000
30. BOUFEKRANE–KHEMISSET	1,000

present on Georges Bank as early as the Late Triassic, and that the basal salt must be Triassic or older. The salt may be correlative with the Osprey Salt (Late Triassic) of Grand Banks. Whereas both interpretations are shown on Figure 3 (that from the rock record and that from the seismic record), this paper relies more heavily on the paleontologic record. Above the palynologically determined Triassic/Liassic interval, abundant Bajocian palynomorphs in core number 4 (4,035–4,051.6 m) mark the Middle Jurassic and constrain the upper age of the postrift unconformity on Georges Bank (Fig. 3).

Although none of the other COST wells drilled off the eastern United States penetrated fossiliferous rocks older than Middle Jurassic, geophysical evidence indicates that as much as 5 km of Triassic (?) strata may lie below the G-2 well (Schlee and Klitgord, this volume; see also Fig. 5).

Figure 1. Early Mesozoic predrift reconstruction of eastern North America and northeast Africa, outlining the broad Variscan-Alleghanian orogenic belt, and the location of Triassic-Liassic basins and lithofacies (from Manspeizer, 1982). For details of the basins in North America, see Figures 4 and 7.

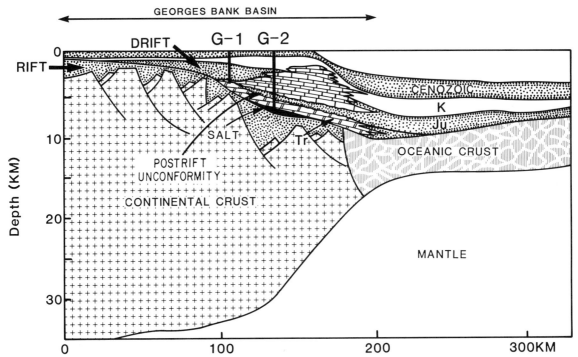

Figure 2. Schematic geologic cross section across the southwestern end of Georges Bank Basin, along USGS multichannel seismic line 5, showing the locations of the G-1 and G-2 wells. Modified from Grow (1981), indicating the new position of the postrift unconformity (see Figs. 3, 5 for details of the 2 wells).

SYNRIFT/POSTRIFT CONCEPT

The tectonic regime that produced the Atlantic passive margin began with rifting of the Variscan-Alleghanian orogen in the Middle to Late Triassic. Geophysical studies across the Atlantic margin (e.g., Grow, 1981; Schlee and Jansa, 1981; and Klitgord and others, 1982), including the Bay of Biscay (Montadert and others, 1979), reveal that the margin consists of two early Mesozoic tectono-stratigraphic facies, an older synrift sequence that is confined to rift basins, and a younger blanket sequence of postrift strata that unconformably overlies the older sequence. The unconformity, marking the change from synrift to postrift is the postrift unconformity. The stratigraphic position of the unconformity varies along the shelf. It embraces most of the Late Triassic on the Scotian Shelf and much of the Lower Jurassic on Georges Bank, and perhaps elsewhere, on the U.S. margin. The unconformity appears to reflect intermittent episodes of rifting and drifting over an interval of about 20 m.y., thus recording the onset of fundamental change (from rifting to drifting) in the tectonic history of the newly evolving passive margin.

The rift stage, involving rapid stretching of the crust, was accompanied by faulting, igneous activity on land, and rapid deposition in deep, elongate, fault-governed troughs. The drift stage, involving the slow cooling of the plate, was accompanied by thermal subsidence, during which time deposition expanded beyond the confines of fault-controlled rift basins, overlapping onto the adjacent basement with successively younger strata. Evolution of the margin thus produced a two-fold depositional couplet of rifting and drifting, yielding in cross section a steer's head depositional configuration (see McKenzie, 1978; Bradley, 1982; and Watts, 1981).

Within this context, the onshore and offshore Triassic-Lower Jurassic strata comprise the synrift deposits, and the offshore wedge of Middle Jurassic-Recent strata comprise the postrift or drift deposits (see Figs. 2, 3, and 5). On the Scotian Shelf, where synrift strata rest on a Meguma-Avalonian terrane, the synrift is underlain by an older passive margin sequence consisting of Upper Devonian to Carboniferous synrift strata, and late Carboniferous to Lower Permian drift strata. Where these terranes occur on the passive margin, as in Georges Bank, the Gulf of Maine, the Scotian Shelf, and Grand Banks, we speculate that the margin may consist of two synrift/drift phases (i.e., a younger Triassic-Jurassic couplet and an older Devonian to Early Permian sequence (see Bradley, 1982; and Fig. 5). In some deep basins, such as the Maine Basin of Georges Bank, where up to 8 km of synrift sediment has accumulated, Triassic strata may rest unconformably on Permo-Carboniferous drift strata as shown by Ballard and Uchupi (1975) for the adjacent basins in the Gulf of Maine.

The concept of multiple rift/drift cycles (successor basins) is shown diagrammatically in Figure 5. Each tectono-stratigraphic

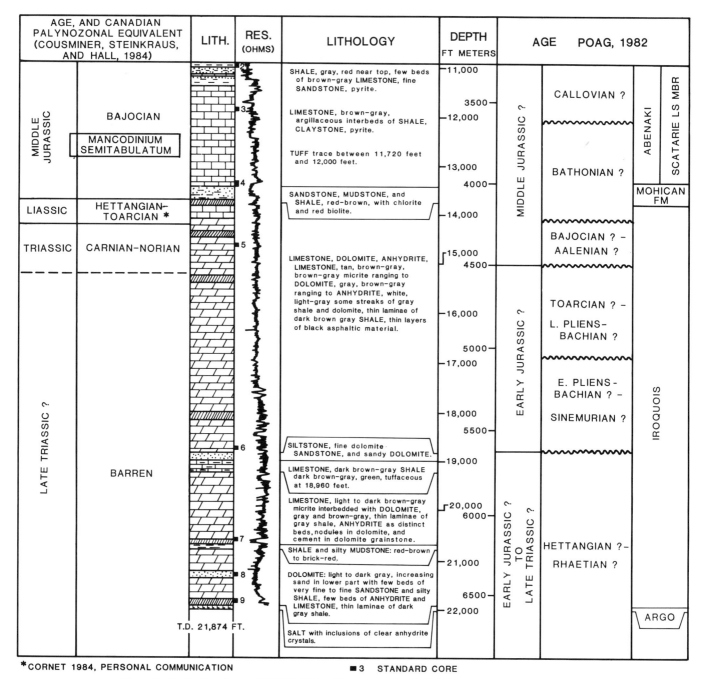

Figure 3. Geologic column of COST G-2 well, core number 2 (3,340.6 m) to total depth (6,667 m). Modified from Poag (1982).

cycle is marked by such first order characteristics as unit geometries, bounding unconformities, lithologies, and structures, as described below. The primary correlation, establishing contemporaneity of onshore and offshore events, is based on biostratigraphic data from the G-2 well. In particular, note that: (a) the onshore volcanic-lacustrine sequence is a time correlative of the offshore post-rift unconformity. (b) the postrift unconformity is diachronous, embracing varying Liassic to perhaps Late Triassic intervals; (c) two (or more) salt formations occur on the shelf, the

Osprey Formation of Carnian-Norian age, and the younger Argo Salt of Hettangian-Sinemurian age; and (d) the oldest Mesozoic marine transgression, documented by dinoflagellates in the G-2 cores, is of Carnian to Norian age.

SYNRIFT BASINS (TRIASSIC LIASSIC)

General Setting

Late Triassic proto-Atlantic rifting extended from the

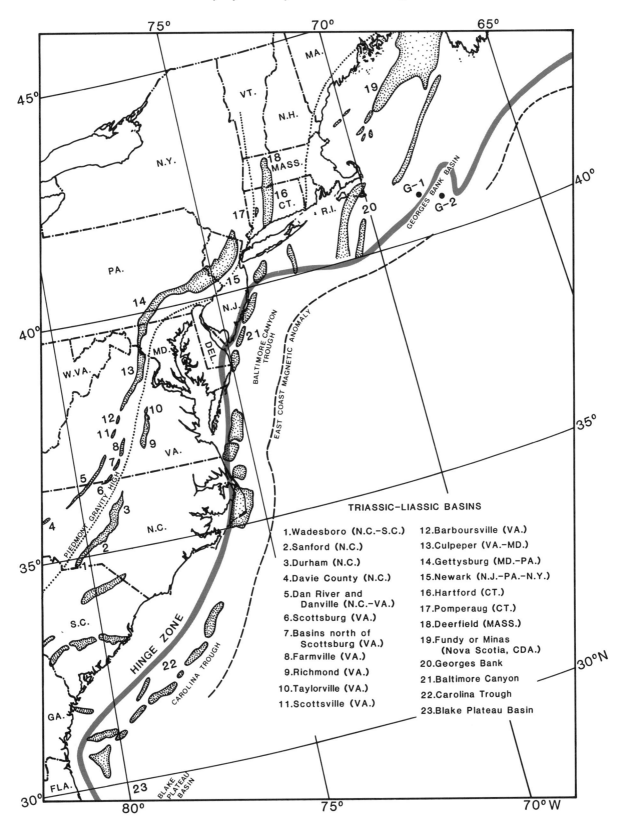

Figure 4. Exposed and concealed Triassic-Liassic basins of eastern North America, delineated by stippled pattern. Also shown: Piedmont Gravity High, East Coast Magnetic Anomaly, Hinge Zone, and the location of the COST G-1 and G-2 wells. Modified after Klitgord, written communication, 1985.

TRIASSIC-LIASSIC BASINS

1. Wadesboro (N.C.-S.C.)
2. Sanford (N.C.)
3. Durham (N.C.)
4. Davie County (N.C.)
5. Dan River and Danville (N.C.-VA.)
6. Scottsburg (VA.)
7. Basins north of Scottsburg (VA.)
8. Farmville (VA.)
9. Richmond (VA.)
10. Taylorville (VA.)
11. Scottsville (VA.)
12. Barboursville (VA.)
13. Culpeper (VA.-MD.)
14. Gettysburg (MD.-PA.)
15. Newark (N.J.-PA.-N.Y.)
16. Hartford (CT.)
17. Pomperaug (CT.)
18. Deerfield (MASS.)
19. Fundy or Minas (Nova Scotia, CDA.)
20. Georges Bank
21. Baltimore Canyon
22. Carolina Trough
23. Blake Plateau Basin

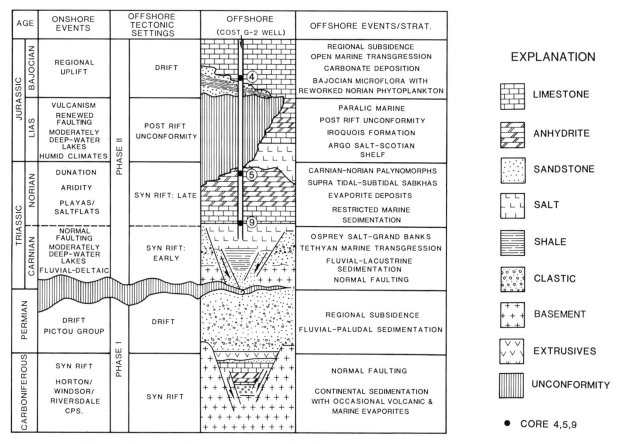

Figure 5. Diagrammatic representation of multiple rift/drift cycles on the North Atlantic borderland.

Cobequid-Chedabucto-Gibralter fracture zone on the north to the Bahamas fracture zone in the south, and across the Variscan-Alleghanian orogen from east to west onto the adjacent bordering cratons of Africa and North America (Fig. 1). Rifting may not have continued south of the Bahamas fracture zone, which apparently was reactivated in the Jurassic as a transform connecting the spreading centers of the Atlantic and Gulf of Mexico (Klitgord and Schouten, 1980). Within this broad terrane lay many northeast-trending, elongate half-grabens and pull-apart basins with intervening horsts that follow the fabric of the orogen. Today, after extensive uplift and erosion, about 20 basins occur in two narrow bands on the American plate where they are separated by the hinge zone (Fig. 4). Several basins, such as the Newark and Hartford, may have been contiguous (see the Broad Terrane Hypothesis as described by Sanders, 1963). One set, the onshore basins, and those beneath the Coastal Plain from Florida to Long Island, follow the core of the main gravity high of the Appalachians; the other set, the offshore basins, are situated west of the East Coast Magnetic Anomaly. The gravity high in the southern Appalachians is located in the Piedmont Province on a transitional crust and the basins are distributed about its axis. The Durham–Deep River–Wadesboro basins, for example, lie southeast of the gravity high and have west-facing border faults with

east-dipping strata, and form a complementary pair with the Dan River Basin to the northwest with its east-facing border fault and west-dipping strata. In the northern Appalachians, one branch of the gravity high extends east of the Newark-Gettysburg Basin and west of the Hartford Basin, whereas another branch extends east to the Funda Basin along the Avalonian-Meguma suture (Figs. 1 and 4).

In cross section, most Newark-type basins are asymmetric half-grabens, constrained on only one side by a system of en echelon normal faults, and on the other side primarily by sedimentary overlap. None of the larger onshore Triassic basins seems to exhibit a classical graben structure with both sides of the basin bounded by normal faults of similar magnitude. In map view, the basins show both right-and-left-offset, and thus appear to be linked to each other by strike-slip faults that have been identified elsewhere as transform or transfer faults (see Bally, 1981, Figs. 21–24.

Where the basins overlie the Meguma-Avalon subplate, the Triassic strata typically overlie a Stephanian-Autunian basement (latest Carboniferous–earliest Permian). To the south the Triassic rests on an older metamorphosed Paleozoic basement. Strata within the basin commonly dip from 10°–15° towards the border fault where they are warped into broadly plunging folds, or com-

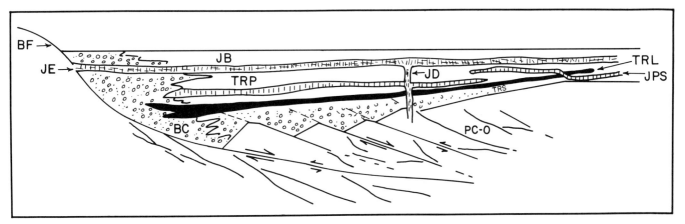

Figure 6. Hypothetical northeast-southwest cross section of the Newark Basin along the Delaware River during deposition of the Early Jurassic Boonton Formation (JB). Abbreviations as follows: BF, border fault; JD, Early Jurassic diabase dikes; JE, Early Jurassic tholeiitic extrusives with interbeds of sedimentary strata; JPS, Early Jurassic Palisades sill and related intrusives; BC, border conglomerate; TrP, Passaic Formation; TrL, Lockatong Formation; TrS, Stockton Formation; PC-O, Precambrian and Cambro-Ordovician rocks of the Taconic and Alleghany thrust sheets; small arrows indicate direction of Taconic and Alleghany thrust movement during the Paleozoic; large arrows indicate movement of re-activated faults during the Mesozoic.

pressed into en echelon folds with axes oriented at acute angles to the border fault, or displaced by small strike-slip and/or normal faults. Whereas some broad warps or folds may have developed from differential slip along normal faults, others, like the tight folds at Lepreau Harbor in the Fundy Basin (Stringer and Lajtai, 1979) and those of the Hammer Creek and Passaic Formations in the Jacksonwald Syncline along the narrow neck of the Newark-Gettysburg Basin display tightly spaced cleavage, and were formed by horizontal compression, probably due to wrench tectonics (Lucas and others, 1985).

Major oblique trending cross faults, with vertical displacement up to 3 km (Van Houten, 1969) and strike-slip displacement up to 20 km (Sanders, 1963), cut the Newark Basin into rhomb-shaped fault blocks. Published geophysical studies of the Newark-Gettysburg Basin (Sumner, 1978), the Durham Basin (Bain and Harvey, 1977), and the Hartford Basin (Wenk, 1984) also postulate the presence of major cross faults creating intrabasinal grabens that formed concurrently with the onset of sedimentation (Cloos and Pettijohn, 1973).

Van Houten (1969) speculated that some of the larger northeast-trending strike-slip faults of the Newark Basin may be part of a transcurrent system including the western border fault, the Ramapo Fault, which according to Ratcliffe (1980) is controlled by reactivation along major semi-ductile faults of Proterozoic and Paleozoic age. The Ramapo border fault, like the border faults of the Culpeper, Taylorsville, and Riddleville basins, is considered to be a listric normal fault that was reactivated along an eastward-dipping imbricated thrust sheet. The geometry of the Newark Basin (Fig. 6) is modeled after the Paleozoic thrust ramp, shown by Ando and others (1984, Fig. 8) to comprise the transitional crust beneath the Early Mesozoic rift basins of Connecticut

and Massachussetts. In the southern Appalachians, the Culpeper, Richmond, and Taylorsville basins also lie along mylonitized fault zones of the Eastern Piedmont fault system, which according to Bobyarchick (1981), is a wrench fault complex that was superimposed over an Alleghanian thrust belt (see also Glover and others, 1980). Seismic reflection profiles on the Coastal Plain reveal that some buried Triassic basins (e.g., the Riddleville Basin) may be bounded to the west by listric normal faults of Mesozoic age that merge eastward into a system of decollements of Alleghanian age (Cook and others, 1981). Listric normal faults apparently provide an important structural mechanism for stretching and thinning the brittle crust of the youthful margin over a ductile lower crust and mantle (Bally, 1981) and are well-documented along other passive margins, such as the Gulf of Biscay (Montadert and others, 1979) and the Red Sea (Lowell and others, 1975).

Newark Supergroup

Extension along and marginal to the proto-Atlantic axis during the Late Triassic, led to the formation of at least 30 clastic and/or evaporite basins (Fig. 1) on both the American and African plates. Recurrent and differential subsidence, due to downwarping of the ductile crust followed by faulting in the brittle crust, provided space to accumulate about 7–9 km of synrift (Late Triassic/Early Jurassic) strata in the Newark-Gettysburg Basin (Olsen and others, 1982), 4 km in the Hartford Basin (Hubert and others, 1978), 4 km in the Durham Basin (Bain and Harvey, 1977), 7 km in the Culpeper Basin (Olsen and others, 1982), 3–4 km in the Dan River Basin (Olsen and others, 1982), and as much as 5–8 km of synrift strata in the Baltimore Canyon, Georges Bank, and Carolina Trough basins. The onshore covered

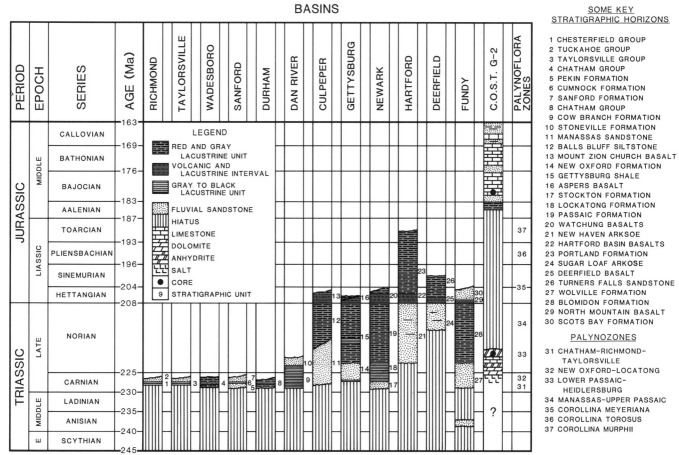

Figure 7. Time-Correlation chart: interbasinal correlation of Newark strata based on palynofloral zones and/or extrusive horizons, data principally from Olsen, written communication, 1985. Correlation is also made with lower Mesozoic strata of the COST G-2 cores, see Figure 12 and Table 1.

basins of Georgia, South Carolina, and Florida are notably thinner, containing about 1 km of strata in Dunbarton, 2.2 km in Riddleville, and 3.5 km in the main South Georgia Rift Basin (Daniels and others, 1983).

Rocks within these basins comprise the Newark Supergroup (Fig. 7) and consist primarily of basal and border petromict conglomerates, arkosic and lithic arenites, gray-to-black siltstones and shale, and red-brown mudstones; interbeds of basaltic lavas occur only from Virgina to Nova Scotia. Limestones, evaporites, dune sands, coal, and kerogen-rich beds are present locally (Fig. 7). Volumetrically, the majority of these beds were deposited in a lacustrine setting and they show a prevalence of laterally extensive gray-black siltstones with fossil fish, and a bulls-eye lithofacies pattern (Olsen, 1980b).

Synrift sedimentation typically began as a diachronous Anisian-to-Carnian event in downsags that subsequently broke into small intrabasinal horsts and grabens, and then into large asymmetric half-grabens along listric normal faults, which were activated along older thrusts. The main fluvial systems drained the flexured margin and flowed along the basin axis, while smaller ephemeral streams drained the uparched and faulted margins. Deposition generally occurred in a closed basin with internal drainage and perched outlets that drained into lower basins (as in historical Lake Magadi of East Africa), and was thus sensitive to subtle changes in climate. Deposition began with fluvial to perhaps lacustrine sandstones and conglomerates that thicken towards the axis of the basin where they interfinger with deeper water alkaline-rich lake deposits (as in the Newark-Gettysburg Basin; see Van Houten, 1969), or with paludal deposits including thin coal seams (as in the Richmond, Sanford, and Taylorsville Basins; see Reinemund, 1955). In some lakes, such as Lake Lockatong of the Newark Basin, sedimentation formed in deep water with perennial anoxic bottom conditions. Varved strata characterize these beds and record the expansion and contraction of the lake, which may have had an areal extent greater than 7,500 km[2]. Besides numerous fish species, these lake beds abound with amorphous algal kerogen, zooplankton, pollen, spores, and plant cuticles (Olsen, 1980b), and thus are potentially important sources of hydrocarbons. As climates became more arid on the valley floor, and lakes began to dry in the succeeding Norian age, extensive playas and mud flats formed along the valley with deposition of clay, including the precipitation of gyp-

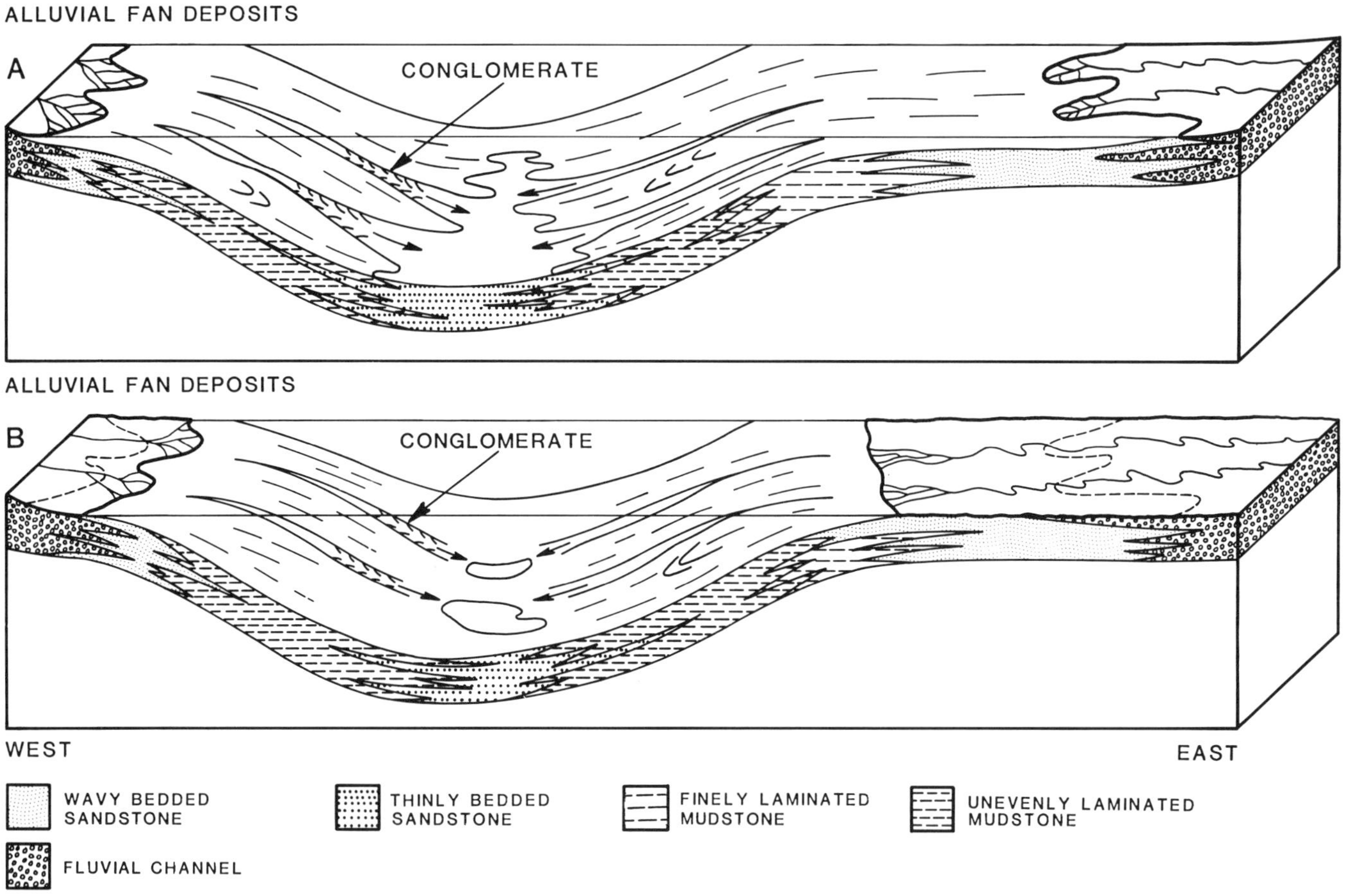

ALLUVIAL FAN DEPOSITS

ALLUVIAL FAN DEPOSITS

WEST   EAST

| | WAVY BEDDED SANDSTONE | | THINLY BEDDED SANDSTONE | | FINELY LAMINATED MUDSTONE | | UNEVENLY LAMINATED MUDSTONE |

| | FLUVIAL CHANNEL |

Figure 8. Reconstructed physiographic profiles of the early Jurassic lakes at (A) high-water level and (B) low-water level, showing the interpreted distribution of lithofacies within the lake basin. The interbedded siltstone and intraclastic sandstone lithofacies are included in the marginal mudflat deposits that occur in embayments between sandy delta plains. Note, (1) exposure and braided stream erosion of sand flats during low-water level, and (2) greater volume of basin-plain turbidite accumulation during high-water level as a result of underflow of sediment charged water from flood-stage feeder rivers (Hentz, 1985).

sum, anhydrite, glauberite, and halite from the Culpeper Basin north through the Hartford, Deerfield, and Fundy Basins (see Hubert and others, 1978). The upland climates generally were humid, and produced an ample supply of clay for Late Triassic red mudstone (e.g., the Passaic, Gettysburg, and Blomiden Formations). Aridity during the Late Triassic (as recorded by the type of sediment accumulating in the basin) was not necessarily caused by a change in the general atmospheric circulation; it may have been caused by regional uplift that created a rain shadow in the leeward valleys (see discussion in Schlee and others, this volume).

A later and distinctively different episode of synrift tectonism started near the beginning of the Jurassic and is marked by plutonic intrusions, basaltic extrusions, and faulting, which was accompanied by uplift of the source terrane with a concomitant increase in rainfall over the watershed and the formation of moderately deep-water oligomictic and meromictic lakes. This phase of tectonism corresponds to the time encompassed by the postrift unconformity (Figs. 2, 3). From the Culpeper Basin north to

the Fundy Basin, plutons (e.g., those of the Narrow Neck in the Newark-Gettysburg Basin) intruded the basin feeding tholeiitic lavas that flowed along the valley floor where they impounded surface runoff, forming pillow lavas in lakes (Manspeizer, 1980). We may infer from the Liassic stratigraphic record that high-discharge ephemeral streams, fed by increased rainfall due to orographic uplift of maritime air, deposited alluvial fan and fan-delta complexes along the upthrown margins of the basins, while clay and silt-bearing perennial streams prograded deltas along the basin's axis and along its flexure margins. Many of the Jurassic lake beds, such as those of the Culpeper, Hartford, and Newark Basins, accumulated in perennially stratified, eutrophic, seasonally expanding and contracting, calcite-precipitating lakes that occupied sub-basins along or near the major border fault (Fig. 8; Hentz, 1985; Olsen, 1980b).

Wherever the Liassic strata are exposed, as in the Culpeper, Gettysburg, Newark, and Hartford Basins, they are marked by a sequence of three to five tholeiitic lava flows that are intercalated in part with lacustrine strata. Each flow is about 50–200 m thick, typically composed of multiple flow units, and intercalated with about 100–300 m of sedimentary rock. The absence of lava flows in the exposed southern basins seems to be the result of deep erosion, since basaltic lavas of similar age are reported in more than 50 wells in South Carolina, Georgia, Alabama, and northern Florida (for details, see Daniels and others, 1983). Paleomagnetic data, in conjunction with geological descriptions of cores, enabled Phillips (1983) to identify 23 different lava flows in the cores at the Clubhouse Crossroad in South Carolina. Although Triassic-Jurassic volcanics are reported from the Gulf of Maine, they have not been found on the adjacent continental margin. The volcanic synrift sequence appears to have been laid down in the narrow time interval of about 500,000 years during the Hettangian epoch (Olsen, personal communication, 1985). Volcanism began at least 20 m.y. after the onset of sedimentation in the Middle Carnian, and then only after 2–6 km of coarse clastics had accumulated in these basins (see Fig. 7). The volcanic emplacements were concurrent with other changes in the geometry and stratigraphy of the basins, thus reflecting fundamental changes in basin tectonics (see Fig. 5). As multiple lava flows, consisting of individual flow units, they record periodic episodes of extension within each basin. It is unlikely that the Triassic border faults, as they break the surface, served as conduits for the egress of lava. Paleoflow studies in the Newark-Gettysburg Basin (Manspeizer, 1980) show that feeder dikes lie along the axes of these basins and neither intersect nor crop out along border faults. Both geochemical and field studies (Puffer and others, 1981) demonstrate that the lava flows of the Newark and Hartford Basins are both rock and time-rock correlatives. As such, they are effective stratigraphic datums (Olsen, written communication, 1985; see Fig. 7).

In addition to the dolerite sills and dikes (e.g., the Palisades Sill in New Jersey, West Rock in Connecticut, and the intrusives of Gettysburg, Pennsylvania, which were emplaced during the depositional history of the basins), a younger igneous event is recorded by those dikes that cut across the lava flows, across the folded and faulted synrift strata, and even across the border faults and marginal highlands. This dike set is most densely concentrated in the Carolinas and is sparse to absent in the northern Appalachians. May (1971) has shown that, on a Bullard reconstruction of the continents, the dike swarm forms a radial pattern centered over the Blake Plateau. The dikes coincide with the Piedmont Gravity High and with the East Coast Gravity Anomaly. DeBoer and Snider (1979) relate the dike swarm to a hotspot located northeast of a triple junction in northern Florida. Swanson (1982) concluded that the dikes occur along a transform system that parallels the orogen and is related to the migration of North America about a pole of rotation in the Sahara. McHone and Butler (1983), recognizing five alkalic igneous rock provinces of Mesozoic age, show that these provinces are located where cross-trending fracture zones intersect Appalachian orogenic structures. Bedard (1985) also has shown that older inherited structures (e.g., the Ottawa-Bonnechere graben and the

Kelvin Fracture Zone) have controlled the emplacement of the Monteregian intrusives and Kelvin Seamounts respectively. Structures localizing the southern White Mountain lineament, however, are less well defined (see McHone and Butler, 1983).

As these late Mesozoic intrusives are younger than the synrift basins, they are discussed by DeBoer and others (this volume).

Basin Origin

For over 100 years, beginning with the studies of Rogers (1842) in Pennsylvania, Newark-type basins were considered graben or fault troughs produced by extension at right angles to the axis of the basin, and thus structurally dominated by dip-slip normal faults (see Sanders, 1963; Thayer and others, 1970; and Marine and Siple, 1974). Today, many of these same basins are observed to have tectonic and sedimentologic features that are characteristic of half-grabens that formed within transforms due to strike-slip or oblique-slip (See Bally, 1981, Figs. 21–24).

Although the accumulation of 5–10 km of sediment in these basins attests to the importance of subsidence in their development, some basins may have formed initially as downwarps along an attenuated crust long before they became fault troughs. Reinemund (1955) suggested that clastic sedimentation in the Deep River Basin was controlled first by downwarping, rather than by faulting. In the absence of a continuous system of faults along the western margin of the Newark-Gettysburg Basin, Faill (1973) concluded that major faulting occurred after sedimentation had ceased, in basins whose basements had been thinned by horizontal extension and subsided by downwarp. Van Houten (1977) and Ratcliffe (1980) have reported strike-slip displacement along the Newark-Gettysburg Basin, while Bain and Harvey (1977), and Ballard and Uchupi (1975) have shown that strike-slip faulting was concurrent with both sedimentation and basin evolution in the Durham and Gulf of Maine Basins respectively. DeBoer and Snider (1979), postulating the presence of a hotspot in the Carolinas, relate the origin of these basins to thermal doming accompanied by crustal extension and rifting, and the injection of the dike swarms to a subsequent phase of strike slip. Manspeizer (1981) has described the Newark–Gettysburg Basin and the Argana Basin of Morocco as pull-aparts resulting from wrench tectonics along east-west-trending transforms (see below). Swanson (1982) speculates that dextral and sinistral shear along an arcuate intracontinental transform system, parallel to the coastline, played a significant role in forming the basins and dike swarms.

The role of transforms has been largely overlooked as a mechanism by which Triassic-Liassic rift basins may originate. Today, where oceanic transforms cross continental margins, as in the Dead Sea Rift (Quennel, 1958; Freund, 1965) and in the San Andreas Rift (Crowell, 1974), the crust has been pulled apart by as much as several hundred kilometers, forming a complex array of graben-like rift basins, whose structures and stratigraphic records serve as models for the origin of Newark-type basins (Manspeizer, 1981). Even the FAMOUS Expedition (Chouk-

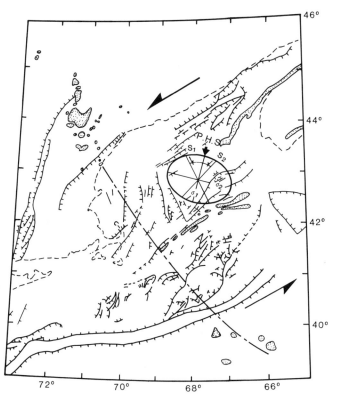

Figure 9. Proposed orientation of left-lateral shear couple and associated strain ellipsoid derived from normal fault pattern in Gulf of Maine and surrounding regions. S_1 and S_2 are probable shear fracture planes; δ_1 and δ_2 are planes of tensional fractures; and P.H.S., principal horizontal stress. Dotted pattern indicates intrusives of Late Triassic to Early Cretaceous age. Dashed line is trend of seismicity belt along which lies the White Mountain magma series and the New England Seamounts chain (Ballard and Uchupi, 1975).

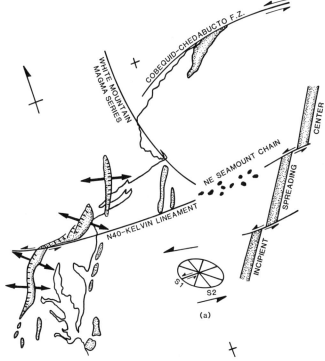

Figure 10. Early Mesozoic tectonic elements of the central Appalachians with proposed east-west–trending, left-lateral shear couple and associated strain ellipsoid (a). Basin trends are controlled by: (1) dip slip along older Paleozoic thrusts within a regional northwest southeast extensional domaine that is perpendicular to the spreading axis; and (2) strike-slip along horizontal shear planes (S_1), as in the Narrow Neck and Fundy basins. In the model (a) the White Mountain magma series, the New England Seamount Chain, and the Cobequid-Chedabucto and N40-Kelvin lineaments all lie along strike-slip shears (S_1 and S_2).

roune and others, 1978), which was taken along young transforms that offset the Mid-Atlantic Ridge axis, has shown that the transform valleys and shoulders are rifted basins, caused by normal faulting within horizontal shear zones.

Thus, the northeast-to-southwest–trending Newark-Gettysburg Basin, for example, which, according to Van Houten (1977), developed along a continental fracture zone, may have formed primarily by extension (northwest-southeast) that resulted from an east-west sinistral shear couple superimposed upon a thrust ramp basement fabric (Manspeizer, 1981; Fig. 6). Continued strike-slip along the transform during deposition seems to have created complex wrench systems distinguished by en-echelon folds and en echelon conjugate strike-slip faults that formed rhomb-shaped sub-basins as in the Newark Basin of New Jersey (Manspeizer, 1981). Where deformation occurred along the transforms between basins (as in the narrow neck between the Newark and Gettysburg Basins), Appalachian type foreland folds formed with axial planar spaced cleavage, clastic dikes, extensional fractures, and numerous basaltic dike intrusions (Lucas and others, 1985). The fact that the dikes are largely confined to the narrowest part of the basin, where compression is greatest, is

difficult to explain in a purely extensional regime. Elsewhere, as in the Hartford Basin, strike-slip seems to have occurred late in the depositional history of the basin, and thus may not be a mechanism related to its origin.

A similar stress field (east-west–trending, left-lateral shear couple) is almost identical to the field inferred by Ballard and Uchupi (1975) to explain the evolution of the offshore basins in the Gulf of Maine (Fig. 9). The same stress field may also explain the origin of the Fundy Basin, which according to Keppie (1982) formed in response to more than 5 km of sinistral strike-slip along the Cobequid-Chedabucto Fracture Zone in the early Mesozoic (Fig. 10). Bedard (1985) suggests that the Triassic tectonic regime may have been more transcurrent than extensional, resembling a pattern of transforms separating en echelon basins (see Figs. 1, 10). Differential rotation of the North American plate during the early opening of the Atlantic, probably created pull-aparts along east-west–trending leaky transforms that were superimposed across a northeast-trending basement fabric (see Le Pichon and Fox, 1971; Sheridan, 1974; Ballard and Uchupi, 1975; Turcotte, 1981; and Bedard, 1985).

Age

Determining the age of the Newark Supergroup has posed a problem because the "red beds" were long thought to be barren of fossils and particularly lacked marine fossils, which are generally used for interregional biostratigraphic correlations. Isotopic age dating and paleomagnetic dating alone are too imprecise to date this stratigraphic sequence.

Froelich and Olsen (1984) note that the Newark Supergroup has been considered to be partly or solely of Early Jurassic age (Rogers, 1842; Lyell, 1847; Redfield, 1856), Late Permo-Triassic age (Emmons, 1857), Jurassic or Late Triassic age (Fontaine, 1883), then solely of Late Triassic, based initially on the presence of rare vertebrates and plant fossils (Ward, 1891; Eastman, 1913), subsequently on vertebrates and plant fossils (Reeside and others, 1957), and finally on radiometric ages of intercalated lava flows (Armstrong and Besancon, 1970). With the use of palynomorphs (Cornet and others, 1973; Cornet and Traverse, 1975; Cornet, 1977; Cornet and Olsen, 1986) and with well-preserved vertebrates (Olsen and others, 1982), workers have shown that some of the basins contain Lower Jurassic beds.

The studies by Cornet and Olsen (1986; see Figs. 7, 11) and their associates have documented that the strata in the Richmond, Taylorsville, Scottsburg, Sandford, Durham, and Dan River basins are Late Triassic, ranging typically from the Middle to Late Carnian. The strata in the Culpeper, Gettysburg, Newark, Deerfield, and Fundy basins, however, range in age from Late Triassic (Carnian to Norian) to Early Jurassic (Hettangian to Toarcian). Rocks in the Fundy Basin may extend downward into the Middle Triassic (Anisian).

The Onshore Paleontologic Record

Fossil organisms of the Newark Supergroup sediments include spores, pollen, plant remains, fresh water fishes, vertebrate skeletal remains and their footprints, insects, conchostrachans (clam shrimp), tadpole shrimp, decapods, *Scoyenia* (arthropod burrows), ostracodes (*Darwinula* type), and fresh water clams (Olsen, 1980a, b). In addition to the palynomorphs, the fish, tetrapods, and reptilian footprints are considered to have biostratigraphic value. The palynostratigraphic data of Cornet (1977), Cornet and others (1973), and Cornet and Traverse (1975), in combination with distributional data on fishes, tetrapods, basalt geochemistry, paleomagnetics, and K/Ar dates have been used by Olsen and his co-workers to build a correlation matrix useful for inter-basinal correlation of Newark strata, and to broadly indicate their relationship to marine type sections and early Mesozoic terrestrial sequences elsewhere (see Olsen and others, 1982; Cornet and Olsen, 1986).

The biostratigraphic value of these non-marine fossil organisms varies considerably, and depends upon documentation of their age equivalence to the established Triassic-Jurassic marine biostratigraphic standard in Europe and North America. The palynomorphs are the only fossils present in Newark strata that are also present in these marine type sections.

Two publications document the ranges of Triassic palynomorphs in the context of ammonoid biostratigraphy. Fisher (1979), based on both sampled exposures and wells drilled on the Canadian Arctic Archipelago, has charted the ranges of 450 palynomorph taxa, and placed them in nine informal assemblage zones of stage range that are correlative with those defined by Tozer (1967; 1973) on ammonoid zones. Another compilation by Visscher and Brugman (1981) charts the ranges of 52 selected palynomorph species having known ranges in Alpine Triassic ammonoid-bearing strata.

Cornet's (1977) extensive study of the palynostratigraphy of the Newark Supergroup recorded the ranges of close to 300 palynomorph species, of which over 100 were considered to have biostratigraphic value. Several of these species commonly occur in ammonoid-bearing strata in both Europe and arctic Canada (see Fig. 12), thus supporting Cornet's dating of the Newark sediments and their external correlation to the worldwide Triassic marine standard.

Commenting on the rapid turnover of species comprising all paleontologic groups in the Triassic portion of the Newark sediments as compared with the Liassic portion, Olsen (written communication, 1985) has concluded that this is linked to more rapid species origination and extinction rates in the Triassic. Paleontologic zonal schemes for the Newark (Cornet and Olsen, 1986) are illuminating in this regard, and indicate that between the mid-Carnian and the end of the Triassic, species replacement rates enable the definition of four palynozones (see Fig. 11), three fossil plant zones, three tetrapod zones, and four fossil fish zones. By contrast the Jurassic portion (from basal Hettangian to the end of the Toarcian) includes only three poorly defined palynozones, one fossil plant zone, one tetrapod zone, and two fish zones. Thus fine-scaled subdivision and inter-basinal correlation of the Triassic portions of the Newark sediments is based on substantial biostratigraphic data.

Particularly well-defined palynologic assemblage zones make possible the separation of strata into a middle Carnian Chatham-Richmond-Taylorsville Zone; a late Carnian New Oxford-Lockatong Zone; an early Norian Lower Passaic–Heidlersburg Zone; and a late Norian (Rhaetian in previous usage) Manassas–Upper Passaic Zone (see Fig. 11). In addition, although the Triassic/Jurassic boundary lacks palynostratigraphic definition in the European and Canadian Arctic marine type localities (Fisher and Dunay, 1981; Cornet, 1977) Cornet and Olsen, (1986) define the Late Norian/Hettangian boundary in the Newark sediments by both a peak occurrence of *Corollina meyeriana* pollen and the first appearance of a number of long-ranging palynomorph species (see Fig. 11).

Earlier interpretations of fossil fish data resulted in a complex Liassic interbasin correlation that placed the extrusive volcanics in these basins at varied stratigraphic levels (Olsen and others, 1982). However, additional fossil fish data indicate that this correlation is invalid and the extrusives are now all considered to be part of an isochronous event of earliest Jurassic age (Olsen, written communication, 1985). Below this datum, paleontologic control depends on a matrix of data points derived from palynologic assemblages, plant megafossils, tetrapod remains, and fossil fishes, and includes all basins with extrusive volcanics (Culpeper, Newark, Pomperaug, Hartford, and Deerfield). Of these only the Hartford and Deerfield Basins are considered to include a Liassic section younger than Hettangian in age, and of these, only the Hartford Basin has been subdivided into three Liassic palynozones; a Hettangian/Sinemurian *Corollina meyeriana* Zone, a Pleinsbachian *Corollina torosa* Zone, and a Toarcian *Corollina murphii* Zone.

Because of the long ranges of the palynomorphs and other fossils of Liassic age, both biostratigraphic subdivision and correlation of strata within the Jurassic are much less precise than in the Triassic.

Of the large number of different kinds of fossil organisms found in the Newark strata, it is only the palynomorphs that have common distribution in both the onshore and offshore synrift basins of eastern North America and the Triassic-Liassic marine basins of arctic Canada and northwest Europe.

The Offshore Paleontologic Record

Based on palynologic studies of a large number of subsurface samples recovered from 67 eastern Canadian offshore wells, Barss and others (1979) have described a Carboniferous to Pleistocene palynozonation that is applicable to both the Grand Banks and Scotian Shelf. They concluded that three of the Grand Bank wells penetrated the entire Late Triassic. Their formal zonation includes terrestrial palynomorphs (spores and pollen) and marine dinoflagellates that are used to date the offshore Canadian strata in all the European Jurassic stage equivalents (Bujak and Williams, 1977). Dinoflagellates are particularly useful because their first appearance in both basins is considered to mark the onset of open marine sedimentation (Pleinsbachian-Aalenian in the Grand Banks, and Aalenian-Bajocian in the Scotian Shelf). In addition, their subsequent diversification in the overlying strata and rapid speciation and extinction rates provide excellent biostratigraphic tops, many of which appear to coincide with those recorded in the standard European Jurassic stages in ammonoid-bearing strata (Woollam and Riding, 1983; Riding, 1984). Also as indicated above, the ranges of many of the Late Triassic palynomorphs recorded from Newark strata (Cornet, 1977; Cornet and Olsen, 1986) have also been documented in strata bearing ammonoids in both arctic Canada (Fisher, 1979), and northwest Europe (Visscher and Brugman, 1981); and many of these species are also present in strata penetrated in several offshore Canadian Grand Banks wells (see Table 1; and Barss and others, 1979).

Considering the geographic proximity of the offshore Canadian basins and the Georges Bank Basin, one would expect that the Canadian palynologic zonation should be useful in the deep tests drilled here off the U.S. coast. However, age diagnostic palynomorphs have not been reported from the COST G-2 well samples for the lower two-thirds of the section (i.e., below 2640 m; see Bebout, 1980; and Poag, 1982).

Cousminer (1983) and Cousminer and others (1984) recovered from the G-2 well (in a sample from core number 5 at 4,440.9–4,441 m) many of the same palynomorph species that were used to identify Triassic sediments in three Grand Banks Canadian wells (Barss and others, 1979) and to mark the age of the rift strata of the onshore Newark Group basins (Table I and Fig. 11). Cornet (personal communication, 1983, 1984) also reported recovering Triassic palynomorphs from cuttings beginning above this cored interval, at 4,344.5 m and extending to 4,634 m; above this, from below 4,055 m, he recovered Hettangian to Toarcian spores and pollen species (Fig. 3). Accordingly, the Liassic/Middle Jurassic contact in the G-2 well is placed at about 4,100 m at the top of the dolomitic anhydrite section that is the dominant lithology to the well's total depth (Fig. 3), and at the base of the Middle Jurassic Mohican Formation (see Poag, 1982, Fig. 24), which Given (1977) suggests represents the first depositional phase of sands after the breakup of North America and Euro-Africa in the Middle Jurassic. The postrift unconformity thus occurs within the condensed Liassic section (less than 300 m), and is operationally placed at the base of the Mohican Formation at 4,153 m.

The palynomorph assemblage recovered in the G-2 core number 5 includes the following species that are also present in both the Grand Banks (Barss and others, 1979) and in the Newark Basins (Cornet, 1977; Robbins; 1981; Cornet and Olsen, 1986): disaccate pollen (cf. *Alisporites* spp.), and the following age restricted species (see Fig. 12): *Camerosporites secatus, Ovalipollis pseudoalatus, Pseudoenzonalasporites summus,* and *Patinasporites densus* (see also Table 1 for complete list of Triassic taxa recovered from the G-2 core sample and the Grand Banks wells).

Although the abundant disaccate pollen of probable coniferous source do not have restricted biostratigraphic significance, they have been reported as frequent constituents of the Late Triassic assemblages described from both the Grand Banks wells (Barss and others, 1979) and the Newark sediments (Cornet, 1977; Robbins, 1981; Cornet and Olsen, 1986). All of the other palynomorphs listed above are age restricted. Thus, all three areas include in their strata sediments of late Triassic, (Carnian to early Norian) age (Fig. 12; Table 1). Of great significance is the fact that the G-2 core number 5 also contains rare marine Triassic elements, noted in some of the Canadian offshore wells (Barss and others, 1979). In the G-2 well these include dinoflagellates (cf. *Noricysta* sp.) and *Tasmanites sp.,* indicating that at least intermittent marine conditions were present in the Georges Bank region as early as the Late Triassic. Reworked Late Triassic dinoflagellates are also present in core number 4 (4,034–4,050.6 m) where they occur with an in situ Bajocian microflora (see Table 1). These include species originally described from the Late Triassic of arctic Canada in ammonite-bearing strata (Bujak and Fisher, 1976).

From the foregoing, it appears probable that the Late Triassic dinoflagellates and other palynomorphs reworked in core number 4, represent strata that were deposited during a Late Triassic marine pulse that entered the Georges Bank region, pos-

Figure 11. Ranges of palynomorph taxa in Newark Supergroup sediments, (from Cornet and Olsen, 1986). The taxa have been rearranged in order of local range tops (last appearances). Note the large decreases in diversity that occur at/or near the palynofloral boundaries. Note also that most of the forms that range into the Liassic are long-ranging, with the exception of a very short-ranging basal Liassic group.

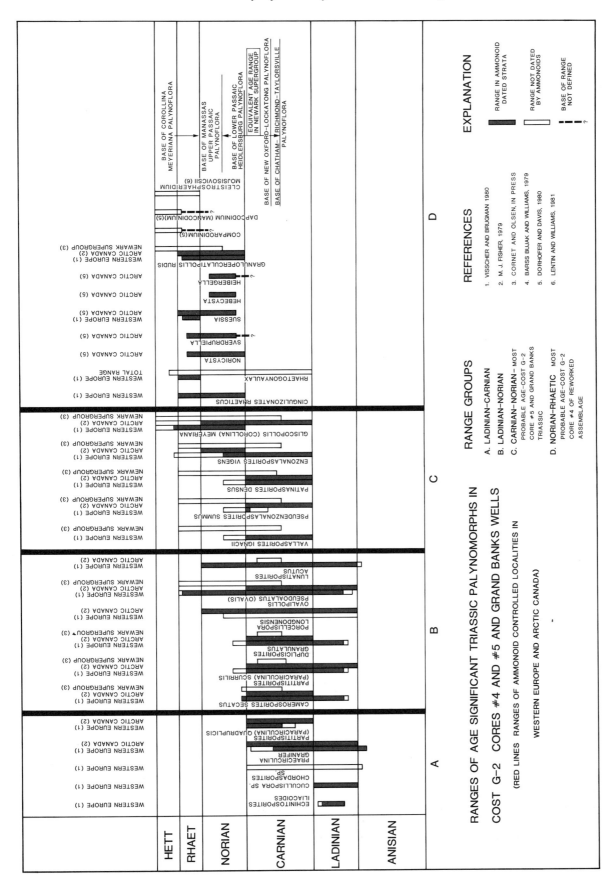

Figure 12. Ranges of age significant Triassic palynomorphs in COST number G-2 well and Grand Banks wells (Barss and others, 1979). The red lines indicate the ranges of these taxa in strata dated by ammonoids in western Europe (Visscher and Brugman, 1981), and arctic Canada (Fisher, 1979).

W. Manspeizer and H. L. Cousminer

TABLE 1. DISTRIBUTION OF LATE TRIASSIC PALYNOMORPHS IN GRAND BANKS WELL, COST G-2 WELL AND NEWARK SUPERGROUP

Well Name	Depth (m)	Cf. Alisporites spp.	Echinitosporites iliacoides	Cucullispora sp.	Chordasporites sp.	Praecirculina granifer	Paracirculina quadruplicis	Camerosporites secatus	Paracirculina scurrilis	Duplicisporites granulatus	Porcellisporites longdonensis	Ovalisporites pseudoalatus	Lunatisporites acutus	Vallasporites ignacii	Pseudenzonalasporites vigens	Patinasporites densus	Enzonalasporites vigens	Gliscopollis meyeriana	Cingulizonates rhaeticus	Cleistrosphaeridium mojsisovics	Granuloperculatipollis rudis	Noricysta spp.	Hebecysta brevicornuta	Heibergella spp.	Suessia swabiana
Grand Banks Wells																									
Coot	3316.4-3563.5																	A							
Sandpiper	2504.5-2713																	A	X						
	2720-2990	A				X	X		X	X		X					*		X						
Osprey	1371-1631																	A							
	1918-2713	A										A			*		*								
	2951-3472			?		X	X		R	X							A*								
Spoonbill	1360-1573																	A							
	2430-2561	A					X							X			*	X							
	2674-2683	A												X											
Georges Bank Cost G-2																									
Core No. 4	to 4051																	A				X	X	X	X R
Core No. 5	to 4441.6	A	R		R	R						R	R		*		*								
Newark Supergroup		A				X		X	X	X		X	X	X	*		*	X			X				

A = Abundant   X = Moderately rare   R = Rare   * = Restricted to late Carnian/early Norian age range

sibly from the Tethys and/or arctic Canada. Judkins and others (1980) concluded, "The most distinguishing stratigraphic feature of the Georges Bank wells is the thick Jurassic section" (4,876.8 m in the G-2 well). However, placing the Triassic top at 4,344 m (at the base of Cornet's Liassic section) reduces the Jurassic section to 2,559 m, which is closer to that in the Canadian basins (north Scotian Basin about 1,676.4 m).

Barss and others (1979) supply palynological datings for 23 wells drilled in the Scotian Shelf. Of 18 wells that have total depths within Jurassic and older rocks, 13 bottom in rocks dated as Middle or Late Jurassic. Five of these wells, the Mohican I-100, Iriquois J-17, Eurydice P-36, Argo F-38, and Hercules G-15, attain Liassic depths. Of these, only two, the Eurydice and Argo wells, penetrated undivided Rhaetian-Hettangian section. In all five wells the late Liassic section is missing, and a considerable hiatus is present between paralic to non-marine strata dated as Sinemurian–early Pleinsbachian, and those of open shelf marine middle Jurassic or younger age (Bajocian to Berriasian). In the Eurydice P-36 well, Berriasian strata at 518 m rest almost directly on those dated as Sinemurian–early Pleinsbachian at 538 m. Although the Mohican I-100 well is shown as having a more or less complete Liassic section, the interval interpreted as being of late Early Jurassic age (3,944–4,259 m) does not carry age restricted palynomorphs to support this interpretation.

Considering the evidence for an extensive Lower Jurassic haitus on the Scotian Shelf, such an unconformity in the Georges

Bank COST G-2 well is not at all anomalous. In summary, the paleontologic data presented here are based on direct study of the rock record, and afford a substantial basis for an interpretation that is alternate to that based solely on seismic stratigraphy (see Poag and Valentine, this volume).

EARLY HISTORY OF THE PASSIVE MARGIN

The Atlantic passive margin evolved through an extended history of recurrent plate activity that accompanied several cycles of Proterozoic and Paleozoic accretionary and then rifting tectonics. Different tectonic regimes were overprinted onto a previously deformed basement, so that by the end of the Paleozoic Era, the basement of the future Atlantic margin was a collage of inherited fabrics that lay along the core of the Variscan-Alleghanian orogen. The final accretionary phase of the orogeny culminated along a 2,000-km dextral shear zone that led to crustal shortening at both ends of the megashear with concomitant thrusting of the Urals and Maurintanides to the east and the Appalachians to the west (Arthaud and Matte, 1977). Subsequent breakup of the orogen and opening of the central Atlantic occurred along these reactivated Paleozoic sutures and along other, inherited structures.

We postulate that wrench and extensional faulting throughout the Late Triassic–Early Jurassic, through clockwise rotation of Gondwanaland with respect to Laurasia, as postulated by Swanson (1982), broke the Variscan-Alleganian orogen along continental extensions of transforms and micro-plates. Sinistral shearing along east-west–trending continental fracture zones (e.g., the Cobequid-Chedabucto and Gibraltar, and the N40-Kelvin lineament/South Atlas Fracture Zone) and extension along older Alleganian thrust faults yielded listric normal faults (e.g., in the Richmond-Taylorsville, Riddleville, and Newark Basins), producing a family of half grabens and pull-apart basins, whose fabrics and geometries largely mimic basement structures. In the Canadian Maritimes, for example, and most probably on the adjacent continental margin, wrench-generated successor Triassic basins lie entirely within older late Paleozoic strike-slip basins. Although the stress field may have changed and/or alternated with time, the primary stress regime throughout the Triassic-Liassic appears to have been an east-west–trending, left-lateral shear couple that resulted from extension along the incipient proto-Atlantic axis. Continued attenuation along listric normal faults, concurrent with strike-slip offset along continental fracture zones, seem to explain the origin of the Atlantic conjugate margins which are markedly asymmetric (see Bally, 1981, Fig. 25; Klitgord and Schouten, 1980).

As the Pangaean Plate migrated north from an equatorial position in the Middle Triassic to a subtropical paleolatitude of about 25°N in the Early Jurassic, it transgressed different climatic regimes, bringing first humid and then arid climates to much of the region (Manspeizer, 1982; see also Hubert and Mertz, 1980). Superimposed on this long-term trend was the annual climatic cycle of the seasonal monsoons. Moderating the trend towards aridity, however, was the opening of the ocean and the transgression of Tethys, which brought water into an otherwise fairly dry continental interior. Finally, the rifting of the continent formed a rift topography, which profoundly affected local climates of the source regions and depositional basins by creating humid climates in mountainous terranes and arid climates in valleys and low-lying playas.

Although a detailed paleogeographic reconstruction has yet to be drawn for the proto-Atlantic during the Triassic-Liassic, it evidently was a time during which a narrow strip of high-standing, coastal and inland ranges with deeply subsiding troughs bordered a rifted continental platform. At times, the platform was transgressed by multiple shallow water tongues from the Tethys seaway to the east, or from Arctic Canada to the north (Fig. 1; see also Schlee and others, this volume).

The first stratigraphic record of a Tethyan transgression is found in upper Middle Triassic rocks (Ladinian Stage) of northern Morocco, along the Gibralter shear zone (Manspeizer and others, 1978). There, andesites are interbedded with carbonates bearing a Lettenkohle *Anoplophora* fauna that is overlain by several hundred meters of massively bedded micrite with molluscan molds and oncolites. A similar marine carbonate and andesite sequence occurs to the southwest, in the Middle High Atlas. Further to the south in the Central High Atlas, Ladinian andesites at the base of the section are overlain by Upper Triassic continental red beds that, according to Biron (1982) are interbedded with sandstones containing Carnian brachiopods and echinoids. As the Carnian Sea transgressed southwestward along the South Atlas transform, it shoaled against the rising Tichka Massif, preventing Tethyan marine waters from entering the embryo Atlantic Basin through the western High Atlas (Fig. 1).

The first marine invasion of the Central High Atlas Basin of Morocco occurred in the Late Triassic (probably Carnian Stage) as shallow hypersaline Tethyan waters trangressed westward through the Gibralter fracture zone (Jansa and Wade, 1975; Manspeizer and others, 1978; Jansa and others, 1980; Lancelot, 1980; and Manspeizer, 1982). As fluvial and lacustrine sedimentation occurred in what is now the onshore and offshore clastic basins of western Morocco and eastern North America, marine carbonates and sulphates were forming in basins from the Alps to southern Spain, and halite with minor amounts of anhydrite and dolomite were forming in evaporite basins in Algeria, Tunisia, and the Aquitaine in southern France (Fig. 1; Busson, 1972; Jansa and others, 1980). Farther west, in the Lusitania Basin on the Iberian Meseta and in the Carson subbasin in eastern Grand Banks, over 2,000 m of pure halite of the Osprey evaporites were precipitated above the continental Kettle red beds, without carbonates or clastics, from hypersaline waters in restricted basins (Jansa and Wade, 1975). Continued marine transgression along the proto-Atlantic Axis, throughout the Late Carnian and into the Norian, brought hypersaline seas and evaporite precipitation to Essaouiara and the western High Atlas Basins of Morocco, as well as to Georges Bank and south to perhaps the Carolina Trough in North America (Fig. 1).

Data from the COST G-2 well show that quiet, nearshore, paralic marine conditions prevailed over much of the platform lying adjacent to the coast, while offshore, dolomite and calcite, with subordinate amounts of anhydrite, algal stromatolites and oolitic sands formed on extensive supratidal sabkhas, that were occasionally flooded by high-energy marine waters (see Arthur, 1982; Poag, 1982). Where hypersaline waters were restricted, as in tectonically active rift basins, thick Carnian salt deposits formed (see Rona, 1982).

That marine conditions prevailed for some time on the newly formed margin at Georges Bank, is documented by the occurrence of Carnian phytoplankton below the postrift unconformity and by reworked late Norian marine palynomorphs within situ Bajocian micro-flora above the unconformity. Similar dinoflagellate assemblages have been described from the subsurface of the Canadian Arctic (Bujak and Fisher, 1976), where they occur with ammonites indicative of a Norian age and marine habitat, and are recorded from the Late Triassic of coastal Israel (Cousminer, 1981). The marine transgression of Georges Bank, which may have extended south to the Carolinas, was part of a much more extensive Late Triassic flooding of the rift basins that

were breaking apart the Pangaean Plate. Marine seas extended south from Arctic Canada along the North Atlantic Rift Zone through eastern Greenland (see Clemmensen, 1982), and west from the Tethys Sea through a complex rift system of western Europe (see Ziegler, 1982).

By the Early Jurassic, the newly evolving margin experienced minor tectonic unrest that was accompanied by regional uplift, marine regression, and erosion. In the onshore basins, it was a time of vulcanism, rifting, and renewed uplift of the source terranes with the concomitant deposition of fan deltas that prograded into moderately deep water lakes. Elsewhere, as in western Europe, the uplift on Georges Bank, which we correlated with the breakup unconformity, is related to the mid-Cimmerian tectonic event. According to Ziegler (1982), it is a major rifting event of Early to Middle Jurassic age that was synchronous with vulcanism in the North Sea, eustatic lowering of sea level, and the rifting episode that preceded the onset of seafloor spreading in the central Atlantic. It also signaled an end to synrift deposition of the U.S. Atlantic margin, and the beginning of its drifting history.

REFERENCES CITED

Ando, C. J., Czuchra, B. L., and 10 others, 1984, Crustal profile of mountain belt; COCORP deep seismic reflection in New England Appalachians and implications for architecture of convergent mountain chains: American Association of Petroleum Geologists Bulletin, v. 68, p. 819–837.

Armstrong, R. L., and Besancon, J., 1970, A Triassic time scale dilemma; K-Ar dating of Upper Triassic mafic igneous rocks, eastern U.S.A. and Canada and post-Triassic plutons, western Idaho, U.S.A.: Ecologae Geologae Helvetiae, v. 63, p. 15–28.

Arthaud, F., and Matte, P., 1977, Late Paleozoic strike-slip faulting in southern Europe ad northern Africa; Result of a right-lateral shear zone between the Appalachians and the Urals: Geological Society of America Bulletin, v. 88, p. 1305–1320.

Arthur, M. A., 1982, Lithology and petrology of COST Nos. G-1 and G-2 wells, in Scholle, P. A., and Wenkam, C. R., eds., Geological studies of the COST Nos. G-1 and G-2 wells, United States North Atlantic Outer Continental Shelf: U.S. Geological Survey Circular 861, p. 11–33.

Bain, G. L., and Harvey, B. W., 1977, Field Guide to the geology of the Durham Triassic basin: Carolina Geological Society Field Trip Guidebook, October, 83 p.

Ballard, R. D., and Uchupi, E., 1975, Triassic rift structure in Gulf of Maine: American Association of Petroleum Geologists Bulletin, v. 59, p. 1041–1072.

Bally, A. W., 1981, Atlantic-type margins, in Bally, A. W., ed., Geology of passive continental margins: American Association of Petroleum Geologists Educational Course Note Series no. 19, p. 1–48.

Barss, M. S., Bujak, J. P., and Williams, G. L., 1979, Palynological zonation and correlation of sixty-seven wells, eastern Canada: Canadian Geological Survey Paper, 78-24, 117 p.

Bedard, J. H., 1985, The opening of the Atlantic, the Mesozoic New England Province, and mechanisms of continental breakup: Tectonophysics, v. 113, p. 209–232.

Biron, P. E., 1982, Le Permo–Trias de la region de l'Ourika [abs.], in Beauchamp, J., ed., Marrakech, Morocco, Le Permo–Trias Marocian, Universite Cadi Ayyad, p. 27.

Bobyarchick, A. R., 1981, The Eastern Piedmont Fault System and its relationship to Alleghanian tectonism in the southern Appalachians: Journal of Geology, v. 89, p. 335–347.

Bradley, D. C., 1982, Subsidence in late Paleozoic basins in the northern Appalachians: Tectonics, v. 1, p. 107–123.

Bujak, J. P., and Fisher, M. J., 1976, Dinoflagellate cysts from the Upper Triassic of arctic Canada: Micropaleontology, v. 22, p. 44–70.

Bujak, J. P., and Williams, G. L., 1977, Jurassic palynostratigraphy of offshore eastern Canada, in Swain, F. M., ed., Stratigraphic micropaleontology of Atlantic basin and borderlands: Elsevier, p. 321–339.

Busson, G., 1972, Principles, methodes et resultats d'une estude stratigraphique du Mesozoique Saharien: Memoires Museum d'Historie Naturell, NS ser. CC, t. 26, 441 p.

Choukroune, P. J., Francheteau, and Le Pichon, X., 1978, In situ structural observations along Transform Fault X, in the FAMOUS area, Mid-Atlantic Ridge: Geological Society of America Bulletin, v. 89, p. 1013–1029.

Clemmensen, L. B., 1982, Tectonic and paleoclimatic aspects of the Triassic sequence in East Greenland [abs.]: Wurzburg, Germany, Geologsche Vereinigung, 72 Jahres Tangung, Julius Maximilians Universitat, p. 26.

Cloos, E., and Pettijohn, F. J., 1973, Southern border of the Triassic basin, west of York, Pennsylvania; Fault or overlap?: Geological Society of America Bulletin, v. 84, p. 523–536.

Cook, F. A., Brown, L. D., Kaufman, S., Oliver, J. E., and Petersen, T. A., 1981, COCORP seismic profiling of the Appalachians orogen beneath the Coastal Plain of Georgia: Geological Society of America Bulletin, v. 92, p. 738–748.

Cornet, B., 1977, The palynostratigraphy and age of the Newark Supergroup [Ph.D. thesis]: Pennsylvania State University, 505 p.

Cornet, B., and Olsen, P. E., 1986, A summary of the biostratigraphy of the Newark Supergroup with comments on early Mesozoic provinciality: Third Latin American Paleobotanical Congress Symposium on Triassic Global Distribution (in press).

Cornet, B., and Traverse, A., 1975, Palynological contribution to the chronology and stratigraphy of the Hartford Basin in Connecticut and Massachusetts: Geoscience and Man, v. 11, p. 1–33.

Cornet, B., Traverse, A., and McDonald, N. G., 1973, Fossil spores, pollen, and fishes from Connecticut indicate Early Jurassic age for part of the Newark Group: Science, v. 182, p. 1243–1246.

Cousminer, H. L., 1981, Palynostratigraphy, thermal alteration index, and kerogen characteristics of the Ga'ash-2 well sequence: Israel Geological Survey Report P2/81, 20 p.

——, 1983, Late Triassic dinoflagellate cysts date Georges Bank deep marine sediments as Rhaeto-Norian [abs.]: San Francisco, California, Proceedings, American Association of Stratigraphic Palynologists, p. 2.

Cousminer, H. L., Steinkraus, W. E., and Hall, R. E., 1984, Biostratigraphic restudy documents Triassic/Jurassic section in Georges Bank COST G-2 well: American Association of Petroleum Geologists Bulletin, v. 68, p. 466.

Crowell, J. C., 1974, Origin of late Cenozoic basins in California, *in* Dickinson, W. R., ed., Tectonics and sedimentation: Society of Economic Paleontologists and Mineralogists Special Publication 22, p. 190–204.

Daniels, D. L., Zeitz, I., and Popenoe, P., 1983, Distribution of subsurface lower Mesozoic rocks in the southeastern U.S., as interpreted from regional aeromagnetic and gravity maps, *in* Gohn, G. S., ed., Studies related to the Charleston, South Carolina, earthquake of 1886; Tectonics and seismicity: U.S. Geological Survey Professional Paper 1313, p. 1–24.

DeBoer, J., and Snider, F. G., 1979, Magnetic and chemical variations of Mesozoic diabase dikes from eastern North America; Evidence for a hotspot in the Carolinas?: Geological Society of America Bulletin, v. 90, pt. 1, p. 185–198.

Eastman, C. R., 1913, Notes on Triassic fishes belonging to the Families Catopteridae and Semionotidae: Annals of the Carnegie Museum, v. 9, p. 139–148.

Emmons, E. E., 1857, American Geology, Part 6: Albany, Sprague and Company, 152 p.

Faill, R. T., 1973, Tectonic development of the Triassic Newark-Gettysburg Basin in Pennsylvania: Geological Society of America Bulletin, v. 84, p. 725–740.

Falvey, D. A., 1974, The development of continental margins in plate tectonic theory: Australian Petroleum Exploration Association Journal, v. 14, p. 95–106.

Fisher, M. J., 1979, The Triassic succession in the Canadian Arctic Archipelago: American Association of Stratigraphic Palynologists Contributions Series, no. 5B, p. 83–100.

Fisher, M. J., and Dunay, R. E., 1981, Palynology and the Triassic/Jurassic boundary: Review of Paleobotany and Palynology, v. 34, p. 129–135.

Fontaine, W. M., 1883, Contributions to the knowledge of the older Mesozoic flora of Virginia: U.S. Geological Survey Monograph 6, 144 p.

Freund, R., 1965, A model for the structural development of Israel and adjacent areas since Upper Cretaceous time: Geological Magazine, v. 102, p. 188–204.

Froelich, A. J., and Olsen, P. E., 1984, Newark Supergroup, a revision of the Newark Group in eastern North America: U.S. Geological Survey Bulletin 1537-A, p. A55–A58.

Given, M. M., 1977, Mesozoic and early Cenozoic geology of offshore Nova Scotia: Bulletin of Canadian Petroleum Geology, v. 25, p. 63–91.

Glover, L., III, Poland, F. B., Tucker, R. D., and Bourland, W. C., 1980, Diachronous Paleozoic mylonites and structural heredity of Triassic-Jurassic basins in Virginia: Geological Society of America Abstracts with Programs, v. 12, no. 4, p. 178.

Grow, J. A., 1981, The Atlantic margin of the United States, *in* Bally, W. A., eds., Geology of passive continental margins; History, structure, and sedimentologic record: American Association of Petroleum Geologists Education Course Note Series, no. 19, article 3, p. 1–41.

Hentz, T. F., 1985, Early Jurassic sedimentation of a rift valley lake, Culpeper basin, northern Virginia: Geological Society of America Bulletin, v. 96, p. 92–107.

Hubert, J. F., and Mertz, K. A., 1980, Eolian dune field of Late Triassic age, Fundy Basin, Nova Scotia: Geology, v. 8, p. 516–519.

Hubert, J. F., Reed, A. A., Dowdall, W. L., and Gilchrist, J. M., 1978, Guide to the red beds of central Connecticut: Society of Economic Paleontologists and Mineralogists, 1978 Field trip, Eastern Section, 129 p.

Jansa, J. F., and Wade, J. A., 1975, Geology of the continental margin off Nova Scotia and Newfoundland, *in* Van der Linden, W.J.M., and Wade, J. A., eds., Offshore geology of eastern Canada: Geological Survey of Canada Paper 74-30, p. 51–105.

Jansa, J. P., Bujak, J. P., and Williams, G. L., 1980, Upper Triassic salt deposits of the western North Atlantic: Canadian Journal of Earth Sciences, v. 17, p. 547–559.

Keppie, J. D., 1982, The Minas Geofracture, *in* St. Julian, T., and Beland, J., eds., Major structural zones and faults of the northern Appalachians: Geological Society of Canada Special Paper 24, p. 263–280.

Klitgord, K. D., and Schouten, H., 1980, Mesozoic evolution of the Atlantic Caribbean and Gulf of Mexico, *in* Pilger, R. H., Jr., ed., The origin of the Gulf of Mexico and the early opening of the central North Atlantic: Proceedings of a symposium, March 3–5, 1980, Baton Rouge, Louisiana State University, p. 100–101.

Klitgord, K. D., Schlee, J. S., and Hinz, K., 1982, Basement structure sedimentation and tectonic history of the Georges Bank basin, *in* Scholle, P. A., and Wenkam, C. R., eds., Geological studies of the COST Nos G-1 and G-2 wells, United States North Atlantic Outer Continental Shelf: U.S. Geological Survey Circular 861, p. 160–186.

Lancelot, Y., 1980, Birth and evolution of the "Atlantic Tethys" (central North Atlantic), *in* Aubouin, J., Debelmas, J., and Letreille, M., eds., Geologie des chaines alpines issues de la Tethys, Memoir de Bureau de Researches Geologiques du Maroc, no. 115, p. 215–225.

Le Pichon, X., and Fox, P. J., 1971, Marginal offsets, fracture zones, and the early opening of the North Atlantic: Journal of Geophysical Research, v. 76, p. 6293–6308.

Lowell, J. D., Genik, G. J., Nelson, T. H., and Tucker, P. M., 1975, Petroleum and plate tectonics of the southern Red Sea, *in* Fischer, A. G., and Judson, S., eds., Petroleum and global tectonics: Princeton, New Jersey, Princeton University Press, p. 129–153.

Lucas, M., Manspeizer, W., and Hull, J., 1985, En echelon folds; A case history from the Jacksonwald syncline, A foreland fold in the Newark Basin of eastern North America [abs.]: American Association of Petroleum Geologists Bulletin, v. 69, p. 1440.

Lyell, C., 1847, On the structure and probable age of the coal field of the James River near Richmond, Virginia: Quarterly Journal of the Geological Society of London, v. 3, p. 261–280.

Manspeizer, W., 1980, Rift tectonics inferred from volcanic and clastic structure, *in* Manspeizer, W., ed., Field studies of New Jersey geology and guide to field trips: New York State Geological Association, p. 314–350.

——— , 1981, Early Mesozoic basins of the Central Atlantic passive margins, *in* Bally, A. W., ed., Geology of passive continental margins; History, structure, and sedimentologic record: American Association of Petroleum Geologists Course Note Series, no. 19, article 4, p. 1–60.

——— , 1982, Triassic-Liassic basins and climate of the Atlantic passive margins: Geologische Rundschau, v. 73, p. 895–917.

Manspeizer, W., Puffer, J. H., and Cousminer, H. L., 1978, Separation of Morocco and eastern North America, A Triassic-Liassic stratigraphic record: Geological Society of America Bulletin, v. 89, p. 901–920.

Marine, I. W., and Siple, G. E., 1974, Buried Triassic basin in the central Savannah River area, South Carolina and Georgia: Geological Society of America Bulletin, v. 85, p. 311–320.

May, P. O., 1971, Pattern of Triassic-Jurassic diabase dikes around the North Atlantic in the context of pre-drift position of the continents: Geological Society of America Bulletin, v. 82, p. 1285–1292.

McHone, J. G., and Butler, J. R., 1983, Tectonic-magnetic origin of Mesozoic alkalie magmas in eastern North America: Geological Society of America Abstracts with Programs, v. 15, p. 640.

McKenzie, D. P., 1978, Some remarks on the development of sedimentary basins: Earth and Planetary Science Letters, v. 40, p. 25–32.

Montadert, L., de Charpal, O., Roberts, D. G., Guennoc, P., and Sibuet, J. C., 1979, Northwest Atlantic passive margins; Rifting and subsidence processes, *in* Talwani, M., Hay, W. W., and Ryan, W.B.F., eds., Deep drilling results in the Atlantic Ocean: American Geophysical Union, Maurice Ewing Series 3, p. 164–186.

Morgan, W. J., 1980, Hotspot tracks and the opening of the Atlantic and Indian oceans, *in* Emiliani, C., ed., The Oceanic lithosphere; The Sea: New York, John Wiley and Sons, v. 7, p. 443–487.

Olsen, P. E., 1980a, Triassic and Jurassic formations of the Newark Basin, *in* Manspeizer, W., ed., Field studies of New Jersey geology and guide to field trips: New York State Geological Association, p. 1–39.

——, 1980b, Fossil great lakes of the Newark Supergroup in New Jersey, *in* Manspeizer, W., ed., Field studies of New Jersey geology and guide to field trips: New York State Geological Association, p. 352–398.

Olsen, P. E., McCune, A. R., and Thompson, K. S., 1982, Correlation of the early Mesozoic Newark Supergroup (eastern North America) by vertebrates, especially fishes: American Journal of Science, v. 282, p. 1–44.

Phillips, J. D., 1983, Paleomagnetic investigations of the Clubhouse Crossroads basalt, *in* Gohn, G. S., ed., Studies related to the Charleston, South Carolina, earthquakes of 1886; Tectonics and seismicity: U.S. Geological Survey Professional Paper 1313, p. C1–C18.

Poag, C. W., 1982, Foraminiferal and seismic stratigraphy, paleoenvironments, and depositional cycles in the Georges Bank Basin, *in* Scholle, P. A., and Wenkam, C. R., eds., Geological studies of the COST Nos. G-1 and G-2 wells, United States North Atlantic Outer Continental Shelf: U.S. Geological Survey Circular 861, p. 11–33.

Puffer, J. H., Hurtubise, D. O., Geiger, F. J., and Lechler, P., 1981, Chemical composition of the Mesozoic basalts of the Newark Basin, New Jersey, and the Hartford Basin, Connecticut; Stratigraphic implications: Geological Society of America Bulletin, v. 92, p. 155–159.

Quennel, A. M., 1958, The structural and geomorphic evolution of the Dead Sea Rift: Quarterly Journal of the Geological Society of London, v. 114, p. 1–24.

Ratcliffe, N. M., 1980, Brittle faults (Ramapo Fault) and phyllonitic ductile shear zones, New York and New Jersey, and their relationship to current seismicity, *in* Manspeizer, W., ed., Field studies of New Jersey geology and guide to field trips: New York State Geological Association, p. 278–311.

Redfield, W. C., 1856, On the relations of the fossil fishes of the sandstone of Connecticut and the Atlantic States to the Liassic and Oolitic periods: American Journal of Science (ser. 2), v. 22, p. 357–363.

Reeside, J. B., Jr., (chairman, Triassic subcommittee) and 16 others, 1957, Correlation chart of the Triassic formations of North America: Geological Society of America Bulletin, v. 68, p. 1451–1514.

Reinemund, J. A., 1955, Geology of the Deep River coal field, North Carolina, U.S. Geological Survey Professional Paper 246, 159 p.

Riding, J. B., 1984, Dinoflagellate cyst range-top biostratigraphy of the uppermost Triassic to lowermost Cretaceous of northwest Europe: Palynology, v. 8, p. 195–210.

Robbins, E. I., 1981, Fossil Lake Danville; The paleoecology of the later Triassic ecosystem on the North Carolina–Virginia border [Ph.D. thesis]: Pennsylvania State University, 396 p.

Rogers, W. B., 1842, Report of the progress of the Geological Survey of the State of Virginia for the year 1841: Richmond, Virginia, 12 p. (reprinted in Geology of the Virginias, 1884, p. 537–546).

Rona, P. A., 1982, Evaporites at passive margins, *in* Scrutton, R. A., ed., Dynamics of passive margins: American Geophysical Union Geodynamic Series, v. 6, p. 116–132.

Sanders, J. E., 1963, Late Triassic tectonic history of northeastern United States: American Journal of Science, v. 261, p. 501–524.

Schlee, J. S., and Jansa, L. F., 1981, The paleoenvironment and development of the eastern North American continental margin, *in* Geology of continental margins: International Geological Congress, 26th, Paris, July 7–17, 1980, Colloque C3, Oceanologica Acts, no. SP, supplement to v. 4, p. 71–80.

Sheridan, R. E., 1974, Conceptual model for the block-fault origin of the North American Atlantic continental margin geosyncline: Geology, v. 2, p. 465–468.

Stringer, P., and Lajtai, E. Z., 1979, Cleavage in Triassic rocks of southern New Brunswick, Canada: Canadian Journal of Earth Sciences, v. 16, p. 2165–2180.

Sumner, J. R., 1978, Geophysical investigation of the structural framework of the Newark-Gettysburg Triassic Basin, Pennsylvania: Geological Society of America Bulletin, v. 8, p. 935–942.

Swanson, M. T., 1982, Preliminary model for an early transform history in central Atlantic rifting: Geology, v. 10, p. 317–320.

Thayer, P. A., Kirstein, D. S., and Ingram, R. L., 1970, Stratigraphy, sedimentology, and economic geology of Dan River Basin, North Carolina: North Carolina Geological Society Guidebook, Field Trip, October, 44 p.

Tozer, E. T., 1967, A standard for Triassic time: Geological Survey of Canada Bulletin 156, 103 p.

Turcotte, D. L., 1981, Rifts-tensional failures of the lithosphere, *in* Papers presented to the Conference on the Processes of Planetary Rifting, December 3–5, 1981 (AGU-NSF-NASA): Houston, Texas, Lunar and Planetary Science Institute, p. 5–8.

Van Houten, F. B., 1969, Late Triassic Newark Group, North-central New Jersey and adjacent Pennsylvania and New York, *in* Subinsky, S., ed., Geology of selected areas in New Jersey and eastern Pennsylvania: New Brunswick, New Jersey, Rutgers University Press, p. 314–347.

——, 1977, Triassic-Liassic deposits, Morocco and eastern North America; A comparison: American Association of Petroleum Geologists Bulletin, v. 61, p. 79–99.

Visscher, H., and Brugman, W. A., 1981, Ranges of selected palynomorphs in the Alpine Triassic of Europe: Review of Paleobotany and Palynology, v. 34, p. 115–128.

Ward, L. F., 1981, The plant-bearing deposits of the American Triassic: Science, v. 18, p. 287–288.

Watts, A. B., 1981, The U.S. Atlantic continental margin; Subsidence history, crustal structure, and thermal evolution, *in* Bally, A. W., ed., Geology of passive continental margins; History, structure, and sedimentologic record: American Association of Petroleum Geologists Education Course Note Series, no. 19, article 2, p. 1–75.

Wenk, W. J., 1984, Seismic refraction model of fault offset along basalt horizons in the Hartford Rift basins of Connecticut and Massachusetts: Northeast Geology, v. 6, p. 168–173.

Woollam, R., and Riding, J. B., 1983, Dinoflagellate cyst zonation of the English Jurassic: Report of the Institute of Geological Sciences, no. 83/2, 42 p.

Ziegler, P. A., 1982, Faulting and graben formation in western and central Europe: Philosophical Transactions of the Royal Society of London, A 305, p. 113–143.

MANUSCRIPT ACCEPTED BY THE SOCIETY JANUARY 2, 1986

ACKNOWLEDGMENTS

The authors wish to thank the management of the Minerals Management Service for authorizing publication of this paper, and for the work of the individuals who contributed to this report. In particular, thanks are due to the cartography section of the Atlantic/OCS offices for preparation of the maps and figures accompanying this report.

The Geology of North America
Vol. I-2, The Atlantic Continental Margin: U.S.
The Geological Society of America, 1988

Chapter 11

Mesozoic and Cenozoic magmatism

J. Z. de Boer
Department of Earth and Environmental Sciences, Wesleyan University, Middletown, Connecticut 06457
J. G. McHone
Department of Geology, University of Kentucky, Lexington, Kentucky 40506
J. H. Puffer
Geology Department, Rutgers State University, Newark, New Jersey 07102
P. C Ragland and D. Whittington
Department of Geology, Florida State University, Tallahassee, Florida 32306

DISTRIBUTION

Introduction

Mesozoic and early Cenozoic magmatism in eastern North America produced distinct provinces with characteristic igneous and tectonic styles. This discussion is arranged in order of age and province size and is primarily descriptive in nature. Some of the igneous events are not well dated and thorough chemical and petrographic analyses may also be lacking. Our aim is to outline present knowledge of the provinces, gathered from many scattered published and unpublished studies, and to synthesize some of the more important aspects of the rocks and events that produced them. More definitive models will eventually be developed, which we believe can show fundamental relationships between lithospheric tectonism and resulting intraplate-to-rift zone magmatism in eastern North America.

"Newark" Rift Basalts

Exposed flood basalts and sills of early Jurassic age are confined mainly to grabens of the "Newark" rift system. The "Newark" rift system encompasses a series of fault-bound basins extending from Florida to Nova Scotia (Fig. 1). Newark-type basins occur farther north (east Greenland, Surlyk, 1978a, b) and east (Morocco-Portugal, Manspeizer, 1981), indicating that Mesozoic rifting followed both the Caledonian and Hercynian sutures (Ziegler, 1975).

"Newark" rift zones consistently step eastward as one proceeds north and south from Virginia. Their regional distribution thus mimics that of the ocean floor anomalies, which show a dominant sense of right lateral offset to the north and left lateral offset to the south of the Bermuda flowline (Sundvik and others, 1984).

The "Newark" rifts originated as synforms and were almost

half filled with clastic sediments before the tholeiitic igneous activity was initiated. Much of the tilting (and faulting) appears to have occurred during and following emplacement of the igneous masses in the early Jurassic. Sedimentation in the southern basins was initiated in the middle Carnian (225–230 Ma) (Olsen, 1980; Olsen and others, 1982). Only Carnian sediments remain. Subsidence in the central basins may have started at about the same time (Gettysburg and Culpeper Basins) and lasted until Pliensbachian (Hartford basin, Olsen and others, 1982). Subsidence in the northern basins (Martins Head, New Brunswick) can be traced back to the Ladinian (230–235 Ma) (Nadon and Middleton, 1984).

The volcanic units are believed to be principally of Hettangian to Sinemurian age (Olsen and others, 1982). The most reliable K-Ar dates range from 170 to 190 Ma (Sutter, 1985). The number and aggregate thickness of exposed lava flows decrease from Virginia northward. The Culpeper basin (Virginia) contains three flow units with a combined thickness of more than 950 m (Lee, 1977; Leavy and Puffer, 1983). The Fundy basin (Nova Scotia) possesses only one flow unit with an aggregate thickness of about 300 m (Wark and Clarke, 1980). Exposed southern basins do not contain flows. Igneous rocks have, however, been reported from buried basins in South Carolina, Georgia, Alabama, and northern Florida.

Drill holes near Charleston (South Carolina) have penetrated a sequence of what are believed to be "Newark" type lava flows. Gottfried and others (1983) distinguished seven flow units, which may consist of as many as 23 flows (Phillips, 1983). Six of the "flows" possess reversed magnetic remanence (Phillips, 1983). This suggests that their emplacement was not contemporary with that of the "Newark" volcanics, which are all normally magnetized. Using mean inclination, Phillips estimated the flows to be of middle Jurassic age. Ar-Ar radiometric ages vary from 167 to 192 Ma An age of 184 Ma is considered most representative for these rocks (Lamphere, 1983).

de Boer, J. Z., McHone, J. G., Puffer, J. H., Ragland, P. C., and Whittington, D., 1988, Mesozoic and cenozoic magmatism; *in* Sheridan, R. E., and Grow, J. A., eds., The Geology of North America, Volume I-2, The Atlantic Continental Margin, U.S.: Geological Society of America.

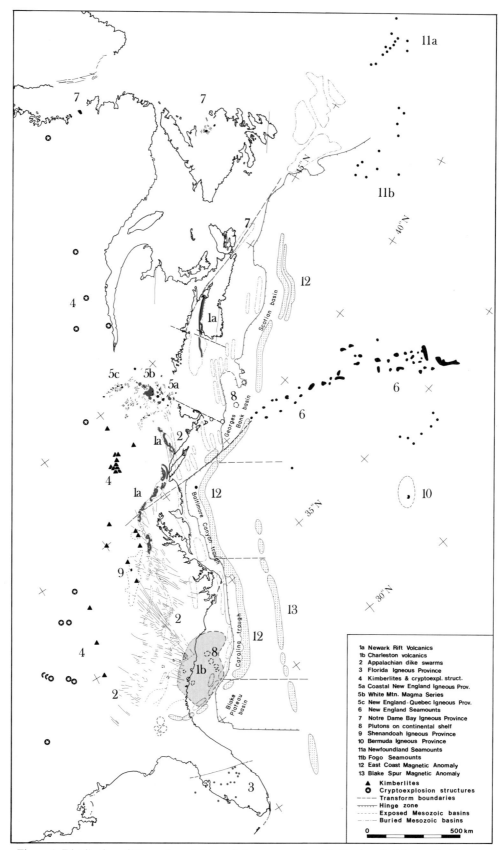

Figure 1. Distribution of Mesozoic and Cenozoic igneous provinces in the Appalachians and adjacent regions. Triassic-Jurassic = red, Cretaceous-Eocene = black.

Seismic studies offshore South Carolina indicate the presence of a strong reflector and high velocity refractor, which are believed to represent the possible eastward extent of these volcanic units below shelf sediments (Dillon and others, 1983) (Fig. 1). The reflector occurs just above the post-rift unconformity. If this is correct the flow units may be chemically of the "Newark" type, but must be younger (probably middle Jurassic). The offshore volcanic rocks cover an area of more than 60,000 km^2, and do not appear solely restricted to rift zones.

Large intrusive bodies (sills) are common in the "Newark" basins of the central Appalachians. Their number, diversity, and aggregate thickness is highest in the Culpeper-Gettysburg-Newark basin sequence. Thicknesses in these basins range from a few hundred to over one thousand meters. The southern intrusives possess relatively thin chilled borders and contact aureoles, while the central Appalachian intrusives are encased by moderately thick borders and aureoles (up to 30% of estimated sheet thickness) (Froelich and Gottfried, 1985). These variations suggest major differences in erosional level along the Appalachian axial zone.

Eastern North American (ENA) Dike Swarms

Triassic(?)-Jurassic dike swarms occur throughout the Appalachians from Alabama to Newfoundland (Fig. 1). Characteristic aeromagnetic anomalies provide evidence for their occurrence below Coastal Plain and shelf sediments. The Brevard-Blue Ridge anticlinorium of the southern Appalachians forms the western border. (Occasional dikes cross this feature, but dike density is significantly reduced.) In the northern Appalachians dikes occur predominantly east of the Connecticut Valley-Gaspé synclinorium. King (1961, 1971) recognized a regional pattern in the orientation of ENA dikes. The subvertical intrusives trend generally NW in the southern, N-S in the central, and NE in the northern Appalachians. Significant variations and overlaps occur, especially in the southern and central Appalachians (de Boer, 1967; Ragland and others, 1983).

Regional variation in dike trend has been attributed to emplacement along gash fractures (mostly reactivated faults) during a period of regional sinistral shear in the late Jurassic (de Boer, 1967). Swanson (1982) expanded this concept and attributed shear to a counterclockwise rotation of North America (relative to Africa) during the middle Jurassic. May (1971) correlated the ENA dikes with similar intrusives in west Africa and northern South America and related their pattern to a pan-Atlantic stress system centered in the present south-central Atlantic. De Boer and Snider (1979) attributed this stress center to a triple (RRR) point, which was located below southern Georgia during the late Triassic–early Jurassic and below the Blake Plateau during the late Jurassic.

The southern Appalachians contain an older NW and younger N-S set of dikes. Both occur at high angle to the Appalachian grain. The NW dikes are on average shorter and more closely spaced than the N-S dikes. The NW dikes are usually clustered (Ragland and others, 1983) and spacing of the clusters is similar to that for offshore transform faults (Schouten and Klitgord, 1982).

NE dikes in the northern Appalachians are predominantly single (often composite) intrusions many times wider and usually longer than the NW oriented southern dikes. They either trend parallel to the Appalachian grain or intersect it at low angle. Philpotts (1985) has chemically correlated the three lava flow units of the Hartford basin with three regional dikes exposed in the metamorphic highlands and rift zone. If these dikes were indeed the principal feeders for the flows, fissures progressively opened from E to W, oblique to the principal basin trend. With the exception of southern New England, dike density and total volume of basalt emplaced at present exposure levels decrease northeastward. This decrease follows that for flows/sills in the volcanic "Newark" grabens and suggests that more crustal extension occurred in the southern Appalachians.

Emplacement of the majority of dikes has been attributed to a single phase of magmatism in the Jurassic (de Boer, 1967; Smith and Noltimier, 1979; Swanson, 1982). More detailed paleomagnetic and radiometric studies suggest, however, that dike emplacement occurred in several episodes (de Boer and Snider, 1979; Dooley and Smith, 1982). Field studies of a swarm of Jurassic olivine-normative diabase dikes in Pennsylvania showed the presence of a fracture cleavage, which was not observed in associated quartz-normative dikes (Smith and others, 1975). This suggests that a compressive tectonic phase followed emplacement of the older olivine-normative set, and was contemporaneous with, or preceded emplacement of the quartz-normative dikes.

Southern New England possesses the highest frequency of Mesozoic dikes. This complexity is probably due to the fact that igneous provinces of various ages intersect in this area. Jurassic dike emplacement in southern New England appears to have been controlled by N to NE trending structures, while Cretaceous dikes preferred W to NW trending zones of weakness (McHone, 1978). Differences in dike orientation can be attributed to changes in the regional stress configurations, which respectively preceded and followed initiation of seafloor spreading (see section D2).

The Florida Province

Florida probably contains one of the most interesting igneous provinces of the East American margin. Unfortunately it is least known as it remains deeply buried. The Cambrian granite complex below central Florida is intruded and overlain by Mesozoic volcanic sequences (Fig. 1) reportedly composed of rhyolites (including ignimbrites), alkali, and tholeiitic basalts (Applin, 1951; Bass, 1969; Barnett, 1975; Mueller and Porch, 1983) (Fig. 1). The tholeiites are believed to overlie the rhyolite-alkali basalt sequence. Radiometric ages range from a questionable 90 to 180 Ma (Milton and Grasty, 1964; Milton, 1972). A rhyolite penetrated at 6 km depth yielded an age of 180 Ma. Drill holes in the Bahama platform have penetrated redbeds containing rhyolite pebbles (Sheridan and others, Chapter 15, this volume) indicating that this province may be quite extensive.

Geologic setting and geochemical variety of the supposed Jurassic rocks resemble that of the Neogene volcanics in the Afar region of Djibouti (Boucarut and others, 1984). The Florida basement can thus be compared with a crustal sliver like the Tadjoura-Ali Sabih horst located at a RRR triple point, which separates three areas of crustal extension in the Afar triangle. Mueller and Porch (1983) concluded that the bimodal igneous suite of southern Florida is related to emplacement of a mantle plume.

Kimberlite Belt

Intrusive occurrences of the melilitite-alnoite-kimberlite-carbonatite association and cryptoexplosion structures occur west of the Appalachians (Sears and Filbert, 1973; Kopecky, 1976; Meyer, 1976; and Jackson and others, 1982) in a broad NE trending zone from Tennessee to Labrador (Fig. 1). From Tennessee to Quebec the intrusives appear aligned along two subparallel, slightly arcuate NE lineaments on either side of the Alabama-New York magnetic lineament, which reflect a major crustal break in the basement beneath the Appalachian basin (King and Zietz, 1978). The magnetic lineament connects the Georgian and Adirondack magnetic (*MAGSAT*) lows (Mayhew and Galliher, 1982), which may represent the locations of early Mesozoic hotspots (de Boer, 1983). Emplacement of the kimberlite suite presumably occurred at intersections of several (Paleozoic) growth faults of the Appalachian basin with major Appalachian cross faults (Parrish and Lavin, 1982).

Radiometric (Zartman and others, 1970; De Journett and Smidt, 1975; Basu and Rubury, 1980) and paleomagnetic (Larochelle, 1961, 1968, 1969; Larochelle and Currie, 1967; Currie and Larochelle, 1969; Robertson, 1967) data suggest Mesozoic ages for the ultramafic intrusives. Crough and others (1980) reviewed ages for kimberlites in the circum-Atlantic region and found maxima at 140 ± 10 and 90 ± 10 Ma.

The westernmost and oldest intrusives of the Notre Dame Bay, New England-Quebec, and Shenandoah igneous provinces originated in this belt; implying a genetic link between the kimberlites and these alkalic provinces (Fig. 1). It appears logical to assume that the crustal flexure between the Appalachian basin and fold belt may have been the site of early Mesozoic rifting. Where this flexure (or its continuation into the Canadian craton) is intersected by major transforms (Newfoundland F.Z.; Monteregian, F.Z.; and Cornwall-Kelvin F.Z.) mantle diapirs may have developed and alkalic volcanism resulted (Notre Dame; New England-Quebec; and Shenandoah provinces).

New England Igneous Provinces

Mesozoic igneous rocks in New England and adjacent areas can be divided into three provinces based upon age, distribution, physical aspects, and petrology (McHone and Butler, 1984).

Coastal New England Province. The coastal New England province is characterized by Triassic intrusive rocks of alkaline affinity that occur in a relatively narrow (60 km) belt (Fig. 1) along the coast of Massachusetts, New Hampshire, and southern Maine (McHone and Butler, 1984). The province is represented chiefly by the Litchfield, Agamenticus, and Abbott plutons, and by a series of NE trending olivine diabase dikes in coastal New Hampshire. Radiometric ages for the intrusives range from middle to late Triassic (240 to 210 Ma) (Burke and others, 1969; Foland and Faul, 1977; Bellini and others (1982). If the Triassic ages are substantiated, this province provides evidence for the earliest post-orogenic magmatism in the Appalachians.

White Mountain Magma Series (WMMS). The White Mountain magma series (WMMS) has often been used as the group named for all Mesozoic intrusives in southern New England. When only plutons of early Jurassic age, contiguous New Hampshire occurrence and characteristic petrology are included, the classic WMMS is more cohesive as a distinct igneous province (McHone and Butler, 1984).

The plutonic belt trends NNW and consists of more than 30 igneous complexes. Because the principal mode of emplacement involved doming and cauldron subsidence, overlapping caldera complexes are common (Creasy and Eby, 1983) and the controlling structure is poorly understood. Chapman (1968) suggested that the arrangement of intrusive centers was related to an intersecting network of NNW and E-W trending fractures. Geometrically, the White Mountains occupy a NNW trending, sinistral transform fault zone (with E-W Riedel gashes) between the Hartford-Deerfield and Fundy basins (Paijitprapapon, 1980).

The White Mountain magma series is alkaline and provides a classic example of intraplate, non-orogenic magmatism. Creasy and Eby (1983) distinguished four petrologic associations: gabbro-diorite-monzonite, syenite-nepheline syenite, alkali syenite-quartz syenite-granite, and subaluminous biotite granite.

The majority of the igneous complexes yield Jurassic radiometric ages varying from 200 to 155 Ma. A younger set of intrusives was emplaced in the middle Cretaceous (125–100 Ma) (Foland and Faul, 1977; Doherty and Lyons, 1980). The latter phase was less important volumetrically and is chemically related to volcanism in the Monteregian Hills and New England seamounts (section A 5C).

A regional (negative) Bouguer anomaly occurs in northern New Hampshire and northeast Vermont. It has been interpreted as a large area of partial melting at the base of the crust. Local gravity and magnetic anomalies over individual intrusives suggest that some plutons possess mafic, others, felsic roots. Two distinct groups of plutons can be recognized geophysically, a Jurassic set of large plutons consisting primarily of granites and a Cretaceous group of smaller plutons composed predominantly of mafic rocks (Sharp and Simmons, 1983). Fission track analyses suggest that the Jurassic plutons were emplaced at depths ranging from 5 to 8 km, while the Cretaceous plutons cooled at depths between 3 and 4 km (Doherty and Lyons, 1980).

The large WMMS plutons appear to be derived from widespread anatectic melting of the lower crust, while the smaller mafic plutons could have originated by emplacement of magmas derived from upper mantle melts of alkali basalt affinity (Loiselle and Hart, 1978; Creasy and Eby, 1983). Intermediate Sr isotope ratios, high La/Yb ratios, and negative Eu anomalies support these hypotheses (Creasy and Eby, 1983). REE data indicate that plagioclase fractionation played an important role (Carr, 1980; Loiselle, 1978).

Paleomagnetic analyses of the White Mountain rocks have provided numerous data with little consistency. Even after magnetic and thermal cleaning, directional scatter remains large and results in poles that do not plot on the American polar paths (Opdyke and Wensink, 1966; Bouley, 1971). Paijitprapapon (1980) proposed that directional scatter in several complexes can be attributed to significant post-emplacement rotation.

New England-Quebec Igneous Province. The New England-Quebec province is the most extensive of the magmatic areas in New England. It includes the Monteregian Hills (Gold, 1967; Philpotts, 1974), early Cretaceous dikes and plutons in western Vermont (Laurent and Pierson, 1973; McHone and Corneille, 1980), and similar intrusives in eastern Vermont and southern New England (Foland and Faul, 1977; McHone, 1978, 1981) that were traditionally grouped with the WMMS (McHone and Butler, 1984).

Most representative for this province are the Monteregian Hills. The Hills consist of a chain of mafic alkalic plutons emplaced in Grenville gneisses and Paleozoic sedimentary rocks. Their geologic setting has been discussed by Gold and Marchand (1969), Philpotts (1974), and Eby (1983). Tectonically the Hills mimic the New England seamount chain. Intrusives of the western segment are aligned E-W; those of the eastern segment NW-SE. Location of individual plutons was apparently controlled by the intersection of E-W and NW fractures with NE-trending normal faults (Philpotts, 1974). Explosive activity (breccia pipes, diatremes, alnoites, kimberlites) was confined to the western segment. More passive emplacement (ring dikes and cone-sheet sills) characterizes plutons in the eastern segment. Silica content of the intrusives increases (consistently) eastward (Gold, 1967; Philpotts, 1974). Three groups of rocks have been recognized by Eby (1983) on the basis of silica saturation: moderately to strongly undersaturated, slightly undersaturated to critically saturated, and oversaturated. The latter are presumed to have had a crustal origin and represent melts generated in response to intrusions of the mafic magma (basanite and alkali picrite), which gave rise to the former two groups (Eby, 1983). Recent petrologic models favor formation of magmas in response to upward transport of metasomatic fluids, possibly coupled with a rise in mantle isotherms (Eby, 1983, 1984a-c).

Early geochronological studies (Lowden, 1961; Fairbairn and others, 1963; Shafiquallah and others, 1970) yielded ages ranging from 95 to 126 Ma. Recent Rb/Sr and fission track data suggest two distinct periods of magmatism: 117–120 and 128–141 Ma (Eby, 1983).

Eby (1983) believes that the two age groups suggest an upward progression of melting beginning with the formation of olivine-rich, alkali picrite magmas in a garnet lherzolite mantle and ending with formation of basanite magma in a spinel lherzolite mantle.

Paleomagnetic analyses by Larochelle (1961, 1968, 1969) and Foster and Symons (1979) are consistent with an age of about 120 ± 4 Ma. The easternmost group possesses normal magnetic polarity, the central and western group (except Mt. Johnston) a reverse remanence. Foster and Symons (1973) proposed that the Hills were emplaced progressively from west to east during four normal (relatively short) and reversed polarity intervals.

New England Seamounts

The New England seamount chain consists of a linear series of about 30 major peaks extending over more than 1300 km from Bear seamount on the continental slope to the Corner seamount complex in the Mid Atlantic Ridge (M.A.R.) foothills (Fig. 1). Its western segment (single chain) trends WNW; the eastern segment (double chain) NNW. Mesozoic seafloor anomalies show differences both in spreading rate and direction on either side of the chain (Barrett and Keen, 1976). Anomalies south of the chain trend NNE, virtually at right angles to the western segment, anomalies north of it NE, orthogonal to the eastern set of seamounts.

Radiometric ages suggest that the seamounts of the western segment (Bear-Gosnold) became progressively younger (125 to 90 m.y.) eastward, while the eastern seamounts were emplaced more or less simultaneously about 90 Ma. (Duncan and Houghton, 1977; Houghton, 1978; Houghton and others, 1977). K-Ar ages, however, are scattered and may reflect low temperature (seawater) alteration (Duncan, 1982). Incremental heating and fusion Ar-Ar ages document an eastward decrease in seamount construction age from 103 Ma (Bear) to 82 Ma (Nashville) (Duncan, 1982).

Magnetic anomalies associated with the seamounts suggest, however, that their principal volcanic mass may have been emplaced in the Jurassic (Mayhew, personal communication). The difficulty of assigning ages is characterized by the Bear seamount. Its magnetic anomaly suggests an early Jurassic age; igneous rocks collected from its scarp provided a Cretaceous age, and the presence of a Miocene reef complex on its summit indicates a rather localized uplift, which may have resulted from Neogene magmatic activity.

In view of these apparent inconsistencies it appears logical to assume that igneous activity in the New England transform zone was episodic. As appears to be the case in other transform zones, magmatic foci, however, appear to have moved east over time.

The New England seamounts consist predominantly of alkali basalt (Sullivan and Keen, 1978; Houghton and others, 1979). Their alkalinity and similarity to oceanic island alkalic volcanics is also indicated by their steep REE patterns and relatively high REE contents (Taras and Hart, 1983).

Dextral offset of anomaly sets M0-M4 and M16-M28 suggests sinistral shear along a major transform fault (Barrett and Keen, 1976; Tucholke and Ludwig, 1982) that became the site of the seamounts. The East Coast Magnetic Anomaly (E.C.M.A.) shows a dextral offset of about 30 to 45 km (Vogt and Einwich, 1979) where intersected by the chain. The offset between the Baltimore and Nantucket troughs suggests that the fault zone bends southwest following 40° latitude to the area of apparent dextral offset between the Gettysburg and Newark troughs. From here it may extend farther west, via a series of E-W faults (Transylvania fracture zone; Root and Hoskins, 1977). Depth to basement north of the chain is from 0.5 to 1.0 km lower than south of it (Vogt, 1973; 1974), indicating that vertical movements accompanied strike slip.

Notre Dame Bay and Orpheus Igneous Provinces

Two gabbroic stocks and more than a hundred lamprophyre dikes occur along the southern shore of Notre Dame Bay in Newfoundland (Fig. 1). The stocks are composed mainly of pyroxenites and alkalic gabbro with some cumulus features (Strong and Harris, 1974). Associated dikes are mainly nepheline-normative monchiquites.

K/Ar ages range from 115 to 144 Ma for the dikes (Wanless and others, 1965, 1967) and from 135 to 155 Ma for the Budgell Harbor stock (Helwig and others, 1974). Paleomagnetic analysis shows that the dikes, which are characterized by relatively strong magnetic anomalies, possess a remanence with normal polarity. Virtual geomagnetic poles suggest a late Jurassic (Deutsch and Rao, 1977) or Cretaceous (LaPointe, 1979) age.

The Notre Dame Bay province probably extends considerably northwest. Mesozoic kimberlites on the east coast of Labrador (King and McMillan, 1975) and west coast of Greenland (Andrews and Emeleus, 1971) formed a single province before opening of the Labrador sea. These complexes may have been connected to the Notre Dame Bay intrusives by swarms of WNW and E-W trending (late Jurassic) gabbroic dikes comparable to those on the southwest coast of Greenland. Here the dikes vary in composition from slightly quartz to slightly olivine and nepheline normative (Watts, 1969). A group of seamounts located SE of the Notre Dame Bay province provides a continuity similar to that of the New England-Quebec province and New England seamounts. The Newfoundland seamounts also trend WNW to E-W. Highly weathered alkali basalts and fresh sodic trachyte were recovered from two seamounts in this group. The trachytes provided Ar/Ar ages of 98 Ma (Sullivan, 1975, 1977).

Alkali-basalt sills or flows and volcano clastic units including basalt, trachybasalt, and quartz-trachyte lithologies were penetrated in four wells in the Orpheus basin (E of Nova Scotia) (Jansa and Pe-Piper, 1985). Radiometric ages plot in the range 102 to 125 Ma and are in good agreement with the ages of interbedded Aptian sediments. The Orpheus graben is believed to connect the Fundy-Cobequid fracture system with the Newfoundland-Gibraltar fracture zone, which intersects the Fogo

seamount chain (Hall and others, 1977). So far no dateable rocks have been collected from the latter. It appears reasonable to correlate the Fogo group with the Newfoundland and eastern New England seamounts, which are of Turonian-Cenomanian age. If this is correct, magmatic foci migrated east from early Jurassic in the Fundy, to early Cretaceous in the Orpheus, and possibly late Cretaceous in the Fogo seamount region.

Mafic Plutons of the Continental Shelf and Southern Georgia

A significant number of apparently mafic "plutons" occur beneath the continental shelf (Fig. 1). They are characterized by positive gravity and magnetic anomalies. "Plutons" in the New England shelf are arranged along an arcuate lineament, which follows the hinge zone. Seismic evidence suggests that at least one of these plutons is of middle Cretaceous age (Klitgord, personal communication).

A relatively large pluton (Great Stone Dome) occurs in the Baltimore trough. This lamprophyric body has been identified in well cuttings, and apparently intruded during the middle Early Cretaceous (Barremian?). After erosion of the dome and deposition of Aptian-Albian(?) clastics, rejuvenation occurred during the late Cretaceous and early Tertiary (Crutcher, 1983).

A third area containing mafic "plutons" is located below the shelf off South Carolina and Georgia. One group trends N, the other NW. The N-S plutons (E of Charleston) appear on line with a major N-S swarm of Jurassic tholeiite dikes in the Carolinas (Ragland and others, 1983). They may be contemporaneous or could reflect rejuvenation of crustal zones of weakness during the Cretaceous.

A cluster of three prominent positive gravity and magnetic anomalies occur in southern Georgia. High seismic velocity, density, and magnetic susceptibility suggest the presence of a gabbroic body with diameter of about 30 km. No topographic relief exists at its contact with the overlying Coastal Plain sediments. Seismic refraction data suggest the presence of sills or flows extending outward from the pluton at this contact. Bridges (1973) suggested that the structure may have been emplaced before deposition of the Cretaceous overburden.

Shenandoah Igneous Province

A relatively large number of rather diverse intrusive units occur in a fairly small area northeast of Monterey (Virginia) (Johnson and others, 1971) (Fig. 1). They include peridotites and pyroxenites (Gardner, 1956), kimberlites, teschenites, camptonites, syenites, and "granitic" felsites (Watson and Cline, 1913) and a bimodal alkali basalt-rhyolite suite (Johnson and others, 1971; Gray and Diecchio, 1985). Ages for the intrusives differ. The province may contain an older teschenite-nepheline syenite sequence, which ranges in age from about 114 to 149 Ma (Zartman and others, 1967) and a younger basalt-rhyolite sequence.

Most intrusives in the latter group appear to be of Middle to

Late Eocene age (42 to 47 Ma, Fullagar and Bottino, 1969; Wampler and Dooley, 1975; Ressetar and Martin, 1980). Virtually all Eocene intrusives exhibit a reverse magnetic remanence (Lovlie and Opdyke, 1974). Short (less than 1 m.y.) reversed intervals characterized the period from 38 to 48 Ma, suggesting that emplacement occurred in a relatively short period.

The Eocene intrusives crop out in the northern segment of a NE trending (negative) Bouguer anomaly of regional extent. This anomaly closely coincides with a geothermal high associated with the zone of thermal springs in Bath County (Dennison and Johnson, 1971).

The specific location of the Shenandoah province may be due to the intersection of this high by the 38th parallel fracture zone, a major cross fault of the Appalachians (Dennison and Johnson, 1971; Dennison and others, 1983).

Gravity and geothermal anomalies are believed to be related to the presence of a slowly cooling granitoid body (Dennison and Johnson, 1971). An upwelling of mantle material, which caused regional anatexis of the lower crust may be responsible for the development of this body and for emplacement of the alkalic intrusives.

Bermuda Igneous Province

The Bermuda rise is a broad high that occupies a major part of the Atlantic seafloor between the New England seamount chain and the Blake-Caicos ridges. A hole drilled on Bermuda provided a section composed of almost equal parts of altered tholeiitic lavas and Oligocene (± 33 Ma) limburgite sills (Reynolds and Aumento, 1974; Aumento and others, 1975, 1976). Recovery of late Eocene and middle Oligocene, mica-bearing volcanogenic turbidites (DSDP 386) indicates that intermittent volcanic activity may have occurred at this site. The relatively large volume of young intrusives suggest that the rise may have formed by massive emplacement of alkalic magmas at shallow crustal levels some 80 to 120 m.y. after formation of the seafloor (Aumento and others, 1976).

AGE RELATIONS

Stratigraphic Ages

Mesozoic and Cenozoic volcanic units that provide stratigraphic markers are rare. The lava flows (and a few pyroclastic units) occurring in "Newark" basins are an exception. The Fundy basin contains one, and the Hartford and Newark-Culpeper basins at least three flow units. Using terrestrial vertebrates, palynomorph assemblages and fish remains, Cornet (1977), Cousminer and Manspeizer (1976, 1977), and Olson and others (1982) were able to correlate stratigraphic levels in various rift zones. Comparison of their biostratigraphy with that in Europe led Olson and others (1982) to the conclusion that volcanism was initiated in the earliest Hettangian and continued episodically until middle or late Sinemurian time.

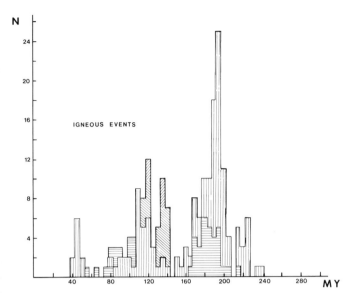

Figure 2. Radiometric ages of Mesozoic-Cenozoic igneous rocks in the Appalachians and adjacent regions. K/Ar ages = vert. lines; Ar/Ar ages = hor. lines; Rb/Sr = diag. lines left; Fission track = diag. lines right. References (3b) to the data used in this compilation are available from the authors.

Radiometric Ages

Wide distribution, relative abundance, and easy accessibility of "young" igneous rocks in the Appalachians have resulted in a large number of radiometric dates. The data base shows Jurassic, Cretaceous, and Eocene maxima, but is heavily weighted towards Jurassic ages (Fig. 2).

Significant age variations exist for each period. K/Ar ages obtained for Jurassic intrusives range from 135 to over 1000 m.y. (Sutter, 1985). Especially disturbing is the variation in apparent ages for single igneous units. Whole rock and mineral K/Ar ages for the Palisades sill (N.J.) vary by 60 m.y. (Erikson and Kulp 1961); Ar/Ar ages for the Hampden (Conn.) flow vary (Conn.) by more than 80 m.y. (Seidemann and others, 1984).

Non-uniform age distribution has been attributed to inhomogeneous distribution of K and Ar (Armstrong and Besancon, 1970); variable loss of ^{40}Ar after crystallization and cooling (Armstrong and Besancon, 1970); incorporation of excess ^{40}Ar during intrusion (Dalrymple and others, 1975; Sutter and Smith, 1979; Seidemann and others, 1984), and a different initial ^{40}Ar/^{36}Ar ratio (Hodych and Hayatsu, 1980). Recent developments offer promise of improved resolution. Post emplacement addition or loss of Ar is addressed by incremental gas spectrum techniques; deviant values for initial Ar isotopic ratios can be recognized by K-Ar isochron techniques (Sutter and Smith, 1977).

Despite potential inaccuracies in reported ages several attempts have been made to distinguish separate episodes within magmatic periods. Foland and others (1971), Foland and Faul (1977), and McHone (1978) distinguish three Mesozoic age groups (240–210; 200–165; and 135–95 Ma) in the data base for

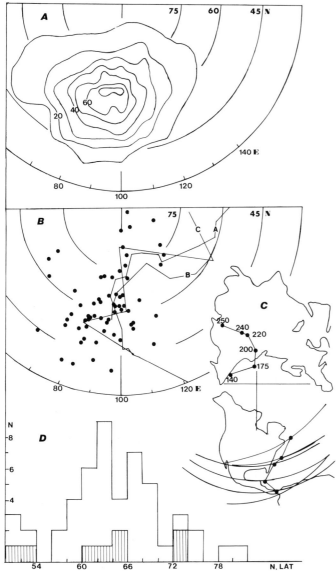

Figure 3. A. Concentration of all presently known paleomagnetic site poles for Triassic-Jurassic igneous units in the Appalachians. B. Selected geomagnetic poles (V.G.P.) for Mesozoic and Cenozoic igneous units in the Appalachians. A, B, C, represent polar wandering paths from Irving 1977, Harrison and Lindh 1982, Van Alstine and de Boer 1978, respectively. C. Southward migration of magnetic equator in the 250 to 140 m.y. period. D. Histogram for mean V.G.P. latitudes for Jurassic flows and sills in "Newark" rifts. Shaded area's represent mean latitudes for respectively Talcott, Holyoke-Watchung, and Hampden events. References (56) to the data used in this compilation are available from the authors.

New England plutons and associated intrusives. The earlier age group is not apparent on the histogram (Fig. 2) because only a few dates are available. Different igneous pulses may be represented within each age group. Two episodes (195 ± 5 and 175 ± 5 Ma) of Jurassic magmatic activity have been recognized in the northern (Hodych and Hayatsu, 1980), central (Sutter and Smith,

1979), and southern Appalachians (Deininger and others, 1975; Dooley and Smith, 1982).

Although significantly fewer dates are available for the Cretaceous, evidence exists for at least two magmatic pulses (135 ± 5 and 110 ± 5 Ma) in the Monteregian Hills (Eby, 1983). Radiometric data for the Shenandoah province suggest a single episode of magmatism in the Eocene.

When the radiometric ages are combined with relative age information obtained from field data (cross cutting relations), paleomagnetism, and geochemistry, at least six magmatic periods can be distinguished. They are late Triassic (Ladinian-Carnian-Norian, 220 ± 15 Ma); early Jurassic (Pliensbachian-Toarcian, 190 ± 10 Ma); middle Jurassic (Bathonian, 175 ± 5 Ma); early Cretaceous (Valanginian, 135 ± 5 Ma); middle Cretaceous (Albian-Cenomanian, 100 ± 10 Ma); and middle Eocene (Lutetian, 45 ± 5 Ma).

Paleomagnetic Ages

Apparent polar wander (APW) paths for North America provide evidence for significant crustal motion in the Mesozoic and Cenozoic. In the late Triassic and early Jurassic the pole shifted more than 15° north (55–71° N. Lat.), and in the late Jurassic more than 35° east (94–130°E) (VanderVoo and French, 1974; Irving, 1977; Van Alstine and deBoer, 1978; Harrison and Lindh, 1982). Because of rapid polar shift and low secular variation rate (Rigotti and Schmidt, 1976) site directions can be used to differentiate and correlate Mesozoic rock units (de Boer, 1968; de Boer and Snider, 1979).

The number of Jurassic Virtual Geomagnetic Poles (V.G.Ps) for Appalachian rocks exceeds that for all other rock units. When site poles are plotted, a large concentration results and differentiation appears impossible (Fig. 3). This scatter may be due to incomplete "cleaning" of samples in the early studies, when alternating current demagnetization was the prevalent method. Thermal demagneetization has shown that a certain percentage of Jurassic rocks carry several remanent components (de Boer and Snider, 1979; Dooley and Smith, 1982; Smith and Hare, 1983) that are difficult to separate. Hydrothermal activity during cooling and/or low grade burial metamorphism may be responsible.

Supposedly "reliable" unit poles provide an Appalachian APW curve similar to the one provided by Irving and other compilers, but offset westward (Fig. 3). The data base used by these compilers for construction of the Mesozoic APW segment for North America is strongly biased by incorporation of a large number of poles obtained from sediments in western North America. Improved resolutions may be obtained using the Appalachians APW curve calibrated by de Boer (1983a).

Virtually all Jurassic dikes, sills, and flows possess a normal remanence. A few dikes with reversed magnetization occur in South Carolina (Thompson, 1968; Bell and others, 1980). Jurassic redbed sequences show mixed polarities (McIntosh, 1976; McIntosh and Hargraves, 1976; McIntosh and others, 1985). The predominant redbed remanence, however, is a CRM (Amerigian,

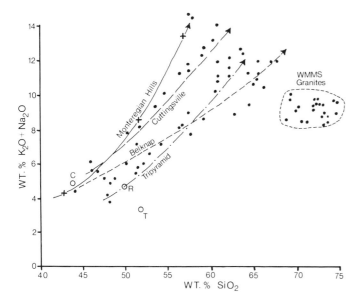

Figure 4. Alkalic-silica fractional diagram, showing relative trends for plutonic rock suites of Monteregian Hills and White Mountains. C = average NEQ camptonite; R = average Rhode Island alkalic dolerite (Pierce, 1976); T = average ENA high-titanium quartz dolerite (Weigand and Ragland, 1970).

1977; Brown, 1979) and the time of its incorporation is unknown. According to Pechersky and Kramov (1973) and Van Hinte (1976), the Mesozoic magnetic field was normally polarized more than 80% of the time. The predominance of normally magnetized igneous units is difficult to understand, however, in view of the more recent data of Steiner (1980) and Kent and Tsai (1978), which suggest that the Jurassic field was predominantly reversed. Channell and others (1980), found a long, normally polarized zone in sediments of Bajocian to Kimmeridgian age (183–152 Ma). If this polarity zone originated earlier (Pliensbachian), it could represent the period in which most Jurassic igneous rocks were emplaced. The magnetic polarity of the Cretaceous and Eocene intrusives is mixed.

GEOCHEMISTRY

Plutonic and Dike Rocks of New England

The geochemical data base for New England is heavily weighted toward Cretaceous rocks. Several hundred major-element, whole-rock chemical analyses are available for Cretaceous intrusions of the New England-Quebec province (NEQ); a few dozen for the Jurassic White Mountain magma series (WMMS), and only very few of the Triassic Coastal New England (CNE) province.

Although a few similar rock types do occur in both the early Jurassic WMMS and early Cretaceous NEQ plutonic units of New England, average rock compositions show clear differences. The large WMMS plutons chiefly consist of granites and syenites with higher $SiO_2/Na_2O + K_2O$ values than the silicic members

of the smaller NEQ plutons (Fig. 4). In general, Cretaceous NEQ plutons are more diverse, comprising mafic to felsic rock series that can be related to differentiation models involving crystal fractionation and/or liquid immiscibility (Philpotts, 1976) (Figs. 5 and 6). Few Jurassic complexes (Belknap and Red Hill) show such fractionation series.

Alkali-SiO_2 diagrams show linear trends for the Cretaceous plutons (mostly NEQ, but including the Belknap complex), whereas WMMS granitoid compositions (mostly Conway granite) form a cluster (Fig. 4). Surface areas of the Monteregian Hills (the Quebec plutons of the NEQ province) and Jurassic White Mountain lithologies may be representative of their actual volumetric proportions. Planimetric measurements of outcrop areas show that granites predominate in the early Jurassic White Mountain batholiths; gabbros and syenites predominate in the Monteregian Hills and form a bimodal assemblage (Fig. 7). Gravimetric anomalies suggest that the volumetric ratio of felsic to mafic rocks must be significantly greater in the White Mountains than in the Monteregian plutons (Sharp and Simmons, 1978; Philpotts, 1974).

The large early Jurassic granitoid-syenitoid plutons of the WMMS province most likely represent products of mass melting in the lower crust of New Hampshire over a period of 40 m.y. (Creasy, 1983). Contemporaneous intrusion of relatively small volumes of mantle-derived basaltic magmas may have produced portions of the associated Red Hill and Belknap complexes. The much smaller early Cretaceous gabbroic-syenitoid plutons of the Monteregian Hills and camptonite dikes of the New England-Quebec province probably represent mantle derived magmas (Eby, 1983). The two distinct magmatic pulses during the early Jurassic and early Cretaceous occurred before and after ocean

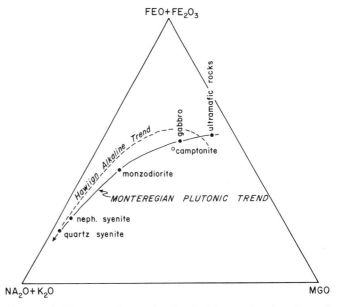

Figure 5. AFM ternary diagram showing the Monteregian plutonic trend and the average Monteregian camptonite, compared with the Hawaiian alkaline trend.

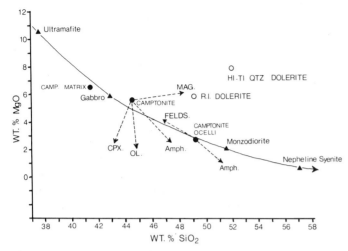

Figure 6. Crystal-liquid fractionation model for Monteregian plutonic rocks, showing liquid compositional trends from removal of specified minerals (dashed lines), and overall fractionation trend for plutonic rocks (solid line). Average camptonite, ocelli, and ocelli matrix compositions are adapted from Philpotts (1976). Open dots; averages for alkalic dolerites from Rhode Island (Pierce, 1976) and high-titanium quartz dolerites from eastern North America (Weigand and Ragland, 1970).

spreading was initiated. The degree of crustal thinning may thus have controlled magmatic processes in the lower crust and upper mantle.

The dikes of New England and adjacent Quebec are mostly mafic. The mafic dikes (e.g., lamprophyres) are often younger and chemically distinct from the last products of plutonic differentiation, and are not arranged in any radial pattern around these complexes (McHone, 1978). Camptonite and dolerite dikes are the most common. Camptonite is compositionally equivalent to a wet basanite, usually with both olivine and nepheline in the norm. The dolerite dikes range from Ne-normative alkali basalts through transitional olivine (± Hy-normative) types to true quartz tholeiites. Their age relations are not well known. The alkalic dolerites may be late Triassic, the high Ti quartz dolerites are early Jurassic, and the camptonites are probably of Cretaceous age. Chemical averages for the dike groups (Fig. 4) illustrate their relative alkalinities.

As the intrusive equivalent of basanite, the Cretaceous camptonite is a good candidate for a mantle-derived magma type. The chemical trends for NEQ plutons could be derived from compositions near the average of NEQ camptonite. The average camptonite plots close to the Monteregian plutonic trend and is similar to the average gabbro (Fig. 6). Ultramafic rocks of the Monteregians are considered to be mostly cumulate in origin (Woussen, 1970; Philpotts, 1974), and so the gabbro/camptonite composition appears to be a reasonable parent for the plutonic fractionation series. Camptonite is a common host for ocelli, small felsic globules interpreted by Philpotts (1976), Eby (1983), and others to be immiscible separates of the alkalic magma. Philpotts (1976) further suggested that immiscible-liquid segregation

could explain much of the differentiation patterns of the Monteregian plutons. Although the average camptonite is similar in SiO_2 content to gabbro, the range of immiscible lamprophyre separates analyzed by Philpotts (1976) overlaps with both the gabbro and syenite of the Monteregian Hills, (Fig. 7), thus immiscibility could be related to the development of the bimodal pattern of plutonic compositions.

The average camptonite, matrix, and ocelli compositions are plotted on an MgO vs SiO_2 diagram (Fig. 6) along with the Monteregian plutonic rock trend and illustrate how immiscible separation can be modeled with crystal fractionation to produce a plutonic composition trend. Accumulation of mafic minerals from a gabbroic magma formed the ultramafic members, while liquid immiscibility led to a mafic-felsic separation within the hypothetical chamber. Accumulation of feldspars and fractionation of amphibole and magnetite could drive the felsic liquid towards syenite. Minor quartz syenite may also result from assimilation of bordering crustal rocks (Philpotts, 1974).

Regional Variations of Appalachian Dike Swarms

A revised five-fold classification has been suggested recently for the early Jurassic ENA (Eastern North America) diabase dikes by Ragland and others (1983), Ragland and Whittington (1983a) and Whittington and others (1983). These authors distinguish two olivine-normative tholeiite groups (low LIL and high LIL), two quartz-normative tholeiite groups (low Ti and high Ti), and a mildly alkalic to transitional olivine diabase group. The latter was added because of common occurrence of these dikes in New England (Pierce, 1976; Trygstad, 1981; McHone and Trygstad, 1982; McHone and Butler, 1984; Pierce and Hermes, 1978; Hermes and others, 1984). A high-Fe quartz-normative group of an earlier classification (Weigand and Ragland, 1970) was abandoned because these rocks represent a differentiate of both olivine- and quartz-normative types. It is difficult to separate the two olivine-normative groups based upon single values for the LIL contents, because many of these elements are mobile and some overlap in their concentrations. Compositional differences become more pronounced with greater degrees of fractionation (Ragland and others, 1983).

Low-LIL, olivine-tholeiite dikes (LLO group) occur primarily as swarms of generally NW-trending dikes in Alabama, Georgia, the Carolinas, and Virginia (Fig. 8). Parental magmas apparently fractionated along a classic Fe-enrichment trend (Fig. 9). M-ratios (mol Mg/Mg + Fe^{2+}) for the least fractionated of these intrusives fall within the range (.70 ± .02) (Fig. 9) frequently considered to indicate primary partial melts of upper-mantle lherzolite (Irvine, 1977). The magmas probably formed at pressures of about 15 kb (Stolper, 1980) in the mantle near the base of the continental crust, and fractionated at about the same level (Fig. 10).

High-LIL, olivine-tholeiite dikes (HLO group) occur mainly in north-trending swarms that extend from below the Coastal plain near Charleston (South Carolina) to central Virginia (Ra-

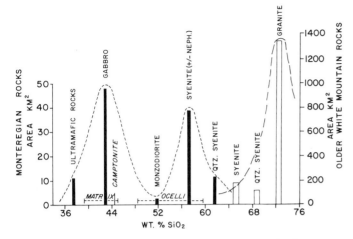

Figure 7. Area vs. SiO₂ content for average rock types of the Cretaceous Monteregian Hills plutons (solid columns) and Jurassic, White Mountain plutons (open columns). The average camptonite composition and ranges of immiscible lamprophyre separates as analyzed by Philpotts (1976) are also shown.

whether the higher content of LILs represents differences in source composition, conditions of fractionation and/or contamination. Paleomagnetic evidence (Smith and Hare, 1983) suggests that the N-S dikes may be younger than the NW-trending swarms. This is consistent with observed cross-cutting relationships.

Low-Ti, quartz-tholeiite dikes (LTQ group) are characterized by TiO_2 <0.9%. Dikes of this group are most common in Georgia and Alabama where they trend generally NW-SE (Fig. 8). They appear to coexist with dikes of the LLO group but are less common. Field relations are insufficiently known to provide relative age estimates. Only a few analyses of this group passed the geochemical and petrographic "screen" used by Ragland and Whittington (1983b), which primarily screened out all altered and phyric rocks. Data suggest Fe + Si enrichment, but no "primary" M ratios are known (Fig. 9). Their LIL-element content is intermediate between that of the LLO and HTQ groups. Weigand and Ragland (1970) suggested that these rocks could have been derived from an olivine-normative magma by relatively shallow-level fractionation. If so, derivation must have been from a HLO type of magma.

High-Ti, quartz-tholeiite dikes (HTQ group) are characterized by TiO_2 >0.9%. The chemical composition of the HTQ types is comparable to that of typical continental basalts (Karroo type). Its incompatable element content is higher than that of the LTQ type, and much higher than that of the olivine normative tholeiites. These dikes are apparently the most common among the early Mesozoic dike swarms of the central and northern Appalachians (Fig. 8). The majority occur from Virginia northward

gland and others, 1983) (Fig. 8). Their parental magmas were similar in some respects to those of the LLO group and apparently originated at a similar depth (approximately 15 Kb of pressure). However, fractionation of the HLO parental magma was probably shallower and characterized by higher oxygen fugacity than proposed for LLO fractionation (Fig. 10). Under the (crustal?) conditions of fractionation the magma evolved along a trend of Fe + Si enrichment (Fig. 9). It is not known at this stage

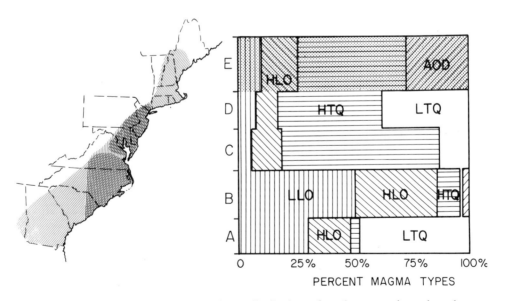

Figure 8. Schematic map showing approximate distributions of northeast-, north-, and northwest-trending dike sets. Ruled patterns on map reflect these orientations; i.e. in the Southeast the dominant strike is to the northwest, in the mid-Atlantic region to the north, and in New England to the northeast. Graph on right shows the relative proportions of magma types within each geographic domain. For convenience, domain boundaries are drawn along state lines: Domain A, Georgia-Alabama; Domain B, North-Carolina-South Carolina; Domain C, Virginia-Maryland; and Domain D, Pennsylvania-New Jersey through Massachusetts. Domain E represents unscreened data from New England.

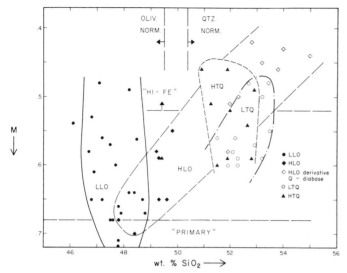

Figure 9. Plot of M (mol% Mg/Mg + Fe^{2+}) vs. SiO$_2$ (wt.%) showing four tholeiitic magma types. Note areas denoting quartz- vs. olivine-normative types and high-Fe types. Field for "primary" magmas based upon M-ratio of .70 ± .02. LLO and HTQ types exhibit Fe enrichment with fractionation; HLO and LTQ, Fe + Si enrichment. All samples plotted are aphyric.

to Nova Scotia and east into Morocco. Dike trends vary from N-S to NE-SW. Most magmas that intruded the "Newark" rift zones, providing the flows and sills, belong to the same type (HTQ). As was the case with the LTQ types, few analyses passed the "screen." The suggested trend is that of Fe enrichment, comparable to fractionation in the Palisades sill (Fig. 9) (see section C 3). The only known dike rock of the HTQ group whose M-ratios approach "primary" values occur in Nova Scotia (Papezik and Hoydich, 1980). The HTQ group might be composed of differentiates from HLO parental magmas that continued to fractionate to "high Fe" dolerites (Weigand and Ragland, 1970). Some evidence exists for magma mixing between more-or-less differentiated end members (Pegram, 1983). Of all tholeiitic ENA basalts, the dikes of the HTQ group are the most enriched in LIL elements.

Dikes of mildly alkalic to transitional olivine diabase (AOD group) occur primarily in New England and typically trend NE. Speculation on the characteristics and origin of this group is limited by the common presence of porphyritic and/or amygdular textures, as well as considerable alteration (Dickenson and Hermes, 1979; McHone and Trygstad, 1982; McHone and Butler, 1983). They appear to have a tendency to be slightly nepheline normative and are marked by relatively high TiO$_2$ values. Rare earth elements and mineral compositions indicate a lack of substantial contamination or crystal fractionation (Hermes and others, 1984). The regional extent and age of these dikes are presently unknown. They may turn out to be of Cretaceous age.

Chondrite-normalized REE patterns for the olivine-normative diabase are slightly depleted relative to quartz-normative basalts and diabases. REE abundances in the olivine normative

type are only about half those of the quartz-normative type (Ragland and others, 1971; Gottfried and others, 1983). Olivine-normative basalts and diabases also contain about half the abundances of the stable incompatible minor and trace elements of the quartz-normative varieties (Gottfried and others, 1983). Among the tholeiitic magma types the order of increasing LREE enrichment as well as incompatible element content generally follows the sequence LLO-HLO-LTQ-HTQ.

Weigand and Ragland (1970) and Ragland and others (1971) suggested that the diabase magmas in the northern Appalachians underwent shallow-level fractionation and selected contamination associated with ponding in the crust, whereas the magmas in the southern Appalachians did so to a (considerably) lesser extent. Although evidence exists against crustal contamination (Pegram, 1983) one generalization can be made. Magmas that intruded NW (cross grain) fractures (most of which are of the LLO variety) were less affected by processes characteristic of shallow-level ponding, such as low pressure fractionation, magma mixing, and perhaps contamination, than magmas intruding NE (parallel grain) faults. Thus structural characteristics may have controlled the degree of magma fractionation during ascent. Such an influence on fractionation, though, is not indicated in areas such as Georgia where compositional variety is observed in dikes exhibiting uniform northwest orientations (Gottfried and others, 1983).

The number of ENA dikes per unit area decreases rapidly from the Carolinas northeastward. This implies that extension was less effective in opening fractures in the northern than in the central and southern Appalachians. Differences in crustal dilation thus may have controlled the rate of magma ascent and thereby the degree of ponding, leading to increased fractionation.

Rift Valley Basalts

Exposed Jurassic flows and sills occur intermittently in "Newark" fault troughs from Virginia to Nova Scotia. Lava flows may be present in similar basins below the sediments of the Coastal Plain and Continental Shelf.

Geochemically all "Newark" basalt flows and the majority of associated sills belong to the HTQ (high-Ti, quartz-normative tholeiite) group of Ragland and Whittington (1983). Puffer and others (1983) and Puffer (1984) distinguished three basalt types within this group. The oldest flow units in the basins containing volcanic rocks are invariably represented by high-Ti basalts. Relatively massive flow units that occur higher in the stratigraphic sequences consist predominantly of high-Fe basalts. These flows dominate both spatially and volumetrically. High Ti-, Fe-, Cu-basalts represent the youngest flow units in the Hartford-Pomperaug and Newark fault troughs (Fig. 11).

The voluminous high-Fe basalts are believed to be fractionation products of the magmas that produced the earlier high-Ti basalts (Puffer and Lechler, 1980; Puffer and others, 1981). The youngest flows in the Newark, Pomperaug, and Hartford basins (Third Watchung and Hampden flow units) differ chemically

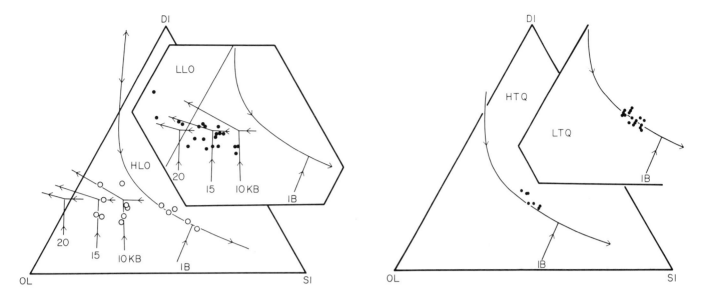

Figure 10. Projections from plagioclase onto the base of the basic tetrahedron (after Walker and others, 1979; Stolper, 1980). Note: (1) LLO and HLO magmas generated at about same depths (15 KB), but LLO magmas fractionated at deep, perhaps upper mantle levels, whereas HLO fractionated at much shallower levels, presumably well up into the crust; (2) HTQ magmas lie along a chord intersecting the cotectic, suggestive of magma mixing (Walker and others, 1979); (3) LTQ magmas lie along the cotectic, fractionated from a more primitive type. All samples plotted are aphyric.

from the older flow units. They contain less SiO_2 and K_2O, but more Ni and Co (Puffer and others, 1981). Magmas that provided these flows cannot have developed by continuation of the high-Ti, high-Fe fractionation trend, but may have evolved from high-Ti magmas via an independent (parallel) fractionation route (Reichenbach and Philpotts, 1983). They may have formed by fractionation of magmas comparable to those of the LLO group in the Carolinas. The high ferric-ferrous ratio, high Ti/Fe ratio, and high Cu content suggest that these fractionating magmas must have been ponded temporarily at relatively shallow crustal levels.

TECTONIC SETTING

Principal Tectonic Phases

Evolution of the Atlantic "passive" margin involved two principal tectonic phases; a rifting phase, involving stretching of the lithosphere and thermal uplift, and a drifting phase, involving separation of continental lithosphere and emplacement-accretion of ocean crust (Bally, 1981). Decoupling in the Appalachians, however, may have been more complex and it has been suggested that a shifting phase separated rifting and drifting (de Boer, 1967; Swanson, 1982).

Rifting predominated in the late Triassic and early Jurassic (Fig. 12). Spreading peaked with development of "Newark" rifts and regional emplacement of relatively large volumes of tholeiitic magmas in the early Jurassic. The majority of early Jurassic flows and sills were emplaced in "Newark" rift zones. The latter gener-

ally paralleled the tectonic grain of the Appalachians and represent classic examples for the rifting phase.

Grain parallel shearing dominated the tectonic evolution in the middle Jurassic (Fig. 12). The sinistral offset can be estimated from the latitudinal differences between "half" basins on opposite continental slopes. Assuming the Blake Plateau, Baltimore Canyon, and Georges Bank basins to represent the western half of the Senegal, Aaiun-Tarfaya, and Essaouira basins of Africa, respectively, the cumulative offset amounts to about 500 km. Much of this offset appears to have been accomplished before the Kimmeridgian (M25). A significant percentage of the offset may have occurred along the ECMA. The sinistral movements may have resulted from a clockwise rotation of Africa with respect to North America (Swanson, 1982). To allow for differential motion between Africa and North America shearing should have occurred following an eastward jump of the spreading axis, possibly to the E.C.M.A. In the Appalachians, this shearing phase culminated with regional emplacement of tholeiitic dike swarms, which (with exception of Nova Scotia and Newfoundland) intersect the Appalachian grain obliquely (Fig. 12). Drifting occurred from late Jurassic to the present (see Klitgord and Schouten, 1986).

Stress Configurations

Early theories on crustal deformation accompanying formation of "Newark" grabens concentrated exclusively on the effects of crustal extension. Sanders (1960, 1962, 1963) drew attention to the structural complexity of these grabens and distinguished an early extensional phase followed by regional arching and three successive compressional events.

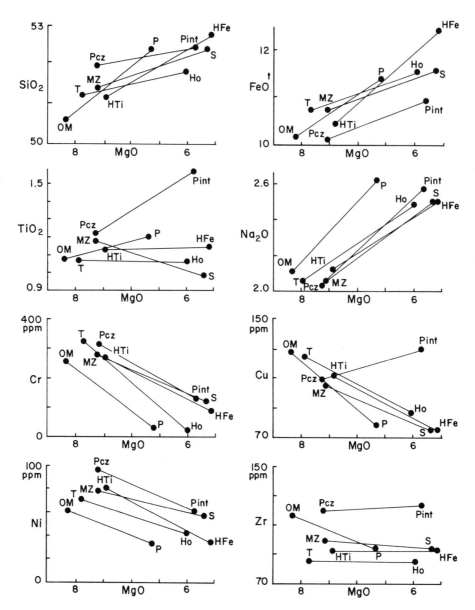

Figure 11. Fractionation of high-Ti type ENA quartz tholeiites. MgO (weight percent) variation diagrams illustrating fractionation of high-Ti tholeiites into high-Fe tholeiites. High-Ti basalts: T= Talcott; OM = Orange Mountain; Pcz = Palisades (Chillzone); MZ = Mount Zion Church. High-Fe basalts: Ho = Holyoke; P = Preakness; Pint = Palisades (Interior); S = Sanders. Note tight clustering of high-Ti basalts and divergent TiO_2 and Cu fractionation trends of Palisades. The latter probably resulted from shallow (in situ) fractionation processes. See Table 1.

Evidence for post "Newark" deformation by sub-horizontal compressive stresses has been accumulating rapidly. Timing and origin of these stresses is still in question. When existing evidence is combined, the following sequence of Mesozoic-Cenozoic stress regimes emerges.

a) Rifting in late Triassic–early Jurassic time resulted from NW (WNW)–SE (ESE) extension (Harrington, 1951; Reinemund, 1955; Randazzo and others, 1965, 1970; Thayer, 1970; Faill, 1973; Piepul 1975; Chandler, 1978; Wise, 1982; Bellini and

others, 1982; Swanson, 1982). The responsible deviatoric stress that acted more or less simultaneously on a longitudinal zone extending from Florida to northern Greenland may have resulted from membrane stresses caused by the relatively rapid northward motion of the Gondwana super-continent (Freeth, 1979, 1980).

Rates of crustal spreading probably peaked in the period when massive volumes of tholeiitic magmas invaded the crust in the southern (NW dikes) and reached the surface in the "Newark" basins of the central and northern Appalachians (flows

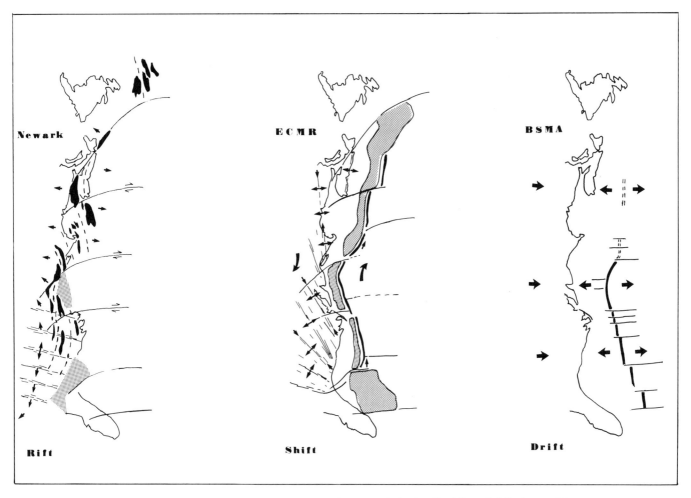

Figure 12. Hypothetical stress configurations during respectively the rift, shift, and drift phases of the Atlantic opening. Diverging arrows; direction of maximal horizontal extensional stress. Converging arrows direction of hypothetical compressive stress in upper crust.

and sills). Upward displacement of the crust above mantle diapirs and/or isostatic compensation following crustal unloading resulted in regional arching in the late stages of the Jurassic deformation. Subvertical stresses predominated in this period (Sanders, 1963).

b) Following regional arching, the principal stress rotated from subvertical to subhorizontal N(NE)–S(SW) (Thayer, 1970; Appel and Fenster, 1977; Goldstein, 1975; Appel and Fenster, 1977; Scharnberger and others, 1979; Houlik and Laird, 1977). The stresses are believed to have resulted from differential motion between Africa and North America due to a clockwise rotation of Africa in the middle Jurassic (Swanson, 1982). Since the pole of rotation was located in the Sahara (Van der Voo and others, 1976) relatively close to the Appalachians, this orogenic belt occupied a small circle of this rotation pole and stress directions varied significantly from N–S in the southern to NE–SW in the central and northern Appalachians. The majority of the references mentioned above assume these stresses to have been compressional. The evidence for a compressive stress is, however,

inconclusive. The shearing probably resulted from a change in the direction of extension with respect to the Appalachian grain. In the early Jurassic the stresses may have been more or less at right angles, in the middle Jurassic oblique to the grain. Such a change may have resulted locally (along transforms) in compression.

c) Stress configurations changed relatively abruptly at the time of initiation of seafloor spreading in the Bathonian. Late Jurassic and younger compressive stresses in the Appalachians were directed NW to W (Metcalf, 1982; Sawyer and Carroll, 1982). Assuming that these compressive stresses have resulted from spreading or "degeoidement" (gravitational slip from geoid highs to lows) (Anderson, 1982, 1984), rotation poles derived from ocean floor anomalies (Klitgord and Schouten, 1986) can be used to determine paleostress directions. As rotational poles changed position over time, variations occurred in plate motion and associated stress configurations. Stress directions changed from WNW to NW in the Hauterivian (± 126 Ma) and from NW to E-W in the late Cretaceous (82 to 60 Ma). Stress configurations changed again in the middle and late Eocene (51 to 37 Ma). Stress directions for different segments of the Appalachians varied by only a few degrees in most periods, but significant divergence (up to 15°) occurred in others. Divergence of stress directions was maximal in the late Jurassic and middle Cretaceous. It may have been responsible for extension along major cross faults and resulting magmatism.

Using the Atlantic reconstructions by Smith and Briden (1977) and Firstbrook and others (1978) and field data on Meso-Cenozoic Appalachian fractures, Metcalf (1982) analyzed the variations of stress over time, distinguishing five tectonic cycles. Each cycle was initiated by a principal stress more or less orthogonal to the Appalachian grain (uplift phase) and terminated with an oblique stress that caused faulting with major (usually dextral) strike slip (shear phase). Major shear phases appear to have occurred in the following periods: 155–147, 120–110, 90–82, 47–38, and possibly 7–0 Ma With exception of the last, each of these periods was characterized by localized magmatic activity along major crossfaults (leaky transforms) in the northern Appalachians and adjacent crustal regions. The absence of contemporaneous magmatism in the southern and central Appalachians (with the possible exception of northern Florida), however, indicates that reactivation of transform faults alone did not lead to magmatism and points to the probability that hotspots played a major role in the localization of igneous provinces. Changes in the stress configuration during the Eocene, for instance, led to dextral shear along grain-parallel faults in the entire Appalachian belt. Magmatism, however, was restricted to the Shenandoah igneous province (Virginia).

Emplacement History: Tectonic Levels

Crustal extension, responsible for subsidence, is presumed to have been attained at shallow lithospheric levels by listric normal faults and at deep levels by ductile flow (McKenzie, 1978; Bott, 1976a, b; Bally and others, 1981). Rather than two tectonic levels, three may have existed in the Appalachians. The upper level extends to depths of about 10 km and is dominated by listric and/or planar normal faults (Cocorp data). The middle level, at depths between about 10 and 35 km, is characterized by the presence of a relatively wide fault zone(s), which provided the principal feeders for magmas, that originated in the third and deepest, ductile upper mantle level. A similar threefold division of tectonic levels was recently proposed for the Rio Grande rift system where listric faults do not appear to extend to depths greater than 5 km (Cape and others, 1983).

The Appalachian gravity high probably provides the surface expression of the feeder zone(s) in the second level. Newark basins are parallel to, and their rockunits dip away from the Appalachian Gravity High (AGH). Because the Newark basins and AGH traverse various litho-tectonic zones of the Appalachians, they are probably manifestations of the same post-Alleghanian tectonic regime. Many models exist for the AGH. None of these can reasonably account for the curvilinear parallelism between anomalies and rift zones and for major amplitudinal variations along strike.

The Hutchinson and others (1982) model appears the most attractive; it incorporates a 20 to 25 km wide SE (45°) dipping zone with density contrast of +0.1 gm/cm^3. This zone probably represents an earlier zone of crustal weakness (thrust-reverse fault) pervasively intruded by tholeiitic dikes, which were the principal feeder for massive sills emplaced in the upper crust at depths generally less than 10 km. Cook (1984) has shown that amplitudinal variation of the AGH along strike is caused by residual anomalies, which can be modeled by crust with a density contrast of + 0.2 gm/cm^3. The anomalously dense crust may reflect emplacement of massive tholeiitic dike/sill complexes in the upper crust. Such anomalously dense crust is concentrated in areas below the four major highs of the Appalachian gravity anomaly. Little anomalous crust exists in the southern Appalachians, where intruding magmas are predominantly olivine-normative, but considerable anomalous crust occurs in the central Appalachians, where the magmas are predominantly quartz-normative. This suggests more ponding in the central Appalachians.

Ragland and Whittington (1983a) estimated that the early Jurassic olivine tholeiite magmas (LLO group) originated at pressures of about 15 Kb (± 50 km). Initial ascent of these magmas (during fractionation in the upper mantle) may have been slow but the rise of magmas in the range from 30 to 10 km must have been relatively fast, because the ratio of magma pressure to vertical stress is greater than unity in this interval (Roberts, 1970) and the horizontal stress was dilational. Geochemical evidence suggests virtually no fractionation (or contamination) at crustal levels, (Ragland and Whittington, 1983b). In the southern Appalachians, LLO magmas rose predominantly along subvertical NW transform fault zones and were emplaced mainly as dike swarms.

Magma ascent in the central and northern Appalachians appears to have followed different routes. Development of grain-parallel rift valleys suggest that the principal fractures in these segments were conjugate sets of normal faults. Focal plane mechanism solutions in the eastern Mediterranean indicate that most major normal faults nucleate at depths of 8 to 12 km and the great majority possess dips of 40° to 45° at their loci (Jackson and McKenzie, 1983). Gravimetric modeling (Hutchinson and others, 1982) implies the presence of a major NE trending (45° SE dipping) zone of weakness at depths from 10 and 30 km below the Newark rift zone. It appears logical to assume that quartz tholeiite magmas (HTQ group) followed this zone to depths of about 10 km and then intruded planar and listric normal faults of the rifts. If listric normal faults predominate in the uppermost crust, ponding should have become more common with shallowing depth as magma following these fractures had to overcome increasingly steeper gravity gradients. As magmas exited faults in the basement complex and entered the late Triassic-early Jurassic sediment wedge in the rift zones, ponding may have occurred frequently by injection along (tilted) bedding surfaces. Retardation of ascent by increased ponding may have facilitated contamination, fractionation, and magma mixing.

The predominance of tholeiitic magma emplacement along NW transform faults in the southern Appalachians and NE normal faults in the central and northern Appalachians resulted from differences in tensile stress. The direction of maximal tensile stress during emplacement of magmatic units in the south was probably NE–SE; in the north NW–SE. This difference may be due to the possible presence of a mantle diapir below Georgia and the Carolinas (de Boer and Snider, 1979).

As a result of differential motion between North America and Africa, the stress configuration changed relatively abruptly in the Middle Jurassic (Swanson, 1982). Sinistral shearing along the Appalachian grain, caused the opening of N- and NE-trending sets of "gash" fractures and emplacement of dominantly quartz tholeiitic dikes. Great extent and remarkable linearity give the impression that the dikes follow newly created fractures only. Detailed studies, however, show that many follow sets of reactivated fractures en echelon or en series (Snyder, 1970; Rastegar, 1973; Koza, 1976; Sawyer and Carroll, 1982).

Dike density decreases and dike width increases northeastward. Measured by cumulative dike width, total extension in the northern Appalachians was considerably less than that in the southern Appalachians in this period.

With the onset of the drifting phase, stress configurations changed again and compressive NW–SE stresses appear to have predominated. Variations in stress amplitude and stress direction may have led to rather localized extension along major cross faults of the Appalachians. Relatively high stress values (resulting from rapid spreading in the Atlantic) and significant stress divergence during the Cretaceous could have led to the emplacement of Cretaceous igneous masses along cross faults in the New England-Quebec and Notre Dame Bay igneous provinces.

Dike orientation varies throughout the New England-Quebec and Notre Dame Bay provinces. In the northwestern part of the NEQ province early Cretaceous lamprophyre dikes trend W (WNM)— E (ESE). They trend predominantly NE-SW in New Hampshire and southern Maine. This suggests that many Cretaceous dikes must follow older faults and that caution should be exercised in using dike trends for stress analysis.

Geotectonics

The development of the Mesozoic "Newark" type fault troughs is related mainly to the tectonic evolution of the Paleo Tethys–Central Atlantic–Gulf of Mexico megarift system, which eventually opened the central Atlantic along the eastern margin of the Appalachians (Dewey and others, 1973; Biju-Duval and others, 1977).

Intracratonic rifting affects wide areas around future plate boundaries. In the Appalachians this is reflected by the emplacement of kimberlites and cryptoexplosion structures along reactivated fault systems in the Appalachian basin and Canadian craton. Early Mesozoic rifting may have occurred as far west as the New Madrid rift zone (Missouri) (Hildenbrand and others, 1977).

With progressive crustal extension, rift systems polarize and peripheral grabens become inactive. Thus, the Appalachian spreading centers may have been consistently abandoned from west to east.

Deactivation of the more westerly basins led to an eastward shift of the spreading foci to the East Coast Magnetic Anomaly (ECMA) zone in the middle Jurassic and again further east to the Blake Spur Magnetic Anomaly (BSMA) zone in the late Jurassic. Thus, a wide (300 km) zone of early Jurassic rifting was stranded in North America by more intense and younger spreading east of the Appalachian orogen. It is not known whether ENA basaltic magmas became younger and/or more voluminous eastward under cover of the Coastal Plain sediments, but evidence recovered near Charleston, South Carolina, suggests such a trend. Eastward deactivation and shift left the majority of the Mesozoic rift systems on the North American plate. In the European crustal assemblage spreading foci shifted west as the majority of Mesozoic rifts are found in European lithosphere (Ziegler, 1982). Transformation from east to west vergence occurred along the Tethys corridor.

North America's fault troughs and basins (both exposed and covered by younger sediments) occur in a wide belt of slightly arcuate, sub-parallel zones. LePichon and Fox (1971) pointed out that there were at least four major transform faults active during the early opening phase. These transforms divide the belt in at least five segments. The dominant sense of apparent offset along the northern transforms (Newfoundland and New England-Kelvi fracture zones) is dextral, as evidenced by the stepwise eastward shift of basin axes. Offsets along the southern transforms (Cape Fear and Bahama fracture zones) appears to have been principally sinistral. The distribution of the Jurassic igneous rocks suggests that thinning may have been greater in the southern than in the northern Appalachians.

Differences in spreading between these segments cannot be accounted for by differential stretching alone. It is inferred (Snider, 1975; de Boer and Snider, 1979) that thermally induced physico-chemical processes affected the upper mantle-lower crust below the Carolinas and caused upward displacement of the crust-mantle boundary, leading to partial melting of the upper mantle, crustal anatexis, and emplacement of relatively large volumes of tholeiitic and/or alkalic magmas. These processes affected crustal ductility and caused additional crustal thinning.

Upwelling mantle currents are believed to cause lithospheric thinning by inducing a thermal upward displacement of the asthenosphere-lithosphere boundary and by exerting tensional stresses (Ziegler, 1982). Many authors believe that doming and regional uplift precede rifting (Cloos, 1939; Burke and Whitman, 1973; Burke and Dewey, 1973, 1974; Kinsman, 1975). Others are of the opinion that rifts develop as a result of horizontal crustal extension with uplift being a later thermal response (Turcotte and Oxburgh, 1973; Molnar and Tapponier, 1975; Scholz and others, 1976; Bott, 1976a, b; Turcotte, 1981). Thus there are two basic models, both involving diapiric emplacement of asthenospheric matter. In the "mantle plume" model emplacement of the diapir results in crustal thinning and subsidence; in the "tensional failure" model relatively high rates of extension thin the lithosphere, resulting in upwelling of the mantle. Density equilibrium is reached as the mass spreads out laterally to form an asthenolith.

The "tensional" failure model fits the database for the late Triassic–early Jurassic rifts and associated magmatism. Contemporaneous magmatism in New England (WMMS) and in the southern Appalachians can be best explained by the "mantle plume" model. WMMS magmatism was characterized by emplacement of granite, derived by anatectic melting of the lower crust; magmatism in Georgia by emplacement of relatively large volumes of tholeiitic magma derived by partial melting from the upper mantle. Differences in the degree of lithospheric spreading may have been responsible for a different development of these provinces. Cretaceous magmatism in the New England-Quebec and Notre Dame Bay provinces has been explained by motion of the North American plate over mantle plumes (Morgan, 1971, 1972; Burke and others, 1973; Vogt, 1973; Coney, 1971; Vogt and others, 1971; Morgan, 1981; and Crough, 1981). Uchupi and others, 1970, LePichon and Fox (1971), Foland and Faul (1977), McHone (1981), and Bedard (1985) on the other hand believe that the linear distribution of igneous rocks in the New England-Quebec province is due to emplacement of magmas along leaky transform faults. Shifts in plate motion effect the stress configurations of adjacent plates and can reactivate old fractures, promoting partial melting of the upper mantle and magmatism (Marsh 1973). It appears logical to assume that magmatic periodicity is related to major changes in spreading direction, and that the basic distribution of alkalic igneous complexes within each province is controlled by transforms. The fact that most provinces are restricted to certain faults and to certain segments of these faults, however, implies that magmatism occurs only where the fault intersects a rather localized hotspot.

Development of several hotspots below the Appalachians could explain Mesozoic and Cenozoic magmatic activity (Burke and others, 1973; Molnar and Francheteau, 1975; de Boer and Snider, 1979; Morgan and Crough, 1979; Crough, 1981; Morgan, 1981; Metcalf, 1982).

As the Pangean plate moved northeast in the late Triassic–early Jurassic eastern North America may have passed over a hotspot, which may have triggered widespread magmatism. Using Smith and Briden's (1977) reconstructions for the opening of the Atlantic, Metcalf (1982) showed that the White Mountains were situated over the hypothetical location of the Freetown (Africa) hotspot during the Triassic (210 Ma). Here the hotspot may have been responsible for anatexis at deep crustal levels and for alkalic magmatism. The Freetown track places the hotspot below the Carolinas at about 190 ± 10 Ma, a period characterized by emplacement of tholeiitic and rhyolitic (Delorey and others, 1982) intrusives at shallow crustal levels. By 180 ± 10 Ma, the Freetown hotspot was located below central Florida and the Bahamas, where it may have been responsible for emplacement of Jurassic rhyolites-basalts and widespread Mesozoic rifting

(Mullins and Lynts, 1977). Thus the late Triassic-early Jurassic hotspot, the track of which virtually paralleled the Appalachians grain for more than 20 m.y., may have aided regional spreading and initiated widespread magmatism from Florida north into the Maritime provinces. Hotspot activity appears to have peaked below New England and the southern Appalachians, significantly weakening the lithosphere and aiding crustal spreading.

As the North American plate moved NW and W during the late Jurassic-Cretaceous, the Appalachians crossed the tracks of two (possibly three) hotspots (Morgan, 1981) (Fig. 13). Earliest magmatic evidence for the northernmost hotspots consists of late Jurassic-early Cretaceous kimberlites and gabbroic intrusives on either shore of the Labrador Sea. This hotspot found itself below Newfoundland about 120 Ma and may have been responsible for emplacement of the Notre Dame Bay alkalic complex. When the trace is reconstructed eastward, using the poles and rates of Klitgord and Schouten (1986) for the Atlantic opening, it ends up below the Azores.

The second (New England) hotspot is believed to have originated below Ontario (Coney, 1971; Morgan, 1972, 1981; Vogt, 1973; Crough, 1981). The earliest geologic evidence for this hotspot, however, occurs in the western Monteregian Hills with the emplacement of alkalic magmas about 140 Ma. The track (using Klitgord and Schouten's poles) brings the hotspot below the White Mountains (and southern New England) around 110 ± 10 Ma and below the New England seamounts in the period from 100 to 70 Ma. Principal evidence for the existence of a Cretaceous hotspot below southern New England is based on fission track ages, which range from 130 to 80 Ma (Crough, 1981).

Morgan (1981) believes that a third hotspot originated below Florida in the late Jurassic. Magmatism in Florida can be attributed to Florida's position over the Bahama-Freetown hotspot in middle-late Jurassic (Metcalf, 1982). No evidence exists in this region for Cretaceous magmatism.

The Bermuda swell resembles that of Hawaii and other oceanic hotspots (Morgan, 1981). Reconstruction of its track westward brings it via the Cape Fear arch below the alkaline Shenandoah igneous province in the Eocene (Fig. 13). The Azores, Great Meteor, and Bermuda hotspots possess tracks that cross the Appalachian grain at high angles. As evidenced by magmatism in the Notre Dame Bay, New England-Quebec, and Shenandoah igneous provinces, their influence was relatively shortlived and spatially restricted.

It is obvious that the occurrence of several of the igneous provinces cannot be explained using "hotspot" models. Neither the Jurassic dike swarms, the grain parallel kimberlite belts, or the chain(s) of Cretaceous plutons along the "hinge" zone can be explained using this concept. Upwelling of upper mantle material following decompression at deep levels below major fracture zones is most likely responsible for emplacement of these provinces. In comparison with semi-permanent "hotspots" one would expect such "leaking" to be relatively shortlived, as rotation poles for the Atlantic opening (and plate stress configurations) changed frequently. Linear chains related to fracture zones should be ex-

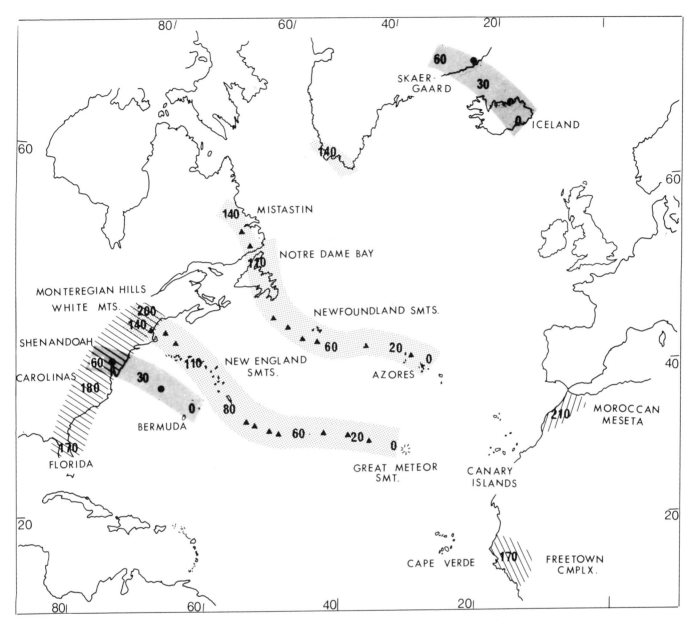

Figure 13. Hypothetical tracks for the White Mountains–Freetown (210–170 Ma)—striped pattern; the Monteregian Hills—Great Meteor and Notre Dame Bay–Azores (140–0 Ma)—light shade; Shenandoah-Bermuda and Skaergaard-Iceland (60–0 Ma)—heavy shade-"hotspots". Tracks obtained by using rotation poles and opening rates provided by Van der Voo and others (1976) and Klitgord and Schouten (this volume). Triangles and dots denote data points for rotations.

tensive, because it is unlikely that "leaking" would occur along certain spatially restricted segments only. Mesozoic and Paleogene igneous provinces in the Appalachians and adjacent crustal zones were most probably emplaced above "hotspots" as well as "leaking" fracture zones. Because of lithospheric extension, "leaking" faults develop over semi-permanent "hotspots" and temporary "hotspots" may develop from initial plumes which followed "leaking" fault zones. Since insufficient data exists, it appears unrealistic to attempt differentiating between these hypotheses at this time.

REFERENCES CITED

Amerigian, C., 1977, Factors controlling the acquisition of primary and secondary magnetizations in sediments [Ph.D. thesis] Kingston, R. I., University of Rhode Island, 209 p.

Anderson, D. L., 1982, Hotspots, polar wander, Mesozoic convection and the geoid: Nature, v. 297, p. 39.

——1984, The earth as a planet: Paradigms and Paradoxes: Science, v. 223, p. 347–355.

Andrews, J. R., and Emeleus, C. H., 1971, Greenland Kimberlites: Rapport, Grønlands Geologiske Undersøgelse, v. 31, 26 p.

Appel, G., and Fenster, D. F., 1977, Deformation of the northern Newark basin: Geological Society of America Abstracts with Programs, v. 9, no. 3, p. 237.

Applin, P. L., 1951, Preliminary report on buried pre-Mesozoic rocks in Florida and adjacent states: United States Geological Survey Circular 91, 28 p.

Armstrong, R. L., and Besancon, J., 1970, A Triassic time-scale dilemma: K-Ar dating of Upper Triassic mafic igneous rocks, eastern U.S.A. and Canada and post-Upper Triassic plutons, western Idaho, U.S.A.: Eclogae Geologicae Helvetiae, v. 63, p. 15–28.

Aumento, F., Ade-Hall, J. M., and Keen, M. J., 1975, 1974—The year of the Mid-Atlantic Ridge: Reviews of Geophysics and Space Physics, v. 13, p. 53–66.

Aumento, F., Mitchell, W. S., and Fratta, M., 1976, Interaction between sea water and oceanic layer two as a function of time and depth—1. Field evidence: Canadian Mineralogist, v. 14, p. 269–290.

Bally, A. W., 1981, Geology of Passive Continental Margins: Atlantic type margins: American Association of Petroleum Geologists, Education course note series 19, p. 1–48.

Barnett, R. S., 1975, Basement structure of Florida and its tectonic implications: Gulf Coast Association of Geological Societies Transactions, v. 25, p. 122–142.

Barrett, D. L., and Keen, C. E., 1976, Mesozoic magnetic lineations, the magnetic quiet zone, and sea floor spreading in the Northwest Atlantic: Journal of Geophysical Research, v. 81, no. 26, p. 4875–4884.

Bass, M. N., 1969, Petrography and ages of crystalline basement rocks in Florida: American Association of Petroleum Geologists, Memoir 11, p. 283–310.

Basu, A. R., and Rubury, E., 1980, Tectonic significance of Kimberlite dikes in central New York: Geological Society of America Abstracts with Programs, v. 12, no. 2, p. 23.

Bedard, J. H., 1985, The opening of the Atlantic, the Mesozoic New England igneous province, and mechanisms of continental breakup: Tectonophysics, v. 113, p. 209–232.

Bell, H., Books, K. G., Daniels, D. L., Hoff, W. E., Jr., and Popenoe, P., 1980, Diabase dikes in the Haile-Brewer area, South Carolina, and their magnetic properties: United States Geologic Survey Professional Paper 1123 A-D, p. C 1–17.

Bellini, F. X., Corkum, D. H., and Stewart, A. J., 1982, Geology of foundation excavations at Seabrook Station, Seabrook, New Hampshire, in Farquhar, D. L., ed., Geotechnology in Massachusetts: University of Massachusetts at Amherst, p. 109–117.

Biju-Duval, B., Dercourt, J., and LePichon, X., 1977, From the Tethys Ocean to the Mediterranean Sea: A plate tectonic model of evolution of the Western Alpine system, in Biju-Duval, B., and Montadert, L., eds., Structural History of the Mediterranean Basin: Technip, Paris, France, p. 143–164.

Bott, M.H.P., 1976a, Mechanisms of basin subsidence—an introductory review: Tectonophysics, v. 36, p. 1–4.

—— 1976b, Formation of sedimentary basins of graben type by extension of the continental crust: Tectonophysics, v. 36, p. 77–86.

Boucarut, M., Clin, M., and Thibault, Cl., 1984, Prolongement du rift Aden en republique de Djibouti; structures superposees dans l'etude des points tectoniques triples: Academie Scientifique. Comptes Rendu, v. 287, p. 679–682.

Bouley, B. A., 1971, Paleomagnetism of the White Mountain Magma series (New Hampshire) and its implications for Atlantic sea-floor spreading [M.S. thesis]: Middletown, Conn., Wesleyan University, 62 p.

Bridges, S. R., 1973, Study of a positive Bouguer gravity anomaly in Tift County (Georgia), Report: Georgia Institute of Technology, 19 p.

Brown, L., 1979, Magnetic observations of the red beds of the northern Connecticut Valley—DRM or CRM? (abstract): Transactions of the American Geophysical Union, v. 60, no. 46, p. 815.

Burke, K., 1976, Development of grabens associated with the initial ruptures of the Atlantic Ocean: Tectonophysics, v. 36, p. 93–112.

Burke, K., and Dewey, J. F., 1973, Plume generated triple junctions: key indicators in applying plate tectonics to old rocks: Journal of Geology, v. 81, p. 406–433.

—— 1974, Two plates in Africa during the Cretaceous?: Nature, v. 249,

p. 313–316.

Burke, K., and Whitman, A. J., 1973, Uplift, rifting, and the break-up of Africa, Tarling, D. H., and Runcorn, S. K., eds., in Implications of Continental Drift to the Earth Sciences, v. 2: New York, Academic Press, p. 735–755.

Burke, K., Kidd, W.S.F., and Wilson, J. T., 1973, Relative and latitudinal motion of Atlantic hot spots: Nature, v. 245, p. 133–137.

Burke, W. H., Otto, J. B., and Denison, R. E., 1969, Potassium-argon dating of basaltic rocks: Journal of Geophysical Research, v. 74, no. 4, p. 1082–1086.

Cape, C. D., McGeary, S., and Thompson, G. A., 1983, Cenozoic normal faulting and the shallow structure of the Rio Grande rift near Socorro, New Mexico: Geological Society of America Bulletin, v. 94, p. 3–14.

Carr, R. S., 1980, Geology and Petrology of the Ossipee Ring-Complex, Carroll County, New Hampshire [M.S. Thesis]: Dartmouth College, 384 p.

Chandler, W. E., 1978, Graben mechanics at the junction of the Hartford and Deerfield basins, Mass. [M.S. thesis]: Amherst, Mass., Cont. no. 33 University of Massachusetts, 151 p.

Channell, J.E.T., Lowrie, W., and Pialli, G., 1980, Gross features of Jurassic geomagnetic stratigraphy from Umbrian land sections: EOS, Transactions, American Geophysical Union, v. 61, p. 944.

Chapman, C. A., 1968, A comparison of the Maine coastal plutons and the magmatic central complexes of New Hampshire, in Zen, E-An, White, W. S., Hadley, J. B., and Thompson, J. B., eds., Studies of Appalachian Geology: Northern and Maritime: New York, Interscience Publishers, p. 385–396.

Cloos, H., 1939, Hebung-spaltung-vulcanism: Geologische Rundschau, v. 30, p. 405–527.

Coney, P. J., 1971, Cordilleran tectonic transitions and motion of the North American plate: Nature, v. 233, p. 462–465.

Cook, F. A., 1984, Geophysical anomalies along strike of the southern Appalachian Piedmont: Tectonics, v. 3, no. 1, p. 45–62.

Cornet, B., 1977, The palynostratigraphy and age of the Newark Supergroup [Ph.D. thesis]: State College, Pennsylvania State University, 527 p.

Cousminer, H. L., and Manspeizer, W., 1976, Triassic pollen date Moroccan High Atlas and the incipient rifting of Pangea as Middle Carnian: Science, v. 191, p. 943–945.

—— 1977, Autunian and Carnian palynoflorules: Contribution to the chronology and tectonic history of the Moroccan pre-Atlantic borderland, in Swain, E. M., ed., Stratigraphic micropaleontology of Atlantic basin and borderlands: Elsevier, p. 185–204.

Creasy, J. W., and Eby, G. N., 1983, The White Mountain batholith as a model of Mesozoic felsic magmatism in New England: Geological Society of America Abstracts with Programs, v. 15, no. 6, p. 549.

Crough, S. T., 1981, Mesozoic hotspot epeirogeny in eastern North America: Geology, v. 9, p. 2–6.

Crough, S. T., Morgan, W. J., and Hargraves, R. B., 1980, Kimberlites: Their relation to mantle hotspots: Earth and Planetary Science Letters, v. 50, p. 260–274.

Crutcher, T. D., 1983, Baltimore Canyon trough Seismic expressions of structural styles, in Bally, A. W., ed., Studies in Geology 15, American Association of Petroleum Geologists, 2.2.3, p. 20–26.

Currie, K. L., and Larochelle, A., 1969, A paleomagnetic study of volcanic rocks from Mistastin Lake, Labrador, Canada: Earth and Planetary Science Letters, v. 6, p. 309–315.

Dalrymple, G. B., Gromme, C., Sherman, and White, R. W., 1975, Potassium-Argon age and paleomagnetism of diabase dikes in Liberia: Initiation of central Atlantic rifting: Geological Society of American Bulletin, v. 86, no. 3, p. 399–411.

de Boer, J., 1967, Paleomagnetic-Tectonic study of Mesozoic dike swarms in the Appalachian: Journal of Geophysical Research, v. 72, no. 8, p. 2237–2250.

—— 1968, Paleomagnetic differentiation and correlation of the Late Triassic volcanic rocks in the central Appalachians (with special reference to the Connecticut Valley): Geological Society of America Bulletin, v. 79, p. 609–626.

—— 1983a, Magnetic and paleomagnetic evidence bearing on hotspot models for

Mesozoic magmatism: Geological Society of America Abstracts with Programs, v. 15, no. 2, p. 91.

—— 1983b, Structural control of Mesozoic magmatism in the Appalachians: Geological Society of America Abstracts with Programs, v. 15, no. 6, p. 554.

de Boer, J., and Snider, F., 1979, Magnetic and chemical variations of Mesozoic diabase dikes from eastern North America: Evidence for a hotspot in the Carolinas: Geological Society of America Bulletin, v. 90, no. 1, p. 185–198.

Deininger, R. W., Dallmeyer, R. D., and Neathery, T. L., 1975, Chemical variations and K-Ar ages of diabase dikes in east-central Alabama: Geological Society of America Abstracts with Programs, v. 7, no. 4, p. 482.

De Journett, J. D., and Schmidt, V. A., 1975, Paleomagnetism of some peridotite dikes near Ithaca, New York (abstract): EOS, Transactions of the American Geophysical Union, v. 56, p. 354.

Delorey, C. M., Dooley, R. E., Ressetar, R., Fullagar, P. D., and Stoddard, E. F., 1982, Paleomagnetism and isotope geochronology of early Mesozoic rhyolite porphyry dikes, eastern southern Appalachian Piedmont: EOS, Transactions, American Geophysical Union, 63-18, p. 309.

Dennison, J. M., and Johnson, R. W., Jr., 1971, Tertiary intrusions and associated phenomena near the thirty-eighth parallel fracture zone in Virginia and West Virginia: Geological Society of America Bulletin, v. 82, p. 501–508.

Dennison, J. M., Parrish, J. B., and Lavin, P. M., 1983, Tectonic model for kimberlite emplacement in the Appalachian plateau of Pennsylvania: A discussion and reply: Geology, v. 11, p. 252–254.

Deutsch, E. R., and Rao, K. V., 1977, Paleomagnetism of Mesozoic lamphrophyres from central Newfoundland (abstract): EOS, Transactions, American Geophysical Union, v. 58, p. 745.

Dewey, J. F., Pitmann, W. C., Ryan, W.B.F., and Bonnin, J., 1973, Plate tectonics and the evolution of the Alpine System: Geological Society of America Bulletin, v. 84, p. 3137–3180.

Dickenson, M. P., and Hermes, O. D., 1979, REE distribution in pre-metamorphic and post-metamorphic diabase dikes from southeast New England (abstract): Geological Society of America Abstracts with Programs, v. 11, no. 1, p. 10.

Dillon, W. P., Klitgord, K. D., and Pauli, C. K., 1983, Mesozoic development and structure of the continental margin off South Carolina: United States Geologic Survey Professional Paper 1313, N 1–16.

Dillon, W. P., Sheridan, E. R., and Fail, J. P., 1977, Structure of the western Blake-Bahamas basin as shown by 24 channel CDP profiling: Geology, v. 4, p. 459–462.

Doherty, J. T., and Lyons, J. B., 1980, Mesozoic erosion rates in northern New England: Geological Society of America Bulletin, 1, v. 91, p. 16–20.

Dooley, R. E., and Smith, W. A., 1982, Age and magnetism of diabase dikes and tilting of the Piedmont: Tectonophysics, v. 90, p. 283–307.

Duncan, R. A., 1982, The New England Seamounts and the absolute motion of North America since mid-Cretaceous time: EOS, Transactions, American Geophysical Union, v. 63, p. 1103–1104.

Duncan, R. A., and Houghton, R. L., 1977, Radiometric age dating of the New England seamounts [Abstract]: American Geophysical Union, Fall Annual Meeting, San Francisco.

Eby, G. N., 1983, Geology, geochemistry and petrogenesis of the Monteregian Hills alkaline province, Quebec: Geological Society of America Abstracts with Programs, v. 15, no. 6, p. 565.

—— 1984a, Monteregian Hills I Petrography, major and trace element geochemistry, and strontium isotope chemistry of western intrusions: Journal of Petrology, v. 25, p. 421–452.

—— 1984b, Geochronology of the Monteregian Hills alkaline igneous province, Quebec: Geology, v. 12, p. 468–470.

—— 1984c, Age, geochemistry, and petrogenesis of lamprophyre pikes from the Monteregian Hills and younger White Mountain igenous province: Geological Society of America, Abstracts with Programs, v. 16, no. 1, p. 14.

Erickson, G. P., and Kulp, J. L., 1961, Potassium-argon measurements in the Palisades Sill, New Jersey: Geological Society of America Bulletin, v. 72,

p. 649–652.

Faill, R. T., 1973, Tectonic development of the Triassic Newark-Gettysburg Basin in Pennsylvania: Geological Society of America Bulletin, v. 84, p. 725–740.

Fairbain, H. W., Faure, G., Pinson, W. H., Hortey, P. M., and Powell, J. L., 1963, Initial ratio of strontium 87 to strontium 86, whole rock age, and discordant biotite in the Monteregian igneous province, Quebec: Journal of Geophysical Research, v. 68, p. 6515–6522.

Firstbrook, P. L., Funnel, B. M., and Hurley, A. M., 1978, Paleoceanic reconstructions 60-0 Ma: Washington, D.C., National Science Foundation, National Ocean Coring Program, 41 p.

Foland, K. A., and Faul, H., 1977, Ages of the White Mountain Intrusives—New Hampshire, Vermont, and Maine, USA: American Journal of Science, v. 277, p. 888–904.

Foland, K. A., Quinn, A. W., and Giletti, B. J., 1971, K-Ar and Rb-Sr Jurassic and Cretaceous ages for intrusives of the White Mountain Magma Series, northern New England: American Journal of Science, v. 270, p. 321–330.

Foster, J., and Symons, D.T.A., 1979, Defining a paleomagnetic polarity pattern in the Monteregian intrusives: Canadian Journal of Earth Science, v. 16, p. 1716–1725.

Freeth, S. J., 1979, Deformation of the African Plate as a consequence of membrane stress domains generated by post-Jurassic drift. Earth and Planetary Science Letters, v. 45, p. 93–104.

—— 1980, Can membrane tectonics be used to explain the break-up of plates? *in* Davies P. A., and Runcorn, S. K., eds., Mechanisms of continental drift and plate tectonics: London, Academic Press, p. 135–149.

Froelich, A. J., and Gottfried, D., 1985, Early Jurassic diabase sheets of the eastern U.S.; A preliminary overview, U.S. Geological Survey Circular 946, p. 79–86.

Fullagar, P. D., and Bottino, M. L., 1969, Tertiary felsite intrusions in the Valley and Ridge province, Virginia: Geological Society of America Bulletin, v. 80, p. 1853–1858.

Gardner, T. E., 1956, The igneous rocks of Pendleton County, West Virginia: West Virginia Geological and Economic Survey, Report of Investigations 12, 26 p.

Gold, D. P., 1967, Alkaline ultrabasic rocks in the Montreal area, Quebec, *in* Wyllie, P. J., ed., Ultramafic and Related Rocks: New York, John Wiley & Sons, p. 288–302.

Gold, D. P., and Marchand, M., 1969, The diatreme breccia pipes and dikes, and the related alnoite, kimberlite, and carbonatite intrusions occurring in the Montreal and Oka areas, Quebec, *in* Pouliot, G., ed., Guidebook for the Geology of Monteregian Hills: Geological Association of Canada, p. 5–42.

Goldstein, A. G., 1975, Brittle fracture history of the Montague basin, north-central Massachusetts: Amherst, University of Massachusetts, Contribution no. 25, Department of Geology and Geography, 108 p.

Gottfried, D., Annell, C. S., and Byerly, G. R., 1983, Geochemistry and tectonic significance of subsurface basalts near Charleston, South Carolina: United States Geologic Survey Professional Paper 1313, A 1–19.

Gray, K. J., and Diecchio, R. J., 1985, Observations of the Eocene dike swarms, Highland County, Virginia, [manuscript]: United States Geological Survey.

Hall, J. M., Barrett, D. L., and Keen, C. E., 1977, The volcanic layer of the ocean crust adjacent to Canada - a review: in Baragar, W.R.A., Coleman, L. C., and Hall, J. M., eds., Volcanic regimes of Canada: Geological Association of Canada Special Paper 16, p. 425–444.

Harrington, J. W., 1951, Structural analysis of the west border of the Durham Triassic basin: Geological Society of America Bulletin, v. 62, p. 149.

Harrison, C.G.A., and Lindh, T., 1982, A polar wandering curve for North America during the Mesozoic and Cenozoic: Journal of Geophysical Research, v. 87, n. B3, p. 1903–1920.

Helwig, J., Aronson, J., and Day, D. S., 1974, A Late Jurassic mafic pluton in Newfoundland: Canadian Journal of Earth Sciences, v. 11, p. 1314–1319.

Hermes, O. D., Rao, J. M., and Dickenson, M. P., 1984, A transitional alkalic dolerite dike suite of Mesozoic age in southeastern New England: Contributions to Mineralogy and Petrology, v. 86, p. 386–397.

Hildenbrand, T. G., Kane, M. F., and Stauder, W., 1977, Magnetic and gravity anomalies in the Northern Mississippi embayment and their special relation in seismicity: United States Geological Survey Miscellaneous Field Studies Map MF-914.

Hodych, J. P., and Hayatsu, A., 1980, K-Ar isochron age and paleomagnetism of diabase along the trans-Avalon aeromagnetic lineament-evidence of Late Triassic rifting in Newfoundland: Canadian Journal of Earth Sciences, v. 17, p. 491–499.

Houghton, R. L., 1978, Petrology and geochemistry of the New England seamount chain [Ph.D. thesis]: Massachusetts Institute of Technology, Woods Hole Oceanographic Institute, 298 p.

Houghton, R. L., Thompson, G., and Bryan, W. B., 1977, Petrological and geochemical studies of the New England seamount chain: American Geophysical Union Transactions, v. 58, p. 530.

Houghton, R. L., Thomas, J. E., Diecchio, R. J. and Tagliacozzo, A., 1979, Radiometric ages of basalts from DSDP Leg 43, sites 382, 384, 385, 386, and 387, in Tucholke, B. E. and Vogt, P. R., eds., Initial report of the Deep Sea Drilling Project, v. 43, p. 739–757.

Houlik, C. W., and Laird, H. S., 1977, Mesozoic wrench tectonics and the development of the Northern Newark Basin: Geological Society of America Abstracts with Programs, v. 9, no. 3, p. 275.

Hutchinson, D. R., Grow, J. A., and Klitgord, K. D., 1982, Crustal structure beneath the southern Appalachians; Nonuniqueness of gravity modeling: Geology, v. 11, p. 611–615.

Irvine, T. N., 1977, Definition of primitive liquid composition for basic magmas: Carnegie Institute, Washington Yearbook, no. 76, p. 454–461.

Irving, E., 1977, Drift of the major continental blocks since the Devonian: Nature, v. 270, p. 304–309.

Jackson, D. E., Hunter, R. H., and Taylor, L. A., 1982, A Mesozoic window into the sub-Appalachian mantle: Kimberlite from the Eastern United States (abstract): Geological Society of America Abstracts with Programs, v. 14, no. 1 & 2, p. 28.

Jackson, J., and McKenzie, D., 1983, The geometrical evolution of normal fault systems: Journal of Structural Geology, v. 5, p. 471–482.

Jansa, L. F., and Pe-Piper, G., 1985, Early Cretaceous volcanism on the northeastern American margin and implications for plate tectonics: Geological Society of America Bulletin 96, p. 83–91.

Johnson, R. W., Milton, C., and Dennison, J. M., 1971, Fieldtrip to the igneous rocks of Augusta, Rockingham, Highland, and Bath counties, Virginia: Virginia Division of Mineral Resources Information Circular 16, 68 p.

Kent, D., and Tsai, L. P., 1978, Paleomagnetism and rock magnetism of Upper Jurassic limestone and basalt from Site 367: Initial Report of the Deep Sea Drilling Project, Supplement to v. 38-41, p. 995–1002.

King, A. F., and McMillan, N. J., 1975, A mid-Mesozoic breccia from the coast of Labrador: Canadian Journal of Earth Science, v. 12, no. 1, p. 44–51.

King, E. R., and Zietz, I., 1978, The New York-Alabama lineament, geophysical evidence for a major crustal break in the basement beneath the Appalachian Basin: Geology, v. 6, no. 5, p. 312–318.

King, P. B., 1961, Systematic pattern of Triassic dikes in the Appalachian region: United States Geologic Survey Professional Paper 424-B, p. B93–B95.

—— 1971, Systematic pattern of Triassic dikes in the Appalachian region-second report: United States Geologic Survey Professional Paper 750-D, p. D84–D88.

Kinsman, D. J., 1975, Rift valley basins and sedimentary history of trailing continental margins, in Fischer, A. G., and Judson, S., eds., Petroleum and Global Tectonics: Princeton, University Press, p. 83–128.

Klitgord, K. D., Dillon, W. P., and Popenoe, P., 1983, Mesozoic tectonics of the southeastern United States Coastal Plain and continental margin: United States Geologic Survey Professional Paper 1313-P, p. 1–15.

Klitgord, K. D., and Schouten, H., 1982, Early Mesozoic Atlantic reconstruction from sea-floor spreading data: EOS, Transactions, American Geophysical Union, v. 63, n. 18, p. 307.

Kopecky, L., 1976, The planetary subcrustal deep fault system in North America: A new concept of a structural control of alkaline magmatism, cryptoexplo-sion structures and related ore deposits, in Podwysocki, M. H., and Barle, J. L., eds., Proceedings of the second international conference on basement tectonics: Denver, Basement Tectonics Committee, Inc., p. 472–483.

Koza, D., 1976, Petrology of the Higganum diabase dike in Connecticut and Massachusetts [M.S. thesis]: University of Connecticut at Storrs, 78 p.

Lamphere, M. A., 1983, $^{40}Ar/^{39}Ar$ ages of basalt from Clubhouse crossroad testhole 2, Charleston, South Carolina: United States Geologic Survey Professional Paper 1313, B 1–8.

Lapointe, P. L., 1979, Paleomagnetism of the Notre Dame Bay lamprophyre dikes, Newfoundland, and the opening of the North Atlantic Ocean: Canadian Journal of Earth Science, v. 16, p. 1823–1831.

Larochelle, A., 1961, Application of paleomagnetism to geological correlation: Nature, v. 192, p. 37.

Larochelle, A., 1968, Paleomagnetism of the Monteregian Hills; New results: Journal of Geophysical Research, v. 73, no. 10, p. 3239–3246.

Larochelle, A., 1969, Paleomagnetism of the Monteregian Hills: Further new results: Journal of Geophysical Research, v. 74, p. 2570–2575.

Larochelle, A., and Currie, K. L., 1967, Paleomagnetic study of igneous rocks from the Manicouagan Structure, Quebec: Journal of Geophysical Research, v. 72, no. 16, p. 4163–4169.

Larson, R. L., and Hilde, T.W.C., 1975, A revised timescale of magnetic reversals for the Early Cretaceous and late Jurassic: Journal of Geophysical Research, v. 80, p. 2586–2594.

Laurent, R., and Pierson, T. C., 1973, Petrology of alkaline rocks from Cuttingsville and the Shelbourne peninsula, Vermont: Canadian Journal of Earth Sciences, v. 10, p. 1244.

Leavy, B. D., 1980, Tectonic and sedimentary structures along trhe eastern margin of the Culpeper basin, Virginia: Geological Society of America Abstracts with Programs, v. 12, p. 182.

Leavy, B. D., and Puffer, J. H., 1983, Physical and chemical characteristics of four Jurassic basalt units in the Culpeper basin, Virginia: Geological Society of America Abstracts with Programs, v. 15, no. 2, p. 92.

Lee, K. Y., 1977, Triassic stratigraphy in the northern part of the Culpeper Basin, Virginia and Maryland: Geological Survey Bulletin, 1422-c, p. C1–17.

LePichon, X., and Fox, P. J., 1971, Marginal offsets, fracture zones, and the early opening of the North Atlantic: Journal of Geophysical Research, v. 76, p. 6294–6308.

Loiselle, M. C., 1978, Geochemistry and petrogenesis of the Belknap Mountains Complex and Pliny Range, White Mountain Magma Series, New Hampshire [Ph.D. thesis]: Cambridge, Massachusetts, Massachusetts Institute of Technology, 302 pp.

Loiselle, M. C., and Hart, S. R., 1978, Sr isotope systematics of the Belknap Mountain Complex, New Hampshire: Geological Society of America Abstracts with Programs, v. 10, p. 73.

Lovlie, R., and Opdyke, N. D., 1974, Rock magnetism and paleomagnetism of some intrusions from Virginia: Journal of Geophysical Research, v. 70, no. 2, p. 343–349.

Lowden, J. A., 1961, Age determinations by the Geological Survey of Canada: Geological Survey of Canada paper 61-17, 69 p.

Manspeizer, W., 1981, Early Mesozoic basins of the central Atlantic passive margins, in A. W., Bally, ed., Geology of Passive Continental Margins: History, Structure and Sedimentologic Record: American Association of Petroleum Geologists Education course notes 19, p. 4-1-60.

—— 1982, Triassic-Liassic basins and climate of the Atlantic passive margins: Geol. Rundschau 71-3: p. 895–917.

—— 1983, Inhereted Appalachian-Hercynian structures, and their impact on Triassic-Liassic "rifting": Geological Society of America Abstracts with Programs, v. 15, no. 3, p. 184.

Marsh, J. S., 1973, Relationships between transform directions and alkaline igneous rock lineaments in Africa and South America: Earth and Planetary Science Letters, v. 18, p. 317–323.

Martello, A. R., Gray, N. H., Philpotts, A. R., and Dowling, J. J., 1984, Mesozoic diabase dikes of southeastern New England: Geological Society of America Abstracts with Programs, v. 16, no. 1, p. 48.

May, P. R., 1971, Pattern of Triassic-Jurassic diabase dikes around the North Atlantic in the context of predrift positions of the continents: Geological Society of America Bulletin, 82, p. 1285–1292.

Mayhew, M. A., and Galliher, S. C., 1982, An equivalent layer magnetization model for the United States derived from Magsat data: Geophysical Research Letters 9, p. 311–313.

McHone, J. G., 1978a, Lamprophyre dikes of New England [Ph.D. thesis]: University of North Carolina at Chapel Hill, 189 p.

—— 1978b, Distribution, orientations, and ages of mafic dikes in central New England: Geological Society of America Bulletin, v. 89, p. 1645–1655.

—— 1981, Comment on Mesozoic hotspot epeirogeny in eastern North America: Geology, v. 9, p. 341–342.

McHone, J. G., and Butler, J. R., 1983, Petrogenesis of lamprophyre dikes and related White Mountain-Monteregien Intrusive complexes (abstract): Geological Society of America Abstracts with Programs, v. 15, p. 75.

—— 1984, Mesozoic igneous provinces of New England and the opening of the North Atlantic ocean; Geological Society of America Bulletin, v. 95, p. 757–765.

McHone, J. G., and Corneille, E. S., 1980, Alkalic dikes of the Lake Champlain valley: Vermont Geology, v. 1, p. 16–21.

McHone, J. G., and Trygstad, J. C., 1982, Mesozoic mafic dikes of southern Maine: Maine Geology, Bulletin 2, p. 16–32.

McIntosh, W. C., 1976, Magnetic reversals in the Brunswick Formation of the Newark Group in New Jersey and Eastern Pennsylvania (abstract): EOS, Transactions, American Geophysical Union, v. 57, p. 238.

McIntosh, W. C., and Hargraves, R. B., 1976, Magnetic reversals in the Brunswick Formation of the Newark Group in New Jersey and eastern Pennsylvania (abstract): EOS, Transactions, American Geophysical Union, v. 57, no. 4, p. 238.

McIntosh, W. C., Hargraves, R. B., and West, C. L., 1985, Paleomagnetism and oxide mineralogy of upper Triassic to lower Jurassic redbeds and basalts in the Newark basin: Geological Society of America Bulletin, v. 96, p. 463–480.

McKenzie, D., 1978, Some remarks on the development of sedimentary basins: Earth and Planetary Science Letters, v. 40, p. 25–32.

Metcalf, T. P., 1982, Intraplate Tectonics of the Appalachians in Post-Triassic Time [M.S. thesis]: Middletown, Conn., Wesleyan University, 238 p.

Meyer, H.O.A., 1976, The kimberlites of the continental United States: A review: Journal of Geology, v. 84, p. 377–404.

Milton, C., 1972, Igneous and metamorphic basement rocks of Florida: Florida Bureau of Geology Bulletin, no. 55, 125 p.

Milton, C., and Grasty, R., 1964, "Basement" rocks of Florida and Georgia: American Association of Petroleum Geologists Bulletin, v. 53, no. 12, p. 2483–2493.

Molnar, P., and Francheteau, J., 1975, The relative motion of "hot spots" in the Atlantic and Indian oceans during the Cenozoic: Geophysical Journal of the Royal Astronomical Society, v. 43, p. 763–774.

Molnar, P., and Tapponier, P., 1975, Cenozoic tectonics of Asia; effects of a continental collision: Science, v. 189, no. 4201, p. 419–426.

Morgan, W. J., 1971, Convection plumes in the lower mantle: Nature, 230, p. 42–43.

—— 1972, Plate motions and deep mantle convection, *in* Shagam, R., and others, Studies in Earth and Space Sciences, Geological Society of America Memoir 132, p. 7–22.

—— 1981, Hotspot traces and the opening of the Atlantic and Indian Oceans, *in* Emiliani, C., ed., The Sea, v. 7, The Oceanic Lithosphere: New York, Wiley and sons, p. 443–487.

Morgan, W. J., and Crough, S. T., 1979, Bermuda hotspots and Cape Fear Arch (abstract): EOS, Transactions, American Geophysical Union, v. 60, no. 18, p. 392–393.

Mueller, P. A., and Porch, J., 1983, Two volcanic provinces in the sub-surface of the Florida Peninsula: Geological Society of America Abstracts with Programs, v. 15, no. 2, p. 63.

Mullins, H. T., and Lynts, G. W., 1977, Origin of the Northwestern Bahama Platform: Review and interpretation: Geological Society of America Bulletin, v. 88, p. 1447–1461.

Nadon, G. C., and Middleton, G. V., 1984, Tectonic control of Triassic sedimentation in southern New Brunswick: Local and regional implications: Geology, v. 12, p. 619–622.

Olsen, P. E., 1980, Triassic and Jurassic formations of the Newark Basin: Annual Meeting of the New York State Geological Association, no. 52, p. 2–39.

Olsen, P. E., McCune, A. R., and Thomson, K. S., 1982, Correlation of the Early Mesozoic Newark supergroup by vertebrates, principally fishes: American Journal of Science, v. 282, p. 1–44.

Opdyke, N. D., and Wensink, H., 1966, Paleomagnetism of rocks from the White Mountain plutonic-volcanic series in New Hampshire and Vermont: Journal of Geophysical Research, v. 71, no. 12, p. 2045–3051.

Paijitprapapon, V., 1980, Paleomagnetism of Mesozoic plutonic and volcanic rocks in the White Mountains (New Hampshire), with emphasis on the influence of tectonic deformations [M.A. thesis]: Middletown, Conn., Wesleyan University, 94 p.

Papezik, V. S., and Hodych, J. P., 1980, Early Mesozoic diabase dikes of the Avalon Peninsula, Newfoundland: Petrochemistry, mineralogy and origin: Canadian Journal of Earth Science, v. 17, p. 1417–1430.

Parrish, J. B., and Lavin, P. M., 1982, Tectonic model for Kimberlite emplacement in the Appalachian Plateau of Pennsylvania: Geology, v. 10, p. 344–347.

Pechersky, D. M., and Khramov, A. N., 1973, Mesozoic paleomagnetic scale of the USSR: Nature, v. 244, p. 499–501.

Pegram, W. J., 1983, Isotopic characteristics of the Mesozoic Appalachian tholeiites: Geological Society of America Abstracts with Programs, v. 15, no. 6, p. 660.

Phillips, J. D., 1983, Paleomagnetic investigations of the Clubhouse Crossroads basalt: United States Geologic Survey Professional Paper 1313, C1–C18.

Philpotts, A. R., 1974, The Monteregian province, *in* Sorenson, H., ed., The alkaline rocks: New York, Wiley & Sons, p. 293–310.

—— 1976, Silicate liquid immiscibility: Its probable extent and petrogenetic significance: American Journal of Science, v. 276, no. 9, p. 1147–1177.

—— 1978, Rift-associated igneous activity in eastern North America, *in* Ramberg, I. B., and Neumann, E. R., eds., Petrology and Geochemistry of Continental rifts: Reidel Dordrcht, p. 133–154.

—— 1985, Recent petrologic studies of Mesozoic igneous rocks in Connecticut: U.S. Geological Survey Circular 946, p. 107–110.

Piepul, R. G., 1975, Analysis of jointing and faulting at the southern end of the Eastern Border Fault, Connecticut: Amherst, University of Massachusetts, Contribution no. 23, Department of Geology, 109 p.

Pierce, T. A., 1976, Petrology of dolerite-metadiorite dikes of southeastern New England [M.S. thesis]: University of Rhode Island, Kingston, Rhode Island.

Pierce, T. A., and Hermes, O. D., 1978, Petrology and geochemistry of diabase and metadiabase dikes in Southeastern New England [abstract]: Geological Society of America Abstracts with Programs, v. 10, p. 80.

Puffer, J. H., 1984, Igneous rocks of the Newark basin: Petrology mineralogy, ore deposits and guide to fieldtrip: Geological Associations of New Jersey 1st annual field conference.

Puffer, J. H., Hortubise, D. O., Geiger, F. J., and Lechler, P., 1981, Chemical composition and stratigraphic correlation of Mesozoic basalt units of the Newark Basin, New Jersey, and the Hartford Basin, Connecticut: Summary: Geological Society of America Bulletin, Pt. 1, v. 92, p. 155–159.

Puffer, J. H., and Lechler, P., 1980, Geochemical cross sections through the Watchung Basalt of New Jersey: Summary: Geological Society of America Bulletin, Pt. 1, v. 91, p. 7–10.

Puffer, J. H., Postana, E. M., and Theoktitoff, G., 1983, Geochemical characteristics of rift related volcanic rocks in the Appalachian orogen: Geological Society of America Abstracts with Programs, v. 15, no. 3, p. 184.

Ragland, P. C., Brunfelt, A. O., and Weigand, P. W., 1971, Rare-earth abundances in Mesozoic dolerite dikes from Eastern United States, *in* Brunfelt, A. O., and Steinnes, E., eds., Activation Analysis in geochemistry and Cosmochemistry: Oslo, Universitetsforlaget, p. 227–235.

Ragland, P. C., Hatcher, R. D., Jr., and Whittington, D., 1983, Juxtaposed Mesozoic diabase dike sets from the Carolinas: A preliminary assessment: Geology, v. 11, no. 7, p. 394–404.

Ragland, P. C., and Whittington, D., 1983a, "Primitive" Mesozoic diabase dikes of eastern North America: Primary or derivative?: Geological Society of America Abstracts with Programs, v. 15, no. 2, p. 92.

——1983b, Early Mesozoic diabase dikes of eastern North America: Magma types: Geological Society of America Abstracts with Programs, v. 15, p. 666.

Randazzo, A. F., Swe, W., and Wheeler, W. H., 1965, The stratigraphy of the Wadesboro Triassic basin in North and South Carolina [M.S. thesis]: University of North Carolina at Chapel Hill, 52 p.

——1970, A study of tectonic influence on Triassic sedimentation—The Wadesboro basin, central Piedmont: Journal of Sedimentary Petrology, v. 40, no. 3, p. 998.

Rastegar, I., 1973, A detailed study of the diabase dikes in the Middle Haddam quadrangle, Connecticut [M.S. thesis]: Wesleyan University, 88 p.

Reichenbach, I., and Philpotts, A. R., 1983, Compositional relations between the Mesozoic basalts of the Hartford basin, Connecticut: Geological Society of America Abstracts with Programs, v. 15, no. 3, p. 173.

Reinemund, J. A., 1955, Geology of the Deep River coal field of North Carolina: United States Geologic Survey Professional Paper 246, 159 p.

Ressetar, R., and Martin, D. L., 1980, Paleomagnetism of Eocene igneous intrusions in the Valley and Ridge Province, Virginia and West Virginia: Canadian Journal of Science, v. 17, p. 1583–1588.

Reynolds, P. R., and Aumento, F. A., 1974, Deep Drill 1972: Potassium-argon dating of the Bermuda drill core: Canadian Journal of Earth Sciences, v. 11, p. 1269–1273.

Rigotti, P., and Schmidt, V. A., 1976, Upper Triassic secular variations as recorded by the Palisades sill New Jersey (abstract) EOS, Transactions of the American Geophysical Union, v. 57, p. 238.

Roberts, J. L., 1970, The intrusion of magma into brittl rocks, *in* Newall, G., and Rast, N., eds., Mechanism of Igneous Intrusion: Liverpool, Seel House Press, p. 287–338.

Robertson, W. A., 1967, Manicouagan, Quebec, paleomagnetic results: Canadian Journal of Earth Sciences, v. 4, p. 1–9.

Root, S. I., and Hoskins, D. M., 1977, Lat 40°N fault zone, Pennsylvania: A new interpretation: Geology, v. 5, p. 719–723.

Ross, M. E., and Reidel, S. P., 1982, Mafic dikes of northeastern Massachusetts: Geological Society of America Abstracts with Programs, v. 13, no. 2, p. 78.

Sanders, J. E., 1960, Structural history of Triassic rocks of the Connecticut Valley belt and its regional implications: New York Academy of Science Transcripts, v. 23, p. 119–132.

——1962, Strike-slip displacement on faults in Triassic rocks in New Jersey: Science, v. 136, no. 3510, p. 40–42.

——1963, Late Triassic tectonic history of northeastern United States: American Journal of Science, v. 261, p. 501–524.

Sawyer, J., and Carroll, S. E., 1982, Fracture Deformation of the Higganum Dike, South-Central Connecticut: U.S. Nuclear Regulatory Commission, no. NUREG/Cr-2479, Washington, D.C., 52 p.

Scharnberger, C. K., Nichols, P. H., and Bria, J. J., 1979, Diabase dikes in Lancaster County, Pennsylvania: Evidence of Late Triassic shear; Geological Society of America Abstracts with Programs, v. 11, no. 1, p. 52.

Scholz, C. H., Koczynski, T. A., and Hutchins, N. G., 1976, Evidence for incipient rifting in southern Africa: Geophysical Journal of the Royal Astronomical Society, v. 44, p. 135–144.

Schouten, H., and Klitgord, K. D., 1982, The memory of accreting plate boundary and the continuity of fracture zones: Earth and Planetary Science Letters, v. 59, p. 255–266.

——1977, Map showing Mesozoic magnetic anomalies; western North Atlantic: United States Geological Survey Miscellaneous Field Studies Map MF-915, scale 1:2,000,000.

Sears, C. E., and Gilbert, M. C., 1973, Petrography of the Mt. Horeb Virginia Kimberlite: Geological Society of America Abstracts with Programs, v. 5, no. 5, p. 434.

Seidemann, D. E., Masterson, W. D., Dowling, M. P., and Turekian, K. K., 1984: K-Ar dates and $^{40}Ar/^{39}Ar$ age spectra for Mesozoic basalt flows of the Hartford Basin, Connecticut, and Newark Basin, New Jersey, Geological Society of America Bulletin, v. 95, p. 594–598.

Shafiquallah, M., Tupper, W. M., and Cole, T.J.S., 1970, K-Ar age of the carbonatite complex, Oka, Quebec: Canadian Mineralogist, v. 10, p. 541–552.

Sharp, J. A., and Simmons, G., 1978, Geologic/Geophysical models of intrusives of the White Mountain magma series: Geological Society of America Abstracts on Programs, v. 10, p. 85.

——1983, Geologic-Geophysical models of intrusives of the White Mountain magma series: Geological Society of America Abstracts with Programs, v. 15, p. 85.

Sheridan, R. E., 1983, Phenomena of pulsation tectonics related to the breakup of the eastern North American Continental margin: Tectonophysics, v. 94, p. 169–185.

——1978, Structural and stratigraphic evolution and petroleum potential of the Blake plateau: Offshore Technical Conference Proceedings, p. 363–368.

Smith, A. G., and Briden, J. C., 1977, Mesozoic and Cenozoic Paleocontinental Maps: Cambridge, England, Cambridge University Press, 63 p.

Smith, R. C., Rose, A. W., and Lanning, R. M., 1975, Geology and geochemistry of Triassic diabase in Pennsylvania: Geological Society of America Bulletin, v. 86, p. 943–955.

Smith, T. E., and Noltimier, H. C., 1979, Paleomagnetism of the Newark trend igneous rocks of the North Central Appalachians and the opening of the Central Atlantic Ocean: American Journal of Science, v. 279, p. 778–807.

Smith, W. A., and Hare, J. C., 1983, Paleomagnetic results from north-south and northwest trending diabase dikes from the Piedmont of North Carolina: Geological Society of America Abstracts with Programs, v. 15, p. 198.

Snider, F. G., 1975, Analysis of magnetic and chemical data from Mesozoic diabase dikes of the Appalachians, with implications for the presence of a Triassic hotspot in the Carolinas [M.A. thesis]: Wesleyan University, 61 p.

Snyder, G. L., 1970, Bedrock geologic and magnetic maps of the Marlborough Quad., east-central Connecticut: United States Geologic Survey Quadrangle Map GQ-791, scale: 1:24,000.

Steiner, M. B., 1980, Investigation of the geomagnetic field polarity during the Jurassic: Journal of Geophysical Research, v. 85, p. 3572–3586.

Stolper, E., 1980, A phase diagram for mid-ocean ridge basalts: Preliminary results and implications for petrogenesis: Contributions in Mineralogy and Petrology, v. 14, p. 13–27.

Strong, D. F., and Harris, A., 1974, The petrology of Mesozoic alkaline intrusives of central Newfoundland: Canadian Journal of Earth Sciences, v. 11, p. 1208–1219.

Sullivan, K. D., 1975, Petrology and geochemistry of volcanic rocks from the Newfoundland Seamounts: Geological Society of America Abstracts with Programs, v. 7, no. 6, p. 866.

——1977, Newfoundland seamounts: Petrology and Geochemistry: *in* Baragar, W.R.A., Coleman, L. C. and Hall, J. M., eds., Volcanic regimes in Canada: Geological Association of Canada Special Paper 16, p. 461–476.

Sundvik, M., Larson, R. L., and Detrick, R. S., 1984, Rough-smooth basement boundary in the western North Atlantic basin: Evidence for a seafloor-spreading origin: Geology, v. 12, p. 31–34.

Surlyk, F., 1978a, Jurassic basin evolution of East Greenland: Nature, v. 274, no. 5667, p. 130–133.

——1978b, Mesozoic faulting in East Greenland: Geologie en Mijnbouw, v. 56, no. 4, p. 311–327.

Sutter, J. F., 1985, Progress on geochronology of Mesozoic diabases and basalts: U.S. Geological Survey Circular 946, p. 110–114.

Sutter, J. F., and Smith, T. E., 1977, $^{40}Ar/^{39}Ar$ incremental release spectra and paleomagnetism of Mesozoic diabase intrusives from Connecticut and Maryland: Geological Society of America Abstracts with Programs, v. 7, p. 1288.

——1979, $^{40}Ar/^{39}Ar$ ages of diabase intrusions from Newark trend basins in Connecticut and Maryland: Initiation of Central Atlantic rifting: American Journal of Science, v. 279, p. 808–831.

Swanson, M. T., 1982, Preliminary model for an early transform history in central

Atlantic rifting: Geology, v. 10, p. 317–320.

Taras, G., and Hart, S. R., 1983, Sr, Nd, and Pb isotopic compositions of the New England Seamount Chain: EOS, Transactions, American Geophysical Union, v. 64, no. 45, p. 907.

Thayer, P. A., 1970, Stratigraphy and geology of Dan River Triassic basin, North Carolina: Southeastern Geology, v. 12, p. 1–31.

Thompson, G. M., 1968, A magnetic study of fault-diabase dike relationships in the Cedar Creek community, South Carolina [M.S. thesis]: University of South Carolina, p. 8–27.

Trygstad, J. C., 1981, The petrology of Mesozoic diabase dikes in southern New Hampshire and Maine: Geological Society of America Abstracts with Programs, v. 13, no. 3, p. 181.

Tucholke, B. E., and Ludwig, W. J., 1982, Structure and origin of the J. Anomaly Ridge, Western North Atlantic Ocean: Journal of Geophysical Research, v. 87, p. 9389.

Turcotte, D. L., 1981, Rifts-tensional failures of the lithosphere, *in* Papers presented to the Conference on Processes of Planetary Rifting, Christian Brothers' Retreat House, Napa Valley, California, 3-5 December 1981: Lunar and Planetary Institution Contributions, v. 451, p. 5–8.

Turcotte, D. L., and Oxburgh, E. R., 1973, Mid-plate tectonics: Nature, v. 244, p. 337–339.

—— 1976, Stress accumulation in the lithosphere: Tectonophysics 35, p. 183–199.

—— 1978, Intraplate volcanism: Philosophical Transactions of the Royal Society of London, Ser. A 288, p. 561–579.

Uchupi, E., Phillips, J. D., and Prada, N. E., 1970, Origin and structure of the New England Seamount chain: Deep-Sea Research, v. 17, p. 483–494.

Van Alstine, D. R., and de Boer, J., 1978, A new technique for constructing apparent polar wander paths and the revised Phanerozoic path for North America: Geology, v. 6, p. 137–139.

Van der Voo, R., and French, R. B., 1974, Apparent polar wandering for the Atlantic bordering continents: Late Carboniferous to Eocene: Earth-Science Reviews, v. 10, p. 99–118.

Van der Voo, R., Mauk, F. J., and French, R. B., 1976, Permian-Triassic continental configurations and the origin of the Gulf of Mexico: Geology, v. 4, no. 3, p. 177–180.

Van Hinte, J. E., 1976a, A Jurassic time scale, American Association of Petroleum Geologists Bulletin, v. 60, p. 489–497.

—— 1976b, A Cretaceous time scale: American Association of Petroleum Geologists Bulletin, v. 60, p. 498–516.

Vogt, P. R., 1972, Evidence for Global synchronism in Mantle Plume Convection, and possible significance for Geology: Nature, v. 240, no. 5380, p. 338–342.

—— 1973, Early events in the opening of the North Atlantic, *in* Tarling, D. H., and Runcorn, S. K., eds., Implications of Continental Drift to the Earth Sciences, v. 2: London, Academic Press, p. 693–712.

—— 1974, Volcano spacing, fractures, and thickness of the lithosphere: Earth and Planetary Science Letters, v. 21, no. 3, p. 235–252.

Vogt, P. R., Anderson, C. N., and Bracey, D. R., 1971, Mesozoic magnetic anomalies, sea-floor spreading, and geomagnetic reversals in the southwestern North Atlantic: Journal of Geophysical Research, v. 76, p. 4796–4823.

Vogt, P. R., and Einwich, A. M., 1979, Magnetic anomalies and seafloor spreading in the western North Atlantic, and a revised calibration of the Keathly (M) geomagnetic reversal chronology, *in* Tucholke, B. E., and Vogt, P. R., eds., Deep Sea Drilling Project, Leg 43: Washington, D.C., U.S. Government Printing Office, v. 43, p. 857–876.

Walker, D., Shibota, T., and Delong, S. E., 1979, Abyssal tholeiites from the Oceanography fracture zone: Contributions in Mineralogy and Petrology, v. 70, p. 111–112.

Wampler, J. M., and Dooley, R. E., 1975, Potassium-argon determination of Triassic and Eocene igneous activity in Rockingham County, Virginia: Geological Society of America Abstracts with Programs, v. 7, no. 4, p. 547.

Wanless, R. K., Stevens, R. D., Lachance, G. R., and Rimsaite, R.Y.H., 1965, Age determinations and geological studies, pt. 1—isotopic ages. Report 5: Geologic Survey of Canada Paper 64-17.

Wanless, R. K., Stevens, R. D., Lachance, G. R., and Edmond, C. M., 1967, Age determinations and geologic studies, K-Ar ages. Report : Geologic Survey of Canada Paper 66-17.

Wark, J. M., and Clarke, D. B., 1980, Geochemical discriminators and the paleotectonic environment of the North Mountain basalts, Nova Scotia: Canadian Journal of Earth Sciences, v. 17, p. 1740–1745.

Watson, T. L., and Cline, J. H., 1913, Petrology of a series of igneous dikes in central-western Virginia: Geological Society of America Bulletin, v. 24, p. 301–334, 682–683.

Watts, W. S., 1969, The coast parallel dike swarm of southwest Greenland in relation to the opening of the Labrador Sea: Canadian Journal of Earth Sciences, v. 6, p. 1320–1321.

Weigand, P. W., and Ragland, P. C., 1970, Geochemistry of Mesozoic dolerite dikes from Eastern North America: Contributions in Mineralogy and Petrology, v. 29, p. 195–214.

Whittington, D., Ragland, P. C., and Hatcher, R. D., 1983, Tectonic implications of an apparent Mesozoic radial diabase dike swarm centered in the vicinity of Charleston-Georgetown, South Carolina; Geological Society of America Abstracts with Programs, v. 15, no. 2, p. 92.

Wise, D. U., 1982, New fault and fracture domains of southwestern New England - hints on localization of the Mesozoic basins, *in* Farquhar, D. C., ed., Geotechnology in Massachusetts: Graduate School, University of Massachusetts at Amherst, p. 447–453.

Woussen, G., 1970, La geologie du complexe igne du Mont Royal (The geology of the igneous complex of Mont Royal), *in* Alkaline rocks; the Monteregian Hills: Canadian Mineralogy, v. 15, no. 2, p. 47–66.

Zartman, R. E., Brock, M. R., Heyl, A. V., and Thomas, H. H., 1967, K-Ar and Rb-Sr ages of some alkalic intrusive rocks from central and eastern United States: American Journal of Science, v. 265, p. 848.

Zartman, R. E., Hurley, P. M., Krueger, H. W., and Giletti, B. J., 1970, A Permian disturbance of K-Ar radiometric ages in New England, its occurrence and cause: Geological Society of America Bulletin, v. 81, p. 3359–3374.

Ziegler, P. A., 1975, Geologic evolution of the North Sea and its tectonic framework: Bulletin of the American Association of Petroleum Geologists, v. 59, 1073–1097.

—— 1978, North Sea rift and basin development, *in* Ramberg, I. B., and Neumann, E. R., eds., Tectonics and Geophysics of Continental Rifts. N.A.T.O. Advanced Study Institute, Series C.: Reidel Publishing Company, p. 249–277.

—— 1982, Geological Atlas of Western and Central Europe: New York, Elsevier Scientific Publishing Company, Inc., 130 p.

MANUSCRIPT ACCEPTED BY THE SOCIETY JUNE 26, 1985

Printed in U.S.A.

The Geology of North America
Vol. I-2, The Atlantic Continental Margin: U.S.
The Geological Society of America, 1988

Chapter 12

Georges Bank Basin: A regional synthesis

John S. Schlee and Kim D. Klitgord
U.S. Geological Survey, Woods Hole, Massachusetts 01543

INTRODUCTION

Georges Bank is a shallow part of the Atlantic Continental Shelf southeast of New England (Emery and Uchupi, 1972, 1984). This bank, however, is merely the upper surface of several sedimentary basins overlying a block-faulted basement of igneous and metamorphic crystalline rock (Figs. 1 and 2). Sedimentary rock forms a seaward-thickening cover that has accumulated in one main depocenter and several ancillary depressions, adjacent to shallow basement platforms of Paleozoic and older crystalline rock. Georges Bank basin contains a thickness of sedimentary rock greater than 10 km, whereas the basement platforms that flank the basin are areas of thin sediment accumulation (less than 5 km). We will discuss the structural, stratigraphic, and tectonic framework of the Georges Bank area, presenting a synthesis of geophysical and geologic data, much of which has been collected over the past decade (Austin and others, 1980; Klitgord and others, 1982; Schlee and Fritsch, 1982; Klitgord and Schlee, 1986). In particular, we shall be concerned with the crustal foundations of Georges Bank, the main stages of sedimentary buildup, and the forces that we think have influenced the evolution of the basin.

Previous Studies

A general outline of the geology of Georges Bank basin has been given by Drake and others (1959), Maher (1971), Emery and Uchupi (1972), Schultz and Grover (1974), Mattick and others (1974), Ballard and Uchupi (1975), and Schlee and Klitgord (1982). More detailed studies of the character and distribution of basement structures have been given by Klitgord and Behrendt (1979), Austin and others (1980), Klitgord and others (1982), and Klitgord and Schlee (1986). The character and deposition patterns of the overlying Mesozoic and Cenozoic sedimentary wedge have been studied in detail by Schlee (1978), Valentine (1981), Poag (1982a, 1982b), Klitgord and others (1982), and Schlee and Fritsch (1982). Quaternary evolution of the bank has been described by Emery and Uchupi (1972), Oldale and others (1974), Lewis and Sylwester (1976), and Schlee and Fritsch (1982).

The Gulf of Maine platform (Fig. 1) has been studied by Drake and others (1954), Uchupi (1966), Emery and Uchupi (1972), Uchupi (1970), Ballard and Uchupi (1975), Hermes and others (1978), and Austin and others (1980). These and other studies have shown that the Gulf of Maine is floored by shallowly buried Paleozoic metasedimentary rocks, middle Paleozoic granitic rock, and Upper Triassic and Lower Jurassic sedimentary rocks (Fundy basin). Isolated outliers of Cretaceous and Tertiary sedimentary rocks also have been mapped in the western Gulf of Maine (Oldale and others, 1973). The entire area is veneered by Pleistocene drift and Holocene fill that is inferred to be as much as 200 m thick (Emery and Uchupi, 1972, Fig. 151).

Data Bases

Our knowledge of Georges Bank is based chiefly on inferences from geophysical data coupled with rock samples obtained from dredge hauls, the dives by submersibles, and from drill holes. Approximately 6,500 km of 48-channel common-depth-point (CDP) seismic-reflection profiles collected by the U.S. Geological Survey (USGS) and by the Bundesanstalt fur Geowissenschaften und Rohstoffe (BGR) of the Federal Republic of Germany (30-km spacing regional grid and a 10-km grid over the outer shelf and upper rise) form the primary data used to map basement structures and sedimentary features. These data are supplemented by a grid of approximately 4,000 km of six-channel seismic reflection profiles collected by the Woods Hole Oceanographic Institution (WHOI); Austin and others, 1980; Fig. 3). Information from these profiles has been used to compile a depth-to-basement contour map (Fig. 4) and sediment thickness contour map (Fig. 5).

Additional geophysical information has been derived from magnetic, gravity, and seismic-refraction surveys. The magnetic-anomaly contour map (Fig. 6) was compiled from a coarse-grid aeromagnetic survey by Project Magnet (U.S. Naval Oceanographic Office, 1966; Taylor and others, 1968) and a detailed aeromagnetic survey by the USGS (Klitgord and Behrendt, 1977; Behrendt and Klitgord, 1980). Magnetic data have provided in-

Schlee, J. S., and Klitgord, K. D., 1988, Georges Bank basin: A regional synthesis; *in* Sheridan, R. E., and Grow, J. A., eds., The Geology of North America, Volume I-2, The Atlantic Continental Margin, U.S. Geological Society of America.

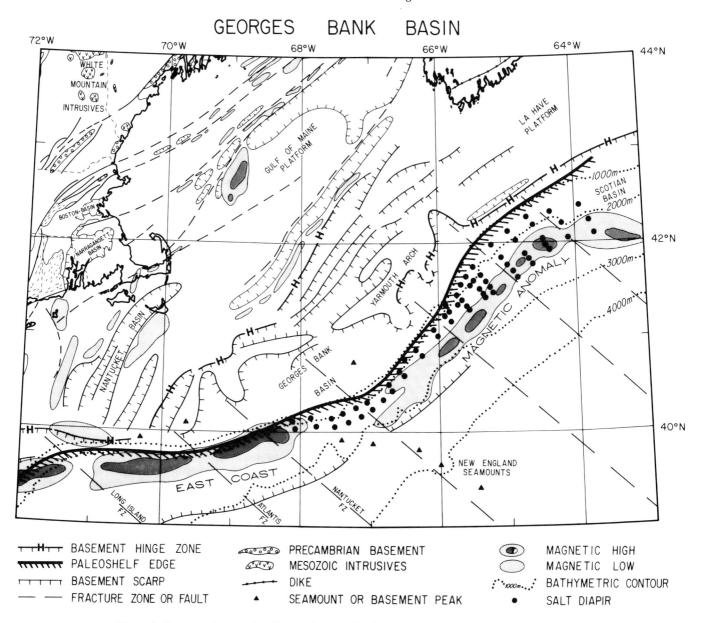

Figure 1. Tectonic elements for Georges Bank–Gulf of Maine region, including faults and grabens structures on the platforms and along the landward edge of Georges Bank basin, ancient shelf edge along seaward edge of Georges Bank basin, salt-diapir province along the seaward edge of Georges Bank, and fracture zones (FZ) and scarps seaward of the bank (from Klitgord and Schlee, 1986, Fig. 12).

Figure 2. Schematic cross section through Georges Bank region, showing distribution of sedimentary wedge bound by sea floor and crystalline basement (from Schlee and Klitgord, 1981). Sediment-facies relationships are indicated. The wavey line above crystalline basement represents postrift unconformity.

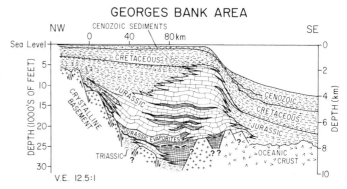

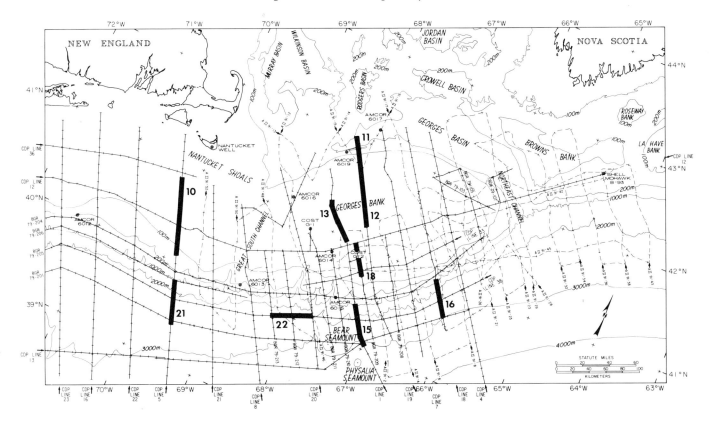

Figure 3. Map showing seismic-reflection and drill-hole data on Georges Bank: black circles, boreholes; light solid lines, USGS and BGR multichannel seismic profiles; dashed lines, WHOI six-channel seismic profiles; heavy solid lines, locations of records figured in this chapter (figure numbers in bold type). Bathymetric contours, in meters, from Uchupi (1965).

formation concerning the geometry and composition of the crystalline basement. Depth-to-basement estimates from magnetic data (Klitgord and Behrendt, 1979) have been used to interpolate basement structure between seismic lines and to infer depth to basement where it is obscured on seismic data (Fig. 4). Data from gravity and seismic-refraction surveys (Grow and others, 1979; Sheridan and others, 1979) have been used to evaluate crustal thicknesses and density structure. The free-air gravity-anomaly contour map (Fig. 7) has also been useful in mapping regions of low- and high-density rock for areas on the shelf where basement is fairly shallow.

Rock samples of offshore subsurface geologic units are sparse. Drill-hole data exist from shallow wells (300 m) on the continental shelf (Hathaway and others, 1979; Poag, 1982a,b), a 102-m well (Coskata) and a 500-m well (AMCOR 6001) on Nantucket Island (Folger and others, 1978), and two deep wells (4,899 and 6,667 m) drilled by consortia of oil companies—COST (Continental Offshore Stratigraphic Test) G-1 and COST G-2 (Amato and Bebout, 1980; Amato and Simonis, 1980; Scholle and Wenkam, 1982; Figs. 3 and 8). On the adjacent Canadian margin, data are publicly available from several ex-

ploratory wells, including the 2,126-m Shell MOHAWK-B-93 well on Browns Bank (Fig. 3) and the 4,393-m Shell MOHICAN-100 well near the Scotian margin shelf edge (Given, 1977; Barss and others, 1979). These data have been supplemented by samples from dredge hauls (Weed and others, 1974) and from dives by submersible (Ryan and others, 1978; Valentine and others, 1980).

STRUCTURAL FRAMEWORK

The primary structural elements of the Georges Bank basin region are typical of passive continental margins: shallow basement platform, deep marginal sedimentary basin, and deep ocean basin. The tectonic evolution of a passive margin creates distinctive crustal structures, basement structures, and sediment-distribution patterns. Passive continental margins form as the result of a continent breaking apart (continental rifting) and then moving apart (continental drift) to create a new ocean basin (sea-floor–spreading). During the rifting phase of margin development, crustal stretching, thinning, and block faulting takes place along the rift zone as the two parts of a plate slowly move

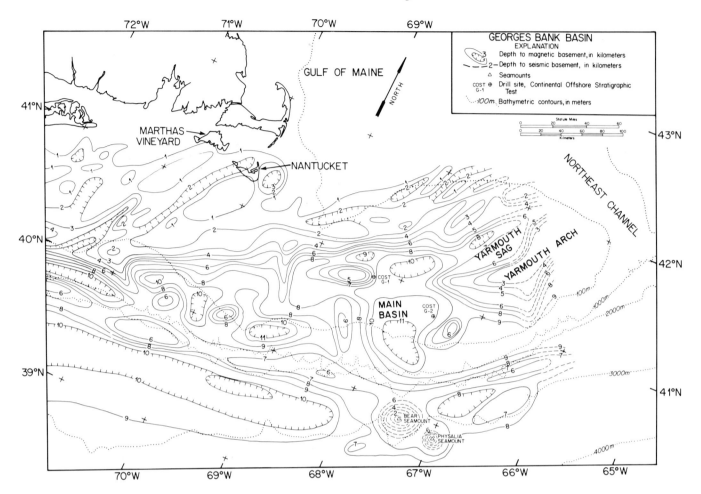

Figure 4. Depth-to-basement map (from Klitgord and Behrendt, 1979, Fig. 7c) determined by using magnetic-depth estimates to interpolate between seismic-basement depth measurements on USGS CDP line numbers 1, 4, 5, 7, 8, 12, and 13. Contour interval is 1 km. Universal Transverse Mercator projection.

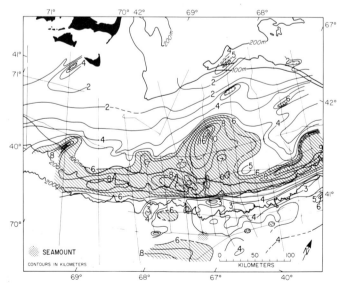

Figure 5. Isopach map of Late Triassic and younger sedimentary rocks in the Georges Bank basin. Thickness in kilometers. Control for thickness was measured along the profiles indicated.

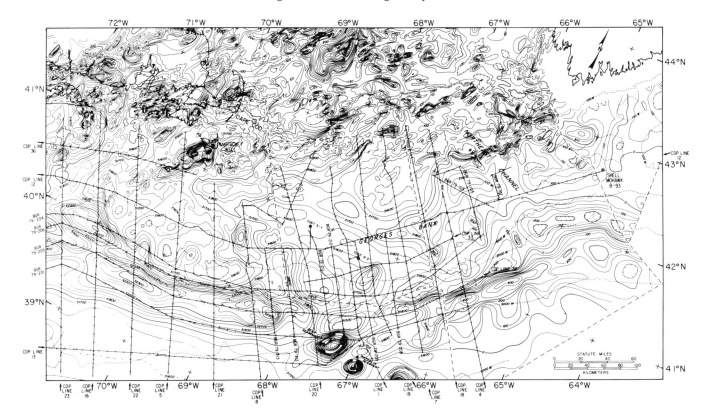

Figure 6. Magnetic-anomaly map compiled from the maps of Klitgord and Behrendt (1977) and Haworth and MacIntyre (1975). Contour interval is 50nT. Grid of USGS and BGR multichannel seismic-reflection profiles is shown. Universal Transverse Mercator (UTM) projection. From Klitgord and others (1982, Fig. 73a).

apart. Extensional tectonic activity in the rift terminates when the extensional plate boundary moves seaward and the sea-floor–spreading process creates new oceanic crust at a midocean ridge. A new deep ocean basin continues to widen during the sea-floor–spreading phase, and margin development is now controlled by thermal and sediment-loading tectonic processes interacting with oceanographic processes.

Margin formation in the Georges Bank region began in early Mesozoic time, when the New England coast was adjacent to the Moroccan coast of Africa, and a depression floored by large fault-bounded blocks of thin continental crust characterized the rift zone. The rate and amount of separation between Africa and North America was sufficient by the Middle Jurassic for the development of a spreading center and the creation of oceanic crust. Sediment accumulation subsequent to rifting in Georges Bank basin on a rapidly cooling and subsiding basement produced a shelf, slope, and rise morphology at the edge of the Atlantic Ocean basin.

Basement Structure

In our study of structural framework, basement structures refer to the structures associated with crystalline basement. *Crys-* *talline basement* is the base of the sediment fill and includes igneous, metamorphic, and metasedimentary rocks. This basement surface is mapped by using both seismic reflection and magnetic data; where appropriate, the surface is identified as either acoustic basement or magnetic basement. *Acoustic basement* refers to the deepest surface observable on a seismic-reflection profile. Depending upon the quality of the seismic data and geologic conditions, acoustic basement can be deeper than crystalline basement or, more commonly, shallower. Where acoustic basement is deeper, crystalline basement is usually well defined by using seismic data, but where acoustic basement is shallower, other information such as magnetic-basement depth estimates must be used. *Magnetic basement* is the upper surface of the source for magnetic anomalies. Magnetic anomaly sources are not unique and depth-to-source estimates are treated with caution. Compared with crystalline rocks, sedimentary rocks are not very magnetic; therefore, magnetic basement provides a rough estimate of crystalline basement. Magnetic depth-to-basement estimates are particularly useful for interpolating basement depths between seismic lines as well as estimating basement depths where acoustic basement is shallower than crystalline basement. For example, the carbonate buildup at the shelf edge, formed

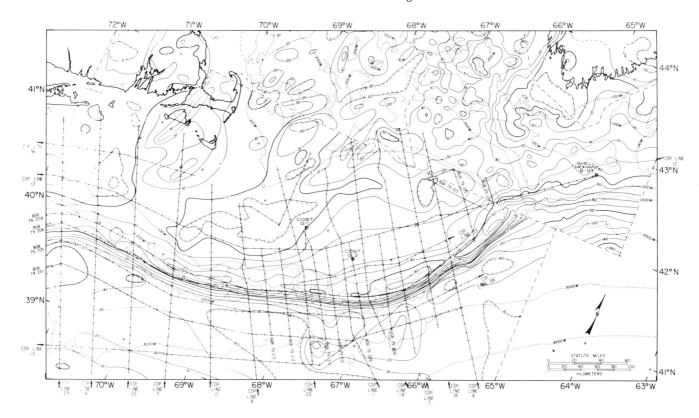

Figure 7. Free-air gravity-anomaly contour data for Georges Bank region and Bouguer gravity-anomaly contour data for Brown's Bank and Gulf of Maine. Contour interval 10 mgals. Grid of USGS and BGR multichannel seismic-reflection profiles is shown. Universal Transverse Mercator projection. From Klitgord and others (1982, Fig. 76a).

during the Jurassic and Early Cretaceous, absorbs most acoustic energy, and little reflected energy is available for imaging structures beneath it. Magnetic-basement depth estimates are used in this region.

In the Georges Bank region, crystalline basement includes Paleozoic and older rocks on the platforms, block-faulted pre-Mesozoic rocks mixed with Triassic and Jurassic igneous rocks beneath the marginal basin, and Jurassic and younger igneous rocks seaward of the marginal basin. Basement samples are few, and most information about basement is inferred from seismic, magnetic, and gravity data. On Georges Bank, graphitic slate, schist, and phyllite (radiometric age of 540–450 Ma) were sampled in the COST G-1 hole at 4,740 m below sea level (Fig. 3; Amato and Bebout, 1980). Weathered basalt (183 ± 8 Ma, Early Jurassic) was encountered in the AMCOR-6001 Nantucket well at a depth of 459 m below sea level (Folger and others, 1978). This basalt is part of a broad flow unit that overlies sedimentary rock in some places and crystalline basement in others (Valentine, 1981). Crystalline rock has not been sampled offshore on the Long Island platform west of Nantucket, but basement is exposed onshore in southern New England and is

formed of various units of Paleozoic-age igneous, plutonic, metamorphic, and metasedimentary rock (Lyons and Brownlow, 1976; Zartman and Naylor, 1984). Numerous dredge samples of middle Paleozoic granitic rock have been obtained from the Gulf of Maine platform (Ballard and Uchupi, 1975; Hermes and others, 1978) and Paleozoic granitic rock was drilled on the LaHave platform at the Shell MOHAWK B-93 well (Fig. 3; Given, 1977). Basaltic igneous rock has been sampled from oceanic crust in the deep sea by the Deep Sea Drilling Project (DSDP), but the closest well is several hundred kilometers away from Georges Bank.

Seismic-reflection profiles show great variability in depth and character of acoustic and crystalline basement. Crystalline basement can be divided into four characteristic zones (Fig. 9): (1) a low-relief zone of Paleozoic or older metamorphic and igneous rocks forming *continental basement* (upper surface of continental crust) on the shallow stable platforms; (2) a *basement hinge zone* (blockfaulted zone of Paleozoic or older metamorphic and igneous rocks that deepens seaward in distinct steps); (3) the *marginal sedimentary basin* where crystalline basement is masked by thick sedimentary units that contain prograding carbonate

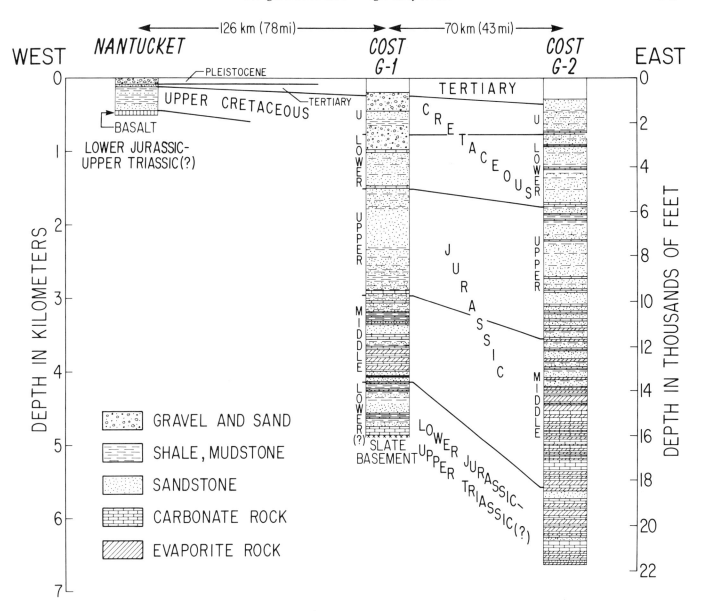

Figure 8. Strata penetrated in Nantucket borehole (AMCOR-6001), COST G-1 well, and COST G-2 well. Siliciclastic deposits in nearshore (west) grade to carbonate deposits offshore (east during the Jurassic when basin subsidence was most rapid); subsequently, siliciclastic deposits predominated. COST G-1 and G-2 stratigraphic information from Amato and Bebout (1980) and Amato and Simonis (1980); system boundaries from Poag (1982a, 1982b); AMCOR-6001 from Folger and others (1978).

deposits at the seaward edge; and (4) a zone in the deep ocean basin of igneous rock forming *oceanic basement* (upper surface of oceanic crust) produced by sea-floor–spreading.

Zone 1. The first basement zone is on the Long Island, Gulf of Maine, and LaHave platforms, where basement depths are generally less than 5 km. Basement is characterized by a strong acoustic reflector that indicates short-wavelength undulating relief (Figs. 10 and 11)—the eroded surface of crystalline rock. Basement structures on the platforms are dominated by half-

grabens that have been partially eroded (Klitgord and Hutchinson, 1985; Hutchinson and others, 1986).

A block-faulted zone on the southeastern side of the Long Island and Gulf of Maine platforms includes three small, elongate subbasins (Nantucket, Atlantis, and Franklin basins) beneath the postrift unconformity (see Basin Fill section below; Figs. 1, 10, and 11). Depth to the tops of these subbasins on the Long Island platform deepens slowly to the south-southeast (e.g., CDP Line 5, Fig. 10), where the seaward ends of the subbasins are obscured by

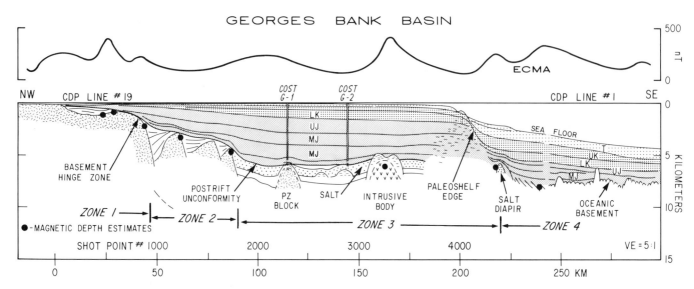

Figure 9. Typical cross section for Georges Bank basin based on USGS CDP seismic line number 19 (modified from Klitgord and others, 1982, Fig. 72), showing different zones of basement type. Dashed lines are basement and sub-basement structures beneath postrift unconformity that have been projected from adjacent seismic profiles. Magnetic-anomaly profile is shown above (East Coast Magnetic Anomaly [ECMA]); locations of COST G-1 and G-2 wells are projected onto profile. Stratigraphic units: Tertiary (T), Upper Cretaceous (UK), Lower Cretaceous (LK), Upper Jurassic (UJ), Middle Jurassic (MJ), Paleozoic (PZ) (from Poag [1982a, 1982a]). Locations of depth-to-magnetic basement estimates from Klitgord and Behrendt (1979) are indicated.

the carbonate buildup near the shelf edge, consisting of Upper Jurassic and Lower Cretaceous rocks. Both the Nantucket and Franklin basins are bounded on their northwest edge by a major border fault (Hutchinson and others, 1986) and on their southeast edge by large magnetic and gravity highs (Figs. 6 and 7).

Zone 2. The basement hinge zone separates the Long Island and Gulf of Maine basement platforms from the deep marginal sedimentary basin. On a few seismic profiles across zone 2, the block-faulted basement and back-tilted sedimentary rocks are distinctive structures beneath the postrift unconformity (Fig. 12). On poorer quality seismic profiles, acoustic basement is shallower than crystalline basement and often coincides with the postrift unconformity (Fig. 13). The nature of crystalline basement somewhere beneath acoustic basement can be inferred only from the character and shape of this unconformity surface.

The block-faulted zone southeast of the Gulf of Maine platform deepens in a series of steps into a broad crustal depression (Yarmouth sag) adjacent to Yarmouth arch (Figs. 1 and 4). Nestled between each "step" is a small subbasin formed by the downdropped block (graben); over these subbasins, the postrift unconformity separates steeply dipping older sedimentary reflectors from more conformable, flat-lying sedimentary units (Fig. 12). The rough character of acoustic basement at the updip side of the half-graben (shot point 1200, Fig. 12) may indicate crystalline basement; on some profiles, this rough basement can be traced beneath the dipping reflectors. On profiles where the

postrift unconformity coincides with acoustic basement (Fig. 13), the undulations of the postrift unconformity over the block-faulted basement structures give acoustic basement a step-like shape.

These graben structures have distinctive magnetic-anomaly patterns associated with them. Magnetic-anomaly lows correlate with deeper basement beneath shallow grabens, whereas magnetic highs are over basement peaks and may be caused by intrusive igneous dikes along the edges of the fault blocks (Klitgord and Hutchinson, 1985; Figs. 6 and 9). Using magnetic-anomaly trends and magnetic-basement information, we infer that crystalline basement forms the bottom of a series of lineated half-grabens beneath the postrift unconformity. Crystalline basement depths at the top edges of the grabens and basins range from less than 2 to more than 6 km.

Yarmouth arch is separated from the Gulf of Maine platform block-faulted zone by the Yarmouth sag (Figs. 1 and 4). These two features form the basement transition between the LaHave platform and the main Georges Bank basin, and at their eastern end they appear to form part of the Gulf of Maine platform block-faulted zone. The Yarmouth sag and arch deepen to the southwest (Figs. 4 and 14) until CDP line 19, where the arch is inferred to be only a small basement high that has a buildup of carbonate rock on top (Poag, 1982a), and the sag is at the seaward edge of the Gulf of Maine platform block-faulted zone. A series of dip lines across this zone (Fig. 14) clearly show this

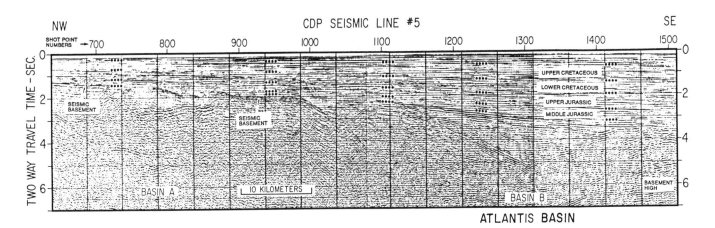

Figure 10. Part of USGS CDP seismic line number 5 across the Atlantis basin (from Klitgord and others, 1982, Fig. 81). Note changes in slope for the postrift unconformity (arrows), and basins beneath the unconformity. See Figure 3 for location. Dots delimit major seismic sequences.

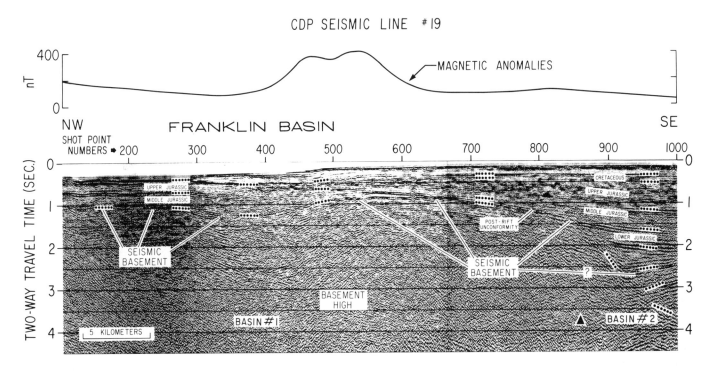

Figure 11. Part of USGS CDP seismic line number 19 across Franklin basin, just landward of Georges Bank basin hinge zone, showing relationship between magnetic-anomaly locations and graben structure (from Klitgord and others, 1982, Fig. 75). Dots delimit major seismic sequences, and magnetic anomaly line is shown above the profile. Triangle at shot point 850 indicates landward edge of half-graben. See Figure 3 for location.

CDP SEISMIC LINE #19

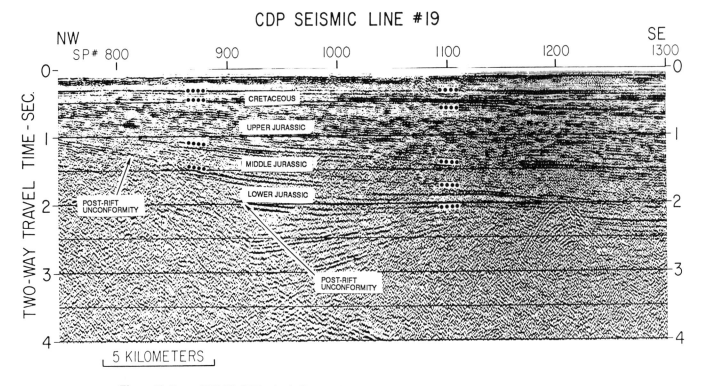

Figure 12. Part of USGS CDP seismic line number 19, showing landward-dipping reflectors beneath postrift unconformity for the landward most of the half-grabens (basin number 2 in Fig. 11) that form the steps in basement (from Klitgord and others, 1982, Fig. 78). Dots delimit major seismic sequences. See Figure 3 for location.

change in character and depth from a platform to a deep basin in the absence of a hinge zone. The southeast side of Yarmouth arch is marked by a sharp drop in acoustic basement, which we infer to be a basement hinge zone at the landward edge of the northeast extension of Georges Bank basin. This hinge zone continues to the northeast (Figs. 1 and 4), and beyond Northeast Channel it becomes the boundary between the LaHave platform and the Scotian basin (Jansa and Wade, 1975). At Northeast Channel, a significant embayment of this hinge zone (Wade, 1977) marks the boundary between the LaHave platform and Yarmouth arch. This embayment is caused by a graben-like basement feature that underlies the seaward part of Northeast Channel (Austin and others, 1980, Fig. 3). Disruptions in the magnetic data indicate that the embayment may continue as a fault zone to the northwest beneath the channel.

Zone 3. The main basin of Georges Bank basin is within zone 3, between the block-faulted zone and the Mesozoic paleoshelf edge. This basin is the region of greatest sediment accumulation (Fig. 5). A few basement peaks can be found on the seismic data, but over most of the basin, acoustic basement is either the postrift unconformity or conformable sedimentary units just beneath it. COST G-1 well was drilled into an uptilted Paleozoic block (Figs. 9 and 13) and several intrusive bodies can be identi-

fied by using magnetic and gravity data (e.g., at shot-point 3100 on CDP line 19, Fig. 9). Crystalline basement beneath the seaward edge of the main basin is marked by the carbonate buildup associated with the Upper Jurassic and Lower Cretaceous rocks of the shelf edge that has been identified along the entire Atlantic margin (Schlee and others, 1979; Jansa, 1981).

The East Coast Magnetic Anomaly (ECMA) is a large linear magnetic anomaly near the shelf edge of the continental margin from Nova Scotia to Georgia (Figs. 1, 6, and 9; Taylor and others, 1968; Klitgord and Behrendt, 1979). It marks the seaward edge of the marginal sedimentary basin and the landward limit (Fig. 9) of oceanic basement (zone 4; Klitgord and Behrendt, 1979). There is a large drop in basement just landward of a shallow ridge in basement (Fig. 4) beneath the landward edge of the ECMA. The paleoshelf edge lies just landward of this basement structure beneath the ECMA and early sediment accumulation patterns may have been localized there by these structures (Klitgord and Behrendt, 1979). Just seaward of the paleoshelf edge, acoustic basement is a series of benches directly beneath the ECMA. These benches have been traced northeastward to the salt ridges off the Scotian margin (Klitgord and Schlee, 1986) and are interpreted as salt diapirs that have trapped sedimentary rock behind them (Fig. 15).

CDP SEISMIC LINE #1

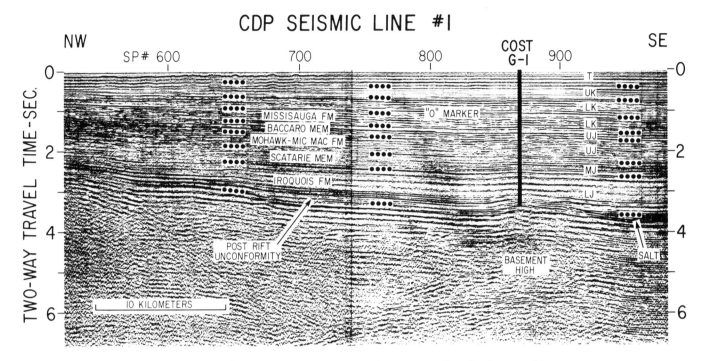

Figure 13. Part of USGS CDP seismic line number 1 over seaward section basement hinge zone (modified from Klitgord and others, 1982; Fig. 79). Notice drops in postrift unconformity of shot points (SP) 550 and 680 that coincide with landward edges of half-grabens. The small basement high that COST G-1 well penetrated is seen as a small rise in the postrift unconformity. Deep reflectors at 6 seconds between shotpoints 650 and 850 may be real (Schlee and others, 1976), indicating a thick sequence of prerift and synrift sedimentary rock in this region, or they may be caused by normal faults in crystalline rock. The stratigraphic nomenclature is after McIver (1972); the Baccaro and Scatarie members are part of the Abenaki Formation. Abbreviations are the same as in Figure 9. See Figure 3 for location.

Zone 4. Acoustic basement in zone 4 has seaward-dipping reflectors at its landward edge and then closely spaced hyperbolic echoes and generally low relief features that typify oceanic basement (Klitgord, 1983). A basement scarp seaward of the ECMA (Fig. 1) separates the zone of seaward-dipping reflectors (Fig. 16) from the basement that looks like typical oceanic crust. The New England Seamount Chain (Uchupi and others, 1970) is within zone 4 and intersects the margin of the Georges Bank at the southwestern edge of the main basin. Bear Seamount and a large gap in the ECMA are at this intersection. These seamounts are in northwest-trending en-echelon groups along oceanic fracture zones (Schouten and Klitgord, 1977), although the gross overall trend of the chain is more west-northwest. Age dating of the seamounts indicates that they become progressively younger to the east (Vogt and Tucholke, 1978; Duncan, 1984) and may have been caused by the North American plate moving over a hot spot. Several basement highs landward of the paleoshelf edge (Figs. 1 and 4) may be Early Cretaceous intrusive bodies associated with the tectonic event that caused the formation of the New England Seamounts.

Crustal Structure

The three primary crustal types in the Georges Bank region are continental, transitional (or rift stage), and oceanic crust: they coincide with the primary structural elements—Gulf of Maine platform, the block-faulted zone and Georges Bank basin, and the deep ocean basin. Continental crust is thickest and underlies the basement platforms. Oceanic crust is the thinnest and underlies oceanic basement in the deep-ocean basin. Crust of the deep marginal sedimentary basin is of intermediate thickness, and its nature is the most speculative. Estimates of crustal thickness and type come from seismic-refraction profiles (Sheridan and others, 1979), gravity models (Grow and others, 1979), and thermal subsidence models (Sawyer and others, 1982, 1983). Gravity model studies indicate that the continental crust under the platform (~35 km thick) thins rapidly at the basement hinge zone and reaches oceanic crustal thickness (~5 km) at the ECMA (Fig. 17). In broader sections of the marginal basin, gravity modeling suggests that the crustal thickness thins rapidly at the hinge zone, becomes more uniform beneath the marginal basin, and

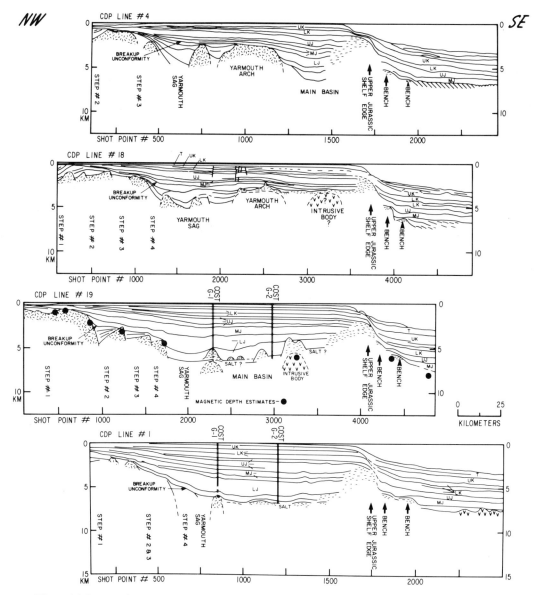

Figure 14. Interpretive cross sections of Georges Bank basin seaward of Gulf of Maine platform along USGS CDP seismic line numbers 4, 18, 19 and 1 (Grow and Schlee, 1976; Grow and others, 1979; Poag, 1982a, 1982b; Klitgord and others, 1982). Letters and lines mark boundaries of seismic sequences. Abbreviations are same as in Figure 9. See Figure 3 for locations.

then thins again at the ECMA (B. A. Swift and others, in preparation). A similar variation in crustal structure has been proposed for the Carolina trough (Hutchinson and others, 1983) and the Baltimore trough (Grow and others, 1983). Subsidence modeling on Georges Bank basin using the COST G-1 and G-2 well data and total tectonic subsidence estimates along CDP line 5 (Sawyer and others, 1982, 1983) are consistent with the crustal thinning estimated from gravity modeling (Sheridan and others, Chapter 9).

Structures within the crust are most readily seen on our seismic profiles across the platform and basement hinge zone.

Crust-cutting faults can be traced downward from the border faults of the half-graben structures at the hinge zone (e.g., Fig. 9) and on the edges of the Nantucket and Atlantis basins (Hutchinson and others, 1986). These features are interpreted as normal faults formed in response to extensional tectonic forces during the rifting phase of margin development; some may be reactivated Paleozoic thrust faults.

BASIN FILL

The Georges Bank sedimentary fill is a prism (Figs. 4 and 5), thickest over several small, irregularly shaped, interconnected

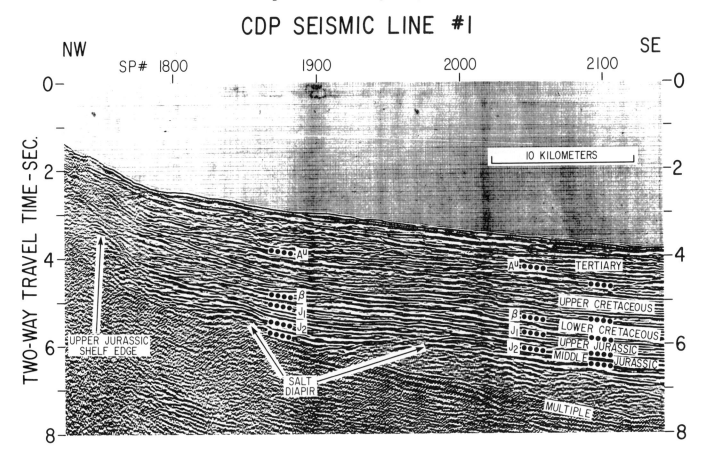

Figure 15. Part of USGS CDP seismic line number 1 across Jurassic shelf edge (from Klitgord and others, 1982, Fig. 82). Identification of deep-sea Jurassic units is from Klitgord and Grow (1980). Age identification of shelf seismic units from Poag (1982a, 1982b). Note how deep reflectors, J_2 and J_1, onlap buried benches labeled salt diapirs and Upper Jurassic shelf edge respectively. See Figure 3 for location. A^u is a conspicuous Tertiary unconformity; β is a horizon that marks the top of the Neocomian; J_1 is a horizon that marks the top of the Cat Gap Formation (~138 Ma); J_2 marks the top of a middle Jurassic sequence; dots mark major sequence boundaries (Klitgord and Grow, 1980).

basins; it thins gradually seaward of the paleoshelf edge and rapidly toward the platforms at the hinge zone. The fill is internally divided top to bottom by a conspicuous unconformity that separates synrift marine and nonmarine deposits from postrift marine deposits; this unconformity is called the *postrift unconformity.* The sedimentary prism is also divided laterally by a buried shelf-edge carbonate platform of Jurassic and Early Cretaceous age that underlies the present Continental Slope. This opaque zone separates sedimentary rock deposited on the shelf from deep-sea sediment deposited during the Jurassic and Early Cretaceous. In the following sections, we use the postrift unconformity and the carbonate platform edge to divide our discussion of basin fill.

Biostratigraphic and lithostratigraphic studies of COST G-1 and G-2 well samples (Fig. 8; Amato and Bebout, 1980; Amato and Simonis, 1980; Arthur, 1982; Poag, 1982a, 1982b; Scholle and Wenkam, 1982; Folger and others, 1978; Hathaway and others, 1976) have provided age, rock type, and environments of deposition (Poag, 1982a, 1982b; Valentine, 1982) for the sedimentary units deposited in the Georges Bank region. These properties have been extrapolated to other parts of the basin by studies of the character of seismic reflections (Sangree and Widmier, 1977; Bubb and Hatlelid, 1977; Schlee and Fritsch, 1982); moreover, seismic-reflection data are also used to map the thickness and distribution of sedimentary units away from the drill sites. We have used a wealth of stratigraphic information from the Canadian margin (Given, 1977; Eliuk, 1978; Wade, 1977) to correlate those deposits with similar type and age deposits of Georges Bank (Judkins and others, 1980; Poag, 1982a).

Postrift Unconformity

The *postrift* (or breakup) *unconformity* (Falvey, 1974; Montadert and others, 1979) is a conspicuous acoustic reflector that

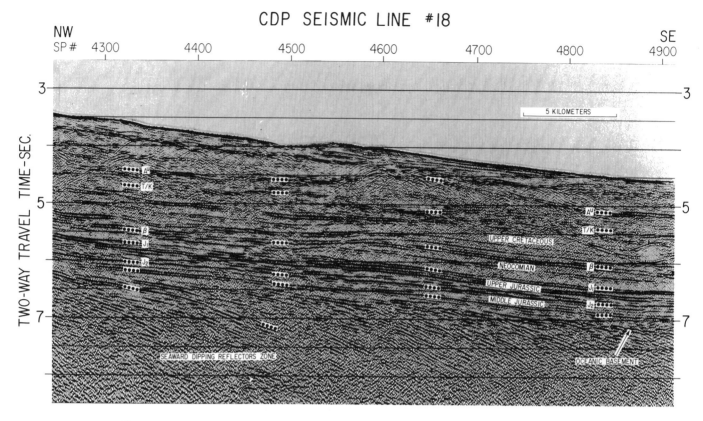

Figure 16. Part of USGS CDP seismic line number 18 across zone of seaward-dipping reflectors. Stratigraphic units above these dipping reflectors are from Klitgord and Grow (1980) and Schlee and others (1985). See Figure 15 for explanation of symbols; T/K marks the Tertiary-Cretaceous boundary. See Figure 3 for location.

marks a major change in deposition pattern during basin evolution. This unconformity separates synrift sedimentary rock from overlying postrift sedimentary rock (Fig. 12). This boundary is caused by a change in tectonic regime from block faulting during rifting to broad crustal subsidence during the passive tectonic phase as the active plate margin shifted seaward. Originally flat-lying strata were tilted and offset by block faulting and rapid flexure of basement during the rift stage of continental breakup. The unconformity is best defined in the block faulted areas at the edge of the platforms where sedimentary rocks within grabens appear as seaward- or landward-dipping reflectors truncated by the postrift subhorizontal reflectors (Fig. 12). Unlike erosion in the basins on the basement platforms, erosion at or seaward of the hinge zone did not cut deeply into crystalline rock; most of the synrift deposits seem to have been preserved in the half-grabens seaward of the top of the hinge zone.

Synrift Deposits (Triassic and Lower Jurassic?)

Sediments that accumulated as continental rifting was taking place (*synrift deposits*) are 400 m of basal red-to-pink poorly sorted nonmarine conglomerates and sandstones and thin mar-

ginal marine limestones in the COST G-1 well (Amato and Bebout, 1980; Arthur, 1982; Poag, 1982a); the redbeds are similar to rocks associated with Triassic and Jurassic grabens onshore (Cornet, 1977; Manspeizer and others, 1978). The COST G-1 hole (Fig. 8) was drilled near a basement high at the western edge of the main basin, in an area where a seismic profile reveals a rapidly thinning sequence marked by alternating continuous high amplitude and discontinuous variable-amplitude reflections (interbedded marine and nonmarine strata). The postrift unconformity is interpreted as the acoustic reflector at the top of a conglomerate layer at COST G-1; it can be traced acoustically to the COST G-2 well, where it coincides with a thin red siltstone-sandstone layer, which caps a 300-m layer of dolomite overlying salt (Amato and Simonis, 1980). Some patches of mounded chaotic reflectors beneath the postrift unconformity have been interpreted as carbonate mounds (Fig. 18; Poag, 1982a; Schlee and Fritsch, 1982). The COST G-2 well penetrated only the upper 12 m of the salt, but seismic evidence suggests that basement is at least 500 m below the top of the salt. The salt has not been dated at the COST G-2 site; it may be as old as Late Triassic. On the Canadian margin, salt interfingers with Upper Triassic redbeds landward of the hinge zone (Jansa, 1981),

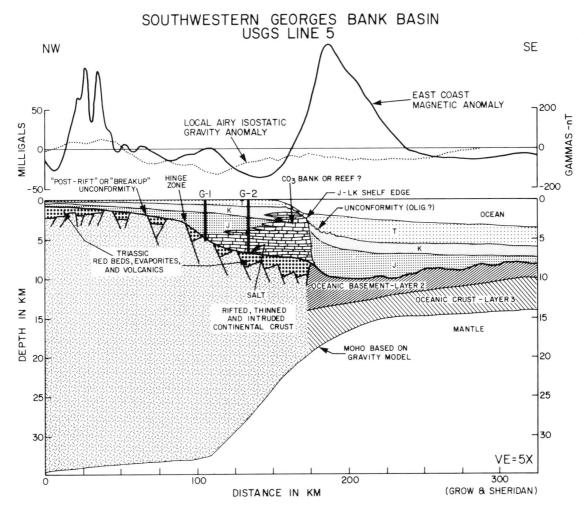

Figure 17. Composite geologic section across southwest end of Georges Bank basin based on USGS CDP seismic line number 5 and drill-hole data projected from the COST G-1 and G-2 wells (Amato and Bebout, 1980; Amato and Simonis, 1980). Mohorovicic discontinuity (Moho) based on gravity model along line number 5 (Grow and others, 1979). Note hinge zone and underlying change in depth to Moho. Abbreviations are same as in Figures 9 and 16.

whereas seaward of the hinge zone, salt is overlain by Lower Jurassic limestone in the Mohican I-100 well (Barss and others, 1979). The restricted seaward limit of the salt-diapir provinces off Nova Scotia–Georges Bank and Morocco (Figs. 1 and 23) (Jansa and Wade, 1975; Uchupi and Austin, 1979; Hinz and others, 1982; Klitgord and others, 1982) indicates that evaporitic conditions did not exist much later than the earliest opening phase of the Atlantic in Early Jurassic time.

Age data for the synrift section is very poor. Cousminer and others (1984) have found Late Triassic dinoflagellate cysts in one sample at a depth of 4.44 km in the COST G-2 well, well above the postrift unconformity (see section on Paleooceanography, chapter 16, for a more complete discussion). Undoubtedly, a Late Triassic phase of salt deposition took place in the North Atlantic (Jansa, 1981), but what distribution did it have seaward of the

basement hinge zone? Tectonic evolution and subsidence studies of the Atlantic margin (Haworth and Keen, 1979; Keen, 1979; Klitgord and Behrendt, 1979; Watts, 1981; Klitgord and others, 1982; Sawyer and others, 1982) indicate large amounts of faulting and subsidence during the rifting phase. Preservation during the rifting phase of thick, broad layers of synrift salt, anhydrite, and dolomite landward of the basement hinge zone is unlikely because of this synrift tectonic activity; this was our primary criterion for identifying the postrift unconformity (Fig. 9). Thus, we would place samples from a depth of 4.44 km at COST G-2 within the postrift sedimentary section. The Late Triassic–age sample of Cousminer and others (1984) and Manspeizer and Cousminer (this volume) is from within the Jurassic postrift sedimentary sequence of Poag (1982a, 1982b), just beneath the carbonate layer correlated with the Scatarie Member of the Abenaki

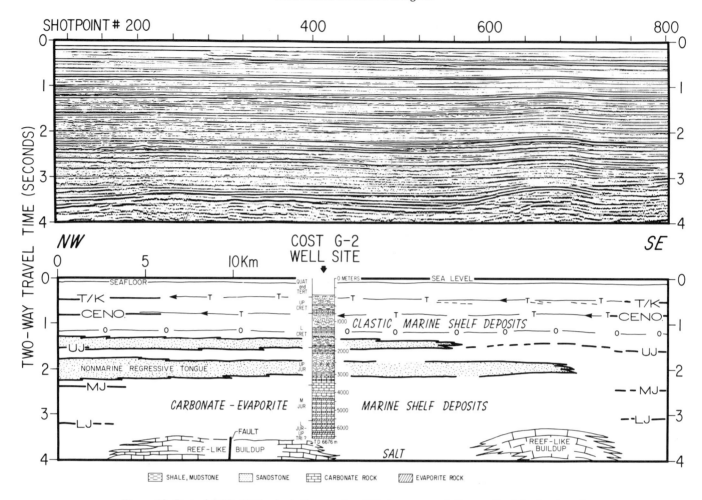

Figure 18. Part of BGR CDP seismic line number 209 across central Georges Bank basin. Actual time-scaled profile is shown; interpretation of stratigraphy is given below. Superimposed on interpretation section is lithology of COST G-2 well at approximate position where profile crossed drill site. T/K, Tertiary/Cretaceous boundary; CENO, Cenomanian; UJ, near top of Upper Jurassic section; MJ, within Middle Jurassic section; LJ, near base of Lower Jurassic(?) section; T, thin transgressive unit; O, correlation with the "O marker"—a zone of limestone within Missisauga Formation beneath Scotian margin (Jansa and Wade, 1975). See Figure 3 for location. From Klitgord and others (1982, Fig. 84).

Formation of Middle Jurassic age on the Scotian margin (Wade, 1977; Poag, 1982a, 1982b); we interpret this sample as reworked material, probably from the Franklin basin.

Normal faults within synrift deposits and possibly crystalline rock can be inferred from acoustic records (Figs. 12 and 13). The maximum thickness of synrift deposits measured on seismic profiles is about 3 km. This thickness may be a minimum because, on most profiles over the main basin, crystalline basement is buried beneath a salt layer that masks returns of acoustic energy. Seaward-dipping reflectors beneath the postrift unconformity along the landward edge of the main basin could be from crystalline basement, overlain by at least 6 km of synrift sediments, or they may be from normal faults offsetting crystalline rock so that only about 3 km of synrift sedimentary rocks are in the main basin.

Beneath the upper continental rise at or landward of the ECMA, possible rift deposits may also be present (Figs. 15 and 16). Basement configuration in the area of the lower slope–upper rise is in the form of linear benches (see previous section on Basement Structure) associated with salt diapirs (Klitgord and Schlee, 1986). Between the benches, in an area of limited areal extent, possible synrift deposits are characterized by moderately continuously reflections of variable amplitude.

Early Postrift Deposits (Jurassic)

The change to a regime of broad basin subsidence was attended by the formation of a broad carbonate bank, as open marine conditions became more prevalent. The carbonate bank may have been built from coalescing carbonate banks above elevated basement blocks or possibly salt swells. Earliest stages were probably marked by patch reefs and banks that were separated by moats and deep-water channels. (See Eliuk [1978] for a description of the Late Jurassic moat-reef complex on the Scotian Shelf.)

The character of the early postrift deposits changes from a dominantly clastic section in the northwest section of the basin (COST G-1) to a dominantly carbonate section in the east (COST G-2; Fig. 8). The lowermost section of postrift Jurassic(?) rocks (lower Pleinsbachian?–Sinemurian?; Poag, 1982a) or lowermost Middle Jurassic? (Poag, this volume) at the COST G-1 is 400 m of dolomite and sandstone interbedded with minor shale and anhydrite. This unit corresponds to a 580-m section of limestone, dolomite, and anhydrite at the COST G-2 site (Fig. 8). Seismic data in the area between the two wells are characterized by zones of very weak discontinuous subparallel reflections and zones of continuous parallel high-amplitude reflections (e.g., Fig. 18)—probably the widespread carbonate section so prevalent in the main basin. The beds of sandstone and shale at the COST G-1 site are not found at COST G-2. Toward the shelf edge and toward the COST G-2 well, reflections become more coherent, continuous, parallel, and have high amplitudes—all characteristics that suggest a broad open-marine—shelf type of depositional environment (Schlee and Fritsch, 1982; Fig. 18). Further, the widespread distribution of high interval velocities (>5 km/sec) associated with these reflections suggests that the limestone and dolomite penetrated in the boreholes cover much of the basin complex. Their uniform lithology in the lower part of the section results acoustically in a loss of continuity at a time-depth of 3 seconds over a half-second part of the record, particularly toward the northwest half of the area.

Middle Jurassic(?) rocks grade from approximately 980 m of interbedded sandstone, shale, and limestone (Mohawk and Mic Mac formations of the Scotian margin [McIver, 1972]) at COST G-1 to approximately 1,350 m of limestone (mainly Scatarie and Baccaro members of the Abenaki Formation–Scotian margin) and minor amounts of sandstone and shale at COST G-2 (Poag, 1982a). The vertical change in the sedimentary section is indicated by an influx of clastic sediments toward the inshore part of the basin (Schlee and Fritsch, 1982; Poag, 1982a), which resulted in an intermix of inner-shelf, marine, and coastal nonmarine paleoenvironments. The age-equivalent sections at COST G-2 are of the upper part of the Iroquois Formation, overlain by a thin Mohican Formation (~100 m thick; Libby-French, 1983); it is in turn overlain by the Scatarie Member (~600 m thick) and separated from the overlying Baccaro Limestone Member (~400 m thick) by the Misaine Member (~100 m thick; Poag, 1982a, Fig. 1).

The Upper Jurassic section in COST G-1 (1,000 m thick and at a depth of 2,000–3,000 m) is mainly shale and sandstone of the Mohawk and Mic Mac formations together with abundant lignite and some coal near the top of the section (Poag, 1982a; Arthur, 1982). The Upper Jurassic rocks are about 1,300 m thick in COST G-2 (depth ~1,700–3,000 m) and, for the first time,

contain abundant sandstone, shale, siltstone, and mudstone of the Mohawk and Mic Mac formations and a few oolitic and algal limestone beds of Oxfordian and early Kimmeridgian age. In the upper Kimmeridgian and Tithonian beds at COST G-2, the section once again becomes predominantly limestone, containing beds of gray shale—stratigraphically equivalent to the Baccaro Member of the Abenaki Formation. During the early Late Jurassic, both well sites were characterized by fluctuation of alluvial, nearshore, and inner-shelf environments (Poag, 1982a; Schlee and Fritsch, 1982). The COST G-2 site returned to an inner-shelf environment by the end of the Late Jurassic. Sediment-accumulation rates were less than during the Early and Middle Jurassic, reflecting the slower subsidence of the margin. Seaward of the COST G-2 well (Fig. 18), reflection profiles indicate that the Upper Jurassic section is characterized by zones of highly continuous parallel reflections interleaved with a few zones of less continuous parallel reflection of moderate amplitude. Toward the present shelf edge, the acoustic character changes as the parallel arrangement of reflections becomes faint and their amplitude variable—this in the main area of carbonate paleoplatform development. (See section below on ancient carbonate platform.)

Late Postrift Deposits (Cretaceous-Tertiary)

Postrift Cretaceous rocks on Georges Bank are less than half as thick as the postrift Jurassic section (in COST G-1, about 1,300 m versus 2,500 m; and in COST G-2, about 1,450 m versus 3,900 m). Subsidence on the margin slowed during the Cretaceous and Tertiary and became more uniform over the entire margin, so that little thickening took place. Further, the Cretaceous marked a changeover to dominantly noncarbonate marine sedimentation.

Lower Cretaceous rocks are 900 m thick in COST G-1 and consist of loosely cemented sandstone and shale containing beds of coal, dolomite, and lignite. The equivalent section in the COST G-2 is about 1,050 m thick; although some thick limestone beds are present, sand containing beds of coal, shale, and mudstone dominates. Strata at both sites were deposited in a shallow nearshore to coastal alluvial environment. The first strata traceable on seismic profiles over the carbonate buildup that formed a paleo-shelf edge (Fig. 2) are within this section (Hauterivian in age; Poag, 1982b). Reefal rocks of Neocomian age were sampled by Ryan and others (1978) in Heezen Canyon at the eastern end of Georges Bank.

The Upper Cretaceous section is thin in COST G-1 (about 450 m) and comprises sand, shale, and gravel, together with glauconite and lignite. In COST G-2, the same section thins to about 350 m and is mainly calcareous sand, claystone, and siltstone; it is marked by several conspicuous unconformities (Poag and Schlee, 1984). Water depths increased to a middle-outer–shelf environment at COST G-2 in low-energy conditions. Seaward, the Upper Cretaceous section thickens rapidly under southwest Georges Bank to more than 1 km, and several units reappear. For example, Campanian and Maestrichtian rocks were

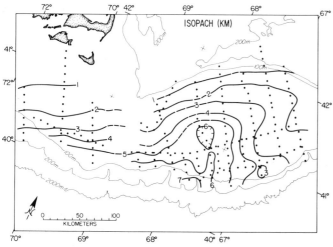

Figure 19. Isopach map of postrift sedimentary section for Georges Bank basin. Heavy contours are in kilometers. Dots indicate data points obtained from spot calculations along grid of seismic-reflection profiles. Light solid lines are bathymetric contours to delimit the Bank in meters. From Schlee and Fritsch (1982).

sampled in various canyons (Ryan and others, 1978; Valentine and others, 1980) and had been deposited in an outer shelf-bathyal environment (Poag, 1982a). At the eastern end of Georges Bank, extensive erosion and outbuilding in the Tertiary cut back the Upper Cretaceous section, so that Middle Eocene chalk unconformably overlies Neocomian limestone in Heezen Canyon (Ryan and Miller, 1981).

The Upper Cretaceous section in the Nantucket AMCOR 6001 hole (Fig. 8) 78 km northwest of COST G-1 is about 340 m thick and is composed of soft shale and unconsolidated clayey sand. At the Nantucket hole, this section of alluvial deposits is separated by the postrift unconformity from an underlying sequence of Jurassic(?) volcanic flows that have been radiometrically dated at 183 ± 8 Ma (Folger and others, 1978). At the COST G-2 well, the same-age section is 350 m thick and consists of poorly consolidated calcareous sandstone shale (Arthur, 1982).

The Tertiary section is estimated to be about 250 m thick in both COST G-1 and G-2. It consists mostly of unconsolidated sand and abundant shale, glauconite, and lignite. Thin Paleocene strata are present in both COST G-1 and G-2 and updip in the Coskata well on Nantucket (10 km northwest of the Nantucket AMCOR 6001 well; Folger and others, 1978). Eocene light-green to dark-green clay and bioclastic limestone are present in COST G-2 and AMCOR hole 6019 (Fig. 3). About 25 m of the Tertiary section in the Nantucket AMCOR 6001 well is probably Eocene. Eocene rocks exposed in Oceanographer and Corsair Canyons (Gibson and others, 1968; Ryan and others, 1978) consist of mudstone, glauconite, and calcarenite.

Oligocene rocks are sparse but may be present in COST

G-2. Samples of Oligocene age were dredged from Oceanographer Canyon (Gibson and others, 1968; Weed and others, 1974). Pleistocene to Miocene rocks are present in the COST G-2 well and updip in the Coskata well at Nantucket. By Early Tertiary time, sediments at the COST G-1 and Nantucket sites were deposited under open-marine conditions in outer-shelf paleodepths. At the COST G-2 site and seaward, deposition took place at bathyal paleodepths for the first time (Poag, 1982a). However, by Oligocene time, inner- to middle-shelf deposits at the COST G-2 site grade updip to nearshore and alluvial deposits. Succeeding Pleistocene sediments were laid down in inner-shelf to nearshore conditions; some deltaic sequences were associated with the outer edge of ice that advanced to the area of Georges Bank during glacial epochs. A thin veneer of Quaternary deposits (~50–100 m) covers the bank.

In summary, synrift sediments (1–3 km thick) accumulated in areally restricted elongate subbasins that trend northeast (sub-parallel to the Appalachian trend) and border the platforms. Thickness of these sedimentary rocks may exceed 3 km in the axis of the main basin. Subsequently, these rocks were tilted and deformed by tectonic activity associated with the opening of the adjacent ocean basin and probably, in part, as a result of igneous intrusions during the Early Jurassic.

Postrift deposits accumulated over a broad area as crustal cooling resulted in subsidence and flexure of the margin adjacent to the opening ocean basin. A thick prism of sediment (6–7 km) accumulated in the Main basin (Fig. 19). These deposits thin outward toward the LaHave and Long Island platforms.

Ancient Carbonate Platform

A conspicuous feature of the Atlantic margin is a buried limestone platform of Jurassic age (Figs. 1 and 20; Schlee and others, 1979; Jansa, 1981). From Nova Scotia to the Gulf of Mexico, the carbonate platform and the slope seaward of it formed a major physiographic province; the platform, banks, and atolls probably flourished as a semicontinuous feature that stretched along most of eastern North America. The platform extends mainly under the southern two-thirds of Georges Bank and is documented on seismic profiles (Fig. 20 and 21; Schlee and others, 1979; Klitgord and others, 1982; Schlee and Fritsch, 1982) and in samples collected by submersibles and taken from drill holes (Austin and others, 1980; Ryan and others, 1978; Ryan and Miller, 1981; Amato and Simonis, 1980; Poag, 1982b).

In the Georges Bank area, the seaward edge of the carbonate platform has two forms (Fig. 20A,B): (1) a rimmed platform (Eliuk, 1978, Fig. 5), and (2) a ramp on which the "shelf break" appears to be in deep water. In the first type of seaward edge (Fig. 20A), the carbonate growth was mainly upward (Jansa, 1981, Fig. 7), giving rise to a topographically steep seaward-facing slope (25–30°) that had a relief of 2 km in the Late Jurassic. The same type of platform edge rimmed the LaHave Platform to the northeast (Eliuk, 1978, Fig. 5) off Nova Scotia. The second type of seaward edge is a broad ramp that prograded into deep water

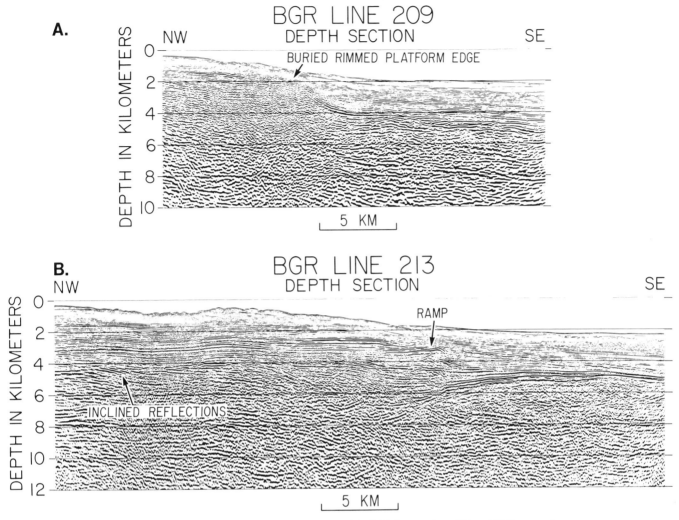

Figure 20. Segments of two seismic depth profiles (BGR CDP line numbers 209 and 213) across slope area seaward of Georges Bank, showing buried carbonate platform transition to deep sea. Upper diagram (A) shows buried steep platform edge; lower diagram (B) shows gradual ramplike transition. Vertical scale in kilometers. Lines were collected and processed by Prakla-Seismos for BGR in 1979. See Figure 3 for locations.

(Fig. 20B). As can be seen from the profile in Figure 20B, the flexure (sudden increase in slope) is lower in the section—2.8 km below sea level, as compared with 2 km below sea level for the rimmed shelf break—and the relief seaward of it is only 1.3 km. The imbricate arrangement of reflections is inferred to represent the migration of an ancient shelf edge into deep water (Mattick, 1982). A similar pattern of reflections along the seaward side of the Scotian basin near Sable Island has been interpreted by Eliuk (1978) on the basis of well data, as representing a progradation and migration by the Abenaki Formation (McIver, 1972) seaward from a basin filling with deltaic sediment. A similar situation may have existed on Georges Bank, as indicated in the COST G-1 and G-2 wells by a sequence of interbedded sandstone, gray and red shale, thin beds of limestone and streaks of coal (Bielak and Simonis, 1980, p. 31; Lachance, 1980, p. 17; Lachance and others, 1980, p. 53–56) of Late Jurassic age. Mohawk and Mic Mac formations (McIver, 1972) are interbedded with the Baccaro Member of the Abenaki Formation at the COST G-2 site. These rocks were deposited under nonmarine to marginal-marine conditions.

Rocks that compose the rimmed platform are crossbedded reef-tract limestone of Early Cretaceous age, as sampled by Ryan and others (1978) from exposures in Heezen Canyon at the eastern end of Georges Bank. As shown by Eliuk (1978, 1981) from drill-hole data beneath the Scotian Shelf, the rimmed platform grades laterally from deep-water shale and limestone mounds to shallow water hexacoral-stromatoporoid-red algal-reef and oolitic shoal deposits (shallow shelf). The deep-water ramp deposits

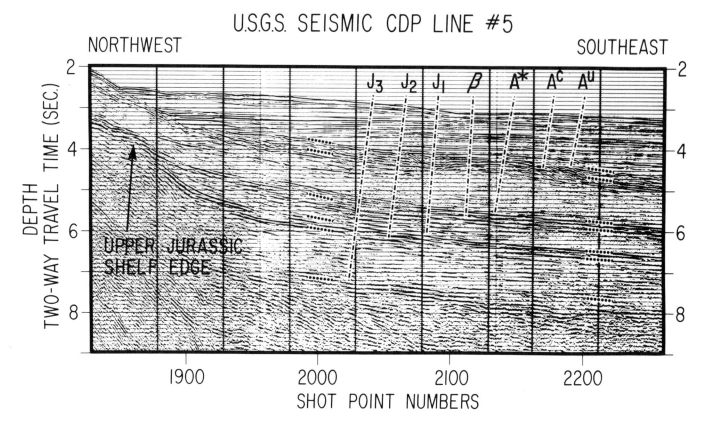

Figure 21. Part of USGS CDP seismic line number 5 across the buried shelf edge and present continental rise (from Klitgord and Grow, 1980, Fig. 7). Note deep reflector J$_3$ that onlaps basement near shotpoint 2300 and lack of the benches seen on seismic line number 1. Most of the horizon symbols are explained in Figures 9 and 15; J$_3$ marks a deep sea horizon at approximately 170 Ma (Klitgord and Grow, 1980); A* marks the top of the Late Cretaceous; and A^c, is a zone of Eocene cherts (Tucholke, 1979, 1981). See Figure 3 for location.

consist of interbedded shale and muddy limestone, which grade upward into biostromal oolitic limestone and eventually into deltaic-marine sandstone, siltstone, and shale (Eliuk, 1981).

Continental Rise Sediments

A continental-rise–type feature did not exist during the rift phase of margin development. The earliest sediments deposited in the new Atlantic basin were probably continental red beds and evaporites. Accumulation of these nonmarine and shallow-water sediments continued into the Early Jurassic late-rift phase of margin development, allowing salt to be deposited on thinned continental (rift-stage) crust (Folger and others, 1979; Uchupi and Austin, 1979).

A continental shelf, slope, and rise formed during the Middle Jurassic, when an open-marine environment became established. A major declivity formed between the carbonate platform (Figs. 2 and 21—paleoshelf edge) at the outer edge of Georges Bank and the ocean basin to the southeast, restricting continental-

rise deposition. Although no drill holes have yet sampled these sediments, the deep-water deposits seaward of the steep carbonate platform are probably reef-flank facies of carbonate debris that slumped from the front and accumulated in fan-shaped aprons. Rapid upbuilding of the carbonate platform on the sinking margin kept it near the sea surface and made it an effective barrier to trap terrigenous siliciclastic sediment (the Mohican, Mohawk, and Mic Mac formations) on the shelf. Thus, one would expect the deep-water Jurassic sedimentary rocks off Georges Bank to be carbonate detritus mixed with pelagic sedimentary deposits.

Three conspicuous acoustic reflectors have been identified within the Jurassic deep-sea sedimentary rocks (Figs. 21 and 22) and traced southward toward Deep Sea Drilling Project (DSDP) sites (Klitgord and Grow, 1980). These reflectors correspond approximately to the top of the Upper Jurassic (J$_1$), top of the Middle Jurassic (J$_2$), and within the lower Middle Jurassic (J$_3$) although only J$_1$ and J$_2$ have been penetrated at DSDP sites. The unit below J$_1$ is the Cat Gap Formation (Jansa and others, 1979),

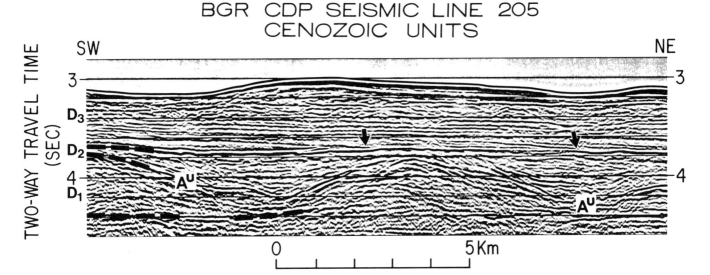

Figure 22. Part of BGR CDP seismic line number 205 along upper continental rise southwest of Georges Bank. Segment shows two conspicuous unconformities that have been used to divide Unit D (Cenozoic sequence) into three subsequences. Lower unconformity (A^u) shows pronounced relief; upper unconformity is nearly flat but marked by downlap of reflections (arrows) in subsequence D_3. See Figure 3 for location.

a Tithonian to Oxfordian red argillaceous limestone. At DSDP site 534A near the Blake Plateau, a lower Oxfordian to middle Callovian mudstone unit was penetrated beneath the Cat Gap Formation (Sheridan and others, 1983); this unit is possibly bounded below by horizon J_2. The great distance of these DSDP sites from Georges Bank, however, makes any direct lithologic correlations tenuous. The Jurassic continental-rise deposits are about 2,500 m thick, whereas the equivalent shelf deposits are about 4,500 m thick in the main basin.

A major shift in sedimentation took place in the Cretaceous when the main depocenter began to shift seaward from Georges Bank to the continental rise and abyssal plain (Fig. 21). The carbonate platform continued to act as an effective sediment trap until Hauterivian time, when clastic sediment overwhelmed it, and material began to be deposited directly in deep water. Carbonate deposition continued on the continental rise until this time (the Blake-Bahama Formation [Jansa and others, 1979]); then the sediment type changed abruptly to clay, claystone, and shale of the Hatteras and Plantagenet formations (Jansa and others, 1979) for the rest of the Cretaceous (Tucholke and Mountain, 1979). These noncarbonate silts and muds that mantled the slope and rise slowly onlapped and buried the steep front of the carbonate platform, forming a gentler slope-rise transition of hemipelagic mud and turbidity-flow deposits (Schlee and others, 1985, Figs. 7–12). Nearly 2,000 m of Cretaceous sediment was deposited on the continental rise in comparison with 1,500 m on the shelf, reflecting the slower subsidence of the shelf and the end of carbonate platform construction.

During the Cenozoic Era, relative sea level fell from the maximum reached in the latest Cretaceous (Pitman, 1978, 1979; Vail and Hardenbol, 1979), and the sediment depocenter completed its shift to the rise (Schlee and others, 1985, Figs. 7–20). In response to a long-term drop in sea level and to short-term shifts of relative sea level, which exposed most of the continental shelf, the continental slope was cut back in several periods of canyon erosion (Ryan and others, 1978), and more erosional debris was deposited on the continental rise. About 1,000 m of Cenozoic sediments mantles the continental rise, thinning on the upper rise and nearly pinching out as these sediments onlap the slope. By comparison, only about 300 m of Cenozoic sediment covers the shelf.

The Cenozoic sediments of the rise wedge contain a conspicuous unconformity A^u (Fig. 22), which is found along the entire margin (Tucholke and Mountain, 1979). Schlee and others (1985) have described this unconformity in the Georges Bank area. Unconformities divide the section into three subsequences (Fig. 22): (1) basin fill below the unconformity A^u but above inferred uppermost Cretaceous reflections; (2) chaotic fill above the unconformity; and (3) widespread parallel-bedded fill that unconformably overlies the chaotic fill. The unconformity A^u is evident as broad channels as much as 15 km wide and having a relief of as much as 0.7 seconds. The channels are most apparent on those profiles oriented parallel to the continental slope and indicate that an ancient slope was cut during the Tertiary and later filled in. We tentatively correlate A^u with a conspicuous shift of the Vail curve of coastal onlap that took place during the late Oligocene. The channels outlined by the profiles over the area do not necessarily line up with the extensions of modern submarine canyons onto the rise.

SUMMARY

The Georges Bank basin shows many similarities in its tectonic setting, basement structure, and sedimentary fill to other marginal sedimentary basins along eastern North America (Schlee and Jansa, 1981; Grow and Sheridan, 1981). Like the adjacent Scotian margin, it is built over a complexly faulted basement (Fig. 9), whose vertical movement during the early stages of basin formation probably led to synrift erosion landward of the hinge zone and to the formation of synrift depocenters seaward of the hinge zone. This synrift basement structure later influenced sedimentary-facies and deposition patterns as the Atlantic Ocean basin formed (Eliuk, 1978; Klitgord and others, 1982; Hutchinson and others, 1986). In common with other eastern North American marginal basins, Georges Bank basin formation spans the interval from Late Triassic(?) to the present.

Basement structure beneath Georges Bank originated as the result of rifting and the subsequent drifting apart of North America and Africa (Fig. 23). Major elements of the basement structure are a shallow continental platform beneath the Gulf of Maine, a deep marginal sedimentary basin beneath Georges Bank, and oceanic crust beneath the continental rise. A series of half grabens, which step down in the seaward direction, marks the hinge zone in the basement that separates the platform from the marginal basin. The trend of these half grabens is parallel to the hinge-zone trend along the entire length of the Atlantic margin, and together the grabens outline the area of Late Triassic–Early Jurassic rifting. The East Coast Magnetic Anomaly marks the seaward edge of the marginal basin, separating it from oceanic crust to the southeast.

Sedimentary rock that accumulated in Georges Bank basin and on the adjacent platform and oceanic crust can be divided into synrift deposits and postrift deposits. The synrift deposits of Early Jurassic and older nonmarine clastic rocks and evaporites accumulated over the broad depression formed by blockfaulting and subsidence in the rift zone. These deposits are probably the source for the narrow zone of salt diapirs beneath the Nova Scotia–Georges Bank continental rise. As the newly forming Atlantic basin widened, postrift-sediment accumulation was partly restricted to the margin, where it was trapped behind elevated basement blocks and the buildup of carbonate rocks in a carbonate platform. The sedimentary deposits that did reach the deep ocean basin formed a continental shelf, slope, and rise. Lower Jurassic anhydrite-dolomite shelf deposits were covered by Middle and Upper Jurassic nonmarine clastic deposits trapped behind the carbonate platform. Near the end of the Early Cretaceous, nonmarine clastic rocks and thin limestone units buried the carbonate platform, resulting in a shift in sediment accumulation to the deep sea. The Cenozoic Era was marked by passing of

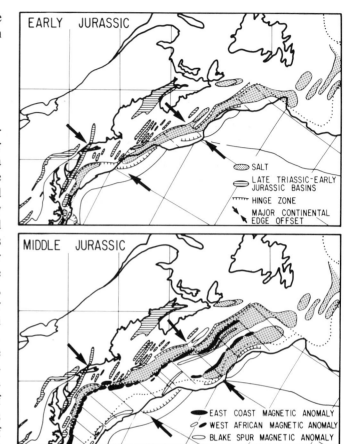

Figure 23. Reconstruction of Atlantic Ocean during the Middle and Early Jurassic (Klitgord and Schouten, 1986). Note possible barriers to water circulation during early period of ocean opening just northeast and just southwest of Georges Bank basin. Lower Jurassic salt deposits are indicated. From Klitgord and others (1982, Fig. 89).

sediments to the slope and rise, periodic cutback of the slope, and a slow rate of sediment accumulation on the bank.

Throughout most of the Jurassic and Early Cretaceous, a carbonate platform that had a steep seaward face was an important feature on the Georges Bank margin. This carbonate platform–shelf-edge complex stretched along the Atlantic margin from Nova Scotia to Florida, controlling margin depositional patterns; nonmarine clastic deposits were partly trapped behind it, and carbonate debris masswasted from its seaward face was an important addition to rise deposits. The form of the platform edge changes laterally from a sharply defined rim to a broad ramp—a change that may be related to the closeness of a terrigenous clastic input. The semicontinuous nature of this feature along much of the Atlantic margin indicates that moderately uniform semitropical conditions prevailed over a wide area.

The U.S. Atlantic continental slope and rise evolved in

response to the buildup of a well-defined carbonate bank. Synchronous siliciclastic sediments accumulated shoreward, but sediment supply during the Jurassic and Early Cretaceous appears to have been inadequate to interfere with the carbonate-sediment accumulation farther offshore. By Hauterivian time (Early Cretaceous), subsidence slowed significantly, and clastic deposits

prograded over the carbonate sediments, finally reaching the open-ocean basin. Except for some fluctuations in the position of the shelf edge, Cenozoic sedimentation was no longer controlled by subsidence of crystalline basement due to crustal cooling, but was controlled mainly by sea-level variations, sediment and ice loading, and continental rebound resulting from melting of the ice.

REFERENCES CITED

Amato, R. V., and Bebout, J. W., eds., 1980, Geologic and operational summary, COST No. G-1 well, Georges Bank area, North Atlantic OCS: U.S. Geological Survey Open-File Report 80-268, 112 p.

Amato, R. V., and Simonis, E. K., eds., 1980, Geologic and operational summary, COST No. G-2 well, Georges Bank area, North Atlantic OCS: U.S. Geological Survey Open-File Report 80-269, 116 p.

Arthur, M. A., 1982, Lithology and petrography of the COST Nos. G-1 and G-2 wells, *in* Scholle, P. A., and Wenkam, C. R., eds., United States North Atlantic Outer Continental Shelf: U.S. Geological Survey Circular 861, p. 11–13.

Austin, J. A., Uchupi, E., Shaughnessy, D. R., III, and Ballard, R. D., 1980, Geology of New England passive margin: American Association of Petroleum Geologists Bulletin, v. 64, p. 501–526.

Ballard, R. D., and Uchupi, E., 1975, Triassic rift structure in the Gulf of Maine: American Association of Petroleum Geologists Bulletin, v. 59, p. 1040–1072.

Barss, M. S., Bujak, J. P., and Williams, G. L., 1979, Palynologic zonation and correlation of sixty-seven wells, eastern Canada: Canada Geological Survey Paper 78-24, 118 p.

Behrendt, J. C., and Klitgord, K. D., 1980, High-sensitivity aeromagnetic survey of the U.S. Atlantic continental margin: Geophysics, v. 45, p. 1813–1846.

Bubb, J. N., and Hatlelid, W. G., 1977, Seismic stratigraphy and global changes of sea level, Part 10; Seismic recognition of carbonate buildups, *in* Payton, C. C., ed., Seismic stratigraphy; Applications to hydrocarbon exploration: American Association of Petroleum Geologists Memoir 26, p. 185–204.

Cornet, B., 1977, Palynostratigraphy and age of the Newark Supergroup [Ph.D. thesis]: University Park, Pennsylvanian State University, 506 p.

Cousminer, H. L., Steinkraus, W. E., and Hall, R. E., 1984, Biostratigraphic restudy documents, Triassic/Jurassic section in Georges Bank COST G-2 well [abs.]: American Association of Petroleum Geologists Bulletin, v. 68, p. 466.

Drake, C. L., Worzel, J. L., and Beckmann, W. C., 1954, Geophysical investigations in the emerged and submerged Atlantic Coastal Plain; Gulf of Maine, Part 9; Geological Society of America Bulletin, v. 65, p. 957–970.

Drake, C. L., Ewing, M., and Sutter, G. H., 1959, Continental margins and geosynclines; The east coast of North America north of Cape Hatteras, *in* Ahrens, L. H., and others, eds., Physics and chemistry of the Earth: London, Pergamon Press, v. 3, p. 110–198.

Duncan, R. A., 1984, Age of progressive volcanism in the New England Seamounts and the opening of the central Atlantic: Journal of Geophysical Research, v. 89, p. 9980–9990.

Eliuk, L. S., 1978, The Abenaki Formation, Nova Scotia shelf, Canada; A depositional and diagenetic model for a Mesozoic carbonate platform: Bulletin of Canadian Petroleum Geology, v. 26, p. 424–514.

——, 1981, Variations along the Mesozoic carbonate shelf, Nova Scotia shelf, Canada, *in* Stoakes, F. A., compiler, Annual Core and Field Sample Conference, January 15–16, 1981: Canadian Society of Petroleum Geologists, p. 15–19.

Emery, K. O., and Uchupi, E., 1972, Western North Atlantic Ocean; Topography, rocks, structure, water, life, and sediments: American Association of Petroleum Geologists Memoir 17, 532 p.

——, 1984, The geology of the Atlantic Ocean: New York, Springer-Verlag, 1050 p.

Falvey, D. A., 1974, The development of continental margins in plate tectonic theory: Australian Petroleum Exploration Association Journal, v. 14, p. 95–106.

Folger, D. W., Hathaway, J. C., Christopher, R. A., Valentine, P. C., and Poag, C. W., 1978, Stratigraphic test well, Nantucket Island, Massachusetts: U.S. Geological Survey Circular 773, 28 p.

Folger, D. W., Dillon, W. P., Grow, J. A., Klitgord, K. D., and Schlee, J. S., 1979, Evolution of the Atlantic continental margin of the United States, *in* Talawani, M., Hay, W., and Ryan, W.B.F., eds., Deep drilling results in the Atlantic Ocean; Continental margins and paleoenvironment: American Geophysical Union, Maurice Ewing Series 3, p. 87–108.

Gibson, T. G., Hazel, J. E., and Mello, J. F., 1968, Fossiliferous rocks from submarine canyons off the northeastern United States: U.S. Geological Survey Professional Paper 600-D, p. D222–D230.

Given, M. M., 1977, Mesozoic and early Cenozoic geology of offshore Nova Scotia: Bulletin of Canadian Petroleum Geology, v. 25, p. 63–91.

Grow, J. A., and Schlee, J. S., 1976, Interpretation and velocity analysis of U.S. Geological Survey multichannel reflection profiles 4, 5, and 6, Atlantic continental margin: U.S. Geological Survey Miscellaneous Field Series Map MF-808.

Grow, J. A., and Sheridan, R. E., 1981, Deep structure and evolution of the continental margin off eastern United States: Proceedings, 26th International Congress, Geology of continental margins, Oceanologica Acta, no. SP, p. 11–19.

Grow, J. A., Bowin, C. O., and Hutchinson, D. R., 1979, The gravity field of the U.S. Atlantic continental margin: Tectonophysics, v. 59, p. 2752.

Grow, J. A., Hutchinson, D. R., Klitgord, K. D., Dillon, W. P., and Schlee, J. S., 1983, Representative multi-channel seismic profiles over the U.S. Atlantic margin, *in* Bally, A. W., ed., Seismic expression of structural styles: American Association of Petroleum Geologists Studies in Geology, series 15, v. 2, p. 2.2.3-1 to 2.2.3-19.

Hathaway, J. C., Schlee, J. S., and 9 others, 1976, Preliminary summary of the 1976 Atlantic Margin Coring Project of the U.S. Geological Survey: U.S. Geological Survey Open-File Report 76-844, 217 p.

Hathaway, J. C., Poag, C. W., and 7 others, 1979, U.S. Geological Survey core drilling on the Atlantic shelf: Science, v. 206, no. 4418, p. 515–527.

Haworth, R. T., and Keen, C. E., 1979, The Canadian Atlantic margin; A passive continental margin encompassing an active past: Tectonophysics, v. 59, p. 83–126.

Haworth, R. T., and MacIntyre, J. B., 1975, The gravity and magnetic field Atlantic offshore Canada: Canada Geological Survey Paper 75-9, 22 p.

Hermes, O. D., Ballard, R. D., and Banks, P. O., 1978, Upper Ordovician peralkalic granites from the Gulf of Maine: Geological Society of America Bulletin, v. 89, p. 1761–1774.

Hinz, K., Dostmann, H., and Fritsch, J., 1982, The continental margin of Morocco; Seismic sequences, structural elements, and geological development, *in* von Rad, U., Hinz, K., Sarnthein, M., and Seibold, E., eds., Geology of the northwest African continental margin: Berlin, Springer-Verlag, p. 34–60.

Hutchinson, D. R., Grow, J. A., Klitgord, K. D., and Swift, B. A., 1983, Deep structure and evolution of the Carolina trough: American Association of Petroleum Geologists Memoir 34, p. 129–152.

Hutchinson, D. R., Klitgord, K. D., and Detrick, R. S., 1986, Rift basins of the Long Island platform: Geological Society of America Bulletin, v. 97,

p. 688–702.

Jansa, L. F., 1981, Mesozic carbonate platforms and banks of the eastern North American margin: Marine Geology, v. 44, p. 97–117.

Jansa, L. F., and Wade, J. A., 1975, Paleogeography and sedimentation in the Mesozoic and Cenozoic, southeastern Canada, *in* Yorath, C. J., Parker, E. R., and Glass, D. J., eds., Canada's continental margins and offshore petroleum exploration: Canadian Society of Petroleum Geologists Memoir 4, p. 79–102.

Jansa, L. F., Enos, P., Tucholke, B. E., Gradstein, F. M., and Sheridan, R. E., 1979, Mesozoic-Cenozoic sedimentary formations of the North American basin; western North Atlantic, *in* Talwani, M., Hay, W., and Ryan, W.B.F., eds., Deep drilling results in the Atlantic Ocean; Continental margins and paleoenvironment: American Geophysical Union, Maurice Ewing Series 3, p. 1–57.

Judkins, T. W., Simonis, E. K., and Heise, B. A., 1980, Correlation with other wells, *in* Amato, R. V., and Simonis, E. K., eds., Geological and operational summary, COST No. G-2 well, Georges Bank Area, North Atlantic OCS: U.S. Geological Survey Open-File Report 80-269, p. 33–36.

Keen, C. E., 1979, Thermal history and subsidence of rifted continental margins; Evidence from wells on the Nova Scotian and Labrador shelves: Canadian Journal of Earth Science, v. 16, p. 505–522.

Klitgord, K. D., 1983, Structure and evolution of the U.S. Atlantic continental margin [abs.]: Inter-Union Commission on the Lithospheric, XVIII General Assembly, International Union of Geodesy and Geophysics, Programme and Abstracts, p. 57.

Klitgord, K. D., and Behrendt, J. C., 1977, Aeromagnetic anomaly map of the U.S. Atlantic continental margin: U.S. Geological Survey Miscellaneous Field Studies Map MF-913, 2 sheets, scale 1:1,000,000.

——, 1979, Basin structure of the U.S. Atlantic margin, *in* Watkins, J. S., Montadert, L., and Dickerson, P. W., eds., Geological and geophysical investigations of continental margins: American Association of Petroleum Geologists Memoir 29, p. 85–112.

Klitgord, K. D., and Grow, J. A., 1980, Jurassic seismic stratigraphy and basement structure of western North Atlantic magnetic quiet zone: American Association of Petroleum Geologists Bulletin, v. 64, p. 1658–1680.

Klitgord, K. D., and Hutchinson, D. R., 1985, Distribution and geophysical signatures of early Mesozoic rift basins beneath the U.S. Atlantic continental margin, *in* Robinson, G. R., Jr., and Froelich, A. J., eds., Proceedings of the second U.S. Geological Survey Workshop on the early Mesozoic basins of the eastern United States: U.S. Geological Survey Circular 946, p. 45–61.

Klitgord, K. D., and Schlee, J. S., 1986, Georges Bank; Subsurface geology, *in* Backus, R. H., ed., Georges Bank: Cambridge, Massachusetts Institute of Technology Press (in press).

Klitgord, K. D., and Schouten, H., 1986, Plate kinematics of the central Atlantic, *in* Tucholke, B. E., and Vogt, P. R., eds., The geology of North America, Volume M, The western North Atlantic region: Geological Society of America (in press).

Klitgord, K. D., Schlee, J. S., and Hinz, K., 1982, Basement structure, sedimentation, and tectonic history of the Georges Bank basin, *in* Scholle, P. A., and Wenkham, C. R., eds., United States North Atlantic Outer Continental Shelf; Geological studies of the COST Nos G-1 and G-2 wells: U.S. Geological Survey Circular 861, p. 160–186.

Lachance, D. J., 1980, Lithology, *in* Amato, R. V., and Bebout, J. W., eds., Geologic and operational summary, COST No. G-1 well, Georges Bank area, North Atlantic OCS: U.S. Geological Survey Open-File Report 80-268, p. 16–21.

Lachance, D. J., Bebout, J. W., and Bielak, L., 1980, Depositional environments, *in* Amato, R. V., and Bebout, J. W., eds., Geological and operational summary, COST No. G-1 well, Georges Bank area, North Atlantic OCS: U.S. Geological Survey Open-File Report 80-268, p. 53–58.

Lewis, R. S., and Sylwester, R. E., 1976, Shallow sedimentary framework of Georges Bank: U.S. Geological Survey Open-File Report 76-874, 14 p.

Libby-French, J., 1983, Petroleum geology of offshore northeastern U.S., *in* Friedman, G. M., ed., Petroleum geology and energy resources of the north-

eastern U.S.: Northeastern Geology, v. 5, p. 119–127.

Lyons, P. C., and Brownlow, A. H., eds., 1976, Studies in New England geology: Geological Society of America Memoir 146, 374 p.

Maher, J. C., 1971, Geologic framework and petroleum potential of the Atlantic Coastal Plain and Continental Shelf: U.S. Geological Survey Professional Paper 659, 98 p.

Manspeizer, W., Puffer, J. H., and Cousminer, H. L., 1978, Separation of Morocco and eastern North America; A Triassic-Liassic stratigraphic record: Geological Society of America Bulletin, v. 89, p. 901–920.

Mattick, R. E., 1982, Significance of the Mesozoic carbonate bank-reef sequence for the petroleum geology of the Georges Bank basin, *in* Scholle, P. R., and Wenkham, C. R., eds., Geological studies of the COST Nos. G-1 and G-2 wells, United States North Atlantic Outer Continental Shelf, U.S. Geological Survey Circular 861, p. 93–104.

Mattick, R. E., Foote, R. Q., Weaver, N. L., and Grim, M. S., 1974, Structural framework of the U.S. Atlantic Outer Continental Shelf north of Cape Hatteras: American Association of Petroleum Geologists Bulletin, v. 58, p. 1179–1190.

Mattick, R. E., Schlee, J. S., and Bayer, K., 1981, The geology and hydrocarbon potential of the Georges Bank–Baltimore Canyon area, *in* Kerr, J. M., and Fergusson, A. J., eds., Geology of the North Atlantic borderlands: Canadian Society of Petroleum Geologists Memoir 7, p. 461–486.

McIver, N. L., 1972, Cenozoic and Mesozoic stratigraphy of the Nova Scotia shelf: Canadian Journal of Earth Sciences, v. 9, p. 54–70.

Montadert, L., de Charpal, O., Roberts, D., Guennx, P., and Sibuet, J. S., 1979, Northeast Atlantic passive continental margins: Rifting and subsidence processes, *in* Talwani, M., Hay, W., and Ryan, W.B.F., eds., Deep drilling results in the Atlantic Ocean; Continental margins and paleoenvironment: American Geophysical Union, Maurice Ewing Series 3, p. 154–186.

Oldale, R. N., Uchupi, E., and Prada, K. E., 1973, Sedimentary framework of the western Gulf of Maine and the southeastern Massachusetts offshore area: U.S. Geological Survey Professional Paper 757, 10 p.

Pitman, W. C., III, 1978, Relationship between eustacy and stratigraphic sequences of passive margins: Geological Society of America Bulletin, v. 89, p. 1389–1403.

——, 1979, The effect of eustatic sea level changes on stratigraphic sequences at Atlantic margins, *in* Watkins, J. S., Montadert, L., and Dickerson, P. W., eds., Geological and geophysical investigations of continental margins: American Association of Petroleum Geologists Memoir 29, p. 453–460.

Poag, C. W., 1982a, Foraminiferal and seismic stratigraphy, paleoenvironments, and depositional cycles in the Georges Bank basin, *in* Scholle, P. A., and Wenkam, C. R., eds., Geological studies of the COST Nos. G-1 and G-2 wells, United States North Atlantic Outer Continental Shelf: U.S. Geological Survey Circular 861, p. 43–92.

——, 1982b, Stratigraphic reference section for Georges Bank basin; Depositional model for New England passive margin: American Association of Petroleum Geologists Bulletin, v. 66, p. 1021–1041.

Poag, C. W., and Schlee, J. S., 1984, Depositional sequences and stratigraphic gaps on submerged United States Atlantic margin: American Association of Petroleum Geologists Memoir 36, p. 165–182.

Ryan, W.B.F., and Miller, E. L., 1981, Evidence of a carbonate platform beneath Georges Bank: Marine Geology, v. 44, p. 213–228.

Ryan, W.B.F., Cita, M. B., Miller, E. L., Hanselman, D., Hecker, B., and Nibbelink, M., 1978, Bedrock geology in New England submarine canyons: Oceanological Acta, v. 1, p. 233–254.

Sangree, J. B., and Widmier, J. M., 1977, Seismic stratigraphy and global changes of sea level; Part 9; Seismic interpretation of clastic depositional facies, *in* Payton, C. E., ed., Seismic stratigraphic applications to hydrocarbon exploration: American Association of Petroleum Geologists Memoir 26, p. 165–184.

Sawyer, D. C., Seift, B. A., Sclater, J. G., and Toksoz, N. M., 1982, Extensional model for the subsidence of the northern United States Atlantic continental margin: Geology, v. 10, p. 134–140.

Sawyer, D. S., Toksoz, N. M., Sclater, J. G., and Swift, B. A., 1983, Thermal evolution of the Baltimore Canyon trough and Georges Bank basin: American Association of Petroleum Geologists Memoir 34, p. 743–762.

Schlee, J. S., 1978, Geology of Georges Bank, *in* Fisher, J. J., ed., New England marine geology; New concepts in research and teaching and bibliography of New England marine geology: National Association of Geology Teachers, New England Section, 26th Annual Meeting, Kingston, Rhode Island, Proceedings, p. 88–92.

Schlee, J. S., and Fritsch, J., 1982, Seismic stratigraphy of the Georges Bank basin, *in* Watkins, J. S., and Drake, C. L., eds., Studies in continental margin geology: American Association of Petroleum Geologists Memoir 34, p. 223–251.

Schlee, J. S., and Jansa, L. F., 1981, The paleoenvironment and development of the eastern North American continental margin: Proceedings, 26th International Congress, Geology of Continental Margins, Oceanologica Acta, no. SP, p. 71–80.

Schlee, J. S., and Klitgord, K. D., 1981, Regional geology and geophysics in the vicinity of Georges Bank basin, *in* Schlee, J. S., ed., Summary report of the sediments, structural framework, petroleum potential, and environmental conditions of the United States middle and northern continental margin area of proposed oil and gas lease sale no. 82: U.S. Geological Survey Open-File Report 81-1352, p. 17–36.

—— , 1982, Geologic setting of the Georges Bank basin, *in* Scholle, P. A., and Wenkam, C. R., eds., Geological studies of the COST Nos. G-1 and G-2 wells, United States North Atlantic Outer Continental Shelf: U.S. Geological Survey Circular 861, p. 4–10.

Schlee, J. S., Behrendt, J. C., Grow, J. A., Robb, J. M., Mattick, R. E., Taylor, P. T., and Lawson, B. J., 1976, Regional framework off northeastern United States: American Association of Petroleum Geologists Bulletin, v. 60, p. 926–951.

Schlee, J. S., Dillon, W. P., and Grow, J. A., 1979, Structure of the Continental Slope off the eastern United States, *in* Doyle, L. J., and Pilkey, O. H., eds., Geology of continental slopes: Society of Economic Paleontologists and Mineralogists Special Paper 27, p. 95–118.

Schlee, J. S., Poag, C. W., and Hinz, K., 1985, Seismic stratigraphy of the Continental Slope and Rise seaward of Georges Bank, *in* Poag, C. W., ed., Geologic evolution of the United States Atlantic margin: New York, Van Nostrand Reinhold, p. 265–292.

Scholle, P. A., and Wenkam, C. R., eds., 1982, Geological Studies of the COST Nos. G-1 and G-2 wells, United States North Atlantic Outer Continental Shelf: U.S. Geological Survey Circular 861, 193 p.

Schouten, H., and Klitgord, K. D., 1977, Map of Mesozoic magnetic anomalies, western North Atlantic, U.S. Geological Survey Miscellaneous Field Studies Map MF-915, scale 1:2,000,000.

Schultz, L. K., and Grover, R. C., 1974, Geology of Georges Bank basin: American Association of Petroleum Geologists Bulletin, v. 58, no. 6, pt. 2, p. 1159–1160.

Sheridan, R. E., and Gradstein, F. M., eds., 1983, Initial Reports of the Deep Sea Drilling Project: Washington, D.C., U.S. Government Printing Office, v. 76, 943 p.

Sheridan, R. E., Grow, J. A., Behrendt, J. C., and Bayer, K. C., 1979, Seismic refraction study of the continental edge off the eastern United States: Tectonophysics, v. 59, p. 1–26.

Taylor, P. T., Zietz, I., and Dennis, L. S., 1968, Geologic implications of aeromagnetic data for the eastern continental margin of the United States: Geophysics, v. 33, p. 755–780.

Tucholke, B. E., 1979, Relationships between acoustic stratigraphy and lithostratigraphy in the western North Atlantic basin, *in* Tucholke, B. E., and Vogt, P. R., eds., Initial Reports of the Deep Sea Drilling Project: Washington, D.C., U.S. Government Printing Office, v. 43, p. 827–846.

—— , 1981, Geologic significance of seismic reflectors in the deep western North Atlantic basin, *in* Warme, J. E., Douglas, R. G., and Winterer, E. L., eds., Deep Sea Drilling Project; A decade of progress: Society of Economic Paleontologists and Mineralogists Special Publication 32, pl 23–37.

Tucholke, B. E., and Mountain, G. S., 1979, Seismic stratigraphy, lithostratigraphy, and paleosedimentation patterns in the North American basin, *in* Talwani, M., Hay, W., and Ryan, W.B.F., eds., Deep drilling results in the Atlantic Ocean; Continental margins and paleoenvironment: American Geophysical Union, Maurice Ewing Series 3, p. 58–86.

Uchupi, E., 1965, Map showing relation of land and submarine topography, Nova Scotia to Florida: U.S. Geological Survey Miscellaneous Geological Investigations Map I-451, 3 sheets, scale 1:1,000,000.

—— , 1966, Structural framework of the Gulf of Maine: Journal of Geophysical Research, v. 71, p. 3013–3028.

—— , 1970, Atlantic Continental Shelf and Slope of the United States; Shallow structure: U.S. Geological Survey Professional Paper 529-I, 44 p.

Uchupi, E., and Austin, J. A., 1979, The geologic history of the passive margin off New England and the Canadian maritime provinces: Tectonophysics, v. 59, p. 53–69.

Uchupi, E., Philips, J. D., and Prada, K. E., 1970, Origin and structure of the New England Seamount Chain: Deep-Sea Research, v. 17, p. 71–79.

U.S. Naval Oceanographic Office, 1966, Total magnetic intensity aeromagnetic survey 1964–1966; U.S. Atlantic coastal region: Washington, D.C., U.S. Naval Oceanographic Office, 15 sheets, scale 1:500,000.

Vail, P. R., and Hardenbol, J., 1979, Sea level changes during the Tertiary: Oceanus, v. 22, p. 71–79.

Valentine, P. C., 1981, Continental margin stratigraphy along U.S. Geological Survey seismic line 5; Long Island platform and western Georges Bank basin: U.S. Geological Survey Miscellaneous Field Studies Map MF-857, 2 sheets.

—— , 1982, Calcareous nannofossil biostratigraphy and paleoenvironment of the COST Nos. G-1 and G-2 wells, in the Georges Bank basin, *in* Scholle, P. A., and Wenkam, C. R., eds., Geological studies of the COST Nos. G-1 and G-2 wells, United States North Atlantic Outer Continental Shelf: U.S. Geological Survey Circular 861, p. 34–42.

Valentine, P. C., Uzmann, J. R., and Cooper, R. A., 1980, Geology and biology of Oceanographer Canyon: Marine Geology, v. 38, p. 283–312.

Vogt, P. R., and Tucholke, B. E., 1978, The New England Seamounts; Testing origins, *in* Tucholke, B. E., and Vogt, P. R., eds., Initial Reports of the Deep Sea Drilling Project: Washington, D.C., U.S. Government Printing Office, v. 43, p. 847–856.

Wade, J. A., 1977, Stratigraphy of Georges Bank basin; Interpolation from seismic correlations to the western Scotian Shelf: Canadian Journal of Earth Sciences, v. 14, p. 2274–2283.

Watts, A. B., 1981, The U.S. Atlantic continental margin; Subsidence history, crustal structure, and thermal evolution, *in* Bally, A., ed., Geology of passive continental margins; History, structure, and sedimentologic record: American Association of Petroleum Geologists Education Course Note Series No. 19, 77 p.

Weed, E.G.A., Minard, J. P., Perry, W. J., Jr., Rhodehamel, E. C., and Robbins, E. I., 1974, Generalized pre-Pleistocene geologic map of the northern United States continental margin: U.S. Geological Survey Miscellaneous Geologic Investigations Map I-861, scale 1:1,000,000.

Zartman, R. E., and Naylor, R. S., 1984, Structural implications of some radiometric ages of igneous rocks in southeastern New England: Geological Society of America Bulletin, v. 95, p. 522–539.

MANUSCRIPT ACCEPTED BY THE SOCIETY MARCH 31, 1986

ACKNOWLEDGMENTS

The authors gratefully acknowledge the comments, ideas, and profiles that have clarified our concepts about Georges Bank through the years. Among our colleagues at the U.S. Geological Survey and Woods Hole Oceanographic Institution, John Grow, Page Valentine, Wylie Poag, Bill Dillon, Elazar Uchupi, Hans Schouten, and Brian Tucholke have been most helpful. We take particular pleasure in thanking our reviewers Bill Ryan, Elazar Uchupi, Deborah Hutchinson, Pete Palmer, and John Grow for their many comments and helpful suggestions. The manuscript was improved through the many suggestions of the USGS editorial personnel in Reston, by the drafting efforts of Patty Forrestel and Jeff Zwinakis, and the patient labor of Peggy Mons-Wengler.

Printed in U.S.A.

The Geology of North America
Vol. I-2, The Atlantic Continental Margin: U.S.
The Geological Society of America, 1988

Chapter 13

Structure and evolution of Baltimore Canyon Trough

J. A. Grow
U.S. Geological Survey, MS 960, Box 25046, Denver Federal Center, Denver, Colorado 80225
K. D. Klitgord and J. S. Schlee
U.S. Geological Survey, Woods Hole, Massachusetts 02543

INTRODUCTION

Baltimore Canyon Trough is the deepest depocenter for Mesozoic and Cenozoic sedimentary rocks along the U.S. Atlantic continental margin, with up to 18 km deposited beneath the continental shelf off New Jersey (Figs. 1, 2, and 3). The landward edge of the trough is marked by a hinge zone where an abrupt increase in depth to basement is observed on seismic reflection profiles. The depth to the top of basement at the hinge zone usually occurs between 4 and 6 km, but secondary hinge zones in places are observed where basement is as shallow as 1 km or as deep as 8 km. The depth to basement increases toward the East Coast Magnetic Anomaly (ECMA), which marks the landward edge of oceanic crust. The width of the trough, as measured by the distance between the hinge zone and the main axis of the ECMA, varies from 60 km off Virginia to 100 km off New Jersey. In the northern half of the trough, the Jurassic paleoshelf edge complex has prograded seaward of the ECMA onto the oceanic crust by as much as 40 km (Figs. 2 and 4). During the late Cenozoic, the shelf edge, as defined by the present 200-m isobath, has retreated by 10 to 20 km from its prior position during Late Jurassic through early Cenozoic. The distance from the hinge zone to the Jurassic shelf edge varies from 50 km off Virginia to 140 km off New Jersey (Fig. 2).

Baltimore Canyon initially formed in response to extensional rifting between North America and Africa during Triassic through Early Jurassic time, which was followed by sea-floor spreading in the early Middle Jurassic (see Klitgord and Schouten, 1986; Klitgord and others, this volume). Gravity models across Baltimore Canyon Trough and other basins along the U.S. Atlantic margin indicate an abrupt change in crustal thickness beneath the hinge zone from 35 to 40 km on the northwest to 25 to 30 km beneath the main part of the basins (Grow and others, 1983; Hutchinson and others, 1983; Sheridan and others, Chapter 9, this volume). The main zone of crustal extension and thinning occurs between the hinge zone and the ECMA and has been referred to as "rift-stage" or transitional crust (Klitgord and Behrendt, 1979; Hutchinson and others, 1983). Volcanism, mafic

intrusions, metamorphism, and partial melting of the lower crust may have accompanied the extension during the formation of the rift-stage crust, but the quantitative effects of these processes are still poorly understood. Following the initiation of sea-floor spreading, the mid-ocean ridge migrated away from the margin. Subsequent lithospheric cooling and sediment loading caused continued subsidence of Baltimore Canyon Trough and further sediment accumulation on the shelf.

The Baltimore Canyon Trough has been the focus of frontier research ever since marine refraction surveys of the continental shelf were conducted off New Jersey over 50 years ago. As new geophysical techniques and tectonic concepts have evolved, the geological interpretation of the continental margin has undergone numerous revisions (Drake, this volume). The combination of deep drill holes (4 to 7 km), modern multichannel seismic-reflection profiles, and a high-resolution aeromagnetic survey, which became available in the last decade, has produced a rapid evolution of concepts. However, the deepest parts of Baltimore Canyon Trough remain untested by any drilling, and geophysical data are our only clues to the deep sedimentary, basement, crustal structure, and early tectonic evolution of the margin.

The purpose of this chapter will be to summarize the major structural and stratigraphic features of Baltimore Canyon Trough and its tectonic evolution, primarily using seismic-reflection profiles. The overall tectonic framework of the U.S. Atlantic margin is reviewed by Klitgord and others (this volume) with comparisons of Baltimore Canyon Trough to the other major basins along the margin. Lithostratigraphic and biostratigraphic studies of Baltimore Canyon Trough exploration wells are covered in Poag and Valentine (this volume), and the results of oil and gas exploration for the trough are reviewed by Mattick and Libby-French (this volume). The paleogeography of the area is covered by Schlee and others (this volume). Seismic refraction and gravity data for the entire margin are summarized in Sheridan and others (Chapter 9, this volume). New Large Aperture Seismic Experiment (LASE) studies across northern

Grow, J. A., Klitgord, K. D., and Schlee, J. S., 1988, Structure and evolution of Baltimore Canyon Trough, *in* Sheridan, R. E., and Grow, J. A., eds., Continental Margin, U.S.: Geological Society of America, The Geology of North America, v. I-2.

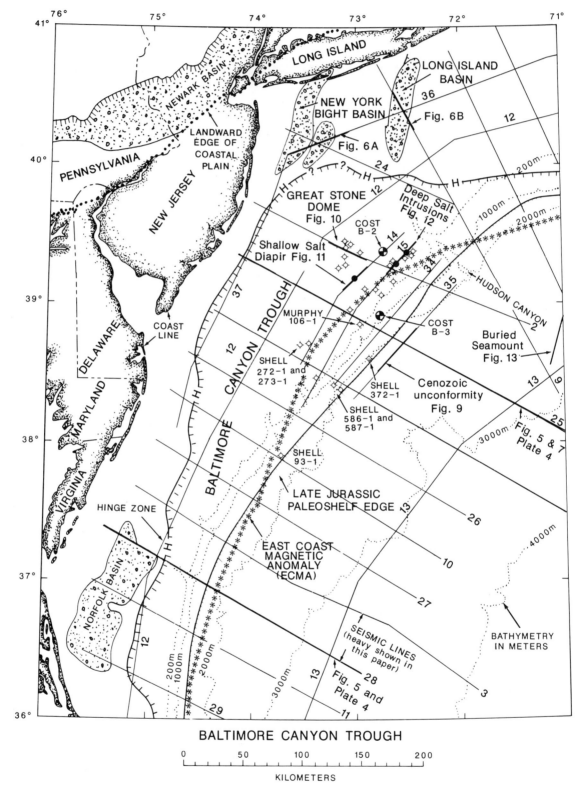

Figure 1. Location map of seismic lines, COST wells, and selected exploration wells in Baltimore Canyon Trough. See Plate 1B for complete map of seismic-reflection profiles.

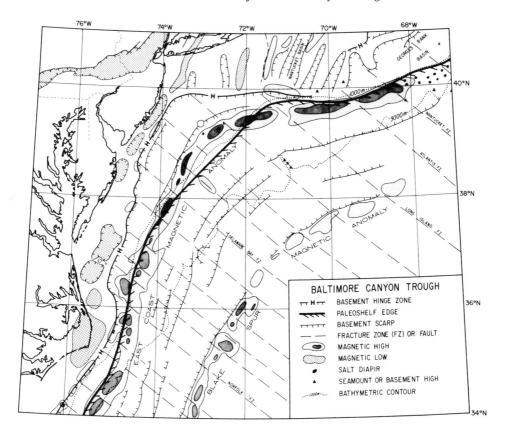

Figure 2. Tectonic map of Baltimore Canyon Trough (simplified from Klitgord and others, 1987).

Baltimore Canyon Trough are presented by Diebold and others (this volume). A seismic stratigraphic interpretation of the New Jersey continental margin with evidence of Cenozoic sea-level fluctuations is discussed in Greenlee and others (this volume).

PREVIOUS INVESTIGATIONS

Early refraction studies of the continental shelf off New Jersey and Long Island by Maurice Ewing and his colleagues established that the coastal plain sedimentary rocks thickened to at least 3 to 5 km beneath the middle continental shelf (Drake and others, 1959; Drake, this volume). The refraction studies also indicated a shallowing of the high-velocity layers (greater than 4.5 km/s) beneath the outer continental shelf and upper continental slope. Although the high-velocity layer beneath the outer continental shelf was originally interpreted as a continuation of the basement from beneath the Coastal Plain, subsequent multichannel seismic-reflection profiles and exploration drilling have shown that the high-velocity horizons beneath outer continental shelf are indurated Jurassic and Lower Cretaceous sedimentary rocks, in part carbonate reef material, which forms a paleoshelf edge system (Figs. 2, 3, 4).

The advent of continuous marine seismic-reflection profiling

using airguns as sources and single-channel streamers as receivers ushered in new studies of the continental slope and rise (Emery and others, 1970; Uchupi, 1970; Emery and Uchupi, 1972). The development of marine multichannel seismic-reflection techniques by the petroleum industry in the 1960s resulted in the acquisition of proprietary data on the continental shelves, which indicated much thicker sedimentary basins than the earlier refraction data had implied. While the early industry data were never published directly, the research community gradually became aware of the thicker sedimentary fill and began to publish interpretations of the margin with up to 10 km of sediment beneath the outer continental shelf (Emery and Uchupi, 1972; Sheridan, 1974; Mattick and others, 1974). In light of this, the high-velocity refraction horizons at 3 to 5 km depth beneath the outer shelf and slope were reinterpreted by Emery and Uchupi (1972) as carbonate bank or reef deposits.

The exploration for oil and gas along the U.S. Atlantic Margin was anticipated as far back as 1962 when the U.S. Geological Survey (USGS) began funding a cooperative research project led by Kenneth O. Emery of Woods Hole Oceanographic Institution (e.g., Emery and others, 1970; Emery and Uchupi, 1972, 1984). Another USGS synthesis of the coastal plain geology with marine geologic and geophysical data along the conti-

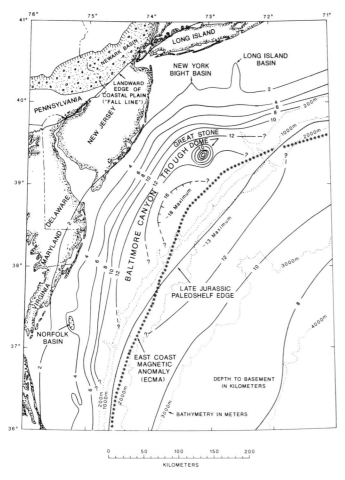

Figure 3. Depth-to-basement map of Baltimore Canyon Trough (modified from Klitgord and Behrendt, 1979; and Benson, 1984).

Large, linear magnetic anomalies occur parallel to the margin and provide distinct markers between crustal types that form the margin. Aeromagnetic and marine magnetic profiling determined that there is a large positive magnetic anomaly, known as the East Coast Magnetic Anomaly (ECMA), which is located along the outer continental shelf and continental slope, from South Carolina to Nova Scotia (Drake and others, 1963; Taylor and others, 1968). High-resolution aeromagnetic maps (Klitgord and Behrendt, 1977, 1979; Behrendt and Klitgord, 1979) and second vertical derivative aeromagnetic maps (Behrendt and Grim, 1983, 1985, and this volume, Plate 2A and 2B) clearly outline the hinge zones and reveal a secondary positive approximately 40 km east of the ECMA in the southern Baltimore Canyon Trough. Keen (1969) initially proposed that the ECMA could be explained as a magnetic edge effect between thick continental crust on the west and thin oceanic crust on the east. Subsequent multichannel seismic-reflection profiles combined with detailed magnetic and gravity modeling of the anomaly have been consistent with Keen's basic "edge-effect" hypothesis (Klitgord and Behrendt, 1979; Grow and others, 1979b; Hutchinson and others, 1983; Alsop and Talwani, 1984; Klitgord and others, this volume). Alsop and Talwani (1984) suggest that one or two intermediate blocks (30 to 40 km wide each) of positively magnetized mafic intrusive rocks may be required between the rifted continental crust and the oceanic crust, although they note that the eastern secondary high of the ECMA and the eastern magnetic block are atypical and not essential characteristics of the ECMA.

DRILLING AND LEASING HISTORY

Offshore drilling began in 1968 along the continental slope with a shallow (305 m maximum penetration), open-hole, drilling program sponsored by four oil companies and executed by the drilling vessel *Caldrill* (Poag, 1978; Poag and Valentine, this volume). Deep Sea Drilling Project (DSDP) Leg 11 drilled DSDP sites 105 through 108 on the continental slope and rise off New Jersey during 1969 (Hollister and Ewing, 1972). The first deep-penetration stratigraphic test well on the U.S. Atlantic shelf was the COST (Continental Offshore Stratigraphic Tests, an industry-funded group) B-2 well, which began in December 1975 and penetrated to 4,891 m into Baltimore Canyon Trough. This well terminated in Upper Jurassic nonmarine shales and sandstones with interbedded coal (Scholle, 1977, 1980; Poag, 1985a; Poag and Valentine, this volume).

The first offshore oil and gas lease on the U.S. Atlantic outer continental shelf was held in Baltimore Canyon Trough in August 1976 (Sale No. 40). After some delays due to litigation in the courts concerning environmental and economic effects on coastal communities, the first wild-cat well was commenced in 1978 (Mattick and Libby-French, this volume).

The COST B-3 well (Fig. 1) on the upper continental slope (819 m water depth) commenced in October 1978 in anticipation of the next Baltimore Canyon Trough lease sale (No. 49).

nental shelf was prepared by Maher (1971), who first proposed the term Baltimore Canyon Trough for this major offshore basin.

The first publicly available multichannel seismic-reflection profiles were obtained by the USGS in 1973: line 1 over Georges Bank and lines 2 and 3 over Baltimore Canyon Trough (Behrendt and others, 1974; Schlee and others, 1976). Between 1973 and 1979, the USGS obtained a total of 35 lines of 24- or 48-channel seismic-reflection profiles in Baltimore Canyon Trough totaling approximately 9,000 km (Fig. 1 and Plate 1B). A large number of additional studies based on the USGS seismic lines in Baltimore Canyon Trough region have been published in recent years (Mattick and others, 1978; Klitgord and Behrendt, 1979; Schlee and others, 1979; Grow and others, 1979a, 1979b, 1983; Grow, 1980; Schlee and Grow, 1980; Klitgord and Grow, 1980; Bally, 1981; Schlee, 1981; Benson, 1984; Poag, 1985a, 1985b; Mountain and Tucholke, 1985; Klitgord and others, this volume). A few industry multichannel seismic lines have also been published recently (Crutcher, 1983; Lippert, 1983; Morgan and Dowdall, 1983).

POST RIFT UNCONFORMITY OCEANIC-BASEMENT SURFACE

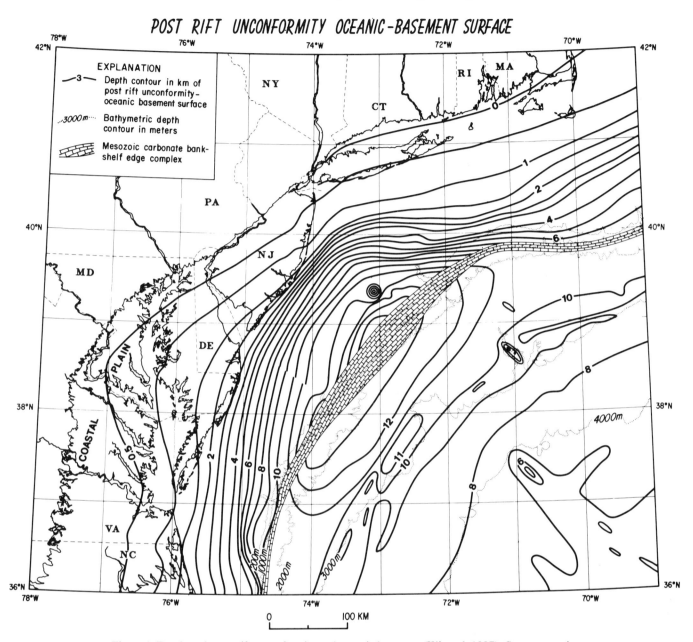

Figure 4. Depth to the postrift unconformity and oceanic basement (Klitgord, 1987). Contours are in kilometers below sea level. The carbonate bank complex behind the paleoshelf edge masks knowledge of the region indicated by the brick pattern.

This lease sale offered tracts down to a depth of 1,000 m on the continental slope and was held in February 1979. The COST B-3 well penetrated to a depth of 4,822 m below sea level, ending in Upper Jurassic interbedded shales, limestones, and sandstones (Scholle, 1980). Lease Sale No. 59 was held in December 1981 and included tracts down to the base of the continental slope (2,200 m). To date, there have been a total of 31 exploration wells in Baltimore Canyon Trough, the deepest penetration being the Murphy Oil No. 106-1 (5,611 m) on the outer continental

shelf edge (Fig. 1). The deepest–water site drilled to date is the Shell No. 587-1 well, a dry hole in 1,966 m of water, which was drilled in 1983 and penetrated to 4,418 m below sea level (Mattick and Libby-French, this volume). This well was one of four wells drilled by Shell Oil Co. to test the petroleum prospects of the Upper Jurassic and Lower Cretaceous sedimentary structures around the paleoshelf edge (Meyer, 1986; Karlo, 1986; Ringer and Patton, 1986; Eliuk and others, 1986; Edson, 1986). Although shows of gas and lesser amounts of oil have been discovered by Texaco and Tenneco at sites approximately 100 km northeast of the Shell deep-water tests, no commercial-size reserves have been found to date.

MAJOR TECTONIC AND STRATIGRAPHIC ELEMENTS

The major tectonic and stratigraphic elements in Baltimore Canyon Trough have been determined primarily from a regional grid of seismic-reflection profiles and a regional high-sensitivity aeromagnetic survey merged with available drill hole data (Klitgord and others, this volume; Poag and Mountain, 1987; Poag and Valentine, this volume). The deepest of the exploration and COST wells in Baltimore Canyon Trough penetrated only approximately one third of the 18-km sedimentary section. The overall structure of the margin in the northern and southern parts of the trough is best displayed by a tectonic feature map (Fig. 2), a depth-to-basement map (Fig. 3), and a depth to postrift unconformity map (Fig. 4) (Klitgord and Schouten, 1986, 1987; Klitgord, 1987; Klitgord and others, 1987) combined with cross-sections along USGS seismic lines 25 and 28, respectively (Fig. 5 and Plate 4). In Baltimore Canyon Trough, the prerift and synrift sediments are separated from the postrift sediments by the postrift unconformity (PRU). This unconformity is the most important depositional break within the sedimentary basin and forms the deepest mappable unconformity over most of the region. Although Baltimore Canyon Trough has a much thicker sequence of sedimentary rocks than Georges Bank Basin (Schlee and Klitgord, this volume), the Carolina Trough, or the Blake Plateau Basin (Dillon and Popenoe, this volume), the separation of the postrift sediments from the synrift sediments by the PRU is common to all four sedimentary basins along the U.S. Atlantic Margin.

In this section, we will first illustrate basement structures, prerift sediments, synrift sediments, the PRU, and postrift sediments on lines 25 and 28. Following the discussion of these two profiles across the entire margin, we will examine shorter seismic profiles over other rift grabens on the Long Island Platform, multiple Cenozoic unconformities on the continental rise, a large volcanic intrusion within shelf sediments, salt intrusions, and buried seamounts, which illustrate the other major tectonic and stratigraphic elements within Baltimore Canyon Trough.

Postrift unconformity

The postrift unconformity separates angularly discordant prerift and synrift sediments affected by extensional rifting of the preexisting continent from the overlying postrift sediments that were deposited after sea-floor spreading commenced.

After sea-floor spreading began, the mid-ocean ridge migrated away from the continent, and lithospheric cooling caused strengthening of the lithosphere and rapid subsidence seaward of the hinge zone. Subsequent to the block faulting at the hinge zone during rifting, the crust became welded together across the hinge

zone. The lithospheric cooling and sediment loading seaward of the hinge zone weighed down the crust west of the hinge zone so that the postrift sediments began a long-term onlap onto the erosional surface landward of the hinge zone as well as seaward onto the newly formed oceanic crust. While worldwide sea-level fluctuation and local variations in sediment supply modulate this long-term coastal onlap of the postrift sediments landward of the hinge zone, the primary driving force was cooling of the lithosphere (Watts, 1981).

The PRU is the deepest surface that can be mapped reliably across the entire margin (Fig. 4). The depth-to-basement map (Fig. 3) is rather speculative because of poor resolution of the crystalline basement and uncertain identification of the prerift-synrift boundary. An earlier depth-to-basement map using magnetic depth-to-source estimates and fewer seismic profiles (Klitgord and Behrendt, 1979) suggested the existence of a basement ridge at depths of 8 to 10 km along the axis of the ECMA, but subsequent higher-power seismic profiles have indicated that no ridge is present in the area of USGS line 25 (Grow, 1980). In most areas, complex topography along the continental slope and strong reflectors in the Jurassic paleoshelf edge prevent adequate seismic penetration to prove or disprove the presence of such a ridge, but the best data available at this time suggest that a ridge may be absent in some areas, such as where fracture zones intersect the ECMA, or smaller and deeper than Klitgord and Behrendt (1979) originally proposed (Klitgord and others, this volume). The character of the PRU varies along the margin as can be seen in the following seismic line examples. Line 28 extends approximately 50 km landward of the hinge zone and has a classically developed postrift unconformity that rises from a depth of 10 km near the ECMA to a depth of 5 km near the hinge zone and to less than 2 km near the coastline (Fig. 5b, Klitgord and Hutchinson, 1985). Landward of the hinge zone, the smooth, warped surface of the unconformity suggests that this part of the unconformity was planed off by subaerial erosion. Seismic line 25 (Fig. 5A) terminated a short distance seaward of the main hinge zone and does not show this onlap of the PRU and younger unconformities onto basement so clearly. However, seismic lines south of Long Island, where the trough narrows again, show a similar sediment onlap pattern to that on line 28. The entire Middle Jurassic and most of the Upper Jurassic section may be missing landward of the hinge zone (Poag, 1985). Because of the large time of nondeposition and erosion, we prefer to use the term "postrift unconformity" rather than the term "breakup unconformity" proposed by Falvey (1974). This is to emphasize the importance of this boundary with respect to sedimentation patterns, rather than its importance to the next step removed in the tectonic interpretations.

Synrift and prerift sediments

Both prerift and synrift sedimentary units may occur between the PRU and the crystalling basement. Both sediment types may be faulted during rifting but they would be expected to have

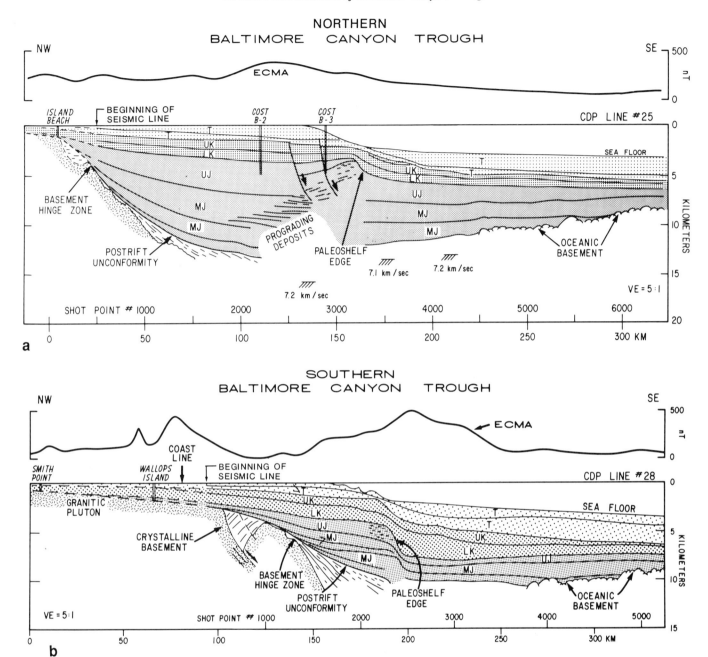

Figure 5. Line drawing summaries of USGS seismic lines 25 and 28. See Plate 4 for actual seismic displays of these two lines (from Grow and others, 1983, Fig. 3; and Klitgord, and Hutchinson, 1985, Fig. 9.7).

distinctly different seismic character. In the simplest cases, the seismic signature of prerift sedimentary units is assumed to be parallel-bedded reflectors that are offset by block-faulting during the rifting phase. In contrast, the signature of synrift sedimentary units is assumed to be fan-shaped reflectors from sedimentary horizons, which were deposited and progressively tilted during the later phases of rifting (Klitgord and Hutchinson, 1985; Schlee and Klitgord, this volume; Hutchinson and Klitgord, 1987).

Along the U.S. Atlantic continental shelf, grabens beneath the postrift unconformity are covered by 2 to 6 seconds of postrift sediments, and the seismic profiles do not always have sufficient resolution to discriminate between synrift and prerift sedimentary units. The best seismic profiles over grabens beneath the postrift unconformity are seen landward of the hinge zone where the postrift sedimentary column is less than 3 km thick (approximately 2 seconds two-way travel time). One of the best examples occurs on line 28 (Fig. 5b; Plate 4) between shotpoints 400 and 1,000; it is referred to as the Norfolk basin (Klitgord and Hutchinson, 1985). This basin is bordered on the west by a listric (?) normal fault with up to 5 km of possible displacement. The sedimentary reflectors within the Norfolk basin dip westward in a subparallel manner, suggesting prerift sediments or the deeply eroded section of synrift sedimentary units.

A series of rift basins have been mapped on the Long Island Platform by Hutchinson and others (1986). Seismic profiles over the New York Bight basin and Long Island basin (Fig. 6) indicate that sedimentary reflectors within these basins resemble the reflectors in Norfolk basin. The Long Island basin occurs at the intersection of USGS seismic lines 36 and 9 (Fig. 1) where the postrift sediments are 1 to 2 km thick. The Long Island basin is a relatively simple graben with a low-angle listric (?) fault on the west side, which dips to the east. Sediments within this graben dip to the west with a hint of fanning, and Hutchinson and others (1986) infer that these are the deepest synrift sediments in the basin and that over half of the graben fill has been eroded away. The New York Bight basin occurs at the intersection of lines 24 and 36 near the entrance to New York harbor (Fig. 1), and it is shallow and has a much more complex internal structure with multiple steps. On line 36, the New York Bight basin appears to have east-dipping sediments with several west-dipping faults that have small offsets.

Rift basins are not detected by the marine seismic lines across the western edge of the northern Baltimore Canyon Trough because the basement hinge zone occurs near the coastline. In the New Jersey area, the Newark basin occurs onshore between 80 and 120 km landward of the main hinge zone (Fig. 1). Proprietary industry data exist in Newark basin and indicate approximately 5 km of Triassic and Lower Jurassic synrift sedimentary rocks composed of nonmarine arkosic sandstones and lacustrine deposits (Manspeizer and Cousminer, this volume). The western border fault dips to the east at approximately 30 degrees (Ratcliffe and others, 1986).

Synrift sediments seaward of the hinge zone and landward of the ECMA dip toward the east on both lines 25 and 28 (Fig. 5 and Plate 4) and are typical of all the profiles in Baltimore Canyon Trough. These dipping reflectors have been attributed by Klitgord and Hutchinson (1985) to a volcaniclastic synrift sedimentary wedge that formed just seaward of the hinge zone over rift-stage crust. A true amplitude seismic display of line 25 near the ECMA shows strong seaward-dipping reflectors between 6 and 7.5 seconds in two-way reflection time or 13 and 17 km in depth (Fig. 7).

A more seaward set of seaward-dipping reflectors are found east of the ECMA offshore Georges Bank (Klitgord and others, this volume) and are probably interbedded basalt and sedimentary layers formed on young oceanic crust. Seaward-dipping reflectors have been observed in other continental margins (Hinz, 1981; Mutter and others, 1982) where the postrift sediment cover is thin enough to allow high-quality seismic data and scientific drilling. These have been shown to be interbedded postrift clastic

sediments and volcanic flows off the Norwegian margin (Hinz, 1981; Mutter and others, 1982). Hinz (1981) has interpreted seaward-dipping reflectors landward of the ECMA in a seismic profile 30 km southwest of line 25 as volcanic sequences that formed immediately after continental breakup and the onset of sea-floor spreading. Klitgord and others (this volume), however, has distinguished two types of seaward-dipping reflector sequences. Seaward-dipping reflectors landward of the ECMA are associated with a synrift clastic wedge just east of the hinge zone, whereas seaward-dipping reflectors seaward of the ECMA off Georges Bank Basin are inferred to be early postrift structures in the oceanic basement similar to those drilled in the Norwegian margin.

Salt deposition appears to be another characteristic of the transition from synrift to postrift deposition. The true amplitude display of line 25 also shows another amplitude anomaly at 6.5 seconds two-way reflecting time or approximately 14 km in depth, which occurs exactly at the axis of the ECMA (shot point 2,300, Fig. 7). Salt diapirs have been found near the ECMA along the edge of the Carolina Trough, Baltimore Canyon Trough, and Georges Bank Basin (Grow and Markl, 1977; Grow, 1980; Dillon and others, 1983; Schlee and Klitgord, this volume), and the amplitude anomaly associated with the ECMA at 14-km depth may be a pocket of evaporites within the upper synrift sediments and volcanics. An alternative interpretation of this deep amplitude anomaly is that it is due to an interbedded dolomite/anhydrite (high-velocity) and shale/mudstone (low-velocity) sequence in the basal postrift sedimentary rocks similar to those drilled at the COST G-1 and G-2 wells on Georges Bank (Taylor and Anderson, 1982).

The top of the synrift sediments on line 25 occurs at approximately 13-km depth near the ECMA (Plate 4). The boundary between the synrift and postrift sediments is difficult to discriminate beneath such a thick sediment accumulation, but the underlying synrift sediments appear significantly more disrupted than the inferred postrift sediments above 13-km depth. If the amplitude anomaly discussed in the previous paragraph was due to interbedded dolomite, anhydrite, and shale in the basal postrift sequence, then the PRU could be over 15-km deep near the ECMA. This alternative deeper PRU is shown on the depth-to-postrift unconformity map in Figure 4. While a PRU deeper than 15 km at the ECMA is possible, the shallower 13 km depth will be the value cited in the remainder of this chapter.

Since no drill holes penetrate the synrift sedimentary units in Baltimore Canyon Trough, their age and composition can only be inferred by comparison with the Newark and Hartford basins onshore (Manspeizer and Cousminer, this volume), with the Canadian margin (Schlee, 1981; Poag, 1985a), and with the Georges Bank basin offshore (Schlee and Klitgord, this volume; Poag and Valentine, this volume), where the COST G-2 well penetrated to a depth of 6,667 m, and can be merged with an overall kinematic model for margin evolution (Klitgord and others, this volume). The Hartford and Newark basins contain Upper Triassic and Lower Jurassic arkosic red,

USGS LINE 36 NEW YORK BIGHT BASIN

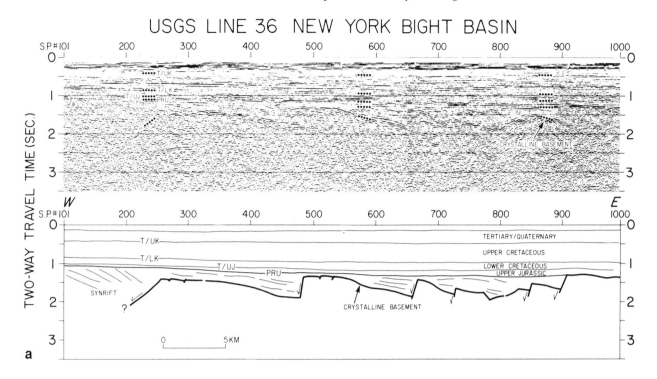

USGS LINE 9 LONG ISLAND BASIN

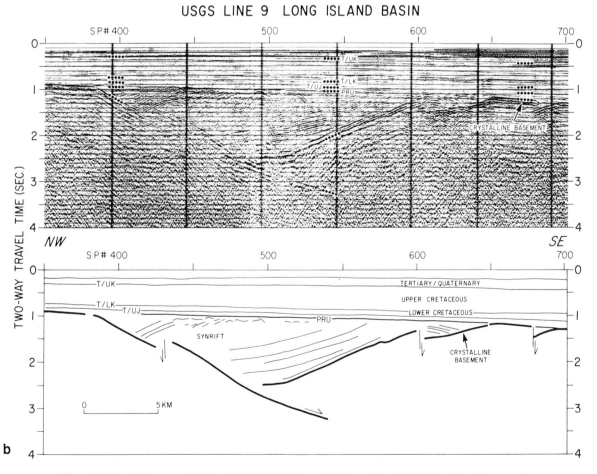

Figure 6. Seismic lines over (a) New York Bight basin and (b) Long Island basin (from Hutchinson and others, 1986). See Figure 1 for line locations.

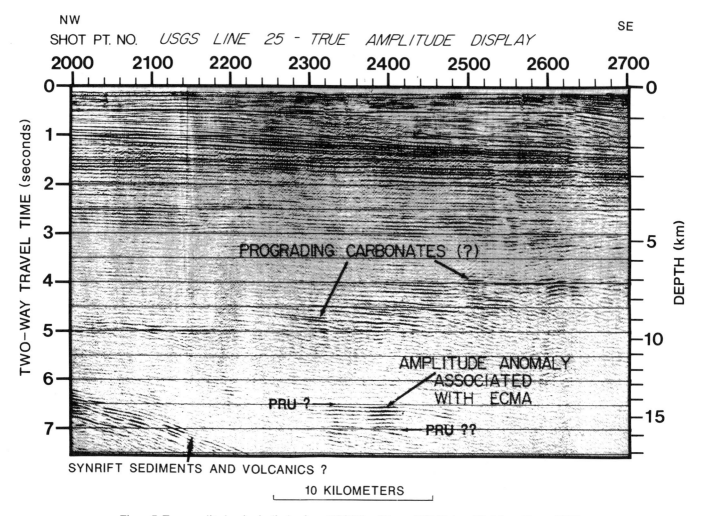

Figure 7. True amplitude seismic display from USGS line 25 near ECMA (modified from Grow, 1980). Note strong reflectors between 13 and 18 km, which are inferred to be synrift deposits including nonmarine red-bed clastics, lacustrine deposits, volcanics, and evaporites. Salt intrusions near ECMA (Figs. 11 and 12) are inferred to rise from synrift deposits.

gray, and black fluvial sandstones and lacustrine deposits with mafic volcanics at the base of the Jurassic (Manspeizer and Cousminer, this volume). The synrift sequence in Baltimore Canyon Trough is also thought to be nonmarine sandstone, volcanic debris and flows, and salt (?), possible equivalents of the Eurydice Formation or the younger Argo Salt of the Canadian margin (Poag, 1985a, Fig. 6-4; Poag and Valentine, this volume). In Georges Bank Basin, the COST G-2 from 4,000 m to the bottom at 6,667 m was dominantly limestones, dolomites, and anhydrites except for halite in the last few meters. The biostratigraphic control in this interval generally is poor, and seismic stratigraphic interpretations suggest that the basal salt horizon occurs precisely at a well-developed postrift unconformity (Poag and Valentine, this volume; Schlee and Klitgord, this volume). A re-examination of the samples from the lower part of the COST G-1 well has indicated that early Middle Jurassic Bajocian sedimentary units

lie just above the PRU (Ascoli, 1983). Seismic correlations to the COST G-1 well on Georges Bank and biostratigraphic ties to Canadian wells lead Schlee and Klitgord (this volume) and Ascoli (1983) to infer a Middle Jurassic age for the base of the postrift sediments at 6,667 m in the COST G-2 well. We infer that the deep synrift reflectors between 13 and 18 km depth on seismic line 25 would also have similar lithologies to the deep sedimentary units of Georges Bank Basin and the Scotian Basin.

Manspeizer and Cousminer (this volume) have analyzed a sidewall core from 4,441 m in COST G-2 containing Triassic palynomorphs and prefer to place the breakup or postrift unconformity at 4,153-m depth. According to their interpretation, the limestones, dolomites, anhydrites, and salt between 4,153 and 6,667 m in the COST G-2 should be considered as synrift Triassic sediments. While there is no dispute that there are Triassic palynomorphs in the sidewall core, the seismic stratigraphic correlation from the COST G-2 to the COST G-1 well by Schlee and Klitgord (this volume) indicates that this horizon is over a kilometer above the Middle Jurassic fauna reported by Ascoli (1983) at the G-1 well. Poag and Valentine (this volume) believe that these palynomorphs were reworked into the Middle Jurassic from Triassic outcrops exposed further to the northwest, and Klitgord and others (this volume) interpret their source as the Franklin Basin in the Gulf of Maine. We favor the latter interpretation with the limestones, dolomites, and anhydrites as representing the early postrift sediments and the halite at 6,667 m as occurring at the top of the synrift sediments.

In summary, while no direct drilling of synrift sediments has occurred in the Baltimore Canyon Trough, we would infer these to be Upper Triassic through Lower Jurassic red arkosic sandstones deposited in fluvial and lacustrine environments with interbedded volcanics and evaporites near the top. Prerift sedimentary rocks, metasediments, and igneous rocks of variable Paleozoic ages and lithologies are inferred to underlie the synrift deposits.

Postrift sediments

Postrift sedimentary units include all of the sedimentary rocks above the postrift unconformity to the landward side of the ECMA and all of the sediments above oceanic basement on the seaward side of the ECMA (Fig. 2, 4, and Plate 4). The thickest accumulation of postrift sediments in Baltimore Canyon Trough is off New Jersey, where the Late Jurassic shelf edge migrated over the oceanic crust. This is the only area along the U.S. Atlantic Margin where migration of the shelf edge over the oceanic crust appears to have occurred. The shingled reflectors between 4 and 10 km depth beneath the outer continental shelf seen on line 25 are inferred to represent a prograding carbonate shelf edge system that migrated during the Late Jurassic (Plate 4).

The lithology of the lower part of this postrift sequence is again based on inferences from multichannel seismic-reflection profiles and the results of exploratory drilling on Georges Bank basin and the eastern Canadian margin. Schlee (1981, Table 1) inferred the early postrift sequence to be an interbedded sedimentary section of nonmarine sandstones, marine shelf limestones, and evaporites, 2 to 5 km thick, with interval velocities in the range of 4.0 to 6.4 km/s. The paleoenvironment of the early postrift sequence has been discussed by Poag (1985a, p. 227–228, Fig. 6-4 and Fig. 6-6), and the seismic sequences have been correlated by him with the Argo Fan, Iroquois Fan, and Mohican Fan of the Scotian margin (see Poag and Valentine, this volume).

Structure contour maps to the postrift unconformity (inferred late Early Jurassic), the top of the Jurassic, and the top of the Cretaceous reveal maximum depths of approximately 13, 4, and 2 km, respectively, beneath the continental shelf (Fig. 8). Approximately 70 percent of the postrift sediments are Middle and Upper Jurassic, while the accumulation of Cretaceous and Cenozoic sedimentary rocks represent about 15 percent each. A remarkable 9 km of Middle and Upper Jurassic postrift sediments

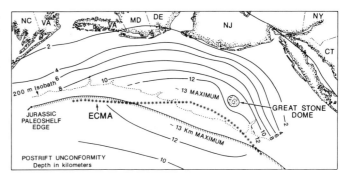

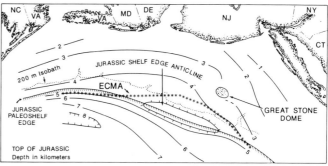

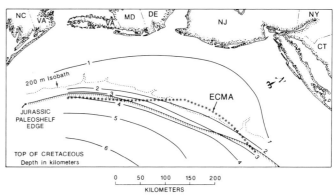

Figure 8. Structure contour maps to postrift unconformity (early Middle Jurassic), top of Jurassic, and top of Cretaceous (modified from Schlee, 1981; and Schlee and Hinz, in preparation). Detailed structure contour maps at a 1:1,000,000 scale of these horizons will be published in late 1987 by Klitgord and Schlee.

accumulated in Baltimore Canyon Trough. The exceptional thickness of the postrift sediments off New Jersey (13 km) is probably due in part to a tremendous influx of clastic material during the Middle Jurassic and may be analogous to modern deltaic margins such as the shelf off the Misssissippi River. These early postrift clastics are inferred to have been derived from youthful Appalachian Mountains and were probably concentrated off New Jersey because Baltimore Canyon Trough was nestled into a major offset of the mountain belt. Eroding mountainous regions surrounded the basins on the north and west sides.

The shingled reflectors within the Middle and Upper Jurassic units on seismic line 25 have been interpreted as as prograding carbonate platform, and significant amounts of back-reef carbonates were found in the Upper Jurassic in the COST B-3 well (Grow, 1980; Libby-French, 1981), and the Shell wells to the south (Meyer, 1986; Karlo, 1986; Ringer and Patton, 1986; Eliuk and others, 1986; Edson, 1986). The extremely thick Middle and Upper Jurassic section behind this platform in Baltimore Canyon Trough is probably much more clastic-prone than the thinner and highly carbonate-prone section drilled at the COST G-2 well in Georges Bank. The imbricate arrangement of reflections seen here is similar to that described by Eliuk (1978) for the Scotian margin and by Schlee and Fritsch (1982) for Georges Bank. Based on drill-hole data on the Scotian shelf, Eliuk speculated that the shingled reflections were caused by the migration of the carbonate bank seaward as the Sable Island delta built into the area and introduced fine-grained clastic sediments. The Upper Jurassic section found in the Shell 273-1 well (5,335 m), approximately 30 km landward of the COST B-3, was dominantly shales and sands (Libby-French, 1981). In contrast to Georges Bank Basin, the Middle and Upper Jurassic carbonate platform in Baltimore Canyon Trough was more restricted to the outer part of the prograding shelf edge, whereas the inner and middle shelf was probably dominated by massive shales and sands eroded from the Appalachians.

Although the Cretaceous and Cenozoic sections of both Baltimore Canyon Trough and Georges Bank Basin have intermittent chalk and limestone units, these periods are dominated by clastic deposition in both basins (Poag and Valentine, this volume). A constructional reef continued to survive into the Early Cretaceous, trapping clastic material behind it. A clastic fan dominated the outer-shelf–slope construction during the large sea-level rise of Valanginian to Hauterivian age. An Aptian oolite bar developed at the shelf edge during the mid-Aptian sea-level drop and represents the final phase of carbonate bank shelf-edge development (Meyer, 1986; Karlo, 1986; Ringer and Patton, 1986). Late Cretaceous and Ceonzoic outer-shelf, slope, and rise construction was dominated by clastic deposition that bypassed the shelf and was transported onto a thickening continental rise.

Cenozoic unconformities on the upper continental rise

Cenozoic sea-level fluctuations have caused extensive erosion on the continental slope, and the shelf edge has retreated by 10 to 20 km in Baltimore Canyon Trough. While erosion has dominated on the continental slope, deposition has dominated on the continental rise. The best-preserved record of these sea-level fluctuations is seen on seismic profiles along the upper continental rise where numerous Cenozoic unconformities have up to 0.3 second of relief locally (Fig. 9). The buried ridges and valleys are generally perpendicular to the margin, as are the modern submarine canyons on the New Jersey continental slope (Robb and others, 1981; Schlee and others, 1985; Poag, 1985b). Recent drilling by the Deep Sea Drilling Project (DSDP) in 1983 on the

New Jersey slope and rise has revealed 12 depositional sequences between upper Campanian (Late Cretaceous) to Quaternary sections (Poag, 1985b, 1987; Poag and Valentine, this volume; Van Hinte, 1985; Schlee and Hinz, in preparation). Periods of nondeposition and erosion represent more than half of this 75-million-year history, and most of the 11 unconformities have counterparts in the worldwide unconformities and sea-level fluctuations inferred by Vail and others (1977). The deepest channeling is now thought to have occurred during the transition from early to middle Eocene, but numerous canyon fillings and erosional episodes can now be documented between the Late Cretaceous and present (Poag, 1985b). The erosion is generally attributed to turbidity currents, debris flows, and slumps in the submarine canyons that were re-excavated numerous times during periods of low sea level.

Construction of the continental rise off Baltimore Canyon Trough is best examined using a detailed grid of USGS and BGR (Bundesanstalt für Geowissenschaften und Rohstoffe) multichannel seismic-reflection profiles (Schlee and others, 1985; Poag, 1987). The rise was constructed in three main phases. The first phase, during the Late Jurassic–Early Cretaceous, involved slope fill-in seaward of the carbonate platform as carbonate fan debris and fine-grained clastic sediments. The second phase was the deposition of a broad sedimentary blanket on the slope-rise area of fine-grained clastic sediments to build a gentle continental slope and rise and completely bury the carbonate paleoshelf edge complex during the Late Cretaceous. In a final phase during the Cenozoic and in response to irregularly falling sea level, the gentle continental slope was cut back in a complicated history of gully erosion, channeling by geostrophic currents, and buildout of submarine fans and turbidite blankets. The cutback of the slope and input from the shelf resulted in deposition of a sedimentary wedge up to 2.3 km thick of slump deposits, hemipelagic ooze, and channel fill.

Volcanic intrusion: Great Stone Dome

Early marine and aeromagnetic anomaly profiles over Baltimore Canyon Trough identified a high-amplitude, circular, positive anomaly in the middle of the continental shelf, which is approximately 20 km in diameter and 40 km landward of the ECMA (Drake and others, 1963; Taylor and others, 1968; Schlee and others, 1976). The positive magnetic anomaly over the dome is 400 nT, while the residual positive gravity residual gravity anomaly is approximately 20 mgal (Grow and others, 1979b). In 1973, the USGS obtained multichannel seismic line 2 (Fig. 10) over this anomaly, which indicated that it was probably a large mafic intrusion that uplifted Lower Cretaceous and older strata (Schlee and others, 1976, Grow, 1980). Excellent large-scale industry seismic displays over Great Stone Dome have also recently been published by Crutcher (1983) and Lippert (1983). The age of the unconformity truncating the dome is now thought to be within the Aptian epoch of the Early Cretaceous (Crutcher, 1983). Crutcher (1983) interpreted the igenous body as ultra-

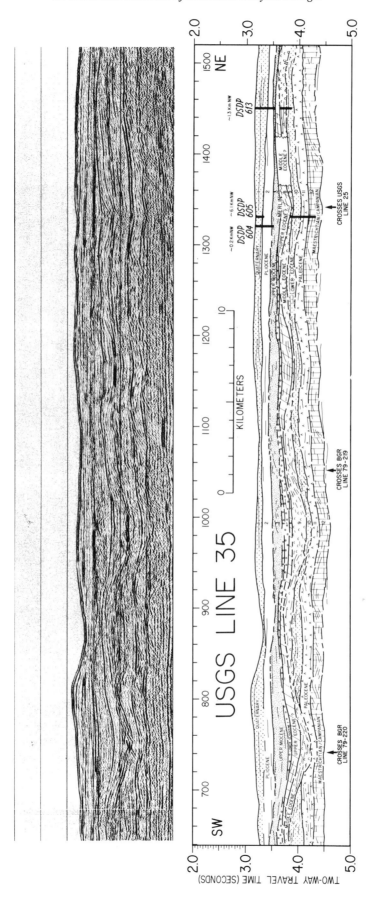

Figure 9. Tertiary unconformities beneath the upper continental rise (modified from Poag, 1985b, Plate 9-3).

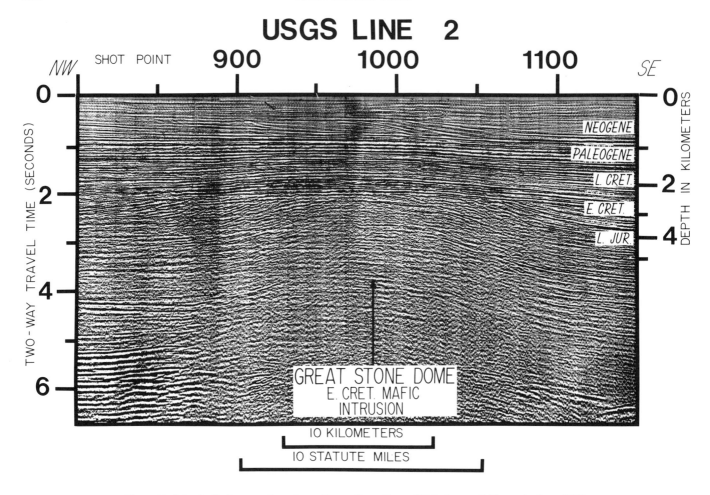

Figure 10. Seismic display over Great Stone Dome (from Grow, 1980; also see Schlee and others, 1976; Crutcher, 1983; and Lippert, 1983).

mafic on the basis of lamprophere cuttings in exploration wells. Lippert (1983) has pointed out that differential compaction of sedimentary units around Great Stone Dome has produced up to 900 ft (274 m) of closure in sediments deposits on top of the dome, and industry bid 650 million dollars for leases on top of this dome. Unfortunately, all seven wildcat wells in Great Stone Dome have turned up dry (Mattick and Libby-French, this volume). Great Stone Dome is the largest igneous event to intrude the postrift sedimentary strata along the U.S. Atlantic margin landward of the ECMA (Grow, 1980). Smaller intrusive bodies are found in Georges Bank Basin (Klitgord and others, 1982; Schlee and Klitgord, this volume; Hurtubise and others, 1987). DSDP sites along the New England seamounts off Georges Bank have encountered volcanic debris within the Cretaceous sediments along the flanks of the seamounts (Tucholke and others, 1979).

Salt intrusions

Salt diapirs are important features along much of the Atlantic margin (Jansa and Wade, 1975; Dillon and others,

1983; Schlee and Klitgord, this volume; Dillon and Popenoe, this volume), but only one shallow diapiric feature has been detected in Baltimore Canyon Trough that appears to be a possible salt structure (Fig. 11). This narrow piercement structure penetrates into Cenozoic strata but has never been drilled by any exploration wells. One exploration well on the southeast flank of the Great Stone Dome did encounter salt at a depth of 3.8 km (Oil and Gas Journal, 1978), but no publicly available seismic profiles exist over this well. Other deeper swells or small domes have also been interpreted as salt intrusions (Fig. 12; Grow, 1980; Mattick and others, 1981). Multiple exploration wells on the structure near SP 2300 on USGS line 14 (Fig. 12) have encountered significant shows of gas, but not commercial amounts (Mattick and Libby-French, this volume). The number of salt diapirs along the Baltimore Canyon Trough is very small compared to the 23 diapiric structures that have been found along the ECMA in the Carolina Trough and are inferred to be salt (Grow and Markl, 1977; Dillon and others, 1983; Dillon and Popenoe, this volume). Salt was also encountered at the bottom of the COST G-2 well (6,667 m) on Georges Bank (Scholle and Wenkam,

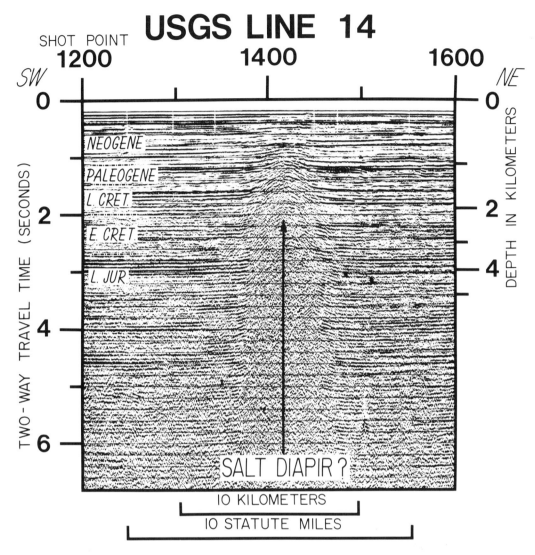

Figure 11. Shallow salt diapir intruding Tertiary sediments near ECMA (from Grow, 1980).

1982). Additional massive diapirs that are thought to be salt occur along the continental slope off Nova Scotia (Jansa and Wade, 1975; Uchupi and Austin, 1979). Upper Triassic salt has been drilled off Newfoundland on the Grand Banks landward of the hinge zone, and lowermost Middle Jurassic sediments have been recovered just above the salt in the Scotian Basin seaward of the hinge zone (Jansa and others, 1980). The age of the salt off the U.S. Atlantic continental margin remains uncertain but probably is of the same age as the Scotian Basin. The salt off the U.S. Atlantic margin could be part of the late synrift or early postrift sedimentary deposits. It is possible that the salt formed during the relatively brief transition from rifting to sea-floor spreading when intermittent intrusions of seawater invaded the rift grabens and evaporated as sills moved up or down and/or sea-level fluctuations created the evaporite environment.

Oceanic basement and buried seamounts

Analysis of seismic profiles immediately seaward of the ECMA or Jurassic shelf edge off Baltimore Canyon Trough generally reveal sediments to depths of 10 to 13 km (Figs. 3–5; Plates 2D and 4). Slightly farther seaward (approximately 50 km), irregular diffractions off the top of a basement surface can be seen; they rise gradually to the southeast. This diffracting surface has been interpreted as the top of the early Middle Jurassic oceanic basement (Klitgord and Grow, 1980) (Fig. 5; Plate 4). Successive units of postrift continental rise sediments progressively onlap this oceanic basement surface in a regular manner farther seaward, indicating that the oceanic crust becomes younger to the east (Klitgord and Grow, 1980).

Buried seamounts of various dimensions have been detected

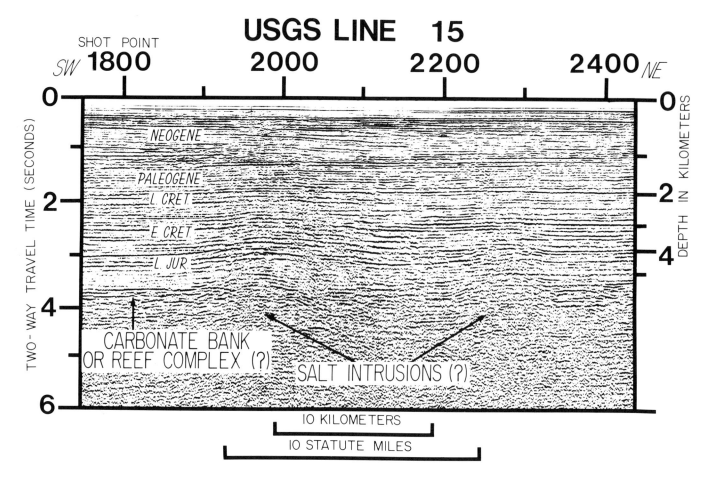

Figure 12. Deeper salt intrusions near ECMA (from Grow, 1980).

on seismic profiles over the continental rise (Schlee and others, 1985, Fig. 7-13; Klitgord and others, 1987). An example of one such seamount (Fig. 13), which rises high up into the sedimentary column, is seen on a profile near Hudson Canyon (Fig. 13). This seamount is the shallowest part of the northwest-trending Hudson Ridge (Klitgord and others, 1987). It rises to within 3 km of sea level, approximately 7 km above the general depth to oceanic basement. The onlapping character of the acoustic reflector J_1 (top of Jurassic) and younger reflectors suggests that the sea-mount formed before the end of the Jurassic. Therefore, this seamount probably formed during or just after the oceanic crust formed (early Middle Jurassic), but before the end of the Jurassic. Other more deeply buried seamounts can be seen, which suggests formation at or near the time that oceanic crust itself was formed (e.g., line 32 off North Carolina; Plate 4; Dillon and Popenoe, this volume).

DEEP CRUSTAL STRUCTURE

The deep crustal structure and evolution of the U.S. Atlantic margin off New Jersey have been investigated by numerous seismic-refraction, magnetic, and gravity techniques for almost half a century (Drake, this volume). The earlier studies interpreted the high-velocity refractors at 3 to 5 km depth beneath the outer continental shelf as a volcanic ridge separating a miogeosyncline and eugeosyncline (Drake and others, 1959, 1963), and early interpretations of the ECMA suggested a mafic igneous intrusion within the crust at a depth of only 6 km (Taylor and others, 1968). Some early studies also interpreted the western edge of oceanic crust at the Blake Spur Magnetic Anomaly (Fig. 2) rather than the ECMA (Rabinowitz, 1974). The advent of improved single-channel and multichannel seismic-reflection techniques showed that the oceanic crust continues westward almost to the ECMA (Emery and others, 1970; Schlee and others, 1976; Grow and Markl, 1977; Klitgord and Behrendt, 1979; Klitgord and Grow, 1980). Long-range sonobuoy refraction measurements also identified a 7.1 to 7.2 km/s layer dipping west-ward beneath the upper continental rise at depths of 13 to 16 km, which was interpreted as oceanic crustal layer 3 (Sheridan and others, 1979).

USGS contract multichannel seismic lines in the Baltimore Canyon Trough region prior to 1978 showed 10 to 12 km of

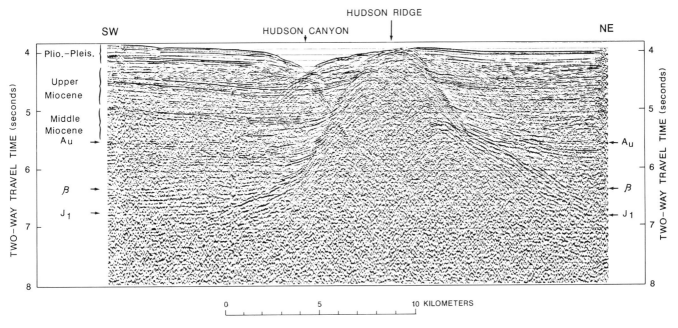

Figure 13. Buried seamount near Hudson Canyon (Klitgord and Schlee, in preparation). This is the largest buried seamount beneath the continental rise seaward of Baltimore Canyon Trough.

sediments beneath the shelf and rise, but seismic penetration beneath the outer shelf and slope over the ECMA was limited on most lines by the carbonate bank shelf-edge complex (Fig. 4) and a strong deep acoustic reflector seaward of the Jurassic paleoshelf edge (sedimentary horizon J_3 of Klitgord and Grow, 1980). Magnetic depth-to-source studies had indicated the source of the ECMA was at a depth of 6 to 10 km within the masked basement region, and that the ECMA may be a magnetic edge effect of a thickened landward edge of the oceanic crust rising to 6 to 8 km depth near the ECMA (Klitgord and Behrendt, 1979; Grow and others, 1979a, b). USGS line 25 was acquired in 1978 with a significantly larger airgun array (2,000 in³) and is the only publicly available line (except for the LASE line across the same location—see LASE Study Group, 1986, and Diebold and others, this volume) to penetrate this Jurassic paleoshelf edge region. Line 25 achieved 18 km of penetration over the ECMA and 13 km penetration beneath the upper continental rise (Grow, 1980). No volcanic ridge is evident in the vicinity of the ECMA on line 25, in spite of the fact that the magnetic depth-to-source studies indicated magnetic sources at 8- to 10-km depths at the ECMA near line 25 (Klitgord and Behrendt, 1979). Grow (1980) interpreted the shallow magnetic sources as small igneous dikes and sills or salt intrusions that were intruded into the postrift sedimentary rocks. A 40-km-wide zone between the eastern flank of the ECMA and the Jurassic paleoshelf edge on line 25 achieved only about 6 km of seismic penetration, due to masking by strong reflectors in the prograded Jurassic paleoshelf-edge carbonates, but the deep reflectors on the east and west sides suggested that they projected through the masked zone. In sum-

mary, line 25 showed up to 18 km of postrift and synrift sediments in the vicinity of the ECMA and a depth-to-oceanic basement of 13 km seaward of the Jurassic paleoshelf edge, with no compelling evidence for a large basement ridge at the landward edge of the oceanic crust (Fig. 14). However, this does not preclude the possibility that significant quantities of mafic extrusive rocks may be interbedded in both the late synrift and early postrift sedimentary rocks.

Klitgord and Schouten (1986) and Klitgord and others (this volume) have mapped a series of oceanic fracture zones that intersect the ECMA, and in each case the amplitude of the ECMA anomaly is reduced. They suggest that line 25 crosses the ECMA at one of the fracture-zone locations and, therefore, is in a region where the amplitude of the ECMA is low and the basement ridge at the landward edge of the oceanic crust is smaller and deeper than average (i.e., 12- to 13-km depth beneath the masked zone between the ECMA and Jurassic paleoshelf edge).

A magnetic model of the ECMA in the Carolina Trough (Hutchinson and others, 1983; Klitgord and others, this volume) indicates that weak positive induced magnetization in the rift-stage crust and strong reversed remnant magnetization in the oceanic crust, both at depths of approximately 11 km, provide an excellent fit to the observed ECMA. The Carolina Trough model was in an area where the ECMA has a single positive peak, making the simplest edge-effect model easily applicable. Although magnetic depth of source studies in the Carolina Trough indicated bodies at 8 to 10 km depth (Klitgord and Behrendt, 1979), the simple edge-effect model was sufficient.

Alsop and Talwani (1984) have modeled the ECMA in

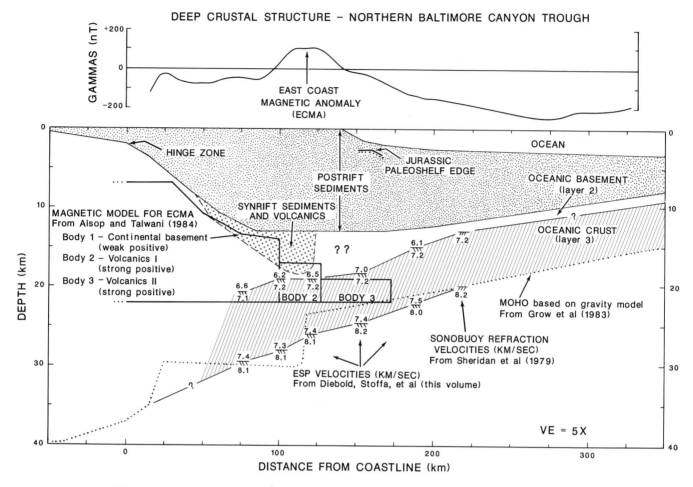

Figure 14. Deep crustal structure beneath northern Baltimore Canyon Trough based on sonobuoy refraction profiles (Sheridan and others, 1979), gravity modeling (Grow and others, 1983) magnetic models (Alsop and Talwani, 1984), and expanding-spread velocity profiles (LASE Study Group, 1986; Diebold, this volume). A summary geologic cross-section for northern Baltimore Canyon Trough along USGS line 25 is shown in Chapter 1 (Fig. 2, Grow and Sheridan, this volume).

the Baltimore Canyon Trough near line 25 and coincident LASE line 6 with three magnetic blocks (Fig. 14): (1) continental basement on the west at a depth of 14 km with weak positive magnetization; (2) a 30-km-wide, strongly positive magnetization block between 17 and 22 km depth directly beneath the ECMA, inferred to be volcanics (VOLCANICS I); and (3) a thin 40-km-wide strongly positive magnetization block between 19 and 22 km depth on the seaward side of the main ECMA peak, beneath a secondary positive, and also inferred to be volcanics (VOLCANICS II). The ECMA along line 25 is much lower in amplitude than for the adjacent regions; it has a double-humped main positive and broader subsidiary positive to the east, and a more complex source zone for the ECMA is required in the Baltimore Canyon Trough than in the Carolina Trough. The Alsop and Talwani (1984) model did not specifically assign a block to the oceanic crust on the southeast, but the inferred sense

of magnetization of the oceanic crust is reversed with respect to the three positive blocks on the northwest, i.e., the same relative transition as interpreted by Hutchinson and others (1983). Therefore, magnetic modeling of the ECMA in Baltimore Canyon Trough is possible with two 30- to 40-km-wide deep positive blocks between the weak positive continental crust on the west and reversed oceanic crust on the east.

Alsop and Talwani (p. 1191, 1984) note that their eastern block (VOLCANICS II) is only required for Baltimore Canyon Trough and is not an essential part of their general model for the ECMA. However, they do emphasize that the central VOLCANICS I body, at a depth of 17 to 22 km directly beneath the ECMA, is "material with higher magnetic intensity (which) arises from highly magnetized material being intruded into a depression caused by early rifting instead of a basement ridge." In both the Hutchinson and others (1983) and the Alsop and Talwani (1984) magnetic models for the ECMA, relatively deep edge-effect sources (11 to 17 km deep) were capable of fitting the general waveform of the ECMA. In summary, the ECMA magnetic models indicate that a single deep edge-effect between rift-stage crust and oceanic crust is sufficient in the Carolina Trough where a single positive anomaly exists, while three deep edges (i.e., with two intermediate positive blocks) are possible between the rift-stage crust and the oceanic crust in Baltimore Canyon Trough where a more complex positive anomaly exists.

Recent wide-angle reflection and refraction experiments (LASE: Large Aperture Seismic Experiment; and ESP: Expanding Spread Profile) along USGS line 25, using three ships, two multichannel streamers, and both airgun and explosive sources, have further resolved the velocity structure beneath Baltimore Canyon Trough (Alsop and Talwani, 1984; LASE Study Group, 1986; Diebold and others, this volume). These studies show that a 7.1- to 7.5-km/s layer, identified as oceanic crustal layer 3 on the seaward side of the ECMA (Sheridan and others, 1979), is found up to 40 km landward of the ECMA and that the depth to mantle increases gradually from 20 km beneath the upper continental rise to 29 km beneath the central part of Baltimore Canyon Trough (Fig. 14). Whether or not this 7.1- to 7.5-km/s horizon is a single lithogenetic structural layer across the ECMA is not certain because of the ambiguous velocity structure beneath the paleoshelf-edge region (a low-velocity zone), and because constant velocity surfaces often do not necessarily coincide with structural surfaces (Trehu and others, 1987). Gravity models across Baltimore Canyon Trough in the vicinity of USGS line 25 have shown a significant thinning of the crust seaward of the hinge zone and a more subtle thinning of the crust seaward of the ECMA (Watts, 1981; Grow and others, 1983).

The most remarkable result of the LASE and ESP experiments is the continuity of the 7.2-km/s seismic layer from beneath the continental rise up to 40 km landward of the ECMA. The LASE Study Group (1986; Diebold and others, this volume) considered several structural models that might explain these observations. While the 7.2-km/s layer could represent a deep metamorphic layer in the continental crust formed during rifting (Falvey and Middleton, 1981), or a remnant of deep Appalachian crust that has similar velocities (Taylor and others, 1980), they prefer an interpretation of plutonic underplating during the transition from rifting to sea-floor spreading. They envision an upper zone of brittle extensional normal faulting overlying a lower crustal layer that is pervasively intruded by mafic magmas. To date, only this one transect of LASE and ESP profiles has been conducted on the U.S. Atlantic margin, and follow-up transects are needed to test these important conclusions in other parts of this margin and other Atlantic-type margins.

SUMMARY

Northern Baltimore Canyon Trough is the thickest accumulation of synrift and postrift sediments along the U.S. Atlantic continental margin with up to 18 km of Mesozoic and Cenozoic sedimentary rocks (Grow and Sheridan, this volume, Fig. 2). Exploratory drilling has only probed the top third of these sediments, and the deeper crustal structure can only be inferred from various geophysical methods. Seismic-reflection profiles indicate that the thickest sediment accumulation occurs between a hinge zone beneath the inner continental shelf and the ECMA, a trough which varies from 60 to 100 km in width. Gravity models and refraction data indicate that the depth to mantle is 35 to 40 km deep west of the hinge zone, 25 to 35 km deep between the hinge zone and the ECMA, and less than 25 km thick seaward of the ECMA. The thick crust landward of the hinge zone is interpreted as relatively undisturbed continental crust, although several onshore and offshore grabens occur landward of the hinge zone and document at least some extension in the upper continental crust (Hutchinson and others, 1985 and 1986). Klitgord and others (this volume) refer to this zone as rifted continental crust. Reflection, refraction, gravity, and magnetic studies indicate that the relatively thin crust seaward of the ECMA is oceanic crust formed by sea-floor spreading during the early Middle Jurassic. The intermediate thickness crust between the hinge zone and the ECMA, which underlies the thickest sediment accumulations, is inferred to have formed by extensional rifting and crustal thinning during Triassic and Early Jurassic time. The crustal extension may have been accompanied by listric normal faulting in the upper crust and mafic intrusives in the lower crust that underplate rift-stage crust.

Comparison with other Atlantic-type continental margins, such as the Bay of Biscay (Montadert and others, 1979; Bally, 1981), where the postrift cover is thin, suggests that the brittle crustal extension with grabens bounded by listric faults occurs in the upper crust while the lower crust deforms in a more ductile manner. In spite of the thin postrift sediment cover and high-quality seismic profiles on the Bay of Biscay, the precise mechanism and amounts of extension in the upper and lower crust are a matter of continuing debate (Chenet and others, 1983; Le Pichon and others, 1983; Le Pichon and Barbier, 1987). Subsidence studies of the COST B-2 and B-3 wells suggest that the rift-stage crust beneath Baltimore Canyon Trough has been stretched and thinned by a factor of 3 to 5 (Watts, 1981; Sawyer and others, 1983).

Synrift deposits in the deepest part of Baltimore Canyon Trough, 13- to 18-km depth near the ECMA, are inferred to be a mixed assemblage of red arkosic sands, lacustrine deposits, volcanics, and evaporites. Salt intrusions near the ECMA and on the southeast flank of Great Stone Dome are inferred to have risen from evaporites within the late synrift sediments. Rapid cooling of the lithosphere and massive sediment loading during the Middle and Late Jurassic resulted in up to 9 km of postrift Jurassic deposits in northern Baltimore Canyon Trough. These deposits prograded 40 km seaward onto the oceanic crust, and, although the Jurassic shelf edge probably contained significant carbonates, the postrift Jurassic sediments in Baltimore Canyon Trough are probably much more dominated by clastic sediments than those of Georges Bank.

Only approximately 4 km of sediments were deposited beneath the continental shelf during the Cretaceous and Tertiary periods, reflecting slower lithosphere cooling and reduced sediment supply. The shelf edge retreated by approximately 20 km during the Tertiary, probably in response to numerous fluctuations of sea level. Eleven major Tertiary unconformities have been identified beneath the upper continental rise that were probably cut during numerous low stands of sea level.

During the late Early Cretaceous, a mafic intrusion penetrated the postrift sediments, forming the Great Stone Dome. Although no definitive mineralogic or geochemical studies are available, we have assumed that this intrusive formed in the upper mantle. Early oil and gas exploration in Baltimore Canyon Trough concentrated on this intrusion, but no significant shows of oil and gas have been discovered. A total of 31 oil and gas exploration wells in Baltimore Canyon Trough have now been completed, including four Shell Oil Company wells on the lower continental slope over the Jurassic paleoshelf edge. Noncommer-

cial quantities of gas were discovered on the deep salt intrusions just landward of the Jurassic shelf edge but other results have been negative, and no exploratory wells have been drilled since 1984.

Except for the extreme thickness of postrift sediments, Baltimore Canyon Trough has had a similar structural history to Georges Bank Basin and the Carolina Trough. All three of these basins were extended during the Late Triassic through Early Jurassic and covered by up to 5 km of synrift sediments. Evidence for evaporites within the late synrift sediments exists in all three basins. Sea-floor spreading was initiated during the early Middle Jurassic, and the mid-ocean ridge migrated away from the margin. This was followed by rapid subsidence due to lithospheric cooling and sediment loading during the remainder of the Jurassic. Subsidence slowed during the Cretaceous and Tertiary periods. Fluctuation of sea levels during the Tertiary, due in part to increasing glacial effects, caused erosion and retreat of the shelf edge by 10 to 20 km in Baltimore Canyon Trough.

REFERENCES CITED

Alsop, L. E., and Talwani, M., 1984, The East Coast magnetic anomaly: Science, v. 266, p. 1189–1191.

Ascoli, P., 1983, Supplementary report on the biostratigraphy (Foraminifera and Ostracoda) and depositional environments of the COST G–1 well, Georges Bank, from 10,000 to 14,800 feet: Geological Survey of Canada Report #PGS–PAL–21–83PA, 6 p.

Bally, A. W., 1981, Atlantic-type margins, *in* Bally, A. W., ed., Geology of passive continental margins: American Association of Petroleum Geologists Education Course Notes, Series no. 19, p. 1-1–1-48.

Behrendt, J. C., and Grim, M. S., 1983, Structural elements of the U.S. Atlantic margin delineated by the second vertical derivative of the aeromagnetic data: U.S. Geological Survey Geophysical Investigations Map GP–956, scale 1:2,500,000.

—— , 1985, Structure of the U.S. Atlantic continental margin from derivative and filtered maps of the magnetic field, *in* Hinze, W. J., The utility of regional gravity and magnetic anomaly maps: Society of Exploration Geophysicists, p. 325–338.

Behrendt, J. C., and Klitgord, K. D., 1979, High-resolution aeromagnetic anomaly map of the U.S. Atlantic continental margin: U.S. Geological Survey Geophysical Investigations Map GP–931, scale 1:1,000,000.

Behrendt, J. C., Schlee, J., and Foote, R. Q., 1974, Seismic evidence of a thick section of sedimentary rock on the Atlantic outer continental shelf and slope of the United States [abs.]: EOS Transactions of the American Geophysical Union, v. 55, p. 278.

Benson, R. N., 1984, Structure contour map of pre-Mesozoic basement, landward margin of Baltimore Canyon Trough: Delaware Geological Survey, Miscellaneous Map Series no. 2, scale 1:500,000.

Chenet, P., Montadert, L., Gairaud, H., and Roberts, D., 1983, Extension ratio measurements on the Galicia, Portugal, and northeastern Biscay continental margins; Implications for evolutionary models of passive continental margins, *in* Watkins, J. S., and Drake, C. L., eds., Studies in continental margin geology: American Association of Petroleum Geologists Memoir 34, p. 703–715.

Crutcher, T. D., 1983, Baltimore Canyon Trough, *in* Bally, A. W., ed., Seismic expression of structural styles, vol. 2: American Association of Petroleum Geologists Studies in Geology, Series no. 15, p. 2.2.3-20–2.2.3-26.

Dillon, W. P., Popenoe, P., Grow, J. A., Klitgord, K. D., Swift, B. A., Paull, C. K., and Cashman, K. V., 1983, Growth faulting and salt diapirism; Their relationship and control in the Carolina Trough, eastern North America, *in*

Watkins, J. S., and Drake, C. L., eds., Studies in continental margin geology: American Association of Petroleum Geologists Memoir 34, p. 21–46.

Drake, C. L., Ewing, M. N., and Sutton, G. H., 1959, Continental margins and geosynclines; The east coast of North America north of Cape Hatteras, *in* Ahrens, L. H., and others, eds., Physics and chemistry of the Earth, v. 3: London, Pergamon Press, p. 110–198.

Drake, C. L., Heirtzler, J., and Hirshman, J., 1963, Magnetic anomalies off eastern North America: Journal of Geophysical Research, v. 68, p. 5259–5275.

Edson, G. M., 1986, Morphology and Evolution of Mesozoic Carbonate Paleoshelf Edge, Eastern Baltimore Canyon Trough [abs.]: American Association of Petroleum Geologists Bulletin, v. 70, p. 586.

Eliuk, L. S., 1978, The Abenaki Formation, Nova Scotia shelf, Canada; A depositional and diagenetic model for a Mesozoic carbonate platform: Bulletin of Canadian Petroleum Geology, v. 28, no. 4, p. 424–514.

Eliuk, L. S., Cearley, S. C., and Levesque, R., 1986, West Atlantic Mesozoic Carbonates—Comparison of Baltimore Canyon and Offshore Nova Scotian Basins [abs.]: American Association of Petroleum Geologists Bulletin, v. 70, p. 586.

Emery, K. O., and Uchupi, E., 1972, Western North Atlantic Ocean; Topography, rocks, structure, water, life, and sediments: American Association of Petroleum Geologists Memoir 17, 532 p.

Emery, K. O., and Uchupi, E., 1984, The geology of the Atlantic Ocean: New York, Springer-Verlag Inc., 1050 p.

Emery, K. O., Uchupi, E., Phillips, J. D., Bowin, C. O., Bunce, E. T., and Knott, S. T., 1970, Continental rise of eastern North America: American Association of Petroleum Geologists Bulletin, v. 54, p. 44–108.

Falvey, D. A., 1974, The development of continental margins in plate tectonic theory: Australian Petroleum Exploration Association Journal, v. 14, p. 95–106.

Falvey, D. A., and Middleton, M. F., 1981, Passive continental margins; Evidence for a prebreakup deep crustal metamorphic subsidence mechanism: Oceanologica Acta, Proceedings 26th International Geologic Congress, Geology of Continental Margins Symposium, Paris, 7–17 July 1980, p. 103–114.

Grow, J. A., 1980, Deep structure and evolution of the Baltimore Canyon Trough in the vicinity of the COST no. B–3 well, *in* Scholle, P. A., ed., Geological studies of the COST no. B–3 well, United States mid-Atlantic continental slope area: U.S. Geological Survey Circular 833, p. 117–125.

Grow, J. A., and Markl, R. G., 1977, IPOD–USGS multichannel seismic reflection profile from Cape Hatteras to the mid-Atlantic Ridge: Geology, v. 5, p. 625–630.

Grow, J. A., Mattick, R. E., and Schlee, J. S., 1979a, Multichannel seismic depth sections and interval velocities over outer continental shelf and upper continental slope between Cape Hatteras and Cape Cod, *in* Watkins, J. S., and others, eds., Geological and geophysical investigation of continental margins: American Association of Petroleum Geologists Memoir 29, p. 65–83.

Grow, J. A., Bowin, C. O., and Hutchinson, D. R., 1979b, The gravity field of the U.S. Atlantic continental margin: Tectonophysics, v. 59, p. 27–52.

Grow, J. A., Hutchinson, D. R., Klitgord, K. D., Dillon, W. P., and Schlee, J. S., 1983, Representative multichannel seismic profiles over the U.S. Atlantic margin, *in* Bally, A. W., ed., Seismic expression of structural styles, vol. 2: American Association of Petroleum Geologists, Studies in Geology, Series no. 15, p. 2.2.3-1–2.2.3-14.

Hinz, K., 1981, A hypothesis on terrestrial catastrophies; Wedges of very thick oceanward dipping layers beneath passive, continental margins; Their origin and significance: Geologisches Jahrbuch, Reihe E, Heff 22, p. 2–28.

Hollister, C. D., Ewing, J. I., eds., 1972, Initial reports of the Deep Sea Drilling Project: Washington, D.C., U.S. Government Printing Office, v. 11, 1,077 p.

Hurtubise, D. O., Puffer, J. D., and Cousminer, H. L., 1987, An offshore Mesozoic igneous sequence, Georges Bank Basin, North Atlantic: Geological Society of America Bulletin, v. 98, p. 430–438.

Hutchinson, D. R., and Klitgord, 1987, Deep structure of rift basins from the continental margin around New England: U.S. Geological Survey Bulletin 2776 (in press).

Hutchinson, D. R., Grow, J. A., Klitgord, K. D., and Swift, B. A., 1983, Deep structure and evolution of the Carolina Trough, *in* Watkins, J. S., and Drake, C. L., eds., Studies in continental margin geology: American Association of petroleum Geologists Memoir 34, p. 129–152.

Hutchinson, D. F., Klitgord, K. D., and Detrick, R. S., 1985, Block Island Fault, a Paleozoic crustal boundary on the Long Island Platform: Geology, v. 13, no. 12, p. 875–879.

—— , 1986, Rift basins of the Long Island Platform: Geological Society of America Bulletin, v. 97, p. 688–702.

Jansa, L. F., and Wade, J. A., 1975, Geology of the continental margin off Nova Scotia and Newfoundland, *in* van der Linden, W.J.M., and Wade, J. A., eds., Offshore geology of eastern Canada; vol. 2, Regional geology: Geological Survey of Canada Paper 74–30, p. 189–197.

Jansa, L. F., Bujak, J. P., and Williams, G. L., 1980, Triassic salt deposits of the western North Atlantic: Canadian Journal of Earth Sciences, v. 17, p. 547–559.

Karlo, J. F., 1986, Results of Exploration in Mesozoic Shelf-Edge Carbonates, Baltimore Canyon Basin [abs.]: American Association of Petroleum Geologists Bulletin, v. 70, p. 605.

Keen, M. J., 1969, Magnetic anomalies off the eastern seaboard of the United States; A possible edge-effect: Nature, v. 222, p. 72–74.

Klitgord, K. D., 1987, Baltimore Canyon Trough; Base of postrift sedimentary deposits; Surface depth contours: U.S. Geological Survey Miscellaneous Field Studies Map MF–XX, scale 1:1,000,000.

Klitgord, K. D., and Behrendt, J. C., 1977, Aeromagnetic anomaly map—U.S. Atlantic continental margin: U.S. Geological Survey Miscellaneous Field Studies Map MF-913, scale 1:1,000,000.

—— , 1979, Basin structure of the U.S. Atlantic margin, *in* Watkins, J. S., Montadert, L., and Dickerson, P. W., eds., Geological and geophysical investigations of continental margins: American Association of Petroleum Geologists Memoir 29, p. 85–112.

Klitgord, K. D., and Grow, J. A., 1980, Jurassic seismic stratigraphy and basement structure of the western north Atlantic magnetic quiet zone: American Association of Petroleum Geologists Bulletin, v. 64, no. 10, p. 1658–1680.

Klitgord, K. D., and Hutchinson, D. R., 1985, Distribution and geophysical signatures of early Mesozoic rift basins beneath the U.S. Atlantic continental margin, *in* Robinson, G. R., Jr., and Froelich, A. J., eds., Proceedings of the Second U.S. Geological Survey Workshop on the early Mesozoic basins of the eastern United States: U.S. Geological Survey Circular 946, p. 45–61.

Klitgord, K. D., and Schouten, H., 1986, Plate kinematics of the central Atlantic, *in* Vogt, P. R., and Tucholke, B. E., eds., The western North Atlantic region: Boulder, Colorado, Geological Society of America, Geology of North America, v. M, p. 351–378.

—— , 1987, Tectonic and magnetic structure; Baltimore Canyon Trough and adjacent magnetic quiet zone: U.S. Geological Survey Miscellaneous Field Studies Map MF–XX, scale 1:1,000,000.

Klitgord, K. D., Schlee, J. S., and Hinz, K., 1982, Basement structure, sedimentation, and tectonic history of the Georges Bank Basin, *in* Scholle, P. A., and Wenkam, C. R., eds., United States North Atlantic outer continental shelf; Geological studies of the COST Nos. G–1 and G–2 wells: U.S. Geological Survey Circular 861, p. 160–186.

Klitgord, K. D., Hutchinson, D. R., Glover, L., Grow, J. A., and Schouten, H., 1987, Baltimore Canyon Trough; Tectonic features and geophysical lineament: U.S. Geological Survey Miscellaneous Field Studies Map MF–XX, scale 1:1,000,000.

LASE Study Group, 1986, Deep structure of the U.S. East Coast passive margin from large aperture seismic experiments (LASE): Marine and Petroleum Geology, v. 3, p. 234–242.

Le Pichon, X., and Barbier, F., 1987, Passive margin formation by low-angle faulting within the upper crust; The northern Bay of Biscay Margin: Tectonics, v. 6, p. 133–150.

Le Pichon, X., Angelier, J., Sibuet, J.-C., 1983, Subsidence and stretching, *in* Watkins, J. S., and Drake, C. L., eds., Studies in continental margin geology: American Association of Petroleum Geologists Memoir 34, p. 731–741.

Libby-French, J., 1981, Lithostratigraphy of Shell 272–1 and 273–1 wells; Implications as to depositional history of Baltimore Canyon Trough, mid-Atlantic OCS: American Association of Petroleum Geologists Bulletin, v. 65, p. 1476–1484.

Lippert, R. H., 1983, The "Great Stone Dome"; A compaction structure, *in* Bally, A. W., ed., Seismic expression of structural styles, vol. 1: American Association of Petroleum Geologists Studies in Geology, Series no. 15, p. 1.3-1–1.3-4.

Maher, J. C., 1971, Geologic framework and petroleum potential of the Atlantic Coastal Plain and continental shelf: U.S. Geological Survey Professional Paper 659, 98 p.

Mattick, R. E., Foote, R. Q., Weaver, N. L., and Grim, M. S., 1974, Structural framework of the U.S. Atlantic outer continental shelf north of Cape Hatteras: American Association of Petroleum Geologists Bulletin, v. 58, p. 1179–1190.

Mattick, R. E., Girard, O. W., Jr., Scholle, P. A., and Grow, J. A., 1978, Petroleum potential of the U.S. Atlantic slope, rise, and abyssal plain: American Association of Petroleum Geologists Bulletin, v. 62, p. 592–608.

Mattick, R. E., Schlee, J. S., and Bayer, K., 1981, The geology and hydrocarbon potential of the Georges Bank; Baltimore Canyon Trough area, *in* Kerr, J. M., and Fergusson, A. J., eds., Geology of the North Atlantic borderlands: Canadian Society of Petroleum Geologists Memoir 7, p. 461–486.

Meyer, F. O., 1986, Facies Specificity of Megaporosity in Mesozoic Shelf-Edge Carbonates, Baltimore Canyon Basin [abs.]: American Association of Petroleum Geologists Bulletin, v. 70, p. 621.

Montadert, L., Roberts, D. G., de Charpal, O., and Guennoc, P., 1979, Rifting and subsidence of the northern continental margin of the Bay of Biscay, *in* Montadert, L., and Roberts, D. G., eds., Initial reports of the Deep Sea Drilling Project: Washington, D.C., U.S. Government Printing Office, v. 48, p. 1025–1060.

Morgan, L., and Dowdall, W., 1983, The Atlantic continental margin, *in* Bally, A. W., ed., Seismic expression of structural styles, vol. 2: American Association of Petroleum Geologists Studies in Geology, Series no. 15, p. 2.2.3-30–2.2.3-35.

Mountain, G. S., and Tucholke, B., 1985, Mesozoic and Cenozoic geology of the U.S. Atlantic continental slope and rise, *in* Poag, C. W., ed., Geologic evolution of the United States Atlantic margin: New York, Van Nostrand Reinhold Co., p. 293–341.

Mutter, J. C., Talwani, M., and Stoffa, P. L., 1982, Origin of seaward dipping reflectors in oceanic crust off the Norwegian margin by "subaerial seafloor spreading": Geology, v. 10, p. 353–357.

Oil and Gas Journal, 1978, Houston oil and minerals plugs third Baltimore Canyon dry hole: Oil and Gas Journal, v. 76, no. 38, p. 72.

Poag, C. W., 1978, Stratigraphy of the Atlantic continental shelf and slope of the United States: Annual Review of Earth and Planetary Science Letters, v. 6, p. 251–280.

——, 1985a, Depositional history and stratigraphic reference section for central Baltimore Canyon Trough, in Poag, C. W., ed., Geologic evolution of the United States Atlantic margin: New York, Van Nostrand Reinhold Co., p. 217–264.

——, 1985b, Cenozoic and Upper Cretaceous sedimentary facies and depositional systems of the New Jersey slope and rise, in Poag, C. W., ed., Geologic evolution of the United States Atlantic margin: New York, Van Nostrand Reinhold Co., p. 343–365.

——, 1987, The New Jersey transect; Stratigraphic framework and depositional history of a sediment-rich passive margin, in Poag, C. W., Watts, A. B., and others, eds., Initial reports of the Deep Sea Drilling Project: Washington, D.C., U.S. Government Printing Office, v. 95, p. 763–817.

Poag, C. W., and Mountain, G. S., 1987, Late Cretaceous and Cenozoic evolution of the New Jersey continental slope and rise; An integration of borehole data with seismic reflection profiles, in Poag, C. W., and Watts, A. B., eds., Initial reports of the Deep Sea Drilling Project: Washington, D. C., U.S. Government Printing Office, v. 95, p. 673–724.

Rabinowitz, P. D., 1974, The boundary between oceanic and continental crust in the western North Atlantic, in Burk, C. A., and Drake, C. L., eds., The geology of continental margins: New York, Springer-Verlag, p. 67–83.

Ratcliffe, N. M., Burton, W. C., D'Angelo, R. M., and Costain, J. K., 1986, Low-angle extensional faulting, reactivated mylonites, and seismic reflection geometry of the Newark basin margin in Pennsylvania: Geology, v. 14, p. 766–770.

Ringer, E. R., and Patten, H L., 1986, Biostratigraphy and Depositional Environments of Late Jurassic and Early Cretaceous Carbonate Sediments in Baltimore Canyon Basin [abs.]: American Association of Petroleum Geologists Bulletin, v. 70, p. 639.

Robb, J. M., Hampson, J. C., Jr., Kirby, J. R., and Twitchell, D. C., 1981, Geology and potential hazards of the continental slope between Lindenkohl and South Toms Canyons, offshore mid-Atlantic United States: U.S. Geological Survey Open-File Report 81–600, 38 p.

Sawyer, D. S., Toksoz, M. N., Sclater, J. G., and Swift, B. A., 1983, Thermal evolution of Baltimore Canyon Trough and Georges Bank Basin, in Watkins, J. S., and Drake, C. L., eds., Studies in continental margin geology: American Association of Petroleum Geologists Memoir 34, p. 743–762.

Schlee, J. S., 1981, Seismic stratigraphy of Baltimore Canyon Trough: American Association of Petroleum Geologists Bulletin, v. 65, p. 26–53.

Schlee, J. S., and Fritsch, J., 1983, Seismic stratigraphy of the Georges Bank Basin complex, offshore New England, in Watkins, J. S., and Drake, C. L., eds., Studies in continental margin geology: American Association of Petroleum Geologists Memoir 34, p. 223–251.

Schlee, J. S., and Grow, J. A., 1980, Seismic stratigraphy in the vicinity of the COST No. B–3 well, in Scholle, P. A., eds., Geological studies of the COST No. B–3 well, United States mid-Atlantic Continental Slope area: U.S. Geological Survey Circular 833, p. 111–116.

Schlee, J. S., Behrendt, J. C., Grow, J. A., Robb, J. M., Mattick, R. E., Taylor, P. T., and Lawson, B. J., 1976, Regional geologic framework off northeastern United States: American Association of Petroleum Geologists Bulletin, v. 60, p. 926–951.

Schlee, J. S., Dillon, W. P., and Grow, J. A., 1979, Structure of the continental slope off the eastern United States, in Doyle, L. J., and Pilkey, O. H., eds., Geology of continental slopes: Society of Economic Paleontologists and Mineralogists Special Publication 27, p. 95–117.

Schlee, J. S., Poag, C. W., and Hinz, K., 1985, Seismic stratigraphy of the continental slope and rise seaward of Georges Bank, in Poag, C. W., ed., Geologic evolution of the United States Atlantic margin: New York: Van Nostrand Reinhold Co., p. 265–292.

Scholle, P. A., ed., 1977, Geological studies on the COST No. B–3 well, U.S. mid-Atlantic outer continental shelf area: U.S. Geological Survey Circular 750, 71 p.

——, ed., 1980, Geological studies of the COST No. B–3 well, United States mid-Atlantic continental slope area: U.S. Geological Survey Circular 833, 182 p.

Scholle, P. A., and Wenkam, C. R., eds., 1982, Geological studies of the COST No. G–1 and G–2 wells, United States north Atlantic outer continental shelf: U.S. Geological Survey Circular 861, 193 p.

Sheridan, R. E., 1974, Atlantic continental margin off North America, in Burk, C. A., and Drake, C. L., eds., The geology of continental margins: New York, Springer-Verlag, p. 391–407.

Sheridan, R. E., Grow, J. A., Behrendt, J. C., and Bayer, K. C., 1979, Seismic refraction study of the continental edge off the eastern United States: Tectonophysics, v. 59, p. 1–26.

Taylor, D. J., and Anderson, R. C., 1982, Geophysical studies of the COST Nos. G–1 and G–2 wells, in Scholle, P. A., and Wenkam, C. R., Geological studies of the COST Nos. G–1 and G–2 wells, United States North Atlantic outer continental shelf: U.S. Geological Survey Circular 861, p. 153–159.

Taylor, P. I., Zeitz, I., and Dennis, L. S., 1968, Geologic implications of aeromagnetic data for the eastern margin of the United States: Geophysics, v. 33, p. 755–780.

Taylor, S. R., Toksoz, M. N., and Chaplin, M. P., 1980, Crustal structure of the northeastern United States; Contrasts between Grenville and Appalachian Provinces: Science, v. 208, p. 595–597.

Trehu, A. M., Klitgord, K. D., Sawyer, D. S., and Buffler, R. T., 1987, Regional investigations of the upper Mantle; Atlantic and Gulf of Mexico continental margins, in Pakiser, L., and Mooney, W., eds., Geophysical framework of the continental United States: Geological Society of America Memoir (in press).

Tucholke, B., and Vogt, P. R., eds., 1979, Initial reports of the Deep Sea Drilling Project: Washington, D. C., U.S. Government Printing Office, v. 43, 1115 p.

Uchupi, E., 1970, Atlantic continental shelf and slope of the United States; Shallow structure: U.S. Geological Survey Professional Paper 529–I, 44 p.

Uchupi, E., and Austin, J. A., Jr., 1979, The geologic history of the passive margin off New England and the Canadian maritime provinces: Tectonophysics, v. 59, p. 53–69.

Vail, P. R., Mitchum, R. M., Jr., and Thompson, S., III, 1977, Global cycles of relative changes in sea level, in Poag, C. W., ed., Seismic stratigraphy; Applications to hydrocarbon exploration: American Association of Petroleum Geologists Memoir 26, p. 83–97.

van Hinte, J. E. and 13 others, 1985, Deep-sea drilling on the upper continental rise off New Jersey, DSDP Sites 604 and 605: Geology, v. 13, p. 397–400.

Watts, A. B., 1981, The U.S. Atlantic continental margin; Subsidence history, crustal structure, and thermal evolution, in Bally, A. W., ed., Geology of passive continental margins: American Association of Petroleum Geologists Education Course Note, Series no. 19, p. 2-1–2-75.

MANUSCRIPT ACCEPTED BY THE SOCIETY AUGUST 26, 1987

ACKNOWLEDGMENTS

Discussions with Bill Dillon, Deborah Hutchinson, Wylie Poag, Page Valentine, Al Uchupi, Brian Tucholke, John Ewing, and John Behrendt over many years have stimulated the interpretations presented in this paper. We thank Manik Talwani, Steve Eittreim, John Behrendt, and John Ewing for critical reviews of the manuscript. We also thank Patricial Worl for typing the manuscript, and Tom Kostick and Patricial Forrestel for drafting the new illustrations.

Printed in U.S.A.

The Geology of North America
Vol. I-2, The Atlantic Continental Margin: U.S.
The Geological Society of America, 1988

Chapter 14

The Blake Plateau Basin and Carolina Trough

William P. Dillon and Peter Popenoe
U.S. Geological Survey, Woods Hole, Massachusetts 02543

INTRODUCTION

Presently, the continental margin of the southeastern United States (Fig. 1) forms a zone of transition between the actively building, steep-fronted carbonate platform of the Bahamas and the typical eastern North American terrigenous clastic-dominated, drowned, shelf-slope-rise configuration. This region of the continental margin is underlain by two major sedimentary basins—the Blake Plateau Basin and the Carolina Trough (Fig. 2)—which are different in shape, basement structure, and history. Indeed, the two southern basins show some of the greatest contrasts of any basins of eastern North America, especially in their early response to rifting and in the change from rifting to drifting. The region has experienced abrupt major changes in geological conditions, most notably the onset of Gulf Stream flow in the early Tertiary.

Morphologically, the area is dominated by the broad, flat Blake Plateau at about 800–1,000 m water depth (Fig. 1). The plateau is bounded to the east by the extremely steep Blake Escarpment, descending to 5,000 m water depths. To the west, a short continental slope rises to a continental shelf. This Blake Plateau morphology characterizes the margin east of Florida and north of the Bahamas. North of Florida the margin merges into the typical shelf-slope-rise morphology. Just north of the Blake Escarpment and its northern projection, the Blake Spur, the Blake Ridge extends away from the continental slope at water depths exceeding 2,000 m (Fig. 1). This broad ridge is a Cenozoic, sedimentary drift deposit controlled by bottom currents. (For the reader who is beginning to wonder why half of the features of this region seem to be named "Blake", the *Blake* was a Coast Survey steamer from which investigations off the southeastern U.S. were carried out in 1877 to 1880. Ferromanganese nodules were discovered on the Blake Plateau at that time [Murray, 1885].)

The data most useful for understanding the deep structure and evolution of the region have been common-depth-point (CDP) multichannel seismic-reflection profiles. Most of such data that are publicly available have been collected by the U.S. Geological Survey (USGS) or for the USGS by Teledyne, Geophysical Service Incorporated (GSI), University of Texas Marine Science Institute (UTMSI), and Institut Français du Pétrole

(Fig. 1; Grow and Markl, 1977; Buffler and others, 1979; Dillon and others, 1976, 1979a, 1979b, 1979c, 1983a, 1983b, 1985; Dillon and Paull, 1978; Schlee and others, 1979; Grow and others, 1983). Other CDP profiles have been collected by UTMSI (Buffler and others, 1978; Shipley and others, 1978) and Lamont-Doherty Geological Observatory (Sheridan and others, 1979, 1981a, 1981b). A two-ship constant offset, very deep penetration multichannel seismic profile was collected by the North Atlantic Transect experiment about midway between profiles FC7 and BT1. This work was done by Lamont-Doherty Geological Observatory, Bundesanstalt für Geowissenschaften und Rohstoffe, University of Texas, and University of Rhode Island and has not been published at this time. Single-channel seismic profiling has been carried out for many years and, in recent years, the line spacing of such surveys commonly has been closer than for CDP profiling, and shallow (subbottom) resolution commonly has been superior. Therefore, the single-channel surveys are generally superior for analysis of structure and seismic stratigraphy of shallow Cenozoic strata (Ewing and others, 1966; Uchupi, 1967; Uchupi and Emery, 1967; Emery and Zarudzki, 1967; Uchupi, 1970; Schlee, 1977; Paull and Dillon, 1980a; Pinet and Popenoe, 1985a, 1985b; Popenoe, 1985). Seismic-refraction studies have been carried out over a long period, but, except for recent work using ocean-bottom seismographs (Anne Trehu, oral communication, 1985), the great depth to basement and the presence of high-velocity layers have prevented the collection of very deep information in the basins (Woollard and others, 1957; Hersey and others, 1959; Bonini and Woollard, 1960; Antoine and Henry, 1965; Bunce and others, 1965; Sheridan and others, 1966; Dowling, 1968; Grim and others, 1980; Dillon and McGinnis, 1983). Deeper structure for the region has been inferred from gravity and magnetics (Taylor and others, 1968; Klitgord and Behrendt, 1977, 1979; Grow and others, 1979; Kent and others, 1979; Behrendt and Klitgord, 1980; Hutchinson and others, 1983; Behrendt and Grim, 1983).

Few wells that provide stratigraphic data have been drilled offshore of the southeastern U.S. (Sheridan and Enos, 1979). On the shelf, a Continental Offshore Stratigraphic Test (COST) well

Dillon, W. P., and Popenoe, P., 1988, The Blake Plateau Basin and Carolina Trough, *in* Sheridan, R. E., and Grow, J. A., eds., The Atlantic Continental Margin, U.S.: Geological Society of America, The Geology of North America, v. I-2.

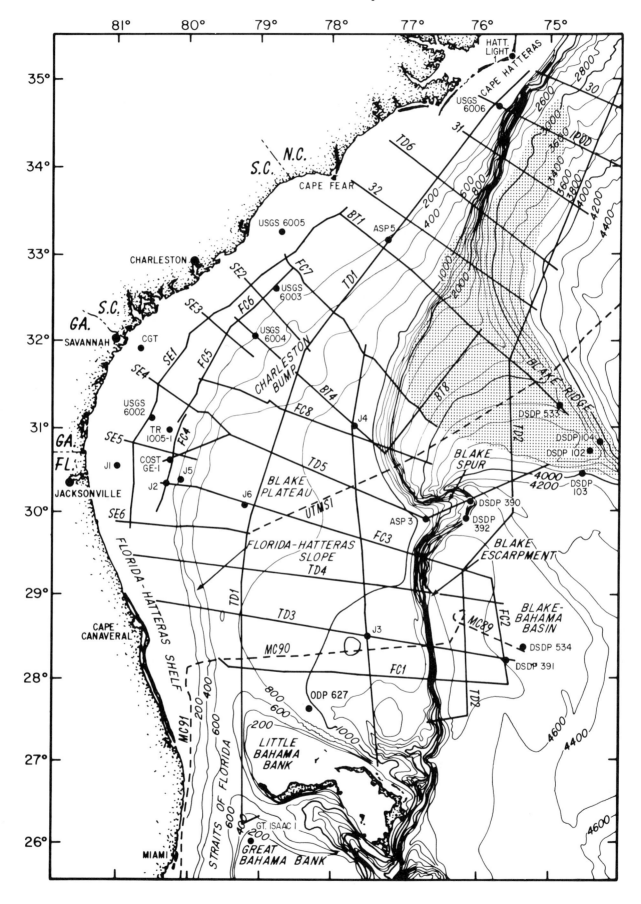

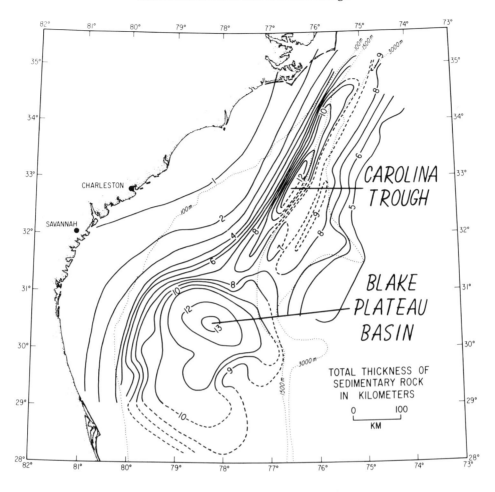

Figure 2. Isopach map showing total thickness in kilometers of sedimentary rock above the inferred postrift unconformity.

Figure 1. Bathymetry of the southeastern U.S. continental margin, locations of deep penetration multichannel seismic reflection profiles and selected wells. Bathymetric contours in meters. The solid track lines represent profiles that were collected for the U.S. Geological Survey. Copies of them are available (except for profiles labeled BT) from the National Geophysical Data Center in Boulder, Colorado. The dashed profile labeled UTMSI was collected by the University of Texas (Shipley and others, 1978) and the dashed profiles labeled MC were collected by Lamont Doherty Geological Observatory (Sheridan, 1981). Other medium penetration multichannel profiles have been collected by the U.S. Geological Survey and many single-channel profiles have been obtained in this region (see Fig. 20). The dot pattern shows the extent of the Bottom Simulating Reflector (BSR) created by gas hydrates in the sediments.

is available (Scholle, 1979) and six other exploration wells were drilled. Several shallow-penetration stratigraphic wells were drilled by Joint Oceanographic Institutions' Deep Earth Sampling Program (JOIDES) (Bunce and others, 1965; Charm and others, 1969), by the USGS (Hathaway and others, 1976, 1979; Poppe, 1981), and by a consortium of oil companies (Atlantic Slope Project—ASP—wells, Poag, 1978; Popenoe, 1985). Some Deep Sea Drilling Project (DSDP) holes were drilled on the Blake Spur (Benson and others, 1978), and some DSDP holes were placed on the ocean floor just seaward of the continental margin (Hollister and others, 1972; Benson and others, 1978; Sheridan and others, 1983). A series of Ocean Drilling Project (ODP) holes recently has been drilled around the northern Bahamas (Leg 101, Scientific Drilling Party, 1985; Leg 101, Shipboard Scientific Party, 1985). Only the COST well, the oil wells, one USGS hole, one ODP hole, and two DSDP wells on the Blake Spur penetrated deep enough to reach Cretaceous rocks in the offshore region north of the Bahamas and south of Cape Hatteras. Additional collection of Cretaceous rocks has come from the Deep Submergence Research Vessel *Alvin* submersible

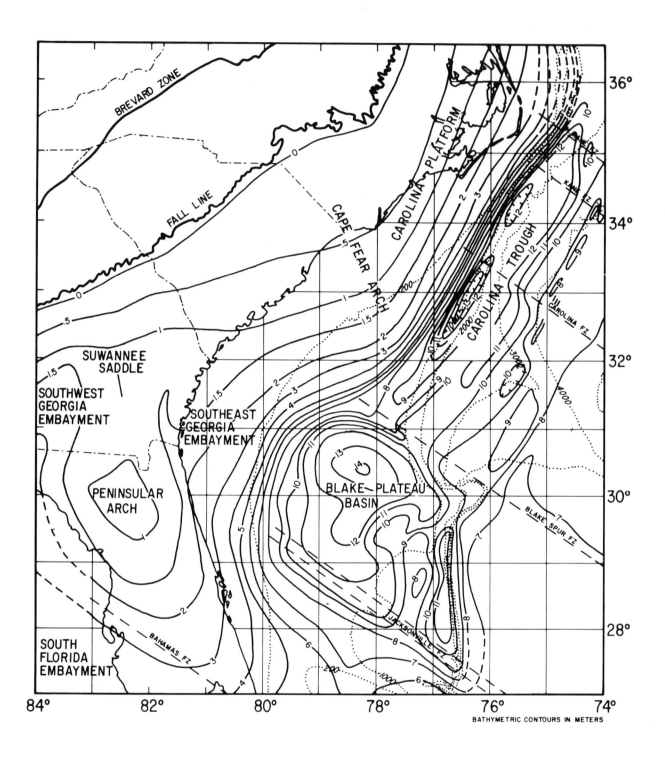

Figure 3. Structure contour map showing depth to inferred postrift unconformity in kilometers and major structural features of the southeastern U.S. continental margin region.

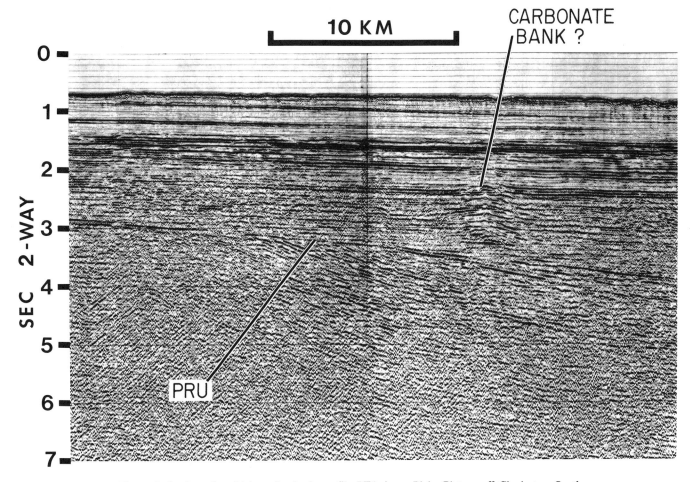

Figure 4. Section of multichannel seismic profile BT4, inner Blake Plateau off Charleston, South Carolina. Note angular unconformity dipping right from about 3 to 4 s. This is the inferred postrift unconformity (PRU) probably overlain by continental and nearshore facies strata and underlain by Paleozoic, low-grade metamorphic rocks. The change from strong discontinuous reflections below 2.2 s to continuous reflections above probably signifies a change from continental and nearshore deposits below to open-shelf deposits above. The irregular sediment surface displays the erosional affects of the Gulf Stream.

dives to the Blake Escarpment (Dillon and others, 1985; Dillon and others, 1987). Dredge hauls also have obtained Cretaceous rocks on the escarpment (Heezen and Sheridan, 1966; Sheridan and others, 1969, 1971).

BASEMENT AND EARLY DEVELOPMENT

Formation of the continental margin off the southeastern United States presumably began with rifting and stretching of the continental block as Africa and North America began to move apart. Thinning of the crustal block in the region of extension may also have involved some deep crustal flow (Bott, 1979). At the end of this episode of *rifting,* when extension of continental basement ceased and continental *drifting* began, with generation of new oceanic crust, an unconformity was cut subaerially across

most of the region. This unconformity, known as the post-rift unconformity (PRU), is the mapped horizon for Figure 3. At shallow basement depths, to about 5 km, on the continental basement of the platforms, the post-rift unconformity commonly appears as an angular unconformity (Crutcher, 1983). In Figure 4, this unconformity dips offshore from about 3 seconds (s) depth on the left of the figure to about 4-s on the right. These basement rocks of the platforms probably are largely Paleozoic, low-grade metamorphic rocks and Paleozoic and older crystalline rocks, based on the evidence from seismics, magnetics, and wells drilled on the Florida-Hatteras Shelf. Broad areas of this unconformity also occur on Jurassic basaltic flows beneath the Florida-Hatteras shelf and inner Blake Plateau from about 31°N to 33.5°N (Dillon and others, 1979a, 1983b; Dillon and McGinnis, 1983). The flows form a high-velocity refractor and strong reflec-

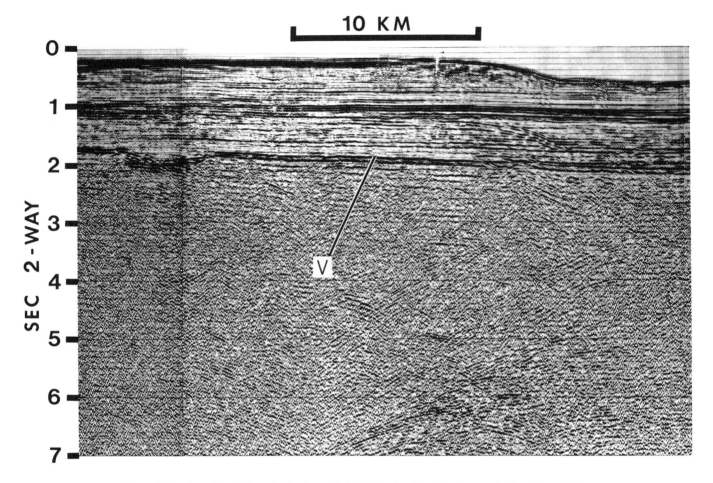

Figure 5. Section of multichannel seismic profile BT4 showing Florida-Hatteras Shelf and inner Blake Plateau off Charleston, South Carolina. The shelf has built out against the flank of the Gulf Stream current. The reflection event labeled V is considered to be a volcanic layer that covered a broad area of the postrift unconformity during the latter part of the rifting period (Dillon and McGinnis, 1983). It has been dated at 184 ±3.3 Ma (Lanphere, 1983).

tor, clearly visible in seismic profiles (Fig. 5, reflector "V"). Beneath the basins, the PRU is recognized as a somewhat stronger reflector than other events at similar depths (5–7 s) and it commonly forms a slight angle to other reflections (Fig. 6).

Nature of rift-stage basement

Rift-stage basement has not been sampled, so its nature must be deduced indirectly. Magnetic intensity values over the basins show an irregular pattern of very low amplitude anomalies with no evidence of a linear, ocean-opening pattern (Klitgord and Behrendt, 1977, 1979). This lack of a pattern of paleomagnetic stripes is especially apparent across the broad Blake Plateau and would seem to confirm that the Blake Plateau is not underlain by oceanic crust.

The crustal structure of the continental margin off southeastern North America has been modeled using seismic and magnetic data to constrain the shape of the sedimentary prism, and gravity to model the base of the crust (Fig. 7; Kent, 1979; Kent and others, 1979; Hutchinson and others, 1983). These crustal models are not unique, of course, and details of structure should not be depended on, especially where steep slopes, such as the Blake Escarpment, produce very large, abrupt, lateral density contrasts near the surface. However, the general structure suggested by these models probably is approximately correct. The crustal models suggest that basement beneath the basins is about 10 to 20 km thick, intermediate between typical oceanic and continental values. The contact between basin crust and oceanic crust seems to be very abrupt, and an abrupt contact between Carolina Trough and continental platform crust also is suggested.

The basement of the continental margin presumably was affected by stretching and fracturing of the pre-existent continental crust, as noted above (Sawyer, 1985). Basins are formed in areas where the crust is more intensively stretched and thinned to

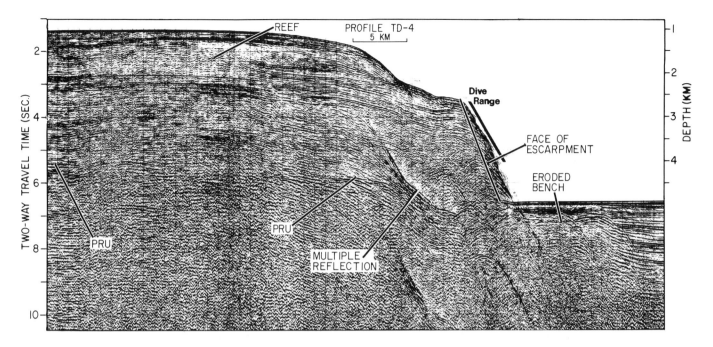

Figure 6. Section of multichannel seismic profile TD4 across the outer Blake Plateau and Blake Escarpment. A well-developed reef of Aptian and Albian age is shown. Erosion of horizon A^u time (Oligocene) and later has cut back the old platform slope, leaving an eroded bench and creating a steep escarpment. Because of its steepness the escarpment is obscured by diffractions. The postrift unconformity (PRU) appears at about 5.5 to 6 s under the plateau. True depth is around 9 to 10 km, much deeper than the sea floor. The PRU surface appears to be shallower than the sea floor in the seismic record because of the high velocity of the Blake Plateau Basin rocks, compared to the velocity of sea water.

form what is referred to as rift-stage basement. The breakup and extension of continental blocks causes extensive intrusion of dikes (De Boer, 1967; May, 1971) and produces rough fracture topography to form rift basins. Such rift basins are common along eastern North America and were filled soon after their formation with continental facies deposits. The rift-stage basement, therefore, must include fragments of continental crust, major intrusions and extrusions of mantle-derived, basaltic igneous rock, and rift-basin fillings of continental, terrigenous, clastic sedimentary rock and evaporites.

That basement beneath the southern Blake Plateau is of oceanic character was proposed by Dietz and others (1970) on the basis of an overlap in a prerift fit of the continents. Recent improved fits are somewhat different (see summary in Klitgord and Schouten, 1986), but the overlap at continental closure still precludes the presence of normal continental basement beneath the Blake Plateau. On the other hand, the thickness of basement and lack of a striped ocean-opening paleomagnetic pattern militate against oceanic basement beneath the Blake Plateau Basin and probably also the Carolina Trough. The composite rift-stage basement would appear to be required, but the conclusion of Dietz and others (1970) probably has some validity in that much of the volume of the basement beneath the Blake Plateau Basin

must have been derived as intrusions from the mantle (or by magmatic underplating). Note that the width of the Blake Plateau Basin is roughly six times that of the Carolina Trough (Figs. 2, 7, 8) and that the basement thickness in the Blake Plateau Basin is about twice as great as for the Carolina Trough. Thus, a swath across the Blake Plateau Basin contains about 12 times the volume of rift-stage basement as the Carolina Trough.

Timing

It is clear that the Carolina Trough took much less time to form than the Blake Plateau Basin. Rifting in the two basins must have begun at the same time, if rigid plate tectonics is assumed, because there is no evidence for a transform fault crossing the North American continent between the two basins, which would have allowed the rifting of one to be isolated from the other, and a propagating rift is structurally improbable. On the other hand, onset of drifting (ocean-opening) occurred much earlier at the Carolina Trough than at the Blake Plateau Basin. This conclusions is reached because there is a much broader zone of oceanic crust between the Carolina Trough and a paleomagnetic time marker in oceanic crust to the southeast than there is between the Blake Plateau Basin and the same time marker. The time marker

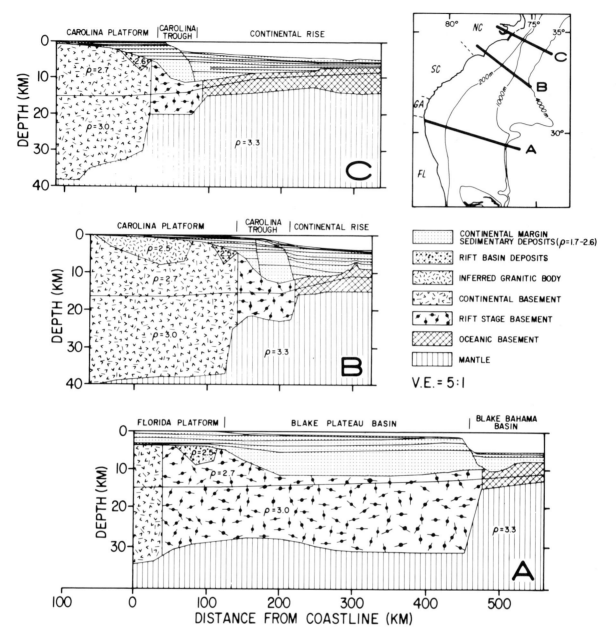

Figure 7. Major structure of the basins of the U.S. southeastern continental margin. Profile A is from Kent (1979) and Kent and others (1979), and profiles B and C are from Hutchinson and others (1983).

referred to is the Blake Spur Magnetic Anomaly (BSMA) and the situation is shown diagrammatically in Figure 8. Because oceanic crust must form at a constant (angular) rate within an ocean basin between two unfractured continental blocks, we must conclude that more time was necessary to form this broader zone of early ocean-basin crust off the Carolina Trough. Therefore, the onset of drift must have occurred earlier there than off the Blake Plateau Basin. The BSMA is not a normal ocean-opening paleomagnetic stripe, but rather it marks the location of a ridge jump (Vogt, 1973). That is, after an initial episode of ocean opening, the

Atlantic opening center (ridge) jumped eastward to the edge of the African continent (Klitgord and Schouten, 1986). Thus, oceanic crust to the west of the BSMA represents the entire ocean basin formed in the earliest episode of basin formation. Klitgord and Schouten (1986) have estimated that 5 m.y. were required to generate the crust between the BSMA and the Carolina Trough (Sheridan, 1983, has proposed a period just half that long). If the Klitgord and Schouten rate is correct, and only one fifth as much oceanic crust is present between the Blake Plateau Basin and the BSMA, then that crust south of the Blake Spur Fracture zone formed in about 1 m.y. The missing 4 m.y. represents an additional period of time during which rift-stage basement continued to be generated in the Blake Plateau Basin.

Rifting seems to have occurred during the late Early to early Middle Jurassic—approximately 195 to 172 Ma, on the basis of the age of basaltic dikes and flows (De Boer, 1967; Milton and Grasty, 1969; Behrendt and Wotorson, 1970; May, 1971; Barnett, 1975; Dalrymple and others, 1975; Van Houten, 1977; Gohn and others, 1978b; Klitgord and others, 1984). Onset of extension probably occurred significantly earlier than this (Van Houten, 1977); some older dates, exceeding 210 Ma and commonly obtained from nonbasaltic rocks, may reflect this earlier extension (Milton and Grasty, 1969; Barnett, 1975). In a broad area around Charleston, South Carolina, basalt flows cover Paleozoic basement rocks and Mesozoic synrift sedimentary rocks (Gohn, 1983). These have been dated at 184 ±3.3 Ma (Lanphere, 1983). The basalts seem to be only slightly faulted (Fig. 5, reflector V), suggesting that the major block-faulting and rift-basin filling of the platforms had nearly ceased by the time these flows were extruded. Probably the stretching and intrusion of the regions that became basins was continuing, however.

An estimate of when rifting ceased must be approached from knowledge of age of the ocean floor, and leads into some divergence of opinion. Sheridan (1983) has suggested that the age of the Blake Spur Magnetic Anomaly (Fig. 8) is 155 Ma, whereas Klitgord and Schouten (1986) conclude that the anomaly age is 170 Ma. The reason for the major part of this discrepancy (perhaps 13 of the 15 m.y.) is the use of different time scales by the different workers. Sheridan (1983) used the Van Hinte (1976) time scale, whereas Klitgord and Schouten (1986) used the more recently published Kent and Gradstein (1985; 1986) time scale. The Kent and Gradstein values have been adopted as the DNAG standard (Palmer, 1983), so these seem to be the preferable ones to use. The small remainder of the discrepancy regarding the age of the Blake Spur Magnetic Anomaly may be due to differences in interpretation of how the age of oldest sediments drilled is related to the age of oldest basement near the drill site. We have accepted the Klitgord and Schouten dates as our working values. The latter authors determine ocean-opening rates that allow extrapolation back to the edge of the continental block (East Coast Magnetic Anomaly), and they estimate 175 Ma to be the age of the oceanic crust adjacent to the continent.

In summary, then, rifting in southeastern North America probably was underway by 195 Ma. By 184 Ma it seems to have become concentrated in the region of the present basins. At about 175 Ma the rifting of the Carolina Trough ended and generation of new oceanic crust commenced. This change from rift to drift was delayed until about 171 Ma at the Blake Plateau Basin. By Blake Spur Magnetic Anomaly time, 170 Ma, when the spreading-center jump took place, the ocean-opening center had established itself to the south end of the Blake Plateau Basin, and rifting had ceased in the region we consider in this paper.

The reasons for differences in structure (thickness) of rift-stage crust between the basins, timing of rift-stage crust develop-

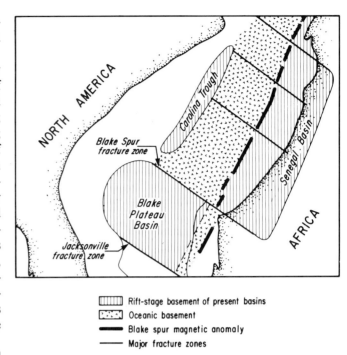

Figure 8. Reconstruction of the Atlantic Ocean at Blake Spur Magnetic Anomaly time (reconstruction provided by K. D. Klitgord, personal communication, 1981). Note difference in width of oceanic basement between the rift stage basement of the U.S. East Coast basins and the time marker of the Blake Spur Magnetic Anomaly. This indicates that the change from rifting to drifting occurred much later off the Blake Plateau Basin than off the Carolina Trough.

ment, and possibly differences in composition, may relate to differences in initial composition of the pre-existing rocks of the basin regions. It has been proposed (summarized in Dillon and Sougy, 1974, p. 339–340) that the basement of Florida and Georgia and their offshore regions, including the Blake Plateau Basin, is an African terrane, part of Africa rather than North America in early to middle Paleozoic time. If so, that might have led to mechanical differences between the two basement types, resulting in different responses to rifting and thus initiating the major offset between the basins that resulted in formation of the Blake Spur Fracture Zone.

EARLY SEDIMENTATION IN BASINS

Prior to the onset of drift, while extensional faulting still was going on, clastic sedimentation rates were high as a result of rapid erosion from the rough, faulted topography of the broad rift zone and from the Appalachian and Mauritanide orogens (Fig. 9, rifting stage; Manspeizer, 1985). These deposits commonly are red beds, including volcaniclastic sediments, where they have been mapped onshore beneath the coastal plain (e.g., Chowns and Williams, 1983) and in one drill site on the northwest corner of Great Bahama Bank (Great Isaac No. 1—Tator and Hatfield,

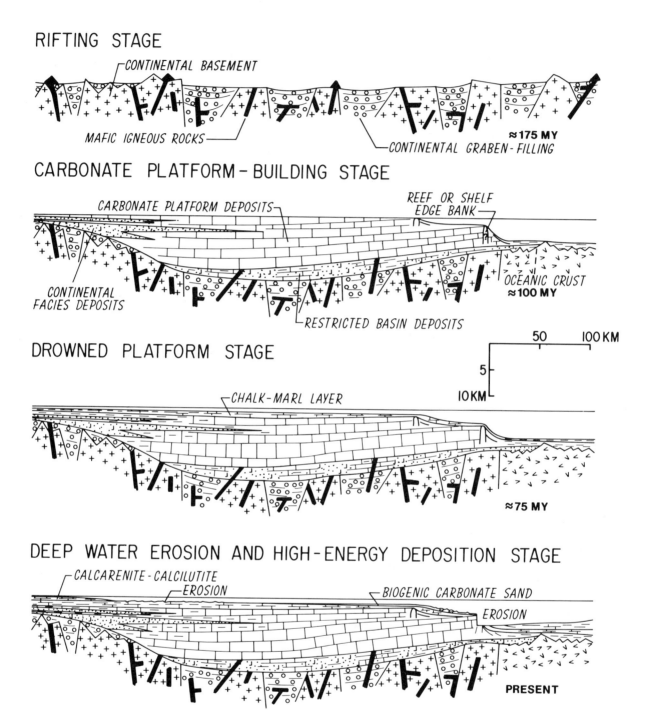

Figure 9. Conceptual stages of continental margin development of the Blake Plateau Basin.

1975; Jacobs, 1977). No sampling of these deposits is available in the deep basins, that is, the Blake Plateau Basin and Carolina Trough; indeed, no rocks older than Early Cretaceous have been sampled from the basin areas (Poag and Valentine, this volume).

Blake Plateau Basin

The zone of rifting was very broad for the Blake Plateau Basin, as indicated not only by its present width, but also by the extent of rift-related red bed deposits that extend westward beneath the coastal plain in Georgia and northern Florida (Daniels and others, 1983; Chowns and Williams, 1983). In the rifting stage, the Blake Plateau Basin also was close to an orogen, the Mauritanides in Africa (Dillon and Sougy, 1974), which presumably was at least a small sediment source. Both from local and distant sources, the Blake Plateau Basin seems to have accumulated more rift-stage deposits than the Carolina Trough prior to the onset of drift, when the newly formed ocean basin would have subsided rapidly and soon would have become a barrier to the input of sediments from Africa. Clastic terrigenous deposits are inferred to dominate in the late rifting–early drifting stage of the Blake Plateau Basin; although some evaporites certainly were formed, as sampled beneath Great Bahama Bank (Jacobs, 1977). some small faults are interpreted to break the deepest layers in the Blake Plateau Basin at 5.5 to 6.5 seconds, just above the postrift unconformity (upper section in Fig. 10). The movement on these faults might be related to minor flow of salt, as has been suggested for similar structures in the Baltimore Canyon Trough (Poag, 1985).

Carolina Trough

The early sedimentation history in the Carolina Trough contrasts markedly with that of the Blake Plateau Basin. In the Carolina Trough, the zone of rifting was very narrow, and evaporite deposition was of major significance (Fig. 11). The presence of salt is apparent from the many diapirs that appear at the seaward margin of the trough (Fig. 12; Dillon and others, 1983b). Although these structures have not been sampled by drilling, evidence for the presence of salt is provided by high salinities (increasing to above that of sea water) that were present in interstitial water of short cores obtained over the diapirs (F. T. Manheim, unpublished data, 1980). We infer that the salt may produce the strong reflection at about 6 to 7.5 seconds (about 11 km) noted in profiles across the Carolina Trough. Profiles of Figure 13 (line BT1) and Plate 4 (line 32) show this reflection; locations are shown in Figures 1 and 12. Interpretations of the lines are shown in Figures 14 and 15.

The presence of salt in the Carolina Trough and its lack (or at least, lack of significant seismic evidence for much salt) in the Blake Plateau Basin may be related to their difference in basement structure. If the crust beneath the Carolina Trough truly is half as thick as that of the Blake Plateau Basin (Fig. 7), the surface of this crust must have floated much lower isostatically than that

of the Blake Plateau Basin in the early period (late rifting–early drifting) of continental margin development. This may have made the difference between the basin's subsidence below sea level, with accompanying flooding by oceanic waters, and lesser subsidence that did not allow such flooding. This conclusion is supported by gravity modeling of eastern North American basins, which suggests the presence of thin basement (about 10 km) beneath the two basins that have significant amounts of salt (the Carolina Trough and Scotian Basin) and thicker crust (about 15 to 20 km) beneath other basins (Dillon and others, 1983a).

MAIN STAGE OF BASIN DEPOSITION

Jurassic

The nature of Jurassic strata must be inferred from seismic profiles and rocks sampled in fairly distant areas. In the southeastern U.S., only a few wells in southern Florida (Klitgord and others, 1984)—the Great Isaac No. 1 well on the northwest corner of Little Bahama Bank (Tator and Hatfield, 1975; Jacobs, 1977) and, questionably, the Hatteras Light No. 1 well at the tip of Cape Hatteras (Brown and others, 1972; Poag, 1979)—are thought to have penetrated Jurassic strata. All oil wells drilled on the shelf in the southeastern region have reached basement below Lower Cretaceous strata (Poag and Valentine, this volume).

Blake Plateau Basin. The Jurassic deposits of the northern East Coast U.S. basins (Baltimore Canyon Trough and Georges Bank Basin) apparently are a mixture of limestone and terrigenous clastic deposits (Poag, 1985; Schlee and Fritsch, 1982). The Blake Plateau Basin is situated farther from sources of terrigenous sediments in the Appalachian Orogen than are the other East coast basins. It also is located farther south, in a presumably warmer Jurassic climate, which should have been more conducive to carbonate deposition. We have interpreted carbonate banks in the upper part of the Jurassic section beneath the Blake Plateau (Fig. 10, upper section). From all the above, we infer that Jurassic deposition was dominantly carbonate with, probably, some terrigenous input to landward (Fig. 9). Similar conclusions have been reached by other workers who have considered the nature of deep Blake Plateau Basin seismic units (Shipley and others, 1978; Dillon and others, 1979a, 1985; Sheridan and others, 1981a, 1981b).

Structure of Blake Plateau Basin Jurassic strata is simple. Subsidence in the middle to inner basin resulted in the present landward dip of the major part of the strata (Fig. 10). Within Jurassic deposits in seismic profiles, broad, semitransparent zones (zones of only very weak reflections) are interpreted as carbonate bank deposits beneath the outer Blake Plateau (Fig. 10, upper section).

Carolina Trough. Jurassic deposits of the Carolina Trough probably are comparable to those of the Baltimore Canyon Trough to its north. The salt deposition most likely ended at or shortly after onset of drift, which began to form an open ocean, so nonevaporite rocks might be expected in the upper Middle and

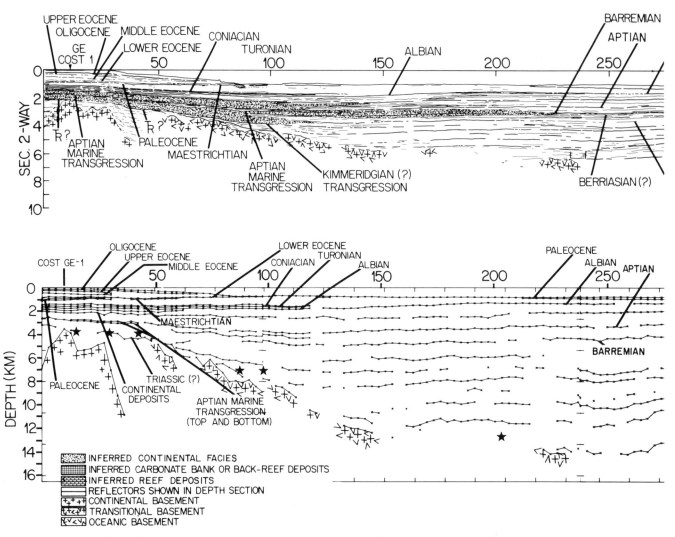

Figure 10 (this and facing page). Interpretation of seismic profile TD5 (location in Fig. 1). Upper section shows stratigraphic and facies interpretation with vertical scale in travel time. Lower section shows a conversion to a vertical scale in depth. Three drill sites were crossed by this profile, the COST GE-1, drilled to Paleozoic basement (Fig. 16), the ASP 3, drilled to Paleocene, and the DSDP 390, drilled to Barremian. This profile is discussed in Dillon and others (1979b) and Dillon and others (1985).

Upper Jurassic sections. Kimmeridgian and Tithonian rocks penetrated by a COST well in the seaward part of the Baltimore Canyon Trough are shallow-marine and coastal deposits of limestone and sandstone and include minor coal (Poag, 1985). In the Baltimore Canyon Trough, the deeper rocks are unsampled, but Poag (1985) has inferred on the basis of seismic profiles that they are a mixed carbonate-terrigenous sequence. We draw a similar conclusion for the Carolina Trough. A change in seismic reflection character, from strong continuous reflections above a depth of about 4 s to discontinuous reflections below, appears in most Carolina Trough profiles (Fig. 13). This change also is apparent landward of the trough on the Carolina Platform (Fig. 4; change occurs at about 2.2 s in this figure). In the Blake Plateau Basin, where well control is available, such a change in reflection charac-

teristics from strong continuous to discontinuous reflections has been interpreted as a change from open-shelf to nearshore-coastal-nonmarine facies (Dillon and others, 1985). Within the deeper strata, some reef- or carbonate bank–type structures are interpreted (Fig. 4). Thus, we suggest that the deeper strata of the Carolina Trough (mainly Jurassic) represent primarily very shallow-water, mixed carbonate and terrigenous clastic deposits. The occasional strong reflectors that cross the trough (Fig. 13) may represent the deposits that accumulated during sea-level rises that spread deeper water conditions across the Jurassic shelf.

Structurally, the Carolina Trough is dominated by a major fault system located at the landward side of the trough and by a series of salt diapirs on its seaward side. The faults are clear on the profiles (Fig. 13 and Plate 4); interpretations of these two profiles

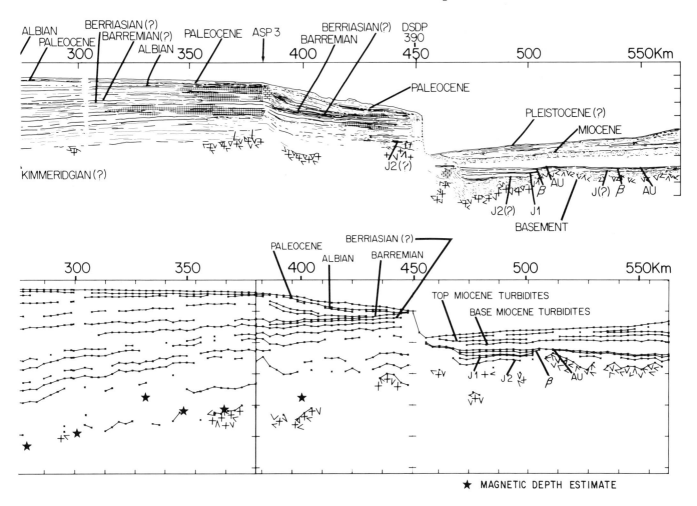

★ MAGNETIC DEPTH ESTIMATE

are shown in Figures 14 and 15. The fault system and diapirs are mapped in Figure 12, for which profiles both of Figure 1 and Figure 20 were employed.

The fault (actually a fault system) is a growth fault, that is, throw increases progressively with depth as a result of continuous movement on the fault during deposition (Dillon and others, 1983a). Furthermore, the movement is continuing, as indicated by offsets of near-bottom reflectors and of the sea floor (Sylwester and others, 1979; Popenoe, 1985). The fault does not flatten significantly with depth, as down-to-basin normal faults ordinarily do, but rather continues at approximately a constant slope into the inferred salt layer (Figs. 13, 14, 15, Plate 4). We conclude that the faulting, rather than resulting from gliding of a block to seaward under the force of gravity, results from withdrawal of material (salt) from beneath the subsiding block (Dillon and others, 1983a).

The salt domes also appear to be active, as indicated by up-warping of the sea floor over the diapirs, collapse due to solution of salt, and resultant slumping of broad areas of the sea floor (slumps are shown as cross-hatched area in Fig. 12) (Dillon and others, 1983a; Cashman and Popenoe, 1985). The diapir locations

are closely associated with the East Coast Magnetic Anomaly (Fig. 12), which has been interpreted as the result of both basement topography and change in crustal composition from rift stage to oceanic basement. Since flowing salt is unaware of magnetics or basement composition, we conclude that the salt, as it rose, probably was guided by a basement inflection at the East Coast Magnetic Anomaly that also may have bounded the initial salt-deposition basin.

There probably is a structural relationship between the salt diapirs and the faults, in that the diapirs result from rising of the salt out of the salt layer that underlies the Carolina Trough, and the faults are due to subsidence of the block of strata overlying the trough as the salt flows out from beneath it. This flow of salt, accompanied by diapirism and faulting, has continued for much of the basin's history (Fig. 11).

Early Cretaceous

Blake Plateau Basin. Lower Cretaceous rocks have been sampled to a limited extent, so we need not depend solely on seismic profiling data to interpret continental margin evolution.

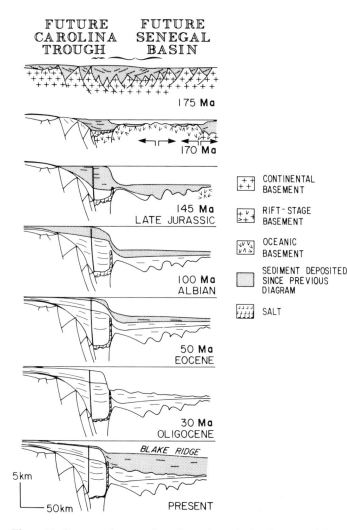

Figure 11. Conceptual stages of continental margin development of the Carolina Trough. In each stage only deposits that accumulated since the previous stage have symbols. Salt always is marked because it is mobile through most of the basin's history. The two pairs of dual arrows in the second stage locate the spreading center that produced the Blake Spur Magnetic Anomaly (left) and a new spreading center against the African continental margin (right). At Blake Spur Magnetic Anomaly time, a spreading-center jump occurred in which the western (left) spreading center was shut down and the new center to the east became the new Atlantic opening center.

In the Blake Plateau Basin, data are available from seven wells that were drilled relative to oil exploration. Unfortunately, all were drilled in a restricted area of the outer continental shelf off northern Florida and southern Georgia between lat. 30°21′N and 31°05′N and long 80°14′W and 80°32′W (data are available from the National Geophysical Data Center, Boulder, Colorado). All of the wells penetrated Lower Cretaceous strata above basement (Poag and Valentine, this volume). We have prepared summaries of data from two of these wells (Fig. 16): the Conti-

nental Offshore Stratigraphic Test, Southeast Georgia Embayment No. 1 (COST GE-1) and the Transco OCS-A-3671, Brunswick Block 1005 No. 1 (TR 1005-1; locations Fig. 1). These wells represent sampling in the far northwestern flank of the Blake Plateau Basin; this is apparent from the location of the COST GE-1 well, shown on the left end of profile TD5 (Fig. 10). On the landward side of the Blake Plateau Basin (Fig. 16), most of the Lower Cretaceous section consists of terrestrial facies, including red sandstone and shale with coal. Probably such a sequence, but of older age, would appear above basement farther out in the basin as well. A general trend toward marine conditions is apparent, and by Albian time, nearshore to midshelf limestones and shales began to accumulate. Paleoenvironmental studies of COST GE-1 samples (Poag and Hall, 1979; and Fig. 16) have allowed us to relate depositional facies to seismic signatures in profile TD5, which passes through the well site (Dillon and others, 1985). Continental-nearshore facies show strong, markedly discontinuous reflections, whereas open-shelf deposits appear as strong continuous reflections, and slope-depth marls (calcareous claystone, argillaceous chalk, and calcareous shale; Fig. 16) produce almost no reflections and thus result in a seismically transparent unit. Such recognition allows the tracing of the boundary between continental-nearshore and open-shelf facies in a profile. When this approach is applied in profile TD-5, we can define the irregular pattern of transgressive-regressive deposits identified by the dot pattern in Figure 10 (upper profile). During the Early Cretaceous, marine transgressions extended westward beyond profile TD5, and regressions resulted in extension of continental deposition out to the central Blake Plateau. Poag and Hall (1979) have found close correspondence between the transgressive-regressive facies pattern as indicated by foraminifera at the COST GE-1 well and the worldwide transgression-regression pattern proposed by Vail and others (1977). Because the transgression-regression curves correspond for the younger part of the section at this location, we have assumed that they are likely to correspond for the older part. Therefore, we have used the transgression-regression pattern deduced from the profile to key to the curve of Vail and others (1977) and to estimate ages of reflectors that are older than Aptian (Fig. 10; Dillon and others, 1985).

On the seaward side of the Blake Plateau Basin, drilling has reached Lower Cretaceous strata, and sampling by submersible and dredge has obtained Lower Cretaceous rocks. Dredge samples have provided Lower Cretaceous rocks at several locations on the Blake Escarpment (Heezen and Sheridan, 1966; Sheridan and others, 1969, 1971). A series of submersible dive transects, extending from the top of the Blake Escarpment down to 4,000 m water depth along profiles TD3, 4, and 5 (locations Fig. 1), has provided shallow-water limestones that show a complete stratigraphic sequence of the Valanginian-Hauterivian to Albian stages (Dillon and others, 1985, 1987). DSDP drill sites 390 and 392 are located on the Blake Spur (Fig. 1), and profile TD5 passed through site 390 (Fig. 10). Barremian limestones that were penetrated at both drill sites had been deposited in very shallow shelf conditions with subaerial intervals (site reports in Benson and

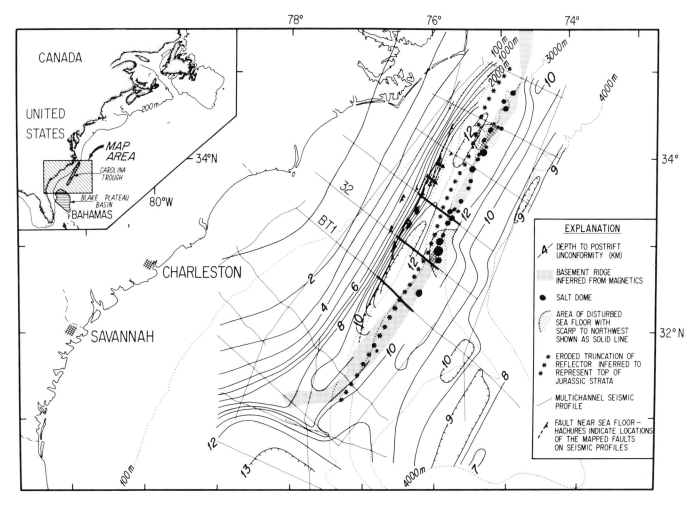

Figure 12. Structure of the Carolina Trough, showing relationship of salt domes and normal growth fault system to basin structure.

others, 1978; Enos and Freeman, 1978). Barremian shallow marine deposits were succeeded by marine clayey nannofossil ooze that accumulated in relatively deep water conditions (several hundred meters) from the Aptian into the early Tertiary (Cenomanian to Santonian strata were not sampled). The structure of the outer Blake Plateau shown in Figure 10 accounts for this sequence at the DSDP sites. Prior to the late Barremian, the entire Blake Plateau was a broad carbonate platform, probably with a reef to the east that is now eroded away. However, at about that time, the reef structure at the platform edge ceased its growth and a new Aptian-Albian reef was initiated to landward. Thus, at the outermost part of the platform, moderately deep-water conditions were instituted. The retreat of reef construction occurred at several locations; it was a maximum—more than 50 km—at the Blake Spur (Fig. 10). This Aptian-Albian reef development is clearly shown in profile TD4 (Fig. 6) and can be observed in most other profiles across the seaward part of the Blake Plateau Basin, including FC1, MC90 (Sheridan and others, 1981a), and

TD3, where the reef is truncated, as well as TD4, TD5, and the UTMSI line (Shipley and others, 1978; see Fig. 1 for location). Sheridan (1981, p. 2574), perhaps more precisely, refers to the reef feature as a "massive reflectorless bank margin complex." One sampling transect of the escarpment by submersible has been made at a location where profile TD3 shows the Aptian-Albian reef to have been truncated by erosion (Dillon and others, 1985, 1987). Samples here include caprinid rudists (Norman F. Sohl, written communication, 1981), suggesting that the feature is, at least in part, a rudist reef.

In summary, sampling of Barremian and older Lower Cretaceous (Neocomian) rocks and seismic profiling data suggest the presence of a carbonate platform, possibly bounded by reefs to the east, and merging westward into terrigenous nearshore and coastal plain deposition beneath the present Florida-Hatteras Shelf. Later in the Early Cretaceous, during the Aptian and Albian, reefs developed landward from the previous platform edge in many cases, and the outermost Blake Plateau was a site of

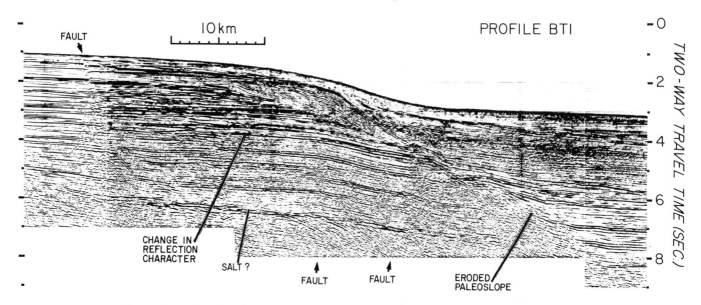

Figure 13. Section of multichannel seismic profile BT1 across the Carolina Trough. A line drawing interpretation of this profile is shown in Figure 14. The result of formation of the outer shelf/slope region by multiple episodes of erosion and deposition is apparent. The history of this profile is shown diagramatically in Figure 11. The major fault on the landward side of the trough (left) is active as shown by progressively greater offsets of reflectors at greater depths. It is thought to result from flow of salt to seaward, which removes support for the major block of strata above the inferred salt horizon. The strong reflector about ½ s below the sea floor and approaching the sea floor on progressing up the slope is not produced by a sedimentary stratum and, indeed, crosses reflections produced by strata. It is known as a bottom simulating reflector, or BSR (note its extent in the interpretation, Fig. 14). The BSR is produced by formation of clathrate compounds (gas hydrates) within the sediments just below the sea floor, as discussed at the end of this paper.

deposition in moderately deep water (Fig. 9). In general, although the Blake Plateau Basin's deposition was continuing to keep up with subsidence in Early Cretaceous, a continuing marine transgression took place on its landward side, and westward retreat of reef building occurred on its seaward side. This may represent the first evidence for environmental stress on the reefs, which shortly thereafter caused their demise and resulted in a major change in depositional character.

No significant structural development took place in the Blake Plateau Basin during Early Cretaceous time, just a continuing (and slowing) subsidence due to cooling and loading of the lithosphere.

Carolina Trough. No offshore sampling is available for the Carolina Trough. However, the Esso Standard Oil Company Hatteras Light No. 1 well (Spangler, 1950; Brown and others, 1972) was drilled at the point of Cape Hatteras, and thus occupies a position on the northwest flank of the Carolina Trough similar to that of the COST and Transco wells in the Blake Plateau Basin (compare locations in Fig. 1 to structure in Fig. 3). Not surprisingly, the lithologies are very similar, the Hatteras Light No. 1 well having recovered Lower Cretaceous continental clastics interbedded with marine sands and shales and minor

limestones. Lignite is common, and sands just above basement are arkosic and gravelly as in the COST well. Lower Cretaceous sediments in the Baltimore Canyon Trough, to the north of the Carolina Trough, are dominantly terrigenous sandstone and shale. Carbonate deposition decreased drastically at this time (Poag, 1978, 1985; Poag and Valentine, this volume). Because the Carolina Trough would appear to be a transitional region between the Baltimore Canyon Trough and the carbonate platform of the Blake Plateau Basin, we suggest that carbonate deposition, if active in the Carolina Trough during the Jurassic, probably weakened considerably in the Early Cretaceous. No clearly defined reefs appear in the profiles of the Carolina Trough, although zones of landward dip and zones of incoherent reflections beneath the unconformity formed by an old eroded continental slope might be considered evidence for reefs or carbonate banks (Grow and Markl, 1977). Such features appear on line 32 (Plate 4) just west of the upturned beds associated with a diapir. We prefer to interpret the northwest dips of beds to be the result of basin subsidence and salt withdrawal, and the zone of incoherent reflections as due to interference by diffractions from the rough unconformity. Thus we conclude that no significant reefs are present within the strata of the Carolina Trough.

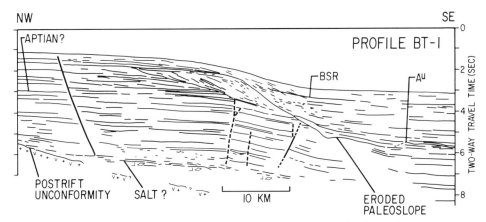

Figure 14. Interpretation of section of profile BT1 shown in Figure 13.

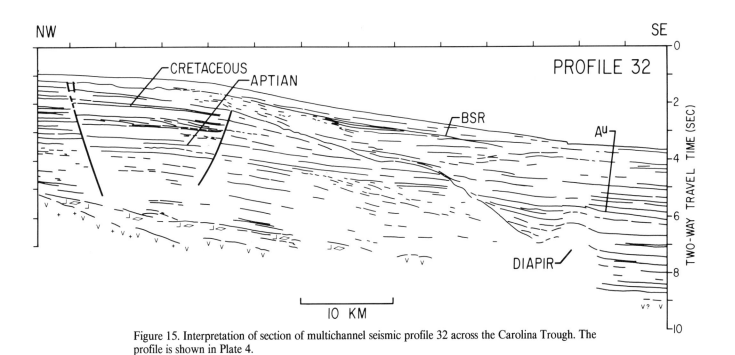

Figure 15. Interpretation of section of multichannel seismic profile 32 across the Carolina Trough. The profile is shown in Plate 4.

During the Early Cretaceous, the structural evolution of the Carolina Trough continued to be dominated by basin subsidence and salt flow (Fig. 11). The result of these movements, which continue to the present, on overall basin structure has been a broad subsidence along the trough. This affect is shown by structure contours on a horizon within the Lower Cretaceous strata (Fig. 17). A broad "shelf-edge anticline" is apparent, now seaward of the shelf edge because of shelf retreat in the Tertiary. This feature results from the northwest dips that are produced by rotation of the outer-shelf strata down toward the trough due to subsidence and salt removal within the trough, and from southeast dips that result from deposition on the paleoslope.

TERMINATION OF CARBONATE PLATFORM DEPOSITION AND INITIATION OF THE BLAKE PLATEAU

Late Cretaceous

Blake Plateau Basin. A radical change in deposition from shallow carbonate-platform deposition to deep-water marl accumulation occurred in the Blake Plateau Basin, approximately in the Cenomanian. This initiated a deep-water plateau off the southeastern United States that has persisted to the present (drowned platform stage—Fig. 9). Initially the Aptian-Albian reef died,

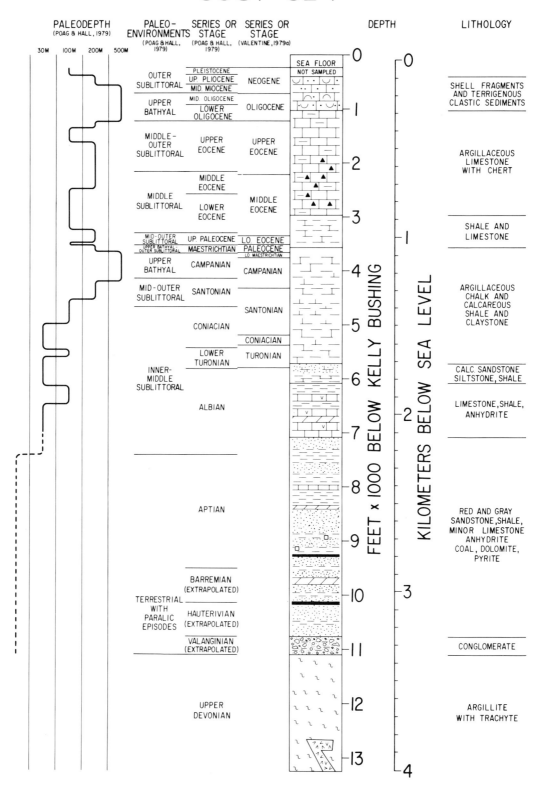

Figure 16. Biostratigraphy, lithology, and paleoenvironments interpreted for COST GE-1 well and Transco (TR) 1005-1 well in the Florida-Hatteras Shelf. See Figure 1 for locations. Information for COST well is reported in Scholle, 1979. Information for Transco well is available from National Geophysical Data Center, Boulder, Colorado. The biostratigraphy of Valentine (1979a) is based on

TR 1005-1

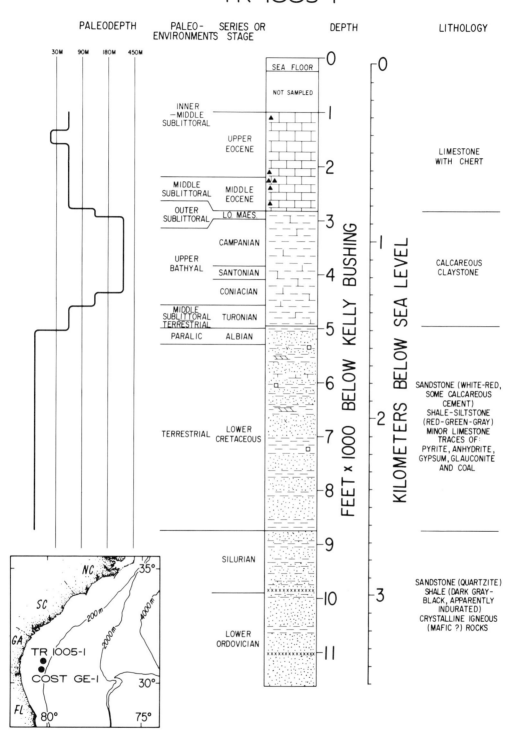

calcareous nannofossils. That of Poag and Hall (1979) is based primarily on planktic and benthic foraminifera back to Albian time, as are the paleodepth and paleoenvironment estimates. Identification of Aptian and older palynomorphs is reported in Poag and Hall (1979) also. Barremian, Hauterivian, and Valanginian upper boundaries are very approximate, determined by extrapolation of sedimentation rates from above. Basement dates are reported by Simonis (1979).

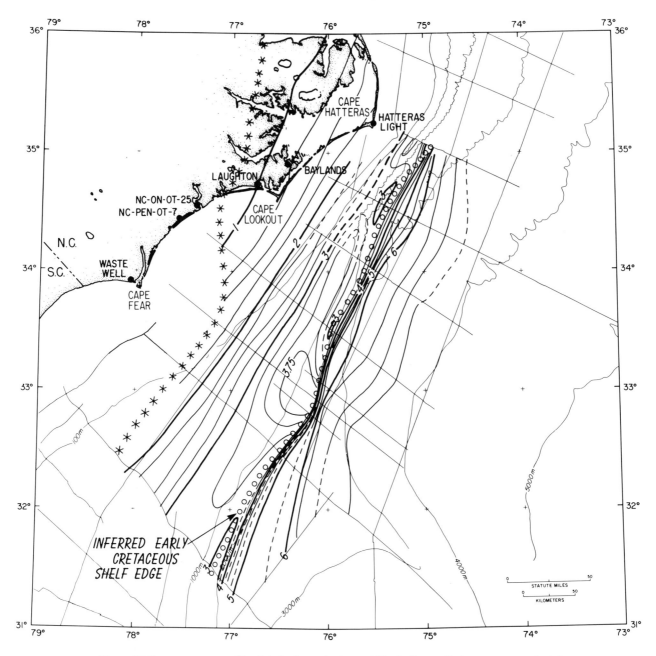

Figure 17. Structure contours (depth in km) on a horizon within the Lower Cretaceous strata, adapted from Paull and Dillon (1982). Line of stars represents location of pinch-out of the selected horizon.

perhaps due to a Cenomanian regression (Vail and others, 1977) that is indicated by lack of Cenomanian strata in the shelf wells of Figure 16. After the Cenomanian, sea level apparently rose markedly and flooded the platform, probably due to worldwide fluctuations (Vail and others, 1977; Pitman, 1978; Watts and Thorne, 1984). The period is marked by fairly deep water conditions, evidenced even at shelf wells on the far western side of the basin, especially in Santonian-Campanian time (Fig. 16). Deep-

water (shelf, then slope-depth) Late Cretaceous marls also appear above shallow-water Albian platform carbonates at the recently drilled ODP site 627, just north of the Bahamas (Fig. 1; Leg 101, Scientific Drilling Party, 1986; Leg 101, Shipboard Scientific Party, 1985). Why the reef that flourished in Aptian-Albian time did not reestablish itself during the rising sea level is not clear, as Schlager (1981) has argued that reefs can keep pace with any sea-level rise. Presumably other environmental stresses may have come into play. The continent had been drifting northward and climates had been changing, so climatic stress may have been the principle reason for the final demise of the Blake Plateau Basin reefs (Jansa, 1981).

A very abrupt transition between the Blake Plateau and the Bahama Banks occurs at the south end of the plateau. The banks are parts of the carbonate platform that continued to grow when construction of the Blake Plateau portion ceased. The present Bahama Banks have maintained themselves near sea level, whereas, just a short distance to the north, the Blake Plateau now lies at about 800 m water depth (Fig. 1). The abruptness of this transition might suggest the presence of tectonic control (Mullins and Lynts, 1977; Sheridan, 1974) or of a dramatic environmental or oceanographic boundary at this location. Such direct control probably is not necessary. Whatever the controlling factors, the bathymetric boundary is most likely to be sharp because the growth of a carbonate platform is an all-or-nothing process. It must maintain itself near sea level to produce significant volumes of limestone. If it sinks, even perhaps to only several tens of meters, the processes of carbonate deposition become much slower. Because the carbonate platform-building process tends to be either on or off, a transition zone will be likely to be abrupt unless clastic sedimentation (of biogenic products) is very active (Leg 101, Shipboard Scientific Party, 1985).

At the shelf wells (and ODP 627), the Upper Cretaceous drowned platform deposits are essentially marls (Fig. 16). Seismically, these deposits are structureless (seismically transparent), a distinctive characteristic that allows them to be traced easily all across the Blake Plateau (Dillon and others, 1985; Shipley and others, 1978). This seismic unit is shown in the interpretation of profile TD5 (upper section, Fig. 10) in which few reflections appear between Albian and Paleocene horizons because of the seismic character of the rocks. The fine-grained nature and apparent lack of bedding of the Upper Cretaceous deposits suggest that the deep-water conditions also were very quiet, with no significant current activity to disrupt deposition of these fine-grained carbonates and clays. We infer that depositional conditions across the Blake Plateau Basin in the Late Cretaceous were quiet, moderately deep water (several hundred meters) in which a gentle rain of fine-grained deposits accumulated on the sea floor. This precludes the presence of a Gulf Stream across the southern Blake Plateau at this time, as it would have prevented deposition or torn up these fine-grained deposits (as it subsequently did, to some extent). Sheridan and others (1981a) have identified a rough surface at the top of Aptian-Albian rocks in the Straits of Florida as possible first evidence of Gulf Stream flow. If so, the flow must have been a temporary pulse, or the erosional effects may simply have been related to the Cenomanian low sea-level stand.

Onshore, the Late Cretaceous transgression extended across the present Coastal Plain, where the oldest sediments are unfossiliferous, coarse-grained, feldspathic sands and clays of continental facies that grade upward into poorly fossiliferous, marginal marine shales and sandstones of Turonian age (Applin and Applin, 1967; Gohn and others, 1980; Valentine, 1984). By late

Santonian time, the transgression that began in the Cenomanian had inundated the Coastal Plain almost to the Fall Line, widely depositing silty clay, carbonaceous clay, and coarse-grained sand. Over the central Carolina Platform (Cape Fear Arch), these beds are dominantly continental, terrigenous clastics (Gohn, 1983; Valentine, 1979b, 1984) that grade laterally southward into marine clastics on the flank of the arch, and light-colored, chalky limestone within the Southeast Georgia Embayment and over peninsular Florida. On the northern flank of the Peninsular Arch of Florida, carbonate sands and gypsiferous dolomite, indicative of a shallow-water carbonate platform, were deposited (Applin and Applin, 1967).

Near the end of the Cretaceous, Campanian and Maestrichtian sediments record complete marine inundation of the Coastal Plain accompanied by the initiation of current flow across southern Georgia through the Suwannee Strait (Fig. 18), which linked the Gulf of Mexico with the Atlantic (Hull, 1962; Chen, 1965; Applin and Applin, 1967; Valentine, 1979b; Pinet and Popenoe, 1985a; Popenoe, 1987). Onshore, the strait has been traced more than 300 km across the panhandle of Florida and southern Georgia as a distinct depression on the top of the Campanian (Tayloran) Stage beds and as an erosional-nondepositional hiatus of Maestrichtian (Navarroan) Stage beds. The current flow through the strait provided a boundary between facies by which sandy calcareous shales of the coastal plain are restricted to its north, and carbonate bank and reef deposits of the Florida peninsula to its south (Chen, 1965). The extension of the Suwannee Strait offshore to the east, under the Florida-Hatteras Shelf, is seen in seismic-reflection records as a broad (100-km-wide) buried channel in which the Campanian beds appear to be thin and chaotic. The south bank of the channel is relatively abrupt, rising about 150 m to an apparent reefal edge, whereas the north bank is gentler and is underlain by a thickened prograded accumulation of Santonian through Paleocene deposits. Maestrichtian deposits are absent across Georgia in the center of the channel (Applin and Applin, 1967) due to erosion or nondeposition. However, beneath the Florida-Hatteras Shelf and inner Blake Plateau, an eastward-prograding buildup of Campanian and Maestrichtian deposits was observed at the southern side of the mouth of the Suwannee strait (Fig. 19, profiles 19 and 24, Campanian-Maestrichtian). Well data (Fig. 16) show that these deposits are fine-grained marls, suggesting that this was a deltaic accumulation from currents that were debouching from the channel into quiet, deeper water (Pinet and Popenoe, 1985a, 1985b). In contrast, seismic profiles show that Campanian-Maestrichtian deposits on the north side of the channel were accumulated in a somewhat stronger current regime in which better-defined bedding and broad progradations are apparent (Fig. 19, line 14).

Clearly, the dominant water flow from the Gulf of Mexico to the Atlantic during the Late Cretaceous and Paleocene was through the Suwannee Strait, rather than the present flow, southward around Florida and then northward as a high-speed current through the Straits of Florida (Fig. 18). The Suwannee

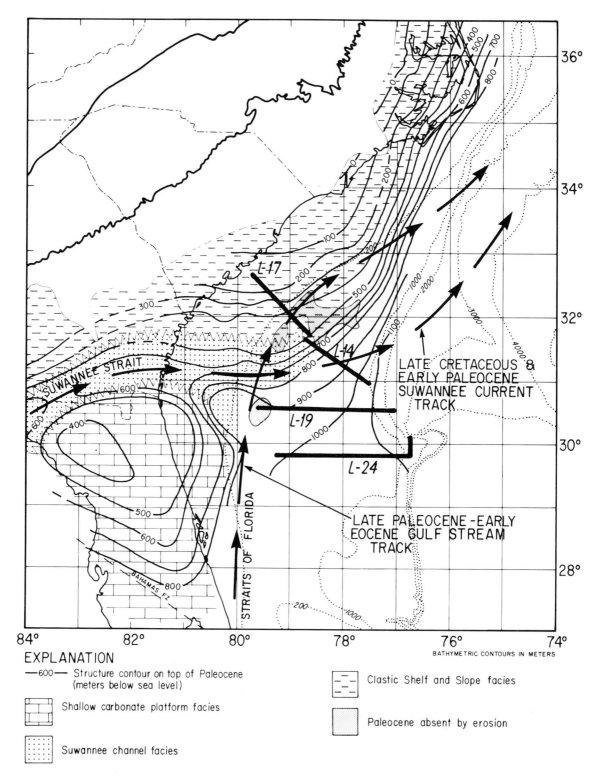

EXPLANATION

—600— Structure contour on top of Paleocene
(meters below sea level)

Shallow carbonate platform facies

Suwannee channel facies

Clastic Shelf and Slope facies

Paleocene absent by erosion

Figure 18. Structure contour map of the top of the Paleocene deposits showing facies of the Paleocene, the track of the early Paleocene Suwannee Current through the Suwannee Strait, and the initial track of the Gulf Stream (Florida Current) through the Straits of Florida in the late Paleocene to early Eocene. Note the late Paleocene–early Eocene erosion where the Gulf Stream impinged on the shelf-slope deposits of the Charleston Bump. Heavy lines indicate locations of seismic profiles shown on Figure 19. Compiled from Chen, 1965; Brown and others, 1972; Paull and Dillon, 1980a; Gohn and others, 1978a, 1980; Popenoe, 1985; and Pinet and Popenoe, 1985a.

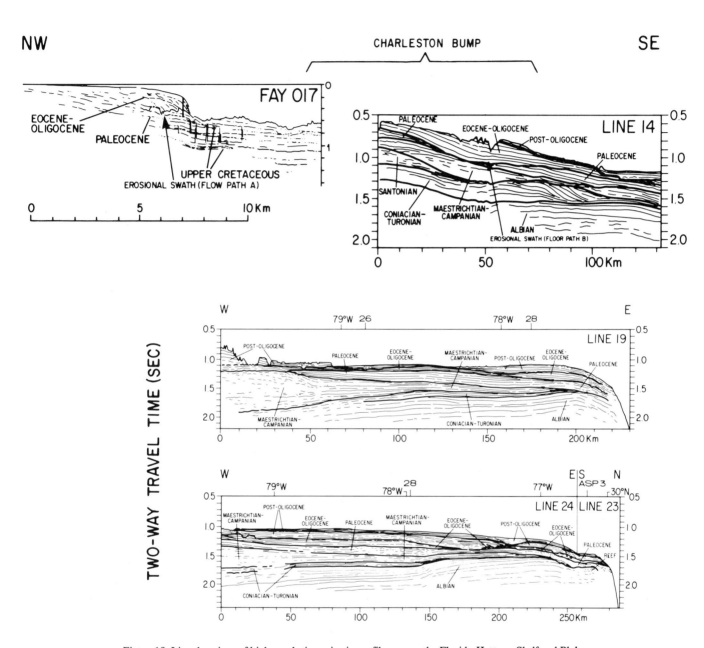

Figure 19. Line drawings of high-resolution seismic profiles across the Florida-Hatteras Shelf and Blake Plateau showing the Late Cretaceous progradational wedges over the Charleston Bump, the thick buildup of Maestrichtian-Campanian strata at the mouth of the Suwannee Strait, and the erosional unconformity on the top of Paleocene strata on the south flank of the Charleston Bump that marks the initiation of the Florida Current (modified from Pinet and Popenoe, 1985a, 1985b). Location of profiles is shown on Figure 18.

current through the Suwannee Strait had a major impact on patterns of sedimentation at that time. The Suwannee current exited from the strait and crossed the continental margin just north of the Blake Plateau Basin, skirting the Charleston Bump. The Charleston Bump (Brooks, 1986) is a topographic feature (Fig. 1) at a location where facies boundaries have occurred in the past. The bump is at the northern corner of the Blake Plateau Basin, where the basement inflectioon bounding the northwestern sides of the Carolina Trough and Blake Plateau Basin forms an abrupt right-angle bend where the Blake Spur Fracture Zone terminates against continental crust (Fig. 3). High-resolution seismic profiling shows that even late in margin development, differential subsidence here at this crustal boundary has controlled transitions in sedimentary facies, perhaps in part through effects on current flow. Pinet and Popenoe (1985a, 1985b) mapped three pronounced Upper Cretaceous facies boundaries where clastic wedges of Turonian-Coniacian, Santonian, and Campanian-Maestrichtian age were shown to have prograded southeastward from the shallow platform corner into the deeper-water carbonate deposits of the Blake Plateau Basin (Fig. 19, Line 14 and Fay 017). This sedimentary ramp of clastic deposits has been modified by additional deposition and subsequent Gulf Stream erosion, but still affects the bathymetry of the Blake Plateau to form the "bump".

Carolina Trough. Characteristic seismic-reflection signatures in the Upper Cretaceous strata of the Carolina Trough suggest that the marl present in the Blake Plateau Basin is not present in the Carolina Trough, but rather that the trough is characterized by normal shelf deposition with occasional episodes of progradation, characteristic of transgressions. This form of deposition is not surprising, as the Suwannee Current, reinforced by the Antilles Current, presumably swept the shelf of the Carolina Trough and would have prevented deposition of very fine sediments. Deposits are inferred to be dominantly sandstones and shales, as in the Baltimore Canyon Trough (Poag, 1985).

Extensive erosion in the Carolina Trough area cut back the continental slope, and this eroded paleoslope then was covered again by Late Cretaceous deposition, which blanketed the slope with a uniform homogeneous layer of sediment during the episode of high sea level (Mountain and Tucholke, 1985). The paleoslope is clearly apparent in Figures 13, 14, and 15, and Plate 4. The erosion that created this unconformity is undated, but may have occurred during the Cenomanian low sea level (although it certainly was submarine, not subaerial, erosion).

Paleocene

During most of the Paleocene, deposition similar to that of the Late Cretaceous continued in the Blake Plateau Basin, and the thermal- and loading-driven subsidence of the basin proceeded at a rate faster than deposition, so the region remained as a deep-water, sediment-starved plateau (Figs. 9 and 18). The Carolina Trough continued to accumulate normal shelf deposits (Fig. 11), while progradational clastic shelf deposits accumulated on the

north flank of the Suwannee Strait and shallow-water carbonate platform deposits were deposited over peninsular Florida (Chen, 1965).

EROSIONAL SHAPING OF
THE CONTINENTAL MARGIN

During the remainder of the Cenozoic (after Paleocene), erosional processes began to reshape the continental margin. In the upper water, erosion was initiated by the onset of a new strong geostrophic flow, the Gulf Stream. In the deep sea, episodes of major erosion occurred, the most notable being the horizon A^u erosion that severely eroded the North Atlantic abyssal floor, especially off the southeastern United States.

Eocene

Toward the close of the Paleocene or in the earliest Eocene, a dramatic strengthening of north-flowing currents through the Straits of Florida heralded the onset of Gulf Stream flow, and this was accompanied by a diminishing of the Suwannee Current (Paull and Dillon, 1980a; Sheridan and others, 1981a; Pinet and Popenoe, 1985a, 1985b). This shift of currents from the Suwannee Strait to the Straits of Florida was probably caused by a major drop in sea level, which would have shallowed the Suwannee Straits and forced the currents south of the carbonate platform of Florida. Vail and Hardenbol (1979) do not show a major sea-level drop at this time; however, Hazel and others (1984) found that the most widespread and longest hiatus of the Paleogene, which produced a major unconformity on both the southeastern Atlantic margin and the Gulf Coast, occurred in the early Eocene.

Within the Straits of Florida the currents inhibited deposition and cut a rugged unconformity (Sheridan and others, 1981a). Farther north, on the inner Blake Plateau, the currents cut an extremely rugged unconformity (Fig. 19, Fay 017 line) on the Paleocene muds of the east-trending former Cretaceous-Paleocene shelf (Paull and Dillon, 1980a; Pinet and others, 1981; Pinet and Popenoe, 1982, 1985b; Popenoe, 1985, 1987). Note in Figure 18 the areas in which Paleocene sediments were removed by this erosion. The track of the new (late Paleocene–early Eocene) Gulf Stream current is shown in Figure 18 by arrows directed north out of the Straits of Florida.* The onset of Gulf Stream flow changed the entire sedimentary regime of the southeastern U.S. Atlantic margin as the stream became a much more effective barrier than the Suwannee Current had been to the transport of sediment from the continent to the northern Blake Plateau.

Apparently, the Suwannee Current had carried much very fine terrigenous sediment derived from the rivers of the northeast-

---

*The Gulf Stream is often termed the Florida Current within and just north of the Straits of Florida, but we will continue to refer to the entire flow as the Gulf Stream to avoid confusion.

ern Gulf of Mexico, whereas the Gulf Stream probably lost most of this sort of material as it detoured around the Florida carbonate banks. When the Gulf Stream was flowing, only very thin, dominantly biogenic sands and authigenic deposits accumulated on the Blake Plateau (Charm and others, 1969; Schlee, 1977). This active current flow and sediment starvation, along with continued subsidence, created the present-day deep-water Blake Plateau. The Florida-Hatteras Shelf formed to land-ward of the flow, with its seaward progradation restricted by the flank of the Gulf Stream current. The shelf break had stepped back over 400 km from the Early Cretaceous shelf edge at the Blake Escarpment (Figs. 10 and 19).

During most of the Paleogene, sea levels were higher than present (Vail and Hardenbol, 1979), although drops to near the present level or below are evident from the stratigraphic record. Major relocations of the Gulf Stream were induced by these sea-level fluctuations. The large network of high-resolution seismic-reflection data off the southeastern margin (Fig. 20) has enabled us to map major vertical and lateral shifts in the uncon-formity cut by bottom currents of the Gulf Stream and the coin-cident shifts of its flanking facies, and to create structure contour and isopach maps of Tertiary units, including onshore data (Pinet and others, 1981; Pinet and Popenoe, 1982, 1985a, 1985b; Popenoe, 1985). The resulting maps show that not only did the position of the Gulf Stream shift more than 150 km across the Blake Plateau (Pinet and others, 1981), but that, during the mid-dle to late Eocene and early to middle Oligocene, sea levels were high enough to reactivate the Suwannee Current across Georgia (Figs. 21 and 22) as originally proposed by Chen (1965). The axis of the Suwannee Current was shifted northwest from the original Late Cretaceous-Paleocene Suwannee Strait to a new location, referred to as the Gulf Trough (Patterson and Herrick, 1971; Gelbaum and Howell, 1982; Fig. 18). This new location may have been occupied because sea levels were higher than during the Late Cretaceous. Although its location is changed, we still refer to the current through the Gulf Trough as the Suwannee Current. Facies boundaries that were mapped onshore by Chen (1965) depict the Gulf Trough as a northeast-trending subsurface depression on the top of the middle and upper Eocene and Oligo-cene strata. The offshore continuation of the Gulf Trough trend is a rugged erosional unconformity, bordered on its northern flank by southeastward-prograding Eocene strata (middle to late Eo-cene Florida-Hatteras Shelf, Fig. 21).

Episodes of deposition across the Straits of Florida and the southern Blake Plateau occurred during Eocene and early Oligo-cene time (Kaneps, 1979; Paull and Dillon, 1980a; Sheridan and others, 1981a; Pinet and Popenoe, 1985a). These can be ac-counted for by rises of sea level, which would have caused the Gulf Stream to have less effect on the bottom of the straits, and to periods of diminished Gulf Stream flow, which would have oc-curred while the Suwannee Strait or Gulf Trough were accepting some flow of water from the Gulf of Mexico. These depositional intervals were punctuated by intervals of strong erosion, when water flow from the Gulf to the Atlantic appears to have been restricted to the Straits of Florida with little or no flow across Georgia, presumably during periods of lowered sea level. We feel that the concept outlined above is a more reasonable explanation for velocity fluctuations of the Gulf Stream than control by plate arrangements, as proposed by Kaneps (1979).

The Eocene and lower Oligocene sediments also record the building of the coast and shelf into the margin of the Carolina Trough at Cape Hatteras. Prior to the Eocene, the coastline was more aligned with Carolina Trough structure (see structure con-tours in the Cape Hatteras area, Fig. 18). During the Eocene and middle Oligocene, the shelf at Cape Hatteras became built out at the merge point of northeast-flowing Gulf Stream and south-flowing shelf currents from the Mid-Atlantic Bight (Figs. 21 and 22; Popenoe, 1985). Another major depocenter for both Eocene and lower Oligocene sediments is within the former Late Cre-taceous–Paleocene Suwannee Channel off Georgia. This sediment accumulation was drilled by both the COST GE-1 and Transco 1005-1 wells (Fig. 16), and over 600 m of micritic, argillaceous limestones and cherts and carbonate muds, deposited in middle-to outer-shelf depths within the former Suwannee Channel, were penetrated.

Late Oligocene and Neogene

Blake Plateau—Continental Shelf. The sedimentation history of the late Oligocene and Neogene constrasts markedly with that of the Paleogene. The Neogene was dominated by sea levels that were mainly lower than at present (Vail and Hardenbol, 1979) and which confined the Gulf Stream to the Straits of Florida. These lower sea levels also removed the former barrier to clastic input to the Florida Peninsula, the Suwannee Straits. Whereas Paleogene sedimentation had been dominantly biogenic or authigenic over Florida and southern Georgia, Neo-gene deposition was distinctly more terrigenous, including clays, silty clays, and sands of Piedmont or Gulf provenance. A decrease in carbonate and increase in terrigenous deposits is universal on the eastern North American margin at this time, but additional regional control by the Suwannee Straits is significant in the southeastern region.

Offshore seismic data indicate a consistent pattern to Neo-gene sedimentation over the Blake Plateau; whenever sea level dropped, the Gulf Stream was forced seaward, resulting in a southward shift of the band of erosion produced by the current and a seaward progradation of the shelf. Such episodes of lowered sea level took place during the late Oligocene (Chattian), early Miocene (Burdigalian), late Miocene (late Tortonian-Messinian), and late Pliocene. Each of these caused a seaward shift of depoition from the Florida-Hatteras shelf out across the bathymetrically higher northern plateau, and a southward shift of the erosional unconformity cut by Gulf Stream currents across the bathymetrically lower southern Blake Plateau (see late Oligo-cene track and late Oligocene depocenter of Fig. 22; Popenoe, 1985). During the most pronounced lowstand, the late Oligo-cene, a band of erosional scour was cut from the Straits of Florida

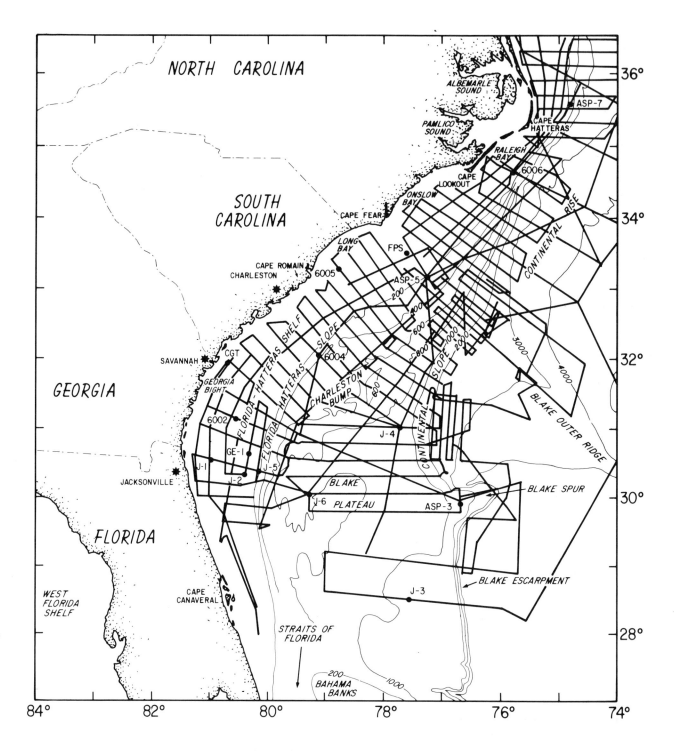

Figure 20. Track chart of USGS high-resolution, single channel, seismic-reflection profiles and offshore wells along the southeastern U.S. Atlantic margin.

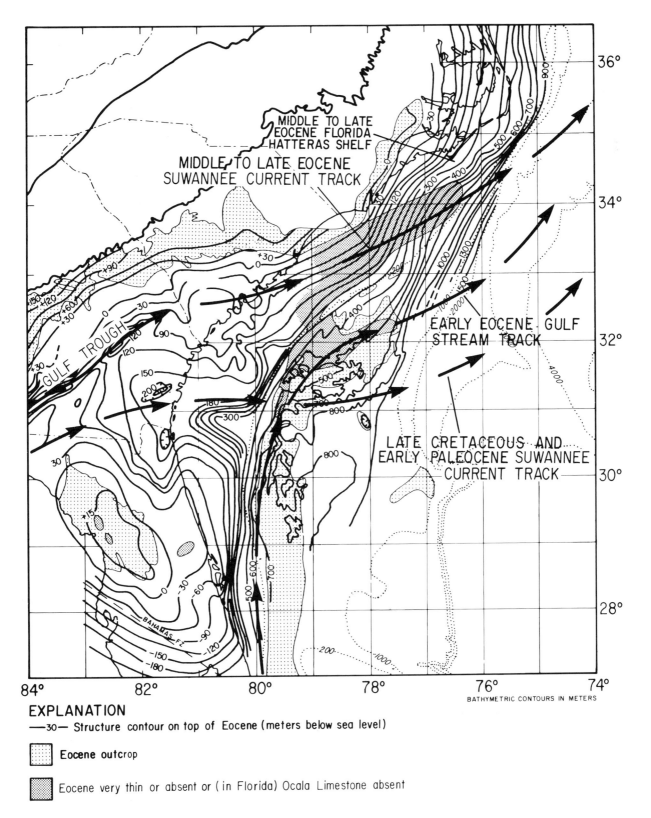

Figure 21. Structure contour map of the top of the Eocene showing Gulf Stream track for the early Eocene and Suwannee current tracks for the Late Cretaceous and middle to late Eocene (from Popenoe, 1987).

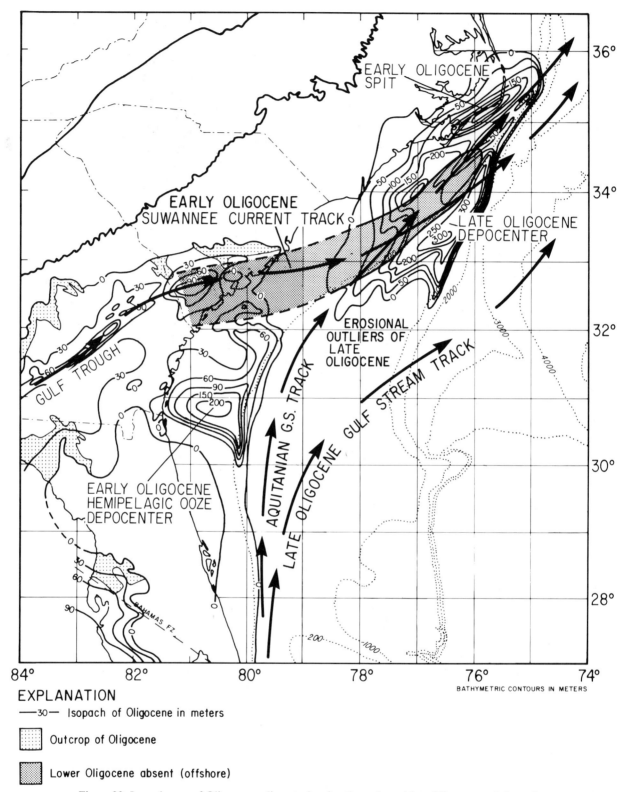

EXPLANATION

—30— Isopach of Oligocene in meters

Outcrop of Oligocene

Lower Oligocene absent (offshore)

Figure 22. Isopach map of Oligocene sediments showing the early and late Oligocene and the early Miocene (Aquitanian) Gulf Stream tracks (modified from Popenoe, 1987). The south side of the Charleston Bump has many buttelike remnant outliers of flat-lying upper Oligocene sediments presumably deposited while the Gulf Stream flowed across the southern Blake Plateau. During the latest Oligocene-Aquitanian transgression these sediments were cut into and largely removed as the Gulf Stream straightened its course.

across the southern Blake Plateau (Pinet and others, 1981) to an area just north of the Blake Spur (Fig. 22). In this area, Dillon and Kent (in Dillon and others, 1986) have mapped a number of deep canyons, eroded into the continental slope, which they believe were cut by turbid water cascading downslope from the Blake Plateau while the Gulf Stream was in this southernmost position (Fig. 23).

Deepening water in the latest Oligocene (late Chattian), early Miocene (Aquitanian), and middle Miocene (late Burdigalian and Langhian) caused the Gulf Stream to straighten its course and override the bathymetrically higher northern Blake Plateau (for instance, the Aquitanian track, Fig. 22). With the Gulf Stream in this position, low-stand shelf sediments were stripped from the south flank of the Charleston Bump and deposited as a leeside, northward-prograding drift on the northern plateau (see mounded lower Miocene strata, Fig. 24). During the maximum transgressions (Aquitanian, Langhian, and early Serravallian) these mounded deposits were overriden and in large part stripped away by Gulf Stream currents. The high stands are expressed as rugged unconformities that can be traced landward to the present Florida-Hatteras Shelf and into coeval sediment packages that formed the shelf and slope (middle Miocene unconformity, Fig. 23; Popenoe, 1985). A concomitant landward shift of shelf sedimentation onto the coastal plain also can be demonstrated (Popenoe, 1987).

Blake Escarpment—continental slope and rise. The onset of erosion on the Blake Plateau at the end of the Paleocene was paralleled by the onset of intense erosion of the Blake Escarpment at the beginning of Oligocene time. This latter occurrence was part of the broad regional erosion of the deep-sea floor of the western North Atlantic produced by a strong flow of deep bottom water moving south out of the Arctic (Mountain and Tucholke, 1985). This flow was initiated when the Greenland Sea opened by plate motions, and a general climatic cooling took place. The entire western part of the basin was scoured and, off the southeastern U.S., all previous Tertiary deposits were removed in an event that produced the horizon A^u unconformity (Tucholke, 1979; Mountain and Tucholke, 1985, Fig. 22). This thermohaline current erosion, which was concentrated at the greatest sea-floor depths and focused on the westernmost part of the basin, apparently tended to steepen the continental slope by intensified erosion at its base, and did so in a most spectacular manner at the Blake Escarpment (Gilbert and Dillon, 1981). At the escarpment, a buried, eroded bench remains where horizon A^u erosion cut into the toe of the carbonate platform (Figs. 6, 10; Paull and Dillon, 1980b). At the Blake Spur, the base of the Blake Escarpment retreated about 15 km (Fig. 10).

In middle Oligocene time, a sea-level drop caused relocation of the Gulf Stream, as discussed above. This apparently resulted in mobilization of shelf and Blake Plateau sediment, creating turbid water flows that cut canyons, now buried, in the slope north of the Blake Spur (Fig. 23).

By early to middle Miocene time, deep circulation in the western North Atlantic basin had become slower, and rapid sed-

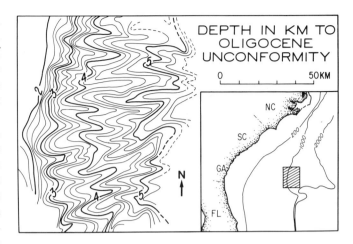

Figure 23. Structure contour map showing depth in kilometers to inferred middle Oligocene unconformity. These canyons or gullies apparently were eroded when a middle Oligocene sea-level drop forced the Gulf Stream to flow across the continental margin at this location and to provide sediments to generate turbid flows down the continental slope.

iment deposition occurred at a confluence of oceanic currents. The sediments accumulated in a major deep-sea sediment drift deposit—the Blake Ridge (Fig. 1)—off the Carolina Trough (Markl and others, 1970; Markl and Bryan, 1983; Mountain and Tucholke, 1985). Probably at this time, sediments accumulated against the base of the previously eroded Blake Escarpment. A subsequent pulse of erosion, apparently in Pliocene time, scoured the latter accumulation, truncating the Miocene strata and further cutting back the Blake Escarpment (Paull and Dillon, 1980b). Although early reports of major erosional retreat of the Blake Escarpment were controversial (Paull and Dillon, 1981b; Sheridan, 1981), there is now considerable evidence for erosional retreat and steepening of the feature (Dillon and others, 1985), including: (1) the buried bench with truncated strata that is continuous with the horizon A^u erosional unconformity (Fig. 10); (2) the truncation of strata at the sea floor, indicating a second major phase of erosion; (3) the steepness of the Blake Escarpment, which in places is nearly vertical over considerable depth ranges, and which is consistently at slopes much greater than the angle of repose and much steeper than could be accounted for by deposition (Gilbert and Dillon, 1981); (4) the truncated platform interior facies now exposed on the Blake Escarpment; and (5) the lack of a talus slope at the foot of the escarpment and the presence of a moat on the sea floor at that location. In making submersible dives on the escarpment, the observations of strong currents (estimated to exceed 4 km/hour), megarippled biogenic sand, exposed, strongly jointed rock faces, and collapsed, angular joint blocks are taken as evidence that mechanical erosion of the escarpment is active. Similar conclusions have been reached by workers regarding the Bahama Escarpment, a comparable feature to the south of the Blake Escarpment (Freeman-Lynde and

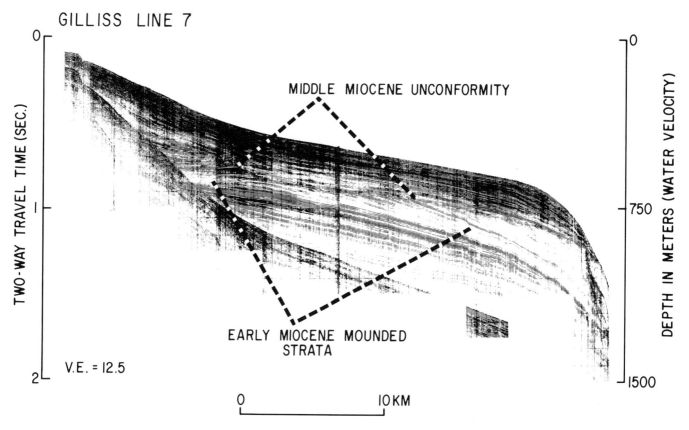

Figure 24. Seismic-reflection profile across the northern Blake Plateau showing mounded sediments of early Miocene age. These mounded beds prograde toward the northeast (into the section) and are interpreted as a drift produced as the Gulf Stream overrode the Charleston Bump during the middle Miocene transgression. The pronounced unconformity is of middle Miocene age and marks the middle Miocene (Langhian-early Serravallian) Gulf Stream position.

others, 1981). The moat and lack of talus slope at the escarpment's base may be evidence for chemical corrosion at great depth, below the carbonate compensation depth and in a location that is similar to the shallower base of the West Florida Escarpment. Recent observations at the Florida Escarpment (Paull and others, 1984) have suggested active chemical weathering processes at the base of the exposed limestone cliff, resulting from localized effects of low pH's produced by emanation of sulfide-rich groundwater from the base of the escarpment and its interaction with organisms.

An unusual feature in seismic profiles of the Blake Ridge is a reflection event that approximately parallels the sea floor and thus is known as the Bottom Simulating Reflector, or BSR (Fig. 25; Markl and others, 1970; Ewing and Hollister, 1972; Lancelot and Ewing, 1972; Bryan, 1974; Tucholke and others, 1977; Shipley and others, 1979; Dillon and others, 1980; Paull and Dillon, 1981a; Markl and Bryan, 1983). Initially, this reflector was referred to as reflector Y by Markl and others (1970). The BSR actually diverges slightly from the sea floor as water depth increases; it is found about 200 m below the sea floor at its

shallow limit in approximately 1,000-m water depth; it deepens to about 800 m below the sea floor at its maximum observed depth on the Blake Ridge in approximately 4,000 m of water (Dillon and Paull, 1983, Fig. 6). The area where the BSR is observed is mapped in Figure 1 (Tucholke and others, 1977; Paull and Dillon, 1981a).

The BSR is believed to be caused by the bottom boundary of a layer in which sediment pore space is filled with an ice-like material known as gas hydrate. A gas hydrate is one of a class of compounds known as clathrates, in which a crystal lattice includes voids or cages that contain a guest molecule and which is not stable unless the guest molecule is present (Hand and others, 1974; Miller, 1974; Davidson, 1983). In gas hydrates, the crystal lattice is formed of water molecules (essentially an expanded ice lattice) and the guest molecule is a gas. About 100 species of molecules are known to form gas hydrates (Davidson, 1983), including many gases that occur naturally in marine sediments, such as the low-carbon-number hydrocarbons (up to isobutane, $i\text{-}C^4$), carbon dioxide, and hydrogen sulfide (Hitchon, 1974). These gas hydrates are stable at the pressure/temperature condi-

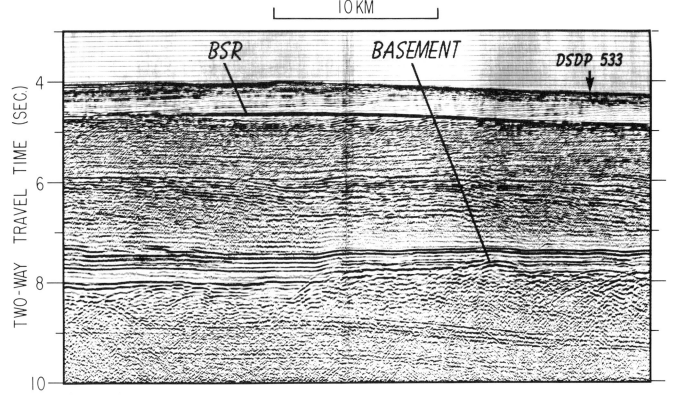

Figure 25. Section of seismic reflection profile BT1 along the axis of the Blake Ridge (Fig. 1). The BSR is obvious and the site of DSDP site 533 is shown (Sheridan and others, 1983; Kvenvolden and Barnard, 1983). Note that the BSR parallels the sea floor and cuts through reflections that arise from strata. The region above the BSR displays much smaller amplitude reflections than below, presumably indicating the effect of cementation of the strata above the BSR by gas hydrates.

tions that exist at the sea floor beyond several hundred meters water depth, and will form within the sediments if gas is present (Claypool and Kaplan, 1974; Kvenvolden, 1983). The gas hydrate would be unstable at lower pressure or higher temperature, and since temperature increases downward into the sediments, the base of the zone of stability is controlled primarily by temperature. The geothermal gradient is reasonably constant over an area, so that depth below the sea floor at which instability occurs is also fairly constant; this accounts for the consistent subbottom depth of the BSR. Its general deepening with greater water depths is a pressure effect.

Gas hydrates have relatively high acoustic compressional wave velocity (compared to water), and a sediment cemented by them has velocities much higher than those of normal shallow subseabed sediments (Stoll and others, 1971; Stoll, 1974; Whalley, 1980; Pandit and King, 1983). Because the depth of the BSR appears to correspond to pressure/temperature conditions that would cause breakdown of methane gas hydrate, and because velocities in the sediment are very high above the BSR and drop abruptly below it (Shipley and others, 1979; Dillon and Paull, 1983), the conclusion is obvious that this geochemical phase

boundary produces the seismic reflection. Actually, some analyses suggest that velocities below the BSR drop below water velocity, which would imply that the gas hydrate–cemented sediment surface layer may be acting as a seal and trapping free gas beneath it (Dillon and others, 1980; Dillon and Paull, 1983).

Direct evidence of gas hydrates in sea-floor sediments was first obtained in 1970 during DSDP drilling on the Blake Ridge at sites 102, 103, and 104 (Fig. 1). The cores at these sites evolved large volumes of gas over long periods (1–2 hours), and high sediment seismic velocities were deduced, leading for the first time to the proposal that gas hydrates were present (Stoll and others, 1971; Lancelot and Ewing, 1972). In 1980, site 533 (Fig. 1) was drilled primarily to analyze the gas hydrate (Sheridan and Gradstein, 1983), which by that time had been studied extensively using seismic data (note location of site 533 on the seismic profile of Fig. 25). This hole allowed direct observation of natural gas hydrate in the Blake Ridge sediments; it appeared as thin, mat-like layers of white crystals (Kvenvolden and Barnard, 1983). Several other lines of evidence discussed by Kvenvolden and Barnard (1983) confirmed that gas hydrates were present. For example, two arguments for presence of gas hydrate are that gas

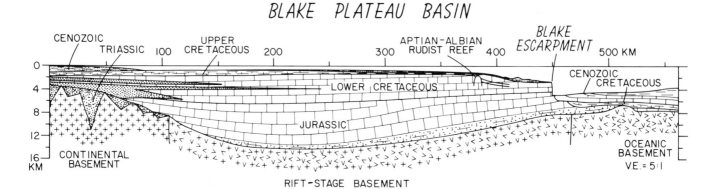

Figure 26. Idealized section across the Blake Plateau Basin.

that was evolved from a sample produced a volume of 20 times the pore space (gas hydrate is a gas concentrator) and that hydrocarbon gases that form hydrates were present, but hydrocarbon gases having large molecules (larger than isobutane) were lacking. The gas within the Blake Ridge sediments is primarily methane (Lancelot and Ewing, 1972; Kvenvolden, 1985).

SUMMARY

The continental margin of the southeastern United States, south of Cape Hatteras, is underlain by two sedimentary basins: the Blake Plateau Basin off Florida and Georgia, and the Carolina Trough off South Carolina and North Carolina. These two basins, although adjacent, show some of the greatest contrasts of any two basins of eastern North America. The Blake Plateau Basin is the widest basin, and the Carolina Trough the narrowest, on this continental margin; the Carolina Trough appears to have the thinnest basement beneath it, and the Blake Plateau Basin has basement as thick as any basin; the Carolina Trough has extensive salt diapirism, whereas the Blake Plateau Basin has none; the Blake Plateau Basin formed as a huge carbonate platform, while the Carolina Trough appears to have been characterized by considerable clastic deposition like the basins to the north.

We identify five stages of continental margin development off the southeastern U.S. These are process-related rather then strictly time-bounded, and the first two show considerable time overlap. They are: (1) a stage of rifting and stretching of continental crust with intrusion of mantle material, (2) an early sedimentation stage of synrift and early postrift deposition, (3) a main period of basin filling in which marine shelf-type deposition occurred with pure carbonate platform development to the south and mixed carbonate-terrigenous clastic deposition to the north, (4) an episode in which the carbonate-platform deposition failed to keep up with continental margin subsidence and a deep-water marginal plateau formed in the southern part of the area, and (5) an episode during which erosional processes were dominant in

shaping the continental margin. This history resulted in two major continental margin basins whose structures are shown in simplified form in Figures 26 and 27.

The basement of the basins is a rift-stage basement that is assumed to consist of a mixture of rifted blocks of continental crust, with fault basins containing synrift sediments, all massively intruded by mafic, mantle-derived, igneous material, especially in the Blake Plateau Basin. The basins owe their initial subsidence to a more intense thinning of the crust beneath them than occurred beneath the platforms. The rift-stage basement of the Blake Plateau Basin appears to be twice as thick, much broader, and to have taken much longer to form than that of the Carolina Trough. Extension of the lithosphere probably began in the Late Triassic, but sufficient rifting and fracturing to induce intrusion and extrusion of basaltic mantle material did not result until about 195 Ma (Early Jurassic). By about 184 Ma, the rifting seems to have become concentrated in the basin areas, and generation of rift-stage basement was well underway. At some time— we suggest about 175 Ma—the Carolina Trough ceased its rifting, and new ocean crust began to form seaward of it. The change from rift to drift occurred later, perhaps about 171 Ma, at the Blake Plateau Basin. The terrestrial deposits (commonly red beds) and rift-stage basalts spread broadly across the basins and the present coastal plain areas, especially in the southern part of the region. The deep strata of the basins are not sampled, and, indeed, no basinal rocks older than Cretaceous have been sampled in the region. Early basinal sedimentary rocks probably consist of terrestrial and volcanogenic deposits with some carbonates and evaporites. A line of salt diapirs off the Carolina Trough shows that the early deposition there included considerable salt. There is no evidence for significant amounts of salt in the Blake Plateau Basin. Presumably the Carolina Trough sank deeper and was flooded with marine waters early in margin history in order to precipitate salt, perhaps because its rift-stage basement was thinner.

During the main stage of basin filling, on the basis of evidence from seismic profiling, Jurassic deposition in the Blake

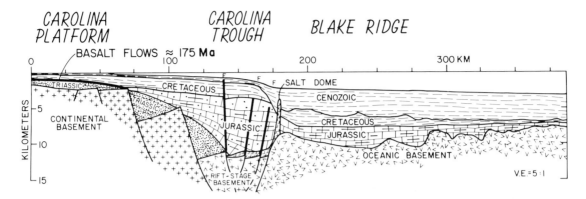

Figure 27. Idealized section across the Carolina Trough.

Plateau Basin consisted dominantly of carbonate platform construction with shelf-edge carbonate banks or reefs (Fig. 26). The Carolina Trough probably was a shallow clastic shelf with mixed carbonate-terrigenous deposition (Fig. 27). Salt began to flow from the depths of the narrow, linear Carolina Trough, resulting in diapirism along its seaward side and a long normal fault system on its landward side. The faulting was the response to subsidence of the block above the salt layer as salt was withdrawn into the diapirs (Fig. 27). This diapirism and related fault movement has continued to the present.

Deposition in the Early Cretaceous seems to have followed the same pattern as in the Jurassic, but for these rocks we have samples obtained by using submersibles on the Blake Escarpment, where the basinal strata crop out, and from oil exploration holes on the landward flanks of the basins. At the end of Neocomian, the outer reefal growth in the Blake Plateau Basin was interrupted, and the succeeding Aptian-Albian reef commonly is located landward of the older carbonate platform edge (Fig. 26). The Carolina Trough is inferred to have had mixed deposition; reef structures, if they existed, are largely eroded away.

The shelf-margin reef of the Blake Plateau Basin expired at the beginning of the Late Cretaceous and never reestablished itself, perhaps due to environmental changes. A sea-level rise created deep-water conditions across the Blake Plateau and resulted in extension of depositional conditions across the present coastal plain. The plateau was covered by a marl layer that accumulated in water depths of several hundred meters. The deposition of this fine-grained sediment precludes the possibility

of a Gulf Stream flow during most of Late Cretaceous and Paleocene. Flow of Gulf of Mexico water to the Atlantic was through the Suwannee Strait, across Georgia, and swinging northward across the Carolina Trough shelf, but bypassing the Blake Plateau. This flow presumably entrained sediment shed from the continent and prevented its deposition on the Blake Plateau or on the Florida platform.

The Gulf Stream flow began in earnest at the end of the Paleocene or beginning of the Eocene, when the flow of water from the Gulf of Mexico no longer could pass through the Suwannee Strait and began instead to flow around the southern tip of the carbonate banks of the Florida peninsula. The Gulf Stream was, and is, a powerful flow that has scoured the surface of the Blake Plateau, and the western flank of the stream has limited the progradation of the Florida-Hatteras Shelf. Episodes of shelf progradation and erosion have been influenced by migration of the Gulf Stream back and forth across the Blake Plateau and renewed flow across Georgia in the Gulf Trough, under the control of sea-level oscillations. At the beginning of Oligocene, a new deep-water flow was concentrated on the western margin of the basin. This flow eroded and steepened the continental slope and, especially off the Blake Plateau Basin, a significant retreat of the lower slope created the steep cliff of the Blake Escarpment (Fig. 26). Another erosional episode after the Miocene cut the escarpment back farther, and erosion is continuing; whereas, to the north off the Carolina Trough, post-Oligocene time was dominated by formation of a major deep-sea sediment drift deposit, the Blake Ridge (Fig. 27).

REFERENCES CITED

Antoine, J. W., and Henry, V. J., 1965, Seismic refraction study of shallow part of continental shelf off Georgia coast: American Association of Petroleum Geologists Bulletin, v. 49, p. 601–609.

Applin, P. L., and Applin, E. R., 1967, The Gulf Series in the subsurface of northern Florida and southern Georgia: U.S. Geological Survey Professional Paper 524-G, 34 p., 8 plates.

Barnett, R. S., 1975, Basement structure of Florida and its tectonic implications: Gulf Coast Association of Geological Societies Transactions, v. 25, p. 122–142.

Behrendt, J. C., and Grim, M. S., 1983, Structural elements of the U.S. Atlantic margin delineated by second vertical derivative of aeromagnetic data: U.S. Geological Survey Geophysical Investigations Map GP-956, scale 1,

2,500,000.

Behrendt, J. C., and Klitgord, K. D., 1980, High-sensitivity aeromagnetic survey of the U.S. Atlantic continental margin: Geophysics, v. 45, p. 1813–1846.

Behrendt, J. C., and Wotorson, C. S., 1970, Aeromagnetic and gravity investigations of the coastal area and continental shelf of Liberia, West Africa, and their relation to continental drift: Geological Society of America Bulletin, v. 81, p. 3563–3574.

Benson, W. E. and Sheridan, R. E., eds., 1978, Initial Reports of the Deep Sea Drilling Project, v. 44: Washington, D.C., U.S. Government Printing Office, 1005 p.

Bonini, W. E., and Woollard, G. P., 1960, Subsurface geology of the North Carolina–South Carolina Coastal Plain from seismic data: American Association of Petroleum Geologists Bulletin, v. 44, no. 3, p. 298–315.

Bott, M.H.P., 1979, Subsidence mechanisms at passive continental margins, *in* Watkins, J. S., Montadert, L., and Dickerson, P. W., eds., Geological and geophysical investigations of continental margins: American Association of Petroleum Geologists Memoir 29, p. 3–9.

Brooks, D. A., 1986, Doing the Charleston Bump: EOS, Transactions of the American Geophysical Union, v. 67, no. 1, p. 3.

Brown, P. M., Miller, J. A., and Swain, F. M., 1972, Structural and stratigraphic framework and spatial distribution of permeability of the Atlantic Coastal Plain, North Carolina to New York: U.S. Geological Survey Professional Paper 796, 79 p.

Bryan, G. M., 1974, *In situ* indications of gas hydrate, *in* Kaplan, I. R., ed., Natural gases in marine sediments: New York, Plenum Press, p. 299–308.

Buffler, R. T., Shipley, T. H., and Watkins, J. S., 1978, Blake continental margin seismic section: American Association Petroleum Geologists, Seismic Section no. 2, 1 sheet.

Buffler, R. T., Watkins, J. S., and Dillon, W. P., 1979, Geology of the offshore Southeast Georgia Embayment, U.S. Atlantic continental margin, based on multichannel seismic reflection profiles, *in* Watkins, J. S., Montadert, L., and Dickerson, P. W., eds., Geological and geophysical investigations of continental margins: American Association of Petroleum Geologists Memoir 29, p. 11–25.

Bunce, E. T., Emery, K. O., Gerard, R. D., Knott, S. T., Lidz, L., Saito, T., and Schlee, J., 1965, Ocean drilling on the continental margin: Science, v. 150, p. 709–716.

Cashman, K. V., and Popenoe, P., 1985, Salt tectonic induced slumping on the continental slope and rise off North Carolina: Marine and Petroleum Geology, v. 2, p. 259–271.

Charm, W. B., Nesteroff, W. D., and Valdez, S., 1969, Detailed stratigraphic descriptions of the JOIDES cores on the continental margin off Florida: U.S. Geological Survey Professional Paper 581-D, 13 p.

Chen, C. S., 1965, The regional lithostratigraphic analysis of Paleocene and Eocene rocks of Florida: Florida Geological Survey Bulletin 45, 105 p.

Chowns, T. M., and Williams, C. T., 1983, Pre-Cretaceous rocks beneath the Georgia Coastal Plain; Regional implications, *in* Gohn, G. S., ed., Studies related to the Charleston, South Carolina, earthquake of 1886; Tectonics and seismicity: U.S. Geological Survey Professional Paper 1313, p. L1–L42.

Claypool, G. E., and Kaplan, I. R., 1974, The origin and distribution of methane in marine sediments, *in* Kaplan, I. R., ed., Natural gases in marine sediments: New York, Plenum Press, p. 299–308.

Crutcher, T. D., 1983, Southeast Georgia Embayment, *in* Bally, A. W., ed., Seismic expression of structural styles: American Association of Petroleum Geologists Studies in Geology No. 15, v. 2, p. 2.2.3-27 to 2.2.3-29.

Dalrymple, G. B., Gromm, C. S., and White, R. W., 1975, Potassium-argon age and paleomagnetism of diabase dikes in Liberia; Initiation of Central Atlantic rifting: Geological Society of America Bulletin, v. 86, p. 399–411.

Daniels, D. L., Zietz, I., and Popenoe, P., 1983, Distribution of subsurface lower Mesozoic rocks in the southeastern United States, as interpreted from regional aeromagnetic and gravity maps, *in* Gohn, G. S., ed., Studies related to the Charleston, South Carolina, earthquake of 1886; Tectonics and seismicity: U.S. Geological Survey Professional Paper 1313, p. K1–K24.

Davidson, D. W., 1983, Gas hydrates as clathrate ices, *in* Cox, J. L., ed., Natural gas hydrates; Properties, occurrence, and recovery: Boston, Butterworth Publishers, p. 1–15.

De Boer, J., 1967, Paleomagnetic-tectonic study of Mesozoic dike swarms in the Appalachians: Journal Geophysical Research, v. 72, p. 2237–2250.

Dietz, R. S., Holden, J. C., and Sproll, W. P., 1970, Geotectonic evolution and subsidence of Bahama platform: Geological Society of America Bulletin v. 81, p. 1915–1928.

Dillon, W. P., and McGinnis, L. D., 1983, Basement structure indicated by seismic-reflection measurements offshore from South Carolina and adjacent areas, *in* Gohn, G. S., ed., Studies related to the Charleston, South Carolina, earthquake of 1886; Tectonics and seismicity: U.S. Geological Survey Professional Paper 1313, p. 01–07.

Dillon, W. P., and Paull, C. K., 1978, Interpretation of multichannel seismic-reflection profiles of the Atlantic continental margin off the coasts of South Carolina and Georgia: U.S. Geological Survey Miscellaneous Field Investigations Map MF-936.

—— , 1983, Marine gas hydrates; II, Geophysical evidence, *in* Cox, J. L. ed., Natural gas hydrates; Properties, occurrences, and recovery: Boston, Butterworth, p. 73–90.

Dillon, W. P., and Sougy, J.M.A., 1974, Geology of West Africa and Cape Verde Islands, *in* Nairn, A.E.M., and Stehli, F. G., eds., The ocean basins and margins, Volume 2, The North Atlantic: New York, Plenum Press, p. 315–390.

Dillon, W. P., Sheridan, R. E., and Fail, J. P., 1976, Structure of the western Blake Bahama Basin as shown by 24 channel CDP profiling: Geology, v. 4, p. 459–462.

Dillon, W. P., Paull, C. K., Buffler, R. T., and Fail, J.-P., 1979a, Structure and development of the Southeast Georgia Embayment and northern Blake Plateau; Preliminary analyses, *in* Watkins, J. S., Montadert, L., and Dickerson, P. W., eds., Geology and geophysical investigations of continental margins: American Association of Petroleum Geologists Memoir 29, p. 27–41.

Dillon, W. P., Poag, C. W., Valentine, P. C., and Paull, C. K., 1979b, Structure, biostratigraphy, and seismic stratigraphy along a common-depth-point seismic profile through three drill sites on the continental margin off Jacksonville, Florida: U.S. Geological Survey Miscellaneous Field Studies Map MF-1090.

Dillon, W. P., Paull, C. K., Dahl, A. G., and Patterson, W. D., 1979c, Structure of the continental margin near the COST GE-1 well site from a common depth point seismic reflection profile, *in* Scholle, P. A., ed., Geological studies of the COST GE-1 well, United States South-Atlantic outer continental shelf area: U.S. Geological Survey Circular 800, p. 97–107.

Dillon, W. P., Grow, J. A., and Paull, C. K., 1980, Unconventional gas hydrate seals may trap gas off southeast U.S.: Oil and Gas Journal, v. 78, no. 1, p. 124, 126, 129, 130.

Dillon, W. P., Popenoe, P., Grow, J. A., Klitgord, K. D., Swift, B. A., Paull, C. K., and Cashman, D. V., 1983a, Growth faulting and salt diapirism, their relationships and control in the Carolina Trough, eastern North America, *in* Watkins, J. S., and Drake, C. L., eds., Studies in continental margin geology: American Association of Petroleum Geologists Memoir 34, p. 21–46.

Dillon, W. P., Klitgord, K. D., and Paull, C. K., 1983b, Mesozoic development and structure of the continental margin off South Carolina, *in* Gohn, G. S., ed., Studies related to the Charleston, South Carolina, earthquake of 1886; Tectonics and seismicity: U.S. Geological Survey Professional Paper 1313, p. N1–N16.

Dillon, W. P., Paull, C. K., and Gilbert, L. E., 1985, History of the Atlantic continental margin off Florida; The Blake Plateau basin, *in* Poag, C. W., ed., Geologic evolution of the United States Atlantic margin: New York, Van Nostrand Reinhold, p. 189–215.

Dillon, W. P., Manheim, F. T., Jansa, L. F., Palmason, G., Tuckolke, B. E., and Landrum, R. S., 1986, Resource potential of the western North Atlantic Basin, *in* Vogt, P. R., and Tucholke, B. E., eds., The Western North Atlantic Region: Boulder, Colorado, Geological Society of America, The Geology of North America, v. M, p. 661–676.

Dillon, W. P., Valentine, P. C., and Paull, C. K., 1987, The Blake Escarpment; A product of erosional processes in the deep ocean, *in* Cooper, R., ed., National Oceanic and Atmospheric Administration, Symposium Series for Undersea Research, v. 2, no. 2, (in press).

Dowling, J. J., 1968, The East Coast onshore-offshore experiment; II. Seismic refraction measurements on the continental shelf between Cape Hatteras and Cape Fear: Seismological Society of America Bulletin, v. 58, no. 3, p. 821–834.

Emergy, K. D., and Zarudzki, E.F.K., 1967, Seismic reflection profiles along the drill holes on the continental margin off Florida: U.S. Geological Survey Professional Paper 581-A, 8 p.

Enos, P., and Freeman, T., 1978, Shallow-water limestones from the Blake Nose, sites 390 and 393, *in* Benson, W. E., and Sheridan, R. E., eds., Initial reports of the Deep Sea Drilling Project, vol. 44: Washington, D.C., U.S. Government Printing Office, p. 413–461.

Ewing, J., Ewing, M., and Leyden, R., 1966, Seismic profiler survey of Blake Plateau: American Association of Petroleum Geologists Bulletin, v. 50, p. 1948–1971.

Ewing, J. I., and Hollister, C. H., 1972, Regional aspects of deep-sea drilling in the western North Atlantic, *in* Hollister, C. D., and Ewing, J. I., eds., Initial Reports of the Deep Sea Drilling Project, v. 11: Washington, D.C., U.S. government Printing Office, p. 951–973.

Freeman-Lynde, R. P., Cita, M. B., Jadoul, F., Miller, E. L., and Ryan, W.B.F., 1981, Marine geology of the Bahama Escarpment: Marine Geology, v. 44, p. 119–156.

Gelbaum, C., and Howell, J., 1982, The geohydrology of the Gulf Trough, *in* Arden, D. D., Beck, B. F., and Morrow, E., eds., Proceedings, Second Symposium on the geology of the southeastern coastal plain: Georgia Geological Survey Information Circular 53, p. 140–153.

Gilbert, L. E., and Dillon, W. P., 1981, Bathymetric map of the Blake Escarpment: U.S. Geological Survey Miscellaneous Field Investigations Map MF-1362, 1:250,000.

Gohn, G. S., ed., 1983, Studies related to the Charleston, South Carolina, earthquake of 1886; Tectonics and seismicity: U.S. Geological Survey Professional Paper 1313, 375 p.

Gohn, G. S., Bybell, L. M., Smith, C. C., and Owens, J. P., 1978a, Preliminary stratigraphic cross sections of the Atlantic Coastal Plain sediments of the southeastern United States; Cenozoic sediments along the South Carolina coastal margin: U.S. Geological Survey Miscellaneous Field Studies Map MF-1015B, 2 sheets.

Gohn, G. S., Gottfried, D., Lanphere, M. A., and Higgins, B. B., 1978b, Regional implications of Triassic or Jurassic age for basalt and sedimentary red beds in the South Carolina Coastal Plain: Science, v. 202, no. 4370, p. 887–890.

Gohn, G. S., Smith, C. C., Christopher, R. A., and Owens, J. P., 1980, Preliminary stratigraphic cross sections of the Atlantic Coastal Plain sediments of the southeastern United States; Cretaceous sediments along the Georgia coastal margin: U.S. Geological Survey Miscellaneous Field Studies Map MF-1015-C, 2 sheets.

Grim, M. S., Dillon, W. P., and Mattick, R. E., 1980, Seismic reflection, refraction, and gravity measurements from the continental shelf offshore from North and South Carolina: Southeastern Geology, v. 21, p. 239–249.

Grow, J. A., and Markl, R. G., 1977, IPOD-USGS multichannel seismic reflection profile from Cape Hatteras to the Mid-Atlantic ridge: Geology, v. 5, p. 625–630.

Grow, J. A., Bowin, C. O., and Hutchinson, D. R., 1979, The gravity field of the U.S. Atlantic continental margin: Tectonophysics, v. 59, p. 27–52.

Grow, J. A., Hutchinson, D. R., Klitgord, K. D., Dillon, W. P., and Schlee, J. S., 1983, Representative multichannel seismic reflection profiles over the U.S. Atlantic continental margin, *in* Bally, A. W., ed., Seismic expression of structural styles: American Association of Petroleum Geologists, Studies in Geology Series, no. 15, p. 2.2.3-1 to 2.2.3-19.

Hand, J. H., Katz, D. L., and Verma, V. H., 1974, Review of gas hydrates with implication for ocean sediments, *in* Kaplan, I. R., ed., Natural gases in marine sediments: New York, Plenum Press, p. 179–194.

Hathaway, J. C., Schlee, J. S., and 9 others, 1976, Preliminary summary of the 1976 Atlantic Margin Coring Project of the U.S. Geological Survey: U.S. Geological Survey Open-File Report 76-844, 217 p.

Hathaway, J. C., and Poag, C. W., and 7 others, 1979, U.S. Geological Survey core drilling on the Atlantic shelf: Science, v. 206, no. 4418, p. 515–527.

Hazel, J. E., Edwards, L. E., and Bybell, L. M., 1984, Significant unconformities and the hiatuses represented by them in the Paleogene of the Atlantic and Gulf Coastal Province, *in* Schlee, J. S., ed., Interregional unconformities and hydrocarbon accumulation: American Association of Petroleum Geologists Memoir 36, p. 59–66.

Heezen, B. C., and Sheridan, R. E., 1966, Lower Cretaceous rocks (Neocomian-Albian) dredged from Blake Escarpment: Science, v. 154, p. 1644–1647.

Hersey, J. B., Bunce, E. T., Wyrick, R. F., and Dietz, F. T., 1959, Geophysical investigation of the continental margin between Cape Henry, Virginia, and Jacksonville, Florida: Geological Society of America Bulletin, v. 70, no. 4, p. 437–465.

Hitchon, B., 1974, Occurrence of natural gas hydrates in sedimentary basins, *in* Kaplan, I. R., ed., Natural gases in marine sediments: New York, Plenum Press, p. 195–225.

Hollister, C. D. and Ewing, J. I., eds., 1972, Initial reports of the Deep Sea Drilling Project, v. 11: Washington, D.C., U.S. Government Printing Office, 1077 p.

Hull, J.P.D., Jr., 1962, Cretaceous Suwannee Strait, Georgia and Florida: American Association of Petroleum Geologists Bulletin, v. 46, p. 118–122.

Hutchinson, D. R., Grow, J. A., Klitgord, K. D., and Swift, B. A., 1983, Deep structure and evolution of the Carolina Trough, *in* Watkins, J. S., and Drake, C. L., eds., Studies in Continental Margin Geology: American Association Petroleum Geologists Memoir 34, p. 129–152.

Jacobs, J. A., 1977, Jurassic lithology in Great Isaac 1 well, Bahamas; Discussion: American Association of Petroleum Geologists Bulletin, v. 61, p. 443.

Jansa, L. F., 1981, Mesozoic carbonate platforms and banks of the eastern North American margin: Marine Geology, v. 44, p. 97–117.

Kaneps, A. G., 1979, Gulf Stream; Velocity fluctuations during the late Cenozoic: Science, v. 204, p. 297–301.

Kent, D. V., and Gradstein, F. M., 1985, A Cretaceous and Jurassic geochronology: Geological Society of America Bulletin, v. 96, p. 1419–1427.

Kent, D. V., and Gradstein, F. M., 1986, A Jurassic to recent chronology, *in* Tucholke, B. E., and Vogt, P. R., eds., The Western Atlantic Region: Geological Society of America, The Geology of North America, v. M, p. 45–50.

Kent, K. M., 1979, Two dimensional gravity model of the Southeast Georgia Embayment–Blake Plateau [M.S. thesis]: Newark, University of Delaware. 89 p.

Kent, K. M., Grow, J. M., and Dillon, W. P., 1979, Gravity studies of the continental margin off northern Florida: Geological Society of America Abstracts with Programs, v. 11, no. 4, p. 184.

Klitgord, K. D., and Behrendt, J. C., 1977, Aeromagnetic anomaly map of the United States Atlantic continental margin: U.S. Geological Survey Miscellaneous Field Studies Map MF-913, 2 sheets, scale 1:1,000,000.

——, 1979, Basin structure of the U.S. Atlantic margin: American Association Petroleum Geologist Memoir 29, p. 85–112.

Klitgord, K. D., and Schouten, H., 1986, Plate kinematics of the central Atlantic, *in* Vogt, P. R., Tucholke, B. E., eds., The Western North Atlantic region: Boulder, Colorado, Geological Society of America, The Geology of North America, v. M, p. 351–378.

Klitgord, K. D., Popenoe, P., Schouten, H., 1984, Florida; A Jurassic transform plate boundary: Journal of Geophysical Research, v. 89, p. 7753–7772.

Kvenvolden, K. A., 1983, Marine gas hydrates; I. Geochemical evidence, *in* Cox, J. L., ed., Natural gas hydrates; Properties, occurrence, and recovery: Boston, Butterworth Publishers, p. 63–72.

——, 1985, Comparison of marine gas hydrates in sediments of an active and passive continental margin: Marine and Petroleum Geology, v. 2, p. 65–71.

Kvenvolden, K. A., and Barnard, L. A., 1983, Gas hydrates of the Blake Outer Ridge, Site 533, Deep Sea Drilling Project Leg 76, *in* Sheridan, R. E., and Gradstein, F. J., eds., Initial Reports of the Deep Sea Drilling Project, v. 76: Washington, D.C., U.S. Government Printing Office, p. 353–365.

Lancelot, Y., and Ewing, J. I., 1972, Correlation of natural gas zonation and carbonate diagenesis in Tertiary sediments from the northwest Atlantic, *in* Hollister, C. D., and Ewing, J. I., eds., Initial Reports of the Deep Sea Drilling Project, v. 11; Washington, D.C., U.S. Government Printing Office, p. 791–799.

Lanphere, M. A., 1983, $^{40}Ar/^{39}Ar$ ages of basalt from Clubhouse Crossroads Test Hole No. 2, near Charleston, South Carolina, *in* Gohn, G. S., ed., Studies related to the Charleston, South Carolina, earthquake of 1886; Tectonics and seismicity: U.S. Geological Survey Professional Paper 1313, p. B1–B8.

Leg 101, Scientific Drilling Party, 1985, Megabank found? Flanks record sea level: Geotimes, v. 30, no. 11, p. 12–15.

Leg 101, Shipboard Scientific Party, 1985, Rise and fall of carbonate platforms in the Bahamas: Nature, v. 315, p. 632–633.

Manspeizer, W., 1985, Early Mesozoic history of the Atlantic passive margin, *in* Poag, C. W., ed., Geologic evolution of the United States Atlantic Margin: New York, Van Nostrand, Reinhold, p. 1–23.

Markl, R. G., and Bryan, G. M., 1983, Stratigraphic evolution of Blake Outer Ridge: American Association of Petroleum Geologists Bulletin, v. 67, p. 666–683.

Markl, R. G., Bryan, G. M., and Ewing, J. I., 1970, Structure of the Blake-Bahama Outer Ridge: Journal of Geophysical Research, v. 75, p. 4539–4555.

May, P. R., 1971, Pattern of Triassic-Jurassic diabase dikes around the North Atlantic in the context of predrift position of the continents: Geological Society America Bulletin, v. 82, p. 1285–1292.

Miller, S. L., 1974, The nature and occurrence of clathrate hydrates, *in* Kaplan, I. R., ed., Natural gases in marine sediments: New York, Plenum Press, p. 151–177.

Milton, C., and Grasty, R., 1969, "Basement" rocks of Florida and Georgia: American Association Petroleum Geologists Bulletin, v. 53, p. 2483–2493.

Mountain, G. S., and Tucholke, B. E., 1985, Mesozoic and Cenozoic geology of the Atlantic continental slope and rise, *in* Poag, C. W., ed., Geologic evolution of the United States Atlantic margin: New York, Van Nostrand Reinhold, p. 293–341.

Mullins, H. T., and Lynts, G. W., 1977, Origin of the northwestern Bahama Platform; Review and reinterpretation: Geological Society of America Bulletin, v. 88, p. 1447–1461.

Murray, J., 1885, Report on the specimens of bottom deposits collected by the U.S. Coast Survey Steamer Blake, 1887-1880: Bulletin of the Museum of Comparative Zoology, v. 12, p. 37–61.

Palmer, A. R., 1983, The Decade of North American Geology 1983 geologic time scale: Geology, v. 11, p. 503–504.

Pandit, B. I., and King, M. S., 1983, Elastic wave velocities of propane gas hydrates, *in* Cox, J. L., ed., Natural gas hydrates; Properties, occurrence, and recovery: Boston, Butterworth Publishers, p. 49–61.

Patterson, S. H., and Herrick, S. M., 1971, Chattahoochee anticline, Appalachicola Embayment, Gulf Trough, and related structural features, southwestern Georgia, fact or fiction: Geological Survey of Georgia Information Circular 41, 16 p.

Paull, C. K., and Dillon, W. P., 1980a, Structure, stratigraphy, and geologic history of the Florida-Hatteras Shelf and inner Blake Plateau: American Association of Petroleum Geologists Bulletin, v. 64, p. 339–358.

—— , 1980b, Erosional origin of the Blake Escarpment; An alternative hypothesis: Geology, v. 8, p. 538–542.

—— , 1981a, Appearance and distribution of the gas hydrate reflection in the Blake Ridge region, offshore southeastern United States: U.S. Geological Survey Miscellaneous Field Studies Map MF-1252, scale 1:1,000,000.

—— , 1981b, Reply *to* Comment *on* Erosional origin of the Blake Escarpment; An alternative hypothesis: Geology, v. 9, p. 339–341.

—— , 1982, Carolina Trough structure contour maps: U.S. Geological Survey Miscellaneous Field Studies Map MF-1402, scale 1:1,000,000.

Paull, C. K., Hecker, B., and 8 others, 1984, Biological communities at the Florida Escarpment resemble hydrothermal vent taxa: Science, v. 226, p. 965–967.

Pinet, P. R., and Popenoe, P., 1982, Blake Plateau; Control of Miocene sedimentation patterns by large-scale shifts of the Gulf Stream axis: Geology, v. 10, no. 5, p. 275–259.

Pinet, P. R., and Popenoe, P., 1985a, Shallow seismic stratigraphy and post-Albian geologic history of the northern and central Blake Plateau: Geological Society of America Bulletin, v. 96, no. 5, p. 627–638.

—— , 1985b, A scenario of Mesozoic-Cenozoic ocean circulation over the Blake Plateau and its environs: Geological Society of America Bulletin, v. 96, no. 5, p. 618–626.

Pinet, P. R., Popenoe, P., and Nelligan, D. F., 1981, Reconstruction of Cenozoic flow patterns over the Blake Plateau: Geology, v. 9, no. 6, p. 266–270.

Pitman, W. C., III, 1978, Relationship between eustasy and stratigraphic sequences of passive margins: Geological Society of America Bulletin, v. 89, p. 1389–1403.

Poag, C. W., 1978, Stratigraphy of the Atlantic continental shelf and slope of the United States: Annual Review of Earth and Planetary Sciences, v. 6, p. 251–280.

—— , 1979, Stratigraphy and depositional environments of Baltimore Canyon Trough: American Association of Petroleum Geologists Bulletin, v. 63, p. 1452–1466.

—— , 1984, Neogene stratigraphy of the submerged U.S. Atlantic margin: Paleogeography, Paleoclimatology, and Paleoecology, v. 47, p. 103–127.

—— 1985, Depositional history and stratigraphic reference section for central Baltimore Canyon Trough, *in* Poag, C. W., ed., Geologic evolution of the United States Atlantic margin: New York, Van Nostrand Reinhold, p. 217–264.

Poag, C. W., and Hall, R. E., 1979, Foraminiferal biostratigraphy, paleoecology, and sediment accumulation rates, *in* Scholle, P. A., ed., Geological studies of the COST GE-1 well, United States Outer Continental Shelf area: United States Geological Survey Circular 800, p. 49–63.

Popenoe, P., 1985, Cenozoic depositional and structural history of the North Carolina margin from seismic stratigraphic analyses, *in* Poag, C. W., ed., Stratigraphy and depositional history of the U.S. Atlantic margin: New York, Van Nostrand Reinhold Publishing Company, p. 125–187, 4 plates.

Popenoe, P., 1987, Paleo-oceanography and paleogeography of the Miocene of the southeastern United States, *in* Burnett, W. C., and Riggs, S. R., eds., World phosphate deposits, Volume 3, Neogene phosphorites of the southeastern United States: Cambridge University Press (in press).

Poppe, L. J., 1981, Data file, the 1976 Atlantic Margin Coring (AMCOR) Project of the U.S. Geological Survey: U.S. Geological Survey Open-File Report 81-239, 96 p.

Sawyer, D. S., 1985, Total tectonic subsidence; A parameter for distinguishing crust type at the U.S. Atlantic continental margin: Journal of Geophysical Research, v. 90, p. 7751–7769.

Schlager, W., 1981, The paradox of drowned reefs and carbonate platforms: Geological Society of America Bulletin, v. 92, p. 197–211.

Schlee, J. S., 1977, Stratigraphy and Tertiary development of the continental margin east of Florida: U.S. Geological Survey Professional Paper 581-F, 25 p.

Schlee, J. S. and Fritsch, J., 1982, Seismic stratigraphy of the Georges Bank Basin complex, offshore New England, *in* Watkins, J. S., and Drake, C. L., eds., Studies in continental margin geology: American Association Petroleum Geologists Memoir 34, p. 223–251.

Schlee, J. S., Dillon, W. P., and Grow, J. A., 1979, Structure of the continental slope off the eastern United States, *in* Doyle, L. J., and Pilkey, O. H., eds., Geology of continental slopes: Society of Economic Paleontologists and Mineralogists Special Publication 27, p. 95–117.

Scholle, P. A., ed., 1979, Geological studies of the COST GE-1 well, United States outer continental shelf area: U.S. Geological Survey Circular 800, 114 p.

Sheridan, R. E., 1974, Atlantic continental margin of North America, *in* Burk, C. A., and Drake, C. L., eds., The geology of continental margins: Berlin, Springer-Verlag, p. 391–407.

—— , 1981, Comment on erosional origin of Blake Escarpment; An alternative hypothesis: Geology, v. 9, p. 338–339.

——, 1983, Phenomena of pulsation tectonics related to the breakup of the eastern North American continental margin, *in* Sheridan, R. E., and Gradstein, F. M., eds., Initial Reports of the Deep Sea Drilling Project, v. 76: Washington, D.C., U.S. Government Printing Office, p. 897–909.

Sheridan, R. E., and Enos, P., 1979, Stratigraphic evolution of the Blake Plateau after a decade of scientific drilling, *in* Talwani, M., Hay, W., and Ryan, W.B.F., eds., Deep drilling results in the Atlantic Ocean; Continental margins and paleoenvironment: Maurice Ewing Series, v. 3, American Geophysical Union Geophysical Monograph, p. 109–122.

Sheridan, R. E., Drake, C. L., Nafe, J. E., and Hennion, J., 1966, Seismic refraction study of continental margin east of Florida: American Association of Petroleum Geologists Bulletin, v. 50, p. 1972–1991.

Sheridan, R. E., Smith, J. D., and Gardner, J., 1969, Rock dredges from Blake Escarpment near Great Abaco Canyon: American Association Petroleum Geologist Bulletin, v. 53, p. 2551–2558.

Sheridan, R. E., Berman, R. M., and Corman, D. B., 1971, Faulted limestone block dredged from Blake Escarpment: Geological Society of America Bulletin, v. 82, p. 199–206.

Sheridan, R. E., Windisch, J. I., Ewing, J. I., and Stoffa, P. L., 1979, Structure and stratigraphy of the Blake Escarpment based on seismic reflection profiles, *in* Watkins, J. S., Montadert, L., and Dickerson, P. W., eds., Geological and geophysical investigations of continental margins: American Association of Petroleum Geologists Memoir 29, p. 177–186.

Sheridan, R. E., Crosby, J. T., Bryan, G. M., and Stoffa, P. L., 1981a, Stratigraphy and structure of the southern Blake Plateau, northern Florida Straits and northern Bahama Platform from multichannel seismic-reflection data: American Association of Petroleum Geologists Bulletin, v. 65, no. 12, p. 2571–2593.

Sheridan, R. E., Crosby, J. T., Kent, K. M., Dillon, W. P., and Paull, C. K., 1981b, The geology of the Blake Plateau and Bahamas region, *in* Kerr, J. W., and Fergusson, A. J., eds., Geology of the North Atlantic borderlands: Canadian Society of Petroleum Geologists Memoir 7, p. 487–502.

Sheridan, R. E., and Gradstein, F. M., eds., 1983, Initial Reports of the Deep Sea Drilling Project, v. 76: Washington, D.C., U.S. Government Printing Office, 947 p.

Shipley, T. H., Buffler, R. T., and Watkins, J. S., 1978, Seismic stratigraphy and geologic history of the Blake Plateau and adjacent western Atlantic continental margin: American Association of Petroleum Geologists, v. 62, p. 792–812.

Shipley, T. H., Houston, M. H., and 5 others, 1979, Seismic evidence for widespread possible gas hydrate horizons on continental slopes and rises: American Association of Petroleum Geologists, v. 73, p. 2044–2213.

Simonis, E. K., 1979, Radiometric age determinations, *in* Scholle, P. A., ed., Geological studies of the COST GE-1 well, United States South Atlantic outer continental shelf area: U.S. Geological Survey, Circular 800, p. 71.

Spangler, W. B., 1950, Subsurface geology of Atlantic Coastal Plain of North Carolina: American Association Petroleum Geologists Bulletin, v. 34, p. 100–132.

Stoll, R. D., 1974, Effects of gas hydrates in sediments, *in* Kaplan, I. R., ed., Natural gases in marine sediments: New York, Plenum Press, p. 235–248.

Stoll, R. D., Ewing, J., and Bryan, G. M., 1971, Anomalous wave velocities in sediments containing gas hydrates: Journal of Geophysical Research, v. 76, no. 8, p. 2090–2094.

Sylwester, R. E., Dillon, W. P., and Grow, J. A., 1979, Active growth fault on seaward edge of Blake Plateau, *in* Gill, D., and Merriam, D. F., eds., Geomathematical and petrophysical studies in sedimentology: Oxford,

Pergamon Press, p. 197–209.

Tator, B. A., and Hatfield, L. E., 1975, Bahamas present complex geology: Oil and Gas Journal, 1: v. 73, no. 44, p. 172–176: 2: v. 73, no. 45, p. 120–122.

Taylor, P. T., Zietz, I., and Dennis, L. S., 1968, Geologic implications of aeromagnetic data for the eastern continental margin of the United States: Geophysics, v. 33, p. 755–780.

Tucholke, B. E., 1979, Relationships between acoustic stratigraphy and lithostratigraphy in the western North Atlantic basin, *in* Tucholke, B. E., and Vogt, P. R., eds., Initial Reports of the Deep Sea Drilling Project, v. 43: Washington, D.C., U.S. Government Printing Office, p. 827–846.

Tucholke, B. E., Bryan, G. M., and Ewing, J. I., 1977, Gas-hydrate horizons detected in seismic-profile data from the western North Atlantic: American Association of Petroleum Geologists Bulletin, v. 61, no. 5, p. 698–707.

Uchupi, E., 1967, The continental margin south of Cape Hatteras, North Carolina; Shallow structure: Southeastern Geology, v. 8, p. 155–177.

——, 1970, Atlantic continental shelf and slope of the United States; Shallow structure: U.S. Geological Survey Professional Paper 529-I, 44 p.

Uchupi, E., and Emery, K. O., 1967, Structure of the continental margin off Atlantic coast of the United States: American Association of Petroleum Geologists Bulletin, v. 51, p. 223–234.

Vail, P. R., and Hardenbol, J., 1979, Sea level changes during the Tertiary: Oceanus, v. 22, no. 3, p. 71–79.

Vail, P. R., Mitchum, R. M., Jr., and Thompson, S., III, 1977, Global cycles of relative changes of sea level, *in* Payton, C. E., ed., Seismic stratigraphy; Applications to hydrocarbon exploration: American Association of Petroleum Geologists Memoir 26, p. 63–97.

Valentine, P. C., 1979a, Calcareous nannofossil biostratigraphy and paleoenvironmental interpretation, *in* Scholle, P. A., ed., Geological studies of the COST GE-1 well, United States South Atlantic outer continental shelf area: U.S. Geological Survey Circular 800, p. 64–70.

——, 1979b, Regional stratigraphy and structure of the Southeast Georgia Embayment, *in* Scholle, P. A., ed., Geological studies of the COST GE-1 well, United States South Atlantic outer continental shelf area: U.S. Geological Survey Circular 800, p. 7–17.

——, 1984, Turonian (Eaglefordian) stratigraphy of the Atlantic Coastal Plain and Texas: U.S. Geological Survey Professional Paper 1315, 21 p.

Van Hinte, J. E., 1976, A Jurassic time scale: American Association of Petroleum Geologists Bulletin, v. 60, p. 489–497.

Van Houten, F. B., 1977, Triassic-Liassic deposits of Morocco and eastern North America; Comparison: American Association Petroleum Geologists Bulletin, v. 61, p. 79–99.

Vogt, P. R., 1973, Early events in the opening of the North Atlantic, *in* Tarling, D. H., and Runcorn, S. K., eds., Implications of continental drift to the earth sciences: London, Academic Press, p. 693–712.

Watts, A. B., and Thorne, J., 1984, Tectonics, global changes in sea level, and their relationship to stratigraphical sequences at the U.S. Atlantic continental margin: Marine and Petroleum Geology, v. 1, p. 319–399.

Whalley, E., 1980, Speed of longitudinal sound in clathrate hydrates: Journal of Geophysical Research, v. 85, p. 2539–2542.

Woollard, G. P., Bonini, W. E., and Meyer, R. P., 1957, A seismic refraction study of the sub-surface geology of the Atlantic Coastal Plain and Continental Shelf between Virginia and Florida: Madison, University of Wisconsin, Department of Geology, Geophysics Section, Technical Report, Contract No. N7onr-28512, 128 p.

MANUSCRIPT ACCEPTED BY THE SOCIETY JANUARY 16, 1987

ACKNOWLEDGMENTS

We wish to thank Margaret Mons-Wengler, Patricia Forrestel, Jeff Zwinakis, and Dann Blackwood for their most professional help in preparing this manuscript. This paper is a very brief summary of many years of study that have involved many fellow scientists at the U.S. Geological Survey in Woods Hole. When we began this work 12 years ago, one of the two major basins of the southeastern U.S. continental margin was completely unknown and many other facts that we now take for granted were undiscovered. Some of those that have joined and aided us in the excitement of discovery are Charles Paull, Kim Klitgord, Page Valentine, Wylie Poag, John Grow, John Schlee, Kathleen Kent, Ann Swift, Lewis Gilbert, Elizabeth Miller, David Reynolds, Deborah Hutchinson,

Anne Trehu, Mahlon Ball, John Behrendt, David Folger, John Hathaway, Robert Sheridan (University of Delaware), Paul Pinet (Colgate University), and Conrad Neumann (University of North Carolina). We thank these and all the others who have joined in the arguments in the halls. This paper was thoroughly reviewed by Wylie Poag, John Schlee, and Elizabeth Winget for the USGS, and we thank them for their efforts. We also thank the reviewers for GSA, Henry Mullins and John Ladd.

The Geology of North America
Vol. I-2, The Atlantic Continental Margin: U.S.
The Geological Society of America, 1988

Chapter 15

Geology and geophysics of the Bahamas

R. E. Sheridan*
Department of Geology, University of Delaware, Newark, Delaware 19716
H. T. Mullins
Department of Geology, Heroy Geology Laboratory, Syracuse University, Syracuse, New York 13210
J. A. Austin, Jr.
Institute for Geophysics, University of Texas, Austin, Texas 78751
M. M. Ball
Office of Marine Geology, U.S. Geological Survey, Woods Hole, Massachusetts 02543
J. W. Ladd
Lamont-Doherty Geological Observatory of Columbia University, Palisades, New York 10964

LOCATION AND PHYSIOGRAPHY

The broad, shallow banks and intervening deep water (800–4000 m) channels of the Bahama Platform are unlike any other topography on the larger Atlantic margin (Fig. 1). The Bahama Platform covers roughly an area of 470,000 km^2, which is nearly equivalent to the area of the exposed Atlantic coastal plain from Cape Hatteras to Florida. The Bahama Platform is divided into a northwest part and a southeast part by a northeast trending line between Rum Cay bank and Oriente, Cuba (Fig. 1). The northwest part of the Platform is characterized by a larger portion of the area being shallow banks and a lesser portion being deep-water channels and basins with water depths of 800–4000 m. In contrast, the southeast portion of the Platform has a greater portion of the area as deep water channels surrounding smaller isolated banks; the depths of the channels are generally deeper than 2000 m. (Uchupi and others, 1971).

The deep channels and basins segmenting the Bahama Platform are of three different types: 1. open seaways with openings to deeper water at both ends; 2. closed seaways of linear proportions but open to deeper water only at one end; and 3. circular and semi-circular basins. Examples of open seaways are the Florida Straits, the Santaren/Nicholas Channels, the Northwest Providence Channel, the Northeast Providence Channel, and the Old Bahama Channel; Tongue of the Ocean (TOTO) and Exuma Sound are closed seaways; and Columbus Basin and Caicos Basin are semi-circular types.

The slopes of the straits are generally planar or convex upward (Malloy and Hurley, 1970; Uchupi, 1969). Off Miami there is the Miami Terrace (200–375 m) with a rough topography interpreted to be karstic in origin (Malloy and Hurley,

1970; Mullins and Neumann, 1979b). At the base of the terrace is a topographic trough separating the Miami Terrace from a sedimentary ridge ("depositional anticline" of Malloy and Hurley, 1970) that parallels the base of slope along the west side of the straits.

The main floor of the straits is shallowest off West Palm Beach, Florida (~700 m); the floor deepens to the north (>800 m) and south (>2000 m). The flatness of the main floor of the straits gives a U-shaped profile of the channel. In the extreme southwest, the Florida Straits become more V-shaped. On the Bahamas side of the Florida Straits there are two hemiconical noses that extend from the northwestern corners of Great and Little Bahama Banks (Mullins and others, 1980a).

On the eastern slopes of the northern Florida Straits, there is an area of small scale roughness, with local relief of 50–100 m, at depths of 600–800 m (Malloy and Hurley, 1970). Neumann and others (1977) made *Alvin* submersible observations of these small topographic features and found them to be numerous north-south striking elongate lithified biohermal mounds, which they defined as "lithoherms."

The bank margins have common geomorphic zones (Mullins and others, 1984; Hooke and Schlager, 1980; Schlager and Chermak, 1979; Crevello and Schlager, 1980). These zones include: 1) a marginal escarpment; 2) an upper gullied slope; 3) a non-gullied lower slope; and 4) a basin or channel floor. Marginal escarpments ring the Bahama Banks as abrupt, precipitous slopes, commonly in excess of 45°, extending from water depths of ~20 m to ~150–180 m (Hine and Mullins, 1983). The upper gullied slopes (4°–15°) form a distinct break with the marginal escarpments where the slopes abruptly decrease into gentler slopes at water depths of ~200 m. Many small submarine canyons with

*Present address: Department of Geological Sciences, Rutgers University, New Brunswick, New Jersey 08903

Sheridan, R. E., Mullins, H. T., Austin, Jr., J. A., Ball, M. M., and Ladd, J. W., 1988, Geology and geophysics of the Bahamas; *in* Sheridan, R. E., and Grow, J. A., eds., The Geology of North America, Volume I-2, The Atlantic Continental Margin, U.S.: Geological Society of America.

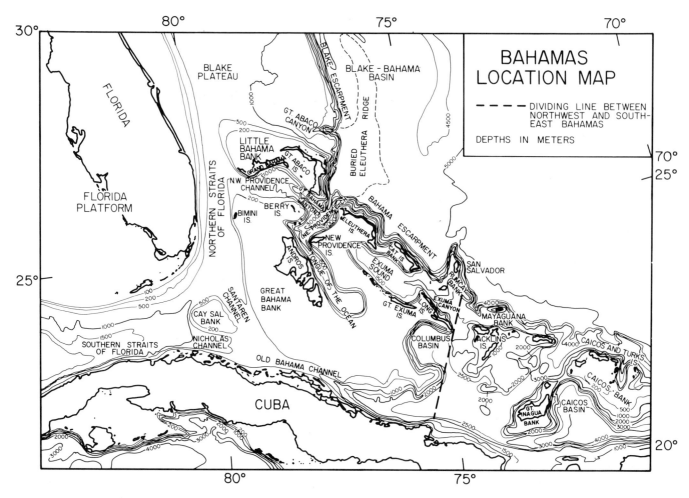

Figure 1. Location map of the Bahama Banks with names of major physiographic features (from Paulus, 1972).

relief of 10 to 150 m notch the slopes (Hooke and Schlager, 1980; Mullins and others, 1984). The gullies appear to terminate at the base of the upper slope, although there is evidence that some cut across the lower slope as broad, shallow channels (Mullins and others, 1984). The gentle (0.5–2.0°) lower slopes are broad, smooth regions interrupted locally by hummocks of less than 50 m relief. The basin, channel, or plateau floor at the base of the lower slope is generally nearly flat and smooth, characterized by a "layered" echo character (Schlager and Chermak, 1979).

The channels opening to the Atlantic Ocean have a distinct morphological zonation, passing downslope from smooth, U-shaped channels to rough, distinctly V-shaped canyons (Athearn, 1962, 1963; Lee, 1981; Andrews and others, 1970). Depths range from 800–1000 m in the U-shaped headward portions of these channels to 2000–4000 m in the V-shaped seaward parts of the channels.

Columbus Basin is a semi-circular Bahama basin with a flat floor at depths of 2000–2600 m. (Bornhold and Pilkey, 1971). The cul-de-sac nature of the basin is emphasized where the slopes bounding the basin are present around approximately 250° of its

perimeter. The Bahama Escarpment, a steep slope, with inclines of more than 45° and ledges with near vertical walls, is an example of an erosional carbonate slope (Mullins and Newmann, 1979a; Freeman-Lynde and others, 1981; Schlager and Ginsburg, 1981). Extending from the marginal escarpments at 80–140 m depths of the banks off Cat and Eleuthera Islands, the Bahama Escarpment plunges continuously to the abyss at depths of 4900–5000 m. There are major canyons cutting through the escarpment, such as Great Abaco Canyon (Mullins and others, 1982a) and Great Bahama Canyon (Andrews, 1970).

PREVIOUS RESEARCH AND INTERPRETATIONS

Marine geological research on the Bahama Platform began in the 1850s with the dredgings by Louis Agassiz and Louis Pourtales. In the 1880s, Agassiz's son Alexander continued the research on the Blake-Bahama area and on the Gulf Stream using the steamer *Blake* (Agassiz, 1888, 1894; Schlee, 1973).

Geological interpretations about the Bahamas were made as

early as the 1850s (Nelson, 1853; Hilgard, 1871, Gabb, 1873; Suess, 1888, 1908). These 19th century ideas are well summarized in an article by Meyerhoff and Hatten (1974). By the early 1900s, some geophysical data in the form of gravity measurements were available that led to interpretations about the crust under the Bahama Platform (Guppy, 1917; Woodring, 1928; Hess, 1933, 1960; Field and others, 1931; Schuchert, 1935; Meyerhoff and Hatten, 1974). During the 1950s and early 1960s, new data were available, such as geological samples of the sediments from the banks and channels and land and aeromagnetic surveys (H. A. Meyerhoff, 1946, 1954; Ericson and others, 1952; Newell, 1955; Drake and others, 1963; Meyerhoff and Treitel, 1959; King, 1959; Meyerhoff and Hatten, 1974).

In the 1960s and early 1970s seismic studies on the Bahama Platform, such as that of Sheridan and others (1966), reported refraction profiles in the deep water areas of the Florida Straits, Santaren Channel, and Providence Channels. The first public seismic reflection data (Ewing and others, 1966) consisted of a line through the Providence channels. This line had limited penetration of about 1 sec. of reflection time and revealed relatively flat-lying sedimentary rocks in the channel. Other reflection profiler studies in the 1960s include the work of Uchupi (1966), Rona and Clay (1966), and Malloy and Hurley (1970) in the northern Florida Straits. All these studies were of single channel, reflection surveys with low energy-sources, thus penetration was less than 1 sec. and only the shallow stratigraphy and structure were observed. Mullins and Lynts (1976) investigated the seismic stratigraphy of the Northeast Providence Channel and northern TOTO. They suggested that parts of the channel were controlled by normal faults that were sporadically active into the Tertiary. Ball and others (1968) surveyed Exuma Sound using single-channel seismic reflection techniques and suggested, along with Lidz (1973), that there were nine "diapiric structures" in the center of the basin. In a seismic refraction study along the Cat Island block, Ball and others (1971) indicated that a 5.5 km–thick layer with a velocity of 4.6 km/sec overlies 17 km of a layer of 5.7 km/sec velocity, much like what Sheridan and others (1966) found in other Bahama profiles. Sheridan (1972) used surface wave dispersion from three earthquakes in Old Bahama Channel north of Hispaniola to determine a regionalized Rayleigh-wave dispersion curve for the Bahamas. Sheridan's studies (1972, 1974; Sheridan and Osburn, 1975) suggested an intermediate crust (7.2 km/sec) formed by a granitic continental crust (6.2 km/sec) pervasively intruded by high velocity ultramafic rocks (8.1 km/sec) exists under the Bahamas.

Talwani and others (1960) compiled a modified Bouguer anomaly map of the Bahamas. The residual gravity anomalies over the shallow banks were small, + 10 mgal, while those over the deep channels were larger and negative, ~–30 to –40 mgal. This suggested that the channels were underlain by anomalously light material at a shallow depth (~1–4 km). Talwani and others (1960) suggested that the channels were down-faulted graben structures. Ball (1967a) also favored this structural interpretation of the gravity anomalies. Interpretations of widely-spaced aero-

magnetic data by Drake and others (1963) implied a continuity between the northwest-southeast anomaly trends of south Florida and the Bahamas. Taylor and others (1968), however, published a map of the Bahamas north of 25°N latitude that showed the prominent Florida anomalies terminating at the Florida Straits.

Well control in the 1960s and 1970s was largely limited to those drilled onshore in Florida (Applin and Applin, 1965; Maher, 1965), four deep wells drilled on islands in the Bahamas (Andros Island—Spencer, 1967; Cay Sal—Paulus, 1972; Long Island—Meyerhoff and Hatten, 1974; and Great Isaac, Tator, and Hatfield—1975), and some shallow holes on Eleuthera, San Salvador, and New Providence Islands (Meyerhoff and Hatten, 1974; Paulus, 1972). One Deep Sea Drilling Project DSDP hole, Site 98, was drilled in Northeast Providence Channel and reported on by Paulus (1972) and Hollister and others (1972). DSDP Sites 99, 100, and 101 were drilled just east of the Bahama Escarpment (Hollister and Ewing, 1972). Some of the deep wells and drill sites in Florida and the Bahamas sampled rocks as old as Late Jurassic age.

Observations from submersibles in the Bahamas began in the 1960s. *Aluminaut* dives in the northern Florida Straits by Neumann and Ball (1970) indicated a southerly flowing, near-bottom counter current along the western margin, confirming the work of Hurley and Fink (1963). Mullins and Neumann (1979a, 1979b) interpreted data on the Miami Terrace collected from the research submersible *Alvin*, and Neumann and others (1977) used *Alvin* to describe the deep-water biohermal buildups (litho-herms) that occur along the eastern margin of the Florida Straits. Schlager and others (1976) used *Alvin* to observe the slopes of the Tongue of the Ocean; Gibson and Schlee (1967) used *Alvin* to find outcrops of Miocene deep sea ooze on the eastern wall of TOTO.

Piston cores and echo soundings were used to study processes of sediment gravity flow and pelagic deposition in the deep areas (Rusnak and Nesteroff, 1964; Pilkey and Rucker, 1966; Rucker, 1968; Lidz, 1973; Bornhold and Pilkey, 1971; Kier and Pilkey, 1971; Bennetts and Pilkey, 1976; Mullins and Neumann, 1979 a & b; Wilber and Neumann, 1977; Schlager and Cher-mak, 1979). On the shallow banks, studies dealt mostly with biogenic and bioclastic deposition, oolite deposits and production and distribution of fine muds over the banks (Purdy, 1963; Traverse and Ginsburg, 1966; Ball, 1967b; Buchanan, 1970; Shinn and others, 1969; Land, 1970; Hoffmeister and Multer, 1968; Enos, 1974; Neumann and Land, 1975).

Basic questions on the long term geologic evolution of the Bahama Platform were debated in the 1970s. Dietz and others (1970) considered the overlap that existed where the Bahama Platform overlies areas of the African continental crust in the reconstruction of the Atlantic continents preferred by Bullard and others (1965). The apparent overlap of the Bahamas would contradict the reconstruction if, as Meyerhoff and Hatten (1974) believed, the Bahamas were underlain by continental crust. Deitz and others (1970) presented three possible alternative models for the Bahamas crust: A) foundered extension of continental basement from North America under all the archipelago; B) thick crust of volcanic origin, which might be oceanic; and C) subsided oceanic crust of normal thickness buried by sediment. Models B and C are compatible with various close-fit reconstructions of west Africa against eastern North America, because these crusts would have been accreted on the North American plate since seafloor spreading began. Model A creates overlap upon reconstruction. Dietz and others (1970) favored model C and explained the evolution of the Bahamas as one of rapid clastic sediment infill of a narrow drift basin on a thin oceanic crust, which subsided with the deposition of more than 5 km of limestone and dolomite.

In contrast, Lynts (1970), and later Mullins and Lynts (1977), favored model A, interpreting the crust under the Bahamas to be underlain by a block-faulted modified continental basement. Lynts (1970) visualized faulting of the Bahamas basement caused by a now inactive southwest dipping subduction zone along the northeastern part of the Bahamas. Such an occurrence would possibly explain the overlap problem in that the subduction might be invoked to allow the Bahamas crust to move northeastward into its present position with respect to North America only after Africa drifted out of the way. A subduction model and continental crust was also favored by Mattson (1973), who concluded that the Bahamas and the North American/North Atlantic Ocean plate were being subducted southwestward beneath Cuba.

Sheridan (1974) accepted the idea of an oceanic origin for the crust under the Bahamas, although in a previous discussion with Dietz and Holden (1971), Sheridan (1971) had favored a crust of intermediate velocity and thickness formed perhaps by basic dike intrusion during the drifting and spreading process. The intermediate type of crust could be thicker than the more dense typical oceanic crust favored by Dietz and others (1970) and was more compatible with both Rayleigh wave dispersion models (Sheridan, 1972) and gravity models (Talwani and others, 1960). As though by compromise, Uchupi and others (1971) interpreted the crust to be oceanic in the southeastern Bahamas and continental in the northwestern part based mainly on gravity data.

Questions persisted on the origin of the deep channels and the segmentation of the Bahama platform into separate, isolated banks. Some researchers favored direct structural control of the deep channels and basins, with formation beginning as grabens (Talwani and others, 1960; Ball, 1967a; Mullins and Lynts, 1977). Such grabens were perceived as originating in the rift stage of the North Atlantic separation, so the relief of the modern channels is thought to have persisted in the same location since the Jurassic (Mullins and Lynts, 1977). On the other hand, Dietz and others (1970) believed that the channels and banks owed their origin to normal sedimentation and erosion processes in carbonate environments, with currents through the channels controlling the delivery of nutrients to the organic components of the bank edges. Normal marine erosional processes, such as turbidity currents, deepened and carved the channels with canyons (Andrews, 1970), while shallow water carbonates built upwards at a

pace equal to subsidence, resulting in a continuous coexistence of banks and channels in a dynamic equilibrium.

In contrast some researchers think that the present channel/bank configuration has only evolved since the Late Cretaceous and that prior to this event there existed a megabank, or large unsegmented platform (Paulus, 1972; Meyerhoff and Hatten, 1974; Schlager and Ginsburg, 1981; Sheridan, 1974). In this scenario, shallow-water bank limestones and dolomites are presumed to pass beneath the channels, where they are overlain by deep-water pelagic oozes and chalks. If the channels began late in the history of the Bahama Platform, or if the channels changed positions drastically in the Cretaceous, why did this occur at that particular time? Some authors suggest a purely environmental control, with erosion and upbuilding being the active processes that created the relief (Paulus, 1972; Dietz and others, 1970; Andrews, 1970; Malloy and Hurley, 1970). Others suggest structural control (Talwani and others, 1960; Ball, 1967a) and invoke graben structures of substantial structural relief. A combination of processes has been proposed in which small-scale fault relief of the sea-floor might localize the erosion and upbuilding processes to produce the configuration of the evolving channels (Sheridan, 1974).

NEW DATA AND INTERPRETATIONS

In the late 1970s and 1980s, the Bahamas were investigated by deep-penetration seismic-reflection profiling and also by concentrated, high-resolution shallow seismic studies. These surveys, complemented with standard piston-coring and dredging studies and detailed stratigraphic sampling from the submersible *Alvin* resolved some of the previous contradictory speculations.

Deep-penetration seismic reflection profiles of Sheridan at al. (1981), Austin (1983, 1985), Ladd and Sheridan (1986), Schlager and others (1984), and Ball and others (1985) now are available in all the deep-water channels and basins of the Bahama Platform (Fig. 2). Basement and sequences as old as Middle Jurassic have been interpreted. New dredging information has been collected in critical locations where direct ties could be made with the new reflection profiles (Corso, 1983; Schlager and others, 1984). Rocks as old as Early Cretaceous age have been sampled.

High-resolution single-channel reflection profiles have been made in grids on many of the bank margins (Mullins and Neumann, 1979a, b; Mullins and others, 1980a, 1981; Van Buren and Mullins, 1983; Lee, 1981, Hine and others, 1981, Schlager and Chermak, 1979) (Fig. 2). A great variety of seismic facies has been interpreted with this data using the seismic sequence and facies analysis techniques of Vail and others (1977a), but little direct calibration by drilling (with the exception of DSDP Site 98) has been done.

Direct sampling of the rocks underlying the Bahama Platform has been done along the erosional Bahama Escarpment from the submersible *Alvin* (Freeman-Lynde and others, 1981). Closely spaced samples in the depth range of 2000–4000 m were

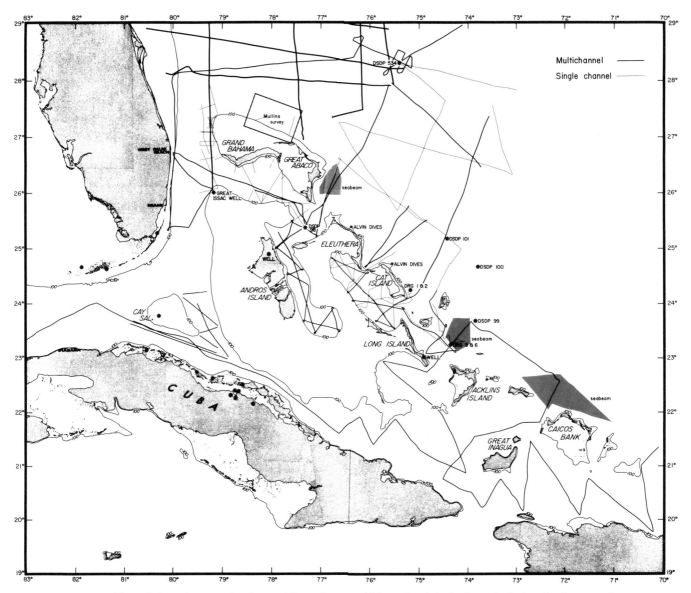

Figure 2. Location map showing tracklines of new multichannel and single channel seismic reflection data, deep exploration wells in the Bahamas, deep wells in Florida and Cuba, DSDP Sites 98, 99, 100, 101, and 534, dredge sites on the Bahama Escarpment, and *Alvin* dive sites. References for these new data are given in the text.

collected near Cat Island (Fig. 2). The ages of the limestones and dolomites range from Late to Early Cretaceous, with some Eocene pelagic ooze covering the older rocks as a drape.

DEEP SEISMIC STRUCTURE

Northern Florida Straits

The base of the drift-phase sediments in the northern Florida Straits is an angular unconformity identified as a horizon above truncated dipping internal reflectors (Fig. 3) (Sheridan and others, 1981). This horizon is projected at a depth of a little more

than 5 km to the nearby Great Isaac well where it is correlated with the major unconformity encountered in that hole between overlying Upper Jurassic limestones, dolomites, evaporites, and underlying red, arkosic rhyolitic volcaniclastics (Tator and Hatfield, 1975) (Fig. 4). These Great Isaac volcaniclastics are of unknown age, but under south Florida where similar volcanic "basement" rocks are drilled they consistently date to the 190–140 m.y. range, and are thus considered to be Early to Middle Jurassic in age (Milton and Grasty, 1969). This major angular unconformity is the "breakup unconformity," as the term is used by Falvey (1974), that separates the drift-phase sediments from the underlying rift-phase sediments.

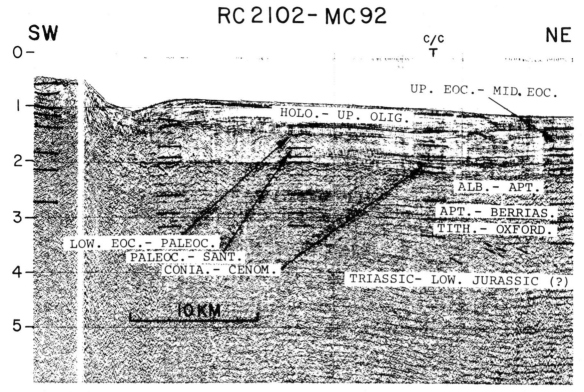

Figure 3. Portion of line MC 92 across Miami Terrace on western margin of Florida Straits. (from Sheridan and others, 1981).

Above this basement breakup unconformity, the throughgoing, continuous high-amplitude reflections are interpreted to be the Upper Jurassic through Lower Cretaceous drift sequences of shallow water limestones, dolomites, and evaporites (Sheridan and others, 1981) (Figs. 3, 4). The Upper Cretaceous and Cenozoic sequences have seismic characteristics correlated with deep-water pelagic oozes, chalks, and limestones (Figs. 3, 4) (Sheridan and others, 1981). On the slopes and margins of the Straits, the Cenozoic sequences are characteristically in the form of sigmoidal clinoform reflectors. Off Key Largo and the Miami Terrace, the westward dipping clinoforms are part of a north-south elongated mounded sequence forming the "depositional anticline" of Malloy and Hurley (1970) (Fig. 3). In the Upper Cretaceous and Tertiary sequences, distinct gullying occurs at the sequence boundaries and restricted zones of hyperbolic reflections also occur there (Figs. 3, 4) (Sheridan and others, 1981). These are interpreted to be erosive effects of the various Florida Current positions through geologic time (Fig. 4).

In several places the seismic facies are transitional horizontally from the transparent, planar stratified deeper-water limestones to the acoustically opaque, reflectorless or hummocky mass of the massively bedded shallow-water limestones of a carbonate platform edge or reef structure (Figs. 3, 4, and 5). This facies transition documents the evolution of the Florida Straits as a deep-water channel between shallow banks at least as early as the early Late Cretaceous (Sheridan and others, 1981).

Providence Channels

On multichannel seismic reflection profiles in the Northwest Providence Channels (Sheridan and others, 1981) (Figs. 6 and Plate 6) the base of the drift-phase sequence is the horizon with truncated dipping reflectors below. This is interpreted to be the breakup unconformity on Triassic-Jurassic volcaniclastics (Fig. 7 and Plate 5) (Sheridan and others, 1981). Under Northeast Providence Channel, the truncated breakup unconformity is no longer seen. Instead, a sedimentary sequence interpreted to be of Middle Jurassic age forms planar, continuous reflections of high amplitude and reverberatory character (Fig. 6). Where this sequence is seen, the underlying basement(?) is only vaguely seen as some hummocky, low frequency events. Because of its hummocky nature, and also because of the projection of the Blake Spur magnetic anomaly (to be discussed later), the vague basement is presumed to be oceanic in nature (Sheridan and others, 1981) (Fig. 6).

The Upper Jurassic and Lower Cretaceous seismic sequences, interpreted to be shallow water carbonates and evaporites, extend throughout Northwest and Northeast Providence Channels (Figs. 6, 7, and Plate 5). These sequences are traceable to Abaco Knoll (between distances 370 and 390 km on Fig. 7) where they are exposed (Fig. 7 and Plate 5) (Sheridan and others, 1981). Dredging on the northeast face of Abaco Knoll has recovered a Lower Cretaceous shallow-water limestone documenting this seismic correlation (Corso, 1983).

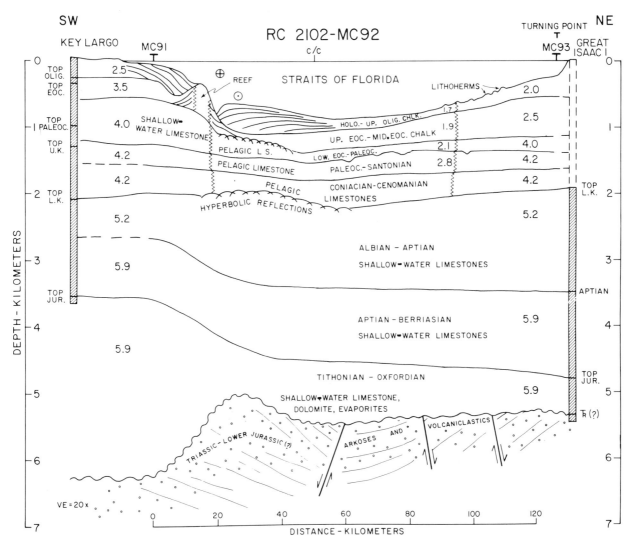

Figure 4. Diagrammatic structural cross-section of northern Florida Straits across Miami Terrace based on depth conversion of MC 92. Velocities used for depth conversion are shown in km/sec. Current directions indicate shallower Florida Current (⊙ out of section) and deeper counter current (⊕ into section). Dashed upper portion of Great Isaac well indicates that no published accounts exist of this section (Tator and Hatfield, 1975) (from Sheridan and others, 1981).

The very distinct Jurassic and Cretaceous reflectors can be traced for hundreds of kilometers through the Northwest and Northeast Providence Channels. However, in one area, these otherwise continuous reflectors are disrupted into a chaotic jumble and faulting is interpreted to offset these horizons by several hundred meters (Fig. 7 and Plate 5). While resolution is poor through this chaotic zone, depth corrections require offsets in sequences as young as Early Cretaceous (Sheridan and others, 1981).

Shallow-water limestones and dolomites of Early Cretaceous age are overlain by continuous layers correlated with Upper Cretaceous and Cenozoic deep-water oozes and chalks (Sheridan and others, 1981). This transition in seismic facies is similar to that interpreted for the Florida Straits; it occurs at approximately the same geological time in both areas (Fig. 7).

Seismic stratigraphic correlations were made from the Bahamas to the Blake-Bahama Basin where correlated seismic sequences have been drilled at DSDP Site 391 (Fig. 7) and more recently at Site 534, 22 km northeast of Site 391 (Sheridan and others, 1981; Sheridan and others, 1983). With this control, reflectors A^u, β, C, and D have been mapped where they impinge on the base of the Bahama Escarpment. These reflectors bound Middle Jurassic and younger sequences composed of deep-sea oozes interbedded with carbonate turbidites and debris flows. Near the base of the escarpment, it is interpreted that the rocks in the A^u–D sequences are interlayers of carbonate turbidites washing into the Blake-Bahama Basin (Fig. 7).

A broad zone of faulting in the Great Abaco fracture zone (Fig. 7) is detected by offsets on the seismic profile affecting the Cretaceous sequences at least up to the Horizon A^u reflector, a

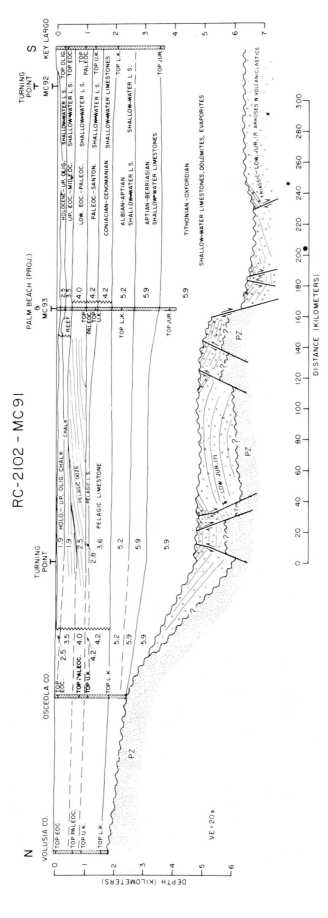

Figure 5. Diagrammatic structural cross-section of Florida shelf based on line MC 91 and projected to the coastal wells. Velocities given in km/sec. (from Sheridan and others, 1981).

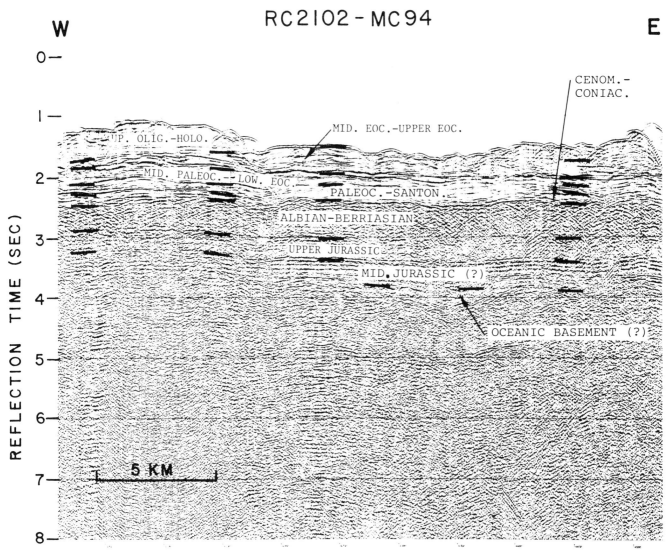

Figure 6. Portion of line MC 94 in Northwest Providence Channel (from Sheridan and others, 1981).

major post-Eocene unconformity. Whether the faults penetrated into the overlying Miocene sediments or actually broke the Tertiary sediments that have been eroded subsequently is uncertain. The significant thing is that extensive faulting has occurred after Cretaceous time (Sheridan and others 1981). The Great Abaco fracture zone is typical in orientation of the western Atlantic fracture zones, which are older Jurassic transform faults. The Great Abaco fracture zone crosses the margin just north of Great Abaco canyon and appears to cross the Blake Plateau just north of Little Bahama Bank. In fact, the pronounced linearity of the northern margin of Little Bahama Bank and its northwest-southeast strike suggests some post-Cretaceous fault control of this margin, and indeed the entire northern structural boundary of the Bahama Platform could be influenced by this post-Cretaceous reactivation of the Great Abaco fracture zone (Sheridan, 1974) and/or other Atlantic fracture zones.

Blake Plateau–Bahama Platform Boundary

Besides one 24 channel flexichoc line (FC 10) of the Institute Francais du Petrole (IFP)–USGS study of the Blake Plateau (Dillon and others, 1979), which crosses the Blake Plateau north to south and terminates on the northern margin of Little Bahama Bank, the only information on the deep structure of this Blake Plateau-Bahama boundary is the magnetic and gravity data. Klitgord and others (1984) have interpreted the magnetic anomalies as reflecting a major fracture zone passing northwest-southeast as far west as the Florida continental shelf off Jacksonville. They call this the Jacksonville fracture zone, which is a continuation of the Great Abaco fracture zone seaward of the Blake Plateau.

Van Buren and Mullins (1983) have correlated single-channel airgun data with the Upper Cretaceous shallow water limestones and dolomites under the Blake Plateau (Ewing and

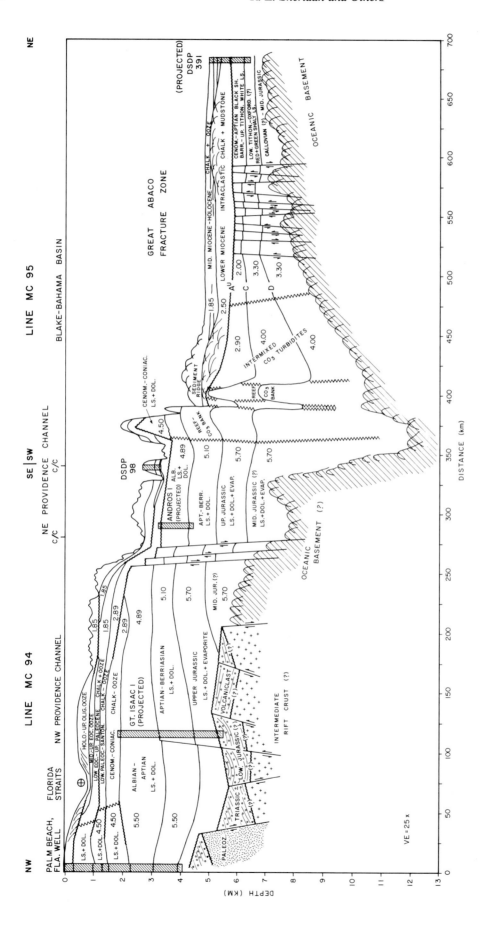

Figure 7. Diagrammatic structural cross-section through Northwest and Northeast Providence Channels and across Bahama Escarpment and Blake-Bahama Basin based on depth conversion of lines MC 94 and 95. Velocities used for the conversion are in km/sec. (from Sheridan and others, 1981).

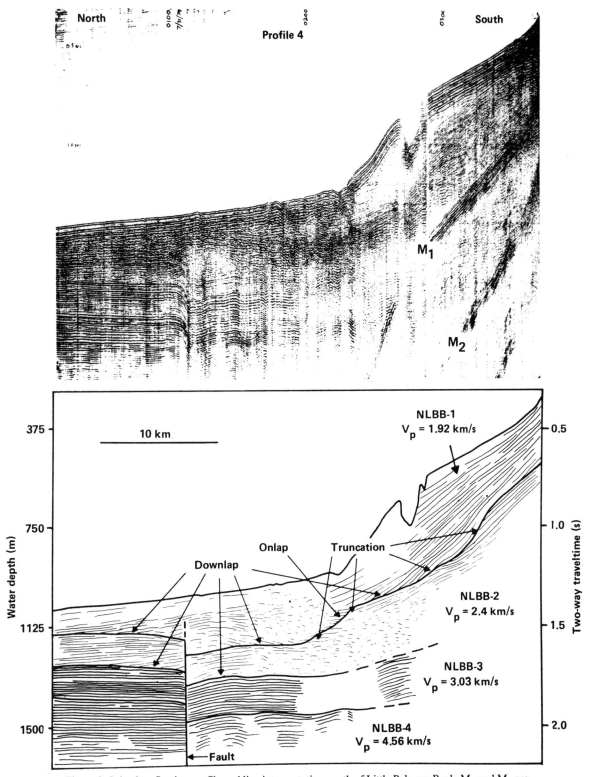

Figure 8. Seismic reflection profile and line interpretation north of Little Bahama Bank. M_1 and M_2 are water bottom multiples. Compressional wave velocities are from Sheridan et al. (1966). Note planar, continuous parallel reflector facies on the upper slope; chaotic hummocky reflector facies at the base of slope; and discontinuous, parallel planar reflectors in the basin facies of the upper two sequences. (from Van Buren and Mullins, 1983).

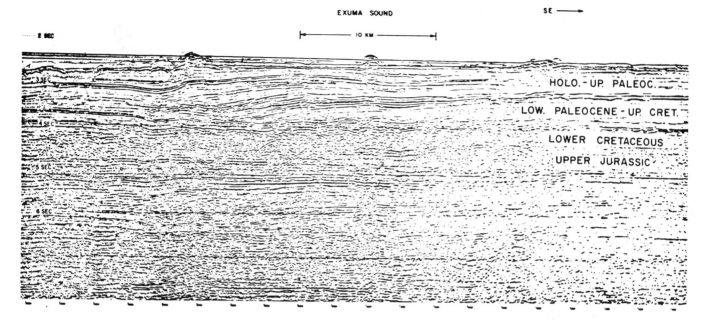

Figure 9. Portion of line MC 361 in Exuma Sound. Note topographic knolls previously interpreted as diapirs and how prograding Tertiary sequences pass beneath these knolls. The primary dips of the prograding sequences were previously interpreted as rim-syncline structures (Ball and others, 1967) (From Ladd and Sheridan, 1986).

others, 1966; Sheridan and others, 1966; Van Buren and Mullins, 1983) (Fig. 8). The overlying layers are uppermost Cretaceous through Cenozoic chalks and oozes of the deeper-water deposition on the Blake Plateau. These seismic profiles reveal that the deepest layers remain relatively flat under the northern Little Bahama Bank lower margin slope, and that the overlying layers are prograding from the bank (Fig. 8). The internal seismic facies of prograding layers have a threefold zonation, consistent along strike: 1) an upper-slope, planar-continuous, low-amplitude reflector facies, 2) a proximal lower-slope, chaotic, hummocky reflector pattern, and 3) a lower slope-basin facies of discontinuous planar reflectors. These can be correlated with and interpreted as a zonation between upper slope, periplatform oozes, a proximal lower slope apron of slump masses and debris flows, and a distal lower slope transition with turbidites intermixed with basinal pelagic deposits (Mullins and others, 1984).

A high-angle, normal(?) fault offsets Cretaceous through possibly Oligocene sequences and may even extend to the present sea floor (Mullins and Van Buren, 1981; Van Buren and Mullins, 1983) (Fig. 8). This fault is mapped as having a northeast-southwest strike, downthrown on the southeast. Projecting the fault bankward, it coincides with a magnetic anomaly trend that suggests a basement fault (Sheridan, 1974; Sheridan and Osburn, 1975). This trend crosses Walker's Cay and hence has been informally named the Walker's Cay fault zone (Mullins and Van Buren, 1981). Although Walker's Cay fault strikes perpendicular to the Little Bahama Bank margin, its association with an apparent major basement boundary reflected in a marked gravity and magnetic anomaly, as well as its apparent youth (Cretaceous to

middle Tertiary movements), suggest that late stage reactivation of basement structures does influence the margin in this area.

Sedimentary processes, such as gravity sliding and slumping, have also had a strong influence on the northern margin of Little Bahama Bank (Van Buren and Mullins, 1983; Mullins and others, 1984). Seismic reflection profiles show that the nose north of Little Bahama Bank is a mounded sediment pile formed as a drift deposited by contour currents since the mid-Tertiary (Mullins and Neumann, 1979a; Sheridan, 1983a).

Exuma Sound

The deepest seismic reflector in Exuma Sound is a rather low relief, continuous event with a characteristic, low-frequency wavelet (Fig. 9 and Plate 5) (Ladd and Sheridan, 1986). It is similar in appearance to the reflector correlated by Sheridan and others (1981) with the top of the Middle Jurassic in the Northwest and Northeast Providence Channels. No reflection events are seen beneath this low-frequency signal. To the southeast, a comparable low frequency reflector can be seen clearly as an undulating, hummocky, hyperbolic reflector with onlapping sediment drape, characteristic of the oceanic basement. Whether this low-frequency reflector under northwest Exuma Sound is basement, or perhaps sedimentary rocks just above basement, is not certain. However, extrapolation of magnetic anomalies and trends from known oceanic crust to the northeast suggests that there may be oceanic basement under Exuma Sound (Klitgord and others, 1984; Sheridan, 1974; Sheridan and Osburn, 1975).

In northwest Exuma Sound two sequences are correlated

with Upper Jurassic to Lower Cretaceous shallow-water limestones and dolomites (Ladd and Sheridan, 1986) (Fig. 9 and Plate 5). Tracing these sequences to southeast Exuma Sound, there is a drastic facies change. Sequences become acoustically transparent and are interpreted to be a deep-water pelagic ooze, chalk, and limestone facies of Late Jurassic to Early Cretaceous age (Lee, 1981; Ladd and Sheridan, 1986). This correlation is supported by recent dredging in the deeply incised canyon at the southeast end of Exuma Sound (Fig. 10) (Corso, 1983; Schlager and others, 1984), where Lower Cretaceous deep-water chalks and limestones have been recovered.

One interpretation of these results is that an embayment of deep water existed in the southeastern Exuma Sound by the Early Cretaceous. Corso (1983) and later Schlager and others (1984) call this Lower Cretaceous embayment of the Atlantic Ocean the Samana Reentrant. Seismic reflection profiles to the south and to the northwest of the Samana Reentrant show that the deep-water facies of Early Cretaceous age does not persist in those directions, that is, into Columbus Basin and into northwest Exuma Sound. Thus, the Samana Reentrant was apparently a dead-end closed seaway channel in the Bahama Platform and did not extend for a great distance across the region.

Sequences that have a distinct sigmoidal clinoform structure of prograding slope facies overlie the Lower Cretaceous in northwest Exuma Sound (Ladd and Sheridan, 1986) (Fig. 9 and Plate 5). Such prograding units are correlated with similar prograding units of Late Cretaceous through Oligocene age found on the Florida Shelf and margins of the Florida Straits (Sheridan and others, 1981). Prograding, clinoform sequences of approximately the same age are also seen in Tongue of the Ocean (discussed below). We interpret these sequences as evidence for a time of extensive drowning of the Bahama Platform. After bank margins stepped back, enough sediments were available on the banks to shed debris into the prograding wedges.

Above these prograding sequences, there are two hummocky sequences (Fig. 9 and Plate 5) (Ladd and Sheridan, 1986). The hummocky reflections of the sequence boundaries are clearly the result of deep-marine erosional processes that have carved interpreted Oligocene through Pliocene sediments. The top of the Pliocene sequence appears to outcrop on the sea floor as a number of distinct topographic knolls 20–50 m high and about 1–2 km across. On 3.5 khz profiles, these knolls have higher reflectivity than the surrounding sediments and have sharp multiple peaks.

The knolls first reported by Ball and others (1968) and later studied by Lidz (1973), were thought to be the tops of diapirs of either undercompacted sediments or salt, which form similar knolls in the Gulf of Mexico. The underlying dipping reflections were interpreted as sediments deformed by flow tectonics into rim syncline structures (Ball and others, 1968; Lidz, 1973). The new multichannel seismic data in Exuma Sound (Fig. 9 and plate 5) reveal that the shallow sedimentary sequences pass beneath the knolls without positive evidence of deformation and the knolls are only surficial features, and definitely not salt diapirs (Ladd

and Sheridan, 1986). The dipping reflections beneath the knolls are prograding paleoslope sequences, not rim-syncline deformation. It is possible that the hummocks are erosional remnants of Pliocene(?) sediments that protrude through a stratified blanket of Pleistocene and Holocene gravity flow and turbidite deposits. Alternatively, deep-water biohermal buildup such as lithoherms might also explain the knolls.

Tongue of the Ocean (TOTO)

The deepest primary reflector in Tongue of the Ocean (TOTO) (Fig. 11 and Plate 5) is similar to the deepest reflector under Exuma Sound in its low-frequency reverberatory nature and its general horizontality (Ladd and Sheridan, 1986). This reflector can be tied with a similar reflector under Northeast Providence Channel (Figs. 6, 7 and Plate 5) where it is correlated with Middle Jurassic limestone, dolomite, and evaporite (Sheridan and others, 1981).

A deeper reflector is seen above tilted, truncated internal reflectors only in the northwest corner of TOTO near Andros Island. The truncation is typical of the breakup unconformity identified in Northwest Providence Channel (Fig. 7). The restricted extent of this horizon implies that this might be the southeast limit of the breakup unconformity and underlying Jurassic rifted, volcaniclastic crust that is interpreted to occur all the way from southern Florida. Farther southeast in TOTO, no similar breakup unconformity is seen on the reflection profiles, and the deepest low-frequency reflector could represent sediments just above the undulating oceanic basement, as is interpreted under Exuma Sound and Northeast Providence Channel (Fig. 7).

Correlation of the low-frequency reflector with Middle Jurassic limestone, dolomite, and evaporite agrees with classical models of rifted passive margins where marine drift-phase sediments onlap the unconformity soon after breakup. If the age of the low-frequency reflector is Middle Jurassic, as proposed by Sheridan and others (1981), then breakup in the area of the northern TOTO began in Middle Jurassic time.

A rough reflection horizon traceable throughout TOTO is correlated to the horizon identified as the top of Lower Cretaceous shallow-water carbonates in the Providence Channels, Exuma Sound, and Columbus Basin (Ladd and Sheridan, 1986; Austin, 1985).

The Upper Cretaceous-Tertiary deep-water carbonate sediments exhibit sigmoidal prograding clinoforms similar to those seen in Exuma Sound (Fig. 11). The buildout of bank margins from the isolated banks that evolved after the mid-Cretaceous rapid sealevel rise is evident in TOTO just as in Exuma Sound and the Florida Straits.

Columbus and Caicos Basins

The seismic stratigraphy in the Columbus and Caicos Basins of the southeastern Bahama Platform is remarkably like that of Exuma Sound (Austin, 1983; 1985). A characteristic low-

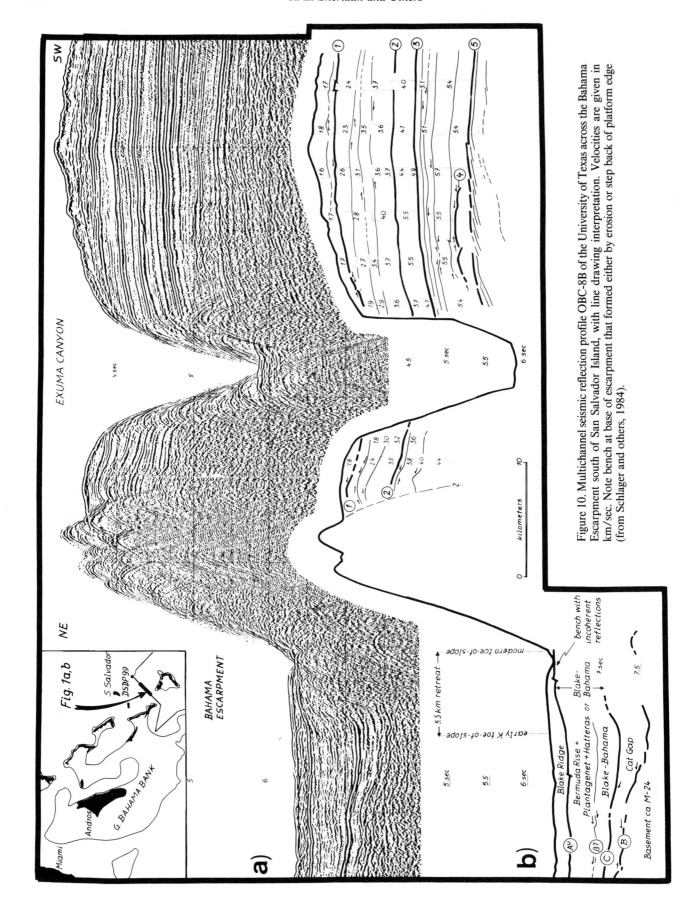

Figure 10. Multichannel seismic reflection profile OBC-8B of the University of Texas across the Bahama Escarpment south of San Salvador Island, with line drawing interpretation. Velocities are given in km/sec. Note bench at base of escarpment that formed either by erosion or step back of platform edge (from Schlager and others, 1984).

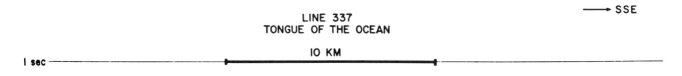

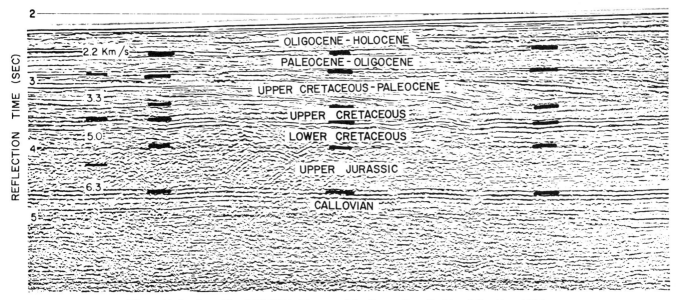

Figure 11. Portion of line MC 337 in Tongue of the Ocean (from Ladd and Sheridan, 1986).

frequency event occurs sporadically as the deepest observed reflector (Fig. 12). This event correlates through profile 8 to MCS line 355 in the southeast Exuma Sound where a reflection at this depth beneath transparent sediments looks like typical oceanic basement.

Above this low-frequency reflector is a sequence tentatively correlated with the Upper Jurassic to Lower Cretaceous shallow-water limestones and dolomites (Fig. 12). The top of the sequence forms a strong, rough reflector that is mappable over a wide area. It is correlated with a similar transition in Exuma Sound between the shallow-water limestones and the overlying deep-water oozes and chalks of Late Cretaceous and Tertiary age (Austin, 1985; Schlager and others, 1984) (Fig. 12). The rough reflector is interpreted to be a major unconformity related perhaps to the exposure of a large area of the Bahama Platform in mid-Cretaceous time, subsequent erosion, then rapid flooding and burial with deep-water carbonates (Sheridan and others, 1981; Austin, 1985). The seismic data reveal some structure on this unconformity in that it deepens generally northwestward in Columbus Basin, from 3.4 sec to 4.2 sec. depth. If this reflector is an unconformity once near sea-level, then this deepening could reflect a gentle post mid-Cretaceous tilting of parts of the Bahama Platform on a large scale. The tilt (0.44°) results in as much as 500 m of structural relief over a distance of 65 km. Similar tilt

and relief are mapped on a correlative surface in Exuma Sound (Ladd and Sheridan, 1986).

In Columbus Basin there are sigmoidal clinoforms in prograding wedges in the lower Tertiary sequences (Fig. 12). Again, the step back of the banks with the major drowning after mid-Cretaceous time allowed the shedding of detritus off-bank into the prograding wedges. Some of the prograding clinoform features might also be due to erosion and deposition by bottom currents similar to that in the Florida Straits (Figs. 3 and 4).

In the Caicos Basin, there are only two multichannel seismic lines (Fig. 2) (Austin, 1983). At present there is little in the way of drilling, dredging, or submersible sampling to correlate the seismic stratigraphy, but the projection of the seismic data to nearby Hispaniola gives some idea of the geology of the area. The basin itself is underlain by well stratified reflectors (Fig. 13), which most likely represent deep-water ooze, chalk, and limestone similar to that found in the other deep Bahama basins. Very likely these sediments include rocks of Cenozoic and Late Cretaceous age.

The deep-water oozes and limestones are displaced by imbricate north-directed thrust faults and are being uplifted along these imbricate thrusts to form the Hispaniola continental slope (Austin, 1983). The compressional tectonics of Cuba and Hispaniola have penetrated the extreme southern basins of the Baha-

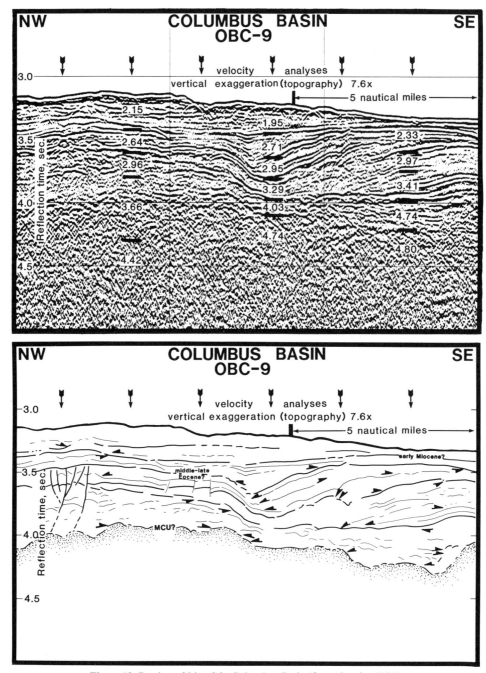

Figure 12. Portion of Line 9 in Columbus Basin (from Austin, 1985).

mas. The orogeny affecting the Greater Antilles lasted from perhaps the mid-Cretaceous through Early Tertiary. Therefore, the involvement of the Caicos Basin section is compatible with the suggested age for the basin sediments of Late Cretaceous through Cenozoic.

Santaren, Nicholas, and Old Bahama Channel

Multichannel seismic reflection data in the deep channels between Great Bahama Bank, Cay Sal Bank, and Cuba (Ball and

others, 1985) show possible interaction between the Bahamas carbonate platform and the Cuban orogenic zone. Meyerhoff and Hatten (1968) present a good summary of the geology of Cuba that involves northward imbricate thrusting. Rocks are largely Mesozoic, including igneous and metamorphic rocks on the south, bordered by volcaniclastics with allochthonous serpentinites and gabbros, with carbonates and evaporites on the north. Flowage of the serpentine and salt northward along thrust zones is indicated. In the northern zone of Cuba, the carbonates are dominantly of Jurassic to Cretaceous age, but some conglome-

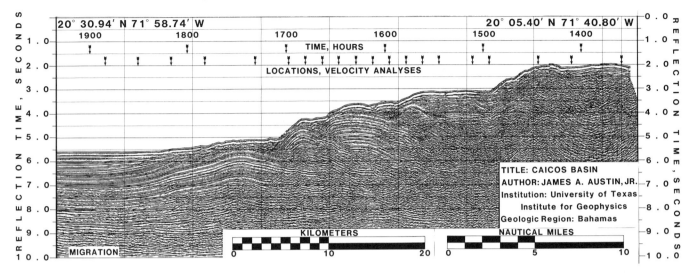

Figure 13. Portion of Line 5 in the Caicos Basin off the northern Hispaniola continental slope. Imbricate thrusting of basin deposits is shown (from Austin, 1983).

ratic limestones are as young as Eocene. These carbonates are thought to have been deposited on shallow banks separated by deep-water tongues and channels. The carbonates are now present as deformed, paraautochthonous rocks in northern Cuba. Some of the Cuban carbonates and other Mesozoic rocks, such as the Middle-Upper Jurassic clastics of the San Cayetano Formation, are in allochthonous thrust slices that have been displaced perhaps hundreds of kilometers north and east. A diapir was interpreted just southeast of Cay Sal Bank from single channel data (Idris, 1975). These diapirs may be an extension of salt diapirs of the Jurassic Punte Allegre Formation along the north coast of Cuba into Old Bahama Channel.

Ball and others (1985) correlate multichannel seismic profiles in Santaren/Nicholas Channels with those of Sheridan and others (1981) some 70 km to the north in the Florida Straits by similarities of reflector depth and character. Above the middle Cretaceous reflector the sequences look similar to those of the Florida Straits and the other deep channels of the Bahamas. Ball and others (1985) interpret these Upper Cretaceous and Tertiary sequences as deep-water carbonates deposited in an off-bank channel.

Tilted blocks, first observed by Idris (1975), are well displayed on new multichannel seismic data (Fig. 14). Ball and others (1985) interpret one of these blocks as a folded, asymmetric anticline that has affected the pre-Tertiary section. Asymmetry in the anticline is apparent in that its north flank has a relief of 1 km, while its south flank has about 500 m (Fig. 14). Ball and others (1985) interpret the discontinuous reflectors on top of the anticline (Fig. 14) as upfolded equivalents of the deep-water upper Cretaceous carbonates to the north of the anticline. A reflectorless, opaque sequence beneath a rough reflector deeper in the anticline is interpreted as Lower Cretaceous shallow water bank carbonates (Fig. 14). Thus, the folding of this anticline has deformed previously continuous flat-lying Upper Cretaceous bas-

inal carbonates that have covered a drowned Lower Cretaceous bank. The flexure apparently was localized at the facies transition of the Lower Cretaceous bank margin. The thinning relationships indicate that the structure was formed near the end of the Cretaceous with maximum relief in the early Tertiary. Younger structural relief might be the result of differential compaction over the bank edge where cemented, harder bank facies give way to porous off-bank facies.

In spite of depth migration of the seismic data (Fig. 14) there remains a troubling crossover of the middle Cretaceous reflector and the dipping overlying Cretaceous sequence (see arrow in Fig. 14). The horizontal event might be the result of reflections from the side out of the plane of the profile or the dipping event might be an internal multiple. The strike of the bank margin and anticline is thought by Ball and others (1985) to be slightly south of west and nearly 90° to the seismic profile, so an oblique crossing is not favored. More work is needed to resolve this problem.

Upper Cretaceous and Cenozoic deep-water carbonates generally onlap and infill the Santaren Channel. On one line there is an abrupt termination of the continuous deep-basinal sequence with a very steep contact. Diffractions and dip changes at the contact led Ball and others (1985) to infer a fault that appears to have affected the entire youngest Cretaceous to middle Cenozoic section.

These faults and anticlines north of Cuba do not resemble the imbricate thrusts found north of Hispaniola (Fig. 13). Rather, they appear to be slightly asymmetrical anticlines and normal faults. Perhaps this difference is due to the presence of a thick, competent Jurassic-Lower Cretaceous carbonate bank near Cay Sal, while less competent, thin-bedded basinal carbonates were deformed off Hispaniola (Ball and others, 1985). Off both Hispaniola and Cuba, deformation is interpreted to affect Cretaceous and Cenozoic rocks, which is compatible with what is known

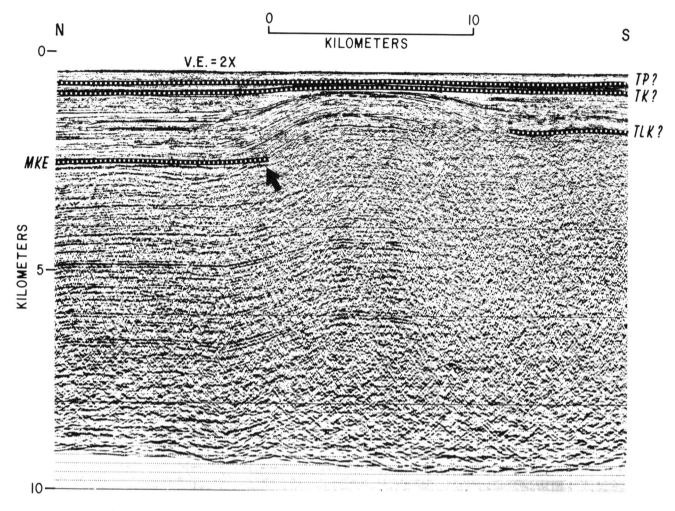

Figure 14. Migrated depth section of a portion of multichannel seismic reflection profile 1 along the axis of Santaren Channel showing the anticline and bank margin structure near the south end of the line. Note arrow pointing to disturbing crossover of reflections (from Ball and others, 1985).

about the Laramide deformation events in Cuba (Meyerhoff and Hatten, 1968). The seismic data suggest that a part of Santaren Channel was a deep-water reentrant into the larger Bahama Platform at least since Early Cretaceous time and that buried shallow platforms of Early Cretaceous age exist under parts of Santaren and Nicholas Channels (Ball and others, 1985).

GRAVITY ANOMALIES

Shipborne data in the deep channel areas and various older marine data on the shallow banks reveal a series of prominent gravity anomalies (Talwani and others, 1960) (Fig. 15) (V. Ewing, personal communication, 1984). Anomaly A, a northwest-southeast trending linear across southern Florida, follows a structural boundary, down-faulted on the south, between Paleozoic crystalline basement under central Florida, and Mesozoic volcaniclastic basement in southern Florida (Fig. 5) (Sheridan and others, 1981).

Anomaly A is abruptly terminated by the north-south trending, linear anomaly B and its continuation anomaly C in the northern Florida Straits. Anomalies B and C reflect partially uncompensated low-density contrasts in the upper 2 km of section, between denser bank carbonates and less consolidated deep-channel deposits. Also, there is apparently a change in basement lithology and perhaps crustal type at this anomaly (Fig. 7) (Sheridan and others, 1981) that must be involved to explain the termination of the deep-sourced anomaly A.

Lower density, possibly unconsolidated deep-water basinal carbonates were deposited under anomaly D (Fig. 15) in a reentrant into the Cretaceous Blake Plateau shallow-water bank from the area of the modern Great Abaco Canyon. This possible deep-water basin may have existed between two shallow-water banks underlain by denser rocks under anomaly E and anomaly F. Perhaps the basin of anomaly D filled in, allowing the coalescence of the two previously isolated banks under E and F to form the modern Little Bahama Bank. Alternatively, a fragment of

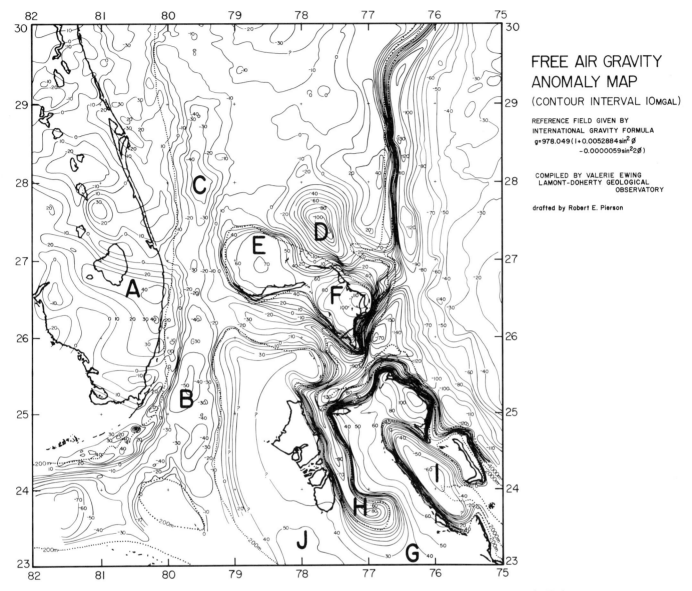

Figure 15. Portions of free-air gravity anomaly map showing northern Bahamas (Valerie Ewing, personal communication, 1984).

lower density igneous crust, such as a continental fragment, might exist under anomaly D. Sheridan and others (1981) present seismic reflection data over the eastern edge of the Blake Plateau showing that a possible splinter of African continental crust could be left under the eastern edge of the Blake Plateau. Such a splinter is explainable by spreading center shifts and plate-tectonic models (Sheridan, 1978).

The sharp termination of anomalies C and D along a northwest-southeast striking line across the southern end of the Blake Plateau is thought to mark a major oceanic fracture zone, called the Jacksonville fracture zone by Klitgord and others (1984), which continues landward of the Great Abaco fracture zone of Sheridan and Osburn (1975).

The large negative free-air anomalies H and I in the deep

channels of Exuma Sound and TOTO could originate in the density contrast between unconsolidated basin deposits and cemented, denser bank carbonates. If so, anomaly G shows a slightly lower positive anomaly under the present bank area to the southeast of TOTO that might be explained as an ancient continuation of this basin.

Another positive anomaly occurs along the southern edge of Great Bahama Bank south of Andros Island (anomaly J). This could be a buried Cretaceous bank, around which Great Bahama Bank coalesced, or alternatively, it might reflect differential uplift and subsidence caused by Cuban tectonism.

All these major anomalies of the Bahamas persist as residual Bouguer anomalies, based on Talwani and others (1960). The removal of the seaward-increasing regional Bouguer anomaly,

TABLE 1.   COMPARISON OF FREE AIR AND RESIDUAL
BOUGUER ANOMALIES*

Anomaly	Free Air (mgls)	Residual Bouguer (mgls)
A	35	30
B	-50	-35
C	-40	-20
D	-100	-70
E	70	20
F	100	10
G	30	-30
H	-80	-40
I	-60	-40
J	50	20

*after Talwani and others (1960)

due to the seaward thinning of the crust under the Bahamas and the shallowing of mantle, leaves pronounced residual Bouguer anomalies as indicated in Table 1.

In general, the bank areas exhibit positive 10–30 mgl anomalies, whereas the channels, and possible buried channels, exhibit -20 to -40 mgl anomalies. These channel anomalies might be explained by 1 to 2 km thick density contrasts between cemented bank rocks and unconsolidated basinal sediments, whereas the smaller bank anomalies could be explained by only 500 m of structural relief uplifting denser basement rock and overlying dense limestones. Both types of structures are observed seismically.

The regional Bouguer anomaly trend across the Bahama Platform increases from 100 to 200 mgls over the Platform itself to more than 300 mgls over the oceanic structure of the Blake-Bahama Basin (Uchupi and others, 1971; Sheridan and Osburn, 1975). Talwani et al., (1960) and Uchupi and others (1971) believe that the 100–200 mgal Bouguer anomaly of the Bahama Platform is caused by a shallower than normal mantle depth, 20–28 km, with a normal density (2.8 g/cc) for a thinner crystalline (continental) crust. On the other hand, Sheridan and Osburn (1975) point out that a crust of intermediate density (3.1 g/cc) under the Bahamas might also cause the observed anomaly. Such a density, 3.1 g/cc, might represent "transitional" continental to oceanic crust, "Icelandic" volcanic crust, a mixture of basic dikes intruded into a fractured and splintered continental crust, or other possible crustal types.

MAGNETIC ANOMALIES

Several publications have reported magnetic data in the Bahamas area (Bracey, 1968; Taylor and others, 1968; Klitgord and Behrendt, 1979; Bryan and others, 1980; and Klitgord and others, 1984). Key features of the airborne and shipborne magnetic data of the area include the steep, short wavelength, high-amplitude anomalies of central Florida that contrast with the long-wavelength, low-amplitude anomalies of the Blake Plateau (Fig. 16). This difference is due to the shallow basement (1–2 km) with Paleozoic igneous rocks under Florida in contrast with

the much deeper (10–12 km) basement and thick sequence of sedimentary rocks under the Blake Plateau (Klitgord and Behrendt, 1979). Presumed Jurassic volcaniclastic basement is shallower (5–6 km) under Little Bahama Bank and Northwest Providence Channel (Sheridan and others, 1981) where the magnetic anomalies are steep and of short wavelength.

The northwest–southeast-striking negative magnetic anomaly across southern Florida corresponds to a positive Bouguer gravity anomaly (A, Fig. 15). South of this anomaly, the magnetic anomalies become longer wavelength and lower amplitude. This reflects the change in basement character (continental to intermediate and oceanic), and the deepening of the basement to more than 6 km (Sheridan and others, 1981; Klitgord and others, 1984). Similarly, anomalies broaden southeast of Little Bahama Bank and Northwest Providence Channel, suggesting that the basement deepens and/or changes character.

In the Blake-Bahama Basin along the base of the Blake Escarpment (Fig. 16), a linear positive anomaly with a northeast-southwest strike is identified as an offset component of the prominent Blake Spur magnetic anomaly (Taylor and others, 1968). This anomaly is offset farther in the area of Great Abaco Canyon where a broader, circular negative anomaly occurs. The strong linear positive continuation of the Blake Spur anomaly is found along the eastern edge of Little Bahama Bank, and it projects into the Bahamas at approximately 77°W.

The Jurassic magnetic quiet zone extends across the Blake-Bahama Basin between the Blake Spur anomaly and the linear marine magnetic anomaly, M25 (Bryan and others, 1980). Projecting the trend of M25 in the Blake-Bahama Basin to the south, it should impinge on the Bahama escarpment just west of San Salvador. However, anomaly M25 is detected 80–90 km east of San Salvador island (Bracey, 1968). This documents the left lateral offset along the Great Abaco fracture zone (Sheridan and Osburn, 1975), also called the Jacksonville fracture zone (Klitgord and others, 1985). A strong linear, southeast trending negative anomaly just seaward of the Bahama Escarpment from southeastward of San Salvador, truncates the M sequence of anomalies north of it (Bracey, 1968). This is the Bahamas fracture zone of Klitgord and others (1984).

The smooth, low amplitude anomalies of the Bahamas contrast markedly with the high amplitude, steeper, short wavelength anomalies of Cuba (Fig. 16) that reflect outcropping igneous and metamorphic rocks and relatively shallow basement. The Cuban anomalies follow the general strike of the Laramide fold belt along the spine of the island. Magnetic lineations and offsets help locate major basement faults and fracture zones (Sheridan, 1974; Sheridan and Osburn, 1975; Klitgord and others, 1984) and indicate the boundaries of the various basement lithologies of the Blake Plateau and central Florida (Fig. 17) (Klitgord and others, 1984).

A fundamental change in crustal type is indicated in southern Florida across the prominent gravity anomaly A (Fig. 15) (Klitgord and others, 1984). This change is interpreted as a thinning of the crystalline crust and a shallowing of mantle from 35

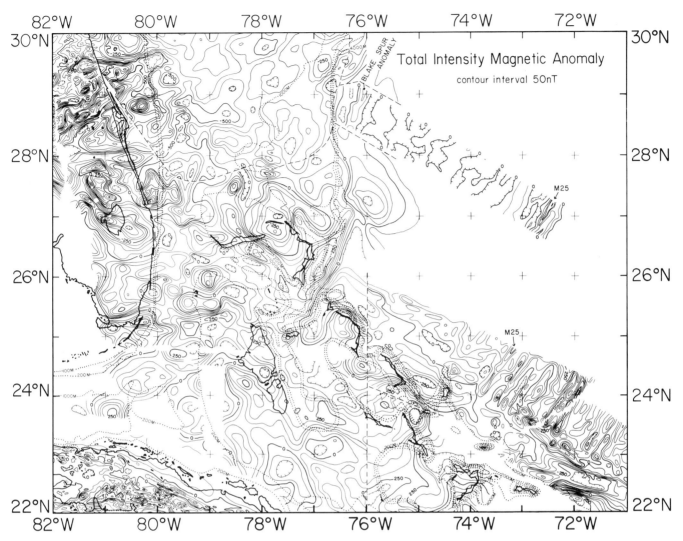

Figure 16. Total intensity magnetic anomaly map of the Bahamas region compiled from the maps of Taylor and others (1968), Bracey (1968), Klitgord and Behrendt (1979), Bryan and others (1980), and Klitgord and others (1984). The dashed lines separate source maps that use different regional datums. No attempt was made to adjust the absolute values of the anomalies.

km to 22 km to compensate for the deeper basement and thicker sediments to the south. This boundary, reflected by both gravity and magnetic anomalies, aligns with the Bahamas fracture zone (Fig. 17). An alternate interpretation is possible, such as one of a denser crust of 3.1 g/cc under the Bahamas (Sheridan and Osburn, 1975), rather than a thinner crust of 2.7–3.0 g/cc density.

The basement type under the Bahamas can be interpreted from seismic reflection data (Sheridan and others, 1981) (Fig. 7). The northwestern Bahamas under the Florida Straits, the Northwest Providence Channel, and a bit of the northern corner of TOTO are underlain by the so-called "intermediate" or "transitional" rift crust, in which Jurassic volcaniclastics are tilted in fault blocks. This crust has a smoother, more circular magnetic anomaly pattern than that of central Florida (Fig. 16). This magnetic and seismic reflection character persists eastward to the projection of the large Blake Spur anomaly in the Northeast Providence Channel (Fig. 17). Here the basement character changes as the tilted, truncated volcaniclastics disappear and only the lower frequency reflector (Middle Jurassic sedimentary rocks) and the hummocky undulating reflector (oceanic crust) are seen (Fig. 7). Using the seismic data, the basement character can be divided in a map view. It is interpreted that transitional crust underlies the northwestern Bahamas to the projected Blake Spur magnetic anomaly, and that oceanic crust underlies the Bahamas farther southeast (Fig. 18).

The nature of the crustal transition from the Bahamas to Cuba is more problematic. Gravity data suggest that the Cuban crust is typical continental density and thickness, an interpretation supported by outcrop of granitic and metamorphic basement. Cuban granitic crust apparently originated in the Mesozoic, and

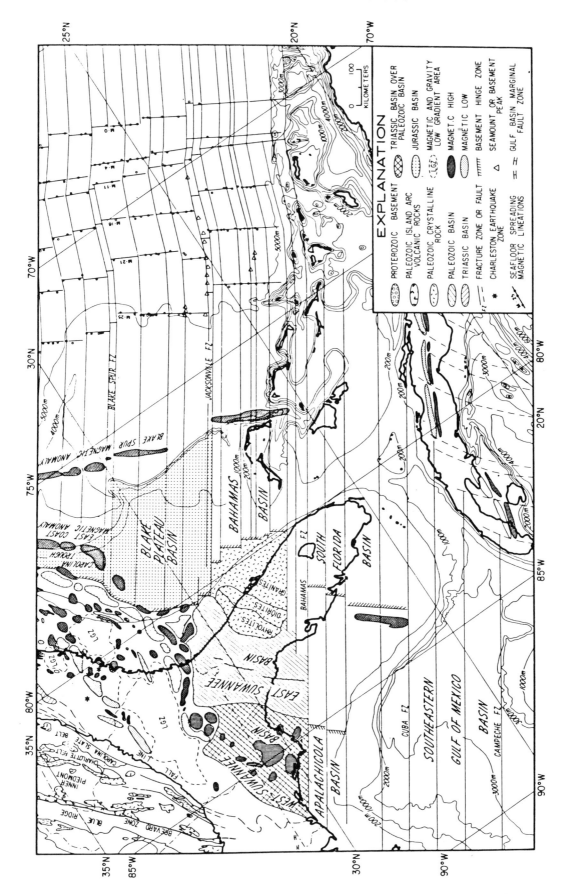

Figure 17. Interpretation based on magnetic anomalies of Atlantic plate fracture zones underlying the Bahamas and southern Florida (from Klitgord and others, 1984).

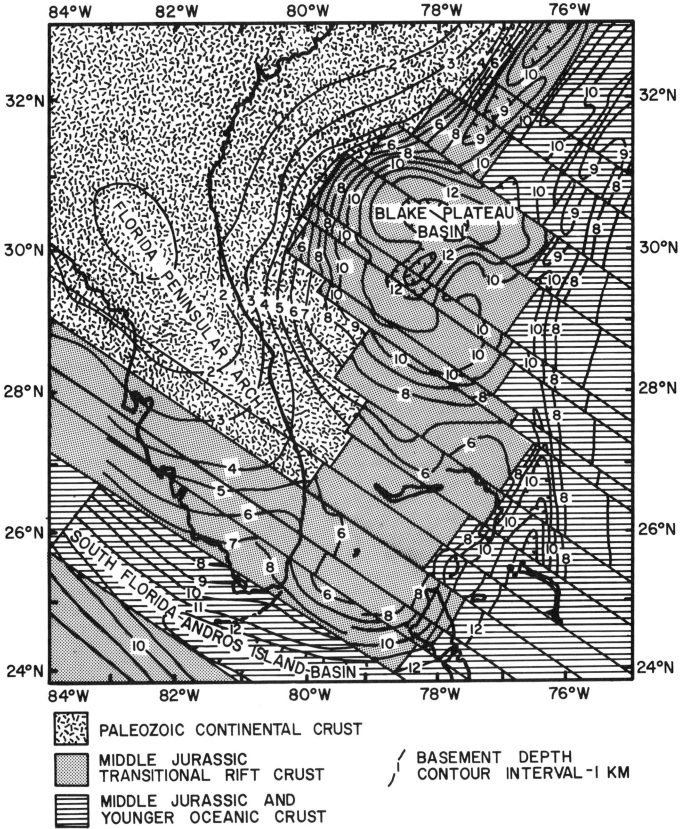

Figure 18. Basement map of the Florida-northern Bahamas region showing the depth in kilometers and type of basement, i.e., continental, oceanic, or transitional, with the possible age ranges.

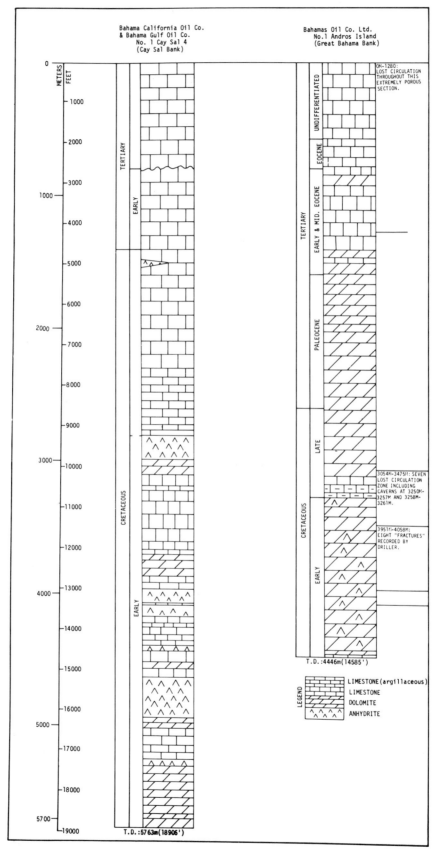

Figure 19. Generalized lithologic logs of the Cay Sal and Andros Island wells (from Paulus, 1972).

was emplaced adjacent to the Bahamas during collision associated with the Laramide Orogeny (Mattson, 1973; Ball and others, 1985).

DEEP STRATIGRAPHY

The deep exploration wells of the Bahamas (Figure 2) include the Cay Sal well (Meyerhoff and Hatten, 1974; Paulus, 1972), the Great Isaac well (Tator and Hatfield, 1975), the Andros Island well (Spencer, 1967; Goodell and Garman, 1969; and Paulus, 1972), and the Long Island well (Meyerhoff and Hatten, 1974). The Cay Sal and Great Isaac wells reportedly penetrated Upper Jurassic limestone, dolomite and evaporite at depths slightly deeper than 5 km. The Andros Island and Long Island wells reportedly bottomed in Lower Cretaceous dolomite (Fig. 19). The Lower Cretaceous sections in the Cay Sal and Andros wells are characterized by several anhydrite units (Paulus, 1972). One explanation of these anhydrites is that during Early Cretaceous time, there existed a broad shallow platform interior with restricted circulation conditions. Consequently, evaporitic events affected an extremely large area. In such an environment, one particular anhydrite unit could cover tens of thousands of square kilometers. The occurrence of the Cay Sal and Andros Lower Cretaceous anhydrites would support the interpretation of a megabank, or wide Bahama Platform during the Early Cretaceous. Moreover, it is compatible with the seismic reflection data now available from the Florida Straits to Exuma Sound, in which Lower Cretaceous and Jurassic seismic reflection horizons of unique shallow water facies character can be traced for hundreds of kilometers.

Deep Sea Drilling Project Site 98 in Northeast Providence Channel (Paulus, 1972) reached a total depth of 3136 m and penetrated deep-water oozes, chalks, and cherts of Tertiary and Late Cretaceous age, thus proving a deep water channel existed at least since Late Cretaceous. By assuming no structural relief between Site 98 and the Andros well where the transition from anhydrite to dolomite coincided with the Early-Late Cretaceous boundary at about 3200 m, Paulus (1972) inferred that a broad Lower Cretaceous shallow water platform was drowned at that time. The disappearance of anhydrite implies segmentation of the megabank into smaller, isolated banks with arms of the ocean reaching well into the interior of what was previously a restricted platform.

The dolomite zones in the Andros and other wells are significant to the sea-level and subsidence history of the Bahamas. Goodell and Garman (1969) equate the dolomite zones in the Andros well to sea-level falls, which could permit dolomitization by reflux action during more evaporitic conditions and shallower water on the banks or even by subaerial exposure and meteoric water effects. The four deep wells have broadly similar total subsidence histories (Schlager and Ginsburg, 1981) (Fig. 20). They exhibit concave upward, exponentially decreasing subsidence curves typical of Atlantic type margins. There are tentative kinks and shifts in the curves between wells based on the present

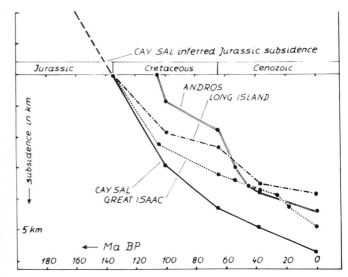

Figure 20. Total subsidence curves for the four deep exploration wells in the Bahamas. It is unclear whether the kinks and differences in the curves reflect vertical tectonic movements or simply errors in age determinations (from Schlager and Ginsburg, 1981).

poor biostratigraphy (Schlager and Ginsburg, 1981), that might reflect tectonic movements on a broad scale. Such deformation would agree with seismic data, which show large post-Cretaceous fault offsets and regional tilts of large blocks causing hundreds of meters of relief (Fig. 7).

Based on many closely spaced samples from the submersible *Alvin,* Freeman-Lynde and others (1981) have constructed a biostratigraphic and sedimentologic column for the depth interval from 2000 to 4000 m on the Bahama Escarpment. Lower Cretaceous rocks occur between 3000 and 4000 m along the Bahamas Escarpment, with "middle Cretaceous" and Upper Cretaceous rocks between 2000 and 3000 m. The bulk of the *Alvin* samples represent restricted platform interior lagoons and tidal flats, whereas only eight horizons contain rudist fragments that indicate platform edge reefs. One important conclusion from this is that the platform-edge reef facies has been largely eroded to expose platform-interior facies.

Controversy about the magnitude and timing of this erosion of the Bahama Escarpment and its northern counterpart, the Blake Escarpment, has arisen. One school favors drastic erosion with up to 25 km of horizontal retreat as late as the Oligocene (Paull and Dillon, 1980, 1981). The alternate view calls for less erosion as a long term continuous process since Cretaceous time (Sheridan, 1981). Freeman-Lynde and others (1981) recovered an eroded middle Cretaceous shallow water limestone with a pelagic drape of latest Cretaceous age, indicating marine erosion of the escarpment as early as Late Cretaceous time. In addition, the *Alvin* section shows that some of the platform-edge facies was indeed preserved. Such an edge facies might only be a kilometer or two in width, acting as a barrier to a quiet lagoon that is only a few kilometers distant from the open ocean, just as occurs in the

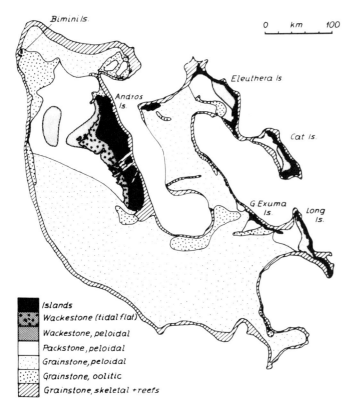

Figure 21. Surface sediment distribution on Great Bahama Bank (from Schlager and Ginsburg, 1981; after Enos, 1974).

Bahamas today. Thus, the magnitude of erosion on the Bahama Escarpment need only be a few km to explain the *Alvin* section.

The dredging along the Bahama Escarpment confirms the *Alvin* work, for the most part. First, Lower Cretaceous shallow-water carbonates generally occur below 3000 m. Second, the bulk of the facies exposed is a platform interior facies (Freeman-Lynde and others, 1981; Corso, 1983). Corso (1983) reported an Early Cretaceous rock from anomalously shallow depths of 2300 to 3300 m on the north face of Abaco knoll. This occurrence could require a fault between the dredge site and Northeast Providence Channel, where Lower Cretaceous shallow water carbonates must be deeper than 3136 m (the T.D. of DSDP 98 which bottomed in Upper Cretaceous chalks). If, however, the dredged rock is of its youngest possible age, Aptian, and if it actually was collected in its deepest possible depth range, 3300 m, no large offset is required. On the other hand, fault offsets of hundreds of meters in this region are not unreasonable.

MODERN CARBONATE SEDIMENTOLOGY

The distribution on the Great Bahama Bank of various kinds and sizes of carbonate particles depends both on their organic origin and on inorganic, physiochemical processes (Fig. 21) (Ginsburg and others, 1963; Enos, 1974; Ginsburg and James, 1974; and Schlager and Ginsburg, 1981). The bank edge is a zone

(1 to 10 km wide) of skeletal sands with reefs and eolianite islands preferentially on the windward side. Corals and mollusk fragments, large Foraminifera, and *Halimeda,* a green algae, are the common particle type.

The largest area of the bank is covered with peloidal sands. These sands are cemented grains ("grapestone"), hardened fecal pellets, and thoroughly micritized particles. Widely scattered patch reefs occur where lithified substrates are available. Oolite sand shoals form locally in areas where island topography focuses currents or at the ends of TOTO and Exuma Sound where large volumes of tidal exchange create high velocity (6 kts) currents (Ball, 1967b; Hine and Mullins, 1983).

The spines of most Bahama islands are large lithified Pleistocene sand dunes that can reach elevations of 100 m (Garrett and Gould, 1984). These structures shelter the banks to the leeward and provide a shadow for quiet water deposition of carbonate silt and clay derived from calcareous algae and shell disintegration (Neumann and Land, 1975). These marine muds grade landward into tidal flats and mangrove swamps (Fig. 21).

Recent shallow coring through the late Pliocene-Pleistocene deposits (Beach and Ginsburg, 1980) has found that while oolites and eolianites occupy the bank edges and peloidal sands occupy the bank interior in the Pleistocene-Holocene, in the Pliocene skeletal grains and reefs dominate the edges on both windward and leeward sides, and the bank interior is dominated by skeletal sands. Schlager and Ginsburg (1981) interpret the Pliocene picture as a reef-encircled atoll structure, but this is not totally proven (Mullins and others, 1982b). This transition from the Pliocene to the Pleistocene may be related to the onset of northern polar glaciation and frequent sea-level changes (Schlager and Ginsburg, 1981).

During sea-level lowstands of the Pleistocene, the exposed bank edge prevented off-bank sand and mud transport, even on the lee side of the bank. Also, during exposure of the bank, the coldwater from the lagoon ceased to drift off the lee side. This allowed fringing reefs to grow (Hine and Neumann, 1977; Hine and others, 1981). With sea-level rise, off-bank sediment transport buried the reef on the leeward side. Consequently, sea-level fluctuations are recorded on the margins of the banks as a welding-on of a fringing reef facies, which in turn causes the lateral and upward accretion of the marginal escarpment (James and Ginsburg, 1979).

Shallow coring has shown that the Pleistocene-Holocene deposits are 24 m thick on Little Bahama Bank and 40 m on Great Bahama Bank (Beach and Ginsburg, 1980). Schlager and Ginsburg (1981) suggest that this difference reflects neotectonism in the Bahamas, with differential subsidence going on between individual banks. Given the presence of post-Cretaceous faults and tilts of major seismic surfaces in the seismic reflection data, differential subsidence continuing into the Holocene is plausible.

The distribution pattern and origin of the deep-water carbonate sediments in the areas surrounding Little Bahama Bank are well documented (Mullins and Neumann, 1979; Cook and Mullins, 1983) (Fig. 22). Off open ocean bank margins, the slope

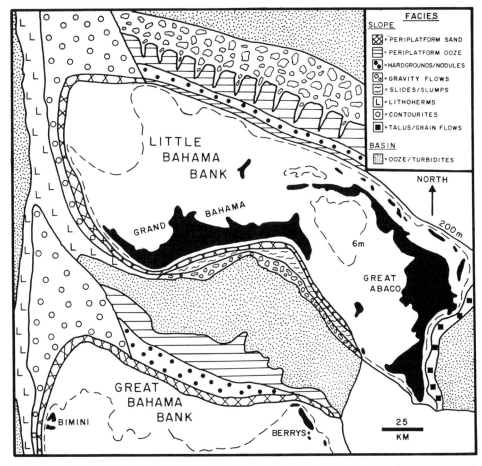

Figure 22. Near-surface sediment distribution map for the deep-water depositional environments around the northern Bahamas (Based on Mullins, 1978; Mullins and Neumann, 1979; from Mullins, 1983).

facies are depth-related zones of periplatform sand, periplatform ooze, gravity-flow deposits, and, finally, interbedded basinal oozes and turbidites. Where along-slope currents prevail, winnowing and cementation may produce hardgrounds and nodules (Mullins and others, 1980a), and carbonate sands accumulate as thick contourite drifts (Mullins and others, 1980b) upon which surficial biogenic lithoherms grow (Neumann and others, 1977). In deep closed-seaway basins like Exuma Sound and TOTO, bottom currents play a reduced role. The facies zonation is again parallel to the margins and dominated by the gravity-transport processes of debris flow and turbidity currents, (Schlager and Chermak, 1979; Crevello and Schlager, 1980). Basinal and periplatform oozes are interlayered.

The only samples of these modern deep-water carbonates are from piston cores. The available samples generally range in age back through part of the Pleistocene (~500,000 years). Unconsolidated periplatform ooze is unbedded, homogeneous fine-grained sediment. Bottom photographs from the ooze areas show that fine-grained muds are being bioturbated, which may explain their seemingly homogeneous nature (Mullins, 1978). Eventually, however, the periplatform ooze does become stratified visually as

well as acoustically. Ooze has a continuous, planar, parallel reflector, and transparent character on seismic profiles (Fig. 8) (Mullins and others, 1979). Where Pleistocene ooze crops out in canyon walls, *Alvin* photographs show good lamination on centimeter and meter scale (Mullins and others, 1982a). The processes of burial and compaction over time with preferential solution and cementation of various oozes of different composition (aragonitic vs. calcitic) apparently leads to the "secondary" bedding in what otherwise appears as homogeneous sediment in fresh cores (Cook and Mullins, 1983).

Periplatform ooze (Schlager and James, 1978) is composed of planktonic coccolith and foraminiferal shells composed of calcite mixed with bank-derived debris such as needles of aragonite and magnesian calcite (Heath and Mullins, 1984). Bank derived aragonite and magnesian calcite components in the ooze decrease with distance from the banks (Heath and Mullins, 1984). Thus during times of increased off-bank transport, percentages of aragonite and magnesian calcite should increase in the periplatform ooze. Pronounced, cyclic aragonite/calcite variations in cores of periplatform ooze are related to sea-level and climate fluctuations (Pilkey and Rucker, 1966; Kier and Pilkey, 1971; Lynts and others, 1973; Droxler and others, 1983). Where greater percentages of calcite occur, normal marine diagenesis has caused preferential cementation, and hence an acoustic layering related to aragonite/calcite variation.

Gravity transport processes (debris flows, grain flows, and turbidity currents) deliver the bulk of carbonate sediments to the deep-water margins and basins of the banks. Bouma (1962) sequences are diagnostic of the carbonate turbidites in piston cores. Several studies in TOTO (Schlager and Chermak, 1979), in Exuma Sound (Crevello and Schlager, 1980), and in Columbus Basin (Bornhold and Pilkey, 1971) employed a close grid of piston cores that permitted the identification of individual turbidite layers produced by distinct flow events. Individual flows from individual canyons as point sources coalesce so the pattern of turbidites persists laterally parallel to the slope, as if a broad apron were deposited from a continuous line source (Schlager and Chermak, 1979; Mullins and others, 1984).

Generally, debris flows are confined to the base of the slope and grade seaward into fine-grained distal turbidites. However, where there is enough bank-margin relief, source area, triggering events, and efficient channeling systems, these debris flow deposits can be found hundreds of kilometers from the banks (Crevello and Schlager, 1980; Bliefnick and others, 1983). A buried deep-sea fan, called the buried Eleuthera Ridge (Fig. 1), is thought to have acted as a conduit for the debris flows, keeping them confined in one of many distributary channels on the fan until the flow reached well out in the Blake-Bahama Basin near DSDP Sites 391 and 534 (Bliefnick and others, 1983).

The possible mechanisms for the generation of gravity flows from the shallow banks include storm waves, slumps, earthquake shocks, and tsunamis. In the Bahamas, no large, damaging earthquakes have occurred in historical times. However, young post-Cretaceous and post-Tertiary faults seen in seismic lines might be reactivated under modern intraplate stresses. Moreover, the occurrence of large earthquakes elsewhere along the Atlantic margin, such as the Charleston earthquake in 1886 and the Grand Banks earthquake in 1929, does suggest that such earthquakes may occur occasionally in the modern Bahamas. Nonetheless the mechanisms of storm waves and slumps due to gravitational instability on the slopes appear to be the most common cause for modern debris flows. During the Pleistocene, with lowering of sea-level below the marginal escarpments of the banks and with harsher climates and larger ocean waves, the setting would have been more favorable for disturbance along the bank margins. With rapid rises in sea-level as glaciers melted, loose sediments rapidly produced on the banks could have been more easily washed downslope, perhaps entraining unconsolidated slope deposits en route to deeper basins.

The nodules of in situ cemented carbonate are confined to a narrow zone parallel to the slope off Little Bahama Bank (Fig. 22) (Mullins and others, 1980b). Mapping by Mullins (1983) shows that these nodular zones are confined under areas where the Antilles Current and an eastward directed spin-off from the Florida Current flow. These warmer water currents

reaching bottom along these slopes might preferentially winnow and cause cementation of the otherwise normal periplatform sediments.

Where the Antilles Current and Florida Current converge (at the northwest corners of Little Bahama Bank and Great Bahama Bank), the lack of constraint of the bank margins causes deceleration and deposition of a thick pile of carbonate sand as contourite drifts (Fig. 22). The contourites form sediment drifts and downlap on a flat reflector, probably of Miocene age (Mullins and Neumann, 1979; Sheridan and others, 1981). On the flanks of these drifts are surficial biohermal buildups (lithoherms) that are mounds of ahermatypic corals facing into the same currents forming the contourite deposits (Neumann and others, 1977).

GEOLOGIC HISTORY

The Bahama crust formed in conjunction with the rifting and drifting of the Atlantic continental margin and the Gulf of Mexico. The early history of the Atlantic constrains that of the Bahamas. Drilling at DSDP Site 534 has given a more definite age on the Blake Spur anomaly as basal Callovian, latest Middle Jurassic (Sheridan, 1983b; Sheridan and Gradstein and others, 1982). This age is some 20 m.y. younger than previously thought (Vogt and Einwich, 1979). This requires that the Jurassic quiet zone had a much higher spreading rate than previously thought (3.8 vs. 1.5 cm/year) (Sheridan, 1983b).

The crustal corridor between the Blake Spur anomaly and the East Coast magnetic anomaly north of the Blake Spur fracture zone (Fig. 17) (Klitgord and others, 1984) is thought to be oceanic in origin and character (Sheridan and others, 1979). This corridor (Fig. 17) is asymmetric with respect to the rest of the central Atlantic and does not have an equivalent oceanic corridor off west Africa (Vogt, 1973; Klitgord and Behrendt, 1979). One explanation of this is that a spreading center shift in the Callovian occurred to form the Blake Spur anomaly, leaving an extinct, double-sided oceanic spreading center in the East Coast-Blake Spur corridor that might thus be called the proto-Atlantic Ocean (Fig. 23 and 24) (Sheridan, 1983b). Extrapolation of the faster spreading rates of the Jurassic quiet zone to the East Coast magnetic anomaly yields an age of middle Bathonian for the ECMA (Sheridan, 1983b; Gradstein and Sheridan, 1983). This implies that the East Coast-Blake Spur corridor formed in only a few million years at double spreading rates of 7.6 cm/yr.

Revised ages for the East Coast (Bathonian) and Blake Spur (Callovian) anomalies (Sheridan, 1983b) agree with ages recently proposed for breakup events in the Gulf of Mexico (Anderson and Schmidt, 1983). Breakup in the Gulf of Mexico began in Bathonian, and a major Callovian shift of the spreading center to the east side of Yucatan Peninsula formed the Jurassic Caribbean crust (which has subsequently been subducted) (Figs. 23 and 24). This implies that the proto-Atlantic and the oceanic Gulf of Mexico are the same age and simplifies the model for the opening of the Gulf of Mexico and the Central Atlantic. South America,

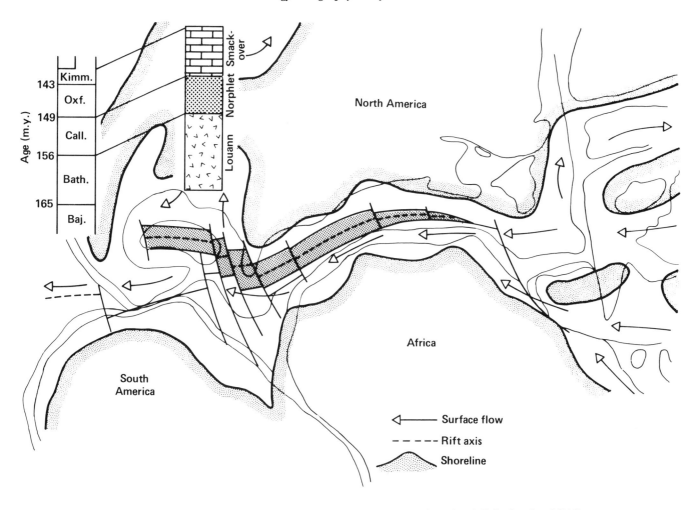

Figure 23. Continental reconstructions at Blake Spur anomaly time (basal Callovian, late Middle Jurassic). Land areas are shown in contrast to epicontinental seas. The proto-Atlantic crust along the eastern North American margin (shaded) is shown to continue into the small oceanic center of the Gulf of Mexico. Surface currents, driven by the Trade Winds, flowed to the west in the near equatorially oriented proto-Atlantic. Also shown is a stratigraphic column for the Gulf of Mexico showing the Louann salt of Bathonian age. The Gulf of Mexico drifting occurred after the Louann deposition, and is now thought to have occurred quickly in the Bathonian, as did the spreading of the proto-Atlantic (from Sheridan, 1983b).

with Yucatan attached, drifted from the place where the oceanic Gulf of Mexico now exists. It was assumed previously that the Atlantic opened first in Early Jurassic, while the Gulf remained closed, and that the Gulf opened later in the Middle Jurassic (Salvador and Green, 1980).

The transitional rift crust under the northwest Bahamas must be of Bathonian age equivalent to the East Coast-Blake Spur oceanic corridor and the oceanic crust in the Gulf of Mexico. Most of the transitional crust is overlain by Upper Jurassic rocks, but possible Callovian sedimentary rocks onlap the rift crust on the eastern side (Fig. 14). Seismic and drilling data indicate that most of the Bathonian rift crust was not covered until Late Jurassic time, so it may have been subaerially exposed, perhaps as a peninsula or isolated islands (Fig. 24). Volcaniclastics were de-

posited in this environment and were tilted during or soon after deposition.

As the Bahama crust accreted into an area equivalent to the East Coast-Blake Spur corridor, any pre-existing continental crust splintered and was injected with igneous rocks similar to the basalts and rhyolites of Jurassic age found under southern Florida. One could visualize a spreading crust with volcanic extrusions building a nearly emergent edifice. An example of such an environment today would be the Afar triangle, where subaerial volcanics exist even at elevations below sea-level. With sea-floor spreading, the large outpouring of volcanics in the Bahama area may have developed a shallow plateau of oceanic basalts similar to Iceland. As in the Iceland-Faeroe Ridge structure, the Bahama volcanic crust could have formed a shallow, linear ridge feature

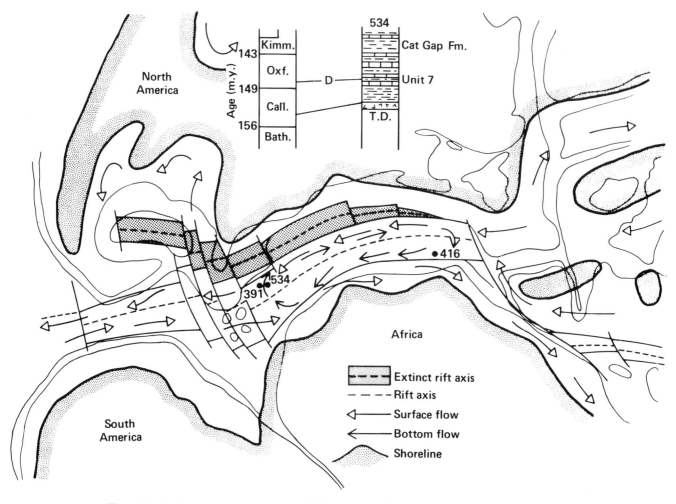

Figure 24. Continental reconstructions at middle Oxfordian time (early Late Jurassic). Note that the extinct spreading center of the Bathonian proto-Atlantic is isolated and attached to the North American plate. After the basal Callovian spreading center shift to the Blake Spur anomaly and southeast Yucatan margin, the modern North Atlantic and Jurassic Caribbean were formed. Hypothetical surface and bottom current flows are shown (from Sheridan, 1983b).

parallel to the general Bahamas fracture zone flow lines (Figs. 17 and 18), similar to the volcanic basement proposed by Dietz and others (1970). Such a volcanic ridge could have formed an effective barrier to circulation between the Bathonian proto-Atlantic and Gulf of Mexico.

After the spreading-center shift in basal Callovian to the Blake Spur magnetic anomaly-Yucatan margin, more oceanic crust continued to form under the accreting Bahama Platform. Callovian drift-phase sediments first onlapped the subsiding transitional crust along its southeast border. Transgression reached southern Florida in Late Jurassic and central Florida by Early Cretaceous time.

Carbonate platform buildup accompanied subsidence of the Bahamas from Late Jurassic onward. The carbonates formed a broad platform, or megabank that included the west Florida shelf, the Florida Platform, the Bahama Platform and the Blake Plateau

(Emery and Uchupi, 1972; Paulus, 1972; Meyerhoff and Hatten, 1974). Presently, drilling and seismic studies of the Bahamas are too limited to document the continuous nature of the Lower Cretaceous megabank. Evidence for deep water reentrants into the megabank in Early Cretaceous time exists in various localities (Schlager and others, 1984; Ball and others, 1985). However, in most of the area surveyed under the present channels and basins, there appears to be Lower Cretaceous shallow-water carbonate facies, which implies that the present channels and basins were not existent at that time.

Bahama deep-water channels and basins appear to have existed generally in their present positions since at least the Late Cretaceous. Drowning of the Blake Plateau also occurred by Late Cretaceous. This resulted in isolation of the Bahama Banks in their present form and allowed access of through-going currents as part of the overall Atlantic and Gulf of Mexico circulation. The

Florida Current of the Gulf Stream began in the Late Cretaceous or Early Tertiary (Sheridan and others, 1981; Crosby, 1980; Paull and Dillon, 1979; Emery and Uchupi, 1972).

Post-Cretaceous faulting resulted in large scale (>500 m) throw and consequent tilting of blocks. The linearity of the bank margins and the alignment of some bank margins with magnetic anomalies and fracture zones have suggested basement fault control of the banks and channels (Ball, 1967a; Moody, 1973; Emery and Uchupi, 1972; Sheridan, 1974; Mullins and Lynts, 1977). Fault control must have been active during the Late Cretaceous, perhaps producing enough small scale sea-floor relief to locate the present bank margins initially, with subsequent depositional-erosional processes then creating the final form and relief of the isolated banks (Sheridan, 1971; 1974). Mullins and Lynts (1977), in contrast, viewed the channels as bounded by faults as old as Jurassic, and that the bank/channel relief was created in its present configuration at that time. The new seismic data suggesting shallow water limestones under the modern channels appear to discount a pre-Late Cretaceous age for large parts of the existing channels.

To explain this late faulting on an otherwise passive Atlantic margin, Sheridan and others (1981) suggest that the Bahama faults were caused by interaction between the Caribbean and North American plates in the Cuban and Antillan orogenies during the Late Cretaceous and Tertiary. Note that seismic data in Old Bahama Channel and Caicos Basin show deformation along the Cuban-Bahamas boundary (Austin, 1983; Ball and others, 1985).

The orientations of Bahama Bank margins seem to fit the preferred directions of fault and fold axes that would result from regional left-lateral shearing (Mullins, 1984; Mullins and Sheridan, 1983). Moody (1973) also included the Bahama Bank margins in a regional wrench fault pattern. The main wrench fault axis for such left-lateral slip would be along the trend of postulated Jurassic transform faults (Klitgord and others, 1984) (Fig. 17). Whatever stresses caused the post-Cretaceous deformation, faulting reactivated older Jurassic lines of weakness.

The stress regime that would produce left-lateral shear in the Bahamas in post-Cretaceous time is compatible with the interaction of the North American and Caribbean plates in oblique subduction (Fig. 25). Thrusting in northern Cuba and the Caicos Basin is towards the northeast, suggesting subduction to the south (Fig. 25). Malfait and Dinkleman (1972), however, indicated subduction to the north. In this latter model the Jurassic Caribbean crust could be destroyed in the subduction zone along a Cuban trench just south of the island. The apparent contradiction between the two models may have been resolved by Mattson's (1979) suggestion of a mid-Cretaceous switch in direction of subduction; northward subduction occurred first followed by southward subduction. This would explain the age and character of Lower Cretaceous serpentinites, ophiolites and metamorphism in Cuba, and the vergence of the later Cretaceous and Tertiary thrusts.

The mid-Cretaceous timing of this important tectonic

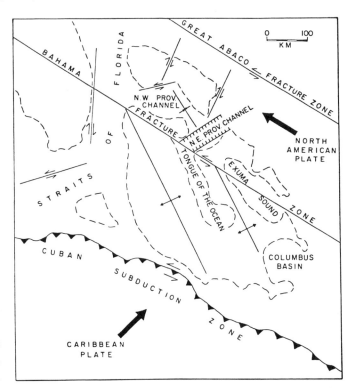

Figure 25. Hypothetical tectonic model for the Bahama Platform resulting from left-lateral wrench faulting caused by the oblique subduction in Cuba of the North American plate under the Caribbean plate. Note that Tongue of the Ocean and Exuma Sound are formed by synclinal warping. (from Mullins and Sheridan, 1983).

change affecting the Bahamas agrees with Ladd's (1976) conclusion that convergence around the rim of the Caribbean began with the separation of South America from Africa. This separation accelerated greatly in mid-Cretaceous time. Malfait and Dinkleman (1972) also indicate Late Cretaceous subduction in Cuba.

During the early Tertiary carbonate sediments prograded along the flanks of the isolated banks into the adjacent deep channels of the Florida Straits, Exuma Sound, and TOTO. Several events of rapid relative sea level rise followed by near still-stands could create these prograding slopes (Vail and others, 1977b). Seismic data show that these prograding processes have led to the filling of pre-existing Lower Tertiary channels and their eventual burial by deep marine carbonates (Fig. 9 and plate 6). Overlap by shallow-water carbonates over a filled deep-channel and the consequent lateral expansion and coalescence of the banks has been suggested (Ball, 1967a).

Bank drowning continued in the Tertiary as the Bahama Platform subsided. In some cases, such as Abaco Knoll, the shallow-water banks persisted until the Latest Cretaceous before drowning. Similar deep knolls and flat topped terraces in the Bahamas are likely remnants of shallow-water carbonate banks that "kept up" with relative sea-level rise for a time, and then "gave up" in the Tertiary (Kendall and Schlager, 1981).

In the later Tertiary, possible acceleration of the Florida Current and Antilles Current influenced the development of the contourite drifts in the Florida Straits (Mullins and others, 1980a). The general later Tertiary cooling and harshening of the global climate is a possible cause for the current acceleration. Additionally, development of the Panama Isthmus by tectono/volcanic processes at about this time may have confined Trade-Wind driven currents and forced more water into the Gulf of Mexico, thus accelerating the Florida Current (Mullins and Neumann, 1979; Mullins and others, 1980a).

During the Quaternary, with generally more frequent and larger sea-level fluctuations and stronger winds, the Tertiary atoll-like configuration of the banks changed to the more asymmetric banks bounded by windward islands that we see today. Sea-level stands higher than the present existed in the Bahamas in the Pleistocene, despite the continued slow tectonic subsidence. At about 125,000 yrs. B.P. coral reefs and beach deposits formed at 5 m above present sea-level on many islands (Neumann and Moore, 1975). Several stages of Late Pleistocene and Holocene eolianite deposits form the dunes of the major islands (Garrett and Gould, 1984).

GEOLOGICAL RESOURCES

There are few proven geological resources in the Bahamas, but the potential for petroleum must be considered. Only 4 deep exploration wells, all dry, have ever been drilled in the Bahamas. A great deal of cavernous porosity was found in the Andros Island well (Spencer, 1967; Meyerhoff and Hatten, 1974). The entire 4.45 km of the Andros Island well section was flushed with sea water. This and a similar occurrence in the Long Island well led Meyerhoff and Hatten (1974) and Tator and Hatfield (1976) to conclude that the seaward margin of the Bahamas is flushed to all reasonable drilling depths and is, therefore, nonprospective for petroleum. The nearness to the erosional Bahama Escarpment, the general history of erosion, and the reentrants cutting deep into the platform that allow sea water to penetrate the individual bank edges even as they evolved, are problems. However, sea water circulation was encountered in the Great Isaac Island well only in the upper 2000 m. In the older section of Lower Cretaceous and Jurassic rocks, both sealed conditions and structural closure were encountered. Anhydrites present in this part of the section, similar to those in Cay Sal and in southern Florida, appear to form effective seals. Thus, this more platform interior facies may be more favorable for petroleum accumulation.

Structural and stratigraphic traps are abundant (Sheridan and others, 1981; Mullins and others, 1978). Seismic data show faults, regional tilting, truncations of tilted beds beneath angular unconformities, and large scale structural relief of the basement between the Blake Plateau Basin on the north and the South Florida-Bahamas Basin on the south. A large number of the faults and regional warpings are of post-Cretaceous age. These young structures may not have been present, however, before the maturation, primary expulsion and migration of any petroleum. In the Bahamas and the Atlantic margin in general, for reasons dis-

cussed below, maturation and origin of petroleum were probably favored prior to Late Cretaceous.

Traps are possible beneath the breakup unconformity where porous volcaniclastics are onlapped by tight Jurassic limestones and anhydrites. The Great Isaac well tested such a possibility and found oil and gas shows with pressures that indicate closure (Tator and Hatfield, 1975). Stratigraphic traps are possible in the Jurassic and Early Cretaceous carbonates well within the larger megabank away from the Atlantic margin. One example is the Sunniland trend in Florida where several tens of millions of barrels of oil are being produced from a reef trend (Meyerhoff and Hatten, 1974).

Little obvious source rock was encountered in the four wells drilled in the Bahamas; there are no occurrences of thick, black organic shales. Some shales were reported and gray, fine grained lagoonal limestones might have source-rock potential. Black shales do occur in the DSDP wells drilled seaward of the Bahamas in the Blake-Bahama Basin (Sheridan and others, 1983), but they are immature and have not acted as source rock. If the equivalents of these DSDP Lower Cretaceous and Jurassic black shales do lie buried somewhere under the Bahamas or the Blake Plateau, there may well have been significant source rock present. Little of the Lower Cretaceous rock under the Bahamas has actually been drilled, and the Middle Jurassic rocks mapped seismically (Sheridan and others, 1981) have never been drilled. Middle Jurassic organic shales were encountered in DSDP Site 534. The occurrence of oil in Florida argues that source rocks may be available somewhere in the similar carbonate environments of the Bahamas. Long range migration might also play a role, in that the Bahama area straddles the structural divide between the Blake Plateau and South Florida basins.

The thermal history of the Bahamas and the timing of petroleum generation is a concern. Thermal gradients in the Andros well were 2.0°C/100 m, reflecting normal heat flow for the area. Thermal gradients taken in the DSDP Sites 533 and 534 off the Bahamas were 3.0-4.0°C/100 m, reflecting normal heat flow for the North American plate of Jurassic age (Sheridan and others, 1983). Landward of the Bahama Escarpment, the Bahama Platform probably underwent a thermal history similar to the rest of the Atlantic margin. Lower Cretaceous and Jurassic rocks are likely to have undergone thermal maturation, while the Upper Cretaceous and younger rocks have not. This means that the Lower Cretaceous and Jurassic rocks are more prospective.

However, the immature Upper Cretaceous and Tertiary rocks might act as reservoirs and traps for petroleum that has migrated upward from the older mature rocks, as in the Gulf of Mexico and the Atlantic margin. The post-Cretaceous structures of the Bahamas have potential, but less than the older rocks and structures.

Large petroleum reserves are found in carbonate rocks analogous to those of the Bahamas, such as the Golden Lane and Poza Rica trends in the Gulf of Mexico (Enos, 1977). Four billion barrels of oil are proven in two distinct occurrences: a long, linear, narrow platform edge facies with secondary porosity and a linear, broad, base of slope, debris flow apron facies. These trends parallel one another, and fine grained facies act as possible seals (Enos, 1977; Mullins, 1983). Such a petroleum reserve might someday be found under the Bahamas.

Other geological resources of the Bahamas include lime-stone used locally for building stone, and oolitic aragonite exported for fine lime. The Department of Lands and Surveys of the Bahamas reports $7.7 million for the value of the aragonite tonnage shipped in 1982.

REFERENCES CITED

Agassiz, A., 1888, Three cruises of the United States Coast and Geodetic Survey steamer "Blake": Harvard Museum of Comparative Zoology Bulletin, v. 14, 314 p. and v. 15, 200 p.

——1894, Reconnaissance of the Bahamas and of the elevated reefs of Cuba in the steam yacht *Wild Duck* January to April 1893: Harvard Museum of Comparative Zoology Bulletin, v. 26, no. 1, p. 1–203.

Anderson, T. H., and Schmidt, V. A., 1983, The evolution of Middle America and the Gulf of Mexico-Caribbean Sea region during Mesozoic time: Geological Society of America Bulletin, v. 94, p. 941–966.

Andrews, J. E., 1970, Structure and sedimentary development of the outer channel of the Great Bahama Canyon: Geological Society of America Bulletin, v. 81, p. 217–226.

Andrews, J. E., Shepard, F. P., and Hurley, R. J., 1970, Great Bahama Canyon: Geological Society of America Bulletin, v. 81, p. 1061–1078.

Applin, P. C., and Applin, E. R., 1965, The Comanche Series and associated rocks in the subsurface in central and southern Florida: U.S. Geological Survey Professional Paper 447, 82 p.

Athearn, W. D., 1962, Bathymetric and sediment survey of the Tongue of the Ocean, Bahamas, I. Bathymetry and sediments: Report submitted to Naval Underwater Ordinance Station, no. 62-65, p. 1–17.

——1963, Bathymetry of the Straits of Florida and the Bahama Islands. II. Bathymetry of the Tongue of the Ocean, Bahamas; Marine Science Bulletin, v. 12, p. 365–377.

Austin, J. A., Jr., 1983, OBC 5-A: Overthrusting in a deep-water carbonate terrane, *in* Bally, A. W., ed.: AAPG Studies in Geology, v. 3, No. 15, p. 3.4.2-167–172.

——1985, Seismic Reflection Survey of Columbus Basin: Bahamas: (Manuscript in preparation).

Ball, M. M., 1967a, Tectonic control of the configuration of the Florida-Bahama Platform: Gulf Coast Association of Geological Societies Transactions, v. 17, p. 265–267.

——1967b, Carbonate sand bodies of Florida and the Bahamas: Journal of Sedimentary Petrology, v. 37, p. 556–591.

Ball, M. M., Dash, B. P., Harrison, C.G.A., and Ahmed, K. O.,1971, Refraction seismic measurements in the northeastern Bahamas (abstract); EOS, Transactions, American Geophysical Union, v. 52, p. 252.

Ball, M. M., Gaudet, R. M., and Leist, G., 1968, Sparker reflection seismic measurements in Exuma Sound, Bahamas (abstract): EOS, Transactions, American Geophysical Union, v. 49, p. 196.

Ball, M. M., and eight others, 1985, Seismic structure and stratigraphy of northern edge of Bahaman-Cuban collision zone: American Association of Petroleum Geologists Bulletin, v. 69, p. 1275–1294.

Beach, D. K., and Ginsburg, R. N., 1980, Facies succession, Plio-Pleistocene carbonate, northwestern Great Bahama Bank: American Association of Petroleum Geologists Bulletin, v. 64, p. 1634–1642.

Bennetts, K., and Pilkey, O. H., 1976, Characteristics of three turbidites, Hispaniola-Caicos Basin: Geological Society of America Bulletin, v. 87, p. 1291–1300.

Bliefnick, D. M., Robertson, A.H.F., and Sheridan, R. E., 1983, Deposition and Provenance of Miocene intraclastic chalks, Blake-Bahama Basin, western North Atlantic, *in* Sheridan, R. E., Gradstein, F. M., eds., Initial Reports of Deep Sea Drilling Project, Vol. 76: U.S. Government Printing Office, Washington, D.C., p. 727–748.

Bornhold, P., and Pilkey, O. H., 1971, Bioclastic turbidite sedimentation in

Columbus Basin: Geological Society of America Bulletin, v. 82, p. 1314–1354.

Bouma, A. H., 1962, Sedimentology of some flysch deposits: Amsterdam, Elsevier, 168 p.

Bracey, D. R., 1968, Structural implications of magnetic anomalies north of the Bahamas-Antilles Islands: Geophysics, v. 33, p. 950–961.

Bryan, G. M., Markl, R. G., and Sheridan, R. E., 1980, IPOD site surveys in the Blake-Bahama Basin: Marine Geology, v. 35, p. 43–63.

Buchanan, H., 1970, Environmental stratigraphy of Holocene carbonate sediments near Frazers, Hog Cay, B.W.I. [Ph.D. thesis]: Columbia University 229 p.

Bullard, E. C., Everett, J. E., and Smith, A. G., 1965, The fit of the continents around the Atlantic: Philosophical Transactions of the Royal Society of London, ser. A, v. 258, p. 41–51.

Cook, H. E., and Mullins, H. T., 1983, Basin margin environment, *in* Scholle, P. A., Bebout, D. G., and Moore, C. H., eds., Carbonate Depositional Environments: American Association of Petroleum Geologists Memoir 33, p. 539–617.

Corso, W., 1983, Sedimentology of rock dredges from Bahamian Platform slopes [Masters thesis]: University of Miami, 75 p.

Crevello, P., and Schlager, W., 1980, Carbonate debris sheets and turbidites, Exuma Sound, Bahamas: Journal of Sedimentary Petrology, v. 50, p. 1121–1148.

Crosby, J. T., 1980, Stratigraphy beneath and geologic origin of the northern Florida Straits from recent multichannel seismic reflection data [Masters thesis]: University of Delaware, 212 p.

Dietz, R. S., and Holden, J. C., 1971, Geotectonic evolution and subsidence of Bahama Platform: Reply: Geological Society of America Bulletin, v. 82, p. 809–812.

Dietz, R. S., Holden, J. C., and Sproll, W. P., 1970, Geotectonic evolution and subsidence of Bahama Platform: Geological Society of America Bulletin, v. 81, p. 1915–1928.

Dillon, W. P., Paull, C. K., Buffler, R. T., and Fail, J. P., 1979, Structure and development of southeast Georgia Embayment and northern Blake Plateau: Preliminary analysis: American Association of Petroleum Geologists Memoir 29, p.27–41.

Drake, C. L., Heirtzler, J. R., and Hirshman, J., 1963, Magnetic anomalies off eastern North America: Journal of Geophysical Research, v. 68, p. 5259–5275.

Droxler, A. W., Schlager, W., and Whallon, C. C., 1983, Quaternary aragonite cycles and oxygen-isotope record in Bahamian carbonate ooze: Geology, v. 11, p. 235–239.

Emery, K. O., and Uchupi, E., 1972, Western North Atlantic Ocean: Topography, rocks, structure, water, life, and sediments: American Association of Petroleum Geologists Memoir 17, 532 p.

Enos, P., 1974, Rock, platforms, and basins of Middle Cretaceous in northeast Mexico: American Association of Petroleum Geologists Bulletin, v. 58, p. 800–809.

——1977, Tambara limestone and the Poza Rica trend, Cretaceous, Mexico, *in* Cook, H. E., and Enos, P., eds., Deep Water Carbonate Environments; Society of Economic Paleontologists and Mineralogists Special Publication 25, p. 273–314.

Ericson, D. B., Ewing, W. M., and Heezen, B. C., 1952, Turbidity currents and sediments in North Atlantic: American Association of Petroleum Geologists

Bulletin, v. 36, p. 481–511.

Ewing, J. I., Ewing, M., and Leyden, R., 1966, Seismic profiler survey of Blake Plateau: American Association of Petroleum Geologists Bulletin, v. 50, p. 1948–1971.

Falvey, D. A., 1974, Continental margin development and plate tectonic theory: Australian Petroleum Explorers Association Journal, v. 14, p. 95–106.

Field, R. M., and collaborators, 1931, Geology of the Bahamas: Geological Society of America Bulletin, v. 42, p. 759–784.

Freeman-Lynde, R. P., Cita, M. B., Jadoul, F., Miller, E. L., and Ryan, W.B.F., 1981, Marine geology of the Bahama Escarpment: Marine Geology, v. 44, p. 119–156.

Gabb, W. M., 1873, Topography and geology of Santo Domingo: Transactions of the American Philosophical Society, v. 15, p. 49–259.

Garrett, P., and Gould, S. J., 1984, Geology of New Providence Island, Bahamas: Geological Society of America Bulletin, v. 97, p. 209–220.

Gibson, T. G., and Schlee, J., 1967, Sediments and fossiliferous rocks from the side of the Tongue of the Ocean, Bahamas: Deep Sea Research, v. 14, p. 691–702.

Ginsburg, R. N., and James, N. P., 1974: Holocene carbonate sediments of continental shelves, *in* Burk, C. A., and Drake, C. L., eds., Geology of Continental Margins: Springer-Verlag, p. 137–155.

Ginsburg, R. N., Lloyd, R. M., Stockman, K. W., and McCullum, J. S., 1963, Shallow water carbonate sediments, *in* Hill, M. N., ed., The Sea, Volume 3: New York, Inter-Science Publishers, p. 554–582.

Goodell, H. G., and Garman, R. K., 1969, Carbonate geochemistry of Superior deep test well, Andros Island, Bahamas: American Association of Petroleum Geologists Bulletin, v. 53, p. 513–536.

Gradstein, F. M., and Sheridan, R. E., 1983, On the Jurassic Atlantic Ocean and a synthesis of results of Deep Sea Drilling Project Leg 76, Sheridan, R. E. and Gradstein, F. M., eds., *in* Initial Reports of Deep Sea Drilling, Vol. 76: U.S. Government Printing Office, Washington, D.C., p. 913–943.

Guppy, H. B., 1917, Plants, seeds, and currents in the West Indies and Azores: London, Williams and Norgate, 531 p.

Heath, K. C., and Mullins, H. T., 1984, Open-ocean, off-bank transport of fine-grained carbonate sediment in the northern Bahamas, *in* Stow, D.A.V., and Piper, D.J.W., eds., Fine-Grained Sediments: Deep-Water Processes and Environments: Geological Society of London, Special Publication 11, p. 199–208.

Hess, H. H., 1933, Interpretation of geological and geophysical observations, *in* U.S. Hydrographic Office-Navy-Princeton Gravity Expedition to the West Indies in 1932: U.S. Hydrographic Office, Washington, D.C., p. 27–54.

—— 1960, Caribbean research project: Progress report: Geological Society of America Bulletin, v. 71, p. 235–240.

Hilgard, E. W., 1871, On the Geological History of the Gulf of Mexico: American Journal of Science, Ser. 3, v. 2, p. 391–404.

Hine, A. C., and Mullins, H. T., 1983, Modern carbonate shelf-slope breaks: Society of Economic Paleontologists and Mineralogists Special Publication No. 33, p. 169–188.

Hine, A. C., and Neumann, A. C., 1977, Shallow carbonate bank margin growth and structure, Little Bahama Bank: American Association of Petroleum Geologists Bulletin, v. 61, p. 376–406.

Hine, A. C., Wilber, R. J., and Neumann, A. C., 1981, Carbonate sand bodies along contrasting shallow bank margins facing open seaways in the northern Bahamas: American Association of Petroleum Geologists Bulletin, v. 65, p. 261–290.

Hoffmeister, J. E., and Multer, G., 1968, Geology and Origin of the Florida Keys: Geological Society of America Bulletin, v. 79, p. 1487–1502.

Hollister, C. D., and Ewing, J. I., eds., 1972, Initial report of the Deep Sea Drilling Project, Vol. 11: U.S. Government Printing Office, Washington, D.C., 1077 p.

Hooke, R. LeB., and Schlager, W., 1980, Geomorphic evolution of the Tongue of the Ocean and the Providence Channels, Bahamas: Marine Geology, v. 35, p. 343–366.

Hurley, R. J., and Fink, L. K., Jr., 1963, Ripple marks show that counter-current exists in Florida Straits: Science, v. 139, p. 603–605.

Idris, F. M., 1975, Reflection seismic measurements in the Old Bahama Channel north of Cuba [Masters thesis]: University of Miami, 41 p.

James, N. P., and Ginsburg, R. N., 1979, The seaward margin of Belize barrier and atoll reefs: International Association of Sedimentology, Special Publication No. 3, 191 p.

JOIDES, 1965, Ocean Drilling on the continental margin: Science, v. 150, p. 709–716.

Kendall, G. St. C., and Schlager, W., 1981, Carbonates and relative changes in sea-level: Marine Geology, v. 44, p. 181–212.

Kier, J. S., and Pilkey, O. H., 1971, The influence of sea level changes on sediment carbonate mineralogy, Tongue of the Ocean, Bahamas: Marine Geology, v. 11, p. 189–200.

King, E. R., 1959, Regional magnetic map of Florida: American Association of Petroleum Geologists Bulletin, v. 43, p. 2844–2854.

Klitgord, K. D., and Behrendt, J. C., 1979, Basin structure of the U.S. Atlantic Margin: American Association of Petroleum Geologists Memoir 29, p. 85–112.

Klitgord, K. D., Popenoe, P., and Schouten, H., 1984, Florida: A Jurassic transform plate boundary: Journal of Geophysical Research, v. 89, p. 7753–7772.

Ladd, J. W., 1976, Relative motion of South America with respect to North America and Caribbean Tectonics: Geological Society of America Bulletin, v. 87, p. 969–976.

Ladd, J. W., and Sheridan, R. E., 1986, Seismic Stratigraphy of the Bahamas: Submitted to American Association of Petroleum Geologists Bulletin.

Land, L. S., 1970, Carbonate mud: Production by epibiont growth on *Thalassia testudinum*: Journal of Sedimentary Petrology, v. 40, p. 1361–1363.

Lee, I., 1981, Shallow seismic stratigraphy of Tongue of the Ocean and Exuma Sound, Bahamas, based on single channel seismic reflection data [Masters thesis]: University of Delaware, 147 p.

Lidz, B., 1973, Biostratigraphy of Neogene cores from Exuma Sound diapirs, Bahama Islands: American Association of Petroleum Geologists Bulletin, v. 57, p. 841–857.

Lynts, G. W., 1970, Conceptual model of the Bahamian Platform for the last 135 million years: Nature, v. 225, p. 1226–1228.

Lynts, G. W., Judd, J. B., and Stehman, C. F., 1973, Late Pleistocene history of the Tongue of the Ocean, Bahamas: Geological Society of America Bulletin, v. 84, p. 2665–2684.

Maher, J. C., 1965, Correlations of Mesozoic and Cenozoic rocks along the Atlantic coast: American Association of Petroleum Geologists cross section 3, 18 p.

Malfait, B. T., and Dinkelman, M. G., 1972, Circum-Caribbean tectonic and igneous activity and the evolution of the Caribbean plate: Geological Society of America Bulletin, v. 83, p. 251–272.

Malloy, R. J., and Hurley, R. J., 1970, Geomorphology and geologic structure: Straits of Florida: Geological Society of America Bulletin, v. 81, p. 1947–1972.

Mattson, P. H., 1973, Middle Cretaceous nappe structures in Puerto Rican ophiolites and their relation to the tectonic history of the Greater Antilles: Geological Society of America Bulletin, v. 84, p. 21–37.

Mattson, P. H., 1979, Subduction, buoyant breaking, flipping and strike-slip faulting in the northern Caribbean: Journal of Geology, v. 87, p. 293–304.

Meyerhoff, H. A., 1946, Tectonic features of the Greater Antilles (abstract): American Association of Petroleum Geologists Bulletin, v. 30, p. 744.

—— 1954, Antillean Tectonics: Transactions of the New York Academy of Sciences, Sec. 2, v. 16, p. 149–155.

Meyerhoff, A. A., and Hatten, C. W., 1968, Diapiric Structures in central Cuba: American Association of Petroleum Geologists Memoir 8, p. 315–357.

—— 1974, Bahamas salient of North America, *in* Burk, C. A., and Drake, C. L., eds., Geology of Continental Margins: Springer-Verlag, p. 429–446.

Meyerhoff, A. A., and Treitel, S., 1959, Geological preliminary evaluation of the Turks and Caicos Islands: La Habana, Min, Ind. unpubl., 10 p.

Milton, C., and Grasty, R., 1969, "Basement" rocks of Florida and Georgia: American Association of Petroleum Geologists Bulletin, v. 53,

p. 2483–2493.

Moody, J. O., 1973, Petroleum exploration aspects of wrench-fault tectonics: American Association of Petroleum Geologists Bulletin, v. 57, p. 449–476.

Mullins, H. T., 1978, Deep carbonate bank margin structure and sedimentation in the northern Bahamas [Ph.D. thesis]: University of North Carolina, Chapel Hill, 166 p.

Mullins, H. T., 1983, Modern carbonate slopes and basins of the Bahamas, *in* Platform Margin and Deep Water Carbonates: Short Course No. 12, Society of Economic Paleontologists and Mineralogists, p.4-1 to 4-138.

Mullins, H. T., 1984, Structural controls of contemporary carbonate continental margins, *in* Platform Margin and Deep Water Carbonates: Short Course No. 12, Society of Economic Paleontologists and Mineralogists, 57 p.

Mullins, H. T., Neumann, A. C., Wilber, R. J., Hine, A. C., and Chinburg, S. J., 1980a, Carbonate sediment drifts in northern Straits of Florida: American Association of Petroleum Geologists Bulletin, v. 64, p. 1701–1717.

Mullins, H. T., and five others, 1982a, Geology of Great Abaco Canyon (Blake Plateau): Observations from the research submersible *Alvin*: Marine Geology, v. 48, p. 239–257.

Mullins, H. T., Heath, U. C., Van Buren, H. M., and Newton, C. R., 1984, Anatomy of a modern open-ocean carbonate slope: Northern Little Bahama Bank: Sedimentology, v. 31, p. 141–168.

Mullins, H. T., Boardman, M. R., and Neumann, A. C., 1979, Echo-character of off-platform carbonates: Marine Geology, v. 32, p. 251–268.

Mullins, H. T., Hine, A. C., and Wilber, R. J., 1982b, Facies succession of Pliocene-Pleistocene carbonates, northwestern Great Bahama Bank: Discussion: American Association of Petroleum Geologists Bulletin, v. 66, p. 103–105.

Mullins, H. T., and Lynts, G. W., 1976, Stratigraphy and structure of Northeast Providence Channel, Bahamas: American Association of Petroleum Geologists Bulletin, v. 60, p. 1037–1053.

—— 1977, Origin of Northwestern Bahama Platform: Review and reinterpretation: Geological Society of America Bulletin, v. 88, p. 1447–1461.

Mullins, H. T., Lynts, G. W., Neumann, A. C., and Ball, M. M., 1978, Characteristics of deep Bahama Channels in relation to hydrocarbon potential: American Association of Petroleum Geologists Bulletin, v. 62, p. 693–703.

Mullins, H. T., and Neumann, A. C., 1979a, Deep carbonate bank margin structure and sedimentation in the northern Bahamas: Society of Economic Paleontologists and Mineralogists Special Publication 27, p. 165–192.

—— 1979b, Geology of the Miami Terrace and its paleoceanographic implications: Marine Geology, v. 30, p. 205–232.

Mullins, H. T., Neumann, A. C., Wilber, R. J., and Boardman, M. R., 1980b, Nodular carbonate sediment on Bahamian slopes: Possible precursors to nodular limestones: Journal of Sedimentary Petrology, v. 50, p. 117–131.

Mullins, H. T., Newton, C. R., Heath, K., and Van Buren, M., 1981, Modern deep water coral mounds north of Little Bahama Bank: Criteria for recognition of deep-water coral bioherms in the rock record: Journal of Sedimentary Petrology, v. 51, p. 999–1013.

Mullins, H. T., and Sheridan, R. E., 1983, Wrench tectonic origin for the northern Bahama Platform: Geological Society of America Abstracts with Programs, v. 15, no. 6, p. 648–649.

Mullins, H. T., and Van Buren, H. M., 1981, Walkers Cay fault, Bahamas: Evidence for Cenozoic faulting: Geo. Mar. Lett., v. 1, p. 225–231.

Nelson, R. J., 1853, On the geology of the Bahamas and on coral formations generally: Geological Society of London Quarterly Journal, v. 9, p. 200–215.

Newell, N. D., 1955, Bahamian platforms; *in* Poldevart, A., ed., The Crust of the Earth; Geological Sosciety of America Special Paper 62, p. 303–315.

Neumann, A. C., and Ball, M. M., 1970, Submersible observations in the Straits of Florida: Geology and bottom currents: Geological Society of America Bulletin, v. 81, p. 2861–2874.

Neumann, A. C., Kofoed, J. W., and Keller, G. H., 1977, Lithoherms in the Straits of Florida: Geology, v. 5, p. 4–10.

Neumann, A. C., and Land, L. S., 1975, Lime mud deposition and calcareous algae in the Bight of Abaco, Bahamas: A budget: Journal of Sedimentary Petrology, v. 45, p. 763–786.

Neumann, A. C., and Moore, W. S., 1975, Sea level events and Pleistocene coral ages in the northern Bahamas: Quaternary Research, v. 5, p. 215–224.

Paull, C. K., and Dillon, W. P., 1979, The subsurface geology of the Florida Hatteras shelf, slope, and inner Blake Plateau: U.S. Geological Survey, Open File Rept. 79-448, 94 p.

—— 1980, Erosional origin of the Blake Escarpment: An Alternative hypothesis: Geology, v. 8, p. 538–542.

—— 1981, Comment and Reply on Erosional origin of the Blake Escarpment: An alternative hypothesis: Geology, v. 9, p. 339–341.

Paulus, F. J., 1972, The geology of Site 98 and the Bahama Platform, *in* Hollister, C. D. and Ewing, J. I., eds., Initial Reports of Deep Sea Drilling Project, Vol. 11: U.S. Government Printing Office, Washington, D.C., p. 877–897.

Pilkey, O. H., and Rucker, J. B., 1966, Mineralogy of Tongue of the Ocean sediments: Journal of Marine Research, v. 24, p. 276–285.

Purdy, E. G., 1963, Recent calcium carbonate facies of the Great Bahama Bank, 2. Sedimentary facies: Journal of Geology, v. 71, p. 472–479.

Rona, P. A., and Clay, C. S., 1966, Continuous seismic profiles of the continental terrace off southeast Florida: Geological Society of America Bulletin, v. 77, p. 31–44.

Rucker, J. B., 1968, Carbonate mineralogy of sediments of Exuma Sound, Bahamas: Journal of Sedimentary Petrology, v. 38, p. 68–72.

Rusnak, G. A., and Nesteroff, W. D., 1964, Modern turbidites: Terrigenous abyssal plain versus bioclastic basin, *in* Miller, R. C., ed., Papers in Marine Geology: New York, Macmillan Co., p. 488–507.

Salvador, A., and Green, A. R., 1980, Opening of the Caribbean Tethys (origin and development of the Caribbean and Gulf of Mexico); Colloque C5, Geology of the Alpine Chains Born of the Tethys, Mem. BRGM, v. 115, p. 224–229.

Schlager, W., and six others, 1984, Early Cretaceous platform retreat and escarpment erosion in the Bahamas: Geology, v. 12, p. 147–150.

Schlager, W., and Chermak, A., 1979: Sediment facies of platform-basin transition, Tongue of the Ocean, Bahamas: Society of Economic Paleontologists and Mineralogists Special Publication no. 27, p. 193–208.

Schlager, W., and Ginsburg, R. N., 1981, Bahama carbonate platforms—The deep and the past: Marine Geology, v. 44, p. 1–24.

Schlager, W., Hooke, R. LeB., and James, N. P., 1976, Episodic erosion and deposition in the Tongue of the Ocean, Bahamas: Geological Society of America Bulletin, v. 87, p. 1115–1118.

Schlee, S., 1973, The edge of an unfamiliar world, a history of oceanography: New York, E. P. Dutton and Co., 398 p.

Schuchert, C., 1935, Historical geology of the Antillean-Caribbean region or the lands bordering the Gulf of Mexico and the Caribbean Sea: New York, Wiley & Sons, 811 p.

Sheridan, R. E., 1971, Geotectonic evolution and subsidence of the Bahama Platform: Discussion: Geological Society of America Bulletin, v. 82, p. 807–809.

Sheridan, R. E., 1972, Crustal structure of the Bahama Platform from Rayleigh wave dispersion: Journal of Geophysical Research, v. 77, p. 2139–2145.

—— 1974, Atlantic continental margin of North America, *in* Burk, C. A., and Drake, C. L., eds., Geology of Continental Margins: New York, Springer-Verlag, p. 391–407.

—— 1978, Structural and stratigraphic evolution and petroleum potential of the Blake Plateau: Offshore Technical Conference Proceedings, p. 363–368.

—— 1981, Comment and Reply on Erosional origin of the Blake Escarpment: An alternate hypothesis: Geology, v. 9, p. 338–339.

—— 1983a, Magnetic, bathymetric, seismic reflection, and positioning data collected underway on *Glomar Challenger*, Leg 76, *in* Sheridan, R. E., Gradstein, F. M., eds., Initial Reports of Deep Sea Drilling Project, Vol. 76: U.S. Government Printing Office, Washington, D.C., p. 341–349.

—— 1983b, Phenomena of Pulsation Tectonics related to the breakup of the eastern North American continental margin, *in* Sheridan, R. E., Gradstein, F. M., eds., Initial reports of Deep Sea Drilling Project, Vol. 76: U.S. Government Printing Office, Washington, D.C., p. 897–909.

Sheridan, R. E., Bates, L. G., Shipley, T. H., and Crosby, J. T., 1983, Seismic stratigraphy in the Blake-Bahama Basin and the origin of Horizon D, *in* Sheridan, R. E., Gradstein, F. M., eds., Initial Reports of Deep Sea Drilling Project, Vol. 76: U.S. Government Printing Office, Washington, D.C., p. 667–683.

Sheridan, R. E., Crosby, J. T., Bryan, G. M., and Stoffa, P. L., 1981, Stratigraphy and structure of southern Blake Plateau, northern Florida Straits,and northern Bahamas from multichannel seismic reflection data: American Association of Petroleum Geologists Bulletin, v. 65, p. 2571–2593.

Sheridan, R. E., Drake, C. L., Nafe, J. E., and Hennion, J., 1966, Seismic refraction study of the continental margin east of Florida: American Association of Petroleum Geologists Bulletin, v. 50, p. 1972–1990.

Sheridan, R. E., Gradstein, F. M., and 24 others, 1982, Early history of the Atlantic Ocean and gas hydrates on the Blake Outer Ridge: Results of the Deep Sea Drilling Project Leg 76: Geological Society of America Bulletin, v. 93, p. 876–885.

Sheridan, R. E., Grow, J. A., Behrendt, J. C., and Bayer, K., 1979, Seismic refraction study of the continental edge off the eastern United States: Tectonophysics, v. 59, p. 1–26.

Sheridan, R. E., and Osburn, W. L., 1975, Marine geological and geophysical studies of the Florida-Blake Plateau-Bahamas Region: Canadian Society of Petroleum Geologists Memoir 4, p. 9–132.

Shinn, E. A., Lloyd, R. M., and Ginsburg, R. N., 1969, Anatomy of a modern carbonate tidal flat, Andros Island, Bahamas: Journal of Sedimentary Petrology, v. 39, p. 1202–1228.

Spencer, M., 1967, Bahamas deep test: American Association of Petroleum Geologists Bulletin, v. 51, p. 263–268.

Suess, E., 1888, Das antlitz der Erda, Zweiter Bank: Vienna, F. Tempsky, 704 p.
—— 1908, Das Antlitz der Erder, Erster Band: Vienna, F. Tempsky, 779 p.

Talwani, M., Worzel, J. L., and Ewing, W. M., 1960, Gravity anomalies and structure of the Bahamas: Transactions 2nd Caribbean Geological Conference, Mayaguez, P. R., p. 156–161.

Tator, B. A., and Hatfield, L. E., 1975, Bahamas present complex geology: Oil and Gas Journal, v. 73, No. 43, p. 172–176 and No. 44, p. 120–122.

Taylor, P. T., Zietz, I., and Dennis, L. S., 1968, Geologic implications of aeromagnetic data for the eastern continental margin of the United States: Geophysics, v. 33, p. 755–780.

Traverse, A., and Ginsburg, R. N., 1966, Palynology of the surface sediments of Great Bahama Bank, as related to water movement and sedimentation: Marine Geology, v. 4, p. 417–459.

Uchupi, E., 1966, Shallow structure of the Straits of Florida: Science, v. 153, p. 529–531.
—— 1969, Morphology of the continental margin off southeastern Florida: Southeastern Geology, v. 11, p. 129–134.

Uchupi, E., Milliman, J. O., Luyendyk, B. P., Bowin, C. O., and Emery, K. O., 1971, Structure and origin of southeastern Bahamas: American Association of Petroleum Geologists Bulletin, v. 55, p. 687–704.

Vail, P. R., Mitchum, R. M., Jr., and Thompson, S. III, 1977a, Seismic stratigraphy and global changes of sea level, Part 4: Global cycles of relative changes of sea level: American Association of Petroleum Geologists Memoir 26, p. 83–97.
—— 1977b, Seismic stratigraphy and global changes of sea level, Part 3: Relative changes of sea level from coastal onlap: American Association of Petroleum Geologists Memoir 26, p. 63–81.

Van Buren, H. M., and Mullins, H. T., 1983, Seismic stratigraphy and geologic development of an open-ocean carbonate slope: The northern margin of Little Bahama Bank, *in* Sheridan, R. E., Gradstein, F. M., eds., Initial Report of Deep Sea Drilling Project, Vol. 76, U.S. Government Printing Office, Washington, D.C., p. 749–762.

Vogt, P. R., 1973, Early events in the opening of the North Atlantic, *in* Tarling, D. H., and Runorn, S. K., eds., Implications of Continental Drift to the Earth Sciences, Vol. 2: New York, Academic Press, p. 693–712.

Vogt, P. R., and Einwich, A. M., 1979, Magnetic anomalies and sea floor spreading in the western North Atlantic, and a revised calibration of the Keathley (M) geomagnetic reversal chronology, *in* Tucholke, B. E. and Vogt, P. R., eds., Initial Reports of Deep Sea Drilling Project, Vol. 43: U.S. Government Printing Office, Washington, D.C., p. 971–974.

Wilber, R. J., and Neumann, A. C., 1977, Porosity controls in subsea cemented rocks from deep-bank environment of Little Bahama Bank: American Association of Petroleum Geologists Bulletin, v. 61, p. 841.

Woodring, W. P., 1928, Tectonic features of the Caribbean region: Proceedings 3rd Pan-Pacific Scientific Congress, Tokyo, v. 1, p. 401–431.

MANUSCRIPT ACCEPTED BY THE SOCIETY MARCH 5, 1985

Printed in U.S.A.

Chapter 16

Paleoenvironments: Offshore Atlantic U.S. margin

John S. Schlee
U.S. Geological Survey, Woods Hole, Massachusetts 02543
Warren Manspeizer
Department of Geological Sciences, Rutgers University, Newark, New Jersey 07102
Stanley R. Riggs
Department of Geology, East Carolina University, Greenville, North Carolina 27834

INTRODUCTION

The U.S. Atlantic continental margin, which stretches 1,850 km from Georges Bank in the north to the Blake Plateau in the south, encompasses an area of 655,000 km². The margin comprises several sedimentary basins of different shapes with platforms in between (see Schlee and Klitgord, this volume). The basins appear to have begun their subsidence at about the same time and to have undergone similar rift and postrift phases of development that resulted in a similar sedimentary section (Schlee and Jansa, 1981). Our objectives in this paper are (1) to portray, at selected intervals, the paleogeography of the margin during the Mesozoic and Cenozoic, (2) to discuss the temporal development of the paleoshelf edge, and (3) to outline the major elements of the several different sedimentary regimes that have prevailed (rift, postrift, carbonate-clastic, and authigenic sediment accumulations).

The main sources of data are interpretations of multichannel seismic-reflection profiles (Dillon 1982; Dillon and others, 1983a; Grow and others, 1979; Schlee, 1981; Schlee and Fritsch, 1982; Schlee and others, 1985), released drill hole data (Scholle, 1977, 1979, 1980; Scholle and Wenkam, 1982; Poag, 1982a, b; Libby-French, 1981, 1984), and Deep Sea Drilling Project (DSDP) data (Hollister and Ewing, 1972; Tucholke and Vogt, 1979; Benson and Sheridan, 1978; Sheridan, and Gradstein, 1983; Van Hinte and Wise, 1987; Poag and Watts, 1987). With few exceptions (Schlee, 1981; Schlee and Fritsch, 1982), published interpretations of offshore basins have been based on a detailed analysis of one or at best a few key profiles (Grow and others, 1979, 1983; Poag, 1982a, b, 1985). Our approach in this chapter is to present interpretations in the form of eight time-slice maps with a brief discussion about data sources, paleogeography, and ties to adjacent areas.

During the construction of the Atlantic continental margin, several major phases of sedimentation have occurred. Simply stated, the Atlantic margin passed through an early rifting stage, an intermediate carbonate-platform stage, and a later clastic infill stage. These stages have resulted in distinctive patterns of sedimentation; thus we discuss each stage separately as a model of sedimentation: its geometry or shape, distinguishing seismic facies, paleogeography, rock types, distinctive unconformities, and rates of subsidence. These models show different parts of the entire margin, both along eastern North America, and along the conjugate Northwest African margin, based on the work of von Rad and others (1982).

PALEOENVIRONMENT TIME INTERVAL MAPS

The collection of paleoenvironmental maps (Triassic and younger, Plate 6A-1 to 7) is modified from the OMD (Ocean Margin drilling) synthesis of the U.S. Atlantic margin (Ewing and Rabinowitz, 1984). The original synthesis was a lithofacies compilation based on analysis of drillhole, dredge, and seismic-reflection data. We have tried to update the maps by including the latest well data (Scholle and Wenkam, 1982). Revision has also involved renaming of the major lithologic units in paleoenvironmental terms as part of the legend and the adding of an inshore paleoenvironmental term (nonmarine) where appropriate. In some areas, such as the Cenozoic of Georges Bank, the sedimentary section is so thin (less than a few hundred meters) that it was not possible to map as many units as Uchupi and others (1984) mapped on their OMD compilation for the Cenozoic.

Triassic–Liassic: Paleogeography

Introduction. Attempts to reconstruct the Triassic-Liassic paleogeography (Plate 6A-1 and 2) of the continental margin off the eastern United States are fraught with much uncertainty and little data. Time-stratigraphic correlations, which provide the bases for paleogeographic constructs, are seriously wanting, and we are left to map this time slice primarily from a series of seismic profiles and inferences from the onshore outcrops.

Schlee, J. S., Manspeizer, W., and Riggs, S. R., 1988, Paleoenvironments: Offshore Atlantic U.S. margin; *in* Sheridan, R. E., and Grow, J. A., eds., The Geology of North America, Volume I-2, The Atlantic Continental Margin, U.S.: Geological Society of America.

Only one well, the COST G-2 (Scholle and Wenkam, 1982) on the United States margin, has penetrated the older synrift sequence. And in that well, even the age and position of the postrift or breakup unconformity (a fundamental tectono-stratigraphic marker on the margin) are uncertain. For example, Poag (1982a, Figs. 21 and 22) and Klitgord and others (1982, p. 163) placed the postrift unconformity at 6,700 m and 6,400 m respectively. However, paleontological data (Cousminer and others, 1984; and Manspeizer and Cousminer, this volume, Figs. 3 and 12), show that the unconformity may be as high as 4,153 m in the core (Fig. 1). The latter are based on the following data: (1) the occurrence of in situ Middle Jurassic Bajocian palynomorphs and reworked Norian dinoflagellates in core No. 4 at a depth of 4,035 m and (2) the presence of 4,440 m of a Late Triassic (Carnian-Norian) palynomorph assemblage, including the index fossils *Patinasporites densus* and *Pseudoenzonalasporites summus,* with the marine phytoplankton, Tasmanites, from core No. 5. These findings were confirmed by Cornet (written and oral communications, 1983 and 1984), who documents the presence of Late Triassic palynomorphs from cuttings at 4,634 m to 4,344 m, and finds Liassic palynomorphs from cuttings at 4,344 m to 4,055 m in the COST G-2 well.

According to these data, the Lower Jurassic stratigraphic section is thus 'condensed' to less than 300 m, and includes the basal sands correlative with the Mohican Formation and the upper part of the underlying, exceedingly thick (more than 7,000 m) evaporite sequence. We locate the unconformity within these basal sands at about 4,153 m. The sands have been correlated (Poag, 1982a, Fig. 22) to the Middle Jurassic Mohican Formation, which Given (1977) identified as a time-transgressive clastic sequence representing the first depositional phase of onlapping sands after the breakup of North America and Europe-Africa. The unconformity embraces a major part of the Early Jurassic time interval, and marks an important tectono-stratigraphic boundary separating the Upper Triassaic synrift clastic-evaporite facies from the overlying Middle Jurassic postrift carbonate-clastic facies. Whereas the upper part of the synrift sequence in the COST G-2 well is dominated by dolomite and anhydrite with minor interbeds of limestone and fine-grained clastic rocks, the lower part, which has not been drilled, probably consists of late Triassic (to perhaps Permian) red beds, lithologically similar to the Eurydice Formation of the Scotian Shelf (see Arthur, 1982).

The COST G-2 section (in this study) thus serves as a template to interpret seismic-reflection profiles across the margin. Its palynomorphs are used to correlate the onshore and offshore basins (see Manspeizer and Cousminer, this volume, Fig. 7) and to establish a correlation network to the Scotian Basin. The Scotian shelf section has been extensively drilled (see McIver, 1972; Jansa and Wade, 1975; Ascoli, 1976; Given, 1977; and Keen, 1982) and is commonly taken as the standard section for the continental margin.

On the Scotian Shelf, Upper Triassic red beds of the Eurydice Formation grade seaward and upward in section to halite and shale; they contain minor anhydrite of the Argo Formation, which has been dated as Early Jurassic (Hettangian-Sinemurian) by Williams (1974) and by Jansa and Wade (1975). Toward the shallower parts of the Scotian Shelf, the "red bed–evaporite sequence is separated from the overlying strata by a conspicuous unconformity," (Jansa and Wiedmann, 1982, p. 233; see also Jansa and Wade, 1975) that may be assigned a post-Sinemurian to pre-Bajocian age. That age designation agrees with the palynological data presented by Barss and others (1979, Fig. 5), who document a major post-Sinemurian to pre-Bajocian unconformity in five of the 18 wells that penetrate Liassic strata on the Scotian Shelf. Although the unconformity is not reported from the deeper parts of the basin, where deposition may have been uninterrupted, an unconformity of comparable age and tectonic setting has been reported as the breakup unconformity in Georges Bank (USGS seismic line 5), in the Baltimore Canyon trough (USGS seismic line 25), and in the Blake Plateau Basin (USGS seismic line FC-3) (see Schlee and Jansa, 1981; Grow, 1981). An unconformity of comparable age and character, the Mid-Cimmerian tectonic event, is also present in western Europe. There, as in North America, the unconformity is associated with volcanism and eustatic changes that Ziegler (1981) asserted are contemporaneous with the rifting that preceded the onset of sea-floor spreading in the central Atlantic.

Environmental Considerations. The geologic record is the result of many factors, the most important of which are tectonism and climate. In this section we examine the effect of a large, northward-migrating and rifting supercontinent on climate and thus on geologic processes.

The final phases of the Variscan-Alleghanian orogeny produced a broadly convex landmass that extended from about paleolatitude 80° S to 70° N. It had an area of about 184×10^6 km^2, a broad central arch standing about 1.7 km high and 720 km across, and an average elevation of more than 1,300 m above the early Mesozoic sea level (Hay and others, 1981).

In the Middle Triassic, most of the future western Atlantic margin lay between the equator and 20° N latitude (Fig. 2), thereby encompassing climates ranging from equatorial rain forest to tropical savannah (Hay and others, 1981; Manspeizer, 1981). As the landmass drifted farther north (Morgan, 1981; and Cornet and Olsen, 1986), transgressing about 10° latitude between the Late Triassic to the Middle Jurassic, the northern rift basins (e.g. those on the Scotian Shelf and Moroccan Meseta) were subjected to increasing aridity as they moved toward the subtropical high pressure cell. The southern basins (e.g., the Dan River and Richmond) stayed humid under the continuing influence of the equatorial low pressure system. These long-term climatic trends are documented in the stratigraphic record; see Hubert and Mertz (1980) for the Fundy Basin; Jansa and Wade (1975) for the Scotian Shelf; and Reinemund (1955) and Olsen and others (1981) for the southern basins. Savannah-like climates, as first suggested by Krynine (1935), seem to have prevailed over much of the region, which was situated between the subtropical desert and equatorial rain forest.

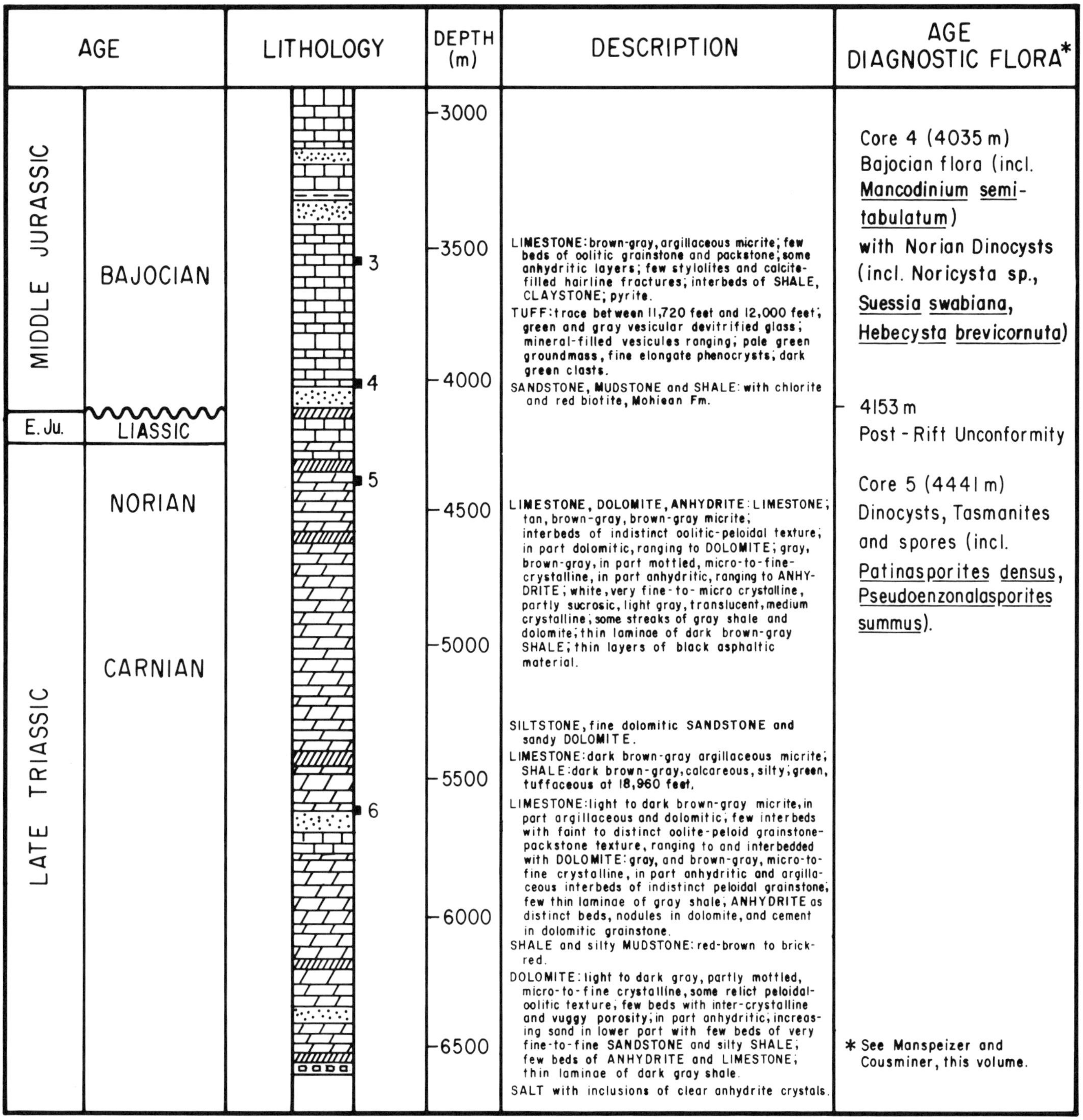

	AGE	LITHOLOGY	DEPTH (m)	DESCRIPTION	AGE DIAGNOSTIC FLORA*
MIDDLE JURASSIC	BAJOCIAN		-3000 -3500 3 -4000 4	LIMESTONE: brown-gray, argillaceous micrite; few beds of oolitic grainstone and packstone; some anhydritic layers; few stylolites and calcite-filled hairline fractures; interbeds of SHALE, CLAYSTONE; pyrite. TUFF: trace between 11,720 feet and 12,000 feet; green and gray vesicular devitrified glass; mineral-filled vesicules ranging; pale green groundmass, fine elongate phenocrysts; dark green clasts. SANDSTONE, MUDSTONE and SHALE: with chlorite and red biotite, Mohican Fm.	Core 4 (4035 m) Bajocian flora (incl. **Mancodinium semi-tabulatum**) with Norian Dinocysts (incl. Noricysta sp., **Suessia swabiana**, **Hebecysta brevicornuta**)
E. Ju.	LIASSIC				4153 m Post-Rift Unconformity
LATE TRIASSIC	NORIAN		5 -4500	LIMESTONE, DOLOMITE, ANHYDRITE: LIMESTONE; tan, brown-gray, brown-gray micrite; interbeds of indistinct oolitic-peloidal texture; in part dolomitic, ranging to DOLOMITE; gray, brown-gray, in part mottled, micro-to-fine-crystalline, in part anhydritic, ranging to ANHYDRITE; white, very fine-to-micro crystalline, partly sucrosic, light gray, translucent, medium crystalline; some streaks of gray shale and dolomite; thin laminae of dark brown-gray SHALE; thin layers of black asphaltic material.	Core 5 (4441 m) Dinocysts, Tasmanites and spores (incl. **Patinasporites densus, Pseudoenzonalasporites summus**).
	CARNIAN		-5000 6 -5500 -6000 -6500	SILTSTONE, fine dolomitic SANDSTONE and sandy DOLOMITE. LIMESTONE: dark brown-gray argillaceous micrite; SHALE: dark brown-gray, calcareous, silty; green, tuffaceous at 18,960 feet. LIMESTONE: light to dark brown-gray micrite, in part argillaceous and dolomitic, few interbeds with faint to distinct oolite-peloid grainstone-packstone texture, ranging to and interbedded with DOLOMITE: gray, and brown-gray, micro-to-fine crystalline, in part anhydritic and argillaceous interbeds of indistinct peloidal grainstone; few thin laminae of gray shale, ANHYDRITE as distinct beds, nodules in dolomite, and cement in dolomitic grainstone. SHALE and silty MUDSTONE: red-brown to brick-red. DOLOMITE: light to dark gray, partly mottled, micro-to-fine crystalline, some relict peloidal-oolitic texture; few beds with inter-crystalline and vuggy porosity; in part anhydritic, increasing sand in lower part with few beds of very fine-to-fine SANDSTONE and silty SHALE; few beds of ANHYDRITE and LIMESTONE; thin laminae of dark gray shale. SALT with inclusions of clear anhydrite crystals.	* See Manspeizer and Cousminer, this volume.

Figure 1. Geologic column of COST G-2 well, modified from Poag, (1982a).

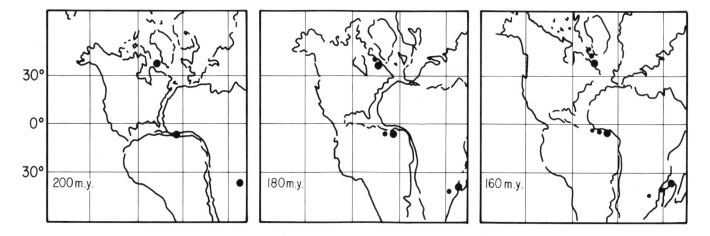

Figure 2. Reconstruction and tracking of the continents around the Atlantic for 200 Ma, 180 Ma and 160 Ma in a fixed hotspot coordinate system. Shown are the Canary, St. Helena, and Crozet hotspots; modified from Morgan (1981) and Cornet and Olsen (1986).

Continents, by virtue of their size, low heat capacity, and topography, modify their climates. Large continents (e.g., Pangaea and Eurasia) have a more profound effect on their climates than do smaller landmasses (e.g., North America). Monsoon circulation, the seasonal alternation of wind direction, was an important factor in sedimentation. The Pangaean winters, for example, were dominated by a subtropical high-pressure cell that carried in warm dry air from the northwest. On the other hand, Pangaean summers were dominated by the Intertropical Convergence Zone that brought in warm moist air from the Tethys Seaway to the east (Manspeizer and others, in preparation). As moist air was uplifted almost 2 km over the mountainous proto-Atlantic dome, it must have cooled adiabatically, owing to decreasing atmospheric pressure, yielding rainwater for streams flowing away from the axis of doming (e.g., the Congo River) and for streams filling high-altitude deep water lakes (e.g., the Rizuzi River that feeds Lake Tanganyika). These high-altitude Mesozoic lakes were thermally stratified and had a well-developed thermocline separating oxygenated surface water from anoxic bottom waters as documented in the stratigraphic record by Olsen (1980) and Manspeizer and Olsen (1981). However, where air descended into low-altitude rift basins along the proto-Atlantic axis, it was warmed adiabatically due to increasing atmospheric pressure. This would have aided precipitation of evaporite minerals wherever marginal marine or fresh water entered rift basins. Similar conditions prevail today in the Red Sea, Lake Magadi, and the Dead Sea. The occurrence of Late Triassic salt flats synchronous with moderately deep lakes, coal swamps, and zeolite-rich playas most eloquently documents the effects of rift topography and elevation on Lower Mesozoic lithofacies (see Manspeizer and Cousminer, this volume, Fig. 7). The stratigraphic record also documents a pervasive and cyclical pattern of wetting and drying, wherein lakes expanded and contracted with the periodicity of 21,000, 42,000, 100,000, and 400,000 years. These intervals agree with the Milankovitch astronomical theory of climates (see Van Houten, 1969; Olsen, 1980; and discussion by Olsen in Manspeizer and others, in preparation).

Late Triassic Paleogeography. Paleogeographic maps (Plate 6A-1 and 2) presented in this study are based on measured sections from the onshore basins and the offshore COST wells (see Manspeizer and Cousminer, this volume, Fig. 7; and Manspeizer, 1985). We infer that during the Late Triassic, the future Atlantic margin was dominated by a basin and range topography whose northeast trend mimicked the fabric of the Variscan-Alleghanian orogen. The rifted terrane extended from the Cobequid-Chedabucto transform on the north to the Bahamas Fracture Zone, and eastward from the American Craton to the African Craton, a distance of about 200 km. Highly elevated, subparallel coastal ranges, which included high-altitude lacustrine and fluvial basins, were situated about 1 km above sea level, and lay adjacent to a narrow, low-lying rifted platform that was situated near sea level and astride the Proto-Atlantic axis. A similar terrane of coastal ranges in the High Atlas of Morocco provided a crude symmetry to the landscape (see Manspeizer and others, 1978; and Hinz and others, 1982). The low-lying continental platform was occasionally transgressed by shallow marine waters from the north and east (Plate 6A-1). We do not envisage an oceanic Atlantic sea floor at this time. Whereas adiabatic cooling of maritime air masses accompanied by rainfall was a dominant process along the high-standing ranges, adiabatic warming and evaporite mineralization were dominant along the near-sea-level basins.

We further infer that where rainfall was excessive, as in the uplands, it fed high-discharge ephemeral streams that flowed from the rifted margins into the basin and built alluvial fans and fan deltas of coarse clastics. Perennial streams, which flowed along the axial or flexured margins of the high-altitude basins, prograded deltas of silt and sand into moderately deepwater lakes and into upland playas. Offshore, away from the immediate influence of coastal ranges, extensive sabkhas, and shallow subtidal platforms were created; the sabkhas were dominated largely by dolomite and limestone and interbeds of lesser anhydrite, fine-grained clastic rocks, and stromatolites. In the Late Triassic, differential subsidence and rifting along the proto-Atlantic axis enabled shallow tongues of hypersaline waters to transgress the narrow platform, perhaps as far south as the Carolina Trough. Where these waters became restricted and where adiabatic warming became excessive, Carnian salt deposits (such as the Osprey Formation) were laid down over continental red beds of Triassic to perhaps Permian age.

Early and Middle Jurassic Paleogeography

The Early Jurassic landscape of eastern North America (Plate 6A-2) was marked by alternating terranes of tholeiitic flood basalt and moderately deep, elongate rift lakes (as inferred from the Culpeper, Gettysburg, Newark, Hartford, and Deerfield Basins) lying adjacent to an erosional platform that was transgressed by shallow marine waters (as inferred from the Scotian shelf and COST G-2 well).

It was a time of major tectonic activity, foretelling the imminence of sea-floor spreading in the Middle Jurassic (Sheridan and Gradstein, 1983). Extension along inverted Alleghanian thrust faults continued to form large asymmetric basins along listric normal faults (see Manspeizer and Cousminer, this volume, Fig. 6). Plutons, confined largely to depositional troughs, fed lavas that flowed along the valley floors and impounded surface runoff to form large stratified lakes. These lacustrine basins became filled with coarse clastic sediments derived from fan-deltas flowing across the active fault margin and with fine detritus derived from deltas prograding from the flexured and axial margins. Concurrently, tholeiitic lavas emanated along Early Mesozoic dike swarms in the Appalachians, forming a broad terrane of flood basalt in the Carolinas and Florida and a smaller flood basalt province in the Northern Appalachians.

Time-stratigraphic correlations (Manspeizer and Cousminer, this volume, Fig. 7) indicate that as a consequence of these thermal events, the landmass stood high, the Late Triassic sea withdrew, and the former marine depositional surface was exposed to erosion. Today that surface, which indicates the end of synrift tectonism on the future margin, is marked by the unconformity in the COST G-2 well. Considerable speculation on the extent of the unconformity and the location of the Early Jurassic shoreline has been advanced in Figure 2 of Plate 6A. The configuration of the hingeline is complicated by the fact that in the Scotian Basin, and perhaps in the deepest parts of the Georges Bank Basin and the Baltimore Canyon Trough, marginal marine conditions may have continued into the Early Jurassic. There, more than 750 m of pure salt of the Argo Formation and almost 580 m of anhy-

dritic dolomite interbedded with red-brown shale (the overlying Iroquois Formation) were deposited in restricted basins.

According to Given (1977), the overlying Middle Jurassic Mohican Formation represents the first postrift deposit of onlapping sands after the breakup of North America and Africa. Shallow-water marine conditions prevailed along the shelf throughout the latter part of the Middle Jurassic, when up to 200 m of oolitic lime packstones to grainstones of the Scatarie Member of the Abenaki Formation was deposited (Poag, 1982a).

TABLE 1.

Reflector	Inferred Age
x	Early Miocene
A^u	Late Eocene-Oligocene
A*	Top of middle Maastrichtian
β	Top of Neocomian
J$_1$	Top of Jurassic
J$_2$	Top of Callovian
J$_3$	Top of Bathonian

Late Jurassic (J$_{2/3}$–J$_1$, Table 1)

In the Late Jurassic (Plate 6A-3), discontinuous carbonate banks formed along the seaward side of the sedimentary basins and platforms, and nonmarine clastic tongues spread from the west and northwest (Schlee and Fritsch, 1982; Poag, 1982a). The carbonate shelf edge, which is described in more detail in a later section, was flanked to the landward side by a transitional carbonate platform of varying width. The broad fan complex off the Baltimore Canyon trough is postulated on the basis of a conspicuous bulge in the sediment isopach between Horizons J$_2$ and J$_1$ (Mountain and Tucholke, 1985). Most of the remaining deepwater areas are thought to be argillaceous limestone and marl that formed as onlapping basin fill south of Georges Bank and as calcareous slope-front fill adjacent to the bank-barrier and the shelf-edge complex (Schlee and others, 1985).

The irregularity of the offshore facies points up the multiplicity of sediment sources—point sources, such as deltas, and linear carbonate sources from reefs and banks. During the late Jurassic, the presence of a delta in the central northern Baltimore Canyon trough (Libby-French, 1984) probably caused the carbonate-shelf edge to migrate seaward 25–30 km (Grow, 1981) over presumably older slope-front deposits (Mattick, 1982). The major leakage through the carbonate barrier appears to have been off the southern Baltimore Canyon trough (offshore Virginia and North Carolina) where a fan complex formed (Van Hinte and Wise, 1987). The main gap in the carbonate platform under Georges Bank is in the south central edge. Here, the gap is in the form of a ramp (Ahr, 1973), characterized by steps in the transition to the deep sea (Schlee and others, 1985). The thickness of deposits seaward of this gap is also greater (range of 3–4 km) than it is to the northeast (1–3 km) and is in elongate isopachs that parallel linear belts of salt(?). The salt occurs as extensions southwestward from the Canadian Sedimentary Ridge Province.

Latest Jurassic–earliest Cretaceous (J₁–β, Table 1)

During the earliest Cretaceous, increasing amounts of clastic sediment entered the northern sedimentary basins at the same time that regressive tongues of sandstone and shale spread southeastward into formerly carbonate areas and gradually restricted the carbonate buildups to the seaward part of the shelf (Plate 6A-4). Libby-French (1984) mapped the buildout of a broad delta system southeast of the New York Bight in the Late Jurassic-Early Cretaceous. Also during this interval, the intrusion of basalt into the Baltimore Canyon trough formed the Great Stone Dome, a broad intrusive complex 12 km across that domed the central part of the delta and created a local unconformity (Schlee and others, 1976).

In the Georges Bank area, the most landward of the New England Seamounts (Bear Seamount) appears to have been intruded during this interval. The broad influx of paralic sediment resulted in the migration of carbonate platform deposits (Grow, 1980) 20 km seaward of the present shelf edge. The carbonate deposits formed a discontinuous set of banks whose shape and paleogeography have been most intricately described by Eliuk (1978) for the Scotian Margin. Onshore, this sequence onlapped the Salisbury Embayment as a nonmarine sequence of unconsolidated silt, clay, sand, and gravel, which was probably deposited by streams and in swamps (Weed and others, 1974).

A seaward gradation from terrigenous inner-shelf deposits to deepwater marls characterized the margin off the southeastern United States. The Blake Plateau was then a shallow-water plateau (Dillon, 1983) that continued south to the Bahamas. Farther offshore, the Blake basin accumulated a thin sequence of deepwater carbonate oozes which to the north became interbedded with terrigenous clastic sediments (Van Hinte and Wise, 1987).

In the deep water south of Georges Bank, onlapping basin fill and contour current deposits (silt and clay) are inferred to have blanketed the area with a 1–1.5 km cover (Schlee and others, 1985).

Top of Neocomian (β) to Near the Top of the Cretaceous (A*, Table 1)

Plate 6A-5 shows a seaward gradation of the paleoshelf edge in the north, from nonmarine clastic rocks on the northwest to slope-rise shale seaward of the edge (as marked by the marl-shale boundary); samples from wells drilled on the Atlantic Coastal Plain for this interval included several nonmarine sandstones, siltstones, and claystones and a few thin intervals that show marginal marine conditions (Schlee, 1981, Fig. 4). Southward, toward the Cape Hatteras area, the marginal marine facies including limestone and marl, are much more prevalent.

Offshore, beneath the shelf, is a broad band of marine and nonmarine terrestrial, fine-grained clastic rocks that stretches from Georges Bank to the Blake Plateau. The sequence (β–A*) was most recently described by Libby-French (1984) as being the result of a major marine transgression that began during the early Late Cretaceous and that covered the area with fine-grained clastics rocks (shale and mudstone) as the deltaic-alluvial system gave way to marine shelf conditions (Libby-French, 1984, Fig. 16). The seismic stratigraphy of the northern half of the margin reveals a blanket of fairly continuous reflections of high to moderate intensity, particularly near the seaward ends of cross-shelf lines. These reflections are indicative mainly of a marine shelf characterized by thin widespread marine limestone and marl interbedded with shale (Schlee and Fritsch, 1982, Fig. 8).

Off the southeastern United States, deep water carbonate rocks accumulated over an ever deepening Blake Plateau (Dillon and others, 1983a; Benson and others, 1978). The area of reef buildup at the seaward edge of the plateau became an area of nondeposition as the reef was abandoned. In the Blake basin and Blake outer ridge, calcareous and siliceous marls accumulated at abyssal depths, whereas less calcareous muds accumulated farther north. In the upper rise south of New England, the deepwater shale is mainly onlapping basin fill (Schlee and others, 1985), except right at the base of the ancestral slope. There the extensively channeled and slumped fill indicates that the discontinuous carbonate banks and reefs, which formerly bordered the shelf edge, had been buried and that sediment from many point sources along the ancestral slope had free access to the deep oceanic basin, there to accumulate as a broad sedimentary blanket.

This interval marked the emergence of the shelf-slope-rise margin we have today. Physiographically, it was probably more subdued than the present one because the slope was much gentler at that time. Partly because of the subdued topography, some of the deep sea reflectors on this interval can be traced through the slope and on to the shelf.

Paleogene (A*–Aᵘ, Table 1)

During this interval, sea level reached a maximum (Poag and Schlee, 1984) as the shoreline, inner shelf, and coastal deposits were all displaced to the northwest (Plate 6A-6). Inner-shelf deposits were laid down in the area of the present Atlantic Coastal Plain; fine-grained clastic and carbonate deposits accumulated farther offshore. The deposits are as much as several hundred meters thick in the mid-Atlantic shelf area (Schlee, 1981) and much thinner off New England (Schlee and Fritsch, 1982).

The paleoenvironmental belts broadly parallel the present trend of the coast: Marl and limestone are more prevalent seaward of New England. The seaward incursion largely covered the Gulf of Maine area where Eocene limestones and porcellanites have been recovered (Hathaway and others, 1979). The Blake Plateau had become a deepwater area, similar to the present area, where the calcareous oozes that accumulated were subjected to scour and perodic movement by the ancestral Gulf Stream (Bunce and others, 1965). Shallow-water limestone was deposited landward of the Florida-Hatteras Slope, under the present day Coastal Plain.

Deepwater areas seaward of the margin comprise mainly marls, siliceous turbidites, and chert. Off New England, the deepwater seismic facies is mainly onlapping basin fill that changes to channel fill and chaotic basin fill near the ancient slope (Schlee and others, 1985). The unit (A*–A^u), which is extensively channeled and eroded beneath the Georges Bank upper rise, points up the late Eocene-Oligocene as a time interval during which the continental slope was cut back and canyons were enlarged.

Early Neogene (A^u–X, Table 1)

The Oligocene-early Miocene interval witnessed a broad series of transgressions accompanied by active sediment input (Vail and Hardenbol, 1979) that resulted in active progradation of the shelf-edge. Perhaps because of a broad uplift of the Appalachians, large amounts of sediment were delivered to the northern Baltimore Canyon trough (Plate 6A-7), there to build a series of deltas out across the shelf (Garrison, 1970; Schlee, 1981). The slope was mainly an area of nondeposition as sediments were carried to the deep sea through an extensive network of channels cut during a major lowstand in the middle Oligocene. The channel system has been mapped over the upper rise south of Georges Bank, where data show channel relief of as much as a 0.7 sec. These channels are 3–15 km across and trend southeast, perpendicular to the present isobaths (Schlee and others, 1985). Associated with them are deposits of chaotic fill and channel basin fill. Non-deposition and erosion resulted in an abbreviated sedimentary section on Georges Bank. The northern Blake Plateau was an area of active bottom scour as a result of the lateral migration of the Gulf Stream, which removed previously deposited sediments and eroded or prevented the deposition of post lower-Miocene sediments (Schlee, 1977). The stream actively scoured the Florida-Hatteras slope and veneered the Plateau with a carbonate ooze (Schlee, 1977). Farther north, where a more typical shelf-slope-rise transition existed, active sediment input from the Baltimore Canyon trough resulted in elongate sediment drifts that covered much of the upper rise and Blake Outer Ridge (Mountain and Tucholke, 1985). Over most of the remaining deepwater area, downslope gravity flows spread a thin sediment blanket over the rise and adjacent abyssal plains. It was during this time interval that the margin began clearly to assume the profile we see at present, a result of the active erosion of what had earlier been a gentle, subdued slope. Though the youngest sedimentary cover is thin and discontinuous in small shelf areas, it has been well described by Jansa and others (1979) and by Poag (1985), for the deep-water area of the Western North Atlantic basin.

TEMPORAL DEVELOPMENT OF THE PALEOSHELF EDGE

The Jurassic paleoshelf edge is a key feature that has been detected all along the Atlantic margin (Grow, 1980). Along

much of the margin (Carolina trough to Georges Bank) it is buried by as much as several kilometers of post-Jurassic sediment; to the south, it is exposed along the Blake-Bahama Escarpment where erosion has cut back the platform several kilometers to exposed back reef facies (Dillon and others, 1981; Freeman-Lynde and others, 1981). The forms of the shelf edge were initially described by Schlee and others (1979) on the basis of a widely spaced grid of multichannel seismic-reflection profiles. Later studies (Schlee and others, 1985) revealed that the buried shelf edge can assume several forms (Plate 6B-1 to 5).

Plate 6B illustrates the several forms of paleoshelf edge as seen on multichannel seismic reflection profiles. On many profiles, the ancient shelf edge is much steeper than the present-day shelf edge, with a relief of 2 km over a horizontal distance of 2.5 km (Plate 6B-1). The declivity is similar to that of the Blake Escarpment, which has a relief of 3.5 km over lateral distance of only a few kilometers. The profile indicates that a major platform edge existed in the Early Cretaceous (Ryan and others, 1978) and that it was eventually buried off the eastern United States in response to a change in the sedimentological regime (see section on "carbonate-clastic infill phase").

The shelf edge seen on the profiles is the final stage in the buildup of a rimmed carbonate platform that existed in a discontinuous fashion along the Atlantic margin (Jansa, 1981). It changed form laterally (Fig. 3 and Plate 6B-2) into a broad ramp beneath the slope off southwestern Georges Bank. The ramp is stepped (as a series of benches that have a relief of 0.5-1 km), and the deep sea transition is 14 km wide. Along the northern edge of the Baltimore Canyon trough, the shelf-edge complex is a buried, upwardly bulbous mass that lacks coherent internal reflectors (Plate 6B-3) and that appears to be a reef-like buildup that persisted into the Early Cretaceous. In the rest of the Baltimore Canyon trough, a rimmed platform (Jansa, 1981) is present in all but the central segment of the outer part of the trough, where it has been intricately block faulted and eroded so that a buried shelf edge is not evident (Plate 6B-4). The platform apparently foundered by faulting, and it was breached and cut back in the late Mesozoic and Cenozoic. A reentrant into the trough, similar to the one described for southwest Georges Bank (Schlee and Fritsch, 1982), may have existed here.

The southern part of the trough is again characterized by a pronounced shelf edge that continued to the Carolina trough, where the buried carbonate platform is presently beneath the upper rise between a row of diapirs and the axis of the trough (Grow and Markl, 1977). Because the shelf edge is back-tilted toward the trough, it in part reflects thermal subsidence and salt-flow movement during intrusion of the zone of diapirs (Dillon and others, 1983b).

The paleoshelf edge trends southeast under the eastern edge of the northern Blake Plateau and emerges from the sedimentary cover in eroded form as the Blake Escarpment. The edge follows the escarpment southward to the Bahamas, where the escarpment forms steep slopes (Freeman-Lynde and others, 1981).

East of Florida, the Blake Escarpment is an eroded cliff

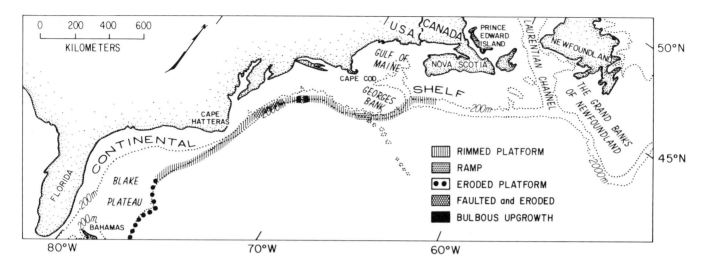

Figure 3. Map showing the distribution of the different types of paleoshelf edge shown on Plate 6B.

3.5 km high and 300 km long (Gilbert and Dillon, 1981). The platform edge was observed at three sites during ten DSDV *Alvin* submersible dives in 1980 (Dillon and others, 1981, 1985). When depths between 1,400 to 4,000 m were investigated and sampled, limestone as old as Early Cretaceous was found. South of the Blake Spur (Plate 7B-5), the escarpment is nearly vertical, current swept, and encrusted by organisms. The dives were also made on a less steep slope of the escarpment 130 km and 200 km south of the Blake Spur. Calcareous nannofossils and sedimentary structures in Cretaceous outcrops revealed that the rocks were deposited in shallow water as a back-reef facies. Talus blocks that contain rudists (Dillon and others, 1981) are the only evidence for the existence of a former reef. The truncation of platform layers and the steep slope of the escarpment clearly indicates the plateau has been extensively eroded, which would have removed most of the reef.

The only places where the intact reef shelf-edge complex has been sampled is under the Scotian margin (Eliuk, 1978), in the deep water seaward of the Baltimore Canyon trough (by Shell Oil Company in exploratory drilling), and at the eastern end of Georges Bank (Ryan and others, 1978). For the Abenaki limestones of the La Have platform (Scotian margin), Eliuk used seismic and well control to map both lateral and vertical carbonate facies in the Bacarro Limestone Member. He found a change from a shale-marl association on the seaward side of the carbonate member to limestone and minor dolomite and shale on part of the offshore bank and—still more landward—to argillaceous limestone and siltstone, part of moat deposits that accumulated landward of the bank. Fauna and flora show a change from a sponge-reef and a coralgal reef on the bank to sparse marine fauna in the moat.

Ryan and others (1978) described massive to medium-bedded limestones and calcareous sandstone (Neocomian age) in the slope seaward of eastern Georges Bank. These were exposed

in cliffs in Heezen Canyon at an interval of 1,334- to 1,257-m water depth and were deposited in a high energy carbonate shelf environment. The COST G-1 and G-2 wells were both drilled some distance from the shelf edge (~150 km from Heezen Canyon), and sampled rocks were of equivalent age (Poag, 1982a). The wells showed coarse to medium-grained sandstone, shale, and mudstone that were deposited in an inner-shelf environment.

Except for Georges Bank and the Canadian Margin, nonproprietary samples of reef rocks are lacking: Shell Oil Company probably sampled the carbonate paleoshelf edge complex in their exploratory deepwater drilling of four wells seaward of the Baltimore Canyon trough during 1983–1985. The Blake Plateau does have regressive reef masses, as sampled by Sheridan and Benson during leg 44 (Benson and Sheridan, 1978) at DSDP sites 390 and 392. These reefs are Early Cretaceous in age and probably originated in a regressive phase well back from the escarpment. Dillon and others (1979a) infer that Albian to Aptian reef and back-reef facies exist under the Outer Blake Plateau and that the platform milieu lasted until the Cenomanian, when deepwater conditions prevailed and the plateau became as deeply submerged as it is at present (Dillon and others, 1979b, 1985).

SEDIMENTARY MODELS

Accumulation of the offshore sedimentary prism has resulted in several distinctive sedimentary associations that warrant further discussion, mainly because their counterparts should exist in folded sedimentary rocks of mountain belts. These associations are the synrift and postrift sediment and volcanic rocks of the block-faulted basement terrane (the broad carbonate platform that acted as a partial barrier to the deep sea) and the authigenic suite of phosphatic sediments that accreted to the southeastern U.S. continental margin in the complex interplay of topography, sea-level changes, and oceanic currents. Both have resulted in distinctive patterns of sedimentation and in distinctive sequences of rocks.

Synrift Model

The separation of Africa and North America in the Middle Jurassic was preceded by a 40-m.y. interval in the Late Triassic and Early Jurassic, marked by rifting, lacustrine and evaporitic sedimentation, and volcanism. As the continental crust began to break apart, continental clastic rocks as much as 8 km thick were deposited in onshore grabens, half-grabens, and pull-apart basins of Morocco and North America (Van Houten, 1969, 1977; Manspeizer and others, 1978). Similar structural basins have been detected offshore on multichannel seismic-reflection profiles (Hutchinson and others, 1983; Grow and others, 1983). They contain evaporites, and are overlain by as much as 4 km of Middle Jurassic and younger postrift strata. In the offshore, the graben fill is separated from the overlying sedimentary wedge by the conspicuous postrift unconformity (see discussion under "Triassic-Jurassic; paleogeography"; and Klitgord and others, 1982).

Virtually all onshore and, presumably, offshore basins seem to lie along reactivated basement structures (Manspeizer and others, in preparation). For example, the Fundy Basin lies within an older late Paleozoic graben that is situated along the Cobequid-Chedabucto transform—a transform that separates the Meguma and Avalon terranes. The Hartford, Culpeper, Taylorsville, and Newark basins all lie along listric normal faults that are reactivated Alleghanian thrusts. Extensional mechanisms, varying from pure northwest-southeast extension to strike-slip along an east trending shear couple, have been advocated by many workers to explain the origin of these basins (see Manspeizer and others, in preparation). In a like manner, sedimentation models span the spectrum from syntectonic to post-tectonic deposition. Recent studies indicate that sedimentation may have begun in crustal downsags that were subsequently broken into small intrabasinal horsts and grabens and then into large, asymmetric, lacustrine troughs, as listric faults formed along inverted thrust faults (see Manspeizer and Cousminer, this volume, Fig. 6). As extension continued in the Triassic, and perhaps because of a change in the stress field at the beginning of the Jurassic, pulses of magma intruded the vast prism of strata, feeding tholeiitic lavas that culminated in rifting onshore. This extensional phase is extraordinary in that the lavas are considered to be isochronous from Virginia to Nova Scotia through a variety of terranes.

Triassic-Liassic synrift sedimentation is notable for its paradoxical occurrence of evaporites offshore and moderately deepwater-stratified lacustrine and fluvial deposits onshore; it records a complex balance between climatic, tectonic, and hydrologic conditions. The distribution of these lithofacies (Pl. 6A-1 and 2; Manspeizer and Cousminer, this volume, Fig. 7) provides a clue to this environmental puzzle (see discussion under "environmental considerations"). Within the late Triassic-Liassic lithosome, three major groups are recognized: (1) anhydrite and dolomite, (2) halite, and (3) K+ and Mg++ salts. Those dominated by anhydrite, reported from Georges Bank (Arthur, 1982), probably formed in a sabkha similar to that of the Persian Gulf. The linear distribution of salt deposits in the Baltimore Canyon Trough and in the Georges Bank and Scotian basins indicate that the evaporites formed in near sea-level rift basins where marine, or perhaps fresh water was restricted (Rona, 1982) and evaporated through adiabatic warming of descending air. The marine evaporite sequence typically stops at the halite stage (Eugster, 1982), but the final concentration products in the early Mesozoic seas were enriched in K+ and Mg++ salts, as recorded in the Triassic-Liassic deposits of Morocco (Salvan, 1972), which bordered the North American Atlantic margin at this time. Because these K+ and Mg++ salts are exceedingly hygroscopic minerals (that can be preserved only under extreme aridity), they probably formed in playas where subsiding brine-filled basins filled with halite (Eugster, 1982).

Almost all of the Triassic-Jurassic lacustrine rocks show a pattern of recurrent lithologies, constituting simple and complex cycles (Van Houten, 1969; Olsen, 1980; and Manspeizer and Olsen, 1981). The Lockatong Formation, the best known of these ancient lake deposits, serves to illustrate Triassic lacustrine conditions. As a huge lacustrine lens of Late Triassic age (Carnian Stage), the formation extended from central New Jersey to Pennsylvania. These strata are about 1,150 m thick and commonly arranged in detrital and chemical cycles that resulted from the expansion and contraction of this ancient lake. According to Van Houten (1962, 1969), their deposition was controlled by the 21,000-year precession cycle.

Compound cycles have been reported by Olsen (in Manspeizer and others, in preparation) with peaks near 42,000, 100,000, and 400,000 years. These large periodic changes are related to cycles in the seasonal variation of sunlight, as prescribed by the Milankovich astronomical theory of climate change (P. E. Olsen, written communication, 1985). Along the Delaware River, each detrital cycle is typically represented by 5 m of sediment comprising a lower black calcareous siltstone that is succeeded upward in the section by dark-gray, calcareous ripple-marked siltstone and fine-grained sandstone. The chemical sequence averages about 3 m thick and contains black and dark gray dolomitic siltstone with lenses of pyritic limestone in the lower part and, in the upper part, massive gray or red analcime and carbonate-rich argillite, disrupted by shrinkage cracks (Manspeizer and Olsen, 1981).

In some respects, the Lockatong Formation is not representative of the other lake deposits. The Newark basin was unique in producing a thick soda-rich silicate chemical facies, although its detrital sequence is similar to that found in other Triassic rift basins (Olsen, 1980). Many of the other lakes were shallow and ephemeral, and evidence shows that they dried completely within one cycle. Also notably different are the thick lacustrine cycles in the Richmond and Taylorsville basins and the laminated limestones of the Scots Bay Formation in the Fundy basin. During the high stand, these deepwater and stratified lakes with anoxic bottom conditions (as in Lake Lockatong) covered an area of about 7,700 km^2 and had a minimum depth of about 100 m (Olsen, 1980). By comparison, Lake Rudolph in the East African Rift has a surface area of about 9,300 km^2 and a maximum known depth of 120 m.

As climates became more arid toward the close of the late Carnian, Lake Lockatong gradually gave way to well-oxygenated mudflats with fringing alluvial fans and braided streams and playas from which glauberite, gypsum, and caliche were precipitated (Van Houten, 1969). Elsewhere in eastern North America (Hubert and others, 1978), the late Triassic Norian Stage was marked by the most extensive development of red beds. It was a time of semi-arid, low-lying source terranes and broad mudflats that slopes gently eastward to a marine sabkha. The sabkha was transgressed by Tethyan waters that drained through the Gibraltar Fracture Zone (see Manspeizer and others, 1978; Jansa and others, 1980; Lancelot, 1980) or by Arctic waters flowing through the North Atlantic rift zone of East Greenland (see Clemmensen, 1982; and Ziegler, 1981).

As Late Triassic clastic sedimentation took place in eastern North America and western Morocco, marine carbonates and sulphate deposits formed in basins from the southern Alps to southern Spain, and halite—together with minor amounts of anhydrite and dolomite—formed in evaporite basins of Algeria, Tunisia, and the Aquataine (Busson, 1972; Jansa and others, 1980). In the Lusitania Basin on the Iberia Meseta and in the Carson Subbasin of eastern Grand Banks, salts were precipitated from hypersaline waters in restricted basins (Manspeizer, 1985). Precipitation of the salt was rapid, but sporadic. Using a precipitation rate of 1 cm/yr, Wade (1980) estimated that 2,000 m of Osprey salt could be precipitated in only 200,000 years. During the Early Jurassic, extreme aridity caused the precipitation of more than 1,000 m of halite with potassium salts in non-rift, non-clastic drift basins (Doukkala, Berrichid, Khemisset) that lay on the Moroccan Meseta and under the subtropical high pressure system (Salvan, 1972; Busson, 1972; Manspeiser and others, 1978). A remarkable balance must have been maintained between the influx of brines from the Tethys and the subsidence of this broad salt flat in Morocco. Inasmuch as the thickness of the Moroccan salt is less than its calculated thickness (see Van Houten, 1977), salt must have dissolved during periods of non-precipitation. On the Scotian Shelf, the synrift sequence is composed of basal red beds (Eurydice Formation), medial salts (Argo Formation), and upper evaporites (Iroquois Formation); there, the sequence is overlain unconformably by the Mohican Formation, a Bajocian sandstone that, according to Given (1977), represents the first drift strata above the postrift unconformity.

The youngest synrift strata drilled to date on the U.S. margin are within the upper part of the thick evaporite sequence in core 5 of the COST G-2 well (Fig. 1), where the sequence is underlain by an equally thick basal synrift section of unknown age(s) and lithology(ies). The dated evaporite section carries the same Carnian-Norian palynomorph assemblage as that reported by Barss and others (1979) in the Osprey Formation from the

Osprey H-84 and the Spoonbill G-30 wells in Grand Banks. We infer that this basal section also consists of Upper Triassic (Carnian-Norian) continental clastic rocks that are overlain by and interbedded with Triassic salt.

The final phase of basin filling onshore took place in the Early Jurassic. It is clearly marked by multiple sequences of tholeiitic lava flows and with interbeds of moderately deepwater lacustrine deposits, as in North America, or with carbonates and evaporite deposits, as in Morocco. Most important, volcanism began almost simultaneously over an area perhaps 2,500 km long and 1,000 km wide, about 20 m.y. after the onset of Triassic sedimentation, after 5–6 km of continental clastic sediments had been deposited. Recurrent movement along the transforms and continental fractures substantially altered the tectonic framework of these basins, so that lacustrine fan deltas were deposited during the Early Jurassic in the same basins where fluvial red beds had accumulated in the latest Triassic. Early Jurassic basins were asymmetric and marked by active listric fault margins and high rates of sedimentation. Subsidence in the Early Jurassic appears to have been almost 2 to 3 times greater during the volcanic-lacustrine phase of synrift sedimentation than during the non-volcanic Triassic red-bed phase, estimated by Van Houten (1969) at about 0.3 m/1,000 years.

Elsewhere, the Early Jurassic was a time of continued rifting, eustatic lowering of sea level, and volcanism in the North Sea. On the newly forming margins of North America and northwest Africa, the final phases of synrift deposition were marked in the offshore by regional uplift, marine regression, tilting, extensive erosion, and perhaps volcanism. In the Middle Jurassic, as the locus of igneous activity moved farther offshore to the axis of future spreading, the inner margin cooled and subsided, ushering in the beginning of postrift sedimentation.

Carbonate—Clastic Infill Phases of Margin Development

From the Grand Banks to the Gulf of Mexico, over a distance of 6,000 km, the form of the main shelf edge has been described in an earlier section ("Temporal development of the paleoshelf edge"). In this section we wish to describe the older buildups on the margin that eventually coalesced to form a broad carbonate platform. Oxley (1981) described isolated patch reef-like bodies deeply buried beneath Georges Bank. These bodies, together with a horseshoe-shaped atoll that borders the southern part of Georges Bank, were mapped on a grid of proprietary multichannel seismic-reflection profiles over the continental shelf. Some of the structures were built on salt ridges and some on elevated basement blocks. These buildups are in the form of broadly arched continuous high amplitude reflectors at greater than 3-seconds depth (two-way travel time). Shallower reflectors at 2-seconds are also slightly bowed in a broad carbonate structure (Fig. 4). Internally, the buildups lack many coherent reflections, and the few they have show reflection terminations at the edge of the structure.

These buildups are best shown in the Georges Bank basin

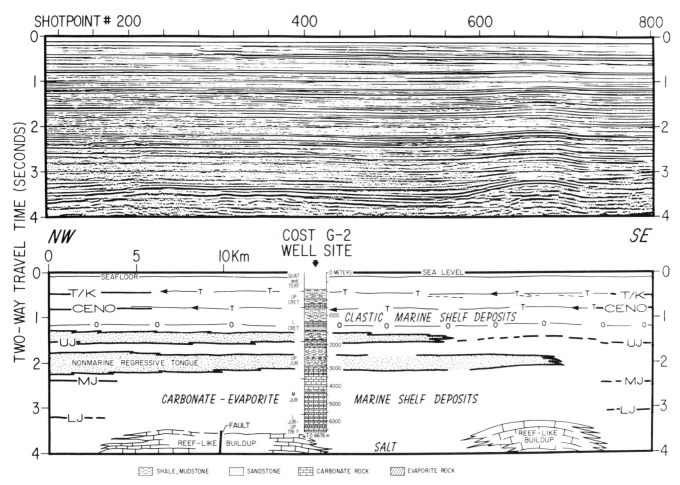

Figure 4. A part of BGR line 209 across central Georges Bank basin. The actual time-scaled profile is shown with interpretation of the stratigraphy below. Also superimposed on the interpretative section is the lithology of the COST G-2 well at the approximate position where the profile crossed the drill site. T/K = Tertiary/Cretaceous boundary; CENO = Cenomanian; UJ = near the top of the Upper Jurassic section; MJ = within the middle Jurassic section; LJ = near the base of the Lower Jurassic (?) section. T marks a thin transgressive unit. O marks correlation with the "O marker"—a zone of limestone within the Missisauga Formation beneath the Scotian Margin (Jansa and Wade, 1975).

area—a part of the Atlantic margin that was extensively fragmented during the continental breakup (Klitgord and others, 1982). The establishment of this uneven basement was probably enhanced by differential subsidence of the cooling blocks after continental separation. Similar reef growth on upraised fault blocks in the western Red Sea has been described by Carella and Scarpa (1962). The high velocity of the rocks and the subdued relief of the buildups strongly indicates a reeflike origin rather than a salt-structure origin.

In the Baltimore Canyon trough, the basement is much more even, and isolated carbonate buildups are lacking. Landward of the hinge zone, half grabens and horsts are present, but the blocks lack the arcuate arrangement of reflectors that could signify a reef buildup. The deep crustal structure of this basin was discussed by Grow and others (1983) and Mattick and others

(1981). Though isolated deposits are missing, the northern part of the trough does appear to have formed a carbonate platform early in its history—a platform that prograded seaward 30 km (Grow, 1980).

An imbricate pattern of older reflections was also described by Schlee and Klitgord (1980) and Schlee and Fritsch (1982, Fig. 4) for the southwest Georges Bank basin. In the same area, Mattick (1982) has described an oblique-progradational seismic facies of older reflections (Middle-Late Jurassic), which he interprets as caused by buildout of the continental platform in the early stages of margin construction.

Mapping of shelf-edge complexes (Mattick, 1982, Fig. 37; Eliuk, 1978, Fig. 5) indicates that after a few million years the isolated buildups filled in and coalesced to create a platform that in the Late Jurassic had either prograded seaward or built upward. In the absence of drill-hole data and using only seismic profiles, we explain the progradational arrangement of reflectors as being caused by clastic wedges in the sedimentary section of central Georges Bank, a section thought to be correlative with the Mohawk–Mic Mac Formations (Poag, 1982a) of the Scotian Margin. The effect of the siliciclastic influxes has been to restrict the carbonate regime to the outer part of Georges Bank and to broaden the pattern of nonmarine facies (Schlee and Fritsch, 1982). Seismically, the clastic influxes show up as tongues of discontinuous reflections of variable amplitude that change laterally into high amplitude continuous reflections that probably represent clastic marine-shelf deposits (Fig. 4). The tongues appear to build into the basin from the northwest and to pinch out to the southeast. The COST G-2 well encountered a sequence of light gray sandstone, gray shale, and light-gray limestone of Late Jurassic age. The upper tongue (Fig. 4) is gray, silty calcareous mudstone and white to light-gray sandstone equivalent to the Missisauga Formation of the Scotian shelf (Poag, 1982a). The fauna indicates that the tongues were deposited under inner shelf to nonmarine paleoenvironments (Poag, 1982a, Fig. 19; Valentine, 1982, Fig. 14). These tongues are major siliciclastic sedimentary influxes that foretell the end of the carbonate regime. Thin beds of widespread shelf limestones (O-marker of the Scotian Shelf) are present in the Georges Bank Cretaceous section, but these are interbedded with gray calcareous mudstone and claystone (shown as zones of weak reflections in the upper second of the profile, Fig. 4). The pattern of reflections is similar to patterns of the earlier carbonate phase in that reflections are horizontally continuous and of moderate to high amplitude. The overlying widespread transgressive limestones are interbedded, with regressive onlapping shales (Schlee and Fritsch, 1982, Fig. 8) reflecting the effect of sea level fluctuations in the Late Cretaceous.

In summary, the carbonate phase of margin construction began, in part, from patch reef growth on elevated basement blocks; later, intervening fill coalesced to build a modest shelf, fringed on the seaward side by a discontinuous reef or ramp. In the late Mesozoic, conditions began to change in all but the southern basins, because the northward drift of the continent resulted in a colder climate (Jansa, 1981). Deltaic influxes of clastic sediment, which at first restricted and then eventually overwhelmed the offshore barriers, began to build a wedge of sediment across the continental rise (Jansa and Wade, 1975; Eliuk, 1978; Schlee and others, 1985; Dillon and others, 1985).

Model for Changing Patterns of Cenozoic Sedimentation on the Southeastern United States Coastal Plain and Continental Shelf

Changing Sediment Patterns in the Cenozoic. This section summarizes the detailed sediment patterns recognized in the upper Paleogene and Neogene rock sequences of the southeastern United States. The interpretations are based upon a well-known coastal-plain section and recent high-resolution seismic and drilling data on the adjacent continental shelf. The sedimentary model

is based on the changing patterns of deposition in Paleogene sediments and distinctive patterns of the Neogene. The Neogene sediments suggest that tectonic factors were decreasing in importance and that major paleoclimatic and paleoceanographic changes had become the dominant factors in controlling Neogene sedimentation.

The rock sequences deposited through the Cenozoic of the southeastern United States display four general types of variations. (1) Shallow-water carbonate sedimentation occurs in tropical to subtropical latitudes and grades northward into temperate siliciclastic deposition (Plate 6A-7; Fig. 5). (2) Siliciclastic sedimentation is dominant on the landward side of the continental margin; carbonate and other authigenic sedimentation (Fig. 5) are dominant toward the seaward side. (3) Both of these sediment types migrate along and across the continental shelf (Plate 6A-6 and 7; Fig. 5) in direct response to changing sea level and oceanographic conditions. (4) The resulting depositional patterns within the upper Cenozoic sediments reflect cyclical deposition on at least three different time scales, representing second-order (greater than 10 m.y. duration), third-order (1 to 10 m.y. duration), and fourth-order (100,000 to 1 m.y. duration) cycles of sea-level fluctuation (Vail and others, 1977). The resulting rock sequence is characterized by gradational sediment patterns and interbedded cyclic sediment units deposited in response to major global changes in paleoclimatological and paleoceanographic conditions. The details of these processes and patterns of upper Cenozoic sedimentation are considered by Riggs and Belknap (this volume).

The global cycles of relative sea level during the Cenozoic (Vail and Mitchum, 1979) display a dramatic change in the patterns of sea-level fluctuation (Fig. 6). The curves depict the second- and third-order sea-level changes during the Paleogene to be generally high and of long duration. After a major regression in the late Oligocene, the Neogene and Quaternary sea-level shifts were generally lower and had rapid third-order fluctuations. On the basis of lithologic cycles characteristic of the Neogene depositional units, Riggs (1984) attributed the coincident changes in patterns of sedimentation at the end of the Oligocene and into the Neogene to a major paleoclimatic change that was dominated by cycles of glaciation and deglaciation. Oxygen isotope analyses on foraminifera from deep-sea sediments (Miller and Fairbanks, 1983; Keigwin and Keller, 1984) suggest that an Antarctic ice cap developed at least 29 m.y. ago, during the late Oligocene. They believe that the opening of Drake Passage established an increased global cooling.

Continental-shelf sediments deposited during the periods of relatively high sea level in the Eocene and early Oligocene were dominated by carbonate deposits characteristic of subtropical to tropical-bank deposition (Plate 6A-7, Fig. 7A and B). Within the Eocene section, siliciclastic sediments were almost nonexistent in the southern part of the shelf and gradually increased in concentration northward, where they formed as thin siliciclastic interbeds and as major diluents within many carbonate facies. In addition to this south/north carbonate/siliciclastic gradient, there

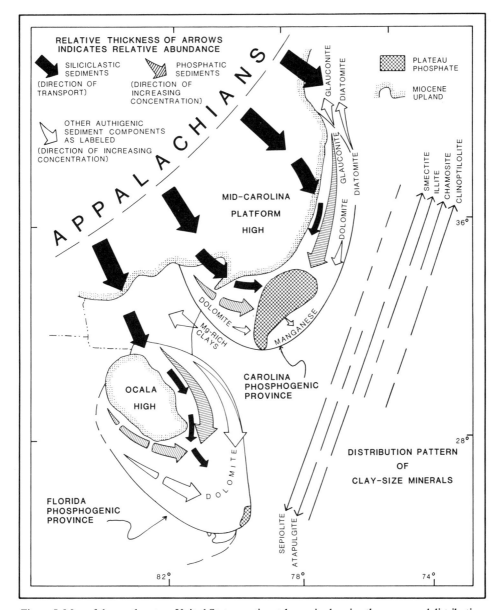

Figure 5. Map of the southeastern United States continental margin showing the source and distribution of major sediment facies associated with the Miocene sediments. Size and direction of arrows reflect the relative magnitude and increasing importance of each sediment facies. Clay mineral distributions are from Lyle (1984).

is a general upsection increase in total siliciclastic sediment and a respective decrease in carbonate sediments throughout the Eocene and Oligocene deposits (Fig. 6). During the late Oligocene and continuing through the Neogene and Quaternary, siliciclastic sediments became the dominant lithology, and carbonate sediments became subordinate; however, the same latitudinal and shoreward relationships still existed between the two lithofacies (Fig. 7C and 7D).

Riggs (1986) found that major episodes of phosphogenesis and the associated aberrant authigenic suite of sediments do not form within extensive sequences of uniform sediments. These

sediments include phosphate, dolomite, diatomaceous muds, organic matter, Mg-rich clays (attapulgite, sepiolite), iron sediments (glauconite, pyrite), manganese, and ferromanganese. Most phosphogenic sequences occur in the transition zone between two distinct lithologic groups of sediments; that is, in the transition zone between thicker and more uniform sections of carbonate and siliciclastic or volcaniclastic sequences. Throughout the geologic column, this transition zone occurs in concert with major changes in sea level and is apparent in large first- and second-order sea-level fluctuations and in the smaller time framework associated with the local facies and unconformities that resulted

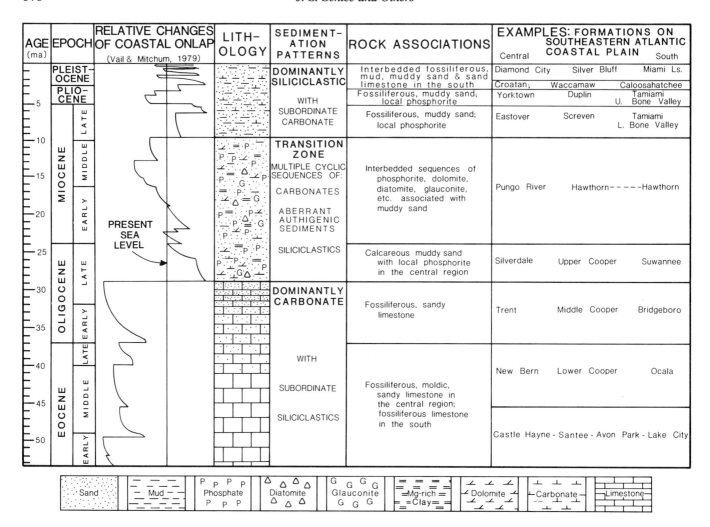

Figure 6. Relative changes of coastal onlap for the last 50 million years of the Tertiary (Vail and Mitchum, 1979) showing changing patterns of dominant sedimentation along the southeastern United States continental margin through time.

from third- and fourth-order fluctuations. Riggs (1986) believes that major episodes of aberrant authigenic sedimentation reflect global changes in tectonism and/or climates that affected and modified the oceanographic processes.

On the southeastern United States continental shelf, a major phosphogenic episode, with its associated aberrant authigenic sediment sequence, took place in the transition zone between limestones (Eocene and Oligocene) and siliciclastic sediment of Pliocene and Quaternary age (Figs. 6 and 7). The phosphogenic transition period began on a local basis during the late Oligocene, when phosphatic sediments formed in the Carolinas around the nose of the Mid-Carolina Platform High (Fig. 8). Carbonate deposition was taking place contemporaneously around the Ocala High. This change coincided with the major sea-level event at 30 Ma and represents the beginning of the transition zone of sedimentation (Fig. 6). During the Miocene second-order sea-

level transgression, which continued to about 13 Ma, an abnormally high concentration and volume of phosphorus was deposited (see Riggs and Belknap, and Riggs and Manheim, this volume) along with contemporaneous facies of glauconite, diatomaceous muds, organic matter, dolomite, and Mg-rich clays (Fig. 8) (Riggs, 1984).

Specific Miocene lithofacies represent periods of increased phosphate formation and accumulation (as much as 75 percent phosphate grains), though minor concentrations of phosphate grains (less than 2 percent) occur within most lithofacies. Miocene phosphate formation (Fig. 6) began by at least the late early Miocene, about 19 Ma, and continued cyclically into late middle Miocene, about 13 Ma (Riggs, 1984; Riggs and others, 1985). It formed around the seaward nose and flanks of major or first-order structural highs (the Ocala and Mid-Carolina Platform Highs), defining the Florida and Carolina phosphogenic prov-

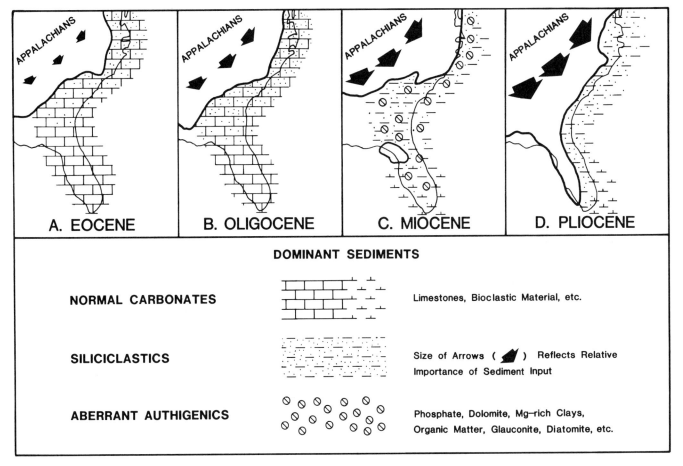

Figure 7. Four-part map series of the southeastern United States continental margin showing changing patterns of dominant sedimentation from the Eocene through the Pliocene.

inces, respectively (Fig. 8). Within each province, second-order or local structural and topographic highs controlled specific sites and amounts of phosphate deposition (Fig. 8) (Riggs, 1984).

By the late Miocene and into the Pliocene and Pleistocene, sedimentation was dominated by siliciclastic deposits in which subordinate carbonates occurred as the southernmost facies or as minor interbeds of carbonate-rich sediments capping a dominantly siliciclastic sediment (Fig. 7). Minor concentrations of primary phosphorite formed locally during the short Pliocene second-order sea-level cycle between 5 and 4 Ma (Riggs, 1984). Preliminary studies suggest that local concentrations of Pleistocene-Holocene phosphorite on the North Carolina shelf are only partly reworked from the Miocene Pungo River Formation; some phosphate seems to be primary and to have formed during one or more of the brief Pleistocene phosphogenic episodes that coincided with the Pleistocene transgressive cycles (Riggs and others, 1985).

Cyclic Depositional Patterns Within the Neogene. By the early Miocene, shelf deposition was characterized by multiple sediment units deposited within established third-order cycles of

eustatic sea-level change (Vail and Mitchum, 1979). These fourth-order sediment units were deposited in response to transgressions and regressions associated with sea-level cycles that ranged between 100,000 and 1 million years in duration (Riggs, 1984; Riggs and others, 1985; Snyder, 1982).

Each fourth-order sea-level event is represented by the following idealized lithologic cycle or some variation of it (Riggs, 1984). Siliciclastic sedimentation was dominant during the early stage of each transgression and within shallower, landward environments; carbonate sedimentation was dominant during the late stage of each transgression and within deeper, seaward environments. If an aberrant sequence of authigenic sediments, including phosphorites, formed, their deposition began slowly on the outer shelf during the early stage and increased to a maximum across the shelf during the mid-stage of transgression. Thus, the phosphorites occur both laterally and vertically within the transition zone between dominantly siliciclastic and dominantly carbonate sedimentation.

Pinet and others (1981), Pinet and Popenoe (1982), and Popenoe (1985) have demonstrated that the western boundary

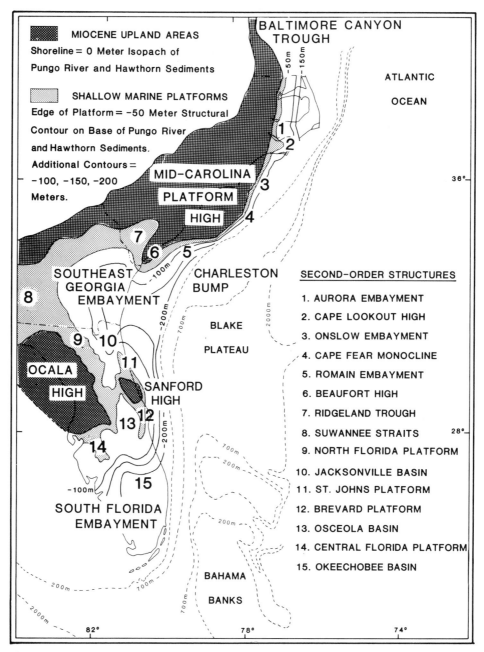

Figure 8. Map of southeastern United States showing the general Miocene shoreline and major first- and second-order structural and topographic features that controlled phosphate sedimentation. Map is from Riggs (1984).

current and the Gulf Stream controlled depositional and erosional processes on the Atlantic continental margin throughout the Neogene. The Gulf Stream axis migrated across the Blake Plateau in response to major fluctuations in global sea level, eroding major channel networks through previously deposited sediments and acting as a major barrier that separated margin sedimentation from deep-sea carbonate deposition on the seaward side of the Stream. Riggs (1984) demonstrated that the interaction of the

Gulf Stream and the continental margin during various stages of sea level played several important roles. First, such interaction determined the type, location, formation, and extent of various authigenic sediments including phosphorites, associated aberrant authigenic sediments, and the subsequent carbonate rocks capping each depositional cycle (Fig. 9). Second, Gulf Stream processes determined the extent and location of subsequent erosion of sediment units.

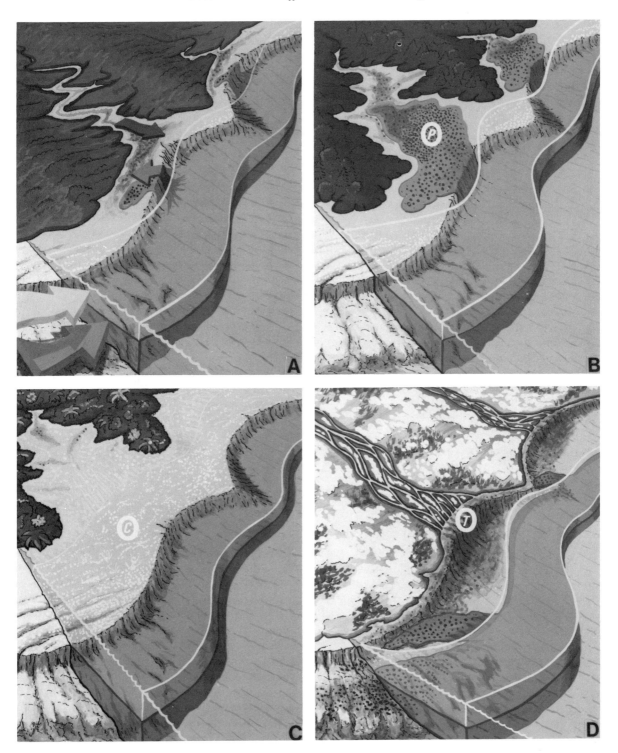

Figure 9. Idealized cycle of Neogene sedimentation on the southeastern United States Atlantic continental margin (from Riggs, 1984). Deposition is a direct response to fluctuations in climates, sea levels, and continental shelf water masses. The first stage (A) reflects cold climates and sea-level low stands of glacial maximums. The next three stages (B through D) reflect warming climates through a sea-level transgression associated with deglaciation. The increased interaction of the Gulf Stream with the configuration of the continental margin accentuates upwelling of nutrient-rich waters and phosphate deposition. P = Phosphate deposition; C = carbonate deposition, and T = terrigenous deposition. Painting by Whiting M. Toler, Washington, North Carolina.

Each transgression could have produced a sediment unit characterized by a complete lithologic sequence. Preservation of all or part of each depositional unit would depend upon the various associated erosional processes. During each transgression, the Gulf Stream eroded portions of the outer continental shelf and slope sediments (Snyder, 1982; Snyder, and others, 1982; Riggs, 1984). During the regression and lowstand part of each sea-level cycle, the exposed portion of each unit is characterized by nondeposition, scouring, and channeling, plus diagenetic alteration of included mineral components. These destructional processes have severely complicated the depositional patterns of the Neogene shelf and slope sediments.

Model of Cyclic Neogene Sedimentation. Neogene sedimentation on the southeastern United States Atlantic continental shelf and slope was cyclic in direct response to fluctuations in the paleoclimates, resulting in multiple glaciation and deglaciation events (Riggs, 1984). These fluctuations caused major changes in vegetation cover on adjacent landmasses. In addition these fluctuations caused changes in rates and types of erosion and resulting sediment and nutrient supplies to the continental margin. Further, they affected variations in eustatic sea levels, current patterns, density stratification, and chemical characteristics of continental shelf and oceanic waters. These processes are summarized in the schematic block diagrams in Figure 9 (Riggs, 1984).

Deglaciation produced the second-, third-, and fourth-order Neogene sea-level events (Vail and others, 1977; Riggs, 1984). Each transgression was characterized by warming climates, an episode of major sediment deposition, and increased interaction of the Gulf Stream with the continental margin bathymetry. This interaction accentuated the Gulf Stream meanders and produced current erosion and topographic upwelling of deeper, nutrient-rich waters (Fig. 9A and 9B). Thus, the seaward extension of first-order structural highs and adjacent flanks became sites of organically rich sedimentation, including phosphorites, dolomites, diatomites, and glauconites. Deposition of these authigenic and biogenic sediments began when large-scale topographic upwelling of nutrient-enriched waters occurred in response to major deflections of the Gulf Stream. Phosphorite sedimentation generally decreased into regional embayments as shelf environments became more remote from the oceanic system (Fig. 8).

Interglacial sea level maximums were dominated by temperate to subtropical carbonate deposition (Fig. 9C). Glacial lowstands were dominated by subarctic to temperate climates and were characterized by erosion and diagenesis of previously deposited sediment units and the development of hardgrounds and unconformity surfaces (Fig. 9D).

The transitional period of aberrant authigenic sedimentation was definitely on the wane by the time of the early Pliocene second-order sea-level transgression. The Pliocene sediments are dominantly siliciclastic and have only minor amounts of authigenic sediments, including phosphorites. The sediments deposited during the remainder of the Pliocene and Quaternary are domi-

nated by temperate to sub-tropical carbonates and siliciclastic sedimentation off much of the southeastern United States. The siliciclastics grade into sub-tropical to tropical carbonate sediments (1) in southern Florida, (2) seaward toward the shelf edge, and (3) vertically into carbonate sediments that formed during the sea-level maximums of deglaciation (Fig. 7D).

The long duration of sea-level highstands and associated warm climates prior to the Neogene may have been instrumental in the development of a stratified ocean: a significant deep-ocean phosphorus reservoir resulted from increased chemical weathering and nutrient supplies. Major global cooling associated with the Neogene glaciation may have brought about a breakdown of oceanic stratification and increased oceanic overturn.

Significant accretion occurred on the continental margin during the 6-m.y. period of rapid deposition associated with the early-to-middle Miocene second-order sea-level transgression (Schlee and others, 1979; Riggs and others, 1985; Snyder, 1982). Most of the volume of Miocene sediments were deposited as slope clinoforms that prograded the shelf seaward, so that only a small volume of actual shelf sedimentation is represented in the Pungo River and Hawthorn Formations. In contrast, the much briefer and more rapid sea-level events associated with the Pliocene and Quaternary ages have produced only a relatively minor amount of shelf/slope progradation. Popenoe (1985) believes that the abrupt pinchout of Miocene sediments on the seaward side is a direct result of Gulf Stream erosion.

Summary. In summary, the Cenozoic section of the southeastern United States continental shelf and slope displays a complete change of depositional mode in response to major changes in the paleoenvironmental conditions through time. The resulting sediment sequence changed from a dominantly carbonate sediment regime during the Eocene and Oligocene (Fig. 7A and 7B), through a transition zone of aberrant authigenic sediment during the Miocene (Fig. 7C), into the dominantly siliciclastic depositional regime characteristic of the post-Miocene (Fig. 7D). Sediment units formed during the Miocene transition zone are characterized by significant concentrations of various types of aberrant authigenic sediments (Fig. 7C) in association with carbonate and siliciclastic components.

These sedimentological patterns form a unique stratigraphic sequence that can be explained by the interaction of at least four environmental factors operating throughout 30 m.y. of geologic time: (1) the size, shape, and structural framework of the continental shelf/slope depositional system; (2) the intensity, stability, and duration of oceanographic conditions and associated climates which controlled the physics, chemistry, and biology of the shelf margin environments; (3) the amount and rate of diluent sedimentation that resulted from siliciclastic and carbonate sources; and (4) additional secondary factors which severely modified the stratigraphic units such as subsequent history of deposition, structural deformation, erosion, burial, diagenetic history, and weathering of the sediments.

REFERENCES CITED

Ahr, W. M., 1973, The carbonate ramp; An alternative to the shelf model: Gulf Coast Association of Geological Society Transactions, v. 23, p. 221–225.

Arthur, M. A., 1982, Lithology and petrology of the COST Nos. G-1 and G-2 wells, *in* Scholle, P. A., and Wenkam, C. R., eds., Geological Studies of the COST Nos. G-1 and G-2 wells, United States North Atlantic outer continental shelf: U.S. Geological Survey Circular 861, p. 11–33.

Ascoli, P., 1976, Foraminiferal and ostracode biostratigraphy of the Mesozoic-Cenozoic, Scotian shelf, Atlantic Canada: Maritime Sediments, Publication 1, pt. B, p. 653–771.

Barss, M. S., Bujak, J. P., and Williams, G. L., 1979, Palynological zonation and correlation of sixty-seven wells, eastern Canada: Geological Survey of Canada Paper 78-24, 118 p.

Benson, W. E., and Sheridan, R. E., 1978, Sites 389 and 390; North rim of Blake Nose, *in* Benson, W. E., Sheridan, R. E., and others, Initial Reports of the Deep Sea Drilling Project: Washington, D.C., U.S. Government Printing Office, v. 44, p. 69–151.

Bunce, E. T., Emery, K. O., Gerard, R. D., Knott, S. D., Lidz, L., Saito, T., and Schlee, J., 1965, Ocean Drilling on the continental margin: Science, v. 150, p. 709–716.

Busson, G., 1972, Principles, methodes et resultats d'une estude stratigraphique du Mesozoique Saharient: Mem Mus. d'Histoire Naturell, NS, Ser. C.T. 26, 441 p.

Carella, R., and Scarpa, N., 1962, Geological results of exploration of Sudan: Beirut, 4th Arab Petroleum Congres, AGIP Mineraria Ltd.

Clemmensen, L. B., 1982, Tectonic and paleoclimatic aspects of the Triassic sequence in East Greenland [abs.]: Wurzburg, Germany, Geologsche Vereinigung, 72 Jahres Tangung, Julius Maximilians Universitat, p. 26.

Cornet, B., and Olsen, P. E., 1986, A summary of the biostratigraphy of the Newark Supergroup with comments on early Mesozoic provinciality: Third Latin American Paleobotanical Congress Symposium on Triassic Global Distribution (in press).

Cousminer, H. L., Steinkraus, W. E., and Hall, R. E., 1984, Biostratigraphic restudy documents Triassic-Jurassic section in the COST G-2 well: American Association of Petroleum Geologist Bulletin, v. 68, no. 4, p. 466.

Dillon, W. P., ed., 1982, Summary of regional geology, petroleum potential, resource assessment, and environmental considerations for oil and gas lease sale area #56: U.S. Geological Survey Open-File Report 82-398, 63 p.

—— , 1983, Regional geology and petroleum potential, *in* Dillon, W. P., ed., Geology report for proposed oil and gas lease sale #90; Continental margin off the southeastern United States: U.S. Geological Survey Open-File Report 83-186, p. 6–84.

Dillon, W., Paull, C. K., Dahl, A. G., and Patterson, W. C., 1979a, Structure of the continental margin near the COST No. GE-1 well, United States South Atlantic Continental Shelf area: U.S. Geological Survey Circular 800, p. 97–107.

Dillon, W. P., Poag, C. W., Valentine, P. C., and Paull, C. K., 1979b, Structure, biostratigraphy, and seismic stratigraphy along a common-depth-point seismic profile through three drill sites on the continental margin off Jacksonville: U.S. Geological Survey Miscellaneous Field Studies Map MF-1090.

Dillon, W. P., Paull, C. K., Valentine, P. C., and others, 1981, Blake Escarpment carbonate platform edge; Conclusions based on observations and sampling from research submersible: American Association of Petroleum Geologists Bulletin, v. 65, p. 918.

Dillon, W. P., Klitgord, K. D., and Paull, C. K., 1983a, Mesozoic development and structure of the continental margin off South Carolina, *in* Gohn, G., ed., Studies related to the Charleston, South Carolina, earthquake of 1886; tectonics and seismicity: U.S. Geological Survey Professional Paper 1313, p. N1–N16.

Dillon, W. P., Popenoe, P., Grow, J. A., Klitgord, K. D., Swift, B. A., Paull, C. K., and Cashman, K. V., 1983b, Growth faulting and salt diapirism; Their relationship and control in the Carolina Trough, eastern North America, *in* Watkins, J. S., and Drake, C. L., eds., Studies in Continental Margin Geol-ogy: American Association of Petroleum Geologists Memoir 34, p. 21–46.

Dillon, W. P., Paull, C. K., and Gilbert, L. E., 1985, History of the continental margin off Florida; The Blake Plateau basin, *in* Poag, C. W., ed., Geologic Evolution of the United States Atlantic Margin: New York, Van Nostrand-Rinehold, p. 189–216.

Eliuk, L. S., 1978, The Abenaki Formation, Nova Scotia shelf, Canada; A depositional and diagenetic model for a Mesozoic carbonate platform: Bulletin of Canadian Petroleum Geology, v. 26, p. 424–512.

Eugster, H. P., 1982, Climatic significance of late and evaporite deposits, *in* Climate in Earth History: Washington, D.C., National Academy Press, p. 105–111.

Ewing, J. I., and Rabinowitz, P. D., 1984, Eastern North American continental margin and adjacent ocean floor, 24° to 41°N and 68° to 78°W: Woods Hole, Massachusetts, Ocean Margin Drilling Program Regional Atlas Series 4, Marine Science International, 32 p.

Freeman-Lynde, R. P., Cita, M. B., Jadoul, F., Miller, E. L., and Ryan, W.B.F., 1981, Marine geology of the Blake Escarpment: Marine Geology, v. 44, p. 119–156.

Garrison, L. E., 1970, Development of Continental shelf south of New England: American Association of Petroleum Geologists Bulletin, v. 54, no. 1, p. 109–124.

Gilbert, L. E., and Dillon, W. P., 1981, Bathymetric map of the Blake Escarpment: U.S. Geological Survey Field Studies Map MF-1362.

Given, M. M., 1977, Mesosoic and early Cenozoic geology of offshore Nova Scotia: Bulletin of Canadian Petroleum Geology, v. 25, p. 63–91.

Grow, J. A., 1980, Deep structure and evolution of the Baltimore Canyon Trough in the vicinity of the COST No. B-3 well, *in* Scholle, P. A., ed., Geological studies of the COST No. B-3 well, United States Mid-Atlantic Continental Slope: U.S. Geological Survey Circular 833, p. 117–126.

—— , 1981, Structure of the Atlantic Margin of the United States, *in* Bally, A. W., ed., Geology of Passive Continental Margins: American Association of Petroleum Geologists Education Course Note Series #19, p. 3-1–3-41.

Grow, J. A., and Markl, R. G., 1977, IPOD-USGS multichannel seismic reflection profile from Cape Hatteras to the Mid-Atlantic Ridge: Geology, v. 5, p. 625–630.

Grow, J. A., Mattick, R. E., and Schlee, J. S., 1979, Multichannel seismic depth sections and interval velocities over Outer Continental Shelf and upper continental slope between Cape Hatteras and Cape Cod, *in* Watkins, J. S., Montadert, L., and Dickerson, P. W., eds., Geological and geophysical investigations of continental margins: American Association of Petroleum Geologists Memoir 29, p. 65–83.

Grow, J. A., Hutchinson, D. R., Klitgord, K. D., Dillon, W. P., and Schlee, J. S., 1983, Representative multichannel seismic profiles over the U.S. Atlantic margin, *in* Bally, A. W., ed., Seismic expression of structural styles; A picture and work atlas: American Association of Petroleum Geologists Studies in Geology Series #15, v. 2, p. 2.2.3.1–2.2.3.19.

Hathaway, J. C., Poag, C. W., Valentine, P. C., Miller, R. E., Schultz, D. M., Manheim, F. T., Kohout, F. A., Bothner, M. H., and Sangree, D. A., 1979, U.S. Geological Survey core drilling on the Atlantic Shelf: Science, v. 206, p. 515–527.

Hay, W. W., Barron, E. J., Sloan, J. L., and Southan, J. R., 1981, Continental drift and the global pattern of sedimentation: Geologische Rundschau, v. 70, p. 34–60.

Hinz, K., Dostmann, H., and Fritsch, J., 1982, The continental margin of Morocco; Seismic sequences, structural elements, and geological development, *in* von Rad, U., Hinz, K., Sarnthein, M., Seibold, E., eds., Geology of Northwest Continental Margin: Berlin, Springer-Verlag, p. 34–60.

Hollister, C. D., and Ewing, J. I., eds., 1972, Initial Reports of the Deep Sea Drilling Project: Washington, D.C., U.S. Government Printing Office, v. 11, 1077 p.

Hubert, J. F., and Mertz, K. A., 1980, Eolian dune field of Late Triassic age, Fundy Basin, Nova Scotia: Geology, v. 8, p. 516–519.

Hubert, J. F., Reed, A. A., Dowdall, W. L., and Gilbert, J. M., 1978, Guide to the redbeds of Central Connecticut (1978 Field Trip, Eastern Section Society of Economic Paleontologists and Mineralogists, contribution no. 32): University of Massachusetts Department of Geology and Geography, 129 p.

Hutchinson, D. R., Grow, J. A., Klitgord, K. D., and Swift, B. A., 1983, Deep structure and evolution of the Carolina trough, *in* Watkins, J. S., and Drake, C. L., eds., Studies in Continental Margin Geology: American Association of Petroleum Geologists Memoir 34, p. 129–152.

Jansa, L. F., 1981, Mesozoic carbonate platforms and banks of the Eastern North American margin: Marine Geology, v. 44, p. 97–117.

Jansa, L. F., and Wade, J. A., 1975, Geology of the continental margin off Nova Scotia and Newfoundland, *in* Van der Linden, W.J.M., and Wade, J. A., eds., Offshore geology of eastern Canada: Canada Geological Survey Paper 74-30, p. 51–105.

Jansa, L. F., and Weidmann, J., 1982, Mesozoic-Cenozoic development of the eastern North American and northwest Africa continental margins; A comparison, *in* von Rad, U. K., Sarthein, M., and Seibold, E., eds., Geology of the northwest African Continental Margin: Berlin, Springer-Verlag, p. 215–269.

Jansa, L. F., Enos, P., Tucholke, B. E., Gradstein, F. M., and Sheridan, R. E., 1979, Mesozoic-Cenozoic sedimentary formations of the North American basin; Western North Atlantic, *in* Talwani, M., Hay, W., and Ryan, W.B.F., eds., Deep drilling results in the Atlantic Ocean; Continental Margins and Paleoenvironment: American Geophysical Union, Maurice Ewing Series 3, p. 1–57.

Jansa, L. F., Bujak, J. P., and Williams, G. L., 1980, Upper Triassic salt deposits of the Western North Atlantic: Canadian Journal of Earth Sciences, v. 17, p. 547–559.

Keen, C. E., 1982, The continental margin of Eastern Canada; A review, *in* Scrutton, R. A., ed., Dynamics of Passive Margins: American Geophysical Union Geodynamic Series, v. 6, p. 45–58.

Keigwin, L., and Keller, G., 1984, Middle Oligocene cooling from Equatorial Pacific DSDP Site 77B: Geology, v. 12, p. 16–19.

Klitgord, K. D., Schlee, J. S., and Hinz, K., 1982, Basement structure sedimentation and tectonic history of the Georges Bank Basin, *in* Scholle, P. A., and Wenkam, C. R., eds., Geological studies of the COST Nos. G-1 and G-2 wells, United States North Atlantic outer continental shelf: U.S. Geological Survey Circular 861, p. 160–186.

Krynine, P. D., 1935, Arkose deposits in the humid tropics; A study of sedimentation in southern Mexico: American Journal of Science, ser. 5, v. 29, p. 353–363.

Lancelot, Y., 1980, Birth and evolution of the "Atlantic Tethys" (Central North Atlantic), *in* Aubouin, J., Debelmas, J., and Letreille, M., eds., Geologie des chaines alpines issues de la Tethys: Memoire du B.R.G.M., no. 115, p. 215–225.

Libby-French, J., 1981, Lithostratigraphy of the Shell 272-1 and 273-1 wells; Implications as to depositional history of the Baltimore Canyon trough, Mid-Atlantic outer continental shelf: American Association of Petroleum Geologists Bulletin, v. 65, p. 1476–1484.

———, 1984, Stratigraphic framework and petroleum potential of the northeastern Baltimore Canyon trough, Mid-Atlantic outer continental shelf: American Association of Petroleum Geologists Bulletin, v. 68, no. 1, p. 50–73.

Lyle, M. E., 1984, Clay mineralogy of the Pungo River Formation, Onslow Bay, North Carolina continental shelf [M.S. thesis]: Greenville, North Carolina, East Carolina University, 129 p.

Manspeizer, W., 1981, Early Mesozoic basins of the central Atlantic passive margins, *in* Bally, A. W., ed., Geology of Passive Continental Margins: American Association of Petroleum Geologists Educational Course Note Series No. 19, p. 4-4–4-60.

———, 1985, Early Mesozoic History of the Atlantic passive margin, *in* Poag, C. W., ed., Geologic evolution of the United States Atlantic Margin: New York, Van Nostrand-Rinehold, p. 1–24.

Manspeizer, W., and Olsen, P. E., 1981, Rift basins of the passive margin; Tectonics, organic-rich lacustrine sediments, Basin Analyses, *in* Hobbs, G. W., ed.,

Field Guide to the Geology of the Paleozoic, Mesozoic, and Tertiary rocks of New Jersey and the Central Hudson Valley, p. 25–103.

Manspeizer, W., Puffer, J. H., and Cousminer, H. L., 1978, Separation of Morocco and eastern North America; A Triassic-Liassic stratigraphic record: Geological Society of America Bulletin, v. 89, p. 901–920.

Mattick, R. E., 1982, Significance of the Mesozoic carbonate bank-reef sequence for the petroleum geology of the Georges Bank basin, *in* Scholle, P. A., and Wenkam, C. R., eds., Geological studies of the COST Nos. G-1 and G-2 wells, United States North Atlantic outer continental shelf: U.S. Geological Survey Circular 861, p. 93–104.

Mattick, R. E., Schlee, J. S., and Bayer, K., 1981, The geology and hydrocarbon potential of the Georges Bank-Baltimore Canyon area, *in* Kerr, J. M., and Ferguson, A. J., eds., Geology of the North Atlantic borderlands: Canadian Society of Petroleum Geologists Memoir 7, p. 461–486.

McIver, N. L., 1972, Cenozoic and Mesozoic stratigraphy of the Nova Scotia shelf: Canadian Journal of Earth Sciences, v. 71, p. 54–70.

Miller, K. G., and Fairbanks, R. G., 1983, Evidence for Oligocene-Middle Miocene abyssal circulation changes in the western North Atlantic: Nature, v. 306, p. 250–253.

Morgan, W. J., 1981, Hot spot tracks and the opening of the Atlantic and Indian Oceans, *in* Emiliani, C., ed., The Sea: New York, John Wiley and Sons, v. 7, p. 443–487.

Mountain, G. S., and Tucholke, B. E., 1985, Mesozoic and Cenozoic geology of the U.S. Atlantic Continental slope and rise, *in* Poag, C. W., ed., Geologic evolution of the United States Atlantic Margin: New York, Van Nostrand-Rinehold, p. 293–342.

Olsen, P. E., 1980, Fossil great lakes of the Newark supergroup in New Jersey, *in* Manspeizer, W., ed., Field Studies of New Jersey Geology and Guide to Field Trips: New York State Geological Association, p. 352–398.

Olsen, P. E., McCune, A. R., and Thompson, K. S., 1981, Correlation of the Early Mesozoic Newark Supergroup (Eastern North America) by vertebrates, especially fishes: American Journal of Science, v. 282, p. 1–44.

Oxley, P., 1981, Exploration potential of Georges Bank: American Association of Petroleum Geologists Bulletin, v. 65, no. 9, p. 1668.

Pinet, P. R., and Popenoe, P., 1982, Blake Plateau; Control of Miocene sedimentation patterns by large-scale shifts of the Gulf Stream axis: Geology, v. 10, p. 257–259.

Pinet, P. R., Popenoe, P., and Nelligan, D. F., 1981, Gulf Stream; Reconstruction of Cenozoic flow patterns over the Blake Plateau: Geology, v. 9, p. 266–270.

Poag, C. W., 1982a, Foraminiferal and seismic stratigraphy, paleoenvironments and depositional cycles in the Georges Bank Basin, *in* Scholle, P. A., and Wenkam, C. R., eds., Geological studies of the COST Nos. G-1 and G-2 wells, United States North Atlantic outer continental shelf: U.S. Geological Survey Circular 861, p. 43–91.

———, 1982b, Stratigraphic reference section for Georges Bank basin; Depositional model for New England Passive Margin: American Association of Petroleum Geologists Bulletin, v. 66, p. 1021–1041.

———, 1985, Depositional history and stratigraphic reference section for Central Baltimore Canyon Trough, *in* Poag, C. W., ed., Geologic evolution of the United States Atlantic margin: New York, Van Nostrand-Reinhold, p. 217–264.

Poag, C. W., and Schlee, J. S., 1984, Depositional sequences and stratigraphic gaps on the submerged U.S. Atlantic margin: American Association of Petroleum Geologists Memoir 36, p. 165–182.

Poag, C. W., and Watts, A. B., eds., 1987, Initial Reports of the DSDP: Washington, D.C., U.S. Government Printing Office, v. 95 (in press).

Popenoe, P., 1985, Seismic stratigraphy and Tertiary development of the North Carolina continental margin, *in* Poag, C. W., ed., Geological Evolution of the U.S. Atlantic Continental Margin: New York, Van Nostrand-Rinehold, p. 125–188.

Reinemund, J. A., 1955, Geology of the Deep River coal field, North Carolina: U.S. Geological Survey Professional Paper 246, 159 p.

Riggs, S. R., 1984, Paleoceanographic model of Neogene phosphorite deposition, U.S. Atlantic continental margin: Science, v. 223, no. 4632, p. 123–131.

——, 1986, Phosphogenesis and its relationship to exploration for Proterozoic and Cambrian phosphorites, *in* Cook, P. J., and Shergold, J. H., eds., Proterozoic and Cambrian Phosphorites: Cambridge University Press, v. 1, p. 352–369.

Riggs, S. R., Snyder, S.W.P., Hine, A. C., Snyder, S. W., Ellington, M. D., Mallette, P. M., and Stewart, T. L., 1985, Phosphate resource potential in Onslow Bay, North Carolina continental shelf: Economic Geology, v. 80, p. 716–738.

Rona, P. A., 1982, Evaporation at passive margins, *in* Scrutton, R. A., ed., Dynamics of passive margins: American Geophysical Union Geodynamic Series, v. 6, p. 116–132.

Ryan, W.B.F., Cita, M. B., Miller, E. L., Hanselman, D., Nesteroff, W. D., Hecker, B., and Nibblelink, M., 1978, Bedrock geology in New England Submarine Canyons: Oceanologica Acta, v. 1, p. 233–254.

Salvan, H. M., 1972, Les saliferes Marocains, leurs caracteristiques et leurs problems, *in* Richter-Burburg, R., ed., Geologie des depots salins: UNESCO, Sco. Terre, no. 7, p. 147–159.

Schlee, J. S., 1977, Stratigraphy and Tertiary development of the continental margin east of Florida: U.S. Geological Survey Professional Paper 581-F, 25 p.

——, 1981, Seismic stratigraphy of the Baltimore Canyon trough: American Association of Petroleum Geologists Bulletin, v. 65, p. 26–53.

Schlee, J. S., and Fritsch, J., 1982, Seismic stratigraphy of the Georges Bank complex, offshore New England, *in* Watkins, J. L., and Drake, C. L., eds., Studies in Continental margin geology: American Association of Petroleum Geologists Memoir 34, p. 223–251.

Schlee, J. S., and Jansa, L. F., 1981, The paleoenvironment and development of the eastern North American continental margin, *in* Geology of continental margins; International Geological Congress, 26th Paris, July 7–17, 1980, Colloque C3: Oceanologica Acta, no. SP, supp. to v. 4, p. 71–80.

Schlee, J. S., and Klitgord, K. D., 1980, Structural Development of Georges Bank: Geological Society of America Abstracts with Programs, v. 12, no. 2, p. 81.

Schlee, J. S., and others, 1976, Regional framework off northeastern United States: American Association of Petroleum Geologists Bulletin, v. 60, no. 6, p. 926–951.

Schlee, J. S., Dillon, W. P., and Grow, J. A., 1979, Structure of the continental slope off the eastern United States, *in* Doyle, L. J., and Pilkey, O. H., eds., Geology of continental slopes: Society of Economic Paleontologists and Mineralogists Special Paper 27, p. 95–118.

Schlee, J. S., Poag, C. W., and Hinz, K., 1985, Seismic stratigraphy of the continental slope and rise seaward of Georges Bank, *in* Poag, C. W., ed., Geologic evolution of the United States Atlantic Margin: New York, Van Nostrand-Rinehold, p. 265–292.

Scholle, P. A., ed., 1977, Geological studies on the COST No. B-2 well, U.S. mid-Atlantic Outer Continental Shelf area: U.S. Geological Survey Circular 750, 71 p.

——, ed., 1979, Geological studies on the COST GE-1 well, United States South Atlantic Outer Continental Shelf area: U.S. Geological Survey Circular 800, 114 p.

——, ed., 1980, Geological studies of the COST No. B-3 well, United States mid-Atlantic Continental Slope area: U.S. Geological Survey Circular 833, 132 p.

Scholle, P. A., and Wenkam, C. R., eds., 1982, Geological studies of the COST G-1 and G-2 wells, United States North Atlantic Outer Continental Shelf area: U.S. Geological Survey Circular 861, 193 p.

Sheridan, R. E., and Gradstein, F. M., eds., 1983, Initial Reports of the Deep Sea Drilling Program: Washington, D.C., U.S. Government Printing Office, v. 76, 943 p.

Snyder, S.W.P., 1982, Seismic stratigraphy within the Miocene Carolina Phosphogenic Province; Chronostratigraphy, paleotopographic controls, sea-level cyclicity, Gulf Stream Dynamics, and the resulting depositional framework [M.S. thesis]: Chapel Hill, University of North Carolina, 183 p.

Snyder, S.W.P., Hine, A. C., and Riggs, S. R., 1982, Miocene seismic stratigraphy, structural framework, and sea-level cyclicity, North Carolina continental shelf: Southeastern Geology, v. 23, p. 247–266.

Tucholke, B. E., and Vogt, P. R., eds., 1979, Initial reports of the Deep Sea Drilling Project: Washington, D.C., U.S. Government Printing Office, v. 43, 1115 p.

Uchupi, E., Sancetta, C., Eusden, J. D., Jr., Bolmer, S. T., McConnell, R. L., and Lambiase, J. J., 1984, Lithofacies maps, *in* Rabinowitz, P. D., and Ewing, J. I., Eastern North American continental margin and adjacent ocean floor, 34° to 41°N and 68° to 78°W: Woods Hole, Massachusetts, Marine Science International, Ocean margin drilling program Regional Atlas Series 4, 32 p.

Vail, P. R., and Hardenbol, J., 1979, Sea-level changes during the Tertiary: Oceanus, v. 22, p. 71–79.

Vail, P. R., and Mitchum, R. M., Jr., 1979, Global cycles of relative changes of sea level from seismic stratigraphy, *in* Watkins, J.S . Montadert, L., and Dickerson, P. W., Geological and Geophysical Investigations of Continental Margins: American Association of Petroleum Geologists Memoir 29, p. 469–472.

Vail, P. R., and others, 1977, Seismic stratigraphy and global changes of sea level, *in* Payton, C. E., ed., Seismic Stratigraphy; Applications to Hydrocarbon Exploration: American Association of Petroleum Geologists Memoir 26, p. 49–212.

Valentine, P. C., 1982, Calcareous nannofossil biostratigraphy and paleoenvironment of two deep stratigraphic wells in the Georges Bank basin, *in* Scholle, P. A., and Wenkam, C. R., eds., Geological studies of the COST Nos. G-1 and G-2 wells, United States North Atlantic Outer Continental Shelf area: U.S. Geological Survey Circular 861, p. 34–42.

Van Hinte, J. E., and Wise, S. W., eds., 1987, Initial Reports of the DSDP: Washington, D.C., U.S. Government Printing Office, v. 93 (in press).

Van Houten, F. B., 1962, Cyclic sedimentation and the origin of analcime-rich upper Triassic Lockatong Formation, west-central New Jersey and adjacent Pennsylvania: American Journal of Science, v. 260, p. 561–576.

——, 1969, Late Triassic Newark Group, North-Central New Jersey and adjacent Pennsylvania and New York, *in* Subitsky, S., ed., Geology of selected areas in New Jersey and eastern Pennsylvania: New Brunswick, New Jersey, Rutgers University Press, p. 314–347.

——, 1977, Triassic-Liassic deposits, Morocco and eastern North America; A comparison: American Association of Petroleum Geologists Bulletin, v. 61, p. 79–99.

Von Rad, U., Hinz, K., Sarnthein, M., and Seibold, E., eds., 1982, Geology of the Northwest African continental margin: Berlin, Springer-Verlag, 703 p.

Wade, J. A., 1980, Geology of the Canadian Atlantic margin from Georges Bank to the Grand Banks; Geology of the North Atlantic Borderlands: Canadian Society of Petroleum Geologists Memoir 7, p. 447–460.

Weed, E.G.A., Minard, J. P., Perry, W. J., Jr., Rhodehamel, E. C., and Robbins, E. I., 1974, Generalized pre-Pleistocene geologic map of the northern United States continental margin: U.S. Geological Survey Miscellaneous Geologic Investigations Map I-861, scale 1:1,000,000.

Williams, G. L., 1974, Report on the Paleontological analyses of Shell Mohican 1-100: Geological Survey of Canada Internal Report EPGS-PA1, p. 9–74.

Ziegler, P. A., 1981, Evolution of sedimentary basins in north-west Europe, *in* Petroleum Geology of the Continental shelf of north-west Europe: London, Institute Petroleum, p. 3–39.

Manuscript Accepted by the Society July 18, 1986

ACKNOWLEDGMENTS

The authors gratefully acknowledge the many helpful comments, ideas, and suggestions of our colleagues at the U.S. Geological Survey that have added to our knowledge about the paleoenvironments of the Atlantic Margin during the Cenozoic and Mesozoic. Among our associates are Elazar Uchupi (Woods Hole Oceanographic Institution), Kim Klitgord, John Grow, W. P. Dillon, Peter Popenoe and C. W. Poag (all of the USGS). Reviewers included L. F. Jansa, A. R. Palmer, Peter Popenoe, Jorn Thiede and scientists from ARCO. The manuscript was improved by the many suggestions given by the Technical Reports Unit in Reston and by the drafting efforts of Patty Forrestel and Jeff Zwinakis and the typing efforts of Peggy Mons-Wengler.

Printed in U.S.A.

The Geology of North America
Vol. I-2, The Atlantic Continental Margin: U.S.
The Geological Society of America, 1988

Chapter 17

A large aperture seismic experiment in the Baltimore Canyon Trough

John B. Diebold
Lamont-Doherty Geological Observatory of Columbia University, Palisades, New York, 10964
Paul L. Stoffa
Institute for Geophysics, University of Texas, Austin, Texas 78712
The LASE Study Group*

INTRODUCTION

A large aperture seismic experiment (LASE) was conducted in the area of the Baltimore Canyon Trough (Fig. 1) by scientists and ships from the Lamont-Doherty Geological Observatory (L-DGO) of Columbia University, the Woods Hole Oceanographic Institution (WHOI), the University of Texas Institute for Geophysics (UTIG), and the Bedford Institute of Oceanography (BIO). Three ships, *Oceanus, Moore,* and *Dawson,* were used to acquire a body of exploration seismic data using a new and innovative method, synthetic aperture Common Depth Point (CDP) profiling, as well as the well established expanding spread profile method. This project was created to obtain information about velocities and structures of deep horizons that had been poorly resolved by previous surveys or by drilling.

The main scientific objectives were defined by questions arising from the analysis of USGS Line 25. These included determining the nature of material underlying Jurassic carbonates and the existence of an underlying basement ridge at the present shelf edge. It was also hoped that any structures associated with the east coast magnetic anomaly might be revealed. It was expected that the large offsets employed in the multi-ship CDP profiling method would allow improved velocity resolution and better rejection of multiple reverberation than in the case of conventional, single ship data.

Since this project involved innovative techniques of acquisition, processing, and analysis, it is necessary to describe them to some extent. The geological results are no less important, however, and provide important benchmarks for future interpretations of the development of the North Atlantic margin.

DATA ACQUISITION

LASE CDP Profiling

A large seismic aperture array was synthesized by using

three ships: the *Fred Moore* from UTIG; the *Oceanus* from WHOI; and the *Dawson* from BIO.

The ships were positioned with the *Dawson* in the lead, approximately 6.5 km in front of the *Fred Moore,* while *Oceanus* followed the tail buoy of the *Moore* seismic array (Fig. 2). Identical source arrays consisting of one 2,000 in³ and one 1,000 in³ air gun were deployed from the *Moore* and the *Dawson.* These seismic sources were fired alternately by the *Dawson* (on the minute) and the *Moore* (on the half-minute). Both the *Moore* and the *Oceanus* received and recorded seismic data for all the shots. When *Moore* fired, it acquired conventional CDP data in the offset range of 0 to 3.6 km and OCEANUS acquired data with source-receiver offsets between 4.0 and 6.5 km. When *Dawson* fired, *Moore* recorded data with source receiver offsets between 6.5 to 11.0 km and *Oceanus* recorded the 11.0 to 14.0 km range.

The distance between the ships at every shot was determined using Mini Ranger. Shot times were determined using identical National Bureau of Standard Geostationary Operational Environmental Satellite clock receivers and systron Donner oscillators. Identical Loran-C units aboard each ship were used to steer the desired course.

Four dip lines and two strike lines, for a total of approximately 900 line km, were acquired with source-receiver offsets of 0–14 km. LASE Line 6, on which we report here, was shot directly over USGS Line 25 (Fig. 1) extending from just offshore Atlantic City to the base of the Continental Rise.

Expanding Spread Profiles

In addition to LASE CDP lines, nine Expanding Spread Profiles (ESPs) were acquired centered along Line 6 (Fig. 1). ESPs 1A ("A" for air gun), 2, and 2A were most landward, located on what was thought to be continental crust; ESPs 3 and 3A were located on the East Coast Magnetic Anomaly (ECMA); ESPs 4 and 4A were located on the Continental Slope above the outer carbonate bank; and ESPs 5 and 5A were located on the upper Continental Rise. For all of the ESPs, *Dawson* provided the sources, while *Moore* received and recorded. At the ESP 1 location, only the airgun profile could be shot because of ecologi-

---

*See acknowledgments.

Diebold, J. B., and Stoffa, P. L. and others, 1988, A large aperture seismic experiment in the Baltimore Canyon Trough; *in* Sheridan, R. E., and Grow, J. S., eds., The Geology of North America, Volume I-2, The Atlantic Continental Margin, U.S.: Geological Society of America.

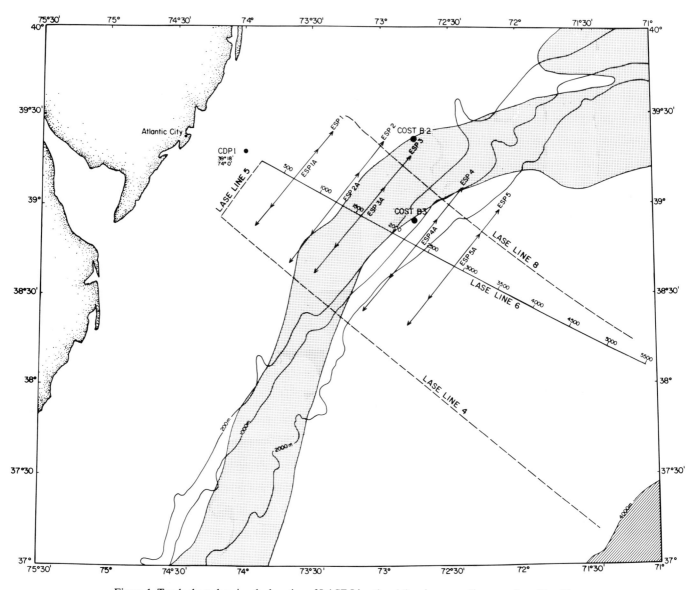

Figure 1. Track chart showing the location of LASE Line 6 and the nine expanding spread profiles. The shaded area defines the location of the East Coast Magnetic Anomaly. LASE Line 6 follows a previously acquired 48-channel profile shot with a 2,000 in (33 liter) tuned airgun array (USGS Line 25: see Grow, 1980; Schlee, 1981; Grow and others, 1983; and Grow, Schlee, and Klitgord, this volume). Previous refraction data in this area has been summarized by Sheridan and others (1979). See Scholle 1977, 1980) for drilling information from the COST B-2 and B-3 wells. Recent studies of the East Coast Magnetic Anomaly include Klitgord and Behrendt (1979), Hutchinson and others (1983), and Alsop and Talwani (1984).

cal considerations. At each of the other four locations, however, two ESPs were shot. One was acquired using the airgun source array, and the other using 25 kg explosive charges. In general, the explosive sources produced deeply penetrating arrivals, while the airguns provided greater resolution in the reflection data from the upper sedimentary column.

During the ESPs, each ship moved at about 5 kts, for a combined ship speed of 18 km/hr. For the explosive profiles, the shooting ship fired 25 kg explosive charges at 10-minute intervals,

providing essentially 100% coverage at the 70 m group spacing of the receiving array. The explosive ESP lines ranged from 70 to 110 km in length. The airgun profiles were acquired using a one-minute shooting schedule. Between shots, the ships separated only 300 m, resulting in a considerable overlap in offset coverage. By combining the data into 50 m source-receiver offset "bins" during processing, the signal-to-noise level improved significantly and deep refracted arrivals were sometimes observed for offsets up to 70 km.

LASE EXPERIMENT CONFIGURATION

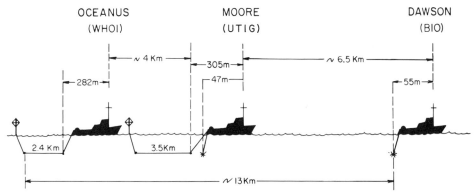

Figure 2. Schematic diagram of the LASE large aperture CDP profiling plan.

DATA ANALYSIS—CDP DATA

The *Moore* recorded 48 channels of seismic data in SEG-B format, using a DFS-IV data recording system and the *Oceanus* recorded 24 channels in a modified SEG-Y format. Shot and recording times, navigation, and the distance between ships were also digitally recorded. During processing, errors in these auxiliary data were found by inspection and corrected.

Because of the unusual profiling configuration, the data were gathered into four separate CDP sub-profiles: *Moore* shooting, *Moore* recording; *Moore* shooting, *Oceanus* recording; *Dawson* shooting, *Moore* recording; and *Dawson* shooting, *Oceanus* recording. The four sub-profiles were then merged by combining the individual gathers at common geographic coordinates. Each CDP was approximately 48 fold and separated by a distance of 50 m along a line running 126° from CDP 1, whose position was 39°18′ north latitude, 74°00′ west longitude.

Figure 3 shows two representative CDP gathers. On the left are data near the ECMA, while on the right are data from the Outer Carbonate Bank area. In both cases, traces from three successive CDP gathers are plotted with the same origin. Each display, then, represents the combination of 142 traces from different shot points acquired over more than one hour of data acquisition. On both examples, wide angle reflections and refractions can be observed, as can deep reflections at approximately 6 to 9 seconds of two-way normal time.

Examination of Figure 3 shows why the entire CDP data field could not be included in the process of determining stacking velocities. Even in relatively deep water, refractions are present at the largest offsets. Some reflected arrivals cross others, a condition that hyperbolic normal moveout cannot satisfactorily correct. Depending on the velocity function, the assumption of hyperbolic T(X) trajectories fails (Stoffa and others, 1981), resulting in a degradation, rather than enhancement of the data after stacking. Therefore, the offending parts of the data had to be muted during the process of selecting stacking velocities. When this was done, standard hyperbolic T(X) semblance velocity scans were made

and used to define stacking velocity functions. Similar mutes were used after normal moveout, and before stacking. The result of the muting was that offsets larger than the standard 4 km were included in the CDP stack only below 4.5 seconds of two-way normal time, on the average.

Before stacking, gain proportional to arrival time and trace equalization were applied to the data. To collapse the source waveform, predictive deconvolution was performed using a 200 msec filter with a 30 msec prediction distance and 1% white noise. After stacking, a zero-phase, time-varying band-pass filter was applied. In Figure 4, a running mix of nine traces was used to improve the definition of the deep structure, followed by spatial decimation by a factor of three. Thus, a CDP spacing of 150 m was used for display but each trace has been averaged over 450 m. Post-stack predictive deconvolution was also applied to reduce both source and water column reverberation.

LASE Line 6 data were migrated (after stacking) using the phase-shift method (Gazdag, 1978; Dubrulle and Gazdag, 1979). The interval velocity structure used was obtained by converting stacking velocities in the upper part of the section, extended to depth using the velocity results from the Expanding Spread Profiles (see below). Figure 4 is a comparison of the migrated and unmigrated data showing a major fault at CDP 1250, just landward of the ECMA. In the migrated record section (right), the long diffraction events present in the stacked section (left) have been substantially reduced. It is now quite clear that at 6.4 seconds of two-way time, CDP 1250 marks the edge of what we interpret as a major normal fault associated with the early rifting of this margin (Alsop and Talwani, 1984). On the lower right of both sections the downward thrown block is clearly observed. This fault can be identified in other lines in the Baltimore Canyon Trough and is always coincident with the ECMA, indicating that both features developed during the early rifting of the margin.

DATA ANALYSIS—ESP DATA

Interval velocities were derived from the LASE ESP data by analyzing seismic traveltime trajectories in both the T(X) and

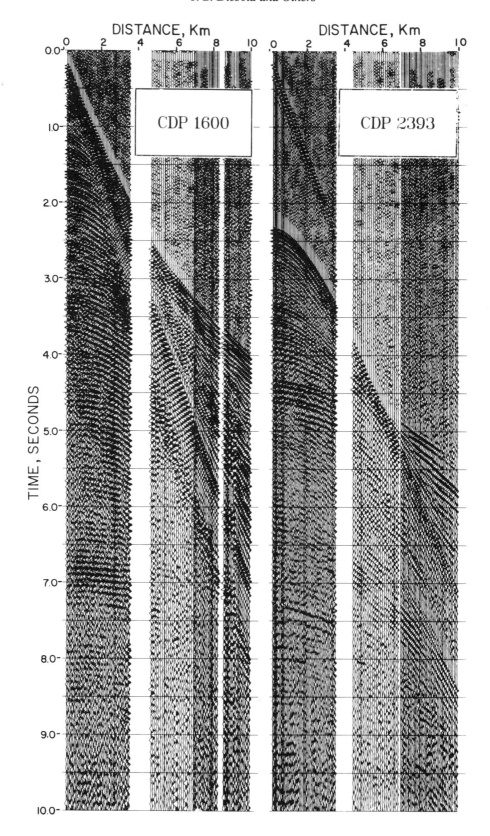

Figure 3. LASE O–10 km CDP gathers plotted at their true source-receiver offset from the area of the East Coast Magnetic Anomaly (left) and outer carbonate bank area (right). Three CDPs are displayed on each panel corresponding to a CDP group spacing of 150 m.

LASE
LINE 6

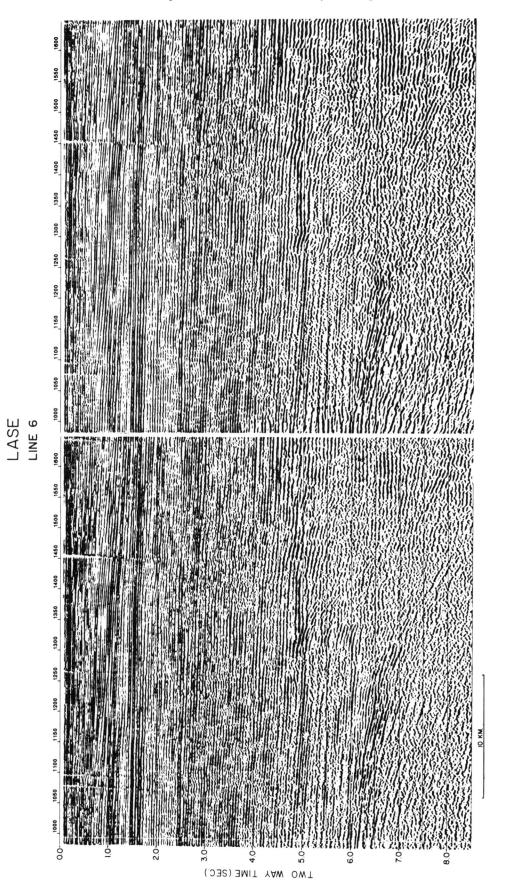

Figure 4. 0–10 km stacked CDP record section (left) and after migration (right). After migration it is clear that CDP 1250 marks the edge of a major normal fault at 6.4 seconds of traveltime.

LASE ESP 3A

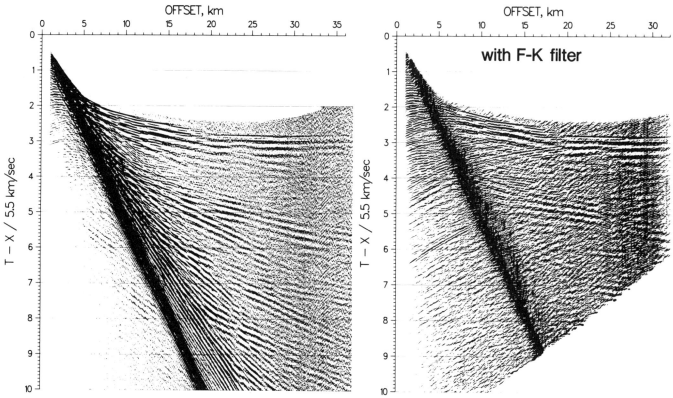

Figure 5A. (Left) LASE Expanding Spread Profile 3A, located at approximately CDP 1620. A time correction corresponding to a reducing velocity of 5.5 km/sec has been applied to separate the arrivals. (Right) LASE Expanding Spread Profile 3A, after f-k filtering, designed to eliminate water-borne energy (low phase velocity arrivals). In this record section it is quite clear that the wide angle reflections from 4 to 6.5 seconds can be followed to the larger offsets.

tau(p) domains. Both near-vertical incidence reflections (for the airgun profiles) and wide angle reflections and refractions were used. The airgun profiles were initially analyzed in the tau(p) domain and the velocity structure was refined by ray tracing and comparison with the observed T(X) data. Starting with these results, which gave more detailed solutions for the sedimentary section, the explosive ESP data were then analyzed in a similar fashion to obtain interval velocities for the crust. In the case of ESP 1, where no explosives were shot, the velocity results are based entirely on the airgun data.

The method of analysis used for the airgun Expanding Spread Profiles is illustrated for ESP 3A (Fig. 5). The T(X) data for source-receiver offsets of 0–35 km were transformed to the tau(p) domain using data with a ray parameter sampling interval of 2 msec/km. The critically refracted and postcritically reflected arrivals were inverted to obtain V(Z) by the slope-intercept or "Tau-Sum" method (Diebold and Stoffa, 1981). By picking closely-spaced tau(p) points, an accurate representation of velocities and gradients in the upper part of the section was obtained. The tau(p) interval velocity model thus derived was smoothed

and then verified by comparing ray traced T(X) arrival times with the original data. In this way, information from precritical reflections and critical points was included in the inversion process.

The final velocity function shown in Figure 5B has a low velocity zone below 5 seconds of two-way time. Therefore, the tau-sum method was unsuitable for analysis below this point, and in any case, the corresponding shadow zone included no deeper postcritical arrivals in the tau(p) data. Consequently, precritical reflections were analyzed to obtain the velocity function in this zone. As a point of departure, the T^2–X^2 method of Le Pichon and others (1968) was used. The data were phase velocity filtered to eliminate arrivals with low phase velocities and to minimize the interference of the seafloor reflection and its multiples. This process, followed by deconvolution, greatly enhanced the wide angle reflections used in the T^2–X^2 analysis (Fig. 5A).

The T^2–X^2 solution consisted of the velocity and thickness for a single, homogeneous layer. Ray tracing subsequently showed that this model was too simple to fit the reflections well. Since no identifiable reflections or refractions could be detected

LASE ESP 3A

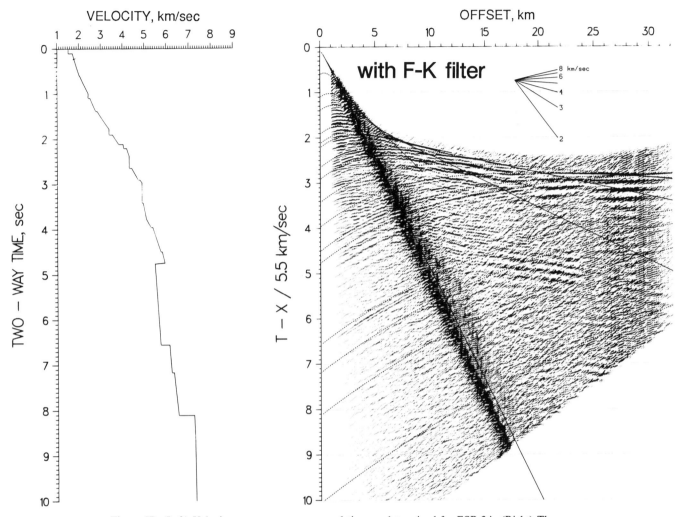

Figure 5B. (Left) Velocity versus two-way normal time as determined for ESP 3A. (Right) The ray traced arrivals predicted for the velocity-depth model superimposed on the f-k filtered data.

from within the layer, a velocity gradient was assumed, and iterative ray tracing was performed to provide a better fit. Velocities below 6.5 seconds for this ESP were obtained entirely from the explosive ESP data (see below).

In general, because of the poor source characteristics of the explosive ESPs and the loss of resolution at short offsets due to interference, the data are of little use for detailed resolution of the upper sedimentary column. The source levels, however, were great enough that arrivals from deeper horizons can be seen at large offsets, thereby complementing the airgun ESP data.

Velocity functions for the explosive ESPs were derived by forward traveltime modeling. Starting with the velocity models obtained from the airgun data for the sedimentary column, the velocity functions for the deep part of the section were developed using ray tracing, layer by layer, from the top down. The ray

tracing was performed assuming laterally homogeneous layers with velocity gradients that were linear with depth, and/or vertically homogeneous layers.

ESP 2 (Fig. 6) is the most landward of the explosive ESPs. The minimum offset in the T(X) data is 6 km, and continuous first arrivals are not seen until 12 km, due to errors in the shot and/or record timing. Therefore, the uppermost part of the velocity model is constrained only by the ESP 2A airgun data. Between 12 and 67 km, the explosive ESP 2 data contain a large number of arrivals with greatly variable phase velocities and amplitudes. The velocity function between 2.5 and 4.5 seconds of two-way time was determined by matching first arrivals. The points at which the modeled arrivals must cross were defined by breaks in slope of the first arrivals, and the velocity gradients were controlled by arrival curvatures, as well as the range of offsets

LASE ESP 2

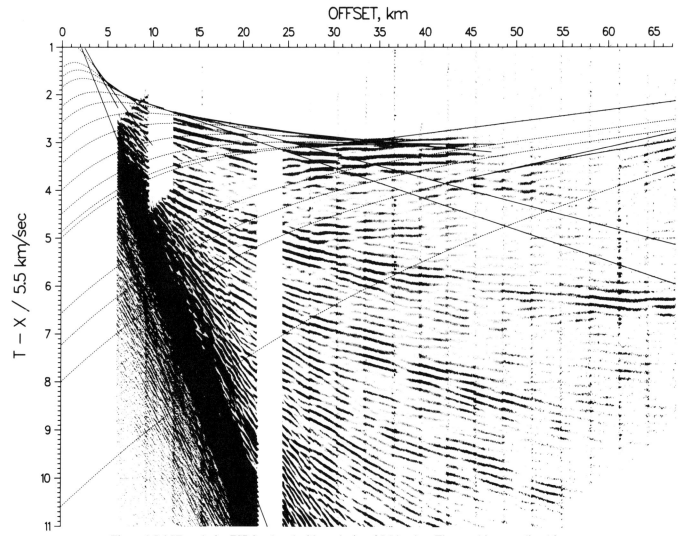

Figure 6. LASE explosive ESP 2 reduced with a velocity of 5.5 km/sec. The traveltimes predicted from the final model are superimposed with precritical arrivals shown as dashed lines; postcritical as solid lines.

over which they are seen with appreciable amplitude. Beyond 30 km, secondary arrivals played an important role in determining the velocity function. In some cases, such as the reflection from the 7.2 km/sec layer just below 8 seconds of two-way time, coherent arrivals can be followed over a range of 30 km or more. The airgun ESP 2A revealed near-vertical reflections between 4.5 and 8.2 seconds that could be correlated with wide angle arrivals seen in explosive ESP 2.

The fact that arrivals can be correlated and timed over such large instances strongly constrains the velocity function. The combined velocity functions derived from ESPs 2, 3, and 4 all contain a low velocity zone (LVZ) below the prograding carbon-

ate sequence (Figs. 7 and 8). Reflected arrivals from above and below this zone are particularly well developed in ESP 3A and between 5 and 6.5 seconds of two-way normal time (Fig. 5B).

In the other ESP data, a reflection could not be detected from above the LVZ, due perhaps to a gradational transition, and this assumption is reflected in the velocity solutions of Figure 7. In these ESPs, the reflection from beneath the LVZ is usually seen, and the velocity can be estimated by the location of critical point, if any.

DISCUSSION

LASE Line 6 was laid out to coincide with USGS Line 25

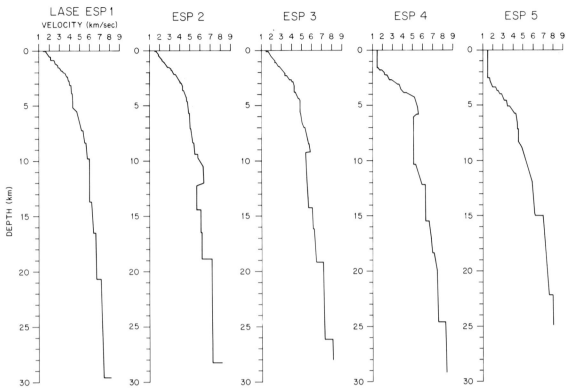

Figure 7. Velocity depth function for LASE ESPs. Note the low velocity zone underlying the carbonate sequence in ESPs 2, 3, and 4 and the deep 7.2 km/sec layer.

for two reasons; to allow a comparison of the results of the new multi-ship synthetic aperture profiling method with the more conventional single ship CDP results, and so the ESP results could be used to aid in the interpretation of both lines. As outlined above, the added offset of the multi-ship method does not come into play during CDP stacking until great depths are reached. This, coupled with the relatively poor source arrays available during LASE, make the USGS line the clear "winner," at least in the post-Jurassic section. The high resolution velocity results obtained from analysis of the ESP and LASE CDP data do not depart significantly from the velocities from Line 25, except in the amount of detail now available. Therefore, aside from its role as the pilot study for the feasibility of the multi-ship method, LASE's main contribution lies in its imaging and velocity information of deeper horizons.

Figure 9 shows the major geologic horizons observed in either the LASE Line 6 stacked record section or inferred from the interpretation of the ESP data. Also shown are the magnetic data. In Figure 9, we see a reflector near ESP 1, at a two-way normal time of approximately 7.5 seconds. Analysis of the ESP 1 data (Fig. 7) indicates an increase in velocity at this two-way normal time from 6.4 to 6.65 km/sec. We interpret this reflector as the top of the continental basement. In the overlying section between 6.2 and 7.2 seconds the velocity function is more transitional and the section probably consists of synrift sediments and volcanics. In ESPs 2 and 3, a 7.2 km/sec reflector is seen with

two-way normal times of 8.0 and 8.1 seconds, respectively. Some reflection energy is present in the stacked section at these times, but the signal level is too low to permit a reliable correlation. Between ESP 3 and ESP 4 no coherent strong reflection events are observed. On the seaward part of the profile, oceanic basement can be visually traced landward to approximately CDP 3300.

Landward of CDP 3300, the ESP velocity results indicate no dramatic arching or basement high beneath the prograding Jurassic carbonates. Though the Moho seems to rise at the location of ESP 3 (Fig. 9) no corresponding elevation is seen in the V(Z) functions (Fig. 7), indicating that the apparent Moho high is an artifact due to the effect of the increasing water depth. The ESP results indicate instead that the Moho rises seaward in an orderly fashion.

Beginning at about CDP 1500, and at 4 seconds, a high velocity zone associated with a prograding carbonate sequence is observed (Gamboa and others, 1985). The prograding carbonate sequence continued seaward to the outer reef complex at the site of ESP 4. This sequence produces a series of high amplitude reflection events which overlie a zone of lower velocity material, possibly salt or shale. No coherent reflections are observed from within this low velocity zone. Analysis and examination of the data from ESPs 3A and 4A seem to indicate that there are small-scale velocity fluctuations within this LVZ, but it is hard to characterize these more definitely. In any case, the LVZ is a

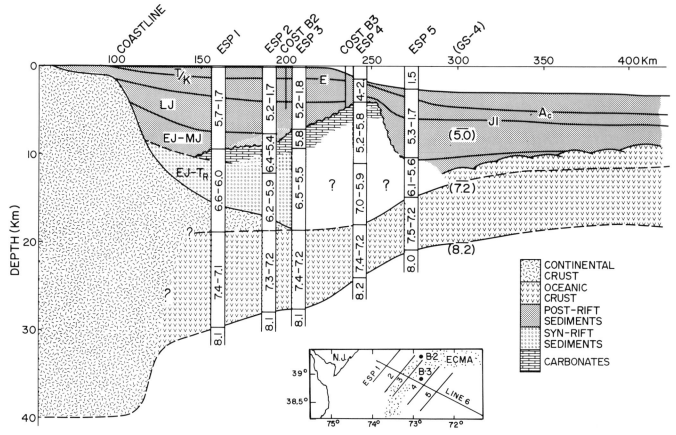

Figure 8. Seismic cross section of the continental margin with inferred geology. Seismic velocity ranges shown in km s⁻¹ under ESPs; velocities in brackets from conventional profile (4). The inset shows the location of the seismic measurements relative to the East Coast Magnetic Anomaly (stippled) and COST wells B-2 and B-3. Profile GS-4 is from earlier sonobuoy refraction data (Sheridan and others, 1979).

consistent feature in each of the four seaward ESP locations. The solution for ESP 1 does not require a low velocity zone, but its presence might have been missed, since no explosive ESP data could be obtained at that location.

CONCLUSIONS

The Large Aperture Seismic Experiment has produced preliminary results (Alsop and Talwani, 1984; and Gamboa and others, 1985) including three outstanding features.

First, the migrated CDP stacked data strongly support the existence of a major fault near CDP 1250. The velocity-depth results of ESPs 1 and 2, which straddle this feature, provide additional information for its interpretation. Landward of CDP 1250, ESP 1 shows a steady increase in velocity, with discontinuities at 5.0 and 6.3 seconds of two-way time, both of which are disturbed by the faulting (Fig. 4). The velocity between these discontinuities is 6.0 km/sec, increasing from 6.2 to 6.4 km/sec beneath the lower horizon. On the seaward side of the fault, corresponding velocities are not seen in ESP 2 until much deeper,

in the interval between 6.6 and 7.2 seconds. This observation, coinciding with the landward edge of the ECMA (Fig. 9), suggests that this fault marks an important change in the rifting process with pre- and synrift sediments landward, and lower velocity postrift sediments seaward.

Second, seaward of this fault, ESPs 2, 3, and 4 all show a similar sequence consisting of a high velocity "cap" overlying sediments with lower velocities (Fig. 7). The high velocity layer correlates with what is interpreted in the reflection profile to be a prograding carbonate sequence, overlying shallow marine shales, sands, and fore reef deposits, which built up as the margin subsided. These velocities are very well determined, as described above by Gamboa and others (1985), largely due to the large offsets and high signal-to-noise ratios of the airgun ESP data.

Third, a deep layer with seismic velocity of about 7.2 km/sec is observed at all ESP locations. This velocity is characteristic of both oceanic Layer 3b and the deep crustal layer observed beneath continental crust at this and other passive margins. The 7.2 km/sec layer appears to be continuous across the entire Baltimore Canyon Trough transect. If it is as ubiquitous as our (unfortunately, coarse) ESP coverage seems to indicate, the im-

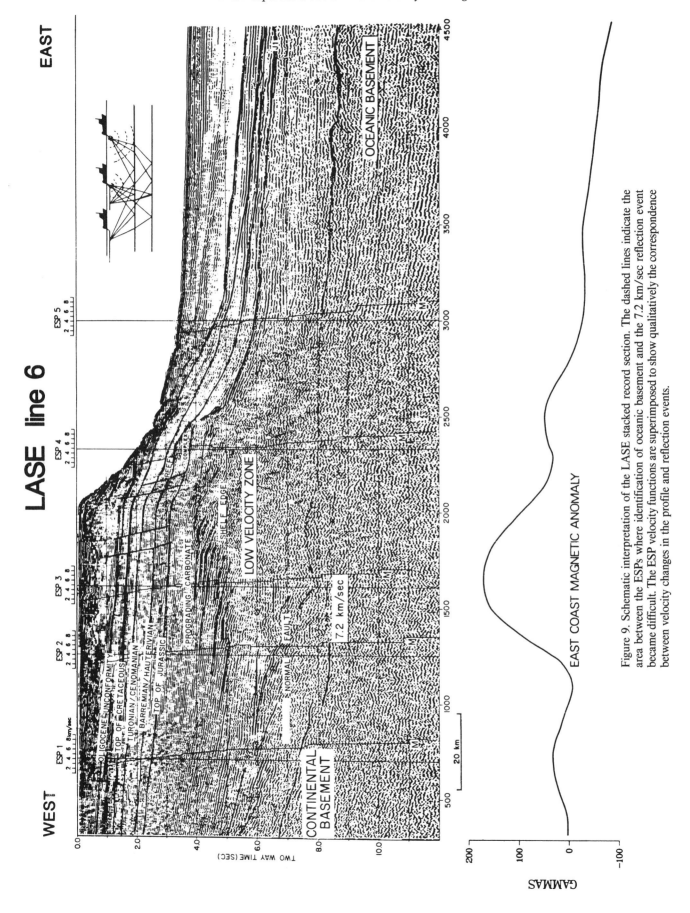

Figure 9. Schematic interpretation of the LASE stacked record section. The dashed lines indicate the area between the ESPs where identification of oceanic basement and the 7.2 km/sec reflection event became difficult. The ESP velocity functions are superimposed to show qualitatively the correspondence between velocity changes in the profile and reflection events.

plication is that the rifting process involves production of this high velocity material in such a way that its top has a velocity increase sharp enough to produce seismic arrivals that appear to be reflections. Velocities in the neighborhood of 7.2 km/sec have been detected at various rifted zones (Keen and others, 1975; Talwani and others, 1978; Weigel and others, 1982), and in crust of unknown origin overlain by continental crust (Harris and Bayer,

1979; Cook and Oliver, 1981). The results of the LASE ESPs indicate continuity of 7.2 km/sec material between rifted and normally created oceanic crust, and suggest (but do not prove) a continuity with similar material within continental crust. The LASE results will, in any case serve as an additional constraint and clue for future determinations of the nature of crustal formation.

REFERENCES CITED

Alsop, L. E., and Talwani, M., 1984, The east coast magnetic anomaly: Science, v. 226, p. 1189–1191.

Buhl, P., Diebold, J. B., and Stoffa, P. L., 1982, Array length magnification through the use of multiple sources and receiving arrays: Geophysics, v. 47, no. 3, p. 311–315.

Cook, F. A., and Oliver, J., 1981, The late Precambrian–early Paleozoic continental edge of the Appalachian orogen: American Journal of Science, v. 218, p. 993–1008.

Diebold, J. B., and Stoffa, P. L., 1981, The traveltime equation, tau-p mapping, and inversion of common midpoint data: Geophysics, v. 45, no. 3, p. 238–254.

Dubrulle, A. A., and Gazdag, J., 1979, Migration of phase shifts; An algorithmic description for array processors: Geophysics, v. 44, p. 1661–1666.

Gamboa, L. A., Truchan, M., and Stoffa, P. L., 1985, Middle and upper Jurassic depositional environments at the outer shelf and slope of the Baltimore Canyon Trough: American Association of Petroleum Geologists, v. 69, p. 610–621.

Gazdag, J., 1978, Wave equation migration with phase shift method: Geophysics, v. 43, p. 1342–1351.

Grow, J. A., 1980, Deep structure and evolution of the Baltimore Canyon Trough in the vicinity of COST No. B-3 well, in Scholle, P. A., ed., Geological studies of the COST No. B-3 well, United States Mid-Atlantic Continental Slope area: U.S. Geological Survey Circular 833, p. 117–126.

Grow, J. A., Hutchinson, D. R., Klitgord, D. D., Dillon, W. P., and Schlee, J. S., 1983, Representative multichannel seismic reflection profiles over the U.S. Atlantic continental margin, in Bally, A. W., ed., Seismic expression of structural styles: American Association of Petroleum Geologists Studies in Geology Series Number 15, p. 2.2.3-1 to 2.2.3-19.

Harris, L. D., and Bayer, K. C., 1979, Sequential development of the Appalachian orogen above a master décollement; A hypothesis: Geology, v. 7, p. 568–572.

Hutchinson, D. R., Grow, J. A., Klitgord, K. D., and Swift, B. A., 1983, Deep structure and evolution of the Carolina Trough, in Watkins, J. S., and Drake, C. L., eds., Studies in continental margin geology: American Association of Petroleum Geologists Memoir 34, p. 129–152.

Keen, C. E., Keen, M. J., Barrett, D. L., and Heffler, D. G., 1975, Some aspects of the ocean-continent transition at the continental margin of eastern North America, in Offshore geology of eastern Canada: Geological Survey of Canada Paper 74-30, p. 189–197.

Klitgord, K. D., and Behrendt, J. C., 1979, Basin structure of the U.S. Atlantic Margin, in Watkins, J. S., Montadert, L., and Dickerson, P. W., eds., Geological and geophysical investigations of continental margins: American Association of Petroleum Geologists Memoir 29, p. 85–112.

Le Pichon, X., Ewing, J., and Houtz, R., 1968, Deep-sea sediment velocity determination made while reflection profiling: Journal of Geophysical Research, Paper 74-30, p. 189–197.

Schlee, J. S., 1981, Seismic stratigraphy of Baltimore Canyon Trough: American Association of Petroleum Geologists Bulletin, v. 65, p. 26–53.

Scholle, P. A., ed., 1977, Geological studies of the COST No. B-3 well, U.S.

Mid-Atlantic Outer Continental Shelf: U.S. Geological Survey Circular 750, 71 p.

—— , ed., 1980, Geological studies of the COST No. B-3 well, U.S. Mid-Atlantic Continental Slope area: U.S. Geological Survey Circular 833, 132 p.

Sheridan, R. E., Grow, J. A., Behrendt, J. C., and Bayer, K. C., 1979, Seismic refraction study of the continental edge off the eastern United States: Tectonophysics, v. 59, p. 1–26.

Stoffa, P. L., and Buhl, P., 1979, Two-ship multichannel seismic experiments for deep crustal studies; Expanded spread and constant offset profiles: Journal of Geophysical Research, v. 84, p. 7645–7660.

Stoffa, P. L., Diebold, J. B., and Buhl, P., 1981, Inversion of seismic data in the tau-p plane: Geophysical Research Letters, v. 8, no. 8, p. 869–872.

Talwani, M., Mutter, J. C., Houtz, R., and Konig, M., 1978, The margin south of Australia; A continental margin paleorift, in Ramberg, I. E., and Newmann, F. R., eds., Tectonics and geophysics of continental rifts: Dordrecht, Holland, D. Reidel, p. 203–219.

Weigel, W., Wissman, G., and Goldflam, P., 1982, Deep seismic structure (Mauritania and central Morocco), in Van Rad, U., Hinz, K., Sarthein, M., and Seibold, E., eds., Geology of the northwest African continental margin: New York, Springer-Verlag, p. 132–158.

MANUSCRIPT ACCEPTED BY THE SOCIETY APRIL 15, 1986
LAMONT-DOHERTY GEOLOGICAL OBSERVATORY CONTRIBUTION NO. 3976.

ACKNOWLEDGMENTS

This work was supported by the following grants from the National Science Foundation: OCE79-22884 and OCE83-19518, and the industrial participants of the Ocean Margin Drilling Program: Atlantic Richfield, Chevron, Cities Service Company, Conoco, Exxon, Mobil, Penzoil, Phillips Petroleum, Sunmark, Union Oil, and Gulf Research and Development Company.

Canadian participation in this study was funded by Energy, Mines and Resources, through an OERD grant to the Geological Survey of Canada.

The Institutions and principal contributors to the LASE Study Group are:

J. Diebold, P. Buhl, J. Mutter, R. Mithal, J. Alsop, Lamont-Doherty Geological Observatory of Columbia University, Palisades, New York 10964.
P. L. Stoffa, J. D. Phillips, T. Stark, Institute for Geophysics, University of Texas, Austin, Texas.
J. Ewing, M. Purdy, Hans Schouten, Woods Hole Oceanographic Institution, Woods Hole, Massachusetts 02543.
D. McCowan, M. Truchan, R. Houtz, Gulf Research and Development Company, Houston, Texas 77236.
C. Keen, I. Reid, J. Woodside, B. Nicholas, Bedford Institute of Oceanography, Atlantic Geoscience Centre, Dartmouth, Nova Scotia, Canada.
T. O'Brien, U.S. Geological Survey, Woods Hole, Massachusetts 02543.

Printed in U.S.A.

The Geology of North America
Vol. I-2, The Atlantic Continental Margin: U.S.
The Geological Society of America, 1988

Chapter 18

Subsidence and basin modeling at the U.S. Atlantic passive margin

Michael S. Steckler
Lamont-Doherty Geological Observatory of Columbia University, Palisades, New York 10964
Anthony B. Watts
Lamont-Doherty Geological Observatory of Columbia University and Department of Geological Sciences, Columbia University, Palisades, New York 10964
Julian A. Thorne
ARCO Resources Technology, 2300 West Plano Parkway, Plano, Texas 75075

INTRODUCTION

It was little more than a decade ago (Sheridan, 1974) that it was realized that the sedimentary thickness at the U.S. Atlantic margin was in excess of 10 km, 2 to 3 times the previous estimates of basement depth. In the years that followed, multichannel seismic (MCS) reflection profile data first became available (Schlee and others, 1976), the first deep offshore well was drilled (COST B-2 in March 1976), and industry and academic scientists rapidly increased their knowledge of the margin.

Concurrent with the increase in data, the first models of continental margin subsidence (Sleep, 1971; Falvey, 1974; McKenzie, 1978) were developed, and techniques for studying basin subsidence were introduced (Watts and Ryan, 1976; Van Hinte, 1978; Steckler and Watts, 1978). As a result, the U.S. Atlantic margin was one of the first to be subjected to quantitative subsidence analysis, and our present view of passive margin development has been greatly influenced by research at this margin. Thus, it is appropriate that we now assess what has been learned and what information may be obtained from future subsidence studies of this margin.

BACKSTRIPPING

The ultimate source of the record of subsidence in a sedimentary basin are the sediments that fill it. The pattern of accumulation of these sediments is dependent on a number of different processes: tectonic, sedimentary, isostatic, and eustatic. The relationships between these factors are complex and are still not yet fully understood. Early subsidence studies therefore approached the problem by working backward from the data, using a technique referred to as backstripping (e.g., Watts and Ryan, 1976; Steckler and Watts, 1978).

Backstripping is an attempt to correct the sedimentary record for some of the geological processes that are active in basins

in order to determine the tectonic subsidence. Tectonic factors, by creating the space for sediments to accumulate, are the primary cause of thick sedimentary sequences. The purpose of backstripping is to calculate and remove the effects of compaction, sediment loading, changing paleobathymetry, and sea-level variations to isolate the tectonic component. The relationships between the factors affecting the sedimentary subsidence are nonlinearly coupled and, therefore, we can only approximate the influence of components of subsidence by backstripping. For example, the thermal state of the lithosphere determines the flexural response to sediment loading. Thus, the tectonic subsidence cannot be accurately deduced without already knowing the thermal state. In practice, reasonable estimates can be made of flexural parameters, and iteration can produce self-consistent models. Backstripping, when carried out carefully, has proven to be a useful tool for understanding the development of sedimentary basins.

Figure 1 illustrates the backstripping technique and some of the difficulties involved with its use on the U.S. Atlantic margin. The first column shows a summary of the stratigraphy at the COST B-2 well (Poag, 1980). Although this well drilled 4.8 km into the sedimentary section in the Baltimore Canyon, it penetrated less than half of the total sediment thickness and provides no information on the subsidence during the first 50 m.y. of the margin's history. Also, the depth to basement is poorly known. In MCS profiles, basement reflectors cannot be traced beneath the outer shelf. Seismic refraction evidence shows an 8- to 10-km-thick, 7.2- to 7.4-km/sec layer beneath the outer shelf (LASE Study Group, 1986). The upper surface of this layer, however, is not necessarily the top of basement.

The first step in backstripping is to reconstruct the sediment thicknesses and densities for previous times during margin history. This is dependent on reconstructing the progressive loss of porosity of the sediments with burial. In practice, this has been

Steckler, M. S., Watts, A. B., and Thorne, J. A., 1988, Subsidence and basin modeling at the U.S. Atlantic passive margin, *in* Sheridan, R. E., and Grow, J. A., eds., The Atlantic Continental Margin, U.S.: Geological Society of America, The Geology of North America, v. I-2.

BACKSTRIPPING

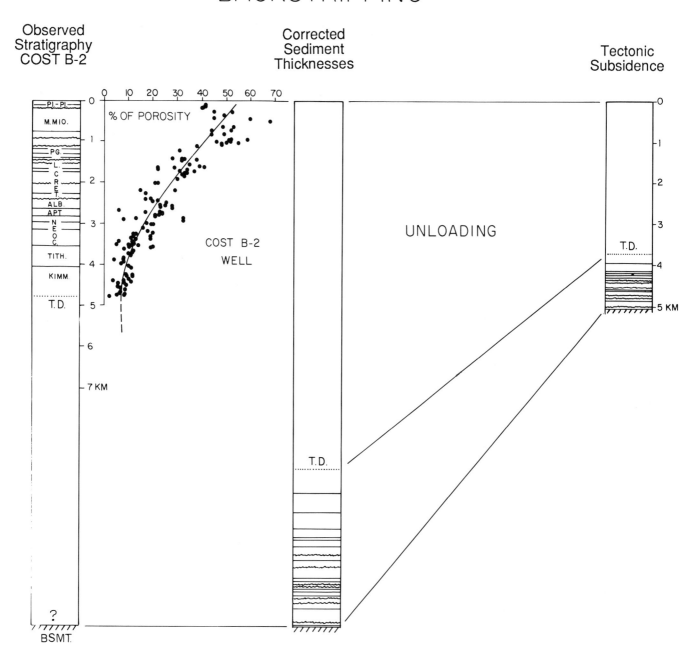

Figure 1. Diagram illustrating backstripping technique using the data from COST B-2 well. The left column indicates the observed sedimentary section (Poag, 1980). The wavy lines note unconformities and the dotted line the total depth of penetration of the well. Next to the stratigraphy, the observed porosity measurements for each depth are indicated. These data are used to calculate the central column, the total sediment thickness through time corrected for compaction of the sediment layers. Stratigraphy is inverted to show depth to basement at each time. The final column is the result of unloading the sedimentary layers according to the varying densities. The result is the net subsidence that would have occurred in the absence of sediment input or tectonic subsidence.

done in two ways. The first, illustrated in Figure 1, is to construct an empirical curve for the porosity decrease based on the data from an individual well (Steckler and Watts, 1978). A drawback of the method is that it does not allow for lithologic variations. An improvement would be to separate the different lithologies and fit a curve to each. This may not, however, give enough of a depth range for each lithology to reconstruct a curve and, furthermore, the compaction properties of shale may change dramatically downhole with compositional changes. An alternative is to use published porosity-depth relationships based on laboratory studies (Sclater and Christie, 1980). In this manner, multiple lithologies are easily accommodated. It is important when using curves determined from other data sets that the porosity data in the well of interest is matched. The variation in porosity-depth from place to place is large, and errors can be propagated by backstripping if the wrong proportion of water and rock is assumed. Bond and Kominz (1984) dealt with this problem by using two sets of porosity curves: one maximizing and one minimizing the effects of compaction.

One solution to the problem of estimating compaction is to use a group of wells within the region of interest in which the same formations and lithologies occur at different depths. In this manner, more complete local compaction curves can be estimated, although local facies variations will still be a limiting factor.

An exponential decrease of porosity with depth is the form that has most generally been used in backstripping studies (Sclater and Christie, 1980). Theoretically, porosity depends on effective overburden pressure. In most cases, the use of depth rather than overburden will make little difference. However, the development of overpressuring can make effective overburden pressure gradients up to an order of magnitude less than normal gradients. Also, if lithologies of very different properties are involved, significant errors, such as a decreasing tectonic subsidence with time, can occur. A more complete solution is to calibrate predictive models of porosity fluid flow and overpressure with a suite of borehole porosity permeability and pressure data throughout the basin of interest (Thorne, 1985). This approach is limited by the large amounts of data required.

The central column in Figure 1 shows the reconstructed basement depths at the times of the dated horizons in the COST B-2 well. Of note is that the basement subsidence for the time period sampled by the well is less than the thickness of sediments accumulated during this time. This is because some of the space for the sediments was created by compaction of the underlying layers. In general, the compaction corrections will increase the percentage of the subsidence that occurred earlier at the expense of the later subsidence.

The second step in backstripping is the isostatic removal of the sediments and other loads. The computationally simplest way to account for sediment loading is using an Airy, or local model, of isostatic adjustment. According to isostatic principles, at the depth of compensation all mass columns balance, yielding the following formula for the tectonic subsidence:

$$T = S^* \left[\frac{\rho_m - \rho_s}{\rho_m - \rho_w} \right] + WD + \left[\frac{\rho_m}{\rho_m - \rho_w} \right] SL$$

where S^* is the corrected sediment thickness, WD is the paleo-water depth, SL is eustatic sea level and ρ_m, ρ_s, and ρ_w are the mantle, average sediment, and water densities, respectively (Steckler and Watts, 1978).

It is known from previous studies (e.g., Watts and Cochran, 1974; Caldwell and others, 1976; Watts, 1978) that the response of the lithosphere to surface loads is generally not that of an Airy mechanism. Due to the strength of the upper lithosphere, the region around a load will deform. This decreases the subsidence due to the sediments directly beneath the load, but increases the subsidence in the adjacent crust. Flexure studies have shown that the shape of the response of the oceanic lithosphere is well described by the deflection of a thin elastic plate overlying a weak substratum, where the thickness of the elastic plate, T_e (and its equivalent flexural rigidity), approximates the depth to the 450°C isotherm (Watts, 1978; Bodine and others, 1981).

The rheology of the continental lithosphere, on the other hand, is only now beginning to be understood. Despite the probably presence of a weak layer in the lower crust (Bird, 1978; Molnar and Tapponnier, 1981; Vink and others, 1984), the highest values of T_e have come from continental studies (Karner and Watts, 1983; Lyon-Caen and Molnar, 1983, 1985; Ahern and Ditmars, 1985). Recent results on modeling continental rheologies (Karner and others, 1983; Willet and others, 1985; Kusznir and Karner, 1985) indicate that the rigidity of young or hot lithosphere will be controlled by the weaker crustal rheology while in old or cold lithosphere the mantle rheology will determine the flexural rigidity, as in the oceans. In addition, a thicker thermal continental lithosphere is needed to reproduce the high observed flexural rigidities (Karner and others, 1983).

A number of workers have applied the oceanic relationship for rigidity/age at continental margins. This is probably valid in the region of greatly thinned crust, but the more landward parts of the margin, if heated, may show a T_e corresponding to a lower isotherm due to weaker crustal rheology (Kusznir and Karner, 1985). This may explain, in part, the low isotherm for T_e needed by Beaumont and others (1982) in their study of the Nova Scotian margin.

The effect of flexure is to cause the sediment thicknesses at each point to depend on the overall geometry of the margin. Thus, in order to backstrip one needs to unload two- or three-dimensional layers of sediments simultaneously. We believe that while the response at a margin can be approximated using a rigidity/age relationship, flexural unloading is best carried out as an iterative process together with forward modeling of the temperature structure.

The other factors that need to be considered in backstripping are the paleobathymetry of the margin and sea-level changes. The water depth represents the margin subsidence that has not been filled with sediments and, therefore, adds directly to the tectonic subsidence. Sea-level changes produce both additional loads of

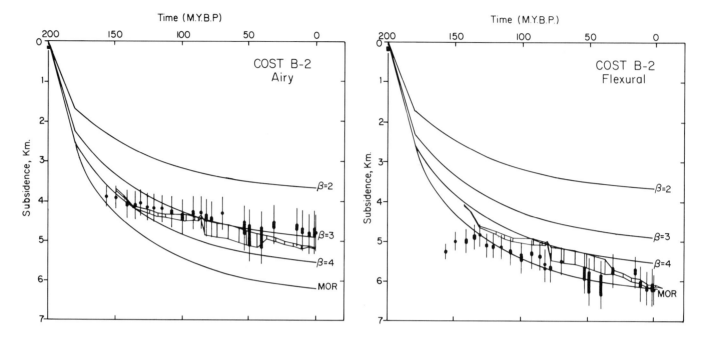

Figure 2. Comparison of backstripping results for the COST B-2 well by Steckler and Watts (1978) and Sawyer and others (1982a, 1982b). The connected lines are from Steckler and Watts (1978) where the height of the line reflects the range of water depth estimates. The heavy lines are from Sawyer and others (1982a, 1982b) where the central thicker part of the line indicates the water depth range and the thinner part an uncertainty estimate. The curves are plotted with theoretical curves from the uniform extension model (McKenzie, 1978). The Airy, or local isostasy, curves underestimate β while the flexural rigidities were not coupled to forward models and overestimating β.

water, which need to be isostatically corrected, and a change in base level.

The column on the right in Figure 1 shows the tectonic subsidence through time resulting from unloading with an Airy mechanism. Since the average density of the sediment column increases with time and thickness, once again the proportion of the subsidence represented by the more recent sediments is decreased. This demonstrates one of the main difficulties in interpreting the subsidence of the U.S. Atlantic margin. Our understanding of the causes of subsidence in the deep basins is based on only the latter one-third of the total margin subsidence.

A quantitative estimate of the errors that accumulate in backstripping at a margin is difficult to make. Sawyer and others (1982a) assumed a 5 percent error for the tectonic subsidence, while Sawyer and others (1982b) estimated a maximum error with a worst-case approach by using extreme values for densities, compaction coefficients, and flexural rigidities. Here, we estimate the uncertainty by comparing the results of two independent efforts at backstripping the COST B-2 well (Fig. 2). These studies were carried out using two alternative methods of compaction correction with different parameters for the compaction properties of the sediments and different biostratigraphic interpretations of the ages and paleowater depth of the sediments. The estimates

of present depth to basement and flexural rigidity also differed. Despite these differences, Figure 2 shows the agreement between the two sets of results is relatively close. Although the curves differ in detail, there is an overall agreement on the shape and amplitude of the curves. The estimates of the stretching factor, β (McKenzie, 1978), derived from each data set would be very similar. Judging from the comparison in Figure 2, an estimate of ± 10 percent of the tectonic subsidence appears to be a reasonable estimate of the resolution of the data.

The form of the tectonic subsidence obtained by Sawyer and others (1982a, 1982b) tends to be flatter than that obtained by Steckler and Watts (1978). This is most likely due to a difference in compaction and indicates a larger compaction effect in the studies by Sawyer and others (1982a, 1982b). The decreasing tectonics subsidence from 160 to 130 Ma in the flexural model of Sawyer and others (1982b) appears to be the result of a problem in the flexural unloading or reconstructed geometry of the margin, since at earlier times the flexural results should be closer to the Airy model and diverge from it as the margin cools. The other, short term, differences are attributable to differences in biostratigraphic interpretation of the well and the current limit of resolution of this type of data. Heller and others (1982) interpreted these sharp changes, which in his data set are entirely due to the bathymetric estimates, as evidence of additional tectonic activity on the margin. We believe that this is probably an overinterpretation of the data.

RIFT MODELS

A number of models have been proposed to explain the processes that occur at rifted continental margins and interpret tectonic subsidence curves. Broadly speaking, these can be divided into two main categories: crustal thinning (or densification) and lithospheric heating. Various authors have identified processes by which the crustal thinning or densification can take place. These include erosion (Sleep, 1971), subcrustal flow (Bott, 1972), ductile necking (Artemjev and Artyushkov, 1971), rotation of listric (de Charpal and others, 1978) or planar faults (LePichon and Sibuet, 1981), pervasive dike intrusion (Royden and Keen, 1980), and crustal phase changes (Falvey, 1974). Additional factors that can modify the crustal thickness have also been proposed such as melt differentiation (Beaumont and others, 1982) and underplating (McKenzie, 1984). Similarly, several means of introducing heat into the lithosphere have been proposed, ranging from hot spots (Burke and Dewey, 1973), to magma injection (Mareshal, 1983), to passive upwelling in response to extension (McKenzie, 1978), to diapiric upwelling at the base of the lithosphere (Neugebauer, 1983), to small-scale convection (Buck, 1984), to depressurization melting (Yoder, 1976; McKenzie, 1984).

McKenzie (1978) proposed that rift-type sedimentary basins form as a result of uniform lithospheric extension (Fig. 3). During rifting, there is uniform thinning of the crust and mantle portion of the lithosphere. The crustal thinning induces rapid isostatic subsidence. The lithospheric thinning produces a passive upwelling of the asthenosphere. This overall advection of heat above the depth of compensation produces uplift. In general, the combination of these two factors produces an initial rapid subsidence during rifting, followed by a slower thermal subsidence after rifting as the lithosphere cools to its original thickness. The form of this subsidence after the first approximately 20 m.y. is that of a decaying exponential. This was first established at the U.S. East Coast margin (Sleep, 1971). In the absence of flexural effects, the asymptotic level to which the margin subsides is controlled by the crustal thinning, and the amplitude of the exponential by the lithospheric heating. Uniform extension of the lithosphere determines the lithospheric heating for a given amount of crustal thinning. This defines the heating that occurs in passive extension (Sengör and Burke, 1978), where rifting is presumably due to inplane forces acting on the lithospheric plates.

While the uniform extension model explains the first-order features of rifts and sedimentary basins, it fails to account for the detailed history of subsidence and uplift during rifting. Additional heat sources must generally be imposed to match the observations (Royden and Keen, 1980; Hellinger and Sclater, 1983; Watts and Thorne, 1984). Models in which extension is distributed differently at depth were developed to account for a deficit in the

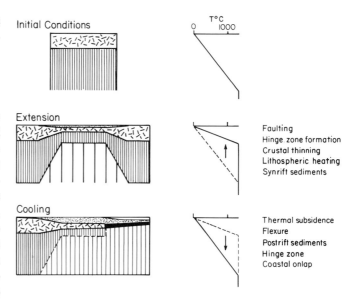

Figure 3. Cartoon illustrating the main phases of continental margin formation. The left side shows schematic cross sections where the random pattern represents continental crust, the solid black represents oceanic crust, the narrow vertical lines represent lithosphere, the wide vertical lines represent asthenosphere, the intermediate spaced lines represent mantle cooling from asthenosphere to lithosphere, and the dotted pattern represents sediments. The central column indicates the evolution of the geotherm during the phases. On the right several characteristic features of these phases are noted.

observed amount of synrift sediment relative to postrift predicted by uniform extension models (Royden and Keen, 1980). In one class of models, the total extension at all depths is constant, but is distributed over different widths at different depths (Vierbuchen and others, 1982; Hellinger and Sclater, 1983; Rowley and Sahagian, 1986). A variation on these models by Wernicke (1985) suggests that the zones of crustal and lithospheric extension may not always overlie each other. While this class of models can explain rift flank uplift, simple calculations show that it is not possible to simultaneously explain the uplift and subsidence in the Gulf of Suez (Steckler, 1985). Alternative models include more extension at greater depth than in the crust. This has been used to account for observations requiring excess heat (Sclater and others, 1980; Royden and Keen, 1980; Steckler and Watts, 1980; Steckler, 1985), particularly landward of hinge zones at continental margins (Royden and Keen, 1980; Watts and Thorne, 1984). In this situation, literal interpretation of the increased extension at depth results in a space problem (e.g., Kligfield and others, 1984). If greater subcrustal stretching than crustal stretching is taken as an indication of excess heat, then these models provide a simple numerical means of reproducing the subsidence and yields an estimate of the magnitude of additional heating necessary. However, it does not address the physics of the cause of the heat and is therefore nonpredictive.

Most means of transferring additional heating to the lithosphere are active; that is, they are based on the introduction of asthenosphere heat sources. Passive extension, until recently, appeared to be unable to explain these observations. Steckler (1985), studying the Gulf of Suez, demonstrated that the large rift flanking uplift did not precede rifting, but apparently developed as a dynamic response to rifting. Calculations by Buck (1983, 1984, 1986) showed that the thermal perturbation caused by extension can drive small-scale convection beneath rifts. In studying how tensional stresses could lead to necking of a viscous lithosphere, Keen (1985) also finds that convection would accompany any localized thinning of the lithosphere. This direct dynamic consequence of rifting significantly enhances the flow of heat into the rift. This additional heating is broader than the rift and can produce rift-flank uplifts of up to 1,000 m. Also, adiabatic depressuring of rising asthenosphere may result in substantial melting and upward advection of heat. An important line of future research is a better quantitative understanding of the mechanisms that introduces this additional heat during rifting.

TECTONIC SUBSIDENCE OF THE EAST COAST

The variation in sediment thicknesses through the margin obtained from seismic and borehole data have delineated the major structural regimes of the U.S. continental margin (Fig. 4). From landward to seaward, one crosses first a thickening wedge of coastal plain sediments. This wedge began to develop in the Late Jurassic and extended landward primarily during the Lower Cretaceous. It is believed to be due to flexural downwarping of the continental lithosphere by sediment loading (Watts and Steckler, 1979).

Seaward of the coastal plain, the basement depth increases rapidly in a series of faults and flexures in a region known as the hinge zone (Jansa and Wade, 1975). Isostatic considerations indicate that this region marks a major transition in crustal thickness and is the boundary between relatively undeformed continental crust and the highly thinned and heated crust that has been extensively modified by rifting (Watts and Steckler, 1979). The hinge zone appears to be about 30 to 50 km wide on the East Coast, and seaward of this region is the locus of the thickest sediment accumulations.

Still farther seaward lies true oceanic crust. The transition from continental to oceanic crust is thought to be marked by the East Coast Magnetic Anomaly (ECMA; Keen, 1969; Grow and others, 1979). On the basis of subsidence and sediment thickness, this boundary is difficult to locate. The contrast between the transitional lithosphere and oceanic lithosphere in terms of crustal thickness and the thermal conditions is relatively minor. Some gravity modeling has identified a change in crustal thickness coinciding with the ECMA (Grow and others, 1979). However, the basement near this region is difficult to identify, and the ocean/continent boundary may well be diffuse.

The major structural elements are clearly delineated in the cross sections shown on Figure 4. The hinge zone is easily identi-

fied and does not correlate with the shoreline or shelf edge. North of the Blake Plateau, the distance from the hinge zone to the ECMA, which represents the width of highly thinned continental crust, averages approximately 100 km.

This overall picture of the continental margin is confirmed by the total tectonic subsidence (TTS) analysis of Sawyer (1985). This is a simplification of backstripping in which the sediment column of a cross section of the margin is unloaded, but no subsidence history is calculated. Because the Atlantic margin is old, the TTS is close to the asymptotic subsidence and can be approximately interpreted for extension and crustal thinning. Without the detailed knowledge of the vertical section and the use of time slices, compaction and flexure corrections can only be made to first order. The hinge zone is marked on TTS profile by a rapid increase in extension and basement depth.

The small change is basement depth and TTS predicted at the ocean/continent boundary makes choosing this boundary more problematic. Although Sawyer (1985) identifies an ocean/continent boundary, which generally is very close to the ECMA, this determination is limited by the available depth-to-basement data (Tucholke and others, 1982). MCS reflection data in this region have difficulty imaging basement, and magnetic depth-to-basement estimates (Klitgord and Behrendt, 1979) yield values that are too shallow beneath the ECMA (Grow and others, 1979; Hutchinson and others, 1983).

It is important to realize that major structural boundaries of the margin may not coincide with the physiographic boundaries, such as the shelf edge. In fact, the position of the shelf edge has migrated through time in the Baltimore Canyon region. Thus, the interweaving of the shelf edge with the hinge zone has produced the use of the terms "basin" and "platform" to describe regions of the shelf underlain by thicker or thinner sediments beneath the continental shelf. These observations are a consequence of the fact that the shelf edge, for reasons of sediment supply and depositional dynamics, is generally straighter than the hinge zone. In all cases, the thickest sediment accumulation is found in the transitional region seaward of the hinge zone, whether this lies beneath the shelf, slope, or rise.

The five COST wells that have been drilled on the east coast are located in different structural provinces of the margin. The

Figure 4. Summary geologic cross sections along the continental margin off North America aligned along the hinge zone (heavy dashed linne). The stratigraphy of the land boreholes is from Brown and others (1972). The stratigraphy of the COST G-1 well is from Amato and Bebout (1980), the COST G-2 well from Amato and Simonis (1980), the COST B-2 well from Scholle (1977), the COST GE-1 well from Amato and Bebout (1978), and DSDP site 391 from Scientific Party (1976). The solid lines in the four sections indicate prominent seismic reflectors identified on nearby multichannel seismic profiles (Schlee and others, 1976; Grow and others, 1979; Grow and Markl, 1977; Buffler and others, 1979; Dillon and others, 1983; Sheridan, 1976). The arrow labeled ECM refers to the position of the East Coast Magnetic Anomaly and BSA to the Blake Spur Anomaly.

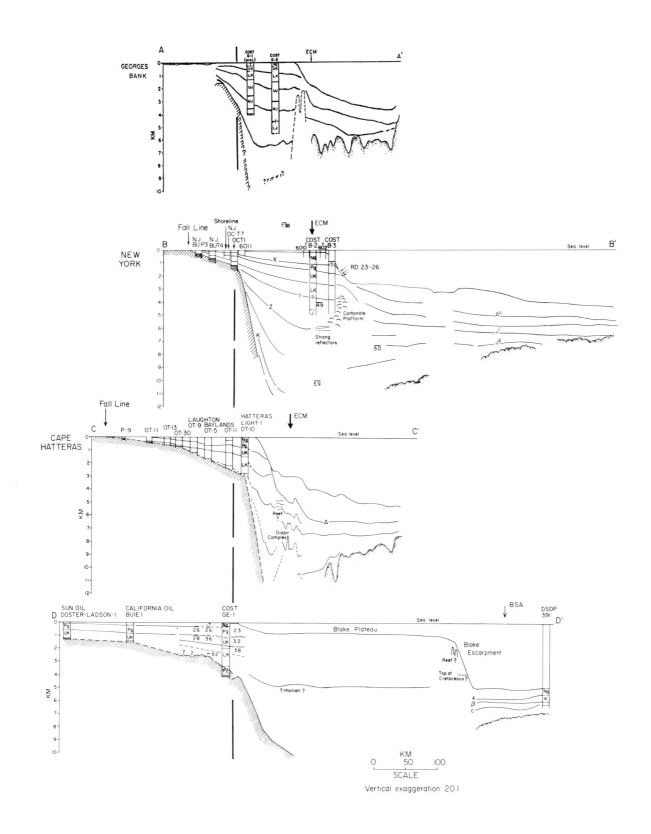

COST GE-1 is landward of the hinge on the coastal plain wedge (despite being offshore). The COST G-1 and G-2 are within the complex hinge zone region of Georges Bank. The COST B-2 lies within the highly thinned transitional crust, and the COST B-3 is thought to overlie oceanic crust. Only the two most landward wells have reached basement. In addition, there are numerous other wells, both scientific and commercial, that have been drilled onshore within the coastal plain wedge.

Since continental breakup and the end of rifting occurred during the Jurassic, the margin has cooled to the point where it is close to equilibrium. The subsidence curves can, therefore, be used to independently determine the amount of crustal thinning. Figure 2 shows the backstripping results from the COST B-2 well discussed earlier compared to theoretical curves generated using a simple extension model. The Airy results suggest a value of β of about 3 to 3.5. This is an underestimate because of flexural effects. The flexural results indicate a larger value of β ranging from $\beta = 5$ up to mid-oceanic ridge subsidence. The rigidities and geometries in these studies were not coupled to forward modeling results. The values used overestimated the flexural effect. A reasonable value between these extremes is crustal thinning by a factor of 4. If some crustal thickening mechanism such as underplating has been operative, then the $\beta = 4$ reflects the net thinning of the crust, and the actual amount of extension was greater. In either case, the crust has been thinned to where it is only slightly thicker than oceanic crust, in agreement with gravity modeling of crustal thicknesses (Grow and others, 1979). The COST B-3 also shows extremely high extension factors, consistent with oceanic crust (Watts, 1981; Sawyer and others, 1982b).

The George's Banks wells have been backstripped by Sawyer and others (1982a, 1982b) who estimate β's of 1.6 and 2.5 for the two wells. Lying at the hinge zone, these wells were probably highly affected by lateral heat transport and flexure. While Sawyer and others' (1982a, 1982b) values were corrected for flexure, there is still probably a fair amount of uncertainty in the actual amount of extension. It is unfortunate that these wells were relatively barren of fossils below the Upper Jurassic, since these wells were located near the edge of the main rift and could have provided much needed constraints on the thermal state of the lithosphere at the time of rifting.

The COST GE-1 is barren of fossils in its lowermost 1,300 m. Heller and others (1982) backstripped only the fossiliferous Aptian to recent sediments; thus no estimate of extension can be made. This well, like others landward of the hinge zone, probably contained little or no Jurassic sediments (Cape Hatteras well, Island Beach well; Brown and others, 1972). The relatively thin sediment cover of this region is indicative of low amounts of extension. In fact, the observed sediment thicknesses of the coastal plain can be accounted for by flexural bending of the lithosphere caused by the thermal subsidence and weight of the sediment prism. Thus, there is little need for synrift extension landward of the hinge zone. Rather, this region was uplifted during rifting. Only during the Late Jurassic and Early Cretaceous did thermal subsidence and flexure succeed in inundating

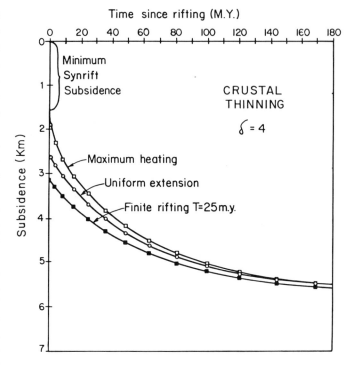

Figure 5. Plot of tectonic subsidence versus time for several theoretical models. In all curves, the crust has been thinned by a factor of 4 consistent with results for the COST B-2 well. The convergence of the curves indicates the difficulty of extrapolating to the conditions of rifting from the later rifting history sampled at the COST B-2 well.

this part of the margin. As will be discussed, the evolution of the coastal plain is a major constraint on thermomechanical models of the margin and sea-level changes.

The magnitude of the tectonic subsidence thus varies systematically across the margin, delineating a major crustal boundary at the hinge zone. Extension seaward of the hinge zone reaches to values as great as 4. The 30- to 50-km-wide hinge zone (average $\beta = 1.5$ to 2.0) and approximately 100-km ($\beta = 4$) wide region of highly thinned crust reconstruct into an approximately 50-km half-width for the initial size of the main rift. This is consistent with the approximately 100-km width commonly observed in rift zones today.

While the record of tectonic subsidence gives good estimations of the crustal thinning at the Baltimore Canyon, most of the heat introduced into the margin has dissipated. It is very difficult to extrapolate back and determine the thermal conditions at the point of rifting and, thereby, distinguish between the various models. Indeed, Cochran (1982) could not do this in the Gulf of Aden where sea-floor spreading began only 10 m.y. ago.

Figure 5 expresses the uncertainty in the conditions at the time of rifting. The central curve shows the expected subsidence for a β of 4 using an instantaneous uniform extension model. This is the standard model, in wide use for interpreting basin subsidence. The lower curve was computed by incorporating a 25-m.y. period of rifting. During the rifting period, heat escapes, resulting in a margin that is much cooler at the conclusion of rifting. The effect is larger than predicted by Jarvis and McKenzie (1980) because lateral heat loss is important when the margin is narrow in the early stages of rifting (Cochran, 1983). The net result is that a greater proportion of margin subsidence occurs during the synrift phase at the expense of the postrift. The margin stratigraphy and effective flexural rigidity will also show considerable difference. Another alternative model is demonstrated by the upper curve. There is evidence that in at least some rifts, the heating that occurs is greater than uniform extension predicts (Royden and Keen, 1980; Steckler, 1985). The upper curve was generated by combining crustal thinning by a factor of 4 with a maximally heated margin, that is, subsidence rates equivalent to a mid-oceanic ridge. The extra heat and thermal expansion yields less synrift subsidence followed by very rapid subsidence immediately following break up.

The uncertainty in the thickness of synrift sediments is a factor of 2. All three curves adequately fit the tectonic subsidence in the COST B-2 well (Fig. 2). Due to the lack of sampling and/or the inability to date the early sediments in the deep basin and the rapid dispersal of heat, the alternative models cannot be distinguished with the current data. Resolution between the various rift models can only be made from the earliest part of margin history.

More landward wells can penetrate farther into the sedimentary package or into basement. By virtue of their thinner sediment accumulation, these wells sample regions less disturbed by the rifting process and crustal thinning. In addition, they are affected to a much greater extent by the various two-dimensional processes at work in margin development such as lateral conduction of heat and flexure (Steckler and Watts, 1980, 1981; Steckler, 1981; Jarvis, 1984) and possibly small-scale convection (Steckler, 1985; Buck, 1984, 1986). Models that include these effects must be used to correctly estimate crustal thinning, lithospheric heating, and other rifting parameters. This is not to say that landward wells are less useful for understanding rifting. Rather, they are critically needed to constrain the influence of these secondary effects that can significantly alter sediment accumulation and the distribution of the tectonic subsidence.

MARGIN RECONSTRUCTION

The limitations of backstripping for determining the early rift history of the U.S. Atlantic coast do not mean that the subsidence data is not of great use. In fact, because all models predict essentially the same thermal subsidence, it can be used to help determine other quantities. Usually, in backstripping sediment thicknesses, ages and water depths are input to determine the tectonic subsidence. However, by using predicted tectonic subsidence together with the sediment thicknesses and ages, the water depth through time can be computed. This is similar to the backtracking of sediments in the deep sea.

In the Baltimore Canyon region, the shelf edge, which prograded seaward throughout the Jurassic, is interpreted to be marked by a major carbonate bank. This carbonate shelf edge reached a maximum seaward extent of approximately 25 km seaward of the present shelf edge. After the Hauterivian the bank became extinct, and by late Miocene the shelf edge was in its present position. Grow and others (1979) hypothesized a major erosional event during the Tertiary that cut the shelf edge back to its current position. In this section, we will show, by reconstruction of the margin history utilizing backstripping techniques, that this event is of much less importance than previously thought.

Figure 7a is a line drawing of the depths to several horizons along U.S.G.S. line 25 by Poag (1985). The oldest horizon shown, the Hauterivian, is the time of extinction of the reef. The younger horizons all appear to be flat lying and show a break in slope at the position of the Hauterivian shelf edge. This suggests that the shelf edge remained in that seaward position until the Eocene and was then cut back. Indeed, Eocene sediments are exposed on the lower continental slope and have been partially eroded. However, the middle Eocene to Holocene section changes dramatically in thickness across the profile. Thus, it is likely that the present flatness of the horizons is the result of recent differential sediment loading and should not be used to interpret their earlier attitude.

In order to reconstruct the earlier geometry of the shelf, the sediment load is backstripped layer by layer taking into account sediment compaction and flexure. Then, the surface is shifted upward to correct for the thermal subsidence that has occurred between that time and the present.

The effect of compaction is quite dramatic and is illustrated in Figure 6. In the first column, a large thickness of sediments, such as is found in the Baltimore Canyon, is shown. The proportion of water and rock is illustrated for each depth and decreases exponentially through the column. If a 750-m-thick sequence of unconsolidated sediments is deposited on this column, then all of the underlying column will compact and expel water due to the additional overburden. If the column maintains its porosity-depth relations, then by conservation of the solid component of the added sediment,

$$S(1 - \phi) = S'(1 - \phi'),$$

where S and ϕ are the sediment thickness and porosity and the primes denote the values at the base of the column. In other words, the amount of rock added at the top equals the volume of rock in the additional thickness at the base of the column. If a 750-m pile of sediment with 50 percent porosity is added to the column and the porosity at the base is 5 percent, then the column height is increased by 395 m and 355 m of water is expelled.

The subsidence due to this sediment load is

$$U = S' \left[\frac{\rho'_s - \rho_w}{\rho_m - \rho_w} \right] = S(1 - \phi) \left[\frac{\rho_{sg} - \rho_w}{\rho_m - \rho_w} \right],$$

where ρ_{sg} is the sediment grain density. For $\rho_{sg} = 2.65$, $\rho_m = 3.33$,

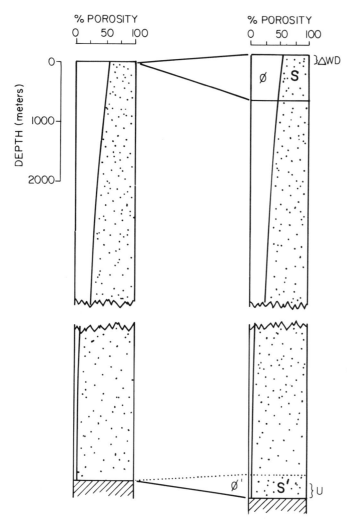

Figure 6. Effect of loading/unloading sediment on a thick sedimentary column. Each column is divided to show the percentage of water and sediment with depth. Details of calculations and meaning of symbols are discussed in the text.

and $\rho_w = 1.03$ there is 264 m of subsidence, and the water depth decreases by 131 m. This means that in backstripping this 750-m layer the water depth would only increase by 131 m. There is an isostatic uplift of 264 m in the basement and 355 m of expansion by decompaction, for a total rebound of 85 percent of the layer thickness. Thus, we can see that unloading the wedge-shaped post-Eocene sediment column will create a change in attitude of the older layers.

If this is carried out for each layer, a sharp break in slope remains in the reconstructed horizons at the site of the present shelf edge. This is the result of assuming Airy isostasy. The unloading was, therefore, carried out flexurally using elastic thicknesses of 25 to 40 km, appropriate for the basement age at each timestep. It was found that both flexure and compaction

needed to be included in order to obtain realistic results. If flexure was incorporated but decompaction not, then the water depths found were unrealistically large. The resulting surfaces were then shifted upward to correct for thermal subsidence of sea floor with an age of 175 Ma. This is a reasonable assumption as the ocean/continent boundary is probably located near shotpoint 2,500 on the seismic profile.

The final reconstruction of the history of the shelf is shown on Figure 7b. The Hauterivian shows the sharp shelf break common to carbonate shelves. The deeper water behind the shelf edge is probably an artifact and shows the resolution of the calculations. The artifact is probably due to not taking into account the lateral variation of lithology or the spatial change in rigidity. After the Hauterivian, the old shelf edge is seen to subside progressively and monotonically. With the extinction of the carbonate bank, clastic sedimentation could not keep pace with subsidence; the entire shelf became deeper, and the shelf edge rolled back. By the Middle Eocene, there is no sharp shelf edge at all, but a gently sloping ramp with considerable (500–700 m) water depth across the shelf.

The present topography was formed during the Miocene when large foreset beds built out the present shelf edge, and erosion by contour currents removed at most a few hundred meters of Eocene at the base of the slope (Farre, 1985). It was the loading due to this outbuilding that backtilted the underlying layers into their present horizontality. The horizontality of the beds beneath the continental rise and the anticlinal structure of the old Cretaceous shelf edge are thus recent features. The earlier carbonate shelf, an apparently promising anticlinal straucture for petroleum exploration, did not become a trap until the middle Miocene.

Thus, a layer-by-layer approach to backstripping, utilizing the smooth thermal subsidence, can provide great help in seismic interpretation of the history of a margin.

FORWARD MODELING

In backstripping, individual wells or observed cross sections are used to isolate the components contributing to basin subsidence. Forward modeling, on the other hand, examines the interaction of the various factors controlling basin formation by constructing synthetic stratigraphic cross sections and comparing them to observed sections. A forward model is based on a specified set of input parameters and assumptions that attempt to describe the physical laws controlling geological processes.

Steckler (1981) and Steckler and Watts (1981) set up a forward model for the evolution of U.S. Atlantic margin basin in which the subsidence, heat flow, temperature structure, and flexural rigidity were computed across the margin and at varying times during its evolution. The effects of both vertical and horizontal heat flow in modifying the temperature structure were included in the model. By summing the flexural effects of individual sediment loads through time, they constructed synthetic stratigraphic cross sections of the passive margin basin.

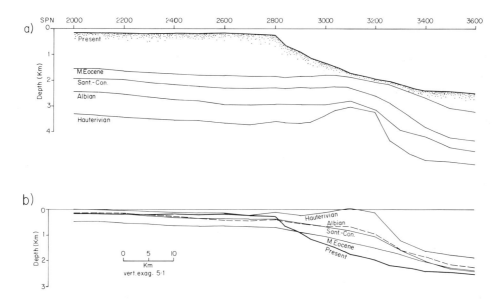

Figure 7. a) Depth section of portion of USGS MCS line 25 near the edge of the continental shelf. Shot point numbers are indicated at top. Depths to horizons and identification of reflectors by Poag (1985). b) Reconstruction of paleobathymetry for indicated times. After extinction of carbonate bank in Hauterivian, clastic sedimentation cannot keep pace with subsidence and the shelf deepens until a Miocene clastic wedge and Tertiary bottom currents shaped the current profile. Details of calculations are in the text.

One of the main results of this early attempt to forward model was to demonstrate that the interaction of flexural loading and thermal contraction of the lithosphere was a major control on the broad "architecture" of basins. In particular, the model was able to explain some of the details of the stratigraphic patterns of onlap, the nature of the transition from a local subsidence over the heated region soon after rifting to a broad downsagging later in basin development, and the presence of stratigraphic "highs" near the shelf break on margins with wide shelves. In a later paper, Watts (1982) showed that some of the details of the synthetic stratigraphic patterns bore a striking resemblance to patterns that Vail and others (1977) had used previously to infer sea-level changes through time. Thus, some of the stratigraphic patterns of onlap mapped by Vail and others (1977) may have tectonic, rather than eustatic, origins.

Beaumont and others (1982) also applied modeling techniques to the study of the U.S. and Canada Atlantic passive margin basins. Their model incorporated most of the features in the Steckler and Watts (1981) model as well as additional physical processes such as variable sediment compaction and conductivity, sediment blanketing, and radiogenic heat production in the sediment. The main difference between their model and that of Steckler and Watts (1981) was in the manner that the sediment load was defined for purposes of the flexure calculation. Steckler and Watts (1981) added sediments to the margin until some specified water depth profile was reached and predicted the stratigraphic thickness in the basin. Beaumont and others (1982), on the other hand, added sufficient sediment to explain observed stratigraphic thicknesses and predicted the water depth changes through time. Thus, in their approach, the results must be compared to paleobathymetric interpretations rather than to observed stratigraphic sections.

Beaumont and others (1982) pointed out the problem of the decreasing sensitivity of the margin to the thermal effects of rifting through time. They also noted that despite this problem, some depth-dependent extension was indicated for the landward parts of the margin.

Watts and Thorne (1984) used the block model of Steckler and Watts (1981) and modified it to include compaction, sea-level changes, and erosion. In their modified model, the boundary between deposition and nondeposition varied during evolution of the basin by an amount depending on the effectiveness of erosion, sea-level changes, and compaction to modify the shape of the margin. The actual position of the shoreline, however, did not vary through time. Thus, their forward models do not incorporate the actual sedimentary dynamics of the basin. Despite this limitation, Watts and Thorne (1984) were able to show that their model was able to reproduce many of the stratigraphic relationships observed on basin cross sections of the Baltimore Canyon Trough.

Watts and Thorne (1984) took advantage of the insensitivity of the later margin subsidence to the initial rifting conditions to investigate more fully the coastal onlap-offlap relationships due to sea-level variations on a subsiding margin. The model was not sensitive to the exact nature of the variations in extension across the margin, due to the smoothing effects of lateral heat flow and flexure. The only substantial difference from uniform extension that could be detected was landward of the hinge zone where considerable lithospheric heating beneath unstretched crust was needed. Even then, the distribution of this heat was not important, and Watts and Thorne (1984) were able to unrealistically extend their subcrustal extension hundreds of kilometers landward of the margin and still correctly predict coastal plain stratigraphy.

In order to examine the applicability of the Watts and Thorne (1984) model, we have modified some of the parameters used in their model to make them more realistic and constructed an additional set of synthetic cross sections for a portion of the U.S. Atlantic margin with a different geometry. Figure 8 summarizes the components of the computations of the forward modeling. Figure 8C illustrates the distribution of extension used in the following models. The hinge zone marks the transition from the zone of no crustal extension to $\beta = 4$. This value for extension is adopted from the calculations presented earlier. Further seaward, oceanic lithosphere is modeled using the two-layered stretching model as calibrated in Steckler (1981). Calculations were performed utilizing both uniform extension throughout the continental lithosphere and with additional thinning of the mantle part of the lithosphere landward of the hinge zone as indicated in Figure 8C. The additional subcrustal thinning decays away from the margin using a Gaussian curve with a standard deviation of $\sigma = 150$ km.

Figure 8B indicates that the thermal subsidence and sediment loading are presumed to take place on an elastic lithosphere whose thickness is determined by the depth to the 450°C isotherm. This isotherm gives the effective elastic thickness T_e in the oceanic lithosphere (Watts, 1982). T_e thereby varies across the margin and through time as the thermal state of the lithosphere evolves.

Beaumont and others (1982) found that for the Nova Scotian margin a 250°C isotherm gave a better fit to the observed gravity anomalies. There are two possibilities for this difference. First, Beaumont and others (1982) did not include sedimentation during rifting when the flexural rigidity would be low. This would bias their result toward a lower value for the isotherm determining T_e. Second, the estimates of the rheology of continental lithosphere indicate that when heated, T_e would follow a lower effective isotherm (Kusznir and Karner, 1985). This effect would be most important landward of the hinge zone while the continental margin was relatively young. Thus, a true estimate for the isotherm could lie between 250° and 450°C and vary with time across the margin.

Figure 8A illustrates the sedimentary-depositional parameters from Watts and Thorne (1984). The curves show the variation in compaction parameters from sand to shale across the continental margin. In addition, sea level varies through time, but the bathymetry shallower than 2000 m remains constant (i.e. fixed relative to sea level). Thus, the shoreline remains fixed, and

sea level variations are accommodated by changes in sedimentation without facies changes. This maximizes the onlap-offlap response to eustatic changes. The sea level curve used is that of Watts and Thorne (1984) with the mid-Oligocene sea level fall adjusted to be in better agreement with its world-wide timing.

Figure 9 shows the results for the coastal plain and shelf region of U.S.G.S. line 25 of the forward model. This is the same section studied by Watts and Thorne (1984). Figure 9a is based on a one-layer stretching model while Figure 9b is based on the two-layer stretching distribution shown in Figure 8. Both models show a well developed coastal plain, hinge zone and inner shelf region underlain by a thick sequence of seaward dipping Mesozoic-Tertiary sediments. The main differences between the models is in the stratigraphy predicted for the coastal plain region. Figure 9a shows a greater thickness, which in the coastal plain regions is almost entirely accounted for by the presence of Jurassic-age sediments. As is well known, the Jurassic is noticeably absent in stratigraphic cross-sections and well data in the U.S. Atlantic coastal plain. In Figure 9b, the absence of Jurassic sediments is the result of extensive heating landward of the hinge zone at the time of rifting, causing emergence of the coastal plain region for about 40 m.y. after initial rifting. Note that this excess heating, indicated by the Jurassic stratigraphy, also improved the fit to the Tertiary sediments which are widely exposed in the coastal plain (Fig. 4).

The necessity for, and the shape of, this additional heating are consistent with the results of Steckler (1985) for the Gulf of Suez. He found that heating was greatly in excess of crustal thinning and that the additional heat was distributed over a broad zone flanking the rift. In fact, within the uncertainties, extension within the rift could not be distinguished from uniform extension. The heating was interpreted by Steckler (1985) as resulting from small-scale convection beneath the rift, driven by the horizontal temperature gradients introduced by rifting (Buck, 1984, 1986). This mechanism effectively "erodes" the corners of the lithosphere bordering the rift, smoothing the isotherms in the lower part of the lithosphere and introducing additional heating over a broad zone. In the Gulf of Suez, the uplift postdates the initiation of rifting and suggests that the uplift is a direct consequence of the dynamics of rifting (Steckler, 1985). However, this cannot be demonstrated for the U.S. Atlantic margin.

The Atlantic border uplift needs to have extended over a larger distance than in the case of Suez. Using $\sigma = 100$ km for the Gaussian decay of the uplift observed at Suez yields deposition of Jurassic sediments at the landward part of the coastal plain. This result is consistent with a strong dependence on β for the vigor of small-scale convection. An alternative possibility is that the uplift is caused by an overall thickening of the crust during rifting due to the addition of low-density melt to its base (McKenzie, 1984). It is not possible to rule out this mechanism on the basis of the models in Figure 9.

Figure 10 shows a computed and observed stratigraphy along USGS line 32 in the Carolina Trough, a narrow continental shelf with less sediment loading seaward of the hinge zone than

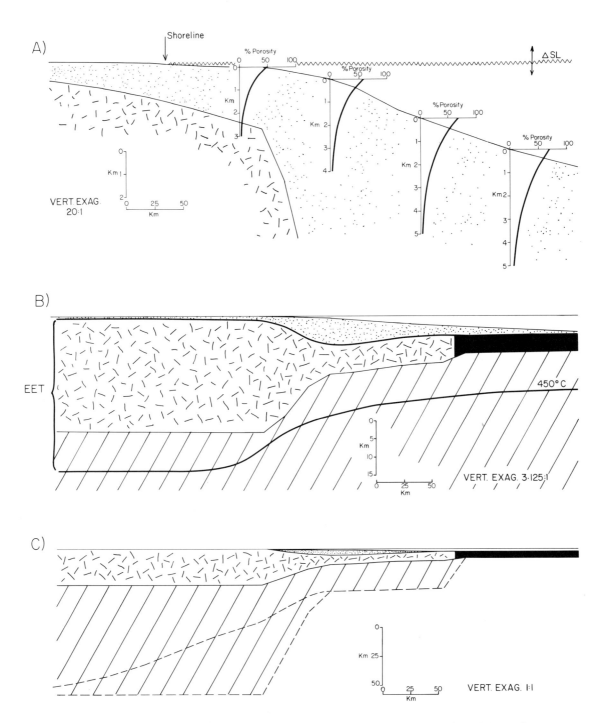

Figure 8. Diagram illustrating the factors included in the forward modeling calculations. A) Inset plots indicate varying sediment compaction with depth along the profiles. Sediments are added at each timestep to maintain constant bathymetric profile while subsidence rates and sea level change. B) Sediments (dots) are located onto an elastic plate (EET) whose thickness varies with the depth to the 450°C isotherm. C) Initial configuration of model at the end of rifting period. Continental crust (random dashes) thins across hinge zone to $\beta = 4$ and thence to oceanic crust (solid). Lower dashed line indicates uniform lithospheric thinning. Upper dashed line indicates additional lithospheric thinning landward of the hinge zone.

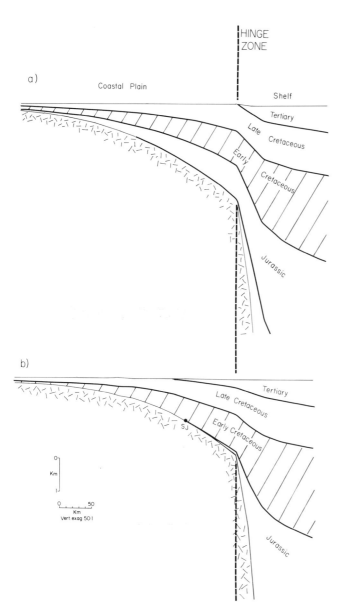

Figure 9. Synthetic stratigraphic cross section of the coastal plain and shelf region of New Jersey in the vicinity of USGS line 25. The synthetic profiles were generated using the flexural loading model for passive margin evolution described in Watts and Thorne (1984). a) "One-layer" model in which the crust and lithosphere have been extended by equal amounts. b) "Two-layer" model in which the crust and lithosphere are thinned by similar amounts seaward of the hinge zone and only the lithosphere is thinned landward of the hinge zone. The shape of the thinning is Gaussian and is illustrated in Figure 8. Note that only b) shows pinch out of the Jurassic beneath the coastal plain and a progressive onlap of individual units in the Early Cretaceous onto basement.

line 25. For line 32, the only changes made were the bathymetric profile and a narrowing of the width of the stretched lithosphere by 20 km. This is indicated by the relative positions of the hinge zone and ECMA. As was the case for line 25 (Fig. 9) the model without additional heating cannot explain the absence of Jurassic beneath the coastal plain.

Celerier (1986) modeled the subsidence of the Carolina Trough using the forward models of Sawyer and others (1982b). His modeling procedure followed that of Beaumont and others (1982) in that he included the observed sediment thicknesses as an input and calculated paleowater depths. Celerier's (1986) uniform extension simulation predicts large unrealistic water depth in the coastal plain early in the margin history. He concluded that extensive heating landward of the hinge zone was needed to create a realistic margin model. This thermal uplift landward of the hinge zone observed in modern rifts and indicated along old margins appears, therefore, to be a general feature of rifting.

The model in Figure 10 shows more stratigraphic horizons than the model in Figure 9, thus revealing more details of the stratigraphy of the margin. Of particular interest is the prediction of an unconformity separating late Oligocene above from Cretaceous sediments below. The hiatus representing the unconformity increases across the slope and shelf from a few million years on the lower slope to about 30 m.y. beneath the inner shelf. In our model, the seaward offlap of the Tertiary is the result of overall falling sea level from the Late Cretaceous to the present. This offlap is well developed in the U.S. Atlantic margin because it is subsiding sufficiently slowly that the effects of sea-level changes dominate the stratigraphy. The unconformity separating the late Oligocene and the Cretaceous is the result of a small amplitude (few tens of meters) change in sea level superimposed on the overall decrease in sea level from the Cretaceous to the present. We note that in the model, the sea-level change is not sufficient to cause it to drop below the shelf break. This suggests that it is possible to produce extensive unconformities on the shelf without dropping sea level below the shelf break.

We caution, however, against too literal an interpretation of the results in Figure 10. The models are limited in their application by the assumption that water depths have remained on the average the same as the present during the geologic past. The shape of the synrift sediments and the apparent outer basement high are artifacts of the constancy of the water depth through time.

Clearly forward modeling would be substantially improved if better models for sediment dynamics could be incorporated. Unfortunately, there is insufficient knowledge at present about the role of processes such as aggradation, starvation, and bypass during the past to incorporate them into forward models. We believe, however, that given better models for sediment dynamics and their response to sea-level changes, it will be possible in the future to use forward models to quantitatively evaluate the relative role of sea level and tectonics in controlling the stratigraphic development of the U.S. Atlantic margin. Since the smoothly varying tectonic subsidence across the margin can be estimated

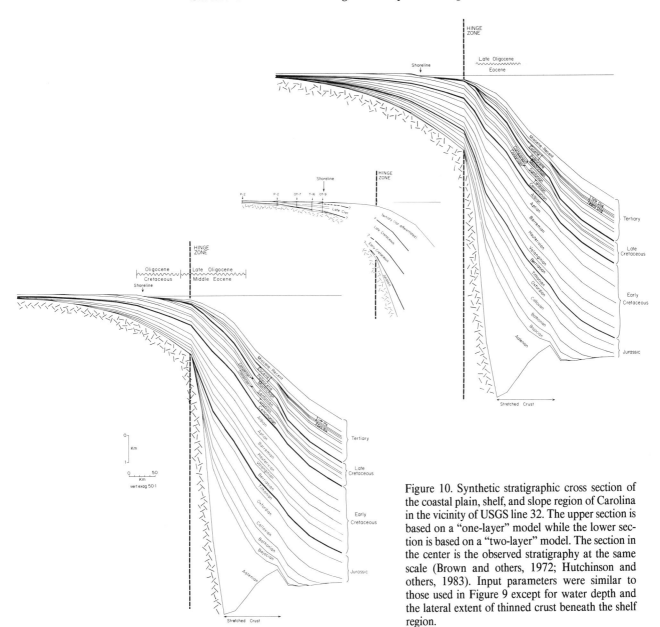

Figure 10. Synthetic stratigraphic cross section of the coastal plain, shelf, and slope region of Carolina in the vicinity of USGS line 32. The upper section is based on a "one-layer" model while the lower section is based on a "two-layer" model. The section in the center is the observed stratigraphy at the same scale (Brown and others, 1972; Hutchinson and others, 1983). Input parameters were similar to those used in Figure 9 except for water depth and the lateral extent of thinned crust beneath the shelf region.

without needing extensive knowledge of the rifting history, the interaction of the tectonic subsidence, sea level, and sedimentation could be accurately modeled throughout Cretaceous–Holocene time if the latter was better known.

CONCLUSIONS

The past decade has seen a significant increase in our understanding of the evolution of the U.S. Atlantic margin. This has been achieved through a combination of improved seismic and well data and new analysis techniques such as backstripping and forward modeling. The margin seaward of the hinge zone has been subjected to extensive net crustal thinning ($\beta = 4$). Because the body of well data in this region only penetrates the last third of the tectonic subsidence, the lithospheric heating during rifting and synrift subsidence are not well constrained. Thus the resolution of present data does not allow firm conclusions as to whether or not extension was uniform seaward of the hinge zone. The lack of Jurassic sediments landward of the hinge zone, on the other hand, provides clear evidence of extensive heating in a zone of little crustal thinning. In contrast to the more seaward parts of the margins, which experienced considerable subsidence and sediment accumulation during rifting, the portions landward of the hinge zone were uplifted and eroded. Small-scale convection driven by the rifting process is a possible mechanism consistent with the observations. Data from this less extended part of the margin is critically needed to constrain models of convective and flexural effects.

Detailed forward modeling is able to simulate the major stratigraphic sequences on the margin. It suggests smaller and more smoothly varying sea-level changes (Steckler, 1984; Watts and Thorne, 1984) than the Exxon group (Vail and others, 1977; Vail and Hardenbol, 1979; Haq and others, 1987) has proposed. At older margins such as this, the smoothly varying thermal subsidence, which is relatively indifferent to details of the rifting history, holds great promise for unraveling the eustatic, tectonic, and sedimentary record at the Atlantic margin. It also makes it possible to use forward modeling and careful backstripping to reconstruct the evolution of the margin bathymetry in Cretaceous and Tertiary times. These types of two-dimensional analyses are necessary in order to obtain correct interpretations of seismic sequences. Thus, a tectonic framework for understanding the development of the U.S. Atlantic margin has been established and provides a basis for future work.

REFERENCES CITED

Ahern, J. L., and Ditmars, R. C., 1985, Rejuvenation of continental lithosphere beneath an intracratonic basin: Tectonophysics, v. 120, p. 21–35.

Amato, R. V., and Bebout, J. W., 1980, Geologic and operational summary, COST No. G-1 well, Georges Bank area, North Atlantic OCS: U.S. Geological Survey Open-File Report 80-268, 112 p.

Amato, R. V., and Simonis, E. K., 1980, Geologic and operational summary COST No. G-2 well, Georges Bank area North Atlantic OCS: U.S. Geological Survey Open-File Report 80-269, 116 p.

Artemjev, M. E., and Artyushkov, E. V., 1971, Structure and isostasy of the Baikal rift and the mechanism of rifting: Journal of Geophysical Research, v. 76, p. 1197–1211.

Beaumont, C., Keen, C. E., and Boutilier, R., 1982, On the evolution of rifted continental margins; Comparison of models and observations for Nova Scotian margin: Geophysical Journal of the Royal Astronomical Society, v. 70, p. 667–715.

Bird, P., 1978, Initiation of Himalayan intracontinental subduction: Journal of Geophysical Research, v. 83, p. 4975–4987.

Bodine, J. H., Steckler, M. S., and Watts, A. B., 1981, Observations of flexure and the rheology of the oceanic lithosphere: Journal of Geophysical Research, v. 86, p. 3695–3707.

Bond, G. C., and Kominz, M. A., 1984, Construction of tectonic subsidence curves for the early Paleozoic miogeocline, southern Canadian Rocky Mountains; Implications for subsidence mechanisms, age of breakup, and crustal thinning: Geological Society of America Bulletin, v. 95, p. 155–173.

Bott, M.H.P., 1972, Subsidence of Rockall Plateau and of the continental shelf: Geophysical Journal of the Royal Astronomical Society, v. 27, p. 235–236.

Brown, P., Miller, J. A., and Swain, F. M., 1972, Structural and stratigraphic framework and spatial distribution of permeability of the Atlantic Coastal Plain, North Carolina to New York: U.S. Geological Survey Professional Paper 796, 79 p.

Buck, W. R., 1983, Convection beneath continental rifts; The effect of cooling and subsidence [abs.]: EOS, American Geophysical Union Transactions, v. 64, p. 838.

—— , 1984, Small-scale convection and the evolution of the lithosphere [Ph.D. thesis]: Cambridge, Massachusetts Institute of Technology, 256 p.

—— , 1986, Small-scale convection induced by passive rifting; The cause for uplift of rift shoulders: Earth and Planetary Science Letters, v. 77, p. 362–372.

Buffler, R. T., Watkins, J. S., and Dillon, W. P., 1979, Geology of the offshore Southeast Georgia Embayment, U.S. Atlantic continental margin, based on multichannel seismic reflection profiles, *in* Watkins, J. S., Montadert, L., and Dickerson, P. W., eds., Geological and geophysical investigations of continental margins: American Association of Petroleum Geologists Memoir 29, p. 11–26.

Burke, K., and Dewey, J. F., 1973, Plume-generated triple junctions; Key indications in applying plate tectonics to old rocks: Journal of Geology, v. 81, p. 406–433.

Caldwell, J. G., Haxby, W. F., Karig, D. E., and Turotte, D. L., 1976, On the applicability of a universal elastic trench profile: Earth and Planetary Science Letters, v. 31, p. 239–246.

Celerier, B., 1986, Models for the evolution of the Carolina Trough and their limitations [Ph.D. thesis]: Cambridge, Massachusetts Institute of Technology, 206 p.

Cochran, J. R., 1982, The magnetic quiet zone in the eastern Gulf of Aden; Implications for the early development of the continental magin: Geophysical Journal, v. 68, p. 171–201.

—— , 1983, Effects of finite extension times on the development of sedimentary basins: Earth and Planetary Science Letters, v. 66, p. 289–302.

de Charpal, O., Guennoc, P., Montadert, L., and Roberts, D. G., 1978, Rifting, crustal attenuation and subsidence in the Bay of Biscay: Nature, v. 275, p. 706–711.

Dillon, W. P., Popenoe, P., Grow, J. A., Klitgord, K. D., Swift, B. A., Paull, C. K., and Cashman, K. V., 1982, Growth faulting and salt diapirism; Their relationship and control in the Carolina Trough, eastern North America, *in* Watkins, J. S., and Drake, C. L., eds., Studies in continental margin geology: American Association of Petroleum Geologists Memoir 34, p. 21–48.

Falvey, D. A., 1974, The development of continental margins in plate tectonic theory: Australian Petroleum Exploration Association Journal, v. 14, p. 95–106.

Farre, J. A., 1985, The importance of mass-wasting processes on the continental slope [Ph.D. thesis]: New York, Columbia University, 200 p.

Grow, J. A., and Markl, R. G., 1977, IPOD-USGS multichannel seismic reflection profile from Cape Hatteras to the Mid-Atlantic ridge: Geology, v. 5, p. 625–630.

Grow, J. A., Mattick, R. E., and Schlee, J. S., 1979, Multichannel seismic depth sections and interval velocities over outer continental shelf and upper slope between Cape Hatteras and Cape Cod, *in* Watkins, J. S., Montadert, L., and Dickerson, P. W., eds., Geological Investigations of continental margins: American Association of Petroleum Geologists Memoir 29, p. 65–83.

Haq, B. U., Hardenbol, J., and Vail, P. R., 1987, Chronology of fluctuating sea levels since the Triassic: Science, v. 235, p. 1156–1167.

Heller, P. L., Wentworth, C. M., and Poag, C. W., 1982, Episodic post-rift subsidence of the United States Atlantic continental margin: Geological Society of America Bulletin, v. 93, p. 379–390.

Hellinger, S. J., and Sclater, J. G., 1983, Some comments on two-layer extensional models for the evolution of sedimentary basins: Journal of Geophysical Research, v. 88, p. 8251–8269.

Hutchinson, D. R., Grow, J. A., Klitgord, K. D., Swift, B. A., 1983, Deep structure and evolution of the Carolina trough, *in* Watkins, J. S., and Drake, C. L., eds., Studies in Continental Margin Geology: American Association of Petroleum Geologists Memoir 34, p. 129–152.

Jansa, L. F., and Wade, J. A., 1975, Geology of the continental margin off Nova Scotia and Newfoundland, *in* Van der Linden, W.J.M., and Wade, J. A., eds., Offshore geology of eastern Canada, Geological Survey of Canada Paper 74-30, v. 2, p. 51–105.

Jarvis, G. T., 1984, An extensional model of graben subsidence; The first stage of basin evolution: Sedimentary Geology, v. 40, p. 13–31.

Jarvis, G. T., and McKenzie, D. P., 1980, Sedimentary basin formation with finite extension rates: Earth and Planetary Science Letters, v. 48, p. 42–52.

Karner, G. D., and Watts, A. B., 1983, Gravity anomalies and flexure of the lithosphere at mountain ranges: Journal of Geophysical Research, v. 88, p. 10449–10477.

Karner, G. D., Steckler, M. S., and Thorne, J., 1983, Long-term thermomechanical properties of the continental lithosphere: Nature, v. 304, p. 250–253.

Keen, C. E., 1985, The dynamics of rifting; Deformation of the lithosphere by active and passive driving forces: Geophysical Journal of the Royal Astronomical Society, v. 80, p. 95–120.

Keen, M. J., 1969, Possible edge effect to explain magnetic anomalies off the eastern seaboard of the United States: Nature, v. 222, p. 72–74.

Kligfield, R., Crespi, J., Naruk, S., and Davis, G. H., 1984, Displacement and strain patterns of extensional orogens: Tectonics, v. 3, p. 577–609.

Klitgord, K. D., and Behrendt, J. C., 1979, Basin structure of the U.S. Atlantic margin, in Watkins, J. S., and Montadert, L., and Dickerson, P. A., eds., Geological and geophysical investigations of continental margins: American Association of Petroleum Geologists Memoir 29, p. 85–112.

Kusznir, N., and Karner, G. D., 1985, Dependence of the flexural rigidity of the continental lithosphere on rheology and temperature: Nature, v. 316, p. 138–142.

LASE Study Group, 1986, Deep structure of the U.S. East Coast passive margin from large aperture seismic experiments: Marine and Petroleum Geology, v. 3, p. 234–242.

LePichon, X., and Sibuet, J. C., 1981, Passive margins; A model of formation: Journal of Geophysical Research, v. 86, p. 3708–3720.

Lyon-Caen, H., and Molnar, P., 1983, Constraints on the structure of the Himalaya from an analysis of gravity anomalies and a flexural model of the lithosphere: Journal of Geophysical Research, v. 88, p. 8171–8191.

——, 1985, Gravity anomalies, flexure of the Indian plate, and the structure, support, and evolution of the Himalaya and Ganga basin: Tectonics, v. 4, no. 6, p. 513–538.

Mareschal, J-C., 1983, Mechanisms of uplift preceding rifting: Tectonophysics, v. 94, p. 51–66.

McKenzie, D. P., 1978, Some remarks on the development of sedimentary basins: Earth and Planetary Science Letters, v. 40, p. 25–32.

——, 1984, A possible mechanism for epeirogenic uplift: Nature, v. 307, p. 616–618.

Molnar, P., and Tapponier, P., 1981, A possible dependence of tectonic strength on the age of the crust in Asia: Earth and Planetary Science Letters, v. 52, p. 107–114.

Neugebauer, H. J., 1983, Mechanical aspects of continental rifting: Tectonophysics, v. 94, p. 91–108.

Poag, C. W., 1980, Foraminiferal stratigraphy, paleoenvironments, and depositional cycles in the Outer Baltimore Canyon Trough, in Scholle, P. A., ed., Geological studies of the COST No. B-3 Well, U.S. Mid-Atlantic Continental Slope area: U.S. Geological Survey Circular 833, p. 44–65.

——, ed., 1985, Geologic evolution of the U.S. Atlantic margin: New York, Van Nostrand Reinhold Company, 383 p.

Rowley, D. B., and Sahagian, D., 1986, Depth-dependent stretching; A different approach: Geology, v. 14, p. 32–35.

Royden, L., and Keen, C. E., 1980, Rifting process and thermal evolution of continental margin of eastern Canada determined from subsidence curves: Earth and Planetary Science Letters, v. 51, p. 343–361.

Sawyer, D. S., 1985, Total tectonic subsidence; A parameter for distinguishing crust type at the U.S. Atlantic continental margin: Journal of Geophysical Research, v. 90, no. 89, p. 7751–7769.

Sawyer, D. S., Swift, B. A., Sclater, J. G., and Toksz, M. N., 1982a, Extensional model for the subsidence of the northern United States Atlantic continental margin: Geology, v. 10, p. 134–140.

Sawyer, D. S., Swift, B. A., Toksz, M. N., and Sclater, J. G., 1982b, Thermal evolution of the Baltimore Canyon Trough and Georges Bank Basin, in Watkins, J. S., and Drake, C. L., eds., Studies in Continental Margin Geology: American Association of Petroleum Geologists Memoir 34, p. 743–764.

Schlee, J., Behrendt, J. C., Grow, J. A., Robb, J. M., Mattick, R. E., Taylor, P. T., and Lawson, B. A., 1976, Regional geologic framework off northeastern United States: American Association of Petroleum Geologists Bulletin, v. 60, p. 926–951.

Scholle, P. A., 1977, Geological studies on the COST B-2 well, United States Mid-Atlantic outer continental shelf area, in Scholle, P. A., ed., Geological studies of the COST B-2 Well U.S. Mid-Atlantic Outer Continental Shelf Area: U.S. Geological Survey Circular 750, p. 103.

Scientific Party, 1976, Deep Sea Drilling Project, Leg 11: Geotimes, v. 21, p. 23–26.

Sclater, J. G., and Christie, P. A., 1980, Continental stretching; An explanation of the post Mid-Cretaceous subsidence of the central North Sea basin: Journal of Geophysical Research, v. 85, p. 3711–3739.

Sclater, J. G., Royden, L., Horvath, F., Burchfiel, B. C., Semken, S., and Stegena, L., 1980, The formation of the intra-Carpathian basins as determined from subsidence data: Earth and Planetary Science Letters, v. 51, p. 139–162.

Sengör, A.M.C., and Burke, K., 1978, Relative timing of rifting and volcanism on Earth and its tectonic implications: Geophysical Research Letters, v. 5, p. 419–421.

Sheridan, R. E., 1974, Atlantic continental margin of North America, in Burk, C. A., and Drake, C. L., eds., The Geology of Continental Margins: New York, Springer-Verlag, p. 391–408.

——, 1976, Sedimentary basins of the Atlantic margin of North America: Tectonophysics, v. 36, p. 113–131.

Sleep, N. H., 1971, Thermal effects of the formation of Atlantic continental margins by continental breakup: Geophysical Journal of the Royal Astronomical Society, v. 24, p. 325–350.

Steckler, M. S., 1981, The thermal and mechanical evolution of Atlantic-type continental margins [Ph.D. thesis]: New York, Columbia University, 261 p.

——, 1984, Changes in sea level, in Trendall, A. F., and Holland, H. D., eds., Patterns of change in Earth evolution, Report of Dahlem Konferenz: Berlin, Springer-Verlag, p. 103–122.

——, 1985, Uplift and extension at the Gulf of Suez; Indications of induced mantle convection: Nature, v. 317, p. 135–139.

Steckler, M. S., and Watts, A. B., 1978, Subsidence of the Atlantic-type continental margin off New York: Earth and Planetary Science Letters, v. 42, p. 1–13.

——, 1980, The Gulf of Lion; Subsidence of a young continental margin: Nature, v. 287, p. 425–429.

——, 1981, Subsidence history and tectonic evolution of Atlantic-type continental margins, in Scrutton, R. A., ed., Dynamics of Passive Margins: American Geophysical Union Geodynamics Series, v. 6, p. 184–196.

Thorne, J. A., 1985, Studies in stratology; The physics of stratigraphy [Ph.D. thesis]: New York, Columbia University, 523 p.

Tucholke, B. E., Houtz, R. E., and Ludwig, W. J., 1982, Sediment thickness and depth to basement in western North Atlantic Ocean basin: American Association of Petroleum Geologists Bulletin, v. 66, p. 1384–1395.

Vail, P. R., and Hardenbol, J., 1979, Sea-level changes during the Tertiary: Oceanus, v. 22, p. 71–79.

Vail, P. R., Mitchum, R. M., Jr., and Thompson, S., III, 1977, Global cycles of relative changes of sea level: American Association of Petroleum Geologists Memoir 26, p. 83–98.

Van Hinte, J. E., 1978, Geohistory, analysis; Application of micropaleontology in exploration geology: American Association of Petroleum Geologists Bulletin, v. 62, p. 201–222.

Vierbuchen, R. C., George, R. P., and Vail, P. R., 1982, A thermal-mechanical model of rifting with implications for outer highs on passive continental margins: American Association of Petroleum Geologists Memoir 34, p. 765–742.

Vink, G. E., Morgan, W. J., and Zhao, W.-L., 1984, Preferential rifting of continents; A source of displaced terranes: Journal of Geophysical Research, v. 89, p. 10072–10076.

Watts, A. B., 1978, An analysis of isostasy in the world's oceans, 1., Hawaiian-Emperor seamount chain: Journal of Geophysical Research, v. 83,

p. 5989–6004.

——— , 1981, The U.S. Atlantic continental margin; Subsidence history, crustal structure, and thermal evolution: American Association of Petroleum Geologists Education Course Note Series 19, 75 p.

——— , 1982, Tectonic subsidence, flexure and global changes in sea level: Nature, v. 297, p. 469–474.

Watts, A. B., and Cochran, J. R., 1974, Gravity anomalies and flexure of the lithosphere along the Hawaiian-Emperor seamount chain: Geophysical Journal of the Royal Astronomical Society, v. 38, p. 119–141.

Watts, A. B., and Ryan, W.B.F., 1976, Flexure of the lithosphere and continental margin basins: Tectonophysics, v. 36, p. 25–44.

Watts, A. B., and Steckler, M. S., 1979, Subsidence and eustasy at the continental margin of eastern North America: American Geophysical Union Maurice Ewing Symposium Series 3, p. 218–234.

Watts, A. B., and Thorne, J., 1984, Tectonics, global changes in sea-level and their relationship to stratigraphic sequences at the U.S. Atlantic continental margin: Marine and Petroleum Geology, v. 1, p. 319–339.

Wernicke, B., 1985, Uniform-sense normal simple shear of the continental litho-sphere: Canadian Journal of Earth Sciences, v. 22, p. 108–125.

Willet, S. D., Chapman, D. S., and Neugebauer, H. J., 1985, A thermo-mechanical model of continental lithosphere: Nature, v. 314, p. 520–523.

Yoder, H. S., 1976, Generation of the basaltic magma: Washington, D.C., National Academy of Sciences, 265 p.

MANUSCRIPT ACCEPTED BY THE SOCIETY FEBRUARY 6, 1987

ACKNOWLEDGMENTS

We would like to thank J. Farre for introducing us to problems in shelf reconstructions. This paper was improved thanks to helpful reviews by G. D. Karner, A. R. Palmer, W. C. Pitman, L. Royden, and especially B. Celerier's detailed comments. This work was supported by NSF grants EAR 81-09473 and OCE 83-09983 and by a grant from Texaco. Lamont-Doherty Geological Observatory contribution number 4132.

The Geology of North America
Vol. I-2, The Atlantic Continental Margin: U.S.
The Geological Society of America, 1988

Chapter 19

Thermal evolution

Dale S. Sawyer
Institute for Geophysics, The University of Texas at Austin, 8701 North Mopac Blvd., Austin, Texas 78759

INTRODUCTION

We can use information about present and past subsurface temperatures in basins of the U.S. Atlantic Continental Margin in several ways. The event that formed the margin—rifting—had a large thermal signature. The decay of that signature led to subsidence and the accumulation of thick sediments. The temperature history of those sediments in part controlled their diagenesis. Therefore, temperature data can help understand the rifting mechanism, the rate and distribution of subsidence, and the modification of sediment after its deposition.

The present tempratures and heat flow on the U.S. Atlantic Margin are of almost no use in constraining the rifting mechanism or the subsidence. The thermal anomaly from the rift that formed the margin is nearly gone. The temperature differences that might be used to distinguish rift mechanisms decay quickly and fall below what is measurable in less than 50 m.y. This margin is at least 150 m.y. old.

The present temperature distribution is more directly useful in predicting parameters like oil and natural gas maturation in sediments. These reactions are sensitive to temperature and may be considered to be most affected by the highest sustained temperature the sediment has reached. Since sediment temperature typically increases monotonically with burial on this margin, the highest temperature reached by sediment is its present temperature.

Remarkably few reliable data are available to constrain the present temperature or heat-flow distribution on the U.S. Atlantic Continental Margin. Within the offshore part of the margin there are tempreature logs and bottom-hole temperature measurements for the five Continental Offshore Stratigraphic Test (COST) wells and one preliminary heat-flow study in the COST B-2 well. The logs and bottom-hole temperatures were not collected under conditions even approaching thermal equilibrium and are therefore suspect. Onshore there are a substantial number of heat-flow measurements (Costain and Speer, this volume). There are also a few heat-flow measurements on the continental rise and abyssal Atlantic Ocean over what is presumed to be normal oceanic crust.

Paleotemperature and paleo–heat-flow data would be much more useful in constraining rifting mechanisms on this margin. They cannot of course be directly observed. There are, however,

at least two physical phenomena that record information about the thermal state of the margin: the thermal alteration of kerogen in sediment (hydrocarbon maturation) and the deposition of sediment. The alteration of kerogen is thought to proceed as a monotonic function of both time and temperature, the rate increasing exponentially with temperature. A measurement of kerogen alteration therefore gives the present value of the integrated temperature history of a packet of sediment. The measurements of kerogen alteration in the COST wells may yield some information about sediment temperatures and conductivities through forward-modeling studies. The second "recorder" of thermal information is the sediment distribution. The subsidence of a basin is the result of cooling of the crust and upper mantle. The rate of subsidence is therefore a measure of how much anomalous heat is being allowed to escape from the system, or in other words, heat flow. The use of well data to estimate subsidence history is described by Steckler and others (this volume). The information in the sediment record can also be extracted using forward-modeling studies.

In this chapter we describe the few thermal measurements that have been made on the U.S. Atlantic Continental Margin and then show some of the forward modeling that has been used to predict paleotemperatures and estimate kerogen alteration there. During the modeling of three Atlantic marginal basins, similarities and differences between them became apparent. Two properties of the basins emerged as important to their large-scale development: the sediment supply and the configuration of the thinned continental or rift-stage crust under the margin. The lateral change in the thickness of rift-stage crust, determined by thermal and gravity modeling, suggests how the crust responded to extension during continental breakup. The models indicate that it may be possible to predict the maximum sediment thickness, temperature, and hydrocarbon source maturity for a marginal basin knowing only the location of the continental shelf edge and of the boundary between rift-stage and oceanic crust.

TEMPERATURE MEASUREMENTS AND HEAT FLOW

Bottom-hole temperatures have been measured and temperature logs run in five COST wells on the U.S. Atlantic Continen-

Sawyer, D. S., 1988, Thermal evolution, *in* Sheridan, R. E., and Grow, J. A., eds., The Atlantic Continental Margin, U.S.: Geological Society of America, The Geology of North America, v. I-2.

tal Margin. All of the temperature measurements were made within the few days following the termination of drilling and mud circulation in the wells, so they do not represent equilibrium temperature measurements. Temperatures recorded at different amounts of time after mud circulation were used to interpolate or extrapolate the temperatures to a standard time of 25 hours after mud circulation (Jackson and Heise, 1980). The reason for their choice of 25 hours is not given. Thermal gradients in the five wells calculated by linear regression of the corrected temperatures (Table 1) are similar, averaging 22°C/km. They are consistently less than average gradients in the Texas Gulf Coast and North Sea basin and similar to those in wells on the Nova Scotian Shelf to the north.

A heat-flow calculation has been attempted in only the COST B-2 well. Von Herzen and Helwig (1981) found few cores from the well suitable for the measurement of thermal conductivity so they measured the conductivity of lithologically segregated drill cuttings. They then used lithology and porosity data in the well to appropriately average the conductivities measured for the distinct lithologies. They found heat-flow values, 1.26 to 1.30 heat flow units (HFU), that they believed to be reliable in the depth interval 1,220 to 4,104 m. This value is reasonable, if not by itself particularly informative. This technique has not yet been applied to the other wells on the U.S. Atlantic Continental Margin.

MODELING METHODS

It is clear that we have measured little of the present temperature field on the U.S. Atlantic Continental Margin; nor can we ever measure the paleotemperatures. If we are to know anything about these temperatures, it will come from modeling other observable parameters on the margin and determining how they constrain the gross features of the temperature fields. The methods that have been employed include methods of forward modeling temperature in a basin and methods of making predictions of hydrocarbon maturation from those temperatures. These are summarized below.

Forward modeling of basin temperature

The objective of modeling basin temperature is to be able to define the temperature field as a function of depth, location, and time in a basin. Any method of doing this is highly dependent on assumptions, and it can be reasonably easy or difficult to make a model according to the type and number of assumptions that are made. Most basin thermal models fall into one of two classes, those that assume a simple heat-flow history for a basin and those that start with simple initial conditions for the basin and simulate its evolution.

Models that assume a simple heat-flow history are frequently one dimensional and are used to predict temperature history at a single location such as the site of a well. Typically, assumed heat-flow histories include a constant heat flow or heat

TABLE 1. GEOTHERMAL GRADIENTS ON THE U.S. ATLANTIC AND OTHER PASSIVE CONTINENTAL MARGINS*

Area	Well	Gradient (°C/km)
Baltimore Canyon	B-2	24
Baltimore Canyon	B-3	22
Georges Bank	G-1	22
Georges Bank	G-2	27
Southeast Georgia Embayment	GE-1	16
Texas Gulf Coast	(1)	29
North Sea Basin	(1)	32
Scotian Shelf	(1)	27

*Table from Arthur (1982). Data from Jackson and Heise (1980), Robbins and Rhodehamel (1976), Smith and others (1976), Amato and Simonis (1979), Amato and Bebout (1978), Heise and Jackson (1980), and Issler (1984).

(1) Average of several wells.

flow decreasing with time after the rifting event that formed a marginal basin. These sort of models ignore many physical processes in marginal basins including lateral heat flow, thermal blanketing by sediments, and flexural response to sediment loading, but they can model the gross temperature field remarkably well. Models of Arthur (1982) and Royden and others (1980), shown below, and those of Keen (1979), Watts (1981), and Angevine and Turcotte (1981) are of this general type.

Temperature simulations are most frequently two dimensional and are used to model a cross section of an evolving passive continental margin basin. Beaumont and others (1982) gave an exceptionally good description of such a model. Finite difference numerical methods are used to predict the temperature field throughout a basin, at a given time, using the temperature field at an earlier time and a model of the physical properties of the material in and under the basin. The input to a model of this sort includes the initial condition of the model and the amount and properties of the sediment that is to be deposited while the basin evolves. The initial condition is usually taken either at the beginning of rifting, if the model is sufficiently sophisticated to account for rifting, or at the end of rifting, if not. The description of the initial condition of the model would include the entire temperature field of the basin and the physical properties of all the material. The strategy employed in using such a model involves iteratively modifying the initial conditions until the model is able to accommodate the observed sediment budget and still predict the correct present and paleobathymetry. Often this means varying the amount of crustal extension applied to each part of the basin. These models vary in the degree to which they account for physical processes that determine the evolution of basins. Most of these models account for lateral and vertical heat flow, some form of time and location varying flexural rigidity, nonuniform physical properties, and initial conditions appropriate to different basin-forming mechanisms. Models of Watts and Thorne (1984), Sawyer (1982), Sawyer and others (1982), and Swift and others (1987) are of this second type.

Hydrocarbon maturation estimation

Most studies of the temperatures in a basin have as their purpose, or include as a by-product, predictions of the degree of hydrocarbon maturation. Most methods for estimating maturation have their roots in a study by Lopatin (1971; in Russian) that has been summarized and expanded (in English) by Waples (1980). He suggested that both time and temperature are important in the chemical reactions that constitute hydrocarbon maturation and that a given degree of maturation may be achieved by either a long time at a low temperature or a short time at a high temperature. Chemical reaction rate theory further suggests that for each 10°C increase in temperature, the reaction rates will double. This idea has been expressed quantitatively in at least two, equivalent ways: the maturation parameter C of Royden and others (1980) and the time temperature index (TTI) of Waples (1980). Each has been estimated for a series of wells in which vitrinite reflectance, a common measure of maturation, is known, and a calibration function derived that associates particular values of C and TTI with the stages of maturation: immaturity (C <10; TTI <15), maturity for oil generation (10 <C <16; 15 <TTI <1,000), maturity for gas generation (16 <C <19; 1,000 <TTI <972,000), and overmaturity (C >19; TTI >972,000). The estimates for C are from Royden and others (1980) and for TTI are from Waples (1980). They are in agreement except for the upper limit of oil maturation, since TTI = 1,000 corresponds to C = 14.2 rather than C = 16. A particularly good study by Issler (1984) evaluates these calibrations for a variety of sedimentary basins including the Scotian shelf. Hydrocarbon maturation for a particular packet of sediment is estimated by constructing its temperature history and then applying the appropriate equation.

Predictions of hydrocarbon maturation are particularly sensitive to some of the thermal parameters in the temperature model upon which they are based. For example, radiogenic heat production in the upper crust, background heat flow from the mantle, and the thermal conductivity of the sediments all directly affect the predicted temperature gradient in the sediment. Hydrocarbon maturation is highly sensitive to temperature (an increase of 10°C doubles the reaction rate), so errors in the choice of these parameters can lead to serious errors in the predictions of maturity. The best way to deal with these uncertainties is to make predictions of hydrocarbon maturity at the location of a well where maturity has been observed. Determine whether your predictions match and if they don't, vary appropriate aspects of your model to improve the fit. Further predictions in the basin should then be more reliable.

BASIN MODELS

Baltimore Canyon trough

The Baltimore Canyon trough is perhaps the best studied and most drilled marginal basin on the U.S. east coast. Seismic data show reflections to depths of as much as 18 km (Grow, 1980). Data from the COST wells allow investigators to date the seismic horizons back into the Jurassic. Gravity data have been used to model the thickness distribution of rift-stage and oceanic crust under the margin (Grow, 1981). Magnetic anomaly analyses have yielded basement-depth and crust-type information (Klitgord and Behrendt, 1979).

The Baltimore Canyon trough model of Sawyer and others (1982) is along USGS seismic line 25 (Fig. 1). The west end of that line is offshore where about 2 km of sediment overlie the basement. The model extends 90 km landward of the west end of line 25, where the sediment pinches out on basement. The oceanward boundary of the model is at shotpoint 5000 of line 25. The width of the model is 360 km, divided into 18 grid columns, each 20 km wide. Horizontal position in the model is expressed as km east of shotpoint 0 of line 25. For additional modeling detail see Sawyer and others (1982) or Sawyer (1982).

The isotherms for the end of the Jurassic in the Baltimore Canyon trough show that the deepest sediments had reached a temperature of about 300°C (Fig. 2). The pull-up of isotherms in the sediments is due to their lower thermal conductivity. About 80 percent of all the sediment in the basin was deposited during the Jurassic, so by the end of that period the basin had taken its present shape. Because the sediment supply was high, the continental rise and slope were shallower than they are now.

The isotherms in the basin changed little after the end of the Jurassic (Fig. 2). The deepest sediments are now only slightly warmer. The water over the continental slope and rise is deeper and the position of the continental shelf edge has changed, migrating, due to erosion, about 30 km landward. That position of the shelf edge is indicated by the location of carbonate buildups that can be observed in the seismic section.

The temperature of sediments in the Baltimore Canyon trough at the location of the COST B-3 well (Fig. 3) increased rapidly during the first 60 m.y. of basin evolution. After that, the increase in temperature was gradual and was due to the slower burial by the recent sediments. Hydrocarbon maturation reactions progress quickly when the temperature is high. The Late Triassic sediments rapidly passed through the zones of oil and gas maturity and into overmaturity. Younger sediment, for example that deposited in the Late Jurassic, took much longer to mature to the oil-generating stage and, 160 m.y. after deposition, is still in the oil-generating stage. The present maturity of sediments is not the only important factor in controlling the generation of hydrocarbons. The Late Triassic sediment, although presently overmature (Fig. 2), may have generated large volumes of oil or gas while in the appropriate maturity window. Those hydrocarbons may have then migrated upward to cooler sediments, rather than remaining in the deeper sediments and being destroyed by overmaturation. It is therefore unlikely that the overmature zone will contain reservoirs of hydrocarbons now, but it may have acted as the source for oil or gas. The zone presently mature for gas generation may contain gas reservoirs or have generated oil in the past. The shallower material may contain reservoirs of any sort,

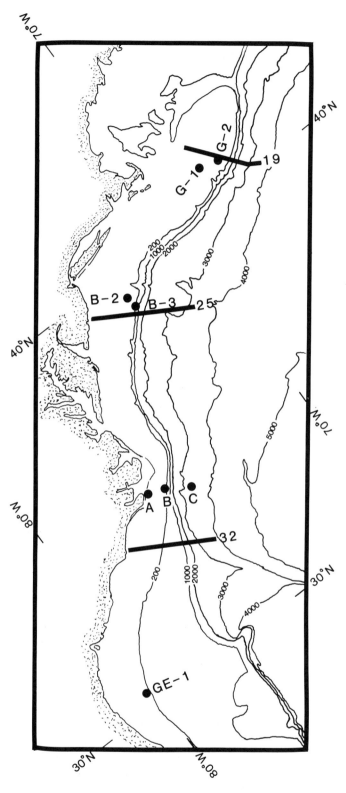

Figure 1. Map showing the location of USGS seismic lines (19, 25, and 32), COST wells (G-1, G-2, B-2, B-3, and GE-1), and locations of thermal models (A, B, and C) from Royden and others (1980).

even though it may lack sufficient thermal maturity to have generated hydrocarbons.

Vitrinite reflectance was observed in samples from the two COST wells in the Baltimore Canyon trough (Claypool and others, 1977; Miller and others, 1980). Those values, converted to the C scale using Waples (1980), are plotted versus depth and compared with the model predictions (Fig. 4). The agreement at the B-2 well location is quite good. Both the slope and absolute value of the predictions are close to those of the data. The fit in the B-3 well is not quite as good. The slope of the group of data points shows a less rapid increase of hydrocarbon maturity with depth than does the model. The absolute values of the model predictions are reasonable; in no case does a data point miss the prediction by more than one unit of maturity.

The Baltimore Canyon trough is predicted to have sediments at all four stages of maturity (Fig. 2). Based on this analysis alone, it would seem likely that hydrocarbons would have been generated in the basin. However, another important parameter is the nature of the sediments deposited in the basin. Seismic stratigraphy can be used to identify depositional sequences and their environment of deposition. All of the sediments considered mature or overmature for hydrocarbon generation in the Baltimore Canyon trough are within depositional sequences A and B of Schlee (1981). He identifies a depositional sequence A, that corresponds roughly to the zones mature for gas and overmature for other hydrocarbons, to be composed of mixed nonmarine and restricted marine sediment of Triassic(?) and Early Jurassic age. Unit B, of Jurassic age, contains the zone of oil maturity and the top portion of the zone of gas maturity. Seismic stratigraphic analysis indicates that it too was deposited in a nonmarine environment with some restricted marine evaporites. Sediments of nonmarine origin are most likely to contain Type III kerogen, the organic precursor to hydrocarbon generation (Tissot and Welte, 1978). This type of kerogen is most often associated with the generation of natural gas. If anoxic conditions exist in restricted marine environments, more oil-rich Type I kerogen-bearing sediment may be deposited.

It seems that the bulk of the sediments in the deep Baltimore Canyon trough are natural-gas prone and have experienced proper thermal conditions for gas maturation. Significant quantities of natural gas may have been generated. On the basis of the sediment type, the generation of oil in a large quantity seems less likely.

As of March, 1981, 25 exploratory wells had been drilled in the Baltimore Canyon trough area. In only five of these wells have significant hydrocarbon shows been reported (Mattick and Ball, 1981). These five wells are close together, about 20 km east of the COST B-2 well. Their locations project to USGS line 25 (Fig. 2) about midway between the two COST wells. Natural gas flowed at low rates from thin sandstone formations, at depths ranging from 3,800 to 4,800 m in all five wells. One well, centrally located among the five, was tested at 2,535 m, and a small amount of oil flowed from a Lower Cretaceous sandstone bed (Mattick and Ball, 1981). All these discoveries are probably

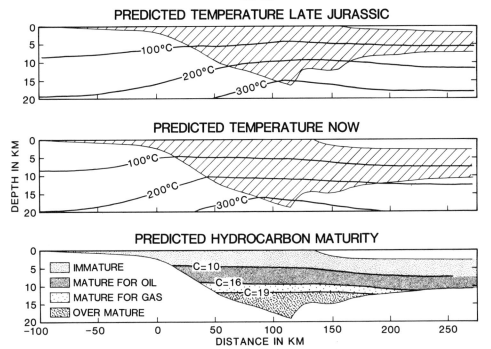

Figure 2. Results of thermal model of the Baltimore Canyon trough along USGS Line 25 by Sawyer and others (1982).

within a single structure, but it has not yet proven to be of commercial size.

The oil discovery is in Lower Cretaceous sediment that is indicated by the thermal simulation of the basin to consist of nearly mature oil-source rock. The gas-bearing horizons occur in sediments of Late Jurassic age, which should be mature for oil generation, but not for gas. It seems likely that all the oil and gas discovered so far were generated at greater depth and migrated to the sandstone reservoirs in which they are found. Their presence supports the contention that some sediment in the Baltimore Canyon trough has experienced appropriate temperature histories for the generation of both oil and natural gas. The apparently greater amount of gas in the reservoir is consistent with the predominance of continentally derived, gas-prone, sediment in the part of the trough that should be mature for natural-gas generation.

Georges Bank basin

The Georges Bank basin lies offshore of Massachusetts and is the site of two COST wells, the G-1 and the G-2 (Fig. 1). Arthur (1982) studied the thermal evolution at the COST G-2 well using the one-dimensional thermal modeling technique described above. Swift and others (1987) used the two-dimensional method to model the whole of USGS seismic line 19, which crosses the margin near the wells.

One-dimensional model. Arthur's (1982) model assumes that the temperature field at the wellsite has varied through time according to the lithospheric attenuation stretching model of Royden and others (1980) appropriate for stretching to $\beta = 2.5$ (Fig. 5). The burial history curves for four horizons (Fig. 5) show that their temperatures increased rapidly until about 110 Ma, after which they were nearly constant. The maximum predicted temperature in the well is 190°C at the base of the well. This sediment is judged to be just past the window of oil maturity and into that for gas maturity (zones from Royden and others, 1980). The sediment deposited at 156 Ma is predicted to have reached oil maturity about 95 Ma but not to have progressed much further since then.

Two-dimensional model. The Swift and others (1987) model is along USGS line 19 in the Georges Bank basin (Fig. 1). The model is 324 km wide, the length of line 19 plus an additional 36 km on each end. It is divided into 18 columns with horizontal grid spacing of 18 km.

At the end of the Jurassic, about 80 percent of the sediment in the basin has been deposited. Sediments in the center of the basin have reached temperatures higher than 100°C (Fig. 6). This is in stark contrast to the temperatures in the Baltimore Canyon

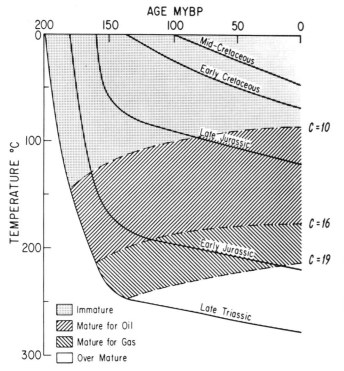

Figure 3. Predicted temperature history curves for sediment of various ages deposited at the location of the COST B-3 well in the Baltimore Canyon trough. Deposition of sediments and their increase in temperature were rapid during the Jurassic. Contours of hydrocarbon source maturity are determined by integrating the temperature history of each packet of sediment. Sediments in the Baltimore Canyon trough have reached all levels of maturity from under-mature to over-mature.

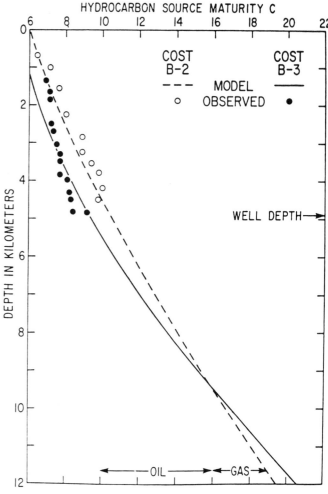

Figure 4. Comparison of predicted hydrocarbon source maturity with observations in the two COST wells in the Baltimore Canyon trough. Vitrinite reflectance was observed in the wells and converted to the parameter C using the relation given by Waples (1980).

trough at the same time. The deepest sediments in that basin had reached 300°C. The difference obviously relates to the differences in maximum thickness of sediment in the two basins. The bathymetry at the end of the Jurassic shows a sharp shelf edge located very close to the present shelf edge. The continental shelf remained within about 100 meters of sea level throughout its development, filling quickly when subsidence was rapid, and more slowly later on.

The isotherms predicted for the present Georges Bank basin (Fig. 6) indicate that the temperature in the basin has changed little since the Late Jurassic. The maximum sediment temperature is still little more than 100°C.

The development of hydrocarbon thermal maturity was slightly more complex in this basin than in the last. The sediments deposited near the edge of the oceanic lithosphere experienced a strong pulse of heating early in their history (Fig. 7C). This was due to the rapid deposition of sediment on very hot, extended crust. The lateral heat flow was high and the sediments became quite hot. Following the initial surge of sediment, the rate decreased and stayed slow. The hot sediments cooled, some by as much as 20°C before sedimentation picked up again in the Ter-

tiary. These Tertiary sediments accumulated quickly enough to bury and heat the older sediment. The temperature within sediments is the result of an interplay between the heat flow from the crust and the rate of sedimentation. The decrease in sediment supply, during the Cretaceous, was probably due to the construction of an effective barrier reef at the shelf edge. This reef was swamped with sediment and killed in the middle Cretaceous, again allowing sediment to reach the continental slope and rise.

Only oil maturation conditions would seem to have been reached along this cross section of the basin. The maturity developed in isolated packets beginning in the Late Jurassic. Later, the region of maturity over rift-stage crust grew. This area is of particular interest because it contains salt and sediments associated with the breakup, when restricted marine conditions probably existed. Sediments deposited in a restricted marine environment may have the proper character to act as oil source rocks. The

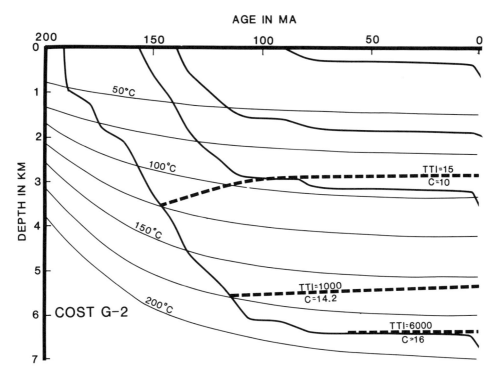

Figure 5. Temperature-depth history for sediments (bold solid lines) at the location of the COST G-2 well as modeled by Arthur (1982). The temperature profiles (light solid lines) are taken from Royden and others (1980) and are for the lithospheric attenuation model with $\beta = 2.5$. Isomaturity lines (dashed lines) correspond to the onset of oil generation, Waple's (1980) upper limit of oil generation, and Royden and others (1980) upper limit of oil generation.

many isolated, fault-bounded, down-dropped blocks in the Georges Bank basin may have concentrated oil-prone sediments during the rifting stage. These sediments reached the proper maturity during the Cretaceous. Oil formed at the time could have migrated into reservoirs in the carbonate banks above. These parts of the basin, which are near the carbonate banks lying over the fault blocks, may have the greatest hydrocarbon potential.

Most of the Jurassic sediments in this basin are probably similar to those of the Baltimore Canyon trough. These would tend to be gas prone and thus are not likely to have generated significant hydrocarbons in a cool basin. A layer of black shales was deposited during the Albian and Aptian over most of the margin (Schlee and others, 1979). Such rocks might be expected to contain oil-source material. However, the thermal model indicates that these sediments never reached temperatures higher than 40°C anywhere on the cross section. Even with errors in the model, it is doubtful that they have reached the approximately 100°C temperature necessary to achieve oil maturity.

Carolina trough

The Carolina trough is perhaps the simplest of the U.S. Atlantic Margin basins. Like the Baltimore Canyon trough, it formed at a part of the margin oriented perpendicular to the oceanic fracture zones, and therefore, probably is a normal pull-apart basin.

One-dimensional model. Royden and others (1980) modeled the thermal history and hydrocarbon maturity at three locations (Fig. 1) in the Carolina trough. They calculated the temperature field using two models of crustal extension, the dike-intrusion model and the lithospheric-attenuation model. The temperature history curves for the sediments were relatively insensitive to the model chosen. The lithospheric-attenuation model curves show that the present temperatures at the bases of the sediment columns are greatest over oceanic crust and decrease landward (Fig. 8). Note that the maximum temperature over oceanic crust (>200°C) is predicted to occur at about 170 Ma. The effects of rapid sedimentation are not included in this simple model and I do not believe that temperatures in the sediment actually got that hot. The hydrocarbon maturity is likewise greatest over oceanic crust where there are sediments predicted to be overmature. However, this is probably an overestimate, with most of this apparent maturation occurring during the short period of very high temperatures mentioned above. At point B (Fig. 1) the predicted temperature history and maturation, mature for oil generation, are reasonable (Fig. 8). At point C the temperature history has been insufficient to mature any hydrocarbons.

Two-dimensional model. Sawyer (1982) and Celerier

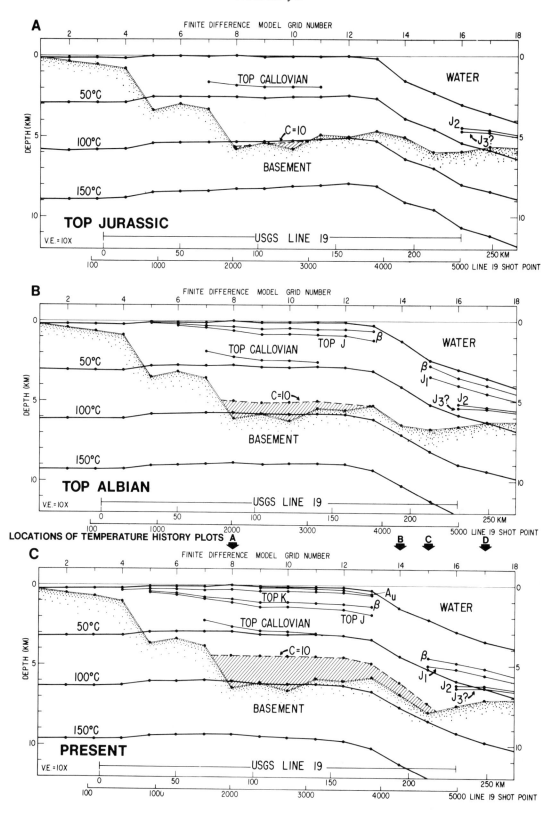

Figure 6. Results of thermal model of the Georges Bank basin along USGS Line 19 by Swift and others (1987). The locations of four time-temperature profiles, A to D in Figure 7, are marked on the lower panel.

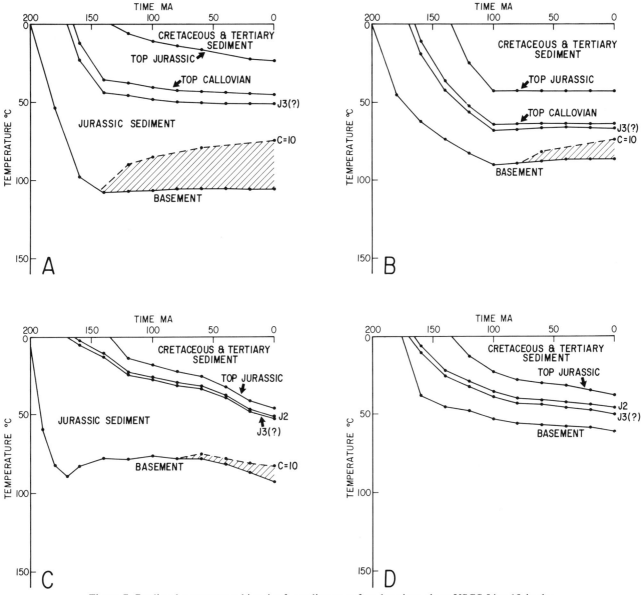

Figure 7. Predicted temperature histories for sediment at four locations along USGS Line 19 in the Georges Bank basin (Fig. 1). The shaded areas indicate sediments mature enough for the generation of oil.

(1986) modeled the Carolina trough along USGS line 32. Sawyer's (1982) finite difference grid extends 40 km westward from the end of that line. The grid is 340 km long extending eastward to shot point 6000. Seventeen horizontal grid columns are distributed over this range at a spacing of 20 km.

The predicted isotherms for the end of the Jurassic show sediments to have reached about 120°C (Fig. 9). The mound of sediment is a carbonate buildup in the Jurassic. The water seems too deep for this to be a normal reef. The predicted present isotherms indicate that sediment deposition has been rapid enough to heat up the sediment during the last 150 m.y. (Fig. 9). The deepest sediment has reached about 165°C.

The temperature history of the deepest sediments in the Carolina trough is appropriate for the generation of oil (Fig. 9). The sediments are just short of the natural gas maturity window. With the range of errors possible in this calculation, gas generation cannot be ruled out. The sediments mature for oil generation are all Early to Middle Jurassic in age.

No seismic stratigraphy estimates of depositional environments have been made in the Carolina trough, nor have any deep wells been drilled, so the nature of these sediments can only by the subject of speculation. It is likely that they are similar to the deep sediments in the Baltimore Canyon trough and therefore, of continental or shallow-marine origin and gas prone. The hydro-

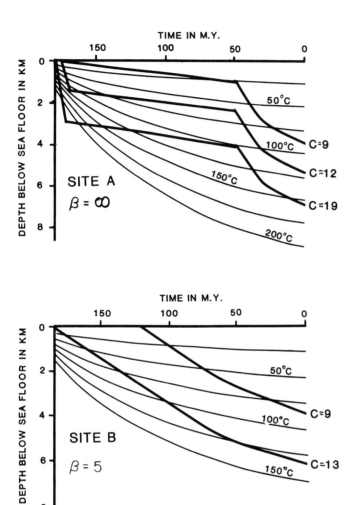

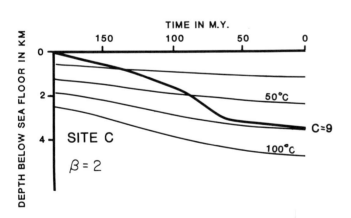

Figure 8. Temperature-depth history curves for sediment deposited at three locations in the Carolina trough area (Fig. 1) as modeled by Royden and others (1980). They are based on the lithospheric attenuation model for different amounts of extension noted on each panel.

carbon generation potential of this basin seems somewhat greater than that of the Georges Bank basin because there are sediments of greater thermal maturity. The salt diapirs and carbonate build-ups, near the rift-stage to oceanic crust boundary, overlie the hydrocarbon-mature sediments and might provide good reservoirs.

EXPLORATION FOR DEEP, HOT, HYDROCARBON-MATURE MARGINAL BASINS

The differences in the sediment thicknesses, shelf widths, hydrocarbon maturities, and other features of marginal basins are often more the result of variabilities in the sediment supply than variation of the crust underneath. The sedimentation can have a more significant influence than is obvious. The maximum values of temperature and hydrocarbon source maturity in the three basins examined occur in the Baltimore Canyon trough. They are there because the maximum sediment thickness, 18 km, on the entire U.S. Atlantic Margin is in that basin. It seems likely therefore that a method to identify areas of maximum sediment accumulation could be used to find potential high-temperature sediments and hydrocarbon source areas.

The extensional model relates the tectonic subsidence of a region to the crustal type underneath (McKenzie, 1978; Sawyer, 1985). The amount of tectonic subsidence determines how much sediment and water can load the crust. The maximum sedimentary thickness in a basin is very sensitive to the relative position of the continental shelf edge and the oceanic crust boundary (Fig. 10). When the shelf edge is located over oceanic crust, as in the Baltimore Canyon trough, the sediment reaches a maximum thickness of about 18 km. When the shelf edge lies continentward, the maximum sediment thickness drops off rapidly. That relationship occurs in the Georges Bank basin and Carolina trough, and the maximum sediment thickness in those basins is significantly less than 18 km. It is easy to observe the present bathymetry of a continental margin. If there is a potential field anomaly, such as the East Coast Magnetic Anomaly, or other feature associated with the location of the transition from rift-stage to oceanic crust (Klitgord and Behrendt, 1979), then it too can be easily observed.

The position of the continental shelf edge is probably largely determined by the sediment supply. If sediment is plentiful, the shelf edge will be built out over oceanic crust. The relation between sediment influx and shelf edge position is not linear. The further out the shelf edge migrates, the more sediment it takes to move it out another unit. This increase seems to stabilize the position of the shelf edge over highly thinned, but not oceanic, crust. The width of the zone of rift-stage crust also influences the amount of sediment required to move the shelf edge over oceanic crust. If the zone is broad, more sediment is required. In the example shown, local isostatic compensation and a particular crust extension geometry are assumed. Flexural isostatic equilibration would not allow the sharp bends in the crust, but the

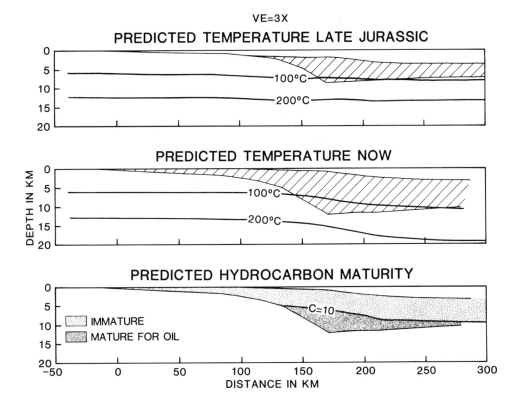

Figure 9. Results of thermal model of the Carolina trough along USGS Line 32 by Sawyer (1982).

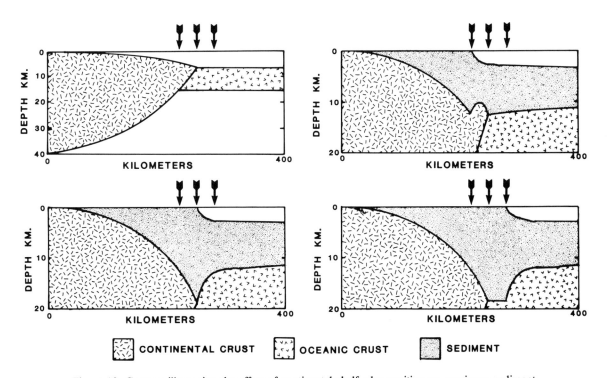

Figure 10. Cartoon illustrating the effect of continental shelf edge position on maximum sediment accumulation in a passive margin basin. The extended continental and oceanic crust, prior to sediment loading, are shown in the upper left. That crust is loaded with sediment to put the shelf edge at three locations (arrows) with respect to the edge of oceanic crust. The sediment is assumed to be compensated locally.

thickest sediment should still accumulate over oceanic crust under the shelf edge.

This approach can be used as a regional exploration tool to identify parts of continental margins likely to contain sediments of high hydrocarbon source maturity. They are most likely to occur when the continental shelf edge is over or very close to oceanic crust. This event is most likely to occur in regions where the sediment influx has been high and the zone of rift-stage crust is narrow.

REFERENCES CITED

Amato, R. V., and Bebout, J. W., 1978, Geologic and operational summary, COST no. GE–1 well, southeast Georgia Embayment area, South Atlantic OCS: U.S. Geological Survey Open-File Report 78-668, 122 p.

Amato, R. V., and Simonis, E. K., 1979, Geologic and operational summary, COST no. G-2 well, Georges Bank area, North Atlantic OCS: U.S. Geological Survey Open-File Report 80-269, 120 p.

Angevine, C. L., and Turcotte, D. L., 1981, Thermal subsidence and compaction in sedimentary basins; Application to Baltimore Canyon trough: American Association of Petroleum Geologists Bulletin, v. 65, p. 219–225.

Arthur, M. A., 1982, Thermal history of the Georges Bank basin, *in* Scholle, P. A., and Wenkam, C. R., eds., Geological studies of the COST nos. G-1 and G-2 wells, United States North Atlantic outer continental shelf: U.S. Geological Survey Circular 861, p. 143–152.

Beaumont, C., Keen, C. E., and Boutilier, R., 1982, On the evolution of rifted continental margins; Comparison of models and observations for the Nova Scotia margin: Geophysical Journal of the Royal Astronomical Society, v. 70, p. 667–715.

Celerier, B., 1986, Models for the evolution of the Carolina trough and their limitations [Ph.D. thesis]: Cambridge, Massachusetts Institute of Technology, 206 p.

Claypool, G. S., Lubeck, C. M., Baysinger, J. P., and Ging, T. G., 1977, Organic geochemistry, *in* Scholle, P. A., ed., Geological studies on the COST no. B-2 well, U.S. Mid-Atlantic outer continental shelf area: U.S. Geological Survey Circular 750, p. 46–59.

Grow, J. A., 1980, Deep structure and evolution of the Baltimore Canyon trough in the vicinity of the COST no. B-3 well, *in* Scholle, P. A., ed., Geological studies of the COST no. B-3 well, United States Mid-Atlantic continental slope area: U.S. Geological Survey Circular 833, p. 117–124.

——, 1981, The Atlantic Margin of the United States: American Association of Petroleum Geologists Education Course Note Series 19, 41 p.

Heise, B. A., and Jackson, D. S., 1980, Geothermal gradient, *in* Amato, R. V., and Simonis, E. K., eds., Geologic and operational summary, COST no. G-2 well, Georges Bank area, North Atlantic OCS: U.S. Geological Survey Open-File Report 80-269, p. 53–55.

Issler, D. R., 1984, Calculation of organic maturation levels for offshore eastern Canada; Implications for general application of Lopatin's method: Canadian Journal of Earth Sciences, v. 21, p. 477–488.

Jackson, D. S., and Heise, B. A., 1980, Geothermal gradient, *in* Amato, R. V., and Bebout, J. W., eds., Geologic and operational summary, COST no. GE-1 well, southeast Georgia Embayment area, South Atlantic OCS: U.S. Geological Survey Open-File Report 78-668, p. 79–81.

Keen, C. E., 1979, Thermal history and subsidence of rifted continental margins; Evidence from wells on the Nova Scotia and Labrador shelves: Canadian Journal of Earth Sciences, v. 16, p. 505–522.

Klitgord, K. D., and Behrendt, J. C., 1979, Basin structure of the U.S. Atlantic Margin, *in* Watkins, J. S., Montadert, L., and Dickerson, P. W., eds., Geological and geophysical investigations of continental margins: American Association of Petroleum Geologists Memoir 29, p. 85–112.

Lopatin, N. V., 1971, Temperature and geologic time as factors in coalification (in

Russian): Izvestiya Akademii Nauk SSSR, Seriya Geologicheskaya, no. 3, p. 95–106.

Mattick, R. E., and Ball, M. M., 1981, Petroleum geology, *in* Grow, J. A., ed., Summary report of the sediments, structural framework, petroleum potential, and environmental condition of the United States middle and northern Atlantic Continental Margin in the area of proposed oil and gas lease sale no. 76: U.S. Geological Survey Open-File Report 81-765, p. 69–82.

McKenzie, D. P., 1978, Some remarks on the development of sedimentary basins: Earth and Planetary Science Letters, v. 40, p. 25–32.

Miller, R. E., and 7 others, 1980, Organic geochemistry, *in* Scholle, P. A., ed., Geological studies of the COST no. B-3 well, United States Mid-Atlantic continental slope area: U.S. Geological Survey Circular 833, p. 85–104.

Robbins, E. I., and Rhodehamel, E. C., 1976, Geothermal gradients help predict petroleum potential of Scotian shelf: Oil and Gas Journal, v. 74, p. 143–145.

Royden, L., Sclater, J. G., and Von Herzen, R. P., 1980, Continental margin subsidence and heat flow; Important parameters in formation of petroleum hydrocarbons: American Association of Petroleum Geologists Bulletin, v. 64, p. 173–187.

Sawyer, D. S., 1982, Thermal evolution of the northern United States Atlantic Continental Margin [Ph.D. thesis]: Cambridge, Massachusetts Institute of Technology, 264 p.

——, 1985, Total tectonic subsidence; A parameter for distinguishing crust type at the U.S. Atlantic Continental Margin: Journal of Geophysical Research, v. 90, p. 7751–7769.

Sawyer, D. S., Toksoz, M. N., Sclater, J. G., and Swift, B. A., 1982, Thermal evolution of the Baltimore Canyon trough and Georges Bank basin, *in* Watkins, J. S., ed., Studies in continental margin geology: American Association of Petroleum Geologists Memoir 34, p. 743–764.

Schlee, J. S., 1981, Seismic stratigraphy of the Baltimore Canyon trough: American Association of Petroleum Geologists Bulletin, v. 65, p. 26–53.

Schlee, J. S., Dillon, W. P., and Grow, J. A., 1979, Structure of the continental slope off the eastern United States: Society of Economic Paleontologists and Mineralogists Special Publication no. 27, p. 95–117.

Smith, M. A., and 8 others, 1976, Geological and operational summary, COST no. B-2 well, Baltimore Canyon trough area, Mid-Atlantic OCS: U.S. Geological Survey Open-File Report 76-774, 79 p.

Swift, B. A., Sawyer, D. S., Grow, J. A., and Klitgord, K. D., 1987, Subsidence, crustal structure, and thermal evolution of Georges Bank basin: American Association of Petroleum Geologists Bulletin (in press).

Tissot, B. P., and Welte, D. H., 1978, Petroleum formation and occurrence: New York, Springer-Verlag, 538 p.

Von Herzen, R. P., and Helwig, J. A., 1981, Geothermal heat flux determined from COST wells on the Atlantic Continental Margin [abs.]: American Association of Petroleum Geologists Bulletin, v. 65, p. 1364.

Waples, D. W., 1980, Time and temperature in petroleum formation; Application of Lopatin's method to petroleum exploration: American Association of Petroleum Geologists Bulletin, v. 64, p. 916–926.

Watts, A. B., 1981, The U.S. Atlantic Continental Margin; Subsidence history, crustal structure, and thermal evolution: American Association of Petroleum Geologists Education Course Note Series 19, p. 2.1–2.75.

Watts, A. B., and Thorne, J., 1984, Tectonics, global changes in sea level and their relationship to stratigraphical sequences at the U.S. Atlantic Continental Margin: Marine and Petroleum Geology, v. 1, p. 319–339.

MANUSCRIPT ACCEPTED BY THE SOCIETY MARCH 31, 1987

ACKNOWLEDGMENTS

I appreciate the efforts of reviewers Charlotte Keen and Charles Angevine whose comments improved the paper. While preparing the paper I was supported by National Science Foundation Grant OCE-8401621 and Office of Naval Research grant N00014-81-K-0728. This is University of Texas Institute for Geophysics contribution no. 687.

Printed in U.S.A.

The Geology of North America
Vol. I-2, The Atlantic Continental Margin: U.S.
The Geological Society of America, 1988

Chapter 20

Sea-level changes and their effect on the stratigraphy of Atlantic-type margins

W. C. Pitman
Lamont-Doherty Geological Observatory of Columbia University, Palisades, New York 10964
X. Golovchenko*
Department of Geophysical Research, Marathon Oil Co., P.O. Box 269, Littleton, Colorado 80160

INTRODUCTION

Passive margins generated by rifting and separation of continents are associated with some of the world's thickest sediment accumulations, of which the east coast of North America is a typical example. The sedimentary wedge that has accumulated there is as much as 15 km thick at the seaward edge and represents an almost continuous succession of shallow marine sediments, ranging in age from probable Triassic-Jurassic at the base of the section to Recent at the surface (e.g., Drake and others, 1959; Bally, 1981). The bottommost part of the section may contain rift-valley sequences (Bally, 1981).

The individual postrift strata are thickest on the ocean side of the margin and systematically thin to zero thickness (either by pinch-out or truncation) toward the landward edge. The horizons between strata are quite planar, disrupted occasionally by faulting related to sediment processes only. The back-stripping technique has been developed to remove the effect of sediment loading as a function of time and thus reveal the driving or tectonic subsidence (Sleep, 1971; Watts and Ryan, 1976; Steckler and Watts, 1978; Steckler, 1981). By use of this technique it has been established that the driving subsidence is caused by deep-seated cooling with a thermal cooling constant usually in the range of 50 m.y. to 100 m.y. (Sleep, 1971; Watts and Ryan, 1976; Steckler and Watts, 1978).

In recent years there has been increased interest in the Atlantic margin and other passive margins both because of the hydrocarbon potential and because the very thick, relatively undisturbed sedimentary section provides a singular opportunity for stratigraphic studies. Vail and his co-workers (see Vail and others, 1977a) have studied in detail the stratigraphic patterns at a number of margins, including the Atlantic margin. They recognize the depositional sequence as the basic stratigraphic unit and

define it as a set of conformable strata bounded at the top and bottom by unconformities or their laterally correlatable conformities. They assume that sedimentary sequences are laid down during periods of relative sea-level rise and that the unconformities at the sequence boundaries are caused by the relative fall of sea level. Vail and his associates have defined a temporal and spatial pattern of depositional sequences that they believe is correlatable on a worldwide basis. They believe this correlatable pattern of sequences is a reflection of eustasy and they have interpreted the pattern to produce a curve of worldwide sea-level change (Vail and others, 1977b).

It is our purpose in this paper to point out some of the difficulties in interpretation of the patterns of depositional sequences at an Atlantic-type margin in terms of eustasy and to show that the correlation of transgressive and regressive events within a basin, between basins, and with a temporal pattern of sea-level change may be subject to errors of up to 3 m.y.

A BASIC MODEL

The origin of the driving subsidence of these passive margin basins is now known to be related to the rift process (McKenzie, 1978; Royden and others, 1980; Royden and Keen, 1980). During the rifting stage, which is usually of short duration (<20 m.y.), the lithosphere (which includes the crust) is stretched by a factor β (and thinned by $1/\beta$). This takes place by faulting at the surface and ductile flow at depth. During rifting, uplift may occur if the ratio of the original crustal thickness to lithosphere thickness is small enough (usually <0.14). However, under most circumstances, this ratio is too large and so synrift subsidence takes place. Subsequent to the rifting, there is a very long period of subsidence as the attenuated lithosphere thickens by the cooling of the underlying aesthenosphere. The cooling is deep-seated; hence, the cooling time constant is large (50 m.y. to 90 m.y.).

*Present address: Lamont-Doherty Geological Observatory of Columbia University, Palisades, New York 10964.

Pitman, W. C., and Golovchenko, X., 1988, Sea-level changes and their effect on the stratigraphy of Atlantic-type margins, *in* Sheridan, R. E., and Grow, J. A., eds., The Atlantic Continental Margin, U.S.: Geological Society of America, The Geology of North America, v. I-2.

Because of the postrift cooling and rethickening of the litho-sphere, postrift movement along basement faults is negligible. The subsidence is persistent but slow, with a maximum rate of several centimeters per 1,000 yrs immediately after rifting, and the rate decreases exponentially with time; the thermal time constants range from about 50 m.y. to 90 m.y. Sedimentation is usually able to keep pace with subsidence. Because the attenuated cooling lithosphere has some finite elastic strength, the lithosphere re-sponds flexurally to the load.

We can use the above concept of postrift subsidence (McKenzie, 1978) to construct a model of the sedimentary and geomorphic processes that take place on a margin to show how these are affected by eustatic sea-level changes. During the rifting of the Atlantic margin, for example, a lithospheric strip between Africa and North America was gradually attenuated until it rup-tured and aesthenosphere came to the surface to form an ocean basin that has continued to grow by continued separation. At the time of the transition from rift to drift, the amount of attenuation of the lithosphere would have been a maximum at the oceanic edge of the rifted zone and would decrease in the continental direction to a point where there had been no attenuation. The rate of thermally driven subsidence is always greater at the most highly attenuated end of the platform (usually the basinward end) and decreases smoothly to zero toward the landward end. Hence, as shown in Figure 1, at any point in time the rate of thermally driven subsidence is a maximum toward the basin and decreases to zero on the landward side. We will call the region of zero thermal subsidence the hinge zone. For the moment we will ignore flexure.

The subsidence rates are slow enough that under usual cir-cumstances there is sufficient sediment to keep pace with subsi-dence. It is observed that in a tectonically quiescent region such as a passive margin, geomorphic processes tend to maintain graded surfaces. In our discussion, we will assume that there is a suffi-cient sediment supply for the erosional and depositional processes to maintain a graded slope on the shelf (the subaqueous part) and coastal plain (the subaerial part). These are different geomorphic provinces, but, for the moment, we will assume they have the same graded slope. The several solid vertical arrows in Figure 1 show the rate of subsidence at several discrete points on the margin. Because the time rate of change of the subsidence rate is very slow we approximate that at each point on the margin the rate of change is constant for periods of several m.y. We have shown the beach to be at point A on the margin, but, under the conditions we set above, the beach is the point at which neither erosion nor deposition is occurring; instead, subsidence is taking place. Therefore, given the above boundary conditions, the only way the beach can remain at point A is if sea level is falling at a rate equal to the local rate of subsidence (Pitman, 1978). We call this the equilibrium point. If this is the case, if the beach is at the equilibrium point, the beach will remain at that point on the margin as long as the rate of sea-level fall equals the rate of subsidence. As the margin rotates downward about the hinge zone, erosion will continue to bevel the coastal plain to a graded

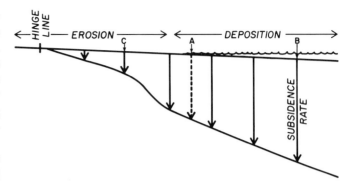

Figure 1. A section across a model passive margin is shown. The vertical solid arrows show the local rate of subsidence. The dashed vertical arrow gives the rate of eustatic sea-level fall. The rate of subsidence is at maximum basinward and decreases smoothly to zero at the hinge zone. It is assumed that the erosional and depositional processes maintain a graded slope for the coastal plain and shelf respectively. For the beach to be stationary on the margin (A) the rate of sea-level fall must equal the rate of subsidence. If the rate of sea-level fall increases, the beach moves seaward (B). If the rate of sea-level fall decreases, the beach moves landward (C).

slope, causing angular truncation of previously deposited strata; deposition will occur on the subsiding shelf, maintaining a graded surface. The consequent strata will be that of a *stillstand*.

If the rate of sea-level fall is increased, then the rate of sea-level fall at point A will exceed the rate of subsidence, expos-ing the shelf. The beach will then migrate seaward until it reaches the new equilibrium point (Pitman, 1978). This is the point on the margin where the new greater rate of sea-level fall is equal to the local rate of subsidence (say at point B); the shoreline will remain there as long as the new rate of sea-level fall and the rate of local subsidence remain equal. If the rate of sea-level fall then de-creases, the shoreline will transgress landward to that point (C) where the new reduced rate of sea-level fall is equal to the local rate of subsidence (Pitman, 1978). This entire sequence of events took place during a time interval when eustatic sea level was constantly falling, but at varying rates.

What is most important is that the same sequence of events would take place at *all* passive margins at which the rate of subsidence increases smoothly from zero at the landward side to a maximum seaward, and where the geomorphic processes main-tain graded slopes (Pitman, 1978). The magnitude of the regres-sion and transgression will vary depending upon the geometry of each particular margin.

Two other variations must be considered. First, if we were to start with the beach at point A during a stillstand and then let the rate of sea-level fall decrease to zero, because of the combined effect of subsidence and the maintenance of graded slopes, the shoreline will slowly migrate toward the hinge zone where the subsidence rate is zero. But because the margin responds flexu-rally to sediment loading, and because as cooling takes place, the

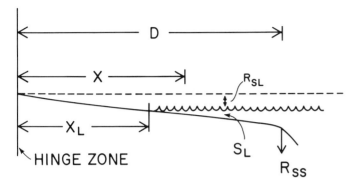

Figure 2. The parameter used to quantify the effect of sea-level changes on the transgressive and regressive events are shown. D, distance from the hinge zone to the shelf edge; X, distance from the hinge zone to any point on the margin; X_L, distance from the hinge zone to the shore line; R_{SL}, rate of sea-level change with respect to a horizontal plane drawn through the craton; S_L, the slope of the shelf and coastal plain; R_{SS}, the rate of subsidence at the shelf edge; R_{SS} decreases linearly to zero at the hinge zone.

flexural rigidity increases, the hinge zone or peripheral bulge will slowly migrate landward and, in the case of zero rate of sea-level change, the beach will follow the flexural bulge, causing a transgression and coastal onlap (Watts, 1982).

Secondly, as discussed above, the rate of thermal subsidence decreases exponentially with time everywhere on a margin. If we start at a stillstand with the beach at point A and sea level falling at a constant rate, the decreasing thermal subsidence rate will cause the shoreline to continuously regress toward its seaward-migrating equilibrium point.

This model has been quantified by Pitman (1978; Fig. 2) to give the position of the beach SL at any time as a function of sea level and sea-level change.

$$(1) \quad X_L/D = R_{SL}/R_{SS} - (R_{SL}/R_{SS} - X_{LI}) \exp(-T \cdot R_{SS}/D \cdot S_L)$$
$$(2) \quad SED(X) = X \cdot R_{SS} \cdot T/D - R_{SL} \cdot T - (\exp(-T \cdot R_{SS}/D \cdot S_L) - 1) \cdot (R_{SL} - X_{LI} \cdot R_{SS}/D) \cdot D \cdot S_L/R_{SS}.$$

R_{SS} is the rate of subsidence at the shelf edge. $X/D \cdot R_{SS}$ is the rate of subsidence at any point X on the margin. X_L = the position of the shoreline zone at the end of the time interval T. R_{SL} is the rate of sea-level change during the time interval T. X_{LI} is the position of the shoreline at the beginning of the time interval. SED(X) is the amount of sediment deposited or eroded (negative SED) at point X during the time interval. For the rest of the discussion we will ignore flexure. We will assume we are dealing with margins several tens of millions of years after rifting so that we may safely assume that at any point on the margin during a period of several tens of millions of years, the rate of change of the rate of thermally driven subsidence is much less than the changes in the rate of sea-level change.

FROM STRATIGRAPHY TO SEA-LEVEL CHANGES

Aggradation

Vail and others (1977a) used the sedimentary sequences to derive curves of relative sea-level change. Their method is illustrated in Figure 3. Figure 3a shows a stratigraphic cross section interpreted in terms of sedimentary sequences (A, B, C, D, and E). The relative sea-level rise represented by sequence A is given by the coastal aggradation as shown in Figure 3c. The coastal aggradation is determined by measuring successive stratal thicknesses above points of pinch-out. Relative sea-level fall is estimated by measuring the vertical drop from the topmost point of transgressive onlap of a sedimentary sequence to the next onlapping strata of the overlying sequence.

However, if our model is correct, then transgressions are not necessarily correlatable with a rise in eustatic sea level; rather, transgressions will take place systematically on any and all passive margins that obey the above boundary conditions and when the rate of sea-level fall is systematically decreasing. If that is the case, then coastal onlap is not necessarily caused by a sea-level rise.

Another of the problems with using the method of onlap to measure "local" sea-level rise is that the reference frame relative to which sea level is being measured is changing. For example, in Figure 4 the aggradation (A–B) measured at point A is the apparent amount of sea-level rise that would be seen by an observer standing at point A. However, if the observer had been at *any* other point on the surface E–A–C–F for the same time interval (Fig. 4), the amount of coastal aggradation measured would have been different because the rate of subsidence is different at each point on the surface. The next element of coastal aggradation in the summation is measured from the same surface from point C (which is subsiding at a different rate than point A) to D. Again, if the measurement had been made from any other point on the surface E–A–C–F, the amount of coastal aggradation observed would be different. Thus, without a fixed reference frame the measurement has no meaning quantitatively. Figure 4 also illustrates the point that a decrease in sea-level fall will cause systematic aggradation, which can be mistaken for sea-level rise.

Coastal Onlap

The consistent observation of continued onlap of "nonmarine" coastal deposits during periods of regression has been difficult to interpret (see Fig. 5, regional super cycle). Coastal deposits are defined as *nonmarine* sediments and do not include lagoonal facies (Vail and others, 1977a). In fact, they often consist of fluvial facies. The littoral zone occupies a narrow zone between the coastal and marine sediments. During a transgressive phase, it has been observed that the point of pinch-out or onlap of coastal deposits and the terminus of the marine deposits move together steadily landward (Fig. 4). However, as regression takes over and the littoral zone moves seaward, the coastal deposits

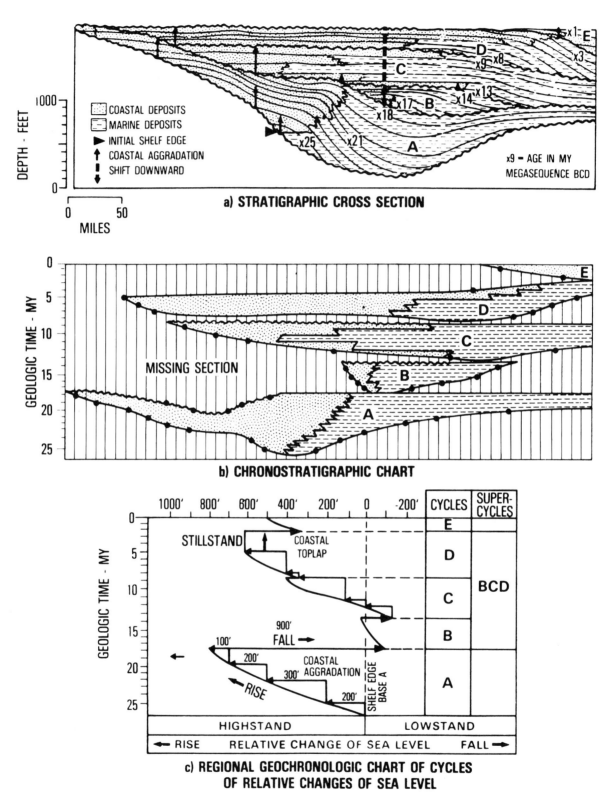

a) STRATIGRAPHIC CROSS SECTION

b) CHRONOSTRATIGRAPHIC CHART

c) REGIONAL GEOCHRONOLOGIC CHART OF CYCLES
OF RELATIVE CHANGES OF SEA LEVEL

Figure 3. From Vail and others, 1977a. The figures illustrate the technique used by Vail and others (1977a) to determine local sea-level change. The technique is explained in the text.

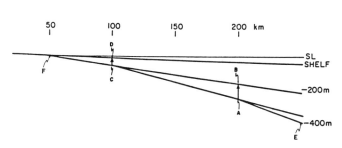

Figure 4. This is a model section through a passive margin. The hinge zone is to the left and the basin to the right. The bottom surface is an unconformity. The overlying onlapping strata are the lower section of a sedimentary sequence. Two methods of interpreting this section are discussed in the text.

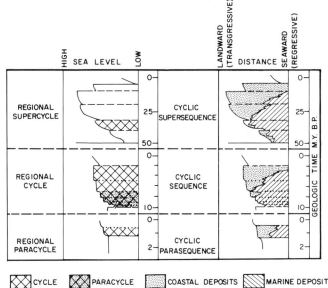

Figure 5. Shows the Vail and others (1977a) interpretation of the relationship between patterns of onlap of coastal sediments (on the right) and sea-level change. Note that as long as the coastal (nonmarine) sediments are onlapping landward, it is assumed that sea level is rising, even during intervals when the littoral is regressing seaward.

continue to onlap landward. At about the time of maximum regression, the coastal deposition ceases precipitously (Fig. 5; Vail and others, 1977a).

There have been several interpretations of this sequence of events. Initially, it was thought that as long as the coastal deposits continued to onlap landward, local sea level must be rising, even though the boundary between the coastal and marine deposits (the littoral zone) continues to move seaward (Vail and others, 1977a). The sudden cessation of coastal deposition was interpreted to mean that sea level had fallen precipitously from the level of the topmost coastal onlap to the landward edge of the marine deposits (Fig. 5, super cycle). Recently Vail and others (1985) concluded that the transition from a transgression to a regression, even though accompanied by continued coastal onlap, indicated a change from sea-level rise to sea-level fall. We reiterate that the transition from a transgression to a regression may equally be caused by an increase in the rate of sea-level fall.

We offer here a mechanism by which the initiation of a regression causes the continuation of coastal onlap (Pitman, 1982). One of the observations of geomorphologists is that tectonically stable regions tend to become beveled to a graded slope, which is maintained by erosion and deposition. This gradient is dependent upon the physical properties of the strata and the environment. In fluvial systems, erosion and deposition take place so as to maintain a graded profile. If the profile over-steepens, water velocity increases and erosion takes place; if the gradient of any part of the system becomes too shallow, water velocity decreases and deposition occurs. In the marine areas, the gradient is controlled by the dynamics of the overlying water mass and sediment supply.

Let us consider a coastal region in which a large graded fluvial system enters the ocean at the coastline and in which the littoral zone has been transgressing slowly for a long period of time. In this case it is likely that the graded slope of the shelf will be less than the gradient of the lower course of the fluvial system

that is being progressively transgressed. In that case, if there is a regression exposing the shelf, subaerial deposition will take place. Because the exposed shelf is at a gentler slope than the lower course of the fluvial system, the rivers crossing this exposed shelf must slow down and hence will deposit a part of the load. These will be entirely nonmarine coastal deposits. This coastal onlap will continue during regression. This sequence of events is shown in Figure 6 (Pitman, 1982). The slopes in this drawing have been exaggerated for illustrative purposes. At 53 Ma, a transgressive phase has ended. At the shoreline, sea level is falling at the same rate as the underlying strata are subsiding; hence the strand line moves neither seaward nor landward. As subsidence occurs, the strata tilt downward in the seaward direction, and erosion of the coastal plain extends landward from the shoreline. Beginning at 52 Ma, the rate of sea-level fall increases, causing a regression. As subsidence occurs, the shoreline now moves seaward toward a new, more seaward, equilibrium position. The now-exposed shelf becomes part of the lower course of the fluvial system. But because its grade is too shallow, the stream velocity decreases and deposition takes place. The wedge of deposits will extend from the new shoreline landward, onlapping the old coastal plain. As the rate of regression increases, the coastal deposits continue to onlap landward, overlapping the previously deposited coastal sediments (51–50 Ma). Now, if in the next interval (50–49 Ma) the rate of the regression decreases, the coastal sediments will still onlap, but the point of onlap maximum will have moved seaward. If at 49 Ma the shoreline reaches a new equilibrium point where the new rate of sea-level fall is equal to subsidence at the shoreline, then the shoreline will remain stationary and erosion and deposition will occur. In this case, *all* deposition of coastal sediment will cease. This surface will be a very distinct unconformity marking the cessation of the seaward migration of the littoral zone. Thus a regression may in fact cause coastal onlap (Pitman, 1982).

THE POSITION OF THE LITTORAL ZONE AS A FUNCTION OF SLOWLY VARYING SEA-LEVEL CHANGE

In the previous discussion we have looked at the stratigraphy of a margin after long periods of sea-level change. We will now attempt to calculate the position of the littoral zone on a margin during periods of time when the rate of sea-level change is changing slowly and smoothly. To do this we have set up a varying rate of sea-level change consisting of a summation of several sine functions (see Figs. 7B and 8B). Using equation 1 (see also Fig. 2) with a time interval of 50,000 years, we have calculated the position of the littoral zone divided by D, the distance from the hinge zone to the shelf edge, as a function of time. If $X_L/D > 1$, the littoral zone is out over the shelf edge. If $X_L/D < 0$, the littoral zone has transgressed landward of the hinge zone. We have made these calculations for two different rates of sea-level change, four different graded slopes, and two values of driving subsidence at the shelf edge. The rate of subsidence at any point on the margin is given by $X/D \cdot R_{SS}$. In each calculation we begin with the rate of sea-level change equal to zero and the littoral zone at the hinge zone, which is its equilibrium position.

Of particular interest here is the varying time lag between changes in direction of the rate of sea-level change and changes from transgressive to regressive or from regressive to transgressive phases. Let us examine the first change, that is, 0 to 5 Ma in Figure 7A. In this case the rate of subsidence at the shelf edge is 1.0 cm/1,000 yrs. The rate of sea-level change is initially zero, and the littoral zone is in each case at the hinge zone. The rate of sea-level fall slowly increases from 0.0 at zero Ma to a maximum of about 0.63 cm/1,000 yrs at 2.65 Ma. At this point in time, if the littoral zones were precisely tracking the rate of sea-level change, the littoral zones would be at $X_L/D = 0.63$. But all of the shorelines are still well landward of the equilibrium point. Hence, as the rate of sea-level fall rolls over and slowly begins to decrease, the regression continues. In fact, the regression continues in each case until that point in time when the position of the shoreline X_L/D equals R_{SL}/R_{SS}. For example, for the shoreline with a gradient of 1:10,000, the regression continues, although at a decreasing rate, until the shoreline is at the point where X_L/D equals 0.4 at 4.2 Ma and at that point in time $R_{SL} = 0.4$ cm/1,000 yrs; now, because the rate of sea-level fall continues to decrease, the 1:10,000-slope shoreline must begin to transgress. Note that the 1:5,000-slope shoreline does not reach its equilibrium point until 4.5 Ma, the 1:2,500-slope shoreline at 5.15 Ma, and the 1:1,250-slope shoreline at 5.5 Ma. The change in direc-

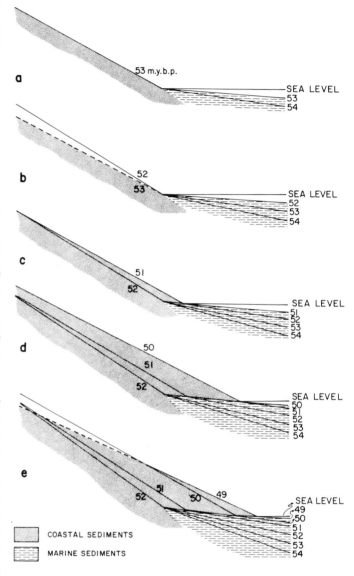

Figure 6. This illustrates the sequence of events expected at a margin in which the slope of the coastal plain is steeper than the slope of the shelf. In this case a regression driven by an increase in the rate of sea-level fall causes simultaneous onlap of coastal sediments.

tion of the 1:1,250-slope shoreline lags the maximum in the R_{SL} curve by almost 2.9 m.y. The difference in lag time of the 1:10,000-slope shoreline and the 1:1,250-slope shoreline is 1.3 m.y.

The next change takes place at 7.2 Ma (Fig. 7A). In this case sea level is rising and the rate of sea-level rise begins to decrease. There is almost a constant time lag of 1.8 m.y., regardless of the slope, between the time of occurrence of the minimum in the rate of sea-level change and the change from a transgression to a regression.

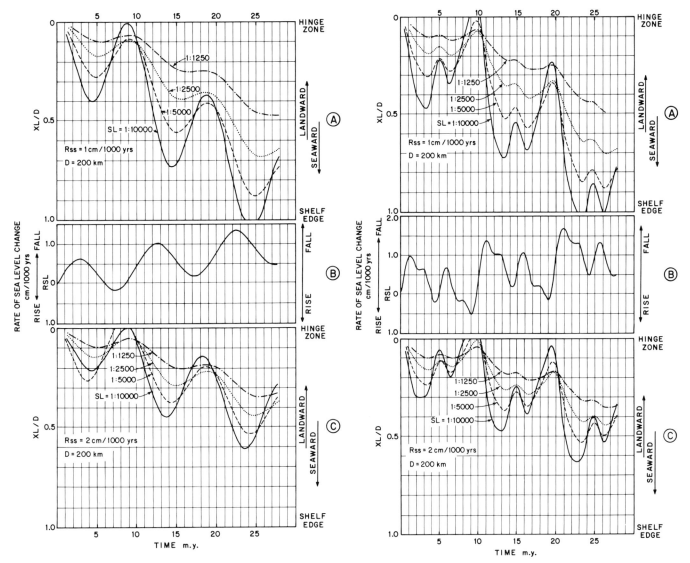

Figure 7. The pattern of transgressive and regressive events in response to sea-level changes and changes in the rate of sea-level change have been calculated for various values of the shelf edge subsidence rate and the shelf–coastal plain slope. The parameters used are those of Figure 2. The rate of sea-level change as a function of time is given in Figure 7B. 7A and 7C show the transgressive and regressive response as a function of time.

Figure 8. Same as Figure 7 except that the rate of sea-level change (8B) differs from that of 7B as it contains some higher frequency components.

The next maximum in the rate of sea-level change occurs at 12.8 Ma, and there is again a rollover from an increasing rate of sea-level fall to a decreasing rate of sea-level fall. The same pattern of time lag as between 0 and 5 Ma emerges. But in this case, the lag ranges from 1.8 m.y. for the 1:10,000-gradient shoreline to 2.8 m.y. for the 1:1,250-gradient shoreline.

We have also made the same calculations using a curve for the rate of sea-level change with some shorter wavelength components (Figs. 8A, B, and C). In this case the rate of sea-level change starts at 0.0 cm/1,000 yrs at zero Ma. Sea level begins to

fall at an increasing rate, reaching a value of 0.95 cm/1,000 yr at 1.2 Ma. At this point the rate of sea-level fall begins to decrease. But, for the margin with a gradient of 1:10,000, the shoreline starts at the hinge zone at zero Ma and gradually regresses as the rate of sea-level fall increases and continues to regress until 3.2 Ma. This is a lag time of 2 m.y.; the lag time for margins with lesser slopes is greater, with the greatest lag time being 2.6 m.y. for gradients of 1:1,250. We have also made the same calculations for margins at which the rate of subsidence at the shelf edge is 2 cm/1,000 yrs (Figs. 7C and 8C). In these cases the lag time is slightly less than at a margin where the subsidence at the shelf edge is 1 cm/1,000 yrs.

Examination of these diagrams reveals that the lag times are quite variable; from near 0 to almost 3.0 m.y. They are not only dependent on the geometry of the margin but also on the prior sea-level history. We conclude that correlation of transgressive and regressive sequences within a basin, between basins, and with a sea-level time scale may be quite difficult to do with a precision as great as one or two faunal zones.

CONCLUSIONS

It is observed that at most passive margins the rate of thermal subsidence varies from zero at the hinge zone on the landward side and increases systematically and smoothly to a maximum somewhere on the basin side at or near the shelf edge (e.g., Steckler, 1981).

If the operative geomorphologic processes of erosion and deposition maintain graded slopes for the coastal plain and the shelf, then:

1. The only way the littoral zone may be maintained on the subsiding platform is if sea level is falling. A stillstand takes place when the rate of sea-level fall is equal to the rate of subsidence minus the rate of sedimentation (if any) at the shore line. A regression will take place when the rate of sea-level fall is increased, and a transgression will take place when the rate of sea-level fall is decreased.

2. Onlap caused by the landward migration of the littoral zone may take place when sea level is falling. Consequently, attempts to measure the amount of sea-level rise by coastal aggradation may lead to error.

3. The continued onlap of coastal deposits during a regression may be explained as the consequence of fluvial systems depositing part of their sedimentary load on the very flat former shelf regions exposed by the regression.

4. Under the above conditions, if sea level is falling, the littoral zone will always be moving toward that point on the margin where the instantaneous rate of sea-level fall is equal to the rate of subsidence minus deposition. But even if the rate of sea-level change varies smoothly, there may be a time lag of up to 3 m.y. between the time of the rate change and the point in time at which there was a change in direction of the movement of the shoreline.

REFERENCES CITED

Bally, A. W., 1981, Atlantic-type margins, *in* Geology of passive continental margins; History, structure, and sedimentologic record: American Association of Petroleum Geologists Course Notes no. 19, 324 p.

Drake, C. L., Ewing, M., and Sutton, G., 1959, Continental margins and geosynclines; The East Coast of North America north of Cape Hatteras: Physics and Chemistry of the Earth, v. 5, p. 110–198.

McKenzie, D., 1978, Some remarks on the development of sedimentary basins: Earth and Planetary Science Letters, v. 40, p. 25–32.

Pitman, W. C., 1978, Relationship between sea level changes and stratigraphic sequences: Geological Society of America Bulletin, v. 89, p. 1389–1403.

——— , 1982, The effects of sea-level changes on the stratigraphy of continental margins: Proceedings of the Joint Oceanographic Assembly, 1982, Dalhousie University, Nova Scotia, p. 22.

Royden, L., and Keen, C. E., 1980, Rifting process and thermal evolution of the continental margin of Eastern Canada determined from subsidence curves: Earth and Planetary Science Letters, v. 51, p. 343–361.

Royden, L., Sclater, J. G., and Von Herzen, R. P., 1980, Continental margin subsidence and heat flow; Important parameters in formation of petroleum hydrocarbons: American Association of Petroleum Geologists, v. 64, p. 173–187.

Sleep, N. H., 1971, Thermal effects of the formation of Atlantic continental margins by continental break-up: Geophysical Journal of the Royal Astronomical Society, v. 24, p. 325–350.

Steckler, M. S., 1981, Thermal and mechanical evolution of Atlantic-type continental margins [Ph.D. thesis]: Columbia University, Department of Earth Science, 261 p.

Steckler, M. S., and Watts, A. B., 1978, Subsidence of the Atlantic-type continental margin off New York: Earth and Planetary Science Letters, v. 41, p. 1–13.

Vail, P. R., Mitchum, R. M., Jr., and Thompson, S., III, 1977a, Relative changes of sea level from coastal onlap, *in* Payton, C., ed., Stratigraphic interpretation of seismic data: American Association of Petroleum Geologists Memoir 26, p. 63–81.

——— , 1977b, Seismic stratigraphy and global changes of sea level, *in* Payton, C., ed., Stratigraphic interpretation of seismic data: American Association of Petroleum Geologists Memoir 26, p. 83–97.

Vail, P. R., Hardenbol, J., and Todd, R. G., 1985, Jurassic unconformities, chronostratigraphy, and sea level changes from seismic stratigraphy and biostratigraphy, *in* Schlee, J. S., ed., Interregional unconformities and hydrocarbon accumulation: American Association of Petroleum Geologists Memoir 36, p. 129–144.

Watts, A. B., 1981, The U.S. Atlantic continental margin; Subsidence history, crustal structure, and thermal evolution, *in* Geology of passive continental margins; History, structure, and sedimentologic record: American Association of Petroleum Geologists course notes no. 19, 324 p.

——— , 1982, Tectonic subsidence flexure and global changes in sea level: Nature, v. 197, p. 469–474.

Watts, A. B., and Ryan, W.B.F., 1977, Flexure of the lithosphere at continental margin basins: Tectonophysics, v. 36, p. 25–44.

Manuscript Accepted by the Society February 6, 1987

ACKNOWLEDGMENTS

We thank the reviewers and editor for their many suggestions which have improved the manuscript. This research was supported by National Science Foundation grants OCE-79-26308 and OCE-79-22884.

Printed in U.S.A.

The Geology of North America
Vol. I-2, The Atlantic Continental Margin: U.S.
The Geological Society of America, 1988

Chapter 21

Seismic stratigraphic and geohistory analysis of Tertiary strata from the continental shelf off New Jersey; Calculation of eustatic fluctuations from stratigraphic data

S. M. Greenlee, F. W. Schroeder, and P. R. Vail*
Exxon Production Research Co., P.O. Box 2189, Houston, Texas 77001

INTRODUCTION

Observations of geologically synchronous basinward shifts in coastal onlap patterns on seismic reflection profiles, from varied tectonic settings and in widespread areas, led Vail and others (1977) to suggest the use of these shifts as a means of global correlation. Sea-level curves based on the interpretation of stratigraphic data are being rigorously tested on the eastern U.S. Continental Margin. The availability of outcrop, well, and seismic data in this region, as well as the general lack of structural overprint over much of the margin, has resulted in its being a focus of research on the effects of sea-level fluctuations on the sedimentary development of basins.

Here we present the results of a study in which we apply seismic stratigraphic and geohistory analysis techniques to data from the Baltimore Canyon trough to interpret sea-level changes during the Tertiary. We first develop a stratigraphic framework for the study area through the interpretation of a regional grid of seismic reflection data tied to available well control. We then use the ages of the major depositional sequence boundaries at the COST B-2 well site to predict the thermotectonic subsidence of the basin since the Early Jurassic and to estimate long-term sea-level changes. Finally, we analyze the Tertiary sedimentary patterns expressed in seismic and well data to derive a detailed curve of eustatic changes of sea level.

BACKGROUND

The geological evolution of the Baltimore Canyon trough has been the subject of numerous investigations and is summarized by Schlee (1981), Grow and Sheridan (1981), Poag (1985b), and other publications. Tertiary development of the Baltimore Canyon trough shelfal area is documented by Poag (1985b) and Poag and Schlee (1984). These authors describe

deep-water conditions prevailing on the shelf until the Neogene when a series of progradational wedges built seaward to the present shelf edge. Unconformities in Tertiary sediments noted on seismic data from the continental slope have been studied by Greenlee (1982), Farre (1985), Thorne and Watts (1984), and Poag (1985c). Unconformities recognized on the continental rise off New Jersey have been studied by Van Hinte and others (1985a, b) and Poag (1985c).

Although most authors recognize the existence of periodic erosion and the development of regional unconformities during the Tertiary, the causal mechanism of this erosion remains a point of controversy. Rapid eustatic falls of sea level as noted by Vail and Hardenbol (1979) are believed to be a factor in the development of these surfaces by several authors (e.g., Poag and Schlee, 1984; Van Hinte and others, 1985a, b). Thorne and Watts (1984), however, after study of the COST B-2, COST B-3, and DSDP site 612 biostratigraphic interpretations and the slope portion of USGS Line 25 suggested problems associated with the correlation of sea-level falls to erosional events on the continental margin. These problems were: (1) lack of documentation of a mechanism for dropping sea level rapidly enough to bring sea level below the shelf edge in a rapidly subsiding margin, (2) inability to recognize all of the unconformities on Vail's cycle chart on all margins, and (3) that seismic reflectors do not necessarily indicate sequence boundaries. Our study addresses the recognition of short-term second- (1–20 m.y.) and third-order (~1–5 m.y.) sea-level fluctuations and the estimation of their magnitude based on stratal surfaces observed on seismic reflection data tied to well information.

STRATIGRAPHIC FRAMEWORK

The initial step in the development of a Tertiary stratigraphic framework for shelfal areas of the Baltimore Canyon trough was the recognition of depositional sequences on seismic sections

---

*Present address: Department of Geology and Geophysics, Rice University, Houston, Texas 77251.

Greenlee, S. M., Schroeder, F. W., and Vail, P. R., 1988, Seismic stratigraphic and geohistory analysis of Tertiary strata from the continental shelf off New Jersey; Calculation of eustatic fluctuations from stratigraphic data, *in* Sheridan, R. E., and Grow, J. A., eds., The Atlantic Continental Margin, U.S.: Geological Society of America, The Geology of North America, v. I-2.

(Mitchum and others, 1977a) and the correlation of these sequences throughout a seismic grid. An interpretation of the depositional environment associated with these sequences was then accomplished using seismic facies analysis techniques as outlined by Mitchum and others (1977b). COST and industry test wells were then tied to the seismic sections using synthetic seismograms where possible, and used to establish the age of each depositional sequence and to calibrate the facies interpretations. Detailed Tertiary biostratigraphy was available for COST B-2, COST B-3, USGS Island Beach no. 1, AMCOR 6011 (Poag, 1980, 1985b). Exxon 684 no. 1 and 902 no. 1, Gulf 857 no. 1, Houston Oil and Minerals 676 no. 1, and Shell 272 no. 1 (M. Crane, personal communication, 1985). An age model based on the physical stratigraphy from seismic profiles and biostratigraphic age-dating from available wells was then established and summarized on a chronostratigraphic chart. Plate 7A shows an interpretation of the key seismic section from our seismic grid that best illustrates Tertiary depositional sequences. Beneath it is a chronostratigraphic chart that summarizes the geographic extent, ages, position of the depositional shelf edge[1] and generalized facies of the depositional sequences observed on this line. Figure 1 is an index map that shows the location of the key seismic line, wells drilled in the Baltimore Canyon area, and the location of wells with available biostratigraphic data.

Paleocene-Eocene

The base of the Tertiary is characterized by a high-amplitude reflection that is downlapped by an overlying prograding wedge. This downlap surface is associated with a major submarine condensed section, which is interpreted to be caused by a rapid eustatic rise of sea level in the late Maastrichtian and early Paleocene. The overlying prograding wedge thins seaward by downlap and erosional truncation where it is dated by offshore wells as late Paleocene in age. Two thin sequences overlie the Paleocene sequence and show local evidence of erosional truncation. These two sequences and the late Paleocene sequence consist of deep-water carbonate-rich sediments. They are dated at COST B-2 and industry wells as early and middle Eocene. During the Paleocene and Eocene the shoreline was located well landward of the study grid. Because of this, downward shifts in coastal onlap were not observed, making the recognition of minor sequence boundaries difficult. Thus, only major sequence boundaries near the top of the Paleocene (55 Ma), near the top of the lower Eocene (49.5 Ma), and near the top of the middle Eocene (39.5 Ma) were recognized.

The seismic data show evidence of onlap of upper Eocene beds at the most landward edge of the seismic grid (Plate 7A). This is consistent with the absence of upper Eocene rocks at the Island Beach Well to the east of the seismic line in Plate 7A

---

[1]The depositional shelf edge refers to the point at which there is a marked change in slope between flat shelfal and coastal plain beds and more steeply dipping slope beds. This may or may not be coincident with the present-day structurally controlled shelf edge.

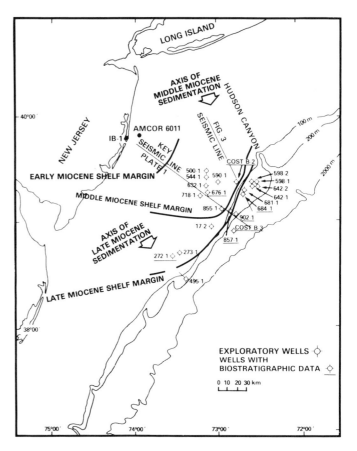

Figure 1. Base map showing location of seismic line and exploratory wells, positions of latest early, middle, and late Miocene depositional shelf margins, and major axis of middle and late Miocene sedimentation.

(Poag, 1985b). Upper Eocene rocks are present further basinward.

Oligocene

One sequence of Oligocene age is identified in offshore well control. This sequence is dated as late Oligocene (Poag, 1980, 1985b). It onlaps an erosional unconformity that cuts out older section basinward where it overlies upper Eocene sediments (Plate 7A). This Oligocene unconformity is correlative with the fall of sea level in the late Oligocene (30 Ma). Seismic data suggest that rocks older than this unconformity and younger than the onlapping upper Eocene exist landward of shot point 4 on Plate 7A. These have been interpreted to be of lower Oligocene or lowermost upper Oligocene age, although the exact age or presence of Oligocene deposits at Island Beach is uncertain (see Poag, 1985b, p. 237–238). A lower Oligocene section recently described by Bybell and others (1986) from a core located in Atlantic County, New Jersey confirms the presence of lower Oligocene rocks updip. The absence or near absence of early

Oligocene sediments on the offshore portion of the key seismic line is believed to be caused by sediment starvation due to a sea-level rise in the early Oligocene combined with erosion associated with the 30 Ma sea-level fall. A downlap surface indicating starvation is present on the landward portion of the key seismic line between the 30 Ma and 39.5 Ma sequence boundaries.

Early Miocene

Four sequences of early Miocene age have been recognized on the seismic sections from our study area. Each sequence shows evidence of onlap in a landward direction and downlap in a seaward direction. The early Miocene sequences are a series of stacked shelf edges just seaward of the late Oligocene shelf margin (Fig. 1). They thin landward by onlap and seaward by downlap. The seaward portions of these sequences are very thin or absent in the area of well control. Exxon paleontologic reports (M. Crane, personal communication, 1985) suggest a thin zone of early Miocene (N6) present over Oligocene deposits at B-2. A point of maximum starvation during the Neogene is present between the 22 Ma and 25.5 Ma sequence boundaries at the downlap surface indicated on the landward side of the section.

Middle Miocene

Four middle Miocene sequences are also recognized. These sequences are strongly progradational, with each successive shelf margin seaward of the previous shelf margin (Plate 7A). Figure 1 shows the position of the depositional shelf edge at the end of the middle Miocene and the axis of maximum thickness of middle Miocene progradational lobes. The middle Miocene depocenter is in the vicinity of the COST B-2 well, which sampled middle Miocene sediment spanning zones N10 to N14 overlying late Oligocene deposits (Poag, 1980, 1985b). Tertiary depositional sequences at COST B-2 are illustrated in Figure 2, which shows a seismic section passing through the COST B-2 well and the Exxon 624-1 well. Miocene chronostratigraphy is shown below the line. Two sequence boundaries are found within sediment dated as zone N10 to N12 by Poag (1980) and are correlated with sea-level falls in the mid-Serravallian (13.8 Ma and 12.5 Ma). A sequence boundary near the top of the middle Miocene (10.5 Ma) overlies sediment dated as N14 (Poag, 1980) and is characterized by a major downward shift in coastal onlap.

Late Miocene-Pliocene

Late Miocene depositional sequences are thickest south of the middle Miocene depocenter (Fig. 1). These deposits thin by onlap against the middle Miocene progradational wedges in a landward direction and along strike lines to the north. On the key seismic line shown in Plate 7A, the uppermost portion of the sequence below the 6.3 Ma sequence boundary extended over the latest middle Miocene shelf margin. This sequence continues onlapping to the north and is absent at the seismic line in Figure 2.

Latest Miocene sediments are widespread in a landward direction as are early Pliocene depositional sequences.

DETERMINING TECTONIC SUBSIDENCE

Geohistory analysis (Van Hinte, 1978) provides a technique for determining total basin subsidence. When total subsidence is corrected for the effects of compaction and sediment loading, the resultant subsidence curve represents the contributions of thermotectonic processes and eustatic changes. Several workers (McKenzie, 1978; Royden and others, 1980; Sclater and Christie, 1980) have developed models of thermal subsidence for passive margins. On passive margins where thermotectonic subsidence is predominantly a result of thermal contraction of the crust and mantle, departures from theoretical curves can be attributed to long-term eustatic fluctuations (Hardenbol and others, 1981).

A geohistory diagram for the earliest Jurassic at the COST B-2 well site is shown in Figure 3A. The well/seismic stratigraphic interpretation used to produce this diagram is included in Table 1. First the effects of compaction were accounted for by restoring each sedimentary unit above the Hettangian to its original thickness using the equations in Hardenbol and others (1981). Total subsidence is obtained by including estimates of paleobathymetry. Note that sea level is not varied in Figure 3A.

An important part of total subsidence is due to sediment load. We used an isostatic-loading model assuming an Airy type crust (Horowitz, 1976) since the variations in thickness of individual sedimentary units near the well site are too minor to warrant using a flexural-loading model. The subsidence due to sediment loading was subtracted from the total subsidence curve. The resultant curve represents subsidence due to thermotectonic and eustatic changes.

To separate thermotectonic from eustatic effects, we compared the resultant subsidence curve with theoretical crustal subsidence curves calculated for various amounts of crustal stretching (equations of Royden and others, 1980). We interpret two thermal cycles. The first started in the Early Jurassic and is associated with the cooling associated with the change from rift to drift of North America and Africa. A second thermal event occurred in the latest Early Cretaceous and is associated with a period of volcanic intrusions (e.g., Great Stone Dome, Grow, 1980). Thermotectonic subsidence slowed and possibly changed to uplift. As a result, eustatic falls were enhanced (most notably in the Barremian and Aptian) and led to localized erosion. The dashed line on Figure 3A indicates our interpretation of the thermotectonic subsidence since Early Jurassic.

The difference between the load-corrected subsidence curve and the interpreted thermotectonic subsidence curve is attributed to eustatic changes. After correcting these estimates of eustasy for changes in water load (Hardenbol and others, 1981), we obtained the long-term sea-level curve shown in Figure 3B. Using this sea-level curve, we replotted total subsidence, subtracted sediment loading effects, and obtained the thermotectonic subsidence curve.

MIOCENE CHRONOSTRATIGRAPHY

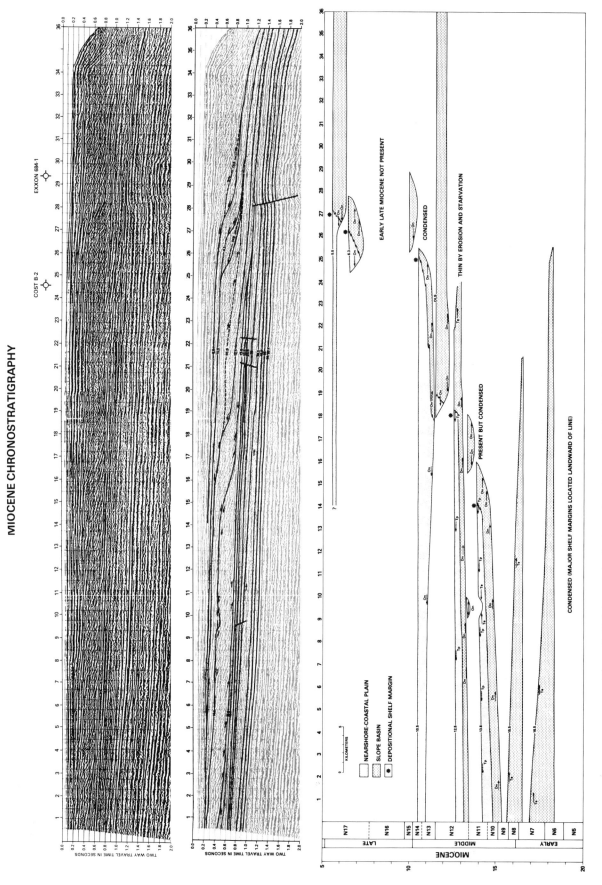

Figure 2. Uninterpreted seismic section, interpreted seismic section, and Miocene chronostratigraphy passing through the COST B-2 and EXXON 684-1 wells. See Figure 1 for the location of the seismic line. Abbreviations in the chronostratigraphic chart are as follows: On = onlap, Dn = downlap, Te = erosional truncation, Tp = toplap, and DLS = downlap surface. (See plate 7 for expanded seismic section.)

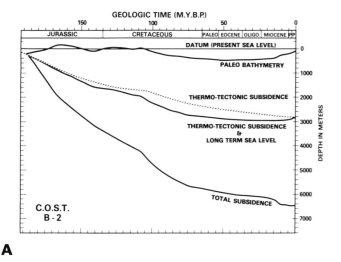

A

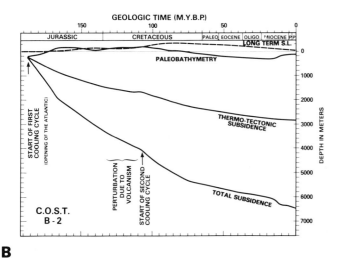

B

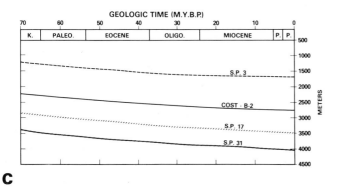

C

Figure 3. Geohistory curves showing (A) thermotectonic subsidence not corrected for long-term eustasy, (B) thermotectonic subsidence corrected for long-term eustatic change, and (C) Tertiary subsidence curves. S.P. 3, 17 and 31 refer to shotpoint locations on the seismic line shown on Plate 7A.

TABLE 1. COST B-2 STRATIGRAPHY

Age (Ma)	Depth Below K.B.[1] (ft)	Paleobathymetry Biostratigraphy[2] (ft)	Model[3] (ft)
2.8	830	100-300	300
3.8	830	100-300	300
5.5	1200	100-300	300
6.3	1510	0-100	300
8.2	1510	0-100	300
10.5	1510	0-100	300
12.6	2900	300-600	800
13.8	3150	0-100	800
15.5	3500	100-300	1000
16.5	3500	100-300	1000
21.0	3640	600-1500	900
39.5	4200	600-1500	900
49.5	4680	600-1500	1100
68.0	5150	0-100	1100
74.0	5300	300-600	900
84.0	6060	0-50	500
88.0	6655	50-100	1000
97.0	7500	50-100	200
102.0	8220	300-600	350
108.0	9110	0	+150[4]
112.0	9775	0	100
131.0	11655	100-300	100
141.0	13130	0	300
156.0	14650	0	+300[4]
167.0	16050	---	+400[4]
174.0	18160	---	200
188.0	21340	---	600

[1] K.B. = Kelly Bushing elevation, which is referenced to sea level.

[2] Poag (1980) and unpublished Exxon reports.

[3] Model data are values used to obtain thermo-tectonic subsidence curves.

[4] + indicates elevations above sea level.

Several sites along the seismic line shown on Plate 7A were analyzed using the same procedure, and comparable results were obtained. Figure 3C illustrates the resultant thermotectonic component of subsidence throughout the Tertiary at four locations. Depths increase downdip as would be expected, but the rate is approximately the same (6.5 m/m.y.) at each location.

ESTIMATION OF EUSTASY FROM STRATIGRAPHIC INFORMATION

Based on the premise that the rate of thermotectonic subsidence in the Baltimore Canyon trough was nearly constant in the Tertiary, we have used abrupt basinward and landward shifts in coastal onlap as observed on the seismic sections to estimate changes in relative sea level (Fig. 4).

The magnitude of a relative sea-level fall is calculated based on the shift of coastline position from the end of the preceeding highstand to the lowest point of coastal onlap associated with the subsequent lowstand deposits. The actual amount of sea-level fall may be somewhat greater because no true coastal onlap occurs during periods of slope bypass and submarine-fan deposition. As

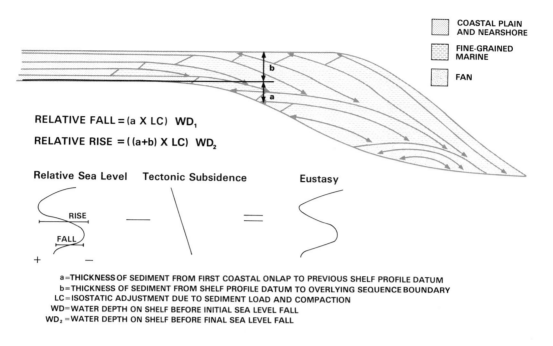

Figure 4. Method used to calculate eustatic change from stratigraphic data.

no coastal sediments are preserved, seismic data cannot resolve the position of sea level at this time. Clearly, it is important to determine whether the observed onlap is true coastal onlap or deep-marine onlap. This is accomplished, where possible, through facies interpretation from well data. The vertical sediment thickness is measured from the point of first coastal onlap to a datum projected out from the previous shelf-margin profile. This sediment thickness must then be decompacted and corrected for sediment-loading effects. Sources of error will arise from an incorrect determination of the first coastal onlap and erroneous compaction and loading functions.

The magnitude of a relative sea-level rise is calculated by measuring the stratigraphic thickness from the first true coastal onlap to the top of the sequence. In the case of a sequence that is restricted to areas seaward of the previous shelf edge, the relative rise will be less than the preceeding relative fall. For sequences extending landward over the previous shelf margin, the relative rise will be greater than the preceeding relative fall. Calculations of sea-level rise may be underestimated using this technique if there has been an unknown amount of erosion associated with the overlying sequence boundary or if insufficient sediment was available to keep pace with the sea-level rise. Subsequent calculations of sea-level rise and fall will then be affected by these errors. Error may also be introduced if compaction and/or loading estimates are incorrect.

Relative sea-level measurements must be corrected for thermotectonic subsidence effects. We used the rate of 6.5 m/m.y. as discussed above. The calculated relative and eustatic sea-level curves derived from the Baltimore Canyon trough are shown on Plate 7B. This plate also illustrates the most recent version of the Tertiary coastal onlap curve, the summarized coastal onlap from the Baltimore Canyon trough study area, and the most recent revision of the short- and long-term eustatic curves (Haq and others, 1987). Because Paleocene and Eocene downward shifts in coastal onlap were not observed in our study area, the sea-level curve for this interval cannot be directly measured using our techniques. Our measurements begin at the mid-Oligocene fall and extend through the Miocene.

DISCUSSION

Sedimentation patterns noted on seismic reflection profiles from the Baltimore Canyon trough and other basins are controlled by relatively long-term episodes of thermotectonic subsidence and lithospheric flexure, short-term tectonism related to salt and shale flowage and diapirism, long and short-term changes of sea level, and changes in the supply of sediment. The Baltimore Canyon trough has undergone two cycles of thermotectonic subsidence. Uplift and erosion in the Early Cretaceous recognized in the geohistory analysis at COST B-2 is in part responsible for the accelerated input of coarse clastic material into the deep sea noted at DSDP Site 603 (Van Hinte and others, 1985b). Superimposed on these long-term tectonic events are much shorter period cycles manifest on seismic profiles as cyclic landward and basinward shifts of coastal onlap.

Although the number and timing of Neogene sea-level cycles interpreted from our regional seismic grid agree well with global cycle charts prepared by researchers at Exxon, significant departures exist in portions of the coastal onlap curve and the eustatic curve (Plate 7B). Specifically, the latter portion of the middle Miocene was estimated to be a period of more widespread coastal onlap and higher eustatic sea level than shown on the global curves. This is most likely a result of the fundamental assumptions made in modeling sea level from stratigraphic data. In estimating changes of sea level using the technique described above, it is assumed that enough sediment is being made available to fill all space being created by sea-level rise and subsidence at the point of measurement. For this reason we attempt to obtain measurements as near to the depocenter as possible. For example, if there were not enough sediment available to aggrade the regressive highstand wedge shown in Figure 4 to sea level at the location where we measured sediment thickness b, then our calculation of relative sea-level rise would be too small. Similarly, if a significant amount of section had been removed during formation of the overlying unconformity, then our measurement of relative rise would be too small by that amount of missing section. Subsequent measurements of relative sea-level rise are then overestimated by the amount of previous underestimations, assuming enough sediment is then available to fill the new space. Calculations of subsequent sea-level falls would be underestimated. It is possible that shelfal erosion during the 13.8 Ma unconformity caused an underestimation of the sea-level rise between the 15.5 Ma and 13.8 Ma unconformities. Subsequent calculations of sea-level rise in the latter portion of the middle Miocene would then have been greater than the actual relative rise.

The possibility of error involved in this and other aspects of modeling sea-level change from stratigraphic data discussed above emphasizes the difficulty of calculating the magnitude of eustatic fluctuations from stratigraphic data. Because our technique involves summing of individual measurements, errors will affect latter portions of the resultant curve. Only through a synthesis of several regional studies can a confident determination of the magnitudes of short- and long-term sea-level changes be made using stratigraphic data. Other methods of estimating magnitudes of Tertiary eustatic change have arrived at values both less than and greater than those derived in this study. Oligocene sea-level falls in the basal and uppermost Chattian have been estimated using isotopic evidence by Miller and others (1985) at 30 and 20 m, respectively, whereas our results suggest falls of 90 and 60 m. Mathews (1984) suggests Paleogene sea-level fluctuations of no more than 50 m based on isotopic evidence. Higher magnitude fluctuations are suggested by Popenoe (1985) based on studies of Gulf Stream paleo–flow patterns. Clearly, a need exists for future multidisciplinary studies on the magnitude of eustatic fluctuations.

Our investigation of Tertiary sediments beneath shelfal areas of the Baltimore Canyon trough suggests that depositional sequences of approximately 1 m.y. duration (third-order) are grouped in larger packages of sequences ("supersequences" of Vail and others, 1977). However, these larger groups of depositional sequences that show many of the attributes of individual

sequences are deposited on time scales of approximately 10 to 20 m.y. In deeper water basins these supersequence boundaries may be the only recognizable unconformities in the otherwise apparently conformable marine section. Such appears to be the case in our study area during the Paleocene and Eocene. Nearer to the depocenter these supersequence boundaries are characterized by one or more basinally restricted overlying third-order sequences (as in the case of the 30 and 10.5 Ma unconformity on Plate 7A). Nearer to the depositional shelf margin, the third-order sequences are noted by downward shifts in coastal onlap. In the absence of extensive erosion associated with these sequence boundaries, the unconformity recorded in the sedimentary section in basinal areas is commonly at or below the limits of biostratigraphic resolution. Such is the case in the Neogene in our study area. The ages of the Oligocene and Miocene supersequence boundaries apparent on our Baltimore Canyon trough grid are 30 Ma and 10.5 Ma (Plate 7B). Second-order lowstands are composed of basinally restricted third-order sequences. During the portion of the supercycle when sea level is generally rising, shelf edges stack vertically or back step. A significant hiatus (as discussed in Vail and others, 1984, p. 134–135) may occur in a basinward direction during this time due to condensed sedimentation followed by minor erosion associated with individual sequence boundaries. This may explain the large hiatuses identified on Plate 7A: early Paleocene, early Oligocene, and early Miocene at COST B-2 (Poag, 1980, 1985b). Early Miocene shelf edges were stacked just seaward of the Oligocene shelf margin, well landward of COST B-2. Following the 15.5 Ma unconformity, shelf edges prograded rapidly seaward (Plate 7A). These prograding third-order sequences make up the second-order highstand, which is bounded by the overlying 10.5 Ma sequence boundary. Thus, a combination of condensed sedimentation during the supersequence sea-level rise and minor erosion during the sea-level fall at 15.5 Ma may have resulted in a large hiatus at COST B-2. Hiatuses may thus be associated with erosion due to supersequence boundaries and nondeposition coupled with minor erosion during supersequence sea-level rises.

CONCLUSIONS

Analysis of a regional grid of seismic data tied to shelf wells in the Baltimore Canyon trough has documented the presence of abrupt basinward shifts in coastal onlap, which are interpreted as a sedimentary response to relative falls in sea level. Geohistory analysis of the COST B-2 well suggests that this area was characterized by slow, continuous thermotectonic subsidence during the Tertiary. Thus, these sedimentary patterns have been used to interpret changes of eustatic sea level. Magnitudes of third-order Neogene sea-level fluctuations have been calculated from the landward and basinward shifts in onlap patterns and range from 110 to 30 m.

Three orders of sea-level change have been recognized. A long-term sea-level fall during the Tertiary (Kominz, 1984, and references therein) was in part responsible for continued seaward movement of shelf edges during the Tertiary and was recognized in the geohistory analysis. Cyclicity on the order of 10 to 20 m.y. appears to have controlled the deposition of supersequences. Biostratigraphically determined hiatuses appear to be related to erosion at the supersequence boundaries as well as sediment star-vation and minor erosion during periods of rising sea level within the condensed section of supersequences. Depositional sequences of approximately 1 m.y. duration further subdivide the stratigraphic section and represent the shortest interval of cyclicity that we were able to recognize on seismic reflection data in the Baltimore Canyon trough.

REFERENCES CITED

Bybell, L. M., Poore, R. G., and Ager, T. A., 1986, Paleogene biostratigraphy of New Jersey Core, ACGS 4 [abs.]: Society of Economic Paleontologists and Mineralogists Annual Midyear Meeting Abstracts, v. 3, p. 17.

Farre, J. A., 1985, The importance of mass wasting processes on the continental slope [Ph.D. thesis]: New York, Columbia University, 227 p.

Greenlee, S. M., 1982, Tertiary seismic stratigraphy of the continental slope off New Jersey between South Toms and Lindenkohl canyons [M.S. thesis]: Kingston, University of Rhode Island, 95 p.

Grow, J. A., 1980, Deep structure and evolution of the Baltimore Canyon trough in the vicinity of the COST no. B-3 well, *in* Scholle, P. A., ed., Geological studies of the COST no. B-3 well, United States and Mid-Atlantic continental slope area: U.S. Geological Survey Circular 833, p. 117–132.

Grow, J. A., and Sheridan, R. E., 1981, Deep structure and evolution of the continental margin off the eastern United States: Oceanologica Acta, Supplement, v. 4, p. 11–19.

Haq, B. U., Hardenbol, J., and Vail, P. R., 1987, Chronology of fluctuating sea levels since the Triassic: Science, v. 235, p. 1156–1167.

Hardenbol, J., Vail, P. R., and Ferrer, J., 1981, Interpreting paleoenvironments, subsidence history, and sea-level changes on passive margins from seismic biostratigraphy: Oceanologica Acta, Supplement, v. 4, p. 33–44.

Horowitz, D. H., 1976, Mathematical modeling of sediment accumulation in prograding deltaic systems, *in* Merrian, D. F., ed., Quantitative techniques for the analysis of sediments; An international symposium: Oxford, Pergamon Press, p. 105–119.

Kominz, M. A., 1984, Oceanic ridge volumes and sea-level change; An error analysis, *in* Schlee, J. S., ed., Inter-regional unconformities and hydrocarbon accumulation: American Association of Petroleum Geologists Memoir 36, p. 108–123.

Mathews, R. K., 1984, Oxygen isotope record of ice-volume history; 100 million years of glacio-eustatic sea-level fluctuation, *in* Schlee, J. S., ed., Interregional unconformities and hydrocarbon accumulation: American Association of Petroleum Geologists Memoir 36, p. 97–107.

McKenzie, D. P., 1978, Some remarks on the development of sedimentary basins: Earth and Planetary Science Letters, v. 40, p. 25–32.

Miller, K. G., Mountain, C. S., and Tucholke, B. E., 1985, Oligocene glacio-eustacy and erosion on the margins of the North Atlantic: Geology, v. 13, p. 10–13.

Mitchum, R. M., Jr., Vail, P. R., and Thompson, S., 1977a, Seismic stratigraphy and global changes of sea level; Part 2, The depositional sequence as a basic unit for stratigraphic analysis, *in* Payton, C. E., ed., Seismic stratigraphy; Applications to hydrocarbon exploration: American Association of Petroleum Geologists Memoir 26, p. 53–62.

Mitchum, R. M., Jr., Vail, P. R., and Sangree, J. B., 1977b, Seismic stratigraphy and global changes of sea level; Part 6, Stratigraphic interpretation of seismic reflection patterns in depositional sequences, *in* Payton, C. E., ed., Seismic stratigraphy; Applications to hydrocarbon exploration: American Association of Petroleum Geologists Memoir 26, p. 117–133.

Poag, C. W., 1980, Foraminiferal stratigraphy, paleoenvironments, and depositional cycles in the outer Baltimore Canyon trough, *in* Scholle, P. A., ed., Geological studies of the COST no. B-3 well, United States Mid-Atlantic continental slope area: U.S. Geological Survey Circular 833, p. 44–65.

—— , 1985a, Cenozoic and Upper Cretaceous sedimentary facies of New Jersey continental slope and rise [abs.]: American Association of Petroleum Geologists Bulletin, v. 69, no. 2, p. 297.

—— , 1985b, Depositional history and stratigraphic reference section for central Baltimore Canyon trough, *in* Poag, C. W., ed., Geologic evolution of the United States Atlantic Margin: New York, Van Nostrand Reinhold, p. 217–264.

—— , 1985c, Cenozoic and Upper Cretaceous sedimentary facies and depositional systems of the New Jersey slope and rise, *in* Poag, C. W., ed., Geological evolution of the United States Atlantic margin: New York, Van Nostrand Reinhold, p. 217–264.

Poag, C. W., and Schlee, J. S., 1984, Depositional sequences and stratigraphic gaps on submerged United States Atlantic margin, *in* Schlee, J. S., ed., Inter-regional unconformities and hydrocarbon accumulation: American Association of Petroleum Geologists Memoir 36, p. 165–182.

Popenoe, P., 1985, Cenozoic depositional and structural history of the North Carolina margin from seismic-stratigraphic analyses, *in* Poag, C. W., ed., Geologic evolution of the United States Atlantic margin: New York, Van Nostrand Reinhold, p. 217–264.

Royden, L., Sclater, J. G., and Von Herzen, R. P., 1980, Continental margin subsidence and heat flow; Important parameters in formation of petroleum hydrocarbons: American Association of Petroleum Geologists Bulletin, v. 64, p. 173–187.

Schlee, J. ., 1981, Seismic stratigraphy of the Baltimore Canyon trough: American Association of Petroleum Geologists Bulletin, v. 65, p. 26–53.

Sclater, J. G., and Christie, P.A.F., 1980, Continental stretching; An explanation of the post-mid-Cretaceous subsidence of the central North Sea basin: Journal of Geophysical Research, v. 85, p. 3711–3739.

Thorne, J., and Watts, A. B., 1984, Seismic reflectors and unconformities at passive continental margins: Nature, v. 311, p. 365–368.

Vail, P. R., and Hardenbol, J., 1979, Sea level changes during the Tertiary: Oceanus, v. 22, p. 71–79.

Vail, P. R., Mitchum, R. M., Jr., and Thompson, S., III, 1977, Seismic stratigraphy and global changes of sea level; Part 4, Global cycles of relative changes of sea level, *in* Payton, C. E., ed., Seismic stratigraphy applications to hydrocarbon exploration: American Association of Petroleum Geologists Memoir 26, p. 83–97.

Vail, P. R., Hardenbol, J., and Todd, R. G., 1984, Jurassic unconformities, chronostratigraphy, and sea-level changes from seismic stratigraphy and biostratigraphy, *in* Schlee, J. S., ed., Inter-regional unconformities and hydrocarbon accumulation: American Association of Petroleum Geologists Memoir 36, p. 129–144.

Van Hinte, J. E., 1978, Geohistory analysis; Application of micropaleontology in exploration geology: American Association of Petroleum Geologists Bulletin, v. 62, p. 201–222.

Van Hinte, J. E., and 13 others, 1985a, Deep sea drilling on the upper continental rise off New Jersey, DSDP Sites 604–605: Geology, v. 13, p. 397–400.

Van Hinte, J. E., and 13 others, 1985b, DSDP Site 603; First deep (>1,000-m) penetration of the continental rise along the passive margin of eastern North America: Geology, v. 13, p. 392–396.

Manuscript Accepted by the Society March 31, 1987

Printed in U.S.A.

Chapter 22

Petroleum geology of the United States Atlantic continental margin

Robert E. Mattick
U.S. Geological Survey, Reston, Virginia 22092
Jan Libby-French
U.S. Geological Survey, Box 25046, Denver Federal Center, Denver, Colorado 80225

INTRODUCTION

Early hydrocarbon exploration on the U.S. Atlantic continental margin was concentrated in the shallow water depths of the Continental Shelf over Georges Bank Basin, Baltimore Canyon Trough, and the Southeast Georgia Embayment (Figs. 1, 2, and 3). Wildcat drilling started in March of 1978 when the Exxon 684-1 well was spudded-in near the shelf edge in the Baltimore Canyon Trough (Fig. 2). To date, industry has drilled 45 deep wildcat wells on the U.S. Atlantic Continental Shelf. Although significant natural gas discoveries and shows were reported from wells drilled in the Baltimore Canyon Trough area (Table 1), a commercial gas field has yet to be established. Crawford (1978) estimated that a daily flow of about 5.7 million m³ (200 million ft³) and reserves of about 34 billion m³ (1.2 trillion ft³) of natural gas would be required to warrant establishing an offshore production platform and pipeline. Since mid-1982, wildcat drilling on the shelf has come to a standstill, and leases for numerous blocks on the shelf have been relinquished by industry.

In 1983–84 exploration shifted from the shelf area to the slope. During this period, deep-water drilling tested the Jurassic-Cretaceous reef trend beneath the mid-Atlantic Continental Slope. It was hoped that the shales here, deposited in a reducing environment, might contain a high percentage of preserved marine organic matter and that reservoirs in backreef, reef, and forereef areas might have been charged with petroleum generated from the shales. To date, industry has completed 4 wells (93-1, 372-1, 586-1, and 587-1) in water depths of about 2,000 m in the Baltimore Canyon Trough area (Fig. 2). Data from the 93-1 well has not bee released; the other three wells were reported to be dry holes.

Some areas of the continental margin, such as the Carolina Trough and the Blake Plateau Basin, remain untested. In addition, with the exception of the Tertiary and uppermost part of the Cretaceous section, the thick wedge of sedimentary rocks (8–9 km) on the lower slope and upper rise remains to be tested. Drilling results from the base of the Continental Rise and Abyssal Plain (Tucholke and Mountain, 1979; Arthur and Natland, 1979) indicate that, with the exception of several limited rock intervals

of Turonian-Hauterivian age, Mesozoic sediments were deposited in generally oxygenated bottom waters. It is unlikely, therefore, that petroleum has been generated from these rocks. On the lower slope and upper rise, however, petroleum may have been generated from Jurassic shales if a strong oxygen minimum zone existed in this area then, a condition dependent upon the nature of the deep ocean currents. In other areas, the deep sedimentary rock section has been only partly tested. For example, the Triassic section—especially on Georges Bank where sedimentary rocks of this age, and possibly Early Jurassic, are contained in an extensive system of buried, synrift grabens—remains to be fully tested.

Despite the large amount of geological and geophysical research that has been done on the U.S. Atlantic margin, many problems related to the petroleum geology remain unresolved. One of the most interesting is the conflict in interpretation concerning maturity of source rocks. Different lines of geochemical evidence give varying results as to the depth of thermal maturity with respect to liquid hydrocarbons; and thermal models of the Atlantic margin that are based on geologic data differ from those based on maturation-depth data.

Geothermal temperature gradients measured on the Canadian and U.S. Atlantic shelves vary from about 16° C/km to about 24° C/km (Jackson and Heise, 1980). Gradients of this magnitude (about average on a worldwide basis) would indicate that maturity of Jurassic rocks, with respect to liquid hydrocarbons, should begin at depths of 2–3 km. Data from geochemical analysis summarized in this paper, however, suggest that maturity does not begin much above depths of 4 km.

The purpose of this chapter is to review exploration efforts, summarize the petroleum geology of the U.S. Atlantic margin, and comment on the potential of yet untested or partly tested rock sections of the margin.

LEASE SALES

Nine lease sales on the U.S. Atlantic continental margin have been held. In August 1976, 93 tracts (214,272 hectares) on

Mattick, R. E., and Libby-French, J., 1988, Petroleum geology of the United States Atlantic continental margin; *in* Sheridan, R. E., and Grow, J. A., eds., The Geology of North America, Volume I-2, The Atlantic Continental Margin: U.S.: Geological Society of America.

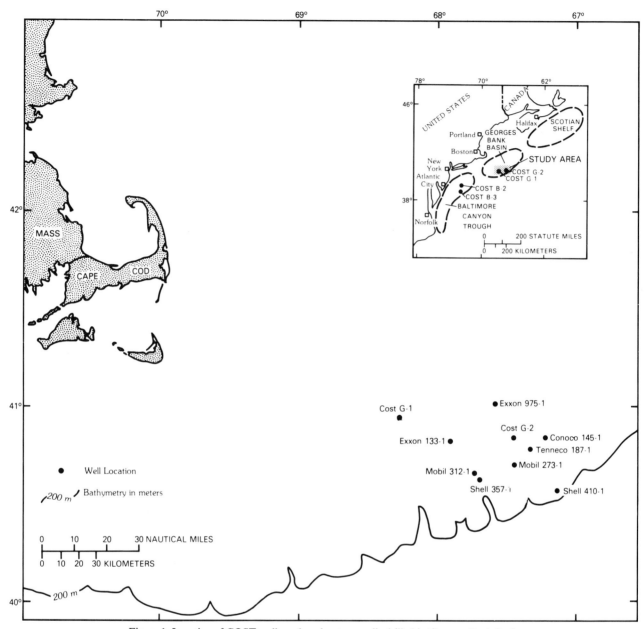

Figure 1. Location of COST wells and exploratory wells drilled in Georges Bank Basin.

the Continental Shelf in the Baltimore Canyon area were leased during Sale 40 for $1.1 billion. During Sale 49 (February 1979), 39 tracts in the same area on the outermost shelf and uppermost slope were leased for $40 million. Sale 43 (March 1979) resulted in the lease of 43 tracts on the Continental Shelf offshore of Georgia for $101 million. During Sale 42 (December 1979), 63 tracts on the Continental Shelf in the Georges Bank area were leased for $816 million. Sales 56 (August 1981) and 59 (December 1981), which chiefly involved tracts on the Continental Slope from New Jersey to South Carolina, resulted in bonus bids of $343 million for 47 tracts and $322 million for 50 tracts,

respectively. More recently, Sales 76 (April 1983) and 78 (July 1983) attracted bids of $68 and $13 million, respectively. At Sale 76, 37 tracts on the Continental Slope in the Baltimore Canyon area were leased; at Sale 78, 11 tracts in the Carolina Trough area were leased. The most recent sale, 82 in October 1984 in the Georges Bank area, attracted no bids.

PETROLEUM GEOLOGY

In the discussion that follows, the petroleum geology of the U.S. Atlantic margin is divided into 4 sections: (1) source and

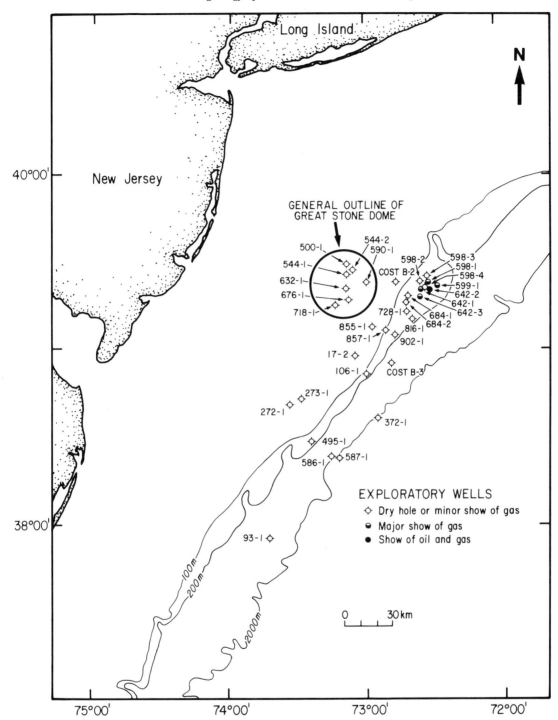

Figure 2. Location of COST wells and deep exploratory wells drilled in the Baltimore Canyon Trough area.

reservoir rocks on the Continental Shelf; (2) traps; (3) potential of buried rift basins; and (4) deep-water potential. In the first two sections, results on the shelf obtained from analyses of rock samples and seismic data are summarized. In the latter two sections, the petroleum potential of yet untested or partly tested areas are discussed; these sections, therefore, are of a more speculative nature.

Source and Reservoir Rocks of Continental Shelf

Southeast Georgia Embayment. Although geochemical studies on the rocks penetrated by exploration wells drilled in the Southeast Georgia Embayment have not been published, organic-carbon content and thermal-maturation properties of rocks penetrated in the COST GE-1 well (Fig. 4) were published by Miller

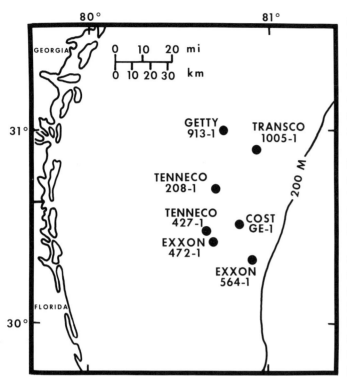

Figure 3. Location of COST GE-1 well and exploratory wells drilled in the Southeast Georgia Embayment.

and others (1979) and Scholle (1979). The highest total organic carbon values occur in Upper Cretaceous shales between depths of 800 and 1,800 m. The average organic-carbon content of this part of the section is 1.24% by weight. A few individual beds between 1,100 and 1,800 m contain in excess of 3% organic carbon. Much of the organic matter in the Upper Cretaceous shales consists of hydrogen-rich, oil-prone kerogens that probably reflect deposition in a marine environment (Miller and others, 1979; Scholle, 1979). Although this part of the section contains sufficient amounts of organic matter to be considered a potential source rock, thermal alteration indices (TAI) of 1+ to 2−, and vitrinite-reflectance values of 0.15 to 0.35 indicate that the Tertiary–Upper Cretaceous section is not mature enough to generate petroleum.

Values of the total organic-carbon content of the upper and middle Lower Cretaceous section, between depths of about 1,800 and 2,800 m, are generally less than 0.5% and TAI and vitrinite-reflectance values average about 2 and 0.35, respectively. Although these values correspond to the generation of immature oil, the oil-generating potential, or convertibility, of this part of the section is assumed to be low because it contains kerogens that are chiefly of terrestrial origin, hydrogen-lean, and gas prone (Miller and others, 1979; Scholle, 1979).

The organic matter in rocks penetrated below a depth of about 2,800 m in the COST number GE-1 well are believed to be thermally mature. TAI values in this interval range from 2 to 2+ and vitrinite-reflectance values vary from 0.6 to 0.75. Values

TABLE 1. DISCOVERIES IN WELLS DRILLED ON THE U.S. CONTINENTAL SHELF

Well name	Depth (m)	Interval Thickness (m)	Natural Gas Flowage* m³/day	(ft³/day)	Oil or Condensate (bbls/day)
[1]Texaco 598-1	>4267	12	210,000	(7,500,000)	
	>3960	12	270,000	(9,400,000)	21
[2]Tenneco 642-2	4017	4.3	340,000	(12,000,000)	100
	3860	?	28,000	(1,000,000)	
	2535	?			630
[3]Tenneco 642-1	>4720	?	160,000	(5,500,000)	18
	>3871	?	535,000	(18,900,000)	
	>3962	?	402,000	(14,200,000)	
[4]Tenneco 642-3	4305	6	103,000	(3,600,000)	
	4357	6	170,000	(6,000,000)	
[5]Exxon 599-1	3770	15	230,000	(8,000,000)	
	>3786	?	230,000	(1,000,000)	

*Flow rates represent initial tests. In addition to the wells above, shows of natural gas were reported in the following wells (see Fig. 2): 857-1, 106-1, 495-1, 816-1, 728-1, and COST B-3.

Sources: [1]Texaco press release (1978); [2]Energy Information (1979); [3]Oil and Gas Journal (1980a); [4]Oil and Gas Journal (1980a); [5]Oil and Gas Journal (1980b).

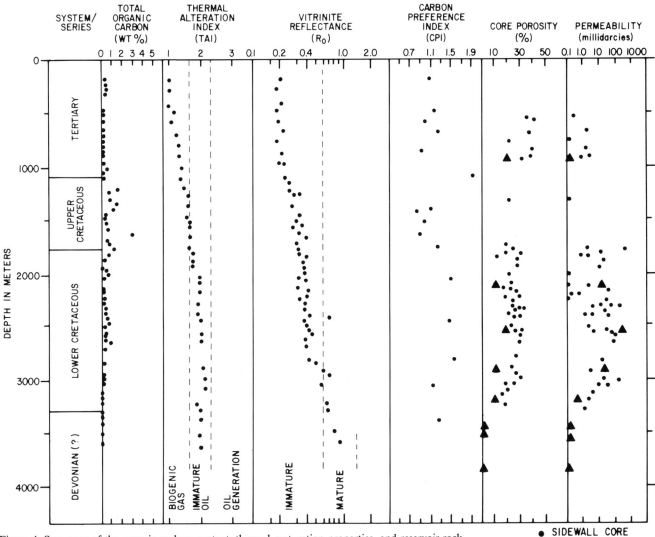

Figure 4. Summary of the organic carbon content, thermal maturation properties, and reservoir-rock properties of rocks penetrated in the COST GE-1 well in the Southeast Georgia Embayment. Figure modified from Miller and others (1979) and Amato and Bebout (1978).

such as these mark peak generation of liquid-hydrocarbons. However, rocks in this part of the section are largely of nonmarine origin and, because of their low organic carbon content, generally less than 0.3% by weight, are believed to have little or no potential as petroleum source rocks (Miller and others, 1979; Scholle, 1979).

Porosity and permeability data (Fig. 4) indicate that the Lower Cretaceous section at the COST GE-1 site contains excellent-quality reservoir rocks. Throughout this 1,500-m-thick interval, porosity measurements average about 25% and permeabilities are greater than 10 millidarcies (md). In the upper part of this interval (1,750–2,200 m), high primary and secondary porosity values occur in shelf carbonate deposits (Halley, 1979). Below the carbonate rocks, sandstone beds in excess of 10 m thick have permeabilities of hundreds of millidarcies (Amato and Bebout, 1978). The probability of high quality reservoir rocks occurring

in the Devonian section below 3,300 m in the vicinity of the GE-1 site is negligible because of loss of porosity and permeability due to metamorphism (Halley, 1979).

Porosity and permeability data from the seven southeast Georgia Embayment wells (Fig. 3) indicate that Lower Cretaceous sandstones have the best reservoir rock potential. Throughout this 300–1,500-m thick section, sidewall-core porosity values range between 15% and 35%, and permeability values range between less than 1 and 700 md, although most are less than 100 md. Lower Cretaceous carbonate rocks, overlying the sandstone beds, have high porosity values, but permeability is generally low. Sidewall-core porosity values range between 15% and 25% and permeability values are mostly less than 2 md. The Upper Cretaceous calcareous mudstone and Cenozoic limestone units also have low permeability values, generally less than 10 and 5 md, respectively. Porosity and permeability are virtually

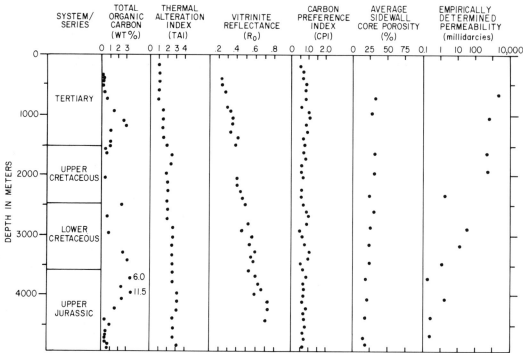

Figure 5. Summary of source rock quality, maturation properties, and reservoir rock characteristics (porosity and permeability) of samples from the COST B-2 well on the Continental Shelf in the Baltimore Canyon Trough. Total organic carbon content, thermal alteration index (TAI), vitrinite reflectance (R_o), and carbon preference index (CPI; Claypool and others, 1977). Porosity and permeability measurements (Smith and others, 1976).

negligible in the Paleozoic section because of induration and metamorphism. These porosity and permeability trends are reflected in the COST GE-1 well data summarized in Figure 4.

Thus, evidence from the COST GE-1 well indicates that mature organic-rich source rocks are absent in the Southeast Georgia Embayment. Although Scholle (1979) has suggested that Upper Cretaceous shales could be excellent source rocks in areas of deeper burial or in areas of higher geothermal gradient, this suggestion must be tempered by the fact that interpretations of seismic-reflection data by Dillon and others (1979a) indicate that Upper Cretaceous rocks are at shallow depths throughout the area shoreward of the 200-m bathymetric contour. Adjacent to the Southeast Georgia Embayment, in deep-water areas where the sedimentary section is thicker, such as the Carolina Trough and the Blake Plateau, porous Lower Cretaceous carbonate rocks and sandstone could provide excellent reservoirs to trap migrating hydrocarbons generated in underlying Jurassic strata or, perhaps, in overlying Upper Cretaceous rocks where they are deeply buried.

Baltimore Canyon Trough. Detailed geochemical studies on the rocks penetrated by the exploration wells in the Baltimore Canyon Trough have not been published. Summarizing data from these wells, however, Smith and others (1981) concluded that the Jurassic through Cenozoic section has marginal to excellent organic richness; but, based on TAI and vitrinite reflectance

data, burial below 3,500 m is necessary for hydrocarbon generation. Cretaceous and Cenozoic rocks are, for the most part, thermally immature. Detailed geochemical studies of the COST B-2 and B-3 wells are in agreement with the work by Smith and others, and they are used in this paper to explain what is known about the source rock potential of the basin (Figs. 5 and 6).

In the Tertiary section, at the COST B-2 well site (Fig. 5) between depths of 600 and 1,500 m, some total organic-carbon values exceed 1%; it is unlikely, however, that Tertiary rocks have been buried deep enough in the Baltimore Canyon area to have reached thermal maturity. In Lower Cretaceous and Upper Jurassic rocks, several units between 3,048 and 4,267 m have abnormally high total organic-carbon values (6%) and contain indigenous solvent extractable hydrocarbons (not shown) with concentrations as high as 5,747 ppm. These high values in the Cretaceous-Jurassic section are believed to be related to abundant coaly material and this part of the section, therefore, is considered to be gas prone. Below 4,267 m, the total organic-carbon values drop dramatically to around 0.1% to 0.2% and the extractable hydrocarbons (not shown) to less than 500 ppm.

Metamorphism of organic matter or maturation studies conducted on rock samples from the COST B-2 well show a general increase in thermal maturity with depth. Interpretations of the maturation parameters in the COST B-2 well have been discussed by Claypool and others (1977). Although some inter-

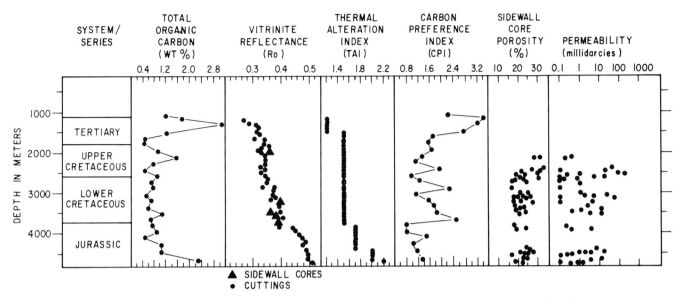

Figure 6. Summary of organic-carbon content, thermal maturation properties, and reservoir-rock properties of rocks penetrated in the COST B-3 well in the Baltimore Canyon Trough. Data from Scholle (1980) and Amato and Simonis (1979).

pretational inconsistencies exist concerning the extent of hydrocarbon-generating reactions, the "threshold or transition temperature of intense oil-generation" is believed to start between 3,048 and 3,535 m, and the zone of peak generation occurs between 3,962 and 4,267 m.

Thus, the results from the COST B-2 well indicate that potential gas-producing source rocks containing predominantly gas-prone, hydrogen-lean kerogen occur in the Lower Cretaceous and Upper Jurassic sections, but fully mature, oil-prone source rocks have not been penetrated. The significant flow of oil recorded from the 642-2 well (Table 1), however, suggests that mature oil-prone source rocks are present locally or, perhaps, deeper in the section.

Studies of the COST B-3, which lies seaward of the COST B-2 well, have been made by Amato and Simonis (1979), Miller and others (1980), and Scholle (1980). Data from these authors are summarized in Figure 6.

Tertiary rocks at the B-3 site down to a depth of about 1,829 m contain an average total organic-carbon content of 1.5%. The solid organic matter in this section is composed predominantly of an amorphous, hydrogen-rich oil-prone type I variety. However, thermal immaturity, as indicated by vitrinite reflectance values of 0.26 to 0.35 and TAI values of 1.2 to 1.5, suggests that the Tertiary section does not contain source rocks that have produced petroleum by thermal generation. Although gas occurred at a depth of about 1,219 m in the well, it was probably methane and of biogenic rather than thermal origin.

The Cretaceous section in the COST B-3 well extends from about 1,841 to 3,780 m. Compared with the overlying Tertiary section, which contains primarily hydrogen-rich kerogens, the Upper and Lower Cretaceous intervals contain solid organic mat-

ter that is hydrogen-lean and predominantly of the type III exinite-vitrinite humic, gas-prone variety with atomic hydrogen-to-carbon (H/C) ratios of 0.6 to 0.8 (Scholle, 1980). Average total organic-carbon values of 0.77 (Fig. 6) and an average extractable hydrocarbon content of 144 ppm (Miller and others, 1980), suggest that the Cretaceous section contains only poor to fair source rocks. According to Miller and others (1980), the Cretaceous section, penetrated in the COST B-3 well, is thermally immature.

The Jurassic section in the COST B-3 well extends from approximately 3,780 m to a total depth of 4,822 m. Total organic carbon values and total extractable hydrocarbons for this interval average 1.06 (weight percent) and 420 ppm, respectively (Miller and others, 1980). Such organic-richness values indicate moderate to good source-rock potential. The kerogens of the Jurassic section are predominantly gas-prone, hydrogen-lean exinite and vitrinite types. According to the same authors, the section below approximately 4,572 m has reached thermal maturity. The zone from 4,572 m to about 4,816 m is believed to be an excellent potential gas source and to be close to thermal maturity (Scholle, 1980). A gas show between 4,799 m to 4,801 m in the B-3 well and the natural gas discoveries on blocks 598, 599, and 642 (Fig. 2) would appear to confirm this evaluation.

Lower Cretaceous and Jurassic sandstone beds have the best reservoir potential. Data on the porosity and permeability of Lower Cretaceous sandstones is available from 12 wells (Fig. 2). Sidewall-core porosity values range between 15% and 30%; permeability values range between 0.1 and 500 md, but most were less than 10 md. The reservoir rock properties of units penetrated in the COST numbers B-2 and B-3 wells are summarized in the last two columns of Figures 5 and 6, respectively. The COST B-2

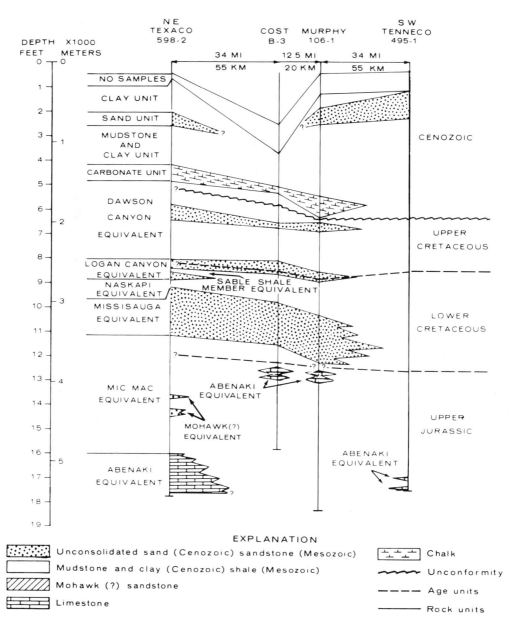

Figure 7. North-south cross section through the eastern Baltimore Canyon Trough wells showing pinchout of sandstone facies (Libby-French, 1984).

well penetrated many thick sandstone units throughout the section. However, the reservoir quality of these rocks deteriorates with depth because of a progressive breakdown of feldspar accompanied by the increase in authigenic clay and silica cement (Scholle, 1977a). As a result, few of the sandstone units penetrated below 3,500 m in the COST B-2 well had more than 1 md permeability. Farther seaward, at the COST B-3 site, time-equivalent sandstone beds have higher permeabilities (tens of millidarcies).

Some sandstone beds pinch out in a seaward direction from the north-northwest part of basin. Sandstone units, as shown in Figure 7, are less abundant at the 495-1, 273-1, and B-3 sites,

than at the 544-1, 718-1 and B-2 sites (Simonis, 1979; Libby-French, 1984). The decrease in sandstone content probably is the result of an increase in distance from the terrigenous sediment source.

The Jurassic section of the basin exhibits a similar depositional pattern. Thick (30–90 m), shaley sandstone and siltstone sequences are present at the northern well sites (590-1, 544-1, COST B-2, and 598-2). At the southern and seaward well sites, however, equivalent sequences are much thinner or are absent. For example, the Jurassic sandstone beds penetrated by the B-3 well are 3–10 m thick and total about 47 m. Porosity values measured on B-3 sidewall cores range from 17% to 25%, and logs

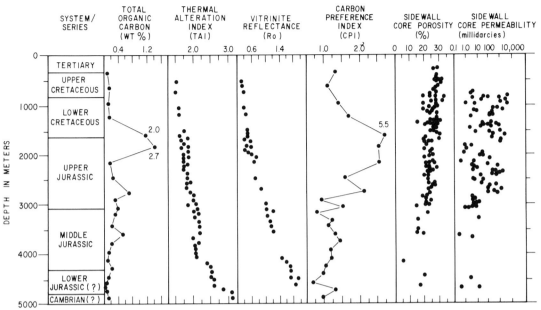

Figure 8. Summary of organic-carbon content, thermal maturation properties, and reservoir-rock properties of rocks penetrated in the COST G-1 well. Data from Scholle and Wenkam (1982) and Lachance and Amato (1980).

indicate that porosities are more than 8% and permeabilities are mostly less than 10 md (Simonis, 1979).

These data suggest that conditions for the entrapment of natural gas in Jurassic sandstone reservoirs are favorable. However, future discoveries are not expected to be large because of the thinness of the reservoir beds and the seaward pinchout of Jurassic sandstone beds. The possibility of vertical migration from deep, thermally mature sources into overlying Lower Cretaceous units, which contain thicker and more porous sandstone units, is enhanced by the presence of growth faults that cut Jurassic and Lower Cretaceous strata along the shelf margin.

Georges Bank Basin. Little geochemical data has been published on the Georges Bank Basin exploration wells. Carpenter and Amato (1984) reported on the source rock potential of the Shell 410-1 well, and they concluded that the source rock potential is poor to fair and the total organic carbon percentages are low. Most of the kerogen is of terrigenous origin and, therefore, gas prone. These conclusions are in general agreement with the COST well data. Studies of the COST G-1 and G-2 wells in the Georges Bank Basin are summarized in Figures 8 and 9.

There is little likelihood that the thin Tertiary section penetrated in the COST G-1 and G-2 wells will contain oil and gas accumulations because of its shallow burial depths and the thermal immaturity of contained kerogens and extractable organic matters.

The Cretaceous section, penetrated between depths of 300–1700 m in the COST G-1 and G-2 wells, has source-rock richness properties in the poor to fair quality range (Miller and others, 1982). According to Miller and others (1982), the H/C ratios of the Cretaceous kerogens range from 0.63 to 0.97 and

these values are comparable with type II and III terrestrially derived, hydrogen-poor, humic gas-prone varieties. However, the extractable Cretaceous organic matter and kerogens are not believed to have experienced a sufficient time/temperature maturation history to have generated liquid-petroleum or gas (Miller and others, 1982; Scholle and Wenkam, 1982).

Source-rock characteristics of the Jurassic section are variable. At the G-2 site, the Jurassic rocks are characterized by abundant shallow-water marine dolomite, limestone, and anhydrite-evaporite facies. In comparison, the Jurassic section penetrated at the G-1 site contains a higher percentage of sands and silts, and less carbonate and argillaceous rocks (Scholle and Wenkam, 1982). This lithologic difference is reflected in the distribution and type of solid organic matter contained in the rocks. The Jurassic kerogens at the G-1 site are predominantly terrigeneous, humic gas–prone, type III; they contain little hydrogen-rich, oil-prone, type I variety. At the G-2 site, the Jurassic kerogens contain some amorphous algal, type I, as well as gas-prone, herbaceous, type II and woody type III kerogens (Miller and others, 1982). The average total organic-carbon values at the G-1 site are 0.73, 0.25, and 0.11, and the average extractable hydrocarbons values are 38, 36, and 39 ppm, respectively, for the Upper (1,800–3,600 m), Middle (3,600–5,400 m), and Lower (below 5,400 m) Jurassic rocks (Figs. 8 and 9). In the COST G-2 well, the average total organic-carbon values (Fig. 9) range from 0.62 (Upper Jurassic) to 0.23 (Lower Jurassic).

Vitrinite reflectance values (Figs. 8 and 9) indicate that the peak generation of petroleum corresponds to depths of about 4–4.5 km (middle Middle Jurassic at the G-2 site). Although there is some indication that thermal maturation occurs at shal-

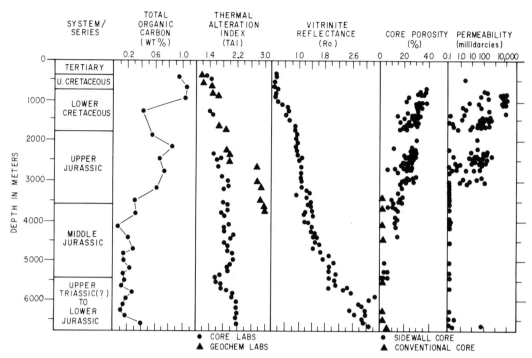

Figure 9. Summary of organic-carbon content, thermal maturation properties, and reservoir-rock properties of rocks penetrated in the COST G-2 well in Georges Bank Basin. Data from Miller and others (1982) and Malinowski (1980).

lower depths, these values may not be reliable because the rocks at these shallower depths may contain older recycled oxidized organic matter (Miller and others, 1982). According to these same authors, there is consistent agreement, however, between the saturated hydrocarbon to total organic-carbon ratio, the C_4 to C_7 gasoline range hydrocarbon composition and concentration, the molecular composition and concentration of the C_{15}^+ hydrocarbons, thermal pyrolysis maximum temperature, and TAI values—all indicate that the peak generation of petroleum occurs at depths of about 4 to 4.5 km in the Georges Bank Basin. Dry gas may be expected to occur below depths of about 5.0 km (Miller and others, 1982).

Even though Middle Jurassic and older rocks penetrated at the G-1 and G-2 sites appear to be thermally mature with respect to the generation of petroleum, the potential of these rocks is not expected to be high. The average total organic-carbon content of this part of the section is less than 0.3% (Figs. 8 and 9) and maximum permeabilities of this predominantly carbonate section on the outer shelf are less than 0.5 md. Higher porosities and permeabilities, however, could be concentrated locally over structural highs and in the vicinity of patch reefs where coarse sand deposition may have occurred. Secondary processes may have resulted in high porosities in reefal limestones if the reefs were exposed to freshwater leaching. The Jurassic carbonate section penetrated by exploratory wells (Fig. 1), however, is not promising. Sidewall- and conventional-core porosity values are mostly less than 10% and permeability values are mostly less than 1 md.

Thick accumulations of Triassic and possibly Lower Jurassic sedimentary rocks fill synrift basins in the Georges Bank area (Poag and Valentine, this volume; Schlee and Klitgord, this volume) and may have been penetrated in the COST G-1 and G-2 wells. Because these rocks could have significant petroleum potential and they occur beneath the Georges Bank and Baltimore Canyon Trough areas in an extensive synrift basin system, they will be discussed in more detail in a later section of this chapter.

The preceeding discussion of the source and reservoir properties of rocks penetrated on the Continental Shelf of the U.S. Atlantic margin indicates that, although Tertiary and some Cretaceous rocks have a relatively high organic-carbon content, these rocks generally are immature with respect to the thermal generation of petroleum. More deeply buried Mesozoic rocks, within the petroleum generation window, have, in general, a low organic-carbon content or are relatively impermeable, or contain gas-prone kerogens. This conclusion may reflect industry's assessment as well. Exploratory drilling on the U.S. Atlantic Continental Shelf has ceased, at least temporarily.

The low-organic carbon content of Jurassic and some Cretaceous rocks penetrated thus far reflects deposition under oxidizing conditions in an open shelf environment—conditions not conducive to the preservation of organic matter, and hence, not conducive to the generation of petroleum. Rocks representing deposition in a more reducing environment, such as in topographic lows within an open shelf basin or in slope environments, could have better potential. The Continental Shelf in the vicinity of Oceanographer Canyon on Georges Bank, perhaps, is an example of an area, on the present Continental Shelf, that could contain sedimentary rocks deposited in a slope environment. There, along the Outer Continental Shelf, deposits of Jurassic and Cretaceous age may have prograded over older Jurassic basinal shales that were deposited in water depths comparable to those of the present slope (Fig. 10).

Traps

Seismic-reflection surveys and industry drilling have revealed a variety of potential traps, some of which have been drilled without success, on the shelf. These include anticlinal structures associated with intrusions, rollover traps, drape structures over basement blocks, and carbonate buildups. In deeper water environments, potential traps include: (1) a structural high related to draping of Cretaceous sediments over the Jurassic shelf edge; (2) pinchouts of Lower Cretaceous and Jurassic sedimentary rocks that were deposited at the base of the paleoshelf; and (3) abandoned canyon-fan complexes. Some of these features are shown on a diagrammatic section across the Baltimore Canyon Trough (Fig. 11).

Intrusions. Dome-shaped structures associated with intrusion of igneous rock and salt, or possibly shale, have been mapped on the shelf and slope. The Great Stone Dome (Fig. 11) is inferred to be a compaction feature associated with an Early Cretaceous intrusion of mafic rock (Mattick, 1981; Lippert, 1983). Seven wells were drilled in the vicinity of the structure, but no reports of oil or gas shows were made. Igneous intrusions may also occur at the shelf margin in the vicinity of the ocean-continent boundary. Here, shallow depths, computed to basement rock on the basis of magnetic data (Klitgord and Behrendt, 1979), may reflect igneous intrusions of Jurassic through Cretaceous age.

Extensive salt diapirism has been reported on the shelf off Nova Scotia by Jansa and Wade (1975); and Schlee and others (1977) inferred that these features continue at least as far south as the Northeast Channel and possibly farther south into the Georges Bank Basin. A linear chain of diapiric structures, thought to be salt related, were mapped on the slope and rise off Cape Hatteras (Grow and Markl, 1977; Dillon and others, 1982), and at least three additional possible salt structures have been mapped at the shelf margin in the Baltimore Canyon Trough area (Grow, 1980).

Rollover traps. On the basis of sparse seismic coverage, it is speculated that the natural-gas discovery on lease blocks 598, 599, and 642 is associated with a rollover structure above a salt intrusion (Fig. 11). In that area, Mattick and others (1981) mapped a listric fault whose throw progressively increases with depth. According to these authors, movement began in Jurassic time and probably continued through Late Cretaceous time. Seismic data indicate that numerous other listric faults occur near the shelf margin in the Baltimore Canyon Trough area (Mattick and others, 1980). Although beds on the downthrown side of some of these faults show dip reversal or rollover, they are usually not associated with salt intrusions.

Drape structures over basement blocks. Schlee and others (1977) reported that large structural highs associated with draping of sedimentary rocks over basement blocks are potential traps in the Georges Bank area. Major relief on these structures is usually restricted to Jurassic horizons; however, near the edge of the shelf, Lower Cretaceous horizons are arched. On seismic-reflection profiles from Georges Bank, these structures appear to be related to block faulting (horst and graben type) of the basement surface, probably during Triassic time. A diagrammatic example of this type of structure on the inner shelf is shown in Figure 11.

Minor carbonate buildups. In a later section of this chapter, the Jurassic-Cretaceous reef trend—a large, chiefly Jurassic carbonate-bank complex beneath the Continental Slope—is discussed. Seismic data indicate that many smaller carbonate buildups may occur on the Continental Shelf shoreward of the larger trend. These are relatively small, subtle anticlinal features that can be detected on numerous seismic profiles from the Georges Bank Basin (Fig. 12). They may represent concentrations of limestone, bioherms(?), or banks in areas of local paleotopographic highs. Although relief on a particular horizon is small, these low-relief structures can persist through thousands of meters of section.

Some of these features have been tested. According to High (1985), the Georges Bank exploratory wells (Fig. 1) were drilled on seismic anomalies interpreted as representing various carbonate environments. The Mobil 312-1 and Shell 357-1 wells tested a prospect that was interpreted as a reef, but proved to be a carbonate mound. The anomaly that the Exxon 975-1 well tested proved to have resulted from a salt bed. The objectives of the Mobil 273-1, Tenneco 187-1, and Conoco 145-1 wells were carbonate banks overlying a salt structure, but only thin oolitic grainstone was penetrated at depth. Likewise, the Shell 410-1 was drilled to test a carbonate bank, but good porosity zones were not present. The Exxon 133-1 well penetrated a volcanic cone, originally interpreted as a patch reef.

Drape over Jurassic shelf edge. Evidence from the COST and exploratory wells in the Baltimore Canyon Trough indicates that the Cenomanian–Lower Cretaceous section consists of relatively thick deltaic sandstone units, many of which have significant reservoir potential (Scholle, 1977b; Scholle, 1980; Amato and Simonis, 1979; Libby-French, 1984). This interval probably represents a delta area in which clastic input overwhelmed the Cretaceous rise in sea level, resulting in the deposition of massive sand sheets out to the shelf edge and beyond. In the Baltimore Canyon Trough and Georges Bank areas, the Cenomanian–Lower Cretaceous sandstones are draped over Jurassic and Lower Cretaceous carbonate rocks of the paleoshelf margin (Fig. 11). This structural high, which is relatively continuous for a distance of several hundred kilometers, has as much as 300 m of local relief (Mattick and others, 1978). This play has been penetrated locally by the COST B-3, 586-1, and 587-1 wells, but without success (Fig. 2).

Pinchouts at base of paleoshelf. At the base of the paleoshelf, coarse-grained Lower Cretaceous and Jurassic sedimentary

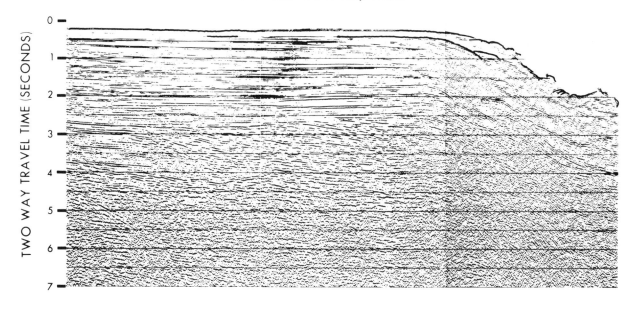

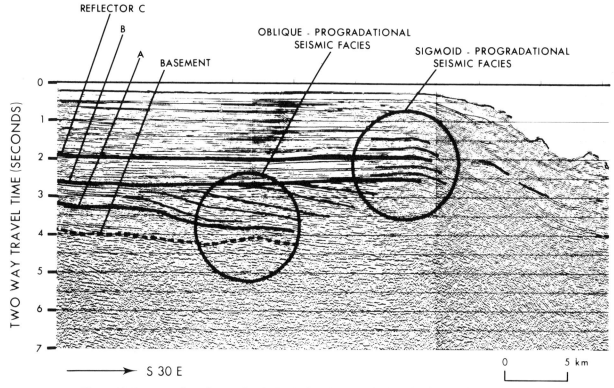

Figure 10. Interpretation of part of U.S. Geological Survey seismic-reflection profile 20 across the Continental Shelf and Slope in the vicinity of Oceanographer Canyon on Georges Bank. For location of seismic profile refer to Schlee and Klitgord (this volume). Reflector A correlates with the base of prograding reflections. Reflector B is at a depth of about 3,000 m, near a contact between a thick Upper Jurassic sandstone unit and an underlying Upper Jurassic limestone unit. Reflector C, at a depth of about 1,600 m, occurs within Lower Cretaceous strata near the top of the Jurassic section. The bottom part, or fondoform zone, of the oblique-progradational reflectors are interpreted as representing basinal facies, probably shale. The sigmoid-progradational seismic facies probably represent sandstone and carbonate rocks that built up over the basinal facies to form the present shelf margin. Oil, possibly generated in the basinal shales, could have migrated into traps in the overlying sandstone and carbonate rocks. Figure from Mattick (1982).

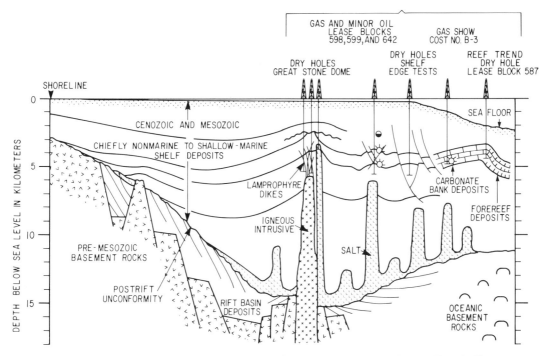

Figure 11. Diagrammatic section showing structural features in the Baltimore Canyon Trough. Figure modified from Benson (Delaware Geological Survey, written communication, 1985) and Benson and Doyle (1984).

rocks could pinchout against finer grained rocks or dense carbonate rocks that make up the paleoshelf. These stratigraphic traps, although difficult to locate by geophysical techniques, should not be overlooked. Migration of hydrocarbons from sediment facies below the present rise and abyssal plain could contribute significant amounts of hydrocarbons to traps near the base of the paleoshelf.

Abandoned canyon-fan complexes. If major regressions of the sea took place during Cretaceous and Jurassic times, as they did during the Pleistocene and Tertiary, the ancestral continental slope may have been traversed by many deeply incised submarine canyons during those periods. These canyons, later abandoned, presumably would have been filled by coarse sediment with good reservoir potential and later abandoned and possibly covered by fine-grained material.

Synrift Basins

In the COST number G-1 well, sedimentary rocks, consisting of sandstones, shales, conglomerates, and interbedded dense limestone and dolomite, were penetrated in the bottom 488 m. In the COST G-2 well, halite-bearing strata were penetrated at a depth of 6,667 m. These units are inferred to be of Late Triassic age on the basis of dinoflagellates (Poag, 1982, p. 53 and 66; Cousminer and others, 1984). From the pink, red, and red-brown colors of the clastic rocks penetrated in the COST G-1 well, Poag (1982, p. 72) inferred that they were deposited in an oxidizing, high-energy, nearshore (alluvial to fluvial) environment; the in-

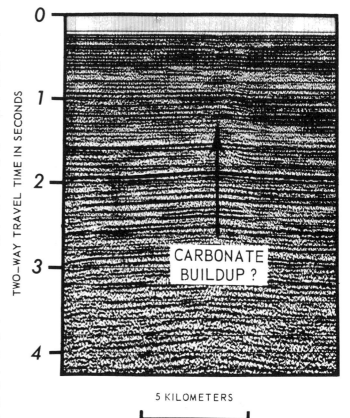

Figure 12. U.S. Geological Survey seismic profile 7 (Plate 1B) from Georges Bank Basin showing low-relief structure—possibly a buildup of carbonate rocks.

terbeds of anhydrite, halite, and dolomite are inferred to indicate occasional flooding by shallow marine waters.

From the interpretation of seismic profiles in the Georges Bank area, Klitgord and others (1982, p. 180) and Schlee and Klitgord (this volume) inferred that Triassic and possibly Lower Jurassic sedimentary rocks are part of a synrift sequence. According to these authors, the synrift sequence represents sediments that filled deep grabens during an early rifting stage prior to continental separation. The top of the synrift sequence (or top of the Triassic [?] section), therefore, correlates with the postrift unconformity that marks the transition from rifting to seafloor spreading. Up to 8 km of synrift deposits (Triassic [?] age sedimentary rocks) underlie the central Georges Bank area beneath a postrift cover of 4–7 km of chiefly Jurassic and Cretaceous age sedimentary rocks (Klitgord and others, 1982, p. 182). Similar thicknesses of synrift sedimentary rocks are believed to underlie the outer shelf area of the Baltimore Canyon Trough (Grow and others, this volume). In this area, however, the synrift sequence is buried beneath a postrift cover more than 10 km thick. According to Benson (Delaware Geological Survey, written communication, 1985), seismic evidence indicates that synrift basins occur at shallow depths on the inner shelf of the Baltimore Canyon Trough, and, by inference, also may occur beneath the Coastal Plain in the same area.

Although the sections penetrated in the COST G-1 and G-2 wells represent deposition in a highly oxidizing environment—an environment not conducive to the preservation of organic matter and, hence, not conducive to the generation of petroleum in other areas, or perhaps deeper in the section—Triassic rocks could represent more favorable conditions. Examples of synrift basins (both marine and nonmarine), which contain large reserves of petroleum, occur worldwide (Mattick, 1984). The Hibernia Field, a Jurassic–Early Cretaceous synrift basin with oil reserves estimated to be in excess of one billion barrels, is located to the north on the Canadian Atlantic margin (Oilweek, 1980). These examples usually involve deep basins with restricted circulation. From the discontinuous character and variable amplitude of reflections from the Triassic section beneath Georges Bank, Klitgord and others (1982, p. 180) concluded that the entire sequence is similar to the section drilled in the lowest part of the COST G-1 well. However, these reflection patterns could just as well represent turbidite deposition in a deep basin. If this proves to be true, then the Triassic section could contain strata rich in organic carbon, and the high temperatures associated with early rifting would have promoted early generation of petroleum; it is possible that the generated gases and some of the fluids would be preserved despite deep burial. Early migration would have filled the pore spaces of reservoirs with fluids and gases, which would tend to locally inhibit the loss of porosity and permeability due to subsequent heat and pressure. The escape of gases and liquids by upward migration would have been prevented by overlying impervious evaporite-bearing strata.

Deep-Water Potential

Jurassic-Cretaceous Reef Trend. In 1983 and 1984 petroleum exploration on the U.S. Atlantic margin was focused on the Jurassic–Lower Cretaceous carbonate-bank buildup beneath the outer-most shelf and upper slope (Fig. 13). Shell Offshore Incorporated, in partnership with Amoco Production Company, drilled 3 unsuccessful exploratory wells (587, 586, and 372) on the Continental Slope (Figs. 1 and 13), presumably to test reef, backreef, and fore reef facies.

Edson (1985) observed bioclastic grainstone, packstone, wackestone, floatstone, rudstone, and possibly boundstone in the conventional cores and thin sections from these wells. He concluded that the limestone represented bioclastic debris buildups and reefal bioherms. The limestone from these three wells contained good porosity (mostly between 10% and 25%, and in the uppermost carbonate rocks penetrated by the 587-1 well, vuggy porosity was observed; however, permeability values of most of the limestone samples were low, mostly less than 0.1 md (Gary Edson, U.S. Minerals Management Service, oral communication). Shell's fourth and final well on the slope, well 93-1, was spudded in 1984, and the well data is proprietary as of publication.

Prior to drilling of exploratory wells on the Continental Slope, some authors (Mattick and others, 1978) had compared the Mesozoic shelf margin complex of the U.S. Atlantic margin to the Golden Lane trend and the Reforma-Campeche shelf complex, a major petroleum producing area of Mexico. The area of Mexico is similar to the Atlantic reef trend in that both may represent Mesozsoic carbonate-bank complexes that contain an associated reef that built upward through time at the shelf-slope break. Major differences in structural style, however, exist between the areas in Mexico and the U.S. Atlantic margin. The Atlantic reef trend represents a relatively undisturbed section of chiefly Jurassic-age carbonate rocks that are buried by Cretaceous and Tertiary sandstones and clays. However, the principal petroleum producing fields of the Reforma area of Mexico overlie a horst and graben system that was rejuvenated repeatedly during Late Jurassic to middle Eocene time (Meyerhoff, 1980). As a result of this tectonic activity, numerous angular unconformities and disconformities are present throughout the section. According to the same author, movement of salt, which underlies the producing strata, caused intense fracturing that produced the huge reservoirs and thick, uninterrupted, effective petroleum columns 500–1,000 m thick in Cretaceous and Paleocene sections of the Mexican fields (Meyerhoff, 1980).

The reef trend on the U.S. Atlantic, perhaps, may be better compared with the Abenaki Formation of the Scotian Shelf offshore of eastern Canada where numerous wells have found rocks of a shelf-edge carbonate complex to be relatively impermeable and undisturbed (Given, 1977). Most of the Scotian Shelf carbonate rocks are impermeable owing to diagenetic cementation, and the same may be true for age-equivalent carbonate rocks on the U.S. margin. On the other hand, secondary porosity and permeability may have been developed during periods of subaerial exposure. Regional unconformities of Jurassic and Cretaceous age, interpreted from seismic records and well data, may be re-

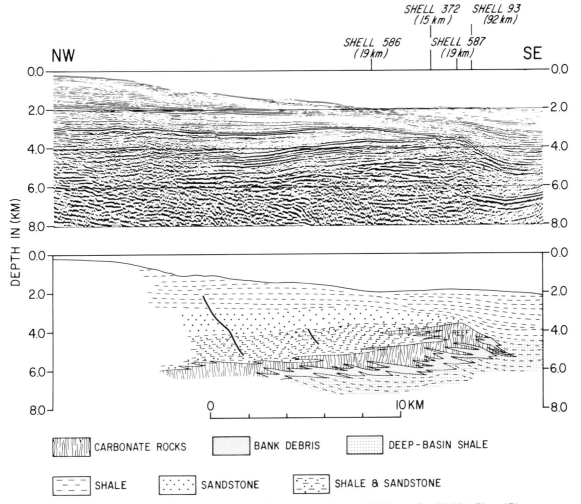

Figure 13. Depth profile through Jurassic-Cretaceous reef trend (BGR profile 79-221, Plate 1B). Uninterpreted seismic record above and interpretation below. Record was converted to depth on the basis of the best available velocity data. Horizontal and vertical scales are approximately equal. Location of industry's four deep-water wells are shown at top. Number in parenthesis beneath well name is the distance of well from seismic profile. Wells 587, 586, and 93 were projected northeast parallel to the strike of the shelf-slope break. Well 372 was projected southwest. Wells 587, 586, 372, and 93 penetrated 2,440, 3,083.5, 1,412, and 3,014 m of sedimentary rock, respectively, and were located in water depths of 1,966, 1,780, 2,119.5, and 1,544.5 m, respectively. Well 372 set the world record for the greatest water depth so far attempted.

lated to periods of subaerial exposure, which may have resulted in meteoric solution and increased porosity of the carbonate rocks (Epstein and Friedman, 1983). Eliuk (1978) attributed solution features in the Scotian Shelf limestone to intermittent subaerial exposure. The Shell Demascota G-32 well, drilled as a shelf-edge test on the Scotian Shelf, indicates that good porosity can be preserved. This well penetrated 168 m of porous dolomite and limestone, including coralgal reef carbonates; porosity values are as large as 14% in the limestone, and fracture porosities of the dolomite are unknown (Given, 1977, p. 74).

On the U.S. Atlantic Continental Slope, in the area of the Jurassic-Cretaceous reef trend, it is difficult to determine struc-

tural closure on the basis of seismic reflection data alone. Because seismic records simply record the travel time of a seismic wave from the ocean's surface to a particular reflecting horizon and back, and depths to that horizon are determined by the interpreter's choice of a velocity function, the final interpreted depth sections can be highly subjective. In most areas of the world, velocity varies slowly in both vertical and lateral directions, so minor errors in the calculation of a velocity function will not affect the overall structural interpretations. During Lease Sale 59, however, of the seven consortia that submitted bids on tracts that overlie the paleoshelf margin, four groups placed their highest bids in the central part of the area; two placed their highest bids in the northern part of the area; and one group placed their highest bids in the southern part of the area. Although some of this variability certainly depended on how each group estimated such factors as reservoir quality, seals, source rock, and maturation, there are indications that some of the differences in estimated tract value depended on how each group interpreted regional structure from their choice of a velocity model.

Blake Plateau Basin and Carolina Trough. According to Dow (1979, p. 432), deep closed bathymetric basins on continental slopes are among the most favorable sites on oceanic margins for the deposition and accumulation of sediments rich in oil-generating organic matter. He attributes this to several interacting conditions such as anoxic bottom water in the oxygen minimum zone, minimum wave and current activity, intermediate rates of deposition, and the deposition of aquatic organic matter.

Much of the Cretaceous-Jurassic section in the Blake Plateau Basin probably represents deposition in a shallow, open-water shelf environment rather than in a deep-water environment. This area probably did not develop into a deep-water basin until relatively late in its geologic history. The Blake escarpment (the eastern edge of the Carolina Trough and Blake Plateau) is marked by intermittent reef growth from Early Jurassic time into Early Cretaceous time (Dillon and others 1979b, p. 107). Behind these reefal deposits, on what is now the Blake Plateau, a broad carbonate platform developed. During this time, the outer Blake Plateau was characterized by shallow-water carbonate-bank deposits and the western part of the plateau by more terrigenous sediments. Deep-water environments, on what is now the Blake Plateau, did not develop until Late Cretaceous time (Dillon and others, 1979b, p. 107).

The early history of the Carolina Trough may have been different from that of the Blake Plateau. According to Dillon and others (1982), the basement rocks beneath the Carolina Trough consist of transitional crust similar to the crustal rocks beneath the Blake Plateau; but, in this area the crust may be thinner than in any other place on the U.S. Atlantic margin. If these hypotheses, which are based on an analysis of gravity data are correct, then the Carolina Trough could have been a deep-water basin since Late Jurassic or, possibly Middle Jurassic time. Thin, underlying crustal rocks that contain a high percentage of relatively dense oceanic material would be expected to cool rapidly and to subside at a faster rate than surrounding areas.

SUMMARY AND CONCLUSIONS

Forty-five wildcat wells and five COST wells have been drilled on the U.S. Atlantic shelf and uppermost slope. Although five of the wildcat wells in the Baltimore Canyon Trough area have yielded significant recoveries of natural gas, a commercial-size field has not been established. Although Tertiary rocks have a high organic-carbon content (average greater than 1%) and contain hydrogen-rich oil-prone kerogens that reflect deposition in a marine environment, they are thermally immature.

Cretaceous rocks, in general, have average organic-carbon contents of less than 1%, contain hydrogen-poor gas-prone kerogens, and are immature. Exceptions include Lower Cretaceous beds in the Southeast Georgia Embayment and the Baltimore Canyon Trough. Basal Lower Cretaceous beds, below depths of about 3,000 m in the former area, are thermally mature; but, because of the low organic-carbon content of these rocks (less than 0.3%), they are not expected to have generated significant quantities of petroleum. In the Baltimore Canyon area, organic-carbon values of basal Lower Cretaceous beds, which contain abundant coaly material at depths of 3,300–3,400 m, reach 3.5%. These rocks, although not fully mature, probably are at or near the threshold for thermal generation of natural gas.

Although peak thermal maturity is reached at depths of 4,000–4,500 m in the Baltimore Canyon Trough and Georges Bank Basin, most of the Jurassic rocks penetrated at or below these depths have a low organic-carbon content and contain hydrogen-lean, gas-prone kerogens. In the Baltimore Canyon Trough, however, some Jurassic intervals are considered to have moderate to good potential for natural gas generation. At the B-3 site, the Jurassic interval below 4,300 m, in particular, is considered to be an excellent potential natural gas source (Scholle, 1980).

The poor source-rock quality of the older shales and carbonate rocks, thus far penetrated on the shelf, probably is a reflection of their deposition in an oxidizing, open-shelf environment and the fact that the little organic matter they contain is of terrestrial origin and, thus, poor for generating oil. Weeks (1958) has stressed the significance of restricted basins and closed deeps within basins where, due to restricted circulation, they provide reducing bottom conditions favorable for the preservation of marine organic matter. With the exception of synrift basins containing chiefly Triassic age rocks, the existence of such restricted basins on the U.S. Atlantic shelf has not been reported.

A related factor that may account, at least in part, for the apparent lack of commercial accumulations of petroleum on the U.S. Atlantic shelf is the relative spatial stability of the shelf-slope break throughout much of its history. Hedberg (1964) has emphasized the relation between petroleum accumulations and transition facies. On the Atlantic margin, however, the shelf-slope break has occupied roughly the same spatial position throughout most of its history; terrestrial, fluvial, and shelf sandstones, therefore, accumulated in a thick wedge that lies far shoreward of basinal shales deposited on the slope and rise. Migration from possible source beds on the slope to reservoir beds on the shelf could be complicated by the relatively impervious Mesozoic carbonate bank complex that separates shelf-derived sediments from fine-grained, deep-water deposits.

Extensive seismic-reflection exploration has revealed a variety of potential structural and stratigraphic traps on the shelf and upper slope. These include anticlinal structures associated with intrusions, rollover traps associated with faults, carbonate buildups, and drape structures over basement blocks. As of this report, many of these structures in Georges Bank Basin, Baltimore

Canyon Trough, and the Southeast Georgia Embayment have been drilled; but no commercial-sized discoveries have been reported, and drilling on the shelf has ceased.

Following exploration of the shelf, industry's efforts in 1982–84 shifted to the slope in water depths of about 2,000 m, where four unsuccessful wells were drilled to test chiefly Jurassic-age backreef, reef, and forereef facies.

Although the thick section of Triassic, and possibly Lower Jurassic, synrift sedimentary rocks of the Atlantic margin are generally thought to consist of highly-oxidized redbeds with little or no petroleum potential, it is possible that some of these rocks represent deep-water lacustrine sedimentation or, perhaps, shallow marine sedimentation during early continental rifting. If this proves to be the case, the Triassic section of Georges Bank, where these sedimentary rocks occur extensively, should be viewed as as potential petroleum target.

Speculation suggests that the Carolina Trough could be another significant target area. This is based on the speculation that subsidence rates in the Carolina Trough during Mesozoic time were high in comparison to other areas of the U.S. Atlantic margin, and that during this time it was a deep, closed basin with restricted circulation.

REFERENCES CITED

Amato, R. V., and Bebout, J. W., 1978, Geological and operational summary, COST No. GE-1 well, Southeast Georgia Embayment area, South Atlantic OCS: U.S. Geological Survey Open-File Report 78-668, 122 p.

Amato, R. B., and Simonis, E. K., 1979, Geological and operational summary, COST No. B-3 well, Baltimore Canyon Trough area, mid-Atlantic OCS: U.S. Geological Survey Open-FIle Report 79-1159, 118 p.

Arthur, M. A., and Natland, J. H., 1979, Carbonaceous sediments in the North and South Atlantic; The role of salinity in stable stratification of Early Cretaceous basins, *in* Talwani, M., Hay, W., and Ryan, W.B.F., eds., Deep drilling results in the Atlantic Ocean; Continental margins and paleoenvironment: American Geophysical Union Maurice Ewing Series 3, p. 375–401.

Benson, R. N., and Doyle, R. G., 1984, Inner margin of Baltimore Canyon Trough; Future exploration play [abs.]: American Association of Petroleum Geologists Bulletin, v. 68, no. 4, p. 454.

Carpenter, G. B., and Amato, R. V., 1984, Geology report, proposed North Atlantic OCS oil and gas Lease Sale 96: U.S. Minerals Management Service OCS Report MMS 84-0062, 19 p.

Claypool, G. E., Lubeck, C. M., Baysinger, J. P., and Ging, T. G., 1977, Organic geochemistry, *in* Scholle, P. A., ed., Geological studies on the COST No. B-2 well, U.S. mid-Atlantic Outer Continental Shelf area: U.S. Geological Survey Circular 750, p. 46–59.

Cousminer, H. L., Steinkraus, W. E, and Hall, R. E., 1984, Biostratigraphic restudy documents Triassic/Jurassic section in Georges Bank COST G-2 well [abs.]: American Association of Petroleum Geologists Bulletin, v. 68, no. 4, p. 466.

Crawford, D., 1978, East Coast 78; A year of initiation: Offshore Journal, v. 38, no. 13, p. 82–87.

Dillon, W. P., Paull, C. K., Buffler, R. T., and Fail, J. P., 1979a, Structure and development of the Southeast Georgia Embayment and Northern Blake Plateau; Preliminary analysis, *in* Watkins, J. S., Montadert, L., and Dickerson, P. W., eds., Geological and geophysical investigations of continental margins: American Association of Petroleum Geologists Memoir 29, p. 27–41.

Dillon, W. P., Paull, C. K., Dahl, A. G., and Patterson, W. C., 1979b, Structure of the continental margin near the COST No. GE-1 drill site from a common depth-point seismic-reflection profile, *in* Scholle, P. A., ed., Geological studies of the COST GE-1 well, United States South Atlantic Outer Continental Shelf area: U.S. Geological Survey Circular 800, p. 97–107.

Dillon, W. P., Popenoe, P., Grow, J. A., Klitgord, K. D., Swift, B. A., Paull, C. K., and Cashman, K. A., 1982, Growth faulting and salt diapirism; Their relationship and control in the Carolina Trough, eastern North America, *in* Watkins, J. S., and Drake, C. L., eds., Studies in continental margin geology: American Association of Petroleum Geologists Memoir 34, p. 21–46.

Dow, Wallace G., 1979, Petroleum source beds on continental slopes and rises, *in* Watkins, J. S., Montadert, L., and Dickerson, P. W., eds., Geological and geophysical investigations of continental margins: American Association of Petroleum Geologists Memoir 29, p. 423–442.

Edson, G. M., 1985, The Mid-Atlantic Mesozoic paleoshelf edge; Carbonate buildup or reef? [abs.]: Americal Association of Petroleum Geologists Bulletin, v. 69, no. 9, p. 1436.

Eliuk, L. S., 1978, The Abenaki, Formation, Nova Scotia, Canada; A depositional and diagenetic model for a Mesozoic carbonate platform: Bulletin of Canadian Petroleum Geologists, v. 26, p. 424–514.

Energy Information, 1979, Gas found in deep zone at Baltimore canyon delineation well: Energy Information published by Petroleum Information Corporation, Denver, v. 8, no. 43, p. 1.

Epstein, S. A., and Friedman, G. M., 1983, Depositional and diagenetic relationships between Gulf of Elat (Aqaba) and Mesozoic of United States East Coast offshore: American Association of Petroleum Geologists Bulletin, v. 67, no. 6, p. 953–962.

Given, M. M., 1977, Mesozoic and early Cenozoic geology of offshore Nova Scotia: Bulletin of Canadian Petroleum Geology, v. 25, no. 1, p. 63–91.

Grow, J. A., 1980, Deep structure and evolution of the Baltimore Canyon Trough in the vicinity of the COST No. B-3 well, *in* Scholle, P. A., ed., Geologic studies of the COST No. B-3 well, United States mid-Atlantic Continental Slope area: U.S. Geological Survey Circular 833, p. 117–126.

Grow, J. A., and Markl, R. G., 1977, IPOD-USGS multichannel seismic reflection profile from Cape Hatteras to the mid-Atlantic Ridge: Geology, v. 5, p. 625–630.

Halley, R. B., 1979, Petrographic summary, *in* Scholle, P. A., ed., Geological studies of the COST GE-1 well, United States South Atlantic Outer Continental Shelf: U.S. Geological Survey Circular 800, p. 42–48.

Hedberg, H. D., 1964, Geologic aspects of origin of petroleum: Bulletin of the American Association of Petroleum Geologists, v. 48, no. 11, p. 1755–1803.

High, L. R., Jr., 1985, Regional geology of Georges Bank Basin—OCS Sale 42 drilling results [abs.]: American Association of Petroleum Geologists Bulletin, v. 69, n. 2, p. 265–266.

Jackson, D. S., and Heise, B. A., 1980, Geothermal gradient, *in* Amato, R. V., and Bebout, J. W., eds., Geologic and operational summary, COST G-1 well, Georges Bank area, North Atlantic OCS: U.S. Geological Survey Open-File Report 80-268, p. 79–81.

Jansa, L. F., and Wade, J. A., 1975, Geology of the continental margin off Nova Scotia and Newfoundland, *in* Van der Linden, W.J.M., and Wade, J. A., eds., Offshore geology of eastern Canada: Geological Survey of Canada Paper 74, v. 2, Regional Geology, p. 51–105.

Klitgord, K. D., and Behrendt, J. C., 1979, Basin structure of the U.S. Atlantic margin, *in* Watkins, J. S., Montadert, L., and Dickerson, P. W., eds., Geological and geophysical investigations of continental margins: American Association of Petroleum Geologists Memoir 29, p. 85–112.

Klitgord, K. D., Schlee, J. S., and Hinz, K., 1982, Basement structure, sedimentation, and tectonic history of the Georges Bank Basin, *in* Scholle, P. A., and Wenkam, C. R., eds., Geological studies of the COST Nos. G-1 and G-2 wells, United States North Atlantic Outer Continental Shelf: U.S. Geological Survey Circular 861, p. 160–193.

Lachance, D. J., and Amato, R. V., 1980, Core descriptions and analyses, *in* Amato, R. V., and Bebout, J. W., eds., Geologic and operational summary, COST No. G-1 well, Georges Bank area, North Atlantic OCS: U.S. Geological Survey Open-File Report 80-268, p. 27–38.

Libby-French, J., 1984, Stratigraphic framework and petroleum potential of northeastern Baltimore Canyon Trough, mid-Atlantic Outer Continental Shelf: American Association of Petroleum Geologists Bulletin, v. 68, no. 1, p. 50–73.

Lippert, R. H., 1983, The "Great Stone Dome"; A compaction structure, *in* Bally, A. W., ed., Seismic expression of structural styles: American Association of Petroleum Geologists Studies in Geology Series No. 15, v. 1.

Malinowski, M. J., 1980, Core descriptions and analyses, *in* Amato, R. V., and Simonis, E. K., eds., Geologic and operational summary COST No. G-2 well, Georges Bank area, North Atlantic OCS: U.S. Geological Survey Open-File Report 80-269, p. 57–76.

Mattick, R. E., 1981, Hydrocarbon potential of the United States Atlantic continental margin: Proceedings of the Southwestern Legal Foundation, Exploration and Economics of the Petroleum Industry, v. 19, p. 163–214.

—— , 1982, Significance of the Mesozoic carbonate bank-reef sequence for the petroleum geology of the Georges Bank Basin, *in* Scholle, P. A., and Wenkam, C. R., eds., Geological studies of the COST Nos. G-1 and G-2 wells, United States North Atlantic Outer Continental Shelf: U.S. Geological Survey Circular 861, p. 93–104.

—— , 1984, Petroleum exploration in Malawi: U.S. Geological Survey Open-File Report No. 84-384, 14 p.

Mattick, R. E., and Hennessy, J. L., 1980, Structural framework, stratigraphy, and petroleum geology of the area of Oil and Gas Lease Sale No. 49 on the U.S. Atlantic Continental Shelf and Slope: U.S. Geological Survey Circular 812, 101 p.

Mattick, R. E., Girard, O. W., Jr., Scholle, P. A., and Grow, J. A., 1978, Petroleum potential of U.S. Atlantic slope, rise, and abyssal plain: American Association of Petroleum Geologists Bulletin, v. 62, no. 4, p. 592–608.

Mattick, R. E., Schlee, J. S., and Bayer, K. C., 1981, The geology and hydrocarbons potential of the Georges Bank–Baltimore Canyon Trough area, *in* Kerr, W., ed., Canadian atlas of North American borderland: Canadian Society of Petroleum Geologists Memoir No. 7, p. 461–468.

Meyerhoff, A. A., 1980, Geology of Reforma–Campeche shelf: Oil and Gas Journal, April 21, 1980, p. 121–124.

Miller, R. E., Schultz, D. M., Claypool, G. E., Smith, M. A., Lerch, H. E., Ligon, D., Gary, C., and Owings, D. K., 1979, Organic Geochemistry, *in* Scholle, P. A., ed., Geological studies of the COST GE-1 well, United States South Atlantic Outer Continental Shelf area: U.S. Geological Survey Circular 800, p. 74–92.

Miller, R. E., Schultz, D. M., Claypool, G. E., Smith, M. A., Lerch, H. E., Ligon, D., Owings, D. K., and Gary, C., 1980, Organic geochemistry, *in* Scholle, P. A., ed., Geological studies of the COST No. B-3 well, United States mid-Atlantic Continental Slope area: U.S. Geological Survey Circular 833, p. 85–104.

Miller, R. E., Lerch, H. E., Claypool, G. E., Smith, M. A., Owings, D. K., Ligon, D. T., and Eisner, S. B., 1982, Organic geochemistry of the Georges Bank Basin, *in* Scholle, P. A., and Wenkam, C. R., eds., Geological studies of the COST Nos. G-1 and G-2 wells, United States North Atlantic Outer Continental Shelf: U.S. Geological Survey Circular 861, p. 105–142.

Oil and Gas Journal, 1979, More gas tested from canyon well: Oil and Gas Journal, v. 77, no. 49, p. 33.

—— , 1980a, More Jurassic gas found off U.S. East Coast: Oil and Gas Journal, v. 78, no. 42, p. 119.

—— , 1980b, Exxon, U.S.A. gauges gas in Baltimore Canyon: Oil and Gas Journal, v. 78, no. 43, p. 183.

Oilweek, 1980, Mobil tells all about Hibernia: Oilweek, December 15, 1980, p. 59.

Poag, C. W., 1982, Foraminiferal and seismic stratigraphy, paleoenvironments and depositional cycles in the Georges Bank Basin, *in* Scholle, P. A., and Wenkam, C. R., eds., Geological studies of the COST Nos. G-1 and G-2 wells, United States North Atlantic Outer Continental Shelf: U.S. Geological Survey Circular 861, p. 43–91.

Schlee, J. S., Martin, R. G., Mattick, R. E., Dillon, W. P., and Ball, M. M., 1977, Petroleum geology on the United States Atlantic Gulf of Mexico margins, *in* Cameron, V. S., ed., Exploration and economics of the petroleum industry; New ideas, methods, new developments: Southwestern Legal Foundation, New York, Matthew Bender and Company, v. 15, p. 47–93.

Scholle, P. A., 1977a, Data summary and petroleum potential, *in* Scholle, P. A., ed., Geological studies on the COST No. B-2 well, U.S. mid-Atlantic Outer Continental Shelf area: U.S. Geological Survey Circular 750, p. 8–14.

—— , 1977b, Geological studies on the COST No. B-2 well, Outer Continental Shelf: U.S. Geological Survey Circular 750, 71 p.

—— , 1979, Data summary and petroleum potential, *in* Scholle, P. A., ed., Geological studies of the COST GE-1 well, United States South Atlantic Outer Continental Shelf area: U.S. Geological Survey Circular 800, p. 18–23.

—— , 1980, Geological Studies of the COST No. B-3 well, United States mid-Atlantic Continental Slope area: U.S. Geological Survey Circular 833, 132 p.

Scholle, P. A., and Wenkam, C. R., 1982, Geological Studies of the COST Nos. G-1 and G-2 wells, United States North Atlantic Outer Continental Shelf: U.S. Geological Survey Circular 861, 193 p.

Simonis, E. K., 1979, Petroleum potential, *in* Amato, R. V., and Simonis, E. K., eds., Geological and operational summary, COST No. B-3 well, Baltimore Canyon Trough area, mid-Atlantic OCS: U.S. Geological Survey Open-File Report 79-1159, p. 100–105.

Smith, M. A., 1979, Geochemical analysis, *in* Amato, R. V., and Simonis, E. K., eds., Geological and operational summary, COST No. B-3 well, Baltimore Canyon Trough area, Mid-Atlantic OCS: U.S. Geological Survey Open-File Report 79-1159, p. 81–99.

Smith, M. A., Amato, R. V., Furbush, M. A., Pert, D. M., Nelson, M. E., Hendrix, J. S., Tamm, L. C., Wood, G., Jr., and Shaw, D. R., 1976, Geological and operational summary, COST No. B-2 well, Baltimore Canyon Trough area, mid-Atlantic OCS: U.S. Geological Survey Open-File Report 76-774, 79 p.

Smith, M. A., Fiorillo, A. R., and Fry, C. E., 1981, Geochemical framework and source-rock evaluation of Mid-Atlantic OCS [abs.]: American Association of Petroleum Geologists Bulletin, v. 65, p. 1671.

Texaco, 1978, Additional potential confirmed in Baltimore Canyon discovery: Denver, News from Texaco Incorporated, 1 p. (press release, August 18, 1978).

Tucholke, B. E., and Mountain, G. S., 1979, Seismic stratigraphy, lithostratigraphy, and paleosedimentation patterns in the North American Basin, *in* Talwani, M., Hay, W., and Rayn, W.B.F., eds., Deep drilling results in the Atlantic Ocean; Continental margins and paleoenvironment: American Geophysical Union Maurice Ewing Series 3, p. 58–86.

Weeks, L. G., 1958, Habitat of oil and some factors that control it, *in* Habitat of oil: American Association of Petroleum Geologists, p. 1–61.

MANUSCRIPT ACCEPTED BY THE SOCIETY APRIL 15, 1986

The Geology of North America
Vol. I-2, The Atlantic Continental Margin: U.S.
The Geological Society of America, 1988

Chapter 23

Hydrogeology of the Atlantic continental margin

F. A. Kohout (deceased)
U.S. Geological Survey, Woods Hole, Massachusetts 02453
Harold Meisler
(U.S. Geological Survey, retired), 32 Winding Way West, Morrisville, Pennsylvania 19067
F. W. Meyer
(U.S. Geological Survey, retired), 6001 SW 70th Ave., Miami, Florida 33143
R. H. Johnston
U.S. Geological Survey, 4311 9th St., East Beach, St. Simons Island, Georgia 31522
G. W. Leve
G. Warren Leve, Inc., 6501 Arlington Expressway, Suite 114, Jacksonville, Florida 32211
R. L. Wait
(U.S. Geological Survey, retired) 3084 Cleethorpe Drive, Lithonia, Georgia 30058

INTRODUCTION

This paper links extensive hydrogeologic investigations on the Atlantic coastal plain of the United States with geologic and oceanographic studies related to oil exploration on the Atlantic continental margin. Changing sea level caused by ice accumulation during the Pleistocene glacial epoch greatly influenced the recharge, storage, and movement of fresh ground water along the entire reach of the Atlantic seaboard. In the south, artesian aquifers of relatively high permeability convey fresh ground water seaward in coastal plain sediments more or less in hydrodynamic equilibrium with the present level of the sea. In the north, low permeability confining beds and rapid overflooding during the last stages of glacial melting have trapped fresh ground water in artesian aquifers underlying the continental shelf. The intent of this synthesis is to integrate research studies on saltwater intrusion and the geographic occurrence of fresh ground water to describe the complex circulation systems that presently exist.

The Atlantic continental margin is defined as extending from the mainland shoreline to the foot of the continental slope. The continental shelf and slope are subdivisions of the continental margin. The coastal plain extends landward from the shore to the fall line, which marks the eastern border of the piedmont province. The continental shelf is defined as extending from the shore seaward to the 200-m isobath. As used in this report, the term coastal plain sediments applies to all sediments overlying crystalline bedrock extending from the piedmont to the foot of the continental slope.

Exploration for fresh ground water on the Atlantic continental margin probably began in 1895 when a test well for water supply was drilled to a depth of 727 m at Key West about 100 km southwest of the Florida mainland. The well tapped Eocene limestone of the Floridan Aquifer System at a depth of about 326 m and produced saltwater with salinity of 30.4 g/kg (about 86 percent sea water). Subsequent test wells in the lower Florida Keys drilled by the Florida East Coast Railroad in the early 1900's also were unsuccessful in locating potable artesian ground water. However, in 1962, a test well by a developer on Key Largo near the Florida mainland produced a large flow of artesian water having a salinity of 4.6 g/kg from a depth of 344 m. The head in terms of the prevailing fluid density was 12 m above sea level, which compares favorably with artesian heads on the mainland and suggests hydraulic continuity of artesian aquifers extending under the continental margin. Similar examples of hydraulic continuity between the mainland and the islands offshore from Georgia, the outer banks of the Carolinas, Maryland, New Jersey, and Long Island, New York, could be drawn, but are not within the scope of this synthesis.

Proof that fresh ground water occurred much farther offshore beneath the Atlantic continental margin was obtained by the U.S. Geological Survey (USGS) during the JOIDES (Joint Oceanographic Deep Earth Sampling) drilling project in 1965. Figure 1 is a photograph taken on the deck of the drilling vessel CALDRILL showing ground water flowing freely from test hole J-1B, located 40 km offshore from Jacksonville, Florida. The relatively fresh ground water flowed vigorously from the drill pipe about 5 m above sea level after penetrating Eocene limestone of the Floridan Aquifer System at a depth of about 250 m below sea

Kohout, F. A., Meisler, H., Meyer, F. W., Johnston, R. H., Leve, G. W., and Wait, R. L., 1988, Hydrogeology of the Atlantic continental margin; *in* Sheridan, R. E., and Grow, J. A., eds., The Geology of North America, Volume I-2, The Atlantic Continental Margin, U.S.: Geological Society of America.

Figure 1. The natural freshwater flow from JOIDES test hole J-1B, 40 km offshore from Jacksonville, Florida, being sampled by G. W. Leve, U.S. Geological Survey. The top of the pipe protrudes through the drilling hole in the center of the Caldrill drilling vessel to about 5 m above sea level. The bottom of the drill pipe was set opposite Eocene limestone about 250 m below sea level at the time of the photograph. Photo by R. L. Wait, U.S. Geological Survey.

level. The chloride content of the water was about 0.7 g/kg. The pressure head at that depth was measured by closing in the well and raising an attached graben hose up the derrick until the flow stopped at about 10 m above sea level (Kohout, 1966; G. W. Leve and R. L. Wait, USGS, oral com., 1965). This probably was the first measurement of freshwater pressure head in a test hole drilled in deep water on the continental margin of the United States that established hydraulic continuity with artesian aquifers on the mainland.

In 1976 and 1977, test wells were drilled through freshwater aquifers to basement rocks on Nantucket Island and Martha's Vineyard Island (respectively, 64 and 32 km off the New England mainland; Fig. 2) as a part of water resources investigations by the USGS. The position of these wells in the offshore area, although technically drilled on the land area of the islands, provided pressure-head measurements for deep aquifers beneath the northeastern segment of the continental margin and will be discussed in detail.

In 1979, Tenneco, Incorporated cooperated with the USGS in hydrologic testing for fresh ground water prior to abandonment of its oil exploratory well about 88 km east of Fernandina Beach, Florida. The work was done by a drill-stem test with packers at about 320 m below sea level after gun-perforating the existing well casing in the Ocala Limestone in the upper part of the Floridan Aquifer System.

The above tests perhaps represent the only measurements of pressure head for freshwater artesian aquifers underlying the Atlantic continental margin. Measurements of this type are sorely

needed for initial and boundary conditions of offshore extensions of aquifers in order that mathematical models may produce reliable simulation of saltwater intrusion into coastal well fields.

In 1976, hydrologic data from 20 test holes (Fig. 2), which were drilled as a part of the USGS AMCOR (Atlantic Margin Coring) Project, for the first time provided an indication of the overall offshore extent of relatively fresh ground water in subsea-floor aquifers. Because the project involved open-hole drilling without blowout preventers, the test holes were limited to a maximum depth of about 310 m below the sea floor. The salinity of interstitial water was determined from pore water that was squeezed from sediment cores with a hydraulic press.

GENERAL HYDROGEOLOGIC FRAMEWORK AND CONSIDERATIONS

In general, sediments of the Atlantic continental margin thicken seaward from coastal plain outcrops on the mainland to the continental slope. Aquifers do not crop out on the shelf. Geologic formations of Jurassic, Triassic, Cretaceous, and Tertiary age, generally called coastal plain sediments, typically dip seaward to form a thick wedge of sediments overlying crystalline basement rocks. The sediments north of Cape Hatteras, North Carolina, are mainly clastic, and those south of the Cape grade into carbonates in Georgia and Florida. Pleistocene sediments that mantle the shelf tend to be fine-grained in the south and have increasing amounts of sand and gravel off the New England coast where glacial outwash influenced deposition.

A zone of diffusion (a gradational transition zone) usually exists between freshwater and seawater in coastal aquifers. Salinity (dissolved solids) in the zone of diffusion ranges from that of fresh ground water (less than 1 g/kg) to that of seawater (about 35 g/kg).

Unconfined (water-table) coastal aquifers are not considered in this report because of their local nature and because the freshwater discharge is generally complete at distances less than 1 km from the shoreline. Artesian (confined) aquifers, on the other hand, are essentially closed hydrogeologic flow systems, confined above and below by relatively low-permeability beds, and can extend many miles offshore beneath the continental shelf. A low stand of sea level during the Pleistocene glacial maximum exposed large areas of the shelf to subaerial erosion and infiltration of rainfall took place over extended periods of time. A rise of sea level started about 18,000 years ago; then, almost cataclysmically, the shelf was overflooded at 100–200 m bpsl (below present sea level) about 8,000 years ago (Emery and Garrison, 1967; Emery and others, 1967; Milliman and Emery, 1968; Dillon and Oldale, 1978). Thus, the physical scale and time factors of seawater encroachment in artesian aquifers underlying the continental margin are more complex than those of unconfined aquifers. Nevertheless, the hydrologic principles of seawater encroachment are similar for both types of aquifers, and details obtained in research studies in the unconfined Biscayne aquifer at Miami, Florida, are presented briefly herein to help explain the occur-

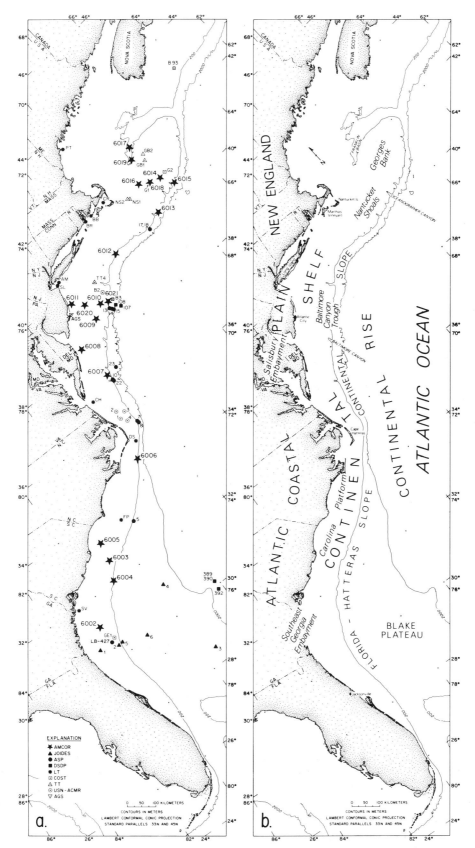

Figure 2. (*a*) Map showing location of drilling sites on the Atlantic continental margin. (*b*) Map showing geographic and geomorphic features.

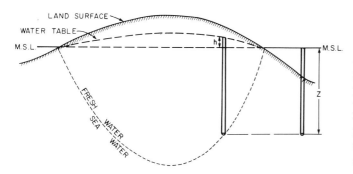

Figure 3. Idealized diagram of an oceanic island showing relation between freshwater and saltwater according to the Ghyben-Herzberg principle.

rence of freshwater in artesian aquifers that underlie the continental shelf from New England to Florida.

The Ghyben-Herzberg Relation

The basic premise of the Ghyben-Herzberg relation is that the position of the freshwater-seawater interface in a coastal aquifer would be governed by hydrostatic equilibrium between the relatively light freshwater and heavier saltwater (Fig. 3). The formulation is obtained from an application of the hydrostatic pressure equations:

$$p = \rho g l$$

where p is the pressure exerted at a point in the aquifer by the overlying column of water, ρ is the density of the water, g is the acceleration due to gravity, and l is the length of water column above the point of interest. Since the pressures at the interface must be equal, the righthand term for the saltwater column z can be equated to the lefthand freshwater column (h + z):

$$\rho_f g (h + z) = g z$$

$$\rho_f h = (\rho_s - \rho_f) z$$

$$z = \frac{\rho_f}{\rho_s - \rho_f} \cdot h$$

where z is the depth to the interface below sea level, h is the freshwater head above sea level, and the subscripts f and s refer to freshwater and saltwater respectively. The densities of freshwater and seawater are commonly assumed to be 1.000 and 1.025 g/cm^3 respectively. This gives rise to the commonly used ratio, z = 40 h, which indicates that for each meter of freshwater head above sea level there will be 40 meters of freshwater in storage below sea level.

The Zone of Diffusion

Other studies (Hubbert, 1940; Cooper and others, 1964; Kohout, 1960a, 1960b; Kohout and Klein, 1967; Henry and Kohout, 1972; Pinder and Cooper, 1970; Segol and others, 1975) have shown that rather than a sharp freshwater-seawater interface, there will be a transition zone (or zone of diffusion) between freshwater and seawater (Fig. 4). Also, it has been recognized that the saltwater in a coastal aquifer is not static as assumed in the Ghyben-Herzberg principle, but that the seawater flows in a cycle inland from the floor of the sea into the zone of diffusion, thence upward and back to the sea. The data presented herein suggest these phenomena are generally operative for artesian aquifers that underlie the shelf. The configuration of the zone of diffusion in the unconfined Biscayne aquifer at Miami, Florida (Fig. 4) is presented to provide background for comparison with Nantucket data, and with data obtained during the AMCOR drilling on the continental margin.

In the Biscayne aquifer, the zone of diffusion is a zone of substantial thickness in which there is a gradation of chloride content from that of freshwater, about 16 ppm (parts per million), to that of seawater (about 19,000 ppm, equivalent to about 35 g/kg salinity, or dissolved solids). The chloride data were obtained from water samples collected from fully cased wells that also provided measurements of pressure head at isolated points in the aquifer (Kohout, 1960a, 1960b). For example, at a distance of 107 m (350 ft) from the shoreline in Figure 4, a drilled well would start in fresh water showing a slow increase of salinity at first, then a rapid increase with depth at 15–18 m (50–60 ft), then a slow increase in the underlying high salinity zone as the chloride content approaches that of seawater (19,000 ppm). Other aquifers—for example, the Floridan aquifer in Florida and the basaltic aquifer underlying Hawaii—have similar vertical salinity profiles (Henry and Kohout, 1972, Fig. 26). Figure 4 provides information for comparison with data from the Nantucket area, the AMCOR drilling project on the continental margin, the New Jersey area, the southeast Georgia area, and southeastern Florida.

RESULTS OF NANTUCKET AND MARTHA'S VINEYARD TEST WELLS

Geologic Setting

Cape Cod and the islands of Nantucket and Martha's Vineyard (Fig. 5) are underlain by a basement complex of crystalline schist, gneiss, and igneous rocks whose surface generally slopes southeastward from about sea level on the New England mainland to as much as 490 m below sea level at Nantucket (Oldale, 1969; Oldale and others, 1974), and to 7.6–9.1 km near the edge of the continental shelf (Sheridan, 1974).

Overlying the basement complex, coastal plain sediments of Cretaceous to Tertiary age form a wedge that thickens seaward from a featheredge along a buried cuesta edge approximately at the position shown in Figure 5 to about 365 m at Nantucket and to more than 2000 m at the edge of the shelf. These sediments are

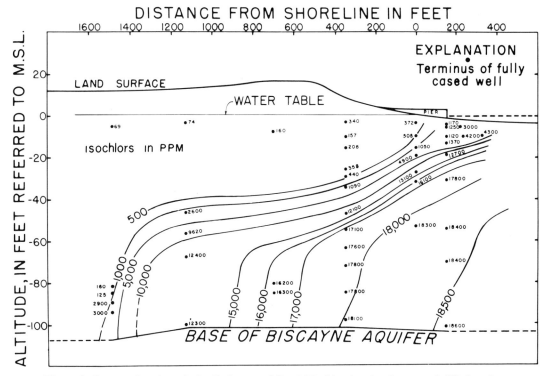

Figure 4. Cross section through the Cutler area, Miami, Florida, showing the zone of diffusion, September 18, 1958. (After Kohout, 1960, Fig. 4).

part of the Atlantic coastal plain sedimentary rock belt extending westward to Long Island, New York, and southward from there to Florida.

During Pleistocene time, glaciers pushed southward across Cape Cod and the islands in three distinct but coalescing lobes (Woodworth and Wigglesworth, 1934; Kaye, 1964; Schafer and Hartshorn, 1965; Oldale, 1969; Oldale and others, 1973). The earlier glacial advances are recorded only in subsurface deposits, but the latest advance (Wisconsin stage glaciation) is recorded in two distinct belts of end moraines and outwash plains forming Cape Cod and the offshore islands (Fig. 5). The north sides of Nantucket and Martha's Vineyard are occupied by end moraines related to a late stillstand of the ice (Oldale and O'Hara, 1984).

Further retreat of the glacier, and a subsequent stillstand of the ice margin, formed the Sandwich moraine at the north side of Cape Cod, while a similar stillstand of the Buzzards Bay lobe formed the end moraine now represented by the Elizabeth Islands (Fig. 5). As the glacier receded, a lake was trapped between the ice mass and the Nantucket-Martha's Vineyard moraines, and a later lake was trapped in Cape Cod Bay behind the Sandwich moraine (Oldale, 1969). These ancestral lakes provided freshwater recharge to the artesian aquifers that underlie the region.

During Pleistocene time, sea level probably fell to about 120 m bpsl (below present sea level) as indicated by recovery of elephant teeth by fishermen and radiocarbon dating of freshwater peat recovered in nets and dredges (Whitmore and others, 1967;

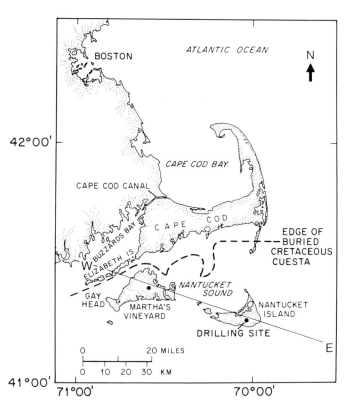

Figure 5. Index map showing drilling sites on Martha's Vineyard and Nantucket Island, and the line of section for Figure 7.

Emery and Garrison, 1967; Emery and others, 1967). Deep submersible expeditions have recently shown geomorphic features on the continental slope that appear to be the result of spring sapping (Robb, 1984). The continental shelf, therefore, would have been exposed during the Pleistocene seaward to near its edge (approximately the 200-m isobath, Fig. 2).

Near the Elizabeth Islands and the landward edge of the cuesta (Fig. 5), Cretaceous sediments are located less than 30 m bpsl (O'Hara and Oldale, 1979). Because of the southeastward dip of the strata, equivalent horizons again reappear downdip in the continental slope 160 km seaward at depths greater than 1000 m (Fig. 2). A wedge of Tertiary sediments overlies the Cretaceous sediments with angular unconformity and serves as a cap over much of the area of the continental shelf.

The above geomorphic-geologic configuration affected the entire continental shelf and is an important consideration in the occurrence of deep freshwater in artesian aquifers, not only at Nantucket and Martha's Vineyard islands, but over the Atlantic continental margin generally.

Data Collection

In 1976, a 514-m test well at Nantucket Island was cored continuously and samples analyzed for mineralogy, texture, paleontology, inorganic and organic chemistry, and porosity and permeability; geophysical logs were run to total depth. These data are discussed in detail by Folger and others, 1978. Subsequent to completion of core drilling, piezometer wells isolated by cement plugs were set at 11, 146, and 436 m, to observe head differences between the several aquifers penetrated (Kohout, Hathaway, and others, 1977).

The salinity of the groundwater was determined by testing interstitial water squeezed from parts of the core not invaded by drilling mud. The technique is well described by Manheim (1966) and Manheim and Sayles (1974) and references therein.

In 1977, a test hole was drilled on Martha's Vineyard Island about 32 km to the west, or toward the shoreline, from Nantucket Island (Delaney, 1980; Kohout and Delaney, 1979). Six piezometers were installed to a depth of 56 m to observe head and collect water samples for chemical analyses. Also, a piezometer was installed just above the base of the aquifer at a depth of 227 m to determine the head at the bottom of the freshwater lens under Martha's Vineyard Island.

The salinity of interstitial water from the Nantucket Island test well is compared with logs of electrical conductivity and lithology in Figure 6. The electrical conductivity log is the reciprocal of the Schlumberger induction resistivity log. The peaks on the conductivity log are reduced in magnitude due to the logarithmic scale used in making the log. However, the comparison shows excellent agreement between electrical conductivity and the salinity of ground water squeezed from the core.

Hydrogeologic Interpretation

Considering first the circumstances of Nantucket, water in the upper part of the test hole to a depth of 158 m has a salinity of less than about 1 ppt (part per thousand). From 158–192 m, the salinity increased irregularly from 0.5 to 29 ppt, compared to the average salinity of ocean water in the vicinity of Nantucket of about 33 ppt (Bigelow and Sears, 1935; Bumpus, 1965).

The rapid increase of salinity in 34 m (from 158 to 192 m) compares closely with observations in the Biscayne aquifer at Miami, Florida (Fig. 4), where the zone of diffusion is 18–31 m thick (Kohout, 1960a; Kohout and Klein, 1967; Henry and Kohout, 1972). The major difference is that at Nantucket the salinity does not increase uniformly from that of freshwater to that of seawater, but is high in sandy zones and low in clay zones (Fig. 6). The difference suggests that, subsequent to the rise in sea level after the last glacial retreat, seawater intruded more rapidly in the permeable sand zone.

The freshwater zone and the zone of diffusion, which comprise the upper 192 m in the well, can be considered to be the Ghyben-Herzberg freshwater lens in the unconfined aquifer beneath Nantucket Island (indicated by the heavy dots in Figure 7).

Below 192 m, the salinity theoretically should have increased to 33 ppt and remained at this concentration at greater depths. Anomalously, the salinity decreased and remained below 0.5 ppt from 222 m to 250 m. This zone is sufficiently permeable to be regarded as an aquifer isolated from the upper unconfined water-table aquifer by tight clay. A coarse sand from 350-457 m also is an aquifer containing water of uniform low salinity (2–3 ppt, Fig. 6).

The occurrence of freshwater in these deeper coastal plain sediments is anomalous because the coastal plain sediments thin to a featheredge westward of Martha's Vineyard Island (Figs. 5 and 7). The offshore aquifers feather out below sea level, or are truncated by the inland edge of a cuesta buried below the sea floor of Nantucket Sound. Thus, there is no evidence of a recharge area in the present-day, above-sea-level part of the New England coastline that could produce sufficient head for hydrostatic or hydrodynamic equilibrium to exist in any artesian aquifer that might underlie the continental shelf. Hydraulic continuity with a freshwater recharge area is lacking and there is no straightforward recharge-discharge explanation for the occurrence of low-salinity water at depths of as much as 460 m below sea level, 64 km offshore, beneath Nantucket Island. The absence of an apparent source of recharge contrasts with the circumstances on Long Island (noted by Collins, 1978) where recharge is known to occur from surficial glacial deposits downward through clay deposits into the Magothy Formation and the Lloyd Sand Member of the Raritan Formation (Lusczynski and Swarzenski, 1962, 1966; Perlmutter and Geraghty, 1963; Perlmutter and others, 1959).

Ground Water Movement Between Nantucket and Martha's Vineyard

Figure 7 is a diagrammatic cross section to facilitate discussion of hypothetical ground water movement in coastal plain sed-

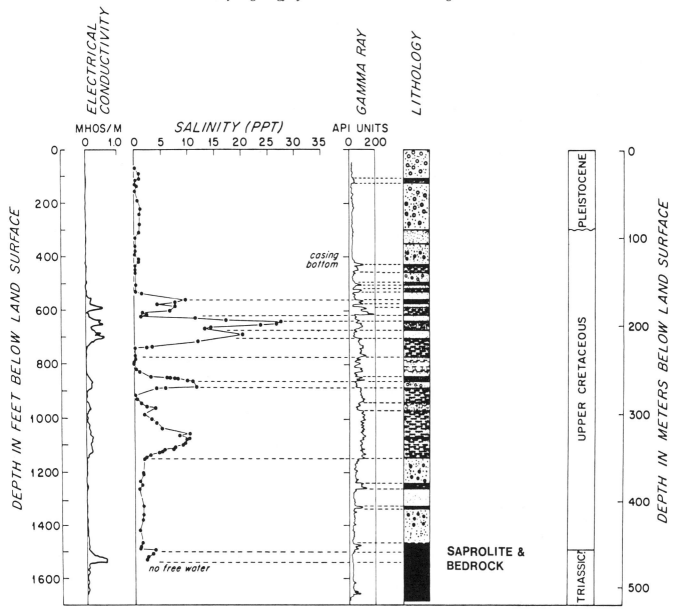

Figure 6. Graph comparing salinity of interstitial water with electrical conductivity and lithology in the Nantucket test well.

iments underlying Nantucket and Martha's Vineyard islands. Three piezometers were installed in the Nantucket test hole to test the following aquifer zones: (1) the shallow water-table aquifer with a piezometer about 11 m deep; (2) the deep part of the freshwater lens with the terminus of a fully-cased piezometer screened at 146 m; and (3) the deepest aquifer above basement rocks, with the terminus of a fully-cased piezometer screened at 436 m. The head in the 146-m piezometer is 12 cm lower than the head in the shallow 11-m well, indicating downward movement of ground water—as is to be expected in this recharge area of the island. The average head of slightly less than 3.6 m in the 146-m piezometer would produce hydrostatic equilibrium to a

depth of only about 146 m bpsl and therefore it is not high enough to account for any of the freshwater occurrences at greater depth in the test hole. In addition, low-permeability clay beds effectively isolate the shallow lens from deeper coastal plain aquifers. Apparently, the permeable sand-and-gravel outwash underlying Nantucket permitted the shallow aquifer to rapidly adjust to flooding of the shelf after the last glacial maximum.

The aquifer tapped by the 436-m piezometer has a head approximately 7 m above msl, some 3.6 m higher than that of the shallow freshwater lens. Theoretically, this head would produce hydrostatic balance for a sharp freshwater-seawater interface in an artesian aquifer at about 299 m below present sea level. The

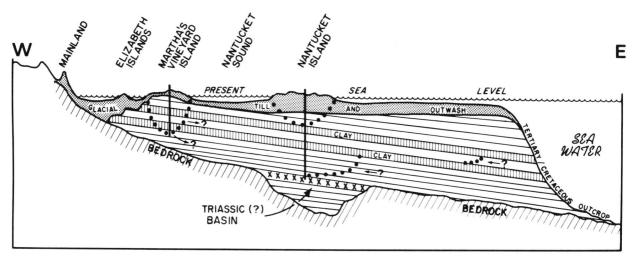

Figure 7. Diagrammatic cross section to facilitate discussion of theoretical hydrologic circumstances involved with occurrence of fresh groundwater in the Nantucket and Martha's Vineyard test wells. (*Heavy dots* indicate possible positions of freshwater-saltwater transition zones.)

existence of relatively fresh ground water (1200 mg/l chloride) in the 436-m piezometer near the base of the aquifer is considered a major departure from theory.

At Martha's Vineyard, the head in the six shallow piezometers installed at various depths to 56 m is about 5.0 m above msl and the head in the deep piezometer at 228 m, just above basement rocks, is about 5.3 m above msl. Shoreward thinning of the coastal plain sediments is such that the aquifer tapped by the 436-m piezometer at Nantucket correlates with the aquifer at 228 m in the Martha's Vineyard well (Hall and others, 1980; Fig. 5). Although hydraulic continuity of the flow system is not proven by these widely spaced piezometers, the head at Nantucket (+7 m msl) is higher than that at Martha's Vineyard (+5 m msl). As the stratigraphic unit has no outcrop equivalent on the mainland, the two heads, +7 m) at Nantucket 64 km offshore and +5 m at Martha's Vineyard 32 km offshore, suggest the existence of a hydraulic gradient associated with an updip landward density circulation. In this concept, seawater would be in transit in the aquifer from outcrop areas deep on the continental slope, driving freshwater ahead of it, toward discharge points in the sea floor of Nantucket Sound, Vineyard Sound, and Buzzards Bay. The concept suggests that a freshwater artesian system developed during a low stand of sea level is now decaying as freshwater is squeezed out by advancing seawater related to the latest sea level transgression. The whole multiaquifer system would be moving slowly toward a condition of lowest potential energy in consonance with the present level of the sea and with sustainable heads in present-day, above-sea-level recharge areas.

Although mathematical models usually assume that equilibrium has already been established, the facts now emerging suggest that many coastal artesian aquifers are still in a transient flow state inherited from the last seawater transgression of the continental shelf that began about 18,000 years ago. The U.S. Geological Survey AMCOR (Atlantic Margin Coring) Project of 1976 has shown that relatively fresh ground water extends as much as 100 km off the New Jersey coast and that the observations at Nantucket are not unique.

RESULTS OF THE AMCOR DRILLING PROJECT

In 1976, the U.S. Geological Survey conducted the Atlantic Margin Coring Project (AMCOR) to obtain information on stratigraphy, mineral resources, hydrology and ground water chemistry, and geotechnical properties of sediments at sites along the continental shelf and slope of the eastern United States, locations shown in Figure 2 (Hathaway and others, 1979). Squeezing techniques (Manheim, 1966; Manheim and Sayles, 1974) were used to collect 175 interstitial water samples from sediment cores after all mud-invaded surfaces had been removed. The salinity values were reported by Hathaway and others (1976).

Figure 8 shows the minimum salinity observed in AMCOR test holes. The presence of freshwater in aquifers underlying the continental shelf was observed as a sharp decrease of salinity through the clay confining bed before the drill penetrated more permeable sand and gravel of an underlying aquifer. The salinity profile for hole 6008 (inset, Fig. 8) is typical of sites where an aquifer containing relatively freshwater underlies and is protected from rapid downward intrusion of seawater by a low-permeability confining bed. Based on the minimum salinity in the observed profile, the value above the line adjacent to the hole location (Fig. 8) is salinity in g/kg; the value below the line is the depth below sea level at which the minimum salinity was observed.

Salinity Transect Off New Jersey

One of the most significant discoveries of the AMCOR test holes was that relatively fresh ground water occurred beneath much of the Atlantic continental margin. Documentation of this low-salinity water was obtained in a transect of five core holes across the shelf east of Barnegat Light, New Jersey. Isochlor lines in Figure 9 show that relatively fresh ground water (5 g/kg) forms a flat-lying lens extending more than 100 km offshore. The estimated position of the lower isochlor lines are based in part on data from onshore observation wells (Gill and others, 1969; Weigle, 1974) and on the distribution of isochlors observed in the transition zones in the Biscayne (water table) aquifer and the Floridan (artesian) Aquifer System of Florida, where extensive studies of saltwater encroachment have been made (Kohout, 1960a, 1960b, 1965, 1967; Cooper and others, 1964; Kohout and Klein, 1967; Henry and Kohout, 1972; Kohout, Henry, and Banks, 1977).

Overlying this lens of relatively low-salinity water, which has a minimum chlorinity of 0.8 g/kg, is an extremely sharp chlorinity gradient increasing toward the chlorinity of seawater in holes 6011, 6020, and 6009 (insets in Fig. 9). This high gradient occurs in the low-permeability clay in the upper part of the Miocene deposits in hole 6011 and in Pleistocene deposits in holes 6020 and 6009. The clay constitutes a confining bed for the underlying permeable beds of the Kirkwood Formation, which were under artesian pressure on the mainland prior to intensive ground water development beginning about 1900. Ground water levels in aquifers underlying the island beaches (equivalent to the "800-foot sand" at Atlantic City) have been drawn down as much as 30 m below sea level by heavy pumping. The integrity of the confining bed is important as it is an impediment to rapid downward infiltration of seawater into the freshwater aquifer of the mainland well fields.

The circumstances off the New Jersey coast also prevail off Ocean City, Maryland. In AMCOR hole 6008 (inset, Fig. 8) chlorinity had decreased sharply during penetration of the offshore extension of the Ocean City-Manokin aquifer (equivalent to the Miocene Kirkwood Formation of New Jersey) when the drill pipe became stuck and the hole lost due to heaving sandy aquifer material.

THEORETICAL EFFECTS OF SEA LEVEL RISE ON FRESHWATER-SEAWATER INTERFACE IN COASTAL PLAIN AQUIFER SYSTEM, NEW JERSEY AREA

As part of a study by the U.S. Geological Survey, entitled "Regional Aquifer System Analysis of the Northern Atlantic Coastal Plain," a finite-difference computer model was used to analyze the effect of Pleistocene sea level changes for the development of the transition zone between fresh ground water and underlying saline water (Meisler and others, 1984).

The model simulates, in cross section, the sedimentary

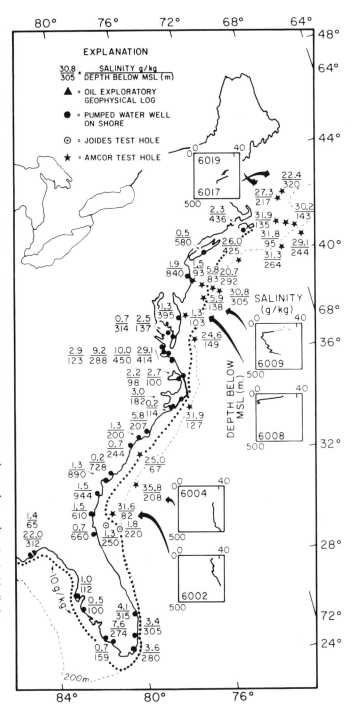

Figure 8. Map showing estimated position of the 10 g/kg isohaline line in ground water underlying the U.S. Atlantic continental shelf. Based on minimum salinity in the observed profile, the value above the line adjacent to the hole location is salinity in g/kg; the value below the line is the depth below mean sea level (msl) at which the minimum salinity was observed. For onshore observation wells, the salinity (above the line) is related to the depth of the well screen (below the line).

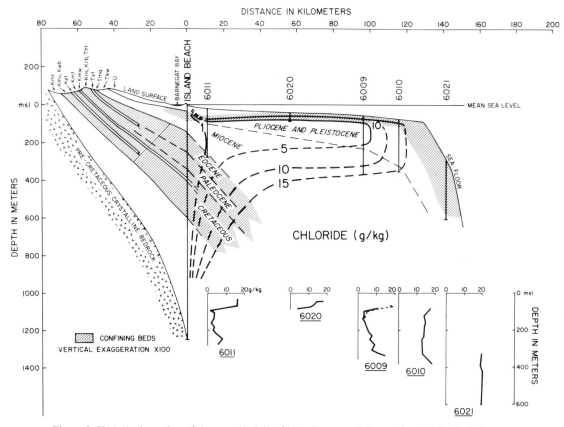

Figure 9. Hydrologic section of the coastal plain of New Jersey and the continental shelf off Barnegat Inlet, New Jersey. Dashed contours show the chlorinity of pore water. Stratigraphic units of the New Jersey coastal plain are: *U,* undifferentiated aquifer; *Tkw,* aquifer in Kirkwood Formation; *Tmq,* Manasquan Formation; *Tvt,* aquifer in Vincentown Formation; *Tht,* Hornerstown Sand; *Krb,* Red Bank Sand; *Kns,* Navesink Formation; *Kmw,* aquifer in Mount Laurel Sand and Wenonah Formation; *Kmt,* Marshalltown Formation; *Ket,* aquifer in Englishtown Formation; *Kwb,* Woodbury Clay; *Kmv,* Merchantville Formation; and *Kmr,* aquifer in Magothy and Raritan formations.

wedge from the Delaware River estuary in New Jersey to the continental slope. The transect (Fig. 10) lies about 30 km south of that shown in Figure 9. Onshore isochlor lines are based on data from water wells, whereas those offshore are extrapolated mainly from the AMCOR data.

Figure 10 compares the position of isochlors in the transition zone with the simulated positions of sharp freshwater-seawater interfaces for eustatic sea levels 0, 15, 30, and 46 m bpsl. Computer solutions capable of modeling the transition zone between freshwater and saltwater (INTERCOMP, 1976; INTERA, 1979) were found to be inapplicable because of the large aspect ratio (i.e., the large horizontal distance compared to the relatively small vertical thickness of the aquifers) involved in the continental shelf sedimentary wedge.

Simulated interface position is sensitive to anisotropy and sea level elevation. Increased anisotropy causes the interface to be shallower and extend farther offshore. Lowered sea level causes the interface to be deeper and to extend farther offshore.

Hydraulic conductivities of the coastal plain aquifers in New

Jersey generally range from less than 3 to about 30 m/day. Vertical hydraulic conductivities of confining beds generally range from 1.5×10^{-3} m/day to 3×10^{-6} m/day (Meisler and others, 1984). The hydraulic conductivities of sediments under the continental shelf and slope are not well known, but the intrinsic lateral permeability of cores from the COST B-2 well located 144 km east of Atlantic City indicates that the sediments become less permeable with depth and with increased distance from the coast (Scholle, 1980). The lateral hydraulic conductivity based on the intrinsic permeability of a sample obtained at a depth of 610 m in COST B-2 is about .76 m/d.

A glacio-eustatic curve based on oxygen isotope data was devised by Zellmer (1979) after Shackleton and Opdyke (1973) for the last 900,000 years (Meisler and others, 1984). The curve indicates that eight large-scale sea level fluctuations of more than 61 m occurred during this period. An additional eight sea level fluctuations had an amplitude of 30 to 60 m.

The most realistic distribution of lateral hydraulic conductivities and a constant anisotropic ratio of 30,000 were used for

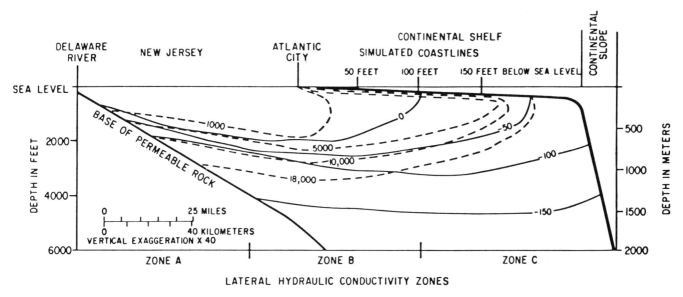

Figure 10. Position of computer-simulated sharp freshwater-seawater interfaces for eustatic sea levels 0, 15, 30, and 46 m (0, 50, 100, and 150 ft) below present sea level compared to observed distribution of saltwater as indicated by dashed lines of chlorinity in milligrams per liter. The simulated lateral hydraulic conductivity (k_l) in zone A, B, and C is 3, 1.5, and 0.8 m (10, 5, and 2.5 ft) per day, respectively. The vertical hydraulic conductivity (K_v) is simulated between 0.10 and 0.025 mm (0.00033 and 0.000083 ft) per day to maintain the anisotropy ratio constant at 30,000.

modeling the interfaces of Figure 10. The observed offshore position of the center of the transition zone (10,000 mg/1 isochlor) coincides approximately with the interface computed for a sea level of 15 m bpsl. Onshore, the interface simulated for a sea level of 30 m bpsl coincides with the 10,000 mg/1 isochlor. These results suggest that the observed freshwater-saltwater transition zone may reflect a sea level that was 15 to possibly 30 m bpsl.

The freshwater-saltwater transition zone is presently moving slowly landward and upward in response to the most recent rise in sea level, starting about 18,000 years ago. Computer estimates were made of the lateral groundwater flow velocity and these suggest that the offshore interface may be moving landward at a rate of about .3 km per 10,000 years.

Cyclic movement of saltwater associated with large-scale eustatic sea level fluctuations caused mixing of saltwater and freshwater to produce the presently observed broad transition zone. Although average sea level for the Quaternary is probably in the range 15 to 30 m bpsl, the calculated lateral flow velocities suggest that the transition zone may reflect sea level conditions for a longer period, possibly extending back to Pliocene time.

The much thinner transition zone, from the sea floor to the underlying fresher water, undoubtedly comes about as a function of (1) protection of the underlying freshwater by relatively low-permeability sediments underlying the sea floor, and (2) the rapid overflooding of the continental shelf at about 100 m bpsl by rising sea level 8,000 years ago.

RESULTS OF THE TENNECO LB-427 DRILL-STEM TEST AND ITS HYDROGEOLOGIC IMPLICATIONS, SOUTHEAST GEORGIA

Construction of water wells specifically for hydrologic testing in offshore areas is generally not feasible because of the prohibitively high cost of leasing offshore drill rigs. An alternative is to take advantage of the presence of a drill rig at a desired location. An opportunity presented itself during the drilling of an exploratory oil-test well offshore from southeast Georgia and northeast Florida, about 88 km east of Fernandina Beach, Florida, on OCS Lease Block 427, reported position 30°34.57' Lat.N and 82°32.05' Long.W. The site was believed to be near the seaward limit of freshwater in the offshore extension of the late Eocene Ocala Limestone, in the upper part of the Floridan Aquifer System. Tenneco permitted the U.S. Geological Survey to conduct hydrologic testing after the well proved to be non-productive of oil and before plugging and abandonment (Johnston and others, 1982).

Wells drilled in the area prior to 1979 were JOIDES J-1, J-2, and J-5, and COST GE-1. The only pressure-head measurement in the area had been made in J-1B, which flowed through the open drill pipe (see Fig. 1 and text). Figure 11 shows the results of the drill-stem test as applied to aquifer evaluation following techniques of Horner (1951) and Bredehoeft (1965).

Based on electric, gamma-ray, and drilling-time logs and

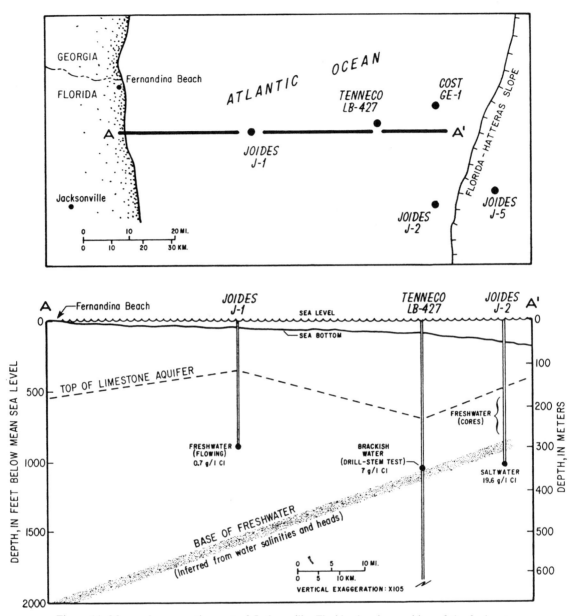

Figure 11. Map and cross section east of Jacksonville, Florida showing position of the freshwater-saltwater transition zone related to the Tenneco LB-427 drill-stem test.

correlation with data from COST GE-1, the casing and grout in the interval 320 to 326 m bpsl were gun-perforated. Packers isolated the section from upper and lower parts of the well. The drill-stem test was only partly successful in that sufficient water could not be produced in the time frame of the test to provide uncontaminated samples of the formation water. Nevertheless, the chloride content of 6–7 g/kg (supported by data of tritium contamination from modern seawater mixed into the presumably dead formation water) indicated good correlation with the 0.7 g/kg chloride in JOIDES J-1B at about the same depth (Manheim and Horn, 1968; Manheim and Paull, 1981). The equivalent freshwater head was estimated at 7 to 9 m above sea level.

Figure 11 shows the inferred position of the freshwater-saltwater transition zone based on chlorinity from JOIDES J-1B (675–1,025 mg/l, Wait and Leve, 1967), J-2 (19,600 mg/l for bottom hole samples, G. W. Leve, written com., 1965), and the Tenneco well (1,000–7,000 mg/l). In summary, the heads and chlorinities in the three wells are compatible with the modern (post-Pleistocene) onshore flow system. Present-day pumping probably has caused small head declines at JOIDES J-1 and the Tenneco sites, and the interface may be transient between positions for predevelopment and present-day heads.

Within 50 to 100 km south of the cross section shown in Figure 11, the Ocala Limestone at the top of the Floridan Aquifer

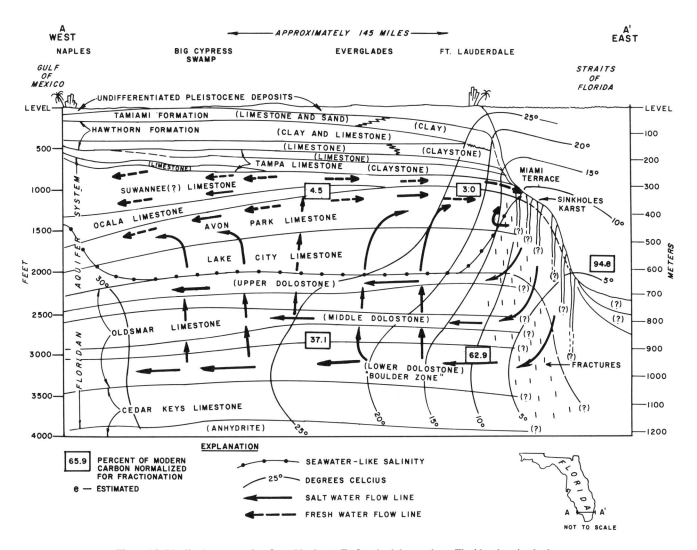

Figure 12. Idealized cross section from Naples to Ft. Lauderdale, southern Florida, showing hydrogeologic circumstances of groundwater flow related to geothermal conditions underlying the Floridan Plateau.

System rises upward to form a regional high (Meisburger and Field, 1976; Popenoe, 1983). This karstic unit, placed in a relatively elevated position under the shelf, has allowed sinkholes to breach the sea floor and form the Crescent Beach submarine spring 4 km off Crescent Beach, Florida (Brooks, 1961; Kohout, 1966; Manheim, 1967), and Red Snapper Sink, a submarine sinkhole about 40 km offshore (Kohout, Leve, and others, 1977; Wilcove, 1975; Popenoe and others, 1984). Breaching of the overlying confining beds by these submarine sinkholes permits ready access of seawater to the underlying aquifer. Saltwater is intruding the Floridan aquifer south of St. Augustine to the Daytona Beach-Cape Canaveral area (Leve, 1968).

HYDROGEOLOGY AND GROUND WATER MOVEMENT IN THE FLORIDAN AQUIFER SYSTEM, SOUTHEAST FLORIDA

Peninsular Florida is the above-sea-level part of a much larger land mass called the Floridan Plateau. The central axis of the Plateau approximates the west coast of Florida. On the west, the Plateau is flanked by abyssal depths of about 3600 m in the Gulf of Mexico—shallowing to less than 1500 m in the southern and eastern Straits of Florida. These cold deep seawater boundaries represent the hydraulic boundaries for the Floridan Aquifer System, which is an extensive group of Tertiary artesian limestone aquifers that supply water to thousands of municipal, industrial, and irrigation wells in Florida and parts of South Carolina, Georgia, and Alabama.

In Florida, recharge of freshwater to the aquifer system occurs chiefly through sinkhole lakes and by direct infiltration through surficial deposits in the karst region of north-central Florida. After moving downward into the aquifer system, the freshwater then moves laterally beneath clay, silt, and marl confining beds toward points of discharge downgradient along the shoreline. The piezometric surface of the aquifer system was first mapped by Stringfield (1936, 1966).

In 1965, Kohout applied the term "Boulder Zone," derived from oil company usage, to identify cavernous limestone and dolomite in the deep saline part of the Floridan Aquifer System (Kohout, 1965, 1967; Kohout, Henry, and Banks, 1977). Since that time, the shallow and deep parts of the Floridan Aquifer System have come under practical use for: (1) injection of industrial and municipal liquid wastes (Vernon, 1970; Garcia-Bengochea and Vernon, 1970; Kaufman and others 1973; Henry and Kohout, 1972); (2) desalination of low-salinity brackish water (Kohout and Sachs, 1969; Kohout, 1970); (3) temporary storage and recovery of freshwater from saline zones (Klein and others, 1975; Merritt and others, 1983); and (4) use of brackish artesian water to promote wildlife survival in Everglades National Park (Meyer, 1971).

Temperature surveys in oil exploratory and waste-injection wells show that the geothermal gradients underlying the Floridan Plateau are affected by the presence of cold seawater in the adjacent deeps of the Gulf of Mexico and the Florida Straits. The geothermal gradient is negative (ground water becomes colder with depth) to depths of about 1220 m near the edge of the plateau (Kohout, Henry, and Banks, 1977, Figs. 10, 11). In the horizontal direction, the temperature increases toward the central axis of the plateau. The horizontal and vertical temperature distributions indicate that cold, dense seawater flows inland through the cavernous dolomite in the deep part of the aquifer system where it becomes progressively warmed and less dense by geothermal heat flow. The decreased density produces upward convective circulation. After mixing with freshwater in the upper part of the aquifer, the diluted saltwater flows seaward to discharge by upward leakage through confining beds or through submarine springs on the continental shelf.

This review of cyclic circulation by geothermal heating and dilution is an extension of earlier studies related to the mechanism of saltwater encroachment in the shallow Biscayne aquifer of southern Florida. In the Biscayne aquifer, the cyclic circulation is caused chiefly by density differences related to hydrodynamic dispersion of salt. Certain aspects provide background and the following reports are recommended to the reader: Cooper, 1959; Cooper and others, 1984; Henry, 1959, 1960; Kohout, 1960a, 1960b; Pinder and Cooper, 1970; Segol and Pinder, 1976; Bear, 1979; Mualem and Bear, 1974. In the thicker Floridan Aquifer System, geothermal heat and dilution are important additional energy sources for the cyclic-flow mechanism (Kohout, 1965,

1967; Kohout and others, 1977b; Henry and Kohout, 1972; Rubin, 1976). Recent studies (Meyer, 1985) suggest rising sea level is another source of energy as stored freshwater in the aquifer system is displaced by inland-moving seawater.

Figure 12 is an idealized hydrogeologic section across southern Florida from Naples to Ft. Lauderdale (Meyer, 1985). The diagram shows the geologic formations that constitute the confining beds (Tamiami, Hawthorn, and, in part, Tampa) and the Floridan Aquifer System (Suwannee, Ocala, Avon Park, Lake City, Oldsmar, and Cedar Keys Limestones). The base of the aquifer system is the low-permeability anhydrite in the Cedar Keys at about 1220 m below sea level. The isotherms in the Florida Straits are adopted from Sverdrup and others (1942). Isotherms underlying the Floridan Plateau are based on temperature measurements in oil-exploratory and waste-injection wells. Also shown in Figure 12 are idealized flow lines for both freshwater (dashed) and saltwater (solid). Carbon-14 activity (in % of modern carbon) in water samples collected during the drilling of waste-injection wells is shown in boxes in proper position in the cross section. Thus, seawater having a carbon-14 activity of about 94% modern carbon in the Florida Straits decreases landward by isotopic decay to about 37% in the deep cavernous boulder zone. An upward-convective-flow component produced by geothermal heat brings the relatively young seawater into contact with relatively old freshwater (because of long transit time from recharge area in central Florida) in the upper part of the aquifer, yielding a blend that is about 4.5% modern carbon. The mixed diluted saltwater (about 3% modern carbon) flows seaward through the upper part of the freshwater-saltwater transition zone to discharge in the Straits of Florida.

Meyer (1985) estimates that municipal and industrial liquid wastes are presently being injected into the boulder zone along the southeast coast of Florida at depths averaging 914 m at a rate of about 100 Mgal/d (million gallons per day; 4,380 l/s); and that oil field brines are presently being injected into the boulder zone in southwest Florida at a rate of about 2 Mgal/d (88 l/s). Currently under consideration are other projects that would use the cold seawater (about 10° C) in the boulder zone for air conditioning and thermally-related power production. The possibility that large-scale injection of warm waste water might rule out the use of the cold water for heat-exchange purposes is a concern. On the other hand, it can be presumed that some of the freshwater being stored might be recovered from the saline aquifer for future use (Kohout, 1970).

The cyclic-circulation mechanism is an important consideration in the practical exploitation of the aquifer system. Interested readers are recommended to previous references and to the references cited therein.

THE 10 g/kg ISOHALINE LINE

The Environmental Protection Agency (EPA) has designated dissolved solids concentration of 10 g/kg (10,000 ppm) as the upper limit for the protection of freshwater aquifers (U.S.

House of Representatives, 1974). Aquifers that contain ground water with higher concentrations of dissolved solids can be used as receptacles of injected domestic and industrial wastes, provided that they do not overlie underground sources of drinking water and they are confined. The 10 g/kg salinity concentration boundary occurs near the middle of the zone of diffusion where maximum salinity gradient takes place. Therefore, the 10 g/kg boundary is hydrologically significant, has quasi-legal status as defined by EPA, and is useful as a criterion for mapping areas in which ground water must be protected.

Figure 8 shows the geographic position of the 10 g/kg isohaline line as determined from minimum ground water salinity observed in test holes drilled through sediments underlying the continental margin.

SUMMARY AND CONCLUSIONS

Exploration for fresh ground water on the Atlantic continental margin began in the late 1800s with the drilling of wells in the Florida Keys and the barrier islands offshore from Georgia, the Carolinas, Delaware, Maryland, New Jersey, and Long Island, New York. Until recently, knowledge of aquifers under most of the continental margin was guesswork based on onshore data.

In 1965, an offshore drilling program for geologic and hydrologic data was initiated by the JOIDES (Joint Oceanographic Institutes Deep Earth Sampling) project. The first test hole, JOIDES J-1B 40 km offshore from Jacksonville, Florida, encountered fresh ground water that flowed above the deck of the drilling ship from a depth of about 300 m below sea level. The finding was not altogether a surprise because Stringfield (1966, p. 161) had calculated there should be enough head remaining at 100 km from shore to cause freshwater to discharge at a depth of 500 m below sea level. This prediction was supported by data from JOIDES J-2 near the edge of the continental shelf. Pore water squeezed from middle Eocene cores decreased in salinity to 1.8 g/kg at 220 m below sea level (Manheim, 1967; Manheim and Horn, 1968). The correlation of low salinity with sediments equivalent to the top of the Floridan Aquifer System on the mainland provides good evidence that relatively fresh ground water is present and that discharge is probably occurring at the edge of the continental shelf 120 km east of Jacksonville.

In 1976, a test hole drilled at Nantucket Island, 64 km offshore from the New England mainland, found fresh groundwater in coastal plain sediments to a depth of 500 m below sea level. Because the coastal plain sediments thin and pinch out toward shore, the New England mainland lacks the recharge potential and hydraulic connection needed to account for the deep freshwater occurrence at Nantucket. During low stand of sea level in the Pleistocene glacial epoch, ice-marginal lakes and streams provided freshwater recharge to aquifers underlying this part of the continental margin. About 8000 years ago, rising sea level almost cataclysmically overflooded the continental shelf at 100–200 m pbsl. Freshwater in the underlying aquifers was effectively trapped and has not had time to be displaced by seawater

advancing inland from outcrops on the continental slope. This contrasts with the situation in the Floridan Aquifer System where the high permeability of the limestone already may have permitted hydrodynamic adjustment to the present level of the sea.

In 1976, as part of investigation of hydrocarbon potential and environmental hazards of oil exploration under the continental margin, the U.S. Geological Survey AMCOR project drilled 20 core holes from offshore Massachusetts to Georgia. A line of five test holes on a transect across the shelf east of the New Jersey coast permitted construction of a cross section through the lens of fresh ground water and the transition zone to seawater. The cross section showed that low-chlorinity water (less than 2 g/kg) extended about 16 km offshore; and that water having a chlorinity of less than 5 g/kg formed a flat-lying lens extending more than 100 km offshore (Fig. 9).

An implication of these recent findings is that saltwater intrusion would have occurred long before now if it had not been for the existence of the offshore freshwater lens—the extent of which was not even guessed in early studies that expressed concern about saline encroachment (Barksdale and others, 1936).

In 1979, Tenneco, Incorporated permitted the U.S. Geological Survey to conduct hydrologic testing after its exploratory well LB-427, 88 km east of Fernandina Beach, Florida, proved to be non-productive of oil and before plugging and abandonment. The site was believed to be near the seaward limit of freshwater in the offshore extension of the late Eocene Ocala Limestone in the upper part of the Floridan Aquifer System. The chloride content of water produced in the drill-stem test (6–7 g/kg) indicated the well was located in the transition zone between freshwater and seawater. Pressure-head and chlorinity measurements provided good correlation with data from other test holes in the vicinity (JOIDES J-1B and J-2).

In this paper, salinity data from recently drilled offshore core holes have been supplemented by data from selected onshore observation wells to construct a map showing distribution of ground water salinity under the U.S. Atlantic continental margin. The EPA criterion of 10 g/kg salinity for protection of freshwater aquifers has been adopted to indicate areas of major water-resources interest. The geographic extent of this interest is indicated by the position of the 10 g/kg isohaline line at distances of 50 to 100 km offshore in Figure 8.

Mathematical models of saltwater encroachment require definition of the initial and boundary conditions of aquifers in order to produce reliable predictions of the intrusion progress. Our present knowledge of the offshore extent of aquifers, and the properties of their overlying confining beds can be appropriately described as primitive. The primary purpose of the AMCOR drilling program was calibration of seismic profiles related to offshore oil and gas production. The data obtained on water-resources aspects were, for the most part, obtained through serendipity. Considering that some coastal well fields have drawdowns well below sea level, the potential for saltwater encroachment is great. Offshore drilling programs designed specifically for water-resources objectives are needed to appraise this threat.

REFERENCES CITED

Barksdale, H. C., Sundstrom, R. W., and Brunstein, M. S., 1936, Supplementary report on the groundwater supplies of the Atlantic City region: State of New Jersey Department of Conservation and Economic Development, Special Report 6, 139 p.

Bear, Jacob, 1979, Hydraulics of ground water: Englewood Cliffs, New Jersey, Prentice-Hall, 604 p.

Bigelow, H. B., and Sears, M., 1935, Studies of the waters on the continental shelf, Cape Cod to Chesapeake Bay, II. Salinity: Papers on Physical Oceanography and Meteorology, v. 4, no. 1, p. 94.

Bredehoeft, J. D., 1965, The drill-stem test: The petroleum industry's deep-well pumping test: Ground Water, v. 3, no. 3, p. 31–36.

Brooks, H. K., 1961, The submarine spring off Crescent Beach, Florida: Quarterly Journal, Florida Academy of Science, v. 24, p. 122–134.

Brown, John S., 1925, A study of coastal ground water, with special reference to Connecticut: U.S. Geological Survey Water Supply Paper 537, p. 101.

Brown, P. M., Miller, J. A., and Swain, F. M., 1972, Structural and stratigraphic framework, and spatial distribution of permeability of the Atlantic Coastal Plain, North Carolina to New York: U.S. Geological Survey Professional Paper 796, 79 p.

Bumpus, D. F., 1965, Residual drift along the bottom on the continental shelf in the Middle Atlantic Bight Area: Limnology and Oceanography, v. 10, Supplement, p. R50–R53.

Collins, M. A., 1978, Comment on "Fresh ground water stored in aquifers under the continental shelf: Implications from a deep test, Nantucket Island, Massachusetts," by Kohout and others, 1977: American Water Resources Bulletin, v. 14, no. 2, p. 484–485.

Cooper, H. H., Jr., 1959, A hypothesis concerning the dynamic balance of fresh water and salt water in a coastal aquifer: Journal of Geophysical Research, v. 64, no. 4, p. 461–467.

Cooper, H. H., Jr., Kohout, F. A., Glover, R. E., and Henry, H. R., 1964, Sea water in coastal aquifers: U.S. Geological Survey Water Supply Paper 1613-C, 84 p.

Delaney, David F., 1980, Ground-water hydrology of Martha's Vineyard, Massachusetts: U.S. Geological Survey Hydrologic Atlas HA-618, 2 sheets.

Dillon, W. P., and Oldale, R. N., 1978, Late Quaternary sea-level curve: Reinterpretation based on glaciotectonic influence: Geology, v. 6, no. 1, p. 56–60.

Emery, K. O., and Garrison, L. E., 1967, Sea levels 7000 to 20,000 years ago: Science, v. 157, no. 3789, p. 684–687.

Emery, K. O., Wigley, R. L., Bartlett, A. S., Rubin, Meyer, and Barghoorn, E. S., 1967, Freshwater peat on the Continental Shelf: Science, v. 158, no. 3806, p. 1301–1307.

Folger, D. W., Hathaway, J. C., Christopher, R. A., Valentine, P. C., and Poag, C. W., 1978, Stratigraphic Test Well, Nantucket Island, Massachusetts: U.S. Geological Survey Circular 773, 28 p.

Garcia-Bengochea, J. I., and Vernon, R. O., 1970, Deep-well disposal of wastewaters in saline aquifers of south Florida, in Saline Water Symposium: Water Resources Research, v. 6, no. 5, p. 1464–1470.

Gill, H. E., Seaber, P. R., Vecchioli, John, and Anderson, H. R., 1969, Evaluation of geologic and hydrologic data from the test-drilling program at Island Beach State Park, New Jersey: New Jersey Department of Conservation and Economic Development, Division of Water Policy and Supply, Water Resources Circular 12, 25 p.

Hall, R. E., Poppe, L. J., and Ferrebee, W. M., 1980, A stratigraphic test well, Martha's Vineyard, Massachusetts: U.S. Geological Survey Bulletin 1488, 19 p.

Hathaway, J. C., Schlee, J. S., Poag, C. W., Valentine, P. C., Weed, E.G.A., Bothner, M. H., Kohout, F. A., Manheim, F. T., Schoen, R., Miller, R. E., and Schultz, D. M., 1976, Preliminary summary of the 1976 Atlantic Margin Coring Project of the U.S. Geological Survey: U.S. Geological Survey Open-File Report No. 76-844, 217 p.

Hathaway, J. C., Poag, C. W., Valentine, P. C., Miller, R. E., Schultz, D. M.,

Manheim, F. T., Kohout, F. A., Bothner, M. H., and Sangrey, D. A., 1979, The U.S. Geological Survey core drilling on the U.S. Atlantic Shelf: Science, v. 206, no. 4418, p. 515–527.

Henry, H. R., 1959, Salt intrusion into fresh-water aquifers: Journal of Geophysical Research, v. 64, no. 11, p. 1111–1119.

—— 1960, Salt intrusion into coastal aquifers: International Association of Scientific Hydrology, Commission of Subterranean Waters, v. 52, p. 478–487.

Henry, H. R., and Kohout, F. A., 1972, Circulation patterns of saline ground water affected by geothermal heating—As related to waste disposal: American Association of Petroleum Geologists Memoir 18, p. 202–221.

Horner, D. R., 1951, Pressure build-up in wells, in Proceedings, Third World Petroleum Congress, Section II: Leiden, Holland, E. J. Brill, p. 503–521.

Hubbert, M. K., 1940, The theory of ground-water motion: Journal of Geology, v. 48, no. 8, pt. 1, p. 785–944.

INTERA Environmental Consultants, Inc., 1979, Revision of the documentation for a model for calculating effects of liquid waste disposal in deep saline aquifers: U.S. Geological Survey Water Resources Investigations 79-96, 73 p.

INTERCOMP Resource Development and Engineering, Inc., 1976, A model for calculating effects of liquid waste disposal in deep saline aquifers: U.S. Geological Survey Water Resources Investigations 76-61, 253 p.

Johnston, R. H., Bush, P. W., Krause, R. E., Miller, J. A., and Sprinkle, C. L., 1982, Summary of hydrologic testing in Tertiary limestone aquifer, Tenneco offshore exploratory well—Atlantic OCS, Lease Block 427 (Jacksonville NH 17-5): U.S. Geological Survey Water Supply Paper 2180, 15 p.

Kaufman, M. I., 1973, Subsurface wastewater injection, Florida: American Society of Civil Engineers Proceedings Paper 9598, Journal of Irrigation and Drainage Division, v. 99, no. IR 1, p. 53–70.

Kaufman, M. I., Goolsby, D. A., and Faulkner, G. L., 1973, Injection of acidic industrial waste into a saline carbonate aquifer: Geochemical aspects: American Association of Petroleum Geologists, Symposium on Under Ground Waste Management and Artificial Recharge, v. 1, p. 526–551.

Kaye, C. A., 1964, Outline of Pleistocene Geology of Martha's Vineyard, Massachusetts: U.S. Geological Survey Professional Paper 501-C, p. C134–C139.

Klein, Howard, Armbruster, J. T., McPherson, B. F., and Freiberger, H. J., 1975, Water and the south Florida environment: U.S. Geological Survey Water Resources Investigation 24-75, 165 p.

Kohout, F. A., 1960, Cyclic flow of salt water in the Biscayne aquifer of southeastern Florida: Journal of Geophysical Research, v. 65, no. 7, p. 2133–2141.

—— 1960b, Flow pattern of fresh water and salt water in the Biscayne aquifer of the Miami area, Florida: International Association of Scientific Hydrology, Committee on Subterranean Waters, v. 42, p. 440–448.

—— 1965, A hypothesis concerning cyclic flow of salt water related to geothermal heating in the Floridan aquifer: New York Academy of Science Transactions, v. 28, no. 2, p. 249–271.

—— 1966, Submarine springs, in Encyclopedia of Earth Sciences Series: New York, Reinhold Publishing Corporation, v. 1, p. 878–883.

—— 1967, Ground-water flow and the geothermal regime of the Floridan Plateau, in Symposium on geological history of the Gulf of Mexico Caribbean Antillean Basin: Transactions, Gulf Coast Association of Geological Societies, v. 17, p. 339–354.

—— 1970, Reorientation of our saline water resources thinking, in Symposium on Saline Water—A Valuable Resource: Water Resources Research, v. 6, no. 5, p. 1442–1448.

—— 1979, Relict fresh ground water of the U.S. Atlantic continental shelf: An unevaluated buffer in present-day saltwater encroachment, in Proceedings, Second Symposium on the Geology of the Southeastern Coastal Plain: Georgia Geological Survey Information Circular No. 53, p. 170–176.

Kohout, F. A., and Delaney, D. F., 1979, Reply to discussion by Michael A. Collins on "Fresh Ground Water Stored in Aquifers Under the

Continental Shelf: Implications from a Deep Test, Nantucket Island, Massachusetts," by Kohout et al., 1977: American Water Resources Bulletin, v. 15, no. 1, p. 252–254.

Kohout, F. A., and Klein, H., 1967, Effect of pulse recharge on the zone of diffusion in the Biscayne aquifer: International Association of Scientific Hydrology, Publication No. 72, Symposium of Haifa, p. 252–270.

Kohout, F. A., and Sachs, M. S., 1969, Modifying the water balance by desalination: American Water Resources Association, Symposium Series No. 7, p. 214–222.

Kohout, F. A., Hathaway, J. C., Folger, D. W., Bothner, M. H., Walker, E. H., Delaney, D. F., Frimpter, M. H., Weed, E.G.A., and Rhodehamel, E. C., 1977a, Fresh ground water stored in aquifers under the continental shelf: Implications from a deep test, Nantucket Island, Massachusetts: American Water Resources Association Bulletin, v. 13, no. 2, p. 373–386.

Kohout, F. A., Henry, H. R., and Banks, J. E., 1977b, Hydrogeology related to geothermal conditions of the Floridan Plateau, *in* Smith, D. L., and Griffin, G. M., eds., The Geothermal Nature of the Floridan Plateau: Florida Bureau of Geology, Special Publication No. 21, p.1–42.

Kohout, F. A., Leve, G. W., Smith, F. T., and Manheim, F. T., 1977c, Red Snapper Sink and ground water flow offshore southeastern Florida, *in* Proceedings, 12th International Congress, Karst Hydrogeology: International Association of Hydrogeologists Memoirs, v. XII, p. 193.

Kohout, F. A., Munson, R. C., Turner, R. M., and Royal, W. R., 1981, Satellite observations of a geothermal submarine spring off Florida west coast, *in* 5th Wm. T. Pecora Memorial Symposium, Satellite Hydrology: American Water Resources Association, Technical Publication Series TPS 81-1, p. 570–578.

Leve, G W., 1968, The Floridan Aquifer in northeast Florida: Ground Water (NWWA), v. 6, no. 2, p. 19–29.

Lusczynski, N. J., and Swarzenski, W. V., 1962, Fresh and salty ground water in Long Island, New York: Proceedings, American Society of Civil Engineers, Journal of the Hydraulics Division 38(HY4), p. 173–194.

——1966, Salt water encroachment in Southern Nassau and Southeastern Queens Counties, Long Island, New York: U.S. Geological Survey Water Supply Paper 1613-F, 76 p.

Manheim, F. T., 1966, A hydraulic squeezer for obtaining interstitial water from consolidated and unconsolidated sediments: U.S. Geological Survey Professional Paper 550-C, p. C256–C261.

——1967, Evidence for submarine discharge of water on the Atlantic continental slope of the United States, and suggestions for further search: New York Academy of Science, Transactions, Series 2, v. 29, no. 5, p. 839–852.

Manheim, F. T., and Hall, R. E., 1976, Deep evaporitic strata off New York and New Jersey—Evidence from interstitial water chemistry of drill cores: U.S. Geological Survey Journal of Research, v. 4, no. 4, p. 697–702.

Manheim, F. T., and Horn, M. K., 1968, Composition of deeper subsurface waters along the Atlantic continental margin: Southeastern Geology, v. 9, no. 4, p. 215–236.

Manheim, F. T., and Paull, C. K., 1981, Patterns of groundwater salinity changes in a deep continental-oceanic transect off the southeastern Atlantic coast of the U.S.A.: Journal of Hydrology, v. 54, p. 95–105.

Manheim, F. T., and Sayles, F. L., 1974, Composition and origin of interstitial waters of marine sediments, based on deep sea drill cores, *in* Goldberg, E. D., ed., The Sea, v. 5: J. Wiley & Sons, p. 527–568.

Meisburger, E. P., and Field, M. E., 1976, Neogene sediments of the Atlantic inner continental shelf off northern Florida: American Association of Petroleum Geologists Bulletin, v. 60, no. 11, p. 2019–2037.

Meisler, Harold, Leahy, P. P., and Knobel, L. L., 1984, The effect of Pleistocene sea-level changes on saltwater-freshwater relations in the northern Atlantic Coastal Plain: U.S. Geological Survey Water Supply Paper 2255, 28 p.

Merritt, M. L., Meyer, F. W., Sonntag, W. H., and Fitzpatrick, D. J., 1983, Subsurface storage of freshwater in south Florida: A prospectus: U.S. Geological Survey Water Resources Investigation Report 83-4214, 69 p.

Meyer, F. W., 1971, Saline artesian water as a supplement: Journal of American Water Works Association, v. 63, no. 2, p. 65–71.

——1974, Evaluation of hydraulic characteristics of a deep artesian aquifer from natural water-level fluctuations: Miami, Florida, Florida Bureau of Geology Report of Investigations No. 75, 32 p.

——1979, Disposal of liquid wastes in cavernous dolostones beneath southeastern Florida: U.S. Geological Survey Open-File Report, 7 p.

——1985, Ground-water movement in the Floridan aquifer system, south Florida: U.S. Geological Survey Professional Paper 1403-G (in press).

Milliman, J. D., and Emery, K. O., 1968, Sea levels during the past 35,000 years: Science, v. 162, no. 3858, p. 1121–1123.

Mualem, Y., and Bear, J., 1974, The shape of the interface in steady flow in a stratified aquifer: Water Resources Research, v. 10, no. 6, p. 1207–1215.

O'Hara, C. J., and Oldale, R. N., 1979, United States Geological Survey research in Massachusetts, Nov. 27–28, 1978: Woods Hole Oceanographic Institution Technical Report 79-40, p. 100–108.

Oldale, R. N., 1969, Seismic investigations on Cape Cod, Martha's Vineyard, and Nantucket, Massachusetts, and a topographic map of the basement surface from Cape Cod Bay to the islands: U.S. Geological Survey Professional Paper 650-B, p. B122–B127.

Oldale, R. N., and O'Hara, C. J., 1984, Glaciotectonic origin of the Massachusetts coastal end moraines and a fluctuating late Wisconsin ice margin: Geological Society of America Bulletin, v. 95, p. 61–74.

Oldale, R. N., Uchupi, Elazar, and Prada, K. E., 1973, Sedimentary framework of the western Gulf of Maine and the southeastern Massachusetts offshore area: U.S. Geological Survey Professional Paper 757, 10 p.

Oldale, R. N., Hathaway, J. C., Dillon, W. P., Hendricks, J. D., and Robb, J. M., 1974, Geophysical observations on northern part of Georges Bank and adjacent basins of Gulf of Maine: American Association of Petroleum Geologists Bulletin, v. 58, no. 12, p. 2411–2427.

Perlmutter, N. M., and Geraghty, J. J., 1963, Geology and ground water conditions in southern Nassau and southeastern Queens Counties, Long Island, New York: U.S. Geological Survey Water Supply Paper 1613-A, 205 p.

Perlmutter, N. M., Geraghty, J. J., and Upson, J. E., 1959, The relation between fresh and salty ground water in southern Nassau and southeastern Queens Counties, Long Island, N.Y.: Economic Geology, v. 54, p. 416–435.

Pinder, George F., and Cooper, Hilton H., Jr., 1970, A numerical technique for calculating the transient position of the saltwater front: Water Resources Research, v. 6, no. 3, p. 875–882.

Popenoe, Peter, 1983, Environmental considerations for OCS development, lease sale 90 call area, *in* W. P. Dillon, ed., Geology report for proposed oil and gas lease sale No. 90, Continental margin off the Southeastern United States: U.S. Geological Survey Open-File Report 83-186, p. 67–106.

Popenoe, Peter, Kohout, F. A., and Manheim, F. T., 1984, Seismic reflection studies of sinkholes and limestone dissolution features on the northeastern Florida shelf, *in* Beck, B. F., ed., Sinkholes: Their geology, engineering and environmental impact; Boston, A. Balkema, p. 43–57.

Robb, J. M., 1984, Spring sapping on the lower continental slope, offshore New Jersey: Geology, v. 12, no. 5, p. 278–282.

Rubin, Hillel, 1976, Onset of thermohaline convection in a cavernous aquifer: Water Resources Research, v. 12, no. 2, p. 141–147.

Schafer, J. P., and Hartshorn, J. H., 1965, The Quaternary of New England, *in* Wright, H. E., Jr., and Frey, D. G., eds., The Quaternary of the United States: Princeton, New Jersey, Princeton University Press, p. 113–128.

Scholle, P. A., 1980, Cost No. B-2 Well, *in* Mattick, R. E., and Hennessy, J. L., eds., Structural framework, stratigraphy, and petroleum geology of the area of oil and gas lease sale No. 49 on the U.S. Atlantic continental shelf and slope: U.S. Geological Survey Circular 812, p. 79–84.

Segol, Genevieve, and Pinder, G. F., 1976, Transient simulation of saltwater intrusion in southeastern Florida: Water Resources Research, v. 12, no. 1, p. 65–70.

Segol, Genevieve, Pinder, G. F., and Gray, W. G., 1975, A Galerkin finite-element technique for calculating the transient position of the saltwater front: Water Resources Research, v. 11, no. 2, p. 343–347.

Shackleton, N. J., and Opdyke, N. D., 1973, Oxygen isotope and palaeomagnetic stratigraphy of equatorial Pacific Core V28-238: Oxygen isotope temperatures and ice volumes on a 105-year and 106-year scale: Quaternary Research, v. 3, p.39–55.

Sheridan, Robert E., 1974, Atlantic continental margin of North America, *in* Burke, C. A., and Drake, C. L., eds., The geology of continental margins: New York, Springer-Verlag, p. 391–407.

Stringfield, V. T., 1936, Artesian water in the Florida Peninsula: U.S. Geological Survey Water Supply Paper 773-C, p. 115–195.

——1966, Artesian water in Tertiary limestone in the southeastern states: U.S. Geological Survey Professional Paper 517, 226 p.

Stringfield, V. T., and Cooper, H. H., Jr., 1951, Geologic and hydrologic features of an artesian submarine spring east of Florida: Florida Geological Survey Report of Investigations, no. 7, Part 2, p. 61–82.

Sverdrup, H. U., Johnson, M. W., and Fleming, R. H., 1942, The oceans: Englewood Cliffs, New Jersey, Prentice-Hall, Inc., 1081 p.

U.S. House of Representatives, 1974, Endangerment of drinking water sources: House Report 93-1185, p. 32.

Vernon, R. O., 1970, The beneficial uses of zones of high transmissivities in the Florida subsurface for water storage and waste disposal: Florida Bureau of Geology Information Circular No. 70, 39 p.

Wait, R. L., and Leve, G. W., 1967, Ground water from JOIDES core hole J-1, *in* Geological Survey research 1967: U.S. Geological Survey Professional Paper 575-A, p. A127.

Weigle, J. M., 1974, Availability of fresh ground water in northeastern Worcester County, Maryland: Maryland Geological Survey Report of Investigations, no. 24, 64 p.

Whitmore, Frank C., Jr., Emery, K. O., Cooke, H.B.S., and Swift, D.J.P., 1967, Elephant teeth from the Atlantic continental shelf: Science, v. 156, no. 3781, p. 1477–1481.

Wilcove, Raymond, 1975, The great Red Snapper Sink: NOAA Magazine, v. 5, no. 2, p. 46–47.

Woodworth, J. B., and Wigglesworth, E., 1934, Geography and geology of the region including Cape Cod, the Elizabeth Islands, Nantucket, Martha's Vineyard, No Mans Land, and Block Island: Harvard University Museum Comparative Zoology Memoir 52, p. 322.

Zellmer, L. R., 1979, Development and application of a Pleistocene sea-level curve to the coastal plain of southeastern Virginia [Master's thesis]: Williamsburg, Virginia, School of Marine Science, College of William and Mary, 85 p.

Manuscript Accepted by the Society March 29, 1985

Chapter 24

Sand and gravel resources: U.S. Atlantic continental shelf

David B. Duane
U.S. Department of Commerce, NOAA, 6010 Executive Boulevard, Rockville, Maryland 20852
William L. Stubblefield
U.S. Department of Commerce, NOAA, 6010 Executive Boulevard, Rockville, Maryland 20852

INTRODUCTION

Background

Sand and gravel epitomize bulk commodity; that is, they are consumed in large volume, are relatively plentiful with low unit cost, and transportation costs from source to market are the chief limiting factors when considering alternate supplies. For the United States Atlantic continental margin, sand and gravel deposits are associated with Pleistocene to Recent fluvial or ice contact processes modified by marine processes. Prospective sources represent a continuum from the limit of the fall line, across the present coastline, to the shelf edge near the limit of the Pleistocene low stand shoreline.

With a few localized exceptions in the New York City vicinity, terrestrial sources have traditionally supplied the market. Along the heavily urbanized sectors—which constitute much of the coastal area from Miami, Florida, north to Boston, Massachusetts—dwindling resources on land, coupled with increasing urbanization and terrestrial environment preservation, have stimulated consideration of marine deposits of the continental shelf as replacement resources for terrestrial supplies. Projecting a near-future time when, for one of several reasons, terrestrial sources are no longer available to this coastal sector, the focus of this review is on marine sand and gravel.

Sand and gravel, but especially the former, are a large volume inferred resource of the U.S. Atlantic continental shelf. The inferred resources from many studies (e.g., Duane, 1968, 1969; Schlee and Pratt, 1970; Schlee, 1973; Schlee and Sanko, 1975; Milliman and others, 1972) are on the order of $5.3 \times 10^{11} \text{m}^3$, with indicated reserves being on the order of $1.5 \times 10^{10} \text{m}^3$ (Tables 1 and 2).

By *resource* we mean identified mineral deposits that may eventually become available (Fig. 1); a *reserve* consists of resources that can be extracted profitably with existing technology and under present economic conditions. The discussion of *inferred resources* covers the area from the present shoreline seaward to the edge of the shelf; for *indicated reserves* the area is the present shoreline seaward to a water depth on the order of 40 meters, the approximate limit for commercial dredges and the surveys conducted by the Corps of Engineers from 1964–1981 (e.g., Duane, 1968).

Uses and Market

Requirements for large volumes of aggregate exist, on the order of $8.1 \times 10^8 \text{m}^3$ for the U.S. as a whole by the year 2000 (Tepordei, 1980). Construction in the Washington, D.C., to Boston corridor consumes much aggregate, and a projected rebuilding of major segments of the interstate highway system could use large quantities, as much as 3 to $6 \times 10^4 \text{m}^3$ per running mile. The continental shelves are also spoken of as a place where huge manmade structures such as deep draft port facilities (Kelly, 1973; U.S. Army Corps of Engineers, 1973), airports (Lerner, 1971), and nuclear generating stations (Nutant, 1973) would be sited. These facilities would require large volumes of aggregate. Protection of coastal areas through beach reconstruction or periodic nourishment can also require large volumes of sand. The Corps of Engineers identified more than 3,200 km (2,000 mi) of the Atlantic and Gulf coasts as areas of critical erosion requiring remedial action (U.S. Army Corps of Engineers, 1971). Much of the remedial action will involve large volumes of sand for beach fill and nourishment.

In addition to the sand or gravel itself, a future market may exist for using dredged holes as a repository for the waste products of society. Research is being conducted to assess the utility of such a method and is initially focused on confining material dredged during maintenance of navigable waterways (MITRE Corporation, 1979; Schubel, 1981). How such a use would impact the economics of the resource is not definitely known, but it may be beneficial (Guyette and Wallace, 1979). Indeed, forces other than sand and gravel markets may dictate such additional uses.

Exploration and Recovery Systems

The principal exploration tool for marine sand and gravel is continuous seismic reflection profiling (CSRP) emitting sound

Duane, D. B., and Stubblefield, W. L., 1988, Sand and gravel resources: U.S. Atlantic continental shelf; *in* Sheridan, R. E., and Grow, J. A., eds., The Geology of North America, Volume I-2, The Atlantic Continental Margin, U.S.: Geological Society of America.

TABLE 1. INFERRED SAND AND GRAVEL RESOURCES FOR SPECIFIC AREAS ON THE U.S. ATLANTIC CONTINENTAL SHELF

Site	Area (km^2)	Estimate of Thickness (m)	Volume (m^3)	Citation
<u>North Atlantic Province</u>		Total:	3.4×10^{11}	
Mass. Bay (Gravel/Sand) Western Mass Bay			2.1×10^7	1
Northern N.J. to Western Scotian Shelf (sand)	112×10^3	3	3.36×10^{11}	2
<u>Mid Atlantic Province</u>		Total:	1.9×10^{11}	
New York Shelf	---	----	6.0×10^9	3
Shrewsburg Rocks N.J. to Rockaway, N.Y.	2.0×10^2	30	6.0×10^9	4
Hudson Shelf Valley Inshore (sand?)	4.5×10^2 9.0×10^2	22 ----	3.3×10^9 6.6×10^9	5
Hudson Shelf Valley Offshore (sand?)	2.5×10^2 7.5×10^2	5-10 ----	1.2×10^9 7.5×10^9	5
Ancestral Hudson Valley (sand?)	1.6×10^2 1.4×10^3	10-20 ----	1.6×10^9 4.0×10^{10}	6
Shoal at mouth of Ancestral Hudson Shelf Valley	6.0×10^2	15	9.0×10^9	6
Nearshore New Jersey	----	----	5.4×10^9	7
Nearshore New Jersey	----	----	2.3×10^9	8
New York Bight	----	----	1.5×10^9	9
Raritan Bay	----	----	3.4×10^6	10
Long Island Inner Shelf	6.9×10^3	3	2.1×10^{10}	11
New Jersey Inner Shelf	1.5×10^4	3	4.5×10^{10}	11
Delaware Inner Shelf	1.4×10^3	3	4.2×10^9	11
Tiger Scarp (N.J. Shelf)	----	----	1.1×10^{10}	12
Ancestral Great Egg Drainage	2.9×10^2 4.4×10^2	10 ----	3.0×10^9 4.4×10^9	13
Ancestral Delaware Valley, Inner (sand?)	$1.4-2.8 \times 10^2$	10-15	4.2×10^9	14
Ancestral Delaware Valley	$1.6-2.2 \times 10^2$	30	6.6×10^9	14
Central DELMARVA peninsula	----	----	1.7×10^9	15
Cape May, N.J.	----	----	1.1×10^9	16
<u>South Atlantic Province</u>	1.1×10^5	2	Total: 2.2×10^{11}	

1. Calculated from Schlee (1964) data.
2. From Manheim (1972).
3. From Williams (1976).
4. Calculated from Williams (1975), Williams and Duane (1975) data.
5. Calculated from Freeland and others (1981) data.
6. Calculated from Knebel and others (1979) data.
7. Calculated from McClennen (1983) data.
8. From Duane (1968).
9. From Williams and Duane (1974).
10. From Bokuniewicz and Frey (1979).
11. From Schlee and Sanko (1975). Note: Inner shelf is defined as 10 km from shore out to 50 m water depth.
12. From Knebel and Spiker (1977).
13. Calculated from McClennen (1973a) data.
14. Calculated from Twichell and others (1977).
15. From Field (1979).
16. From Meisburger and Williams (1980).

TABLE 2.   INDICATED RESERVES OF COMMERCIAL QUALITY SAND ON THE
U.S. ATLANTIC CONTINENTAL SHELF

Site	Volume (m^3)	Annual Beach Fill Requirements[**] (m^3)	Citation
North Atlantic Province Total: 7.7×10^8			
Western Massachusetts Bay			
Cat Island	8.5×10^5		1,2
Nahant	5.9×10^5	2.3×10^4	1,2
Nantasket	2.3×10^6	2.3×10^4	1,2
New Inlet	2.1×10^6	3.8×10^3	1,2
Cape Cod Bay			
Duxbury Beach	8.1×10^6		2
Sagamore Beach	3.1×10^7		2
Plum Island, Massachusetts	4.1×10^7		2
Saco Bay, Maine	9.4×10^7		2
Port Judith, Rhode Island	1.1×10^8		2
Bridgeport, Connecticut	9.9×10^7		2
Long Island Sound	1.9×10^8		3
Long Island Sound			
Matinicock Pt. Shoal, N.Y.	1.0×10^6		3
Stamford Harbor Shoal	6.0×10^6		3
Cable and Anchor Reef	4.0×10^6		3
Easton Neck, N.Y.	7.0×10^6		3
Norwalk Isles Shoal, Conn.	4.0×10^6		
Bridgeport, Conn.	3.7×10^7		3
Crane Neck Shoal, N.Y.	5.0×10^6		3
Port Jefferson Shoal, N.Y.	5.0×10^6		3
Milford Shoal, Conn.	1.0×10^7		3
North Shore Beach, N.Y.	7.0×10^6		3
Townshend Ledge, Conn.	3.0×10^6		3
Roanoke Point Shoal, N.Y.	4.0×10^7		3
Mattituck Shoal, N.Y.	5.4×10^7		3
Long Island Shoal, Conn.	6.0×10^6		3
Mid Atlantic Province Total: 1.3×10^{10}			
South Shore Long Island (area: 2×10^3 km^2)	6.0×10^9	4.7×10^7	4
Inner Long Island Shelf[*]	7.0×10^6		5
New Jersey Shelf[*]	1.5×10^7		5
Delaware Shelf[*]	1.4×10^7		5
Inner N.Y. Bight			
Rockaway Beach	7.9×10^8	3.5×10^5	6
Sandy Hook to Mammouth, N.J.	7.8×10^8	1.8×10^5	6
Central New Jersey Shelf			
Barnegat to Towsend Inlet	1.7×10^8		7
Sandy Hook, N.J.	3.6×10^8		2
Manasquan, N.J.	4.6×10^7		2
Barnegat, N.J.	3.4×10^8		2
Little Egg Harbor, N.J.	1.4×10^8		2
Cape May, N.J.	1.4×10^9		2
Delmarva Peninsula	1.7×10^9		8
Thimble Shoals in Chesapeake Bay (sand and gravel)	1.5×10^7		9
Thimble Shoals in Chesapeake Bay (fine sand)	1.4×10^9		9
South Atlantic Province Total: 1.6×10^9			
Cape Fear, N.C.	1.1×10^9		10
Northern Fl.	2.3×10^8	6.3×10^5	11
Cape Canaveral, Fl.	8.9×10^7	5.7×10^5	12
Fort Pierce, Fl.	7.5×10^7	2.9×10^5	13
Miami to Palm Beach, Fl.	1.2×10^8	6.3×10^5	14

[*]10 km from shore to 50 m of water depth.
[**]data listed where given.

1. From Meisburger, 1976.
2. From Duane, 1969.
3. From Williams, 1981.
4. From Williams, 1976.
5. From Schlee and Sanko, 1975.
6. From Williams and Duane, 1974.
7. From Meisburger and Williams, (1982).
8. From Field, 1979.
9. From Meisburger, 1972.
10. From Meisburger, 1977.
11. From Meisburger and Field, 1975.
12. From Field and Duane, 1974.
13. From Meisburger and Duane, 1971.
14. From Duane and Meisburger, 1969.

TOTAL RESOURCES

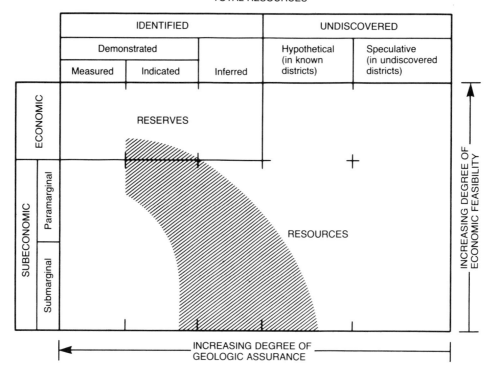

Figure 1. Classification for mineral resources. On this conceptual graph all sand and gravel deposits discussed have been placed subjectively within the shaded envelope. Depending upon use, some deposits discussed could move in or out of the economically feasible region. Modified from Brobst and Pratt, 1973.

pulses in the range of 20Hz to 12kHz. Depth of penetration is on the order of a few meters to as much as 30 m. An important adjunct to CSRP is side scan sonar, where the pulsed beam is directed at an angle to the seafloor. Sidescan sonar is particularly sensitive to surface roughness (i.e., degree of reflectivity) and therefore to surface sediment textures and bedforms. With the advent of microcomputers, modern side scan sonar systems are now corrected for slant-range distance and can provide a "map" of the seafloor in real time.

Confirmation of the character of material comprising the seafloor and shallow subbottom deduced from sonograms is done best with cores. Gravity, or free-fall corers, provide limited penetration of granular materials. Electric, pneumatic, or hydraulic vibrating-type corers have been successful in obtaining coarse granular material in excess of 10 m length (Duane and Meisburger, 1969).

To mine the aggregate, a variety of types of dredges exist. The one selected for use will reflect characteristics that best suit it for a particular job. Of existing conventional dredges, the hydraulic hopper dredge seems to be a universal type suited for work in open marine waters.

The availability of the sand and gravel resource on the continental shelves for the U.S. was put into a different perspective when on March 10, 1983, President Reagan declared an Exclu-

sive Economic Zone (EEZ) for the U.S. Extending from the marine base line of the coast seaward 200 nautical miles, it provides exclusive jurisdiction over the vast resources within an area approximately 1.7 times larger than the land acreage of the U.S. (USGS, 1984).

Environmental and Economic Considerations

Major obstacles or deterrents to marine mining have been a combination of aesthetic and environmental concerns, development of technology, cost-benefit factors, and the lack of clear government regulations. The first three of these were considered by Schlee and Sanko (1975) when they limited their estimate of reserves in the New York Bight area to that region between 10 km from shore and 50 m of water depth. However, environmental concerns, real or imagined, are also major obstacles. The environmental impact of marine dredging has been discussed by Padan (1977) for the Massachusetts Bay; Grant (1973) for offshore Rhode Island; Schlee and Sanko (1975) and Kastens and others (1978) for the New York Bight. Ecological monitoring and evaluation of several offshore borrow (dredging) projects have been conducted (e.g., Courtenay and others, 1980; Reilly and Bellis, 1983; and Naqvi and Pullen, 1982). Where reefs are present, as in south Florida, damage to them as a consequence of fine sand and silt deposition occurs (Courtney and others, 1980); however, species diversity and number were noted to have recovered in other offshore areas as well as in the borrow areas (Courtenay and others, 1980; Turbeville and Marsh, 1982). Removal of large volumes of sediment changes the bottom topography and may change the sediment character of the exposed seafloor. Ecological perturbation therefore will occur and immediately affect nonmobile communities. Depending upon local circumstances, offshore dredging activities will exert a range of impacts that may be detrimental, beneficial, or have no determinable effect on fauna. Careful planning of a sand recovery operation will minimize ecological effects. Guidance and factors to consider for most nearshore areas are provided by Naqvi and Pullen (1982).

The payback for marine mining of sand and gravel is tied to the local availability (or lack) of onshore deposits. A cost analysis for the New York area by Dehais and others (1981) suggests that at 1978 prices ($2.20/m^3) the total payback of costs will be achieved in less than two years provided annual production exceeds 2.8×10^6m^3. Offshore sources for beach nourishment are economically competitive for that purpose as demonstrated on the south shore of Long Island, where 5.1×10^6m^3 of sand was placed along 10 km of beach at an aggregated cost of $2.80/m^3 (Nersesian, 1977).

REGIONAL RESOURCE ASSESSMENT

Background

The U.S. Atlantic continental shelf is typical of shelves on the trailing edges of migrating continents. Characteristically its surface is broad, dipping gradually seaward to a pronounced break-in-slope, usually measured as transitional from <1° to >3° (Inman and Nordstrom, 1971). Depth of water at the shelf break varies from 20 m off the Florida Keys to more than 160 m on Georges Bank and the Scotian Shelf. Width of the shelf varies from less than 3 km off southern Florida to more than 200 km on the Scotian Shelf (Emery and Uchupi, 1972).

From the standpoint of discussing sand and gravel resources, (basic exploration schemes, processes of deposit formation and subsequent modifying marine processes, and general considerations for recovery), the scheme of Emery and Uchupi (1972) is followed here (Fig. 2). Provinces discussed are: (1) North Atlantic Province (Scotian Shelf to Long Island Sound, New York, (2) Middle Atlantic Province (Long Island south shore to Cape Lookout, North Carolina); and (3) the South Atlantic Province (Cape Lookout to Miami, Florida). A combination of processes is responsible for the gross characteristics of the sand resource in each province; their boundaries are transitional. The most dominant process of a region (glacial, outwash, subaerial, or marine processes) makes an overriding, persistent stamp on the resource, which governs, in a general but predictable way, its composition, dimension, and volume.

Sand and gravel deposits on the shelf are comprised of a host

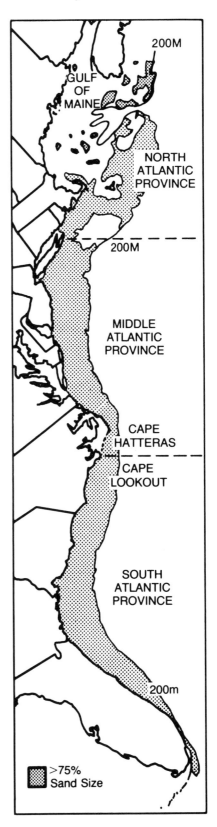

Figure 2. Index map of the eastern North American continental shelf showing the three geological provinces discussed in the text. Figure is modified from Knebel (1981).

of types such as blanket (the surficial sand sheet), plano-convex bodies (finger shoals and relict beach ridges), lenticular (channel fill), or plano-concave (scour or kettle holes). Consequently, an estimate of the average sand thickness produced by "integrating" the dimensions of these sand bodies from mapped areas to the rest of the shelf area creates a resource volume subject to considerable error.

North Atlantic Province

Processes and Sediment Character. The entire region is within the limits of Pleistocene glacial advance, which was chiefly responsible for the land forms subsequently drowned and reworked by marine processes. Ice contact depositional and erosional features are responsible for the character and location of most of the sand and gravel deposits. Study of geophysical records show bedrock with an undulating surface. Blanket-like till has been deposited on much of the bedrock. Thickness varies greatly from several hundred meters to places where bedrock is exposed in the seafloor. In some nearshore areas, glacial depositional features remain and influence normal marine processes but marine winnowing and other reworking processes have been instrumental in producing the present seafloor sediment distribution. Surficial distribution varies between restricted areas such as the Gulf of Maine and Long Island Sound and that of the exposed continental shelves.

Gulf of Maine. Marine processes, following glacial activity in what is now the Gulf of Maine, resulted in very patchy distribution of sediments. Basins within the Gulf are floored by thick accumulations of silts and clays except for the more southern Georges and Franklin basins, which are sand (Uchupi, 1965; Schlee, 1973). Other than these two basins, sand and gravel are restricted to areas of strong currents or elevated topography such as ledges, shallow banks, and hummocky topography between basins. Much of the sand and gravel is poorly sorted, and the sand-size fraction is quartz and feldspar with abundant rock fragments and heavy minerals. The greatest concentrations of gravels are found offshore of western Nova Scotia and off Massachusetts, in the vicinity of Jeffreys Ledge, Stellwagen Bank (in Massachusetts Bay) and the inshore shelf in western Massachusetts Bay between Hull and Plymouth (Fig. 3; Schlee and Pratt, 1970; Schlee and others, 1973). Size distribution of these gravels is polymodal or unimodal. Polymodal deposits were formerly till deposits whereas the unimodal gravels were deposited as outwash or postglacial alluvium later reworked by marine processes (Schlee and Pratt, 1970).

The greatest concentration of sand in the Gulf is around the fringes, with specific areas being along the coast and in Cape Cod Bay extending northward through western Massachusetts Bay to Cape Ann, Massachusetts; a secondary area exists off Portland, Maine, in northwestern Gulf of Maine. Much of this material is reworked glacial outwash but some is a result of erosion of rocky headlands (Cunningham and Fox, 1974; Aubrey and others,

1982) and reworking of sandy beach deposits (Schlee and others, 1973). Throughout the Gulf of Maine the variety of sediment types is large and lateral changes between types are rapid.

Long Island Sound. The sound is a partially stratified estuary, approximately 150 km long, with a maximum width of 40 km (Bokuniewicz and others, 1977). In the deepest area, which is in the central part of the sound near the southern shoreline, the water depth is approximately 45 m. At the eastern mouth of the sound the water depth varies from 7 to 55 m over a series of large sand ridges.

Long Island Sound was probably in existence in late Tertiary or early Pleistocene time but much of the present morphology is the result of glacial erosion and deposition during late Wisconsinan time (Fuller, 1914). During glacial advances, ice lobes overrode the present sound and gouged deep channels into bedrock. Between 15,000 and 13,500 years B.P. the sound was occupied by a glacial lake formed by blockage of the existing channels with glacial till (Newman, 1977). Lake sedimentation resulted in fine-grain lacustrine deposits overlaying the glaciated surface. Glacial outwash was deposited on the lacustrine deposits and in turn was locally covered by modern estuarine deposits (Williams, 1981).

In addition to clays and silts, some sand and gravel continues to be introduced into the sound. Bokuniewicz and others (1977) used the large asymmetric sand waves at the eastern entrance of the sound to suggest that coarse-grained material is being carried into the eastern sound from the adjacent shelf. Along Long Island's north shoreline, sands are most abundant where glacial moraines and outwash fans are exposed to wave action and littoral processes (Williams, 1981). Along the shoreline, strong currents remove the fine sediment, leaving clay and silt to cover much of the southern sound and some of the central portions. Fine to medium grain sands are also abundant along the Connecticut coast. Coarse sands and poorly sorted gravels are localized in small pockets and are found near the shore of Connecticut where the wave activity and littoral currents are strongest. These sands are composed of quartz, feldspar, rock fragments, and abundant heavy minerals.

Continental Shelf. Even more than the Gulf of Maine and Long Island Sound, the continental shelf, including Georges Bank, is a product of glacial activity modified by modern marine processes. The exact positioning and timing of the ice advances is a matter of conjecture. Among others, McMaster and Ashraf (1973a, b) suggest that the ice advances were probably guided by preglacial fluvial topography. As suggested by Figure 3, the ice front extended across Long Island, and probably along the northern edge of Georges Bank, to the outer Scotian Shelf. In the vicinity of Georges Bank, however, the ice advance may have been as far south as the shelf's edge (Emery and others, 1967). For the Scotian Shelf, Shepard and others (1934) and Emery and Uchupi (1972) used relict fauna and the absence of glacial cobbles and pebbles to suggest that the ice front did not extend as far southeast as Sable Island. In contrast, Schlee and Pratt (1970) suggest the ice covered the entire shelf. By 11,000 years B.P. the

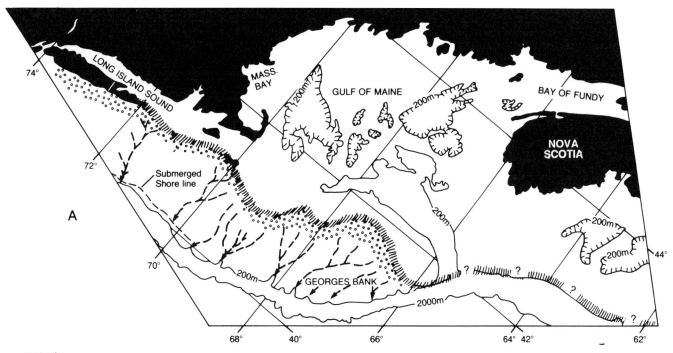

|||| Limit of Ice Advance

°°°°° Outwash, Moraine

◄ ─ ─ Surface and Subsurface Channels

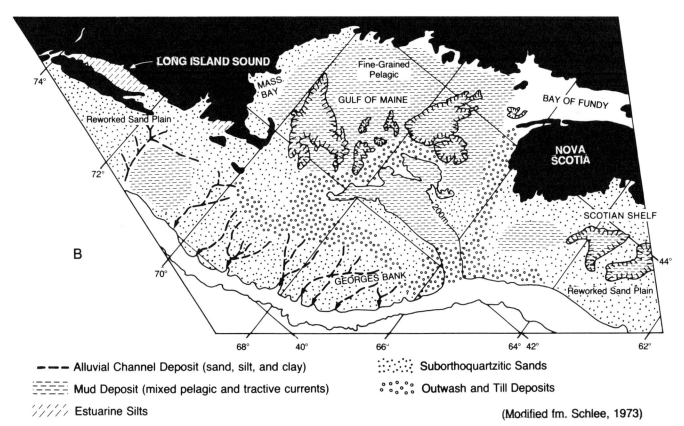

─ ─ ─ Alluvial Channel Deposit (sand, silt, and clay)

▭▭▭ Mud Deposit (mixed pelagic and tractive currents)

///// Estuarine Silts

∴∴∴ Suborthoquartzitic Sands

°°°°° Outwash and Till Deposits

(Modified fm. Schlee, 1973)

Figure 3. A reconstruction of the: (*A*) Pleistocene paleoenvironment; and (*B*) Lithofacies of the continental shelf in the North Atlantic Province. Figure modified from Schlee (1973).

shelves were submerged (Curray, 1965; Tucholke and Hollister, 1973).

The glacial advance over the Scotian Shelf resulted in topography typical of glaciated coasts; an inner irregular zone, a central basin zone, and an outer bank zone. The inner zone is an irregular belt about 30 km wide and bordered seaward by an approximate 50 km wide depression zone of basins. The outer banks extend out to the average shelf width of 200 km (Uchupi, 1969). These shallow banks are in water depths from 28 to 100 m and are entirely covered with sands and gravels (Berger and others, 1966; King, 1967, 1970; Kontopoulos and Piper, 1982).

Berger and others (1966) suggest that these banks are related to Tertiary sedimentation, but most workers agree with Johnson (1925) that the sediment is derived from glacial outwash associated with Pleistocene glaciation and is the result of both fluvial and littoral processes (King, 1967; Stanley and Cok, 1968; James and Stanley, 1968). However, minor amounts of sediment may have come from erosion of the substrate during low stands of the sea or by ice rafting, both before and during the Holocene (James and Stanley, 1968).

When viewed in concert with the Gulf of Maine, Georges Bank is physiographically and sedimentologically similar to the outer Scotian Shelf. Shepard and others (1934) broke with the thinking of earlier workers, who had viewed Georges Bank as either a purely morainal deposition or as an erosional remnant of coastal plain strata, by suggesting that the area was a coastal-plain cuesta that had been separated from the mainland by stream and glacial erosion of the Gulf of Maine and slightly modified by glacial deposition atop the bank. This suggestion has been supported by Emery and Uchupi (1965), Uchupi (1966), Knott and Hoskins (1968), Garrison (1970), and Ballard and Uchupi (1974).

Subsequent to ice advances, Georges Bank experienced four to five cycles of glacial outwash deposition deposited and dissected by rain and melt water (Knott and Hoskins, 1968). The subaerial topography was subsequently smoothed and the shallower channels filled by reworking of the sediments during the last marine transgression (Knott and Hoskins, 1968; Garrison, 1970).

The result of this reworking is a medium- to coarse-grained, well sorted sand with some resistant heavy minerals. The surficial sediments in the southern half of the bank have been redistributed to form a smooth surface that slopes gently (0° 03′) toward the south and is broken only by faint impressions of terraces resulting from lower sea level stands (Emery and Uchupi, 1972). Some of the more pronounced terraces near the outer edge of the shelf, such as Nichols and Franklin shorelines, can be traced south of Hudson Canyon (Fig. 4).

As was true for the Sable Island area, there are two types of modern shelf deposits in the vicinity of Georges Bank: (1) mud south of Nantucket and Martha's Vineyard; and (2) reworked surficial sand on the shoals. Fine grain sediments are accumulating on the inner shelf immediately east of Block Island and in Long Island Sound (McMaster, 1960; Bokuniewicz and others,

1976; Williams, 1981; Knebel and others, 1982). Similar deposition is probably occurring in a 13,000 km² area on the southern New England coast (Twichell and others, 1981; Bothner and others, 1981). Transport of clays and silts from Georges Bank to this area is supported by long term current meter observations by Butman and others (1983) who report a mean vector flow to the southwest throughout the full water column. If these authors are correct, the 13 m of silt and clay accumulation south of Martha's Vineyard is perhaps the only area on the mid to outer continental shelf of eastern U.S. where present day sedimentation is occurring. Other workers, notably Garrison and McMaster (1966), McMaster and Garrison (1966), and Garrison (1970) question the modern accumulation and contend that these sediments are relict.

Of much greater interest to resource aspects of this paper are the large-scale asymmetrical sand ridges and smaller sand waves on Georges Bank and Nantucket Shoals. As reported by many researchers (e.g., Shepard and others [1934], Jordan [1962], Stewart and Jordan [1964], Uchupi [1968], and Mann and others [1981]), the crests of some of these shoals are within four m of the water's surface. On Georges Shoal, one of several such shoals on Georges Bank, the ridges consist of coarse sand with the crest coarser than the flanks. Gravel often floors the troughs (Stewart and Jordan, 1964).

During a 28 year period Stewart and Jordan (1964) observed that the ridges migrated. They postulated the ridge migration was due to seasonal storms or hurricanes that produce currents in excess of 1000 cm/sec on Georges Bank (Hunter and Horton, 1960; Jordan, 1962). Seaward of these active sand ridges are a series of ridges that show no evidence for migration. Recently Twichell (1983) suggested the tidal currents were in fact transporting sand on Georges Bank in concert with storm induced currents. Fine-grained sediment is being removed from the sand ridges for deposition along the northern and southern sides of Georges Bank and the shelf west of Nantucket Shoals.

For the neighboring Nantucket Shoals, Mann and others (1981), Twichell and others (1981), and Moody and others (1984) observed near-bottom tide-driven currents capable of moving sand size material. These currents are strongest on the shallow crest of the bank, weaker on the northern and southern flanks, and extremely weak (less than 5 cm/sec) south of Cape Cod.

Inferred Resources and Indicated Reserves. *Gulf of Maine.* Critical to evaluating the magnitude of the resource is calculating sediment character with sediment depth; but little data exist. Available information indicates that extensive variation in thickness is the rule. Aubrey and others (1982) report sand thickness off eastern Cape Cod to be less than 2 m, whereas off Monomoy Point, Massachusetts, there are 15 m of sand with traces of gravel, and south of Hyannis Port, Massachusetts, the surficial sand measures 3 m (Manheim, 1972). Off Portland, Maine, the sand varies from 0 to 3 m in thickness (Manheim, 1972). Based on limited information from these data and that of Schlee and others (1973), Manheim (1972) estimated a sand

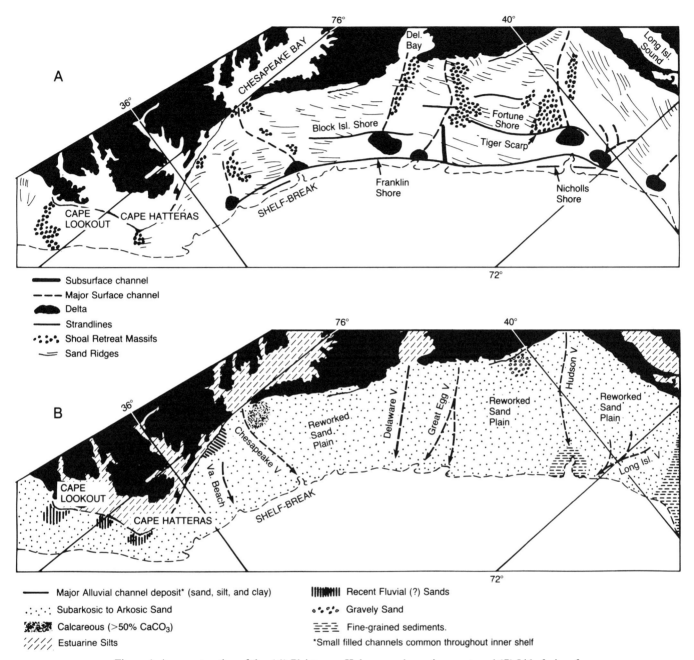

Figure 4. A reconstruction of the: (*A*) Pleistocene-Holocene paleoenvironment; and (*B*) Lithofacies of the continental shelf in the Middle Atlantic Province. Data are from Uchupi, 1968; Emery and Uchupi, 1965; Milliman, 1972a; and Swift, 1976.

resource of $1 \times 10^9 m^3$ available from Cape Cod Bay to Cape Ann, Massachusetts (Table 1). A major program by the National Oceanic and Atmospheric Administration (NOAA), the New England Offshore Mining Environmental Study (NOMES), centered on a shoal portion of Stellwagen Bank due east of Boston Harbor. This indicated a sand and gravel reserve of $5 \times 10^6 m^3$ (Padan, 1977) composed of 25%–30% feldspar, 20%–30% quartz, 15%–20% hornblende, 5%–10% mica, and 5% carbonate.

The most complete assessment of the resources in this area is a result of a study by the Corps of Engineers. Duane (1969) and Meisburger (1976) report several sites along the western side of the Gulf of Maine, from Cape Cod to Portland, Maine, where the grain size is suitable for beach renourishment; quantities may be sufficient to warrant commercial exploitation. These sites, which are all in less than 20 m of water, are included in Table 2.

Long Island Sound. No estimates are available for the quantity of sand and gravel available throughout Long Island Sound but Duane (1969) and Williams (1981) have used data from the

Corps of Engineers' program to identify a series of sites where sufficient reserves of these deposits may support commercial exploitation. Fourteen such sites, in water depths less than 20 m, vary in thickness from 2.1 to 4.3 m. These sites are estimated to contain $1.9 \times 10^8 m^3$ of sand and gravel (Table 2), but based on the known geology and hydrology, larger volumes are likely.

Continental Shelf. Schlee and Pratt (1970) delineated various areas on the shelf where the surficial sediment consisted of 75% or more sand and gravel. A part of this shelf, which falls within our grouping of the North Atlantic Province, totals approximately $75 \times 10^3 km^2$. The data used by Schlee and Pratt only included the extreme southwestern portion of the Scotian shelf; when the other portions of this shelf are included, as is done here, the area totals approximately $112 \times 10^3 km^2$. For the Scotian shelf, King (1967) estimated a thickness of 10 m of sand on the outer shelf. This sand thickness is projected to a resource of $3.6 \times 10^{11} m^3$. For Georges Bank, where Lewis and Sylwester (1976) estimated a minimum thickness of 20 m of sand, the resource totals $1.3 \times 10^{12} m^3$. If the total North Atlantic Province, minus those areas of active mud accumulation on the inner shelf east of Block Island and the outer shelf off southern New England, is assumed to be covered with a 15 m thick layer of sand (average between the 10 m and 20 m estimates) the estimated supply is $1.6 \times 10^{12} m^3$. Such an estimate would be ample to meet the needs of northeastern North America for several thousand years.

Probably much of the sand and gravel resources on the shelf south of Cape Cod is suitable for commercial exploitation; however, only one area has been identified as an indicated reserve. Duane (1969) reports on an offshore area 2 to 10 km south of Point Judith, Rhode Island, where gravel overlies coarse- and medium-grained sand. Average thickness of the sand is 2 m, which provides an estimated volume of $1.1 \times 10^8 m^3$ of sediment suitable for beach restoration and nourishment projects.

Sand ridges on Georges Bank and Nantucket Shoals may also be an indicated resource. These features range in height from 12 to 20 m and may achieve lengths of 20 km and widths of 2 to 3 km (Stewart and Jordan, 1964; Mann and others, 1981). Composed of generally well sorted, medium-coarse sands, these ridges lie in water depths generally less than 10 m. Using the minimum dimensions for a ridge, the volume of sand for a single ridge exceeds $4.8 \times 10^8 m^3$. This volume, when multiplied by the many ridges within this area, indicates an impressive resource of commercial grade sand.

Middle Atlantic Province

This province, which consists of repeated sequences of headlands, barrier spits, and barrier islands, lies south of maximum Pleistocene ice advance but received large volumes of coarse detritus from meltwater streams. Coastal processes and large volumes of detritus strongly influence the character of the sand deposits. The shelf is marked by various topographic and subbottom features inherited from the subaerial processes modified by the marine transgression. These include filled channels, retreat paths

of estuaries and capes, inner shelf shoals, and old shorelines. Each of these is a potential source for large quantities of sand and gravel.

Processes and Sediment Character. Holocene sea level rise was probably interrupted by a series of stillstands or near stillstands followed by a rapid rise. This process produced a series of shore parallel breaks-in-slope on the shelf, which in some instances are subtle, and are interpreted as old shorelines (Fig. 4). Where sampled, these old shorelines consist of sand (Knebel and Spiker, 1977).

A large number of channels also can be recognized on the shelf from either bathymetry or shallow seismic studies. The most pronounced of these features is the Hudson Shelf Valley, which extends the full width of the shelf from the inner New York Bight apex to the Hudson Canyon. Similar large relict drainage systems include the Great Egg Valley (possibly the ancestral Schuylkill River), Delaware Valley, Chesapeake Valley, Virginia Beach Valley, and Albemarle Valley (Fig. 4). In addition to these major channels are numerous smaller or discontinuous channels. Williams (1975, 1976) identified 16 channels emanating off the south coast of Long Island; Freeland and Swift (1978) cite at least 5 in the vicinity of Little Egg Inlet, New Jersey; Sheridan and others (1974) report on 2 filled channels offshore of northern Delaware; and Field (1980) identified at least 24 small channels on the inner Maryland shelf. Some of these channels are largely filled with sand or gravel (Williams, 1976) whereas others contain mostly lagoonal muds (Knebel and others, 1979). In all probability a similar distribution of filled channels exists over much of the Middle Atlantic Province.

Schlee (1964) mapped a $1,780$ km^2 terrace of gravel and pebbly sand off central New Jersey in water depths of 10 to 40 m and attributed the deposit to a southern migration meander of the ancient Hudson River. Knebel and others (1979) noted a similar southerly ancestral path that extended for over 80 km to a termination in an abandoned estuary near Tiger Scarp (Fig. 4). Over this distance the width of the channel varies from 2 to 17 km with a relief of 3 to 15 m. This channel is filled with heterogeneous fluvial sediments capped by 10 to 30 m of shelf sediments (Knebel and others, 1979). The valley fill contains cut and fill structures reminiscent of channels. The upper 0.1 to 2.4 m consists of fine-to-coarse sand with scattered gravel. Underlying this surficial sand, to the limit of the short cores, are interbedded layers of sand and clay.

Other ancestral drainage systems are comparably impressive in size. Paleochannels on the Long Island shelf may be from 30 to 200 m deep and filled with a mixture of fine to coarse sand and gravel (Williams, 1976). Widths of these valleys range up to several kilometers but the lengths are not known. The Great Egg Valley, where mapped between 50 and 79 m of water depth, varies in width from 4 to 6 km for at least 75 km along its length and is filled with 10 to 15 m of bedded sediments (McClennen, 1973a). Twichell and others (1977) mapped the Delaware Valley across much of the shelf and noted widths of 3 to 8 km with 10 to 30 m of infilled sediments. Off the coast of Delaware, Sheridan and others (1974) found ancestral valleys with reliefs of 27 to 30 m. In the latter studies some of these were filled with sands and capped by muds whereas others were filled with muds and capped by sand. For the smaller channels on the Maryland shelf, Field (1980) observed widths of 50 to 500 m in water depths of 16 to 30 m.

Another dimension to shore-normal distributary channels is given by the cut-and-fill "gorge" features of paleotidal inlets. These readily discernable nearshore features have been discussed by Field and Duane (1976), Williams (1976), Meisburger (1979), and Field (1980). Cut and fill features further seaward from the present shoreline are less easily identified as paleoinlets. The nearshore features, and possibly the more seaward features, are a reservoir of sand and gravel acting as a deposition site through the interruption of longshore transport. Additionally, a study of modern inlets (e.g., Fitzgerald, 1984) has demonstrated that currents associated with these inlets (rip and ebb tide currents) carry sand seaward beyond the shoreface, onto the inner continental shelf.

Large constructional sand features seaward of nearshore depositional centers have been identified on the shelf by Duane and others (1972), Swift and others (1972a), and Swift (1976). The features, referred to as shoal retreat massifs (Fig. 4), are believed to represent retreat paths of littoral drift convergences at estuary mouths or off cuspate forelands during sea level rise. In the Middle Atlantic Province these massifs are best developed on the northern side of the ancestral shelf valleys as a result of deposition during the retreat of the estuary mouth shoals. In several places the sand has filled a former subaerial river-cut valley.

Large portions of the Middle Atlantic Province's surficial bodies are molded into a series of sand ridges. This ridge and swale topography dominates the inner shelf (Fig. 5); average ridge spacing varies from 1.6 to 6 km, wave length is approximately 2 km, amplitude is 2 to 10 m, and length is 9 to 56 km (Veatch and Smith, 1939; Uchupi, 1968; Emery and Uchupi, 1972; Duane and others, 1972; Swift and others, 1972a). In the nearshore the ridges are aligned at a fairly ubiquitous 20 to 30 degrees to the coastline; further offshore on the middle shelf this trend becomes coast parallel (Stubblefield and others, 1983a, b). Unlike the shoals on Georges Bank and the Nantucket Shoals these ridges were not formed by tidal currents (Swift and others, 1972a; Stubblefield and others, 1975); consequently, other models have been invoked to explain the origin of these ridges.

On the Long Island shelf, McKinney and Friedman (1970) contend that the Holocene shelf topography reflects a Pleistocene coastal plain fluvial drainage topography that has not been obscured by recent sediment flux. Similarly, Swift and others (1972a) interpret portions of the bathymetry on the northern New Jersey and inner Long Island shelf as bearing the imprint of subaerial tributaries. Many workers, however, have interpreted the shelf topography of the U.S. Middle Atlantic Bight and adjacent coastal waters as being indicative of overstepped coastlines (e.g., Emery and others, 1967; Shephard, 1973; Uchupi, 1970; McClennen and McMaster, 1971).

A more recent origin for the sand ridges has also been proposed. In a detailed investigation of the nearshore Atlantic shelf of the United States, extending from Long Island to Florida, Duane and others (1972), Swift and others (1972a), and Field and Duane (1976) concluded that the "positive" shelf topography reflects post-transgressive processes. With the exception of an imprint of fluvial drainage in isolated areas, these authors suggested that most of the "positive" features formed in nearshore waters adjacent to the retreating shoreface; either as shoreface-connected ridges or ridges associated with a retreating estuary or cape. Nearshore wave and current motion combine to produce and maintain these features.

In their examination of shoreface-connected and similar nearshore ridges, Duane and others (1972) used the frequent 20°–30° intersection angle with the coastline trend as an argument against the idea of the ridges being submerged coastlines (Fig. 5). More than 200 of these features occur on the inner shelf of the Middle and South Atlantic Province, with the greatest number in the northern sectors. Plano-convex in cross section, they trend northeast and make an acute angle with the adjacent coast. They are composed generally of well-sorted, medium-grained sands similar to adjacent beaches. Sediments of these shoals contrast with the underlying strata and the interface creates a strong sonic reflection facilitating computation of reserve estimates (Fig. 5).

Recently Stubblefield and others (1983a, b) viewed the ridges on the New Jersey shelf as having dual origins. These authors contend that the nearshore ridges and certain outer shelf ridges are post-transgressive features as suggested by Duane and others (1972) and Swift and others (1972a), but that many of the shore parallel mid-shelf ridges, in water depths of 20 to 30 m, are degraded barriers subsequently modified by shelf currents.

Under modern conditions most fluvial sediments are being trapped within the estuaries and little is escaping to the shelf (Meade, 1969). However, some new sediment is being introduced to the nearshore by erosion of the shoreface as the sea level rises (Swift, 1972b), and some of the shoreface sediment may be originally derived from the inner shelf itself (McMaster, 1954). In the Middle Atlantic Province the extent of the introduction of sizeable quantities of reworked sand-size material remains in question.

The reworking of the surficial sands is better documented. Twichell (1979) notes present-day reworking of the sand ridges near the head of Wilmington Canyon by storm-driven bottom currents or breaking internal waves. On the mid-shelf of New Jersey, Stubblefield and others (1975) and Stubblefield and Swift (1976, 1981) document not only reworking but nearly 0.7 m of a ridge aggradation within the last 100 years. These authors also speculate that this aggradation is accompanied by trough erosion, which adds some new material to the shelf's sediment budget. Reworking of surficial sediments is also documented with numerous sidescan surveys (summarized by Swift and Freeland, 1978). Bottom currents capable of reworking the sand-size material have been observed during summer and winter storms

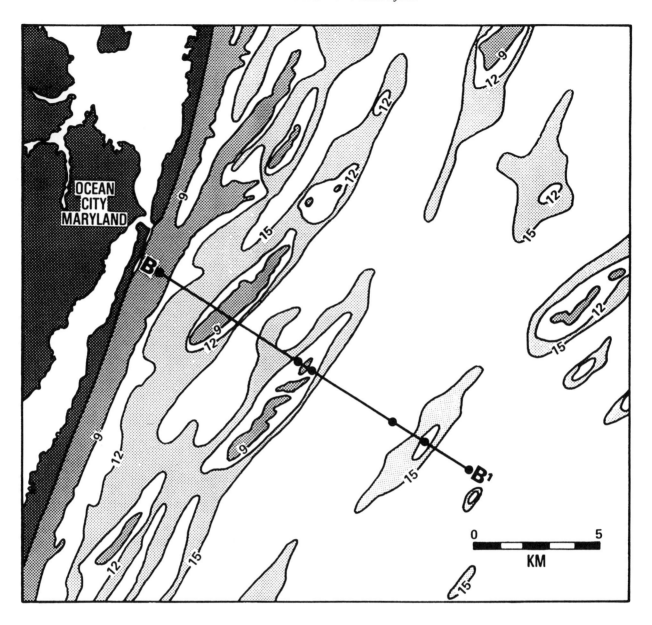

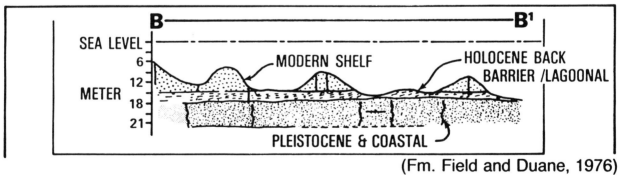

(Fm. Field and Duane, 1976)

Figure 5. Plan view and representative cross-sectional profile from the Maryland inner continental shelf. Depths are in meters. The plano-convex nature of the linear shoals along line *B-B'* is typical of all linear shoals on the middle and south Atlantic shelf. Large dots along profile *B-B'* indicate location of vibratory cores (from Field and Duane, 1976).

(McClennen, 1973b; Butman and others, 1979; Allen and others, 1983).

Gravels are spotty in distribution. In addition to the gravel deposit off the central New Jersey coast (Schlee, 1964), small deposits are found seaward of the Delaware and Chesapeake bays (Milliman, 1972a, b). Gravels floor the troughs in some of the sand ridge complexes.

Inferred Resources and Indicated Reserves. There have been numerous local and a few regional studies that provide a measure of the surficial sediment thickness and thus an indication of the reserves and resources (Tables 1, 2). Sand thickness can be estimated for the total shelf and for areas of local concentrations such as the ancestral drainage systems. For the middle and outer New Jersey shelf Knebel and Spiker (1977) and Folger and others (1978) used short cores and seismic data to estimate a surficial thickness between 0.5 to 6.0 m. Comparable surficial sediment thicknesses were estimated for similar areas from seismic surveys by McClennen (1973a) who observed a thickness of 1 to 5 m over much of the New Jersey shelf; Shideler and others (1972) who found the Virginia shelf to have a thickness between 0 to 9 m with an average of 3 m; and Shideler and Swift (1972) who with CSRP data measured thicknesses up to 12 m with an average of 5 m on the shelf between Cape Henry, Virginia, and Cape Hatteras, North Carolina. The few long cores in the area confirm at least these thicknesses. The USGS Atlantic Margin Coring Project (Hathaway and others, 1979) cored 40 m of sand and gravel interlayered with mud on the inner New Jersey shelf and 15 m of sand on the outer Maryland shelf. Assuming a sand blanket with a thickness of 3 m for the Middle Atlantic Province, a resource on the order of $2.6 \times 10^{12} \text{m}^3$ is obtained.

However, the above figure can be increased by inclusion of volumes in deposits of other shapes that, for many areas of the shelf, produce a sand thickness much greater than 3 m. For example, relict shorelines are areas of thickened sand (and perhaps gravel) deposits. Knebel (1981) reports that these accumulations vary from 2 to 10 m in thickness whereas McClennen (1973a) estimates such features on the outer shelf are as much as 16 m thick. In contrast, filled ancestral channels contain appreciably more sediment but the grain size of the material is less known. The surficial sediment in these relict drainage systems is usually sand or gravel (Schlee, 1964; Schlee and Sanko, 1975; Freeland and others, 1981; Williams, 1981) but the underlying channel fill may consist of interlayered sand, gravel, and mud (Knebel and Spiker, 1977). Without long cores through these channels there is no way to estimate with confidence the amount of sand and gravel; but if the assumption is accepted that the bulk of the material is useable in terms of aggregate mining, a resource of $6.1 \times 10^9 \text{m}^3$ to $2.7 \times 10^{10} \text{m}^3$ exists (Table 1). This quantity is restricted to the mapped areas of the four major relict channels (Hudson Shelf Valley, ancestral Hudson Valley, Great Egg Valley, and Delaware Valley). When the unmapped portions of the major channels and the numerous small channels are considered, the estimated quantity of sand and gravel is substantially increased.

The "beach" ridge complexes and the massifs contain large quantities of well sorted fine to coarse sand. Knebel and colleagues mapped two such features. The area around Tiger Scarp on the central New Jersey shelf (Fig. 4) contains an estimated total of $1.1 \times 10^{10} \text{m}^3$ of sand (Knebel and Spiker, 1977). For a large shoal at the mouth of the ancestral Hudson River, Knebel and others (1979) mapped dimensions that equate to approximately $3.5 \times 10^9 \text{m}^3$. By similar reasoning, a resource of $1.1 \times 10^7 \text{m}^3$ to $1.3 \times 10^9 \text{m}^3$ is available for each ridge on the shelf (using ridge dimensions of Swift and others, 1972a; Duane and others, 1972). When all the ridges on the shelf are considered, a staggering quantity results.

For the Middle Atlantic Province, some estimates of indicated reserves are available. Schlee and Sanko (1975) considered the limitation of the operating dredges and established their outer boundary at 50 m of water depth, and they introduced aesthetic considerations by imposing an inner boundary of 10 km from shore so that dredging operations would not be readily observed from the shore. Within these boundaries, Schlee and Sanko estimated reserves of $6.9 \times 10^6 \text{m}^3$ for the Long Island shelf, $1.5 \times 10^7 \text{m}^3$ for the New Jersey shelf, and $1.4 \times 10^7 \text{m}^3$ for the Delaware shelf (Table 2). Using McClennen (1983) data for the New Jersey shelf, out to water depths of 20 m, quantities of $5.4 \times 10^9 \text{m}^3$ are determined.

Several workers used the results of the Corps of Engineers' program to establish reserves. Williams (1976) reports that $6 \times 10^9 \text{m}^3$ of sand suitable for beach restoration and nourishment on the Long Island shelf is in water depths suitable for dredging. Duane (1969) identified five major areas along the nearshore New Jersey shelf where mining is feasible with a total volume of $1.0 \times 10^9 \text{m}^3$ (Table 2). In a follow-on study Meisburger and Williams (1982) report comparable values to those of Duane for central New Jersey but identify 15 potential borrow sites. To the south, 18 sites are identified around Cape May, New Jersey, with an estimated $1.1 \times 10^8 \text{m}^3$ of useable sands (Meisburger and Williams, 1980). Field (1979) conservatively estimates $1.7 \times 10^9 \text{m}^3$ of sand suitable for beach nourishment along the central DELMARVA peninsula. Near the mouth of Chesapeake Bay $1.4 \times 10^9 \text{m}^3$ of fine sand is found in water depths of 15 m or less with another $1.5 \times 10^7 \text{m}^3$ of coarse and gravelly sand available from Thimble Shoals within the bay (Meisburger, 1972).

South Atlantic Province

From the vicinity of Cape Lookout, North Carolina, southward, the coastline is characterized by chains of short, low relief barrier islands separated by a few large capes (Fear, Romaine, and Canaveral) with associated shoals (Fig. 6). Being far removed from glacial action and the associated large volumes of sediment runoff, the inner shelf of this province is topographically subdued and composed for the most part of irregularly outlined highs separated by broad topographic depressions, not dissimilar from the coastal plain morphology. Relief is commonly less than 6 m. Absent are the large northeast trending sand shoals of the Middle Atlantic region, except for the area of Fort Pierce, Florida (midway between Miami and Cape Canaveral).

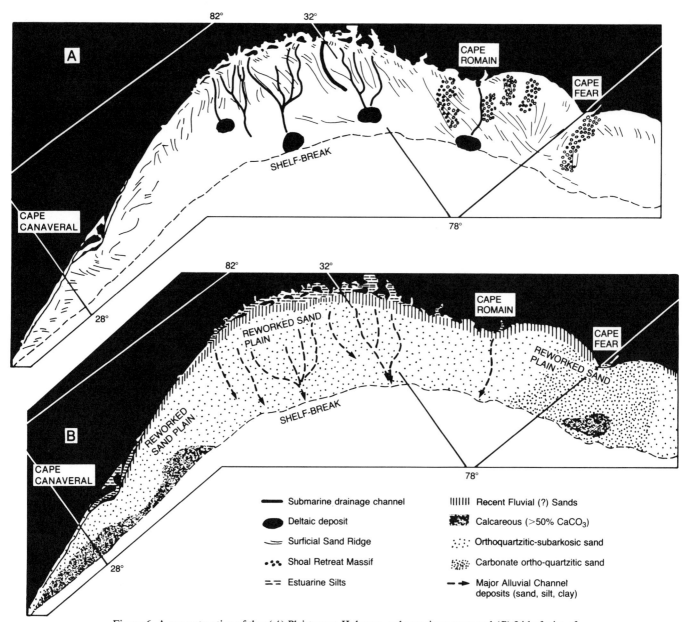

Figure 6. A reconstruction of the: (*A*) Pleistocene-Holocene paleoenvironment; and (*B*) Lithofacies of the continental shelf in the South Atlantic Province. Data are from Uchupi, 1968; Emery and Uchupi, 1965; Milliman, 1972a; and Swift, 1976.

Unlike the continental shelf north of Cape Lookout, the shelf of the South Atlantic Province is neither traversed by shelf channels nor incised by the heads of submarine canyons. A series of terraces are found, some of which are sufficiently deep to cut into the Florida-Hatteras slope below the shelf break (Emery and Uchupi, 1972). Outcropping over much of the shelf are hardgrounds, which are generally interpreted to be pre-Pleistocene reefs (Emery and Uchupi, 1972).

The surficial sand sheet is thin and discontinuous with outcrops of Cretaceous, Tertiary, and Pleistocene age material common (Luternauer and Pilkey, 1967; Meisburger, 1979; Meis-

burger and Field, 1975; Field and Duane, 1974; Riggs, 1984). For the most part the inner shelf sediment is fine-grained, largely quartz, sand to the vicinity of Cape Canaveral. South of Cape Canaveral the grain size and carbonate (biogenic) content increases. Milliman (1972a) ascribed the fine sand of the inner shelf as a modern fluvial deposit and believed it continuous between Cape Fear and Cape Canaveral. Seaward of approximately the −18 m MLW contour, coarser iron-stained Pleistocene sands occur on the surface (Pilkey and Frankenberg, 1964). The line is abrupt and is referred to as the relict-recent boundary. The supposed fluvial origin of the nearshore sediment is debatable. As was mentioned previously in this paper, several kinds of ocean currents and processes may be responsible for transporting sediment seaward from the coast. The large extent of surface exposures and near surface occurrence of Pleistocene and Tertiary strata provides localized sources for the surficial sand sheet.

Processes and Sediment Character. Recently Pilkey and others (1980, 1981) addressed the question of relict versus recent deposition. They note that most of the non-carbonate sand was deposited on the shelf under conditions different from those prevailing today; however, based on mineralogy and incorporated shell material they believe that the surficial sediments, down to depths of 5 m, are being reworked by modern storms. They based their suggestions for reworking on the scarcity of mud in the surficial sands. In such a context, these sediments, like the non-carbonate surficial sediments in the Middle Atlantic Province, are relict in terms of original deposition but recent in terms of reworking.

Calcareous material is largely indigenous. The bulk of the non-carbonate surficial sands were fluvially-derived during lower sea levels. Based on heavy minerals, several workers suggest the sands were carried directly to the shelf by rivers draining the metamorphic and igneous source rocks of the Piedmont region (Pilkey, 1964; Giles and Pilkey, 1965; Doyle and others, 1968; Judd and others, 1970). In addition to the fluvial input, sediments especially on the Florida shelf, have been produced by erosion of underlying Tertiary rock outcrops (Luternauer and Pilkey, 1967; Meisburger and Field, 1976; Pilkey and others, 1980).

As expected, the carbonate fraction of the surficial sands is younger than the non-carbonate material. The carbonate is restricted to the most recent transgression with the oldest shell material being less than 15,000 years B.P. (Pilkey and others, 1981). The absence of any shells older than very late Pleistocene age implies that: (1) older carbonate shells were destroyed by subaerial leaching, mechanical abrasion, and biological degradation during sea level regression; and (2) the carbonate fraction is renewed by the secretion of shells during succeeding transgressions (Pilkey and others, 1981; Knebel, 1981).

Surficial sediments of the middle to outer shelf have been influenced by modern oceanographic processes. A suite of cross-shelf cores studied by Pilkey and others (1980, 1981) show that the upper few meters of the sediments were reworked during the Holocene. Moreover, large scale sand ridges, common over much of the area between central Florida and Cape Lookout, are being formed and modified by present shallow water processes (Swift and others, 1972a; Field, 1974; Field and Duane, 1976).

The orthoquartzitic-subarkosic sands in this province (Fig. 6) are medium- to coarse-grained on the central and outer shelf and generally fine-grained in the nearshore (Milliman, 1972a; Pilkey and others, 1981). Near capes, the grain size increases to medium-to-coarse (Duane and others, 1972). Shell hash constitutes upwards to 20% of the surficial sediment and locally is higher. Phosphate grains are abundant in two distinct phosphogenic provinces (from Cape Lookout to Charleston, South Carolina, and from northern Florida to Miami) and are believed to

have been derived from erosion of the underlying Tertiary formations (Pilkey and Luternauer, 1967; Luternauer and Pilkey, 1967; Riggs, 1984).

Gravels are less common on the shelves of this province and are found in concentrations exceeding 25% only seaward of Cape Fear, Charleston, South Carolina, and Cape Canaveral (Milliman, 1972a). These gravels, which may be shelly, are associated with ancestral estuaries exhumed by nearshore currents.

South of Palm Beach, Florida, the shelf acquires unique characteristics as it changes from a broad, drowned, coastal plain covered with terriginous sediments to a narrow markedly stepped shelf where a "reef like" mass forms each riser (Fig. 7). It is in the vicinity of Hillsboro Beach, 48 km (30 mi) north of Miami, where the change becomes most pronounced. In that sector the step-like shelf is buried beneath a continuous seaward sloping sediment blanket that begins at the shoreface and ends behind a reef-like mass at approximately 20 m water depth. This sediment blanket is predominantly a grey medium-fine-grained calcareous quartz sand (55%–65% quartz; Duane and Meisburger, 1969).

South of Hillsboro Beach the step-like inner shelf is marked by three persistent steps, characteristically at –6 m, –12 m, and –20 m below the sea surface. A thin veneer of sand comprised of biogenous debris (<10% quartz) covers the flats. However, some sectors of the second and third flats contain 3–4 m of sediment, the potential source of sand for aggregate. The reef-like masses forming the risers have been identified by Shinn and others (1977) to be cemented drowned calcareous dunes.

The two distinct shelf sediment facies distinguished in the surface sediments of the southern Florida area persist with increasing depth. The unconsolidated Holocene sediments in the region of Miami, and 48–64 km (30–40 mi) northward are carbonate skeletal sands and gravels composed largely of hard parts of the biota living in the warm shallow waters covering the shelf. About one-third of the skeletal fragments are sufficiently complete to be readily recognized, the remainder is a hash. Quartz is rare and generally is less than 10% by weight. The wide range of sizes, poor sorting, and the character and condition of particles suggest in-situ formation. Furthermore, seismic profiles show a shoreward asymmetry in the accumulation of sediments lying in the flats behind the reef-like risers.

Further north toward Cape Canaveral, most of the sediment recovered in cores is a clean, homogeneous, fine- to medium-grained calcareous quartz sand (55%–65% quartz). The quartz is angular and subrounded clear grains that are mixed with rounded to subrounded gray, black, and brown calcareous particles. Differences in sediments throughout this sector are not large although the more southerly sections are somewhat coarser with higher carbonate content, representing a transition from the reef-like area of the south to the more typical shelf environment of the north. In all samples in this northern sector sediments of the top few centimeters are finer than material below.

Inferred Resources and Indicated Reserves. The recent study by Pilkey and others (1981) provided a reasonable estimate of sediment thickness over much of the shelf from the border

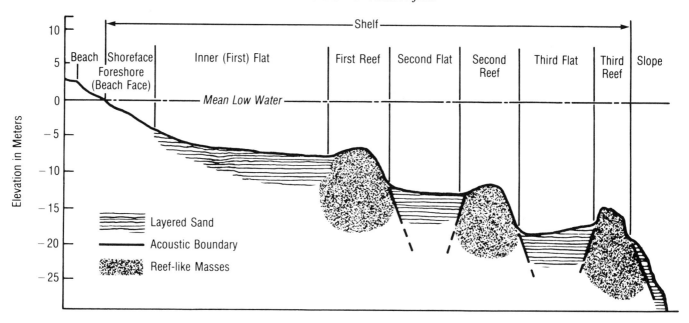

Figure 7. West (left) to east schematic profile of southeastern Florida shore and shelf morphology and interpretation of the stratigraphy (from Duane and Meisburger, 1969).

between North and South Carolina to just north of Cape Canaveral. These studies are complemented, and in some areas expanded, by the Corps of Engineers' coverage of nearshore North Carolina and Florida. Data from these programs show that, except where there are sand ridges, the surficial sediment is uniformly thin over much of the shelf. From North Carolina to northern Florida the thickness averages less than 5 m and thickens slightly offshore (Pilkey and others, 1981). Specifically, off the coast of northern South Carolina the sand is discontinuous in water depths of 12 m and thickens to 6 m in 45 m water depth. Offshore of Charleston, South Carolina, the sand varies in thickness from less than 1 m to 5 m in water depths between 15 to 65 m. The nearshore sands are similarly thin. From Palm Beach, Florida, to southern Georgia the sand thickness varies from 0 to 4 m. The exception is under ridges where the thickness may exceed 12 m and off Ft. Pierce, Florida, where 9 m of sand is found in 12 to 18 m of water (Meisburger and Duane, 1971; Field and Duane, 1974; Meisburger and Field, 1975). As a general rule, the sand is thickest under the ridges, discontinuous in depressions, and averages less than 1.5 m in the flat areas. In the Cape Fear to Cape Lookout area the surface sand is thin to discontinuous except under ridges where thickness of 12 m may exist (Meisburger 1977, 1979).

Based on the estimated thickness of the surficial sand sheet, an approximation can be made for the sand resource on these shelves. If an estimate of 2 m is used for the average thickness of sand, the inferred resource from Cape Lookout to Miami is $2.2 \times 10^{11} m^3$. Indicated reserves are abundant in the areas of the sand ridges. For example, Meisburger and Duane (1971) cite three nearshore ridges (Bethel, Capron, and Indian River) near Fort Pierce, Florida, as sources for constructional grade sand whose

volume is $7.5 \times 10^7 m^3$. Similarly Field and Duane (1974) calculate that $8.9 \times 10^7 m^3$ of sand is available from the Bull, Ohio-Hetzel, and Chester shoals (ridges) off Cape Canaveral, Florida (Table 2). Thirty-two potential borrow areas were identified by Meisburger and Field (1975) in the nearshore zone off northern Florida. Many of these sites are associated with linear shoals. The total reserves are unknown, but from 6 of these sites the indicated reserves exceed $2.3 \times 10^8 m^3$. Comparable volumes can be expected from the ridges seaward of Cape Fear and Cape Hatteras. In the nearshore between Miami and Palm Beach, Florida, Duane and Meisburger (1969) estimate $1.2 \times 10^8 m^3$ sand is available, some of which is of quality suitable for beach nourishment.

Density of data is adequate for sand volume calculations only in the Miami region; only estimates of the reserves to the north are possible (Table 1). Approximately $5.4 \times 10^7 m^3$ of sand size material exist in a 3 km × 3 km grid offshore Miami. Of that volume, approximately $3.7 \times 10^7 m^3$ are contained within the confines of the plateau area at the –12 m water depth. Because of the nature of the accumulation, and the bottom and subbottom nature of the region, similar areas of accumulation can be forecast. In fact, $2.7 \times 10^5 m^3$ of sand were pumped from the –12 m flat offshore of Hillsboro Beach (48 km [30 mi] north of Miami) in 1972 to nourish the beach there.

Although median diameters of beach and offshore sediment are not too dissimilar, compositional differences raise the question of the suitability of the offshore sand for placement on the beach. However, recent beach nourishment projects at Miami, which used offshore sources, have demonstrated stability.

For the North Carolina region $1.1 \times 10^9 m^3$ of sand suitable for beach nourishment is available in the inner shelf off Cape

Fear, in 4 borrow areas (Table 1). In a recent study of inlets of South Carolina, Fitzgerald (pers. com., 1984) observed, seismically, numerous filled channels extending seaward of present day inlets. Some of these channels at shallow depths (2–18 m) are interpreted to be filled tidal inlets while channels at greater depths are interpreted as resulting from estuarine and fluvial conditions during lowered stands of sea level. Similar seismic stratification was observed off the North Carolina coast by Meisburger (1977). Data indicate the largest sand sources are the shoals off Capes Fear and Lookout, and filled channels seaward from major inlets such as the Cape Fear and Newport rivers. Seismic profiles show that the surficial sand sheet is thin and discontinuous. Essentially quartzose sand and some calcareous granular sediment was cored. Indicated reserves are on the order of $1.6 \times 10^6 \text{m}^3$ (Meisburger, 1977). It is probable that analogous deposits occur offshore of the Santee River delta and inlets of South Carolina.

SUMMARY

The resources of sand and gravel on the U.S. Atlantic continental shelf are at best a rough approximation because of the limited and sparsely spaced seismic and core data. Reasoned estimates, however, suggest that for the North Atlantic Province $2.0 \times 10^{12} \text{m}^3$ of sand and gravel are available; for the Middle Atlantic Province $2.6 \times 10^{12} \text{m}^3$; and for the South Atlantic Province $2.2 \times 10^{11} \text{m}^3$. Most of this sediment is sand size material with mineralogy varying between arkosic and suborthoquartzitic. In the South Atlantic Province the carbonate content in the surficial sands increases from trace amounts north of Cape Lookout,

North Carolina, to in excess of 75% south of Palm Beach, Florida. Gravels are common on the glaciated shelves of the North Atlantic Province and in a few relict fluvial channels off the New Jersey coast. Elsewhere gravel deposits are rare. With notable exceptions, surficial sediments contain less than 5% clay and silt. The exceptions are areas of present deposition including the basins north of Georges Bank, the inner shelf east of Block Island, and the outer shelf off southern New England. Modern estuaries are trapping most of the river-borne fine grain sediments.

Obviously not all the $8.1 \times 10^8 \text{m}^3$ annual U.S. needs of sand and gravel by year 2000 (Tepordei, 1980) will come from marine resources. However, due to demand for increased quantities of sand and gravel and fewer onshore mining operations near metropolitan areas, more use will be made of these marine resources, especially in populous coastal areas. Goodier (1972) reported that in one year Connecticut used 14% of the sand and gravel available on land. Massachusetts has demands that exceed supply, and several producers in Long Island have ceased operations due to depleted reserves (Kastens and others, 1978; Schubel, 1981).

Recent research efforts assessing the impact of marine aggregate recovery have shown dredging and rehandling operations produce impacts ranging from detrimental, to beneficial, or may have no discernable effect. Moreover it is apparent that consequences now are predictable and that careful planning prior to operations can minimize undesirable ecological effects.

Where a large volume of sand is required, as in beach nourishment or coastal protection projects, inshore deposits are now commercial resources.

REFERENCES CITED

Allen, J. C., and 12 others, 1983, Physical oceanography of continental shelves: Reviews of Geophysics and Space Physics, U.S. National Report to International Union of Geodesy and Geophysics 1979–1982, v. 21, no. 5, p. 1149–1181.

Aubrey, D. G., Twichell, D. C., and Pfirman, S. L., 1982, Holocene sedimentation in the shallow nearshore zone off Nauset Inlet, Cape Cod, Massachusetts: Marine Geology, v. 47, p. 243–259.

Ballard, R. D., and Uchupi, E., 1974, Geology of the Gulf of Maine: American Association of Petroleum Geologists Bulletin, v. 58, p. 1156–1158.

Berger, J., Cok, A. E., Blanchard, J. E., and Keen, M. J., 1966, Morphological and geophysical studies on the eastern seaboard of Canada: The Nova Scotian Shelf, *in* Garland, G. D., ed., Continental drift: Royal Society of Canada and University of Toronto Press, p. 102–113.

Bokuniewicz, H. J., and Frey, C. T., 1979, The volume of sand and gravel resources in the lower bay of New York Harbor: Stony Brook, New York, State University of New York, Marine Science Research Center, Special Report 12, Reference 79-16, 34 p.

Bokuniewicz, H. J., Gerbert, J., and Gordon, R. B., 1976, Sediment mass balance of a large estuary, Long Island Sound: Estuarine and Coastal Marine Science, v. 4, p. 523–536.

Bokuniewicz, H. J., Gordon, R. B., and Kastens, K. A., 1977, Form and migration of sand waves in a large estuary, Long Island Sound: Marine Geology, v. 24, p. 185–199.

Bothner, M. H., Spiker, E. C., Johnson, P. P., Rendigs, R. R., and Aruscavage, P. J., 1981, Geochemical evidence for modern sediment accumulation on the shelf off southern New England: Journal of Sedimentary Petrology, v. 51,

p. 281–292.

Brobst, D. A., and Pratt, W. P., 1973, United States mineral resources, introduction, *in* Brobst, D. A., and Pratt, W. P., eds., United States mineral resources: U.S. Geological Survey Professional Paper 820, p. 1–8.

Butman, B., Vermersch, J., Beardsley, R., and Noble, M., 1978, Long-term current observations on Georges Bank: EOS (Transactions, American Geophysical Union), v. 59, p. 302–303.

Butman, B., Noble, M., and Folger, D. W., 1979, Long-term observations of bottom currents and bottom sediment movement on the Mid-Atlantic continental shelf: Journal of Geophysical Research, v. 84, p. 1187–1205.

Butman, B., Noble, M., Chapman, D., and Beardsley, R., 1983, An upper bound for the tidally rectified currents at one location on the southern flank of Georges Bank: Journal of Physical Oceanography, v. 13, p. 1452–1460.

Courtenay, W. R., Jr., Hartig, B. C., and Losiel, G. R., 1980, Ecological evaluation of a beach nourishment project at Hallandale (Broward County), Florida: U.S. Army Corps of Engineers, Coastal Engineering Research Center, MR 80-1 (1), 23 p.

Cunningham, R. W., and Fox, W. T., 1974, Coastal processes and dispositional patterns on Cape Ann, Massachusetts: Journal of Sedimentary Petrology, v. 44, p. 522–531.

Curray, J. R., 1965, Late quaternary history, continental shelves of the United States, *in* Frey, G. D., and Wright, H. E., eds., The quaternary of the United States: Princeton, New Jersey, Princeton University Press, p. 723–737.

Dehais, J. A., Guyette, P. L., and Wallace, W. A., 1981, Onshore pressures make offshore mining viable: Rock Products, June 1981, p. 72–76.

Doyle, L. J., Cleary, W. J., and Pilkey, O. H., 1968, Mica: Its use in determining shelf-depositional regimes: Marine Geology, v. 6, p. 381–389.

Duane, D. B., 1968, Sand deposits on the continental shelf: A presently exploitable resource, *in* Proceedings of the national symposium on ocean sciences and engineering of the Atlantic Shelf, March 19–20, 1968, Philadelphia: Washington, D.C., Marine Technology Society, p. 289–297.

—— 1969, Sand inventory program, A study of New Jersey and northern New England coastal waters: Shore and Beach, p. 12–16.

Duane, D. B., and Meisburger, E. P., 1969, Geomorphology and sediments of the nearshore continental shelf, Miami to Palm Beach, Florida: U.S. Army Corps of Engineers, Coastal Engineering Research Center, Technical Memo 29, 47 p.

Duane, D. B., Field, M. E., Meisburger, E. P., Swift, D.J.P., and Williams, S. J., 1972, Linear shoals on the Atlantic inner continental shelf, Florida to Long Island, *in* Swift, D.J.P., Duane, D. B., and Pilkey, O. H., eds., Shelf sediment transport: Process and pattern: Stroudsburg, Pennsylvania, Dowden, Hutchinson, and Ross, p. 447–498.

Emery, K. O., and Uchupi, E., 1965, Structure of Georges Bank: Marine Geology, v. 3, p. 349–358.

—— 1972, Western North Atlantic Ocean: Topography, rocks structure, water, life, and sediments: American Association of Petroleum Geologists Memoir 17, p. 532.

Emery, K. O., Wigley, R. L., Barlett, A. S., Rubin, M., and Barghoorn, E. S., 1967, Freshwater peat on the continental shelf: Science, v. 158, p. 1301–1307.

Field, M. E., 1974, Buried strandline deposits on the central Florida inner continental shelf: Geological Society of America Bulletin, v. 84, p. 57–60.

—— 1979, Sediments, shallow subbottom structure and sand resources of the inner continental shelf, central DELMARVA Peninsula: Fort Belvoir, Virginia, U.S. Army Corps of Engineers, Coastal Engineering Research Center, Technical Paper 79-2, 122 p.

—— 1980, Sand bodies on coastal plain shelves: Holocene record of the U.S. Atlantic inner shelf off Maryland: Journal of Sedimentary Petrology, v. 50, p. 505–528.

Field, M. E., and Duane, D. B., 1974, Geomorphology and sediments of the inner continental shelf, Cape Canaveral, Florida: Fort Belvoir, Virginia, U.S. Army Corps of Engineers, Coastal Engineering Research Center, TM-42, 87 p.

—— 1976, Post-Pleistocene history of the United States inner continental shelf: Significance to origin of Barrier Islands: Geological Society of America Bulletin, v. 87, p. 691–702.

Fitzgerald, D. M., 1984, Interactions between the ebb-tidal delta and landward shoreline: Price Inlet, South Carolina: Journal of Sedimentary Petrology, v. 54, p. 1303–1318.

Folger, D. W., Butman, B., and Knebel, H., 1978, Environmental hazards on the Atlantic continental shelf of the United States: Proceedings, Offshore Technical Conference, 10th, Houston, Texas, p. 2293–2306.

Freeland, G. L., and Swift, D.J.P., 1978, Surficial sediments: Albany, New York, MESA New York Bight Atlas Monography 10, New York Sea Grant Institute, 93 p.

Freeland, G. L., Stanley, D. J., Swift, D.J.P., and Lambert, D. N., 1981, The Hudson shelf valley: Its rule is shelf sediment transport: Marine Geology, v. 42, p. 399–427.

Fuller, M. L., 1914, Geology of Long Island, New York: Reston, Virginia, U.S. Geological Survey Professional Paper 82, 231 p.

Garrison, L. E., 1970, Development of continental shelf south of New England: American Association of Petroleum Geologists Bulletin, v. 54, p. 109–124.

Garrison, L. E., and McMaster, R. L., 1966, Sediments and geomorphology of the continental shelf off southern New England: Marine Geology, v. 4, p. 273–289.

Giles, R. T., ailkey, O. H., 1965, Atlantic beach and dune sediments of the southern United States: Journal of Sedimentary Petrology, v. 35, p. 900–910.

Goodier, J. L., 1972, Reclaiming sand and gravel from the sea: Proceedings, New England Coastal Zone Management Conference, 2nd, p. 27–33.

Grant, M. J., 1973, Rhode Island's ocean sands: Management guidelines for sand and gravel extraction in state waters: Kingston, Rhode Island, Marine Technical Report No. 10, University of Rhode Island, 51 p.

Guyette, P. L., and Wallace, W. A., 1979, A cost analysis of offshore mining and dumping operations in the greater New York metropolitan area: Rensselaer Polytechnic Institute Technical Report, 65 p.

Hallermeier, R. J., 1980, Sand motion initiation by water waves: Two asymptotes: Journal of Waterway, Port, Coastal, and Ocean Division, American Society of Civil Engineers, v. 106, WW3, p. 299–318.

Hathaway, J. C., and eight others, 1979, U.S. Geological Survey core drilling on the Atlantic shelf: Science, v. 206, p. 515–526.

Hunter, A. D., and Horton, C. H., 1960, Underwater structures: Cathodic protection checks corrosion: Underwater Engineering, v. 1, no. 3, p. 40-41, Sheffield Publishing Company, Washington, D.C.

Inman, D. L., and Nordstrom, C. E., 1971, On the tectonic and morphologic classification of coasts: Journal of Geology, v. 79, p. 1–21.

James, N. P., and Stanley, D. J., 1968, Sable Island bank off Nova Scotia: Sediment dispersal and recent history: American Association of Petroleum Geologists Bulletin, v. 52, p. 2208–2230.

Johnson, D. W., 1925, The New England-Acadian shoreline: New York, John Wiley and Sons, 608 p.

Jordan, G. F., 1962, Large submarine sand waves: Science, v. 136, p. 839–848.

Judd, J. B., Smith, W. C., and Pilkey, O. H., 1970, The environmental significance of iron-stained quartz grains on the southeastern United States Atlantic shelf: Marine Geology, v. 8, p. 355–362.

Kastens, K. A., Fray, C. T., and Schubel, J. R., 1978, Environmental effects of sand mining in the lower bay of New York Harbor: Stony Brook, New York, Marine Sciences Research Center, State University of New York, Special Report 15, Reference 78-3, 139 p.

Kelly, J. L., 1973, Offshore terminals and the national perspective: Marine Technology Society, Proceedings 9th Annual Conference, p. 9–18.

King, L. H., 1967, Use of a conventional echo sounder and textural analysis in delineating sedimentary facies, Scotian Shelf: Canadian Journal of Earth Sciences, v. 4, p. 691–708.

—— 1970, Surficial geology of the Halifax-Sable Island map area: Ottawa, Marine Sciences Branch, Paper No. 1, Department of Energy, Mines, and Resources, 16 p.

Knebel, H. J., 1981, Processes controlling the characteristics of the surficial sand sheet, U.S. Atlantic outer continental shelf: Marine Geology, v. 42, p. 349–368.

Knebel, H. J., and Spiker, E., 1977, Thickness and age of surficial sand sheet, Baltimore Canyon Trough area: American Association of Petroleum Geologists Bulletin, v. 61, p. 861–871.

Knebel, H. J., Wood, S. A., and Spiker, E. C., 1979, Hudson River; Evidence for extensive migration on the exposed continental shelf during the Pleistocene: Geology, v. 7, p. 254–258.

Knebel, H. J., Needell, S. W., and O'Hara, C. J., 1982, Modern sedimentary environments on the Rhode Island inner shelf, off the eastern United States: Marine Geology, v. 49, p. 241–256.

Knott, S. T., and Hoskins, H., 1968, Evidence of Pleistocene events in the structure of the continental shelf of the northeastern U.S.: Marine Geology, v. 6, p. 5–43.

Kontopoulos, N., and Piper, D.J.W., 1982, Storm graded sand at 200 m water depth, Scotian Shelf, East Canada: Geo-Marine Letters, v. 2, p. 077–081.

Kraft, J. C., and John, C. J., 1979, Lateral and vertical facies relations of Transgressive Barrier: American Association of Petroleum Geologists Bulletin, v. 63, p. 2145–2163.

Lerner, L., 1971, New York offshore airport feasibility study: Executive Briefing, Saphier, Lerner, Schindler, Environetics, New York, 25 p. (unpublished).

Lewis, R. S., and Sylwester, R. E., 1976, Shallow sedimentary framework of Georges Bank: Reston, Virginia, U.S. Geological Survey Open File Report 76-874, 12 p.

Luternauer, J. L., and Pilkey, O. H., 1967, Phosphorite grains, their application to study of North Carolina shelf sediments: Marine Geology, v. 5, p. 315–320.

Manheim, F. T., 1972, Mineral resources off the northeastern coast of the United States: U.S. Geological Survey Circular 669, 28 p.

Mann, R. G., Swift, D.J.P., and Perry, R., 1981, Size classes of flow-transverse bedforms in a subtidal environment, Nantucket Shoals, North American shelf: Geo-Marine Letters, v. 1, p. 39–43.

McClennen, C. E., 1973a, Nature and origin of New Jersey continental shelf topographic ridges and depressions [Ph.D. thesis]: University of Rhode Island, 94 p.

—— 1973b, New Jersey continental shelf near bottom current meter records and recent sediment activity: Journal of Sedimentary Petrology, v. 43, p. 371–380.

—— 1983, Middle Atlantic nearshore seismic survey and sidescan sonar survey: Potential geologic hazards off the New Jersey coastline: U.S. Geological Survey Open File Report 83-824, Reston, Virginia, 30 p.

McClennen, C. E., and McMaster, R. L., 1971, Probably transgressive effects on the geomorphic features of the continental shelf off New Jersey, United States: Maritime Sediments, v. 7, p. 69–72.

McKinney, T. F., and Friedman, G. M., 1970, Continental shelf sediments of Long Island, New York: Journal of Sedimentary Petrology, v. 40, p. 213–248.

McMaster, R. L., 1954, Petrography and genesis of the New Jersey Beach sands: New Jersey Department of Conservation and Economic Development, Geological Survey Bulletin, no. 63, 239 p.

—— 1960, Sediments of Narragansett Bay system and Rhode Island Sound, Rhode Island: Journal of Sedimentary Petrology, v. 30, p. 249–274.

McMaster, R. L., and Ashraf, A., 1973a, Subbottom basement drainage system of inner continental shelf off southern New England: Geological Society of America Bulletin, v. 84, p. 187–190.

—— 1973b, Extent and formation of deeply buried channels on the continental shelf off southern New England: Journal of Geology, v. 81, p. 374–379.

McMaster, R. L., and Garrison, L. E., 1966, Mineralogy and origin of southern New England shelf sediments: Journal of Sedimentary Petrology, v. 36, p. 1131–1142.

Meade, R. H., 1969, Landward transport of bottom sediments in estuaries of the Atlantic coastal plain: Journal of Sedimentary Petrology, v. 39, p. 222–234.

Meisburger, E. P., 1972, Geomorphology and sediments of the Chesapeake Bay entrance: Fort Belvoir, Virginia, U.S. Army Corps of Engineers, Coastal Engineering Reseach Center, Technical Memorandum-38, 61 p.

—— 1976, Geomorphology and sediments of western Massachusetts Bay: Fort Belvoir, Virginia, U.S. Army Corps of Engineers, Coastal Engineering Research Center, Technical Paper 76-3, 77 p.

—— 1977, Sand resources on the inner continental shelf of the Cape Fear region, North Carolina: Fort Belvoir, Virginia, U.S. Army Corps of Engineers, Coastal Engineering Research Center, Miscellaneous Report 77-11, 20 p.

—— 1979, Reconnaissance geology of the inner continental shelf, Cape Fear region, North Carolina: Fort Belvoir, Virginia, U.S. Army Corps of Engineers, Coastal Engineering Research Center, Technical Paper 79-3, 135 p.

Meisburger, E. P., and Duane, D. B., 1971, Geomorphology and sediments of the inner continental shelf Palm Beach to Cape Kennedy, Florida: Fort Belvoir, Virginia, U.S. Army Corps of Engineers, Coastal Engineering Research Center, Technical Memorandum 34, 111 p.

Meisburger, E. P., and Field, M. E., 1975, Geomorphology, shallow structure, and sediments of the Florida inner continental shelf, Cape Canaveral to Georgia: Fort Belvoir, Virginia, U.S. Army Corps of Engineers, Coastal Engineering Research Center, Technical Memorandum 54, 117 p.

—— 1976, Neogene sediments of Atlantic inner continental shelf off northern Florida: American Association of Petroleum Geologists Bulletin, v. 60, p. 2019–2037.

Meisburger, E. P., and Williams, S. J., 1980, Sand resources on the inner continental shelf of Cape May region, New Jersey: Fort Belvoir, Virginia, U.S. Army Corps of Engineers, Coastal Engineering Research Center, Miscellaneous Report 80-4, 40 p.

—— 1982, Sand resources on the inner continental shelf off the central New Jersey coast: Fort Belvoir, Virginia, U.S. Corps of Engineers, Coastal Engineering Research Center, Miscellaneous Report No. 82-10, 48 p.

Milliman, J. D., 1972a, Atlantic continental shelf and slope of the United States— petrology of the sand fraction of sediments, northern New Jersey to southern Florida: U.S. Geological Survey Professional Paper 529-J, 39 p.

—— 1972b, Marine Geology, *in* Coastal and offshore environmental inventory, Cape Hatteras to Nantucket Shoals: Marine Publication Series, University of Rhode Island, p. 10-1–10-91.

Milliman, J. D., Pilkey, O. H., and Ross, D. A., 1972, Sediments of the continental margin of the eastern U.S.: Geological Society of America Bulletin, v. 83, p. 1315–1334.

MITRE CORPORATION, 1979, Disposal of dredged material within the New York District: Fort Belvoir, Virginia, U.S. Army Corps of Engineers, Coastal Engineering Research Center Technical Report MTR-7808, 2 Vols.

Moody, J. S., and eleven others, 1984, Atlas of tides and tidal currents observations on the northeast continental shelf and slope of North America: U.S. Geological Survey Bulletin, no. 1611, 122 p.

Naqvi, S. M., and Pullen, E. J., 1982, Effects of beach nourishment and borrowing on marine organisms: Fort Belvoir, Virginia, U.S. Army Corps of Engineers, Coastal Engineering Research Center, MR 82-14, 43 p.

Nersesian, G. K., 1977, Beach fill design and placement at Rockaway Beach, New York, using offshore ocean borrow sources: Coastal sediments: American Society of Civil Engineers/Charleston, South Carolina, November 2–4, 1977, p. 228–247.

Newman, D. S., 1977, Late Quaternary paleoenvironmental reconstruction: Some constructions from northwestern Long Island, New York: Annals of the New York Academy of Sciences, v. 288, p. 545–570.

Nutant, J. S., 1973, An environmental assessment of floating nuclear plants: Marine Technology Society Proceedings 9th Annual Conference, p. 367–374.

Padan, J. W., 1977, New England offshore mining environmental study (NOMES): Boulder, Colorado, Special Report, U.S. Department of Commerce, National Oceanic and Atmospheric Administration, 140 p.

Pilkey, O. H., 1964, Heavy minerals of the U.S. south Atlantic continental shelf and slope: Geological Society of American Bulletin, v. 74, p. 641–648.

Pilkey, O. H., and Field, M. E., 1972, Onshore transportation of continental shelf sediment: Atlantic southeastern United States, *in* Swift, D.J.P., Duane, D. B., Pilkey, O. H., eds., Shelf sediment transport: Stroudsburg, Pennsylvania, Dowden, Hutchinson, and Ross Incorporated, p. 429–446.

Pilkey, O. H., and Frankenberg, D., 1964, The Recent-relict sediment boundary on the Georgia continental shelf: Bulletin of the Georgia Academy of Science, v. 22, p. 73–78.

Pilkey, O. H., and Luternauer, J. L., 1967, A North Carolina shelf phosphate deposit of possible commercial interests: Southeastern Geology, v. 8, p. 33–51.

Pilkey, O. H., Blackwelder, B. W., Knebel, H. J., and Ayers, M. W., 1981, The Georgia embayment continental shelf: Stratigraphy of a submergence: Geological Society of America Bulletin, v. 92, p. 52–63.

Reilly, F. J., and Bellis, V. J., 1983, The ecological impact of beach nourishment with dredged materials on the intertidal zone at Boque Banks, North Carolina: Fort Belvoir, Virginia, U.S. Army Corps of Engineers, Coastal Engineering Research Center, MR 83-3, 74 p.

Riggs, S. R., 1984, Paleoceanographic model of Neogene phosphorite deposition, U.S. Atlantic continental margin: Science, v. 223, no. 4632, p. 123–131.

Schlee, J. S., 1964, New Jersey offshore gravel deposit: Pit and Quarry, v. 57, p. 80–81.

—— 1973, Atlantic continental shelf and slope of the United States—Sediment texture: U.S. Geologic Survey Professional Paper 529-L, 64 p.

Schlee, J. S., and Pratt, R. M., 1970, Atlantic continental shelf and slope of the United States—Gravels of the northeastern part: U.S. Geological Survey Professional Paper 529-H, 39 p.

Schlee, J. S., and Sanko, P., 1975, Sand and gravel: Albany, New York, New York Sea Grant Institute, MESA New York Bight Atlas Monograph 21, 26 p.

Schlee, J. S., Folger, D. W., and O'Hara, C. J., 1973, Bottom sediments on the continental shelf of the northeastern United States, Cape Cod to Cape Ann, Massachusetts: U.S. Geological Survey, Miscellaneous Geologic Investigations, Map I-746, 2 Sheets.

Schubel, J. R., 1981, Coastal resources: Exploring our shelf, *in* Gateway to the sea: Albany, New York, National Park Service and New York State Sea Grant Office, 13 p.

Shepard, F. P., 1973, Submarine Geology (third edition) New York, Harper and Rowe, 517 p.

Shepard, F. P., and Cohee, G. V., 1936, Continental shelf sediments off the Mid-Atlantic states: Geological Society of America Bulletin, v. 47, p. 441–458.

Shepard, F. P., Trefethen, J. M., and Cohee, G. V., 1934, Origin of Georges Bank: Geological Society America Bulletin, v. 45, p. 281–302.

Sheridan, R. E., Dill, C. E., Jr., and Kraft, J. C., 1974, Holocene sedimentary environment of the Atlantic inner shelf off Delaware: Geological Society of America Bulletin, v. 85, p. 1319–1328.

Shideler, G. L., and Swift, D.J.P., 1972, Seismic reconnaissance of Post-Miocene deposits, Middle Atlantic continental shelf—Cape Henry, Virginia to Cape Hatteras, North Carolina: Marine Geology, v. 12, p. 165–185.

Shideler, G. L., Swift, D.J.P., Johnson, G. H., and Holliday, B. W., 1972, Late Quaternary stratigraphy of the inner Virginia continental shelf: A proposed standard section: Geological Society of America Bulletin, v. 83, p. 1787–1804.

Shinn, E. A., Hudson, J. H., Halley, R. B., and Litz, B., 1977, Topographic control and accumulation rates of some Holocene coral reefs: South Florida and Dry Tortugas, *in* Proceedings, International Coral Reef Symposium, 3rd: Rosensteil School of Marine and Atmospheric Sciences, Miami, Florida, University of Miami, p. 7.

Stanley, D. J., and Cok, A. E., 1968, Sediment transport by ice on the Nova Scotian shelf, *in* Ocean sciences and engineering of the Atlantic shelf: Transactions of the Maritime Technical Society, p. 109–125.

Stewart, H. B., Jr., and Jordan, G. F., 1964, Underwater sand ridges on Georges Shoal, *in* Miller, R. L., ed., Papers in marine geology, F. P. Shepard Commemorative Volume: New York, Macmillan Company, p. 102–114.

Stubblefield, W. L., and Swift, D.J.P., 1976, Ridge development as revealed by subbottom profiles on the central New Jersey shelf: Marine Geology, v. 20, p. 315–334.

—— 1981, Grain size variations across sand ridges, New Jersey continental shelf, U.S.A.: Geo-Marine Letters, v. 1, p. 45–48.

Stubblefield, W. L., Lavelle, J. W., McKinney, T. F., and Swift, D.J.P., 1975, Sediment response to the present hydraulic regime on the central New Jersey shelf: Journal of Sedimentary Petrology, v. 45, p. 337–358.

Stubblefield, W. L., McGrail, D. W., and Kersey, D. G., 1983a, Recognition of transgressive and post-transgressive sand ridges on the New Jersey continental shelf *in* Tilliman, R. W., and Siesmers, C. T., eds., Ancient elastic shelf sediments: Society of Economic Paleontologists and Mineralogists Special Publication, p. 1–23.

Stubblefield, W. L., Kersey, D. G., and McGrail, D. W., 1983b, Development of middle continental shelf sand ridges: New Jersey, U.S.A: American Association of Petroleum Geologists Bulletin, v. 67, p. 817–830.

Swift, D.J.P., 1976, Continental shelf sediment, *in* Stanley, D. J., and Swift, D.J.P., eds., Marine sediment transport and environmental management: New York, Wiley and Sons, p. 311–350.

Swift, D.J.P., and Freeland, G. L., 1978, Current lineations and sand waves on the inner shelf, Middle Atlantic Bight of North America: Journal of Sedimentary Petrology, v. 48, p. 1257–1266.

Swift, D.J.P., Kofoed, J. W., Saulsbury, F. P., and Sears, P., 1972a, Holocene evolution of the shelf surface central and southern Atlantic shelf of North America, *in* Swift, D.J.P., Duane, D. B., and Pilkey, O. H., eds., Shelf sediment transport: Process and pattern: Stroudsburg, Pennsylvania, Dowden, Hutchinson, and Ross, p. 499–574.

Swift, D.J.P., Holliday, B., Avignone, N., and Shideler, G., 1972b, Anatomy of a shoreface ridge system, False Cape Virginia: Marine Geology, v. 12, p. 59–84.

Swift, D.J.P., Freeland, G. L., Gadd, P. E., Han, G., Lavelle, J. W., and Stubblefield, W. L., 1976, Morphologic evolution and coastal sand transport, New

York–New Jersey Shelf: American Society of Limnology and Oceanography, Special Symposium 2, p. 69–89.

Tepordei, V. V., 1980, Sand and gravel, *in* Mineral facts and problems (1980 edition): U.S. Bureau of Mines, Bulletin 671, p. 781–791.

Tucholke, B. E., and Hollister, C. D., 1973, Late Wisconsin glaciation of the southwestern Gulf of Maine: New evidence from the marine environment: Geological Society of American Bulletin, v. 84, p. 3279–3296.

Turbeville, D. B., and Marsh, G. A., 1982, Benthic fauna of an offshore borrow area in Broward Country, Florida: Fort Belvoir, Virginia, U.S. Army Corps of Engineers, Coastal Engineering Research Center, MR. 82-1, 42 p.

Twichell, D. C., 1979, Medium-scale potentially mobile bedforms on the Mid-Atlantic continental shelf, *in* Middle Atlantic outer continental shelf environmental studies, 3: Gloucester Point, Virginia, Virginia Institute of Marine Science, p. 5-1 to 5-16.

—— 1983, Bedform distribution and inferred sand transport on Georges Bank, United States Atlantic continental shelf: Sedimentology, v. 30, p. 695–710.

Twichell, D. C., Knebel, H. J., and Folger, D. W., 1977, Delaware River, evidence for its former extension to Wilmington Submarine Canyon: Science, v. 195, p. 483–485.

Twichell, D. C., McClennen, C. E., and Butman, B., 1981, Morphology and processes associated with the accumulation of the fine-grained sediment deposit on the southern New England shelf: Journal of Sedimentary Petrology, v. 51, p. 269–280.

Uchupi, E., 1965, Basins of the Gulf of Maine: U.S. Geological Survey Professional Paper 525-D, p. D175–D177.

—— 1966, Structural framework of the Gulf of Maine: Journal of Geophysical Research, v. 71, p. 3013–3028.

—— 1968, Atlantic continental shelf and slope of the United States—Physiography: U.S. Geological Survey Professional Paper 529-C, 30 p.

—— 1969, Marine geology of the continental margin off Nova Scotia, Canada: New York Academy Science Transactions, ser. 2, v. 31, p. 56–65.

—— 1970, Atlantic continental shelf and slope of the U.S.—Shallow structure: U.S. Geological Survey Professional Paper 529-I, 44 p.

U.S. Army Corps of Engineers, 1971, Report on the national shoreline study: Washington, D.C., U.S. Army Corps of Engineers, 59 p.

—— 1973, Shore protection manual: Washington, D.C., U.S. Army Corps of Engineers, U.S. Government Printing Office, 1160 p.

United States Geological Survey, 1984, Symposium proceedings—A national program for the assessment and development of the mineral resources of the United States Exclusive Economic Zone; November, 1983: U.S. Geological Survey Circular 929, 308 p.

Veatch, A. G., and Smith, P. A., 1939, Atlantic submarine valleys of the United States and the Congo Submarine Valley: Geological Society of America Special Paper 7, 101 p.

Williams, S. J., 1975, Anthropogenic filling of the Hudson River (shelf) channel: Geology, v. 3, p. 597–600.

—— 1976, Geomorphology, shallow subbottom structure, and sediments of the Atlantic intercontinental shelf off Long Island, New York: Fort Belvoir, Virginia, U.S. Army Corps of Engineers, Coastal Engineering Research Center, Technical Paper 76-2, 123 p.

—— 1981, Sand resources and geological character of Long Island Sound: Fort Belvoir, Virginia, U.S. Army Corps of Engineers, Coastal Engineering Research Center, Technical Paper 81-3, 65 p.

Williams, S. J., and Duane, D. B., 1972, Regional shelf studies a guide to engineering design: Marine Technology Society, Proceedings 8th Annual Conference, Washington, D.C., p. 227–236.

—— 1974, Geomorphology and sediments of the inner New York Bight continental shelf: Fort Belvoir, Virginia, U.S. Army Corps of Engineers, Coastal Engineering Research Center, Technical Memorandum 45, 81 p.

—— 1975, Construction in the coastal zone: A potential use of waste materials: Marine Geology, v. 18, p. 1–15.

MANUSCRIPT ACCEPTED BY THE SOCIETY APRIL 26, 1985

Printed in U.S.A.

The Geology of North America
Vol. I-2, The Atlantic Continental Margin: U.S.
The Geological Society of America, 1988

Chapter 25

Mineral resources of the U.S. Atlantic continental margin

Stanley R. Riggs
Department of Geology, East Carolina University, Greenville, North Carolina 27834
Frank T. Manheim
Marine Geology Branch, U.S. Geological Survey, Woods Hole, Massachusetts 02543

INTRODUCTION

Most geologic materials may be usable resources in some form and at some time, whether it be for general land fill and aggregate, beach replenishment, construction material, or as a source of metals and fuels. Thus, most natural materials occurring within the Atlantic continental margin are resources, defined as "materials, including those only surmised to exist, that have present or anticipated future value" (U.S.G.S., 1980). Whether a resource becomes a reserve or not (an economically recoverable commodity) depends upon the properties and economic values of that material, which are determined by the following factors: (a) availability, concentration, and occurrence of the material; (b) methods of recovering and processing the commodity; (c) transportation costs of ore and beneficiated products; and (d) environmental setting and costs of permitting and mitigation. Also, the economics of a given mineral resource change dramatically in response to new technological advances, discoveries of new deposits, or as industrial and social demands change through time. An increasingly critical component associated with the economics for development of any mineral commodity, is a good geologic knowledge of the resource base.

At present, most hard mineral extraction industries of the Atlantic continental margin are located on the coastal plain (Fig. 1). Consequently, these resources are fairly well known and they will encompass a large portion of this section. Consideration of offshore resources must deal with the potential; since the continental shelves are geologically only submerged portions of the continent, it is logical to assume that the mineral potential should be comparable to that found on land. However, detailed geologic investigation, exploration, and research in the marine environment is extremely expensive, technologically difficult, and generally a relatively "new" science. Present limiting economic and technological restraints for recovery of marine minerals are rapidly being overcome by major technological advances resulting from offshore petroleum exploration and development. Also, the United States Department of the Interior has become actively involved in developing regulations for the actual leasing and min-

ing of undersea minerals within the U.S. Exclusive Economic Zone (EEZ) and Outer Continental Shelf (OCS) regions, underscoring the increased interest and anticipation of future development of important marine mineral resources.

Those resources considered in this chapter (Fig. 1) include minerals that either have been recovered, are presently being recovered, or are known to occur in significant concentrations to be considered potential future resources. This chapter excludes petroleum, water, and sand and gravel resources (these resources are discussed by Mattick and Libby-French, Kohout and others, and Duane and Stubblefield, this volume).

PHOSPHATES

Agricultural resources, followed closely and now intimately by energy resources, have become the most critical resources and basic ingredients of modern society. As world population continues its exponential explosion, society becomes increasingly more dependent upon a high fertilizer- and energy-based agriculture. The United States produced 54 million metric tons of phosphate in 1980, 88% of which was utilized for fertilizer (U.S. Bureau of Mines, 1980). About 87% of this production came from Atlantic coastal plain deposits in Florida and North Carolina, which represent the single most important world phosphate resource, accounting for 40% of total world production in 1980.

The phosphate resource and reserve potential for the southeastern U.S. is extremely large. However, mounting land use and environmental pressures cause much speculation about the future availability of land-based phosphate reserves, particularly in Florida. This expanding pressure on the world's most important deposits is generating increased interest in deeper and lower grade deposits of other portions of the southeast system, including extensive phosphate resources on the continental shelf and Blake Plateau.

The total volume of phosphorus that was precipitated from the ocean and occurs in Neogene and Pleistocene sediments of the

Riggs, S. R., and Manheim, F. T., 1988, Mineral resources of the U.S. Atlantic continental margin; *in* Sheridan, R. E., and Grow, J. A., eds., The Geology of North America, Volume I-2, The Atlantic Continental Margin, U.S.: Geological Society of America.

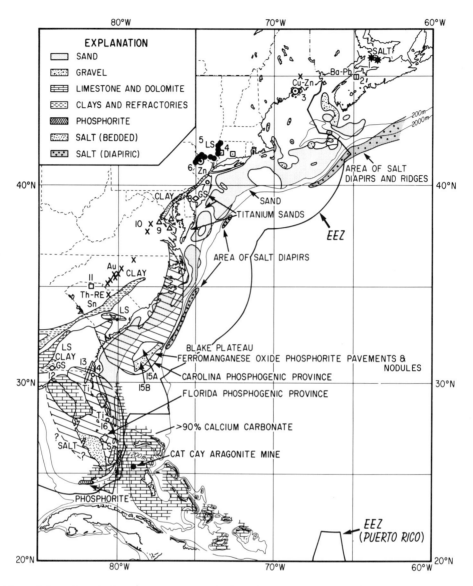

Figure 1. Location of mineral resources along the Atlantic continental margin from Nova Scotia to the Bahamas. Principal references for locations of deposits are Gerlach (1970), Manheim (1972), Manheim and others (1980), and Riggs (1984).

Key to Symbols

EEZ. The U.S. Exclusive Economic Zone (200 nautical miles).

Wide diagonal banding. Carolina and Florida phosphogenic provinces contain phosphate-bearing sediments including commercial and potentially commercial deposits, large areas on the shelf that are poorly known, and some areas with low economic potential.

1. Pugwash and Malagash salt deposits of Mississippian age, Nova Scotia.
2. Walton barite and lead deposits, Nova Scotia.
3. Callahan copper-lead-zinc deposit, Penobscot Bay, Maine.

4. Garnet and mica deposits, Connecticut.
5. Paleozoic limestone, dolomite, and marble deposits, New York, western Connecticut, northern New Jersey, and eastern Pennsylvania.
6. Franklin Furnace zinc deposit, New Jersey.
7. Titanium sand deposits, Lakehurst, New Jersey.
8. Cohansey glass sands, southern New Jersey.
9. Chesapeake Bay low-grade diatomite deposits, Virginia.
10. Gold outcrops and placers, Virginia, North Carolina, and South Carolina.

11. Catawba River thorium-rare earth deposits, South Carolina.
12. Glass sands and attapulgite clays, Attapulgus, Georgia.
13. Trail Ridge and Green Cove Springs titanium and zircon heavy mineral sand deposits, Florida and Georgia.
14. Titanium sand deposits, Jacksonville, Florida.
15. Blake Plateau submarine ferromanganese oxides and phosphorites: (A) nodule province; (B) pavement, slab, and nodule province.
16. Titanium sand deposits, Melbourne, Florida.

Atlantic continental margin is staggering. Much of this phosphorus occurs in sediment facies in minor concentrations and therefore will never be of economic interest. However, within the total stratigraphic section of interest, there are numerous units in many different locations containing significant concentrations of phosphate sediment that must be included in any resource evaluation. The phosphate resource classification is based upon the U.S. Bureau of Mines and U.S. Geological Survey classification scheme and includes all degrees of identification and quantification of resource data (U.S.G.S., 1980, Table 5).

Details of the structural and stratigraphic framework and patterns and processes of sedimentation of Neogene phosphorites on the southeastern U.S. Atlantic continental margin are discussed in chapters 8 and 15 in this volume.

Florida Phosphate Province

Structural and Stratigraphic Framework.
Major economic phosphate sediments within the Florida Phosphate Province are restricted to the Hawthorn and Bone Valley formations of Florida and Georgia. Extensive drilling and stratigraphic work throughout this province by the phosphate companies and other workers (including Altschuler and others, 1964; Bergendahl, 1956; Carr and Alverson, 1959; Cathcart, 1968a, 1968b; Cathcart and McGreevy, 1959; Clark, 1972; Olson, 1966; Puri and Vernon, 1964; Riggs, 1979a, 1979b; Scott, 1982; Sever and others, 1967; Williams, 1971; etc.) have established that the major and most extensive phosphorites and phosphatic sediments of Neogene age are primarily of early to middle Miocene age, with significantly lesser amounts occurring locally in the Pliocene Bone Valley Formation and Pleistocene sediments, respectively.

The Miocene Hawthorn sediments underlie much of the southeastern U.S. coast (Sheridan and others, Chapter 15, this volume, Fig. 4) extending from the southern tip of Florida, through southeast Georgia, and into southern South Carolina. The Hawthorn Formation also extends eastward under the entire Atlantic Continental Shelf from the Pourtales Terrace (Gorsline and Milligan, 1963), northward along the Miami Terrace (Mullins and Neumann, 1979) and the Florida, Georgia, and South Carolina Continental Shelf (Charm and others, 1969; Emery and Uchupi, 1972; Furlow, 1969; Hathaway and others, 1970, 1976, 1979; Poag, 1978; Popenoe, 1983; Popenoe and Meyer, 1983; Woolsey, 1976; etc.). The very broad west Florida Continental Shelf is also underlain by phosphatic sediments of the Hawthorn Formation (Riggs, 1984); however, we are only now beginning to learn about the Tertiary sediments in this shelf area (Mullins, unpublished data).

The Ocala High is a northwest-southeast–trending first-order feature (Fig. 2) with a core of Eocene Ocala Limestone that affected the deposition of all Neogene sediments (Puri and Vernon, 1964; Riggs, 1979b). Two major first-order sediment basins occur in association with the Ocala High (Fig. 2). The southern portion of the Florida Peninsula is occupied by the large Okeechobee Basin, which accumulated more than 225 m of phosphatic carbonate sediments during the Miocene (Freas, 1968). To the north, the Ocala High is terminated in Georgia by the Southeast Georgia Embayment and Gulf Trough or Suwanee Straits. Prior to the Neogene, the Gulf Trough was open between the Gulf of Mexico and Southeast Georgia Embayment, preventing southward transport of terrigenous sediments to peninsula Florida. However, as the basinal trough filled by the end of the Oligocene, a flood of terrigenous sediments began to move across the straits, diluting the inner shelf phosphorites of South Georgia and Florida (Riggs, 1979b).

Miocene phosphorites were deposited in association with the topographically or structurally produced second-order features around the perimeter of the Ocala High (Fig. 2). The southern end of the Ocala High forms the broad and extensive Central Florida Platform that plunges gently into the subsurface; the crest and adjacent flanks of this structure contain the world's largest producing phosphate districts, the Central and South Florida deposits (Altschuler and others, 1964; Cathcart, 1968a, 1968b). Along the eastern side of the Ocala High is a broad, irregular, and shallow continental shelf system referred to as the North Florida Platform. The southern half of the North Florida Platform contains only a thin sequence of Miocene sediments with minor amounts of phosphate due to the shadow effect of the Sanford High upon major oceanographic current systems. The northern half of the North Florida Platform contains an irregular to moderately thick section of Miocene sediments, which are up to 25 m thick with high concentrations of phosphate within the second-order embayments. Seaward, off the flank of the platform, Miocene phosphatic sands and clays increase to more than 150 m in thickness in the Jacksonville Basin (Freas, 1968).

Another positive element of extreme importance to Neogene sedimentation is the Sanford High. This narrow northwest-southeast trending second-order feature is separated from the Ocala High by the Osceola Basin. Long and narrow plunging platforms off the north and south ends of the Sanford High contain a thick and extensive sequence of phosphate-rich sediments that extends down the flanks of the platforms into adjacent Okeechobee, Osceola, and Jacksonville basins. These phosphorites also extend east off the Sanford High and associated platforms onto the Florida Continental Shelf. Core holes off Jacksonville, Florida show thick and relatively phosphate-rich downdip facies of the Miocene stratigraphic sequence (Charm and others, 1969; Hathaway and others, 1970).

The western margin of the Ocala High occurs on the western Florida Continental Shelf in the Gulf of Mexico. Little is known about the Tertiary geology in this area; however, the southwestern limit to the Ocala High is probably coincident with the northwest trending Bahama Fault Zone as delineated by Klitgord and Popenoe (1984). The Neogene phosphorites on the Gulf of Mexico Shelf should have a similar distribution and relationship to first- and second-order structures as elsewhere within the Florida phosphogenic province (Sheridan and others, Chapter 15, this volume, Fig. 4; Riggs, 1984). Birdsall (1978) found minor concentrations of phosphate in the surface sediments throughout

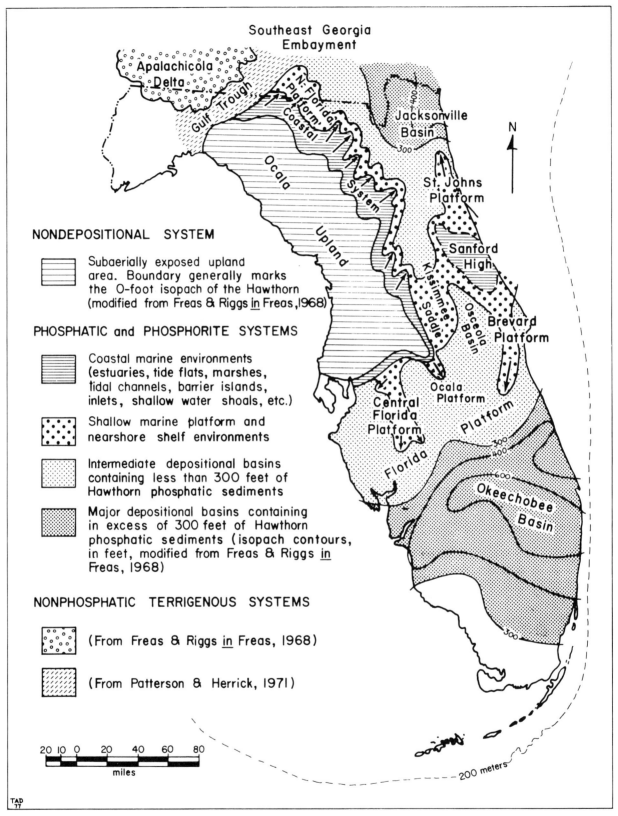

Figure 2. Structural framework and the resulting Miocene depositional environments for the Florida Phosphogenic Province (from Riggs, 1979b).

much of the western Florida continental margin suggesting that major Neogene phosphorites might exist in the shallow subsurface.

The occurrence of extensive phosphate pavements and slabs on the upper Miami Terrace (Sheridan and others, Chapter 15, this volume, Fig. 4) with abundant phosphate nodules on the lower Miami Terrace has been described by Mullins and Neumann (1979). They present evidence suggesting multiple episodes of phosphatization associated with the deposition of the Hawthorn sediments during the lower to middle Miocene. Similar phosphorites have been described on the Pourtales Terrace (Gorsline and Milligan, 1963). Also, Manheim and Mullins have recently found Miocene phosphorites cropping out on the continental slope along the west Florida margin (unpublished data).

Depositional Sequences. Hawthorn sediments form a wedge-shaped unit that ranges from a feather edge on the flanks of the Ocala and Sanford highs to more than 225 m in some downslope basins. Riggs (1979b) subdivided the Hawthorn into the Miocene Arcadia and Noralyn formations and the Pliocene Bone Valley Formation on the basis of lithology, stratigraphic distribution, phosphate petrology, and sedimentary structures. The Arcadia and Noralyn are partially contemporaneous Miocene formations and represent the time of major phosphate formation. The Noralyn Formation is a complex coastal-marine and inner-shelf sequence dominated by terrigenous sediments with a large phosphate component deposited around the Ocala High and into northern basins. The Arcadia Formation is dominantly a dolomite with varying amounts of phosphate and interbedded fine terrigenous sediment deposited as a middle- and outer-shelf facies; it occurs below and down the depositional slope to the south and east of the Noralyn Formation. The Pliocene Bone Valley Formation consists of an irregular concentration of phosphate that occurs in complex fluvial facies in the upslope area and grades downslope into estuarine, open bay, and shallow marine facies containing minor but variable concentrations of phosphate and bearing other stratigraphic names.

The Miocene section of Florida is dominated by three major sediment components (phosphate, terrigenous sediment, and carbonate) that display several significant regional patterns (Riggs, 1979b). Minor concentrations of phosphate formed everywhere within the marine environment; areas of optimum formation were associated with shallow-water platforms that projected onto the continental shelf. The phosphate grains, which were transported as clastic particles, were subsequently deposited and accumulated on the platforms or in adjacent entrapment basins. Some phosphate was transported offshore into the major depositional basins where it was significantly diluted by terrigenous and carbonate components.

Distribution patterns of terrigenous sediment suggest a major source from the north across the Suwannee Straits and into shallow-coastal and inner-shelf environments around the Ocala High (Riggs, 1979b). Sands and clays were then transported east and southeast into the Jacksonville Basin and adjacent areas on the north side of the Sanford High with only minor amounts

moving into southern and eastern Florida. Regional distribution patterns of the carbonate component, predominantly dolosilt, are opposite that of the terrigenous sediments. The carbonate is an authigenic/diagenetic sediment formed in slope to middle-shelf marine environments in southern and eastern Florida, where it is the dominant component. Carbonate rapidly decreases in abundance into the inner-shelf environments around the Ocala High and north of the Sanford High, where it becomes a subordinant component.

Some phosphate formed in most portions of the continental margin around the first-order structural high; however, there are definite regional patterns involving different phosphate grain types within the Hawthorn sediments (Riggs, 1979a, 1979b, 1980). Phosphate grains on the inner-shelf are dominantly medium to fine sand-sized intraclasts (rounded but irregularly shaped grains). Locally, laminae and beds of microcrystalline phosphate mud (microsphorite) and phosphate intraclast gravels occur on the shallow inner-shelf platforms associated with hiatus surfaces characterized by sediment bypass around the nose of first-order structural highs (i.e., the Central Florida Platform in Fig. 2). In outer-shelf and deeper basin environments, phosphate is dominated by well-sorted, fine to very fine sand-size pellets (very regularly shaped geometric grains). The dominant phosphate component on the Blake Plateau and Miami Terrace (Fig. 1), are thick and extensive pavements of indurated microcrystalline phosphate mud and coarse intraclastic plates and pebbles of pavement material (Manheim and others, 1980; Mullins and Neumann, 1979). The latter phosphate types are similar to classic pebble deposits in the Central Florida Phosphate District, around the southern nose of the Ocala High.

Carolina Phosphate Province

Structural and Stratigraphic Framework. The Mid-Carolina Platform High, a relatively broad structural feature extending from Cape Romain, South Carolina, to central Onslow Bay, North Carolina, controlled sedimentation throughout the Tertiary (Riggs and others, 1985, 1982a, 1982b; Snyder, 1982; Snyder, Hine and Riggs, 1982). Deposition of Miocene sediments off the east flank of the Mid-Carolina Platform High took place in the Aurora and Onslow embayments separated by the second-order Cape Lookout erosional high (Grow and Sheridan, this volume, Fig. 1; Riggs and Belknap, this volume, Fig. 1; Snyder and others, 1982; Riggs and others, 1982a, 1982b). These embayments contain the well known Aurora Phosphate District on the North Carolina Coastal Plain and the newly discovered Northeast Onslow Bay and Frying Pan Phosphate districts on the continental shelf, respectively (Fig. 3). Miocene sediments dip east off the Mid-Carolina Platform High as a series of apparent offlapping erosional units which thicken rapidly to over 300 m. The Miocene phosphorite continues in shallow subcrop around the outer nose of the Mid-Carolina Platform High where it was described in a drill hole on the outer shelf edge (Blackwelder and others, 1982).

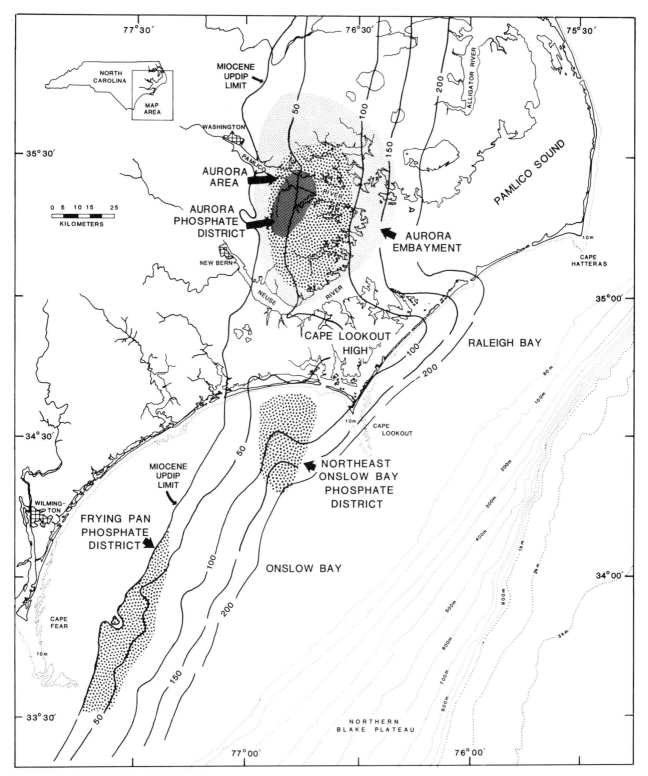

Figure 3. Structural contour map (in meters) on the base of the Miocene Pungo River Formation along the North Carolina Coastal Plain and Continental Shelf. The map also delineates (1) the Cape Lookout High, a paleotopographic feature on top of Oligocene sediments that defines the Miocene Aurora Embayment on the north and Onslow Embayment on the south; and (2) the location of the areas of resource determinations in Tables 2 and 3 including the Aurora, Northeast Onslow Bay, and Frying Pan phosphate districts, the Aurora area, and the Aurora and Onslow embayments. Modified from Riggs and others, 1985.

Phosphate-rich sediments occur within three stratigraphic units of the North Carolina continental margin (Riggs and Belknap, ths volume, Fig. 4). (1) The lower to middle Miocene Pungo River Formation is very extensive and occurs throughout the Aurora and Onslow embayments of the North Carolina continental margin. (2) The Pliocene lower Yorktown Formation occurs in the Aurora district. (3) The Pleistocene/Holocene surface sediments occur in the Frying Pan district of Onslow Bay. The Pungo River Formation is the most important phosphorite unit with successively decreasing amounts of phosphate occurring within the lower Yorktown and Pleistocene/Holocene sediments, respectively (Riggs, 1984; Riggs and others, 1985).

Extensive drilling and stratigraphic work throughout the coastal plain–continental shelf system by many workers (including Brown, 1958; Brown and others, 1972; Cathcart, 1968a, 1968b; Gibson, 1967, 1983; Katrosh and Snyder, 1982; Miller, 1971, 1982; Riggs, 1984; Riggs and others, 1982a, 1982b, 1985; Snyder, 1982; Snyder, Riggs and others, 1982; Waters, 1983; Woolsey, 1976; etc.) have established that the Pungo River Formation is contemporaneous and contiguous with the Hawthorn Formation to the south and the Calvert Formation to the north. Hawthorn sediments extend from southern Florida into southeastern South Carolina. In northern South Carolina and southern North Carolina, the Hawthorn sediments are displaced seaward to the outer edge of the continental shelf by the pre-Tertiary Mid-Carolina Platform High (Grow and Sheridan, this volume, Fig. 1). The same sediments continue northward across the North Carolina Continental Shelf and Coastal Plain as the Pungo River Formation and extend into Virginia with minor and decreasing concentrations of phosphate and increasing concentrations of glauconite and diatomite (Riggs, 1984). In Virginia the sediments are called the Calvert Formation of the Chesapeake Group.

The Pliocene Yorktown Formation consists of a lower phosphorite gravelly sand unit and an upper very fossiliferous clayey sand unit without any phosphate (Riggs and others, 1982b). The lower Yorktown is a prominent unit within the Aurora Phosphate District where it is a potential resource. The regional distribution outside of the mining area is not known and, even though some Pliocene sediments do occur on the continental shelf, this facies has not been recognized to date. Lower Yorktown phosphorites are thought to be stratigraphic equivalents to the Pliocene Bone Valley Formation in the Central Florida Phosphate District (Riggs, 1979a).

High concentrations of phosphate (from 2% to 75%) occur in Pleistocene and Holocene sediments of the Frying Pan Phosphate District in southern Onslow Bay. The thin and irregularly distributed Holocene surface sediments were first described by Pilkey and Luternauer (1967) and Luternauer and Pilkey (1967) and subsequently by Riggs and others (1982a, 1982b, 1983, 1985).

Some phosphate formed in most portions of the continental margin around first-order structural highs; however, there are definite regional patterns involving different phosphate grain types within the Pungo River (Riggs and others, 1982a, 1982b; Lewis and others, 1982; and Scarborough and others, 1982).

Phosphate grains on the inner shelf are dominantly medium to fine sand-size intraclasts (rounded but irregularly shaped grains). In the outer-shelf environments and deeper basins, phosphate is dominated by well-sorted, fine to very fine sand-size pellets (very regularly shaped geometric grains). The dominant phosphate component of the deeper water Blake Plateau deposits off the North Carolina shelf are thick and extensive pavements of indurated microcrystalline phosphate mud and coarse intraclastic plates and pebbles of pavement material (Manheim and others, 1980).

Phosphate Resources

Florida Phosphate Province. Since phosphate mining began in the 1880s, there have been innumerable estimates of the phosphate resouces within the Florida Phosphogenic Province. The latest resource estimates are summarized by Cathcart and others (1984) in Table 1 by economic resource category as defined by the U.S. Geological Survey (1980) and by geographic region within Florida. Identified phosphate resource estimates range from a low of over 4 billion metric tons of phosphate concentrate to a high of over 9 billion metric tons that are either presently economic, marginally economic, or sub-economic (Table 1).

Mayberry (1981) and Cathcart and others (1984) are the only two authors whose estimates realistically consider the hypothetical resource potential within Florida; both authors estimate about 5 billion additional metric tons of phosphate concentrate (Table 1). It should be noted that these hypothetical estimates do not include most of the vast amounts of phosphate that occur in discrete beds within deeper sediments of the Okeechobee, Osceola, and Jacksonville basins (Fig. 2), nor any potential resources that occur on the continental shelf around Florida. Extrapolating on the basis of the minimal data that exist for these areas, it is safe to conclude that the "speculative resource" estimates would probably increase the phosphate resources by at least an order of magnitude. None of this hypothetical or speculative resource is available with conventional mining. However, it does represent future potential that will become available as shallow reserves are depleted and, more important, as competing land-use and environmental pressures increasingly conflict with present surface mining techniques. It will require new mining technology, such as slurry mining procedures presently under development, to ever realize this vast resource.

North Carolina Coastal Plain. Table 2 summarizes the estimated phosphate resources within the Aurora Phosphate District, Aurora Area, and the Aurora Embayment (Fig. 3). Within the Aurora Phosphate District, Cathcart and others (1984) estimated 1+ billion metric tons of "recoverable phosphate" with a 30% P_2O_5 content as "identified resources" based upon company reserve data. Zellers and Williams (1978) calculated the identified resources in a slightly larger Aurora Phosphate District to be 1.6+ billion metric tons of recoverable phosphate product. This category represents phosphate that has been extensively drilled

TABLE 1. SUMMARY OF ESTIMATED PHOSPATE RESOURCES BY VARIOUS AUTHORS FOR FLORIDA
IN MILLIONS OF METRIC TONS
(modified from Cathcart and others, 1984)

	Zellers and Williams, 1978		Fountain and Hayes, 1979		Mayberry, 1981		Cathcart and others, 1984	
	Tons Phos. Conc.	Tons P_2O_5	Tons Phos. Conc.	Tons P_2O_5	Tons Phos. Conc.	Tons P_2O_5	Tons Phos. Conc.	Tons P_2O_5
Identified Economic Resources**								
North Florida	845	260	1056*	320*			900	280
Central Florida	780	250	884	280	1270	400	800	260
South & East Florida	1000	300	3052	920			2000	600
Identified Marginally Economic Resources								
North Florida	404	120	512*	160*			450	140
Central Florida	25	10	31	10	1820	560	20	Tr.
South & East Florida	527	160	1600	490			1000	300
Identified Sub-economic Resources								
North Florida	20	Tr.	33*	Tr.*			20	Tr.
Central Florida	32	Tr.	10	Tr.	1270	390	10	Tr.
South & East Florida	166	50	500	150			400	120
Total Identified Resources	3979	1210	7650*	2240*	9060	2770	5600	1700
Average P_2O_5	30.4%		30.6%		30.6%		30.3+%	
Total Hypothetical Resources**								
North Florida							900	250
Northeast Florida							650	200
South Florida							1425	400
East Florida							2200	650
Total Florida	180	60	no data		4700	1420	5175	1500
Grand Total: Resource Potential	4159	1270	7650	2240	13760	4190	10775	3200

*Resource numbers include south Georgia.

**Identified resources are resources whose tonnage and grade have been determined by drilling and chemical analysis and in which the phosphate product can be recovered using existing technology (Cathcart and others, 1984). The basis for subdividing identified resources into economic, marginally economic, and subeconomic categories is discussed in Cathcart and others (1984).

***Hypothetical resources are resources for which drilling information is sparse (less than one drill hole per section), only lithologic logs of drill holes are available, or where geologic inference indicates that deposits are present (Cathcart and others, 1984).

and controlled by mining companies; therefore these figures represent "demonstrated resources."

Within the Aurora Area (Fig. 3), Cathcart and others (1984) conservatively estimated an additional 8+ billion metric tons of phosphate with a 30% P_2O_5 concentrate as "hypothetical resources." These resources were considered to be subeconomic because of problems related to deep mining. Since most of this latter category has been extensively drilled by major phosphate companies, these should be included in the category of "inferred subeconomic resources."

The category of "hypothetical undiscovered resources" of the Pungo River Formation within the Aurora Embayment on the North Carolina Coastal Plain is vast. Miller (1971, 1982) produced a series of structural contour maps delineating the distribution of the Pungo River Formation throughout the coastal plain of northern North Carolina. A few deep drill holes with associated gamma-ray logging in downdip directions from the present mining district suggest that several additional phosphate-rich beds occur within deeper parts of the stratigraphic section than those presently being mined. Throughout the large and relatively deep mid-basin portion of Aurora Embayment (Fig. 3), poorly known phosphate-rich beds constitute vast potential future resources, particularly with development of new technology such as borehole slurry mining. Fountain and Hayes (1979) evaluated

TABLE 2. SUMMARY OF ESTIMATED PHOSPHATE RESOURCES BY VARIOUS AUTHORS FOR THE NORTH CAROLINA COASTAL PLAIN, CONTINENTAL SHELF, AND ADJACENT BLAKE PLATEAU
(Resources reported in billions of metric tons of phosphate concentrate)

	Zellers and Williams, 1978	Fountain and Hayes, 1979	Manheim and others, 1980, 1982	Cathcart and others, 1984	Riggs and others, 1985	Resource Category*
North Carolina Coastal Plain (based on deep core drilling)						
Aurora Phosphate District	1.62×10^9 540 km^2			$1+ \times 10^9$ 250 km^2		demonstrated
Aurora Area				$8+ \times 10^9$ 1550 km^2		inferred
Aurora Embayment		21×10^9 3600 km^2				hypothetical
North Carolina Continental Shelf (based on shallow vibracores and seismic data)						
Frying Pan Phosphate District					3.75×10^9 722 km^2	inferred
Northeast Onslow Phosphate District					0.78×10^9 614 km^2	inferred
Onslow Embayment				very large		hypothetical
Blake Plateau (based on surface samples and remotely sensed data)						
Charleston Bump Area			$2+ \times 10^9$ 22,000 km^2			inferred

*Resource category based on USGS (1980)--
   Demonstrated: Resources identified on the basis of drilling and chemical analyses, and the phosphate product can be recovered using existing technology.
   Inferred: Postulated extensions of demonstrated resources based largely on knowledge of geologic character with scattered drill holes and chemical analyses.
   Hypothetical: Resources that are similar to known deposits and may reasonably be expected to exist in the same region with analogous geologic conditions; sparse drill hole information exists.

the potential minable phosphate in North Carolina as a base for establishing the uranium resource potential in phosphates. Their resource numbers reflect the entire Aurora Embayment (Fig. 3). Within this area they calculated a "total potential phosphate resource" of 21 billion metric tons of phosphate product with concentrate grades of 29% to 32% P_2O_5. Thus, by subtracting the demonstrated and inferred resources of the Aurora Phosphate District and Aurora Area of Cathcart and others (1984), there is a "hypothetical resource" of about 12 billion metric tons of phosphate concentrate.

North Carolina Continental Shelf. Sediments of the Pungo River Formation crop out in a 150-km-long by 40-km-wide belt that trends southwest from Bogue Banks to Frying Pan Shoals, covering 6,000 km^2 of Onslow Bay and extending into the subsurface to the east and south of this outcrop belt (Riggs and Belknap, this volume, Fig. 2). Riggs and others (1982b, 1985) developed a preliminary evaluation of the phosphate resource potential in Onslow Bay based on an extensive network of high-resolution subbottom seismic profiles in combination with 9-m vibracores. They defined five beds that contain anomalous concentrations of phosphate in two different areas within Onslow Bay; concentrations are significant enough in these beds to be

considered potential future resources. Figure 3 shows the location of the Frying Pan and Northeast Onslow Bay phosphate districts. Table 3 summarizes the preliminary phosphate resource potential on the portions of Onslow Bay for which vibracore, seismic, and analytical data on the five known phosphorite beds have been obtained and analyzed to date.

In considering the phosphate resource potential of Onslow Bay, several important points must be remembered. Work to date (Riggs and others, 1985) represents first-stage exploration. Resource estimates are based upon the following: (1) shallow vibracores penetrate only the uppermost portions of outcropping seismic units on an irregular and fairly wide-spaced pattern; (2) lithologic and analytic core data are extrapolated into the subsurface on the basis of interpreted seismic sections; and (3) subsurface resource interpretations are based upon the assumption that phosphate distribution is at least constant down-dip to mid-basin within any bed. The latter assumption is based on known distribution of phosphate in other basins containing the Pungo River–Hawthorn formations (Riggs and others, 1982a, 1985). Thus, the entire resource potential of Onslow Bay at this point in time can only be considered as "hypothetical resources" (U.S.G.S., 1980).

TABLE 3. PRELIMINARY SUMMARY OF PHOSPHATE RESOURCES OF ONSLOW BAY,
NORTH CAROLINA, CONTINENTAL SHELF
(from Riggs and others, 1985)

Parameter	FRYING PAN PHOSPHATE DISTRICT			NORTHEAST ONSLOW BAY PHOSPHATE DISTRICT		
	PUNGO RIVER FORMATION		HOLOCENE SEDIMENT	PUNGO RIVER FORMATION		
	FPF-1 Outcrop	FPF-1 Subsurf.**	FPH	BBF-1L**	BBF-1U**	BBF-6**
Lithology	Muddy Sands	Muddy Sands	Gravelly Sands	Sli. Muddy Sands	Very Sandy Muds	Muddy Sands
Water Depth	24 to 26 m	24 to 26 m	24 to 26 m	13 to 23 m	21 to 35 m	13 to 17 m
Area	118 km²	704 km²	161 km²	271 km²	289 km²	54 km²
Phosphate Thick.						
Ave.	2 m	4 m	0.5 m	2.5 m	3 m	4 m
Range	0 to 3.5 m	3 to 5 m	0 to 3 m	1 to 4 m	2 to 5 m	2 to 6 m
Overburden Thick.	0 to 2.4 m	0 to 32 m	0 m	0 to 20 m	0 to 17 m	0 to 12 m
Grade: Total Sed.						
Ave.	12.4% P₂O₅	12.4% P₂O₅	6.3% P₂O₅	3.0% P₂O₅	6.1% P₂O₅	3.8% P₂O₅
Range	4.8% to 22.9%	4.8% to 22.9%	3.2% to 21.7%	2.0% to 4.6%	2.9% to 9.8%	2.9% to 6.1%
Grade: Conc.*	29.2% P₂O₅	29.2% P₂O₅	26.5% P₂O₅	31.0% P₂O₅	30.3% P₂O₅	29.7% P₂O₅
Tons Ore (metric)	700 mill.	8,000 mill.	200 mill.	1,900 mill.	2,500 mill.	600 mill.
Tons Conc (metric)	300 mill.	3,400 mill.	50 mill.	200 mill.	500 mill.	80 mill.

SUMMARY OF RESOURCE IN MILLIONS OF METRIC TONS OF PHOSPHATE CONCENTRATE		* % P₂O₅ corrected to 0% insol. to establish a
Frying Pan Phosphate		basis for comparison of chemical analyses.
District	3,750	** Area and thickness figures are based upon seismic
Northeast Onslow Bay		interpretations; the grades are based upon
Phosphate District	780	analyses from the shallow vibracores and the
TOTAL RESOURCE	4,530 MILL. METRIC TONS OF PHOSPHATE CONCENTRATE	assumption that they remain constant with depth.

In the Frying Pan Phosphate District, Pungo River Formation unit FPF-1 contains a conservative estimate of 3.7 billion metric tons of phosphate concentrate with an average P_2O_5 grade of 29.2% (Table 3). This unit presently has the highest potential as a future phosphate resource in Onslow Bay. Superimposed on top of Miocene sediment units is the thin and irregular Holocene sand sheet (unit FPH) that contains an estimated 50 million metric tons of phosphate concentrate with an average grade of 26.5% P_2O_5; the low grade is due to very high shell content.

In the Northeast Onslow Bay Phosphate District, the BBF depositional sequence of the Pungo River Formation contains three beds with significant concentrations of phosphate (Table 3). Phosphate concentrations of the total sediment range up to 10% P_2O_5 with concentrate grades up to 31% P_2O_5. These shallow water, low-grade beds that contain an estimated 780 million metric tons of phosphate concentrate are, at present, only long-term future resources with a low economic potential. However, the BBF sediment sequence represents an extremely widespread, thick, and complex sequence of depositional units (Riggs and Belknap, this volume, Fig. 2), for which there is minimal information.

This 4.5 billion tons of phosphate concentrate within complex Miocene depositional sequences is probably a conservative estimate, and probably represents only the surficial expression of numerous major deposits that occur at depth. Deeper drilling through the formation is the essential next phase in exploration to critically evaluate the total resource potential of the continental shelf; this would have a high potential to significantly expand the "hypothetical resources" of the continental shelf, which are thought to be in tens of billions of tons, and upgrade large portions of them to "inferred" and "demonstrated" resources.

Blake Plateau. Extensive phosphorite pavements, pebbles, and pellets occur in conjunction with manganese sediments on the Blake Plateau off South Carolina (Manheim and others, 1980, 1982). These phosphates (Fig. 1) occur on the nose and flanks of the Charleston Bump (Dillon and Popenoe, this volume), a major topographic wedge of Cretaceous sediments (Paull and Dillon, 1980) extending from the continental slope southeast across the Plateau. Manheim and others consider these primary phosphates to be of late Oligocene to middle Miocene age. This 22,000 km² area on the northern Blake Plateau contains an estimated 2 billion metric tons of phosphate nodules that average 22% P_2O_5 (Manheim and others, 1980, 1982).

Based upon unpublished cruise results in 1982, Manheim believes these earlier estimates are conservative. The new data suggest that virtually the entire Blake Plateau area outlined in Plate 1 may be underlain by a phosphorite pavement, except for local windows where the crust has been breached by erosion. He believes that the phosphate crust averages at least 10 cm in thickness in order to maintain its integrity and continuity over distances of tens of kilometers. These estimates are considered to be "inferred resources" based primarily upon a fairly uniform geologic character and distribution determined by photography and various other remote sensing techniques and a chemical data base derived from scattered surface sampling. See the section on manganese for discussions on the associated resources.

Byproduct Resources

Trace Elements. Sedimentary phosphates of the Atlantic margin are uncommonly rich in numerous elements (Altschuler, 1974, 1980; Altschuler and others, 1967, 1958; Cathcart, 1956, 1978; Dolfi, 1983; Ellington, 1984; Fountain and Hayes, 1979) that may be of economic significance in two different ways. First, some of the elements may themselves have important economic byproduct potential. Second, high abundances of other elements may have significant impacts upon the environment and human

TABLE 4. TRACE ELEMENT COMPOSITION OF HAND-PICKED PHOSPHATE GRAINS FROM THE NORTH CAROLINA
CONTINENTAL SHELF AND COASTAL PLAIN

Onslow Bay (Riggs and others, 1985)*				Aurora District (Dolfi, 1983)**		
Element	MIOCENE Average Range	MIOCENE Mean (n=32)	HOLOCENE Mean (n=11)	Element	MIOCENE AND PLIOCENE Range	Mean (n=76)
Zn	113 to 252 ppm	143 ppm	61 ppm	La	5 to 134 ppm	30 ppm
Cr	112 to 150 ppm	120 ppm	125 ppm	Ce	16 to 229 ppm	64 ppm
Mo	103 to 130 ppm	109 ppm	17 ppm	Nd	7 to 149 ppm	23 ppm
Pb	22 to 24 ppm	22 ppm	29 ppm	Sm	<1 to 20 ppm	5 ppm
As	18 to 24 ppm	19 ppm	122 ppm	Eu	<1 to 4 ppm	<1 ppm
V	17 to 22 ppm	18 ppm	18 ppm	Yb	<1 to 14 ppm	2 ppm
Ni	17 to 20 ppm	19 ppm	10 ppm	Lu	<1 to 2 ppm	<1 ppm
Cu	10 to 19 ppm	17 ppm	15 ppm			
Cd	6 to 38 ppm	13 ppm	1 ppm	U	12 to 124 ppm	57 ppm
Co	5 to 7 ppm	5 ppm	6 ppm	Th	1 to 13 ppm	3 ppm
Se	4 to 9 ppm	8 ppm	6 ppm			
F2	3.45 to 3.55%	3.50%	2.94%***			

*Analyses by inductively coupled argon plasma emission spectroscopy.
**Analyses by neutron activation.
***Analyses by standard wet chemical techniques; n=13 and 3, respectively.

health through the wastes of processing or through the use of fertilizer products. These effects are complex and poorly known; some of them may be good and some bad. Thus, in some cases it may be beneficial to recover certain elements to alleviate the environmental and human health risks or to add to the resource value.

Altschuler (1980) listed 11 elements occurring in world phosphates that are significantly enriched compared to average marine shales (Cd = 60×; U = 30×; Ag = 30×; Y = 10×; Se = 8×; Yb = 5×; Mo = 4×; La = 4×; Sr = 2×; Pb = 2×; and Zn = 2×). Table 4 summarizes detailed trace element data based upon studies of hand-picked phosphate grains representing specific grain types and individual stratigraphic horizons within the North Carolina phosphate section. Many of these elements occur in anomalous concentrations in most phosphate deposits and warrant further investigations as potential byproduct resources.

Uranium. Central Florida Phosphate District is presently a major uranium producer as byproduct recovery from phosphoric acid production. In 1980, the district had a total annual recovery capacity of over 3 million pounds of U_3O_8 (Reaves, 1984). The substantial but diminishing uraniferous phosphate resource of central Florida is expected to be depleted gradually over the next 15 to 25 years (Fountain and Hayes, 1979).

However, Fountain and Hayes (1979) established an extensive uranium potential within the phosphate resources of southeastern United States (Table 5). The 7.16 million metric tons of uranium resource represents over 47 billion tons of potential phosphate concentrate without any recovery factors. Considering present mining limitations and applying major recovery factors, they calculate that only 1.5 million metric tons of uranium is actually recoverable, representing 20 billion tons of phosphate concentrate (average uranium concentration ranges from 60 ppm to

110 ppm). The actual amount of this resource that is potentially recoverable is not fixed; it is dependent upon many rapidly changing economic and technological factors.

Several other factors must also be considered. For example, the uranium resource for the Aurora Embayment of North Carolina was based on the phosphate concentrate grade of 60 ppm uranium in the Aurora Phosphate District (Cathcart, 1975). Data in Table 6 suggest that the actual grade for North Carolina may be somewhat higher. Including extensive continental shelf deposits and the higher grades of the Onslow Bay phosphates, the data indicate that the actual uranium resource as a phosphate byproduct is considerably greater than presented in Table 5.

Phosphogypsum. Phosphogypsum is a byproduct of the wet-process fertilizer production which consists of digestion of phosphate concentrate with sulfuric acid to produce phosphoric acid and phosphogypsum in a ratio of 1:2.5 acid to gypsum (F.I.P.R., 1983). About 50% of the 54 million metric tons of phosphate produced in the U.S. in 1980 (U.S.B.M., 1980) was converted into phosphoric acid. This equates to a very large production of phosphogypsum per year; this byproduct resource and associated potential minor and trace element byproducts (Table 7) cannot be ignored. Because of the high fluorine and silica content, commercial quantities of fluosilicic acid (H_2SiF_6) are presently recovered as a byproduct from some wet-process phosphoric acid plants.

Phosphogypsum is a valuable soil amendment since it contains significant concentrations of the primary nutrient, phosphorus, in combination with secondary nutrients, calcium, and sulfur. Consequently, some tonnage is used as soil conditioner. Locally it is used as a base for road building and as a mix with mine spoils for reclamation purposes. However, the greater bulk of gypsum is presently unused and ends up being stockpiled.

TABLE 5. ESTIMATED URANIUM RESOURCES IN METRIC TONS OCCURRING
IN THE PHOSPHATE RESOURCES IN THE SOUTHEASTERN UNITED STATES
COASTAL PLAIN
(from Fountain and Hayes, 1979)

Phosphate District	Total Resource U (tons)	Recoverable U (tons)	Ave. U Conc. Phos. Product
North Carolina* (Aurora Embayment)	2,153,000	566,000	60 ppm
East Florida	1,305,000	272,000	102 ppm
North Florida- South Georgia*	1,223,000	123,000	70 ppm
South Florida*	931,000	336,000	110 ppm
East Georgia- South Carolina	914,000	142,000	60-90 ppm
Central Florida+	634,000	124,000	110 ppm
Totals	7,160,000	1,563,000	

*Phosphate producing districts.
+Uranium producing districts.

TABLE 6. URANIUM ANALYSES ON SELECTED
HAND-PICKED PHOSPHATE GRAINS FROM NORTH
CAROLINA COASTAL PLAIN AND
CONTINENTAL SHELF

Aurora Phosphate District

Pungo River and Yorktown Fm.

Source	Range	Mean	n
I	5 to 286 ppm	92 ppm	154
II	12 to 124 ppm	57 ppm	76

Frying Pan Phosphate District

Pungo River Fm.

Source	Range	Mean	n
III	54 to 146 ppm	85 ppm	6

Holocene Phosphate

Source	Range	Mean	n
III	58 to 195 ppm	119 ppm	20

I. Indorf, 1982; Indorf and others, 1983.
II. Dolfi, 1983.
III. Burnett, in Riggs and others, 1982c and 1983.

TABLE 7. MAJOR, MINOR, AND TRACE ELEMENT ANALYSIS OF
PHOSPHOGYPSUM FROM THE CENTRAL FLORIDA PHOSPHATE DISTRICT*

$CaSO_4 \cdot H_2O$	95.59%	Ti	4020 ppm	Cu	8 ppm
F	1.20%	Na	252 ppm	Ba	7 ppm
P_2O_5	0.65%	Sb	111 ppm	Cd	7 ppm
SiO_2	0.50%	As	42 ppm	Sn	4 ppm
Al_2O_3	0.14%	W	29 ppm	B	3 ppm
MgO	0.12%	V	19 ppm	Co	2 ppm
Fe_2O_3	0.10%	Mo	16 ppm	Ni	2 ppm
		Mn	15 ppm	Ta	2 ppm
		Re	11 ppm	Yt	2 ppm
		K	11 ppm	Bi	1 ppm
		Sr	10 ppm	Be	1 ppm
		Zr	10 ppm	Pb	1 ppm
		Zn	9 ppm	Ag	<1 ppm
				Pt	<1 ppm

Ra-226 from 20.5 to 33.0 pCi/g

Sources: Kouloheris (1980), May and Sweeney (1981),
Kaufman and Bliss (1977).

METALLIC DEPOSITS

Manganese and Associated Metal Oxides

Geological Occurrence and Genesis. United States and
Canada are essentially "have not" nations with respect to commercial manganese ores on the continent. However, large potential sources of manganese and other important metals occur on the Blake Plateau (Fig. 1), a major morphological feature occurring within the U.S. Exclusive Economic Zone (Commeau and others, 1984).

The Blake Plateau is a large terrace floored by Tertiary and Mesozoic rocks, that forms an extension of the southeast Atlantic Continental Shelf. It extends from the Bahama Banks on the south, where it is up to 270 km wide, northward to where it wedges out against the Florida-Hatteras Slope off North Carolina (Fig. 1). Ferromanganese oxide deposits occur (a) as extensive pavements, slabs, and nodules scoured by the Gulf Stream; (b) in water depths of 600 to 1000 m; and (c) between latitudinal zones of 30.5° and 32.2° N (Dillon and others, 1981).

The ferromanganese oxides occur primarily as encrustations on eroded and recrystallized products of earlier mid-Tertiary phosphorites that occur in a semi-continuous belt along the Atlantic continental margin (Fig. 1). The Blake Plateau phosphorites are carbonate fluorapatite, which occurs as mammal bones, phosphatized middle Tertiary sediments, and frequently as replaced foraminifera and molluscs (Manheim and others, 1980, 1982). Manheim and others believe that primary phosphatization was coeval with the continental shelf and coastal plain phosphorites previously discussed and that the episodic phosphatizing environment was characterized by reducing conditions with high contents of organic matter. Between periods of phosphatization, the plateau was dominated by the Gulf Stream producing high energy bottom environments with associated winnowing of fine sediment and erosion producing extensive areas of coarse lag gravels of phosphorite, phosphatized limestone, and silicified carbonate. These exposed surfaces were subsequently penetrated and encrusted by ferromanganese oxides to form pavements. Smaller nuclei formed slabs and round nodules resembling abyssal manganese nodules (Glasby, 1977).

Polished and thin sections of ferromanganese oxide layers show two general habits of oxide formation: (1) replacement and impregnation zones that are from mm to 10 cm or more thick; and (2) an overlying layer-by-layer accretional zone. Accretional zones are interrupted by transverse and vertical fractures filled with carbonates that are often aragonitic in proximity to phosphates and to low and high magnesian calcite away from the phosphates. Manganese carbonate has not been found, underscoring the highly oxidizing conditions that prevailed during formation of manganese-rich phases.

The Blake Plateau ferromanganese deposits demonstrate that such phases can form at relatively shallow ocean depths, from normal ocean water, and in close proximity to major land masses without an apparent linkage to vulcanism.

Resource Potential and Byproducts. Manganese resources on the Blake Plateau were initially estimated to be between 10 and 100 million tons of dry nodules (Manheim and others, 1982). A 1982 U.S. Geological Survey cruise (Manheim, 1982) produced photographic, seismic, and sampling data suggesting that nodule tonnages are nearer 10 million while ferromanganese pavements are more widespread than previously thought, and should exceed 200 million tons distributed over some 14,000 km^2.

Concentrations of other valuable metals and components in the Blake Plateau ferromanganese nodules are presented in Table 8, along with comparisons to other areas. The Blake deposits resemble world ocean averages for manganese nodules (McKelvey and others, 1983) in terms of manganese, cobalt, and nickel concentrations; however, they are distinctly lower in copper, nickel, and zinc, and much higher in platinum and arsenic than prime Pacific nodules (Haynes and others, 1982). In terms of current commodity prices, molybdenum, vanadium, and cerium have higher potential values in the Blake nodules than copper does.

It is difficult to sample unbroken pavements with normal oceanographic equipment; consequently, numbers of analyses are more sparse. Analyses of large slabs suggest that composition is highly variable; the top layered crusts have similar composition to nodules, whereas, lower phosphatic and partially impregnated portions of pavements have substantially reduced concentrations of manganese-associated metals. Relatively thick layers have lower manganese concentrations by a few percent, as well as lower concentrations of associated metals.

Plateau type deposits cannot be considered in the same terms as standard land ores; they represent entirely new resource potentials with new recovery, processing, and marketing problems. Considered on the basis of manganese alone, they would be distinctly submarginal. However, in contrast to most manganese deposits on land, which have few valuable byproducts, ocean-floor ferromanganese oxides have a long list of associated metals including: nickel, cobalt, copper, molybdenum, vanadium, cerium, and platinum. Concentrations of the latter four elements in the Blake Plateau ferromanganese are considerably higher than in Pacific abyssal nodules heretofore considered to be the principal ocean manganese resource.

Kaufman (1976) compared catalytic properties of Blake nodules to other nodules and found that catalytic hydrogenation capacity (in terms of conversion of benzene to cyclohexane) of the Blake nodules was higher. Use of nodules as catalytic absorbers of vanadium and nickel in petroleum refining would further enhance the spent catalyst in these metals (M. Cruikshank, personal communication, 1983).

In summary, the Blake Plateau represents a large resource of manganese, phosphate, and associated ferromanganese crust metals within the U.S. EEZ (Halbach and Manheim, 1984). Recovery of other byproduct metals from the nodules must be included in future evaluations. Also, the problems of developing new mining, processing, and marketing systems, and overcoming

TABLE 8. COMPARATIVE VALUES OF SELECTED METAL CONSTITUENTS IN BLAKE
PLATEAU MANGANESE NODULES[1], WITH PACIFIC ABYSSAL MANGANESE NODULES[2],
AND THE WORLD AVERAGE DATA FOR MANGANESE NODULES[3]
(Values are in weight percent oven dried material. Standard deviation
is in parantheses.)

Element	Blake Plateau	Pacific Nodules	World average
Si	2.0 (1.9)	6.5 (0.016)*	7.7 (4.1)
Ti	0.2 (0.14)	0.3 (0.002)*	0.69 (0.47)
Ca	11.9 (4.7)	2.19 (0.012)*	2.2 (0.12)
Fe	11.1 (2.7)	5.78 (0.031)*	12.5 (5.7)
Mn	16.3 (4.2)	29.14 (0.080)*	18.6 (7.2)
Ni	0.62 (0.14)	1.34 (0.006)*	0.66 (0.44)
Co	0.33 (0.05)	0.22 (0.001)*	0.27 (0.19)
Cu	0.10 (0.03)	1.15 (0.005)*	0.45 (0.40)
Zn	0.045 (0.005)	0.16 (0.0006)*	0.12 (0.41)
Pb	0.092 (0.0169)	0.055 (0.0006)*	0.093 (0.067)
Mo	0.036 (0.003)	0.076 (0.0004)*	0.038 (0.022)
V	0.058 (0.015)	0.052 (0.001)*	0.052 (0.039)
Ce	0.072 (0.070)	0.029 (0.001)	0.072 (0.059)
As	0.046 (0.071)	0.0039 (0.00096)	0.014 (0.009)
Pt (ppb)	326 (110)	123	---

[1]Manheim and others (1982).
[2]Flanagan and Gottfried (1980).
[3]McKelvey and others (1983).

*Standard error of the labratory means.

environmental constraints are critical initial phases essential prior to the future development of the vast resource base on the Blake Plateau.

Heavy Mineral Placers

Mineral components of placer deposits are chemically and physically stable with high specific gravities. The extremely varied mineralogy includes the following groups, defined in Table 9: titanium, rare-earths, zirconium, aluminosilicates, and valuable metals including gold, silver, tin, platinum, etc. These minerals occur concentrated in placer deposits in modern alluvial/fluvial systems draining the piedmont, coastal plain fluvial, and modern beach systems. Because of the vast number of complex sea-level fluctuations during and since the Neogene, similar deposits occur within the many ancient fluvial channel deposits and ancient beaches on both the coastal plain and continental shelf (see Riggs and Belknap, this volume). On the Atlantic Coastal Plain, placer deposits generally occur in unconsolidated or poorly consolidated Pleistocene shoreline sediments such as the titanium deposits of Trail Ridge in Florida and Georgia (Fig. 1) or the modern alluvial/fluvial sediments such as the Fishing Creek placer gold deposits of North Carolina (Riggs, 1977).

Heavy minerals are commonly mined from deposits on the coastal plain, but have not been successfully mined in the offshore zone. Heavy minerals have been studied in marine sediments in most coastal states, many done in connection with a program initiated by the U.S. Geological Survey in the late 1960s (McKelvey and others, 1968). Recently, the Department of Interior's Hard Mineral Task Force Report (U.S.D.I., 1981; Manheim,

1981) updated the heavy mineral data. New U.S. Geological Survey initiatives began utilizing in 1982 the extensive collection of Atlantic inner-shelf vibracores of the U.S. Army Corps of Engineers (Grosz and Escowitz, 1983). Table 9 presents a summary of onshore and offshore placer compositions for the Atlantic continental margin.

The Atlantic Continental Shelf contains an extensive Holocene surface sand sheet that consists largely of relict and extensively reworked sands (Riggs and Belknap, this volume, Fig. 1; Duane and Stubblefield, this volume, Fig. 4). These shelf sands, plus the extensive and complex network of Pleistocene fluvial channels, have significant future potential for containing major placer deposits. In spite of many reviews of coastal plain placers and reconnaissance studies on the distribution of heavy minerals (Martens, 1935; Pilkey, 1963; McKelvey and others, 1968; Ross, 1970), the realistic potential of offshore mineral placers is largely unknown. The U.S. Atlantic shelf spans 391×10^3 km^2 with a resource estimate of 1300×10^6 m^3 of heavy mineral bearing sand with variable compositions and grades; these shelf sands range from submarginal to high potential (McKelvey and others, 1968). Heavy mineral bearing sands have been located on the shelf off Virginia's Delmarva Peninsula (Fig. 1; Grosz and Escowitz, 1983). These deposits may rival the Florida Trail Ridge deposits in size and grade.

Gold placers were drilled in Nova Scotia during the late 60s, one location in Lunenburg Bay was briefly mined, and drilling has traced gold enrichments at least 3 mi offshore from Isaacs Harbour (Burns, 1979). The eastern edge of the Piedmont from Virginia to Alabama (Fig. 1) contains extensive gold mineralization (Simons and Prinz, 1973); these deposits were among the

TABLE 9. HEAVY MINERAL COMPOSITION OF TYPICAL PLACERS ALONG THE ATLANTIC
COASTAL PLAIN AND CONTINENTAL SHELF

Region	Location	Ti	Zr	Al	Mo	Total*	Ref.
Coastal Plain	New Jersey:						
	Lakehurst, Ocean City						
	(Miocene-Pleistocene)	87	1.4	1+?	1.1	----	1
	South Carolina:						
	Hilton Head (Modern)	40.5	12	----	1.3	----	1
	Georgia:						
	Chatham Co. (Modern)	59	14	4	1	----	1
	Jekyll Is. (Modern)	51	14	11	1.2	----	1
	Florida:						
	Trail Ridge (Pleistocene)	45	15	29	----	----	1
	Jacksonville (Pleistocene)	51	11	----	0.5	----	1
	Atlantic Coast (17 Spls. Modern)	18-53	2-17	6-33	1-8	5	1
Continental Shelf	Long Island Sound to Nova Scotia	----	----	----	----	2.9	2
	New York Bight	27.8	10.7	11.5	1.4	2.5	3
	Virginia	17.6	2.9	23.2	tr	0.7	4
	South Carolina	18.3	3.5	11.4	tr	1.8	4
	Florida (Atlantic)	17.7	4.4	15.6	tr	1.3	4

Notes: All minerals are in weight percent of heavy mineral fraction.

*Total refers to total heavy minerals as a percentage of sand fraction (.062 mm to 2.0 mm).

Column headings: Ti = titanium minerals (ilmenite, titanite, rutile, leucoxene, etc.); Zr = zircon;
Al = aluminum oxides (staurolite, sillimanite, kyanite, andalusite, etc.); Mo = rare earth
minerals (monazite, etc.).

References: 1. McKelvey and others, 1968.   3. Drucker, 1983.
2. Ross, 1970.   4. Grosz and Escowitz, 1983.

largest producers of gold in the U.S. prior to the 1848 California gold strike. Considering weathering and transport processes in association with complex sea-level fluctuations of the Neogene, there should be a significant potential for gold placer deposits in both modern and ancient streams and barrier islands draining these gold-bearing areas and crossing the Atlantic Coastal Plain and Continental Shelf.

Should an economic climate favorable for mineral extraction industries emerge in the future, the combined development of sand and gravel with heavy mineral separation, both on the coastal plain and continental shelf, deserves attention. Bokuniewicz (1981) explored the possibilities of combined waste disposal and sand recovery in the New York Bight. If placer mineral concentrations similar to those reported by Drucker (1983) for the inner New York Bight (Table 9) coincide with potential sand and gravel development, recovery of heavy minerals could be undertaken prior to sale of the sand.

OTHER NONMETALLIC DEPOSITS

Surficial deposits of sand and gravel (including aggregate, high silica sand for glass, and high feldspar sand for potash), limestone, shells, clay, peat, and total sediment for land fill represent the most widely exploited group of mineral resources within the Atlantic continental margin today. For the most part these deposits require modest, if any, beneficiation or preparation prior to use. Also, because of their low unit value, the commodities have limited and often local markets that are dictated by very high transportation costs. Consequently, most operations are very small scale, low budget, and temporary, depending on the highly variable local markets and economies. Duane and Stubblefield (this volume) discuss sand and gravel resources.

Clay Deposits

The Atlantic continental margin contains enormous quantities of clay raw materials, primarily in a belt along the western edge of the coastal plain extending from Georgia northward through North Carolina (Table 10; Fig. 1). Clay deposits fall into four principal categories: kaolins (including ball and fire clay), fuller's earth, common clay, and bentonite. The first three categories are economically very significant mineral resources on the Atlantic margin; bentonite has not been identified in commercial quantities in this province (Ampian, 1981). Production of clays on the Atlantic margin is shown in Table 10.

Kaolin. Kaolin clays occur along the southeast side of the "fall line," where the mining district extends from Aikin, South Carolina to Macon, Georgia, with some mines extending into the Eufaula district in Alabama. This 300 mi-long district accounted for 90% of the U.S. production and 40% of the world's total kaolin output in 1980 (Ampian, 1981). The southeastern U.S. kaolin deposits have hundreds of millions of tons of reserves and billions of tons of resources (Hosterman, 1973).

The kaolin clays are tabular lenses and discontinuous beds

TABLE 10. PRODUCTION OF CLAYS IN 1980 FROM THE U.S. ATLANTIC COASTAL PLAIN
(Ampian, 1981)
(Resources reported in thousands of short tons)

State	No. of Mines	Kaolin	Ball Clay	Fire Clay	Fuller's Earth	Common Clay	Total Tons	Total Value 10^6 $
Maine	5	x	x	x	x	57	57	$ 156
Massachussets	3	x	x	x	x	259	259	$ 1,322
New York	11	x	w	x	x	597	597	$ 2,310
New Jersey	4	x	x	11	x	52	63	$ 553
Maryland	10	x	w	x	x	597	597	$ 1,984
Virginia	15	x	x	x	x	502	502	$ 2,015
North Carolina	59	w	x	x	x	2110	2110	$ 6,838
South Carolina	52	724	x	x	w	907	1632	$ 28,600
Georgia	93	6236	x	x	584	1209	8029	$ 554,000
Florida	15	32	x	x	518	181	731	$ 35
Totals	267	6992	x	11	1102	6671	14777	$ 598,287

w = data withheld by U.S.B.M.
x = no reported production.

within the late Cretaceous to mid-Eocene sediments that lie unconformably on the erosional surface of pre-Cretaceous Piedmont rocks (Patterson and Buie, 1974; Buie and Schrader, 1982). The kaolin deposits are composed mainly of the mineral kaolinite, a hydrated aluminum silicate, with varying amounts of impurities including quartz, muscovite, and anatase. The kaolins may be either partially or completely bauxitized (gibbsite); however, bauxite does not occur in adequate concentrations to consider this an aluminum ore. Tops of individual kaolin bodies are often pisolitic and bauxitized suggesting multiple sea-level fluctuations with alternating episodes of deposition and alteration (Schrader and others, 1983).

The southeastern U.S. kaolin deposits occur in sediments at least partially, if not totally derived from the weathering of crystalline Piedmont rocks (Patterson and Buie, 1974; Buie and Schrader, 1982). Intense weathering under tropical, humid conditions during the late Cretaceous and early Tertiary supplied kaolinites, quartz, muscovite, K-feldspar, and Ti-rich heavy minerals to the fluvial systems. These sediments were transported to the coast and deposited in shallow water fluvial/deltaic/coastal marine environments. Buie (1964) introduced the theory that airborne volcanic ash of a similar composition was deposited in shallow brackish to marine environments and subsequently altered to montmorillinite and then to kaolinite.

Factors that dictate the economics of kaolin clays include the following (Schrader and others, 1983): (1) percent of kaolin group clay minerals (kaolinite + nacrite + dickite, etc.); (2) brightness values or color of the clay; (3) percent grit or sand; (4) chemical impurities such as TiO_2, Fe_2O_3, and SiO_2 polymorphs; (5) particle size distribution of the clay fraction; and (6) physical properties such as viscosity and abrasion. Pure kaolinite has had iron and other impurities leached away, and is used as a raw material for porcelain and other white ceramic material. Fire clay, used to make refractory products, may contain diaspore or boehmite, which increases the alumina content to 40% or more.

Fuller's Earth. The entire early to middle Miocene section extending from Florida through North Carolina, contains anomalous concentrations of magnesium-rich clay minerals including attapulgite (or more properly palygorskite), sepiolite, and montmorillonite (Riggs, 1984; Dillon and others, this volume). Lyle (1984) and Weaver and Beck (1977) demonstrated that concentrations of palygorskite and sepiolite clays in Miocene sediments increase southward from North Carolina to maximums in Florida and they are particularly high in the southern part of the first-order Southeast Georgia Embayment, where there is minimum amount of dilution by other sediments. Fuller's earth deposits thicken into the second-order Gulf Trough that trends northeast across the Florida panhandle and into southern Georgia and is mined in the Meigs, Attapulgus, and Quincy areas along the Georgia-Florida border (Fig. 1; Patterson and Buie, 1974).

Fuller's earth clay occurs as large lenses and discontinuous beds within the upper part of the Miocene Hawthorn Formation (Patterson and Buie, 1974). The clay deposits are composed dominantly of palygorskite, with lesser amounts of sepiolite, both complex magnesium-rich hydrous silicate clay minerals with fibrous crystal habits. Other components of fuller's earth include montmorillonite, some kaolinite, and variable amounts of terrigenous silts and sands, bioclastic calcite hash, phosphate, dolomite, diatoms, and opal concretions (Patterson and Buie, 1974).

Many geologists believe that the magnesium-rich clay minerals that constitute fuller's earth are derived from chemical alteration of volcanic ash; however, no volcanic detritus has ever been found associated with these deposits. Weaver and Beck (1982) believe that palygorskite formed in shallow, brackish-water coastal lagoons by the alteration of montmorillonite in humid, subtropical to tropical climatic conditions. However, this clay mineral assemblage occurs throughout the entire Miocene section of the coastal plain, continental shelf, and Blake Plateau (Riggs, 1984, 1979b). Riggs (see Schlee and others, this volume) believes that this magnesium-rich clay sequence is an authigenic component that formed over much of the continental margin contemporaneously with other "aberrant authigenic" sediments including phosphate, dolomite, diatomaceous muds, organic-rich muds, etc. Highly fluctuating sea levels in response to glaciation and deglaciation produced alternating oceanographic conditions that included episodes of cold, nutrient-rich waters with high organic productivity and extensive dysaerobic oxygen-minimum zones in broad shelf and slope environments.

Fuller's earth, or "swelling clay" has strong adsorption capabilities with marked cation exchanges for taking H_2O and organics into interlayer sites. These clays were initially used for "fulling wool." Now they are used for many aspects of cleaning, purifying, and decolorizing mineral, vegetable, and animal oils.

Common Clay. Common clay and shale resources constitute the largest group (45%) of all clay resources mined on the Atlantic continental margin in 1980 (Table 10). North Carolina was the largest producer (32%) followed by Georgia (18%) and South Carolina (14%; Ampian, 1981). These deposits supply low unit cost commodities critical to the construction industries and therefore are related to the metropolitan markets. Thus, land values and transportation costs are critical parameters in dictating economics. Common clay is primarily used for building bricks, drainage tiles, sewer pipe, Portland cement, plastic and paper filler, construction aggregate, etc.

Common clay is very abundant and has extremely variable characteristics. Important physical properties include plasticity, transverse strength, shrinkage, and fusability. Many of the common clays are dominated by illite mineral assemblages with little to no contaminant material such as organic matter. Along the western edge of the Atlantic Coastal Plain and eastern Piedmont, the extensive red-beds in the Triassic basins—such as the Sanford-Durham-Wadesboro and Dan River basins in North Carolina—supply large volumes of industrial common clay. Further east on the coastal plain, Pleistocene clays are of local importance.

Carbonates

Carbonate resources are abundant, extensive, and quite varied within the Atlantic continental margin (Fig. 1). Limestone and dolomite of Tertiary age are extensively quarried for crushed rock aggregate from relatively low-grade deposits from North Carolina through Florida. Pleistocene coquinas and coral reefs are quarried for ornamental building stone along the Florida east coast from St. Augustine to the Keys. High quality limestones for chemical industries occur in some Tertiary units such as specific facies of the Eocene Ocala and Castle Hayne limestones in Florida and North Carolina, respectively. Modern aragonitic oolites from shallow banks of the Bahama Islands are exceptionally pure ($CaCO_3$ = 94% to 97%; MgO = 0.5% to 1.5%; SiO_2 = 0.021% to 0.08%; Al_2O_3 = 0.02% to 0.15%; SO_3 = 0.10% to 0.20%; SrO = 0.30% to 1.25%; $NaCl$ = 0.06% to 0.25%; and Fe_2O_3 = 0.01% to

0.02%; Manheim, 1981). Oyster and other shells were widely recovered as sources of high quality lime from estuaries in Florida beginning in the 1920s, but such recovery ceased in the late 1970s as a consequence of environmental regulations (Whitfield, 1975).

Diatomites

Pure diatomites are composed of siliceous diatoms, and consequently are uniform and porous. Thus, diatomites make excellent material for absorbents, filtration, fillers, and substrates for catalysts. The U.S. is a world leader in diatomite production, but principally from Miocene deposits on the west coast. There are a few significant, but nonproducing deposits on the Atlantic margin. Low-grade diatomite deposits in the Miocene Calvert and Choptank formations crop out in the Virginia and Maryland cliffs along Chesapeake Bay; these units thicken and become enriched in diatoms into the subsurface (Glaser, 1971; Fig. 1). The lower portion of the Calvert Formation is defined as the Fairhaven Diatomaceous Earth Member (Gibson, 1983). This unit varies from muddy diatomite with diatoms constituting as much as 65% of the total sediment to diatomaceous muds (Glaser, 1971). Similar Miocene sediments in the Pungo River Formation of North Carolina contain extensive and thick facies of diatomaceous mud (Miller, 1971, 1982). However, even the best of these beds constitute low-grade deposits and, except for some minor workings for filler grade materials, are not commercially viable.

Evaporites

The largest salt deposit in the U.S. occurs in deep Jurassic strata in southern Florida (Fig. 1). These deposits contain bromine solutions of 5,000 ppm (Manheim and Horn, 1968) but are too deep to consider solution mining by boreholes.

During the mid- to late-70s, offshore evaporite diapirs were discovered along the U.S. Atlantic coast continuing trends of late Triassic to early Jurassic age on the Canadian continental margin (Fig. 1; Dillon and others, 1986). Chemical composition of offshore salt bodies is not known, however, they may contain late-stage evaporite elements like potassium and bromine.

Peats

Peat deposits of the Atlantic continental margin fall into five general geologic settings: (1) large coastal fresh water swamps or pocosins related to coastal marine processes during multiple sea-level events of the Pleistocene; (2) river floodplains and associated deltaic interdistributaries; (3) intertidal salt marshes in modern estuaries; (4) wetlands resulting from glacial terrains; and (5) Carolina Bays (ellipitical depressions) and small fresh water wetlands. These low-energy transitional environments from water to land, represent areas with poor or immature surface and subsurface drainage, high organic productivity, and subsequent accumulation of organic matter in reducing conditions.

The U.S. Department of Energy initiated an extensive peat survey program in the late 1970s to determine the amount and location of fuel-grade peat in the U.S. that may be mined in an environmentally acceptable manner. The Atlantic continental margin has large deposits of peat that extend from Maine to Florida with the following current state estimates of fuel-grade resources in millions of dry tons (D.O.E., 1982): Maine, 700; Massachusetts, 260; North Carolina, 630; South Carolina, 90; Georgia, 800; and Florida, 760. The total resource is in excess of 3.24 billion dry tons of fuel-grade peat (minimum values: 8000 Btu/lb dry basis, depth 1.5 m, and 80 acres/mi^2; maximum ash level: 25% dry basis). The total estimated resources of all peats for the Atlantic margin exceeds 8.75 billion dry tons (D.O.E., 1982). The greatest future potential importance of peat is as a fuel source and as feedstock for gasification. For example, fuel values for the fuel-grade resources of the Atlantic margin are estimated at about 65 quads (10^{15} Btu), equivalent to 10.8 billion barrels of oil (D.O.E., 1982). Presently, only small volumes of peat are locally mined, primarily for horticultural soil improvement.

REFERENCES CITED

Ampian, S. G., 1981, Clays in 1980: U.S. Bureau of Mines, Mineral Industry Surveys, p. 1–42.

Altschuler, Z. S., 1974, The weathering of phosphate deposits; Environmental and geochemical aspects, in Griffith, E. J., Beeton, A., Spencer, J. M., and Mitchell, D. T., eds., Environmental phosphorus handbook: New York, John Wiley and Sons, p. 33–96.

——1980, The geochemistry of trace elements in marine phosphorites, Part I; Characteristic abundances and enrichments, in Bentor, Y. K., ed., Marine phosphorites: Society of Economic Paleontologists and Mineralogists Special Publication 29, p. 19–30.

Altschuler, Z. S., Clarke, R. S., and Young, E. J., 1958, Geochemistry of uranium in apatite and phosphorite: U.S. Geological Survey Professional Paper 314-D, p. 45–90.

Altschuler, Z. S., Cathcart, J. B., and Young, E. J., 1964, The Geology and Geochemistry of the Bone Valley Formation and its Phosphate Deposits, West Central Florida: Geological Society of America Guidebook, field trip 6, 68 p.

Altschuler, Z. S., Berman, S., and Cuttitta, F., 1967, Rare earths in phosphorites; Geochemistry and potential recovery: U.S. Geological Survey Professional Paper 575-B, p. B1–B9.

Bergendahl, M. H., 1956, Stratigraphy of parts of DeSoto and Hardee counties, Florida: U.S. Geological Survey Bulletin 1030-B, p. 65–98.

Birdsall, B. C., 1978, Eastern Gulf of Mexico, continental shelf phosphorite deposits [M.S. thesis]: St. Petersburg, University of South Florida, 87 p.

Blackwelder, B. W., MacIntyre, I. G., and Pilkey, O. H., 1982, Geology of the continental shelf, Onslow Bay, North Carolina, as revealed by submarine outcrops: American Association of Petroleum Geologists Bulletin, v. 66, p. 44–56.

Bokuniewicz, H. J., 1981, Submarine borrow pits as containment sites for dredged sediment, in Kester, D. R., Ketchum, B. H., Duedall, I. W., and Park, P. K., eds., Dredged-material disposal in the ocean; Waste in the Ocean: v. 2: New York, John Wiley, Interscience, p. 215–227.

Brown, P. M., 1958, The relation of phosphorites to ground water in Beaufort County, North Carolina: Economic Geology, v. 53, p. 85–101.

Brown, P. M., Miller, J. A., and Swain, F. M., 1972, Structural and stratigraphic framework and spatial distribution of permeability of the Atlantic Coastal Plain, North Carolina to New York: U.S. Geological Survey Professional Paper 796, 79 p.

Buie, B. F., 1964, Possibility of volcanic origin of the Cretaceous sedimentary koalin of South Carolina and Georgia [abs.], in Bradley, W. F., ed., Clays and Clay Minerals; Proceedings of the Twelfth National Clay Conference: New York, Macmillan Co., p. 195.

Buie, B. F., and Schrader, E. L., 1982, South Carolina koalin, in Nystrom, P. G.,

and Willoughby, R. H., eds., Geological Investigations Related to the Stratigraphy in the Koalin Mining District, Aiken County, South Carolina: South Carolina Geological Society field trip guidebook, 1982, South Carolina Geological Survey, p. 1–20.

Burns, V. M., 1979, Marine placer minerals: Reviews in Mineralogy, v. 6, p. 347–380.

Carr, W. J., and Alverson, D. C., 1959, Stratigraphy of middle Tertiary rocks in part of west-central Florida: U.S. Geological Survey Bulletin 1092, 111 p.

Cathcart, J. B., 1956, Distribution and occurrence of uranium in the calcium phosphate zone of the land-pebble phosphate district of Florida: U.S. Geological Survey Professional Paper 300, p. 489–494.

——, 1968a, Florida-type phosphate deposits of the United States; Origin and techniques for prospecting, in Seminar on sources of mineral raw materials for the fertilizer industry in Asia and the Far East: Proceedings of the United Nations ECAFE, Mineral Resources Development Series 32, p. 178–186.

——, 1968b, Phosphate in the Atlantic and Gulf Coastal plains, in Brown, L. F., ed., Proceedings 4th Forum of Geology of Industrial Minerals: Austin, University of Texas Press, p. 23–24.

——, 1975, Uranium in phosphate rock: U.S. Geological Survey Open-File Report 75-321, 20 p.

——, 1978, Uranium in phosphate rock: U.S. Geological Survey Professional Paper 988-A, 6 p.

Cathcart, J. B., and McGreevy, L. J., 1959, Results of geologic exploration by core drilling, 1953, Land-pebble phosphate district, Florida: U.S. Geological Survey Bulletin 1046-K, p. 221–298.

Cathcart, J. B., Sheldon, R. P., and Gulbrandsen, R. A., 1984, Phosphate-rock resources of the United States: U.S. Geological Survey Circular 888, 48 p.

Charm, W. B., Nesteroff, W. D., and Valdes, S., 1969, Detailed stratigraphic description of the JOIDES cores on the continental margin off Florida: U.S. Geological Survey Professional Paper 581-D, 13 p.

Clark, D. S., 1972, Stratigraphy, genesis, and economic potential of the southern part of the Florida land-pebble phosphate field [Ph.D. thesis]: Rolla, University of Missouri, 182 p.

Commeau, R. F., Clark, A., Johnson, C. F., Manheim, F. T., and Lane, C. M., 1984, Ferromanganese crust resources in the Pacific and Atlantic oceans: Washington, D.C., Proceedings of OCEANS 84 Conference, p. 421–430.

Dillon, W. P., Paull, C. K., Valentine, P. C., Ball, M. M., Arthur, M. A., Shinn, E., and Kent, K. M., 1981, The Blake Escarpment carbonate platform edge; Conclusions based on observations and sampling from a research submersible: American Association of Petroleum Geologists Bulletin, v. 65, p. 918.

Dillon, W. P., Manheim, F. T., Jansa, L. F., Palmason, G., Tucholke, B. E., and Landrum, R. S., 1986, Resource potential of the western North Atlantic basin, in Vogt, P. R. and Tucholke, B. E., eds., The western North Atlantic region: Boulder, Colorado, Geological Society of America, The Geology of North America, v. M, p.

Dolfi, R. M., 1983, Geochemical facies analysis and stratigraphic correlation of Miocene phosphorite beds in the Aurora district of North Carolina [M.S. thesis]: Raleigh, North Carolina State University, 88 p.

Drucker, B. S., 1983, Distribution and analysis of selected economic heavy mineral species within the inner New York Bight: Houston, Texas, Proceedings, Offshore Technology Conference 15, v. 1, p. 427–436.

Ellington, M. D., 1984, Major and trace element composition of phosphorites of the North Carolina continental margin [M.S. thesis]: Greenville, North Carolina, East Carolina University, 93 p.

Emery, K. O., and Uchupi, E., 1972, Western North Atlantic Ocean, Topography, rocks, structure, water, life, and sediments: American Association of Petroleum Geologists Memoir 17, 532 p.

F.I.P.R., 1983, A review of the potentially hazardous characteristics of phosphate wastes: Bartow, Florida Institute of Phosphate Research, 39 p.

Flanagan, F. J., and Gottfried, D., 1980, U.S. Geological Survey Rock Standards III; Manganese-nodule reference samples USGS–NOD–A–1 and USGS–NOD–P–1: U.S. Geological Survey Professional Paper 1155, 39 p.

Fountain, R. C., and Hayes, A. W., 1979, Uraniferous phosphate resources of the southeastern United States, in DeVoto, R. H., and Stevens, D. N., eds.,

Uraniferous phosphate resources, United States and free world: U.S. Department of Energy Publication GJBX-110(79), p. 65–122.

Freas, D. H., 1968, Exploration for Florida phosphate deposits, *in* Seminar on sources of mineral raw material for the fertilizer industry in Asia and the Far East: Proceedings of the United Nations ECAFE Mineral Resources Development Series 32, p. 187–200.

Furlow, J. W., 1969, Stratigraphy and economic geology of the eastern Chatham County phosphate deposit: Georgia Geological Survey Bulletin 82, 40 p.

Gerlach, A. C., ed., 1970, The national atlas of the United States of America: Washington, D.C., U.S. Geological Survey, 417 p.

Gibson, T. C., 1967, Stratigraphy and paleoenvironment of the phosphatic Miocene strata of North Carolina: Geological Society of America Bulletin, v. 78, p. 631–649.

——, 1983, Stratigraphy of Miocene through lower Pleistocene strata of the United States Central Atlantic Coastal Plain, *in* Ray, C. E., ed., Geology and paleontology of the Lee Creek Mine, North Carolina, Volume I: Smithsonian Contributions to Paleobiology, no. 53, p. 35–80.

Glasby, G. P., ed., 1977, Marine manganese deposits: Amsterdam, Elsevier, 523 p.

Glaser, J. E., 1971, Geological and mineral resources of southern Maryland: Maryland Geological Survey Report of Investigation 15, 84 p.

Gorsline, D. S., and Milligan, D. B., 1963, Phosphatic deposits along the margin of the Pourtales Terrace, Florida: Deep-Sea Research, v. 10, p. 259–262.

Grosz, A. E., and Escowitz, E. C., 1983, Economic heavy minerals of the U.S. Atlantic Continental Shelf: Geological Society of America Abstracts with Programs, v. 15, no. 2, p. 703.

Halbach, P., and Manheim, F. T., 1984, Potential of cobalt and other metals in ferromanganese crusts on seamounts of the Central Pacific Basin: Marine Mining, v. 4, p. 319–336.

Hathaway, J. C., McFarlan, P. F., and Ross, D. A., 1970, Mineralogy and origin of sediments from drill holes on the continental margin off Florida: U.S. Geological Survey Professional Paper 581E, 26 p.

Hathaway, J. C., Poag, C. W., and Valentine, P. C., 1979, U.S. Geological Survey core drilling on the U.S. Atlantic shelf: Science, v. 206, no. 4418, p. 515–527.

Hathaway, J. C., Schlee, J. S., and Poag, C. W., 1976, Preliminary summary of the 1976 Atlantic margin coring project of the U.S.G.S.: U.S. Geological Survey Open-File Report, no. 76–844, 218 p.

Haynes, B. W., Law, S. L., and Barron, D. C., 1982, Mineralogical and elemental description of Pacific manganese nodules: U.S. Bureau of Mines Information Circular 8906, 60 p.

Hosterman, J. W., 1973, Clays, *in* Brobst, D. A., and Pratt, W. P., eds., United States mineral resources: U.S. Geological Survey Professional Paper 820, p. 123–132.

Indorf, M. S., 1982, Uranium-phosphorus determinations for selected phosphate grains from the Miocene Pungo River Formation, North Carolina [M.S. thesis]: Greenville, North Carolina, East Carolina University, 80 p.

Indorf, M. S., Riggs, S. R., and Bray, J. T., 1983, Uranium-phosphorus determinations for selected phosphate grains from the Neogene phosphorites in southeastern Beaufort County, North Carolina: Geological Society of America Abstracts with Programs, v. 15, no. 2, p. 105.

Katrosh, M. R., and Snyder, S. W., 1982, Diagnostic foraminifera and paleoecology of the Pungo River Formation, central coastal plain of North Carolina: Southeastern Geology, v. 23, p. 217–232.

Kaufman, R., 1976, Offshore hard mineral resource potential and mining technology, *in* Symposium on marine resource development in the middle Atlantic states: Society of Naval Architects and Marine Engineering, Chesapeake and Hampton Roads Section, 24 p.

Kaufman, R. F., and Bliss, J. D., 1977, Effects of phosphate mineralization and the phosphate industry on radium-226 in ground water of central Florida: Las Vegas, Nevada, U.S. Environmental Protection Agency, Office of Radiation Programs, EPA/520-6-77-010.

Klitgord, K. D., and Behrendt, J. C., 1979, Basin structure of the U.S. Atlantic continental margin, *in* Watkins, J. S., Montadert, L., and Dickerson, P. W., eds., Geological and geophysical investigations of continental margins:

American Association of Petroleum Geologists Memoir 29, p. 85–112.

Klitgord, K. D., and Poponoe, P., 1984, Florida; A Jurassic transform plate boundary: Journal of Geophysical Research, v. 89, p. 7753–7772.

Koulores, A. P., 1980, Chemical nature of phosphogypsum as produced by various wet process phosphoric acid processes, *in* Borris, D. P., and Boody, P. W., eds., Proceedings of the International Symposium on Phosphogypsum, Lake Buena Vista, Florida: Bartow, Florida Institute of Phosphate Research, v. 1, p. 8–35.

Lewis, D. W., Riggs, S. R., Hine, A. C., Snyder, S.W.P., Snyder, S. W., and Waters, V. J., 1982, Preliminary stratigraphic report on the Pungo River Formation of the Atlantic Continental Shelf, Onslow Bay, North Carolina, *in* Scott, T. M., and Upchurch, S. B., eds., Miocene of the southeastern United States: Southeastern Geological Society and Florida Bureau of Geology Special Publication no. 25, p. 122–137.

Luternauer, J. L., and Pilkey, O. H., 1967, Phosphorite grains; Their application to the interpretation of North Carolina shelf sedimentation: Marine Geology, v. 5, p. 315–320.

Lyle, M. E., 1984, Clay mineralogy of the Pungo River Formation, Onslow Bay, North Carolina continental shelf, [M.S. thesis]: Greenville, North Carolina, East Carolina University, 129 p.

Manheim, F. T., 1972, Mineral resources off the northeastern coast of the United States: U.S. Geological Survey Circular 669, 28 p.

——, 1981, Potential hard mineral and associated resources on the Atlantic and Gulf continental margins, Phase II Task Force Report: National Technical Information Service Report PB81-192643, 41 p.

——, 1982, Cruise Report, GYRE cruise 11-82: U.S. Geological Survey, Office of Marine Geology, Atlantic-Gulf Branch, unpublished administrative document, 6 p., and maps.

Manheim, F. T., and Horn, M. K., 1968, Composition of deeper subsurface fluids along the Atlantic margin: Southeastern Geology, v. 9, p. 215–236.

Manheim, F. T., Pratt, R. M., and McFarlin, P. F., 1980, Composition and origin of phosphorite deposits on the Blake Plateau, *in* Bentor, Y. K., ed., Marine phosphorites: Society of Economic Paleontologists and Mineralogists Special Publication 29, p. 117–137.

Manheim, F. T., Popenoe, P., Siapno, W. D., and Lane, C. M., 1982, Manganese-phosphorite deposits of the Blake Plateau (Western North Atlantic Ocean), *in* Halbach, P., and Winter, P., eds., Marine mineral deposits; New research results and economic prospects: Essen, Verlag Glueckauf, v. 6, p. 9–44.

Martens, J.H.C., 1935, Beach sands between Charleston, South Carolina and Miami, Florida: Geological Society America Bulletin, v. 46, p. 1563–1596.

May, A., and Sweeney, J. W., 1981, Assessment of environmental impacts associated with phosphogypsum in Florida, *in* Borris, D. P., and Boody, P. W., eds., Proceedings of the International Symposium on Phosphogypsum, Lake Buena Vista, Florida: Bartow, Florida Institute of Phosphate Research, v. 2, p. 481.

Mayberry, R. C., 1981, Phosphate reserves, supply and demand, southeastern Atlantic coastal states, 1980–2000 A.D.: Preprint, Denver, Colorado, Society of Mining Engineers, American Institute of Mining Engineering, 21 p.

McKelvey, V. E., Wang, F. H., Schweinfurth, S. P., and Overstreet, W. C., 1968, Potential mineral resources of the United States Outer Continental Shelf: U.S. Geological Survey, unpublished administrative document, 151 p.

McKelvey, V. E., Wright, N. A., Bowen, R. W., 1983, Analysis of the world distribution of metal-rich subsea manganese nodules: U.S. Geological Survey Circular 886, 55 p.

Miller, J. A., 1971, Stratigraphic and structural setting of the middle Miocene Pungo River Formation of North Carolina [Ph.D. thesis]: Chapel Hill, University of North Carolina, 82 p.

——, 1982, Stratigraphy, structure, and phosphate deposits of the Pungo River Formation of North Carolina: North Carolina Department of Natural Resources and Community Development, Geological Survey Bulletin 87, 32 p.

Mullins, H. T., and Neumann, A. C., 1979, Geology of the Miami Terrace and its paleo-oceanographic implications: Marine Geology, v. 30, p. 205–231.

Olson, N. K., 1966, Geology of the Miocene and Pliocene series in the north Florida–south Georgia area: Southeastern Geological Society, 12th Annual

Field Conference, Guidebook, 94 p.

Patterson, S. H., and Buie, B. F., 1974, Field conference on kaolin and fuller's earth: Georgia Geological Survey Guidebook 14, 53 p.

Patterson, S. H., and Herrick, S. M., 1971, Chattahoochee Anticline, Apalachicola Embayment, Gulf Trough, and related structural features, southwestern Georgia; Fact or fiction: Georgia Geological Survey Information Circular 41, 16 p.

Paull, C. K., and Dillon, W. P., 1980, Structure, stratigraphy, and geologic history of Florida; Hatteras Shelf and inner Blake Plateau: American Association Petroleum Geologists Bulletin, v. 64, p. 339–358.

Pilkey, O. H., 1963, Heavy minerals of the U.S. South Atlantic Continental Shelf and Slope: Geological Society of America Bulletin, v. 74, p. 641–648.

Pilkey, O. H., and Luternauer, J. L., 1967, A North Carolina shelf phosphate deposit of possible commercial interest: Southeastern Geology, v. 8, p. 33–51.

Poag, C. W., 1978, Stratigraphy of the Atlantic Continental Shelf and Slope of the United States: Annual Review of Earth and Planetary Science, v. 6, p. 251–280.

Popenoe, P., 1983, High-resolution seismic reflection profiles collected August 4–28, 1979, between Cape Hatteras and Cape Fear and off Georgia and north Florida (Cruise GS-7903-6): U.S. Geological Survey Open-File Report 83-512, 4 p.

Popenoe, P., and Meyer, F. W., 1983, Description of single channel high-resolution seismic reflection data collected from continental shelf/slope and upper rise between Cape Hatteras, North Carolina and Norfolk, Virginia; and Vero Beach to Miami, Florida (Cruise 80-G-9): U.S. Geological Survey Open-File Report 83-515, 4 p.

Puri, H. S., and Vernon, R. O., 1964, Summary of the geology of Florida and guidebook to classic exposures: Florida Geological Survey Special Publication 5, 312 p.

Reaves, M. J., 1984, The importance of by-product uranium to phosphate rock producers, *in* Harben, P. W., and Dickson, E. M., eds., Phosphates; What prospect for growth?: New York, Metal Bulletin Incorporated, p. 247–252.

Riggs, S. R., 1977, The extractive industries in the coastal zone of continental United States, *in* Estuarine pollution control and assessment study: Environmental Protection Agency, Office of Water Planning and Standards, U.S. Government Printing Office, v. 1, p. 121–138.

——, 1979a, Petrology of the Tertiary phosphorite system of Florida: Economic Geology, v. 74, p. 195–220.

——, 1979b, Phosphorite sedimentation in Florida; A model phosphogenic system: Economic Geology, v. 74, p. 285–314.

——, 1980, Intraclast and pellet phosphorite sedimentation in the Miocene of Florida: Journal of Geological Society of London, v. 137, p. 741–748.

——, 1984, Paleoceanographic model of Neogene phosphorite deposition, U.S. Atlantic continental margin: Science, v. 223, no. 4632, p. 123–131.

Riggs, S. R., Hine, A. C., Snyder, S. W., Lewis, D. W., Ellington, M. D., and Stewart, T. L., 1982a, Phosphate exploration and resource potential on the North Carolina Continental Shelf, *in* Proceedings, 14th Offshore Technology Conference, Houston: Dallas, Texas, Offshore Technology Conference, v. 2, p. 737–748.

Riggs, S. R., Lewis, D. W., Scarborough, A. K., and Snyder, S. W., 1982b, Cyclic deposition of Neogene phosphorites in the Aurora area, North Carolina, and their possible relationship to global sea-level fluctuations: Southeastern Geology, v. 23, no. 4, p. 189–204.

Riggs, S. R., Snyder, S. W., Ellington, M. D., Burnett, W. C., and Beers, M., 1982c, Pleistocene/Holocene phosphorite formation; North Carolina continental margin: Geological Society of America Abstracts with Programs, v. 14, p. 77.

Riggs, S. R., Ellington, M. D., and Burnett, W. C., 1983, Geologic history of the Pleistocene/Holocene phosphorites on the North Carolina continental margin: Geological Society of America Abstracts with Programs, v. 15, p. 105.

Riggs, S. R., Snyder, S.W.P., Hine, A. C., Snyder, S. W., Ellington, M. D., and Mallette, P. M., 1985, Geologic framework of phosphate resources in Onslow Bay, North Carolina Continental Shelf: Economic Geology, v. 80,

p. 716–738.

Ross, D. A., 1970, Atlantic Continental Shelf and Slope; Heavy minerals of the continental margin from southern Nova Scotia to northern New Jersey: U.S. Geological Survey Professional Paper 529-G, 40 p.

Scarborough, A. K., Riggs, S. R., and Snyder, S. W., 1982, Stratigraphy and petrology of the Pungo River Formation, central coastal plain, North Carolina: Southeastern Geology, v. 23, no. 4, p. 205–216.

Schrader, E. L., Long, A. L., Muir, C. H., Quintus-Bosz, R., and Stewart, H. C., 1983, General geology and operations of kaolin mining in the "Southeastern Clay Belt"; A perspective from Huber, Georgia: Macon, Georgia, J. M. Huber Corporation, Clay Division, 25 p.

Scott, T. M., 1982, A comparison of the cotype localities and cores of the Miocene Hawthorn Formation in Florida, *in* Scott, T. M., and Upchurch, S. B., eds., Miocene of the southeastern United States: Southeastern Geological Society and Florida Bureau of Geology Special Publication 25, p. 237–246.

Sever, C. W., Cathcart, J. B., and Patterson, S. H., 1967, Phosphate deposits of south-central Georgia and north-central peninsular Florida: Georgia Division of Conservation, South Georgia Minerals Project Report 7, 62 p.

Simons, F. S., and Prinz, W. C., 1973, Gold, *in* Brobst, D. A. and Pratt, W. P., eds., United States mineral resources: U.S. Geological Survey Professional Paper 820, p. 263–276.

Snyder, S. W., Riggs, S. R., Katrosh, M. R., Lewis, D. W., and Scarborough, A. K., 1982, Synthesis of phosphatic sediment-faunal relationships within the Pungo River Formation; Paleoenvironmental implications: Southeastern Geology, v. 23, no. 4, p. 233–246.

Snyder, S.W.P., 1982, Seismic stratigraphy within the Miocene Carolina Phosphogenic Province; Chronostratigraphy, paleotopographic controls, sea-level cyclicity, Gulf Stream dynamics, and the resulting depositional framework [M.S. thesis]: Chapel Hill, University of North Carolina, 183 p.

Snyder, S.W.P., Hine, A. C., and Riggs, S. R., 1982, Miocene seismic stratigraphy, structural framework, and sea-level cyclicity; North Carolina Continental Shelf: Southeastern Geology, v. 23, no. 4, p. 247–266.

U.S. Bureau of Mines, 1980, Phosphate; Mineral commodities profile: U.S. Bureau of Mines, 19 p.

U.S. Department of the Interior, 1981, Executive summary, program feasibility document, OCS hard minerals leasing (with 23 appendices): U.S. Department of Interior, National Technical Information Services, PB 81-192551, 180 p.

U.S. Geological Survey, 1980, Principles of a resource/reserve classification for minerals: U.S. Geological Survey Circular 831, 5 p.

Waters, V. J., 1983, Foraminiferal paleoecology and biostratigraphy of the Pungo River Formation, southern Onslow Bay, North Carolina Continental Shelf [M.S. thesis]: Greenville, North Carolina, East Carolina University, 186 p.

Weaver, C. E., and Beck, K. C., 1977, Miocene of the S. E. United States; A model for chemical sedimentation in a peri-marine environment; *in* Developments in Sedimentology, v. 22: Amsterdam, Netherlands, Elsevier, 234 p.

——, 1982, Environmental implications of polygorskite (attapulgite) in Miocene of the southeastern United States, *in* Arden, D. D., Beck, B. F., and Morrow, E., eds., Proceedings on the Geology of the Southeastern Coastal Plain: Georgia Geological Survey Information Circular 53, p. 118–125.

Whitfield, W. K., 1975, Mining of submerged shell deposits; History and status of regulation and production of the Florida industry: Florida Marine Research Publication 11, 49 p.

Williams, G. K., 1971, Geology and geochemistry of the sedimentary phosphate deposits of northern peninsular Florida [Ph.D. thesis]: Tallahassee, Florida State University, 124 p.

Woolsey, J. R., 1976, Neogene stratigraphy of the Georgia coast and inner Continental Shelf [Ph.D. thesis]: Athens, University of Georgia, 222 p.

Zellers, M. E., and Williams, J. M., 1978, Evaluation of the phosphate deposits of Florida using minerals availability system: U.S. Bureau of Mines Open-File Report 112-78, 106 p.

MANUSCRIPT ACCEPTED BY THE SOCIETY JANUARY 2, 1986

Printed in U.S.A.

Chapter 26

Heat flow and geothermal resource potential of the Atlantic Coastal Plain

John K. Costain
Regional Geophysics Laboratory, Department of Geological Sciences, Virginia Polytechnic Institute and State University, Blacksburg, Virginia 24061
J. Alexander Speer
Orogenic Studies Laboratory, Department of Geological Sciences, Virginia Polytechnic Institute and State University, Blacksburg, Virginia 24061

INTRODUCTION

The geothermal resources in the eastern United States are liquid-dominated, low-temperature systems, oriented toward nonelectric power applications such as space heating and industrial processes (Toth, 1980; John Hopkins University, 1981). Evaluation of the geothermal resource potential in the eastern U.S. takes place in a geologic framework quite unlike that of the western U.S., where geothermal energy is primarily used in the generation of electric power. The resource in the East must be both large and favorably located with respect to potential utilization; fortunately, a large fraction of our energy consumption currently is devoted to space heating. The high costs associated with drilling for, pumping, and circulating warm water make it important to locate the highest temperatures at the shallowest depths.

Systematic efforts to estimate the geothermal resources of the United States have been made by the U.S. Geological Survey (Muffler, 1979; Sammel, 1979; Reed, 1983). Muffler and Cataldi (1978) proposed the use of a consistent terminology for geothermal resource assessment. The "geothermal resource base" is defined as all of the thermal energy in the Earth's crust under a given area, measured from the mean annual temperature. The "accessible resource base" is that part of the resource base shallow enough to be tapped by production drilling, and it is divided into "useful" and "residual" components. The "useful" component is defined as thermal energy that could be extracted at costs competitive with other forms of energy at some specified future time. This useful component is the subject of this section.

Geothermal resources in the Appalachian Mountain system and the Atlantic Coastal Plain can be grouped into four types: (1) Heat-producing granitoids in crystalline basement buried beneath the seaward-thickening wedge of Atlantic Coastal Plain sediments of low thermal conductivity; (2) Normal geothermal gradient resources, including those areas on the Atlantic Coastal Plain over crystalline basement nearly devoid of heat-producing granites; (3) Warm water emanating from fault zones (hot springs); and (4) Hot-dry-rock in regions of abnormally high geothermal gradient.

Resources 1 and 2 are the principal subject of this section. Resource 2 is widely available throughout much of the United States (Sammel, 1979). Resource 3 has been recognized in the Valley and Ridge Province since before 1884 (Rogers, 1884; Perry and others, 1979). Resource 4 is described by Pettitt (1979).

THE RADIOGENIC MODEL

Optimum sites for the development of Resource 1 in the eastern United States are associated with the heat-producing granitoids beneath the seaward-thickening wedge of relatively unconsolidated water-saturated sediments of the Atlantic Coastal Plain. Drill holes in these sediments yield large quantities of water at temperatures that, at any given depth, are higher in some locations than in others, depending on the local value of the geothermal gradient. The "radiogenic model" (Costain and others, 1980; Costain and Glover, 1982) is shown in Figure 1. The geothermal gradient is directly proportional to the local value of heat flow and inversely proportional to sediment thermal conductivity. Conductive heat flow is defined as

$$q = K \, d\theta/dz \qquad (1)$$

where q is heat flow, K is thermal conductivity, and $d\theta/dz$ is the geothermal gradient. Equation (1) applies only to the transport of heat by conduction. Heat transport by fluid convection is more

Costain, J. K., and Speer, J. A., 1988, Heat flow and geothermal resource potential of the Atlantic Coastal Plain, *in* Sheridan, R. E., and Grow, J. A., eds., The Atlantic Continental Margin, U.S.: Geological Society of America, The Geology of North America, v. I-2.

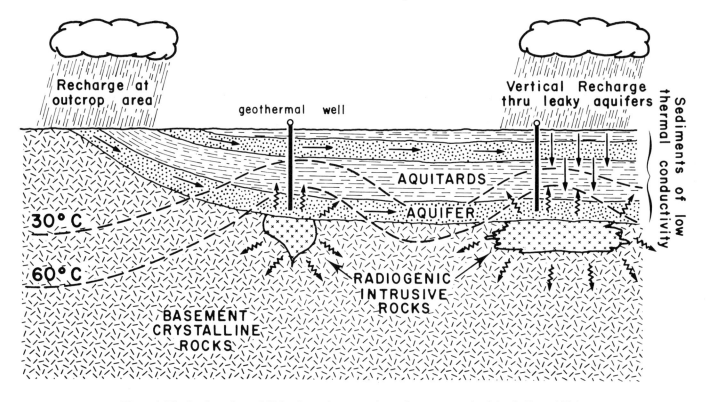

Figure 1. The "radiogenic model" for the optimum geothermal resource on the Atlantic Coastal Plain (after Costain and others, 1980).

efficient, and any successful geothermal system must incorporate both conduction and convection. Heat flow is often reported in "heat flow units," HFU (1 HFU = 1 heat flow unit = 10^{-6}cal/cm^2-sec = 41.84×10^{-3} W/m^2).

A hole drilled into the sediments of the Atlantic Coastal Plain will usually penetrate different lithologies, each with a different thermal conductivity. Because q is constant at any depth of interest in this discussion (we assume no significant heat generation in the sediments and no hydrologic disturbances), a change in K will be accompanied by an opposite change in the geothermal gradient, $d\theta/dz$, in order to keep the product $Kd\theta/dz$ constant. Simply stated, for any given value of heat flow, the highest temperature gradients occur in rocks with the lowest thermal conductivities. For example, mudstones of low thermal conductivity in the Valley and Ridge Province of the Appalachian mountain system will have temperature gradients that may be several times higher than those found in holes drilled into highly conductive dolomite. For the same value of heat flow, q, higher temperatures will be reached at shallower depths in thick sequences of mudstone rather than dolomite. Mudstone is the better insulator. The unconsolidated sediments of the Atlantic Coastal Plain are even better insulators than the mudstones. For the development of a geothermal resource, the combination of high heat flow and low thermal conductivity will result in the highest subsurface temperatures at the shallowest depths.

THE HEAT SOURCE

In a series of extremely important papers, Birch and others (1968), Lachenbruch (1968), and Roy and others (1968) showed that the local heat flow has a well-defined relationship to the concentration of uranium (U) and thorium (Th) in the underlying rocks. The immediate implication of their observation is that the distribution of U and Th in the upper 10 to 20 km of the Earth's crust is primarily responsible for the observed lateral variations in surface heat flow in the northeastern U.S. A similar relationship between surface heat flow and heat generation was reported by Costain and others (1986) for the Piedmont of the southeastern U.S.

The generation of heat in rocks by the process of radioactive decay is one of conversion of mass to energy according to $E = mc^2$. Only three elements have isotopes with long enough half lives to be important for radioactive heat generation in rocks. These are isotopes of uranium (U), thorium (Th), and potassium (K), with half lives on the order of 1 billion years or more. The heat contribution, H, in calories/gram per year from the radioactive decay of these isotopes is given by the equation

$$H = 0.72U + 0.20Th + 0.27K \qquad (2)$$

where U is in parts per million (ppm) uranium, Th is in ppm

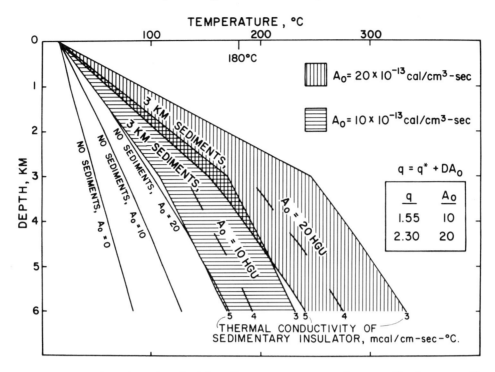

Figure 2. Effect of low thermal conductivity sediments on temperatures for a crystalline basement with different concentrations of heat-producing isotopes.

thorium, and K is in percent potassium (Birch, 1954; Rybach, 1976). Moderate amounts of the heat-producing elements occur in the crystalline basement rocks, concentrated primarily in unmetamorphosed granitoids. In granite, thorium is usually more abundant than uranium, the Th/U ratio falling between 3 and 4. Most of the heat produced comes from U and Th; about 10 to 15 percent comes from K. Heat generations of 15 HGU (1 HGU = 1 heat generation unit = 10^{-13} cal/cm^3-sec = 0.4184×10^{-6} W/m^3) have been determined in granite bodies exposed in the Piedmont of the southeastern U.S. (Costain and others, 1986), and are known to occur in the crystalline basement beneath the Atlantic Coastal Plain. In New England, exposed granites have heat productions of 24 HGU (Birch and others, 1968). The heat generated from radioactive decay is in addition to the normal heat flow around and beneath the granites and can be greater by more than a factor of 2 than the background flux characteristic of the normal heat leaving the Earth. In some parts of New England, for example, heat from radioactive granites constitutes two-thirds of the total heat flow (Birch and others, 1968). In South Carolina, the U, Th, and K concentrations of the Cuffytown Creek Granite are about 10 ppm U, 33 ppm Th, and 3.7% K. Thus, this granite generates about 15 calories per gram per year, or 12×10^{-13} cal/cm^3-sec, or about 20,000 watts per cubic mile. In fact, most of the heat flowing toward the surface in the Cuffytown Creek Granite comes from the heat generated by the radioactive decay of isotopes of U, Th, and K. An understanding of the distribution of granites and of U and Th in the host basement rocks is there-

fore important to define locations where the highest temperatures will occur at the shallowest depths.

One of the principal geological targets in the southeastern U.S. is therefore the relatively young (330 Ma and younger) syn- and postmetamorphic U- and Th-bearing, heat-producing granitoids in the crystalline basement beneath the sediments of the Atlantic Coastal Plain. Granitic rocks relatively enriched in U and Th (concentrations as high as the Cuffytown Creek) occur locally in the basement rocks beneath the sediments of the Atlantic Coastal Plain. The concentrations of U and Th in the granites are low, 2–15 ppm; but these concentrations are higher than the 1–3 ppm in adjacent rocks in the basement. Large volumes of granite with U and Th will substantially increase the subsurface temperature; thus, higher temperatures will be found at shallow depths within the overlying sediments. The leftmost curve in Figure 2 is the temperature-depth profile in basement crystalline rocks that contain no heat-producing elements and are not overlain by sediments. As U and Th are added to produce first 10 HGU and then 20 HGU, the subsurface temperature (and therefore the geothermal gradient) increases. Finally, if U- and Th-bearing granite is blanketed by sediments that have a relatively low thermal conductivity, the subsurface temperature is increased further.

Granite bodies are exposed in the Piedmont Province northwest of the Atlantic Coastal Plain (Fig. 3). To the southeast, they are concealed by a seaward-thickening wedge of sediments beneath the Atlantic Coastal Plain. Few deep holes have been

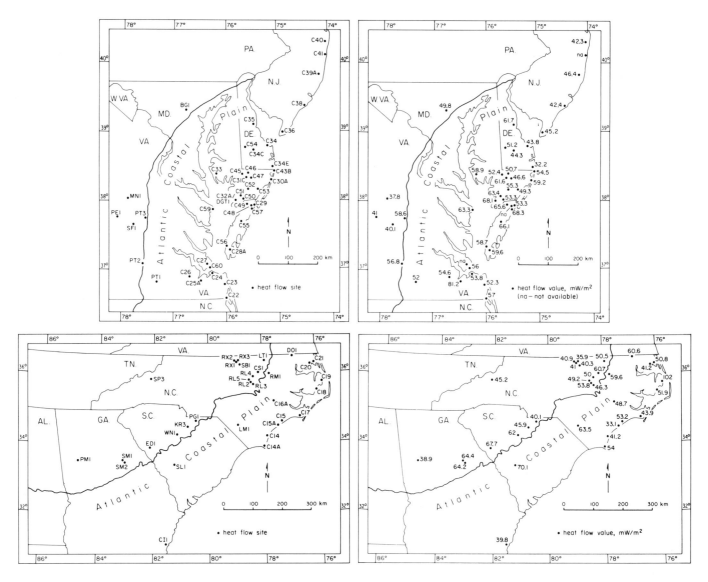

Figure 3. Heat flow sites and heat flow values in the Piedmont and Atlantic Coastal Plain (from Costain
and others, 1986).

drilled through the sediments to basement; but the sediment
thickness is known to be about 3 km at Cape Hatteras, North
Carolina. The relatively unconsolidated sediments of the Atlantic
Coastal Plain that overlie granitoids in the basement are the
geologic framework of high geothermal resource potential in the
southeast U.S. Heat flow and the geothermal gradient should be
higher over these granites if they contain modest concentrations
of U and Th.

The granitoid rocks of the southern Appalachians can be
grouped into several petrographic types (Speer and others, 1980,
1981; Speer, 1981) and characterized by heat production. Petrographic types include:

1. Coarse-grained, amphibole ± pyroxene + biotite granitoids

with heat productions between 1.3 and 3.8 $\mu W/m^3$ that are
characteristic of sites at PT-1, -2, and -3. RL-2, -4, and -5,
CS1, RM1, DO1, KR3, and PG1.

2. Coarse-grained, biotite granitoids with heat productions between 3 and 5 $\mu W/m^3$ include sites SL1, PM1, and SM1.

3. Coarse-grained, biotite granitoids that are evolved facies of
composite plutons include sites C25A (Portsmouth, Virginia)
and ED1 (Cuffytown Creek Granite), where the heat production was found to be between 3.4 and 5.9 $\mu W/m^3$.

4. Fine- to medium-grained, biotite granitoids with heat productions between 2.5 and 5.9 $\mu W/m^3$ at sites WN1 and SM2.

5. Medium-grained, muscovite + biotite granitoids with heat
productions less than 1.7 $\mu W/m^3$. No heat-flow values are yet

available in the low-heat-production muscovite + biotite granitoids.

6. Coarse-grained, cordierite + biotite granitoids with heat productions between 1.7 and 2.3 $\mu W/m^3$, such as at site SU near Stumpy Point, North Carolina (site C19).

Heat-flow determinations have been made from holes drilled into granites at the above sites except at Stumpy Point, North Carolina (SU). The Stumpy Point core was recovered from basement by others (a few tens of kilometers from heat-flow site C19 at Stumpy Point). In addition, unlike the others, the Stumpy Point Granite is a Precambrian granite.

Heat-flow determinations (Fig. 4) have been made at sites located on the Atlantic Coastal Plain (Fig. 4). Most of the drill-core sites for the granitoids were located at the centers of gravity minima and, presumably, are at or near the centers of the granite bodies. The geological settings of these holes are fairly well understood, and the heat production and heat flow are believed to be representative of that section of the crust. By contrast, the drillcores in metamorphosed rocks have generally unknown geological settings; and it is not clear whether they are representative of that section of the crust. For example, the Lumberton, North Carolina (LM1), drillhole penetrates metavolcanic rock but may be underlain by a granite, as indicated by the large gravity minimum and by seismic definition of a granite below the metavolcanics encountered in LM1 (Pratt and others, 1985). The Portsmouth, Virginia, granitoid (Russell and others, 1985) is an excellent example of the "radiogenic model" (Fig. 1) of Costain and others (1980). The Portsmouth Granite is a buried, confirmed, heat-producing granite. The overlying coastal plain sediments are about 600 m in thickness. The geothermal gradient in the sediments over the negative gravity anomaly is about 42 °C/km, but only 27 °C/km in a hole drilled nearby (12 km at site C26) at Isle of Wight off the anomaly in the same lithostratigraphic sequence. The heat flow over the Portsmouth Granite is about 81 mW/m². At Isle of Wight the heat flow is about 55 mW/m².

THE INSULATING COVER

Thermal conductivity is a function of bulk composition and is related directly to mineralogy and water content. In the sediments beneath the Atlantic Coastal Plain, the major mineralogical components are quartz, clay minerals, and calcium carbonate. Virtually all of the coastal plain sediments are water-saturated, and water content contributes significantly to the lower values of thermal conductivity. It is possible to consider a four-component system comprising water, quartz, clay minerals, and calcium carbonates, and quantify the relation between that system and thermal conductivity. This relationship can be used to broadly characterize the thermal conductivity of lithostratigraphic units. Average vertical thermal conductivity can be estimated for each formation, as well as changes in intraformational thermal conductivity. It is not widely appreciated that thermal conductivity differences, as derived directly from geophysical logs of the

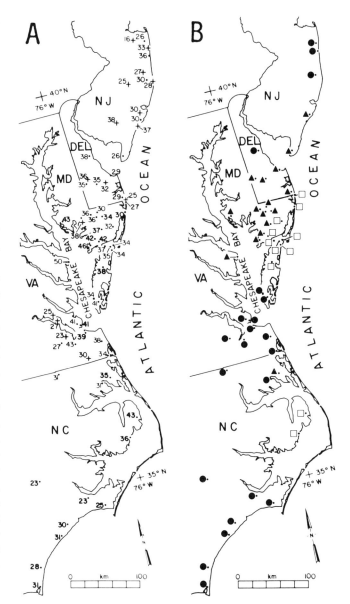

Figure 4. A. Geothermal gradients (°C/km) in the Atlantic Coastal Plain. •, exploratory holes drilled during 1978; +, older data. B. Maximum temperatures expectable at the base of Coastal Plain sediments. •, <55°C; ▲, 55°C to 75°C; □ >75°C (from Lambiase and others, 1980).

geothermal gradient, are useful for lithostratigraphic correlation and interpretation of lithologies. For example, many intervals encountered in Atlantic Coastal Plain drillholes have segments within which the geothermal *gradient* progressively decreases down hole, followed by a rapid increase in gradient and another progressive decrease. A decrease in gradient corresponds to an increase in the quartz/clay ratio; thus, the pattern described above corresponds to a series of relatively thin, fining-upward sequences. There are many such sequences recognizable in the Lower Cretaceous lithologic units throughout much of the Atlantic Coastal Plain. The sequences are equivalent to those generated by fluvial systems and are recognized in a number of modern and ancient examples. Generally, most of the Lower Cretaceous is regarded as nonmarine (Brown and others, 1972). Thus, depositional facies can be recognized on logs of the geothermal gradient. Other depositional characteristics can also be interpreted from geothermal gradients. Vertical changes in quartz/clay ratios are often indicative of changes in depositional environment. Gradual vertical changes in quartz/clay ratio often are representative of transitions between nonmarine, marginal marine, and marine facies. Generally, coastal plain sediments become more clayey as they become more marine. Also, changes in the thickness and abundance of clay beds relative to sand beds often reflect facies changes; nonmarine deposits tend to have fewer, thinner, less abundant clay beds than marine or marginal marine deposits. Thus, a general depositional history often can be reconstructed from an interpretation of geothermal gradient logs.

Lambiase and others (1980) summarized values of the geothermal gradients obtained in the holes drilled on the Atlantic Coastal Plain (Fig. 4). With the exception of two sites in New Jersey, gradients along the coastline of New Jersey, Delaware, and Maryland are low, with values between 16°C/km and 30°C/km. Higher gradients occur throughout most of the Delmarva Peninsula, with values between 30°C/km and 46°C/km. Near Chesapeake Bay, temperature gradients are the highest of any in the five-state area, with values between 39°C/km and 50°C/km. One gradient of 43°C/km was determined in northeastern North Carolina; all other shallow (<300 m) gradients in this area are between 31°C/km and 35°C/km. In southeastern North Carolina, gradients are lower and vary from 22°C/km to 32°C/km.

Geothermal gradients in the upper 300 m can be used to estimate the temperature at the base of and within the deeper coastal plain sediments. Depths to crystalline basement can be estimated by extrapolating between known depths from wells, with adjustments for regional trends in basement topography. For example, near Gardi, Georgia, basement was encountered in a drill hole at a depth of 1.33 km. The temperature predicted at this depth from the average gradient over the upper 296 m is 60°C and is equal to the 60°C basement surface temperature actually measured. Temperatures estimated in this manner may be useful for determining approximate values of basement surface temperature or for identifying regional trends in subsurface temperatures.

However, temperatures estimated from shallow (300 m) geothermal gradients are most likely maxima. There are factors that lower the geothermal gradient at depth in the northern Atlantic Coastal Plain, and consequently temperatures at basement are less than predicted from the shallow gradients. For a constant crustal heat flux, an increase in thermal conductivity of the sediments will cause a decrease in the geothermal gradient; the basement surface temperature will be less than that predicted by extrapolating a shallow geothermal gradient. Thermal conductivity varies with sediment type; abrupt lateral and vertical facies changes occur within the coastal plain sediments (Brown and

others, 1972), making prediction of lithology, and therefore thermal conductivity, difficult. Generally, the sediments in the deeper sections of the northern coastal plain are nonmarine deposits that are more quartz-sand rich than the upper 300 m; this tends to increase thermal conductivity in the deeper part of the section relative to the upper part and thus lowers the geothermal gradient. Also, thermal conductivity increases with compaction, which increases grain-to-grain contact by decreasing porosity and, therefore, water content. Introduction of cement into sediments will also increase thermal conductivity by decreasing porosity and replacing water with material of higher conductivity. Available data from deep holes indicate that a reasonable estimate for the minimum temperature at the base of the sediments may be as much as 20 percent lower than the temperature obtained by extending a shallow geothermal gradient. There are no data to suggest that hydrothermal convection or lateral fluid flow within the sediments of the Atlantic Coastal Plain significantly affects geothermal gradients with depth. The diminished gradients are the result of increasing thermal conductivity.

Results from a hole that was drilled to basement (540 m) near Portsmouth, Virginia (site C25A), and a test hole at Crisfield, Maryland (near site C32A), verify that in some locations the thermal conductivity of Atlantic Coastal Plain sediments does increase with depth, resulting in a decrease in the geothermal gradient with increasing depth. In the Portsmouth hole, the average geothermal gradient in the upper 300 m is 43°C/km. If this remained constant with depth to the top of crystalline basement, the temperature would be 39°C; however, the measured temperature at that depth is 36°C because of a 35 percent decrease in the geothermal gradient from 46 to 28°C/km for the lower section of the hole. The temperature at the surface of crystalline basement in the deep hole at Crisfield, Maryland, was 13 percent less than that predicted by extension of the shallow gradient because the gradient of 46°C/km in the upper 300 m decreased to 30°C/km in the lower part of the hole.

In southern Virginia, estimated temperatures at the basement surface are relatively low despite high geothermal gradients because basement occurs at relatively shallow depths (Lambiase and others, 1980). In New Jersey, estimated temperatures are low both because geothermal gradients are low and the basement surface is shallow.

Generally, temperatures at basement on the Delmarva Peninsula are predicted to be high. Along the Atlantic Coast, geothermal gradients are relatively low; but the thick sequence of coastal plain sediments (up to 2.34 km) results in relatively high basement surface temperatures. On the western margin of the peninsula, gradients are higher; but the coastal plain sediments are thinner (average thickness is approximately 1.52 km). The estimated basement surface temperatures on the western margin of the peninsula are slightly lower than those along the Atlantic coast. Extrapolation of geothermal gradients and estimated temperatures (Lambiase and others, 1980) between and below drill holes must be done with caution. The distribution of radioactive heat sources in the basement is poorly known. This, coupled with variations in thermal conductivity, makes the value of the geothermal gradient strictly site specific, and continuity between exploratory holes can be uncertain.

One area suggested for geothermal development in the northern Atlantic Coastal Plain is on the eastern shore in the area between Crisfield in southern Maryland and Oak Hall in northern Virginia. A test site was located at Crisfield, Maryland, because of the known high geothermal gradients there and the moderate depth to basement. Higher gradients (50°C/km) were found elsewhere—for example, on the flank of the large negative gravity anomaly in the vicinity of Chesapeake Bay (site C59)— but the depth to basement there is less. Temperature at the top of crystalline basement at Crisfield at a depth of 1.36 km was found to be approximately 58°C. Limited pump tests were made to estimate potential fluid production. Three zones in the coastal plain sediment were perforated. Zone 1 was perforated between 1,262 m and 1,285 m. The temperature of the water flowing from the zone was 57.2°C. Zone 2 was perforated from 1,187 m to 1,227 m. Water pumped from this zone for 48 hours at an average rate of 119 gallons per minute (gpm) produced a head drawdown of 84 m (275 ft). The temperature of water at the level of perforation was 56°C; at the surface the discharge temperature was 51°C. Zone 3 was perforated between 1,155 m and 1,170 m. A low-volume pump was set at a depth of 125 m and produced an averaged discharge of 32 gpm for 36 hours, resulting in a static drawdown of 30 m. Down-hole water temperature was 54°C, and surface discharge temperature reached 35°C.

LIFETIME

Limited hydrologic and heat-flow data now available make it possible to estimate the thermal lifetime of a geothermal resource within the sediments beneath the Atlantic Coastal Plain. Laczniak (1980) modeled the response of a leaky aquifer system to a single dipole (pumping plus injection well). The model was run for a simulated period of 15 years or until steady-state thermal and fluid flow was reached. A doublet system (dipole) with direct injection back into the reservoir was shown to be a feasible method of extracting heat in the low-temperature, liquid-dominated geothermal systems of the Atlantic Coastal Plain.

Important conclusions of Laczniak's study were: (1) Direct injection back into the reservoir may be necessary to maintain sufficient fluid pressure at the production well for systems with a low permeability; (2) Temperature distribution within the system is only slightly affected by changes in permeability in the range of 10 to 100 millidarcies; (3) Resting the system for periods of 6 months does not result in a significant recovery of heat at the production well; and (4) A doublet system with thermal and hydrologic conditions similar to those encountered at Crisfield, Maryland, a well spacing of 1,000 m, a permeability of 100 md, and a pumping-injection stress of 500 gpm (injection temperature 44°C) could produce 5.5 million Btu's per hour over a period greater than 15 years.

Assuming a value of 70,000 Btu's per hour for energy expenditure to heat an average insulated home, Laczniak estimates that the simple doublet system described above would support more than 75 households. If pumped for only 6 months a year, the thermal life span of the system would be at least 30 years.

SUMMARY

Geothermal energy may yet become an important energy resource for the Atlantic Coastal Plain. High heat flow and low thermal conductivity yield the highest temperatures at the shallowest depths. An understanding of the distribution of heat-producing granites and of U and Th in the host crystalline basement rocks is important to define locations where the highest temperatures will occur at the shallowest depths in the overlying sediments. One of the principal geological targets in the southeastern U.S. is the relatively young (330 Ma and younger) syn- and postmetamorphic U- and Th-bearing, heat-producing granitoids in the crystalline basement beneath the sediments of the Atlantic Coastal Plain. The concentrations of U and Th in the granites are low, a few parts per million; but these concentrations are higher than those in adjacent rocks in the basement, and large volumes of such granite can substantially increase the subsurface temperature. The relatively unconsolidated, water-saturated sediments of the Atlantic Coastal Plain that overlie these heat-producing granitoids in the basement are the geologic framework of high geothermal resource potential in the southeast U.S. Heat flow and the geothermal gradient should be higher over these granites.

Limited hydrologic and heat flow data now available make it possible to estimate the thermal lifetime of a geothermal resource within the sediments beneath the Atlantic Coastal Plain. A doublet system (dipole) with direct injection back into the reservoir appears to be one feasible method to extract heat from the low-temperature, liquid-dominated geothermal systems of the Atlantic Coastal Plain.

REFERENCES CITED

Birch, F., 1954, Heat from radioactivity, *in* Faul, H., ed., Nuclear geology: New York, John Wiley and Sons, p. 148–174.

Birch, F., Roy, R. F., and Decker, E. R., 1968, Heat flow and thermal history in New England and New York, *in* Zen, E., White, W. S., Hadley, J. B., and Thompson, J. B., Jr., eds., Studies of Appalachian geology: New York, Interscience, p. 437–451.

Brown, P. M., Miller, J. A., and Swain, F. M., 1972, Structural and stratigraphic framework and spatial distribution of permeability of the Atlantic Coastal Plain, North Carolina to New York: U.S. Geological Survey Professional Paper 796, 79 p.

Costain, J. K., and Glover, L., III, 1982, Evaluation and targeting of geothermal energy resources in the southeastern United States: Final Report VIP&SU-78ET-17001-12 to the U.S. Department of Energy, 218 p.

Costain, J. K., Glover, L., III, and Sinha, A. K., 1980, Low-temperature geothermal resources in the eastern United States: EOS American Geophysical Union Transactions, v. 61, p. 1–3.

Costain, J. K., Speer, J. A., Glover, L., III, Perry, L., Dashevsky, S., and McKinney, M., 1986, Heat flow in the Piedmont and Atlantic Coastal Plain of the southeastern United States: Journal of Geophysical Research, v. 91, no. B2, p. 2123–2135.

Johns Hopkins University, 1981, Geothermal energy development in the eastern United States: Final Report JHU-APL, QM-81-130, 135 p.

Lachenbruch, A. H., 1968, Preliminary geothermal model of Sierra Nevada: Journal of Geophysical Research, v. 73, p. 6977–6989.

Laczniak, R. J., 1980, Analysis of the relationship between energy output and well spacing in a typical Atlantic Coastal Plain geothermal doublet system [M.S. thesis]: Blacksburg, Virginia Polytechnic Institute and State University, 98 p.

Lambiase, J. J., Dashevsky, S. S., Costain, J. K., Gleason, R. J., and McClung, W., 1980, Moderate-temperature geothermal resource potential of the northern Atlantic Coastal Plain: Geology, v. 8, p. 447–449.

Muffler, L.J.P., ed., 1979, Assessment of geothermal resources of the United States, 1978: U.S. Geological Survey Circular 790, 163 p.

Muffler, L.J.P. and Cataldi, R., 1978, Methods for regional assessment of geothermal resources: Geothermics, v. 7, no. 2-4, p. 53–89.

Perry, L. D., Costain, J. K., and Geiser, P. A., 1979, Heat flow in western Virginia and a model for the origin of thermal springs in the folded Appalachians: Journal of Geophysical Research, v. 84, p. 6875–6883.

Pettitt, R. A., 1979, Hot-dry-rock program in the eastern U.S., *in* A Symposium of Geothermal Energy and its Direct Uses in the eastern United States: Davis, California, Geothermal Resources Council Special Report no. 5, p. 43–47.

Pratt, T., Costain, J. K., Coruh, C., Glover, L., III, and Robinson, E. S., 1985, Geophysical evidence for an allochthonous Alleghanian(?) granitoid beneath the basement surface of the Coastal Plain near Lumberton, North Carolina: Geological Society of America Bulletin, v. 96, p. 1070–1076.

Reed, M. J., ed., 1983, Assessment of low-temperature geothermal resources of the United States, 1982: U.S. Geological Survey Circular 892, 73 p.

Rogers, W. B., 1884, On the connection of thermal springs in Virginia with anticlinal axes and faults, *in* A reprint of annual reports and other papers on the geology of the Virginias: New York, p. 577–597.

Roy, R. F., Blackwell, D. D. and Birch, F., 1968, Heat generation of plutonic rocks and continental heat flow provinces: Earth and Planetary Science Letters, v. 5, p. 1–12.

Russell, G. S., Speer, J. A., and Russell, C. W., 1985, The Portsmouth Granite, a 263 Ma postmetamorphic biotite granite beneath the Atlantic Coastal Plain, Suffolk, Virginia: Southeastern Geology, v. 26, p. 81–93.

Rybach, L., 1976, Radioactive heat production in rocks and its relation to other petrophysical parameters: Basel, Birkhauser Verlag, Pageoph, v. 114, p. 309–317.

Sammel, E. A., 1979, Occurrence of low-temperature geothermal waters in the United States, *in* Muffler, L.J.P., ed., Assessment of geothermal resources of the United States, 1978: U.S. Geological Survey Circular 790, p. 86–1310.

Speer, J. A., 1981, Petrology of cordierite- and almandine + cordierite-bearing biotite granitoid plutons of the southern Appalachian Piedmont, U.S.A.: Canadian Mineralogist, v. 19, p. 35–46.

Speer, J. A., Becker, S. W., and Farrar, S. S., 1980, Field relations and petrology of the postmetamorphic, coarse-grained granites and associated rocks in the southern Appalachian Piedmont, *in* Wones, D. R., ed., The Caledonides in the U.S.A.: Blacksburg, Virginia Polytechnical Institute and State University, Department of Geological Sciences Memoir 2, p. 137–148.

Speer, J. A., Solberg, T. N., and Becker, S. W., 1981, Petrology of the U-bearing minerals of the Liberty Hill pluton, South Carolina; Phase assemblages and migration of U in granitoid rocks: Economic Geology, v. 76, p. 2162–2175.

Toth, W. J., 1980, Definition of markets for geothermal energy in the northern Atlantic Coastal Plain: The Johns Hopkins University Applied Physics Laboratory Report JHU/APL GEMS-002, QM-80-075, 221 p.

MANUSCRIPT ACCEPTED BY THE SOCIETY FEBRUARY 6, 1987

Printed in U.S.A.

Chapter 27

Geologic hazards on the Atlantic continental margin

David W. Folger
U.S. Geological Survey, Woods Hole, Massachusetts 02543

INTRODUCTION

Since 1975, an extensive program funded primarily by the U.S. Department of Interior's Bureau of Land Management has been carried out by various government, academic, and private research organizations to evaluate the potential impact of drilling and hydrocarbon exploitation on the continental margins of the United States. The study programs were developed in response to Federal statutes, such as the Outer Continental Shelf Lands Act of 1953 (OCSLA) and its 1978 amendments, the National Environmental Policy Act (NEPA), and the Clean Water Act, and to provide new data for environmental impact statements.

The U.S. Geological Survey began geologic environmental studies programs in 1973. As leasing became imminent on the Atlantic margin, the survey developed a Memorandum of Understanding with the Bureau of Land Management in 1975 to evaluate geologic hazards on the margin that might cause or distribute oil spills, and/or of other pollutants associated with drilling operations. Additional studies funded by the bureau were conducted in cooperation with private companies and academic institutions through contracts. The problems investigated include such phenomena as sandwave movement; surface sediment transport; effects of tides, waves, and residual circulation on sediment mobility; geotechnical properties of sediments; gas in sediments; slope stability and mass wasting processes; faulting; and distribution of trace metals and hydrocarbons. As a result of such studies, potential hazards or constraints can be known prior to drilling operations, and may be reduced by proper engineering or rig siting. However, at the beginning of these studies, few data were available on the Atlantic margin to document the magnitude, location, and variability of stresses—geologic and oceanographic—that might be encountered, nor were there many measurements of the levels of various contaminants such as trace metals and hydrocarbons.

This chapter is based mainly on data and analyses contained in a large number of papers and reports resulting from the extensive USGS field studies. Many of these are included in final reports to the Bureau of Land Management or papers derived therefrom (Folger, 1977b; Knebel, 1979, 1984; Popenoe, 1980, 1981; Popenoe and others, 1982; Aaron, 1980; Aaron and others, 1980; Carpenter, 1981a; Robb, 1982; O'Leary, 1982; McGregor,

1983). Initial results of studies in the Georges Bank and Baltimore Canyon areas were summarized by Folger and others (1978) in a similar report.

This chapter will focus mainly on the results of geologic-hazard and sediment-dispersal studies conducted in the three areas leased and drilled on the Atlantic continental margin, namely Georges Bank, the Baltimore Canyon Trough, and the Southeast Georgia Embayment (Fig. 1). In addition, it will emphasize hazards in areas where drilling may soon take place, such as the Carolina Trough and Blake Plateau Basin, and on the continental slope (200–2000 m water depth) along the entire margin (Fig. 1).

Lease Areas

The chronology of leasing and drilling on the Atlantic margin is summarized in Table 1; lease areas and blocks are shown in Figure 1. Environmentally-based lawsuits delayed initial leasing in both the Georges Bank and the Baltimore Canyon Trough areas. In part because of the extensive fishery on Georges Bank, lawsuits have been most intensively pursued there and have successfully blocked further leasing in both Sales 52 and 76.

In addition to the exploratory wells shown in Table 1, two COST (Continental Offshore Stratigraphic Tests) holes were drilled on Georges Bank, two in the Baltimore Canyon Trough, and one in the Southeast Georgia Embayment.

A total of 46 exploratory wells were drilled on the U.S. Atlantic margin by September 1984. Few indications of major reserves were found, but this initial exploratory drilling effort probably did not fully evaluate the potential of the margin. For example, Sherwin cautioned in 1975 that the 99 wells drilled off Canada's east coast were not an adequate number to evaluate the petroleum potential of such a large area. His warning was well-founded because the next 50 holes revealed commercial reserves of gas off Sable Island (Oil & Gas Journal, 1982), discovered the Hibernia and related structures off Newfoundland that may have reserves of 2.5 billion barrels (McKenzie, 1981), and discovered gas and distillate off Labrador (Oil and Gas Journal, 1981).

The 40 exploratory holes drilled on Georges Bank and the

Folger, D. W., 1988, Geologic hazards on the Atlantic continental margin; *in* Sheridan, R. E., and Grow, J. A., eds., The Geology of North America, Volume I-2, The Atlantic Continental Margin, U.S.: Geological Society of America.

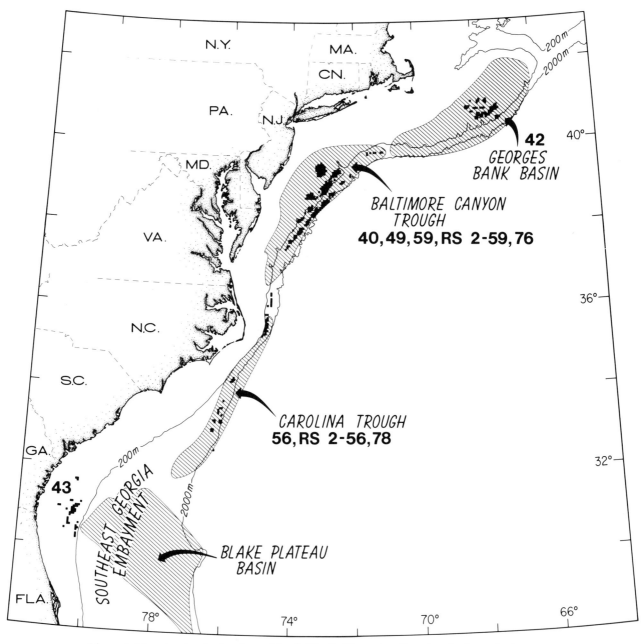

Figure 1. Map showing the locations of the 4 Atlantic Mesozoic sedimentary basins, *striped pattern*; leased blocks, *black*; and lease sales, *numbers*.

Baltimore Canyon Trough constitute one well drilled for each 5,000 km^3 of sediment, whereas according to Sherwin (1975) the average for the continental U.S. is one well drilled for each 0.6 km^3 of sediment.

Based on the Canadian experience, drilling will probably continue on the U.S. Atlantic margin for at least another decade despite the present slowdown. Most of it will be in deeper waters of the continental slope and farther offshore in the Carolina Trough or Blake Plateau Basin, where engineering will be more difficult and where oil spills will be harder to control. There, environmental hazards or constraints will have to be assessed with even greater thoroughness than on the shallow continental shelves.

Nature of Geologic Hazards

Hazards to or derived from offshore exploration and development span a broad spectrum of problems. Some can be alleviated by judicious rig or pipeline siting or proper engineering at relatively low cost; for example, modification of an anchoring design to account for a particular bottom type or current regime. Others are more difficult to alleviate; for example, slumping in

TABLE 1. CHRONOLOGICAL SUMMARY OF ATLANTIC MARGIN LEASING AND DRILLING

Area	Sale	Date	Tracts	Bids Accepted (Million $)	Exploratory Wells Drilled
North Atlantic	42	18 Dec. 1979	63	816	8
(Georges Bank)	52	---	---	---	
	76	---	---	---	
Middle Atlantic	40	17 Aug. 1976	93	1100	32
(Baltimore Canyon	49	28 Feb. 1979	39	40	
Trough)	59	8 Dec. 1981	51	324	
	RS2-59	5 Aug. 1982	18	3.9	
	76	26 Apr. 1983	37	68	
South Atlantic	43	28 Mar. 1978	43	100	6
(Southeast Georgia	56	4 Aug. 1981	47	342	
Embayment and	RS2-56	5 Aug. 1982	8	2.8	
Carolina Trough)	78	27 Jul. 1983	11	13	

deep-water sites. The former have been referred to as constraints because they can be reduced or eliminated by conventional engineering practices. The latter have been referred to as hazards because existing techniques may not eliminate their potential to cause structural damage (Cardinell and others, 1982).

In this chapter, two kinds of hazards will be addressed. The first includes those phenomena that are hazardous to the integrity of a structure (conventional hazard or constraint) such as sandwave migration, faulting, and mass-wasting. The second includes those phenomena (oceanographic or geologic) that might distribute pollutants from a rig or pipeline and hence represent a hazard not to a structure but to biota, ranging from plankton to humans. In short, hazards are those phenomena that may spill or distribute pollutants. Because the focus of this chapter is on geologic rather than meteorologic or oceanographic hazards, no detailed discussion will be devoted to the stresses imposed on structures by winds, currents, and waves; but, rather, to their interaction with sediments that could result in substrate instability and mobility.

The first category of hazards responds to external and internal stresses that cause sediment instability or mobility. External stresses include such phenomena as tidal currents that scour and erode the bottom, waves that cause oscillatory water motion, or ground motion due to earthquakes. (Earthquakes are not the only result of faulting. The presence of faults has resulted in loss of rigs and platforms whose boreholes crossed existing fault planes and penetrated deeper high-pressure gas [Danenberger, 1980]). Internal stresses may reduce the shear strength of sediment to a level at which gravitational forces, even at extremely low slope angles, may cause sediment to move. Such stresses include high pore pressures resulting from rapid sedimentation of fine-grained particles (silts and clay), and the evolution of gas from decomposition of incorporated organic matter.

A combination of both external and internal stresses probably initiates many substrate failures. For example, ground motion due to earthquakes imparts an acceleration to the sediment, but it also may increase the pore pressure by "shaking down" the component particles into a smaller volume. Repeated shaking can cumulatively reduce shear strength (Hampton and others, 1978).

The more common cyclic loading due to waves (or any other repetitive phenomenon) may result in similar loss of shear strength. As a worst case model for susceptibility to failure are fine-grained sediments, high in organic content, that have been rapidly deposited in an area of common earthquakes, such as the Copper River Delta, Alaska (Hampton and others, 1978).

In the second category are processes that entrain, transport, and bury pollutants. Residual, tidal, or wave-induced bottom currents may erode and transport natural and pollutant detritus. Studies of the sources and sinks of natural detritus provide clues to the potential mobility, transport routes, and depositional sites of pollutants that may have similar hydrodynamic properties. If the natural sedimentary systems can be documented, realistic models can then be constructed to predict pollutant residence times, dilution rates, and most likely sites of accumulation and concentration.

SETTING

The nearly flat continental shelf of the eastern United States varies in width from about 200 km off New England to a few kilometers off south Florida (Fig. 1). The continental slope (between 200 and 2000 m water depth) is comparatively narrow (about 35 km) and will become increasingly accessible to petroleum development over the next 5 to 10 years. Much of the shelf is veneered with residual and relict coarse-to-medium–grained sand that transitions with depth to the Pleistocene and Holocene clay and silt that mantles the slope (Duane, this volume). However, Tertiary to Cretaceous strata are exposed in some canyons or areas where slumps incise the slope (Ryan and others, 1978; Robb and others, 1981).

Most pertinent to surficial sediment mobility are the current speeds and their duration, and wave frequency and amplitude. The dynamics of the currents along the Atlantic margin are just starting to be unraveled (Butman and others, 1979; Butman, Beardsley, and others, 1982; Butman, Noble, and Moody, 1982; Mayer, 1982; Mayer and others, 1979, 1982; Ou and others, 1981; Boicourt and Hacker, 1976; Flagg and others, 1982). A

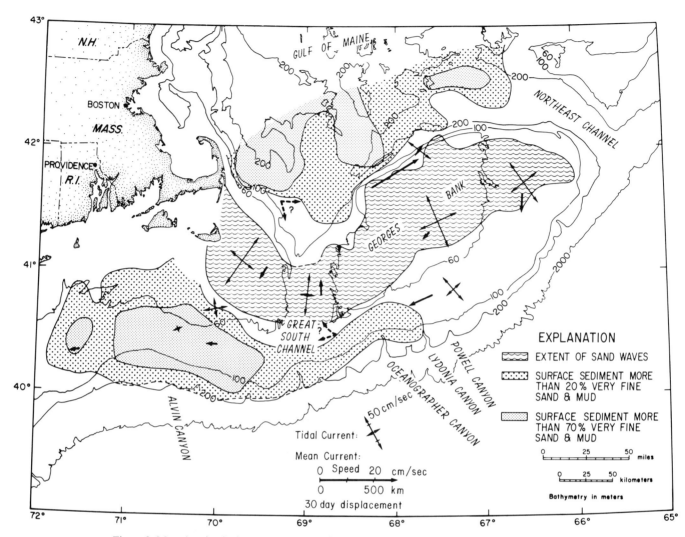

Figure 2. Map showing bathymetry, current regimes, sediment types, and bottom characteristics in the Georges Bank area. On the bank top where waters are as shallow as 5 m, tidal currents are most vigorous and sand-waves are common. Residual or mean currents are strongest on the northwest side of the bank. Both classes of currents are weak south of Nantucket where mud, probably winnowed mainly from the bank top, is accumulating. After Butman and others, 1982a; Twichell and others, 1981; Twichell, 1982.

major contribution to the physical oceanography of the margin between Cape Hatteras and southern Florida is contained in 15 papers published in the Journal of Geophysical Research, v. 88, no. C8, 1983. Tidal currents are dominant on Georges Bank, storm-driven currents in such areas as the Baltimore Canyon Trough (Butman and others, 1979), and boundary currents in the Southeast Georgia Embayment. The Gulf Stream impinges on the Blake Plateau and the continental shelf in the Southeast Georgia Embayment and has, over time, eroded major areas of the margin (Pinet and others, 1981a, b, c). Waves are most often affected by large North Atlantic storms that move out to sea in the area of Cape Hatteras and move northeast, parallel or subparallel to the coast. During late summer and fall, tropical hurricanes move northeast in the North Atlantic at varying distances from shore along the entire margin.

GEOLOGIC HAZARD ASSESSMENT

Georges Bank Basin

The name of this feature, both bank and basin, indicates why more controversy has surrounded petroleum exploration and development here than in any other Atlantic margin area. This deep Mesozoic sedimentary basin is capped by a large positive feature, a bank, roughly the area of West Virginia. (Fig. 2). Oil is generated in basins of this dimension, and abundant fish accumulate in the shallow waters overlying banks of this dimension. Thinning of continental crust, subsidence due to cooling, and subsequent sediment loading all contributed to the collapse of the margin off New England and to the accumulation of a thick section (about 14 km) of Triassic, Jurassic, and Cretaceous rocks

(Grow, 1980; Schlee and Klitgord, 1982; Klitgord and others, 1982; Poag, 1982; Folger and Schlee, 1983; see Schlee, this volume). As subsidence slowed, a thin section of Tertiary rocks formed a cuesta that was subsequently capped by as much as 80 m of detrital sediment deposited on the seaward edge of the continental glaciers during Pleistocene time (Lewis and others, 1980; see Duane, this volume). Thus, two geologic processes—massive subsidence and continental glaciation—resulted in the potential for oil and, in part at least, for the presence of fish.

The bank is a shallow barrier across the Gulf of Maine, breached only by two deep channels—one to the south, the 75-m-deep Great South Channel, and one to the north, the 240-m-deep Northeast Channel (Fig. 2). The tidal range in the northeastern Gulf of Maine is as high as 4 m, much higher than the low 1-m tidal ranges on the seaward side of the bank (J. Moody and B. Butman, pers. com., 1983). Large tidal currents thus flow across Georges Bank into and out of the Gulf of Maine. In addition, the bank lies along the tracks of many major winter storms and tropical hurricanes, and the combination of meteorologic and oceanographic phenomena often results in strong currents and high, confused seas.

Currents. High-speed, semidiurnal tidal currents flow northeast-southwest across the bank. The tidal ellipse has its long axis oriented northwest, at right angles to the bank axis (Fig. 2). Semidiurnal tidal current speeds as high as 75 cm/sec on the Bank top, decrease on the flanks to 40 cm/sec, and to 10 cm/sec over the continental slope (Moody and Butman, 1980). Data from moored current meter arrays confirm the persistent clockwise flow around the Bank depicted by Bigelow (1927) and Bumpus (1973) based on drift bottle tracks. The residual circulation is strongest (30 cm/sec) on the northwest flank of the Bank, and weaker on the southeast flank (5–15 cm/sec) (Butman and others, 1982a). These measurements quantify the flow and show that the clockwise circulation intensifies during spring and weakens during winter (Butman and others, 1982a; Butman and Beardsley, 1984; Flagg and others, 1982). One current drifter (Drogue #620), tracked over the bank top by satellite during the summer and fall of 1979, illustrates well the clockwise circulation around the Bank, and also the variability of the circulation (Fig. 3, upper; Butman and others, 1982a); whereas another drifter (Drogue #433), tracked in the spring and summer of 1979, moved from the northeast end of Georges Bank along the southeast flank. It slowed down and meandered around an area south of Nantucket for about 20 days. This indicates that some materials may not circle around the Bank, but may be transported along-shelf, and redeposited to the southwest (Fig. 3, lower; Butman and others, 1982a).

Continental Shelf Sediment Movement. The vigorous tidal circulation, coupled with storm waves and the residual circulation, exerts a strong influence on the transport of bottom sediment on the bank. The high mobility of sand on Georges Bank, for example, has been known for at least 50 years. Early sailing ships crossing the bank during storms reported sand on the decks after being hit by large waves (Emery and Uchupi, 1972).

Movement of sand waves was first documented by Stewart and Jordan (1964) who, based on sequential bathymetric measurements, estimated that the waves moved at a rate of 12 m per year. The near collapse of a Texas Tower on the bank was attributed to sediment movement (Emery and Uchupi, 1972), but may have been due mainly to structural failure.

Twichell (1983) has described the spectrum and distribution of sand waves on Georges Bank in detail and, based on sidescan sonar imagery, suggests that sand waves move at considerably greater rates. In fact, migration is extremely complex with crest-lines moving as much as 50 m over a 3 month period; but, because the waves move back and forth, net movement over longer time periods is less. Net sand volume movement is about 3000 m^3/km^2/year; however, the total volume of sand in motion probably is in excess of 300,000 m^3/km^2/year (D. Twichell, pers. com., 1983).

Wave lengths of sand waves on the bank vary considerably (Fig. 4). The characteristics of ridges 15–90 km long with wave lengths of 10–15 km and amplitudes of 10–35 m are very different from the smaller more abundant waves that are several hundred meters in wave length and about 10 m in wave height. Smaller, even more common, waves or megaripples have wave lengths of about a meter and wave heights of less than half a meter (Fig. 5). All of the sand waves apparently respond to tidal forcing. I have, for example, watched from a submersible the smaller ones changing their direction of movement over a boulder pavement from north to south in Great South Channel during a tidal change. Obviously, structures such as pipelines should not be located where sand wave movement is most intense if they cannot be buried below the level of sediment migration.

The effect of storms on this already dynamic environment has only recently been well documented. Data from almost seven years of instrument deployments on Georges Bank show that the major sediment mobility is associated with storms (Butman and Folger, 1979; Aaron and others, 1980; Butman and Moody, 1983; Figs. 6, 7). A turbid layer, at least 10 m thick, of sediment resuspended from the bottom, has been observed during intense storms; the sediment in the turbid layer is transported over a large area of the bank by tidal currents, storm flow, and residual flow (Fig. 2). Estimates of water particle excursions east or west during a single storm range from 10–20 km (Twichell and others, 1981).

New work has shown that fine-grained sediments are presently accumulating south of Nantucket where the tidal currents are weak (<5 cm/sec; Fig. 2; Twichell and others, 1981; Bothner and others, 1981). Clearly, this area represents a potential sink for various pollutants associated with drilling operations on the bank. The mobility of the substrate and the dispersal of pollutants in the water column and in the bottom sediments appear to be the two most important problems on the bank itself. The key question is whether pollutants will be effectively diluted in this system, or whether such wide distribution will cause them to interfere with various food and reproductive cycles over much of the bank.

Upper Continental Slope Sediment Movement. The nature and extent of sediment transport on the adjacent upper con-

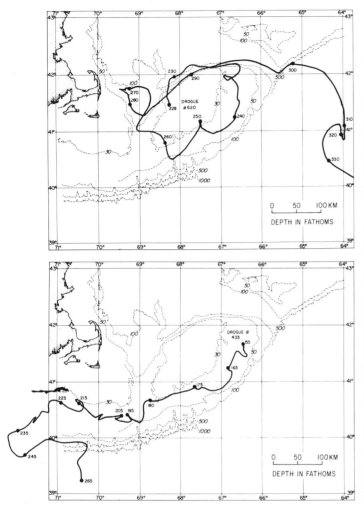

Figure 3. Maps showing tracks of window-shade surface current drifters tracked by satellite in 1979. *Upper,* Drogue #620 confirms the clockwise circulation around the Bank in some seasons. *Lower,* Drogue #433 shows that some water does not recirculate but flows southeast into the Mid-Atlantic Bight. Numbers are in Julian days. After Butman and others, 1982a.

undercutting (Ryan and others, 1978), earthquakes, storm waves, internal waves, or gas may occur now.

Abundant morphological evidence of slumping is present on the continental slope off Georges Bank (Roberson, 1964; McIlvaine, 1973; McGregor, 1979; McIlvaine and Ross, 1979; Aaron and others, 1980; Embley and Jacobi, 1986). Aaron and others (1980) have estimated that as much as 37% of the continental slope between 200 and 2,500 m of water depth, is underlain by slump deposits. High resolution reflection seismic and sidescan sonar data reveal what appear to be slump scars that often resemble the morphology of the heads of slumps or debris flows on land (D. O'Leary, pers. com., 1983; Fig. 9). Ryan and others (1978) state that "A principal agent of erosion is mass-wasting from cliffs which are oversteepened and undercut by currents accelerated within the narrow thalwegs of a dentritic-type drainage network and which are weakened by biological borings and solution diagenesis." And further, "Rock debris and talus blocks and slump blocks in the canyon axes . . . indicate that mass-wasting and gravitational collapse of bedrock is an active process along the steep walls of the present-day canyon." Recent geotechnical measurements on sediments taken from apparent slump scars on the slope adjacent to Alvin Canyon (Fig. 2) suggest that the sediments underlying the scars are overconsolidated, which would indicate that significant overburden of sediment may have been removed by sliding or slumping (Booth and others, 1981). One C_{14} date obtained on sediment from a slide near Alvin Canyon is Holocene in age. However, near the base of Powell Canyon (Fig. 2), evidence of mass wasting has been dated as latest Pleistocene (J. Booth, pers. com., 1983). Before extensive oil and gas exploration proceeds in this area, the potential for failure of sediments must be evaluated in more detail.

Baltimore Canyon Trough

The physiography and hydrology of the Baltimore Canyon Trough area are less complex than those in the Georges Bank area (Fig. 10). The sand-covered continental shelf between New York and Cape Hatteras lies in a large embayment (the New York Bight). The shelf averages about 100 km wide and dips gently seaward (1.5 m/km; Fig. 10). A thin Holocene sand veneer (0–20 m; Knebel and Spiker, 1977) overlies a thick Pleistocene section (as much as 450 m near the shelf edge; Grow, 1980).

Currents. Currents are dominated mostly by tides, but low-frequency fluctuations in speed and direction are associated with wind stress, topographic waves, and deep-sea circulation near the shelf-break; high-frequency fluctuations are associated with internal and surface waves (Butman and others, 1982b). The semidiurnal tidal currents in the Mid-Atlantic Bight are rotary. The major axis of the tidal ellipse is oriented roughly at right angles to the isobaths. Maximum surface tidal currents are about 16 cm/sec, and minimum surface currents, parallel to the isobaths, are about 10 cm/sec. Near the bottom, tidal currents range from a high of 7 cm/sec to a low of about 4 cm/sec (Moody and

tinental slope is not as clear as on the shelf. Despite evidence indicating that the 1929 Grand Banks earthquake initiated a massive slump and turbidity flow on the continental slope south of the Grand Banks (Heezen and Ewing, 1952; Heezen, 1960; Fig. 8), no available data prove that slumping of equal magnitude has taken place in historic time off the U.S. eastern continental slope—despite the occurrence of magnitude 7 to 8 earthquakes in the Cape Ann, Massachusetts (1755) and Charleston, South Carolina (1886) areas. However, conditions such as rapid sedimentation, oversteepening of the slope, and low sediment shear strengths must have been common during glacial times when low sea levels caused debris-filled rivers to flow into the ocean close to the present continental shelf edge (Fig. 2). Instabilities due to these conditions must have been common for at least several thousand years after sea level began to rise and some mass movement, possibly triggered by such phenomena as erosional

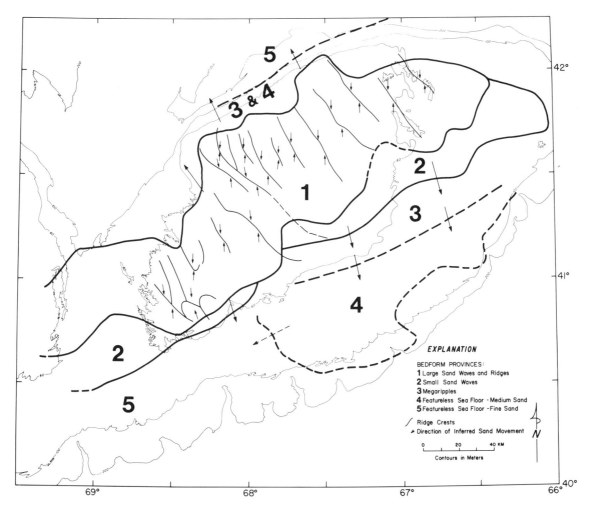

Figure 4. Diagram depicting various scales of sand-waves that are in motion on Georges Bank. Largest sand-waves are located mostly in waters less than 60 m deep. After Twichell, 1982.

Butman, 1980). Thus the tidal currents are much weaker than on Georges Bank.

Mean or residual bottom currents flow southwest, subparallel to the isobaths, 25° offshore, at speeds of 1.7 to 5.4 cm per second. This results in a 30-day long-shelf water-particle displacement of 45–150 km, and cross-shelf displacement of approximately 30 km. However, mean currents are occasionally modified or even reversed for periods as long as 20 days by wind-driven currents associated with storms and, at the shelf-edge, by Gulf Stream rings (Butman and others, 1982b; Mayer and others, 1979; Beardsley and others, 1976; Beardsley and Boicourt, 1981).

Continental Shelf Sediment Movement. Initial studies of prospective lease areas on the outer continental shelf (OCS) off the Mid-Atlantic states revealed several features similar to those described off Georges Bank. Among them are increased turbidity during storms and a few areas with sand waves.

Continuous observations of bottom flow, wave height, turbidity, and bottom sediment motion in the Middle Atlantic Bight clearly show that oscillating water motion due to waves and enhanced storm-driven currents winnows and suspends fine detritus (silt and clay) from the bottom. This sediment may subsequently be transported long distances by storm-associated currents. This situation is similar to that on Georges Bank, but the daily average flow associated with storms is most often parallel to the local isobaths (to the northeast or to the southwest), depending on the character of the storm (Butman and others, 1979; Butman and Folger, 1979).

Winter observations at three stations during 1975 to 1976 showed that average bottom current speeds were less than 15 cm/sec; whereas during the passage of three storms, currents exceeded 30 cm/sec and reached a maximum of 43 cm/sec. During each storm, bottom sediment was resuspended by the storm-generated currents and wave-induced bottom flow. Currents may be expected to move the sediment resuspended during these storms on the order of 10 km cross-shelf and 100 km along-shelf during a winter season (Butman and others, 1979). The rapid development of small asymmetrical ripples on the

Figure 5. Bottom photograph of common, mobile wave ripples 1–2 meters in wavelength and <0.5 meters in wave height near the top of Georges Bank (20 m water depth).

bottom during storms indicates that coarse bottom sediment is put in motion by traction transport, but no estimate has been made of its thickness, volume, or displacement (Butman and Moody, 1983).

Detailed studies of seven areas where earlier reconnaissance surveys had shown wave-like bottom morphology revealed sand waves only in one area, near the head of Wilmington Canyon (see Fig. 10). This field contains about 250 sand waves oriented northwest, and are 100 to 200 m in wave length and 1 to 9 m in wave height. They are asymmetrical to the southwest (Twichell, 1979). The morphology and internal structure of the waves were consistent with those presently in motion on Georges Bank, as well as those described close to the mouth of Delaware Bay (Jordan, 1962; Sheridan and others, 1974). However, subsequent observations of bottom currents revealed no velocities sufficient to maintain sand waves of these dimensions (Twichell, 1979). In addition, a faint reflector about 1 m below the bottom that extends across all the sand waves, suggests that a mobile, reworked layer covers the area but does not penetrate the internal structure of the large waves. Twichell (1979) therefore concludes that the waves are relict and are not in motion and thus pose no threat to structures such as pipelines laid in the area. Despite the generally weak (semidiurnal) tides and mean current flow, enormous volumes of fine sediment move across and along the shelf in response to storm activity; thus, pollutants could be distributed widely but also perhaps greatly diluted, especially during winter.

Measurements were carried out to evaluate the geotechnical characteristics of clay and silt horizons that underlie the surficial sand sheet. These finer-textured materials were first described by Donahue and others (1966); later observed in troughs between the common sand ridges and exposed at the bottom at several localities by observers aboard submersibles (Folger, 1977a); and subsequently described in detail from geophysical data and cores by Knebel and Spiker (1977). Geotechnical studies revealed that most of these sediments have high shear strengths, probably due to subaerial exposure, but site-specific tests should be run before any structures are set up on them (Sangrey and Knebel, 1979).

Several faults have been identified beneath the shelf. One growth fault located about 15 km south of Long Island is at least 30 km long, is oriented almost north-south, and has a down-to-the-west sense of displacement. The fault offsets basement as much as 85 m. It also offsets Cretaceous strata, and may have been active as late as Quaternary time (Hutchinson and Grow, 1982). Another set of faults with the same sense of displacement has been observed on high-resolution seismic profiles farther offshore. These faults definitely offset sediments as young as Holocene but apparently have no expression at the sea bottom (Sheridan and Knebel, 1976). Other faults near the shelf edge and under the slope have the opposite sense of movement and are typical of down-to-the-basin faults in large prograding sediment wedges (shown diagrammatically in Fig. 11; Schlee and Grow, 1980).

In summary, scour and bottom sediment mobility are common during storms. The fine sediment resuspended at these times can scavenge pollutants from the water column and transport them to the bottom. This consideration will be important even if hydrocarbon production is found only on the slope where interest is now focused, because pipelines may transit the shelf area. Faulting of Tertiary and Quaternary strata on the shelf suggests that seismicity, which could serve as a forcing mechanism to initiate slumping on the adjacent continental slope, may be present in the area.

Continental Slope Sediment Movement. Lease Sales 59 and Resale 2-59 were specifically designed to include a significant area of the continental slope off the Mid-Atlantic states mainly because it is underlain by a Jurassic/Cretaceous reef or carbonate buildup that may contain hydrocarbons (Folger and others, 1979; Grow and others, 1979; Schlee and Grow, 1980; Mattick and others, 1981). One exploratory hole, which was dry, penetrated the reef axis. All others appear to be in back-reef facies (W. Poag, pers. com., 1985). Though no commercial hydrocarbon accumulations apparently have been found, future additional exploratory drilling in this area, as noted above, is inevitable.

The problem, as on the slope off Georges Bank, is how stable is the surface? Are slumps and mass movement taking place? And, if so, where and why? Most intensive study has been devoted to this problem in the Mid-Atlantic area. About 450 m of Pleistocene sediment is draped over the upper slope of the Baltimore Canyon Trough area. Tertiary sediments are exposed on the mid-slope (Fig. 11). Geotechnical tests of both Pleistocene

TRIPOD OBSERVATIONS GEORGES BANK STA. A
5 DEC. 1976 – 23 FEB. 1977

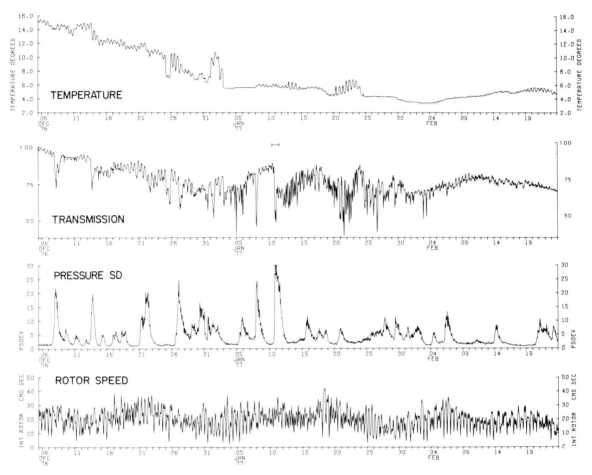

Figure 6. Diagram showing the changes in temperature, water turbidity (light transmission), pressure standard deviation (waves), and current speed on the south flank of Georges Bank between 5 December 1976 and 23 February 1977. Photographs in Figure 7 were taken before, during, and after the major storm indicated by the bar (transmission) when increased water turbidity accompanied the passage of large waves (10–13 January 1977) (Butman and Folger, 1979).

and Tertiary sediments reveal that they are mostly well-consolidated although, in a few areas, they may be susceptible to failure (Booth and others, 1982; Olsen and others, 1982). During glacial times, failures due to rapid sediment deposition in local areas on the upper slope were probable. Now, however, with little new sediment accumulating, instability due to loading is probably rare (Cardinell and others, 1982). Sangrey (1986) concluded from geotechnical studies of cores taken in the Baltimore Canyon Trough that neither waves nor earthquakes have probably caused significant slumping in the area. The slumping evident on seismic records may have been caused mostly by oversteepening due to erosion, especially on steep canyon slopes.

Yet the question of post-Pleistocene mass movement of slope sediments is still controversial. (See Uchupi, 1967b; Keller and others, 1979; McGregor and Bennett, 1977, 1979; Robb and

others, 1981; Farre and others, 1983; Prior and others, 1984; Stanley and others, 1984).

Several studies (Rona and Clay, 1967; Embley and Jacobi, 1977; McGregor and Bennett, 1977; McGregor, 1979; Knebel and Carson, 1979; Malahoff and others, 1980; Cardinell and others, 1982; Embley, 1982; Ryan, 1982; Farre and others, 1983; Stanley and others, 1984) present evidence that large areas of the slope have moved, some of it recently. One large slump near Hudson Canyon (Fig. 10) appears to be associated with a fault that has near-surface expression (Cardinell and others, 1982). Ryan (1982) concludes that "In the upper Slope, more than 75% of the seafloor forms part of the canyon drainage system. There are few sites spared from mass-wasting." And Farre and others (1983) show amphitheatre-shaped scars (their Figures 9 and 10) similar to the one portrayed in Figure 9 (this chapter) near Alvin

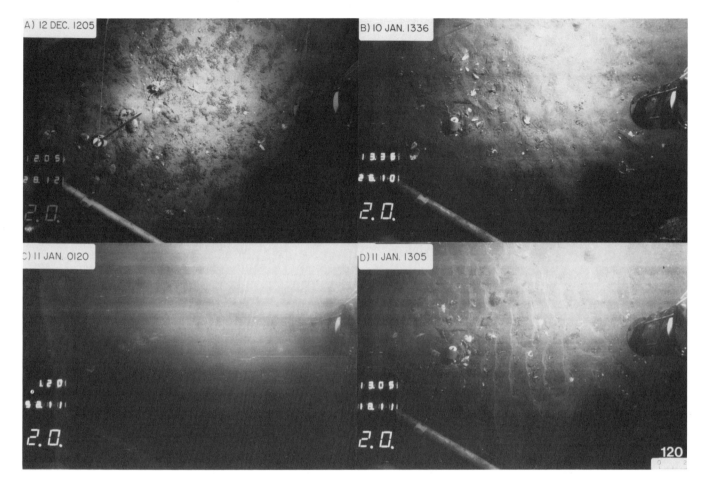

Figure 7. Bottom photographs taken from a tripod on the south flank of Georges Bank (see Butman and Folger, 1979 for location) before, during, and after the major winter storm on January 10–13, 1977 (see Fig. 6). Note the flocky layer (organic and fine terrigenous detritus) on the bottom in Photo *A* one month before the storm; the decrease in flocky material as the storm began in Photo *B*; the greatly increased turbidity in the water during the storm in Photo *C*; and the well-developed asymmetrical ripples, shells, and lack of flocky material on the bottom in Photo *D*.

Canyon. Other studies, however, have led the authors to the opposite conclusion; that is, that recent slumping is sparse (Robb and others, 1981; Twichell, 1986) or non-existent (Prior and others, 1984).

Robb and others (1981) and Robb (1982), for example, concluded from a detailed study of the south Toms, Carteret, and Lindenkohl canyon areas (see Area A, Fig. 10) that identifiable slumps or slides, which are located mainly on steep canyon heads and walls, comprise less than 5% of the slope. And Prior and others (1984), as a result of an even more detailed survey of the Carteret Canyon area (see Area B, Fig. 10), included within Robb's area, went further by concluding that [the slope] "is an ancient, relict landscape largely unmodified by modern slope processes." Clearly, additional studies are essential to resolve this controversy. Different levels and styles of mass-wasting apparently have taken place in different areas of the slope (McGregor, 1986). These include such phenomena as "rockslides and rockfall along bedding and joint planes from an oversteepened slope,

erosion and transport of blocks by turbidity currents, biological attack of cliff faces . . ." (Farre and Ryan, 1985), and even by spring sapping (Robb, 1984). How active these processes are today and what level of hazard they impose over the time frame of hydrocarbon exploration and development remain to be resolved.

More complex coverage with the new mid-range sidescan sonar systems and various deep-towed instruments will help acquire the data necessary to map slope morphology and structure in more detail. Perhaps most important is the need for long cores in deep water to evaluate *in situ* shear strength and gas contents of the strata (Sangrey, 1986).

Southeast Georgia Embayment

Both the physiography and the hydrology of the Southeast Georgia Embayment differ significantly from those of the two areas described above, which lie further north. From the shore to

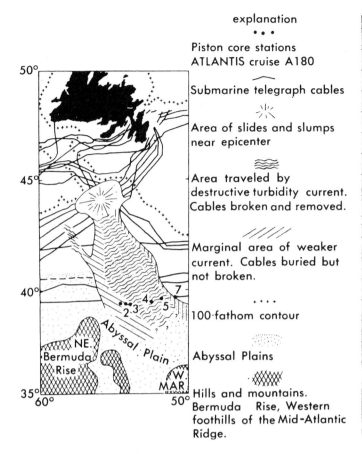

explanation

• • •

Piston core stations
ATLANTIS cruise A180

⌒

Submarine telegraph cables

⁎

Area of slides and slumps
near epicenter

≋

Area traveled by
destructive turbidity current.
Cables broken and removed.

////

Marginal area of weaker
current. Cables buried but
not broken.

• • • •

100-fathom contour

⠐⠐⠐⠐

Abyssal Plains

⧀⧀⧀

Hills and mountains.
Bermuda Rise, Western
foothills of the Mid-Atlantic
Ridge.

Figure 8. Diagram depicting the area of slumps and slides on the continental slope of Newfoundland caused by the Grand Banks earthquake of 1929. After Heezen, 1960.

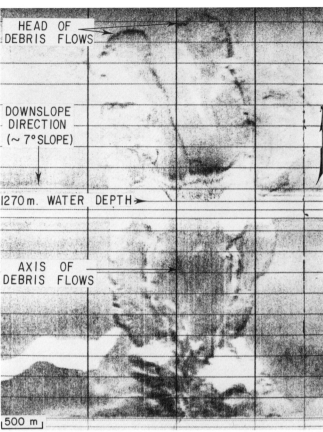

Figure 9. Sidescan sonogram of a slump or debris flow on the continental slope seaward of Georges Bank (approximately 10 km west of Alvin Canyon (see Fig. 2). Courtesy of O'Leary, D. W., unpublished data, 1983.

the 2,000-m isobath, the area of the continental shelf and Blake Plateau between Cape Hatteras and Little Bahama Bank covers 260,000 km^2, which is equivalent to the entire area from Cape Hatteras northward to the Canadian border. South of Cape Hatteras, the continental slope is split into two steps or terraces over much of the area (Fig. 12). The slope dips down only from 200 to 600 m from the seaward edge of the broad (as wide as 130 km), flat Florida-Hatteras shelf. At 600 m, the gradient flattens out and only attains water depths of 1100 m as much as 300 km from the outer shelf edge. The broad intermediate-depth flat terrace, the Blake Plateau, terminates seaward at the Blake Escarpment, which has an even steeper gradient (75°) than the slope north of the plateau. A large cone of sediments on the continental rise extends over 300 km southeast from the foot of the slope (at 2000 m) to abyssal depths at 4000 m. This feature is known as the Blake Ridge (Ewing and others, 1966). North of 31° N, the Blake Plateau is narrow and deepens gradually from about 300 m at the foot of the Florida-Hatteras slope off Charleston, South Carolina, to depths of 1,000 m at the continental slope. South of 31° N, the top of the Blake Plateau is wide and flat, and seaward it drops off abruptly at the Blake Escarpment to the Blake/Bahama Abyssal

Plain. This presents a very different physiographic setting from the continental margin to the north (for greater detail, see Heezen and others, 1959; Pratt and Heezen, 1964; Uchupi, 1967a; Emery and Uchupi, 1972). In part, the morphology has evolved in response to vigorous water circulation in the area that is mainly driven by Gulf Stream flow.

The Gulf Stream (Fig. 13) has played a major role in the shaping and evolution of the Blake Plateau area at least since Paleocene time (Pratt and Heezen, 1964; Bunce and others, 1965; Ewing and others, 1966; Emery and Uchupi, 1972; Schlee, 1977; Kaneps, 1979; Pinet and Popenoe, 1981a, b, c; Popenoe and others, 1982). The high-speed currents have been a barrier to the seaward dispersal of sediments derived from the continent and have eroded and scoured the bottom and prevented significant Tertiary sediment accumulation. This has led, in association with subsidence, to development of the Blake Plateau (128,000 km^2).

The mean position of the modern Gulf Stream front lies just seaward of the 200-m isobath; its normal onshelf fluctuations extend to the 200-m isobath off South Carolina and just landward of the 100-m isobath off Florida. Extreme lateral excursions (50 km) of the front affect a large area of the continental shelf

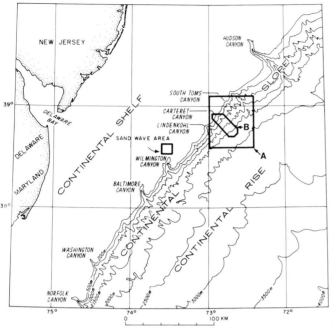

Figure 10. Map showing physiography of the margin off the Middle Atlantic states.

(Waddell, 1982). Surface current speeds of 4 to 6 knots are common off southern Florida, and of 3 knots off Jacksonville, Florida (Richardson and others, 1969; Schmitz and Richardson, 1968; Fig. 13). On the shelf side of the Gulf Stream, complex eddy/ring/filaments associated with strong current shear can reverse the flow along the shelf and slope (Richardson, 1980; Richardson and others, 1973; Richardson and others, 1977). Wind-driven, tidal, and residual currents are superimposed on the complex flow associated with the Gulf Stream. (For detailed references, see the Journal of Geophysical Research, v. 88, 1983.)

Because of the different physiographic, hydrologic, and geologic settings, several of the hazards in the Southeast Georgia

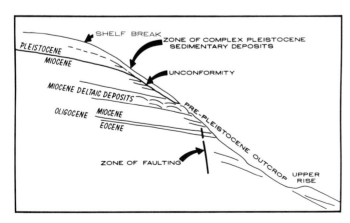

Figure 11. Diagrammatic cross section, based on a seismic line crossing the slope in area A (Fig. 10), that shows exposed Tertiary strata on the middle to lower slope. After Robb and others, 1981.

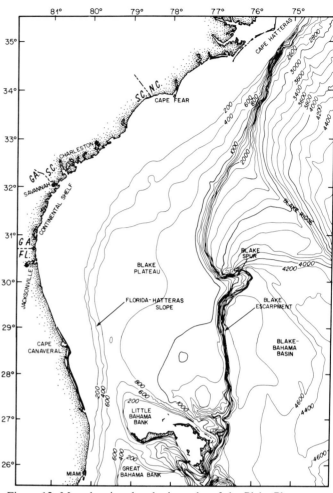

Figure 12. Map showing the physiography of the Blake Plateau area. After Pratt and Heezen, 1964.

Embayment are unique to the area. These include currents and associated scour, limestone solution, and clathrates.

Currents. Experience gained by drilling in the area of high Gulf Stream flow with the R/V GLOMAR CONCEPTION (Hathaway and others, 1978) indicates that special equipment for maintaining position, and perhaps strengthened and streamlined risers may be necessary. In addition, the strong flow makes anchoring of structures more difficult in the hard bottom that is sometimes armoured by manganese or phosphate (Pinet, Popenoe, Otter, and McCarthy, 1981c). Perhaps greatest stress on any structure will occur when hurricanes traverse the area augmenting Gulf Stream flow with storm tides and waves. In areas where the Gulf Stream flow is not strong, normal tidal flow and wind-driven currents do not appear to be a serious problem although some bed-load transport takes place during storms (See Journal of Geophysical Research, v. 88, 1983; Butman and Pfirman, 1980). New studies funded by the Minerals Management Service are underway and will provide valuable new measurements of Gulf Stream flow over deeper parts of the Blake Plateau.

Limestone Solution. Because of repeated, intensive scour

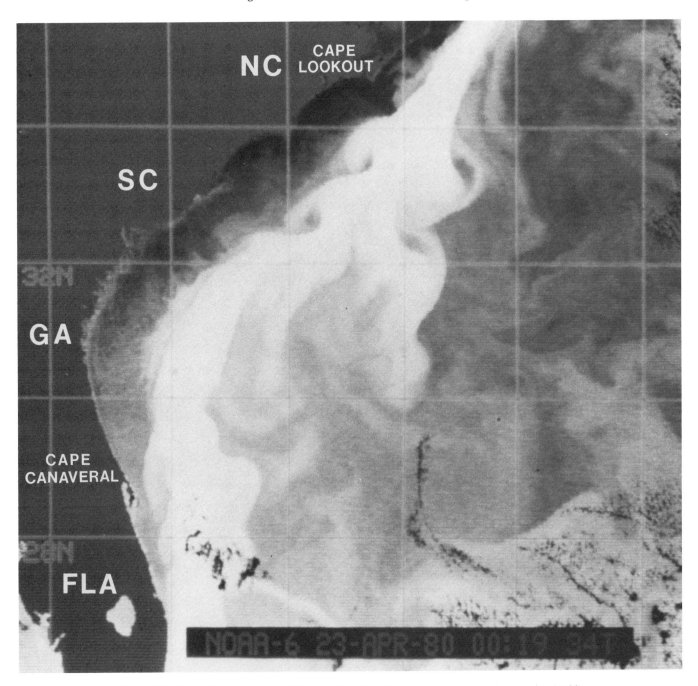

Figure 13. Photo showing the axis (light gray) of Gulf Stream flow and complex associated eddy structure. The high-speed flow may be a hazard to exploration and development in this area. Photograph courtesy of Otis Brown, NOAA, 1983.

across the Plateau throughout Tertiary time, many unconformities exist. Whether or not textural variations of material at these unconformities will cause serious problems for drillers remains to be seen. What is obvious, however, is that thick sections of carbonate rocks of late Cretaceous and Paleogene age occur over wide areas. During subaerial exposure of the shelf, mainly during late Oligocene and early Miocene time, extensive limestone solution resulted in the development of karst topography and large

sinkholes in the shelf area (see Popenoe and others, 1982; Fig. 14). Similar features on mainland Florida have been a serious hazard to rigs, several of which have actually fallen through the thin ceilings of large caverns (R. Worrell, pers. com., 1962). Deeper cavernous zones have also resulted in drilling delays because of lost circulation, pipe jamming, and lost drill strings. Another important consideration is the potential contamination or loss of fresh water from aquifers that are present in

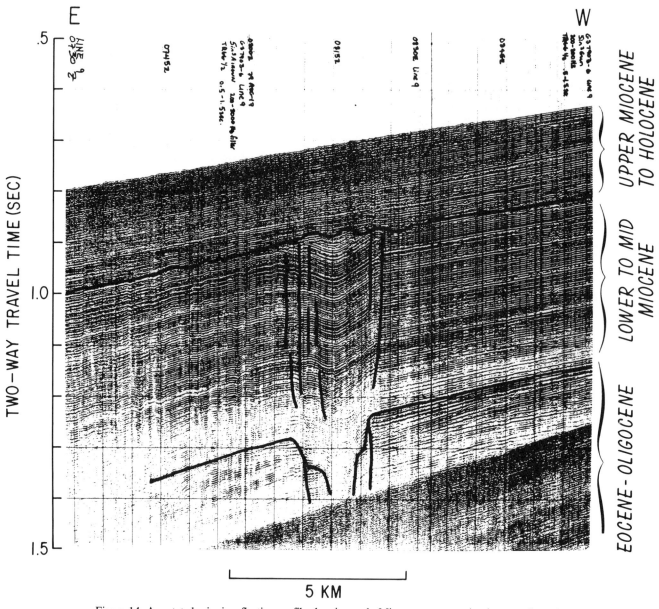

Figure 14. Annotated seismic reflection profile showing early Miocene strata sagging into a collapsed area of cavernous Eocene and lower Oligocene limestone. Solution probably occurred during subaerial exposure of the shelf during late Oligocene and early Miocene time, which caused the overlying beds to collapse during the middle Miocene. After Popenoe and others, 1982.

these zones. Vigorous freshwater flow is common in many areas on the Florida shelf where the sea bottom intersects such solution zones (Brooks, 1961; Kohout and others, 1977; Manheim, 1967).

Clathrates. In much of the area where water depths are over 800 m, frozen gas hydrates or clathrates (Fig. 15) are common (Shipley and others, 1979; Dillon and others, 1980; Paull and Dillon, 1980, 1981). Anomalous bottom simulating reflectors that cut across stratal reflectors, indicative of their presence, were described by Markl and others as early as 1970 on the Blake Ridge. More recently, actual samples of clathrates have been recovered from the deep-sea (Dillon and Kvenvolden, 1982). The

clathrates, which consist of an ice-like crystal lattice of water molecules and entrapped gas molecules, form under low-temperature, high-pressure conditions. A lower phase boundary that causes a bottom-simulating reflector (BSR; Fig. 15), is usually located about 0.4 to 0.6 seconds (2-way travel time) below subbottom where the hydrate becomes unstable due to the geothermal gradient (Tucholke and others, 1977). Penetration of free gas under clathrates in permafrost off the McKenzie Delta has resulted in the loss of several rigs (Peter Day in Popenoe and others, 1982, p. 61), but whether or not gas trapped beneath gas hydrates at great water depths will be of sufficient volume and at

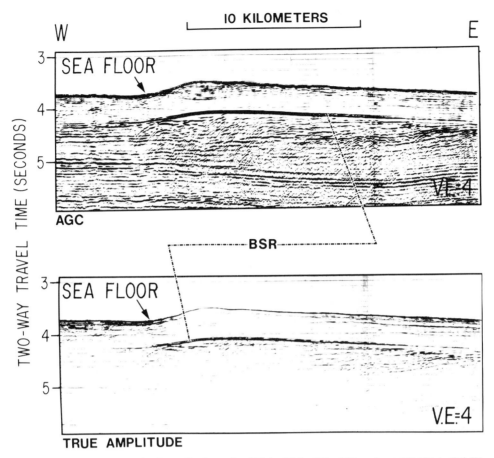

Figure 15. Multichannel seismic profile (on axis of Blake Ridge [Fig. 12] at about 32° N lat., 76° W long.) showing a bottom-simulating reflector (BSR) due to the lower boundary of a clathrate (gas hydrate) in the sediment. Free gas trapped below the clathrate causes the "bright spot" below the boundary. The clathrate causes subdued reflections above the boundary. After Paull and Dillon, 1980.

sufficient pressure to cause potential blowouts is unknown. Free gas, trapped under the hydrate, often produces a "bright spot" on seismic records due to the impedence contrast (Fig. 15). Because clathrates and "bright spots" beneath them are usually easy to recognize, drilling operators can be prepared to penetrate the free gas. Carpenter (1981a,b, 1986) has presented evidence suggesting that a clathrate is related to a slumped mass of sediment located approximately 185 km south of Cape Lookout (Fig. 16). It is possible that clathrates and the underlying free gas, in addition to being possible hazards, may constitute a large unconventional energy source if they formed in sediments that are sufficiently porous and permeable to permit recovery of hydrocarbons.

Faulting. Other hazards in the area are similar to those previously described. They include such phenomena as earthquakes, faulting, slumping, and sand waves. The area has a record of historic seismicity: in 1886, a magnitude earthquake shook the Charleston region (Behrendt and others, 1981). This is one of the largest historic earthquakes in the eastern United States. Faulting is not obvious in the Shelf area, with the exception of the Helena Banks Fault off Cape Romain and the White Oak Lineament

(Snyder and others, 1982), which may or may not be a fault (Popenoe and others, 1982; Fig. 16). Offshore, however, a long (380 km) growth fault parallels the coast at the eastern edge of the Blake Plateau between 32°N. and 35° N. The fault displaces the bottom and has a throw of 1 m at a depth of 10 m below bottom and 450 m at 5 km below bottom. It is a steep, easterly-dipping fault that apparently has been active for a long time in association with movement of Jurassic salt at about 11 km depth (Sylwester and others, 1979; Dillon and others, 1983; Fig. 16). Associated with it are swarms of antithetic faults of small displacement. Karstic collapse structures noted off North Carolina (Fig. 14) are associated with these faults (Popenoe and others, 1982).

Slope Stability. The declivity of the continental slope varies greatly along the eastern edge of the Blake Plateau. It is at least 3.5° where the slope joins the Blake Outer Ridge, greater than 16° off Cape Hatteras, and steepest, about 75°, on the Blake Escarpment (Fig. 12). Off Cape Hatteras and on the Blake Escarpment, deep-sea erosion has exposed Cretaceous to Pleistocene rocks. Between these two areas, only occasional slumps scar

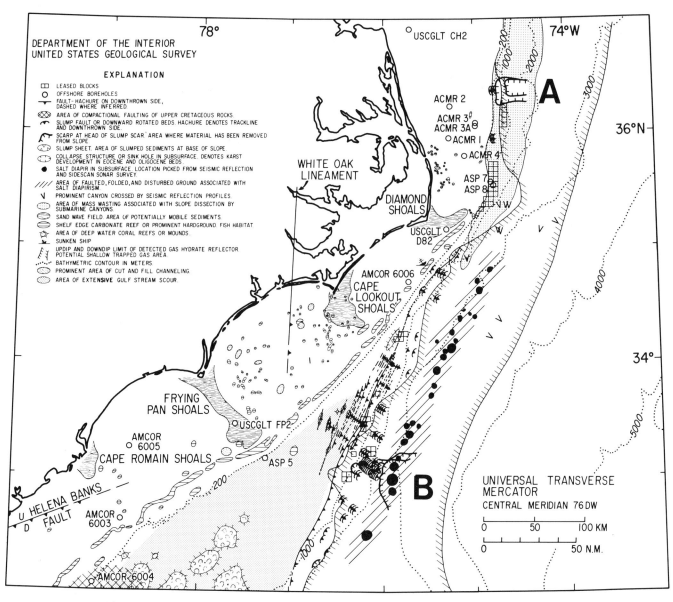

Figure 16. Summary map of hazards in the Southeast Georgia Embayment (Blake Plateau area). Modified after Popenoe and others, 1982.

a smooth surface. Areas where large slumps have been discovered are shown in Figure 16. One large scar is shown at Site A; another, which is about 60 km across, is shown at Site B. At site B, a large block 80 m high, 2 km deep, and of unknown length, apparently is slowly failing, as listric faults above the block offset the bottom several meters (Popenoe and others, 1982). Most submarine canyons are located near and north of Cape Hatteras, where mass wasting and slumping are maintained by head and sidewall erosion. Coring to determine important failure limits should be carried out, and areas determined to be in a failure mode should be avoided as drilling sites.

Sand wave fields off the southeastern U.S. are most common near the capes, such as Cape Romain, Cape Fear, and Cape Lookout, although some occur far offshore in several hundred meters of water (Fig. 16). Commonly, wave lengths are about 800 m and wave heights are about 10 m, similar to the dimensions of sand waves on Georges Bank. In deep water, the steeper faces lie to the north and Popenoe and others (1982) suggest that the waves are being moved by the Gulf Stream.

Drilling in the Carolina Trough and Blake Plateau areas will encounter a suite of geologic hazards that differ from but are about equal in severity to those on Georges Bank. The deeper water of both prospective areas, especially where vigorous and highly variable Gulf Stream flow occurs, will require special precautions. Precaution also should be exercised in the siting of structures in areas of cavernous limestone, and aquifer zones should be cemented to avoid contamination. Innovative engineering may be required to avoid the danger of blowouts by gas trapped under clathrates, especially in areas near the locus of antithetic faults and faults associated with salt diapirs, which could serve as conduits for the escape of gas from deep horizons. Stability of the highly dissected continental slope north of Cape Hatteras may be a problem, but elsewhere mass wasting does not seem to be common. Because of the hard limestone bottom in most areas of the Blake Plateau, scour or sand-wave motion will probably not be a significant problem.

SUMMARY

Geologic and hydrologic hazards to exploitation of the U.S. Atlantic continental margin vary: the most severe conditions are found on Georges Bank followed by the Carolina Trough and Blake Plateau, and the least severe conditions in the Baltimore Canyon Trough area.

Problems on Georges Bank are mainly associated with the winter storms and high tidal current speeds that continuously move small and large sand waves and, mainly during major storms, resuspend abundant fine-textured detritus in sufficient concentrations to scavenge pollutants from the water column. A major depositional site where fine material is presently accumulating occurs south of Nantucket. Slumping has taken place and is

a potential problem on the continental slope southeast of Georges Bank, but the frequency of recent failures has not been determined.

Except for the passage of hurricanes and severe winter storms, which resuspend abundant fine detritus from the bottom, the continental shelf in the Baltimore Canyon Trough area contains few major hazards. Some slumping on the continental slope takes place, probably occurring mainly in canyon heads and walls. And in other local areas, shear strength may be low enough to fail, due to such phenomena as gas, with little external stress from, for example, earthquakes. More studies are needed to determine the frequency, level, and magnitude of mass-wasting phenomena in this area.

In the Southeast Georgia Embayment, the most hazardous areas are in the deep waters of the Carolina Trough and Blake Plateau Basin. Gulf Stream flow and its associated eddies will impose severe stress on any structures or moored facilities, particularly in view of the known phosphate- or manganese-armoured hard bottom that will make mooring or burial difficult. In addition, the substrate in some areas may be unstable because of the cavernous porosity immediately below the bottom that developed during several stages of the Tertiary when sea level was low. Offshore, clathrates are common in water depths between 800 and 2,000 meters. Drilling, particularly in the Carolina Trough, may therefore encounter clathrates and underlying free gas, which is, as yet, a hazard of unknown magnitude.

REFERENCES CITED

Aaron, J. M., 1980, Environmental geologic studies in the Georges Bank area, U.S. Northeastern Atlantic outer continental shelf, 1975–1977: U.S. Geological Survey Open-File Report 80-240, 5 chapters.

Aaron, J. M., Butman, Bradford, Bothner, M. H., and Sylwester, R. L., 1980, Maps showing environmental conditions relating to potential geologic hazards on the United States northeastern Atlantic continental margin: U.S. Geological Survey Map MF-1192.

Beardsley, R. C., and Boicourt, W. C., 1981, On estuarine and continental-shelf circulation in the Middle Atlantic Bight, *in* Warren, B. A., and Wunsch, Carl, eds., Evolution of Physical Oceanography: Cambridge, Massachusetts, MIT Press, p. 198–233.

Beardsley, R. C., Boicourt, W. C., and Hansen, D. V., 1976, Physical oceanography of the Middle Atlantic Bight, *in* Gross, M. Grant, ed., Middle Atlantic continental shelf and the New York Bight: Limnology and Oceanography, Special Symposium, v. 2, p. 20–34.

Behrendt, J. C., Hamilton, R. M., Ackermann, H. D., and Henry, V. J., 1981, Cenozoic faulting in the vicinity of the Charleston, South Carolina, 1886 earthquake: Geology, v. 9, p. 117–122.

Bigelow, H. B., 1927, Physical oceanography of the Gulf of Maine: U.S. Fisheries Bulletin, v. 40, no. 2, p. 511–1027.

Boicourt, W. C., and Hacker, P. W., 1976, Circulation on the Atlantic continental shelf, United States, Cape May to Cape Hatteras: Liege, Society of Royal Scientists Memoir, 6(10), p. 187–200.

Booth, J. S., Farrow, R. A., and Rice, T. L., 1981, Geotechnical properties and slope stability analyses of surficial sediments on the Georges Bank continental slope: U.S. Geological Survey Open-File Report 81-566, 86 p.

——1982, Geotechnical properties and slope stability analysis of surficial sediments on the Baltimore Canyon continental slope, *in* Robb, J. M., ed.,

Environmental geologic studies in the Mid-Atlantic outer continental shelf axis—Results of 1978-1979 field seasons: Unpublished Final Report submitted to the U.S. Bureau of Land Management, p. 7-1 to 7-40.

Bothner, M. H., Spiker, E. E., Johnson, P. P., Rendigs, R. R., and Aruscavage, P. J., 1981, Geochemical evidence for modern sediment accumulation on the continental shelf off southern New England: Journal of Sedimentary Petrology, v. 51, p. 281–292.

Brooks, H. K., 1961, The submarine spring off Crescent Beach, Florida: Florida Academy of Science, Quarterly Journal, v. 24, p. 122–134.

Bumpus, D. F., 1973, A description of the circulation on the continental shelf of the east coast of the United States: Progress in Oceanography, v. 6, p. 111–157.

Bunce, E. T., Emery, K. O., Gerard, R. D., Knott, S. T., Lidz, Louis, Saito, Tsuenemasa, and Schlee, J. S., 1965, Ocean drilling on the continental margin: Science, v. 150, p. 709–716.

Butman, Bradford, and Folger, D. W., 1979, An instrument system for long-term sediment transport studies on the continental shelf: Journal of Geophysical Research, v. 84, p. 1215–1220.

Butman, Bradford, Noble, M. A., and Folger, D. W., 1979, Long-term observations of bottom current and bottom sediment movement on the Mid-Atlantic continental shelf: Journal of Geophysical Research, v. 84, p. 1187–1205.

Butman, Bradford, and Pfirman, S. L., 1980, Observations of bottom current and bottom sediment movement in the Southeast Georgia Embayment, *in* Popenoe, P., ed., Final report, Environmental studies, Southeastern United States, Atlantic OCS 1977—Geology: U.S. Geological Survey Open-File Report 80-140, p. 104–137.

Butman, Bradford, and eight others, 1982a, Recent observations of the mean circulation on Georges Bank: Journal of Physical Oceanography, v. 12,

p. 569–591.

Butman, Bradford, Noble, M. A., and Moody, J. A., 1982b, Observations of near-bottom currents at the shelf break near Wilmington Canyon, *in* Robb, J. M., ed., Environmental studies in the Mid-Atlantic outer continental shelf area—Results of the 1978–1979 Field Seasons: Final Report to the U.S. Bureau of Land Management, p. 3-1 to 3-50.

Butman, Bradford, and Moody, J. A., 1983, Observations of bottom currents and sediment movement along the U.S. east coast continental shelf during winter, *in* McGregor, B. A., ed., Environmental geologic studies on the United States Mid- and North Atlantic outer continental shelf area, 1980–82, v. 3: Woods Hole, Massachusetts, Unpublished U.S. Geological Survey Final Report to the Minerals Management Service, p. 7-1 to 7-60.

Butman, Bradford, and Beardsley, R. C., 1984, Long-term current observations on the southern flank of Georges Bank: Seasonal cycle of current, temperature, stratification, and wind stress: Journal of Physical Oceanography, in press.

Cardinell, A. P., Keer, F. R., and Good, L. K., 1982, Hazard analysis on the Mid-Atlantic Continental Slope, OCS lease sale 59 area: Offshore Technology Conference paper 4173, p. 89–95.

Carpenter, G. B., 1981a, Potential geologic hazards and constraints for blocks in proposed South Atlantic OCS oil and gas lease sale 56: U.S. Geological Survey Open-File Report 81-019, 325 p.

——1981b, Coincident sediment slump/clathrate complexes on the U.S. Atlantic continental slope: Geomarine Letters, v. 1, no. 1, p. 38–42.

——1986, The relation of clathrates and sediment stability, *in* Folger, D. W., and Hathaway, J. C., eds., Conference on continental margin mass-wasting and Pleistocene sea-level changes: U.S. Geological Survey Circular, in press.

Danenberger, E. P., 1980, Outer continental shelf oil and gas blowouts: U.S. Geological Survey Open-File Report 80-101, 15 p.

Deis, J. L., Kurz, F. N., and Porter, E. O., 1982, South Atlantic summary report 2: U.S. Geological Survey Open-File Report 82-15, 50 p.

Dillon, W. P., Grow, J. A., and Paull, C. K., 1980, Unconventional gas hydrate seals may trap gas off southeast United States: Oil and Gas Journal, v. 78, p. 124, 126, 129–130.

Dillon, W. P., and Kvenvolden, K. A., 1982, Gas hydrates in seafloor sediments off southeastern United States: Evidence from seismic reflection and drilling data, *in* Malone, R. D., ed., Technical Proceedings, Methane Hydrates Workshop: Morgantown, West Virginia, U.S. Department of Energy, Morgantown Energy Technology Center, p. 78–81.

Dillon, W. P., and six others, 1983, Growth faulting and salt diapirism: Their relationship and control in the Carolina Trough, in Watkins, J. S. and Drake, C. L., eds., Studies in continental margin geology: American Association of Petroleum Geologists Memoir 34, p. 31–46.

Donahue, J. G., Allen, R. C., and Heezen, B. C., 1966, Sediment size distribution profile on the continental shelf off New Jersey: Sedimentology, v. 7, p. 155–159.

Embley, R. W., 1982, Anatomy of some Atlantic margin sediment slides and some comments on ages and mechanisms, *in* Saxov, S. , and Nieuwenhuis, J. K., eds., Marine slides and other mass movements: New York, Plenum Press, p. 189–213.

Embley, R. W., and Jacobi, R., 1977, Distribution and morphology of large submarine sediment slides and slumps on Atlantic continental margins: Marine Geotechnology, v. 2, p. 205–228.

——1986, Mass wasting in the western North Atlantic: *in* Vogt, P. R. and Tucholke, B. E., eds., The Geology of North America, Volume M. the Western North Atlantic Region: Geological Society of America, in press.

Emery, K. O., and Uchupi, E., 1972, Western North Atlantic Ocean: Topography, Rocks, Structure, Water, Life, and Sediments: American Association of Petroleum Geologists Memoir 17, 532 p.

Ewing, J. D., Ewing, M., and Leyden, R., 1966, Seismic-profile survey of Blake Plateau: American Association of Petroleum Geologists Bulletin 50, p. 1948–1971.

Farre, J. A., McGregor, B. A., Ryan, W.B.F., and Robb, J. M., 1983, Breaching the shelfbreak: Passage from youthful to mature phase in submarine canyon evolution, *in* Stanley, D. J., and Moore, G. T., eds., The shelfbreak: Critical interface on continental margins: Society of Economic Paleontologists and Mineralogists Special Publication No. 33, p. 25–39.

Farre, J. A., and Ryan, W.B.F., 1985, Comment on spring sapping on the lower continental slope, offshore New Jersey: Geology, v. 13, p. 91–92.

Flagg, C. W., Magnell, D. F., Cura, J. J., McDowell, E. E., Scarlet, R. I., 1982, Interpretation of the physical oceanography of Georges Bank: Unpublished Final Report prepared for the U.S. Bureau of Land Management, 8 Chapters.

Folger, D. W., 1977a, Submersible observations of the bottom in and near petroleum lease areas and two dumpsites on the continental shelf off the Middle Atlantic states, *in* Folger, D. W., ed., Middle Atlantic outer continental shelf environmental studies, Vol. III, Geologic Studies: Gloucester Point, Virginia, Virginia Institute of Marine Science, p. 8-1 to 8-40.

Folger, D. W., ed., 1977b, Middle Atlantic outer continental shelf environmental studies, v. III: Geologic studies, final report, 1 June 1975 – 30 June 1976: Gloucester Point, Virginia, Virginia Institute of Marine Science, 12 Chapters.

Folger, D. W., Butman, B., Knebel, H. J., and Sylwester, R. J., 1978, Environmental hazards on the Atlantic outer continental shelf of the United States: Offshore Technology Conference Paper 3313, p. 2293–2297.

Folger, D. W., Dillon, W. P., Grow, J. A., Klitgord, K. D., and Schlee, J. S., 1979, Evolution of the Atlantic continental margin of the United States, *in* Talwani, M., Kay, W., Ryan, W.B.F., eds., Deep drilling results in the Atlantic Ocean: Continental margins and paleoenvironment, Maurice Ewing Series 3, American Geophysical Union, p. 87–108.

Folger, D. W., and Schlee, J. S., 1983, The geologic evolution and petroleum potential of Georges Bank, *in* Tiffney, W. N., and Hill, R. F., eds., Proceedings of the Conference—Georges Bank Hydrocarbon Exploration and Development: Hanover, New Hampshire, The American Society for Environmental Education, p. 1–27.

Grow, J. A., 1980, Seismic stratigraphy in the vicinity of the COST No. B-3 well, *in* Scholle, P. A., ed., Geological studies of the COST No. B-3 well, United States Mid-Atlantic continental slope area: U.S. Geological Survey Circular 833, p. 177–132.

Grow, J. A., Mattick, R. E., and Schlee, J. S., 1979, Multichannel seismic depth sections and interval velocities over outer continental shelf and upper continental slope between Cape Hatteras and Cape Cod: American Association of Petroleum Geologists Memoir 29, p. 65–83.

Hampton, M. A., Bouma, A. H., Carlson, P. R., Molnia, B. F., Clukey, E. C., and Sangrey, D. A., 1978, Quantitative study of slope instability in the Gulf of Alaska: Offshore Technology Conference Paper 3314, p. 2307–2318.

Hathaway, J. C., Poag, C. W., Valentine, P. C., Miller, R. E., Schultz, D. M., Manheim, F. T., Kohout, F. A., Bothner, M. H., and Sangrey, D. A., 1978, U.S. Geological Survey core drilling on the Atlantic shelf: Science, v. 206, p. 515–527.

Heezen, B. C., 1960, Turbidity currents: McGraw-Hill Encyclopedia of Science and Technology, v. 146–147.

Heezen, B. C., and Ewing, M., 1952, Turbidity currents and submarine slumps, and the 1929 Grand Banks earthquake: American Journal of Science, v. 250, p. 849–873.

Heezen, B. C., Tharp, M., and Ewing, M., 1959, The floors of the Ocean—I, The north Atlantic: Geological Society of America Special Paper 65, 122 p.

Hutchinson, D. R., and Grow, J. A., 1982, New York Bight fault: U.S. Geological Survey Open-File Report 82-208, 8 p.

Jordan, G. F., 1962, Large submarine sand waves: Science, v. 136, p. 839–848.

Kaneps, A. G., 1979, Gulf Stream velocity fluctuations during the late Cenozoic: Science, v. 204, p. 297–301.

Keller, G. H., Lambert, P. N., and Bennett, R. H., 1979, Geotechnical properties of continental slope deposits—Cape Hatteras to Hydrographer Canyon, *in* Doyle, L. J., and Pilkey, O. H., eds., Geology of the continental slopes: Society of Economic Paleontologists and Mineralogists Special Publication No. 27, p. 131–151.

Klitgord, K. D., Schlee, J. S., and Hinz, K., 1982, Basement structure, sedimentation, and tectonic history of the Georges Bank Basin, *in* Scholle, P. A., and Wenkham, C. R., eds., Geological studies of the COST Nos. G-1 and G-2

Wells, United States north Atlantic outer continental shelf: U.S. Geological Survey Circular 861, p. 4–10.

Knebel, H. J., 1979, Middle Atlantic outer continental shelf environmental studies, v. III, Geologic studies, final report, 1 October 1976–30 September 1977: Gloucester Point, Virginia, Virginia Institute of Marine Science, Unpublished Administrative Report, 8 Chapters.

—— 1984, Sedimentary processes on the Atlantic continental slope of the United States: Marine Geology, v. 61, p. 43–74.

Knebel, H. J., and Carson, B., 1979, Small-scale slump deposits, middle Atlantic continental slope of eastern United States: Marine Geology, v. 29, p. 221–236.

Knebel, H. J., and Spiker, E., 1977, Thickness and age of surficial sand sheet, Baltimore Canyon Trough area: American Association of Petroleum Geologists Bulletin, v. 61, p. 861–871.

Kohout, F. A., Leve, G. W., Smith, F. T., and Manheim, F. T., 1977, Red Snapper Sink and groundwater flow, offshore northeastern Florida [abs.]: International Association of Hydrologists, 12th International Congress, Alabama, 1975, Abstracts with Program, p. 60.

Lewis, R. S., Sylwester, R. E., Aaron, J. M., Twichell, D. C., and Scanlon, K. M., 1980, Shallow sedimentary framework and related potential geologic hazards of the Georges Bank area, *in* Aaron, J. M., ed., Environmental geologic studies in the Georges Bank area, United States northeastern Atlantic outer continental shelf, 1975–1977; U.S. Geological Survey Open-File Report 80-240, p. V-1-V-25.

Manheim, F. T., 1967, Evidence for submarine discharge of water on the Atlantic continental slope of the United States, and suggestions for further search: New York Academy of Science Transactions, Series 2, v. 29, p. 839–852.

Markl, R. G., Bryab, G. M., and Ewing, J. B., 1970, Structure of the Blake-Bahama Outer Ridge: Journal of Geophysical Research, v. 75, p. 4539–4555.

Mattick, R. E., Schlee, J. S., and Bayer, K. C., 1981, The geology and hydrocarbon potential of the Georges Bank–Baltimore Canyon Trough area: American Association of Petroleum Geologists Memoir 7, p. 461–486.

Mayer, D. H., 1982, The structure of circulation: MESA physical oceanographic studies in New York Bight, 2: Journal of Geophysical Research, v. 87, p. 9579–9888.

Mayer, D. H., Hansen, D. V., and Ortman, D. A., 1979, Long-term current and temperature observations on the Middle Atlantic shelf: Journal of Geophysical Research, v. 84, p. 1776–1792.

Mayer, D. H., Han, G. C., and Hansen, D. V., 1982, Circulation in the Hudson Shelf Valley: MESA physical oceanographic studies in New York Bight, 1: Journal of Geophysical Research, v. 87, p. 9563–9578.

McGregor, B. A., 1979, Variations in bottom processes along the U.S. Atlantic continental margin, *in* Watkins, J., Mantadert, L., and Dickerson, P., eds., Geological and geophysical investigations of continental margins: American Association of Petroleum Geologists Memoir 29, p.139–149.

——ed., 1983, Environmental geologic studies on the United States Mid- and North Atlantic outer continental shelf area, 1980-82: Woods Hole, Massachusetts, Unpublished U.S. Geological Survey Final Report to the Minerals Management Service, 3 volumes.

—— 1986, Diversity of processes and morphology on the U.S. Atlantic continental slope and rise, *in* Folger, D. W., and Hathaway, J. C., eds., Conference on continental margin mass wasting and Pleistocene sea-level changes: U.S. Geological Survey Circular, in press.

McGregor, B. A., and Bennett, R. H., 1977, Continental slope sediment instability northeast of Wilmington Canyon: American Association of Petroleum Geologists Bulletin, v. 61, p. 918–928.

Malahoff, A., Embley, R. W., Perry, R. B., and Fefe, C., 1980, Submarine mass-wasting of sediments on the continental slope and upper rise south of Baltimore Canyon: Earth and Planetary Science Letters, v. 49, p. 1–7.

McIlvaine, J. C., 1973, Sedimentary processes on the continental slope off New England [Ph.D. thesis]: Massachusetts Institute of Technology and Woods Hole Oceanographic Institution, 211 p.

McIlvaine, J. C., and Ross, D. A., 1979, Sedimentary processes on the continental

slope of New England: Journal of Sedimentary Petrology, v. 49, p. 563–574.

McKenzie, R. M., 1981, The Hibernia . . . a classic structure: Oil and Gas Journal, v. 79, no. 38, p. 240–246.

Moody, J. A., and Butman, Bradford, 1980, Semi-diurnal bottom pressure and tidal currents on Georges Bank and in the Mid-Atlantic Bight: U.S. Geological Survey Open-File Report 80-1137, 22 p.

Oil and Gas Journal, 1981, Canada okays two tests off East Coast: v. 79, no. 42, p. 137.

—— 1982, Development of structure off Nova Scotia eyed: v. 80, no. 27, p. 51.

O'Leary, D. W., 1982, Environmental geologic studies in the Georges Bank Area, United States northeastern Atlantic continental margin, 1978–1979, Final report submitted to the Bureau of Land Management: Woods Hole, Massachusetts, U.S. Geological Survey Unpublished Report, 13 Chapters.

Olsen, H. W., McGregor, B. A. Booth, J. S., Cardinell, A. P., and Rice, T. L., 1982, Stability of near-surface sediment on the mid-Atlantic upper continental slope: Offshore Technology Conference 4303, p. 21–27.

Ou, H. W., Beardsley, R. C., Mayer, D., Boicourt, W. C., and Butman, Bradford, 1981, An analysis of subtidal current fluctuations in the Mid-Atlantic Bight: Journal of Physical Oceanography, v. 11, p. 1383–1392.

Paull, C. K., and Dillon, W. P., 1980, The appearance and distribution of the gas hydrate reflector in the Blake Ridge region, offshore southeastern United States: U.S. Geological Survey Open-File Report 80-88, 24 p.

—— 1981, Appearance and distribution of the gas hydrate reflector in the Blake Ridge region, offshore southeastern United States: U.S. Geological Survey Miscellaneous Field Studies Map MF-1252.

Pinet, P. R., Popenoe, P., McCarthy, S. M., and Otter, M. L., 1981a, Seismic stratigraphy of the northern and central Blake Plateau, *in* Popenoe, P., ed., Environmental geologic studies on the southeastern Atlantic outer continental shelf, 1977–78: U.S. Geological Survey Open-File Report 81-582-A, p. 7-1 to 7-91.

Pinet, P. R., Popenoe, P., and Nelligan, D. F., 1981b, Gulf Stream: Reconstruction of Cenozoic flow patterns over the Blake Plateau: Geology, v. 9, p. 266–270.

Pinet, P. R., Popenoe, P., Otter, M. L., and McCarthy, S. M., 1981c, An assessment of potential geologic hazards of the northern and central Blake Plateau, *in* Popenoe, P., ed., Environmental geologic studies on the southeastern Atlantic outer continental shelf, 1977–78; U.S. Geological Survey Open-File Report 81-582-A, 8 Chapters.

Poag, C. W., 1982, Stratigraphic reference section for Georges Bank basin: A depositional model for New England passive margin: American Association of Petroleum Geologists, v. 66, p. 1021–1041.

Popenoe, P., ed., 1980, Environmental studies, southeastern United States Atlantic outer continental shelf, 1977: U.S. Geological Survey Open-File Report 80-146, 11 Chapters, 3 Appendices.

—— 1981, Environmental geologic studies on the southeastern outer continental shelf, 1977–78; U.S. Geological Survey Open-File Report 81-582-A, 8 Chapters and 5 Appendices.

Popenoe, P., Coward, E. L., and Cashman, K. V., 1982, A regional assessment of potential environmental hazards to and limitations on petroleum development of the southeastern United States Atlantic continental shelf, slope, and rise, offshore North America: U.S. Geological Survey Open-File Report 82-136, 67 p.

Pratt, R. M., and Heezen, B. C., 1964, Topography of Blake Plateau: Deep Sea Research, v. 11, p. 721–728.

Prior, D. B., Coleman, J. M., and Doyle, E. H., 1984, Antiquity of the continental slope along the middle-Atlantic margin of the United States: Science, v. 223, p. 926–928.

Richardson, P. L., 1980, Gulf Stream ring trajectories: Journal of Physical Oceanography, v. 10, p. 90–104.

Richardson, P. L., Cheney, R. E., and Mantini, L. A., 1977, Tracking a Gulf Stream ring with a free-drifting surface buoy: Journal of Physical Oceanography, v. 7, p. 580–590.

Richardson, W. S., Schmitz, W. J., Jr., and Nuler, P. P., 1969, The velocity structure of the Florida Current from the Straits of Florida to Cape Fear:

Deep Sea Research, v. 16, p. 220–231.

Richardson, P. L., Strong, A. E., and Krauss, J. A., 1973, Gulf Stream eddies: Recent observations in western Sargasso Sea: Journal of Physical Oceanography, v. 3, p. 297–301.

Robb, J. M., 1982, Environmental geologic studies in the Mid-Atlantic outer continental shelf area—Results of 1978–1979 field seasons, final report submitted to the Bureau of Land Management: Woods Hole, Massachusetts, U.S. Geological Survey, 10 Chapters.

—— 1984, Spring sapping on the lower continental slope, offshore New Jersey: Geology, v. 12, p. 278–282.

Robb, J. M., Hampson, J. C., and Twichell, D. C., 1981, Geomorphology and sediment stability of a segment of the U.S. continental slope off New Jersey: Science, v. 211, p. 935–937.

Rona, P., and Clay, C. W., 1967, Stratigraphy and structure along a continuous seismic reflection profile from Cape Hatteras, North Carolina to the Bermuda Rise: Journal of Geophysical Research, v. 72, p. 2107–2130.

Roberson, M. I., 1964, Continuous seismic profile survey of Oceanographer, Gilbert, and Lydonia submarine canyons, Georges Bank: Journal of Geophysical Research, v. 69, p. 4779–4789.

Ryan, W.B.F., 1982, Imaging of submarine landslides with wide-swath sonar, *in* Saxov, S., and Niewenhuis, J. K., eds., Marine slides and other mass movements: NATO Conference Series, IV, Marine Series, v. 6, p. 175–188.

Ryan, W.B.F., Cita, M. B., Miller, E. L., Hanselman, D., Nesteroff, W. D., Hecker, B., 1978, Bedrock geology of some New England submarine canyons: Oceanologica Acta, v. 1, no. 2, p. 233–254.

Salzmann, M. A., 1980, Outer continental shelf oil and gas information program: U.S. Geological Survey Open-File Report 80-1202, 193 p.

Sangrey, D. A., 1986, Hindcasting analysis of slope stability, *in* Folger, D. W., and Hathaway, J. C., eds., Conference on continental margin mass-wasting and Pleistocene sea-level changes: U.S. Geological Survey Circular, in press.

Sangrey, D. A., and Knebel, H. J., 1979, Geotechnical engineering studies in the Baltimore Canyon Trough area, *in* Knebel, H. J., ed., Middle Atlantic outer continental shelf environmental studies, v. III, Geologic studies: Gloucester Point, Virginia, Virginia Institute of Marine Science, Unpublished Final Report, p. 7-1 to 7-21.

Schlee, J. S., 1977, Stratigraphy and Tertiary development of the continental margin east of Florida: U.S. Geological Survey Professional Paper 581-F, p. F-1 to F-25.

Schlee, J. S., and Grow, J. A., 1980, Buried carbonate shelf-edge beneath the Atlantic continental slope: Oil and Gas Journal, v. 78, no. 8, p. 148–159.

Schlee, J. S., and Klitgord, K. D., 1982, Geologic setting of the Georges Bank basin, *in* Scholle, P. A., and Wenkham, C. R., eds., Geological studies of the COST Nos. G-1 and G-2 wells, United States North Atlantic outer continental shelf: U.S. Geological Survey Circular 861, p. 4–10.

Schlee, J. S. and Pratt, R. M., 1970, Atlantic continental shelf and slope of the United States—Gravels of the northeastern part: U.S. Geological Survey Professional Paper 529-A, 39 p.

Schmitz, W. J., Jr., and Richardson, W. S., 1968, On the transport of the Florida Current: Deep Sea Research, v. 15, p. 679–693.

Sheridan, R. E., and Knebel, H. J., 1976, Evidence of post-Pleistocene faults on New Jersey Atlantic outer continental shelf: American Association of Petroleum Geologists Bulletin, v. 60, p. 1112–1117.

Sheridan, R. E., Dill, C. E., and Kraft, J. C., 1974, Holocene sedimentary environment of the Atlantic inner shelf off Delaware Bay: Geological Society of America Bulletin, v. 85, p. 1319–1328.

Sherwin, D. F., 1975, Canada's east coast petroleum potential only been

scratched: Oil and Gas Journal, v. 73, no. 17, p. 100–104.

Shipley, T. H., Houston, M. H., Buffler, R. T., Shaub, F. J., McMillen, K. J., Ladd, J. W., and Worzel, V. L., 1979, Seismic evidence for widespread possible gas hydrate horizons on continental slopes and rises: American Association of Petroleum Geologists Bulletin, v. 63, p. 2204–2213.

Snyder, S.W.P., Hine, A. C., and Riggs, S. R., 1982, Miocene seismic stratigraphy, structural framework, and sea-level cyclicity: North Carolina continental shelf: Southeastern Geology, v. 23, no. 4, p. 247–266.

Stanley, D. J., Nelsen, T. A., and Stuckenrath, R., 1984, Recent sedimentation on the New Jersey slope and rise: Science, v. 226, p. 125–133.

Stewart, H. B., Jr., and Jordan, G. F., 1964, Underwater sand ridges on Georges Shoal, *in* Miller, R. L., ed., Papers in Marine geology, Shephard commemorative volume: New York, Macmillan Company, p. 102–114.

Sylwester, R. E., Dillon, W. P., and Grow, J. A., 1979, Active growth fault on the seaward edge of Blake Plateau, *in* Gill, D., and Merriam, D. F., eds., Geomathematical and petrophysical studies in sedimentology: New York, Pergamon Press, p. 197–209.

Tucholke, B. E., Bryan, G. M., and Ewing, J. D., 1977, Gas hydrate horizon detected in seismic profile data from the western North Atlantic: American Association of Petroleum Geologists Bulletin, v. 61, p. 698–707.

Twichell, D. C., 1979, Medium-scale potentially mobile bed forms on the Mid-Atlantic continental shelf, *in* Knebel, H. J., ed., Middle Atlantic outer continental shelf environmental studies, vol. III; Gloucester Point, Virginia, Virginia Institute of Marine Science, Unpublished Administrative Report, p. 5-1 to 5-15.

—— 1983, Bed-form distribution and inferred sand transport on Georges Bank, United States Atlantic continental shelf: Sedimentology, v. 30, p. 695–710.

—— 1986, Geomorphic map of the U.S. Atlantic continental slope and upper rise between Hudson and Baltimore Canyons, *in* Folger, D. W., and Hathaway, J. C., eds., Conference on continental margin mass-wasting and Pleistocene sea-level changes: U.S. Geological Survey Circular, in press.

Twichell, D. C., McClennen, C. E., and Butman, B., 1981, Morphology and processes associated with the accumulation of the fine-grained sediment deposit on the southern New England shelf: Journal of Sedimentary Petrology, v. 51, p. 269–280.

Uchupi, E., 1967a, The continental margin south of Cape Hatteras, North Carolina: Shallow structure: Southeastern Geology, v. 8, p. 155–177.

—— 1967b, Slumping on the continental margin southeast of Long Island, New York: Deep Sea Research, v. 14, p. 635–639.

Waddell, E., 1982, South Atlantic OCS physical oceanography, draft final progress report (year Four), Volumes 1, 2, 3; Prepared for the Minerals Management Service by Science Applications, Incorporated, 306 p.

MANUSCRIPT ACCEPTED BY THE SOCIETY FEBRUARY 4, 1985

ACKNOWLEDGMENTS

I am indebted to Margaret Hempenius who typed the manuscript, to Patricia Forrestel and her staff who drafted the illustrations, and to Peter Popenoe and Bradford Butman who kindly reviewed the manuscript. H. Knebel, B. Butman, D. Twichell, J. Robb, P. Popenoe, D. O'Leary, and J. Booth, and K. Scanlon of the U.S. Geological Survey, and H. Benton, J. Roberts, and F. Gray of Minerals Management Service, all provided published and unpublished material used in the compilation. Support for the majority of the work came from the Bureau of Land Management, in large part through the tireless and enlightened efforts of E. Imamura.

The Geology of North America
Vol. I-2, The Atlantic Continental Margin: U.S.
The Geological Society of America, 1988

Chapter 28

Coastal geologic hazards

Orrin H. Pilkey
Department of Geology, Duke University, Durham, North Carolina 27706
William J. Neal
Department of Geology, Grand Valley State College, Allendale, Michigan 49401

INTRODUCTION

Coastal hazards are an inevitable result of human occupation of one of the earth's most dynamic environments, the shore zone. Storms and the associated destruction by winds, waves, and floods have exacted their toll since the earliest coastal settlement. The result of these processes, together with mass wasting and subsidence, is shoreline retreat or, in the vernacular of the property owner, "shoreline erosion." Shakespeare noted in a sonnet of having "seen the hungry ocean gain advantage on the kingdom of the shore," and he surely was aware of the disappearance of coastal villages in Great Britain; for example, along the Lincolnshire coast (King, 1972). Yet each generation in each coastal country seems to discover the hazards anew, as if they are becoming more noticeable or severe in effect. This observation may actually be true because of changing climate and patterns of coastal use; most specifically because of increased density of human habitations. Humans are even reshaping the oceanographic setting of the adjacent shelves, adding to the potential severity of coastal hazards.

North America's Atlantic seaboard has a relatively short record of coastal settlement, but the disappearance or abandonment of entire communities such as Broadwater, Virginia; Diamond City, North Carolina; and Lafayette and Edingsville Beach, South Carolina attests to storm impact, shoreline retreat, and barrier island dynamics. Historic reports from most Atlantic states include examples of storm damage and significant shoreline erosion (e.g., Tuomey, 1848; Vincent, 1870). Loss of life and property due to hurricanes and northeasters along the Atlantic coast has been common since colonial times, but the accelerated coastal development of the twentieth century has led to increasing property losses (e.g., Herbert and Taylor, 1980; Simpson and Lawrence, 1971).

Much of the second half of this century has been relatively free of damaging Atlantic hurricanes, but the threat of damage and loss of life from future storms has increased (Simpson and Riehl, 1981). A recurrence of a hurricane such as that of 1938 which struck the northeastern U.S., or the 1928 storm that killed more than 1,800 people in the Lake Okeechobee, Florida region,

or the 1962 Ash Wednesday northeaster that struck the U.S. East Coast will now result in a vastly higher economic loss, while the threat of high fatalities remains. The rapid rate of growth in the coastal zone since the 1950s has created a series of disasters waiting to happen. The National Hurricane Center commonly predicts potential death tolls ranging from hundreds to thousands, and average Gulf Coast hurricanes now routinely cause damage in excess of $1 billion per storm.

In view of the continued growth in coastal zone population and infrastructure, the failure of the engineered structural approach to eliminate or effectively mitigate coastal hazards, and the more recent complicating factor of an accelerated sea level rise, it appears that the stage is set for at least a partial withdrawal of development from the shore (Howard and others, 1985).

Storms and shoreline retreat are not the only environmental hazards of the coastal zone. Two of the more publicly visible hazards are discussed in other chapters of this volume: waste disposal, by H. Palmer, and the oil spill–associated problems, by D. Folger.

Many of the uses of the coastal zone (summarized by Ross, 1978) are of a conflicting nature, virtually assuring continuing and accelerating environmental crises in the future. Recreational and commercial fishing, military use, transportation, mineral extraction, and waste disposal are all human activities that cannot co-exist entirely peacefully.

Because of the breadth of coastal environmental hazards, "solutions" to environmental problems are certainly not a purely scientific endeavor. As in all of society's environmental problems, strong overtones of politics and economics exist. A society must establish its own limits and priorities regarding coastal zone use and management, and no two societies will view environmental problems in the same fashion. One fact is certain, however; without strong and effective coastal management policies that cross state and national boundaries, the environmental crisis at the shore can only worsen, to the detriment of all segments of society.

STORM ASSOCIATED HAZARDS

The immediate storm threats to coastal property are high

Pilkey, O. H., and Neal, W. J., 1988, Coastal geologic hazards; *in* Sheridan, R. E., and Grow, J. A., eds., The Geology of North America, Volume I-2, The Atlantic Continental Margin, U.S.: Geological Society of America.

TABLE 1. SAFFIR-SIMPSON HURRICANE SCALE*

Category	Winds (mph)	Storm Surge (ft)	Central Pressure (in)	Damage	Examples
1	74-95	4-5	>28.94	minimal	Agnes (1972), Diane (1955), David (1979)
2	96-110	6-8	28.50-28.91	moderate	Cleo and Dora (1964), GA-SC-NC (1940)
3	111-130	9-12	27.91-28.47	extensive	Carol (1954), Connie (1955), NE U.S. (1938, 1944)
4	131-155	13-18	27.17-17.88	extreme	Hazel (1954), Donna (1960), SE Florida (1926, 1928)
5	>155	>18	<27.17	catastrophic	Allen** (1980), Camille** (1969), Florida Keys (1935)

*After Simpson and Riehl (1981) and Herbert and Taylor (1980).

**Gulf Coast Storm.

winds, waves, wave runup, storm surge, and high levels of precipitation and runoff combining to generate coastal flooding and erosive currents. In a longer-term sense, coastal erosion by storms is a threat to virtually all shorefront development.

The greatest physical threat to property along the Atlantic seaboard is the hurricane (Simpson and Riehl, 1981). The annual probability that a hurricane will strike any given 80-km coastal segment ranges from 1% to 16%. Such storms are a threat to the Atlantic coast as far north as Massachusetts.

A typical east coast hurricane forms off the east coast of Africa as a low pressure area (tropical depression) when the surface waters of the ocean reach temperatures of approximately 27°C or more. As warm, moist air rises, it cools and condenses, releasing stored heat that warms the surrounding air, increasing the vertical air flow and intensifying the storm center. The Coriolis effect deflects the air flow, and the counterclockwise rotating air mass takes on the familiar hurricane spiral shape with the "eye," a chimneylike column of rising air, at its center. Surface air rushes in to replace the rising air mass, and the sea provides an endless reservoir of moisture.

When sustained wind velocities reach a minimum of 119 km/hr (74 mph), the storm is classed a hurricane (Table 1). Hurricane winds may exceed 320 km/hr and the diameter of a storm can range from 100 to 1,610 km.

Such hurricanes track north/northwest, and the forward movement of the storm system may reach velocities in excess of 100 km/hr (60 mph). If the hurricane makes landfall, the coast will be subjected to high winds and storm surge, topped by wind-driven waves (Fig. 1). These effects tend to be most severe to the right of the eye in the direction of storm migration because of the counterclockwise air flow producing onshore winds (Fig. 1).

The National Weather Service uses the Saffir-Simpson Scale (Table 1) to describe and rank hurricanes. The scale is based on three storm variables: wind velocity, storm surge level, and baro-metric pressure. Although correlation exists between these variables, the configuration of the shelf and shoreline also affects storm surge levels and wave behavior. Likewise, the associated damage tends to increase with storm rank; however, the level of development in the area hit by a hurricane also will determine the actual dollar loss. Hurricane Agnes, 1972, a category 1 storm, did more than $2 billion worth of damage, in large part because of high rainfall and consequent inland flooding. More recently, Hurricane David, 1979, also a category 1 storm, generated losses approaching $400 million because of its "long distance" impact as it ran along much of the U.S. southeastern Atlantic coast.

Storm surge flooding is the hurricane hazard that accounts for most fatalities as well as widespread destruction—often nearly total for great hurricanes. Storm surge is the rise in sea level, above the normal still water level, that is induced by the combination of the intense low pressure at the center of the storm and the rise of the water column as water is mounded beneath the storm in response to the counterclockwise winds. As the crest of the storm surge moves onshore into shallow water, it may be heightened more, reaching 4 to 6 m or more above sea level. This effect is augmented by wind-generated waves that will add still more height and velocity to the associated flood. Water pushed into sounds, embayments, and estuaries by storm surge and wind may come flooding back seaward to cause additional erosion, as in the case of inlet formation.

Northeasters, like hurricanes, can also cause great wind and flood damage. Frequently northeasters are more damaging than hurricanes from the standpoint of shoreline erosion because they may remain in place for a longer period of time, sometimes through several tidal cycles. These intense winters storms begin as subtropical depressions. As the low pressure center tracks north along the eastern seaboard, strong northeast winds may create storm surge flooding accompanied by heavy wave activity. Such storms may move slowly, resulting in high-water wave and flood damage over more than one tidal cycle. Shoreline retreat is temporarily accelerated as beaches narrow and dunes are eroded and breached.

The Ash Wednesday Storm of March 5–9, 1962, came on top of a spring tide so that the storm surge flooding and wave activity inflicted heavy destruction from the Carolinas to New England. The New Jersey shore was particularly hard hit; the first, and sometimes second, row of houses, boardwalks, and other structures were destroyed. Extensive overwash of the barrier islands also occurred.

While hurricanes and northeasters generate the immediate forces of destruction, the effects of storms are magnified by the facts that much of the coast consists of low-lying barrier islands and that sea level is rising.

SEA LEVEL RISE

The rapid post-glacial rise in sea level may account for the origin and cross-shelf migration of the Atlantic seaboard's extensive barrier island system (Swift, 1975; Niedoroda and others, 1985). Along the glaciated coast the eustatic flooding of river mouths and low-lying mainland valleys produced embayments and fjords. Present eustatic sea level rise along the Atlantic Coast may approach 22 cm/century. Sea level rise is variable because of subsidence (and locally uplift) due to post-glacial isostatic adjustment, compaction, of barrier island sands due to groundwater extraction, loading of compactable sediments such as peat marsh deposits buried by migrating barrier islands, similar loading by the water mass of the encroaching sea, and tectonism. Subsidence may greatly magnify the problem of sea level rise.

Within the last one-half century, the rate of global sea level rise has accelerated (Hoffman and others, 1983; Revelle, 1983; Seidel and Keyes, 1983), probably as a result of increased CO_2 input into the atmosphere. The CO_2, along with other gases such as methane, has induced a global warming, the so-called "greenhouse effect," which has resulted in the accelerated rise in global sea level through increased discharge from melting polar ice (Etkins and Epstein, 1982) and thermal expansion of the ocean surface (Gornitz and others, 1982).

This sea level rise may account, in part, for the apparent increase in the rate of coastal erosion and increased flood potential as a result of storms (Barth and Titus, 1984; Pilkey and Howard, 1981). A continuation of this trend is expected (Hoffman and others, 1983; Revelle, 1983).

Titus and Barth (1984) suggest a number of sea level-rise scenarios based on projections of many highly complex factors, including future trends in world production of carbon dioxide. For example, their most conservative estimate of post-1980 worldwide sea level rise by the year 2000 is 4.8 cm, compared to a high of 17.1 cm. By the year 2100 the rise figure projections are 56.2 cm to 345.0 cm. On coastal plain coasts, as well as low-elevation, low-sloping valleys of the New England and Maritime province glaciated coast, a sea level rise of this magnitude is

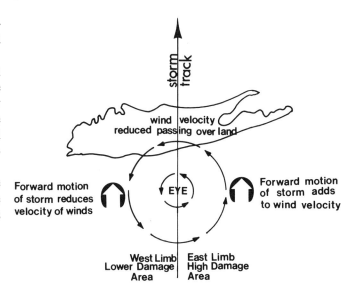

Figure 1. A hypothetical hurricane approaching Long Island, New York, (from McCormick and others, 1984).

accompanied by a lateral shoreline retreat that is orders of magnitude greater than the vertical sea level rise.

The major environmental problem associated with sea level rise is shoreline retreat (Fig. 2). In addition, sea level rise will have the effect of pushing storm waves, storm surges, and flooding further inland. A number of models have been devised to describe and predict the response of shorelines to changes in sea level. The most widely used model is the "Bruun rule" (Bruun, 1962). Basically, Bruun suggests that the shoreface profile is a constant for a given set of conditions of energy grain size and sand supply input. Thus the response to sea level rise is a landward and upward adjustment of the profile. Sand from the eroding shoreface is deposited on the inner shelf. The Bruun Model assumes no gain or loss of sand from longshore drift, an assumption more or less incorrect everywhere.

Dean and Maurmeyer (1983) suggest an important modification of the Bruun rule that applies to barrier islands. Their model takes into account sand loss by overwash across barrier islands and also the presence of "incompatible" sediment, often mud of lagoon origin, on the mid to lower shoreface.

Because of the wide variability of barrier island processes and our still rudimentary understanding of shoreface processes (Niedoroda and others, 1985), none of the models of shoreline response are entirely satisfactory. This is an area of much-needed future research.

Where large seawalls are in existence, such as on portions of the long-developed New Jersey shoreline, retreat is less of a problem than is the increased frequency of wall-topping waves by large storms (Leatherman, 1984). In two case studies, Charleston, South Carolina (Kana and others, 1984) and Galveston, Texas (Leatherman, 1984), other environmental hazards associated with sea level rise are discussed. These hazards include salt water intrusion and inundation or exposure to storm flooding of haz-

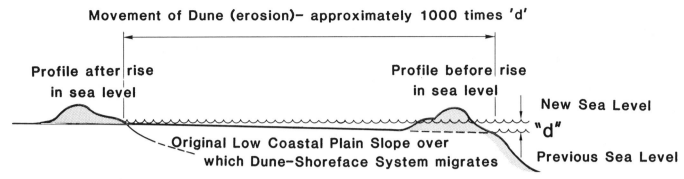

Figure 2. A small sea level rise should produce a relatively large horizontal shoreline retreat, a principal dictated by the slope of the Lower Atlantic Coastal Plain.

ardous waste sites. Salt water intrusion is considered to be a relatively minor problem for the two areas in question. In the Charleston study area, however, five hazardous waste disposal sites would be in the 100-year floodplain by the year 2075 (using the high sea level rise scenario) in addition to five such sites already in the 100-year floodplain. Twenty-two hazardous waste sites would be in the floodplain of a 2075 high sea level rise scenario at Galveston, Texas. Loss of important habitat such as salt marsh is an additional problem (Titus and others, 1984).

SHORELINE EROSION

Retreat of the shoreline, whether related to sea level rise or not, is a widespread phenomenon on the U.S. Atlantic coast (U.S. Army Corps, 1971; Shepard and Wanless, 1971). Shoreline retreat, usually termed erosion by beachfront property owners (Figs. 3 and 4), is a major coastal hazard, particularly since a building perched on the edge of the sea, about to fall in, has very high public visibility. Such situations usually are widely reported in the media.

Causes of erosion are highly variable and are dependent upon local geologic and oceanographic conditions. Fundamentally, erosion occurs on open ocean coasts because wave energy created over large areas of ocean surface is expended in the surf zone. Other than the aforementioned rise in sea level, sand supply (lack or abundance of) is often a major cause of shoreline retreat or advance. Beach sand removed during storms may be deposited on overwash fans on barrier island surfaces, it may end up in temporary storage on the shoreface to be returned to the beach by fair weather wave systems, or it may be lost permanently on the inner shelf seaward of the 10-m depth contour. The major processes of seaward movement of sand during storms include rip current fall-out and bottom currents resulting from the downwelling of water pushed ashore by storm winds (see Niedoroda and others, 1985).

Shoreline retreat is frequently but not always related to storm activity. Certainly storms are usually responsible for the major "jumps" in shoreline retreat rate.

Along most of the Atlantic coast of North America, espe-cially where estuaries exist, no beach sand is being contributed from adjacent fluvial sources. Fluvial sand loads of rivers traversing the Atlantic coastal plain are dumped in the upper reaches of estuaries, often far from the coastal zone. Estuaries also act as sinks for beach sand moving in under the influence of longshore currents in the littoral zone. Such sand is moved into the back barrier environment through tidal inlets forming flood tidal deltas. Without fresh sand supplies, barrier islands, in the short term at least, inevitable retreat.

In the last two decades or so, man probably has been a principal cause of Atlantic shoreline erosion. Many types of shoreline "protection" structures trap or remove the beach sand supply or interfere in various ways with the longshore transport system, thus hastening erosion. The many Atlantic coast jetties, emplaced to enhance navigation through inlets or into harbors, have been a particularly important cause of enhanced erosion rates (Komar, 1983).

Long-range erosion rates are now available for most stretches of Atlantic-facing shorelines. These data come from a variety of sources such as maps, aerial photos, and shoreline surveys. It is important to differentiate long-term (40 to 100 years) from short-term (less than 10 years) data, because in all likelihood short-term erosion rates are accelerating due to a combination of the impact of manmade coastal structures and the accelerating sea level rise. Because erosion is sporadic in nature, short-term erosion rates may be either unreasonably high or low depending on recent storm history of a given shoreline stretch.

Dolan and others (1983) have summarized U.S. Atlantic coastal erosion rates based (usually) on 40 years of aerial photography). Erosion rates of east coast barriers average 0.8 m (2.6 ft) per year. Glacial-induced rebound plays an important role in more northerly regions. For example, in Maine, rebound is responsible for an average shoreline advance of 1 m (3 ft) per year, although the sandy beaches of southern Maine are retreating at rates comparable to most of the Atlantic barrier system.

SHORELINE "PROTECTIVE" STRUCTURES AS ENVIRONMENTAL HAZARDS

The natural inclination of individuals owning buildings on a

Figure 3. The eroding shoreline on Wassau Island, Georgia. Here there is no heralded erosion problem because no man-made structures are threatened by the shoreline retreat. Photo by Bob Frey.

Figure 4. Long Beach, North Carolina, after Hurricane Hazel, 1954. Although the beach changed its position in space, it remained a wide strip of sand suitable for recreation. Thus shoreline retreat is not necessarily a threat to the beach; a point not well understood by the public. Photo furnished by Steve Benton.

retreating shoreline is to stabilize the shoreline in an effort to protect the structure. Methods of shoreline stabilization fall into three categories: (1) beach nourishment; (2) sand trapping structures, such as groins, built perpendicular to the shoreline or offshore breakwaters paralleling the shoreline; and (3) various types of walls built on a parallel to the shoreline. The latter two approaches are "shore hardening" procedures, whereas beach replenishment is considered to be a "soft" solution to the erosion problem. Walls and groins are structural solutions; beach nourishment is a nonstructural solution. The various procedures, techniques, and costs of shoreline stabilization are summarized in the U.S. Army Corps of Engineers' *Shore Protection Manual* (1984).

Shore hardening is usually carried out to protect buildings and community infrastructure from storm waves, flooding, and a retreating shoreline (Fig. 5). In this latter respect, the numerous groins and miles of seawalls along the Atlantic Coast have been quite successful. Thousands of buildings probably owe their very existence to adjacent shore hardening structures, although most such structures cannot be depended upon for protection in major storms. It is now clear, however, that halting the receding shoreline with protective structures benefits the property owners of beachfront property, but at the same time seriously degrades or destroys the natural beach and the value that it holds for the general public.

Shore-parallel structures such as sea walls or revetments tend to (1) increase backwash during storms by wave reflection, and (2) increase the intensity of longshore currents. Both processes cause removal of sand from the recreational beach system. Even if shore parallel structures did not actively cause sand removal, the very act of stabilization of the upper portion of a retreating shoreface system assures eventual long-range loss of the beach as the equilibrium shoreline retreats landward.

Shore-perpendicular structures, such as groins, trap sand moving in the longshore transport system and thus contribute to a diminishing sand supply system. Larger jetties may shunt the sand offshore, entirely out of the longshore drift system.

Some evidence is developing that extensive shoreface steepening is occurring in front of long-stabilized shorelines in New Jersey and New York. Steepened shorefaces result in increased surf zone wave energy, increased rates of overtopping of seawalls by storm waves, and the consequent need for walls of ever-larger dimensions. The history of seawall construction in Sea Bright and Monmouth Beach, New Jersey bears out this scenario of the need for larger and larger walls.

At present the most environmentally acceptable approach to beach stabilization is beach nourishment: the addition of large quantities of compatible sand to build beaches seaward. Beach nourishment, unlike hard stabilization, usually improves the quality of the recreational beach. Furthermore, the presence of a broad beach in front of a community, whether natural or artificial, provides a significant degree of storm or hurricane protection.

Most beach nourishment projects have provided benefits over only short time periods (3–10 years), however. In the Miami Beach project, the nourished beach has been remarkably stable. In most areas where replenished beaches have eroded rapidly, a significant portion of the eroded sand undoubtedly is transported along shore, thereby benefiting adjacent beaches.

The costs of beach replenishment are high, typically of the order of $600,000/km, but the total cost of a Miami Beach type project in the early 1980s was up to $4.8 million/km by 1985. Frequently, serious environmental issues must be resolved, such as destruction of marine fauna and flora in the dredged, sand-source area, or destruction of turtle nests or nesting sites. Often, suitable sand supplies are not available. Beach nourishment is

Figure 5. A shoreline without a recreational beach on Casey Key, Florida. Here can be seen groins, a small wooden seawall, limestone boulders, and rubber tires, all emplaced in what appears to be a failing attempt to halt shoreline retreat. Photo by Dinesh Sharma.

Figure 6. A new inlet formed on Harvey Cedars on Long Beach, New Jersey, during the 1962 Ash Wednesday northeastern storm. Water rushing seaward from the bay apparently was funneled across the island by the bulkhead on the lagoon side of the island. Photo courtesy of the U.S. Army Corps of Engineers, Philadelphia District.

most viable economically and environmentally for areas of dense development, where large sand supplies are available, at sites where wave energy is low, and where environmental issues are reconcilable. Many developed areas are not fortunate enough to have all factors that justify beach nourishment as a long-range solution.

Given the rising sea level, beach replenishment is not likely to be the erosion 'solution' for the long-range future of the Atlantic coastal zone, but nourishment is far superior to the shore hardening solution from the standpoint of beach preservation. The states of North Carolina and Maine now prohibit *all* kinds of "hard" stabilization on open-ocean sandy beaches. Beach replenishment is permissible in these states but frequently is not economically feasible. In effect, these states have declared a retreat from the shoreline, and 37 buildings were moved back from the beach in North Carolina during 1985.

NATURAL BARRIER ISLAND PROCESSES VERSUS DEVELOPMENT

One of the most pressing coastal zone environmental problems is the development, frequently overdevelopment, of barrier islands. The problem involves construction of buildings in locations that are likely to result in damaging interaction with natural barrier island processes. The processes involved vary widely from island to island and region to region depending on such diverse factors as wave energy, tidal amplitude, sand supply, island orientation, climate and its control of vegetation, dominant wind directions and intensities, and man-caused alterations to the island. The problem from a coastal zone manager's viewpoint would be much simpler if all barrier islands were subject to the same processes and responded to the processes in similar and predictable

ways. Such is not the case. The extreme diversity of barrier islands and barrier island processes (Nummendal, 1983; Leatherman, 1979) forms the basis of a regulator's nightmare.

One of the major hazard zones on barriers are areas near inlets. The hazard is due to the propensity of inlets to migrate and to change shape and size. Changes in shoreline position, both seaward and landward, tend to occur at relatively high rates near inlets, due to changes in refracted wave patterns. Wave refraction patterns change in response to the changing shape of the ebb tidal delta.

Formation of new inlets usually results from storm surge return and breakthrough from the lagoon side of the island (Fig. 6). New inlet formation is naturally a hazard to buildings in the immediate vicinity, but such events also destroy roads, and power and water lines. Inlets tend to form only during major storms and are most likely to occur across low, narrow portions of barriers that are backed by a large lagoon. Human modification of barrier islands may enhance the chances for inlet formation as in the case of the construction of finger canals. Unfortunately, prediction of likely sites for new inlet formation with a useful degree of accuracy is not possible with our present state of knowledge.

Overwash is a common storm event on most low elevation transgressive islands. The possibility of overwash is one of the principal reasons for construction of houses on stilts (FEMA, 1984). Unlike the location of new inlets, potential sites of future overwash can be mapped and avoided.

Migration of unvegetated coastal dunes is at least a nuisance and often a genuine hazard to barrier island development. This is especially true when dune vegetation is removed during initial stages of construction.

Shoreline erosion is also a problem on both sides of many Atlantic barrier islands. However, backside erosion is generally much less of a hazard than open ocean shoreline retreat because shoreline stabilization is economically feasible and relatively successful in low wave-energy environments.

HAZARD MITIGATION

In order to avoid or reduce hazards of barrier island development, siting of buildings, roads, and other island infrastructure should be carried out in the context of known barrier island processes. The aim of this approach to development should be two-fold: (1) to reduce hazards to buildings and people, and (2) to allow island processes to continue unabated to the degree that such is feasible and compatible with development (Nordstrom and McCluskey, 1985). The latter is essential to preserve environmentally and economically critical areas such as the open ocean recreational beach and salt marshes on the lagoon side. Coastal hazard mapping is an important first step in mitigation (e.g., Baker, 1980; Komar, 1983; Monday, 1983; Neal and others, 1984).

Where development already exists, the impact of hazards can be reduced through improved construction techniques for both new and existing structures (e.g., FEMA, 1981, 1984; Pilkey and others, 1983). Developed areas also should have evacuation plans, utilizing resources such as NOAA (National Oceanographic and Atmospheric Administration) evacuation maps. Increasingly, mitigation is achieved through legislative regulation at all levels. Examples include the National Flood Insurance Program, the Federal Coastal Resources Act, various state and province coastal zone management programs, and local zoning and building code ordinances. The future will see increasing regulation (e.g., Baker, 1980; Monday, 1983; Howard and others, 1985).

Howard and others, 1985, argue that retreat from the shoreline is an inevitable management strategy for portions of the Atlantic Coast. These authors note: "Sea level is rising and the American shoreline is retreating. We face economic and environmental realities that leave just two choices: (1) plan a strategic retreat now, or (2) undertake a vastly expensive program of armoring the coastline and, as required, retreating through a series of unpredictable disasters."

REFERENCES CITED

Baker, E. J., ed., 1980, Hurricanes and coastal storms: Gainesville, University of Florida, Florida Sea Grant College, Marine Advisory Program, Report No. 33, 219 p.

Barth, M. C., and Titus, J. G., eds., 1984, Greenhouse effect and sea level rise; A challenge for this generation: New York, Van Nostrand Reinhold Company, 325 p.

Bruun, P., 1962, Sea level rise as a cause of shore erosion: Journal of the Waterways and Harbors Division, v. 88 (WW1), p. 117–130.

Dean, R. G., and Maurmeyer, E. M., 1983, Models for beach profile response, *in* Komar, P. D., ed., Handbook of coastal processes and erosion: Boca Raton, Florida, CRC Press, Incorporated, p. 151–166.

Dolan, R., Hayden, B., and May, S., 1983, Erosion of the United States shoreline, *in* Komar, P. D., ed., Handbook of coastal processes and erosion: Boca Raton, Florida, CRC Press, Incorporated, p. 285–299.

Etkins, R., and Epstein, E. S., 1982, The rise of global mean sea level as an indication of climate changes: Science, v. 215, no. 4530, p. 287–289.

Federal Emergency Management Agency, 1981, Design and construction manual for residential buildings in coastal high hazard areas: Washington, D.C., U.S. Government Printing Office, Publication 722-967/545, 189 p.

—— , 1984, Elevated residential structures: Washington, D.C., U.S. Government Printing Office, Publication O-438-116, 4, 137 p.

Gornitz, V., Lebedeff, S., and Hansen, J., 1982, Global sea level trend in the past century: Science, v. 215, no. 4540, p. 1611–1614.

Herbert, P. J., and Taylor, G., 1980, The costliest and deadliest hurricanes, United States 1900–1979: Miami, Florida, U.S. Department of Commerce, National Hurricane Center National Oceanic and Atmospheric Administration Technical Memorandum NWS NHC 7 (amended), 24 p.

Hoffman, J. S., Keyes, D., and Titus, J. B., 1983, Projecting future sea level rise; Methodology, estimates to the year 2100, and research needs: Washington, D.C., U.S. Environmental Protection Agency, Strategic Studies Staff, Office of Policy and Resource Management, 121 p.

Howard, J. D., Kaufman, W., and Pilkey, O. H., Jr., conveners, 1985, National strategy for beach preservation; Second Skidaway Institute of Oceanography Conference on America's eroding shoreline: Savannah, Georgia, Skidaway Institute of Oceanography, 11 p.

Kana, T. W., Michel, J., Hayes, M. O., and Jensen, J. R., 1984, The physical

impact of sea level rise in the area of Charleston, South Carolina, *in* Barth, M. P., and Titus, J. G., eds., Greenhouse effect on sea level rise; A challenge for this generation: New York, Van Nostrand Reinhold Company, p. 105–150.

King, C.A.M., 1972, Beaches and coasts: New York, Saint Martin's Press, 2nd edition, 570 p.

Komar, P. D., ed., 1983, CRC handbook of coastal processes and erosion: Boca Raton, Florida, CRC Press, Incorporated, 305 p.

Leatherman, S. P., ed., 1979, Barrier islands from the Gulf of Saint Lawrence to the Gulf of Mexico: New York, Academic Press, 325 p.

Leatherman, S. P., 1984, Coastal geomorphic responses to sea level rise; Galveston Bay, Texas, *in* Barth, M. P., and Titus, J. G., eds., Greenhouse effect and sea level rise: A challenge for this generation: New York, Van Nostrand Reinhold Company, p. 105–150.

McCormick, C. L., Pilkey, O. H., Jr., Neal, W. J., and Pilkey, O. H., Sr., 1984, Living with Long Island's South Shore: Durham, Duke University Press, 157 p.

Monday, J., ed., 1983, Preventing coastal flood disasters; The role of the states and federal response, *in* Proceedings of a National Symposium of the Association of State Floodplain Managers: Boulder, Colorado, Natural Hazards Research and Applications Information Center Special Publication 7, 386 p.

Morton, R. A., 1976, Effects of Hurricane Eloise on beach and coastal structures, Florida Panhandle: Geology, v. 4, p. 277–280.

Neal, W. J., Blakeney, W. C., Jr., Pilkey, O. H., Jr., and Pilkey, O. H., Sr., 1984, Living with the South Carolina shore; Durham, North Carolina, Duke University Press, 205 p.

Niedoroda, A. W., Swift, D.J.P., and Hopkins, T. S., 1985, The shoreface *in* Davis, R. A., Jr., ed., Coastal sedimentary environments: New York, Springer-Verlag, p. 533–624.

Nordstrom, R. F., and McCluskey, J. M., 1985, The effects of houses and sand fences on the eolian sediment budget at Fire Island, New York: Journal of Coastal Research, v. 1, p. 39–46.

Nummendal, D., 1983, Barrier islands, *in* Komar, P. D., ed., Handbook of coastal processes and erosion: Boca Raton, Florida, CRC Press, Incorporated, p. 77–121.

Pilkey, O. H., Jr., and Howard, J. D., conveners, 1981, Saving the American beach; A position paper by concerned coastal geologists: Skidaway Institute of Oceanography Conference on America's Eroding Shoreline, Savannah, Georgia (reprinted in Geotimes, v. 26, no. 12, p. 18–22).

Pilkey, O. H., Sr., Pilkey, W. D., Pilkey, O. H., Jr., and Neal, W. J., 1983, Coastal design; A guide for builders, planners, and homeowners: New York, Van Nostrand Reinhold Company, 224 p.

Revelle, R. R., 1983, Probable future changes in sea level resulting from increased atmospheric carbon dioxide, *in* Report of the Carbon Dioxide Assessment Committee, Changing climate: Washington, D.C., National Academy Press, p. 433–447.

Ross, D. A., 1978, Opportunities and uses of the ocean: New York, Springer-Verlag, 320 p.

Seidel, S., and Keyes, D., 1983, Can we delay a greenhouse warming?: Washington, D.C., U.S. Environmental Protection Agency, Strategic Studies Staff, Office of Policy and Resources Management, 127 p.

Shepard, F. P., and Wanless, H. R., 1971, Our changing coastlines: New York, McGraw-Hill, Incorporated, 579 p.

Simpson, R. H., and Lawrence, M. B., 1971, Atlantic hurricane frequencies: Washington, D.C., U.S. Department of Commerce, National Oceanic and Atmospheric Administration Technical Memorandum NWS SR-58, 14 p.

Simpson, R. H., and Riehl, H., 1981, The hurricane and its impact: Baton Rouge, Louisiana State University Press, 398 p.

Swift, D.J.P., 1975, Barrier island genesis; Evidence from the Central Atlantic Shelf, eastern U.S.A.: Sedimentary Geology, v. 14, no. 1, p. 1–43.

Titus, J. G., and Barth, M. C., 1984, An overview of the causes and effects of sea level rise, *in* Barth, M. C., and Titus, J. G., Greenhouse effect and sea level rise; A challenge for this generation: New York, Van Nostrand Reinhold Company, p. 1–56.

Titus, J. G., Henderson, T. R., and Teal, J. M., 1984, Sea level rise and wetlands loss in the United States: National Wetlands Newsletter, v. 6, no. 5, p. 3–6.

Tuomey, M., 1848, Report on the geology of South Carolina: Columbia, South Carolina, A. S. Johnston, 293 p.

U.S. Army Corps of Engineers, 1971, Report on the National Shoreline Study: Washington, D.C., Department of the Army, Corps of Engineers, 59 p.

—— , 1984, Shore protection manual: Washington, D.C., U.S. Government Printing Office Publication 008-022-00218-9, (2 volumes), 1,269 p.

Vincent, F., 1870, A history of the State of Delaware: Philadelphia, Pennsylvania, John Campbell, 478 p.

MANUSCRIPT ACCEPTED BY THE SOCIETY MAY 8, 1986

This chapter is a contribution of the Duke University Program for the Study of Developed Shorelines, supported by a grant from the William H. Donner Foundation Incorporated of New York.

Printed in U.S.A.

The Geology of North America
Vol. I-2, The Atlantic Continental Margin: U.S.
The Geological Society of America, 1988

Chapter 29

Cretaceous and Cenozoic tectonism on the Atlantic coastal margin

David C. Prowell

U.S. Geological Survey, Suite B, 6481 Peachtree Industrial Boulevard, Doraville, Georgia 30360

INTRODUCTION

Regional tectonism on the Atlantic coastal margin is expressed in a variety of ways such as uplift, subsidence, tilting of the landmass, geomorphic features, seismicity, and faulting. Of these features, faulting probably is the most definitive evidence of crustal deformation. Major episodes of faulting such as the ductile shearing associated with dynamic metamorphism in the exposed Appalachians and rift faulting associated with the formation of early Mesozoic basins along the Atlantic seaboard have long been a part of the geologic knowledge. However, faulting related to more subtle events, such as the uplift of the Blue Ridge Mountains or post-rift downwarp of the Atlantic continental margin, has received far less attention even though it is an important element of modern geology. Prior to 1970, the eastern United States was generally considered devoid of faults of post-Jurassic age even though evidence of such faulting was available in the late 1800's. McGee (1888) and Darton (1891) recognized faults in Virginia, Maryland, and Washington, D.C. that involved the juxtaposition of crystalline "basement" and Cretaceous or younger Coastal Plain strata. These observations and the local linearity of the inner margin of the Coastal Plain (Fall Line) lead them to postulate a tectonic control for the updip limit of sedimentation. Confirmation of widespread post-Jurassic tectonism, however, was not readily available and arguments favoring passive warping of the continental edge dominated geologic thought.

In the 1970's, the construction of nuclear power plants, large dams, and other large structures generated a need to understand Cenozoic tectonism and seismicity in eastern North America. The resulting studies of fault activity in the eastern United States provided evidence of many large, previously unrecognized, Cretaceous and Cenozoic fault zones. For example, Jacobeen (1972), Mixon and Newell (1977), Prowell and O'Connor (1978), Dischinger (1979), Reinhardt and others (1984), and Dischinger (in press) have described faults with vertical displacements of 30 to 76 m and mapped lengths as great as 100 km (Fig. 1).

Early efforts to inventory documented occurrences of post-Jurassic faulting resulted in publications by York and Oliver (1976) and Howard and others (1978), but the detailed information available to these authors was minimal. Recent studies have significantly increased the fault data base and a more complete inventory of faults is now available. Prowell (1983) reports more than 130 fault localities east of the Mississippi River and approximately 80 of these localities are along the Atlantic coastal margin. The localities described by Prowell (1983) are based almost entirely on geologic information and most of the data are from field observation of fault exposures. His data show a predominance of reverse faults along the Atlantic coastal margin that were not predicted by general plate tectonic theory. These observations have generated considerable scientific interest because of the concern over possible seismic hazards and its implications to plate tectonics. The focus of this presentation will be to summarize the more important features of the major post-Jurassic faults in the eastern United States with emphasis on faults in the emerged Atlantic coastal margin.

Figure 1. Cambrian phyllite (right) faulted over Late Cretaceous sediments (left) along a southeast-dipping reverse fault (marked by right hammer) in the Belair fault zone near Augusta, Georgia. Total vertical displacement at this locality is 20 m. (Small fault splay and drag folding marked by left hammer—view looking northeast).

Prowell, D. C., 1988, Cretaceous and Cenozoic tectonism on the Atlantic coastal margin; *in* Sheridan, R. E., and Grow, J. A., eds., The Geology of North America, Volume I-2, The Atlantic Continental Margin, U.S.: Geological Society of America.

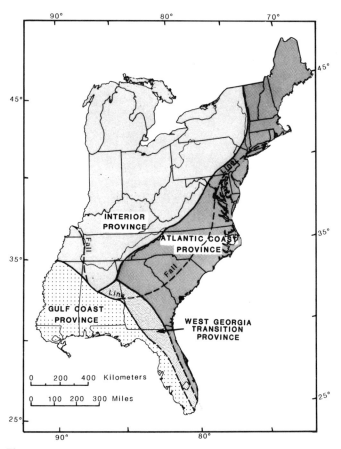

Figure 2. Map of the eastern United States showing Cretaceous and Cenozoic fault provinces defined by type of fault motion (data from Prowell, 1983).

REGIONAL DISTRIBUTION OF FAULTS

The onshore faults along the Atlantic coastal margin reported in Prowell (1983) are shown on Plate 8A along with the Coastal Plain, Piedmont, Blue Ridge, and Valley and Ridge physiographic provinces. Several offshore faults reported here from recent geophysical studies (see Behrendt and others, 1981; Hutchinson and Grow, 1985) and the seismic Ramapo fault (see Aggerwal and Sykes, 1978) have also been included on Plate 8A. The recognition of these Cretaceous and younger faults is generally dependent on the observed displacement of geologic units or contacts of an appropriate age range. Hence, most of the observed faults shown on plate 8A are within the Coastal Plain geologic province, where datable, subhorizontal sedimentary horizons are most abundant. The occurrence of datable Cretaceous and Cenozoic materials in the Appalachians, however, is far more limited and undoubtedly is the primary reason for the small number of reported faults in this region. However, close similarity of the fault style recognized in the Appalachians to that in the Atlantic Coastal Plain allows fault characteristics to be extrapolated across this entire region. In addition, seismologic evidence and seismic reflection profiling in conjunction with limited geologic data have been used to assign an age and location to a few faults shown on

Plate 8A. These criteria are far less definitive than field observations of faulting but they offer evidence of tectonism in otherwise unevaluated areas.

The abundance of fault data along the inner margin of the Coastal Plain could be taken as an indication of a concentration of tectonic activity, but it may only be a reflection of opportune geologic conditions. Reports of faults along the inner edge of the Coastal Plain typically describe crystalline rocks faulted against Cretaceous or younger sediments. Contacts of this sort are far more obvious than those involving the juxtaposition of similar rock types and, therefore, the local geologic conditions may be a major factor in the grouping of faults shown on Plate 8A. Accordingly, areas showing no faults should not be taken to represent a lack of tectonic activity, but rather a lack of sufficient information.

Some of the fault data shown on Plate 8A and in Prowell (1983) were used by Prowell (1976) and Howard and others (1978) to characterize regions of the eastern United States on the basis of fault style. An updated version of their interpretations is shown in Figure 2. The eastern United States is herein subdivided into four tectonic provinces and each can be described by the style of faults within it. These provinces are: (1) the Atlantic Coast province, (2) the Gulf Coast province, (3) the West Georgia Transition province, and (4) the Interior province. A brief description of the fault styles in these provinces provides a framework for understanding the tectonic processes operating during the late Mesozoic and Cenozoic.

Atlantic Coast Province

The Atlantic Coast province is characterized by Cretaceous and younger northeast-trending reverse fault zones and fault systems up to 100 km long. Vertical displacements as great as 76 m have occurred since the Early Cretaceous and progressively smaller offsets have been recognized in rocks spanning the Cenozoic. Although a component of lateral slip has been reported for many reverse faults, dip-slip reverse motion is dominant. The strikes of the fault zones tend to be more northerly in the northern part of the province and more easterly in the southern part of the province, but the strikes are typically within 45 degrees of north. The dips of the fault zones range from 40 to 85 degrees, and the dip of an individual zone may vary depending on the mechanical properties of the rocks in the adjacent fault blocks. Deformation associated with the faulting is extremely brittle in hard rocks and slip surfaces consist of coarse breccias and soft gouge. Coastal Plain strata are typically less sheared and drag folding is well developed. Secondary thermal mineralization is not observed, which indicates that heating and recrystallization (dynamic metamorphism) are not part of this process.

Gulf Coast Province

The Gulf Coast province is characterized by large normal faults that trend subparallel to the present Gulf coastline. Therefore, in the region east of the Mississippi River these normal faults characteristically strike northwestward. Fault zones and fault systems extend for as much as several hundred kilometers in arcuate patterns. Vertical displacements in the lower Mississippi Embayment exceed 400 m in Lower Cretaceous strata and displacements of lesser magnitude are recognized in overlying Tertiary strata as young as Miocene. Faults in the lower Gulf Coastal Plain have larger displacements than those further inland and exhibit mostly down-to-the-coast movement. Faults in the upper Gulf Coastal Plain are commonly in mirror-image pairs defining long, narrow grabens. Vertical displacements in the upper Coastal Plain generally are less than 200 m and lateral displacements are either not reported or are considered minimal.

West Georgia Transition Province

In western Georgia and central Florida(?), a zone of transition between the reverse fault and normal fault provinces of the Atlantic and Gulf regions can be recognized. The faults in this West Georgia Transition generally are east-west-trending vertical faults or fault zones as much as 30 km long. Vertical displacements in early Tertiary strata are as great as 60 meters and lateral displacement is apparently minimal. These near-vertical faults are commonly flanked by smaller secondary reverse faults (Reinhardt and others, 1984) suggesting that compression is a factor in the deformation. The West Georgia faults, like those of the Atlantic Coast province, exhibit only brittle deformation in crystalline rocks with pronounced drag folding in the adjacent Coastal Plain strata.

Interior Province

Verified faults of Cretaceous or younger age are extremely rare in the Interior province due to the lack of post-Jurassic strata. The only available information suggests that faults in this area have northward trends and are approximately vertical (see Prowell, 1983). The greatest reported vertical displacement is about 6 m in river terrace alluvium of probable early Tertiary age near Clifton Forge, Virginia (White, 1952). This fault has been traced laterally for a few hundred meters but nothing is known about its true length. Regional mapping in the area suggests that this fault may parallel a regional joint orientation associated with the broad, open folds in the Paleozoic foldbelt.

CHARACTERISTICS OF ATLANTIC MARGIN FAULTS

The Atlantic coastal margin as recognized in this DNAG volume is largely contained within the Atlantic Coast fault province (this report). In addition, structural elements of the West Georgia Transition province are pertinent to discussions of the Atlantic margin. The following characterization of faulting on the Atlantic coastal margin is based on a synthesis of reported faults in these two fault provinces.

Fault Geometry

The geometry of a "fault" in the Atlantic coastal margin is far more complex than most scientific reports would suggest. These structural features are zones of parallel, closely-spaced (less than 0.5 km), en echelon shear planes. Detailed investigations of this type of fault geometry have been described by Mixon and Newell (1977, 1978), Newell and others (1978), and Prowell and O'Connor (1978). Prowell and O'Connor found that discrete individual fault planes (fault "strands") are approximately 5 to 8 km long with vertical displacement diminishing toward the ends of each fault strand. At least 8 of these closely spaced fault strands form the structure that they named the Belair "fault zone" (Fig. 3). They also found that vertical motion is transferred between fault strands so that movement occurs along the entire length of the fault zone. Mixon and Newell (1977) found that fault zones can group in staggered patterns to form what they call the Stafford "fault system" in northern Virginia (Fig. 4). Fault zones are approximately 25 to 40 km long and they are known to form fault systems as long as 100 km. Reinhardt and others (1984) further recognized that smaller "secondary" faults form at acute angles to the primary fault zone to accommodate localized stress within the upthrown and downthrown blocks. Many of the faults reported in Prowell (1983) are probably secondary faults flanking major fault zones.

The arrangement of fault systems can generate regional structural features such as horsts or grabens. Mixon and Newell (1977) recognized that a long, narrow graben is formed between the Stafford fault system and the mirror image Brandywine fault system (Jacobeen, 1972) 25 km to the east (see Fig. 4). A geologic section showing the subsurface configuration of these fault zones (Fig. 5) is constructed from data found in Mixon and Newell (1978, Fig. 4) and Jacobeen (1972). Both the Stafford and Brandywine fault systems are steeply-dipping reverse fault zones typical of the Atlantic Coast fault province. Offsets of the unconformity between Cretaceous sediments and pre-Cretaceous rocks show the magnitude of vertical displacement. The movements that formed the narrow fault-bounded graben have also affected the Coastal Plain stratigraphic section from the Early Cretaceous through the late Cenozoic.

Structural Orientation

Orientation is an important factor in the analysis of fault zones because it establishes the general configuration of the crustal stress field during faulting. Eleven of the fault zones reported in Prowell (1983, index numbers 19, 28, 29, 31, 33, 35, 60, 62, 69, 75, 77) have been well studied and their orientations are considered diagnostic of other fault zones in the Atlantic coastal margin. A compilation of strike, dip, and dip direction of these faults is shown on the stereographic projection in Figure 6. The solid lines crossing the diameter of the circle represent the strike of each fault zone. The dots on the figure are the points on the great circles that represent the measured dip of each fault zone. The direction of each dot from the solid strike line indicates the direction of inclination of the dipping fault plane.

The large fault zones in the Atlantic coastal margin tend to strike subparallel to the inner margin (Fall Line) of the Coastal Plain (see Plate 11). The more northerly strike lines shown in Figure 6 are fault zones in Virginia and Maryland, and the more

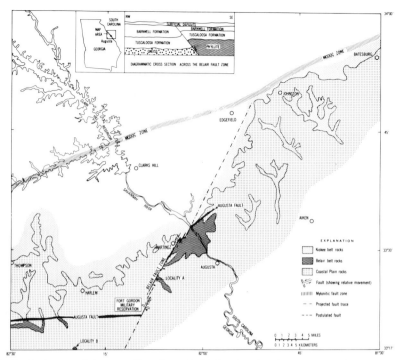

Figure 3. Geologic map of Augusta, Georgia, and vicinity showing the en echelon faults that form the Belair fault zone (from Prowell and O'Connor, 1978, Fig. 1).

northeast-trending faults are found in eastern Georgia and the Carolinas. The two nearly east-west strike lines are fault zones in the West Georgia Transition province and the proximity of the dip symbols to these lines is an indication of their near-vertical dips. The coincidence of major fault zones with the Fall Line on the Atlantic coastal margin suggests that faulting and sedimentation are closely associated; however, the orientations of these features share a common similarity to the regional fabric of the Appalachian crystalline rocks.

The orientation of the reverse faults in the Atlantic Coast fault province and of the vertical faults in the West Georgia Transition province is strongly influenced by the prominent layered fabric of the crystalline (Appalachian-type) basement rocks as well as preexisting structures. Most of the Cretaceous and younger faults reported by Prowell (1983) trend parallel or sub-parallel to the local Paleozoic and Precambrian rock fabric, and reactivation of pre-Cretaceous faults has been reported. Prowell and O'Connor (1978) and Bramlett and others (1982), for example, showed that a Cretaceous and early Cenozoic reverse fault zone in eastern Georgia is a reactivated late Paleozoic tear fault formed during ductile regional overthrusting on the Augusta thrust fault. Similarly, Behrendt and others (1981) and Hamilton and others (1983) presented evidence suggesting that a subsurface normal fault at the edge of an early Mesozoic rift basin near Charleston, South Carolina, has been reactivated with reverse movement during the Cretaceous and early Cenozoic. In addition, Aggerwal and Sykes (1978) have reported seismicity suggestive of reverse motion along the Ramapo fault, which forms the western edge of the Newark Mesozoic basin in New York.

Although the possibility of reactivation of pre-Cretaceous fracture zones has been documented in some cases, most presently known Cretaceous and Cenozoic structures seem to have no obvious connection with preexisting faults or folds. The relationship between fault orientation and basement rock fabric is much more consistent; some faults, however, crosscut all existing rock fabric (for example, Prowell, 1983, fault no. 62) suggesting that rock fabric is not a requirement in fault orientation.

The predominant dip-slip reverse motion of faults in the Atlantic Coast fault province strongly suggests that the faults strike nearly perpendicular to the direction of maximum compressive stress. The small strike-slip component observed on many of the reverse faults (for example, the Balair fault zone) suggests that some faults are not exactly perpendicular to the compressive stress field. The influence of pre-existing rock fabric and rock inhomogeneity on fault orientation can easily explain this misalignment. However, the collective implication of all of the fault zones is that a northwest-southeast compression is responsible for their existence.

Geologic History of Faulting

The detailed studies mentioned in this report consistently indicate that faults in the eastern United States have a history of protracted movement. Most of the faults shown on Plate 8A are reported to have vertical displacements of 3 m or less; however, in many instances, this is only a minimum value for displacement because the condition of the exposure and (or) offset horizons provided incomplete information. Of the fault zones that have been studied in detail, vertical displacement is always many tens of meters. The longevity of this fault movement is demonstrated by the diminishing amount of displacement in progressively younger geologic units which cut the faults. For example, Mixon and Newell (1977, 1978) and Newell and others (1978) found that some of the reverse fault movement along the Stafford fault system in Virginia is confined to the lower Cretaceous Potomac Formation. They also report that successively smaller displacements were found in strata of Paleocene, Eocene, Miocene, and Plio-Pleistocene age in the same general area. Prowell and O'Connor (1978), Reinhardt and others (1984), and Dischinger (1979; in press) report similar sequential displacements on other Atlantic coastal margin fault zones.

The oldest reported displacements along reverse faults on the Atlantic coastal margin are found in early Cretaceous strata in Virginia and Maryland. Nothing is presently known about initiation of compression and reverse faulting on the Atlantic coastal margin, but some inferences can be made from other types of Mesozoic regional tectonism. Triassic rifting, Triassic and Jurassic rift basin sedimentation, and lower and middle Jurassic diabase intrusions are commonly associated with extensional stresses created during continental separation. An extensional stress field would generally prohibit the formation of compressional reverse faults and would therefore place a lower age limit on the propagation of reverse faults. This suggests that the reverse faults could have formed as early as the late Jurassic, if no significant amount of time was required for reversal of the stress field from extension to compression.

The geologic evidence of late Cenozoic fault movement is poor, largely because of the limited distribution of well-defined late Cenozoic materials and the small amounts of fault movement. Late Cenozoic fault movements have been reported by Mixon and Newell (1977, 1978), Pavlides and others (1983), Prowell (1983), and Reinhardt and others (1984) in Virginia, Maryland, and Georgia. These reports show clear evidence of relatively young tectonism as opposed to near-surface gravity slides, which some authors (see Prowell, 1983) report as recent tectonism. The fault described by Pavlides and others (1983) is the youngest known reverse fault involving crystalline basement. The fault is located proximal to the Paleozoic Mountain Run

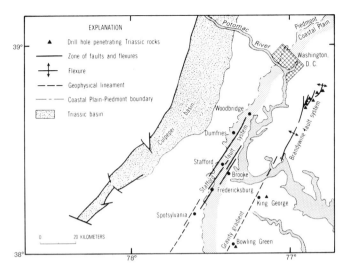

Figure 4. Geologic map of northern Virginia and southern Maryland showing the fault zones that form the Stafford and Brandywine fault systems (modified from Mixon and Newell, 1977, Fig. 1).

fault zone and offsets the base of Pleistocene(?) colluvium about 1.5 m. The location of the faulting within the mylonitic rocks of the older Mountain Run fault zone has tentatively been attributed to reactivation of a zone of weakness. This is not to imply, however, that the same tectonic processes that formed the Paleozoic Mountain Run fault zone also caused the late Cenozoic faulting.

Fault Slip Rates

Comparison of amounts of offset in different chronostratigraphic horizons provides a basis for calculating fault slip rates over geologic time. Wentworth and Keefer (1983) compiled data published by Mixon and Newell (1978), Prowell and O'Connor (1978), and Behrendt and others (1981) for three fault zones in the Atlantic coastal margin and concluded that the average rate of vertical displacement is 0.9 meters/million years (m/m.y.). New and more detailed data have been used to construct the slip rate curves shown in Figure 7. These data imply that the conclusions of Wentworth and Keefer (1983) regarding displacements on the Brooke fault zone (Stafford fault system) and consequently their

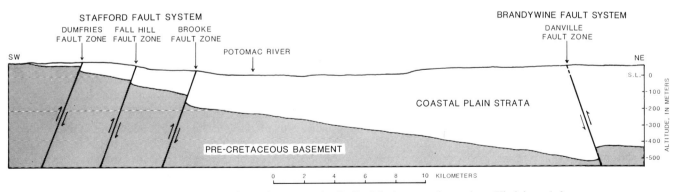

Figure 5. Schematic geologic section across the Stafford fault system in northern Virginia and the Brandywine fault system in southern Maryland showing offset of the base of Cretaceous Coastal Plain strata (data from Mixon and Newell, 1978, and Jacobeen, 1972).

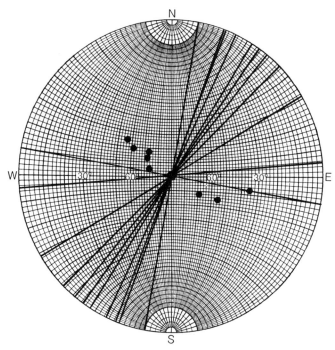

Figure 6. Comparison of strike and dip of eleven well-defined fault zones on the Atlantic coastal margin (data from Prowell, 1983, index nos. 19, 28, 29, 31, 33, 35, 60, 62, 69, 75, 77).

slip rate curves were over-estimated. In addition, recent subsurface borings cast doubt on the geophysical evidence supporting the Cooke fault of Behrendt and others (1981; G. S. Gohn, U.S. Geological Survey, oral comm., 1985). The new slip-rate curves imply that fault movement in the eastern United States ranged from about 0.3 to 1.5 m/m.y. with an average of about 0.5 m/m.y. throughout geologic time. This observation is an important element in the evaluation of recent faulting in the eastern United States and the assessment of the seismic potential of these faults. The consistency of fault movement through geologic time indicates that the compressive stress responsible for the deformation was relatively uniform and unidirectional.

RELATIONSHIP TO OTHER TECTONIC FEATURES

Other types of Cretaceous and Cenozoic tectonism have been recognized in the eastern United States and compare favorably with the faults found in the Atlantic coastal margin. The relatively small number of seismic events recorded in eastern North America in the last 250 years (see York and Oliver, 1976) may be explained by the small slip rates of regional reverse faults. Where localized seismic networks have provided focal plane solutions, the resulting fault plane solutions have typically been attributed to reverse fault movement. Seismicity consistent with reverse faulting has been recognized along the Ramapo fault zone in New York (Ratcliffe, 1971; Aggerwal and Sykes, 1978; Yang and Aggerwal, 1981), near Charleston, South Carolina (Tarr and Rhea, 1983), and at the North Anna Reservoir in central Virginia

(Dames and Moore, 1976). Seismological data in the eastern United States, however, are far from conclusive proof of recent reverse fault activity because many seismologists disagree over the interpretation of seismic evidence. In addition, no observed fault displacements can presently be attributed to historical seismic events.

The crustal stress in the pre-Cretaceous rocks beneath the Atlantic coastal margin is probably responsible for the origin and orientation of the reverse faults. Zoback and Zoback (1980) summarized the state of stress in the conterminous United States and they described the eastern United States as a northwest-southeast compressional regime. However, very few actual stress measurements were available to them and they relied heavily on the fault orientations from Prowell (1983) to determine the direction of principal stress. New stress measurements reported by Zoback and others (1984) suggest that compressional forces are present in the crust of eastern North America but that the maximum compressive stress has a northeast-southwest orientation. This suggested stress orientation is not substantiated by the orientation and slip directions of Pleistocene(?) faults such as the one reported by Pavlides and others (1983) in central Virginia.

Uplift and subsidence of the continental edge is further evidence that regional tectonism has affected the Atlantic coastal margin during the Cretaceous and Cenozoic. Various geomorphic, geophysical, and chronostratigraphic studies have suggested that relative changes in elevation are common occurrences in the geologic past (see Walcott, 1972; Denny, 1974; Isachsen, 1975; Zimmerman, 1977; Hack, 1979, 1982; Lyttle and others, 1979; Brown and others, 1980; and Cronin, 1981). Hack (1979) emphasizes that the uplift and subsequent erosion of the exposed Appalachian crystalline rocks controls sediment deposition on the Atlantic coastal margin. Sediment distribution on the coastal margin is also affected by the development of nearshore archs and embayments (Owens, 1983; Owens and Gohn, 1985) and offshore basins or troughs (Sleep, 1971; Sheridan, 1974, 1976; Watts and Ryan, 1976; Klitgord and Behrendt, 1979; and Owens, 1983). The subsidence of the landmass has been attributed to processes such as extensional rifting (Sheridan, 1974, 1976), lithospheric cooling and contraction (McKenzie, 1978; Keen, 1979), and depositional loading (Walcott, 1970); whereas uplift of the landmass has been attributed to lithospheric bulging adjacent to subsidence (Beaumont, 1978, 1979), glacial unloading (Walcott, 1972), and mantle hotspot migration (Crough, 1981). All of these proposed mechanisms are probably active in the earth's crust, but they fail to explain all of the regional vertical changes in the landmass, especially uplift and tilting, outlined by Hack (1979). Similarly, Heller and others (1982) recognized that the mobility of offshore depocenters and the variability of deposition in different basins could not be explained by simple lithospheric cooling or loading. These observations in conjunction with the regional fault distribution discussed in this paper imply that deep crustal tectonism, locally affected by erosion and sediment loading, is primarily responsible for the present configuration of the Atlantic coastal margin.

SUMMARY AND CONCLUSIONS

This paper briefly summarizes the nature of faulting and associated tectonism in the eastern United States from the Cretaceous to the present. Studies of reverse faults indicate that regional compression has existed in the crust from the Early Cretaceous through the Pleistocene, but evidence of earlier movements is not available. Intrusion of diabase dikes and rift basin sedimentation from the Triassic to the Middle Jurassic, however, suggests that tensional forces associated with rifting were still present at that time. The reversal of the maximum horizontal stress direction from tension to compression probably was not a rapid process and must have resulted in a period of little or no applied stress. This period of transition probably occurred between the Middle Jurassic and the Early Cretaceous, and likely is associated with a period of non-deposition on the continental shelf.

Cretaceous and younger faults in the eastern United States may form along preexisting Paleozoic or Mesozoic faults but they generally parallel local rock fabric. The location of the plane of shearing in the regional geologic framework seems to depend on both the local orientation of the maximum compressive stress and the orientation and mechanical properties of planar geologic elements (for example, faults, foliations, and cleavage). Shear zones that are not oriented at right angles to the maximum compressive stress will have oblique-slip reverse movement.

The long-term compression since the Early Cretaceous is apparently responsible for rather uniform fault movement for the last 110 m.y. Slip rates may vary locally from 0.3 to 1.5 m/m.y., but over geologic time, fault motion closely approximates 0.5 m/m.y. The presence of compressive stress and reverse fault movement in the recent geologic record is inferred by seismic focal plane solutions and modern stress measurements. However, offsets in Holocene strata have not been recognized, probably because of the lack of significant amounts of fault movement.

Other evidence of possible crustal tectonism is uplift, tilting,

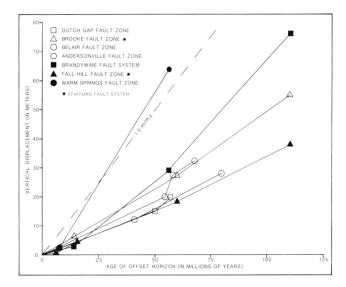

Figure 7. Movement histories of faults on the Atlantic coastal margin showing age of offset horizon versus greatest vertical displacement of horizon (Dutch Gap fault zone from Dischinger, in press; Brooke and Fall Hill fault zones from Mixon and Newell, 1978; Belair fault zone from Prowell and O'Connor, 1978; Andersonville fault zone from H. E. Cofer, Georgia Southwestern College, oral comm., 1983; Brandywine fault system from Jacobeen, 1972; Warm Springs fault zone from Reinhardt and others, 1984).

and subsidence manifested as coastal arches, embayments, and troughs, which are further modified by sediment loading. Similarly, periodic uplift and erosion of source areas for coastal sediment such as the mountainous terrane in the folded Appalachians is also suggestive of regional tectonism. This tectonism is of such regional extent that smaller scale processes such as lithospheric loading or hotspot activity cannot account for its presence. Therefore, faulting and related tectonism must be subtle evidence of previously unrecognized plate tectonics in the eastern United States.

REFERENCES CITED

Aggerwal, Y. P., and Sykes, L. R., 1978, Earthquakes, faults, and nuclear power-plants in southern New York and northern New Jersey: Science, v. 200, no. 4340, p. 425–429.

Beaumont, C., 1978, The evolution of sedimentary basins on a viscoelastic lithosphere: Theories and examples: Royal Astronomical Society, Geophysical Journal, v. 55, p. 471–497.

——— 1979, On the rheological zonation of the lithosphere during flexure: Tectonophysics, v. 59, p. 347–365.

Behrendt, J. C., Hamilton, R. M., Ackermann, H. D., and Henry, V. J., 1981, Cenozoic faulting in the vicinity of the Charleston, South Carolina, 1886 earthquake: Geology, v. 9, p. 117–122.

Bramlett, K. W., Secor, D. T., Jr., and Prowell, D. C., 1982, The Belair fault: A Cenozoic reactivation structure in the eastern Piedmont: Geological Society of America Bulletin, v. 93, p. 1109–1117.

Brown, L. D., and Oliver, J. E., 1976, Vertical crustal movements from leveling data and their relation to geologic structure in the eastern United States: Review of Geophysics and Space Physics, v. 14, no. 1, p. 13–35.

Brown, L. D., Reilinger, R. E., and Citron, G. P., 1980, Recent vertical crustal

movements in the United States: Evidence from precise levelling, *in* Morner, Nils-Axel, ed., Earth rheology, isostasy and eustasy: New York, Wiley and Sons, p. 389–405.

——— 1981, Rates and possible causes of neotectonic vertical crustal movements of the submerged southeastern United States Atlantic coastal plain: Geological Society of America Bulletin, v. 92, Part 1, p. 812–833.

Crough, S. T., 1981, Mesozoic hotspot epeirogeny in eastern North America: Geology, v. 9, no. 1, p. 2–6.

Dames and Moore, Inc., 1976, Summary report on the in-progress seismic monitoring program at the North Anna site, January 21, 1974, through May 1, 1976: Report submitted to Virginia Electric and Power Company (VEPCO), representing results of the seismographic study required by VEPCO by the Nuclear Regulatory Commission in 1973 as Question 2.28 to the Preliminary Safety Analysis Report for North Anna 3 and 4 power reactors, 117 p.

Darton, N. H., 1891, Mesozoic and Cenozoic formations of eastern Maryland and Virginia: Geological Society of America Bulletin, v. 2, p. 431–450.

Denny, C. S., 1974, Pleistocene geology of the northeast Adirondack region, New York: U.S. Geological Survey Professional Paper 786, 50 p.

Dischinger, J. R., Jr., 1979, Stratigraphy and structure of the faulted Coastal Plain near Hopewell, Virginia: Geological Society of America Abstracts with Programs, v. 11, no. 4, p. 177.

——, in press, Late Mesozoic and Cenozoic stratigraphic and structural framework near Hopewell, Virginia: U.S. Geological Survey Bulletin 1567.

Hack, J. T., 1979, Rock control and tectonism—Their importance in shaping the Appalachian highlands: U.S. Geological Survey Professional Paper 1126-B, p. B1–B17.

——1982, Physiographic divisions and differential uplift in the Piedmont and Blue Ridge: U.S. Geological Survey Professional Paper 1265, 49 p.

Hamilton, R. M., Behrendt, J. C., and Ackermann, H. D., 1983, Land multichannel seismic-reflection evidence for tectonic features near Charleston, South Carolina, *in* Gohn, G. S., ed., Studies related to the Charleston, South Carolina, earthquake of 1886—Tectonics and seismicity: U.S. Geological Survey Professional Paper 1313, p I1–I18.

Heller, P. L., Wentworth, C. M., and Poag, C. W., 1982, Episodic post-rift subsidence of the United States Atlantic continental margin: Geological Society of America Bulletin, v. 93, no. 5, p. 379–390.

Howard, K. A., Aaron, J. M., Brabb, E. E., Brock, M. R., Gower, H. D., Hunt, S. J., Milton, D. J., Muehlberger, W. R., Nagata, J. K., Plafker, George, Prowell, D. C., Wallace, R. E., and Witkind, I. J., 1978, Preliminary map of young faults in the United States as a guide to possible fault activity: U.S. Geological Survey Miscellaneous Field Studies Map MF-916.

Hutchinson, D. R., and Grow, J. A., 1985, New York Bight Fault: Geological Society of America Bulletin, v. 96, p. 975–989.

Isachsen, Y. W., 1975, Possible evidence for contemporary doming of the Adirondack Mountains, New York, and suggested implications for regional tectonics and seismicity: Tectonophysics, v. 29, p. 169–181.

Jacobeen, F. H., Jr., 1972, Seismic evidence for high-angle reverse faulting in the Coastal Plain of Prince Georges and Charles counties, Maryland: Maryland Geological Survey Information Circular 13, 21 p.

Keen, C. E., 1979, Thermal history and subsidence of rifted continental margins: American Association of Petroleum Geologists Memoir 29, p. 85–112.

Klitgord, K. D., and Behrendt, J. C., 1979, Basin structure of the U.S. Atlantic margin, *in* Watkins, J. S., Montadert, Lucien, and Dickerson, P. W., eds., Geological and geophysical investigations of continental margins: American Association of Petroleum Geologists Memoir 29, p. 85–112.

Lyttle, P. T., Gohn, G. S., Higgins, B. B., and Wright, D. S., 1979, Vertical crustal movements in the Charleston, South Carolina-Savannah, Georgia, area: Tectonophysics, v. 52, p. 183–189.

McGee, W. J., 1888, The geology of the head of the Chesapeake Bay: U.S. Geological Survey Annual Report 7, p. 537–646.

McKenzie, D., 1978, Some remarks on the development of sedimentary basins: Earth and Planetary Science Letters, v. 40, p. 25–32.

Mixon, R. B., and Newell, W. L., 1977, Stafford fault system—Structures documenting Cretaceous and Tertiary deformation along the Fall Line in northeastern Virginia: Geology, v. 5, no. 7, p. 437–440.

——1978, The faulted Coastal Plain margin at Fredericksburg, Virginia, *in* Mixon, R. B., and Newell, W. L., eds., Guidebook for Tenth Annual Virginia Field Conference, October 13–14, 1978: Virginia Academy of Sciences, Reston, Virginia, 50 p.

Newell, W. L., Prowell, D. C., and Mixon, R. B., 1978, A Piedmont-Coastal Plain fault contact; Stafford fault system, northeastern Virginia, *in* Mixon, R. B., and Newell, W. L., eds., Guidebook for Tenth Annual Virginia Field Conference, October 13–14, 1978: Virginia Academy of Sciences, Reston, Virginia, p. 34–40.

Owens, J. P., 1983, The northwestern Atlantic Ocean margin, *in* Moullade, M., and McNairn, A. E., eds., The Phanerozoic of the world, no. 11, The Mesozoic, Part B: Elsevier Scientific Publishing Co., p. 33–60.

Owens, J. P., and Gohn, G. S., 1985, Depositional history of the Cretaceous series in the United States Atlantic coastal plain: Stratigraphy, paleoenvironments, and basin evolution, *in* Poag, C. W., ed., Geological evolution of the United States Atlantic margin: Stroudsburg, Van Nostrand-Reinhold, Pennsylvania, Chapter 2, p. 25–86.

Pavlides, Louis, Bobyarchick, A. B., Newell, W. L., and Pavich, M. J., 1983, Late Cenozoic faulting along the Mountain Run fault zone, central Virginia Piedmont [abs.]: Geological Society of America Abstracts with Programs, v. 15, no. 2, p. 55.

Prowell, D. C., 1976, Implications of Cretaceous and post-Cretaceous faults in the eastern United States: Geological Society of America Abstracts with Programs, v. 8, no. 2, p. 249–250.

——1983, Index of faults of Cretaceous and Cenozoic age in the eastern United States: U.S. Geological Survey Miscellaneous Field Studies Map MF-1269.

Prowell, D. C., and O'Connor, B. J., 1978, Belair fault zone—Evidence of Tertiary fault displacement in eastern Georgia: Geology, v. 6, p. 681–684.

Ratcliffe, N. M., 1971, The Ramapo fault system in New York and adjacent northern New Jersey; A case of tectonic heredity: Geological Society of America Bulletin, v. 82, no. 1, p. 125–142.

Reinhardt, Juergen, Prowell, D. C., and Christopher, R. A., 1984, Stratigraphy and structure of sediments near Warm Springs as indicators of Cenozoic tectonism in the southwest Georgia Piedmont: Geological Society of America Bulletin, v. 95, p. 1176–1187.

Sheridan, R. E., 1974, Conceptual model for the block fault origin of the North American Atlantic continental margin geosyncline: Geology, v. 2, p. 465–468.

——1976, Sedimentary basins of the Atlantic margin of North America: Tectonophysics, v. 36, p. 113–132.

Sleep, N. H., 1971, Thermal effects of the formation of Atlantic continental margins by continental breakup: Royal Astronomical Society Geophysical Journal, v. 24, p. 325–350.

Tarr, A. C., and Rhea, Susan, 1983, Seismicity near Charleston, South Carolina, March 1973 to December 1979, *in* Gohn, G. S., ed., Studies related to the Charleston, South Carolina, earthquake of 1886—tectonics and seismicity: U.S. Geological Survey Professional Paper 1313, p. R1–R17.

Walcott, R. I., 1970, Flexural rigidity, thickness, and viscosity of the lithosphere: Journal of Geophysical Research, v. 75, p. 3941–3954.

——1972, Late Quaternary vertical movements in eastern North America—Quantitative evidence of glacial-isostatic rebound: Reviews of Geophysics and Space Physics, v. 10, no. 4, p. 849–884.

Watts, A. B., and Ryan, W.B.F., 1976, Flexure of the lithosphere and continental margin basins: Tectonophysics, v. 36, p. 25–44.

Wentworth, C. M., and Keefer, M. M., 1983, Regenerate faults of small Cenozoic offset—probable earthquake sources in the southeastern United States, *in* Gohn, G. S., ed., Studies related to the Charleston, South Carolina, earthquake of 1886—Tectonics and seismicity: U.S. Geological Survey Professional Paper 1313, p. S1–S20.

White, W. A., 1952, Post-Cretaceous faults in Virginia and North Carolina: Geological Society of America Bulletin, v. 63, no. 7, p. 745–748.

Yang, Jih-Ping, and Aggerwal, Y. P., 1981, Seismotectonics of the northeastern United States and adjacent Canada: Journal of Geophysical Research, v. 86, no. B6, p. 4981–4998.

York, J. E., and Oliver, J. E., 1976, Cretaceous and Cenozoic faulting in eastern North America: Geological Society of America Bulletin, v. 87, p. 1105–1014.

Zimmerman, R. A., 1977, The interpretation of apatite fission track ages with an application to the study of uplift since the Cretaceous in eastern North America [Ph.D. thesis]: Philadelphia, University of Pennsylvania, 223 p.

Zoback, M. L., and Zoback, M. D., 1980, State of stress in the conterminous United States: Journal of Geophysical Research, v. 85, no. B11, p. 6113–6156.

Zoback, M. L., Zoback, M. B., and Schiltz, M. E., 1984, Index of stress data for the North American and parts of the Pacific Plate: U.S. Geological Survey Open File Report 84-157, 62 leaves.

MANUSCRIPT ACCEPTED BY THE SOCIETY NOVEMBER 4, 1985

Chapter 30

Seismicity along the Atlantic Seaboard of the U.S.; Intraplate neotectonics and earthquake hazard

Leonardo Seeber and John G. Armbruster
Lamont-Doherty Geological Observatory of Columbia University, Palisades, New York 10964

INTRODUCTION

The wide acceptance of plate tectonics two decades ago caused earth scientists interested in neotectonics and seismicity to focus their attention along plate boundaries with resulting great advances in our understanding of interplate deformation. In contrast, crustal deformation and earthquake hazard in the interior of presumably rigid plates have been considered second-order phenomena and have received relatively little attention. Although the concept of a tectonically quiescent intraplate environment may be justified for some areas, in others, as in parts of eastern North America, the historic record of damaging earthquakes indicates that neotectonics is quite significant, at least from the seismological viewpoint, and needs to be better understood.

Overall earthquake damage potential: Eastern versus western US

Certainly more earthquakes and, particularly, a greater number of damaging earthquakes occur in interplate western North America than in the intraplate rest of the continent. However, the largest known earthquakes in the conterminous U.S. have occurred in the East: the three 1811 to 1812 events in New Madrid, Missouri, and the 1886 event in Charleston, South Carolina. In this discussion the "size" of an earthquake refers to its damage potential, or the size of the area with damaging intensities (which is here arbitrarily taken as the area with Modified Mercalli intensity VIII or greater). Damage potential depends on the spectrum and attenuation of seismic waves and is not a reliable measure of the total radiated seismic energy or the tectonic strain, which are expressed by the moment or magnitude. Thus, the 300-km-long 1906 rupture of the San Andreas fault produced a smaller area of damaging intensities (Fig. 1), but probably a much larger moment than the 1811 to 1812 earthquakes in the East. Such a systematic difference between eastern and western earthquakes has been attributed in part to differences in seismic attenuation properties (e.g., Nuttli, 1973). High stress–drops may also contribute to this difference.

The relatively rare large eastern earthquakes have produced destructive effects over such large areas that they have compensated for the relative quiescence during the intervening periods. When all the known earthquakes are considered, the average rate at which area of intensity VIII or greater has been produced by earthquakes during the historic period is similar east and west of the Rockies (Fig. 1; Seeber, 1983). Thus, if the historic record of seismicity is representative of the long-term activity, the overall level of earthquake hazard is similar in the eastern and western halves of the U.S. The contrast between rates of moment release versus inferred overall levels of hazard in eastern and western U.S. implies that eastern earthquakes are more effective in causing damage than western earthquakes.

Since the 1886 earthquake is among the largest and the 1811 to 1812 earthquakes are the largest historic intraplate earthquakes worldwide (Coppersmith and others, 1986), they may also be unusual in the eastern U.S., and the long-term rate of intensity $\geq$ VIII in this area may be lower than the rate in the historic period. Recent paleoseismic data suggest that large earthquakes have occurred in Charleston, South Carolina, and New Madrid, Missouri, during the Holocene with repeat times of one to a few thousand years (e.g., Talwani and Cox, 1985; Obermeier and others, 1985). If these were the only possible sources of large earthquakes in the eastern U.S., an unusually high rate of seismicity in the nineteenth century would be inferred for that area. If, however, earthquakes similar to the one in 1886, such as the 1929 Grand Banks earthquake (M = 7.2) and the 1934 Labrador earthquake (M = 7.3), can occur elsewhere along the Atlantic Seaboard (e.g., Basham and Adams, 1983), prehistoric and historic rates of damaging intensities may be similar. A uniform distribution of M > 7 earthquakes along 3,000 km of the North American Atlantic Seaboard at an historic rate of three events each 200 years would imply one event in each 100 km portion of the Seaboard every 2,000 years. Such a rate is consistent with paleoseismic data. Thus, it seems reasonably conservative to assume that historic seismicity produced a representative overall long-term rate of damaging intensities.

Seeber, L., and Armbruster, J. G., 1988, Seismicity along the Atlantic Seaboard of the U.S.; Intraplate neotectonics and earthquake hazard, *in* Sheridan, R. E., and Grow, J. A., eds., The Atlantic Continental Margin, U.S.: Geological Society of America, The Geology of North America, v. I-2.

of a series of collision events spanning most of the Paleozoic. This orogen is responsible for the last penetrative phase of deformation to affect rocks of the upper crust in this area. The structural belt of the Appalachians is strongly asymmetric, verging toward the interior of the continent, and is remarkably uniform along strike (Plate 8B). To accommodate convergence across the Appalachians, the Precambrian cratonic basement of North America and part of its sedimentary cover were overthrust by a slab of crystalline rocks comprising metamorphosed Paleozoics and slivers of basement transported hundreds of kilometers northwest from the internal part of the orogen. The Appalachian master detachment at the base of this slab is the fundamental structural feature responsible for its emplacement. This detachment dips shallowly to the southeast and has been mapped for large distances down to midcrustal depths below the Blue Ridge and Piedmont along a few widely spaced seismic reflection profiles (e.g., Cook and others, 1981). The northwestern limit of the crystalline slab tends to be a sharply defined boundary that coincides with a major splay of the master detachment and is named herein the Appalachian Front.

Mesozoic rifting coincided in large part with the Paleozoic collision zone and was strongly controlled by preexisting structure, presumably reactivating in a back-slip mode some of the Paleozoic thrusts (e.g., Ratcliffe and others, 1986), including the master detachment (Swanson, 1986). Rifting occurred over a wide band, only part of which is now exposed along the margin (e.g., Daniels and others, 1983; Klitgord and others, 1983). Mesozoic rifting is dramatically manifested by a set of large basins where synrift sediments are preserved. These basins are only found east of the Blue Ridge, but extension and related faulting were certainly not confined to the preserved portion of the basins. In particular, normal slip is often identified as the last phase of deformation on Paleozoic thrusts throughout the Appalachians (e.g., Boyer and Elliott, 1982) and Mesozoic extensional faulting may have been active west of the Piedmont. Mesozoic rifting is the last plate-boundary event and the last phase of large deformation along the margin.

Even accepting the possibility of some unreal patterns of epicenters generated by an uneven coverage of the seismicity in the catalogs, it is clear from Plates 8A and 8B that earthquakes are not evenly scattered along the Appalachians and Atlantic Seaboard. Broad zones of relatively high seismicity are recognized in the Adirondacks, eastern New England, southern New York–New Jersey, Virginia, and eastern Tennessee. Many authors have attempted to account for the observed patterns of seismicity with systematic relationships between seismicity and geology (e.g., Long, 1986; Alexander and Lavin, 1982; Talwani and others, 1986; additional references in following four paragraphs).

Relationships between structure and seismicity are likely to offer unique insights of the neotectonic process. Any generalizations, however, must be carefully evaluated against the possibility of time-dependent seismicity changes. For example, seismicity is clearly controlled by structure associated with the Newark Basin.

Epicenters are concentrated around the northern half of the basin, but do not occur within it (Fig. 2). Although seismicity is clearly associated with some of the basins (Dewey and Gordon, 1984), it is not correlated with all of the basins (Plates 8A and 8B). For example, the Gettysburg Basin, the southwestern extension of the Newark Basin, is not associated with historic seismicity. This apparent lack of systematics in the relation between seismicity and Mesozoic basins may reflect temporal changes and may disappear in a sufficiently long sample of the seismicity.

On the basis of the correlation between seismicity and Mesozoic basins, rift faults predominate among preexisting structural features suspected to control neotectonics along the Atlantic Margin (e.g., Wentworth and Mergner-Keefer, 1983). These normal faults were thought to be reactivated in a reverse sense in the current stress field on the basis of fault-plane solutions (e.g., Yang and Aggarwal, 1981), or displacement of Neogene sediments (e.g., Prowell, this volume; Hutchinson and Grow, 1985). A normal displacement, however, was found to be the last deformation recorded by small structural features along the Ramapo fault (Ratcliffe, 1980), a Mesozoic border fault suspected of neotectonic reactivation (Aggarwal and Sykes, 1978). Recent results suggest that at least some of the earthquakes are generated by secondary faults trending across the strike of the basins (Quittmeyer and others, 1985; earthquake sequences in Lancaster, Pennsylvania, and in Ardsley, New York, discussed below).

A concentration of seismicity and postrift tectonic activity along the continental extensions of fracture zones and other features on the oceanic lithosphere adjacent to passive margins was proposed by Sykes (1978). The correlation between the Blake Spur Fracture Zone and the seismic source near Charleston, South Carolina, and between the New England Seamount Chain and the Boston-Ottawa Seismic Zone are often quoted examples. Abundant new data on structural features in the oceanic basement, particularly on the location of oceanic fracture zones (e.g., Plate 8B), suggest that this correlation, if it exists, is less systematic than originally envisioned (e.g., Klitgord and others, 1983).

A correlation between seismicity and plutonic bodies has been repeated proposed (e.g., McKeown, 1978; Long, 1976; Campbell, 1978). This correlation may be accounted for by cooling stresses or by stress amplification near boundaries between rocks of contrasting material properties. A correlation between seismicity and zones of postrift subsidence was proposed by Barosh (1981). Following the realization that a series of early Paleozoic sutures separates a number of terranes that accreted to the margin of proto–North America (e.g., Williams and Hatcher, 1982), a correlation was recently proposed between these sutures or distinct terranes and seismicity (Wheeler and Bollinger, 1984). None of these or other hypotheses can individually account for all of the seismicity. Many of them, however, are likely to account for some of the seismicity. Certainly they are of value because they open new points of view and stimulate ideas. Accordingly, some recent results are discussed below intending more to be provocative than to avoid controversial subjects.

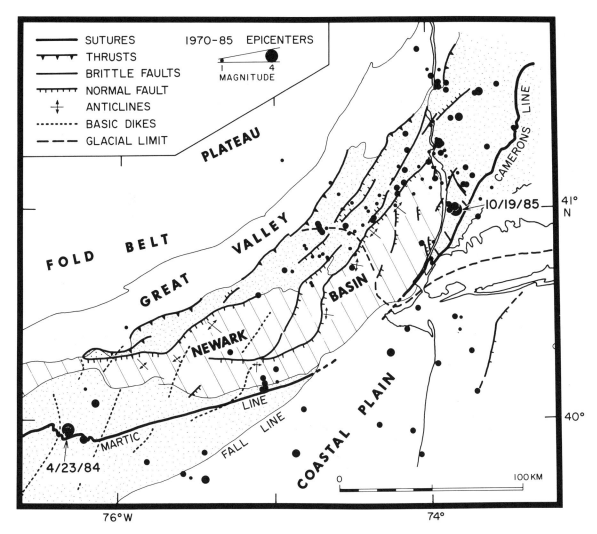

Figure 2. Seismicity detected by the regional network and the Newark Triassic-Jurassic Basin. Location uncertainties for most of the epicenters are smaller than the size of the symbols. Seismicity is concentrated in crystalline rocks adjacent to the basin, but is absent within and below the basin. In particular, epicenters correlate with the trace of the Ramapo border fault, but not with the subsurface extension of this southeasterly dipping fault. The 4/23/84 earthquake in Lancaster County southwest of the Newark Basin and the 10/19/85 earthquake in Westchester County east of the basin are the two largest (m = 4) earthquakes in the period covered by these data (1970 to 1985). Both these events occurred near the Early Paleozoic suture (Cameron's–Martic Line), but they ruptured faults transverse to this suture and to the trend of the basin.

THE APPALACHIANS AND THE ATLANTIC SEABOARD: A DISTINCT SEISMO-NEOTECTONIC PROVINCE?

The northwestern limit of the crystalline Appalachians, the Appalachian Front (see above), is superimposed on epicenters obtained from the regional networks in Figure 3. Although the regional coverage is biased by the uneven distribution of networks, hypocenter accuracy allows resolution of detailed features. These hypocenters were selected to have confidence limits on

depth of 5 km or less and can be expected to contain information about the three-dimensional distribution of seismicity in the crust. The same data are shown in two sections across the Appalachian Front, one for the northern (Fig. 4) and the other for the southern Appalachians (Fig. 5). In these sections, hypocenters are projected on planes perpendicular to the front and plotted at their shortest distance to this front.

A discontinuous zone of seismicity is associated with the Appalachian Front in Figure 3. This broad active zone consists of two distinct parallel belts separated by a narrow zone of low seismicity a few tens of kilometers wide, which follows the front.

L. Seeber and J. G. Armbruster

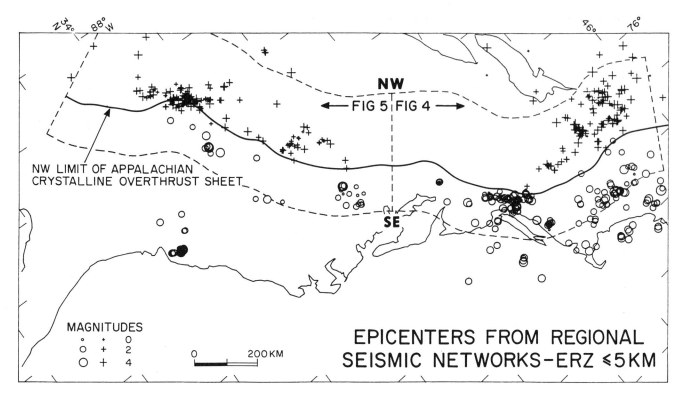

Figure 3. Earthquake epicenters from regional seismic networks along the Appalachians/Atlantic Seaboard. Only earthquakes with depth uncertainties of 5 km or less are included. The thick line is the northwest limit of crystalline rocks in the Appalachian allochthon—the trace of a system of thrust splays branching off the master detachment and named the Appalachian Front (see Fig. 4). Two seismic zones are recognized: one in the cratonic basement below the detachment (crosses) and one in the allochthonous crystalline slab above the detachment (circles).

In both these sections (Figs. 4 and 5), seismicity seems to follow the crystalline rocks. West of the front, seismicity is in the Precambrian basement below the Paleozoic sediments and follows the basement southeastward as it plunges below the foredeep, but only as far as the front. East of the front, seismicity seems to be confined to the allochthonous slab of crystalline rocks above the Appalachian master detachment (e.g., Bollinger and others, 1985a). Also in this zone, seismicity near the front deepens toward the east, possibly reflecting the thickening of the crystalline slab. Further east this trend is reversed, particularly in New England across the Connecticut valley, which is thought to correlate with the buried margin of pre-Appalachian North America (Brown and others, 1983). The detachment and the sedimentary layer wedged between the crystalline basement and allochthonous slab seem to be mostly aseismic.

The control that structure along the Appalachian Front appears to exert on the spatial distribution of seismicity may be interpreted in different ways. Lithologic contrast across faults may affect rheology and strain mechanisms, so that seismicity is spatially associated with these faults, but is not generated by them (Seeber and Coles, 1984). Thus, the Appalachian Front may

separate zones of different seismic regimes without being itself active (e.g., Bollinger and others, 1985a). The opposite end-member hypothesis is that the Appalachian detachment is still active (but not necessarily seismogenic) and is acting as a decoupling layer (Behrendt and others, 1981; Seeber and Armbruster, 1981). The pattern of stress may be useful to distinguish between these two hypotheses. A zone of decoupling would be expected if the Appalachians/Atlantic Seaboard east of the front was found to be a distinct stress province with a characteristic pattern of stress and deformation.

The seismic zones east of the Appalachian Front, in eastern New England, around the Newark Basin, and in central Virginia show a mixed pattern of fault-plane solutions (Fig. 6). In contrast, a relatively uniform pattern of fault-plane solutions characterizes the seismicity in the craton, west of the Appalachian Front. While west of the front the P-axes of the solutions are consistently directed east-northeast, east of the front P-axes are more scattered, including both northeast- and northwest-directed axes. Thus, earthquake data seem to us more consistent with a distinct stress province along the Appalachians and Atlantic Seaboard. Others, however, prefer to fit the earthquake data to a uniform

1970 — 1984   NORTHEAST US NETWORKS HYPOCENTERS

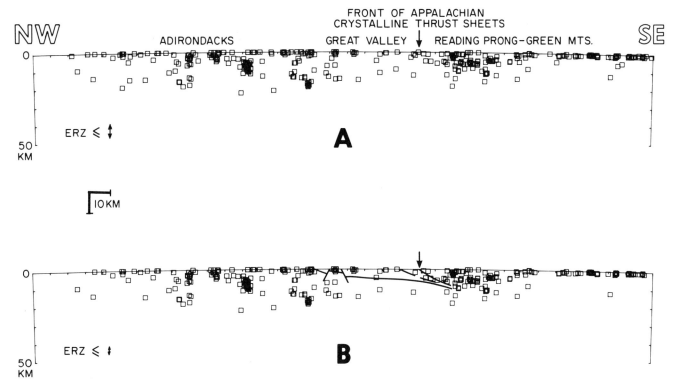

Figure 4. Hypocenters across the northern Appalachian Front; same data as in Figure 3. Hypocenters are projected along the front and plotted on a vertical plane perpendicular to the front. The maximum vertical error is 5 km in A and 3 km in B. Some of the major features from the Adirondacks–Green Mountains profile by COCORP (Brown and others, 1983) are shown in B. Two distinct zones of seismicity occupy the cratonic basement and the allochthonous crystalline slab; little or no seismicity occurs in the sedimentary wedge between these zones (undulations of this wedge along the front can account for some of the discrepancies).

field of east-northeast–directed maximum horizontal compression affecting the entire eastern U.S. (e.g., Zoback and Zoback, 1986; Gephard and Forsyth, 1985).

Maximum horizontal in situ stress from hydrofracture and from borehole elongation along the Appalachians and the Atlantic Seaboard (Zoback and others, 1984; Plumb and Cox, 1987) is generally directed northeast with few exceptions, such as in Moodus, Connecticut (Rundle and others, 1985). A uniform-stress province including both the cratonic interior and the Atlantic Seaboard has been proposed on the basis of these data (e.g., Zoback and Zoback, 1986). The directions of maximum horizontal stress, however, seem better correlated to the local direction of strike in the Paleozoic fabric than to a uniform stress field (Seeber and Armbruster, 1985). Thus, in situ stress directions in the shallow crust may also be consistent with a stress regime that is unique to the Appalachians/Atlantic Seaboard.

Whether or not the Appalachians and the Atlantic Seaboard compose a distinct stress province is a fundamental question for which available data cannot provide a definitive answer. If it is a

distinct province, the Appalachian Front seems a likely boundary between this province and the craton. Yielding to a bias in favor of this hypothesis, the Appalachian Front is chosen in this paper as a convenient boundary to the seismicity relevant to the continental margin.

THE 1886 CHARLESTON, SOUTH CAROLINA, EARTHQUAKE

The 1886 Charleston, South Carolina, earthquake is the largest known along the Appalachians/Atlantic Seaboard and may be a crucial event for understanding the nature of neotectonics and determining the level of hazard in this province. Among large eastern earthquakes it is also the one with the most detailed data on surface effects. These include intensity data from much of the eastern U.S. (Bollinger, 1977), far-field long-period effects (Armbruster and Seeber, 1981), and near-field large-strain effects (Seeber and Armbruster, 1983). These data have been examined to characterize focal parameters (e.g., Evernden, 1975; Nuttli,

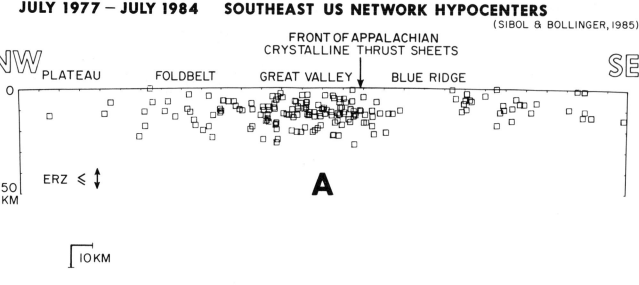

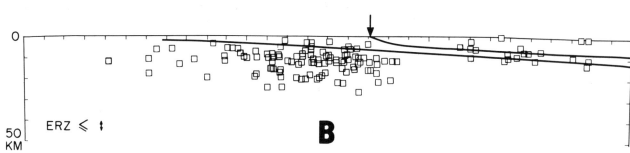

Figure 5. Hypocenters across the southern Appalachian Front; same data as in Figure 3; A and B as in Figure 4. Some of the major features of the Savannah river profile (Cook and others, 1981) are shown in B. Comments in caption of Figure 4 apply also here.

1983). In the last decade, geologic and geophysical studies in the epicentral area explored the tectonic environment to test the hypothesis that this environment is unique (e.g., Rankin, 1977; Gohn, 1983). Although much has been learned about the 1886 earthquake, its significance in terms of intraplate neotectonics and earthquake hazard remains controversial.

Particularly puzzling is the contrast between, on the one hand, the relatively high rate of moment release indicated by the large size of the 1886 main shock and the relatively short repeat time of one to few millennia obtained from paleoseismic data (Obermeier and others, 1985; Talwani and Cox, 1985), and, on the other hand, the lack in the epicentral area of any clearly recognized Cenozoic fault displacement larger than a few tens of meters (Stewart, 1986). Cenozoic faults have been described in relation to the 1886 earthquake (e.g., Talwani, 1982; Hamilton and others, 1983; Behrendt and others, 1983), but the significance of all these features in relation to the 1886 source remains controversial. In any case, Cenozoic faults along the Atlantic Margin are not unique to the 1886 mesoseismal area (e.g., Prowell, this volume). Thus, the structural characteristics of the 1886 source

remain largely unknown and the epicentral area cannot be associated with any clearly identified geologic feature unique along the Atlantic Seaboard.

Seismicity before and after the 1886 main shock: Is intraplate seismicity time-stationary?

Large earthquakes are often associated with regional changes in seismicity. Conversely, unique information on the nature of a large event is often derived from the space-time distribution of related seismicity. In order to improve constraints on the 1886 event, the distribution of southeastern seismicity in the nineteenth century was studied by systematically reexamining archival data. In particular, seismicity patterns preceeding and following this event were compared (Armbruster and Seeber, 1984). If the source of the 1886 earthquake is unique, the seismicity generated by this source should stand out from the background, not only after, but also prior to 1886. From results obtained so far, the level of seismicity in the epicentral area for at least 80 years before 1886 resembles the general level elsewhere

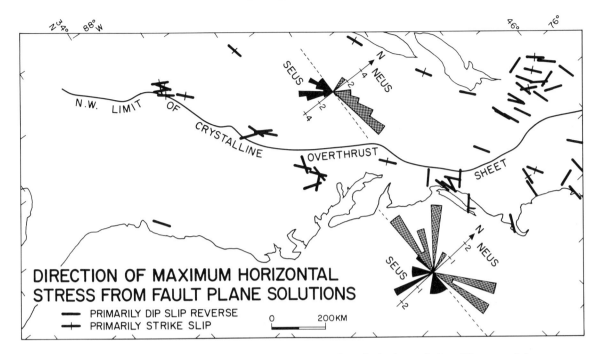

Figure 6. Maximum horizontal stress directions (P-axes) from fault-plane solutions. The trace of the Appalachian Front (see Fig. 3) separates two domains: a cratonic domain to the west where P-axes are predominantly east-northeast and an Appalachian/Atlantic Seaboard domain to the east where P-axes are more randomly directed. Data are shown separately for the northeastern U.S. (NEUS) and for the southeastern U.S. (SEUS). Data are from: Bollinger and others, 1985b; Houlday and others, 1984; Gephard and Forsyth, 1985; Ebel and McCaffrey, 1984; Armbruster and Seeber, 1987; Quittmeyer and others, 1985; Talwani, 1982; Liebermann and Thurber, 1986; Herrmann and others, 1982; Johnston and others, 1985; Wahlström, 1987.

along the southeastern Seaboard (Fig. 7). Thus, seismicity prior to 1886 would probably not have identified the source area of the Charleston event as being unique, except for the immediate fore-shocks starting five days before the main shock.

According to the new data from the systematic archival search, the burst of seismicity that followed the 1886 main shock affected a very large area (hundreds of kilometers) and marked the onset of a pattern of seismicity that differed from the previous pattern (Fig. 7) and has persisted to the present. Thus, seismicity in the southeastern U.S. is not stationary at the time scale of the historic period. A change in the pattern of seismicity has also been recognized in the Cape Ann area of the northeastern U.S. which had been significantly more active in the eighteenth century than in the subsequent period (e.g., Veneziano and Van Dyke, 1986). These results are consistent with the recent discovery that the Meers fault in Oklahoma slipped by several meters during the Holocene, even though very little, if any, seismic energy release can be ascribed to this fault during the historic period (Brocoum and others, 1986).

The hypothesis that large-scale patterns of seismicity may vary drastically over periods comparable to the historic record is

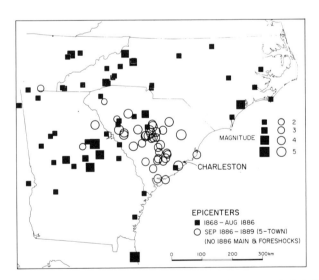

Figure 7. Seismicity before (filled squares) and after (open circles) the August 31, 1886 Charleston, South Carolina, earthquake. Epicenters and magnitudes of these preinstrumental earthquakes were obtained from felt reports in contemporary newspapers. The seismicity after the main shock is represented by the larger aftershocks (i.e., reported by at least 5 towns; Seeber and Armbruster, 1987). They indicate a widespread aftershock zone measured in hundreds of kilometers. Except for immediate fore-shocks, the level of seismicity prior to the main shock in the epicentral zone near Charleston was low and similar to other areas of the southeastern Coastal Plain.

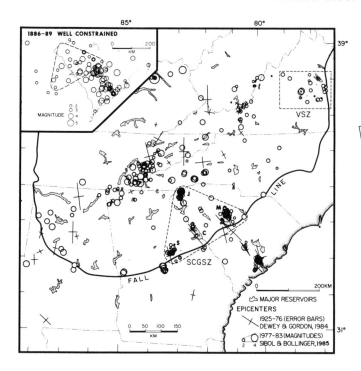

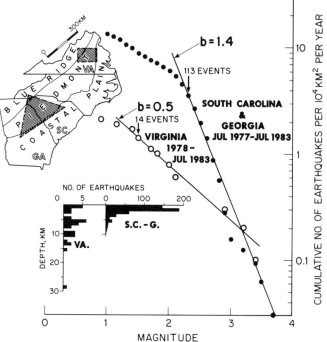

Figure 8. Instrumental epicenters and reservoirs in the southeastern U.S. Induced seismicity has been documented only within the dashed triangular area where clusters of epicenters are spatially and temporally related to reservoirs (J, Jocassee; C, Clark Hill; M, Monticello; S, Sinclair). This peculiar distribution of induced seismicity cannot be accounted for by the distribution of reservoirs. The triangular area of induced seismicity corresponds to the portion of the South Carolina–Georgia Seismic Zone in the Piedmont and is also the locus of 1886 to 89 aftershock activity (inset). From Seeber and Armbruster, 1987.

Figure 9. Size and depth distributions of instrumental hypocenters in the two prominent active zones in the southeastern Piedmont: the Virginia Seismic Zone and the South Carolina–Georgia Seismic Zone (data from Sibol and Bollinger, 1985). These two zones differ markedly in both depth and size distribution. From Seeber and Armbruster, 1987.

crucial for the correct interpretation of earthquake data in the estimate of hazard. Further tests of the hypothesis in the eastern U.S. seem unlikely given the relatively low rate of seismicity and the short time of data coverage in this area. Comparative analysis of intraplate seismicity worldwide may offer the best opportunity for studying large changes of seismicity patterns (Coppersmith and others, 1986; Dewey, 1983).

Induced seismicity and the 1886 to 1889 aftershock zone: Which is the chicken and which is the egg?

During this century, seismicity in the southern Piedmont has been dominated by the South Carolina–Georgia Seismic Zone (SCGSZ) and by the Virginia Seismic Zone (VSZ; Fig. 8). The seismicity in the VSZ is midcrustal and characterized by a low ratio of small to large events; in contrast, the seismicity in the SCGSZ is shallow and characterized by a high ratio of small to large events (Fig. 9). The SCGSZ includes the earthquake-inducing Jocassee, Monticello, Clark Hill, and Sinclair reservoirs and is the only area in the southeastern U.S. where reservoir-

induced seismicity has been unequivocally recognized (e.g., Talwani, 1981). In fact, current seismicity in that area is considered to be for the most part induced (e.g., Sibol and Bollinger, 1985). This peculiar distribution of induced seismicity begs an explanation since reservoirs within the triangular area of Figure 8 resemble, in basic characteristics, reservoirs elsewhere in the Piedmont.

The SCGSZ mimics the 1886 to 1889 aftershock zone (Fig. 8) in spatial distribution with approximately the level of seismicity reached by the aftershock sequence three years after the main shock (Seeber and Armbruster, 1987). Both aftershocks and SCGSZ occupy a triangular portion of the Piedmont approximately 300 km along and 200 km across strike of the Appalachians. The 1886 to 1889 aftershock zone then, is still characterized by a mechanical condition peculiar to that zone and currently manifested by seismic response to reservoir loading. The remarkable spatial correlation between 1886 to 1889 aftershocks, and recent induced seismicity suggests a common cause for these two phenomena.

The correlation between the aftershock zone and induced seismicity may indicate that the perturbation in the state of stress (and/or strength) that caused the burst of aftershocks in the 1880s has persisted to the present and is still manifested locally by

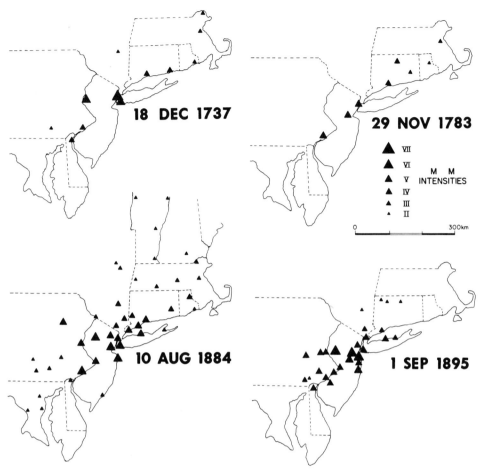

Figure 10. The four largest known earthquakes in the New York City Seismic Zone. These earthquakes are similar in size (magnitudes 4.0–5.0), but are probably from distinct sources.

near-failure conditions. Alternatively, both the 1886 to 1889 aftershocks and the recent induced seismicity in South Carolina and Georgia reflect a peculiar area of the Piedmont where the shallow crust is permanently near failure and can be activated seismically by slight changes in mechanical conditions. In this case, this sensitive area in the Piedmont would probably be distinguishable by other peculiar geological and geophysical characteristics. None have so far been proposed.

SEISMIC ZONES IN THE NEW YORK CITY AREA AND IN LANCASTER COUNTY, PENNSYLVANIA: THE CASE FOR CHARACTERISTIC EARTHQUAKES?

Seismic sources with characteristic "maximum" earthquakes have been postulated and observed along plate boundary zones, such as Japan and California, where earthquake sources are commonly associated with known faults. The "characteristic" size of these earthquakes is determined by the dimensions of the

faults, or portions thereof, that can rupture in a single earthquake (e.g., Coppersmith and others, 1980; Wesnousky and Scholz, 1982). The concepts of characteristic earthquake and seismic gap are fundamental for hazard analysis along plate boundaries.

Lower rates of seismic-energy release and lack of knowledge on the properties of active faults in intraplate regions make it difficult to test whether the concept of characteristic earthquake applies to these regions. The Atlantic Seaboard is no exception. Some of the well-known seismic zones in this province, however, have produced similar earthquakes, which are maximum for the historic period. The 1737, 1783, and 1884 earthquakes in the New York City Seismic Zone, for example, appear to be similar in size to estimated magnitudes 4.5 to 5.0. The 1895 earthquake was probably smaller (Fig. 10).

An ongoing search of newspapers in the northeastern U.S. has identified another persistent zone of seismicity in Lancaster County, Pennsylvania (Armbruster and Seeber, 1987). The new preinstrumental data indicate that four earthquakes of similar size ($M \approx 4$) have occurred in this zone since 1800 and are the largest

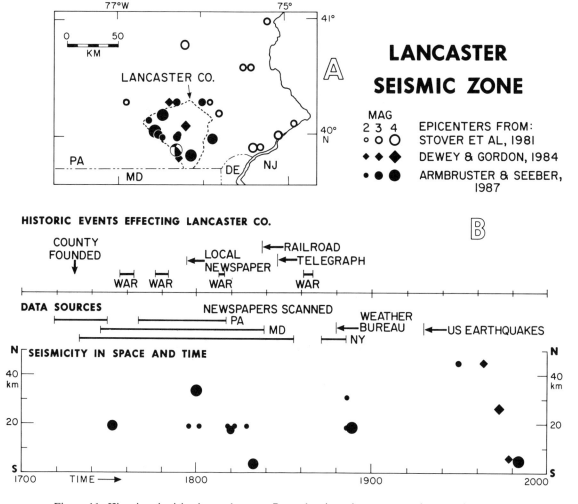

Figure 11. Historic seismicity in southeastern Pennsylvania: epicenter map (A); and time-latitude diagram (B). Most epicenters before 1960 are from intensity data. A recent search of felt reports from newspapers (coverage shown in B) has revealed a persistent seismic zone in Lancaster County and little seismicity in the rest of Pennsylvania. Several events in the historic period, including the April 23, 1984 Martic earthquake, have magnitudes nearly 4, the maximum size of known events in the Lancaster Seismic Zone. The tectonic setting of this zone is shown in Figure 2. From Armbruster and Seeber, 1987.

known for eastern Pennsylvania (Fig. 11). The last one of these, the April 23, 1984, Martic earthquake, has a well-defined aftershock zone about 4.5 km deep and adjacent to the Martic Line (Fig. 2), a Paleozoic suture that appears to bound the Lancaster seismic zone on the south.

In both the New York City and in the Lancaster seismic zones the "maximum" events seem to occur in a small area when viewed in the scale of the eastern Seaboard, but at least some of these occur at distinct locations and are probably from different sources (Armbruster and Seeber, 1987). Thus, the concept of characteristic earthquakes, if at all applicable in the eastern U.S., might apply to zones where seismogenic faults have similar char-

acteristics and tend to produce earthquakes of similar size, in contrast to the type case where a single fault, or a portion of a fault, slips in similar characteristic ruptures.

The concept of characteristic earthquakes applied to seismic zones that have been persistently active during the historic period is in apparent contradictions to the hypothesis that patterns of seismicity may change drastically during the historic period, as suggested by the seismicity in South Carolina (see above). Both concepts may be applicable, but at different temporal and spatial scales. The most satisfactory tests for the general applicability of either of these hypotheses would be to find physical bases for them in the intraplate earthquake generating process.

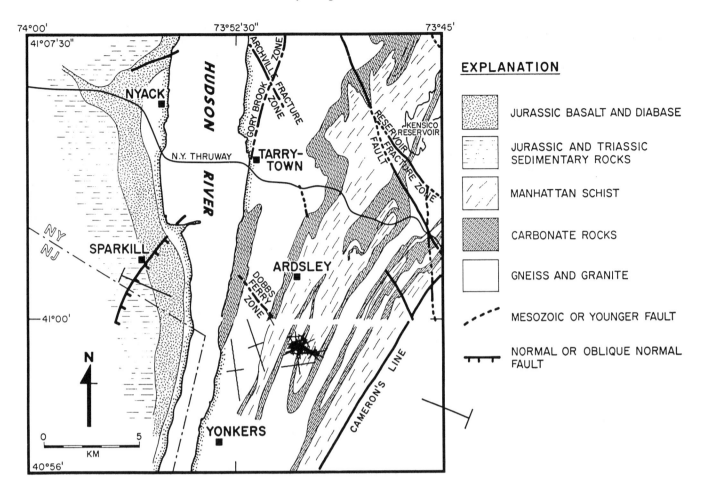

Figure 12. Brittle faults and rock types in the area of southeastern New York and northeastern New Jersey (after Ratcliffe, 1980, 1981; Hall, 1981). This area can be seen in a broader structural context in Figure 2. The Manhattan Prong between the Hudson River and Cameron's Line contains Precambrian North American basement (Fordham Gneiss), which is unconformably overlayed by Lower Paleozoic metasediments. Penetrative Paleozoic multiphase deformation of these high-grade rocks is dominated by thrusting with vergence to the west. Cameron's Line is an Early Paleozoic suture. West-dipping Triassic sediments of the Newark Basin and the Palisades mafic sill outcrop west of the Hudson River. Several steeply dipping brittle faults characterized by both normal and strike-slip displacements have been recognized in the area. Most of these displacements are thought to be Mesozoic, but later displacements are possible. Epicenters are indicated by HYPOINVERSE errorbars. The three epicenters closest to the Hudson River are earthquakes between 1976 and 1980. The tight cluster of epicenters 3-4 km south of Ardsley are from a swarm that started January 1985 and was still active January 1987. The Mblg = 4.0 event on October 19, 1986 has been the largest so far in that swarm. The main fault active in this swarm strikes west-northwest and slipped left-laterally (see Fig. 13). The Dobbs Ferry fault zone is close to the surface extrapolation of the active fault and may be associated with it.

THE 1985 TO 1986 EARTHQUAKE SEQUENCE IN WESTCHESTER COUNTY, NEW YORK

On January 1985, a small (M ~ 2) earthquake was felt near the town of Ardsley in southern Westchester County. This was the first in a series of earthquakes that included the Mblg = 4.0 event on October 19, 1985, which was widely felt from eastern Pennsylvania to western Massachusetts. So far (January 1987), this sequence includes about 28 detected earthquakes. Eleven of

these were felt (M >1), the others were very small events (M <0). The most recent event was on January 13, 1987, with no indication yet that this sequence is terminated.

The Ardsley earthquake sequence (Fig. 12), has been closely monitored, and its source parameters are particularly well constrained. The source of this seismicity is small, about 1 km in the largest dimension, and is centered at a depth of 4.5 km (Fig. 13). The hypocenter zone includes several very small aftershocks of the M = 4 earthquake in October 1985, when a number of

SOUTHERN WESTCHESTER CO. SEQUENCE, 1985

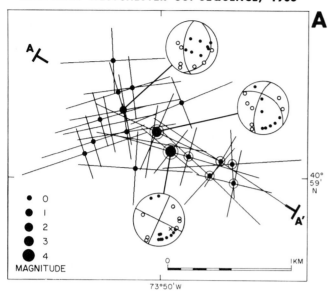

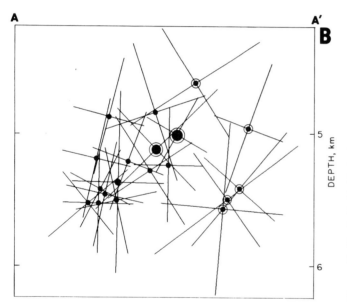

high-gain temporary seismic stations were operated near the epicenter. The aftershocks during the first week after the main shock, were confined to the outer edge of a vertical zone striking westnorthwest (Fig. 13). The planar surface defined by the early aftershocks coincides with the plane of left-lateral slip in the fault-plane solution and may represent the rupture in the main shock. The tendency for aftershocks to be concentrated at the outer edges of the rupture has been noticed in other closely monitored sequences in the eastern U.S. (e.g., Seeber and Coles, 1984).

The surface extrapolation of the inferred 19 October 1985 rupture approximately coincides with the Dobbs Ferry zone, one of the few brittle faults recognized in the Manhattan Prong (Hall, 1981; Fig. 12). This fault is in a family of northwest brittle faults, presumably of Mesozoic age, which are antithetic to the Appalachian trend and to the border faults controlling the Newark Basin. The 125th Street fault in Manhattan is another brittle fault in this family and has a complex history involving both normal and strike-slip displacement (Merguerian, 1986).

Distinct hypocenters can be resolved for about half the earthquakes in the Ardsley sequence, including most of the small ones. They depict a complex source about 1 km across with at least two active faults (Fig. 13). Most of the other events in this sequence produced nearly identical seismograms, except for scale, and are part of a multiplet (Fig. 14). Thus, they originate from the same small volume, not more than a few hundred meters across, they have similar slip geometries, and probably represent slip on the same fault. Surprisingly, this multiplet comprises the larger earthquakes in the sequence, ranging in size from M = 3.0 to M = 0.8. The waveform of the largest event (M = 4.0) can be constructed by several smaller events with the same geometry of slip as events in the multiplet (Carter and others, 1986). The Ardsley multiplet has persisted 1.5 years, including both the first and the most recent event in the sequence, and it is considered still active in January 1987. The significance of this observation is unclear. One possible interpretation is that protracted aseismic slip on the active fault around the seismogenic zone keeps reloading the same asperity.

SUMMARY

Seismicity along the Atlantic Seaboard is moderately high for an intraplate region and includes large damaging earthquakes. Although the rate of moment release in this and other areas of the eastern U.S. is low when compared to the moment release along plate boundaries, this rate seems high when contrasted with the low rate of intraplate deformation inferred from geologic data. In one of the best-documented examples, several large earthquakes seem to have occurred during the Holocene in the Coastal Plain of South Carolina, but structural relief on a Jurassic surface in the same area is less than a few tens of meters. Where faults are found to displace Cenozoic sediments of the Coastal Plain, inferred cumulative rates tend to be less than 1 m per million years (Wentworth and Mergner-Keefer, 1983).

Figure 13. Detailed views of the complex source zone that generated the 1985 to 1986 swarm south of Ardsley in southern Westchester (see Fig. 12). A is a map view. B is a sectional view from the south-southwest (located in A). Data were obtained in part from a network of portable seismographs. Hypocenters (circled) of aftershocks during the first week after the M = 4 event on October 19, 1985, define a vertical zone about ¾ km across that coincides with the left-lateral plane in the fault-plane solution. This zone probably represents the rupture in the main shock. The fault-plane solution indicating reverse faulting pertains to a M = 1.6 event on October 26, 1985. Eight events with magnitudes 0.3 < M < 3.0 and spanning the entire sequence, have almost identical waveforms at several stations (see Fig. 14) and must share almost identical hypocenters and faulting geometry. The largest in this multiplet, the M = 3 event on October 21, 1985, is the only one represented in this figure. The fault-plane solution for this multiplet is similar to the one for the main shock.

STATION : AMNH
Z – COMP

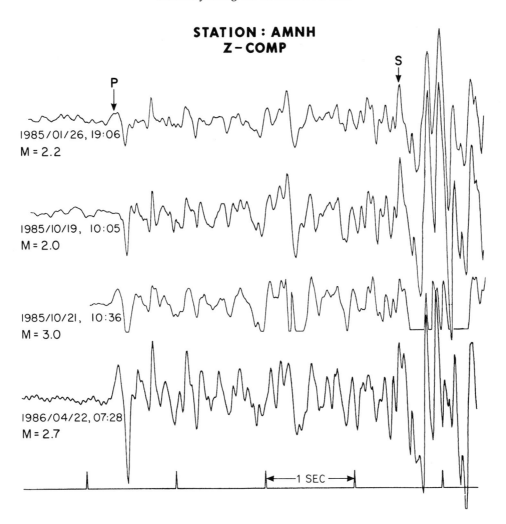

Figure 14. Vertical component seismograms for four earthquakes in the southern Westchester swarm recorded at the Museum of Natural History in Manhattan (station AMNH), about 10 km south of the source. These seismograms are very similar, particularly after allowing for some noise and distortion. Most wiggles can be recognized in each seismogram. These four events are part of a multiplet that comprises 8 of the 11 events in the Westchester swarm with magnitudes M > 0. This multiplet has been active for over a year. The waveforms of the main shock can be modeled by superposing several closely spaced events of this multiplet. Multiplets are symptomatic of sources located within a fraction of a wavelength of the resolvable frequencies, in this case about 100 to 200 m, and are characterized by uniform geometry of rupture.

The apparent contradiction between earthquake and geologic data in intraplate eastern U.S., however, stems from a comparison with seismicity and strain rates in interplate western U.S. While seismic moment release along the San Andreas plate boundary is concentrated on few major faults, in the East the long-term moment release may be distributed on many faults (Anderson, 1986). Moreover, earthquakes with similar macroseismic effects have generally smaller moment release in the East than in the West.

Estimates of fault ruptures of recent earthquakes in the East were obtained from detailed resolution of aftershock zones. These ruptures tend to be small—1 to 2 km across for M ≈ 5 earthquakes—and may be characterized by high stress-drops. Thus, damaging earthquakes in the East can be generated by small faults that do not reach the surface and that geological and geophysical investigations may not detect. A low rate of seismic moment release in the East, then, may be associated with a high rate of damaging intensities, and techniques for earthquake hazard analysis developed for the West may not be effective in the East. However, the contribution of these effects to the apparent discrepancy between high seismicity and little deformation in the East has yet to be evaluated.

The Appalachian Front, defined as the northwestern limit of the allochthonous crystalline slab above the master detachment, appears to control the spatial distribution of seismicity on a large scale. Hypocenters are concentrated in the autochthonous basement below the sedimentary wedge and in the allochthonous crystalline slab above this wedge. The sedimentary wedge and the detachment are relatively aseismic. Stress indicators are ambiguous about a distinct stress province along the Appalachians/Atlantic Seaboard. If this stress province exists, the Appalachian Front and the master detachment are likely boundaries of this province.

At an intermediate scale, seismicity is also often controlled by preexisting structure. The Newark Basin, for example, is itself aseismic while seismicity is concentrated around the border of this basin. Although a similar correlation between Mesozoic basins and seismicity has been noticed elsewhere, this correlation is not universal, since many of the other basins are not associated with seismicity. The apparent lack of a one-to-one correlation between seismicity and preexisting structural features may be the result of changes in the patterns of seismicity that occur over periods longer than the available historic record. Seismicity might be more uniform if it could be integrated over a time longer than the historic record in the U.S.

Similar patterns of seismicity along the Atlantic Seaboard are seen in the historic data, which are dominated by felt events, and in recent instrumental data, which include much smaller earthquakes. In both time and magnitude distribution, then, the pattern of seismicity appears to be stationary through the historic period, for the most part. Significant changes, however, have been identified. The most prominent coincided with the 1886 Charleston, South Carolina, earthquake. For almost a century before this event, seismicity in its epicentral area was low, similar to the level in other portions of the Coastal Plain. After this event, the South Carolina–Georgia Seismic Zone became active. This zone was manifested, first by an intense three-year-long burst of aftershocks, then by a persistent seismicity, which has been primarily reservoir-induced in the last few decades. Although an understanding of tectonic processes that could cause coherent changes of seismicity over large areas is still lacking, nonstationarity in the seismic regime needs to be incorporated into realistic estimates of earthquake hazard in the eastern U.S.

REFERENCES CITED

Aggarwal, Y. P., and Sykes, L. R., 1978, Earthquakes, faults, and nuclear power plants in southern New York and northern New Jersey: Science, v. 200, p. 425–429.

Alexander, S. S., and Lavin, P. M., 1982, Further evidence on the block-tectonic structural fabric of the eastern United States and its present significance [abs.], Earthquake Notes, v. 53, no. 3, p. 11.

Anderson, J. G., 1986, Seismic strain rates in the central and eastern United States: Seismological Society of America Bulletin, v. 76, p. 273–290.

Armbruster, J. G., and Seeber, L., 1981, Intraplate seismicity in the southeastern United States and the Appalachian detachment, in Earthquakes and earthquake engineering; The eastern United States: Ann Arbor, Michigan, Ann Arbor Science Publications, p. 375–396.

—— , 1984, Low seismicity in South Carolina prior to the 1886 earthquake [abs.]: EOS, v. 65, p. 241.

—— , 1987, The April 23, 1984, Martic earthquake and the Lancaster seismic zone in eastern Pennsylvania: Seismological Society of America Bulletin, v. 77, p. 877–890.

Barosh, P. J., 1981, Cause of seismicity in the eastern United States; A preliminary appraisal, in Beavers, J. E., ed., Earthquakes and earthquake engineering; The eastern United States: Ann Arbor, Michigan, Ann Arbor Science Publications, v. 1, p. 397–417.

Barstow, N. L., Brill, K. G., Jr., Nuttli, O. W., and Pomeroy, P. W., 1981, An approach to seismic zonation for siting nuclear electric power generating facilities in the eastern United States: Washington, D.C., U.S. Nuclear Regulatory Commission Report NUREG CR1577, 143 p.

Basham, P. W., and Adams, J., 1983, Earthquakes on the continental margin of eastern Canada; Would future large events be confined to the locations of large historical events?: U.S. Geological Survey Open-File Report 83-843, p. 456–467.

Behrendt, J. C., Hamilton, R. M., Ackerman, H. D., and Henry, V. J., 1981, Cenozoic faulting in the vicinity of the Charleston, South Carolina, 1886 earthquake: Geology, v. 9, p. 117–122.

Behrendt, J. C., Hamilton, R. M., Ackermann, H. D., Henry, V. J., and Bayer, K. C., 1983, Land multichannel seismic-reflection evidence for tectonic fea-

tures near Charleston, South Carolina: U.S. Geological Survey Professional Paper 1313-J, 29 p.

Bollinger, G. A., 1977, Reinterpretation of the intensity data for the 1886 Charleston, South Carolina, earthquake: U.S. Geological Survey Professional Paper 1028, p. 27–32.

Bollinger, G. A., Chapman, M. C., Sibol, M. S., and Costain, J. K., 1985a, An analysis of earthquake focal depths in the southeastern U.S.: Geophysical Research Letters, v. 12, p. 785–788.

Bollinger, G. A., Teague, A. G., Monsey, J. W., and Johnston, A. C., 1985b, Focal mechanism analyses for Virginia and eastern Tennessee earthquakes (1978–1984): U.S. Nuclear Regulatory Commission Report NUREG CR4288, 83 p.

Boyer, S. E., and Elliott, D., 1982, Thrust systems: American Association of Petroleum Geologists Bulletin, v. 66, p. 1196–1230.

Brocoum, S. J., Slemmons, D. B., and Ramelli, A. R., 1986, The Meers fault, Oklahoma; Its implication to our understanding of eastern U.S. seismicity [abs.]: Earthquake Notes, v. 57, no. 1, p. 7.

Brown, L., Ando, C., Klemperer, S., Oliver, J., Kaufman, S., Czuchra, B., Walsh, T., and Isachsen, Y. W., 1983, Adirondack–Appalachian crustal structure; The COCORP Northeast Transverse: Geological Society of America Bulletin, v. 94, p. 1173–1184.

Campbell, D. L., 1978, Investigation of the stress-concentration mechanism for intraplate earthquakes: Geophysical Research Letters, v. 5, p. 477–479.

Carter, J. A., Barstow, N., and Sutton, G. H., 1986, Complex faulting characteristics of the 19 October 1985 Ardsley, N.Y., earthquake [abs.]: EOS, v. 67, p. 1111.

Chiburis, E. F., 1981, Seismicity, recurrence rates, and regionalization of the northeastern U.S. and adjacent Canada: U.S. Nuclear Regulatory Commission Report NUREG/CR-2039, 76 p.

Coffman, J. L., and von Hake, C. A., eds., 1973, Earthquake history of the United States: Washington, D.C., U.S. Department of Commerce, p. 208.

Cook, F. A., Brown, L. D., Kaufman, S., Oliver, J. E., and Petersen, T. A., 1981, COCORP seismic profiling of the Appalachian orogen beneath the crustal plane of Georgia: Geological Society of America Bulletin, v. 92, p. 738–748.

Coppersmith, K. J., Cluff, L. S., and Patnardhan, A., 1980, Estimating the probability of occurrence of surface faulting earthquakes on the Wasatch fault zone, Utah: Seismological Society of America Bulletin, v. 70, p. 1431–1462.

Coppersmith, K. J., Johnson, A. C., and Arabasz, W. J., 1986, Assessment of maximum earthquake magnitudes in the eastern United States [abs.]: Earthquake Notes, v. 57, no. 1, p. 26.

Daniels, D. L., Zeitz, I., and Popenoe, P., 1983, Distribution of subsurface lower Mesozoic rocks in the southeastern United States, as interpreted from regional aeromagnetic and gravity maps: U.S. Geological Survey Professional Paper 1313K, 24 p.

Devine, J. F., Nov. 1982, Letter to R. E. Jackson, U.S. Nuclear Regulatory Commission, Washington, D.C.

Dewey, J. W., 1983, A global search of continental midplate seismic regions for specific characteristics bearing on the 1886 Charleston, South Carolina, earthquake, in A workshop on the 1886 Charleston, South Carolina, earthquake and its implications for today: U.S. Geological Survey Open-File Report 83-843, p. 391–426.

Dewey, J. W., and Gordon, D. W., 1984, Map showing recomputed hypocenters of earthquakes in the eastern and central United States and Canada, 1925–1980: U.S. Geological Survey Miscellaneous Field Studies Map, MF1699, scale 1:2,500,000.

Ebel, J. E., and McCaffrey, J. P., 1984, Hypocentral parameters and focal mechanisms of the 1983 earthquake near Dixfield, Maine: Earthquake Notes, v. 55, no. 2, p. 21.

Evernden, J. F., 1975, Seismic intensities, "size" of earthquakes, and related parameters: Seismological Society of America Bulletin, v. 65, p. 1287–1313.

Gephard, J. W., and Forsyth, D. W., 1985, On the state of stress in New England as determined from earthquake focal mechanisms: Geology, v. 13, p. 70–72.

Gohn, G. S., ed., 1983, Studies related to the Charleston, South Carolina, earthquake of 1886; Tectonics and seismicity: U.S. Geological Survey Professional Paper 1313, 375 p.

Hall, L. M., 1981, Reconnaissance map of fracturing in the White Plains Quadrangle, scale 1:24,000. (Unpublished).

Hamilton, R. M., Behrendt, J. C., and Ackerman, H. D., 1983, Land multichannel seismic-reflection evidence for tectonic features near Charleston, South Carolina: U.S. Geological Survey Professional Paper 1313I, 18 p.

Herrmann, R. B., Langston, C. A., and Zollweg, J. E., 1982, The Sharpsburg, Kentucky, earthquake of 27 July, 1980: Seismological Society of America Bulletin, v. 72, p. 1219–1239.

Houlday, J., Quittmeyer, R., Mrotek, K., and Statton, C. T., 1984, Recent seismicity in north and east-central New York State: Earthquake Notes, v. 55, no. 2, p. 16–20.

Hutchinson, D. R., and Grow, J. A., 1985, New York Bight fault: Geological Society of America Bulletin, v. 96, p. 975–989.

Johnston, A. C., Reinbold, D. J., and Brewer, S. I., 1985, Seismotectonics of the southern Appalachians: Seismological Society of America Bulletin, v. 75, p. 291–312.

Klitgord, K. D., Dillon, W. P., and Popenoe, P., 1983, Mesozoic tectonics of the southeastern U.S. Coastal Plain and continental margin: U.S. Geological Survey Professional Paper 1313P, 15 p.

Liebermann, R. C., and Thurber, C. H., 1986, Stony Brook seismic network on Long Island, New York; Operation and maintenance: U.S. Nuclear Regulatory Commission Report, NUREG/CR4580, 101 p.

Long, L. T., 1976, Speculations concerning southeastern earthquakes, mafic intrusions, gravity anomalies, and stress amplification: Earthquake Notes, v. 47, no. 3, p. 29–35.

—— , 1986, The classification and testing of hypotheses for the 1886 Charleston earthquake [abs.]: Earthquake Notes, v. 57, p. 17.

McKeown, F. H., 1978, Hypothesis: Many earthquakes in the central and southeastern United States are causally related to mafic intrusive bodies: U.S. Geological Survey Journal of Research, v. 6, p. 41–50.

Merguerian, C., 1986, The bedrock geology of New York City [abs.], in The geology of southern New York: New York, Hofstra University Press, p. 8.

Nuttli, O. W., 1973, Seismic wave attenuation and magnitude relations for eastern

North America: Journal of Geophysical Research, v. 78, p. 876–885.

—— , 1983, Average seismic source-parameter relations for midplate earthquakes: Seismological Society of America Bulletin, v. 73, p. 519–535.

Obermeier, S. F., Gohn, G. S., Weems, R. E., Galinas, R. L., and Rubin, M., 1985, Geologic evidence for recurrent moderate to large earthquakes near Charleston, South Carolina: Science, v. 227, p. 408–411.

Plumb, R. A., and Cox, J. W., 1987, Stress directions in eastern North America determined to 4.5 km from borehole elongation measurements: Journal Geophysical Research, v. 92, p. 4805–4816.

Quittmeyer, R. C., Statton, C. T., Mrotek, K. A., and Houlday, M., 1985, Possible implications of recent microearthquakes in southern New York State: Earthquake Notes, v. 56, no. 2, p. 35–42.

Rankin, D. W., ed., 1977, Studies related to the Charleston, South Carolina, earthquake of 1886; A preliminary report, U.S. Geological Survey Professional Paper 1028, 204 p.

Ratcliffe, N. N., 1980, Brittle faults (Ramapo fault) and phillonitic ductile shear zones in the basement rock of the Ramapo seismic zone, in Mannspeizer, W., ed., Field study of New York geology—Guide to field trips: 52nd Annual Meeting of New York State Geological Association, p. 278–312.

—— , 1981, Reconnaissance map of fracturing in the Ossining Quadrangle, scale 1:24,000. (Unpublished).

Ratcliffe, N. M., Burton, W. C., D'Angelo, R. M., Costain, J. K., 1986, Low-angle extensional faulting, reactivated mylonites, and seismic reflection geometry of the Newark basin margin in eastern Pennsylvania: Geology, v. 14, p. 766–770.

Rundle, T. A., Singh, M. M., and Baker, C. H., 1985, In situ stress measurements in the earth's crust in the eastern United States: U.S. Nuclear Regulatory Commission Report NUREG/E1-1126, 75 p.

Seeber, L., 1983, Earthquake hazard and risk in the eastern U.S. versus the western U.S. [abs.]: Earthquake Notes, v. 54, no. 3, p. 31.

Seeber, L., and Armbruster, J., 1981, The 1886 Charleston, South Carolina, earthquake and the Appalachian detachment: Journal of Geophysical Research, v. 86, p. 7874–7894.

—— , 1983, Large strain effects of the 1886 South Carolina earthquakes: U.S. Geological Survey Open-File Report 83-843, p. 132–136.

—— , 1985, Spatio-temporal distribution of seismicity along the Appalachians and the case for detachment reactivation [abs.]: Earthquakes Notes, v. 56, no. 3, p. 66.

—— , 1987, The 1886–89 aftershocks of the Charleston, South Carolina, earthquake; A widespread burst of seismicity: Journal of Geophysical Research, v. 92, p. 2663–2696.

Seeber, L., and Coles, K. S., 1984, Seismicity in the central Adirondacks with emphasis on the Goodnow, October 7, 1983, epicentral zone and its geology, in Potter, D. B., ed., New York State Geological Association Field Trip Guidebook, 56th Annual Meeting, Sept. 20–23, 1984, Hamilton, College, Clinton, New York: p. 334–352.

Shakal, A. F., and Toksoz, M. N., 1977, Earthquake hazard in New England: Science, v. 195, p. 171–173.

Sibol, M. S., and Bollinger, G. A., eds., 1984, Hyopocenter listing from southeastern U.S. seismic network: Bulletins 1–12, Blacksburg, Virginia, 44 p.

Sibol, M. S., and Bollinger, G. A., eds., 1985, Hypocenter listing from southeastern U.S. seismic network: Blacksburg, Virginia, Bulletin of Seismicity of the Southeastern United States, no. 1–14, 46 p.

Stover, C. W., Reagor, B. G., and Algermissen, S. T., 1981, Seismicity map of the state of Pennsylvania: U.S. Geological Survey Miscellaneous Field Studies Map MF-1280, scale 1:1,000,000.

Stewart, R. M., 1986, Review of geological, geophysical, and seismological data relevant to seismotectonics of the Charleston, South Carolina, area [abs.]: Earthquake Notes, v. 57, no. 1, p. 16.

Swanson, M. T., 1986, Preexisting fault control for Mesozoic basin formation in eastern North America: Geology, v. 14, p. 419–422.

Sykes, L. R., 1978, Intraplate seismicity, reactivation of preexisting zones of weakness, alkaline magmatism, and other tectonism postdating continental fragmentation: Reviews in Geophysics and Space Physics, v. 16, p. 621–688.

Talwani, P., 1981, Earthquake prediction studies in South Carolina: American Geophysical Union Maurice Ewing Series, v. 4, p. 381–393.

—— , 1982, An internally consistent pattern of seismicity near Charleston, South Carolina: Geology, v. 10, p. 654–658.

Talwani, P., and Cox, J., 1985, Paleoseismic evidence for recurrence of earthquakes near Charleston, South Carolina: Science, v. 229, p. 379–381.

Talwani, P., Hinze, W. J., Barstow, N., Voight, B., 1986, Intersecting ideas; Or the intersection model for intraplate earthquakes [abs.]: Earthquake Notes, v. 57, no. 1, p. 13.

Veneziano, D., and Van Dyke, J., 1986, Spatial variation of seismicity; Modeling, estimation, and application to the eastern United States [abs.]: Earthquake Notes, v. 57, no. 1, p. 26.

Wahlström, Rutger, 1987, Focal mechanisms of earthquakes in southern Quebec, southeastern Ontario, and southeastern New York with implications for regional seismotectonics and stress field characteristics: Seismological Society of America Bulletin, v. 77, p. 891–924.

Wentworth, C. M., and Mergner-Keefer, M., 1983, Regenerate faults of small Cenozoic offset; Probable earthquake sources in the southeastern United States: U.S. Geological Survey Professional Paper 1313S, p. 20.

Wesnousky, S. G., and Scholz, C. H., 1982, Deformation of an island arc; Rates of moment release and crustal shortening in intraplate Japan determined from seismicity and quaternary fault data: Journal of Geophysical Research, v. 87, p. 6829–6852.

Wheeler, R. L., and Bollinger, G. A., 1984, Seismicity and suspect terranes in the southeastern U.S.: Geology, v. 12, p. 323–326.

Williams, H., and Hatcher, R. D., Jr., 1982, Suspect terranes and accretionary history of the Appalachian orogen: Geology, v. 10, p. 530–536.

Yang, J. P., and Aggerwal, Y. P., 1981, Seismotectonics of northeastern United States and adjacent Canada: Journal of Geophysical Research, v. 86, p. 4981–4998.

Zoback, M., Zoback, M. D., and Schiltz, M. E., 1984, Index of stress data for the North America and parts of the Pacific plate: U.S. Geological Survey Open-File Report 84-157, 62 pp.

Zoback, M. D., and Zoback, M. L., 1986, In-situ stress, lower-crustal strain, and intraplate seismicity [abs.]: Earthquake Notes, v. 57, no. 1, p. 13.

MANUSCRIPT ACCEPTED BY THE SOCIETY MARCH 31, 1987

ACKNOWLEDGMENTS

We are grateful to David Simpson, Klaus Jacob, and Paul Pomeroy for a helpful review of the manuscript. Kazuko Nagao drafted the figures. This work was supported by Nuclear Regulatory Commission grant 04-85-113-02, NSF grant 83-16589, and U.S. Geological Survey grant 14-08-0001-A0261. Lamont-Doherty Geological Observatory contribution number 4103.

The Geology of North America
Vol. I-2, The Atlantic Continental Margin: U.S.
The Geological Society of America, 1988

Chapter 31

Waste disposal in the Atlantic continental margin

Harold D. Palmer
FEMTO Associates, 6436 Bannockburn Drive, Bethesda, Maryland 20817

INTRODUCTION

Perhaps no other endeavor undertaken by geologists presents the challenge that accompanies analyses of the fate and effects of wastes discharged into the terrestrial and marine environments. Not only are geological disciplines called into play, but social, political, international, and emotional factors impinge upon the assessment of disposal options and the decisions leading to site selection and disposal operations. There is justifiable concern surrounding the assessment of waste disposal practices. As a recent report by the National Research Council (1986, p. 40) states:

Anthropogenic changes in the planetary environment are a unique feature in that they may well exceed the limits of natural regulation. There are many indications that human-induced changes are substantive enough to affect the survival of other organisms, both directly and indirectly, and to pose a threat to mankind itself.

Within the Atlantic continental margin, a wide variety of waste-disposal methods are practiced by the industrial, municipal, state, and federal communities. The complexity in dealing with waste-management practices is not confined to scientific uncertainty regarding the behavior of wastes in alluvium (soil), rock, the atmosphere, or nearshore to abyssal sea-floor substrates and water masses, but to a large degree it may be attributed to conflicting regulations, policies, permit requirements, and licenses—all controlled by authorities whose jurisdictions may overlap.

Terminology is in part to blame for much of the confusion surrounding waste-disposal practice. The very word "disposal" is misused when considering the first definition of the term: ". . . an orderly or systematic placement, distribution, or arrangement . . .," whereas the fourth meaning is stated as " . . . a discarding or throwing away" (Grove, 1971, p. 654). As this author has stated previously (Hard and Palmer, 1976), we cannot consider burial in landfills on the coastal plain or ocean dumping as "throwing away" because the waste materials remain, in some form, within the soil, groundwater, marine and estuarine sediments, biota, and/or waters. Throwing away implies no return, but our growing knowledge of global cycles for various elements suggests that "disposed" matter is merely stored in some fashion, and ultimately available for reintroduction into terrestrial systems. The

majority of waste "disposal" practices are neither orderly nor systematic. They may comply with regulations and thus conform to a system that designates a location for disposal, but in the case of ocean dumping in particular, techniques in use are far from orderly, as solids and liquids discharged into the marine environment are immediately affected by oceanographic agents and processes, many of which are still poorly understood. The marine environment poses significant problems in waste assessment not encountered in land-based disposal options. For centuries the "out-of-sight, out-of-mind" rationale pervaded concepts of disposal in the ocean. Concern over the number of nations using the ocean as a dumping ground led to a "Convention on the Prevention of Marine Pollution by Dumping of Wastes and Other Matter," which met in London in 1972. The policies proposed and adopted at this historic meeting represent the Articles of the "London Dumping Convention" (LDC) to which the United States became a signatory. Within these articles lies a definition of ocean dumping (IMCO, 1976):

Article III.1.(a) 'Dumping means'; (1) any deliberate disposal at sea of wastes or other matter from vessels, aircraft, platforms or other man-made structures at sea.

The use of the word "disposal" again creates uncertainty in the definition. For example, some nations have argued that emplacement of high-level radioactive wastes by embedding them in abyssal sediments is not "dumping" and hence outside the context of the articles. The resolution of many domestic and international questions regarding the propriety of the LDC terms awaits clarification, and the dialogue continues between those nations engaged in, or contemplating, the release of waste materials into the marine environment.

The role of the geologist in matters pertaining to the management of wastes is crucial in the assessment of fate and effects predicted for specific disposal practices. Multidisciplinary applications of the earth sciences are essential in the study of groundwater hydrology, petrology, marine geology, sedimentology, geophysics, geochemistry, and other aspects of waste management often assumed by the engineering fraternity. Geologists have a clear obligation to participate in the decision-making processes

Palmer, H. D., 1988, Waste disposal in the Atlantic continental margin, *in* Sheridan, R. E., and Grow, J. A., eds., The Atlantic Continental Margin, U.S.: Geological Society of America, The Geology of North America, v. I-2.

that determine waste-management policy since their perspectives of rates and scales of change exceed short-term cause, effect, and remedy scenarios often sought as expedients in dealing with specific situations.

THE COASTAL PLAIN

The eastern coast of the United States is one of the most densely populated areas of North America. The major urban centers and outlying suburbs within the coastal plain account for perhaps 35 million persons within the 163 coastal counties identified in a recent survey undertaken by the National Oceanic and Atmospheric Administration (Basta and others, 1982). Under the Coastal Zone Management Act of 1972, (Public Law 92-583), coastal counties are included in shorelands that strongly influence, and may have a direct and significant impact on, coastal waters. These East Coast counties contain 52 percent of the nation's population residing in the "coastal zone."

The result of such concentrations of industry, commerce, urban, and suburban endeavors places enormous demands on waste-disposal practices. For example, metropolitan New York City processes 1.5 billion gallons of water in daily activities and discharges this amount as a waste product into adjacent estuaries and the nearshore marine environment bordering this megalopolis (Squires, 1983). Whether or not the content of such wastes, and those deposited or stored in landfills in the coastal plain, in storage tanks, impounded ponds, injection wells, or other methods of waste management is classified as a "hazardous waste" is determined by the Resource Conservation and Recovery Act (RCRA) of 1976. Later, the passage of the Comprehensive Environmental, Response, Compensation, and Liability Act of 1980 (CERCLA) clarified jurisdictions and policies for many older sites containing toxic wastes—an act that is now referred to as "Superfund."

In a 1983 report to the Congress of the United States, the Office of Technology Assessment (OTA, 1983) notes that from 255 to 275 million metric tons of hazardous waste—as defined by either Federal or State criteria—were generated annually. In the general definition adopted by the OTA, "hazardous waste" means a solid waste (or combination of wastes) that may cause mortality, illness, or present a substantial present or potential hazard to human health or the environment when improperly treated, stored, transported, or disposed of or otherwise managed.

In the Atlantic coastal states, roughly 35 percent of such wastes occupy landfills, 22 percent are in storage containers, 13 percent lie in surface impoundments, and the remaining third are dealt with through other means as noted above. In a study conducted by the Environmental Protection Agency (EPA, 1984), approximately 6,300 hazardous waste generators were identified in the four EPA regions bordering the Atlantic Ocean. Of these, 40 percent had reported some form of contamination of at least one medium (soil, air, groundwater, or surface water) resulting from disposal practice.

An analysis of water uses in the United States as of 1980

(Solley and others, 1986) reveals a 15 percent increase in consumption of water between 1975 and 1980. The Atlantic Coastal Plain is heavily dependent upon groundwater, which accounts for perhaps one-third of the public supply. This is water withdrawn from surface and groundwater sources for domestic, public, and industrial use. Contamination of aquifers is a major concern where injection wells are either in use or contemplated. Brown and others (1976) have identified three major subsurface waste-storage reservoirs in the northern Atlantic Coastal Plain that have porosity and permeability attributes meeting proposed disposal criteria—thickness of 6 m or greater, an impermeable layer of at least equal thickness bordering the upper and lower contacts, and a depth below mean sea level of at least 450 m.

One study of waste impact assessment may serve as an example of the application of geological, hydrological, and geochemical disciplines to real situations that threaten residents of coastal plain counties. The Army Creek landfill in New Castle County, Delaware, covers approximately 24 hectares and contains more than 1.5 million m^3 of solid and liquid industrial wastes and municipal refuse. A detailed study of the site by Baedecker and Apgar (1984) identified the paths of leachate migration downgradient from the site, and the attempts to intercept and remove contaminated groundwater through recovery wells, which extract the anerobic waters tainted with gasses, metals, and refractory organic compounds. Unfortunately, continued pumping of these recovery wells has resulted in a lowered head in the underlying aquifer and significant waste of uncontaminated groundwater. In addition, the reversal of the gradient has led to the threat of saltwater intrusion in the eastern portion of the aquifer.

Saltwater incursion of coastal aquifers is a major concern as the rate of groundwater withdrawal increases in all coastal states. Meisler (1986) examines the vertical and horizontal distribution of fresh, brackish, and saltwater within northern Atlantic Coastal Plain aquifers and the effects of fluctuations in relative sea level and concludes that as a result of the most recent rise in sea level, saltwater is slowly encroaching landward at a rate of perhaps 1 m per 35 years. This rate is greatly accelerated when significant withdrawals of groundwater reverse the natural head within permeable strata.

As the population of coastal counties continues to grow, the generation of sewage sludge accelerates. Data available from NOAA (Basta and others, 1982) provide these data for 1980 and projections to 2000 (in parentheses):

Coastal counties, eastern U.S.	163	
Publicly-owned municipal treatment plants (POTW's)	880	(1,190)
Sludge produced (10^3 tons/yr, dry solids)	855	(2,251)
POTW's discharging directly to ocean	95	(143)
Sludge produced (10^3 t/y)	110	(335)
All other POTW's in coastal counties	785	(1,047)
Sludge produced (10^3 t/y	743	(1,914)

In 1980, 36 percent of sludge generated in East Coast counties was discharged into ocean waters, 51 percent was placed in landfills, and the remaining 13 percent was disposed of through other practices. Most sludge consists of water used in the treatment of wastes. A typical solids content ranges from 5 to 8 percent; thus, this waste is usually discharged as a fluid from outfalls—some with diffusers to assist in the mixing of effluent with receiving waters—or barges, which release the liquid while underway in designated offshore dumping areas.

Another waste that has required the attention of geologists results from the use of radioactive materials in a variety of industrial and research applications. These substances are termed "low-level" wastes to distinguish them from spent fuel removed from nuclear power plants, or highly active wastes generated in the reprocessing of such materials, or the "transuranic" wastes resulting from other activities involving radioactive materials. Low-level wastes are currently being disposed of by land burial (storage) at several sites, one lying in the Atlantic Coastal Plain at Barnwell, South Carolina. These materials consist of hospital and research laboratory wastes such as glassware, protective clothing, reagent liquids, and other materials requiring little shielding. All states are currently involved in establishing "compacts" that will designate repositories for low-level wastes serving participating states. At least four compacts are under consideration within Atlantic coastal states, and site evaluations for potential repositories are underway. By law, all states must declare their affiliations in a particular compact before 1 January, 1988, or establish their own repositories within state boundaries.

Natural agents and processes deliver large quantities of dissolved and suspended material to the coastal Atlantic continental margin. These are the "non-point sources" that cannot be traced to a specific point of entry, and as such are difficult to identify and nearly impossible to manage. Natural and anthropogenic substances arrive in coastal and offshore waters through atmospheric washout, from runoff of surface waters in streams, rivers, and estuaries, and from stormwater overflows from municipal sewage systems (see Kneip and others, 1982). Through these channels pass pesticides, herbicides, petroleum products, toxic metals, and nutrients, which can enrich the shallow shelf waters to a point where harmful concentrations are present. Management strategies for non-point source control begin with monitoring and regulation of specific point sources and by instituting policies for best management practices to protect local onshore water quality.

OCEAN DUMPING

Most of the world's major cities have grown to prominence as a result of waterborne commerce—the traffic of vessels on rivers, estuaries, and coastal waters that bear goods to and from the ports and harbors supporting large centers of population. Records kept of the growth of these towns as they became cities over centuries and, in some cases millenia, provide a fascinating account of man's conflict with natural forces of erosion, deposition, sea-level change, and catastrophic events that have shaped the fortunes of nations (see Legget, 1973).

The vitality, in every sense of the word, of the populace gathered in these centers depends upon the collection, removal, and disposal of wastes generated either by man's activities or by nature. For cities lying on the shores of estuaries or the open sea, the logical receptical for such unwanted materials was, and remains, the adjacent water mass. It is convenient, apparently limitless in its capacity to accommodate wastes, and continues to provide the benefits that prompted colonization of its margins. Coastal states in the Atlantic continental margin have used the ocean as such a sink for centuries. The location and nature of ocean dumpsites off the East Coast of the United States is shown in Figure 1.

Dredged materials

In order to ensure continued access to port and harbor facilities and promote increased trade, coastal cities must maintain, and in some cases deepen, the dredged channels that lead shipping to their waterfront facilities. On the basis of total foreign waterborne commerce delivered to the U.S. (total short tons), eight of the twenty cities that constitute the nation's largest ports lie in the Atlantic continental margin. The magnitude of the effort necessary to keep these ports open to deep-draft shipping or to modify existing channels to greater depths is significant.

Data from the U.S. Army Corps of Engineers (Hurme, personal communication, 1986) provide the following volumes for ocean dumping of dredged materials in the Atlantic Ocean:

Year	Volume (million m^3)
1976	18.6
1977	11.5
1978	15.2
1979	10.0
1980	10.9
1981	5.4
1982	9.2
1983	9.6
1984	11.7

Approximately 90 percent of the solid matter dumped in the nation's estuaries and nearshore and offshore waters consists of sediment excavated or removed by dredging activity. Assessment of the fate and effects of dredged material in these environments continues as a controversial issue with policy makers, scientists, engineers, and concerned laymen. The major concern associated with dredged-material dumping surrounds the potential for release of toxic elements and/or compounds that adhere to fine-grained sediments comprising the bulk of dredged materials. It is widely known that clays exhibit highly charged surfaces that scavenge cations and organic compounds from solution (Pravdic, 1970; Brannon and others, 1978; Carter and Wilde, 1972). In some cases the adsorbed materials may reach concentrations that have demonstrated toxic or lethal affects on marine biota. The EPA is currently preparing guidelines to identify levels of elements and compounds that define sediments as "contaminated" wastes on the basis of concentration limits established by labora-

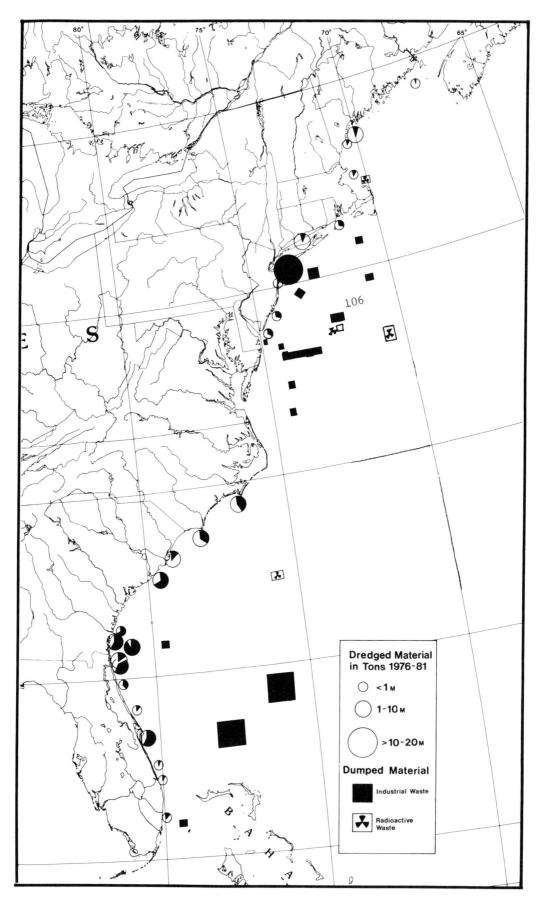

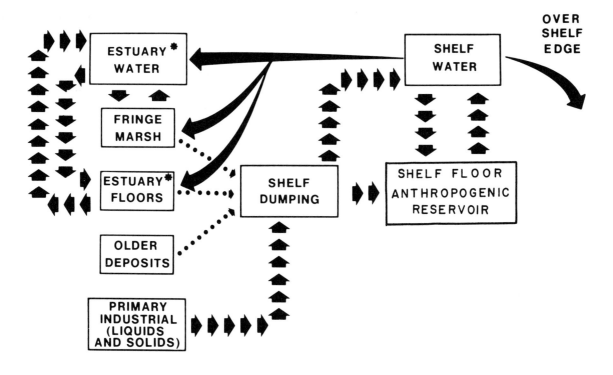

DOTTED LINE PATHS ARE DREDGE ACTIVITIES.
✳ ESTUARIES INCLUDE HARBORS.

Figure 2. Transport and dispersal pathways for anthropogenic wastes dumped on the continental shelf (from Hard and Palmer, 1976).

tory bioassay tests. As noted by the Marine Board of the National Research Council, "The best strategy for the disposal of contaminated dredged material is one that contains the particles, confines the contaminants to the particles, and isolates the deposit from plants and animals, and particularly from man" (1985, p. 130).

The various pathways by which dredged materials may reenter marine and estuarine environments are shown in Figure 2. Techniques of isolating dredged materials dumped in open waters offshore include burial (Bokuniewicz, 1983); "capping" techniques in which noncontaminated and generally coarser-grained materials are employed as a "cap" or blanket deposit overlying the finer-grained contaminated materials (Morton, 1983; O'Connor, 1982); or deep-water disposal at sites where dispersion and ultimate settling of the materials dilutes the waste.

In 1980, federally-funded maintenance dredging programs for major Atlantic ports included 12 projects that required the removal of 43 million cubic meters of sediment at a total cost of

Figure 1. Waste types and disposal sites in the offshore Atlantic continental margin. No radioactive dumpsites have been used since 1967, and only the deep-water (106) and acid dumpsites in the New York Bight are in use today (1986). Modified from NOAA (1980) and U.S. Army Corps of Engineers (1982). Circle diameters represent millions of tons.

$73.1 million. A sedimentological study of the materials dumped in the New York Bight southeast of the Hudson/Ratitan estuaries indicates that perhaps 850 million cubic meters of dredged sediment, fly ash, and other solid industrial and municipal wastes has accumulated in this area as a result of nearly a century of waste-disposal practices (Williams, 1979). The rates of accumulation of bottom "sediment" resulting from the activities far exceed normal depositional regimes associated with annual sediment discharge related to inputs from coastal drainage (Meade, 1972). Comparisons of historical bathymetric maps from as early as 1845 to the late 1970s indicate that accumulations in excess of 15 m over original depths have been verified as resulting from decades of ocean dumping (Williams, 1979). Most of this material was dumped during the period 1936 to 1973, when a sedimentation rate of up to 27 cm/yr was determined by coring studies and isobath analyses (Fuhrman and Dayal, 1982). From Figure 1, it is clear that the New York Bight area represents the most heavily exploited offshore area for waste-disposal activities in the offshore waters of the Atlantic Continental Margin.

Industrial Wastes

By-products and effluents from industrial processes that are not discharged from nearshore outfalls or disposed of through

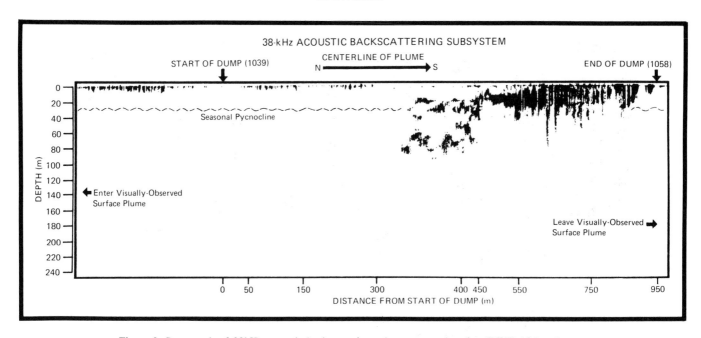

Figure 3. Sonograph of 38kHz acoustic backscattering subsystem employed at DWD-106 to image approximately 1,400 tons of coal ash discharged by a bottom-dumping hopper barge underway on a southerly course. At about 320 m from the initial release point (time 1039) the coarser particles are seen to have penetrated the seasonal pycnocline (zone of maximum density gradient). Settling velocities of the coarsest fraction of ash particles is estimated at 50 m/min (from Rose and others, 1985).

coastal plain options onshore are carried to ocean disposal sites designated by the EPA. Two sites in the offshore Atlantic Continental Margin are in use by a number of firms holding dumping permits, which specify the criteria under which wastes may be released into the marine environment. Scows, barges, hopper dredges, and other craft transport the wastes from coastal generators to the ocean sites.

The "acid waste" dumpsite, which lies some 25 km off the New Jersey coast, receives high acidic materials and a variety of other industrial by-products generated by manufacturing activities in the New York–New Jersey area. The second, and much larger site, is located approximately 195 km southeast of Ambrose Light (roughly 106 nautical miles); this location has been used extensively for a variety of chemical wastes, sludge, and fly ash. It is commonly referred to as "the 106 site" or "DWD-106" for "Deep Water Dumpsite" (Fig. 1). Water depths at this site above on the continental slope range from 1,750 to 2,700 m. These two sites constitute the only industrial waste dumpsites in the waters of the Atlantic Continental Margin now receiving wastes on a regular and controlled basis.

A detailed study of the behavior of dumped particulate matter at DWD-106 was undertaken by Rose and others (1985). They employed a high-frequency acoustic back-scattering system to track the discharged ash plume as it settled through the water column. Concentrations of particular matter ("ground truth" for calculations of mass dispersion) were obtained by water samplers operated in the insonified plume. The results, shown in Figures 3

and 4, demonstrate the ability of the acoustic system to monitor the descent of particular matter, which ranged from 54 to 98 percent solids upon release. Rapid dispersion of the plume is evident, and penetration of the seasonal thermocline by the waste bolus suggests that these particles ultimately settle to the sea floor in a fashion similar to other fine solids introduced through natural processes. However, visual inspection of the sea floor from the research submersible *Alvin* at the site of sediment trap deployments under and near the DWD-106 site (Gardner and others, 1985) revealed no identifiable accumulation of particulate matter that might be attributed to dumping practice.

Numerous other sites formerly used as dumping locales appear in Figure 1. As a result of more stringent criteria for waste-disposal practice in the sea, they are not now receiving industrial wastes from industry or municipalities. However, valuable information regarding the fate and effects of dumped materials on physical, biological, and chemical environments on the Atlantic shelf and upper slope can be assessed through study of these sites. Devine and Simpson (1985) describe such an assessment in their examination of the old Philadelphia sludge dumpsite some 65 km southeast of the mouth of Delaware Bay. During the final three years of sludge dumping (1977–1979), approximately 20,000 metric tons of sewage sludge solids were released within the designated 172-km^2 area. From the onset of dumping in May 1973, approximately 4.1 million tons were dumped in water depths from 50 to 70 m before dumping ceased in November 1980. These investigators found no quantifiable evidence of im-

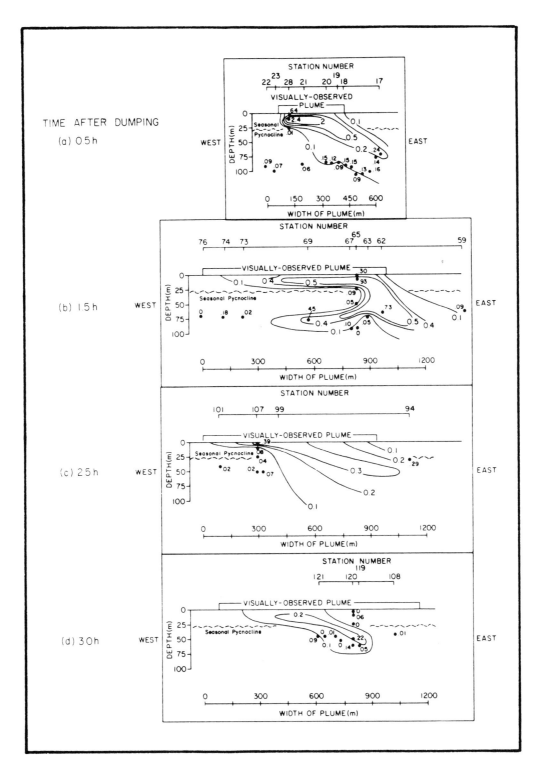

Figure 4. Ash concentrations (in mg/l) at four intervals after discharge of coal ash at DWD-106. Based upon initial concentrations of particulate matter at the time of release, it is estimated that 98 percent of the ash had settled below 100 m by 0.5 hr after dumping.

pact and no signs of alteration that could be construed as a threat to either human health or to the vitality of most benthic organisms.

These conclusions were borne out in earlier manned submersible operations in this same area reported by Folger and others (1979). These first-hand observations, photography, and sampling, together with other findings in a variety of waste-disposal areas of the Atlantic margin at midshelf depths, suggest that the rapid dispersion of particulate wastes released into the water column tends to provide a dilution that quickly diminishes the potential toxicity, smothering aspect, and/or harmful effect(s) of most wastes released at shallower offshore sites. More recent studies comparing the effects of dumping in an acid-waste dumpsite in the New York Bight and an unaffected control site nearby (Phoel and others, 1985) support the conclusion that, at least at these locations where known quantities and volumes were discharged, minimal impact on marine biota was observed.

Radioactive wastes

Disposal of radioactive wastes in the ocean is controlled by Annexes to the London Dumping Convention (see above), two of which bear upon low-level radioactive wastes (LLRW). High-level wastes, consisting of spent fuel assemblies and reprocessing wastes containing high concentrations of fission products and actinides, are prohibited from disposal in the ocean. Low-level materials, those wastes having activities of less than ten nannocuries/gm/sec (Park and others, 1983) were routinely dumped at various marine sites during the period 1946 to 1982. Nearly all of the wastes were contained in steel drums capped with concrete.

Seven sites in the Atlantic continental margin received LLRW until 1970 when the U.S. ceased all ocean dumping of such materials. Three of the sites, Massachusetts Bay, Sandy Hook 1 (EPA's 2,800 m site), and Sandy Hook 2 (EPA's 3,800 m site) received 99 percent of the activity (in curies at time of packaging) and 97 percent of the containers. The shallow Massachusetts Bay site was utilized between 1952 and 1959, and 4,008 steel drums containing the LLRW were dumped in depths averaging 92 m (see Lockwood and others, 1982). The deeper Sandy Hook sites were used between 1951 and 1962, a period in which 28,000 drums of waste were dumped (NACOA, 1984).

Although dumping by the United States has not occurred since 1970, the possibility of renewed use of deep-water sites at depths exceeding 4,000 m (a requirement under LDC regulations) has prompted continued research in the areas of fate and effects of packaged materials previously dumped, containment design, and site selection criteria for designation of possible future sites. In order to comply with and administer regulations that may be proposed for future disposal of LLRW, the EPA has supported research at the three major disposal sites. These efforts have included recovery of waste packages for assessment of container integrity and possible leakage (Colombo and others, 1983), and geological observations and sampling at the two deep-water sites using the research submersible DSRS ALVIN (Rawson and

Ryan, 1983; Hanselman and Ryan, 1983). One waste container encountered in these dives is shown in Figure 5.

Ocean incineration

Certain classes of liquid wastes containing significant quantities of chlorinated hydrocarbons and other toxic organic compounds are best disposed of through combustion, which reduces the components of these substances to water, hydrochloric acid, carbon dioxide, and other nonhazardous products. The concept of incineration at sea far from land and populated areas is, for some, an attractive option to land-based incineration in areas where exhaust plumes might affect terrestrial life. The practice generates negligible particulate matter and is thus devoid of geologic significance.

Drilling Activity and Rig Discharges

The area of submerged land within the Atlantic Continental Margin is nearly equally divided between continental shelf seabed (83 million acres shoaler than 200 m) and the upper slope to a depth of 2,500 m (87 million acres), a depth generally considered a practical limit for resource exploitation in the (OCS) Outer Continental Shelf (American Petroleum Institute, 1984). Five large sedimentary basins—Georges Bank in the northeast; Baltimore Trough in the mid-Atlantic; and the Southeast Georgia Embayment, Carolina Trough, and Blake Plateau Basin in the southeastern Atlantic—prompted extensive geological and geophysical exploration, which led to the leasing of 410 blocks during the 9 OCS sales held between August 1976 and July 1983. At the close of 1984, only 182 leases were active. The exploration drilling that occurred during this period led to 41 dry holes, most in the Baltimore Trough area. Only 5 of the total of 46 holes had any measurable flow of hydrocarbons (mostly gas) in a zone just landward of the buried shelf-edge carbonate reef complex. A thorough summary of Atlantic OCS activity through 1984 is provided by Rudolph and Havran (1985).

Exploration drilling activity produces a variety of waste by-products. On the basis of volume, most are small (runoff water, sanitary and galley wastes, testing fluids, etc.) and are largely fluids quickly dispersed by natural marine processes. Two wastes, however, have significant volume and are generally released into the marine environment at the drillsite. One, cuttings removed from the hole as the bit advances through formation material, consists of sediment particles and rock fragments that are considered natural materials. The other, drilling fluids, consists primarily of mudlike slurries which may carry many chemical additives that have demonstrated toxic effects on test organisms exposed to varying concentrations of the additive. The principal constituents of drilling muds are clay minerals, water, barite (as barium sulfate—used to increase the specific gravity of the fluid), chrome lignosulfonate, lignite, and sodium hydroxide.

The quantity of drilling fluids and cuttings dumped on the sea floor adjacent to an exploration drilling rig depends on the

Figure 5. Photo from DSRV *Alvin,* July 1978, at the 3,800 m low-level radioactive waste disposal site. The 55-gallon drum lies on a firm substrate that displays scour effects and traces of corrosion products (rust) in a down-current (southwesterly) direction. A boulder of Eocene chalk lies at the right; a rattail fish (*Coryphaenoides armatus*) appears in upper left (EPA photograph courtesy R. Dyer).

target depth of the well and the frequency at which conditions require changing the composition of the mud. A typical discharge is cited by Ayres and others (1980) for an exploration well drilled to 4,970 m in a mid-Atlantic exploration well. The total of solids discharged was 2,160 tons, which included 752 tons of barite and 45 tons of chrome lignosulfonate. The accumulation of this material on the sea floor is a function of water depth, local currents, and the manner of release of wastes. In general, physical alteration of the bottom by accumulation of particulate matter is confined to a radius of perhaps 200 m around the drill site. Chemical effects, as revealed by elevated trace metal concentrations in sea floor sediments surrounding the site, may extend for greater distances from the rig.

The documentation on fate and effects of drilling operations in the Atlantic continental margin is voluminous. A succinct presentation of most aspects of the impact of drilling discharges in the marine environment will be found in a report by the National Research Council (1983), which addresses drilling procedures, fate, and effects of discharges, and biological effects of offshore drilling. More specific studies may be located by examination of the most recent annual Atlantic OCS Region Study Index prepared by the Minerals Management Service (1985).

SUMMARY

The term, "assimilative capacity" is often employed in arguments both favoring and decrying the use of the sea as a waste receptacle. Many definitions of this term may be proposed, but the analogy drawn by Goldberg (1981) comparing the introduction of a waste into the marine environment as a titration exercise, with the end point yielding the value for assimilative capacity, provides a focus for the argument. His definition (Goldberg, 1981, p. 8) is "The amount of a given material that can be contained within it (a body of seawater) without producing an unacceptable impact on living organisms or nonliving resources." A companion paper by Kamlet (1981) raises obvious questions of scientific and political (regulatory and policy) definitions of what constitutes an unacceptable impact. The level of concern expressed by commercial fishermen, coastal Chambers of Commerce, public, and private organizations concerned with the serious questions of environmental impacts of waste disposal, and others who have legitimate questions regarding the quality of marine and estuarine waters has become a major issue in coastal communities.

Projections by the U.S. Census Bureau suggest an 8 percent gain in population of all states bordering the Atlantic Ocean by the year 2000. The greatest growth will be in the central and southern states, and estimates actually show a decline of 15 percent during this period in New York, the most populous of Atlantic coastal states.

These data suggest that by 2000, the 71.6 million inhabitants of these coastal states will make further demands upon inland waters and estuaries, the coastal zone, and offshore marine environments. Increased population guarantees increased waste production, and the need for prudent and timely decisions will be a paramount concern of state and federal legislators. As pointed out by the author and his colleagues (Palmer and Gross, 1979), geologists, and especially marine geologists and sedimentologists, have a clear responsibility to become involved in decision-making processes affecting the choices of waste-disposal alterna-

tives. Their concepts of rates and scales of natural agents and processes lend a perspective reaching far beyond terms of elected or appointed policy makers and indeed beyond the lifetimes of coastal inhabitants. Their assessment of waste-management practice thus becomes a critical factor in making rational choices which minimize the impact(s) of disposal practice.

REFERENCES CITED

American Petroleum Institute, 1984, Oil, gas, and deepwater technology; The hidden frontier: Washington, D.C., American Petroleum Institute, 24 p.

Ayers, R. C., Jr., Sauer, T. C., Jr., Meek, R. P., and Bowers, G., 1980, An environmental study to assess the impact of drilling discharges in the mid-Atlantic, *in* Proceedings of a Symposium on Research on Environmental Fate and Effects of Drilling Fluids and Cuttings: Washington, D.C., Courtesy Associates, p. 382.

Baedecker, M. J., and Apgar, M. A., 1984, Hydrogeochemical studies at landfill in Delaware, *in* Groundwater contamination, Studies in Geophysics, National Research Council: Washington, D.C., National Academy Press, p. 127–138.

Basta, D. J., Chambers, B. P., Ehler, C. N., and LaPointe, T. F., 1982, Identifying and evaluating alternative ocean dump sites: National Ocean Service, National Oceanic and Atmospheric Administration, 17 p.

Bokuniewicz, H. J., 1983, Submarine borrow pits as containment sites for dredges sediment, *in* Kester, D. R., Ketchum, B. H., Duedall, I. W., and Park, P. K., eds., Wastes in the ocean, v. 2, Dredged material disposal in the ocean: New York, John Wiley and Sons, p. 215–227.

Brannon, J. M., Plumb, J. H., Jr., and Smith, I., 1978, Long-term release on contaminants from dredged material: Vicksburg, Mississippi, U.S. Army Corps of Engineers, D-78-49, p. 65.

Brown, P. M., and Reid, M. S., 1976, Geologic evaluation of waste storage potential in selected segments of the Mesozoic aquifer system below the zone of fresh water, Atlantic Coastal Plain, North Carolina through New Jersey: U.S. Geological Survey Professional Paper 881, 47 p.

Carter, R. C., and Wilde, P., 1972, Cation exchange capacity of suspended material from coastal sea water off central California: Marine Geology, v. 13, p. 107–122.

Colombo, P., Neilson, R. M., Jr., and Kendig, M. W., 1983, Analysis and evaluation of a radioactive waste package retrieved from the Atlantic Ocean, *in* Park, P. K., Kester, D. R., Duedall, I. W., and Ketchum, B. H., eds., Wastes in the ocean, v. 3, Radioactive wastes in the ocean: New York, John Wiley and Sons, p. 237–268.

Devine, M. F., and Simpson, D., 1985, Ocean dumping at the Philadelphia site; A case study, *in* Proceedings, Coastal Zone '85: New York, American Society of Civil Engineers, p. 613–628.

EPA, 1984, Assessment of hazardous waste mismanagement damage case history: U.S. Environmental Protection Agency, Office of Solid Waste, Publication no. EPA/530-SW-84-002, 369 p.

Folger, D. W., Palmer, H. D., and Slater, R. A., 1979, Two waste disposal sites on the continental shelf off the middle Atlantic states; Observations made from submersibles, *in* Palmer, H. D., and Gross, M. G., eds., Ocean dumping and marine pollution: Stroudsburg, Pennsylvania, Dowden, Hutchinson and Ross, p. 163–184.

Furhman, M., and Dayal, R., 1982, A sedimentological study of the dredged material deposit in the New York Bight: Environmental Geology, v. 4, p. 1–14.

Gardner, W. D., Southard, J. B., and Hollister, C. D., 1985, Sedimentation, resuspension, and chemistry of particles in the northwest Atlantic: Marine Geology, v. 65, p. 199–242.

Goldberg, E. D., 1981, The oceans as waste space; The argument: Oceanus, v. 24, no. 1, p. 2–9.

Grove, P. B., 1971, Webster's 3rd New International Dictionary: Springfield, G.&C. Merriam, 2662 p.

Hanselman, D. H., and Ryan, W.B.F., 1983, 1978 Atlantic 3,800-meter radioactive waste disposal site survey; Sedimentary, micromorphological, and geophysical analyses: Office of Radiation Programs, U.S. Environmental Protection Agency Report EPA 520/1-83-017, 36 p.

Hard, C. G., and Palmer, H. D., 1976, Sedimentation and ocean engineering; Ocean dumping, *in* Stanley, D. J., and Swift, D.J.P., eds., Marine sediment transport and environmental management: New York, John Wiley and Sons, p. 557–577.

IMCO, 1976, Convention on the prevention of marine pollution by dumping of wastes and other matter: London, (United Nations), 36 p. plus Amendment (Publication no. 76.14E), 6 p.

Kamlet, K. S., 1981, The oceans as waste space; The rebuttal: Oceanus, v. 24, no. 1, p. 10–17.

Kneip, T. J., Cutshall, N. H., Field, R., Hart, F. C., Lioy, P. J., Mancini, J., Mueller, J. A., Sobotowski, C., and Szeligowski, J., 1982, Management of non-point sources, *in* Mayer, G. F., ed., Ecological stress and the New York bight: Columbia, South Carolina, Science and management, Estuarine Research Foundation, p. 145–161.

Legget, R. F., 1973, Cities and geology: New York, McGraw-Hill, 624 p.

Lockwood, M., Grunthal, M. C., and Curtis, W. R., 1982, Side-scan sonar of the Massachusetts Bay low-level radioactive waste disposal site, *in* Proceedings, Oceans '85 Conference: Washington, D.C., Marine Technology Society/Institute of Electrical and Electronics Engineers, p. 1150–1155.

Meade, R. H., 1972, Sources and sinks of suspended matter on continental shelves, *in* Swift, D.J.P., Duane, D. B., and Pilkey, O. H., eds., Shelf sediment transport; Processes and patterns: Stroudsburg, Pennsylvania, Dowden, Hutchinson and Ross, Incorporated, p. 247–262.

Meisler, H., 1986, Northern Atlantic Coastal Plain regional aquifer-system study, *in* Sun, R. J., ed., Regional aquifer-system analysis program of the U.S. Geological Survey; Summary of Projects, 1978–1984: U.S. Geological Survey Circular 1002, p. 168–194.

Minerals Management Service, 1985, Atlantic OCS region study index, 1985: Reston, Virginia, U.S. Department of Interior, Minerals Management Service, 209 p.

Morton, R. W., 1983, Precision bathymetric study of dredged-material capping experiment in Long Island Sound, *in* Kester, D. R., Ketchum, B. H., Duedall, I. W. and Park, P. K., eds., Dredged-material Disposal in the Ocean: Wastes in the Ocean, volume 2: New York, John Wiley & Sons, p. 99–124.

NACOA, 1984, Nuclear waste management and use of the sea: Washington, D.C., National Advisory Committee on Oceans and Atmosphere, Special report to the President and Congress, 113 p.

NOAA, 1980, Eastern United States coastal zones data atlas: Washignton, D.C., National Oceanic and Atmospheric Administration, p. 4.27.

National Research Council, 1983, Drilling discharges in the marine environment: Washington, D.C., National Academy Press, 180 p.

—— , 1985, Dredging coastal ports; An assessment of the issues: Washington, D.C., National Academy Press, 212 p.

—— , 1986, Global change in the geosphere-biosphere; Initial priorities for an international geosphere-biosphere program: National Academy Press, Washington, D.C., 91 p.

O'Connor, J., 1982, Evaluation of capping operations at the experimental mud dump site, New York, Bight area, 1980: U.S. Army Corps of Engineers, New York District Final Report no. DACW 39-82-2544, 72 p.

OTA, 1983, Technologies and management strategies for hazardous waste control: Washington, D.C., Office of Technology Assessment, U.S. Government Printing Office, 55 p.

Palmer, H. D., and Gross, M. G., 1979, Ocean dumping and marine pollution; Geological aspects of waste disposal: Stroudsburg, Pennsylvania, Dowden, Hutchinson and Ross, Incorporated, 268 p.

Park, P. K., Kester, D. R., Duedall, I. W., and Ketchum, B. H., 1983, Radioactive wastes in the ocean; Wastes in the ocean, v. 3: New York, John Wiley and Sons, Incorporated, 522 p.

Phoel, W. C., Reid, R. N., Radosh, D. J., Kube, P. R., and Fromm, S. A., 1985, Studies of the water column, sediments, and biota at the New York bight acid waste dumpsite and a control area, *in* Proceedings, Ocean '85 Conference, Ocean engineering and the marine environment, San Diego, v. 2: Marine Technology Society and Institute of Electrical and Electronic Engineers/Ocean Engineering Society, p. 945–948.

Pravdic, V., 1970, Surface charge characterization of sea sediments: Limnology and Oceanography, v. 15, p. 230–233.

Rawson, M. D., and Ryan, W.B.F., 1983, Geologic observations at the 2,800-meter radioactive waste disposal site and associated deepwater dumpsite (DWD-106) in the Atlantic Ocean: Office of Radiation Programs, U.S. Environmental Protection Agency Report EPA 520/1-83-018, 54 p.

Rose, C. D., Ward, T. J., and de Pass, V. E., 1985, Ecological assessment for coal ash dumped at deepwater dumpsite 106, *in* Park, P. K., Kester, D. R., Duedall, I. W., and Ketchum, B. H., eds., Energy wastes in the ocean; Wastes in the ocean, v. 4: New York, John Wiley and Sons, 818 p.

Rudolph, R. W., and Havran, K. J., 1985, Atlantic Summary Report, December 1984: Outer Continental Shelf Oil and Gas Information Program, U.S. Department of the Interior/Minerals Management Service, 74 p.

Solley, W. B., Chase, E. B., and Mann, W. B., IV, 1986, Estimated use of water in the United States in 1980: U.S. Geological Survey Circular 1001, 56 p.

Squires, D. F., 1983, The ocean dumping quandary; Waste disposal in the New York Bight: Albany, State University of New York Press, 226 p.

U.S. Army Corps of Engineers, 1982, 1980 report to Congress on administration of ocean dumping activities: Water Resources Support Center Pamphlet 82-P1, 41 p.

Williams, S. J., 1979, Geologic effects of ocean dumping on the New York bight inner shelf, *in* Palmer, H. D., and Gross, M. G., eds., Ocean dumping and marine pollution; Geological aspects of waste disposal: Stroudsburg, Pennsylvania, Dowden, Hutchinson and Ross, Incorporated, p. 51–72.

MANUSCRIPT ACCEPTED BY THE SOCIETY FEBRUARY 6, 1987

ACKNOWLEDGMENTS

A. Hurne, U.S. Army Corps of Engineers, Water Resources Support Center (Dredging) provided current data on dredged material volume. D. Basta, NOAA National Ocean Service, Strategic Assessment Branch furnished recent compilations of outfall volumes and projections for future waste disposal. R. Dyer, Office of Radiation Programs, Environmental Protection Agency, and C. Rose, ERCO/A Division of ENSCO, Incorporated, kindly provided figures and photos.

Printed in U.S.A.

The Geology of North America
Vol. I-2, The Atlantic Continental Margin: U.S.
The Geological Society of America, 1988

Chapter 32

Synthesis and unanswered questions

Robert E. Sheridan
Department of Geological Sciences, Rutgers University, New Brunswick, New Jersey 08903
John A. Grow
U.S. Geological Survey, Box 25046, Denver Federal Center, Denver, Colorado 80225

INTRODUCTION

This chapter summarizes the major findings of the foregoing chapters and highlights the critical factors in the geologic development of the U.S. Atlantic Continental Margin. In addition, we point out the limitations of the syntheses as a warning that much of what is written is speculative, based on inconclusive data, and/or somewhat dated. In most cases, individual authors have identified uncertainties of interpretations within the chapters themselves, but the reader might fail to recognize these caveats and qualifications.

During the course of the compilation of this volume, some of the chapters were submitted as early as two years prior to final publication. Consequently, some aspects of these chapters are dated. We know of data being released in 1986 and 1987 that are very pertinent to the syntheses of the Atlantic Continental Margin. Unfortunately, these data were not available in time for inclusion in this volume, and, thus, there are some deficiencies. However, on the whole the syntheses are current and represent up-to-date summaries of thinking on the subjects.

In some instances, there are marked differences in the interpretations of the same data presented in one or more chapters. These conflicts and inconsistencies are identified in this chapter as an aid to the reader. Such inconsistencies represent valid differences of opinion and are generally justified because of lack of publicly available data.

While a large amount of geophysical and drilling data have been obtained on the U.S. Atlantic Margin in the last two decades, a plethora of unanswered questions and uncertainties about the geology and geophysics of the region remains. In this chapter we will identify some of the remaining problems that await solution. Finally, we will make recommendations on the needs for future research in this region and propose the means to address the remaining questions.

SUMMARY OF MAJOR POINTS

Tectonics

The five major Mesozoic and Cenozoic basins of the U.S. Atlantic Margin (Georges Bank, Baltimore Canyon trough, Caro-lina trough, Blake Plateau, and Bahamas/South Florida basins) are the dominant features of first-order geological importance. Huge sediment volumes accumulated in the basins, where basement depths sometimes exceed 16 km. The hinge zone, for the most part, is near the current coast line, and landward of this line occur the much less voluminous sediments of the Coastal Plain. Coastal Plain embayments (Southeast Georgia, Albemarle, Salisbury, e.g.) with 1 to 2 km of sediments are interrupted by transverse basement highs (Peninsular and Cape Fear arches, and the New Jersey high, e.g.). These lesser relief structures appear to be controlled by preexisting structural grain of the sutured Paleozoic basement (Klitgord and others, this volume).

The miogeocline of the U.S. Atlantic Continental Margin extends from the Coastal Plain to the continental rise and straddles three very different kinds of crust. Under the deep Mesozoic basins, rifted and possibly intruded, "transitional" crust exists between the unstretched continental crust, landward of the hinge zone, and the normal oceanic crust seaward of the continent/ocean boundary.

Plate reconstructions (Klitgord and others, this volume) lead to interpretations of the U.S. Atlantic Margin as one typical of passive margins, with breakup between the North American and African continents in Middle Jurassic, some possible spreading-center shifts and adjustments in the early drift stage related to the origin of the Gulf of Mexico and Caribbean oceanic crusts, and a later normal drift stage evolution and maturation. Some plate interactions with Cuban orogenies in the Mesozoic and early Cenozoic are indicated (Sheridan and others, Chapter 15, this volume).

Stratigraphy and sedimentary processes

Drilling and seismic data on the continental shelf and slope identify several regional hiatuses: in the middle Cenomanian, Coniacian, Maastrichtian/Paleocene, Oligocene, and Miocene (Poag and Valentine, this volume). These widespread events persist in the region and may be of global extent (Olsson and others; Greenlee and others, this volume).

Sheridan, R. E., and Grow, J. A., 1988, Synthesis and unanswered questions, *in* Sheridan, R. E. and Grow, J. A., eds., The Atlantic Continental Margin, U.S.: Geological Society of America, The Geology of North America, v. I-2.

From south to north, there is a time-transgressive transition, or facies change, between the increasingly more southern occurrence of carbonate sedimentary rocks versus the more northern occurrence of siliciclastic sedimentary rocks. The possible cause of this important paleoenvironmental change is the movement of the Atlantic Margin north of 30° N latitude with drift of the North America plate (Schlee and others, this volume).

During certain times (Maastrichtian-Eocene), deeper water on the shelf and associated transgression of the Coastal Plain permitted deposition of largely authigenic chemical deposits (Olsson and others, this volume). Later, in the Oligocene-Miocene, extensive and thick deposits of prograding wedges of sand and gravel built the shelf off the mid-Atlantic states (Schlee and others, this volume), while the Gulf Stream intensified and finished sculpturing the Blake Plateau and Bahama channels (Poag and Valentine; Dillon and Popenoe; Sheridan and others, Chapter 15, this volume).

The Coastal Plain wedge accumulated as the postrift unconformity tilted seaward by flexure (Steckler and others, this volume), and there were varying amounts of coastal onlap during late Jurassic and later times (Olsson and others; Gohn, this volume). Sea-level fluctuations (Olsson and others; Greenlee and others, this volume) superimposed on this continuous seaward flexure produced the recurring migration of the strandline (transgression and regression) across the Coastal Plain (Steckler and others; Pitman and Golovchenko, this volume).

Frequently occurring sea-level changes in the Miocene through Holocene, caused by glacial eustasy, have created fine-scale sequence development mappable on the continental shelf and in the Coastal Plain (Riggs and others, this volume). Neogene deposits of chemical authigenic sediment were influenced by currents and the small-scale structural warping of the sea floor of the Atlantic shelf. The sea-floor warpings oriented the currents (Riggs and Belknap, this volume). Modern currents and wind patterns have clearly influenced sediment distribution on the slopes of the Bahama Banks and on the banks themselves (Sheridan and others, Chapter 15, this volume).

Basin syntheses

Based on regional seismic reflection lines, there are many places where prerift and synrift sedimentary rocks are found beneath the drift stage sediments of the Atlantic Margin. In contrast, there are only two places where these synrift rocks were penetrated by the drill in deep basins seaward of the hinge line: in the COST G2 well on Georges Bank (Manspeiser and Cousminer, this volume) and in the Great Isaac well in the Bahamas (Sheridan and others, Chapter 15, this volume). In these two wells, the ages of the rocks below the breakup unconformity are not well documented. In the Great Isaac well, the arkose and volcaniclastics below the unconformity are unfossiliferous (Sheridan and others, Chapter 15, this volume), and in the COST G2 well, the biostratigraphy and ages are controversial, as is the depth of the breakup unconformity (to be discussed later). Consequently, direct knowledge of the character of the offshore

synrift and prerift deposits is still very limited.

Seismically, these synrift and prerift basins imitate the onshore basins in structural style and development. Both seaward-dipping and landward-dipping border faults are detected offshore and onshore. Both landward-dipping and seaward-dipping reflectors in the synrift basins are detected. By analogy to the onshore basins, these reflectors are thought to be interbedded nonmarine sedimentary rocks and volcanics.

Volcanics are present extensively along the Atlantic Continental Margin, and were emplaced both during the rift stage, as plate margin effects, and during the drift stage, as intraplate effects. Interestingly, many volcanics of various Mesozoic and Cenozoic ages on the Atlantic Margin can be related to "hot spot" tracks (DeBoer and others, this volume).

At the end of the rift stage, the first incursions of the sea into the subsiding rift generally leave a salt deposit that immediately underlies the drift-stage sequence. Evidence for such salt is found in three of the basins: Georges Bank, Baltimore Canyon trough, and Carolina trough (Poag and Valentine; Schlee and Klitgord; Dillon and Popenoe; Grow and others, this volume). The strongest diapirism is found under the continental slope off the Carolina trough.

The history of the Blake Plateau during the early drift stage is similar to the basins to the north, with rapid subsidence and thick carbonate accumulations in the Jurassic and Early Cretaceous. In the middle Cretaceous, the Blake Plateau deepened with continued subsidence, and the Florida Current of the Gulf Stream eroded the surface in the Cenozoic (Dillon and Popenoe, this volume). Erosion by various means sculpted the Blake Escarpment into a sheer cliff and beveled a bench at the base.

The Blake Outer Ridge formed as a thick pile of fine-grained sediments deposited from contour-following currents. Rapid sedimentation of organic-rich sediments in the proper depth range resulted in the development of extensive gas hydrate deposits on the ridge.

In the Bahamas, the Jurassic and Lower Cretaceous carbonates cover the transitional and oceanic crust, and are coalesced into a "megabank" that combined with the carbonate bank of Florida and the Blake Plateau (Sheridan and others, Chapter 15, this volume). Late-stage tectonism affected the Bahamas in the Late Cretaceous and Early Tertiary. Extensive faulting of this age is found. These faults are thought to be related to the collision of the Bahamas with the Cuban orogen (Sheridan and others, Chapter 15, this volume).

Deep crustal structure

Gravity modeling of the Atlantic Margin indicates that there is an abrupt shallowing of mantle beneath the hinge zone seaward of which the mantle boundary rises to depths of 20 km. A second abrupt shallowing of mantle occurs at the continent/ocean boundary seaward of which the mantle occurs at depths of 15 km.

Recent magnetic modeling of the East Coast Magnetic Anomaly (ECMA) has been consistent with an edge effect of the boundary of oceanic crust, although intermediate blocks west of the oceanic crust are necessary in some areas (Grow and others, this volume). Seismic reflection data do not reveal any ridge structures as had been proposed for the anomaly in the past, but some normal faulting is suggested in the Baltimore Canyon trough, which would bring oceanic basement against nonmagnetic synrift sediments (Grow and others, this volume).

The Brunswick Magnetic Anomaly (BMA) can be modeled as a border effect of a narrow, linear synrift basin, again with magnetic basement in contact with nonmagnetic sedimentary rocks. The Blake Spur Magnetic anomaly (BSMA) can be attributed to a structural oceanic basement ridge marking the position of a major spreading center shift in Middle Jurassic time. It also marks the edge effect between Jurassic transition crust beneath the Bahamas and oceanic crust east of the Blake-Bahama Escarpment (Sheridan and others, Chapter 15, this volume).

One of the most recent seismic experiments on the Atlantic Margin, the Large Aperture Seismic Experiment (LASE), revealed that the distinctive 7.1 to 7.5 km/sec layer under the continental rise and slope of New Jersey continues well landward under the shelf and west of the ECMA (Diebold and others, this volume). This intriguing result demands new thinking on the origin of this layer. Offshore, under the slope and rise, it is reasonably interpreted as oceanic layer 3 (Grow and others, this volume). Landward of the ECMA, its origin is debatable, but its continuity with the same seismic layer east of ECMA favors an origin related to Mesozoic breakup processes. Deep crustal metamorphism or "underplating," with massive basaltic intrusions injected laterally into the rifted continental crust adjacent to the continent/ocean boundary (ECMA), might explain this.

Theoretical studies

Modern theoretical studies involving forward modelling have been applied to the Atlantic Margin. Extension during rifting and breakup appears to have involved two layers with independent extensional behavior (β–factors; Steckler and others, this volume). Backstripping and backtracking of geohistory curves indicated that only in the late stage of sedimentation (Oligocene-Miocene) did the pronounced reversal in dip of the Jurassic/Lower Cretaceous shelf edge evolve (Steckler and others, this volume). If this youthful age for this important closure is correct, there are significant petroleum potential implications. However, the recent drilling on this shelf edge indicates the presence of Lower Cretaceous wedges, which suggests tilting was going on even in the Cretaceous and that some structural relief is quite old.

The thermal modelling of the Atlantic Margin, utilizing extension data, indicates that the basal sediments just at the continent-ocean boundary have reached the hottest temperatures (Sawyer, this volume). The deeply buried sedimentary rocks in this location have gone through the oil maturation window.

Other modelling of the process of strandline migrations on a tectonically tilting margin, with superimposed sea-level changes, reveals that coastal onlap continues in the early stages of sea-level

fall (Pitman and Golovchenko, this volume). The continental shelf sequences are well determined seismically and drilled in several places, permitting the application of seismic stratigraphic techniques to determine sea-level changes for the Atlantic Margin. The resulting curve correlates well with the global curve developed recently by the EXXON group (Greenlee and others, this volume).

Geologic resources and hazards

After drilling approximately 50 exploration wells in the offshore and Bahama areas of the Atlantic Margin, no commercial deposits of oil and gas have been found (Mattick and Libby-French; Sheridan and others, Chapter 15, this volume). Deltaic sands and shales and carbonate rocks have been some of the reservoir types tested in a variety of structural traps. Most of the traps were of drift-stage or late drift-stage origins, and thus quite young relative to the optimum thermal maturation interval for the margin.

Ground water studies reveal that there are important connections between the Coastal Plain aquifers and those offshore. Artesian wells have been discovered on the continental shelf off Florida. In addition, relict fresh-water lenses, now somewhat brackish, have been trapped in the more impermeable shallow units under the continental shelf of the mid-Atlantic region (Kohout and others, this volume).

Hard mineral resources which are currently exploited on the Coastal Plain, such as sand and gravel, phosphate, and heavy-mineral placers, are well documented as known resources offshore. Maganese nodules are abundant on the relatively shallow Blake Plateau (Riggs and Manheim, this volume).

Heat-flow measurements on the Coastal Plain show variations of more than 1 HFU. Higher values are apparently over buried granitic bodies. The theoretical distribution of heat-flow values relates well to late Paleozoic granites beneath the Coastal Plain sediments (Costain and Speer, this volume).

Hazards and environmental problems on the Atlantic Margin are of natural geologic origin, but are aggravated by the more intense use of this province by the large population. Shifting sands endanger offshore drilling platforms (Folger, this volume); coastal erosion endangers seashore homes (Pilkey, this volume); potentially damaging earthquakes are known to occur (Seeber and Armbruster, this volume); and gravity slumps can disturb radioactive wastes dumped on the continental slope (Palmer, this volume).

Conflicts and contradictions

Several chapters in this volume present conflicting and contradictory interpretations of the same data. For example, Manspeiser and Cousminer interpret the age of the anhydritic dolomites below 4,000 m in the COST G2 well on Georges Bank as Late Triassic. If so, these rocks should be beneath the breakup unconformity, accepting the age of breakup proposed by Klitgord and others, and Sheridan and others, of approximately Bathonian (Middle Jurassic). However, the most pronounced seismic "breakup or postrift unconformity" in Georges Bank occurs at the base of the COST G2 well at 6.7 km (Schlee and Klitgord). This contrasts with the unconformity above the contentious Triassic sedimentary rocks, which is not the traditional angular unconformity more usually associated with breakup, but a rather subtle sequence boundary. Could this more subtle unconformity be the actual breakup unconformity, while the deeper, more angular unconformity might be a manifestation of a slightly older Triassic(?) rifting event that did not lead to breakup? Such "stacked" breakup unconformities are seen on other margins, such as off Newfoundland, where several rifting events have reoccupied the same region prior to final breakup.

On some of the cross sections and figures of the Poag and Valentine chapter, they indicate, albeit with question marks, a possible Early Jurassic age for some of the earliest drift-stage sediments. This conflicts with interpretations that the age of the ECMA and, therefore, breakup is of Middle Jurassic age (Sheridan and others, Chapter 15; Klitgord and others). The Poag and Valentine interpretation is based on extrapolating downward the ages of successively deeper sequences below the Late Jurassic sequences documented by drilling. This is admittedly speculative on their part.

Another conflict is the interpretation of the opening of the Atlantic with respect to the opening of the Gulf of Mexico. Sheridan and others suggest that these openings were of the same age, that is, breakup in Bathonian (ECMA) and spreading center shift in Callovian (BSMA) time. In contrast, Klitgord and others place the Gulf of Mexico opening events slightly younger, that is, breakup in Callovian (BSMA) and spreading center shift in Tithonian (M21) time. The conclusion by Klitgord and others is based on plate reconstructions involving the South American plus Yucatan blocks, and assumes that the shapes of these blocks back in Jurassic time were the same as they are today. Very little is known about the basement configuration under the Yucatan block, and much stretching and postbreakup deformation could have occurred there. How wide was the Yucatan block in the Jurassic? If the Gulf of Mexico was spreading up to M21 time, where are the linear sea-floor magnetic anomalies of M21 to M25 age?

Unanswered questions and recommendations for future research

The synthesis chapters reveal many interpretations that are still debatable, and still in need of documentation to confirm current ideas. For example, are the abrupt shallowings of the MOHO at the hinge zones and continent/ocean boundary real? The LASE data seem to imply that the gravity determined "steps" in the MOHO are actually more gradual. While the origin of the BMA is rift related in interpretations in this book, it could be suggested that the anomaly marks a Paleozoic crustal suture (Klitgord and others, this volume). Is this so? Could the ECMA also be a Paleozoic suture reoccupied by the Mesozoic breakup?

What is the transitional crust? Is it more basic in composition, with many gabbroic intrusives (underplating) that form the 7.1 to 7.5 km/sec layer? What are the ubiquitous seaward-dipping reflectors just landward of the continent/ocean boundary? Are these synrift phenomena, or Icelandic-type volcanic wedges overlapping the breakup unconformity and thus of early drift stage origin? Is there more than one breakup unconformity on the Atlantic Margin? Were there earlier Triassic rifting events that did not lead to breakup, and are these "stacked" beneath the true Jurassic breakup unconformity? What causes the small-scale (1 to 2 km) differential relief structures along strike on the Coastal Plain? Magnetically interpreted Paleozoic sutures do rim these Coastal Plain structures. Are they the controlling factors? Are the magnetic anomalies indeed sutures? What was the origin of the Charleston earthquake? Could another such large, damaging temblor occur on the Atlantic Margin? Would this occur in the Charleston area, or could this occur anywhere on the Atlantic Margin? What is the impact of the older Jurassic transforms forming offshore fracture zones (such as the Blake Spur and Bahamas fracture zones) where they cross the continental crust? What kind of crust is under the South Florida–Bahamas basin southeast of the Bahamas fracture zone?

These questions reveal the need for further exploration and interpretation of the geology and geophysics of the U.S. Atlantic Margin. Below we list some projects that should address the remaining questions:

1. Deep-penetration, large-offset, vertical seismic reflection profiling across the margin, from the fall line to the continental rise. (These profiles should have enough penetration to record reflections from the upper mantle below the MOHO.) Deep seismic imaging of the MOHO and the lower crust (transitional crust) is critical. Especially important is the role that the lower crust and MOHO play in the origin of anomalies such as ECMA, BMA, and BSMA.

2. Deeper penetration in Ocean Drilling Project (ODP) drillsites. Drilling into the sequence below the reflector J3 on the continental rise will give an age on the basal drift stage sediments. Seismic ties from these offshore ODP sites into the continental margin will document the basal drift sequence above the breakup unconformity.

3. Deep ODP drilling to sample inner Jurassic Quiet zone (IJQZ) basement. The interpretation of the ECMA as an edge effect requires that the IJQZ be oceanic crust with reversed magnetic polarity. This needs verification.

4. Deeper ODP drilling on the Blake Spur magnetic anomaly. Interpretations of this anomaly as a structural escarpment related to a spreading center shift could be tested. Sediments and basement of different ages should be found on either side of the anomaly, and basaltic oceanic basement should occur landward of the BSMA under the IJQZ.

5. Deeper *Alvin* submersible dives on the Blake-Bahama Escarpment. Hypotheses about erosional cutting back of the escarpment have been proposed. The amount of erosion proposed depends on the age of the sedimentary rocks below 4,000 m. If

Jurassic rocks are found between 4,000 and 5,000 m on the Blake Escarpment, then erosion would have been more extensive than if only Lower Cretaceous rocks are recovered.

6. Direct ties to Coastal Plain and Bahama wells. Land vibroseis or high resolution land seismic profiles should connect offshore seismic lines to onshore wells. In the Bahamas, the deep-water seismic reflection profiles in the channels should be tied to shallow-water profiles on the banks; and then as close to the island wells as possible.

7. Further documentation of local relative sea-level cycles. Studies combining high-resolution seismic reflection profiles tied to well data are needed in all the basins. Region-wide correlations must be established, with publicly available data, and then compared with similar cycles in other basins around the world, to establish a global sea-level cycle curve.

8. High-resolution and deep-penetration seismic studies on the Coastal Plain to: (a) establish sea-level cycles and sequence correlations, (b) locate buried Mesozoic rift basins and structures, (c) document intrabasement Paleozoic structures as continuations of those exposed in New England and Canada and those exposed in Georgia and the Carolinas, and (d) investigate the sutured nature of the basement and the lower crustal and MOHO involvement in those sutures.

9. Further analysis of the biostratigraphy of the older portions (Jurassic/Triassic) of the existing wells drilled on the Atlantic Margin, either COST wells or released data on commercial wells. Atlantic-wide correlations of nonmarine and shallow-marine forms such as dinoflagellates, ostracodes, large foraminifera, pollen, and fossil fish are needed to resolve the conflicting interpretations on the age of the synrift to drift transitions.

Printed in U.S.A.

Index

[Italic page numbers indicate major references]

Typeset by WESType Publishing Services, Inc., Boulder, Colorado
Printed in U.S.A. by Malloy Lithographing, Inc., Ann Arbor, Michigan